AF598621

UNIVERSITY OF BRADFORD Library

REFERENCE ONLY

This book must not be removed from the library

BD 034056212 X

International Encyclopedia of
Composites

Editorial Board

Donald F. Adams, *University of Wyoming*
A. R. Bunsell, *Ecole des Mines*
Tsu-wei Chou, *University of Delaware*
C. K. H. Dharan, *University of California, Berkeley*
James Economy, *University of Illinois*
Timothy Gutoski, *Massachusetts Institute of Technology*
Naohiro Igata, *Science University of Tokyo*
Norman J. Johnston, *NASA—Langley Research Center*
Paul A. Lagace, *Massachusetts Institute of Technology*
Herman Mark, *Polytechnic University*
Luigi Nicolais, *University of Naples*
George Springer, *Stanford University*

International Encyclopedia of Composites

Volume 3

Stuart M. Lee
Editor

VCH
New York

Stuart M. Lee
3718 Cass Way
Palo Alto, California 94306

Library of Congress Cataloging-in-Publication Data
(Revised for volumes 3 and 4)

Encyclopedia of composites.

Includes bibliographical references.
Contents: v. 1. Acetal resins and composites—cyanate ester. — v. 3. Laminates, ceramic—mold, short-fiber composites — v. 4. Natural composites—pultrusion.
1. Composite materials—Encyclopedias. I. Lee, Stuart M.
TA418.9.C6E53 1989 620.1′18′03 89-24893
ISBN 0-89573-290-4 (set)
ISBN 0-89573-733-7 (v. 3)
ISBN 0-89573-734-5 (v. 4)

British Library Cataloguing in Publication Data

International encyclopedia of composites.
Vol. 3
1. Composite materials
I. Lee, Stuart M.
620.118

ISBN 3-527-27949-0

© 1990 VCH Publishers, Inc.

This work is subject to copyright.

All rights are reserved, whether the whole or part of the material is concerned, specifically those of translation, reprinting, re-use of illustrations, broadcasting, reproduction by photocopying machine or similar means, and storage in data banks.

Registered names, trademarks, etc. used in this book, even when not specifically marked as such, are not considered to be unprotected by law.

Printed in the United States of America

ISBN 0-89573-733-7 (volume 3) VCH Publishers
ISBN 0-89573-290-4 (set) VCH Publishers
ISBN 3-527-27949-0 (volume 3) VCH Verlagsgesellschaft
ISBN 3-527-26852-9 (set) VCH Verlagsgesellschaft

Printing History:
10 9 8 7 6 5 4 3 2 1

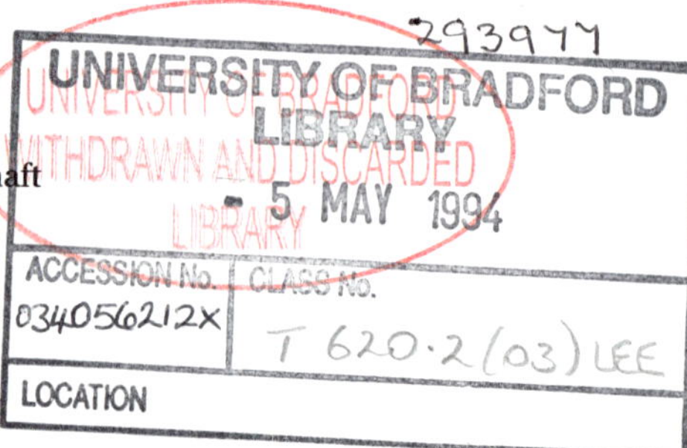

Published jointly by:

VCH Publishers, Inc.
220 East 23rd Street
Suite 909
New York, NY 10010

VCH Verlagsgesellschaft mbH
P.O. Box 10 11 61
D-6940 Weinheim
Federal Republic of Germany

VCH Publishers (UK) Ltd.
8 Wellington Court
Cambridge CB1 1HZ
United Kingdom

Contents

Main Entries

Contributors

Suresh G. Advani, University of Delaware
Molding, Short-Fiber Composites; Flow Processing

Denise M. Aylor, Naval Ship Research and Development
Metal Matrix Composites, Corrosion

M. F. Amateau, Pennsylvania State University
Laminates, Ceramic

D. Wayne Becker, Wichita State University
Molding, Producibility for Polymer-Fiber Composites

Walter L. Bradley, Texas A&M University
Micromechanisms of Delamination in Composite Materials

William I. Childs, Quantum Composites
Molding, Compression

David A. Conti, British Aerospace (Ret.)
Microwave Composites

L. R. Dharani, University of Missouri
Metal Matrix Composites, Continuous Fiber, Fracture

J. F. Dolowy, Jr., DWA Composite Specialties, Inc.
Metal Matrix Composites, Aluminum

Renée G. Ford, Editor, Materials and Processing Report
Molding, Ceramic Injection

Randall M. German, Rensselaer Polytechnic Institute
Molding, Metal Injection

James G. Goree, Clemson University
Metal Matrix Composites, Continuous Fiber, Fracture

Timothy G. Gutowski, Massachusetts Institute of Technology
Manufacturing, Economic Considerations

Terence F. W. Hall, Northrop Corporation
Manufacturing Automation, Polymer Composites

W. C. Harrigan, Jr., DWA Composite Specialties, Inc.
Metal Matrix Composites, Aluminum

Carl T. Herakovich, University of Virginia
Lamination Theory

Siegfried H. Herliczek, Libbey–Owens–Ford Co.
Laminates, Glass

Peter D. Hilton, Arthur D. Little, Inc.
Market Evaluation

A. I. Isayev, University of Akron
Liquid-Crystalline Composites

Walter F. Jones, Clemson University
Metal Matrix Composites, Continuous Fiber, Fracture

Peter W. Kopf, Arthur D. Little, Inc.
Market Evaluation

Julia T. Lee, Michigan Molecular Institute
Literature of Composites

Shaw Ming Lee, CIBA-GEIGY Corporation
Matrix-Controlled Failure Modes of Polymeric Composites

T. Limtasiri, The Siam Cement Co., Ltd.
Liquid-Crystalline Composites

I. E. Locci, NASA Lewis Research Center
Metal Matrix Composites, Rapid Solidification Processing

David L. McDanels, NASA Lewis Research Center
Metal Matrix Composites, Tungsten Fiber Reinforced Copper

Geoffrey F. Meades, British Aerospace (Ret.)
Microwave Composites

G. L. Messing, Pennsylvania State University
Laminates, Ceramic

B. A. Mikucki, Dow Chemical Company
Metal Matrix Composites, Magnesium

John J. Morena, American Composite Education, Inc.
Mold Fabrications

Andreas Mortensen, Massachusetts Institute of Technology
Metal Matrix Composites, Infiltration

Beebhas C. Mutsuddy, Michigan Technological University
Molding, Ceramic Injection

R. D. Noebe, NASA Lewis Research Center
Metal Matrix Composites, Rapid Solidification Processing

S. L. Ogin, University of Surrey
Laminates, Matrix Cracking in Composite Laminates under Static and Fatigue Loading

Donald W. Petrasek, NASA Lewis Research Center
Metal Matrix Composites, Tungsten Fiber Reinforced Superalloys

Pradeep Rohatgi, University of Wisconsin—Milwaukee
Metal Matrix Composites, Casting Processes

D. V. Rosato, Rhode Island School of Design
Materials Selection, Polymeric Matrix Composites

Abdul B. Sadat, California State Polytechnic University
Machining of Composites

Jacques E. Schoutens, MMCIAC, Kaman Sciences Corporation
Metal Matrix Composites, Design Methodology

S. O. Shook, Dow Chemical Company
Metal Matrix Composites, Magnesium

Robert A. Signorelli, NASA Lewis Research Center
Metal Matrix Composites, Tungsten Fiber Reinforced Superalloys

P. A. Smith, University of Surrey
Laminates, Matrix Cracking in Composite Laminates under Static and Fatigue Loading

A. Brent Strong, Brigham Young University
Manufacturing

A. Brent Strong, Brigham Young University
Molding

Stephen R. Swanson, University of Utah
Laminates, Static Strength

S. A. Umar-Khitab, Defense Research Establishment, Valcartier
Laminated Plate Analysis Using the Finite Element Method

Mark R. van den Bergh, DWA Composite Specialties, Inc.
Metal Matrix Composites, Aluminum

K. K. Wang, Cornell University
Molding, Polymer Injection

Franklin E. Wawner, University of Virginia
Metal Matrix Composites, Discontinuously Reinforced, Microstructure and Mechanical Property Correlations

Dexter White, Dow Chemical Company
Molding, Resin Transfer

Conversion Table from SI Units

Property	From	To	Multiply By
Dimension			
	cubic meter (m^3)	cubic inch (in^3)	6.102×10^4
	cubic meter (m^3)	cubic foot (ft^3)	35.315
	cubic meter (m^3)	cubic yard (yd^3)	1.308
	cubic meter (m^3)	gallon (U.S. dry)	227.02
	cubic meter (m^3)	gallon (U.S. liquid)	264.17
	cubic meter per second (m^3/s)	cubic foot per minute (cfm)	2.119×10^3
	meter (m)	inch (in)	39.37
	meter (m)	foot (ft)	3.281
	meter (m)	micron (μm)	1.0×10^6
	meter (m)	mil	3.937×10^4
	meter (m)	mile (statute)	6.214×10^{-4}
	meter (m)	yard (yd)	1.094
	micron (μm)	microinch	39.37
	micron (μm)	angstrom (Å)	1.0×10^4
	newton (N)	kip	2.248×10^{-4}
	square meter (m^2)	square inch (in^2)	1.55×10^3
	square meter (m^2)	square foot (ft^2)	10.76
	square meter (m^2)	square mile (mi^2)	3.861×10^{-7}
	square meter (m^2)	square yard (yd^2)	1.196
Electrical			
	joule (J)	electron volt	6.241×10^{18}
	joule (J)	erg	1.0×10^7
	megajoule (MJ)	kilowatt hour (kW·h)	0.2778
	newton (N)	dyne	1.0×10^5
	newton per meter (N/m)	dyne per centimeter (dyne/cm)	1.0×10^3
Dielectric Strength	millivolts per meter (mV/m)	volts per mil ($V/10^{-3}$ in)	2.539×10^{-2}
Force			
	joule (J)	foot-pound force (lbf)	0.7376
	kilogram per square meter (kg/m^2)	ounce mass per square yard	29.412
	meter per second (m/s)	mile per hour (mph)	2.237
	meter per square sec (m/s^2)	gal	1.0×10^2
	newton (N)	ounce-force	3.597
	newton-meter (N·m)	inch-pound (in-lb)	8.857
	newton-meter/meter (N·m/m)	inch-pound/inch (in-lb/in)	0.2248
	weber (Wb)	maxwell	1.0×10^8
Mass/Volume			
	cubic meter (m^3)	liter (for fluids only)	1.0×10^3
	cubic meter (m^3)	ounce (U.S. fluid)	3.381×10^4
	cubic meter (m^3)	quart (U.S. liquid)	1.057×10^3
	gram per cubic centimeter (g/cm^3)	pound per cubic inch (lb/in^3)	0.0361
	kilogram (kg)	ounce (avoirdupois)	35.27
	kilogram (kg)	ounce (troy)	32.15
	kilogram (kg)	pound (avoirdupois)	2.205
	kilogram (kg)	pound (troy)	2.679
	kilogram (kg)	ton (long, 2240 pounds)	9.842×10^{-4}
	kilogram (kg)	ton (metric)	1.0×10^{-3}
	kilogram (kg)	ton (short, 2000 pounds)	1.102×10^{-3}
	kilogram per cubic meter (kg/m^3)	pound-mass (gallon, U.S. liquid)	8.345×10^{-3}
	milliliter (ml)	cubic inch (in^3)	0.0610
	newton (N)	poundal	7.233
	newton (N)	pound-force (lbf)	0.2248
Core Density	kilogram per cubic meter (kg/m^3)	pounds per cubic foot (lb/ft^3)	6.243×10^{-2}
Density	milligram per cubic centimeter (mg/cm^3)	pound per cubic foot (lb/ft^3)	6.243×10^{-2}
Specific Volume	cubic meter per milligram (m^3/mg)	cubic inches per pound (in^3/lb)	2.768×10^{10}

Conversion Table from SI Units

Property	From	To	Multiply By
Mechanical			
	kilogram per meter (kg/m)	denier	9.0×10^{6}
	kilogram per meter (kg/m)	tex	1.0×10^{6}
	newton per tex (N/tex)	gram force per denier	11.33
	tex	denier	9
Fracture Energy (G_{IC})	joules per square meter (J/m^2)	inch pounds per square inch (in lbs/in^2)	5.710×10^{-3}
Impact Strength Resistance	kilojoule per square meter (kJ/m^2)	foot pounds force per square inch (ft lb/in^2)	0.4758
Izod Impact	kilojoules per meter (kJ/m)	foot pounds per inch (ft lb/in)	18.73
Modulus	gigapascal (GPa)	pound force per square inch (psi)	1.450×10^{5}
Strength	megapascal (MPa)	pound force per square inch (psi)	1.450×10^{2}
Pressure			
	newton (N)	kilogram-force (kgf)	0.1020
	newton (N)	ton-force short	1.124×10^{-4}
	pascal (Pa)	atmosphere (normal)	9.869×10^{-6}
	pascal (Pa)	inch of mercury (32°F)	2.953×10^{-4}
	pascal (Pa)	inch of mercury (60°F)	4.019×10^{-3}
	pascal (Pa)	inch of water (39°F)	4.015×10^{-3}
	pascal (Pa)	millibar	1.0×10^{-2}
	pascal (Pa)	pound force per square inch (psi)	1.450×10^{-4}
	pascal (Pa)	torr (mm Hg, 0°C)	7.501×10^{-3}
Radiation			
	coulomb per kilogram (c/kg)	roentgen	3.876×10^{3}
	joule per kilogram (J/kg)	rad	1.0×10^{2}
Thermal			
	joule (J)	Btu (International Table)	9.478×10^{-4}
	joule (J)	Btu (mean)	9.471×10^{-4}
	joule (J)	Btu (thermochemical)	9.485×10^{-4}
	joule (J)	calorie (International Table)	0.2388
	joule (J)	calorie (mean)	0.2387
	joule (J)	calorie (thermochemical)	0.2390
	joule per kilogram kelvin (J/Kg·K)	calories per gram per °C (cal/g/°C)	2.389×10^{-4}
	joule per square meter (J/m^2)	Btu per square foot (Btu/ft^2)	8.805×10^{-5}
	kilojoule (kJ)	kilocalorie (kcal)	0.2388
	watts per square centimeter °C (w/cm^2 °C)	gram calories per second square centimeter (gcal °C/sec·cm^2 °C)	0.2388
Heat Capacity	joule per kilogram kelvin (J/kg·K)	Btu (International Table) per pound °F	2.389×10^{-4}
	joule per kilogram/kelvin (J/kg·K)	Btu (thermochemical) per pound °F	2.390×10^{-4}
Linear Coefficient of Thermal Expansion	millimeter per millimeter K (mm/mm·K)	inch per inch °F	0.5556
Specific Heat	kilojoule per kilogram K (kJ/kg·K)	Btu per pound °F	0.2388
Temperature	kelvin (K)	centigrade (°C)	subtract 273.
	centigrade (°C)	Fahrenheit (°F)	multiply by 1.80 then add 32
Thermal Conductance	watt per meter kelvin (w/m·K)	Btu·foot per hour·square foot·°F	6.933
	watt per square meter kelvin (w/m^2·K)	Btu per hour·square foot·°F	0.176
Viscosity			
	cubic meter per second (m^2/s)	centistoke	1.0×10^{6}
	pascal second (Pa·s)	centipoise	1.0×10^{3}
	pascal second (Pa·s)	poise (absolute viscosity)	10.0

Laminate Analysis

See Bending Analysis; Finite Element Analysis; Interfacial Analysis; Laminated Plate Analysis Using the Finite Element Method; Laminates, Matrix Cracking in Composite Laminates under Static and Fatigue Loading; Laminates Static Strength; Lamination Theory

Laminated Plate Analysis Using the Finite Element Method

Fiber Composites

The traditional approach to the design of engineering structures has been to start with a given material and design stress level, then detail appropriate members to carry the specified loads without exceeding the material strengths. These strengths are usually defined as a certain percentage of the material's tensile ultimate (F_{tu}) or yield strength (F_{ty}). The advent of advanced fiber composites has given the impetus for a major change in the normal design philosophy. Rather than being constrained to design a structure with a given set of material properties, it is now possible to tailor the required properties into the material itself, thereby rendering the finished product much more efficient in the role for which it is required.

The term *composite* refers to a mixture of two or more materials that are microscopically distinct but together form a single macroscopic entity. In current engineering terms, the word has come to refer to materials made of stacked layers of a fibrous reinforcement material held in place by a rigid binder. Such materials have a long history, and in the past two decades they have found widespread usage in aerospace applications in the form of long fibers embedded in a carrier matrix to make thin plates or shells. Combinations of such materials have several advantages in that they tend to exhibit the best qualities of their constituents as well as other qualities that neither possesses. This has resulted in a broad investigation of these materials using various techniques.

The usual elasticity approach to solving engineering problems is to assume an unknown stress function over the region of interest, and substitute this into the compatibility equations [1]. Application of the boundary conditions then determines the necessary constants for a correct representation of the solution. Unfortunately, this procedure is cumbersome for all but the simplest of problems. It is practically impossible to apply this method to complex geometries.

The most common approach used in the structural analysis of laminated composites involves some variation of classical laminated plate theory. In this case, the laminate displacements are represented through the thickness by a linear function in terms of the midplane displacements and curvatures [2]. Unfortunately, this approach is limited to thin plates, since the effects of shear deformations are not taken into account.

The utilization of laminated composites is not, however, limited to thin plates. Applications also exist for thick laminates, in which case the classical theory cannot be applied. Many higher order laminated plate theories have been proposed to overcome the difficulties encountered as a result of the thickness effects [3–6]. These theories attempt, with varying degrees of success, to take care of the zero shear stress requirement at the top and bottom surfaces of the laminate. Unfortunately, they do not take care of the free-edge effects at the longitudinal and lateral extremes of the plates. These effects become important in considerations of failure for plates fastened along their edges.

Current design procedures for laminated fiber composites require a number of independent steps to arrive at a viable design. One must evaluate the lamina properties, perform a laminate analysis, test for failure, and decide whether any modifications are necessary. Often it is difficult to predict the deflection response from a knowledge of the stresses within the laminate.

It would be advantageous to be able to predict the deflection response of thick laminated fiber composites simply from a knowledge of the lamina orientations, the laminae constituents, and the applied loading. The finite element method allows one to integrate the above steps into one easily mastered process.

The Finite Element Method

The underlying concept of the displacement-based finite element method is the division of the region into a series of subregions or elements. The behavior over each element is described by a set of assumed functions for the displacements. The form of these assumed functions is such that the displacement continuity is satisfied at the element boundaries. The local behavior is described by the use of an appropriate function valid for the region of interest. The advantages of this method lie in its application to complex geometries. In addition, the method may be applied to 3D analysis with much greater ease than the previous methods.

The finite element literature commonly refers to the subregions into which the domain is divided as *elements*. The displacements in these elements and the forces acting on and within the elements are related to one another through what is known as the *element stiffness matrix*. Following the evaluation of the individual element matrices and their subsequent assembly into the global stiffness matrix, standard techniques such as Gauss elimination or Cholesky square root decomposition are used to determine the resulting displacements. These displacements are then used to evaluate the strains and stresses within each lamina, and a failure theory is applied to determine overloads. The material properties in the area of the overload are modified as required, and the process repeated to convergence.

The finite element method is not without its limitations. The plate or shell elements currently available cannot be used to predict the three-dimensional stress fields encountered in problems involving thick plates. Formulations for thick plate and shell elements found in the literature use an increased number of variables to overcome some of the limitations inherent in thin plate theory; however, they still represent the entire laminate thickness by one element.

To overcome the limitations imposed by the use of plate and shell elements, several solid elements may be used to represent the laminate thickness. The use of these elements allows inclusion of all interlaminar stresses as well as end effects in the analysis of thick laminates. The properties of each ply are assigned to one element, and a stack of these represents the plate behavior in the thickness direction. Unfortunately, this leads to an excessive number of degrees of freedom, and some method must be found whereby those degrees of freedom not germane to the problem are eliminated.

Strain Energy Formulation

The finite element solution to an engineering problem represents the behavior of a structure by dividing the structure into a number of subregions over which simple models may be used. Each subregion is bounded by a set of nodes. The usual form of the element stiffness matrix follows from the displacement-based formulation of the finite element equilibrium equations [7]. The elastic strain energy in a loaded element is given by

$$\mathbf{U}_e = \frac{1}{2}\int \sigma_{ij}\varepsilon_{ij}\, dVol \tag{1}$$

where the integration is performed over the volume of the element. The stresses and strains in an element are related through Hooke's law, which may be written as

$$\{\sigma_{ij}\} = [D]\{\varepsilon_{ij}\} \tag{2}$$

The strains in an element are defined in terms of the displacements as

$$\{\varepsilon_{ij}\} = [\mathscr{L}]\{u\} \tag{3}$$

where $\mathscr{L}$ is a suitable linear operator. However, the displacements themselves are usually defined in terms of a set of basis functions valid only within the element. It is customary to write

$$u = \sum_{i=1}^{k} N_i \mathrm{u}_i \tag{4}$$

where u_i represents constant displacements at discrete points or nodes on the element, and N_i represents the shape functions associated with each node. So the strain-displacement matrix may be written as

$$\{\varepsilon_{ij}\} = [B]\{u_i\} \tag{5}$$

Since u_i are constants, they may be taken outside of the integral sign. The strain energy expression then becomes

$$\mathbf{U}_e = \frac{1}{2}\{u_i\}^T\left[\int [B]^T[D][B]\, dVol\right]\{u_i\} \tag{6}$$

It is readily verified that the term in the square brackets represents the element stiffness matrix.

Strain-Displacement Matrix

The strains in an element are defined in terms of the displacements within the element through the use of a suitable operator. In our case, this operator is Green's 3D strain tensor to the second order. This is given by

$$[\mathcal{L}] = \begin{bmatrix} \left(\frac{\delta}{\delta x}\right) + \frac{1}{2}\left(\frac{\delta}{\delta x}\right)^2 & \frac{1}{2}\left(\frac{\delta}{\delta x}\right)^2 & \frac{1}{2}\left(\frac{\delta}{\delta x}\right)^2 \\ \frac{1}{2}\left(\frac{\delta}{\delta y}\right)^2 & \left(\frac{\delta}{\delta y}\right) + \frac{1}{2}\left(\frac{\delta}{\delta y}\right)^2 & \frac{1}{2}\left(\frac{\delta}{\delta y}\right)^2 \\ \frac{1}{2}\left(\frac{\delta}{\delta y}\right)^2 & \frac{1}{2}\left(\frac{\delta}{\delta y}\right)^2 & \left(\frac{\delta}{\delta y}\right) + \frac{1}{2}\left(\frac{\delta}{\delta y}\right)^2 \\ \left(\frac{\delta}{\delta y}\right) + \left(\frac{\delta}{\delta x}\right)\left(\frac{\delta}{\delta y}\right) & \left(\frac{\delta}{\delta x}\right) + \left(\frac{\delta}{\delta x}\right)\left(\frac{\delta}{\delta y}\right) & \left(\frac{\delta}{\delta x}\right)\left(\frac{\delta}{\delta y}\right) \\ \left(\frac{\delta}{\delta y}\right)\left(\frac{\delta}{\delta z}\right) & \left(\frac{\delta}{\delta x}\right) + \left(\frac{\delta}{\delta z}\right)\left(\frac{\delta}{\delta z}\right) & \left(\frac{\delta}{\delta y}\right) + \left(\frac{\delta}{\delta y}\right)\left(\frac{\delta}{\delta z}\right) \\ \left(\frac{\delta}{\delta z}\right) + \left(\frac{\delta}{\delta x}\right)\left(\frac{\delta}{\delta z}\right) & \left(\frac{\delta}{\delta x}\right)\left(\frac{\delta}{\delta z}\right) & \left(\frac{\delta}{\delta x}\right) + \left(\frac{\delta}{\delta x}\right)\left(\frac{\delta}{\delta z}\right) \end{bmatrix} \tag{7}$$

In contrast to the usual elasticity approach, in which an unknown stress function is assumed over the region of interest, the finite element method represents the same function in a piecewise manner by a set of shape functions valid at discrete intervals within the domain. These functions are used in conjunction with constant values at given points in the subdomain to interpolate the unknown function within the element.

There is no mathematical restriction on the type of basis functions used for the interpolation of the unknown. However, it has been customary to use polynomial interpolation because of the ease with which these functions may be developed and differentiated. The latter is especially helpful when computing the element strains. Such functions have been used to develop a large number of elements for use in various engineering problems.

The approximation of the unknown function may be accomplished through the use of either Lagrange or Hermitian interpolation. In the former, a polynomial is sought that passes through a given set of points within the domain while satisfying the values at those points. In the latter, the polynomial must also satisfy the derivative values at those points. For both cases, the general form of the polynomial is given by

$$N_i(x) = \sum_{i=0}^{n} \prod_{j=0}^{n} \left(\frac{x - x_i}{x_i - x_j}\right) u_i \qquad \text{for } j \neq i \tag{8}$$

where the symbol Π denotes a product of the indicated binomials over the indicated range.

Because of the nature of the shape functions, they may be used to interpolate many other variables within the element. They may be used not only in the evaluation of the stiffness matrix, but also in the representation of the applied loading, the evaluation of the mass and damping matrices for dynamic problems, and the stability matrix for buckling problems. It is therefore useful to isolate the evaluation of these shape functions from the requirements of the problem.

To this end, these functions are written in a set of natural coordinates that range from -1 to 1. Such coordinates represent a mapping of the physical coordinates into a nondimensionalized system of parent coordinates. The element stiffness matrix may be easily evaluated in this parent (r, s, t) coordinate system, then mapped to the physical (x, y, z) coordinate system through the use of an appropriate transformation.

For the 20-node solid element considered here (Fig. 1), the shape functions for the corner nodes are given by

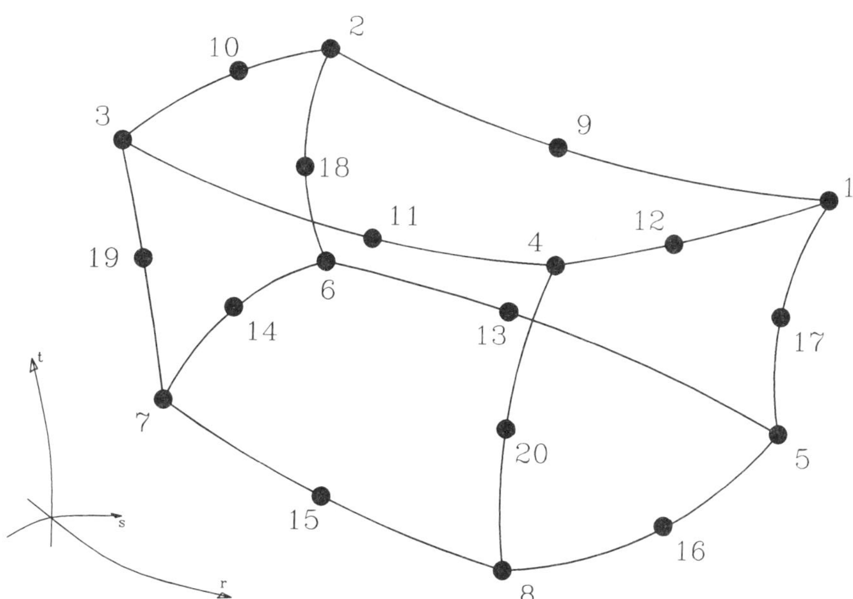

FIGURE 1 **Node numbering for 20-node cuboid.**

$$
\begin{aligned}
N_1 &= \frac{1}{8}(1 + r)(1 + s)(1 + t)(-2 + r + s + t) \\
N_2 &= \frac{1}{8}(1 - r)(1 + s)(1 + t)(-2 - r + s + t) \\
N_3 &= \frac{1}{8}(1 - r)(1 - s)(1 + t)(-2 - r - s + t) \\
N_4 &= \frac{1}{8}(1 + r)(1 - s)(1 + t)(-2 + r - s + t) \\
N_5 &= \frac{1}{8}(1 + r)(1 + s)(1 - t)(-2 + r + s - t) \\
N_6 &= \frac{1}{8}(1 - r)(1 + s)(1 - t)(-2 - r + s - t) \\
N_7 &= \frac{1}{8}(1 - r)(1 - s)(1 - t)(-2 - r - s - t) \\
N_8 &= \frac{1}{8}(1 + r)(1 - s)(1 - t)(-2 + r - s - t)
\end{aligned}
\tag{9}
$$

The shape functions for the mid-side nodes on the top and bottom faces of the element are given by

$$
\begin{aligned}
N_9 &= \frac{1}{4}(1 - r^2)(1 + s)(1 + t) \\
N_{10} &= \frac{1}{4}(1 - r)(1 - s^2)(1 + t) \\
N_{11} &= \frac{1}{4}(1 - r^2)(1 - s)(1 + t) \\
N_{12} &= \frac{1}{4}(1 + r)(1 - s^2)(1 + t) \\
N_{13} &= \frac{1}{4}(1 - r^2)(1 + s)(1 - t) \\
N_{14} &= \frac{1}{4}(1 - r)(1 - s^2)(1 - t) \\
N_{15} &= \frac{1}{4}(1 - r^2)(1 - s)(1 - t) \\
N_{16} &= \frac{1}{4}(1 + r)(1 - s^2)(1 - t)
\end{aligned}
\tag{10}
$$

Finally, the shape functions for the mid-side nodes on the edges parallel to the t direction are given by

$$
\begin{aligned}
N_{17} &= \frac{1}{4}(1 + r)(1 + s)(1 - t^2) \\
N_{18} &= \frac{1}{4}(1 - r)(1 + s)(1 - t^2) \\
N_{19} &= \frac{1}{4}(1 - r)(1 - s)(1 - t^2) \\
N_{20} &= \frac{1}{4}(1 + r)(1 - s)(1 - t^2)
\end{aligned}
\tag{11}
$$

Writing the displacements in terms of the shape functions then gives the strain-displacement matrix as

$$[B] = [\mathcal{L}]\{N\}^T\{u_i\} \tag{12}$$

Clearly $[B]$ consists of first-order derivatives of the shape functions with respect to the natural coordinates. The matrix may be written in partitioned form, with the number of partitions equal to the number of nodes used to define the element. Thus a typical partition is given by

$$
[B]_i = \begin{bmatrix}
\frac{\delta N}{\delta x} & 0 & \frac{1}{2}\left(\frac{\delta N}{\delta x}\right)^2 \\
0 & \frac{\delta N}{\delta y} & \frac{1}{2}\left(\frac{\delta N}{\delta y}\right)^2 \\
0 & 0 & \frac{\delta N}{\delta z} + \frac{1}{2}\left(\frac{\delta N}{\delta z}\right)^2 \\
\frac{\delta N}{\delta y} & \frac{\delta N}{\delta x} & \frac{\delta N}{\delta x}\frac{\delta N}{\delta y} \\
0 & \frac{\delta N}{\delta z} & \frac{\delta N}{\delta y} + \frac{\delta N}{\delta y}\frac{\delta N}{\delta z} \\
\frac{\delta N}{\delta z} & 0 & \frac{\delta N}{\delta x} + \frac{\delta N}{\delta z}\frac{\delta N}{\delta x}
\end{bmatrix}_i
\tag{13}
$$

for the ith shape function.

If deflections are assumed to be small, it is customary to introduce the von Karman small deflection assumptions [8] and ignore the second-order contributions. These imply that the derivatives of u and v with respect to x, y, and z are small.

Jacobian Transformation

Since the derivatives of the shape functions are given in terms of the parent coordinates, it is necessary to transform these into the global coordinate system in order to obtain the global element stiffness matrix. In terms of the parent rst coordinate system, the shape function derivatives in the transformed xyz-coordinate system are given by

$$
\begin{Bmatrix}
\frac{\delta N}{\delta x} \\
\frac{\delta N}{\delta y} \\
\frac{\delta N}{\delta z}
\end{Bmatrix}
= [J]^{-1}
\begin{Bmatrix}
\frac{\delta N}{\delta r} \\
\frac{\delta N}{\delta s} \\
\frac{\delta N}{\delta t}
\end{Bmatrix}
\tag{14}
$$

where $[J]$ is known as the Jacobian operator. This operator relates the transformed coordinate derivatives to the parent coordinate derivatives, and is a measure of the amount of distortion between the two coordinate systems.

Since the displacement interpolations are given by the sum of the products of the shape functions and their respective nodal displacements, the Jacobian operator is given by

$$[J] = \begin{bmatrix} \sum_{i=1}^{n}\left(\frac{\delta N}{\delta r}\right)_i x_i & \sum_{i=1}^{n}\left(\frac{\delta N}{\delta r}\right)_i y_i & \sum_{i=1}^{n}\left(\frac{\delta N}{\delta r}\right)_i z_i \\ \sum_{i=1}^{n}\left(\frac{\delta N}{\delta s}\right)_i x_i & \sum_{i=1}^{n}\left(\frac{\delta N}{\delta s}\right)_i y_i & \sum_{i=1}^{n}\left(\frac{\delta N}{\delta s}\right)_i z_i \\ \sum_{i=1}^{n}\left(\frac{\delta N}{\delta t}\right)_i x_i & \sum_{i=1}^{n}\left(\frac{\delta N}{\delta t}\right)_i y_i & \sum_{i=1}^{n}\left(\frac{\delta N}{\delta t}\right)_i z_i \end{bmatrix} \tag{15}$$

The evaluation of the $[B]$ matrix is done in the natural coordinates of the basis functions, so the integration of the element stiffness matrix extends over the parent volume. Therefore the differential volume element must be written in terms of the parent coordinates. In terms of these coordinates, the volume integration is given by

$$\int dVol = \int\int\int \det[J]\, dr\, ds\, dt \tag{16}$$

where $\det[J]$ refers to the determinant of the Jacobian matrix.

Material Property Matrix

The material property matrix $[D]$ is based upon the engineering constants of the material making up the element. These elastic constants are determined by performing a series of mechanical tests [9]. In general, these tests involve a measurement of the deformation that a material undergoes when subjected to a known force. This allows the compliance matrix $[C]$ to be readily determined.

The most general form of the compliance matrix for an anisotropic material contains 36 elastic constants [10]. Grouping these constants in terms of the technical constants gives

$$[C] = \begin{bmatrix} \frac{1}{E_{11}} & -\frac{\nu_{21}}{E_{11}} & -\frac{\nu_{31}}{E_{11}} & \frac{\eta_{12,1}}{E_{11}} & \frac{\eta_{23,1}}{E_{11}} & \frac{\eta_{31,1}}{E_{11}} \\ \frac{\nu_{12}}{E_{22}} & -\frac{1}{E_{22}} & -\frac{\nu_{32}}{E_{22}} & \frac{\eta_{12,2}}{E_{22}} & \frac{\eta_{23,2}}{E_{22}} & \frac{\eta_{31,2}}{E_{22}} \\ \frac{\nu_{13}}{E_{33}} & -\frac{\nu_{23}}{E_{33}} & -\frac{1}{E_{33}} & \frac{\eta_{12,3}}{E_{33}} & \frac{\eta_{23,3}}{E_{33}} & \frac{\eta_{31,3}}{E_{33}} \\ \frac{\eta_{1,12}}{G_{12}} & \frac{\eta_{2,12}}{G_{12}} & \frac{\eta_{3,12}}{G_{12}} & \frac{1}{G_{12}} & \frac{\mu_{23,12}}{G_{12}} & \frac{\mu_{31,12}}{G_{12}} \\ \frac{\eta_{1,12}}{G_{23}} & \frac{\eta_{2,12}}{G_{23}} & \frac{\eta_{3,12}}{G_{23}} & \frac{\mu_{12,23}}{G_{23}} & \frac{1}{G_{23}} & \frac{\mu_{31,23}}{G_{23}} \\ \frac{\eta_{1,12}}{G_{31}} & \frac{\eta_{2,12}}{G_{31}} & \frac{\eta_{3,12}}{G_{31}} & \frac{\mu_{12,31}}{G_{31}} & \frac{\mu_{23,31}}{G_{31}} & \frac{1}{G_{31}} \end{bmatrix} \tag{17}$$

where

- E_{ii} = Young's modulus in the i direction
- G_{ij} = shear modulus in the ij plane
- ν_{ij} = Poisson's ratio for strain in the j direction when stressed in the i direction
- $\eta_{ij,k}$ = coefficients of mutual influence of the first kind relating normal strain in the k direction due to a shear stress in the ij plane
- $\eta_{i,jk}$ = coefficients of mutual influence of the second kind relating shearing strain in the jk plane due to a normal stress in the i direction
- $\mu_{ij,kl}$ = Chentsov coefficient relating shearing strain in the kl plane due to a shearing stress in the ij plane

Since deformations are isothermal and limited to the elastic regime, it is possible to write a strain energy potential function. During deformation, the incremental work done per unit volume deformed is given by

$$\begin{aligned} \delta W &= \sigma_{ii}\delta\varepsilon_{ii} \\ &= C_{ij}\varepsilon_{jj}\delta\varepsilon_{ii} \end{aligned} \tag{18}$$

Integrating for all the stresses then gives the work done per unit volume as

$$W = C_{ij}\varepsilon_{jj}\varepsilon_{ii} \tag{19}$$

whereupon

$$\frac{\delta^2 W}{\delta\varepsilon_{ii}\delta\varepsilon_{jj}} = C_{ij} \tag{20}$$

Similarly, we may obtain

$$\frac{\delta^2 W}{\delta\varepsilon_{jj}\delta\varepsilon_{ii}} = C_{ji} \tag{21}$$

Since the result must be independent of the order of differentiation, we have

$$C_{ij} = C_{ji} \tag{22}$$

Therefore, only 21 of the 36 elastic constants are truly independent. Such materials are known as *triclinic* since there are no planes of material property symmetry.

If the internal composition of a material possesses symmetry of any kind, it will be observed in the elastic properties. If there are two orthogonal planes of material property symmetry, then there will be symmetry relative to a third mutally orthogonal plane. This will reduce the number of independent elastic constants to nine. Thus

$$[C] = \begin{bmatrix} \frac{1}{E_{11}} & \frac{-\nu_{21}}{E_{11}} & \frac{-\nu_{31}}{E_{11}} & 0 & 0 & 0 \\ \frac{-\nu_{12}}{E_{22}} & \frac{1}{E_{22}} & \frac{-\nu_{32}}{E_{22}} & 0 & 0 & 0 \\ \frac{-\nu_{13}}{E_{33}} & \frac{-\nu_{23}}{E_{33}} & \frac{1}{E_{33}} & 0 & 0 & 0 \\ 0 & 0 & 0 & \frac{1}{G_{12}} & 0 & 0 \\ 0 & 0 & 0 & 0 & \frac{1}{G_{23}} & 0 \\ 0 & 0 & 0 & 0 & 0 & \frac{1}{G_{31}} \end{bmatrix} \tag{23}$$

This case of material isotropy defines an orthotropic material, in which there is no interaction between the normal stresses and the shearing strains or between the shearing stresses and the normal strains. In the case of a fiber composite, there is everywhere a plane in which the mechanical properties are equal in all directions. The compliance matrix then has only five independent constants. If the 2-3 plane is the special plane of isotropy, then

$$[C] = \begin{bmatrix} \frac{1}{E_{11}} & \frac{-\nu_{21}}{E_{11}} & \frac{-\nu_{31}}{E_{11}} & 0 & 0 & 0 \\ \frac{-\nu_{21}}{E_{11}} & \frac{1}{E_{22}} & \frac{-\nu_{32}}{E_{22}} & 0 & 0 & 0 \\ \frac{-\nu_{31}}{E_{11}} & \frac{-\nu_{32}}{E_{22}} & \frac{1}{E_{22}} & 0 & 0 & 0 \\ 0 & 0 & 0 & \frac{1}{G_{12}} & 0 & 0 \\ 0 & 0 & 0 & 0 & \frac{1}{G_{23}} & 0 \\ 0 & 0 & 0 & 0 & 0 & \frac{1}{G_{12}} \end{bmatrix} \tag{24}$$

Such materials are said to be *transversely isotropic*.

The material property matrix $[D]$ relates the strains in a material to the stresses, and so is the algebraic inverse of the compliance matrix. In terms of the technical constants, the $[D]$ matrix is given by Eq. (25) at the bottom of the page,

where

$$\mathbf{D}_n = \frac{1 - \nu_{12}\nu_{21} - \nu_{23}\nu_{32} - \nu_{31}\nu_{13} - 2\nu_{13}\nu_{32}\nu_{21}}{E_{11}E_{22}E_{33}}$$

Laminae Micromechanics

Since a fiber composite is made up of two distinct materials, its mechanical behavior will be a function of the properties of the lamina constituents and their relative abundances. In the case of laminates, the behavior will also be affected by the lamination sequence. It is impractical to experimentally determine the elastic constants for all of the permutations and combinations that may be used. It is much more advantageous to mathematically derive the lamina properties on the basis of the constituent materials. The study of the detailed interaction of the constituents for the purposes of investigating the behavior of a heterogeneous material is known as *micromechanics*. The prediction of lamina properties is an essential adjunct to the design process, and several methods are available to do this [11–16].

These methods involve a basic set of simplifying assumptions consistent with the physical situation and based upon the principles of solid mechanics. The first set of these assumptions is that the lamina can be considered to be macroscopically homogeneous, linearly elastic, and generally orthotropic. Additionally, the fibers and binder are assumed to be homogeneous, linearly elastic, and free of voids. The fibers are assumed to be regularly spaced and perfectly aligned, with complete bonding at the fiber–matrix interface. The differences in the various methods arise from the degree to which each of these basic assumptions is relaxed.

In all but the mechanics of materials approach, the solution for the laminae engineering constants is given by a fairly complicated set of relationships among the constituent properties. Furthermore, variations in manufacturing will always yield variations in the laminate geometry. This makes precise prediction of the required constants a difficult task. Over the last 25 years, the mechanics of materials approach has been used to derive equations for many different properties. Though of simple form, they are scattered throughout the literature.

A unified set of simple working equations has been provided by Chamis [17] with a view towards experimental guidelines for maximum benefit and minimum test-

$$[D] = \begin{bmatrix} \frac{1-\nu_{23}\nu_{32}}{E_{22}E_{33}\mathbf{D}_n} & \frac{\nu_{21}-\nu_{31}\nu_{23}}{E_{22}E_{33}\mathbf{D}_n} & \frac{\nu_{31}-\nu_{21}\nu_{32}}{E_{11}E_{33}\mathbf{D}_n} & 0 & 0 & 0 \\ \frac{\nu_{12}-\nu_{32}\nu_{13}}{E_{11}E_{33}\mathbf{D}_n} & \frac{1-\nu_{13}\nu_{31}}{E_{11}E_{33}\mathbf{D}_n} & \frac{\nu_{32}-\nu_{12}\nu_{31}}{E_{11}E_{33}\mathbf{D}_n} & 0 & 0 & 0 \\ \frac{\nu_{13}-\nu_{12}\nu_{23}}{E_{11}E_{22}\mathbf{D}_n} & \frac{\nu_{23}-\nu_{21}\nu_{13}}{E_{11}E_{22}\mathbf{D}_n} & \frac{1-\nu_{12}\nu_{21}}{E_{11}E_{22}\mathbf{D}_n} & 0 & 0 & 0 \\ 0 & 0 & 0 & G_{12} & 0 & 0 \\ 0 & 0 & 0 & 0 & G_{23} & 0 \\ 0 & 0 & 0 & 0 & 0 & G_{31} \end{bmatrix} \tag{25}$$

ing. The ply properties defined by these equations are given with respect to the material axis orientation in terms of the properties of the constituents and their respective volume fractions. The mechanical properties are computed using

$$E_{11} = \mathbf{V}_f E_{f11} + \mathbf{V}_b E_b \qquad E_{22} = \frac{E_b}{1 - \sqrt{\mathbf{V}_f}\,[1 - (E_b/E_{f22})]}$$

$$\nu_{12} = \mathbf{V}_f \nu_{f12} + \mathbf{V}_b \nu_b \qquad G_{12} = \frac{G_b}{1 - \sqrt{\mathbf{V}_f}\,[1 - (G_b/G_{f12})]}$$

$$\nu_{23} = \frac{E_{22}}{2G_{23}} - 1 \qquad G_{23} = \frac{G_b}{1 - \sqrt{\mathbf{V}_f}\,[1 - (G_b/G_{f23})]} \tag{26}$$

where

$\mathbf{V}_b$ = volume fraction of binder
$\mathbf{V}_f$ = volume fraction of fibers
E_b = elastic modulus of binder
E_{f11} = elastic modulus of fiber in the 1 direction
E_{f22} = elastic modulus of fiber in the 2 direction
G_b = shear modulus of binder
G_{f12} = shear modulus of fiber in the 1-2 plane
G_{f23} = shear modulus of fiber in the 2-3 plane
ν_b = Poisson's ratio of the binder for strain in the 2 direction due to a stress in the 1 direction
ν_{f12} = Poisson's ratio of the fiber for strain in the 2 direction due to a stress in the 1 direction.

However, since it is easier to measure the weight fractions of the fiber and binder, the equations for determining the volume fractions from the weight fractions are also provided. These are

$$\mathbf{V}_f = \frac{1 - \mathbf{V}_v}{1 + (\rho_f/\rho_b)[(1/\lambda_f) - 1]}$$

$$\mathbf{V}_b = \frac{1 - \mathbf{V}_v}{1 + (\rho_b/\rho_f)[(1/\lambda_b) - 1]} \tag{27}$$

where

$\mathbf{V}_v$ = volume fraction of voids
λ_b = weight fraction of binder
λ_f = weight fraction of fiber
ρ_b = density of binder
ρ_f = density of fiber

These equations provide a detailed quantitative insight into the strength and stiffness behavior, and are useful in parametric studies for the evaluation of various constituent parameters. They are ideally suited for use in the evaluation of the stiffness matrix for the modified element being described here.

Lamina Orientation

Laminated fiber composite structures are usually constructed by stacking several unidirectional layers of laminae in a specified sequence of orientations with respect to a reference system of coordinates. Therefore, in order to perform an engineering analysis, the principal directions of material orthotropy must be referenced to a common geometric axis.

From elementary mechanics of materials, the transformation equations required to express the material stresses in a coordinate system inclined to the material axis is given by

$$\{\sigma\} = [T]^T[D][T]\,\{\varepsilon\} \tag{28}$$

where $[T]$ is the transformation matrix relating the strains in the ply principal directions to those in the global reference axis. For the three-dimensional case, the transformation matrix is given by a fourth-order tensor transformation in terms of the directions cosines of the unit vectors in the respective coordinate systems [18]. See Eq. (29) at the bottom of the page,

where

$$\begin{array}{lll} l_1 = \cos(e_x, e_1) & m_1 = \cos(e_y, e_1) & n_1 = \cos(e_z, e_1) \\ l_2 = \cos(e_x, e_2) & m_2 = \cos(e_y, e_2) & n_2 = \cos(e_z, e_2) \\ l_3 = \cos(e_x, e_3) & m_3 = \cos(e_y, e_3) & n_3 = \cos(e_z, e_3) \end{array}$$

If we use tensorial shear strains rather than engineering shear strains, the transformation equations for strain are similarly given by

$$\{\varepsilon\}_{xyz} = [T]\{\varepsilon\}_{123} \tag{30}$$

However, it is usually more convenient to use engineering shear strains. This requires premultiplication of both strain vectors by the inverse of Reuter's transformation matrix [19]. The elements of Reuter's transformation ma-

$$[T] = \begin{bmatrix} l_1l_1 & m_1m_1 & n_1n_1 & 2l_1m_1 & 2m_1n_1 & 2n_1l_1 \\ l_2l_2 & m_2m_2 & n_2n_2 & 2l_2m_2 & 2m_2n_2 & 2n_2l_2 \\ l_3l_3 & m_3m_3 & n_3n_3 & 2l_3m_3 & 2m_3n_3 & 2n_3l_3 \\ l_1l_2 & m_1m_2 & n_1n_2 & l_1m_2 + l_2m_1 & m_1n_2 + m_2n_1 & n_1l_2 + n_2l_1 \\ l_2l_3 & m_2m_3 & n_2n_3 & l_2m_3 + l_3m_2 & m_2n_3 + m_3n_2 & n_2l_3 + n_3l_2 \\ l_3l_1 & m_3m_1 & n_3n_1 & l_3m_1 + l_1m_3 & m_3n_1 + m_1n_3 & n_3l_1 + n_1l_3 \end{bmatrix} \tag{29}$$

trix are given by

$$R_{ij} = \begin{Bmatrix} 1 \text{ for } i = j, \ i \in \{1,2,3\} \\ 2 \text{ for } i = j, \ i \in \{3,4,5\} \\ 0 \text{ for } i \neq j \end{Bmatrix} \tag{31}$$

Substitution of the above into the stress-strain relations then gives

$$\{\sigma\}_{xyz} = [T][D]_{123}[R][T]^{-1}[R]^{-1}\{\varepsilon\}_{xyz} \tag{32}$$

It is readily verified that $[R][T]^{-1}[R]^{-1}$ is equal to the transpose of $[T]$, so the inversion implied by the above equation is not required. The material property matrix transformed to the global directions is then given by

$$[D]_{xyz} = [T][D]_{123}[T]^T \tag{33}$$

Since the laminae rotations are confined to the xy plane, the directions cosines are given by

$$\begin{matrix} l_1 = +\cos\theta & m_1 = -\sin\theta & n_1 = 0 \\ l_2 = +\sin\theta & m_2 = +\cos\theta & n_2 = 0 \\ l_3 = 0 & m_3 = 0 & n_3 = 0 \end{matrix} \tag{34}$$

so the transformation matrix simplifies to

$$[T] = \begin{bmatrix} \cos^2\theta & \sin^2\theta & 0 & -2\cos\theta\sin\theta & 0 & 0 \\ \sin^2\theta & \cos^2\theta & 0 & 2\cos\theta\sin\theta & 0 & 0 \\ 0 & 0 & 1 & 0 & 0 & 0 \\ \cos\theta\sin\theta & -\cos\theta\sin\theta & 0 & \cos^2\theta - \sin^2\theta & 0 & 0 \\ 0 & 0 & 0 & 0 & \cos\theta & \sin\theta \\ 0 & 0 & 0 & 0 & -\sin\theta & \cos\theta \end{bmatrix}$$

where θ is the counterclockwise in-plane rotation angle from the global reference axis to the material reference axis. In terms of the double angle formulas.

$$[T] = \begin{bmatrix} \cos^2\theta & \sin^2\theta & 0 & -\sin 2\theta & 0 & 0 \\ \sin^2\theta & \cos^2\theta & 0 & \sin 2\theta & 0 & 0 \\ 0 & 0 & 1 & 0 & 0 & 0 \\ \cos\theta\sin\theta & -\cos\theta\sin\theta & 0 & \cos 2\theta & 0 & 0 \\ 0 & 0 & 0 & 0 & \cos\theta & \sin\theta \\ 0 & 0 & 0 & 0 & -\sin\theta & \cos\theta \end{bmatrix} \tag{35}$$

Numerical Integration

The evaluation of the element stiffness matrix involves integration of a function over the domain represented by the element. From the above discussion, this integral is given by

$$[K] = \int\int\int [B]^T[T]^T[D][T][B] \det[J]\, dr\, ds\, dt \tag{36}$$

An explicit evaluation of this integral is generally impractical. Exact solutions of this equation are obtainable in the parent coordinate system for only the simplest of these expressions. In general, cases may arise in which such closed-form integration is not possible. In practice, the integrals are evaluated numerically using Gauss quadrature [20], the integration order used generally being dependent upon the degree of the interpolating polynomials used for the element shape functions and the particular matrix to be constructed.

This Gauss technique substitutes the integration by a summation over the domain of the product of a substitute function evaluated at discrete points within the doman and a weighting factor. The number of points at which the function is evaluated is not necessarily the same in all three dimensions of the domain. If W_i represents the weighting factor for a particular point, then the stiffness matrix is evaluated as a triple summation over the domain given by

$$[K] = \sum_{t=1}^{gpt}\sum_{s=1}^{gps}\sum_{r=1}^{gpr} (\mathscr{F}_{(r,s,t)})\, W_r W_s W_t \tag{37}$$

where

gpr = number of integration points in the r direction

gps = number of integration points in the s direction

gpt = number of integration points in the t direction
$\mathcal{F}_{(r,s,t)} = [B]^T[T]^T[D][T][B] \det[J]$

The number of operations required is equal to N^d, where N is the integration order and d corresponds to the number of dimensions of the element. It is therefore essential to choose as small a value of N as is practical. Zienkiewicz [21] suggests that the minimum requirement is that which would integrate the determinant of the Jacobian operator accurately.

The stiffness matrix for each element of the domain is easily evaluated using this technique by substitution of the appropriate matrices into the above equation.

Composite Elements

For homogeneous isotropic materials, the $[D]$ matrix is a constant throughout the element volume. Therefore, the evaluation of this matrix is usually performed only once. In the analysis of fiber composite laminates, the variation in the material properties through the element domain is accounted for by assigning a different $[D]$ matrix for each individual element. Theoretically one can model fiber composite structures by breaking up each ply into the usual finite element mesh, with the thickness of each element representing the thickness of each individual ply.

Practically, this leads to numerical difficulties because of the limitation in the element aspect ratio. The limitation springs from the resulting relative magnitudes of the terms in the element stiffness matrix. Large differences in these magnitudes create problems in accuracy resulting from ill-conditioning [22]. The magnitudes of these differences are greatly dependent upon the aspect ratios of the elements used to model the problem.

For the sake of argument, we may set a maximum value of the aspect ratio at 1:20. Since the usual thickness of a fiber composite ply is more or less 0.013 cm, this limits the physical dimensions of an element with square in-plane sides to a length of no more than 0.254 cm on a side. A typical problem such as a square plate 30 cm on a side with 80 plies through the thickness would thus require well over a million elements for a moderately accurate solution. If an eight-node cuboid element is employed in the modeling, the problem would require almost $3\frac{1}{2}$ million degrees of freedom. This large number of degrees of freedom results in an excessive computational demand, and is generally not considered to be economically justifiable.

One obvious way to overcome this difficulty is to use the method of substructuring to reduce the total number of degrees of freedom [23]. The global stiffness matrix is first partitioned into a part representing an external boundary region, another part representing a region within this boundary, and two other parts representing the connection between them. The internal degrees of freedom are then condensed out, leaving behind a representation of the model in terms of its overall stiffness and equivalent loads. The condensed set of equations is then solved for the external degrees of freedom, and those results are used to obtain the solution in the interior of the structure.

This procedure is identical to performing Gauss elimination on the internal degrees of freedom and results in a large savings in computation if repeated solutions are required for the same structure with different boundary loadings. The disadvantage is that the unknown function used in the derivation of the finite element equilibrium equations is now represented by a series of piecewise approximations inside the region of interest.

In the case of laminated fiber composite analysis, unless the deflections are known to lie within the elastic limit of the material, it is necessary to evaluate the stresses and strains in the interior region to determine whether or not failure has occurred. This requires the solution of the displacements in the interior of the structure. Since such an a priori knowledge cannot be assumed, this method offers no advantages.

An alternative approach is to represent several plies of the structure within one element, as is done in classical laminated plate analysis. If, for example, 10 plies were represented through the element thickness, the above-mentioned problem would require a little over a thousand elements and about 4500 degrees of freedom. Though still large, this is a reduction of over 99% in the number of elements and nodes. For the case of a 20-node cuboid element, the displacement would be represented by a single quadratic function rather than by several quadratics through the thickness.

This approach requires that the material property matrix not be treated as a constant throughout the element volume, since the material properties of any two given plies need not be the same. The variation of the material properties within the element need not be caused only by the differences in the ply constituents, their relative volume fractions, and the ply orientations with respect to the global reference axis, but may also be caused by local material failures. Since these are all parameters in the calculation of the lamina properties, the $[D]$ matrix must be computed at each integration point within any one given ply.

Because of the nonconstant nature of the material property matrix throughout the element volume, it is no longer possible to perform the volume integration using a minimum number of Gauss points. Rather, the integration must be performed using several points through the thickness. The procedure adopted is to divide the element thickness into a number of sections equal to the number of plies through the element thickness, with the ply boundaries delineating each section.

The general expression for the element stiffness matrix then becomes

$$[K] = \sum_{n=1}^{p} \left(\sum_{t=1}^{gpt} \sum_{s=1}^{gps} \sum_{r=1}^{gpr} (\mathcal{F}_{(r,s,t)}) W_r W_s W_t \right)_n \tag{38}$$

where p is the number of plies through the element thickness. The stiffness matrix is evaluated over each ply and the results summed over the element volume. Since Gauss quadrature is used through the ply thickness, the

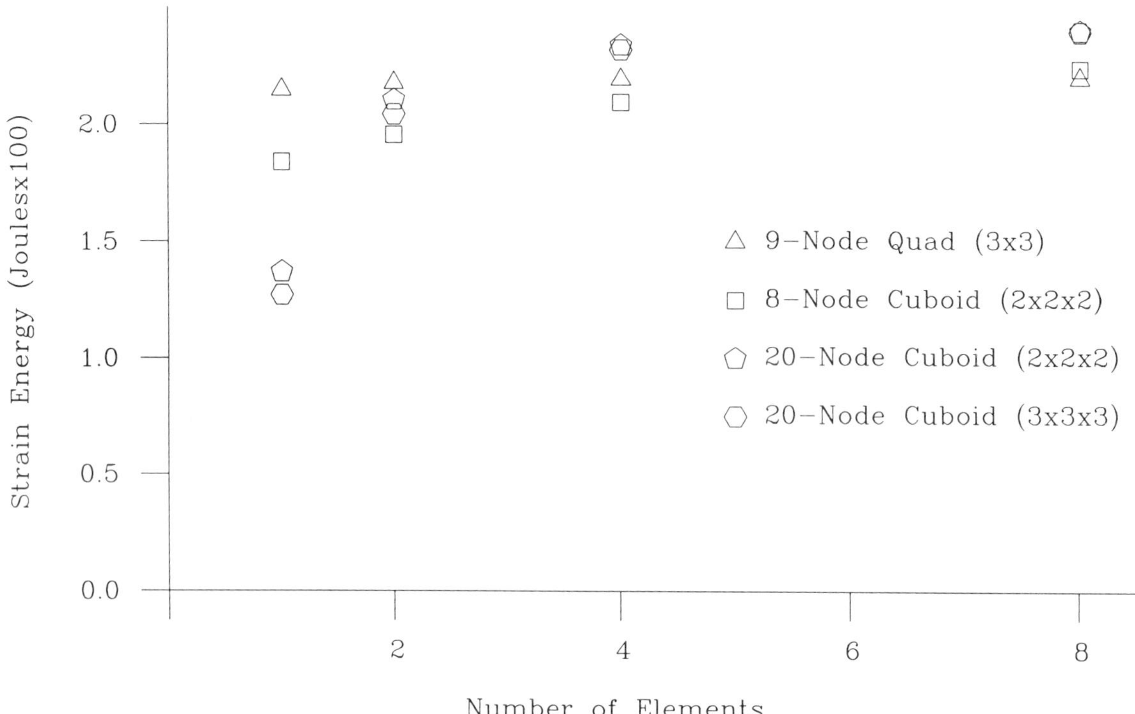

FIGURE 2 **Convergence test results.**

integration points and weights must be corrected for the nonstandard limits. If the upper and lower limits of integration are denoted by L_u and L_1, the new sampling point S_p and weight S_w are given by

$$S_p = \frac{L_1 - L_u}{2} + \frac{L_u - L_1}{2} S_p^o \qquad S_w = \frac{L_u - L_1}{2} S_w^o \quad (39)$$

where

S_p = sampling point based on interval of −1 to 1
S_w = sampling weight based on interval of −1 to 1

The values of the sampling points and weights based upon the parent interval of −1 to 1 have been computed and published in the open literature [24].

Concluding Remarks

The formulation described in this article was tested using a nine-node quadrilateral with 3 × 3 integration and an eight-node cuboid with a ply integration point distribution in the *rst* directions of 2 × 2 × 2. The 20-node cuboid was also tested with 2 × 2 × 2 and 3 × 3 × 3 integration. The problem solved was that of a bimetallic strip built in at one end, with a uniformly distributed load applied at the free end. The strip was 20.32 cm in length by 2.54 cm in width, and composed of two 1.27 cm thick layers of aluminum and steel. The applied load intensity was 6.89 MPa. All elements indicated good convergence (Fig. 2).

It was not too long ago that the computing power available on today's desktops was beyond the reach of the average engineer. The advent of the microcomputer revolution has made commonplace the application of sophisticated numerical analysis techniques to problems not amenable to solution by any other method. Finite element analysis is a case in point, where problems involving complex geometries are now tackled as a matter of course. Unfortunately, the models generally in use are based upon continuum mechanics, and are not properly suited to applications involving fiber composite materials. The composite element presented here is a step in the direction of improving that situation.

S.A. Umar-Khitab

References

1. S. P. Timoshenko and J. N. Goodier, *Theory of Elasticity*, 3d ed., McGraw-Hill, New York, 1970, pp. 90–95.
2. A. B. Basset, *Phil. Trans. Roy. Soc. (London)*, *12*, 433 (1890).
3. E. Reissner and Y. Stavsky, *J. Appl. Mech.*, *28*, 402 (1961).
4. J. M. Whitney and N. J. Pagano, *J. Appl. Mech.*, *37*, 1031 (1970).
5. R. B. Nelson and D. R. Lorch, *J. Appl. Mech.*, *41*, 177 (1974).
6. K. H. Lo, R. M. Christensen, and E. M. Wu, *J. Appl Mech.*, *44*, 633 (1977).

7. R. H. Gallagher, *Finite Element Analysis Fundamentals*, Prentice-Hall, Englewood Cliffs, NJ, 1975, pp. 139–143.

8. Y. C. Fung, *Foundations of Solid Mechanics*, Prentice-Hall, Englewood Cliffs, NJ, 1965, p. 457.

9. Y. M. Tarnoploskii and T. Kincis, *Static Test Methods for Composites*, G. Lubin, Trans., Van Nostrand Reinhold, New York, 1985, pp. 1–281.

10. S. G. Lekhnitskii in J. J. Brandstatter, Ed., *Theory of Elasticity of an Anisotropic Body*, P. Fern, Trans., Holden Day Inc., San Francisco, 1963, pp. 14–15.

11. Z. Hashin and B. W. Rosen, *J. Appl. Mech., 31*, 223 (1964).

12. T. T. Wu, *J. Appl. Mech., 32*, 211 (1965).

13. R. Hill, *J. Mech. Phys. Solids, 13*, 189 (1965).

14. H. Frolich and B. Sack, *Phil. Trans. Roy. Soc.* (*London*), *A185*, 415 (1946).

15. J. M. Whitney and M. B. Riley, *AIAA J., 4*, 1537 (1966).

16. L. R. Hermann and K. S. Pister, *Proceedings of the ASME Annual General Meeting*, PN 63 WA-239, 1963.

17. C. C. Chamis, *SAMPE Q., 15*(3), 14 (1984).

18. K. J. Bath, *Finite Element Procedures in Engineering Analysis*, Prentice-Hall, Englewood Cliffs, NJ, 1982, pp. 258–259.

19. R. C. Reuter, *J. Compos. Mater., 5*, 270 (1971).

20. H. Grandin, Jr., *Fundamentals of the Finite Element Method*, Macmillan, New York, 1986, pp. 189–194.

21. O. C. Zienkiewicz, *The Finite Element Method*, McGraw-Hill, New York, 1977, pp. 201–204.

22. S. D. Conte and C. deBoor, *Elementary Numerical Analysis*, McGraw-Hill, New York, 1965, pp. 150–157.

23. B. Irons and S. Ahmad, *Techniques of Finite Elements*, Wiley, New York, 1986, pp. 185–186.

24. A. N. Loxan, N. Davids, and A. Levenson, *Bull. Amer. Math. Soc., 48*, 739 (1942).

Laminates, Ceramic

The demand for components that can withstand higher temperatures and corrosive environments can often be met with monolithic ceramics such as alumina, silicon carbide, and mullite. However, the low fracture toughness of monolithic ceramics relative to that of metals is a major impediment in applications where reliability is of primary importance. The low fracture toughness of these ceramics can be significantly diminished either by inducing surface compressive stresses [1] or by introducing reinforcing second phases, such as particles, continuous ceramic fibers, or discontinuous ceramic whiskers [2,3]. The concept of laminated ceramic composites allows both strengthening mechanisms to be simultaneously incorporated into a structure.

Surface compressive stresses have been induced in ceramics by surface grinding of transformation-toughened monolithic ceramics such as ZrO_2 [4], quenching from high temperatures [5], coating with low expansion glazes [6], and applying low expansion layers by diffusion treatments [7]. These techniques have inherent limitations on the depth, magnitude, and form of the compressive layers [8].

The use of continuous fibers has allowed some degree of control of the orientation and placement of fibers in the ceramic matrix, therefore permitting tailoring of the strength, elastic, and thermoelastic properties. However, continuous-fiber-reinforced ceramics have been limited to relatively low melting matrices, such as glass and glass–ceramics [9,10]. For oxide matrix ceramics, silicon carbide is a commonly used reinforcement material [11]. Continuous SiC fibers are derived from polymers and contain both oxygen and excess carbon, which limits their thermal stability and ability to be processed [12]. The whisker form of SiC, on the other hand, is more stable than the fiber form. Although whiskers can be introduced easily into the matrix, the existing processing techniques for whisker-reinforced ceramics permit very little control over the orientation and placement of the whiskers [13]. The application of laminate processing techniques to the fabrication of whisker-reinforced composites circumvents these limitations and thus expands the design possibilities for whisker-reinforced ceramic composites.

The concept of a laminated composite is used effectively in the design of polymer matrix composites to achieve a high degree of tailoring of strength, elastic, and thermoelastic properties. Polymer matrix composites reinforced with continuous fibers are designed and fabricated by controlling the stacking sequence of laminae with specific characteristics and orientations. Tape casting techniques, which are commonly used in the ceramics industry, can be adapted to produce tape-cast, whisker-reinforced composite layers in a manner analogous to the prepreg laminae used in manufacturing polymer composites. These tape-cast laminae can be arranged layer by layer to achieve a wide range of mechanical and chemical properties that are unattainable by conventional methods of fabrication. In this article, we first discuss some of the property tailoring and design features of laminated ceramic matrix composites as well as some of the issues involved in their processing, then we demonstrate the applicability of composite laminate theory to the strengthening of SiC_w (silicon carbide whisker)–Al_2O_3 laminates.

Design Concepts

The variety of potential material configurations, and hence mechanical properties, for ceramic laminates is very great. As with polymer composites, the properties of ceramic composites may be calculated from the constituent properties. The procedure for material design is illustrated in Figure 1. Starting with the elastic modulus of the fibers and matrix, E_f and E_m, respectively, the orientation factor f, the volume fraction of the fiber V_f, the fiber aspect ratio l/d, and the coefficients of thermal expansion for the fibers and matrix, α_f and α_m, the elastic properties of the lamina can be calculated using theoret-

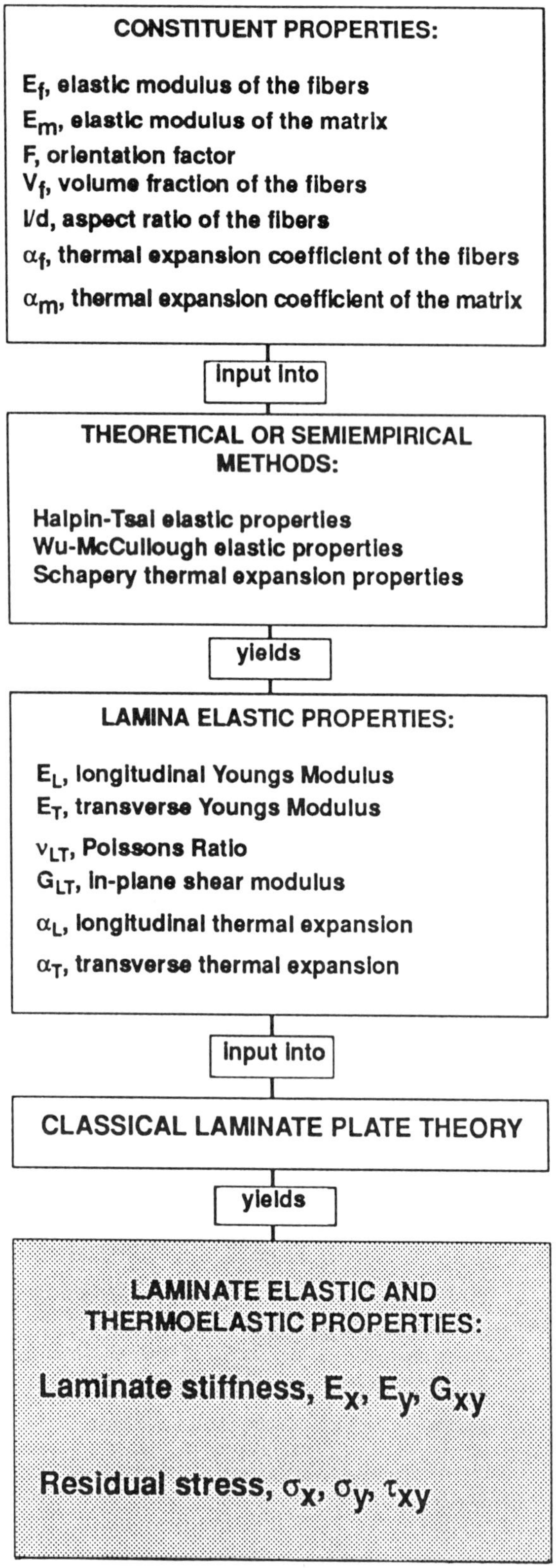

CONSTITUENT PROPERTIES:

E_f, elastic modulus of the fibers
E_m, elastic modulus of the matrix
F, orientation factor
V_f, volume fraction of the fibers
l/d, aspect ratio of the fibers
α_f, thermal expansion coefficient of the fibers
α_m, thermal expansion coefficient of the matrix

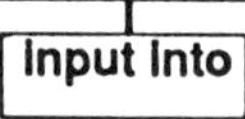

THEORETICAL OR SEMIEMPIRICAL METHODS:

Halpin-Tsai elastic properties
Wu-McCullough elastic properties
Schapery thermal expansion properties

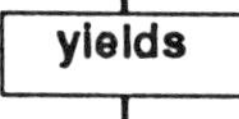

LAMINA ELASTIC PROPERTIES:

E_L, longitudinal Youngs Modulus
E_T, transverse Youngs Modulus
ν_{LT}, Poissons Ratio
G_{LT}, in-plane shear modulus
α_L, longitudinal thermal expansion
α_T, transverse thermal expansion

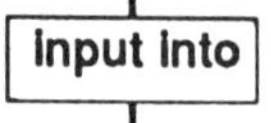

CLASSICAL LAMINATE PLATE THEORY

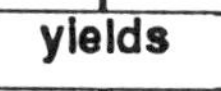

LAMINATE ELASTIC AND THERMOELASTIC PROPERTIES:

Laminate stiffness, E_x, E_y, G_{xy}

Residual stress, σ_x, σ_y, τ_{xy}

FIGURE 1 Design procedure for laminated ceramic composites.

ical and semiempirical methods [14–17]. By selection of the sequence of ply orientations and compositions, various elastic, thermoelastic, strength, physical, and chemical characteristics can be produced. Many, but not all, of these properties may be accurately predetermined. The use of classical laminate plate theory [18–19] has been found to accurately predict the elastic and thermoelastic properties of the ceramic laminate. The strength properties, on the other hand, cannot be readily determined by commonly used laminate failure criteria, since fracture of these laminates is still strongly controlled by the presence of flaws. Modified laminate fracture theories must be employed.

Ceramics can be strengthened by the introduction of surface compressive stresses [20–21]. Virkar et al. [20] fabricated three-layer composites by dry-pressing powders to yield surface layers consisting of unstabilized ZrO_2 and an oxide (e.g., MgO or Al_2O_3), with a core composition of the matrix oxide and stabilized ZrO_2. On cooling, the unstabilized ZrO_2 transformed to monoclinic ZrO_2, and the outer surface expanded in dimension relative to the core. In this manner, they were able to decrease the fracture sensitivity to large surface flaws as long as the surface layer was thicker than the flaw size.

Examples of material designs that can make use of laminated-composite concepts to improve performance are illustrated in Figure 2. The magnitude of the surface compressive stress can be calculated from laminate theory. Figure 2*a* shows a laminate design as an alternative and more general approach to inducing a surface compressive stress; in this design, the layers toward the midplane successively increase in coefficient of thermal expansion. This design can be effected by loading the outer layers with increasing amounts of low expansion materials so that compressive residual stresses will develop in the surface layers as a result of the differential contraction of the core and outer laminae during cooling after the high temperature densification process. A major advantage of laminated-composite processing is that it provides the engineering flexibility to use myriad material and property combinations that would be impossible with traditional methods involving thermal or chemical tempering. This technique also allows the use of non-equilibrium compositions for a greater degree of stress profile variation. For instance, the depth and magnitude of the stress gradient can be independently controlled by proper selection of lamina composition and properties. Maximizing the stress gradient by introducing a high expansion material in the interior of the composite would be impossible by conventional chemical tempering but is quite feasible by lamination.

Strengthening can also be achieved by rendering surface flaws ineffective through the introduction of a tougher ceramic layer below the surface (Fig. 2*b*). This design mitigates surface damage in the outer layers by blunting the cracks when they reach the underlying toughened layer. This layer may contain either whiskers, a toughened ceramic, or metallic particles. The use of a toughened ceramic layer as the outer layer would not be as effective, since abrasion or impact could produce flaws through its entire depth, permitting the crack to

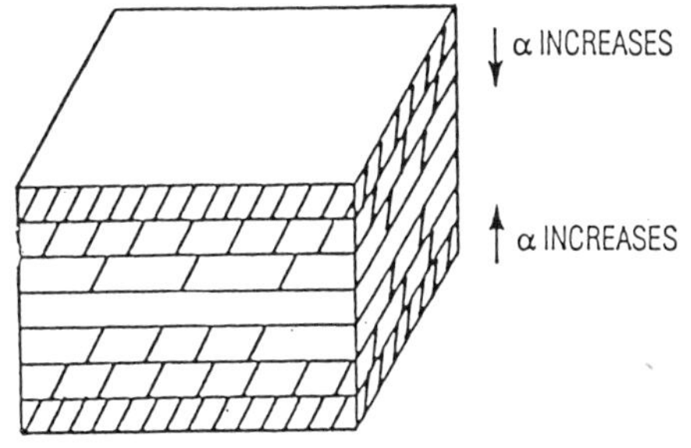

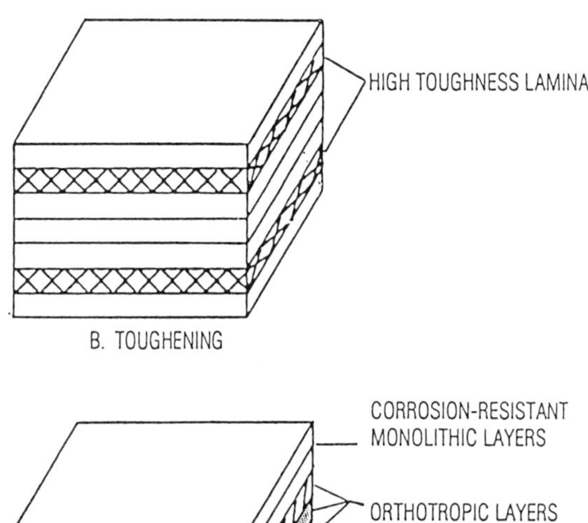

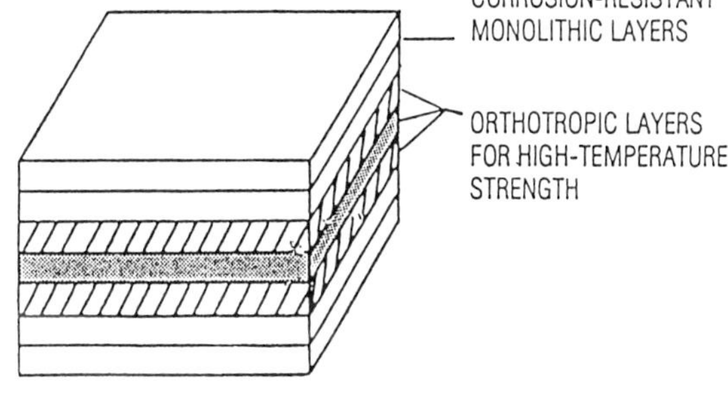

FIGURE 2 **Tailoring concepts in ceramic composite laminates.**

propagate through the lower toughness interior layers with minimum resistance.

In addition to increased strength and toughness, high temperature corrosion resistance can be designed into a composite material by using a corrosion-resistant layer on the exterior surface (Fig. 2*c*) and layers tailored for high temperature strength in the interior. A similar concept may be employed for a material designed as a high temperature heat exchanger by grading the interior layers for high thermal conductivity. Using composite laminate theory, a materials designer can tailor the grading to minimize the deleterious residual tensile stresses that are likely in such a construction.

Applications

Laminated ceramic composites can be effectively used in a number of applications for both high and low temperature environments. Heat exchangers and heat engine components are two of the more important high temperature applications for this concept. Room temperature brittleness and high temperature creep are significant limitations to ceramic structures. Laminated ceramic composite construction can provide effective solutions to these problems. Components often consist of complex geometric forms that are not easily manufactured by normal ceramic composite fabrication processes. Even simple tubular forms can be a problem if densification techniques other than sintering are required. Laminated forms built up from flexible tapes can be conveniently wrapped, shaped, and formed on mandrels similar to these used in the polymer composite industry.

Lamination techniques are also quite effective in developing special structural characteristics, such as controlled porosity. This technique could provide interior air or gas passages for cooling, heating, or fluid transport within the structure. Layer-by-layer fabrication can be used to develop the precise geometric control required for such components.

This technique has great potential for ballistic armor applications. Ceramic materials make effective armor because of their light weight, high hardness, high stiffness, and high compressive strength. Low fracture toughness, brittleness, high cost, and sensitivity to surface flaws currently limit their use. The ability to impart both intrinsic toughening and high compressive prestress on both the impact and back surfaces is a powerful material tailoring technique not available with conventional ceramic composite processing.

Ceramic composites are currently being used in cutting tool bits because of their hardness and high temperature strength. These tool bits are very expensive and still susceptible to impact damage. Laminated composite designs would permit economical use of the high cost reinforcement phases as well as improving the fracture strength and impact resistance.

Processing

Laminated multilayer ceramic structures, such as capacitors and multilevel chip carriers, are routinely fabricated by tape casting. Individual tapes of 100–500 μm in thickness and up to 1 m in width are deposited continuously with a doctor blade onto a carrier substrate, from which they can be separated after drying. The tapes are cast from a formulated slurry composed of a liquid, ceramic powder, organic polymer, plasticizer, and dispersant. These additives provide control over rheology during casting, flexibility during handling and shaping of the dried tape, and plasticity of the tape during consolidation and bonding.

Laminates with a wide variety of architectures and compositional ranges can be produced by stacking the individual layers and then consolidating them. For more complex shapes, the flexible tapes can be wrapped, shaped, and formed on mandrels similar to those used in the polymer composite industry. Because as much as 20 vol % of the unfired laminate is composed of organic additives, organic thermolysis at relatively low temperature (<500°C) is a critical process step for the production of defect-free laminates. Although tape casting provides maximum process flexibility, alternative forming processes such as coextrusion, roll compaction, and sequential colloidal deposition may have advantages for specific laminate structures.

For whisker-reinforced composites, tape casting provides more control over whisker alignment than do alternative forming processes because the rheological conditions during tape casting can enhance whisker alignment in the casting direction. In recent studies of tape-cast and laminated SiC_w–mullite composites, we have determined from x-ray diffraction of cast tapes that, on average, only 30% of the SiC whiskers are aligned in the casting direction [22]. The degree of alignment is limited by the broad distribution of aspect ratios in commercial SiC whiskers. If uniform whisker lengths are used, it should be possible to tailor whisker-reinforced composite structures with a level of control that is now possible only in continuous-fiber-reinforced composites.

Processing of fiber-reinforced composites has received considerable attention because continuous and staple fibers inhibit catastrophic failure better than particles and whiskers in reinforced composites. Well-established fiber winding techniques can be used with composite tapes to tailor fiber-reinforced laminated structures with a range of compositional variation. To date, two processing methods for fiber-reinforced composites predominate: infiltration, which is preferred with glass or glass–ceramic matrices because these materials are able to infiltrate fiber tows during elevated-temperature consolidation, and chemical vapor infiltration (CVI), which is used to deposit nonoxide matrices such as SiC into the fiber tows. Unfortunately, there are numerous problems associated with CVI, including extremely long processing times (for example, several days are required to infiltrate a 3 cm thick section) and difficulties in achieving uniform composition and full density.

Sol-gel processing with either polymeric or colloidal precursors offers opportunities for two infiltration with oxide and nonoxide compositions that are unattainable by either viscous infiltration or CVI. For example, Hindman et al. [23] examined infiltration of alumina and aluminoborosilicate fiber tows with alumina precursor sols. Fiber-reinforced laminated structures could be fabricated by the continuous winding of sol-gel-infiltrated fiber tows to yield a wide variety of composite structures and compositional gradations.

While a range of shapes and laminate architectures is possible with either tape or fiber winding, the presence of reinforcing phases introduces a constraining effect on densification. Laminate structures with different compositions but containing no second phases in the individual laminae can densify at conventional temperatures as long as there is little difference in shrinkage of the individual laminae. However, the presence of as small as 15 vol % of whiskers can significantly reduce densification under conditions that normally allow the matrix to densify fully. At higher volume fractions of whiskers, densification is almost completely retarded. Consequently, whisker- and fiber-reinforced composites must be densified by applying pressure during sintering. For simple shapes, uniaxial hot pressing at up to 20 MPa is sufficient, but complex shapes require hot isostatic pressing (HIP). Hot isostatic pressing increases the process complexity by requiring an impermeable surface around the shaped part before it can be pressed; however, the recent development of a glass encapsulation technology for hot isostatic pressing by ASEA, Inc., promises to reduce the complexity of this processing step [24]. Alternatively, if the composite can be sintered to 95% relative density, at which density the remaining porosity is no longer connected to the surface, then the component can be further densified by hot isostatic pressing without encapsulation.

Thermoelastic Tailoring and Mechanical Properties

In this section, we demonstrate how laminate theory can be used to tailor SiC_w–Al_2O_3 composites for high strength. The elastic moduli and coefficients of thermal expansion of laminae containing different volume fractions of silicon carbide whiskers were obtained from tape-cast and hot-pressed samples [25].

Differences in elastic moduli and coefficients of thermal expansion for laminae containing different volume fractions of silicon carbide whiskers have been used to generate favorable residual stress patterns in fabricated laminates. The thermal stresses σ_x^T, σ_y^T, and τ_{xy}^T in each layer of the laminate at any position through the thickness z, measured from the midplane, caused by the restraint of the neighboring layer can be determined by Hooke's law,

$$\begin{Bmatrix} \sigma_x^T \\ \sigma_y^T \\ \tau_{xy}^T \end{Bmatrix} = \begin{vmatrix} \bar{Q}_{11} & \bar{Q}_{12} & \bar{Q}_{16} \\ \bar{Q}_{12} & \bar{Q}_{22} & \bar{Q}_{26} \\ \bar{Q}_{16} & \bar{Q}_{26} & \bar{Q}_{66} \end{vmatrix} \begin{Bmatrix} \varepsilon_x^o + zK_x \\ \varepsilon_y^o + zK_y \\ \gamma_{xy}^o + zK_{xy} \end{Bmatrix}$$

where ε_x^o, ε_y^o, and γ_{xy}^o are the mid-ply strains in the arbitrary directions of interest and K_x, K_y, and K_{xy} are the curvatures for the laminate. Q_{ij} are the transformed reduced stiffness coefficients for each layer in the laminate.

A symmetric cross-ply laminate is the most practical design for many applications. In this case, the constitutive equation for midplane laminate strains reduces to

$$\begin{Bmatrix} \varepsilon_x^o \\ \varepsilon_y^o \\ \gamma_{xy}^o \end{Bmatrix} = \begin{bmatrix} A'_{11} & A'_{12} & A'_{16} \\ A'_{12} & A'_{22} & A'_{26} \\ A'_{16} & A'_{26} & A'_{66} \end{bmatrix} \begin{Bmatrix} N_x^T \\ N_y^T \\ N_{xy}^T \end{Bmatrix}$$

where A'_{ij} are the in-plane extension flexibility coefficients and N_x^T, N_y^T, and N_{xy}^T are thermal forces generated in the laminate as a result of cooling from the temperature where stress relieving is insignificant, ΔT. For ceramic composites, this may be very close to the hot-pressing temperature. For a design in which all of the whiskers are aligned in one direction, the thermal forces are

$$N_x^{\mathrm{T}} = \sum_{k=1}^{n} (Q_{11}\alpha_1 + Q_{12}\alpha_2) t_k \, \Delta T$$

$$N_y^{\mathrm{T}} = \sum_{k=1}^{n} (Q_{12}\alpha_1 + Q_{22}\alpha_2) t_k \, \Delta T$$

$$N_{xy}^{\mathrm{T}} = 0$$

where t_k is the thickness of the kth layer, Q_{ij} are the untransformed stiffness coefficients, and α_i are the coefficients of the thermal expansion in the principal material directions. The thermal moments M_x, M_y, and M_{xy} are zero. The residual stresses are

$$\sigma_L = Q_{11}(A'_{11}N_x^{\mathrm{T}} + A'_{12}N_y^{\mathrm{T}}) + Q_{12}(A'_{12}N_x^{\mathrm{T}} + A'_{22}N_y^{\mathrm{T}})$$

$$\sigma_T = Q_{12}(A'_{11}N_x^{\mathrm{T}} + A'_{12}N_y^{\mathrm{T}}) + Q_{22}(A'_{12}N_x^{\mathrm{T}} + A'_{22}N_y^{\mathrm{T}})$$

$$\tau_{LT} = 0$$

Using this method, the residual stress patterns for three laminate designs are calculated and presented in Figure 3. Design A, Figure 4*a*, illustrates the effect of a large tensile stress in the core of the laminate. The case layers contain 20 vol % whiskers, while the core is completely unreinforced. The high tensile stress in the core, Figure 3*a*, is generated by the large difference in thermal expansion between the core ($\alpha = 7.7 \times 10^{-6}$ °C^{-1}) and the outside layers ($\alpha_L = 7.18 \times 10^{-6}$ °C^{-1} and $\alpha_T = 7.07 \times 10^{-6}$ °C^{-1}). The thickness as well as the coefficient of thermal expansion difference contributes to the large tensile stress. A micrograph of the as-fabricated laminate cross section is also shown in Figure 4*a*. The high residual tensile stress in the core of the laminate results in cracks that are regularly spaced at ~ 1 mm intervals and are normal to the plane of the laminate. These cracks close just after they enter the surface layer, which is in a state of compressive stress.

To reduce the tensile stress in the core, design B (Fig. 4*b*) uses a reduced thickness for the exterior layers. This design modification results in a calculated increase in the compressive stress in the exterior layers from 100 to 300 MPa (Fig. 3*b*) and a decrease in the tensile stress from ~ 300 to < 200 MPa. Because of these changes, the crack frequency is significantly reduced, and the crack spacing is over 3 mm. Furthermore, the cracks terminate quickly when they enter the strong compressive field.

Design C (Fig. 4*c*) utilizes layers containing 10 vol % of whiskers ($\alpha_T = 7.58 \times 10^{-6}$ °C^{-1} and $\alpha_L = 7.48 \times 10^{-6}$ °C^{-1}) in the core to reduce the thermal expansion difference between the core and the surface layers. In addition, the tensile stress in the core is minimized by reducing the thickness of the exterior layers. The tensile stress is reduced to ~ 70 MPa, and the surface compressive stress is 250 MPa (Fig. 3*c*). The favorable state of stress in this design has completely eliminated the cracking in the core as a result of differential thermal expansion.

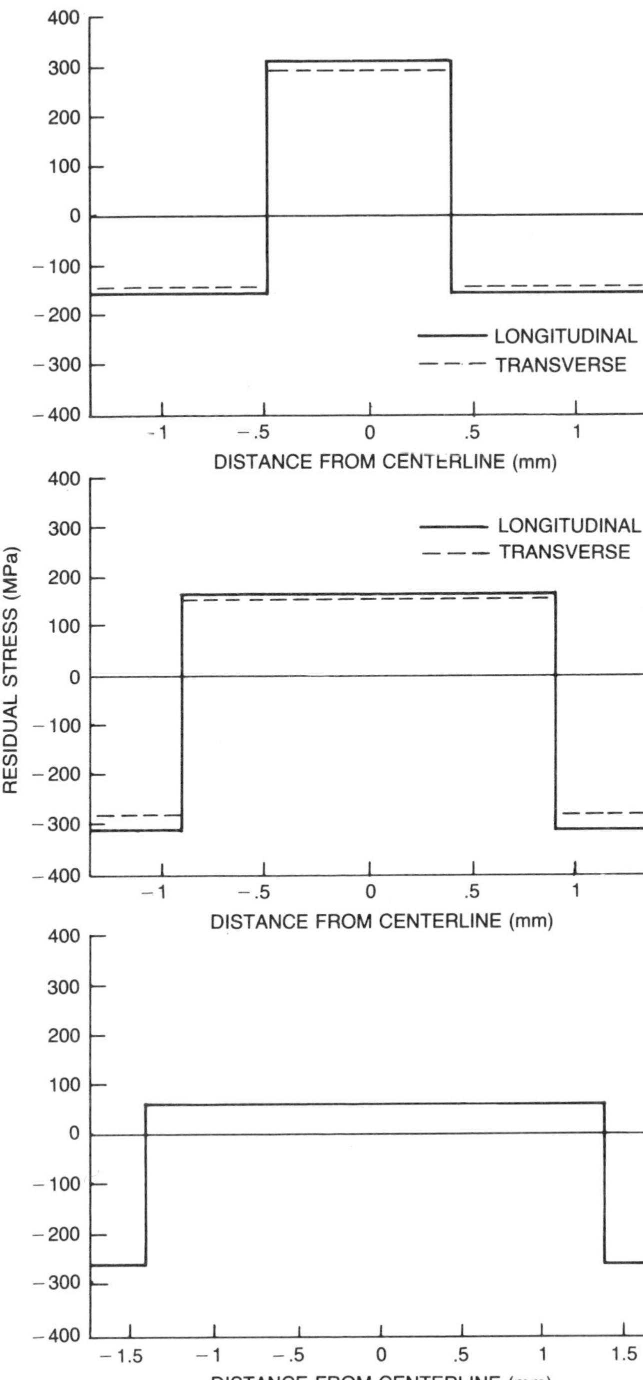

FIGURE 3 **Residual stress profiles for three laminate designs.**

Since fracture of ceramics is controlled principally by the size and distribution of flaws, strengths often are expressed in terms of fracture toughness and flaw size. A common test technique used to measure the strength of ceramics is the indented flexural test, which imparts a known and consistent flaw into the material.

The flexural strengths of unreinforced alumina, SiC-whisker-reinforced alumina, and a thermoelastically tai-

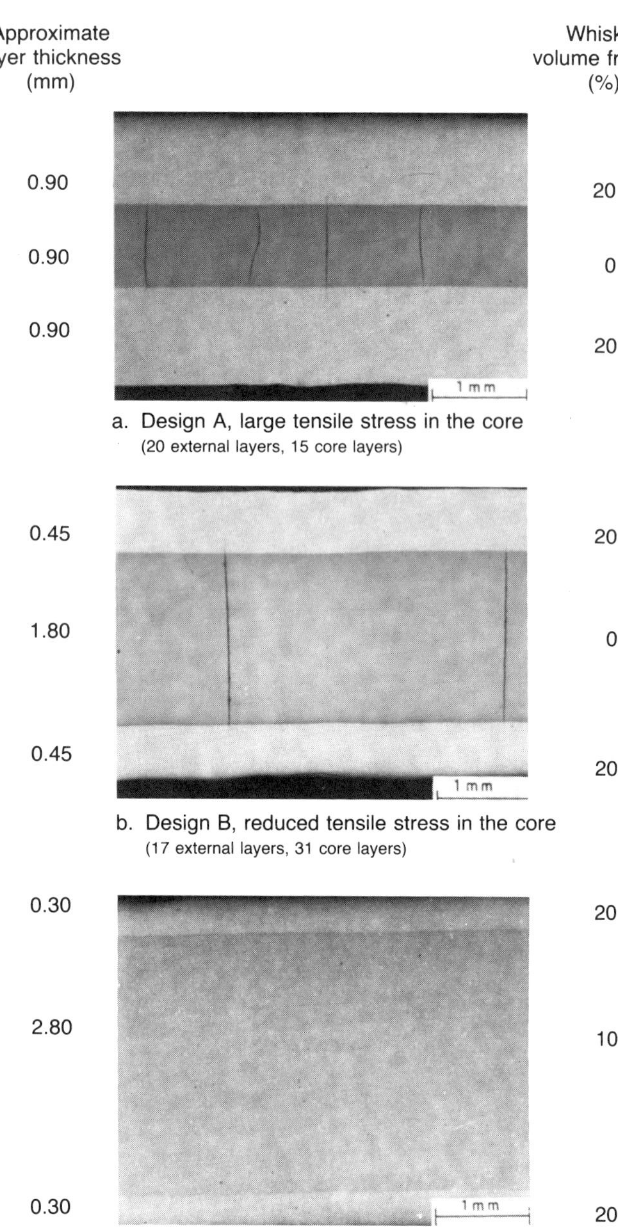

a. Design A, large tensile stress in the core (20 external layers, 15 core layers)

b. Design B, reduced tensile stress in the core (17 external layers, 31 core layers)

c. Design C, optimized for strength (9 external layers, 48 core layers)

FIGURE 4 **Examples of thermostatic tailoring in laminate design, cross sections after hot pressing.**

lored laminate (design C), after indentation, are shown in Figure 5. The strength, σ_F, of the unreinforced, reinforced, and thermoelastically tailored material can be predicted using fracture toughness K_{Ic}, crack length a, and residual compressive strength σ_r:

$$\sigma_F = Y\frac{K_{Ic}}{\sqrt{a}} - \sigma_r$$

where Y is the crack geometry factor. The measured strength of the unreinforced alumina is 120 MPa, which

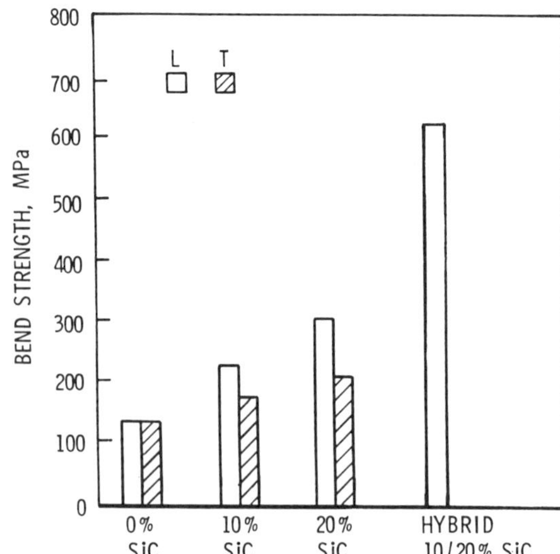

FIGURE 5 **Bend strength of indentation-flawed, tape-cast SiC–alumina composites.**

agrees with the calculated strength using accepted values of fracture toughness. Increasing the whisker content to 10 vol % raises the strength to 225 MPa in the longitudinal direction but to only 180 MPa in the transverse direction. These measured strength values are also in good agreement with values calculated from the flaw size a, the desired residual stress σ_r, and the fracture toughness. Increasing the whisker content to 20 vol % raises the strength to 300 MPa in the longitudinal direction. The discrepancy between longitudinal and transverse strength also increases. A marked increase in strength, however, is produced by the thermoelastically tailored material, design C. The favorable residual stresses in this design result in a doubling of the fracture strength to 620 MPa.

Summary

Laminated ceramic composites offer significant potential for tailoring materials to meet specific application requirements. This results in a significant improvement in strength over monolithic and conventional ceramic composites by combining the effects of whisker toughening and residual compressive surface stresses. Ceramic tape casting is currently the most suitable method for producing flexible preform laminae, which can subsequently be stacked and bonded to form the laminated structures.

M. F. Amateau and G. L. Messing

References

1. H. P. Kirchner, *Strengthening of Ceramics, Treatments, Tests, and Design Application*, Marcel Dekker, New York, 1979.

2. P. F. Becher and G. C. Wei, *J. Am. Ceram. Soc.*, *67*(12), C267 (1984).
3. G. C. Wei and P. F. Becher, *Am. Ceram. Soc. Bull.*, *64*(2), 298 (1985).
4. N. Claussen, "Design of Transformation-Toughened Ceramics," in A. H. Heuer and L. W. Hobbs, Eds., *Advances in Ceramics, Vol. 12, Science and Technology of Zirconia*, American Ceramic Society, Columbus, OH, 1981, pp. 137–163.
5. C. J. Phillips and S. DiVita, *Am. Ceram. Soc. Bull.*, *43*(1), 6 (1964).
6. H. P. Kirchner, R. M. Gruver, and R. E. Walker, *Am. Ceram. Soc. Bull.*, *47*(9), 798 (1968).
7. R. F. Gruszka, R. E. Mistler, and R. B. Runk, *Am. Ceram. Soc. Bull.*, *49*(6), 575 (1970).
8. D. J. Green, F. F. Lange, and M. R. Jones, "Residual Compressive Stresses in Al_2O_3-ZrO_2 Composites," in A. H. Heuer and L. W. Hobbs, Eds., *Advances in Ceramics, Vol. 12, Science and Technology of Zirconia*, American Ceramic Society, Columbus, OH, 1981, p. 250.
9. R. A. Sambell, D. H. Bowen, and D. C. Phillips, *J. Mater. Sci.*, *7*(7), 663 (1975).
10. K. Prewo and J. Brennen, *J. Mater. Sci.*, *15*(2), 463 (1980).
11. J. Homeny, W. L. Wallace, and M. K. Ferber, *Am. Ceram. Soc. Bull.*, *67*(2), 333 (1987).
12. T. Clark, R. Arons, J. B. Stamatoff, and J. Rabe, *Ceram. Eng. Sci. Proc.*, *6*(7–8) 576 (1985).
13. S.-T. Buljan, A. E. Pasto, and H. J. Kim, *Am. Ceram. Soc. Bull.*, *68*(2), 387 (1989).
14. J. C. Halpin and J. L. Kardos, *Polym. Eng. Sci.*, *16*(5), 344 (1976).
15. R. L. McCullough, C. T. Wu, J. C. Seferis, and P. H. Lindenmeyer, *Polym. Eng. Sci.*, *16*(5), 371 (1976).
16. R. A. Schapery, *J. Compos. Mater.*, *2*(3), 280 (1968).
17. B. Paul, *Trans. Met. Soc., AIME*, February 1960, p. 36.
18. K. S. Pister and S. B. Dong, *J. Eng. Mech. Div., ASCE*, October 1959, p. 1.
19. E. Reissner and Y. Stavsky, *J. Appl. Mech.*, September 1961, p. 402.
20. A. Virkar, J. L. Huang, and R. A. Cutler, *J. Am. Ceram. Soc.*, *70*(3), 164 (1987).
21. A. Virkar, J. Jue, J. Hansen, and R. A. Cutler, *J. Am. Ceram. Soc.*, *71*(3), C148 (1988).
22. M. Wu, G. L. Messing, and M. F. Amateau, *Proceedings of the 91st Annual Meeting of the American Ceramic Society*, 1989, paper 51-SI-89.
23. D. Hindman, J. E. Snyder, W. Liang, and M. Schreiner, "Development of Fiber Reinforced Ceramic Tubers for Radiant Tube Furnaces," in P. Vincenzini, Ed., *High Tech Ceramics*, 1987, pp. 2451–2460.
24. H. Larker, "Dense Ceramic Parts Hot Pressed to Shape by HIP," in R. F. Davis, H. Palmour and R. L. Porter, Eds., *Emergent Process Methods for High Technology Ceramics*, Plenum Press, New York, 1984, pp. 571–582.
25. E. D. Kragness, "Processing and Characteristics of Laminated SiC Whisker/Alumina Composites," M.S. thesis, The Pennsylvania State University, University Park, PA, 1988.

Laminates, Glass

The most common glass composites are those made by laminating together two pieces of transparent glass with a sheet of polyvinyl butyral plastic. Such composites are of high optical quality and combine the safety features of the plastic with the durability of glass, and are therefore widely used for automotive and architectural windows. Applications include solar energy control glazings for buildings, windshields for automobiles and airplanes, vision blocks for military vehicles, and burglar- or bullet-resistant glasses for high-security uses. Glass composites for these and other applications may contain one or more pieces of colored, coated, annealed, or tempered glasses; rigid and flexible plastics; electric wires; and metallic or ceramic coatings. Approximately 65% of all the laminated glass made today is used in automotive products, 30% for architectural applications, and 5% for specialty products.

An Englishman named Arthur T. Fullicks made glass composites as early as 1885, when he bonded several pieces of glass together [1]. However, the idea of laminated glass for improved safety was first published by another Englishman, John Crane Woods [2]. He received both British (1905) [3] and United States (1906) [4] patents for bonding sheets of transparent cellulose nitrate, using Canadian balsam, between two sheets of glass. The product was of poor quality and had a high cost, and there was little demand; thus Woods's venture was unsuccessful, and his patents were allowed to lapse. In 1910, a French chemist named Eduard Benedictus obtained both French [5] and British [6] patents for the manufacture of laminated safety glass. His procedure was similar to Woods's except that instead of using Canadian balsam, he used gelatin and other adhesives. He called his product Triplex, and a French company, La Société du Verre Triplex, manufactured the product. In 1912, the Triplex Safety Glass Company, Ltd., of England purchased production rights from the French company and began the manufacture of laminated safety glass at Willesden, England, in the middle of 1913 [7,8].

The merits of laminated glass were well established during World War I, when it was used for gas mask lenses and for windshields in military motor vehicles and aircraft. After the war, however, little progress was made in the glass lamination industry because of the high prices and low quality of the product. Starting about 1924, the development of laminated safety glass closely followed the automotive industry; in that year laminated safety glass was first used commercially in automobiles. By 1928, laminated windshields were featured as the standard by Ford. The trend in the automobile industry of going from open to enclosed cars increased the use of glass for glazing motor vehicles, and since the safety merits of laminated glass were well established, it became the accepted standard for automobile windshields in the United States. In Europe, however, annealed glass was replaced with tempered glass [9].

The first material used to manufacture laminated windshields was transparent cellulose nitrate plasticized with camphor. This was not a durable product, since it became brittle at low temperatures and gradually yellowed with sunlight. Thus, during the late 1920s, a large number of modifications were made to improve both the adhesive and the manufacturing process. About 1932, cellulose acetate started to be used as the interlayer for safety glass. This product was superior in clarity, visibility, and stability, and by 1936 it was used in most laminated safety glass. The two greatest drawbacks to the use of cellulose acetate were the need to use an adhesive to give a dependable and permanent bond between the glass and plastic layers and the need to edge-seal the laminates. This was required because cellulose acetate is hydroscopic, and thus had to be protected from weathering and loss of plasticizer. Even though such composites were stable for the life of the car, they were still subject to serious criticism as being brittle and therefore unsafe, especially at low temperatures.

Since the automotive industry was growing rapidly, it was obvious that a new product was needed. Thus, in the early 1930s, U.S. glass and plastic manufacturers joined forces to develop a new safety glass [10]. In 1937, the new product was tested; it consisted of two sheets of 3.0 mm glass bonded together with 0.38 mm thick polyvinyl butyral. This vinyl resin, when properly plasticized, offered improved safety over a wider temperature range. The self-bonding plastic required no adhesive or edge seal. The elimination of these manufacturing operations offset the higher cost of the plastic, resulting in an improved safety glass at no increased cost to the public [11]. It became the industry standard until 1966.

After World War II, as the number of highway accidents and their severity increased with the number and speed of automobiles, it became apparent that a large number of the injuries involved the windshield [12,13]. In the early 1960s, G. Radloff [14] in Germany suggested that the penetration resistance of laminated safety glass could be greatly improved by reducing the glass–plastic adhesion level through increased water content of the polyvinyl butyral interlayer. Unfortunately, excess moisture created delamination problems with the safety glass. It became apparent that adhesion control with a chemical agent was needed to solve the problem. In 1966, Libbey-Owens-Ford was granted Patent 3,231,461 [15] for a new adhesion control agent. The materials now in common use are potassium or sodium salts of organic acids, formates, or acetates, or a mixture of the two [16]. At the same time, the interlayer thickness was increased from 0.38 mm to 0.76 mm. Thus, the new high penetration resistant (HPR) product improved the penetration velocity of the windshield from about 21 km/h to approximately 35 to 47 km/h [17,18]. The HPR windshield almost immediately became the new standard for the United States and has been used in all automobiles since 1966. A large number of reports [19,20,21] outlining its superior performance in penetration resistance and laceration protection [22,23] have convinced many countries to adopt the HPR windshield.

To further improve the laceration protection properties of the HPR windshield, a large amount of work has been done to find the ideal glass–plastic combination [24,25]. Chemically tempered glass was evaluated in the United States [26,27,28,29], and heat-strengthened glass was produced in England [30,31]. Neither method was commercially successful. Several companies in both the United States and Europe succeeded in commercially producing windshields with an additional coating of plastic on the interior surface of the glass. During an impact, this layer totally eliminates all laceration of the car occupant. Even though this is not a new idea [32], the technology to produce such a product was not available until the 1970s. Such antilacerative windshields using polyurethane were evaluated by several companies [33,34], and first became commercially available in Europe in 1977 [35]. A similar product was developed in the United States jointly by Libbey-Owens-Ford and DuPont [36–38]; it consisted of 0.38 mm polyvinyl butyral (PVB) and 0.1 mm polyethylene terephthalate (PET) [39,40]. A hard coating on the PET provided both abrasion and chemical resistance. It was used commercially on some General Motors cars in 1987 [41,42]. These products are somewhat more expensive than the standard HPR windshields and require some care when cleaning the interior plastic surface of the windshield [43–45]. In recent years, a new version of the antilacerative windshield has been developed; it uses one piece of glass and is commonly called the bi-layer windshield [46–49]. This product, which is not yet commercially available, has only one piece of plastic that performs both functions, penetration resistance and laceration protection. Even though this product has some economic advantages, it has one possible drawback: the interlayer is no longer protected from physical or environmental damage by the two pieces of glass and may become unreliable with time or misuse.

The first standard for automotive glass was written in 1938 and outlined specifications and methods for testing safety glazing materials. This ANSI Z26.1 Code [50] has been routinely modified to keep it abreast of improvements and changes. In 1968, the U.S. government included this code in the U.S. Federal Motor Vehicle Safety Standard 205 for glazing materials [51]. Conformance with the federal standard is done through self-certification by the glazing manufacturers. Conformance with state and provincial regulations is normally achieved by submitting reports from approved testing laboratories to the American Association of Motor Vehicle Administrators (AAMVA) [52]. Automotive safety glass is tested according to its specific use or location in the vehicle and must be labeled as to the DOT (Department of Transportation) Code [53], manufacturer, item classification, and model number. The United States and Canada have almost identical laws governing the use of automotive glazings. Mexico adopted safety standards for windshields in 1975. Laminated windshield safety glass is also required in Denmark, Sweden, Norway, Switzerland, Finland, Italy, Belgium, the Netherlands, and Japan. A large number of European countries have adopted ECE Regulation No. 43 [54] for the certification of motor vehicle glazings. Certification of laminated glass for architectural applications in the United States is governed by federal, state, and local building codes [55,56]. Federal regulations are certified to Federal Standard 16CFR1201 by the Safety Glazing Certification Council (SGCC) [57]. Local and state codes usually require certification to ANSI Z97.1 [56] or by the Consumer Product Safety Commission [58,59].

There are basically two methods of producing laminated glass composites. One is by casting a curable liquid resin between the glass sheets, followed by a heat or UV treatment to produce a solid laminate. This procedure is mainly used for a few specialty applications, but has the advantage of not requiring either expensive equipment or high technology [60]. The other is the most common process presently in use; it involves placing a sheet of thermoplastic PVB between two sheets of glass and bonding these together with high heat and pressure. PVB is a tough transparent polymer that is somewhat tacky and adheres well to glass. Even though other plastics, such as polyurethanes, have been evaluated for some composite applications, PVB has almost exclusively been used for the manufacture of laminated glass for over 50 years [61]. To manufacture laminates by this method, two identical-sized pieces of glass are cut and cleaned, then a piece of PVB is placed between the pieces of glass. The sandwich is heated and pressed together to eliminate excess air and seal the edges of the unit. This is normally done by one of three methods: the use of a vacuum ring, a vacuum bag, or a set of de-airing rolls. To complete the manufacture, the composite is placed in an autoclave and heated to 120–150°C under pressure of 10–15 kg/cm^2 for 20–40 min. Air autoclaves are most commonly used for this purpose. The high temperatures are needed to soften the PVB and allow it to flow and adhere to the glass. The pressure helps to dissolve any remaining traces of air into the plastic. After the glass is cooled in the autoclave, the pressure is removed and the composites are seamed, cleaned, and inspected before being shipped to the customer.

The PVB used to make these products is manufactured by several chemical companies (Table 1) and is normally shipped as 250 and 500 m rolls that are sealed into water-resistant bags. It comes in three forms: refrigerated, interleafed, or dusted with sodium bicarbonate. This is to prevent the plastic from blocking when stored for extended periods of time. After the plastic is unwound, and washed if dusted, it is cut to size and stored in a clean room area that has a temperature- and humidity-controlled environment. Low temperatures are needed to prevent the sheets of PVB from sticking together. The humidity is carefully controlled because the PVB rapidly picks up moisture (Fig. 1), which affects the adhesion level of the plastic to the glass (Fig. 2). The higher the moisture level in the plastic, the lower its adhesion to the glass. If the adhesion is too low, separations in the composite may be produced over time. On the other hand, a low moisture level in the plastic can increase the adhesion level to the point where the penetration resistance of the composite becomes unacceptably low (Fig. 3), especially for windshield applications, where it is an important safety feature.

Glass composites are manufactured in a variety of constructions depending on the requirements of the specific applications. The automobile windshield is by far the most common use of laminated glass; its excellent penetration resistance is a safety feature that has greatly reduced injuries and saved many lives. Laminated windshields are constructed from a variety of glass thicknesses, the most common construction being two pieces of 2.2 mm glass and 0.76 mm of special adhesion-controlled PVB. The penetration resistance of such PVB composites varies with temperature (Fig. 4), but at the present time no other plastic has been found that overcomes this and still offers all the advantages of PVB (i.e., cost, clarity, safety, and weatherability). Aircraft windshields are more complex than their automotive counterparts. They are constructed using multiple glass–plastic layers that include a specially formulated low-plasticizer PVB. The glass is usually coated with a conductive tin oxide coating so that the unit can be electrically heated. These special features, including mechanically fastening the windshield and PVB to the aircraft, are used to give maximum impact resistance. This is required because one of the hazards confronting modern high-speed aircraft is in-flight collisions with birds during takeoffs and landings.

Architectural glass is usually laminated with a special high adhesion type of PVB and, depending on the application, can contain a variety of glass and plastic thicknesses, with 0.38 mm PVB being used in most applications. Laminated architectural glass, unlike ordinary monolithic glass, has the safety characteristic that broken pieces of glass will adhere to the plastic interlayer

TABLE 1
Manufacturers of PVB Film, Trade Name, and Plasticizer

Manufacturer	PVB Trade Name	Plasticizer
DuPont	Butacite	4G7, Tetraethyleneglycol di-*n*-heptanoate
Monsanto	Saflex	DHA, Dihexyl adipate
Sekisui	S-Lec	3GH, Triethyleneglycol di-2-ethyl butyrate
Dynamit-Nobel	Trosifol	3G7, Triethyleneglycol di-*n*-heptanoate

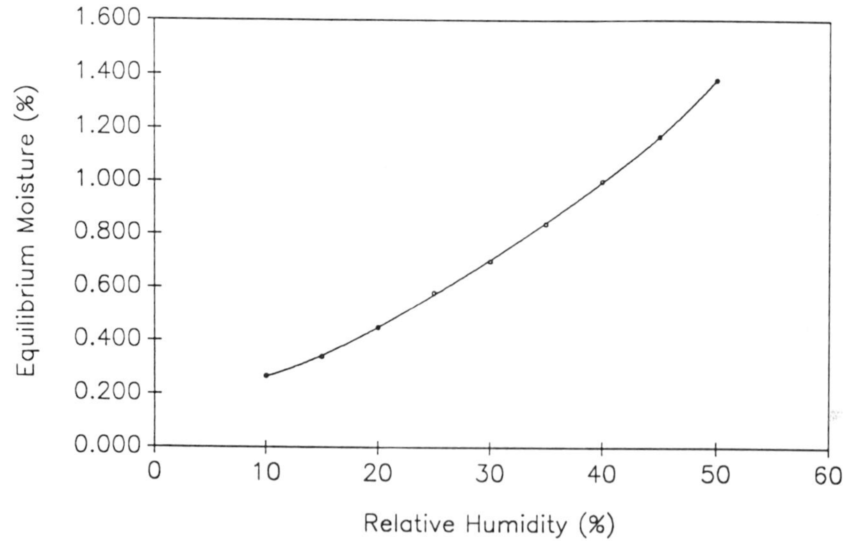

FIGURE 1 Typical effect of humidity on PVB moisture. Equilibrium moisture in 0.76 mm sheet. (Courtesy of E.I. DuPont Company.)

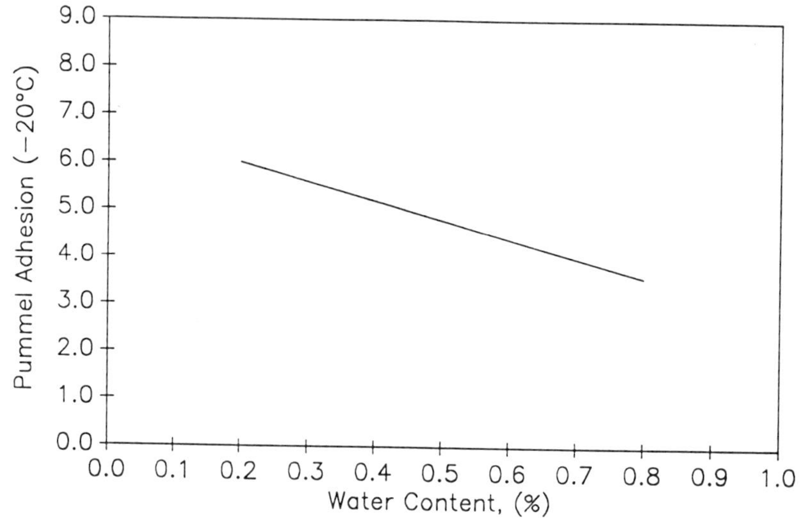

FIGURE 2 Typical effect of water content on PVB adhesion. (Courtesy of Sekisui Chemical Company.)

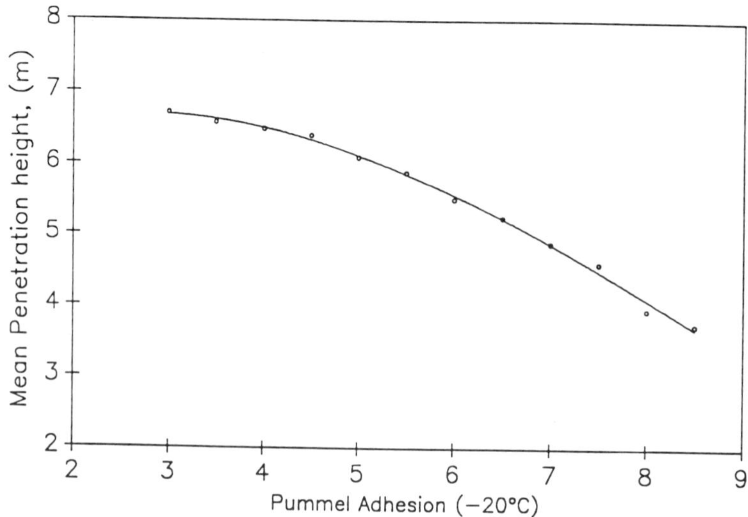

FIGURE 3 Typical variation of mean support level versus adhesion level. Composites (305 × 305 mm) impacted with the 2.27 kg steel ball at 20°C (0.76 mm PVB, 2.4 mm glass). (Courtesy of Monsanto Company.)

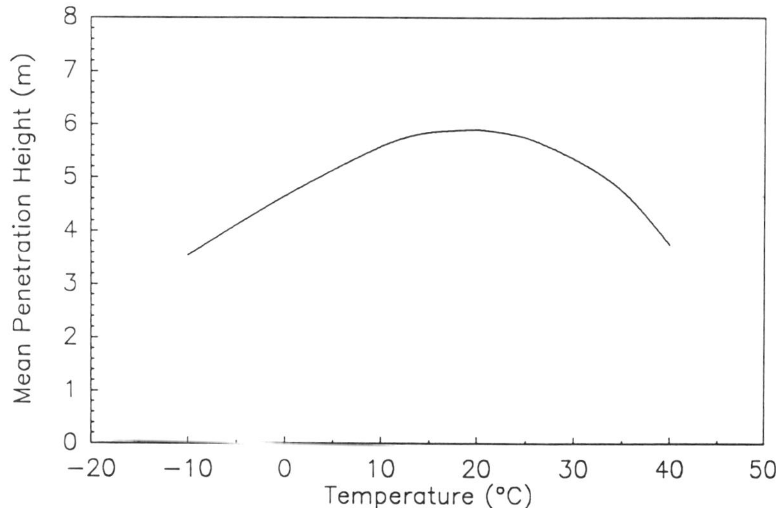

FIGURE 4 **Typical mean penetration resistance of laminated glass versus temperature. Data on composites constructed from two pieces of 2.2 mm glass (305 × 305 mm) and 0.76 mm HPR PVB and evaluated with the 2.27 kg steel ball.** (Courtesy of Libbey-Owens-Ford Company.)

and prevent injuries resulting from fall-out; this also leaves the glazed opening secure as a physical barrier against weather or entry. Bullet-resistant and security glass is usually made with two or more layers of annealed or strengthened glass and an extra-thick layer of PVB. Such products may also contain a layer of rigid plastic, such as polycarbonate, to act as a spall shield. Polycarbonates are usually laminated to glass with an elastomeric polyurethane. This is required because PVB is not compatible with polycarbonate. Polyurethanes also maintain their elastomeric properties over a greater temperature range, thus producing a more stable product between two materials with relatively large differences in their thermal expansions. Laminated glass also offers the advantage of improved sound transmission reduction over regular glass [62] (Fig. 5) and thus is used extensively in buildings where high levels of noise are undesirable, such as in libraries, airport buildings, factories, office buildings, and broadcasting studios.

The light transmittance of laminated glass is different from that of ordinary glass. It absorbs more ultraviolet (UV) and infrared (IR) radiation than does glass alone (Fig. 6). When solar radiation falls on any glass, portions are reflected, absorbed, and transmitted. The absorbed energy heats the glass. The energy is then dissipated by re-radiation and convection from both surfaces of the

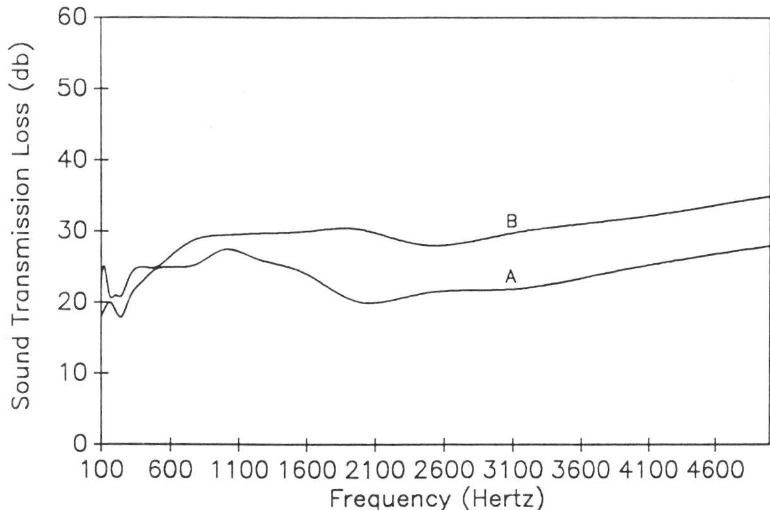

FIGURE 5 **Sound transmission loss versus frequency. Curve A shows 6 mm monolithic glass (STC = 28). Curve B shows 6 mm laminated glass (STC = 34) made with 3 mm glass and 0.76 mm PVB.** (Courtesy of Monsanto Company.)

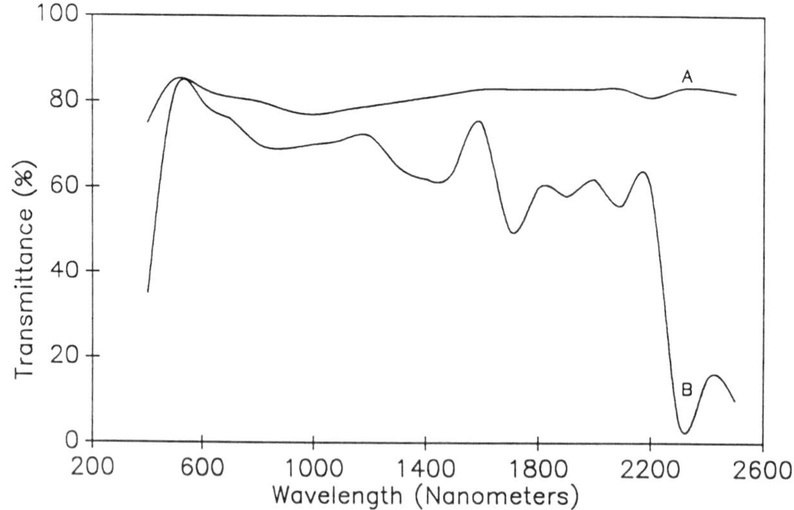

FIGURE 6 **Typical solar transmittance curves of (*A*) 3 mm clear glass and (*B*) 6 mm laminated glass (3 mm glass, 0.76 mm PVB).** (Courtesy of Libbey-Owens-Ford Company.)

glass. Colored interlayers and glass are used to reduce the transmitted solar heat by absorbing part of the solar radiation in the UV, visible, and IR ranges. Specially coated IR reflective glass can be used to reduce the heat load in buildings or automobiles and still have visible light transmissions above 70%. UV radiation (290–380 nm) can cause degradation of dyes, pigments, and polymers, resulting in decomposition of plastic parts and fading of carpets, drapes, art, and other interior fixtures. To prevent this, a specially formulated PVB interlayer is available that selectively filters any radiation below 380 nm (Fig. 7). Such glass laminates are primarily used for art galleries, museums, and shop windows.

PVB plastic is produced via a condensation reaction between polyvinyl alcohol and *n*-butyraldehyde that produces a polymer that has approximately 25% by weight free vinyl alcohol groups [63].

$$\left[CH_2-\underset{OH}{\underset{|}{CH}} \right]_x + C_3H_7\underset{O}{\underset{\|}{C}}H \rightarrow \left[CH_2-\overset{\overset{CH_2}{\frown}}{\underset{O}{\underset{|}{CH}}}\quad\underset{O}{\underset{|}{CH}} \;(\text{O}-\underset{C_3H_7}{\underset{|}{CH}}-\text{O}) \right]_y \left[CH_2-\underset{OH}{\underset{|}{CH}} \right]_z + H_2O$$

The hard resin is next plasticized with one of several high molecular weight esters (Table 1) to produce a tough, flexible polymer that is extruded to produce a continuous sheet. Some of the common properties of PVB are listed in Table 2. The sheeting also contains small amounts of UV stabilizers, antioxidants, and adhesion control agents, the amount and type of each depending on the manufacturer and intended end usage. Since PVB readily absorbs water, water is also present in small amounts; Figure 1 shows the typical equilibrium

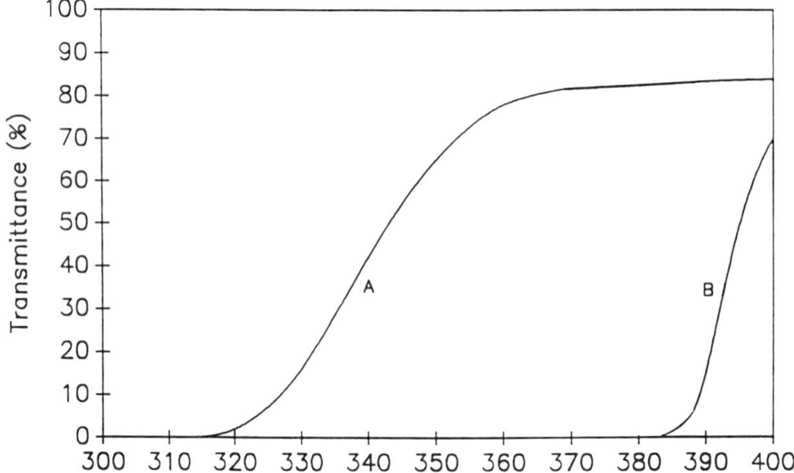

FIGURE 7 **Typical ultraviolet transmittance curves of (*A*) 6 mm clear glass and (*B*) 6 mm laminated glass (3 mm glass, 0.76 mm PVB with UV absorber).** (Courtesy of Dynamit Nobel Company.)

TABLE 2
Typical Properties of PVB Interlayer

Physical Property	Test Conditions	Value
Tensile strength	23°C, 51 cm·min^{-1}	209 kg·cm^{-2}
Tensile elongation	23°C, 51 cm·min^{-1}	373%
Density	25°C	1.059 g·cm^{-3}
Specific heat	65°C	2250 J·kg^{-1}·K^{-1}
Thermal conductivity	65°C	0.20 W·m^{-2}·C^{-1}·m
Poission ratio	—	~0.5
Refractive index	25°C	1.479

Courtesy of Monsanto Company.

between PVB and moist air. PVB is soluble in the common oxygen-containing solvents, such as alcohol, ketones, and esters; it is stable under alkaline conditions, but unstable in the presence of acids.

Special sheeting is made for each of the automotive, architectural, and aircraft markets. The automotive PVB includes clear and gradient band tinted plastic. The shadeband located in the upper part of the windshield is either printed or co-extruded onto the plastic film, with blue and green being the most common colors used for windshields. The architectural plastic is available in a large number of colors, transmission levels, thicknesses, and widths (up to 3.5 m). The aircraft type, made only by Monsanto, is a low plasticized product. The majority of the film is made by four manufacturers. Estimates put the world market at approximately 72,000 metric tons per year [64], with over 80% being used in North America, western Europe, and Japan. The present price for PVB sheeting is approximately $8.40 (U.S.)/kg or $7.07/m^2 for gradient automotive grade; thus approximately $600,000,000 (U.S.) is spent annually on PVB film to produce glass composites. Worldwide, approximately 65% of all PVB is used in automotive applications, while 35% goes into architectural and specialty glass composites.

With the right combination of glass, interlayer, and coating, a glass composite can be designed that has almost any color, safety objective, energy control, or aesthetically pleasing feature.

Siegfried H. Herliczek

[*See also* CIRCUIT BOARD LAMINATES; GLASS MATRIX COMPOSITES.]

References

1. A. T. Fullicks, British Patent 15,303 (1885).
2. A. F. Randolph, *Mod. Plast.*, *18*(10), 31 (1941).
3. J. C. Woods, British Patent 9972 (1905).
4. J. C. Woods, U.S. Patent 830,398 (1906).
5. E. Benedictus, French Patent 405,881 (1910).
6. E. Benedictus, British Patent 1790 (1910).
7. *The Hidden Difference,* E. I. du Pont de Nemours, Inc., Polymer Product Dept., Wilmington, DE, 1989.
8. Anon., *Glass, 51*(1), 9 (1974).
9. S. Kay, *Chem. and Ind., 23,* 1086 (1973).
10. G. O. Morrison, F. W. Skirrow, and K. G. Blaikie, U.S. Patent 2,036,092 (1936).
11. Anon., *Am. Ceram. Soc. Bull., 18*(5), 176 (1939).
12. S. Schwimmer and R. Wolf, *Leading Causes of Injury in Automobile Accidents,* Cornell, A.C.I.R., 1962.
13. R. D. Lester, *Automobile Engineer, 51*(9), 341 (1961).
14. G. Radloff, *Automobiltechnische Zeitschrift, 64*(6), 1979 (1962).
15. P. T. Mattimoe, U.S. Patent 3,231,461 (1966), to LOF Glass Co.
16. J. R. Huntsberger, *J. Adhesion, 13,* 107 (1981).
17. L. M. Patrick, K. R. Trasien, and F. T. DuPont, *Safety Performance Comparison of 30 Mil HPR Laminated and Monolithic Differentially Tempered Windshields,* Society of Automotive Engineers Publishers, Warrendale, PA, SAE Reprint 700427, 1970.
18. L. M. Partick and R. Daniel, "Comparisons of Standard and Experimental Windshields," The Eighth Stapp Air Crash and Field Demonstration Conference, L. M. Patrick, Ed., Wayne State University Press Publishers, Detroit, 1966, pp. 147–166.
19. D. F. Huelke, *Striking the Windshield,* University of Michigan Publishers, Ann Arbor, 1967.
20. G. M. Machay, A. W. Siegel, and P. V. Hight, *Tempered versus HPR Laminated Windshields: A Comparative Study of United Kingdom and United States Collisions,* Society of Automotive Engineers Publishers, Warrendale, PA, 1970, SAE Reprint 70091.
21. R. B. Faro, *Windshield Glazing as an Injury Factor in Automobile Accidents,* Cornell Aeronautical Lab., Inc. CAL No. VJ-1823-R25, 1968.
22. D. F. Hueke et al., *Plastic and Reconstructive Surgery, 41*(6), 1968.
23. R. L. Morrison, *Influence of Ambient Temperatures on Impact Performance of HPR Windshields,* The Fifteenth Stapp Car Crash Conference, SAE Publishers, New York, 1972, p. 603.
24. R. G. Rieser and J. Chabel, *Safety Performance of Laminated Glass Structures,* Society of Automotive Engineers Publishers, Warrendale, PA, SAE Reprint 700481, 1970.
25. H. M. Alexander, P. T. Mattimoe, and J. J. Hofmann, *An Improved Windshield,* Society of Automotive Engineers Publishers, Warrendale, PA, SAE Reprint 700482, 1970.

26. L. M. Patrick, K. R. Trasien, and F. T. DuPont, *Safety Performance of a Chemically Strengthened Windshield*, Society of Automotive Engineers Publishers, SAE Reprint 690485, 1969.
27. J. R. Blizard and J. S. Howitt, *Glass Ind., 50*, 573 (1969).
28. J. R. Blizard and J. S. Howitt, *Glass Ind., 51*, 16 (1970).
29. J. R. Blizard and J. S. Howitt, *Development of a Safer Nonlacerating Automobile Windshield* (in ref. 26), SAE Reprint 690474, 1969.
30. S. E. Kay, V. J. Osla, J. Pickard, and P. A. Breretsn, *Automobiltechnische Zeitschrift, 79*, 389 (1977).
31. S. E. Kay, J. Pickard, and L. M. Patrick, *Improved Laminated Windshields with Reduced Laceration Properties*, The Seventeenth Stapp Car Crash Conference, SAE Publishers, New York, 1973, pp. 127–169.
32. E. W. Ried, U.S. Patent 2,120,628 (1938).
33. R. G. Rieser and J. Chabel, U.S. Patent 3,808,077 (1974).
34. W. Schafer and H. Raedish, U.S. Patent 3,979,548 (1976).
35. *Securiflex, The Anti-Lacerative Windshield*, Saint-Gobain Vitnage, 1984.
36. P. T. Mattimoe, T. J. Motter, J. J. Hofmann, and S. H. Herliczek, U.S. Patent 3,900,673 (1975).
37. P. T. Mattimoe, T. J. Motter, J. J. Hofmann, and S. H. Herliczek, U.S. Patent 4,059,469 (1977).
38. N. W. Johnston, S. H. Herliczek, and C. E. Ash, *Anti-Lacerative Windshield: An Overview* (in ref. 26), 1984, SAE Reprint 840388.
39. J. Dolenga, *Glass Magazine, 35*(1), 44 (1985).
40. D. Zoia, *Ward's Auto World, 20*(8), 44 (1984).
41. Anon., *DuPont Magazine, 80*(5), 20 (1986).
42. Anon., *Chemical Week, 135*(6), 22 (1984).
43. Anon., *SAE Automotive Eng., 92*(7), 48 (1984).
44. E. P. Doolittle, T. B. Horton, and H. P. Blom, *Anti-Lacerative Windshield Materials; Field Evaluation by General Motors* (in ref. 26), SAE Reprint 840391, 1984.
45. D. Winter *Ward's Auto World, 24*(2), 47 (1988).
46. J. Cherrk, W. I. Frey, J. D. Kelly, and L. S. Sokol, U.S. Patent 4,277,299 (1981).
47. Anon., *Glass Magazine, 35*(7), 50 (1985).
48. J. Dolenga, *Glass Magazine, 35*(5), 38 (1985).
49. Anon., *Machine Design, 60*(3), 16 (1988).
50. *Safety Code for Safety Glazing Materials for Glazing Motor Vehicles Operating on Land Highways*, American National Standards Institute, New York, Z26.1-1983.
51. Motor Vehicle Safety Standard No. 205, *Federal Register, 49*(37), 3732 (1984).
52. American Association of Motor Vehicle Administrators, 1201 Connecticut Ave., N.W., Suite 910, Washington, DC.
53. Prime Glazing Material Manufacturers, *Federal Register, 45*(151), 51569 (1980).
54. *Uniform Provisions Concerning the Approval of Safety Glazing and Glazing Materials*, Economic Commission for Europe, Regulation No. 43, United Nations Economic and Social Council, New York.
55. Architectural Glazing Safety Standard, *The Building Official and Code Administrator, 15*(4), 40 (1981).
56. *Safety Performance Specifications and Methods of Test for Safety Glazing Material Used in Buildings*, American National Standards Institute, New York, Z97.1-1984.
57. *Certified Products Directory, Safety Glazing Material Used in Buildings*, Safety Glazing Certification Council, c/o ETL Testing Lab., Inc., Cortland, NY.
58. K. D. Mann, *U.S. Glass, Metal and Glazing, 16*(2), 34 (1981).
59. Consumer Product Safety Commission, 5401 Westband Ave. Bethesda, MD.
60. J. R. St. Clair, *Glass Ind., 65*(11), 28 (1984).
61. R. C. McCoy, *U.S. Glass, Metal and Glazing, 23*(4), 68 (1988).
62. A. R. Shaw, *Glass Digest, 61*(11), 54 (1982).
63. N. G. Gaylord, Ed., *Encyclopedia of Polymer Science and Technology, Vol. 14*, Interscience Publishers, New York, 1971, p. 208.
64. R. Turner, *Chemical Week, 142*(18), 8 (1988).

Laminates, Matrix Cracking in Composite Laminates under Static and Fatigue Loading

In most applications, polymer matrix composites are used in the form of laminates, which have fibers in the individual laminae (or plies) in different directions. Consider a simple laminate configuration such as $(0/\pm45/90)_s$, and imagine that a load is applied parallel to the 0° ply direction. The first form of easily observable damage to develop in the laminate under either static or fatigue loading is cracking in the off-axis (i.e., 90° and 45°) plies. Some authors make a distinction between this type of damage occurring in "subcritical" laminate elements and damage to the "critical" elements, the 0° plies, since damage to the latter leads directly to component fracture. However, this does not mean that matrix cracking is unimportant; in stiffness-limited designs, the stiffness reduction associated with matrix cracking can be sufficient to cause component "failure," and in the chemical industry, matrix cracking may cause a structure to fail by seepage. It has also been demonstrated frequently that matrix cracking can lead to stress concentrations, which may enhance fiber failure in adjacent 0° plies.

This article is divided into two major parts. In the first part we introduce some experimental observations on the initiation and growth of matrix cracks under static and fatigue loading and then look briefly at work concerned with modeling this cracking and predicting residual stiffness properties. For simplicity, we shall concentrate on cross-ply [i.e., (0/90)] laminates. In the second part we outline a recent theoretical model for matrix cracking and comment briefly on the implications for design.

Survey of Matrix Cracking Observations

Development of Transverse-Ply Cracks under Static Loading

The mechanisms of transverse-ply crack initiation and growth have been extensively studied in glass fiber reinforced plastic (GFRP) and carbon fiber reinforced plastic (CFRP) laminates. Observation of transverse-ply cracks is easier in GFRP than in CFRP because of their optical transparency. Techniques used to observe matrix cracking include optical microscopy, edge replication, and X-ray radiography. Under static loading, glass, carbon, and Kevlar (a trade name of the DuPont Company) composites containing 90° plies develop an increasing density of transverse-ply cracks, with the cracks, in general, spanning the thickness of the transverse ply and the width of the laminate. Figure 1 shows the development of cracking in a $(0/90)_s$ GFRP laminate under increasing load. Bailey and Parvizi [1] performed in situ tests inside a scanning electron microscope on a cross-ply GFRP laminate. Viewing on the edge, they confirmed that fiber debonding (assisted by strain magnification in the matrix between the fibers—see the work of Kies [2] used by Garrett and Bailey [3]) is the origin of transverse-ply cracking. At a higher strain, the debonds link up to form a flaw, which subsequently develops into a transverse-ply crack spanning the full thickness of the transverse ply and width of the laminate.

Harrison and Bader [4] and Schulte [5] report tests on cross-ply CFRP laminates. In CFRP, transverse-ply cracks are generally believed to initiate exclusively at the laminate edge because of the local stress field there arising from the free-edge effect. (In GFRP, experimental evidence suggests that cracks initiate predominantly at the edge but also in the bulk.) Harrison and Bader [4] found from optical microscopy that, as in GFRP, the transverse cracks initiated from fiber–matrix debonds. Schulte [5] and Bader and Boniface [6] showed that the cracks initiate at the edge and grow across the laminate width as the applied strain is increased.

In both GFRP and CFRP, as the applied strain is increased beyond that necessary to form the first crack, an increasing number of cracks form. The laminate stress–crack spacing relation in GFRP and CFRP has been predicted by Bader, Bailey, Curtis, and Parvizi [7] on the basis of a shear-lag approach, giving reasonable agreement with experimental data. Eventually an apparent saturation is reached when the cracks are spaced about one ply thickness apart. Reifsnider has termed this pattern the characteristic damage state or CDS (e.g., Reifsnider and Masters [8]) and has suggested that while it is dependent on the transverse-ply thickness and the orientation of the neighboring plies, it is independent of such variables as load history and environment.

Bailey, Curtis, and Parvizi [9] present data for the strain to first cracking, as a function of the transverse-ply thickness, in cross-ply CFRP and GFRP laminates. The cracking strain decreases as the ply thickness increases. They predicted the strain to first failure on the basis of energetics arguments, making use of the multiple failure theory of Aveston and Kelly [10]. Agreement with theory was good for small ply thicknesses, provided that ther-

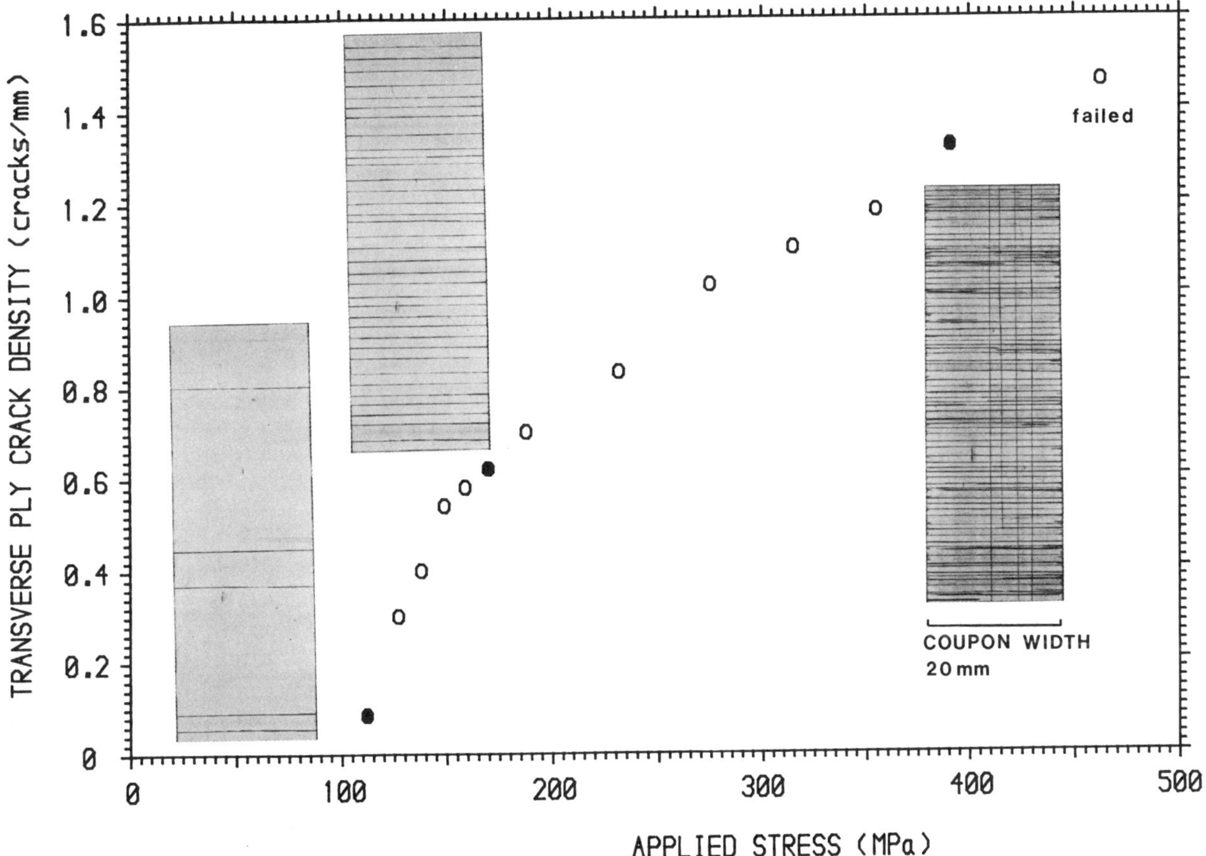

FIGURE 1 Development of cracking in a $(0/90)_s$ GFRP laminate under increasing static loading. (Courtesy of L. Boniface.)

mal strains arising from the curing process were included, but poor at large ply thicknesses, where, they argued, cracking is "mechanism-controlled." This discrepancy was clarified by Flaggs and Kural [11], who pointed out that the predictions of Bailey et al. are based on the existence of a crack spanning the entire width and thickness of the transverse ply, thus precluding the possibility of a critical flaw size smaller than the ply thickness. One of the most important features of the work of Bailey et al. is the implication that by making the transverse plies sufficiently thin, it should be possible to suppress transverse cracking completely.

Another method for predicting the load for first cracking was developed by Wang and Crossman [12]. They used a two-dimensional finite element analysis to model the growth of a transverse-ply crack across the ply thickness. Using a crack closure technique, they evaluated the strain energy release rate as a function of crack length, $2a$, in a transverse ply of thickness $2d$, and found that it had a maximum value at about $a/d = 0.85$. They postulated that the first transverse-ply cracks form when the available maximum energy release rate equals the material energy release rate. They compared their numerical predictions with experimental data from a CFRP $(\pm 25/90)_s$ laminate and found good agreement.

Weibull statistics have also been used to explain the decreasing strain to first cracking with increasing ply thickness. This has now become quite a popular technique in analyzing features of transverse cracking. Manders et al. [13] appealed to a statistics argument to explain the appreciable experimental scatter in load/crack spacing data in GFRP laminates at large crack spacings, and concluded that the strain to first failure is dependent on the specimen volume. At small crack spacings, they argued, the controlling factor was the decreasing stress in the transverse ply—as indicated by a shear-lag model. Other workers [14,15] have used Weibull statistics to predict both the load/crack spacing relation and the dependence of the strain to first failure on ply thickness and neighboring ply orientation. However, although good agreement was obtained, this approach does obscure the mechanisms of crack development, as noted by Peters [16] in a subsequent paper. He concluded that describing the laminate behavior by means of Weibull was impractical since the constraint exerted on flaws close to the ply interface means that they are less likely to propagate than those towards the center of the ply; hence the Weibull shape parameter itself becomes a function of the ply thickness.

As an aside, it is important to note that the flaws involved in a Weibull analysis are not necessarily pre-existent cracks (although thermal stresses can sometimes lead to debonds and cracks, even in the absence of mechanical loading). They may simply be regions of closely spaced fibers, giving rise to high local strains.

Transverse-Ply Cracking under Fatigue Loading

Under fatigue loading, transverse-ply cracking can occur at stress levels much lower than those required to initiate cracking under static loading. Harrison and Bader [4] made a detailed study of the development of transverse-ply cracks in CFRP at fatigue stress levels below the static threshold. They proposed that the fatigue effect arises from the opening and closing of fiber–resin debonds, bringing about plastic deformation at the crack tip and thus assisting crack propagation. They also suggested that the cracks take a finite number of cycles to grow across the laminate width, with those cracks that form later in the test taking longer than those that form first. This was attributed to crack interaction resulting in a lower cyclic stress in the transverse ply, and consequently a slower growth rate for subsequent cracks. This reduction in stress is obviously an important feature to incorporate into a fatigue model. Reifsnider and coworkers have suggested that, in both CFRP and GFRP, at practical stress levels (say, 60% of the ultimate static strength of the laminate), the final transverse-ply crack pattern (which they propose is more or less the same as the CDS seen under static loading and should therefore have no effect on the residual strength of the laminate) forms early in the fatigue life. They have also observed that transverse cracks can act as local stress raisers, promoting fiber breakage and delamination (e.g., Reifsnider and Jamison [17]). A review of the above work is given in Reifsnider [18].

During static loading of cross-ply laminates, or during fatigue loading with stresses of about 60% of the tensile strength or more, matrix cracks appear very rapidly in a coupon specimen. A saturation "steady state" is reached, with the final crack spacing approximately equal to the transverse-ply thickness. However, with much lower fatigue stresses, about 20 to 50% of the tensile strength, it is possible to observe the slow growth and accumulation of matrix cracks.

The fatigue growth of matrix cracks in a transparent cross-ply GFRP laminate is shown in Figure 2. The coupon was cycled with a peak stress of about 20% tensile strength and a stress ratio $\sigma_{min}/\sigma_{max}$ of 0.1. The photographs show a number of interesting features. First, most of the cracks initiate at the coupon edge and grow across the laminate, although some internally initiated cracks can be seen at 150,000 cycles. Second, the number of growing cracks varies throughout the test; new cracks are initiated, while existing cracks either grow completely across the specimen and are "lost," or two crack tips growing towards each other on the same plane merge when they meet, so that both growing crack tips are lost. Third, there is a variety of crack growth rates. Cracks that closely overlap do not grow at all, whereas those crack tips that are spaced further apart from the next nearest crack (measured longitudinally along the specimen) grow rapidly. Fourth, the fatigue growth rate of an individual crack is independent of the length of that crack.

The crack growth rate dependence on crack spacing but independence of crack length is illustrated by Figure 3. The lengths of the cracks labeled A to D in Figure 2 are shown in Figure 3 as a function of the number of cycles. The longitudinal spacing from the crack tip to the next nearest crack is also indicated. It is clear that, within experimental error, the growth rates are independent of crack length but depend strongly on the crack spacing.

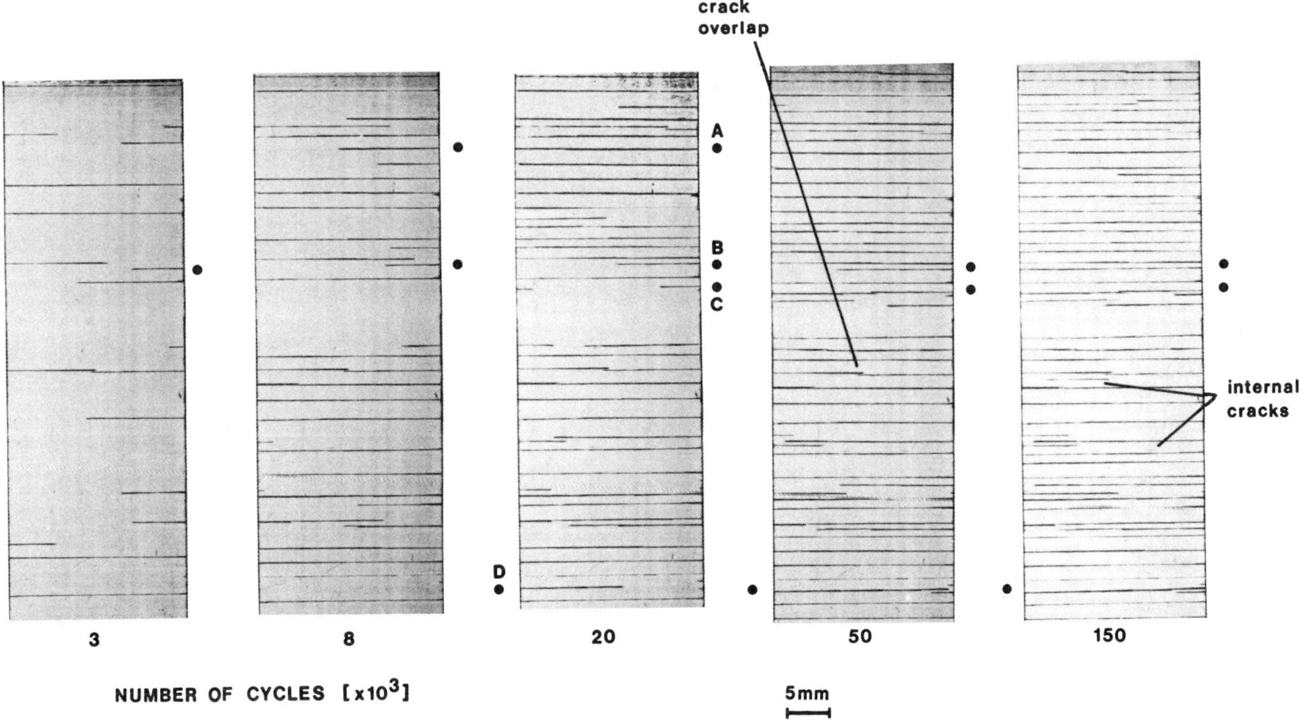

FIGURE 2 The growth of matrix cracks in a $(0/90)_s$ GFRP laminate under fatigue loading: (*a*) 3000 cycles, (*b*) 8000 cycles, (*c*) 20,000 cycles, (*d*) 50,000 cycles, (*e*) 150,000 cycles. (Courtesy of L. Boniface.)

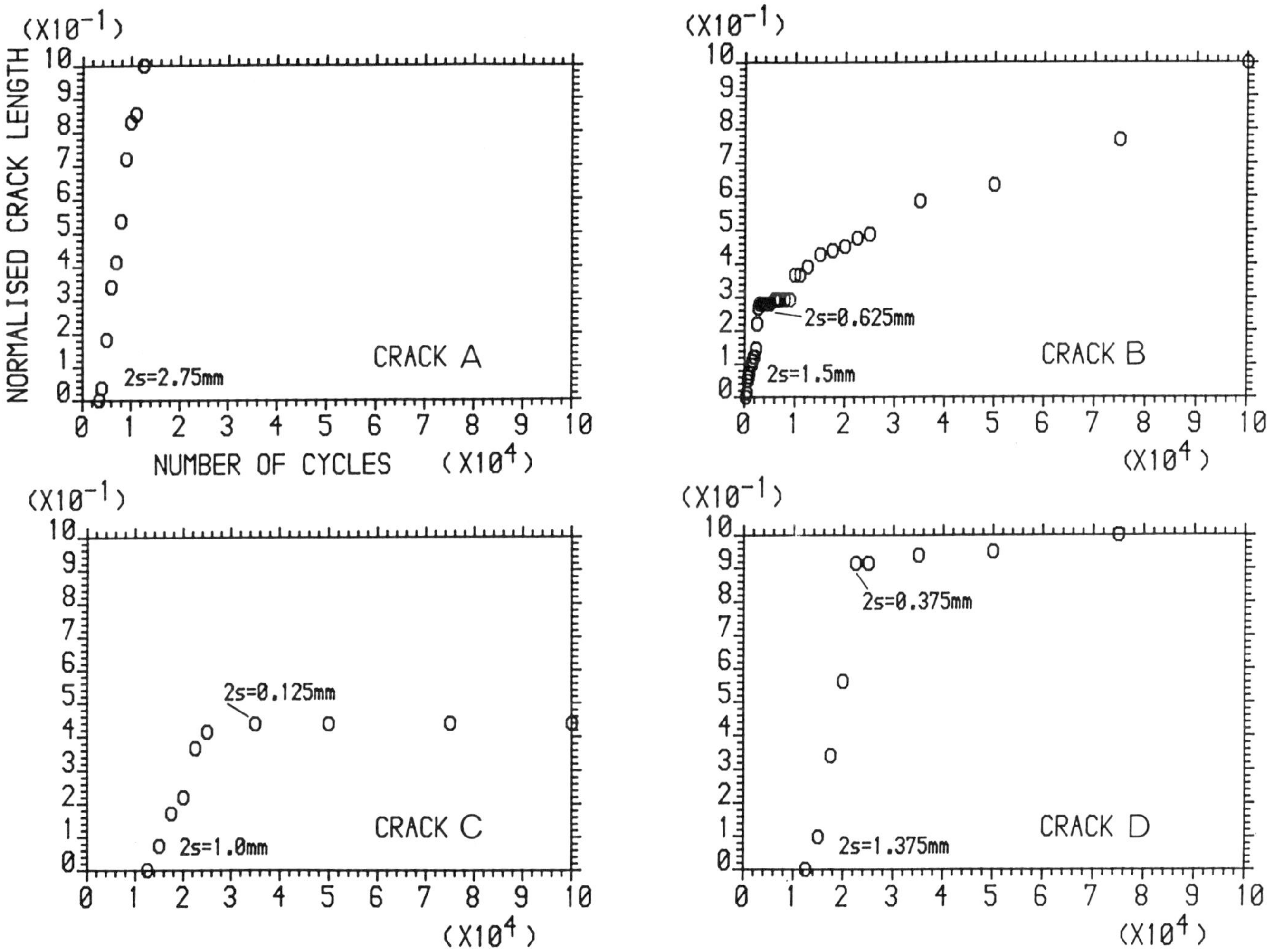

FIGURE 3 Lengths of individual transverse-ply cracks in Figure 2 as a function of number of fatigue cycles. (Courtesy of L. Boniface.)

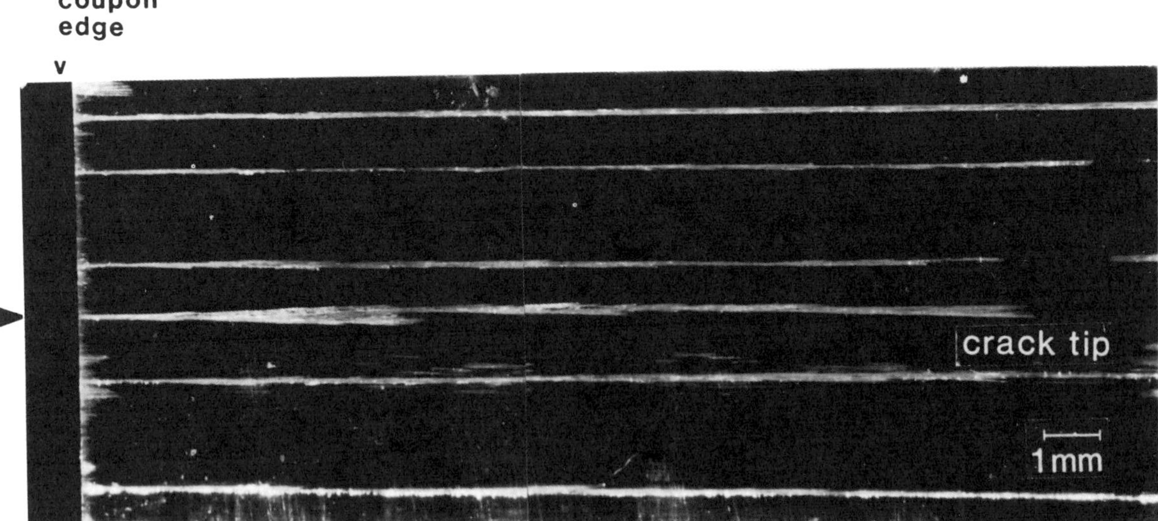

FIGURE 4 Close-up of transverse-ply cracks showing bifurcation at the crack tip. (Courtesy of L. Boniface.)

Similar results have been reported [19] on commercial laminates with a thin transverse ply (0.25 mm). In this case, the number of growing cracks was substantially increased and approached one growing crack tip per 2 mm^2 of laminate area.

Detailed observations on the crack tips themselves show them to be irregularly shaped, in general; in some cases, crack tips can bifurcate (see Fig. 4). If this occurs, the crack growth rate is substantially reduced until one branch has grown beyond the bifurcation point. Viewing crack tips using polarized light enables the stress intensification at the crack tip to be observed with the laminate under load.

Detailed studies [20,21] of the growth rates of individual cracks in CFRP have been made with the aid of penetrant enhanced X-radiography. Figure 5 illustrates the effect of different crack spacings on the growth rate in a $(0/90_3)_s$ CFRP coupon cycled at $\sigma_{\max} = 125$ MPa, $R = 0.1$. At small crack spacings (as experienced by crack *a*) the growth rate is low, but for larger crack spacings (as experienced by crack *b*) the growth rate is much higher. Cracks such as *c* and *d*, which are initially well separated but later overlap, show a dramatic reduction in growth rate.

Similar results have been found in work by Boniface et al. [21] on individual crack growth in $(0/90_2)_s$ CFRP cycled at three different peak stress levels. Using the X-ray technique, it is possible to monitor the number of cracks at the coupon edges and also those that extend beyond the center of the laminate. Figure 6 shows the difference between the number of cracks at both edges and the number crossing the center line of the laminate as a function of the number of cycles. At 2×10^5 cycles, the numbers of cracks at the laminate edges are roughly comparable, but those crossing the center line differ by about a factor of 3. Clearly, if cracks always grow by fast fracture across the laminate width from one edge to the other, or if they develop from flaws that extend the width

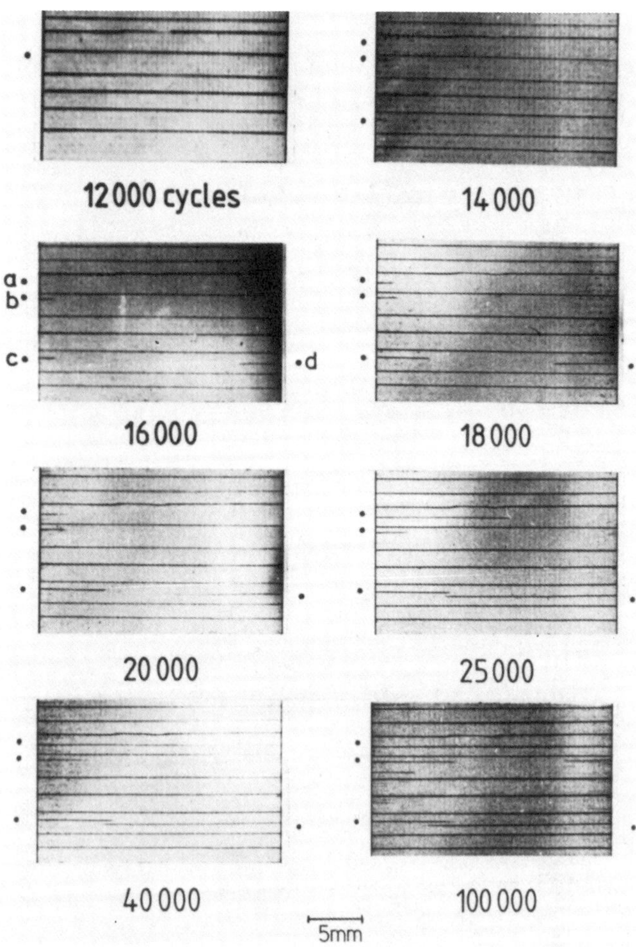

FIGURE 5 Development of cracks in $(0/90_3)_s$ CFRP laminate showing rapid growth across the width at large crack spacings (cycles 12,000 and 14,000), slow crack growth, and overlapping. (Reprinted from Ref. 32 with permission.)

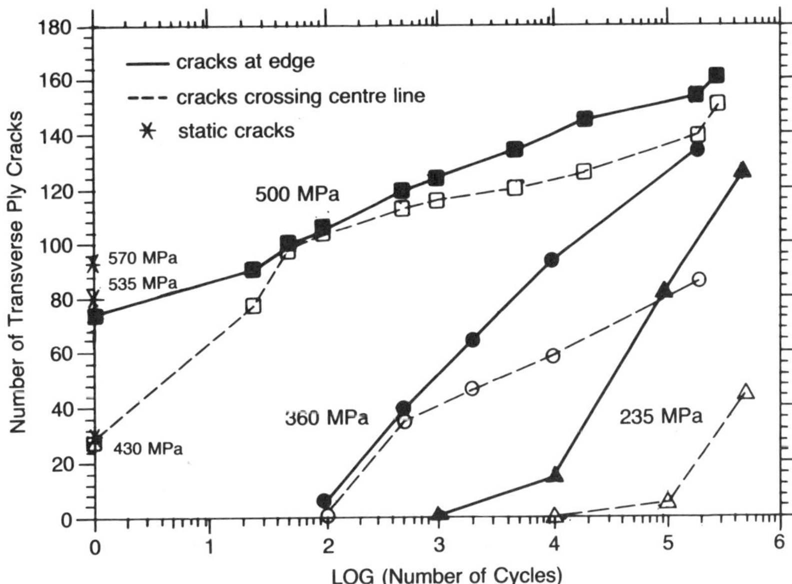

FIGURE 6 **Number of transverse-ply cracks at the edge and centerline against cycles for fatigue at σ_{max} = 500 MPa, 360 MPa, and 235 MPa.** (Reprinted from Ref. 21 with permission.)

of the laminate, we would not expect there to be differences in the number of cracks.

Although many good qualitative descriptions of crack development under fatigue loading have been given, there have been comparatively few attempts to quantify the crack growth and the associated stiffness reduction.

The first possible line of approach appears to have been proposed by Poursartip [22], who suggested that matrix cracking could be characterized by using the well-known expression for the strain energy release rate G: $G = (P^2/2B)dC/da$, where P is the laminate load, B the transverse-ply thickness, and dC/da the change in laminate compliance with transverse-ply crack length. Although he was unable to obtain a satisfactory compliance–crack length relation for his $(0/90)_{2s}$ GFRP laminate, he demonstrated a correlation between the slope of the matrix cracking regime of the stiffness reduction with cycles curve and the peak stress in the fatigue cycle. (The compliance expression for the strain energy release rate has also been used more recently by Caslini et al. [23] to analyze matrix cracking under static loading.)

The fatigue development of transverse-ply cracks has also been analyzed by Wang et al. [24] using a simulation procedure. They supposed that multiple cracking of a transverse ply occurs by the initiation and propagation of a distribution of effective flaws (as in their earlier work, these are through-width flaws growing across the transverse-ply thickness). They assumed that the variation of inherent flaw sizes and their spacings could be represented by normal distributions. If the worst flaw has a size $2a_0$, then, under static loading, the first transverse crack forms when the available energy (determined from a finite element analysis) equals the critical energy release rate. Once a flaw has become a crack, it reduces the energy release rate of neighboring flaws (by stress shielding). This is incorporated in the model by multiplying the energy release rate of the remaining flaws by an energy release rate retention factor, which takes a value between 0 and 1, dependent on the spacing. Once this has been done, the applied load is increased until some other flaw becomes critical, the energy release rates of the neighboring flaws are modified, and the load is increased again. This process is repeated until saturation is reached. To model the crack growth under fatigue loading, they assume that the flaws grow under the applied fatigue load until they reach a critical size, at which point they become transverse-ply cracks. A simple power law dependence of the crack growth rate on the strain energy release rate is assumed, and, as in the static model, the energy release rate is reduced by crack interactions. Good agreement with experimental data has been obtained using this model. An alternative model using fracture mechanics is outlined in the second part of this article.

Laminate Stiffness Properties in the Presence of Transverse-Ply Cracks

When analyzing the effect of matrix cracks on laminate stiffness and internal stress distributions, most workers idealize the cracked laminate as shown in Figure 7. The transverse ply is considered to contain a regular array of cracks, all of which span the full thickness of the transverse ply and extend across the width of the laminate, although in reality crack distribution can be quite different. The first workers to relate the presence of matrix cracks to a reduction in the longitudinal stiffness of the laminate appear to have been Hahn and Tsai [25]. They

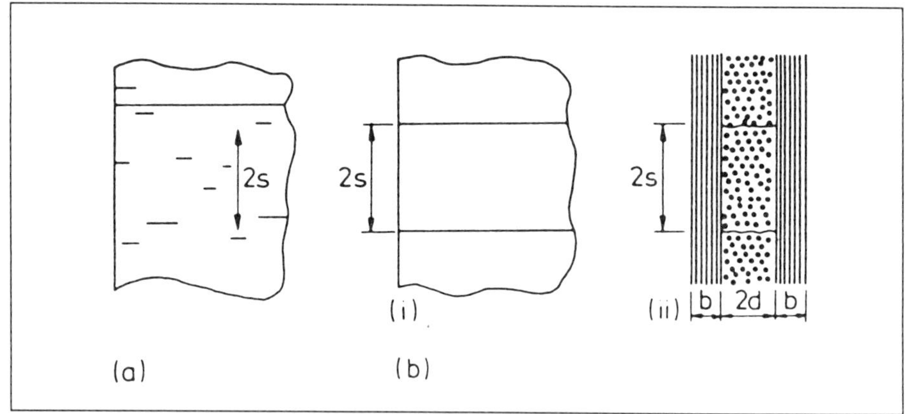

FIGURE 7 Real and idealized geometry of a portion of the cracked laminate: (*a*) real crack distribution; (*b*) idealized crack distribution, (i) front view, (ii) edge view. (Reprinted from Ref. 30 with permission.)

assumed that the presence of a transverse-ply crack led to an ineffective length over which the transverse ply carried no load and derived an expression for the stiffness of the laminate as a function of the number of cracks per unit length. The most common form of analysis used subsequently has been "shear-lag," in which load is considered to be shed into the 90° plies by shear of the constituent plies of the laminate, reaching a maximum value midway between two existing cracks.

There are also rather more sophisticated theory of elasticity approaches in the literature to dealing with the residual stiffness properties of cracked laminates [26]. Talreja [27] used a continuum mechanics model to characterize all four elastic constants of a cracked laminate. He obtained good agreement with published experimental data and also compared his results with the ply discount scheme. The ply discount scheme [28] suggests that the modulus of a heavily cracked laminate can be estimated by setting the transverse and shear stiffnesses of a cracked off-axis ply to zero and recalculating the laminate stiffnesses on that basis. Talreja [27] found that this technique overestimated the stiffness reduction.

A Model for Matrix Crack Growth

In this section a model for transverse-ply crack initiation and growth is outlined. First, the average crack spacing is related to the current laminate stiffness. Second, an approximate expression is derived for the stress intensity factor associated with an individual transverse-ply crack. Third, using the Paris law to describe the growth of individual cracks, the macrobiotic stiffness reduction rate is related to the applied fatigue stress. Finally, the approach is used to derive an expression for the laminate strain to first ply failure under static loading, and a generalized diagram for first ply failure is described.

Stiffness–Crack Spacing Relation

Figure 8 shows transverse-ply cracks in a section of a commercial $(0/90)_s$ GFRP coupon after fatigue loading. A shear-lag model describes the stiffness reduction due to matrix cracking reasonably well and can be reduced to an approximate linear relationship [29]:

$$\frac{E}{E_0} = 1 - \frac{c}{2s} \tag{1}$$

where E is the stiffness of the cracked laminate and E_0 is the stiffness of the uncracked laminate. $2s$ is the average crack spacing, and c is a constant for given material properties and specimen geometry (see Fig. 7), given analytically by:

$$c = \frac{E_0}{E_1}\left(\frac{b + d}{b} - \frac{E_1}{E_0}\right)\frac{2}{\lambda} \tag{2}$$

with:

$$\lambda^2 = \frac{3G(b + d)E_0}{d^2bE_2E_1} \tag{3}$$

FIGURE 8 Transverse-ply cracks in a 20 mm width of a $(0/90)_s$ GFRP coupon viewed in transmitted light after (*a*) 2000 cycles, (*b*) 200,000 cycles. (Courtesy of D. Lloyd-Thomas.)

E_1, E_2, and G are the longitudinal modulus of the longitudinal plies, the longitudinal modulus of the transverse plies, and the shear modulus in the longitudinal direction of the transverse plies, respectively.

Following Poursartip [22], we can define the total crack length in the transverse ply, *a*, by:

$$a = \frac{WL}{2s} \tag{4}$$

where *W* is the laminate width and *L* the gauge length. Equation (1) may then be written:

$$\frac{E}{E_0} = 1 - \left(\frac{c}{WL}\right) a \tag{5}$$

A Stress Intensity Factor for a Transverse-Ply Crack

A stress intensity factor *K* is generally a function of stress and a characteristic length. To estimate *K* for a transverse-ply crack, an idealization of the crack is used, as shown in Figure 9. The crack is considered to be flat and to extend across the thickness (but not across the width) of the transverse ply. A three-dimensional model (Fig. 9) shows the expected load trajectories around the crack. Away from the crack tip, load is transferred into the 0° plies by shear at the 0/90 interfaces and therefore does not build up at the crack tip. There is, however, a stress intensity due to the localized stress disturbance at the crack tip. With a ply thickness of 2*d*, the characteristic length of this disturbance is about *d*. Since the crack is not a through-thickness crack, but is bounded (and restrained) by the 0° plies, the stress intensity factor is less than $\sigma\sqrt{\pi d}$ and is taken to be:

$$K = \sigma_t\sqrt{2d} \tag{6}$$

where σ_t is the stress in the transverse ply acting on the crack tip. This stress is a function of the vertical separation between the cracks in the direction of the applied loading and decreases as the spacing decreases, in accordance with the qualitative suggestion of Harrison and Bader [4].

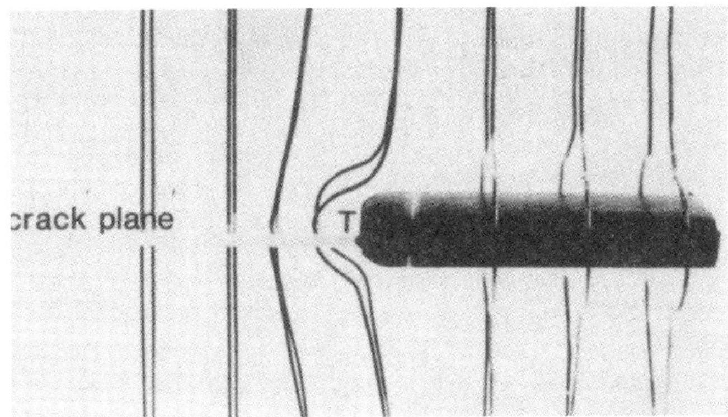

FIGURE 9 A view of a three-dimensional model of a growing transverse-ply crack showing the stress distribution along the crack and near the crack tip (at T). The buildup of stress at the crack tip is independent of the crack length. (Reprinted from Ref. 32 with permission.)

The average stress in the transverse-ply can be written approximately as [30]:

$$\sigma_t = \sigma\left(\frac{E_2}{E_0}\right)\sqrt{2\Omega(s)} \tag{7}$$

$\Omega(s)$ is a dimensionless parameter that describes the fall in the stored elastic energy in the transverse ply as the average vertical crack spacing 2*s* decreases and is a function of the laminate moduli and the thicknesses of the plies. σ is the stress applied to the laminate. When the cracks are far apart, $\Omega(s) \rightarrow 0.5$ and the average stress in the transverse ply tends to the laminated plate theory value; for closely spaced cracks (i.e., spacings less than about one-half of the transverse-ply thickness), $\Omega(s) \rightarrow 0$. Figure 10 shows a plot of the function $\Omega(s)$ for a $(0/90)_s$ GFRP laminate. This function is very similar to the energy release rate retention factor of Wang et al. [24]. So, by combining Eqs. (6) and (7), the stress intensity factor for a transverse-ply crack can be written as:

$$K = \sigma\left(\frac{E_2}{E_0}\right)\sqrt{2\Omega(s)}\sqrt{2d} \tag{8}$$

In deriving Eq. (8), no account has been taken of the heterogeneous and anisotropic nature of the transverse ply. Equation (8) can be simplified by noting that over a range of stiffness reduction, corresponding to an average crack spacing of about 1 to 3 times the ply thickness, the

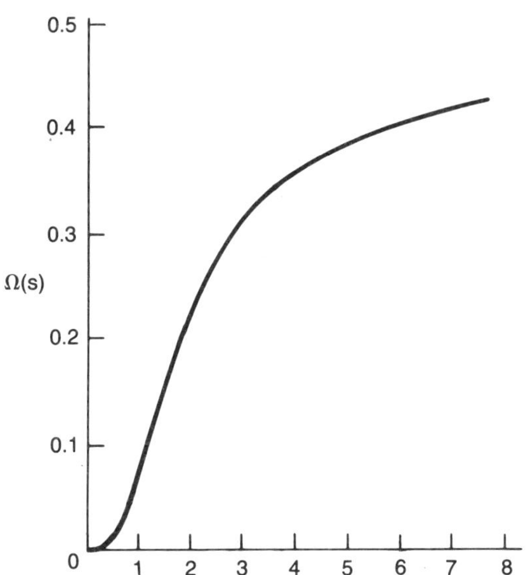

FIGURE 10 $\Omega(s)$ function that describes the fall in stored elastic energy in the transverse ply. (Reprinted from Ref. 32 with permission.)

function $\Omega(s)$ is approximately linearly related to the average crack spacing. Hence Eq. (8) becomes:

$$K = \sigma \left(\frac{E_2}{E_0}\right) \sqrt{\frac{2s}{k}} \sqrt{2d} \tag{9}$$

where k is a constant for a given laminate geometry.

Application of Stress Intensity Factor to Fatigue Behavior

The stress intensity factor is related to the total crack growth rate as follows: The growth rate of an individual transverse-ply crack of length $2z$ is governed by the Paris relation, that is:

$$\frac{dz}{dN} = B'K_{max}^m \tag{10}$$

where the stress intensity factor K_{max} is given by Eq. (9).

Observations and measurements of the growth of individual cracks in commercial GFRP have suggested that over a large range of stiffness reduction, the number of active crack tips in the transverse ply p is very approximately constant. The total crack growth rate is therefore given by:

$$\frac{da}{dN} = pB'K_{max}^m \tag{11}$$

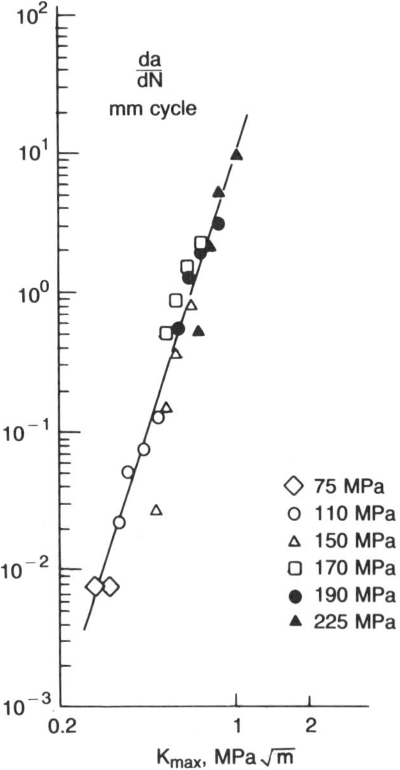

FIGURE 11 Plot of inferred total crack growth rate in the transverse ply against the maximum stress intensity factor for a $(0/90)_s$ GFRP laminate. (Reprinted from Ref. 30 with permission.)

By using Eq. (5) and applying the chain rule for differentiation, the total crack growth rate can be expressed in terms of the stiffness reduction rate:

$$\frac{da}{dN} = \left(\frac{WL}{c}\right)\left(-\frac{1}{E_0}\frac{dE}{dN}\right) \tag{12}$$

where $(-1/E_0)(dE/dN)$ is simply the slope of the normalized stiffness with cycles curve.

The right-hand side of Eq. (11) may be expressed in terms of the current value of the stiffness by rewriting K_{max} (as given by Eq. 9), using Eq. (1), as:

$$K_{max} = \sqrt{\frac{c}{k}} \cdot \left(\frac{E_2}{E_0}\right) \sigma_{max} \sqrt{\frac{1}{1 - E/E_0}} \cdot \sqrt{2d} \tag{13}$$

where σ_{max} is the maximum stress in the fatigue cycle.

From Eq. (11) we would expect to obtain a correlation between da/dN and K_{max} from a log-log plot, and this has been found [27]. Figure 11 shows a log-log plot of the inferred total crack growth rate (from Eq. 12) against the approximate maximum stress intensity factor, for different values of σ_{max} (Eq. 13), for data from the fatigue of a $(0/90)_s$ GFRP laminate. There is a good correlation with the exponent m about 6.

It is also possible to relate the slope of the normalized stiffness with cycles curve to the applied stress level and the current value of the laminate modulus. Combining Eqs. (11) to (13), it can be shown that:

$$-\frac{1}{E_0}\frac{dE}{dN} = A\left[\frac{\sigma_{max}^2}{E_0^2(1 - E/E_0)}\right]n \tag{14}$$

where $n = m/2$ and A is a constant. Integration of this expression enables predictions of stiffness with cycles to be made (see the second section of this article).

Application of Stress Intensity Factor to 90° Ply Failure under Static Loading

The expression for K can be used to find the laminate strain to first cracking of the 90° ply for different transverse-ply thicknesses; first ply failure is taken to be the strain to form the first transverse-ply cracks that span the thickness of the transverse ply and the width of the laminate.

When the crack spacing is very large (i.e., $\Omega(s) \to 0.5$), Eq. (8) becomes:

$$K = \sigma \left(\frac{E_2}{E_0}\right) \sqrt{2d} \tag{15}$$

Assuming that the fracture toughness K_c and the toughness G_c for cracks growing parallel to the fibers in the transverse-ply are related by:

$$K_c = \sqrt{E_2 G_c} \tag{16}$$

then Eq. (15) can be rewritten:

$$\varepsilon_f = \sqrt{\frac{G_c}{E_2}} \frac{1}{\sqrt{2d}} \tag{17}$$

Here ε_f is the laminate strain for the formation of the first transverse-ply crack. This expression should, of course, be modified to include the effect of the (tensile) thermal stresses in the 90° ply. However, this effect is small in GFRP and has been neglected.

The prediction of Eq. (17) can be compared with the experimental results and predictions of Parvizi et al. [31] for a GFRP laminate (see Fig. 12). There is good agreement between the two predictions and the experimental results for ply thicknesses of less than about 0.5 mm. At thicknesses greater than about 0.5 mm, the laminate strains to the first cracking are reasonably constant and about equal to the failure strain of the 90° ply tested alone. Bailey et al. [9] have observed that cracks nucleate from fiber debonds and that at large ply thicknesses the cracks propagate instantaneously. Crack propagation, they suggest, is controlled essentially by a thermodynamic condition (equivalent to Eq. 17 above) at small ply thicknesses, but at larger ply thicknesses it is mechanism controlled.

The origin of the mechanism-controlled growth can be illuminated using the stress intensity model. At a critical applied laminate strain ε_{deb}, fiber debonds appear in the transverse ply, assisted by the strain magnification in the matrix around the fibers. At a higher strain ε_{crit}, some of these debonds link up to form a flaw [1]. The extent of the flaw parallel to the fibers is not known, but will depend, presumably, on the variation in the fiber spacing through the laminate width.

The flaw can now be treated as an enclosed crack with dimensions $2a_d$ by $2l_0$ perpendicular and parallel to the fibers, respectively (see Fig. 13). Whether or not the enclosed crack leads to fast fracture depends on the thickness of the transverse ply.

THICK TRANSVERSE PLIES. The enclosed crack grows by the linking of debonds, at the strain ε_{crit}, to a critical size $2a_d^{crit}$ by $2l'$ (where $2l' > 2a_d^{crit}$), such that the condition:

$$\varepsilon_{crit} E_2 \sqrt{\pi a_d^{crit}} > K_c \tag{18}$$

is satisfied. The enclosed crack then grows by fast fracture across the laminate thickness and width.

THIN TRANSVERSE PLIES. The transverse-ply thickness is smaller than the critical flaw width for fast fracture, $2a_d^{crit}$. A flaw that has grown across the ply thickness is therefore stable. The stress intensity factor perpendicular to the fiber direction is now given by Eq. 15. Fast fracture occurs when the laminate strain ε is increased such that:

$$\varepsilon E_2 \sqrt{2d} > K_c \tag{19}$$

The above approach gives predictions consistent with the experimental results of Parvizi et al. [11]. From their data for a GFRP laminate (Fig. 12), the strain to first failure for thick transverse plies is constant at $\varepsilon_f \sim 0.6\%$. Taking $G_c = 240\ \mathrm{J/m^2}$, then from Eq. (18):

$$a_d^{crit} \sim 0.15\ \mathrm{mm}$$

and the transition from thin to thick transverse plies occurs at $2d = \pi a_d^{crit}$, that is, $2d \sim 0.5$ mm. The experimental results show that the transition occurs at about this value.

This approach enables a generalized diagram for first ply failure to be constructed for laminates containing transverse plies of different thickness and toughness. Equation (17) can be rewritten:

$$\varepsilon_f \sqrt{\left(\frac{E_2}{G_c}\right)} = \frac{1}{\sqrt{2d}} \tag{20}$$

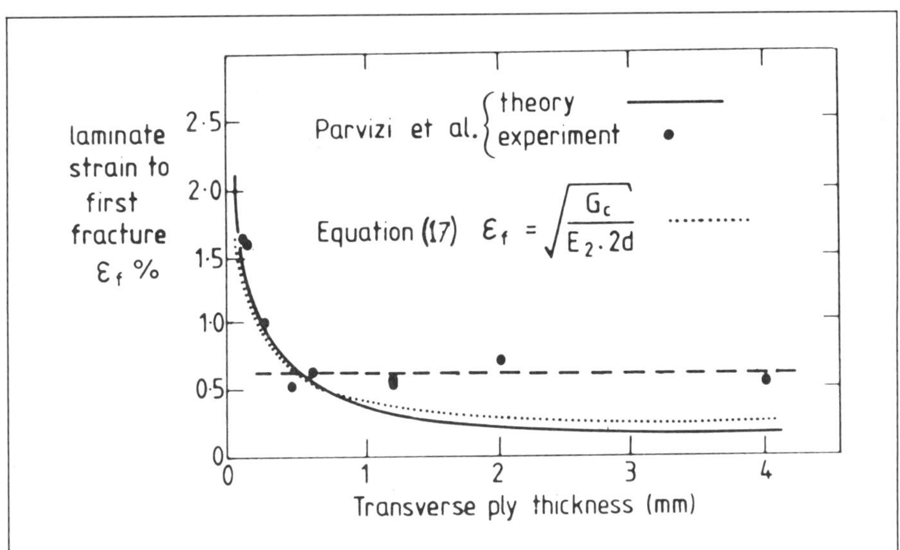

FIGURE 12 **Comparison between the theory and experimental results of Parvizi et al. [11] and the prediction of Eq. (17).** (Reprinted from Ref. 32 with permission.)

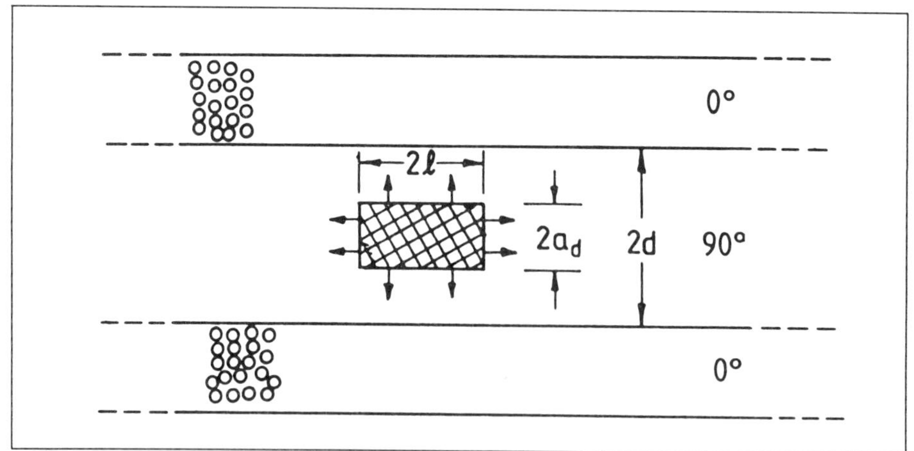

FIGURE 13 Enclosed crack in the transverse ply. (Reprinted from Ogin and Smith, *Scorpta Met.*, 19, 779 (1985) with permission from Pergamon Press plc.)

This suggests that a plot of $\varepsilon_f\sqrt{E_2/G_c}$ against $2d$ should fall onto one curve for 90° plies in any laminate (ignoring any effect of the orientation of neighboring plies), providing that thermal strains are included. This of course will apply only over the region of small transverse-ply thicknesses. The cutoff point for a given fiber–resin system will occur at $\varepsilon_c\sqrt{(E_2/G_c)}$, where ε_c is the unconstrained failure strain of the 90° ply, and therefore there is a range of values over which the cutoff occurs. Data are shown plotted in Figure 14 for a variety of glass and carbon laminates.

Implications for Design against Fatigue

Equation (14) can be used to reduce data (in the form of stiffness reduction with cycles curves) from tests at different peak cyclic stresses onto one plot. Figure 15 shows an example for $(0/90)_s$ GFRP, for which $A = 5.65 \times 10^4$ and $n = 2.83$. Integration of this equation gives a relationship between the current value of the modulus E/E_0, the cycle number N, and the peak stress $\sigma_{\max}$:

$$\frac{E}{E_0} = 1 - \left[25.3 \times \left(\frac{\sigma_{\max}}{E_0}\right)^{1.48} \times N^{0.26}\right]$$

Figure 15, which is a plot of this equation for the same $(0/90)_s$ GFRP laminate, shows the number of cycles required to reach 95%, 90%, and 85% of the initial stiffness. For example, the diagram predicts that with a cyclic peak stress of 170 MPa and with $R = 0.1$ (i.e., $\sigma_{\max} = 0.34$; σ_{UTS} for this system is about 500 MPa), the number of

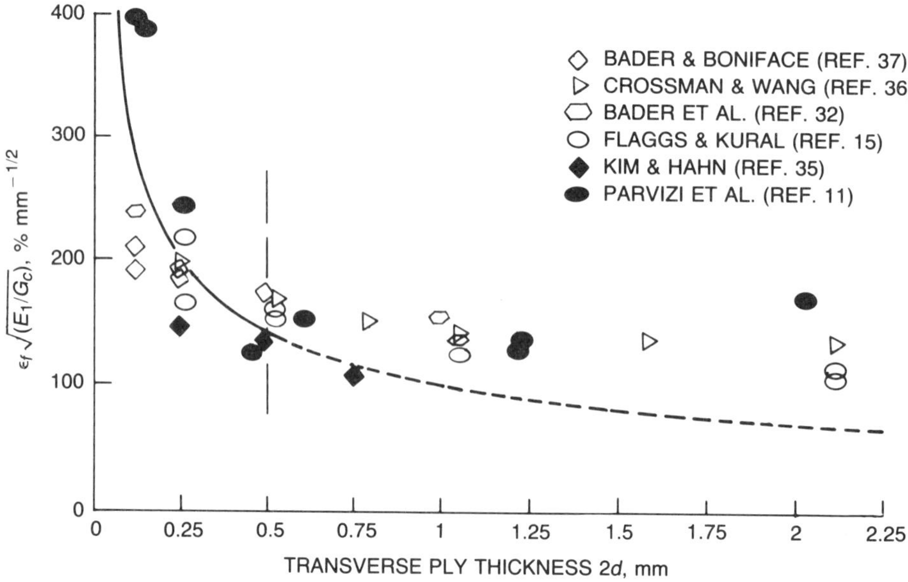

FIGURE 14 Generalized diagram for the total transverse-ply strain to failure for cross-ply and angle-ply laminates. The vertical line shows the approximate position of the transition from thin- to thick-ply behavior.

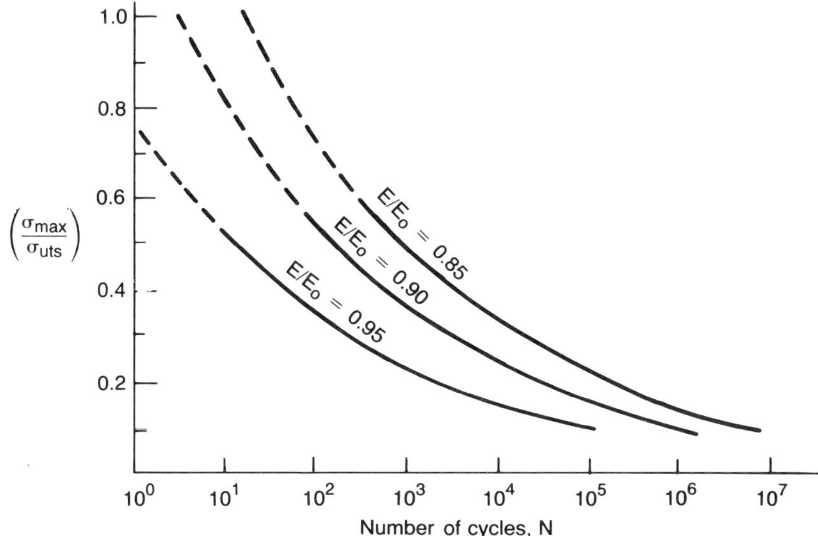

FIGURE 15 **Diagram relating the number of cycles required to reach a given stiffness reduction to the maximum stress in the fatigue cycle expressed as a fraction of the tensile strength.** (Reprinted from Ref. 29 with permission.)

cycles required to produce a 10% stiffness reduction is about 1500. The development of these diagrams could be of practical use in designs where stiffness is a failure criterion [32].

S. L. Ogin and P. A. Smith

References

1. J. E. Bailey and A. Parvizi, *J. Mater. Sci., 16,* 649 (1981).
2. J. A. Kies, "Maximum Strains in the Resin of Fiberglass Composites," U.S. Naval Research Laboratory, Washington, D.C., Report No. 5752, 1962.
3. K. W. Garrett and J. E. Bailey, *J. Mater. Sci., 12,* 2189 (1977).
4. R. P. Harrison and M. G. Bader, "The Micromechanics of Carbon Fibre Composites Damage Initiation and Development under Quasi-Static and Fatigue Loading," Department of Metallurgy and Materials Technology, University of Surrey, Final Report A93B/25, 1981.
5. K. Schulte, in T. Feest, Ed., *Proceedings of the Int. Conference on Testing, Evaluation and Quality Control of Composites,* Surrey University, Butterworths Press, 1983, pp. 233–242.
6. M. G. Bader and L. Boniface, *Proceedings of ICCM-V,* San Diego, 1985, pp. 221–232.
7. M. G. Bader, J. E. Bailey, P. T. Curtis, and A. Parvizi, *Proceedings of the Third Int. Conference on Mechanical Behaviour of Materials,* Vol. 3, Cambridge, England, 1979, pp. 227–239.
8. K. L. Reifsnider and J. E. Masters, "Investigation of Characteristic Damage States in Composite Laminates," American Society of Mechanical Engineers (ASME) Winter Annual Meeting, San Francisco, 1978.
9. J. E. Bailey, P. T. Curtis, and A. Parvizi, *Proc. R. Soc. Lond., A366,* 599 (1979).
10. J. Aveston and A. Kelly, *J. Mater. Sci., 8,* 352 (1973).
11. D. L. Flaggs and M. H. Kural, *J. Compos. Mater., 16,* 103 (1982).
12. A. S. D. Wang and F. W. Crossman, *J. Compos. Mater., 14,* 71 (1980).
13. P. W. Manders, T.-W. Chou, F. R. Jones, and J. W. Rock, *J. Mater. Sci., 18,* 2876 (1983).
14. H. Fukunaga, T.-W. Chou, P. W. M. Peters, and K. Schulte, *J. Compos. Mater., 18,* 339 (1984).
15. H. Fukunaga, T.-W. Chou, K. Schulte, and P. W. M. Peters, *J. Mater. Sci., 19,* 3546 (1984).
16. P. W. M. Peters, *J. Compos. Mater., 18,* 545 (1984).
17. K. L. Reifsnider and R. Jamison, *Int. J. Fatigue, 4,* 187 (1982).
18. K. L. Reifsnider, "Fatigue and Creep of Composite Materials," *Proceedings of the 3rd Riso Int. Symposium of Metallurgy and Materials Science,* 1982, pp. 125–136.
19. S. L. Ogin, *Proceedings of the European Symposium on Damage Development and Failure Processes in Composite Materials,* Leuven, Belgium, 1987, pp. 56–61.
20. S. L. Ogin, L. Boniface, and M. G. Bader, *Proceedings of the Conference on Fibre Reinforced Composites,* Liverpool, I. Mech. E., 1986, pp. 173–178.
21. L. Boniface, P. A. Smith, S. L. Ogin, and M. G. Bader, in: *Proceedings of Sixth International Conference on Composite Materials, ICCM-IV,*F. L. Matthews, N. C. R. Buskell, J. M. Hodgkinson and J. Morton, Eds., Elsevier Applied Science, 1987, pp. 3.156–3.165.
22. A. Poursartip, "Aspects of Damage Growth in Fatigue of Composites," Ph.D Thesis, Cambridge University Engineering Department, 1983.
23. M. Caslini, C. Zanotti, and T. K. O'Brien, *J. Compos. Tech. and Res., 9,* 121 (1987).
24. A. S. D. Wang, P. C. Chou, and S. C. Lei, *J. Compos. Mater., 18,* 239 (1984).
25. H. T. Hahn and S. W. Tsai, *J. Compos. Mater., 8,* 288 (1974).

26. G. J. Dvorak, N. Laws, and M. Hajazi, *J. Compos. Mater., 19,* 216 (1985).

27. R. Talreja, *J. Compos. Mater., 19,* 355 (1985).

28. A. L. Highsmith and K. L. Reifsnider, "Stiffness Reduction Mechanisms in Composites," ASTM STP 775, 103 (1982).

29. S. L. Ogin, P. A. Smith, and P. W. R. Beaumont, *Compos. Sci. and Tech., 22,* 23 (1985).

30. S. L. Ogin, P. A. Smith, and P. W. R. Beaumont, *Compos. Sci. and Tech., 24,* 47 (1985).

31. A. Parvizi, K. W. Garrett, and J. E. Bailey, *J. Mater. Sci., 13,* 195 (1978).

32. S. L. Ogin and P. A. Smith, *European Space Agency (ESA) Journal, 11,* 45 (1987).

Laminates, Static Strength

Advanced fiber composites have excellent strength-to-weight properties and are often used in strength-critical applications. This article presents procedures for relating the strength of laminates to the strength of the constituent materials. Laminates can have three distinct failure modes: delamination, matrix failure, and fiber failure. In general, the existence of multiple failure modes even under static loading complicates the assessment of strength relative to isotropic materials (like metals). However it is possible to establish a rational procedure to assess laminate strength. This article considers the strength of laminates under conditions of uniform static stress.

Basic Approach

Two choices are available in doing a strength assessment of composite laminates. In one approach the laminate being considered is treated as an individual material, and the strength properties are established through tests on that laminate. Because of the many possible laminates of interest, this procedure is not feasible in most cases. The alternative is to consider the properties of the individual plies that make up the laminate, and assess the strength of the laminate on a ply-by-ply basis. This approach is used here. In this approach, however, the particular mode of failure of the individual plies must be carefully considered. In general, matrix cracking may or may not lead directly to failure, while fiber failure usually corresponds to the ultimate failure of the laminate. Examples will be given to clarify these points.

The stress analysis of the laminate furnishes an overall state of strain in the laminate, from which the strains in each individual ply can be calculated. The strains in each ply are then transformed into the directions of the fibers, using standard formulas for rotation of strain components with rotation of coordinate directions. Finally, the in-plane stresses in each ply can be obtained by using the stress–strain relationship for the ply. The stresses and strains will then be known for each ply in the laminate. In practice, this procedure can be most easily accomplished with the aid of a micro or other computer, since the matrix multiplications are tedious to do by hand. The equations employed to obtain the ply stresses and strains are given as follows.

The laminate strains are obtained from the laminate applied loads (in the form of stress resultants and moments) from the standard classical lamination theory (CLT) relationships, given by

$$\begin{Bmatrix} \varepsilon_0 \\ \kappa \end{Bmatrix} = \begin{bmatrix} A & B \\ B & D \end{bmatrix}^{-1} \begin{Bmatrix} N \\ M \end{Bmatrix} \tag{1}$$

where formation of the A, B, and D matrices is as defined by Jones [1], for example. The strains in each ply can then be obtained from the laminate centerline strains ε_0 and curvatures κ by

$$\{\varepsilon\} = \{\varepsilon_0\} + z\{\kappa\} \tag{2}$$

The strains in each ply can then be rotated into the fiber directions by the rule for transformation of coordinates, if the engineering strains are first converted into tensor components. Again, following Jones, [1], this can conveniently be done by multiplying by the R matrix:

$$R = \begin{bmatrix} 1 & 0 & 0 \\ 0 & 1 & 0 \\ 0 & 0 & 2 \end{bmatrix} \tag{3}$$

The transformation of the strains given with respect to the fiber directions is then given by

$$\begin{Bmatrix} \varepsilon_1 \\ \varepsilon_2 \\ \gamma_{12} \end{Bmatrix} = [R][T][R]^{-1} \begin{Bmatrix} \varepsilon_x \\ \varepsilon_y \\ \gamma_{xy} \end{Bmatrix} \tag{4}$$

where

$$T = \begin{bmatrix} \cos^2\theta & \sin^2\theta & 2\sin\theta\cos\theta \\ \sin^2\theta & \cos^2\theta & -2\sin\theta\cos\theta \\ -\sin\theta\cos\theta & \sin\theta\cos\theta & \cos^2\theta - \sin^2\theta \end{bmatrix} \tag{5}$$

The ply stresses can then be obtained by multiplying the ply strains by the lamina stress–strain matrix.

$$\begin{Bmatrix} \sigma_1 \\ \sigma_2 \\ \tau_{12} \end{Bmatrix} = \begin{bmatrix} E_{11}/d & \nu_{12}E_{22}/d & 0 \\ & E_{22}/d & 0 \\ \text{sym.} & & G_{12} \end{bmatrix} \begin{Bmatrix} \varepsilon_1 \\ \varepsilon_2 \\ \gamma_{12} \end{Bmatrix} \tag{6}$$

where $d = 1 - \nu_{12}^2 E_{22}/E_{11}$ and the material constants are those of the orthotropic ply. That is, E_{11} is the modulus in the fiber direction, E_{22} is the modulus transverse to the fibers, G_{12} is the in-plane shear modulus, and ν_{12} is the negative ratio of strain in the transverse direction to strain in the fiber direction, when a uniaxial stress is applied in the fiber direction. Note that the subscripts 1 and 2 are conventionally used in lamination theory to stand for the fiber direction and normal to the fiber, respectively, not for principal stresses. This procedure is just the usual method of obtaining ply stresses, which has been explained in detail in many references, for example, in Ref. 1.

The final product of these manipulations consists of the ply stresses and strains, referred to the direction of the fiber and transverse to the fiber, in each ply of the laminate. The remaining step in the strength assessment is to compare the calculated ply stresses and strains with allowable values, using material properties in conjunction with suitable failure theories. This important step is discussed next.

Ply Failure Theories

Before discussing the various possible failure theories that can be applied to the individual plies of a laminate, it is useful to consider the physical processes involved in failure. In typical polymer-based composites, the resin has sufficient elongation capability that fiber failure occurs before resin failure in the usual unidirectional tension coupon test with the load parallel to the fibers. On a more detailed level, it is believed that the matrix plays a significant role in bridging around the individual fiber breaks that occur at weak points in the fibers [2–4] and that the ultimate fiber failure occurs when these individual fiber breaks are coupled together in sufficient amount. From a macroscopic, ply-level viewpoint, fiber failure is characterized by either the tensile stress or the tensile strain at failure in the unidirectional specimen. The situation is more complicated in compression because both fiber and matrix play a role in determining the strength [2,4,5], in a complicated manner that is not well understood.

When tested in a direction transverse to the fibers, the composite typically fails in the matrix at a transverse strain level often significantly less than the failure strain of neat resin, and also much lower than the fiber failure strain under axial loading. The higher modulus fibers are thought to serve as stress concentration points so that the ply transverse strain is much lower than the neat matrix failure strain. As a consequence of this lower transverse strain to failure, a laminate may exhibit matrix failure in the transverse plies (relative to the major loading axis) well before failure of the fibers that are in the loading direction. Thus, for example, matrix cracking will occur in a [0/90] laminate in the 90° plies, if the loading is in the zero direction. This matrix cracking, the first manifestation of laminate failure, has been studied extensively [6–9]. As the loading is increased, further matrix cracking will occur, forming a more or less regular spacing [7]. While transverse strain (or stress) is often used to characterize the propensity to produce matrix cracks, energetic approaches have also been applied [8]. The energy approach indicates that the "in situ" matrix strength will actually depend on the thickness of adjacent ply groups, which has been reported in experiments [6–9].

The matrix cracking has a number of effects on the laminate. For example, it increases permeability to moisture. It also tends to reduce the effective transverse properties of the ply. A number of studies have in fact attempted to model the reduction in transverse stiffness with matrix cracking, using averaged continuum properties [10–12]. It has been shown that the matrix cracks can serve as initiation sites for delamination under fatigue loading [13].

It is important to note, however, that matrix cracking may or may not lead directly to ultimate laminate failure. If the laminate loading is carried primarily by the matrix, matrix cracking and subsequent softening can lead directly to failure, as for example in the shear loading of a [0/90] laminate. However in most practical situations the laminate is designed so that the load is carried by fibers, to take advantage of the strength of the fibers relative to the weak matrix. In these cases the ultimate strength of the laminate may be several times that of the load corresponding to the initiation of ply cracking, and cracking is caused by the failure of fibers. In typical carbon–epoxy laminates under static tension loading, the stress–strain response may be quite linear up to failure (say within 5%), with the softening of the transverse properties due to matrix cracking being counterbalanced by the stiffening of the fiber with loading. Thus we can conclude that matrix cracking and ultimate laminate failure are typically separate events and must be considered separately. This means that separate failure criteria must be established for matrix cracking and fiber failure [14,15].

Although a large number of possible failure criteria have been suggested for use with composite materials, almost all are derived from applications to materials other than laminated composites and, as a consequence, do not differentiate between the modes of failure. As such, they cannot be applied in a rational manner to composite laminates on a ply basis, where it is necessary to distinguish between matrix and fiber failure.

Before considering specific failure criteria, it is useful to review some experimental evidence on ply and laminate failure. Figure 1 gives the transverse failure properties of AS4/55A and IM7/8551-7 carbon–epoxy systems [16]. The first system is filament wound, while the second is preimpregnated with a "high toughness" resin system. Both are unidirectional, and both are tested in the form of tubes subject to torsion combined with axial tension or compression. The results show that the matrix-dominated transverse strength properties are dependent on both the transverse normal stress and the transverse shear stress. Furthermore, there is a strong interaction between these stress components. Thus any criterion applicable to transverse failure by matrix cracking must account for this interaction of stress components.

Experimental evidence on ultimate laminate failure is given in Figures 2 and 3 for AS4/3501-6 carbon/epoxy [17] and Figure 2 for IM7/8551-7 carbon–epoxy [16]. The laminates for both materials are quasi-isotropic $[0/\pm45/90]_s$ and were tested in the form of 96 mm (3.8 in.) i.d. cylinders with the applied loading being internal pressure and axial tension or compression. Additional data of this type have been obtained on other laminates and materials, including $[0/\pm60]$, quasi-isotropic laminates loaded at an angle to the fibers for biaxial tension loading [18–20], and [0/90] laminates in compression loading [21].

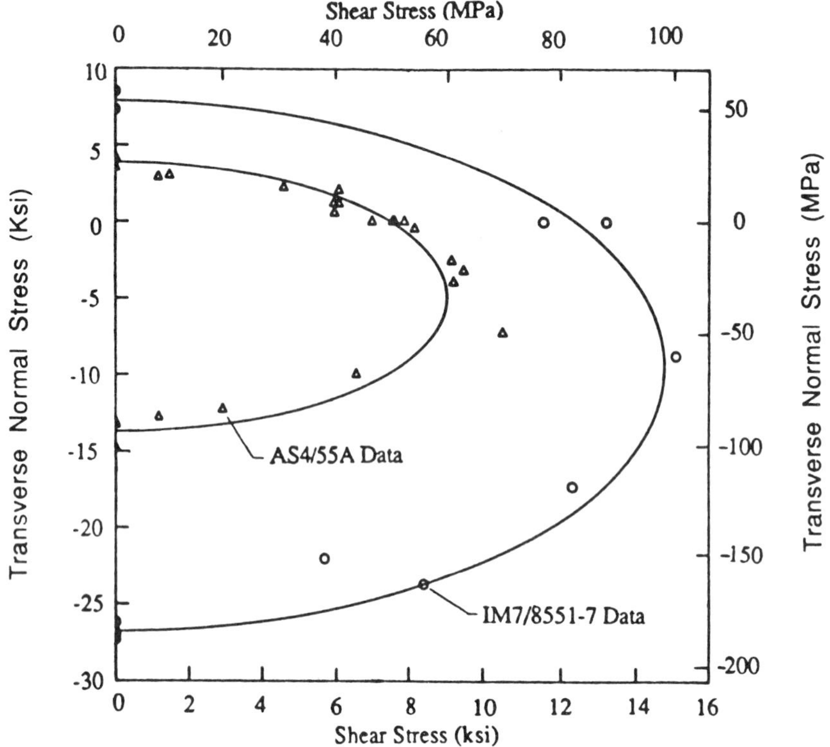

FIGURE 1 Transverse ply failure properties for AS4/55A and IM7/8551-7 carbon–epoxy lamina. (From Ref. 16.)

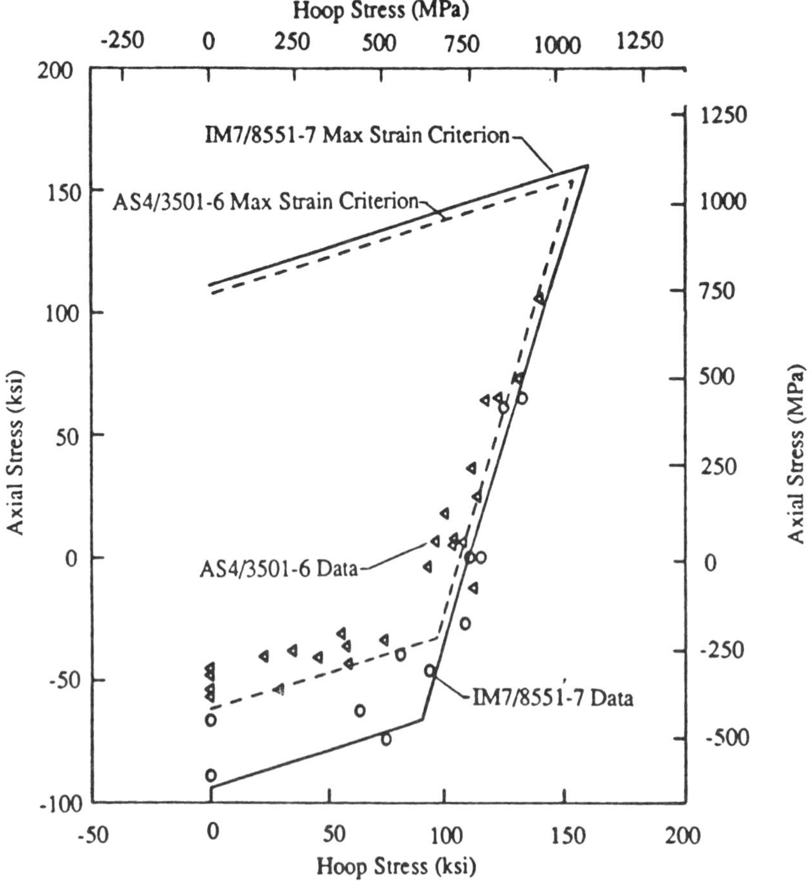

FIGURE 2 Laminate failure stresses for AS4/3501-6 and IM7/8551-7 carbon–epoxy laminates in quasi-isotropic [0/±45/90] configuration, under biaxial stress loading. (From Ref. 16.)

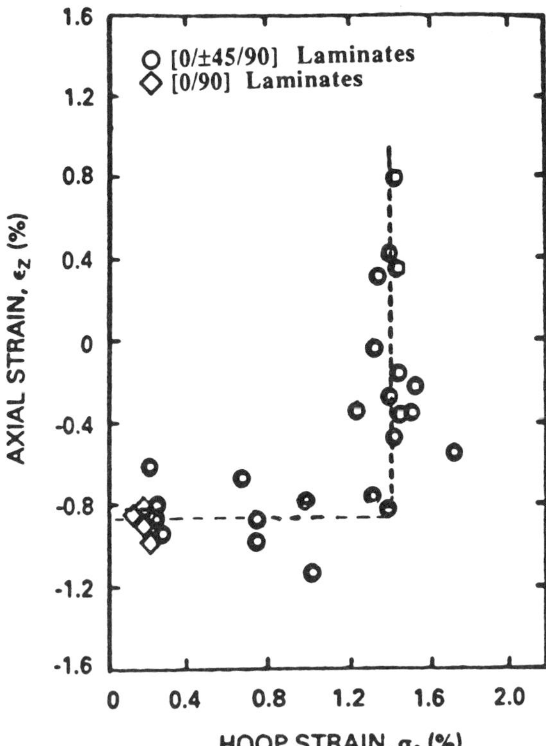

FIGURE 3 **Laminate failure strains in AS4/3501-6 carbon–epoxy laminates in quasi-isotropic [0/±45/90] and [$0_{12}/90_2$] configurations.** (From Ref. 16.)

Several failure criteria have been suggested for direct application to composites. The stress polynomial due to Tsai and Wu [22] is given in the usual quadratic form as

$$F_1\sigma_1 + F_{11}\sigma_1^2 + F_2\sigma_2 + F_{22}\sigma_2^2 + 2F_{12}\sigma_1\sigma_2 + F_{66}\tau_{12}^2 = 1 \quad (7)$$

where the F terms are material constants, and the stresses are the in-plane ply stresses. The various F terms can be easily related to experimental data on ply failure by evaluating Eq. (7) for the various simple material property test conditions. This gives

$$F_1 = 1/X_t + 1/X_c \qquad F_{11} = -1/(X_tX_c) \qquad F_{66} = 1/S^2$$
$$F_2 = 1/Y_t + 1/Y_c \qquad F_{22} = -1/(Y_tY_c) \quad (8)$$

where X_t and X_c are unidirectional strength parallel to the fibers in tension and compression, respectively, Y_t and Y_c are transverse strengths, and S is the shear strength. The F_{12} term must be obtained from multiaxial tests on a lamina, and is usually taken as either

$$F_{12} = 0 \quad (9)$$

or

$$F_{12} = -0.5\sqrt{F_{11}F_{22}} \quad (10)$$

This polynomial can be used directly to predict first ply failure, which may correspond to ultimate laminate failure if the major load is in compression and will usually correspond to matrix cracking if the major load is tension. This criterion does not directly differentiate between matrix and fiber failure, but it can be interpreted as doing so indirectly by assuming that first ply failure corresponds to matrix failure, and last ply failure corresponds to ultimate failure of the laminate [23].

Two additional criteria available for ply failure prediction are based on separating the foregoing polynomial into two parts, one describing matrix failure and the other describing fiber failure. Hahn, Erikson, and Tsai [15] recommend the following:

Fiber failure:

$$F_1\sigma_1 + F_{11}\sigma_1^2 = 1 \quad (11)$$

Matrix failure:

$$F_2\sigma_2 + F_{22}\sigma_2^2 + F_{66}\tau_{12}^2 = 1 \quad (12)$$

A similar proposal has been made by Hashin [14], who recommends the following:

Fiber failure:

$$F_1\sigma_1 + F_{11}\sigma_1^2 + F_{66}\tau_{12}^2 = 1 \quad (13)$$

Matrix failure:

$$F_2\sigma_2 + F_{22}\sigma_2^2 + F_{66}\tau_{12}^2 = 1 \quad (14)$$

In either form, there is no ambiguity about what type of failure is being predicted.

A failure criterion that is widely used for predicting fiber failure on a ply basis is that of maximum fiber direction strain. Because composites are often notably weaker in compression than in tension, two material property values are needed, and the criterion becomes

$$\varepsilon_{1c} < \varepsilon_1 < \varepsilon_{1t} \quad (15)$$

where ε_{1c} and ε_{1t} are fiber direction failure strains in compression and tension respectively.

Comparison with Experimental Data

Matrix Failure

The matrix mode of failure shown in Figure 1 can be represented by either the Tsai–Wu criterion (Eq. 7), the expression for matrix failure by Hahn et al. (Eq. 12), or the equivalent expression for matrix failure by Hashin (Eq. 14). In fact these criteria all reduce to the same expression for the state of stress of Figure 1, which represents combinations of transverse normal and in-plane shear stresses. The correlation of these three criteria with the data is quite good, since they represent the interaction of the transverse normal and shear stresses seen in the experimental results.

As mentioned above, experimental evidence suggests that the matrix failure strength is different within the laminate and in unidirectional laminas due to the restraint offered by the adjacent plies [6,8,9]. It has been suggested that the lamina strength values can be adjusted for this "in situ" effect. The alternative is to abandon the concept of a stress-based failure criterion and use energy release rates in conjunction with the assumption of inherent flaws to determine matrix cracking [6]. This latter approach is of course much more complex.

Ultimate Laminate Failure

The data given in Figures 2 and 3 indicate that the failure of laminates can be correlated within the accuracy of the experimental data by the maximum fiber strain criterion. This is a fundamental conclusion of the experimental studies reported in Refs. 16–21. Since fiber strain has long been used as a laminate failure criterion in practical applications, the results shown in Figures 2 and 3 and in Refs. 16–21 are not particularly surprising. However this carefully controlled experimental work does add credibility to a criterion that is sometimes regarded as being "too simple." The laminate ultimate stress predicted on the basis of the maximum fiber direction strain criterion is shown compared with the data for AS4/3501-6 in Figures 2 and 4, while Figure 3 compares the laminate failure strains with the fiber strain allowable values ε_{1c} and ε_{1t}.

The value of the fiber failure strain in the laminates composed of AS4/3501-6 appears to be nearly the same as that measured in unidirectional tensile coupons. However, the laminate fiber strain value for IM7/8551-7 appears to be about 20% lower than the values measured in tensile coupons. Thus it may be necessary in general to establish allowables from laminate tests, rather than simply from using tensile coupon values. This creates a fundamental difficulty in that it is difficult to test laminates without introducing free-edge effects. In compression, it is not clear whether there is a reduction in strain allowables due to a laminate effect. The compressive failure strains of the laminates shown in Figure 3 are lower than those determined in unidirectional compression specimens, but it is possible that this difference is related to the difficulty of obtaining accurate values with the latter type of test specimen. Because of the presumed complex nature of the compression failure, the present data do not really establish that the compressive fiber strain can be used as a laminate failure criterion with full confidence. The data of Figure 3 do indicate that the strain to failure under compression loading of quasi-isotropic [0/±45/90] laminates is the same as in [0/90] laminates. Clearly more work is required in this area to establish whether there is any laminate dependence of the ultimate compressive failure strain.

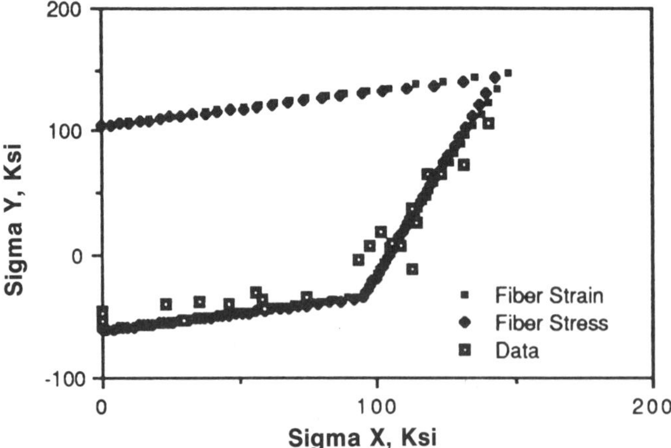

FIGURE 4 Comparison of maximum fiber stress and strain failure theories for ultimate laminate failure with data for AS4/3501-6 carbon–epoxy laminates in a quasi-isotropic [0/±45/90] configuration.

Stress values can also be used as a laminate failure criterion. The use of a maximum fiber direction ply stress criterion can give accuracy equal to that of the strain criterion, if used properly. While it is true that the value of stress in the fiber direction depends not only on the fiber direction strain but also on the transverse strain, in fact this dependence on transverse strain is quite small due to the very low value of minor in-plane Poisson's ratio characteristic of fiber composites. Over the range of variables shown in the experimental data of Figures 2 and 4, the difference between a maximum fiber stress and strain criterion is only a few percent, and is within the scatter of the data.

A problem exists in the nonlinearity of the composite laminate stress–strain response. While the stress–strain curves of laminates appear to be reasonably linear under tension loads, say within 5 or 10% to failure, in fact this is in part due to two counterbalancing sources of nonlinearity. In detail, carbon fibers show a stiffening behavior in tension such that the final secant modulus may be on the order of 15% higher than the initial modulus. Conversely, matrix cracking and nonlinearity in shear can soften the off-axis plies so that the overall laminate response appears linear. If either an accurate nonlinear laminate theory is used, or adjustment is made to the allowable fiber stress values to account for the nonlinearity in an approximate manner, the fiber direction stress can be used with equal accuracy to that of fiber strain. Figure 4 also compares the maximum fiber direction stress criterion used in this manner.

The fiber failure criterion of Hahn, Erikson, and Tsai given in Eq. (11) is equivalent to the use of two separate values for tensile and compressive fiber direction stress and is thus identical to the fiber direction stress criterion discussed above. Thus if proper stress allowables are used, this criterion gives excellent agreement with the experimental data, essentially equivalent to the use of the strain criterion. An easy way to account for the laminate nonlinearity discussed above is to use fictitious values of stress that are taken as the initial modulus multiplied by the strain at failure.

The Hashin fiber failure criterion of Eq. (13) has another shear term, in addition to the fiber normal stress terms. For conditions under which this shear stress is not large, this criterion is equivalent to the maximum

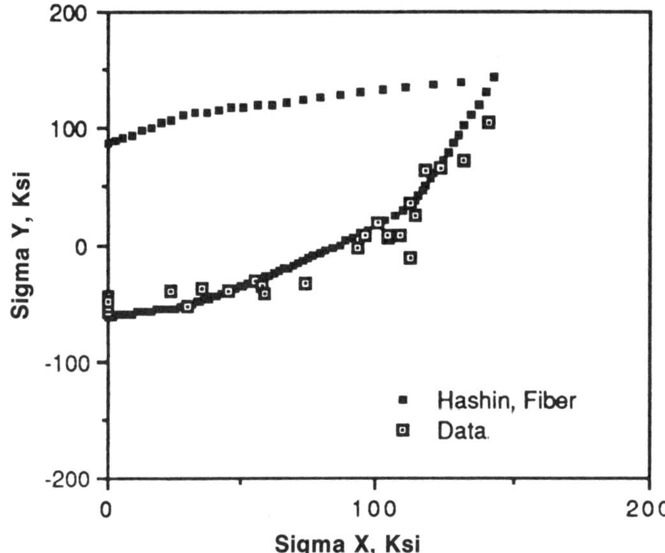

FIGURE 5 Comparison of Hashin fiber failure mode theory for ultimate laminate failure with data for AS4/3501-6 carbon–epoxy laminates in a quasi-isotropic [0/±45/90] configuration.

fiber stress criterion and thus gives acceptable agreement with the data, as shown in Figure 5. However for other conditions, the shear term is apparently overly conservative and agreement with the data is less good [20].

As mentioned above, the Tsai–Wu criterion of Eq. (7) can be used as a fiber failure criterion by making special assumptions. In particular, the first ply to fail is assumed to be transverse, matrix-dominated failure, while the last ply failure is considered to coincide with ultimate laminate failure. Figure 6 compares this approach with the laminate failure data. It can be seen that the Tsai–Wu approach for ultimate laminate failure is conservative by large factors under conditions of multiaxial laminate tensile stress, and nonconservative for multiaxial laminate compressive stress. The inherent problem is that the matrix and fiber failure modes are not clearly differentiated. The transverse stress terms are overly weighted with respect to fiber failure by being based on matrix failure. Fiber failure may indeed be influenced by the transverse stresses, but the magnitude of these stresses is essentially limited by the ability of the matrix to transmit these stresses into the fiber. In general, a transverse stress that is large with respect to matrix allowables can still be small with respect to fiber allowables.

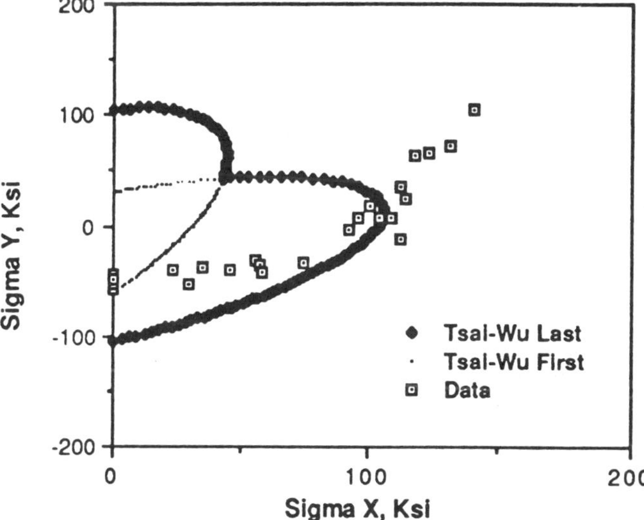

FIGURE 6 Comparison of Tsai–Wu first- and last-ply failure theories with data for ultimate strength of AS4/3501-6 carbon–epoxy laminates in a quasi-isotropic [0/±45/90] configuration.

Because the Tsai–Wu criterion is sensitive to the transverse stresses and overly conservative under conditions of multiaxial tensile stress, any reduction in the calculated transverse stresses may serve to improve the accuracy of the criterion when applied to tensile stress states. For example, using the presumed reduction of effective transverse properties with matrix cracking and nonlinear shear response of Ref. 12 gives a small improvement in the comparison of the Tsai–Wu criterion with experiment. An alternative procedure is suggested in Ref. 23, in which the transverse properties E_{22}, G_{12}, and ν_{12} are multiplied by a degradation factor (DF), usually taken as 0.3, and the criterion of Eq. (7) is used as a "first ply" failure criterion. As shown in Figures 7 and 8, this empirical approach does improve the predictive capability, although it is still not as good as using either maximum fiber direction stress or strain. This procedure is quite empirical, as the degradation factor essentially becomes a free constant. Also, the failure is predicted to be controlled by transverse plies rather than the plies in the loading directions, contrary to the usual interpretation of the experimental evidence.

Discussion

The major result of the evidence presented is that laminate ultimate failure can be represented on a ply-level analysis, which is thus applicable to all laminates. The most accurate failure criterion appears to be either maximum fiber direction stress or strain. These criteria have been used extensively in practical applications, and thus it is reassuring that the laboratory data also verify their validity. To a greater or lesser extent, criteria that include transverse normal or shear stress effects on fiber failure appear to be less applicable. While in general we would expect the transverse stresses to have an effect on fiber strength, the difficulty is in establishing how to accurately calculate these stresses. Clearly the maximum transverse stresses are limited by the strength of the matrix or matrix–fiber interphase. This stress level may be quite low with respect to the fiber strength of carbon fibers. Thus failure theories that include the effect of these stresses appear to be conservative, perhaps by as much as a factor of 4 as shown above. It may be that the

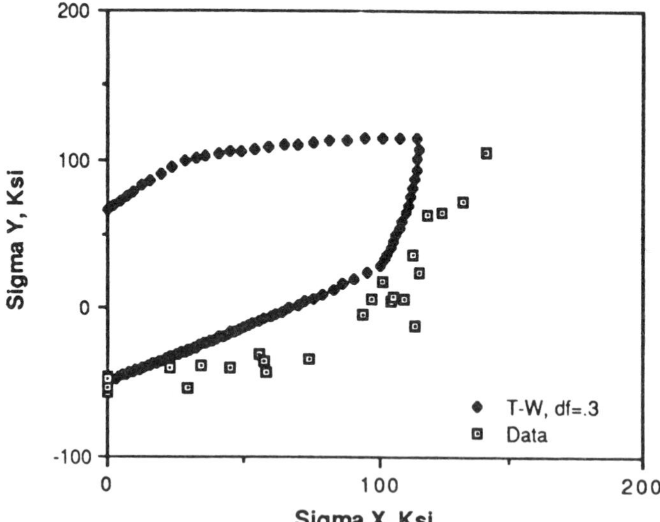

FIGURE 7 Comparison of Tsai–Wu first-ply failure theory with reduced transverse stiffness (DF = 0.3) with stress data for ultimate strength of AS4/3501-6 carbon–epoxy laminates in a quasi-isotropic [0/±45/90] configuration.

transverse stresses are more important in the failure of aramid fiber laminates, because of the weaker transverse strength of that fiber relative to carbon fibers.

The use of a ply-level criterion is of course desirable in that the criterion is presumed to apply to all (fiber-dominated) laminates of interest. While the number of laminates for which valid failure properties are available is not large, the data that exist appear to support this contention [16–21]. Certainly more data are needed for compressive stress states, since the complicated failure mechanisms involved in laminate compressive failure might need additional treatment.

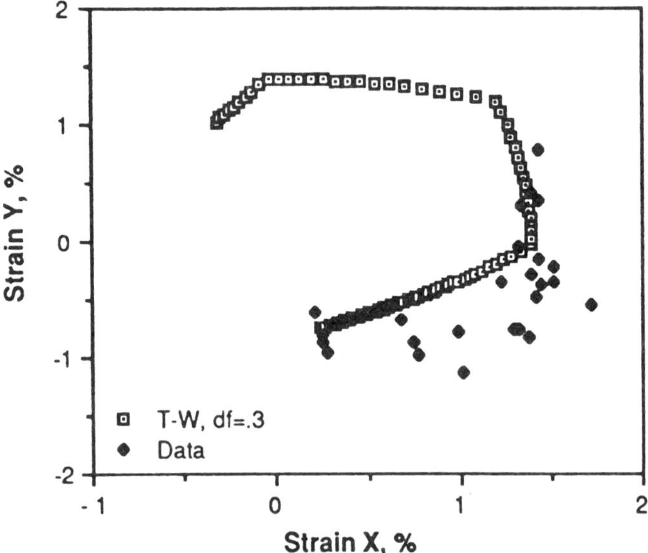

FIGURE 8 Comparison of Tsai–Wu first-ply failure theory with reduced transverse stiffness (DF = 0.3) with strain data for ultimate strength of AS4/3501-6 carbon–epoxy laminates in a quasi-isotropic [0/±45/90] configuration.

The question of how fiber strength or strain capability, say as measured in a unidirectional coupon test, translates into a "delivered" value in a laminate is not settled. While with the AS4/3501-6 carbon–epoxy system the fibers in the laminate appeared to have nearly the same strength properties as in a coupon, there was an apparent loss of strength of approximately 20% with the IM7/8551-7 carbon–epoxy system. Thus it appears that at least two laminate tests are necessary to determine tension and compression "in situ" delivered fiber strengths.

The difference in delivered fiber strengths may be related to the fiber–matrix interphase properties. It has been conjectured that ply cracks in adjacent off-axis plies have a stress–concentration effect and thus reduce the strength of the load-carrying fibers. A mechanism for relieving this stress concentration has been proposed by Cook and Gordon [24], in which a microdelamination around the fiber at the tip of the crack relieves the stress. It is possible that the higher matrix strength and toughness of the 8551-7 system may tend to suppress this delamination, and thus reduce the delivered fiber strength. Similar effects are often observed in brittle matrix systems.

Complicating Effects

The case of statically applied uniform stress discussed above is of course the simplest possible situation involving failure of laminates. Obvious complications are those of fatigue, in which matrix cracking is believed to lead to delamination and thus can directly influence failure [13].

Another complicating effect is that of free edges. Stress-free edges in composite laminates (such as the edges of a laminate coupon) produce interlaminar stress distributions that die out a short distance from the edge [25]. This well-known phenomenon has been studied extensively [25–27]. In many cases the stresses cause delamination and the associated stress concentration in the fibers causes premature failure of the laminate. There is some evidence that tough resins may suppress this failure mode, in which case the failure strength of laminate coupon specimens should be the same as that of laminate tubular specimens (which don't have edges).

Another complication is that associated with nonuniform stresses around stress concentrations. Typical cases are stresses around cutouts and pin-loaded holes. Although composites tend to be brittle in their overall stress–strain response, simply using elastic stress concentration factors has been shown to be overly conservative. The "point" and "average" stress methods have been introduced as empirical ways to use stress values lower than the peak stresses [28]. This approach has been applied on a ply level, using the fiber strain criterion [29]. This procedure was shown to give reasonable comparisons with experiment for pin-loaded holes if delamination adjacent to the hole was suppressed by lateral loads in the experiments [30].

In some laminates and loadings, it is not clear whether ultimate laminate failure can be produced by matrix failure alone or whether fiber failure is required. This difficult area has been addressed by a number of investigators [10,12,31,32]. A typical approach has been to formulate a nonlinear response model that includes softening due to matrix degradation, coupled with appropriate matrix and fiber failure properties and criteria. The nonlinear model then can predict the mode of failure, which may be either excessive deflection due to matrix softening or fiber failure.

Summary and Conclusions

Recent experimental evidence has shown that the ultimate failure of carbon–epoxy laminates can be approached on a ply-level basis, so that a ply criterion can presumably be used for all laminates. If the laminate and loading is "fiber dominated," fiber failure is required to cause ultimate laminate failure. The ply-level failure criterion that best represents the experimental data on the failure of a number of carbon–epoxy laminates is either fiber direction strain or stress. Allowable values for compression are lower than for tension. Criteria that include interactions with transverse stresses are conservative, in some cases by large factors.

Stephen R. Swanson

References

1. R. M. Jones, *Mechanics of Composite Materials*, McGraw-Hill, New York, 1975.
2. B. W. Rosen, "Mechanics of Composite Strengthening," in *Fiber Composite Materials*, American Society for Metals, Metals Park, Ohio, 1965, Ch. 3.
3. C. Zweben and B. W. Rosen, *J. Mech. Phys. Solids*, *18*, 189 (1970).
4. D. Hull, *An Introduction to Composite Materials*, Cambridge University Press, London, 1981.
5. M. R. Piggott, in *Developments in Reinforced Plastics*, Vol. 4, Elsevier Applied Science, London, 1984, p. 131.
6. A. S. D. Wang, *Compos. Tech. Rev.*, *6*, 45 (1984).
7. J. E. Master and K. L. Reifsnider, "An Investigation of Cumulative Damage Development in Quasi-Isotropic Graphite/Epoxy Laminates," in *Damage in Composite Materials*, ASTM STP 775, 1982, pp. 40–62.
8. D. L. Flaggs and M. H. Kural, *J. Compos. Mater.*, *16*, 103 (1982).
9. A. Parvizi, K. W. Garrett, and J. E. Bailey, *J. Mater. Sci.*, *13*, 195 (1978).
10. R. J. Nuismer and S. C. Tan, "The Role of Matrix Cracking in the Continuum Constitutive Behavior of a Damaged Composite Ply," in *Mechanics of Composite Materials, Recent Advances, Proceedings of the IUTAM Symposium on Mechanics of Composite Materials*, Pergamom Press, Elmsford, NY, 1983, pp. 437–448.
11. N. Laws, G. J. Dvorak, and M. Hejazi, *Mech. Mater.*, *2*, 123 (1983).
12. S. R. Swanson and A. P. Christoforou, *J. Eng. Mater. Technol.*, *109*, 12 (1987).
13. K. L. Reifsnider, K. Schulte, and J. C. Duke, "Long-Term Fatigue Behavior of Composite Materials," in *Long Term Behavior of Composites*, ASTM STP 813, 1983, pp. 136–159.
14. Z. Hashin, *J. Appl. Mech.*, *102*, 329 (1980).
15. H. T. Hahn, J. B. Erikson, and S. W. Tsai, "Characterization of Matrix/Interface-Controlled Strength of Unidirectional Composites," in G. Sih and V. P. Tamuzs, Eds., Fracture of Composite Materials, Nijhoff, The Hague, 1982, pp. 197–214.
16. G. E. Colvin, and S. R. Swanson, "Characterization of the Failure Properties of IM7/8551-7 Carbon/Epoxy Under Multiaxial Stress," in ASME AD-Vol. 13, *Recent Advances in the Macro- and Micro-Mechanics of Composite Materials Structures*, 1988, pp. 235–241.
17. S. R. Swanson and M. Nelson, "Failure Properties of Carbon/Epoxy Laminates Under Tension-Compression Biaxial Stress," *Proceedings of the Third Japan–U.S. Conference on Composite Materials*, Tokyo, 1986.
18. S. R. Swanson and A. P. Christoforou, *J. Compos. Mater.*, *20*, 457 (1986).
19. S. R. Swanson and B. Trask, "Biaxial Tests of Off-Axis Quasi-Isotropic Laminates," *Proc. American Society of Composites Second Symposium*, 1987, pp. 225–234.
20. S. R. Swanson and B. C. Trask, *Composites 19*, 400 (1988).
21. S. R. Swanson, *Proc. Pan American Congr. Appl. Mech.* 67 (1989).
22. S. W. Tsai and E. M. Wu, *J. Compos. Mater.*, *5*, 58 (1971).
23. S. W. Tsai, *Composites Design*, 3rd ed., Think Composites, Dayton, OH, 1987.
24. J. Cook and J. E. Gordon, *Proc. R. Soc. London A*, *282*, 508 (1964).
25. R. B. Pipes and N. J. Pagano, *J. Compos. Mater.*, *4*, 538 (1970).
26. S. S. Wang and I. Choi, *J. Appl. Mech.*, *49*, 549 (1982).
27. C. T. Herakovich, *J. Compos. Mater.*, *15*, 336 (1981).
28. J. M. Whitney and R. J. Juismer, *J. Compos. Mater.*, *8*, 253 (1974).
29. C. C. Poe, Jr., *Eng. Fracture Mech.*, *17*, 153 (1983).
30. S. R. Swanson and J. S. Burns, *Proceedings of the 31st National SAMPE Symp.*, 1986, pp. 1078–1086.
31. R. J. Nuismer, "Predicting the Performance and Failure of Multidirectional Polymeric Matrix Composite Laminates: A Combined Micro-Macro Approach," *Proc. ICCM 3*, 1980, pp. 436–452.
32. R. S. Sandhu, R. L. Gallo, and G. P. Sendeckyj, "Initiation and Accumulation of Damage in Composite Laminates," in *Composite Materials: Testing and Design* (6th Conf.), ASTM STP 787, 1982, pp. 163–182.

Lamination Theory

Materials and structures fabricated from fibrous composites can be studied at various levels of analysis ranging from micromechanics to structural mechanics. At the micromechanics level, the individual fibers and surrounding matrix are the fundamental constituents or components of the system. At the level of structural mechanics, the actual goemetry of the structural element under consideration is the fundamental component. An intermediate level of analysis is called lamination theory. In lamination theory, the properties of the fiber and the matrix are smeared to give effective homogeneous properties of individual layers (or lamina), which are bonded together to form laminates. These laminated materials may then be used to form structural components. There are a numer of assumptions associated with lamination theory:

- Individual layers are assumed to be homogeneous, orthotropic, elastic materials.
- Individual layers are assumed to be in a state of plane stress.
- Displacements follow a restrictive class according to Kirchhoff assumptions.
- Individual layers are perfectly bonded to adjacent layers.

The fundamental equations of lamination theory relate in-plane forces and bending moments per unit length to the midplane strains and curvatures of the laminate through a series of coefficients that are defined in terms of the layer material properties, layer fiber orientations, layer thicknesses, and stacking sequence of the layers. Thus, lamination theory describes the response of laminated composites to in-plane forces and bending moments. After solution of these equations describing the response of the laminate, stresses in individual layers can be obtained from the midplane strains and curvatures with the aid of the layer constitutive equations. The individual layer stresses obtained in this manner are valid only in interior regions of a laminate, away from free edges. In the vicinity of the free edge the stress state is three-dimensional, and thus the plane stress assumption of lamination theory is no longer valid. Since the basic equations of lamination theory are stated in terms of in-plane forces and moments per unit length, it is possible that zero force and moment conditions can be satisfied without satisfying identically zero stresses on all layers of the laminate. Thus, stress-free boundary conditions generally will not be satisfied by lamination theory.

This discussion will review:

- Orthotropic elastic constitutive equations.
- Kirchoff assumptions for plate deflections.
- Development of the equations of lamination theory.
- Retrival of layer stresses from lamination theory.
- Prediction of the effective laminate engineering constants.
- Special laminates (symmetric, balanced, cross-ply, angle-ply, quasi-isotropic).

Textbook treatments of lamination theory can be found in Refs. 1–8. All results presented in this paper are for T300/5208 graphite–epoxy composite. This designation indicates that the fiber is T300 (a graphite fiber) and the matrix is 5208 (an epoxy resin).

Laminae

Stiffness and Compliance

Consider a layer of a continuous fiber composite with the fibers aligned parallel to the 1 direction, the 2 direction in the plane of the layer, and the 3 direction perpendicular to the layer (Fig. 1). The 1-2-3 directions are called the material principal coordinate axes. It is common practice in the study of composites to use a reduced notation for stress and strain. Thus, we define the in-plane stresses $\{\sigma\}_1$ and strains $\{\varepsilon\}_1$ in terms of the components of the stress and strain tensors as:

$$\{\sigma\}_1 \equiv \begin{Bmatrix} \sigma_1 \\ \sigma_2 \\ \tau_{12} \end{Bmatrix} \equiv \begin{Bmatrix} \sigma_{11} \\ \sigma_{22} \\ \sigma_{12} \end{Bmatrix} \tag{1}$$

$$\{\varepsilon\}_1 \equiv \begin{Bmatrix} \varepsilon_1 \\ \varepsilon_2 \\ \gamma_{12} \end{Bmatrix} \equiv \begin{Bmatrix} \varepsilon_{11} \\ \varepsilon_{22} \\ 2\varepsilon_{12} \end{Bmatrix} \tag{2}$$

In the above, γ_{12} is the engineering shear strain and ε_{12} is the tensor shear strain.

The individual layers (laminae) of a laminate are assumed to be homogeneous, elastic, orthotropic materials with a constitutive equation in material principal coordinates of the form:

$$\{\varepsilon\}_1 = [S]\{\sigma\}_1 \tag{3}$$

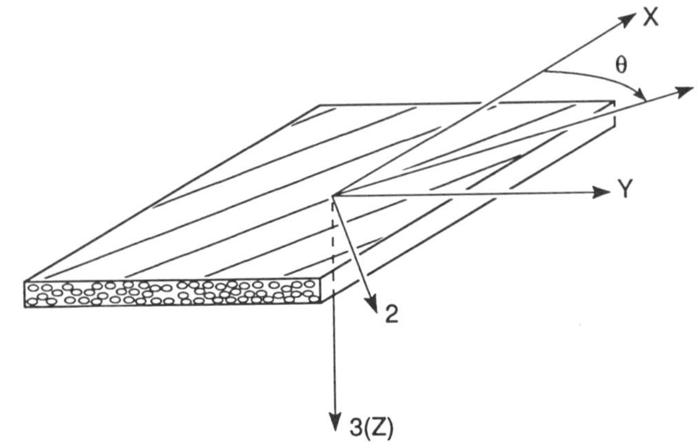

FIGURE 1 **Lamina coordinate systems.**

where $[S]$ is the compliance matrix in material principal coordinates and has the form:

$$[S] = \begin{bmatrix} S_{11} & S_{12} & 0 \\ S_{12} & S_{22} & 0 \\ 0 & 0 & S_{66} \end{bmatrix} \tag{4}$$

(The use of the 6 notation in composite mechanics is traditional in order to accommodate an easy transition to three-dimensional notation.) This equation can be inverted to yield:

$$\{\sigma\}_1 = [Q]\{\varepsilon\}_1 \tag{5}$$

where $[Q] = [S]^{-1}$ is the *reduced* stiffness matrix for plane stress, which has the form:

$$[Q] = \begin{bmatrix} Q_{11} & Q_{12} & 0 \\ Q_{12} & Q_{22} & 0 \\ 0 & 0 & Q_{66} \end{bmatrix} \tag{6}$$

The individual terms of the compliance matrix $[S]$ can be expressed in terms of the engineering constants E_1, E_2, ν_{12}, and G_{12} as:

$$\begin{aligned} S_{11} &= \frac{1}{E_1} \\ S_{12} = S_{21} &= \frac{-\nu_{12}}{E_1} \\ S_{22} &= \frac{1}{E_2} \\ S_{66} &= \frac{1}{G_{12}} \end{aligned} \tag{7}$$

where E_1 and E_2 are the elastic moduli in the fiber direction and transverse to the fibers, respectively, ν_{12} is the Poisson's ratio associated with loading in the 1 direction, i.e.:

$$\nu_{12} = -\frac{\varepsilon_2}{\varepsilon_1} \tag{8}$$

and G_{12} is the shear modulus in the 1-2 plane. Symmetry of the compliance matrix further requires that:

$$E_1\nu_{21} = E_2\nu_{12} \tag{9}$$

where ν_{21} is the Poisson ratio associated with loading in the 2 direction. Components of $[Q]$ can be expressed in terms of the engineering constants by appropriate substitutions.

The stresses, strains, and stiffness coefficients can be transformed to an arbitrary coordinate system corresponding to a fiber orientation θ with the global x axis (Fig. 1) through the transformation equations:

$$\{\sigma\}_1 = [T_1]\{\sigma\}_x \tag{10}$$

and

$$\{\varepsilon\}_1 = [T_2]\{\varepsilon\}_x \tag{11}$$

where

$$[T_1] = \begin{bmatrix} m^2 & n^2 & 2mn \\ n^2 & m^2 & -2mn \\ -mn & mn & (m^2 - n^2) \end{bmatrix} \tag{12}$$

and

$$[T_2] = \begin{bmatrix} m^2 & n^2 & mn \\ n^2 & m^2 & -mn \\ -2mn & 2mn & (m^2 - n^2) \end{bmatrix} \tag{13}$$

The transformation matrices $[T_1]$ and $[T_2]$ differ because engineering shear strain is being used. The transformation matrices would be identical if tensor shear strain were used. It is noted that as used here, positive θ is measured from the positive x axis to the positive 1 axis. Substituting the above transformation equations into the constitutive equation in material principal coordinates gives the constitutive equation in the arbitrary coordinate system as:

$$\{\sigma\}_x = [\bar{Q}]\{\varepsilon\}_x \tag{14}$$

where

$$[\bar{Q}] = [T_1]^{-1}[Q][T_2] \tag{15}$$

In the above equations, $[\bar{Q}]$ is the transformed reduced stiffness matrix. Throughout this presentation the notation $\{\ \}_1$ is used to denote quantities in the material principal 1-2 coordinate system, and the notation $\{\ \}_x$ is used to denote quantities in the global x-y coordinate system.

This equation can be inverted to yield:

$$\{\varepsilon\}_x = [\bar{S}]\{T\}_x \tag{16}$$

The expanded form of the constitutive equation is:

$$\begin{Bmatrix} \sigma_x \\ \sigma_y \\ \tau_{xy} \end{Bmatrix} = \begin{bmatrix} \bar{Q}_{11} & \bar{Q}_{12} & \bar{Q}_{16} \\ \bar{Q}_{12} & \bar{Q}_{22} & \bar{Q}_{26} \\ \bar{Q}_{16} & \bar{Q}_{26} & \bar{Q}_{66} \end{bmatrix} \begin{Bmatrix} \varepsilon_x \\ \varepsilon_y \\ \gamma_{xy} \end{Bmatrix} \tag{17}$$

The expanded inverted form of the constitutive equation in terms of $[\bar{S}]$ has the same general appearance. It is noted here that for off-axis orientations, both the stiffness and compliance matrices are fully populated, i.e., the 16 and 26 terms are nonzero.

The dependence of the $[\bar{Q}]$ terms on fiber orientation θ for graphite–epoxy is demonstrated in Figures 2 and 3. These normalized figures show that $\bar{Q}_{11}$, $\bar{Q}_{22}$, $\bar{Q}_{66}$, and $\bar{Q}_{12}$ are even functions of θ, whereas $\bar{Q}_{16}$ and $\bar{Q}_{26}$ are odd functions of θ. The maximum values for $\bar{Q}_{11}$ and $\bar{Q}_{22}$ are approximately four times those of the other stiffness co-

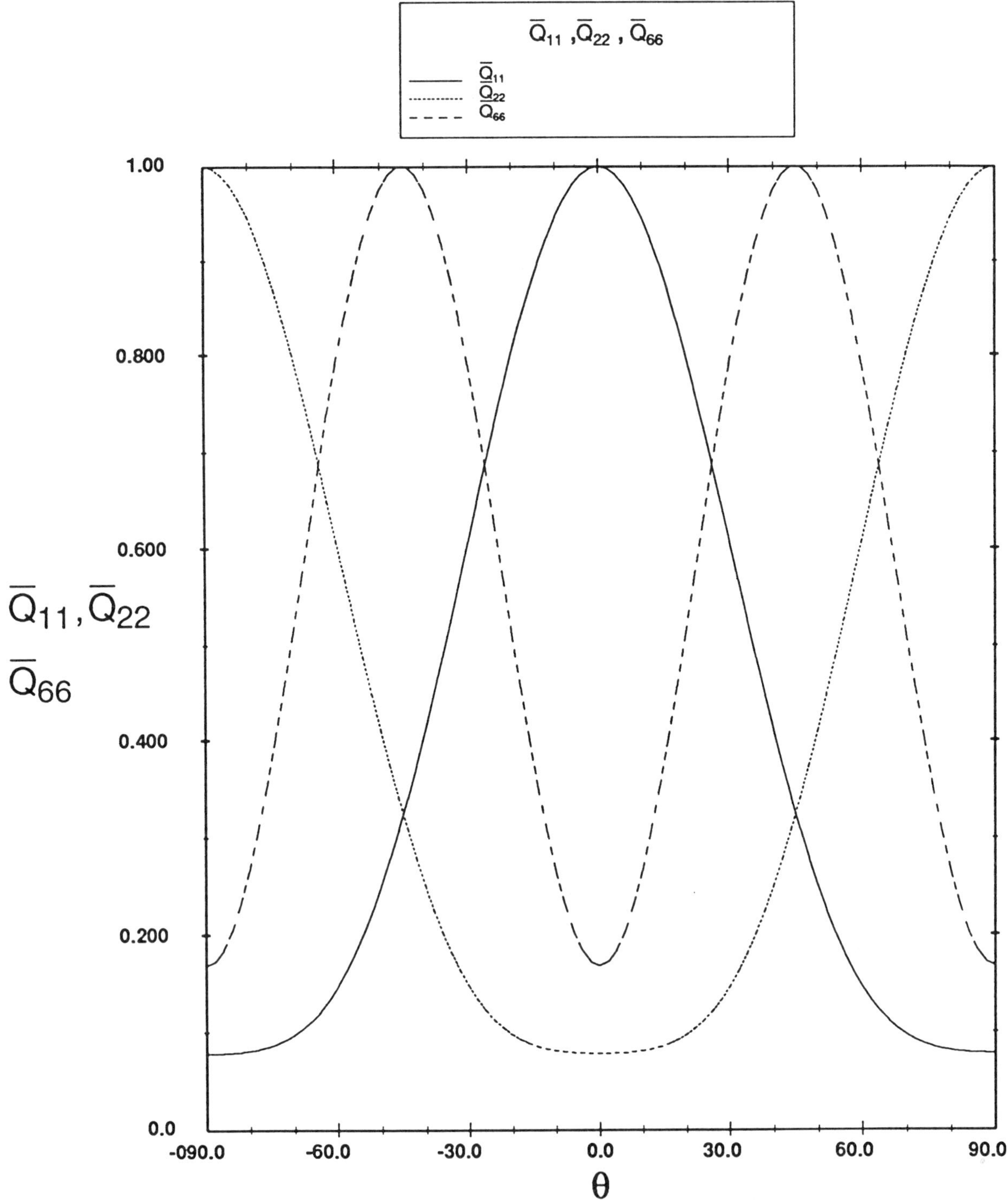

FIGURE 2 Normalized $\overline{Q}_{11}$, $\overline{Q}_{22}$, $\overline{Q}_{66}$ for off-axis lamina.

efficients. Maximum and minimum values of the coefficients correspond to different fiber orientations depending upon the particular coefficient.

Lamina Engineering Constants

The individual terms of the compliance matrix can be expressed in terms of engineering constants through consideration of specific loading conditions. For example, if the lamina is subjected to a uniform loading σ_x with all other stresses zero, we have:

$$\varepsilon_x = \overline{S}_{11}\,\sigma_x \tag{18}$$

and from Hooke's law:

$$\varepsilon_x = \frac{\sigma_x}{E_x} \tag{19}$$

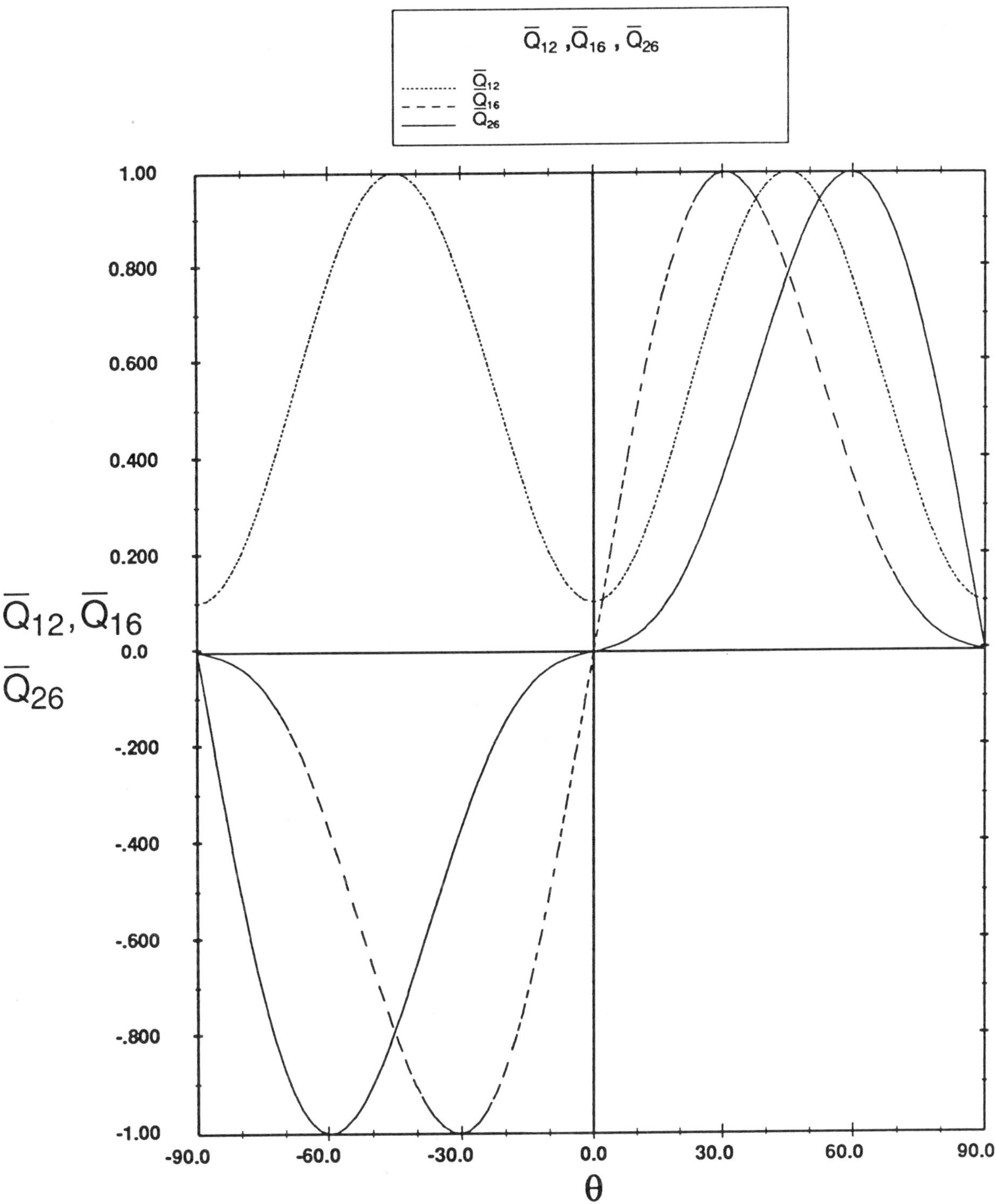

FIGURE 3 Normalized $\overline{Q}_{12}$, $\overline{Q}_{16}$, $\overline{Q}_{26}$ for off-axis lamina.

Thus:

$$\overline{S}_{11} = \frac{1}{E_x} \tag{20}$$

Similar thought experiments lead to the results:

$$\overline{S}_{22} = \frac{1}{E_y} \tag{21}$$

$$\overline{S}_{12} = \overline{S}_{21} = \frac{-\nu_{xy}}{E_x} \tag{22}$$

$$\overline{S}_{66} = \frac{1}{G_{xy}} \tag{23}$$

$$\overline{S}_{16} = \overline{S}_{61} = \frac{\eta_{xy,x}}{E_x} \tag{24}$$

$$\overline{S}_{26} = \overline{S}_{62} = \frac{\eta_{x,xy}}{G_{xy}} \tag{25}$$

where E_x and E_y are the axial and transverse Young's moduli, G_{xy} is the in-plane shear modulus, ν_{xy} is the Poisson's ratio:

$$\nu_{xy} = \frac{-\varepsilon_y}{\varepsilon_x} \tag{26}$$

and $\eta_{xy,x}$ and $\eta_{x,xy}$ are the coefficients of mutual influence:

$$\eta_{xy,x} = \frac{\gamma_{xy}}{\varepsilon_x} \tag{27}$$

$$\eta_{x,xy} = \frac{\varepsilon_x}{\gamma_{xy}} \tag{28}$$

The coefficients of mutual influence are similar to Poisson's ratios in that they are the ratio of the shear strain associated with axial strain (or the inverse). Physically, they are a measure of the coupling between shear and axial strain. They are nonzero if the 16 or 26 terms in the compliance matrix are nonzero.

The variation of engineering constants with off-axis fiber orientation for unidirectional graphite–epoxy is shown in Figures 4 and 5, where all values have been normalized with respect to their maximum value. The Young's modulus E_x is largest in the fiber direction but decreases very rapidly as the off-axis fiber orientation is increased. The ratio of axial to transverse modulus of this typical graphite–epoxy is greater than 12.0. The shear modulus G_{xy} varies over a much smaller range and attains its maximum value at $\theta = \pm 45°$.

Poisson's ratios and shear-axial coupling coefficients of the type $\eta_{xy,x}$ vary over a wide range. Poisson's ratio is always positive, but $\eta_{xy,x}$ is positive or negative depending upon fiber orientation. For the graphite–epoxy under consideration, Poisson's ratio varies from a maximum of 0.372 to a minimum of 0.019. The axial-shear coupling coefficient $\eta_{xy,x}$ varies over a wide range, with maximum and minimum values of ± 2.14 at $\pm 12°$, respectively.

Laminates

Kirchhoff Displacements

The Kirchhoff assumptions for bending of thin plates state that normals to the midplane remain straight and normal to the deformed midplane, and the normals to the midplane do not change length. With these assumptions, the strains $\{\varepsilon\}_x$ at any distance z from the midplane are written:

$$\{\varepsilon\}_x = \{\varepsilon^0\}_x + z\{\kappa\}_x \tag{29}$$

where $\{\varepsilon^0\}_x$ are the midplane strains and $\{\kappa\}_x$ are the plate curvatures, that is:

$$\{\kappa\}_x = \begin{Bmatrix} \kappa_x \\ \kappa_y \\ \kappa_{xy} \end{Bmatrix} = \begin{Bmatrix} -w_{,xx} \\ -w_{,yy} \\ -2w_{,xy} \end{Bmatrix} \tag{30}$$

Forces and Moments

The stresses at a specific location z^k (in the kth layer) are obtained from the constitutive equations and the strain provided by the Kirchhoff assumptions:

$$\{\sigma\}_x^k = [\bar{Q}]^k[\{\varepsilon^0\}_x + z^k\{\kappa\}_x] \tag{31}$$

The forces $\{N\}$ and moments $\{M\}$ per unit length acting over the thickness of a laminate can be written:

$$\{N\} = \int_{-H}^{+H} \{\sigma\}^k \, dz \tag{32}$$

and:

$$\{M\} = \int_{-H}^{+H} z\{\sigma\}^k \, dz \tag{33}$$

Laminate Equations

Integration of the force and moment equations provides the fundamental equations of lamination theory:

$$\begin{Bmatrix} N \\ M \end{Bmatrix} = \begin{bmatrix} A & B \\ B & D \end{bmatrix} \begin{Bmatrix} \varepsilon^0 \\ \kappa \end{Bmatrix} \tag{34}$$

where $[A]$, $[B]$, and $[D]$ are defined for an N-layer laminate as:

$$[A] = \sum_{k=1}^{N} [\bar{Q}]^k (z_k - z_{k-1}) \tag{35}$$

$$[B] = \frac{1}{2}\sum_{k=1}^{N} [\bar{Q}]^k (z_k^2 - z_{k-1}^2) \tag{36}$$

$$[D] = \frac{1}{3}\sum_{k=1}^{N} [\bar{Q}]^k (z_k^3 - z_{k-1}^3) \tag{37}$$

with z_k being the z coordinate to the bottom of the kth layer (Fig. 6). These equations are also referred to as the equations of "classical lamination theory (CLT)." The equation of lamination theory can be inverted and expressed:

$$\begin{Bmatrix} \varepsilon^0 \\ \kappa \end{Bmatrix} = \begin{bmatrix} A^* & B^* \\ C^* & D^* \end{bmatrix} \begin{Bmatrix} N \\ M \end{Bmatrix} \tag{38}$$

where the * quantities are defined appropriately.

The fundamental equations of lamination theory show that, in general, there is a coupling between *bending* and *stretching* through the $[B]$ matrix.

Special Laminates

For the purpose of the following discussion, it is assumed that all layers of the laminate are made with the same composite material and have the same thickness.

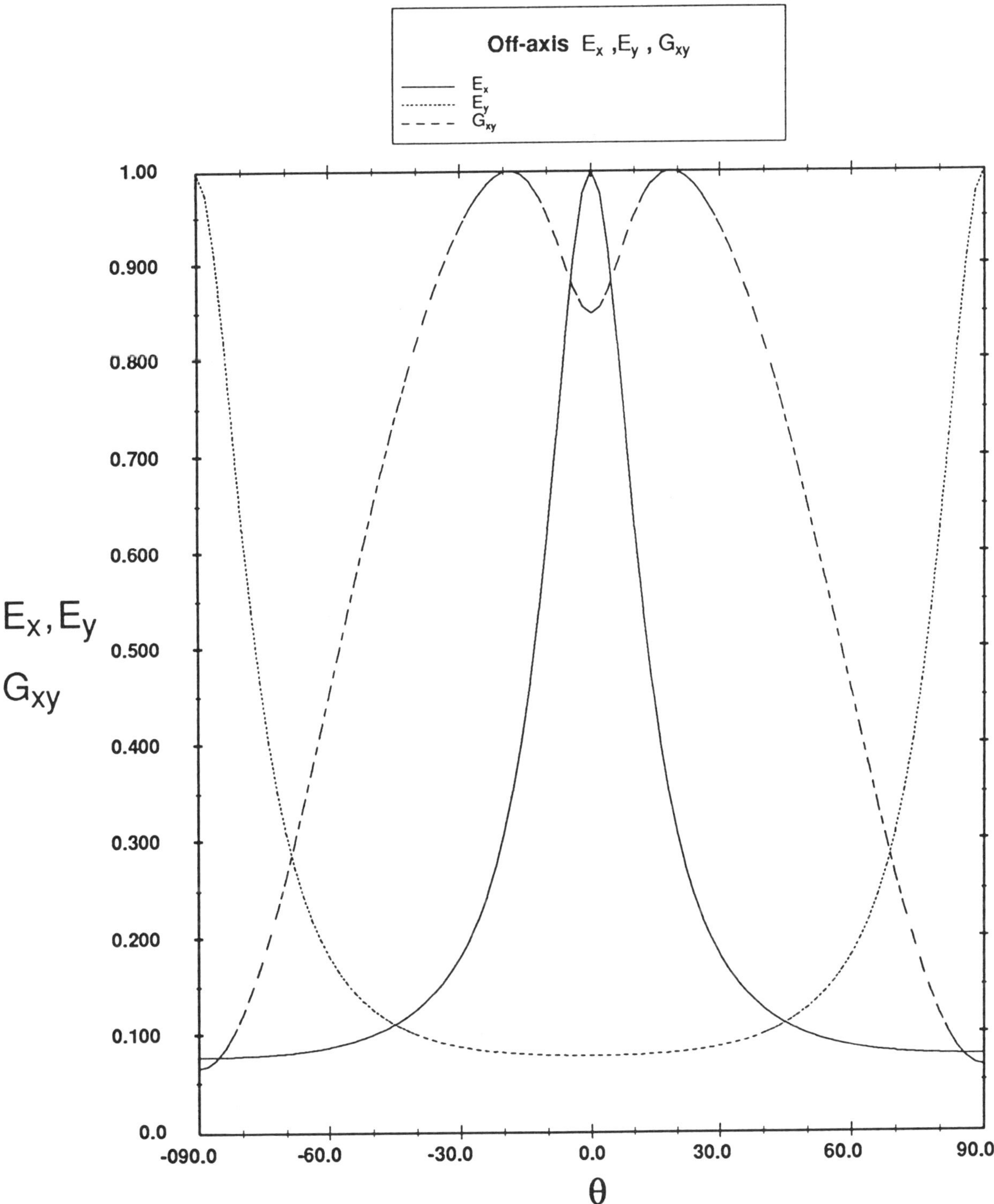

FIGURE 4 **Normalized E_x, E_y, G_{xy} for off-axis lamina.**

These assumptions are made only to demonstrate some consequences of lamination theory. They are not required by lamination theory, which is valid for hybrid laminates with variable layer thickness and arbitrary stacking sequence of the layers.

Symmetric Laminates

The fundamental equations of lamination theory indicate that, in general, in-plane response is coupled with bending response through the $[B]$ matrix. However, if the laminate stacking sequence is symmetric about the midplane, the B_{ij} are identically zero and there is no bending-stretching coupling. Thus, for symmetric laminates

$$\{N\} = [A]\{\varepsilon^0\} \tag{39}$$

and

$$\{M\} = [D]\{\kappa\} \tag{40}$$

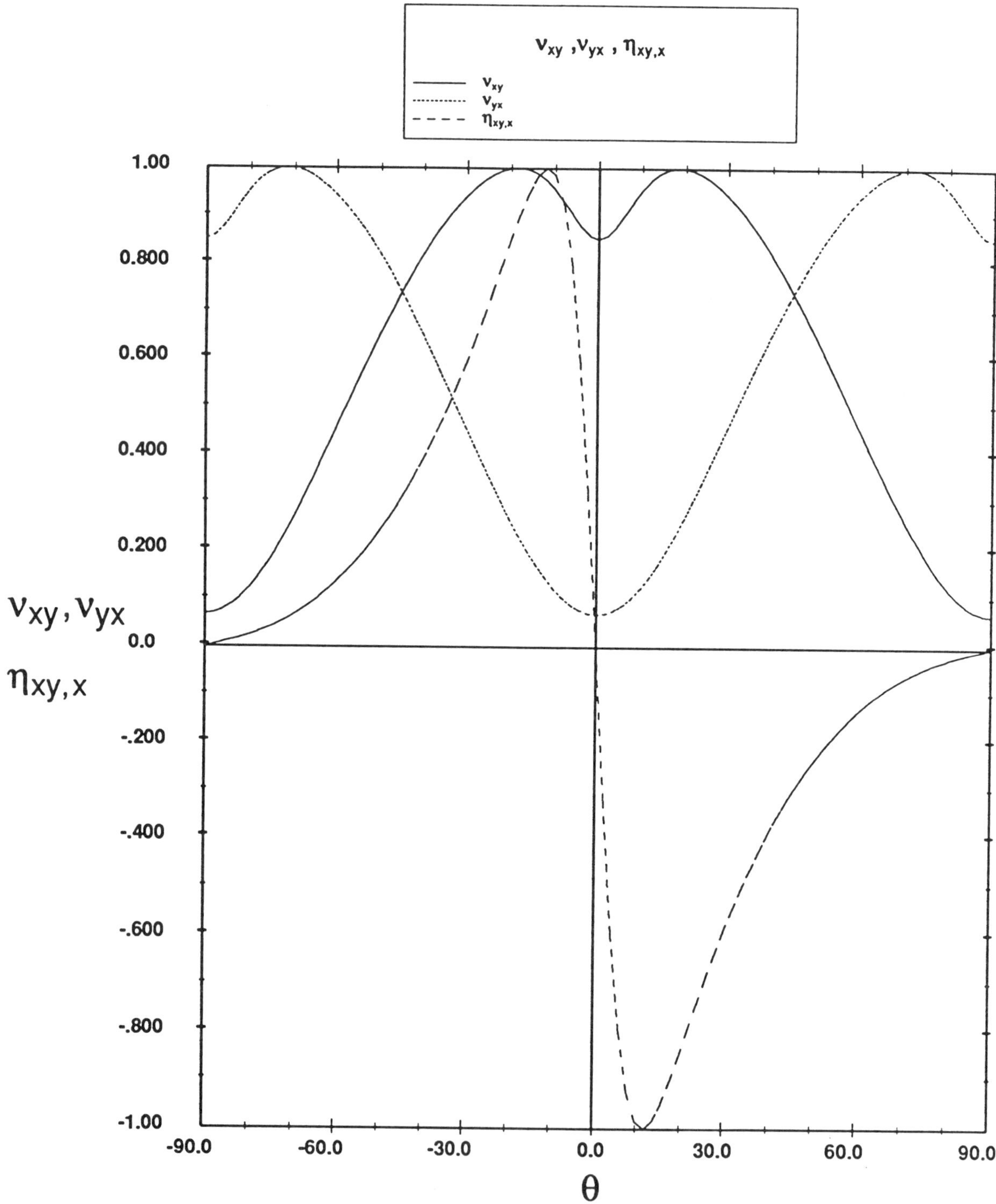

FIGURE 5 **Normalized ν_{xy}, ν_{yx}, $\eta_{xy,x}$ for off-axis lamina.**

The two preceding equations clearly show that the $[A]$ and $[D]$ matrices correspond to the laminate in-plane and bending stiffnesses, respectively.

Engineering Constants

For a symmetric laminate, the in-plane response can be written in inverted form as:

$$\{\varepsilon^0\}_x = [a^*]\{\bar{\sigma}\}_x \tag{41}$$

where a^* are defined:

$$[a^*] = 2H[A]^{-1} \tag{42}$$

and $\{\bar{\sigma}\}_x$ are the laminate stresses averaged through the $2H$ thickness of the laminate.

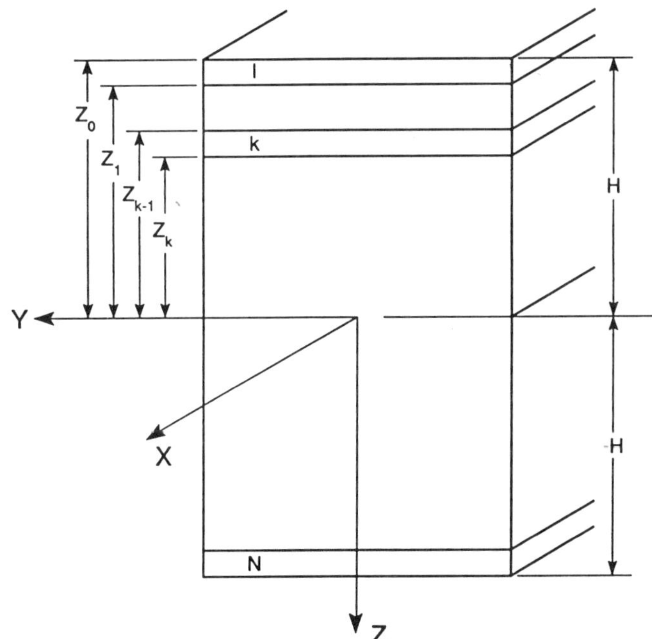

FIGURE 6 Laminate Geometry and Notation.

A series of thought experiments, as were described for laminae, provides expressions for the laminate engineering constants in the form:

$$E_x = \frac{1}{a_{11}^*} \tag{43}$$

$$E_y = \frac{1}{a_{22}^*} \tag{44}$$

$$\nu_{xy} = \frac{-a_{12}^*}{a_{11}^*} \tag{45}$$

$$G_{xy} = \frac{1}{a_{66}^*} \tag{46}$$

$$\eta_{xy,x} = \frac{a_{16}^*}{a_{11}^*} \tag{47}$$

$$\eta_{xy,y} = \frac{a_{26}^*}{a_{22}^*} \tag{48}$$

$$\eta_{x,xy} = \frac{a_{16}^*}{a_{66}^*} \tag{49}$$

$$\eta_{y,xy} = \frac{a_{26}^*}{a_{66}^*} \tag{50}$$

Balanced Laminates

If a laminate contains a $+\theta$ ply for every $-\theta$ ply, it is called a balanced laminate. In this case the A_{16} and A_{26} coefficients are zero, and coupling between in-plane extension and shear is zero. A_{16} and A_{26} are zero in this case because of the summation of $\bar{Q}_{16}$ and $\bar{Q}_{26}$, which are odd functions of θ (Fig. 3). A balanced laminate may also contain any number of 0° and 90° layers, since the 16 and 26 terms are identically zero for these fiber orientations.

Cross-Ply Laminates

Cross-ply laminates are those consisting of all layers with fiber orientations of 0° and 90°. All 16 and 26 coefficients in the [A], [B], and [D] matrices are zero in this case, and the response of the laminate is orthotropic for both bending and stretching. The coefficient of mutual influence $\eta_{xy,x}$ of cross-ply laminates is also zero because of the zero 16 and 26 terms in [A].

ANGLE-PLY LAMINATES. Angle-ply is a term reserved for laminates composed of *only* an equal number of layers at $+\theta$ and $-\theta$ orientations. They are an interesting and important class of laminates because they clearly show the influence of fiber orientation on the laminate properties. Figures 7 to 9 show the variation of engineering properties of angle-ply laminates with fiber orientation. It is evident from these figures that the properties of composite laminates may vary over a wide range. It is particularly noteworthy that Poisson's ratio ν_{xy} (Fig. 9) can vary from near zero for a [90] laminate to 1.27 for a $[\pm 26]_s$ graphite–epoxy laminate. Thus the lateral strain is actually *greater* than the applied axial strain for some fiber orientations of angle-ply laminates. Indeed, close examination of the values indicates that Poisson's ratio is greater than 1.0 for angle-ply laminates with fiber orientations in the range 17° to 37°. These values can be compared with those for the unidirectional off-axis lamina as shown in Figure 8 and typical Poisson's ratios for structural metals (0.25–0.50).

It is also evident from Figure 8 that the shear modulus is a minimum for a unidirectional laminate and is a maximum for a $[\pm 45]_s$ laminate. Poisson's ratios for angle-ply laminates vary by a factor of 4 over the complete range of fiber orientations, whereas the range for unidirectional off-axis laminae is only about 1.5.

Quasi-isotropic Laminates

The term *quasi-isotropic* refers to the fact that the in-plane stiffness [A] of the laminate exhibits isotropic properties. The response is *quasi*-isotropic because the bending stiffness of these same laminates is not isotropic. Any symmetric laminate with equally spaced fiber orientations at intervals of π/N with $N \geq 3$ exhibits quasi-isotropic response.

As examples, the engineering properties of T300/5208 graphite–epoxy $\pi/3$ and $\pi/4$ laminates are given in Table 1.

It is noted that the elastic constants of quasi-isotropic laminates are the same independent of the number of fiber orientations (equal to or greater than 3).

TABLE 1
T300/5208 Lamina and Laminate Properties

Laminate	E_x (GPa)	E_y (GPa)	G_{xy} (GPa)	ν_{xy}
Unidirectional	132.4	10.8	5.65	0.238
$\pi/4$	52.3	52.3	20.1	0.30
$\pi/3$	52.3	52.3	20.1	0.30

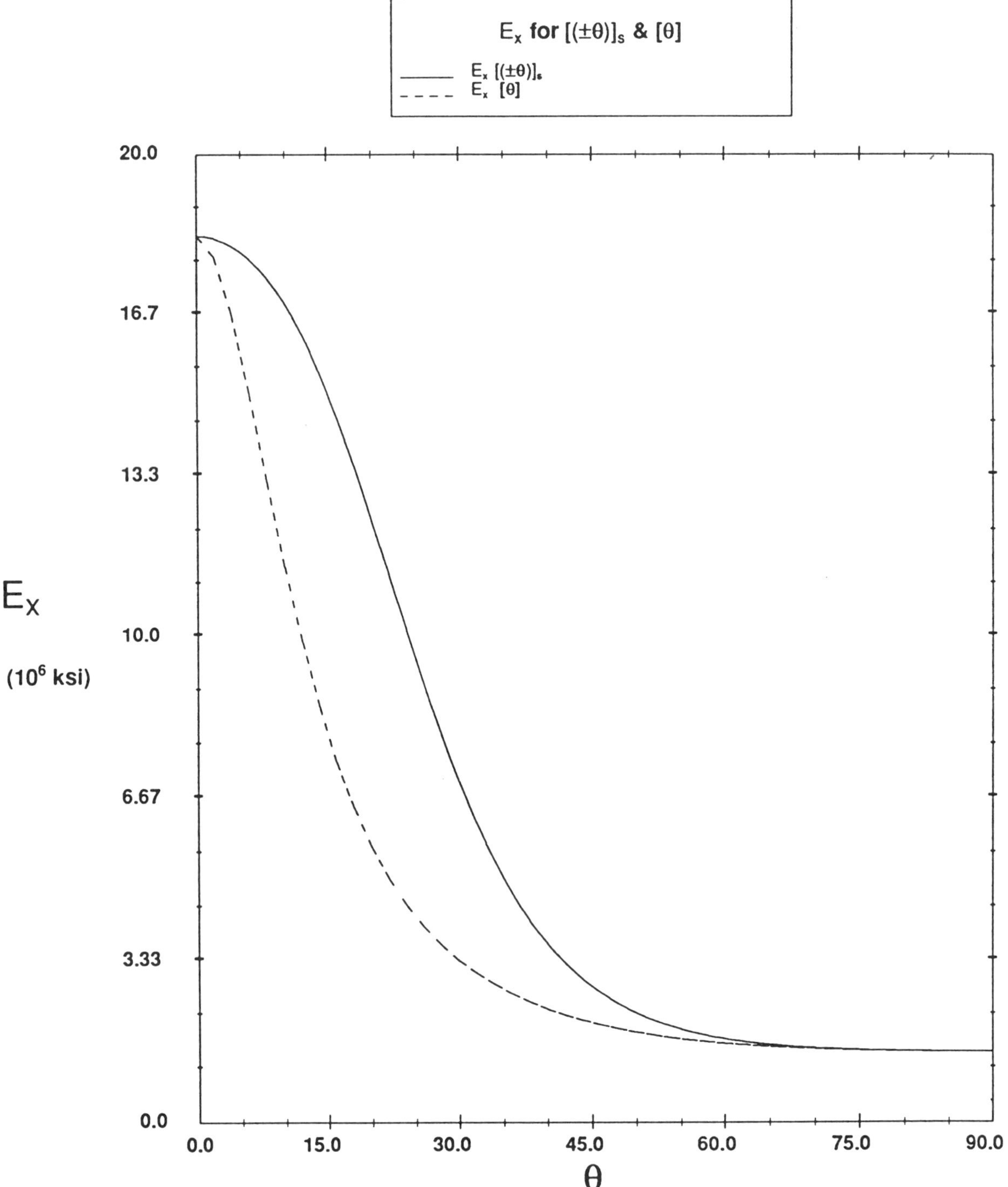

FIGURE 7 E_x for off-axis and angle-ply laminates.

Unsymmetric Laminates

If the stacking sequence of the layers is not symmetric about the midplane, the $[B]$ matrix will not be identically zero. This results in coupling between bending and stretching. As an example, consider an unsymmetric laminate subjected to the force loading $\{N\}$. The resulting curvatures are:

$$\{\kappa\} = [C^*]\{N\} \tag{51}$$

In a similar fashion, moment loading $\{M\}$ results in midplane strains:

$$\{\varepsilon^0\} = [B^*]\{M\} \tag{52}$$

Stress Distributions

The equations of lamination theory as presented in the previous sections are based upon the assumption of a

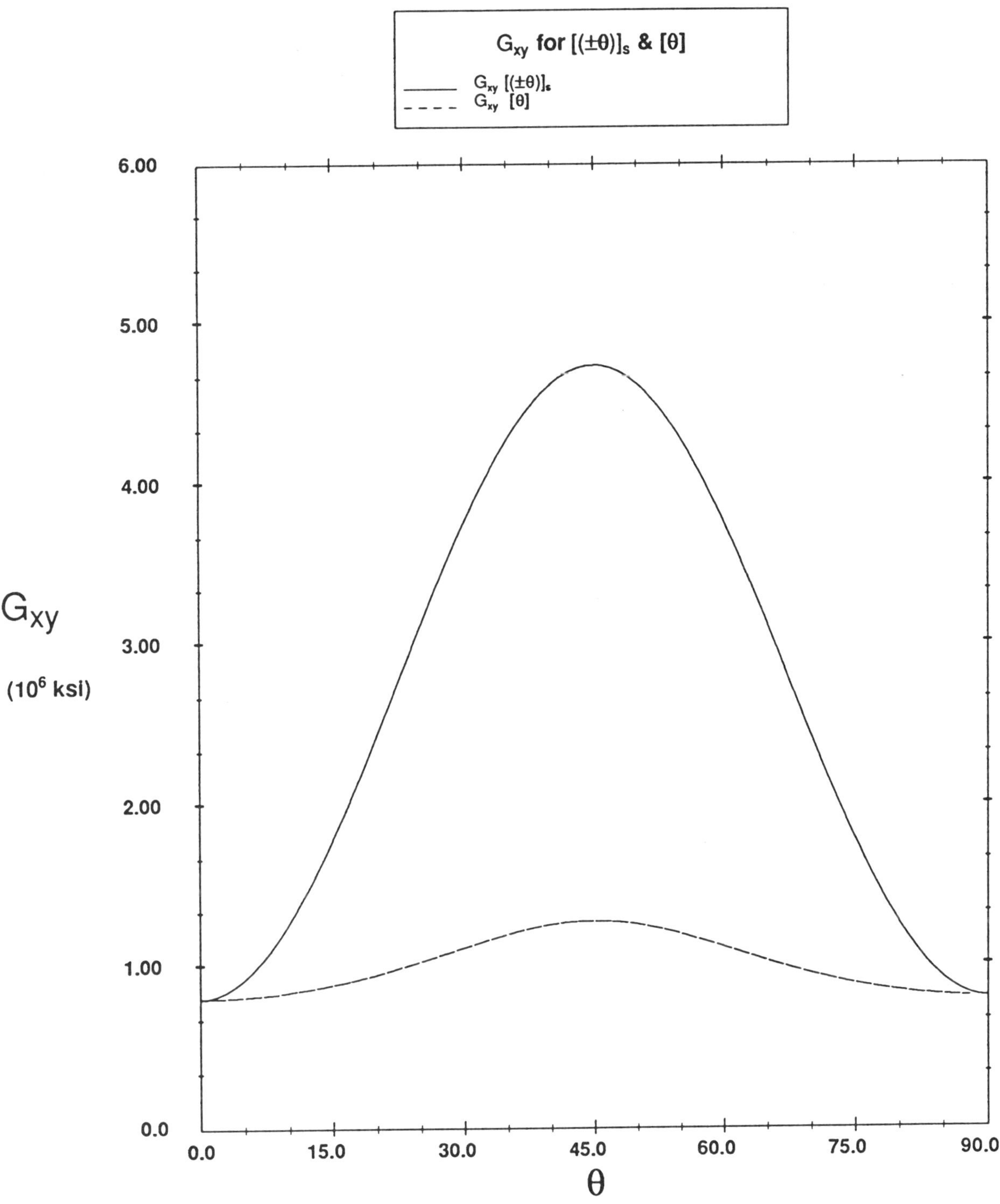

FIGURE 8 G_{xy} for off-axis and angle-ply laminates.

plane state of stress in each layer and a linear distribution of strains through the thickness of the laminate. Since the properties vary from layer to layer, the stresses also vary from layer to layer. This is consistent with the theory of elasticity, as the in-plane components of stress are not required to be continuous from layer to layer. Stress distributions for the pure force loading $N_x = 18$ N/cm on a $[0/\pm45/90]_s$ graphite–epoxy laminate are summarized in Table 2.

The stresses predicted by lamination theory indicate that the stresses are constant in each layer, but discontinuous between layers. The σ_y components are equivalent to a force $N_y = 0$, and the shear components τ_{xy} are equivalent to a zero N_{xy} force with the shear stresses in

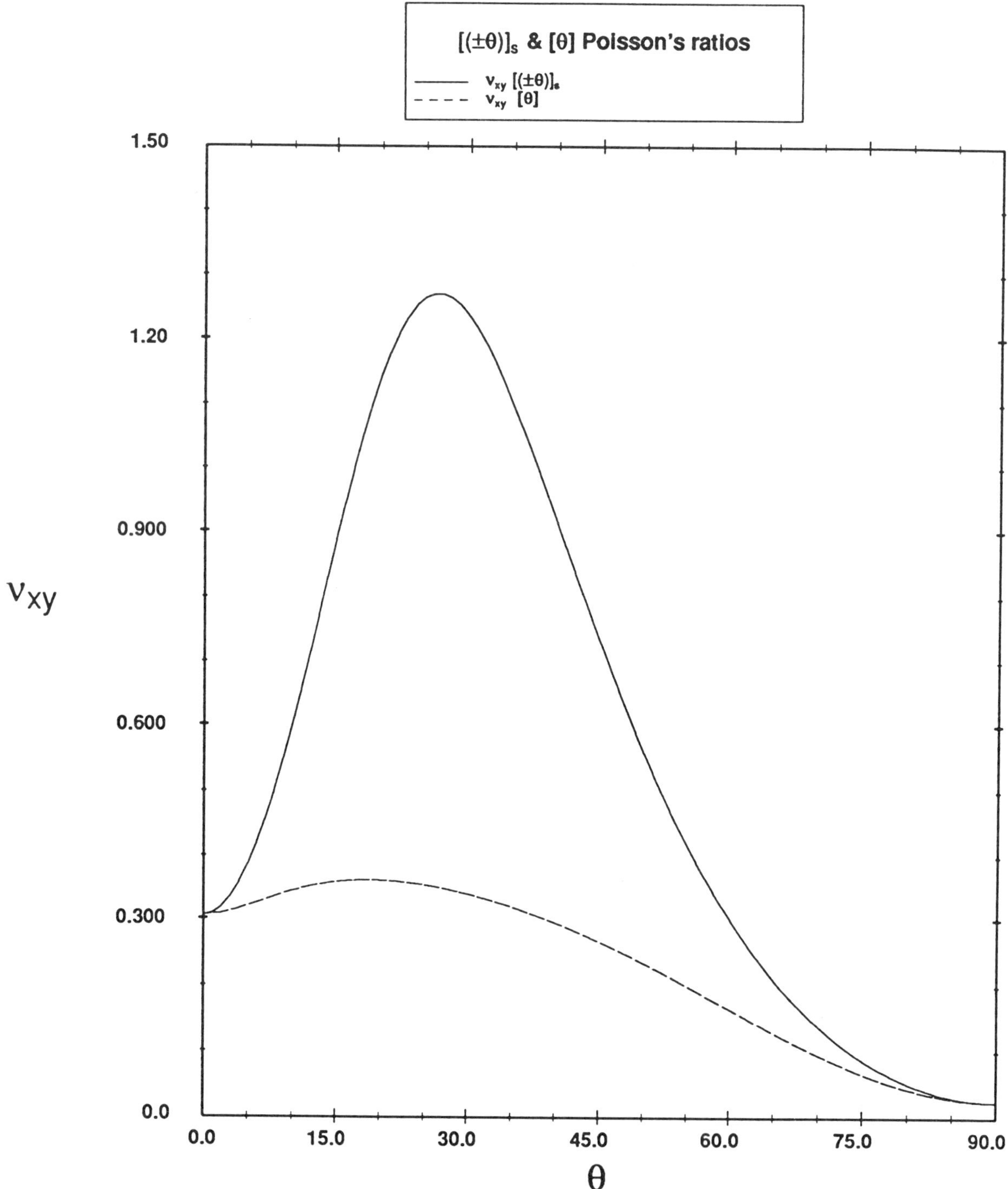

FIGURE 9 ν_{xy} for off-axis and angle-ply laminates.

TABLE 2
Stresses in a $[90/45/0/-45]_s$ Laminate under Axial Load

Layer	σ_x (MPa)	σ_y (MPa)	τ_{xy} (MPa)
0	4.3	−0.02	0
45	1.1	0.62	0.7
−45	1.1	0.62	−0.7
90	0.3	−1.22	0

the +45 and −45 layers equal in magnitude and opposite in sign. Further, the shear stresses are zero in the orthotropic 0° and 90° layers. The midplane strains for this loading are

$$\{\varepsilon^0\} = \begin{Bmatrix} 0.0033\% \\ -0.001\% \\ 0\% \end{Bmatrix} \tag{53}$$

Conclusions

The fundamental equations describing the elastic response of homogeneous orthotropic laminae and laminates made from these laminae have been formulated. The engineering constants for lamina and laminates have been presented and compared as a function of fiber orientation. Special classes of laminates have been defined. Finally, it has been shown that the state of stress in the individual layers of a laminate is biaxial even for the simplest of loadings.

Acknowledgement

The author is indebted to all those former students whose questions have led him to a better understanding of lamination theory and to the various funding agencies whose support provided him the opportunity to study mechanics of composites.

Carl T. Herakovich

References

1. J. E. Ashton, J. C. Halpin, and P. H. Petit, *Primer on Composite Materials: Analysis*, Technomic, Westport, CT, 1969.
2. L. R. Calcote, *The Analysis of Laminated Composite Structures*, Van Nostrand Reinhold, New York, 1969.
3. R. M. Jones, *Mechanics of Composite Materials*, Scripta Book Company, Washington, DC, 1975.
4. S. W. Tsai and H. T. Hahn, *Introduction to Composite Materials*, Technomic, Westport, CT, 1980.
5. B. D. Agarwal and L. J. Broutman, *Analysis and Performance of Fiber Composites*, John Wiley, New York, 1980.
6. R. M. Christensen, *Mechanics of Composite Materials*, John Wiley, New York, 1979.
7. J. R. Vinson and R. L. Sierakowski, *The Behavior of Structures Composed of Composite Materials*, Martinus Nijhoff, Boston, 1987.
8. J. M. Whitney, *Structural Analysis of Laminated Anisotropic Plates*, Technomic, Lancaster, PA, 1987.

Lightening Protection

See Conductive Composites; Electrical and Electronic Composites; Static Electrical Agents

Liquid-Crystalline Composites

While the whole field of engineering thermoplastics is virtually exploding with new grades and families of materials, nothing can beat the thermotropic liquid-crystalline polymers (LCPs). Their outstanding performance has stimulated considerable interest in high-performance specialty applications from aerospace and military to electronics. The main advantage of these polymers is that they can be processed in the melt state. LCPs are capable of forming highly oriented crystalline structures when subjected to shear and elongation above their melting point. Because of their rodlike molecular conformation and the stiffness of their backbone chains, LCPs form fibrous chains in the final product, with their fracture surfaces being similar to those of wood. They also give the neat resin its "self-reinforcing" properties comparable to or exceeding those of fiber-reinforced conventional thermoplastics: flexural moduli up to 16.7 GPa, tensile strengths as high as 137.9 MPa, and heat distortion temperatures up to 355°C at 1.8 MPa [1,2]. Other properties of LCPs are outstanding resistance to chemicals, solvents, UV light, and ionizing radiation, as well as extremely low combustibility and smoke generation without using additives.

Liquid-Crystalline Polymers

General

The terms *meso-phase, anisotropic,* and *liquid-crystalline* as applied to polymers are used interchangeably. Liquid-crystalline polymers may be thermotropic or lyotropic. Thermotropic material implies a material, small-molecule or polymer, that forms an ordered melt on fusion. A lyotropic material, small-molecule or polymer, is one that forms an ordered solution when dissolved in an appropriate solvent. In both thermotropic and lyotropic materials, this order is usually classified as nematic, smectic, or cholesteric (Fig. 1) [3]. The majority of lyotropic and thermotropic polymers being described today are relevant to the nematic structure. This phase structure is characterized by molecules showing parallel, one-dimensional order. The solutions or melts tend to be turbid and of relatively low viscosity. An important quality, low viscosity will aid the processing of polymeric nematic solutions and melts. Smectic molecules align parallel in layers showing two-dimensional order. These solutions or melts, while also turbid, are quite viscous. The molecules of cholesteric order have one or more optically active centers. This structure is characterized by nematic layers arranged in a helical conformation. Cholesteric materials form highly viscous solutions or melts, as do the smectic type. Polymers with smectic and cholesteric structures have been reported, although fabrication and physical property data are scant.

Polymer anisotropy is a quality to be exploited for the development of very high polymer chain extension and chain orientation in a fiber-spinning process (Fig. 2). Flexible-chain polymers are assumed to be in a random coil conformation in the melt, resulting in low chain extension during spinning. This leads to low mechanical properties. However, an LCP starts with a high degree of order in the melt or solution. Melt spinning of such polymers produces fibers with a high degree of nematic order "frozen" in. This also develops high chain continuity,

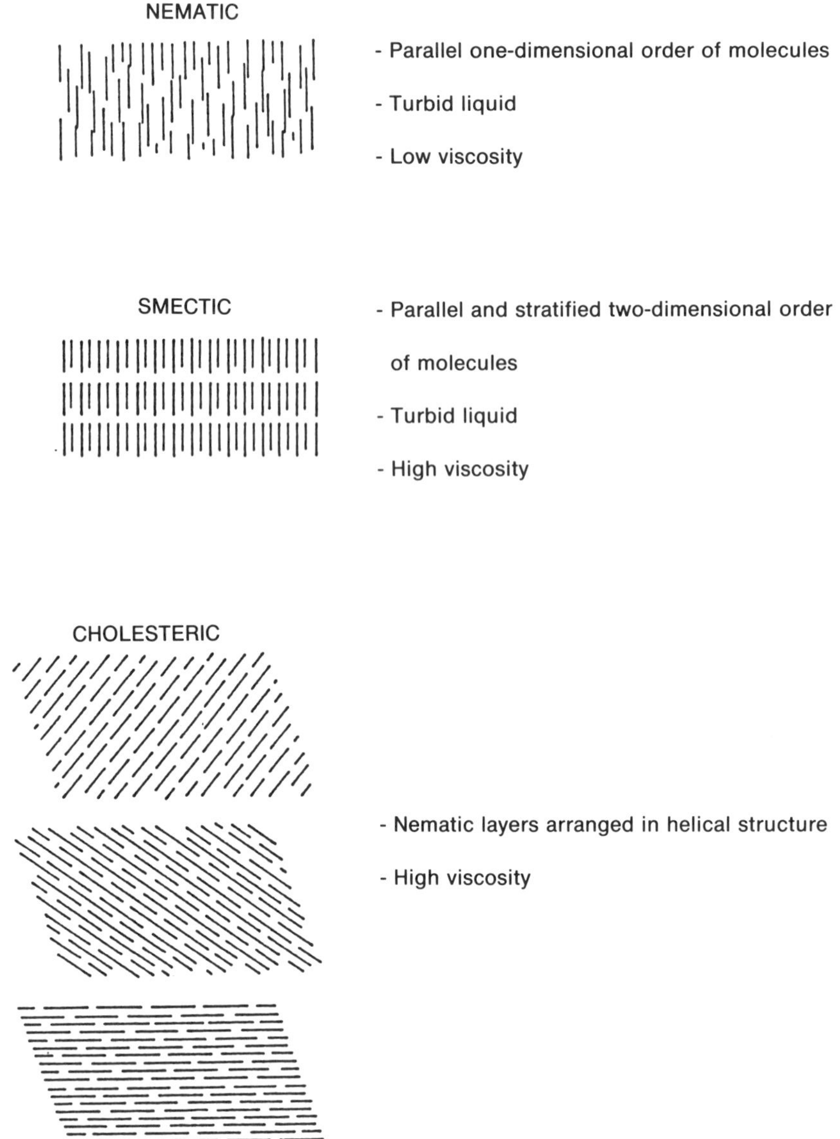

FIGURE 1 **Structure characteristic of thermotropic and lyotropic liquid crystals.** (From Ref. 3.)

highly extended chain structures, and high mechanical properties, especially a high tensile modulus. This comes from certain types of structures, i.e., aromatic polyamides (solutions) and thermotropic wholly aromatic polyesters (melts). Melts of wholly aromatic polyesters are the major concern of this article.

Types of LCPs and Their Structure

LCPs can be characterized depending on the sequential location of rigid liquid-crystalline segments (mesogens), so the main-chain LCP is usually mentioned rather than side-chain copolymers. The latter are also called comb liquid crystals. There are, however, many possibilities, and several of them are discussed in Ref. 4. A classification scheme that enables a more precise definition of a kind of LCP is presented in Table 1 [5].

CLASS A: LONGITUDINAL LCPs. This class is now called main-chain polymers. A new name is necessary to distinguish them from class B. There are numerous examples of LCPs in class A. Jackson and Kuhfuss [6] first studied a copolyester of PET (polyethylene terephthalate)/HBA (poly *p*-hydroxybenzoic acid), and this copolymer is now commercialized. Lenz [7] studied longitudinal polyesters in which the rigid liquid-crystalline (LC) groups are connected together by either rigid or flexible non-LC spacers. Kricheldorf [8] performed broad studies on the synthesis and properties of polyester imides. Polyester amides [9] and polyurethanes [10] also form longitudinal structures.

CLASS B: ORTHOGONAL LCPs. LCPs in this class also contain LC groups in the main chain, but these groups are approximately perpendicular to the back-

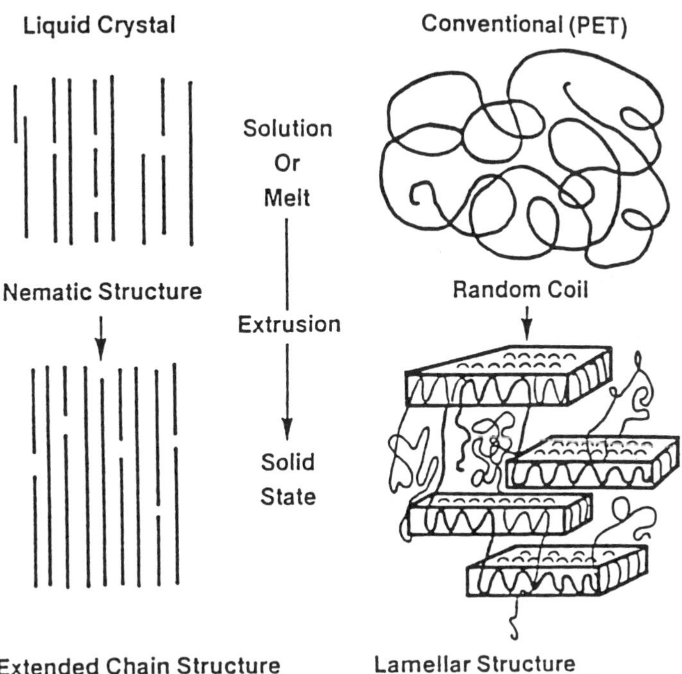

FIGURE 2 Fiber formation model. (From Ref. 32.)

bone. Ringsdorf [4] has extensively studied two kinds of polymers in this class.

CLASS C: CROSS LCPs. Polymers in this class are enantiotropic, whereas polyesters in class B are monotropic. Ringsdorf [4] also synthesized these polymers for the first time.

CLASS D: DISCOTIC POLYMERS. A variety of molecules have been synthesized in this class, including polysiloxanes, polyamides, and polyesters. These materials exhibit low molecular mobility. There are two subclasses, which are distinguished by the function of the rigidity of spacers between discs. Subclass DS contains soft or flexible spacers, and subclass DR contains rigid spacers. Wenz [11] proposed two possible packing structures expected to coexist with each other for subclass DS. Wendorff et al. [12,13] suggested a sanidic (broadlike) packing structure for subclass DR (Fig. 3).

CLASS E: COMBS OR E-SHAPED STRUCTURES. A flexible spacer is introduced between the backbone and the LC segment [14,15]. There are two subclasses in this class. Subclass EO consists of combs with one row of side chains, and subclass EP consists of combs with a palisade of side chains [16]. A large number of LCPs belong to the first subclass, including azo [17], azoxy [18], and phenylazo-azobenzene [19] LCPs.

CLASS F: DISC-COMB STRUCTURES. Kreuder and Ringdorff [20] synthesized disc structures with triphenylene derivatives.

CLASS G: PARALLEL STRUCTURES. LC groups are at side chains and oriented approximately along the chain backbone. Hessel and Finkelmann [21] and Zhou and coworkers [22] synthesized these type of polymers.

CLASS H: INVERSE COMBS. The inverse-comb polymers consist of LC groups. Ballauf [23,24] synthesized polymers in this class with higher molecular masses than those of Weissflow and Demus [25,26].

CLASS I: BIAXIAL STRUCTURES. Nematic structures in this class were obtained by Finkelmann and coworkers [27].

CLASS J: DOUBLE LCPs. These polymers contain LC groups in both the main chain and branches. Such polyesters were synthesized by Reck and Ringsdorf [28].

CLASS K: LCP NETWORKS WITH ELASTOMERIC PROPERTIES. These polymers were first synthesized by Finkelmann and coworkers [29] in 1981. Later, many more materials were prepared by Finkelmann [30] and Zentel and Benalia [31].

Thermotropic LCPs

KEY DEVELOPMENTS. Table 2 lists key discussions and developments that have led to the very high level of research activity in both academia and industry, as initially outlined by Calundann [32] and updated by us. Reinitzer, in 1888, first observed thermotropic cholesteric in small molecules. Polymeric liquid crystallinity

TABLE 1
Classification of LCPs

Type	Structure	Designation
A		Longitudinal
B		Orthogonal
C	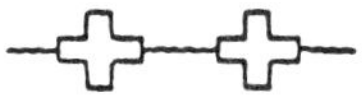	Cross
DS		Soft disc
DR		Rigid disc
EO	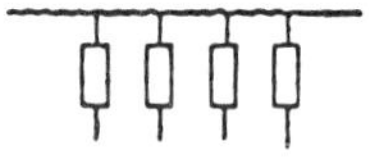	One-comb
EP	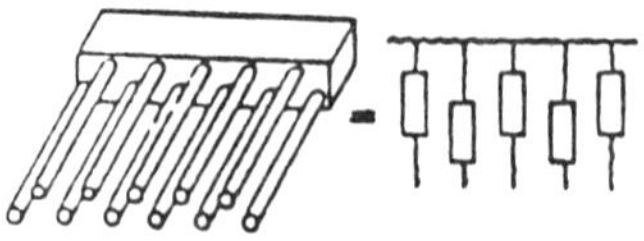	Palisade-comb
F	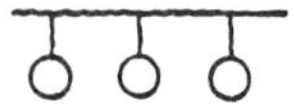	Disc comb
G		Parallel
H		Inverse comb
I	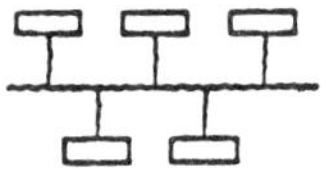	Biaxial
J	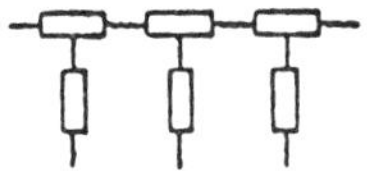	Double
K	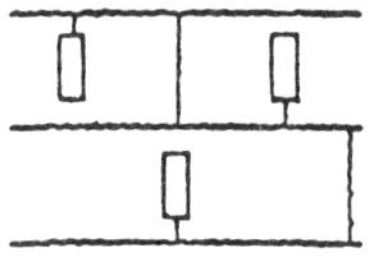	Network

From Ref. 5.

was noted and widely researched in the 1940s and 1950s. Natural and synthetic lyotropic biopolymers such as tobacco mosaic virus, collagen, and poly(γ-benzyl-1-glutamate) were mostly studied. Flory proposed in 1956 that a solution of a rodlike synthetic polymer of suitable rigidity at the appropriate concentration would form an anisotropic phase. Kwolek of the DuPont Company demonstrated this theoretical prediction by observing that certain wholly aromatic polyamides gave anisotropic solutions in alkylamide and alkylurea solvents. This observation led ultimately to the development of Kevlar aramid fiber [33]. Cottis et al. [34,35] patented a high melting aromatic copolyester consisting of the copolyester of *p*-hydroxybenzoic acid, terephthalic acid, and 4,4′-biphenol in 1972, later commercialized as Ekkcel I-2000. This polyester is injection moldable at temperatures in the vicinity of 400°C. Jackson et al. [36,37] and McFarlane [38] reported the first well-characterized thermotropic polymer in 1974. It is an aromatic-aliphatic copolyester,

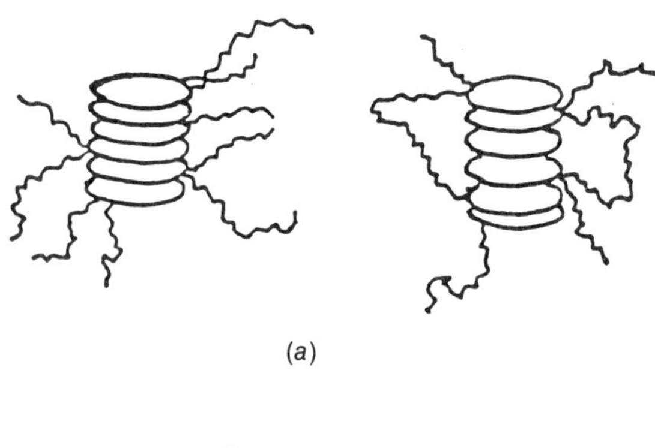

(*a*)

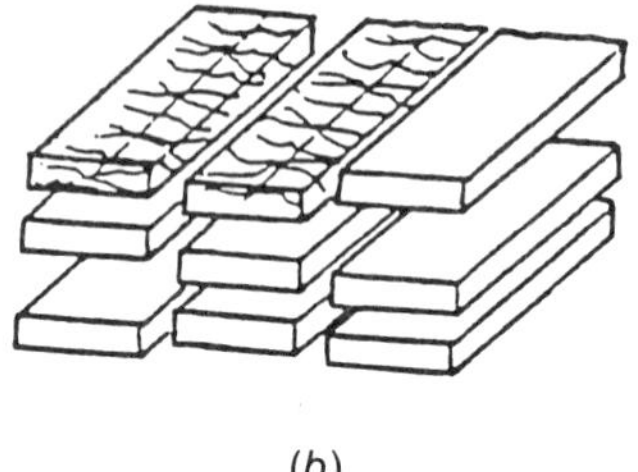

(*b*)

FIGURE 3 **Packing of (*a*) DS (discotic soft) and (*b*) DR (discotic rigid) polymers.** (From Refs. 11–13.)

marketed under the code name X7G. This polymer had several shortcomings, and serious commercial interest did not develop at that time. At the same time, Celanese Company developed a family of wholly aromatic thermotropic copolyesters [32]. These polyesters had many of

TABLE 2
Research Milestones and Commercialization of Anisotropic Polymers

1888	Liquid crystal recognition
1940–1956	Lyotropic biopolymers
	—Tobacco mosaic virus
	—Collagen
	—Poly(γ-benzyl-L-glutamate)
1956	Theory—P.J. Flory
	—"Rigid rod" → anisotropic solution
1965	Lyotropic aromatic polyamides
	—Kevlar
1972	Melt-processable wholly aromatic polyester
	—EKKCEL I2000
1974	Thermotropic polyester
	—Aromatic/aliphatic copolymer—X7G
1975–1976	Wholly aromatic polyesters, polyazomethines
1975	"Linkageless" lytotropic polymers
	—Poly(p-Phenylenebenzobisoxazole)
	—Poly(p-Phenylenebenzobisthiazole)
	Intense patent activity
1985	Dartco Manufacturing Co. (Xydar)
	Celanese Corporation (LCP-2000)
1986	Unitika (LC-2000 series, LC-6000 series)
	Mitsubishi Corporation (E-6000 series)
1987	Bayer AG (KV 9211, KV 9221)
	BASF Corporation (Utrax KR 4002, Utrax KR 4003, Utrax KR 4004)
	ICI (Victrex SRP1, Victrex SRP2)
1988	Granmont Inc. (Granlar LCP)

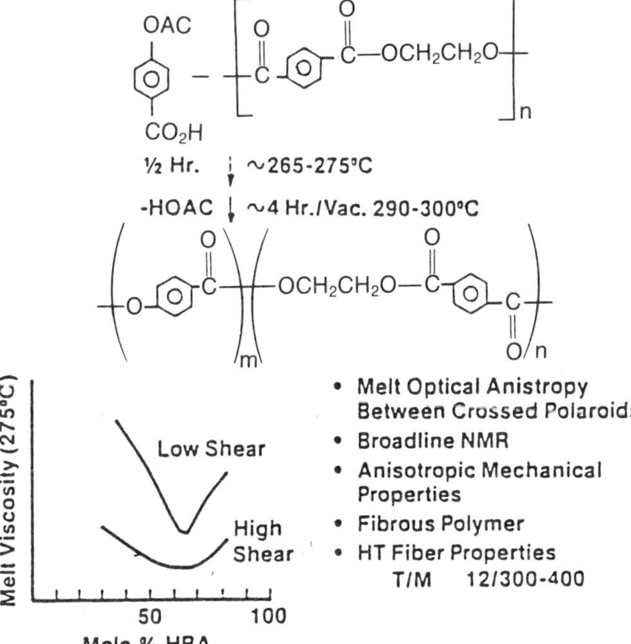

FIGURE 4 **Preparation and characteristics of thermotropic polyesters derived from PET.** (From Ref. 32.)

the same bulk and melt appearance characteristics as X7G; however, they produced superior physical properties. Thermotropic polyazomethines also appeared at this time [39]. Poly-*p*-phenylenebenzobisoxazole and poly-*p*-phenylenebenzobisthiazole, which were initially prepared at the Air Force Materials Laboratory at Wright Patterson, had mechanical properties superior to the aromatic polyamides and esters [40].

The amount of industrial and academic work on liquid-crystalline polymers has piled up since 1975. The variety of anisotropic polymers known is now immense. Several structural categories in addition to the rodlike aromatic polymers are included. The class containing the high performance thermotropic polyesters is the primary subject of this work.

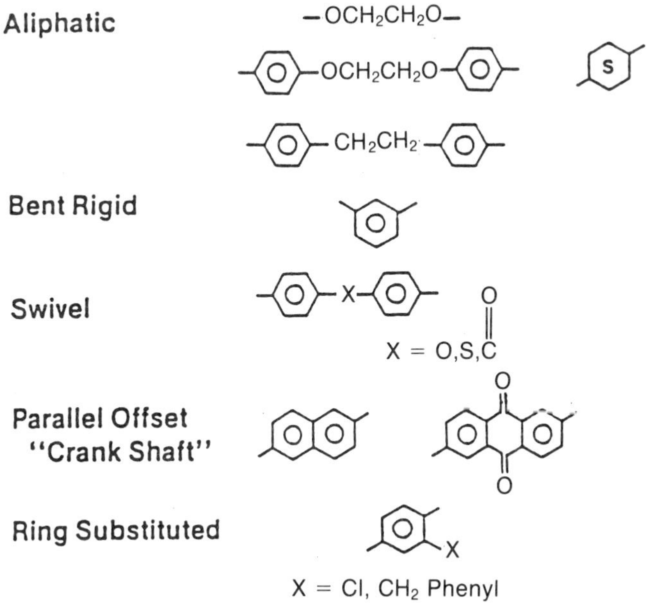

Aliphatic — $-OCH_2CH_2O-$; OCH_2CH_2O; CH_2CH_2

Bent Rigid

Swivel — X = O,S,C

Parallel Offset "Crank Shaft"

Ring Substituted — X = Cl, CH_2 Phenyl

FIGURE 5 **Structures used for tractability in aromatic thermotropic polyesters.** (From Ref. 32.)

Figure 4 summarizes the well-characterized series of thermotropic aliphatic-aromatic polyesters (X7G polymers), a product of copolymerization of *p*-acetoxybenzoic acid with polyethylene terephthalate (PET). The characteristic melt optical anisotropy typical of nematic small molecules is showed in some of these compositions. The melt viscosities of the X7G copolyesters with about equivalent molecular weights were further plotted as a function of *p*-hydroxybenzoyl content under low and high shear conditions (Fig. 4). Melt viscosities were found to be lower for polymers containing 40 to 70 mole % of *p*-hydroxybenzoyl content. The minimum melt viscosities were noted at about 60–70 mole % HBA content with supporting evidence of nematic structure. The fiber properties reported for Eastman's hydroxybenzoyl–PET copolyesters are not particularly impressive, with tensile moduli in the range of 37.2–49.7 GPa and maximum tensile strengths rarely exceeding 1.49 GPa. The most obvious characteristic of thermotropic polyesters is their bulk appearance. These polyesters—both aliphatic-aromatic, as the X7G type, and aromatic structures—have a fibrous, almost woodlike texture. An opalescent, almost metallic surface sheen is another visible characteristic of thermotropic polyesters.

The molecular structures used to promote polyester melt anisotropy at reasonable temperatures are summarized in Figure 5 [32]. The Celanese, Eastman, and DuPont companies have outlined most of the key approaches to a lower polymer melting point via crystalline order disruption. The polyesters are derived typically from symmetric monomers to preserve the inherent melt anisotropy, such as *p*-hydroxybenzoic acid, terephthalic acid, hydroquinone, 4,4′-biphenol, and the like. Eastman's approach involved introduction of aromatic monomers into aliphatic-aromatic polyesters such as PET.

Introduction of bent units derivable, for example, from isophthalic acid or resorcinol certainly is an obvious route towards increasing polymer tractability. However, there are problems associated with the meta linkage. DuPont described a wide variety of tractabilizing molecules leading to thermotropic polyesters [41]. Their research has focused for the most part on ring-substituted monomers such as chloro-, methyl-, or phenyl-substituted hydroquinone and "swivel" or linked ring molecules, examples of which are 3,4′ or 4,4′ functionally (hydroxy or carboxy) disubstituted diphenyl ether, sulphide, or ketone monomers.

Celanese has defined families of thermotropic polyesters based on 2,6-naphthalene dicarboxylic acid (NDA), 2,6-dihydroxynaphthalene (DHN), and 6-hydroxy-2-naphthoic acid (HNA) (Fig. 6). For simplification, many polyesters or polyester amides possible from these monomers will be referred to as naphthalene thermotropic polymers, or NTPs [32].

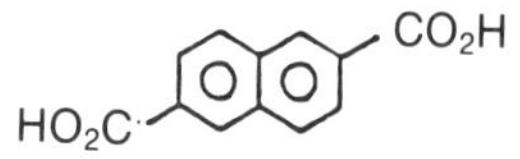

2, 6 - Naphthalene Dicarboxylic Acid

NDA

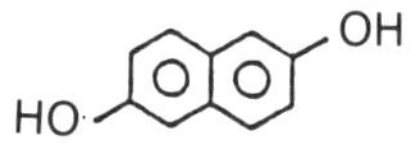

2, 6-Naphthalene Diol

DHN

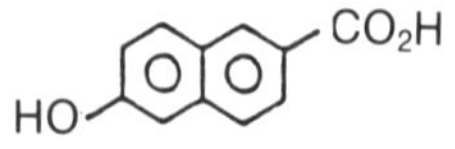

6-Hydroxy-2-Naphthoic Acid

HNA

FIGURE 6 NTP monomers. (From Ref. 32.)

HNA HBA TA HQ

I II

340 320 300 280 260 240

T_m, °C

20 40 60 80

Mole % HNA

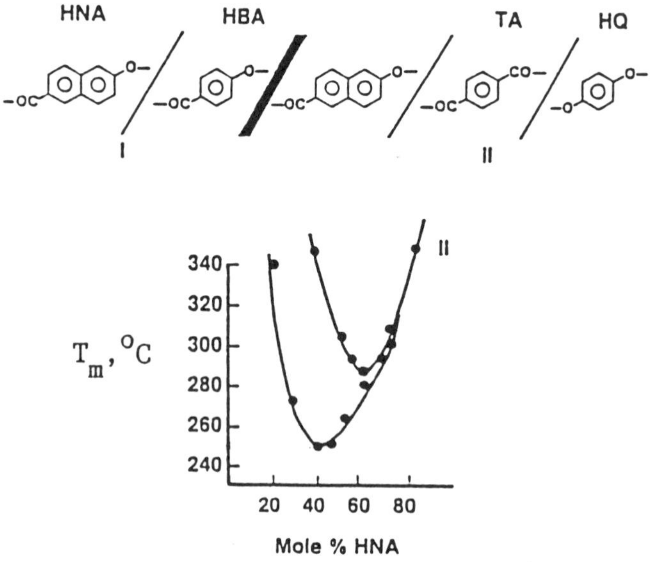

FIGURE 8 Effect of composition on HNA-based NTP T_m. (From Ref. 32.)

The relationship between copolyester composition and the melting points of NDA- [42], DHN- [43], and HNA- [32] based polymers and HBA/TA/HQ [33] is shown in Figure 7. Replacement of TA with 2,6-naphthalene dicarboxylic in the polymerization charge thus gives a series of copolyesters with a melting point minimum at about 325°C, near 60 mole % HBA. Replacing hydroquinone by 2,6-naphthalene diol provides a series with a polymer melting point low near 280°C at about 50 mole % HBA.

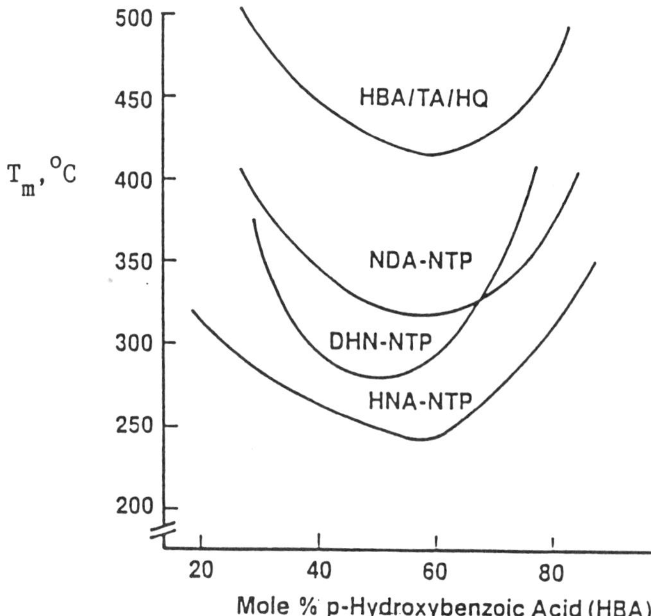

FIGURE 7 Effect of composition on NTP T_m. (From Ref. 32.)

Much of the composition range of the two-component polyester of *p*-hydroxybenzoic acid and 6-hydroxy-2-naphthoic acid (HBA-HNA) falls within the industrially convenient melt temperature zone of 250–310°C. Replacement of the *p*-oxybenzoyl fraction with *p*-phenylene terephthaloyl units in the two-component polyester resulted in a more symmetric, higher melting copolyester series shown as polymer II (HNA/TA/HQ) in Figure 8 [32]. All the compositions of II were thermotropic and ultimately gave high strength and high modulus fibers.

A more complex system of higher tractability over the NDA-NTP copolymer is, for example, interpolymerization of isophthalic acid with HBA, NDA, and HQ (a four-component terpolyester). Figure 9 shows contour plots representing groups of these terpolyester compositions with the same crystalline melting point. The small inner contour with compositions melting near the minimum is about 255°C. Spinning behavior, melt rheology, and fiber properties were similar across this diagram. The shaded area encloses those terpolyesters with close to the optimum processability–fiber property profile, i.e., systems melting at 300°C or less, with equal to or less than 15 mole % isophthaloyl units and from 55–75 mole % *p*-oxybenzoyl units. Stable fiber spinning becomes difficult at above 300°C.

The introduction of a mixed linkage does not appear to offer any gain in system tractability, specifically, an increase in the range of compositions melting below 320°C. Figure 10 shows the relationship between melting point and composition of the HNA/TA/HQ copolyester and of an analogous polyester amide in which *p*-acetoxyacetanilide replaced hydroquinone diacetate in the initial polymerization charge. Amide linkage incorporation resulted in a reduced tractable zone, although minimum melting points for both systems were nearly the same at about 275–280°C (about 60 mole % HBA). The maximum amide linkage was about 25 mole % [43].

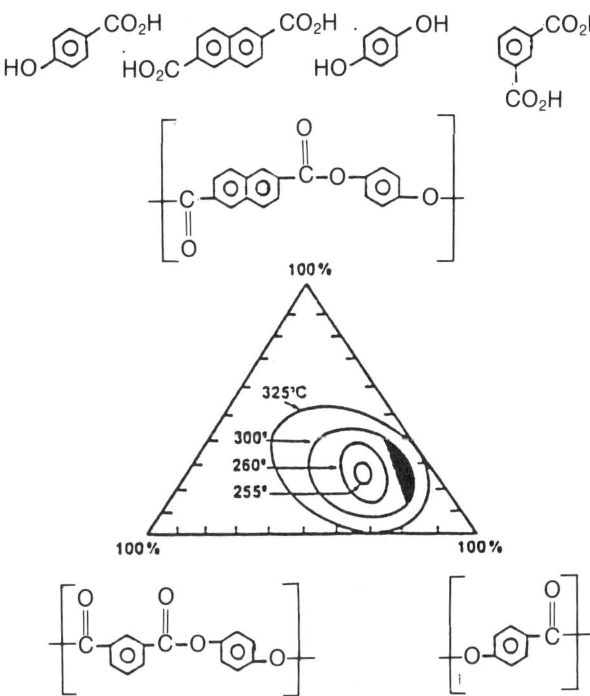

FIGURE 9 **Composition/T_m relationship of NDA-based terpolyester.** (From Ref. 32.)

In summary, the important factors in the molecular design of commercially thermotropic polymers for fiber–resin applications are: (a) processability—polymers should ideally be processed in the range of 250–350°C; (b) melt anisotropy—a careful balance of molecular symmetry is required; (c) end-use properties—optimization of structure property is required; (d) minimum monomer cost; and (e) easy preparation. All of the polymers described so far may be made in a conventional melt acidolysis process starting with the acetoxy derivatives of the hydroxyl-containing monomers used. Figure 11 shows a typical polymerization scheme. This preparation of the two-component polyester is derived from the acetylated hydroxybenzoic and hydroxynaphthoic acids with or without added catalysts. Alternative synthetic methods include acidolysis in inert heat exchange media, a useful approach for preparation of very high melting polymers. Polymers not incorporating *p*-oxybenzoyl units may be made by direct esterification of diacids and diols, with or without HNA, in the presence of a suitable catalyst. So far, only a few thermotropic LCPs have been commercialized (Table 2). Dartco Manufacturing Co. (now belonging to AMOCO) and Celanese Plastics & Specialities Co. introduced LCPs based on wholly aromatic copolymers within a span of a few months under the trade names of Xydar and Vectra, respectively. These have shown properties equal or superior to those of other engineering thermoplastics, and have the distinct advantage of being melt processable. The most recent additions to commercial LCPs are the E-6000 series of Mitsubishi Chemical Industries (1986), the KV series of Bayer AG, the Utrax KR series of BASF, and the Victrex SRP series of ICI, all introduced in 1987.

SOLID-STATE STRUCTURE. Thermotropic LCPs may be melt processed by extrusion and injection molding to form highly oriented rods, strands, and molded products. Extrudates have molecular orientation developing as a result of the effect of the flow field on the

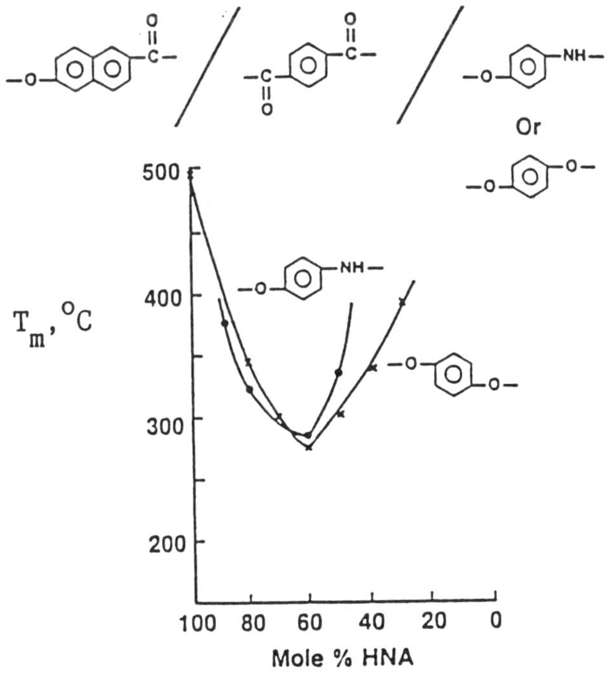

FIGURE 10 **Composition/T_m relationship of HNA-based polyester amide.** (From Ref. 32.)

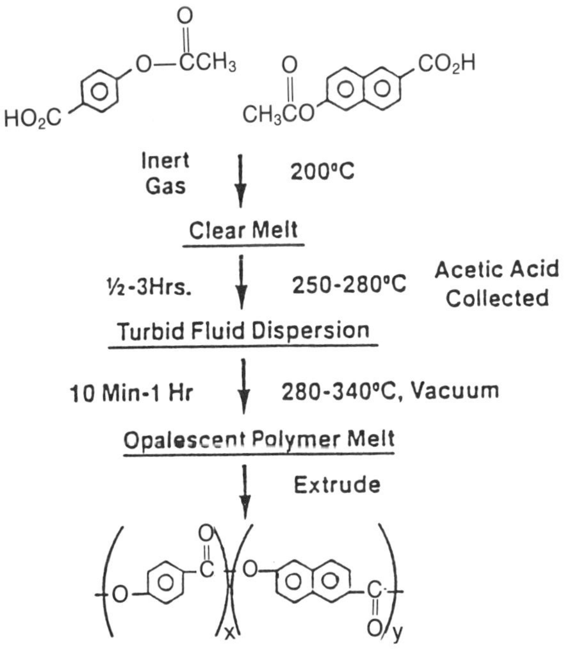

FIGURE 11 **Typical NTP synthetic scheme.** (From Ref. 32.)

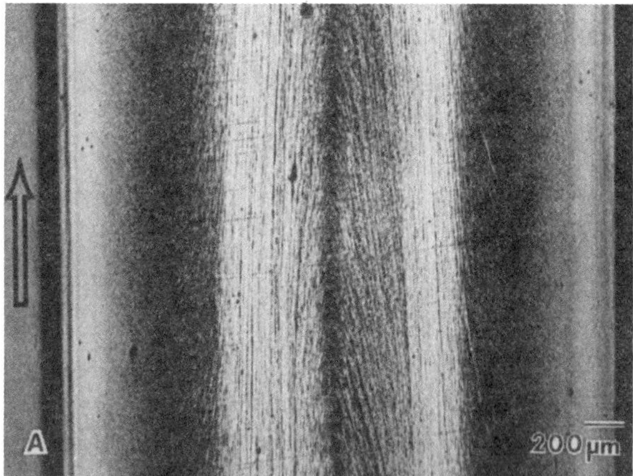

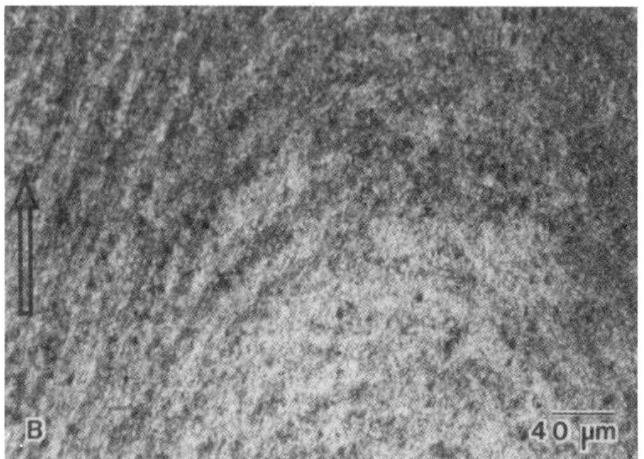

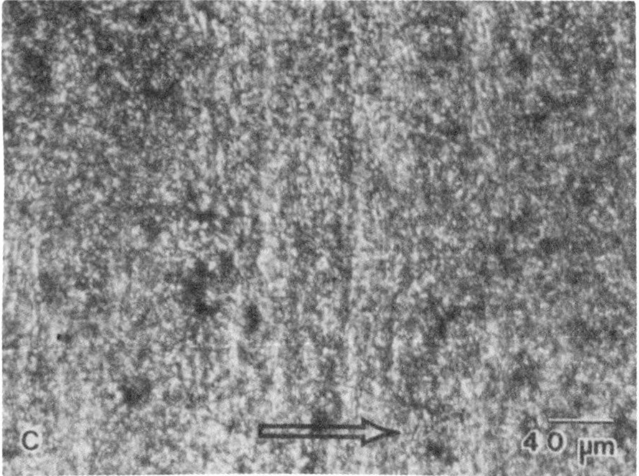

FIGURE 12 (*A*) A reflected light micrograph shows the layered structures in a molded bar aligned parallel to the flow direction (arrow). Variations in density and color reflect variations in orientation from layer to layer. (*B*) Lateral, curved flow patterns are seen in a polarized light micrograph overview of a thin section. (*C*) The flow layers are nearly normal to the flow direction (arrow) in the center of the bar. (From Ref. 51.)

oriented extended chain molecules [44–46]. Properties of molded thermotropic LCPs are better than those of short fiber reinforced composites. Therefore, thermotropic LCPs are termed "self-reinforcing composites" [47,48]. Highly anisotropic structures—i.e., layers (normal to the flow direction), bands (parallel to the flow), and a skin-core structure—give rise to highly anisotropic physical properties. The skin-core structure observed for LCPs is similar to the structure of typical thermoplastic materials with maximum orientation at the surface (skin), because of elongational flow, and minimum at the core, because of shear flow [44,48–50]. These layered structures may be observed by eye owing to the color variations associated with local orientation differences between layers. The layered structure is highly anisotropic and process dependent, affecting mechanical properties.

Molded bars of thermotropic LCPs exhibit layered structures, as shown by reflected light microscopy (Fig. 12*a*) of a cut and polished bar. Thin sections of a molded bar show fine, nematic domains with superimposed flow lines (Fig. 12*b*), especially near the center of the bar (Fig. 12*c*). Skin-core morphologies appear in injection-molded bars and extrudates with domains aligned in the flow direction. Scanning electron microscopy (SEM) of fractured injection-molded bars provides an overall view of the layered structure (Fig. 13*a*), the surface skin (Fig. 13*b*), the internal fibrillar structures in the inner skin (Fig. 13*c*), and the core (Fig. 13*d*). Complex skin-core and banded textures may be observed in extrudates where the orientation is a function of the draw ratio and the final diameter, with higher orientation in finer strands. Figure 14*a*, taken with the flow axis parallel to the polarizer, and Figure 14*b*, at 45° from this position, show the orientation and incomplete extinction in the skin. Nematic domains are seen in the core, and to some extent in the skin; however, the extinction regions in the skin (Fig. 14*a*) reflect higher orientation.

The SEM micrograph (Fig. 15) shows the common woody or fibrillar fracture of extrudates. The orientation of the strand controls this fracture morphology to some extent. Micrographs of polished thin sections of an unoriented NTP rod (Fig. 16*a*) and a slightly oriented rod (Fig. 16*c*) photographed in circularly polarized light reveal a nematic structure banding and evidence of skin orientation (Fig. 16*c*). Secondary electron imaging (SEI) of fracture surfaces reveals a more coherent fibrillar structure in the slightly oriented rod (Fig. 16*d*) compared with the unoriented rod (Fig. 16*b*).

The fine structure of the NTP extrudates may be revealed by etching experiments. Plasma etching of a polished extrudate was monitored by high-resolution SEI. An overview (Fig. 17*a*) shows flow lines of the polymer in a middle region between the skin and core of the strand, and at higher magnification, grainlike domains compose parallel structures (Fig. 17*b* and *c*) not parallel with adjoining domains. A more detailed fine fibrillar texture of these local regions is shown in Figure 17*d*. The etching experiments reveal domain structures that have a high degree of nematic order aligned by the flow process.

Microscopy techniques reveal structural models describing LCP extrudates and moldings [47,51–54]. Thapar

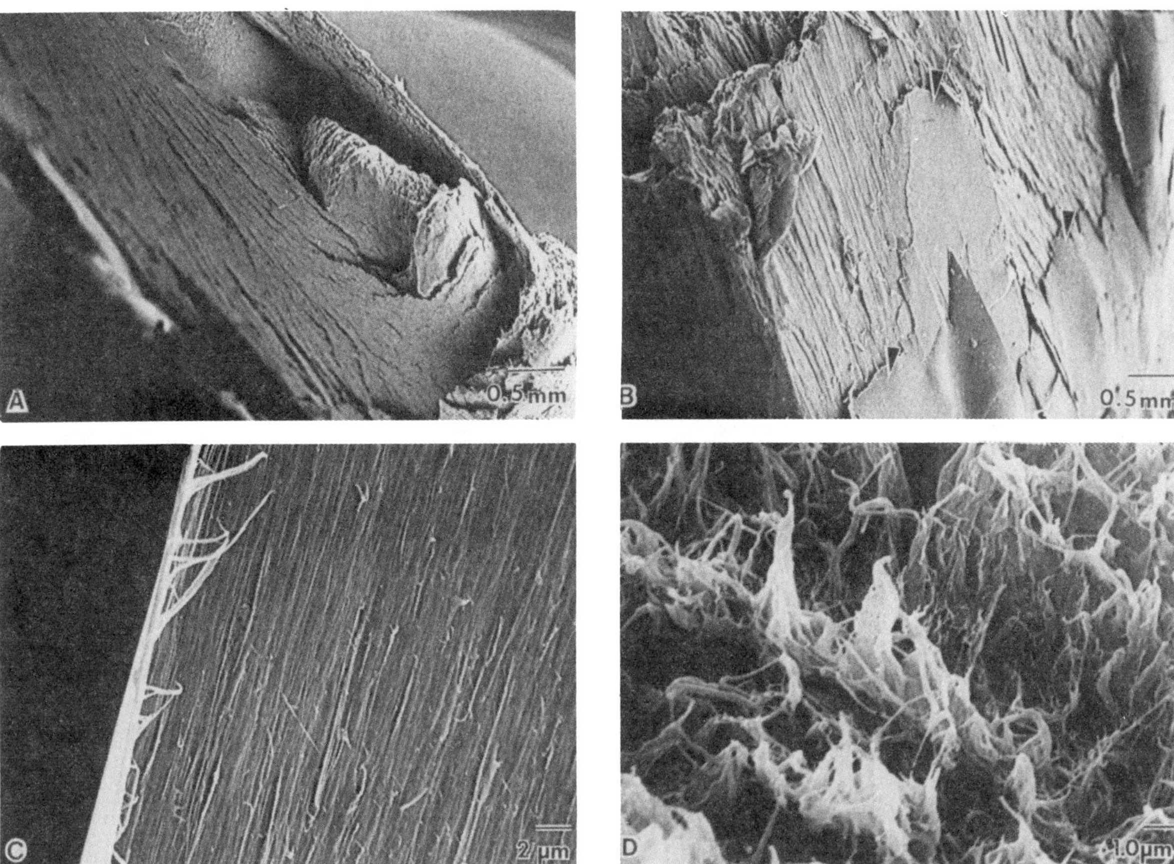

FIGURE 13 The layered structure of molded bars is shown in SEM of fractured NTP specimens. (*A*) An overview shows that the concentric layers are thin, whereas (*B*) the outer, coherent skin (arrows) is layered or sheetlike in structure. Fibrillar structures are observed in (*C*) the skin and (*D*) the core. (From Ref. 51.)

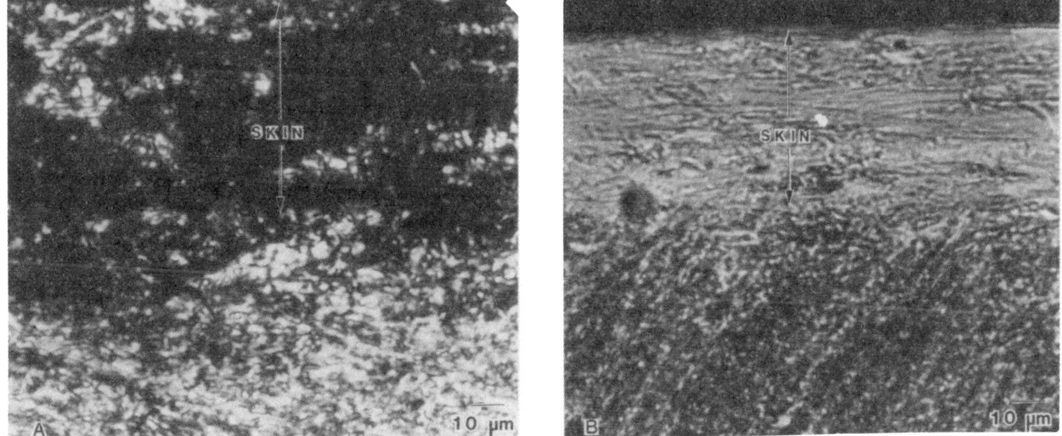

FIGURE 14 (*A*) Skin–core structures shown in more detail in highly magnified polarized light micrographs with the specimen in the orthogonal position. The fine domain textures are partially extinct near the skin as a result of its orientation, although domains are observed. (*B*) With the specimen at 45°, the skin is seen clearly to be more oriented than the core. (From Ref. 51.)

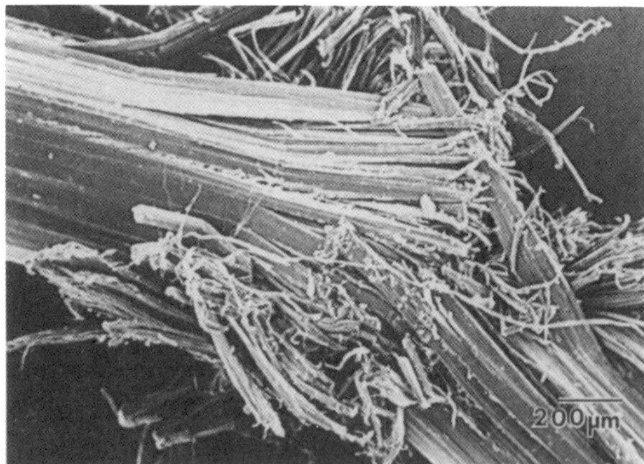

FIGURE 15 **Fractured extrudates appear woody in texture when viewed in the SEM, and coarse and fine fibrils are observed.** (From Ref. 51.)

and Bevis [47] displayed a schematic of the skin-core and layered structures of injection-molded thermotropic LCPs by polishing and etching methods for SEM. Baer et al. [53,54] developed a similar model by SEM fracture methods for an NTP extrudate. Sawyer and Jaffe [52] applied a wide range of microscopy techniques to study LCP extrudates and moldings as well as highly oriented fibrous materials and developed a general structural model. This model, shown in Figure 18, exhibits skin-core, layered, and banded macrostructures and fine, hierarchical fibrillar microstructures ranging from macrofibrils, on the order of 5 μm across, to microfibrils, on the order of 50 nm wide and about 5 nm thick [52].

CHARACTERIZATION OF THERMOTROPIC LCPs. *General.* When a new polymer is prepared in laboratories, a set of characterization methods is usually applied to determine whether the polymer forms a thermotropic melt, what type of LC phase is formed (nematic or smectic), and at what temperatures the phase changes occur (including melting, smectic to nematic, anisotropic to isotropic, and recrystallization). For specific polymers of particular interest, additional methods may be applied, such as small-angle light scattering, SEM, and transmission electron microscopy (TEM) to investigate the morphology of the LC state; wide-angle X-ray diffraction to more fully characterize the smectic and nematic states; dynamic mechanical and temperature transition measurements; mechanical properties such as tensile strength, modulus, elongation, and impact strength; melt viscosity and rheological behavior for investigations on homopolymers and for observing effects of variations

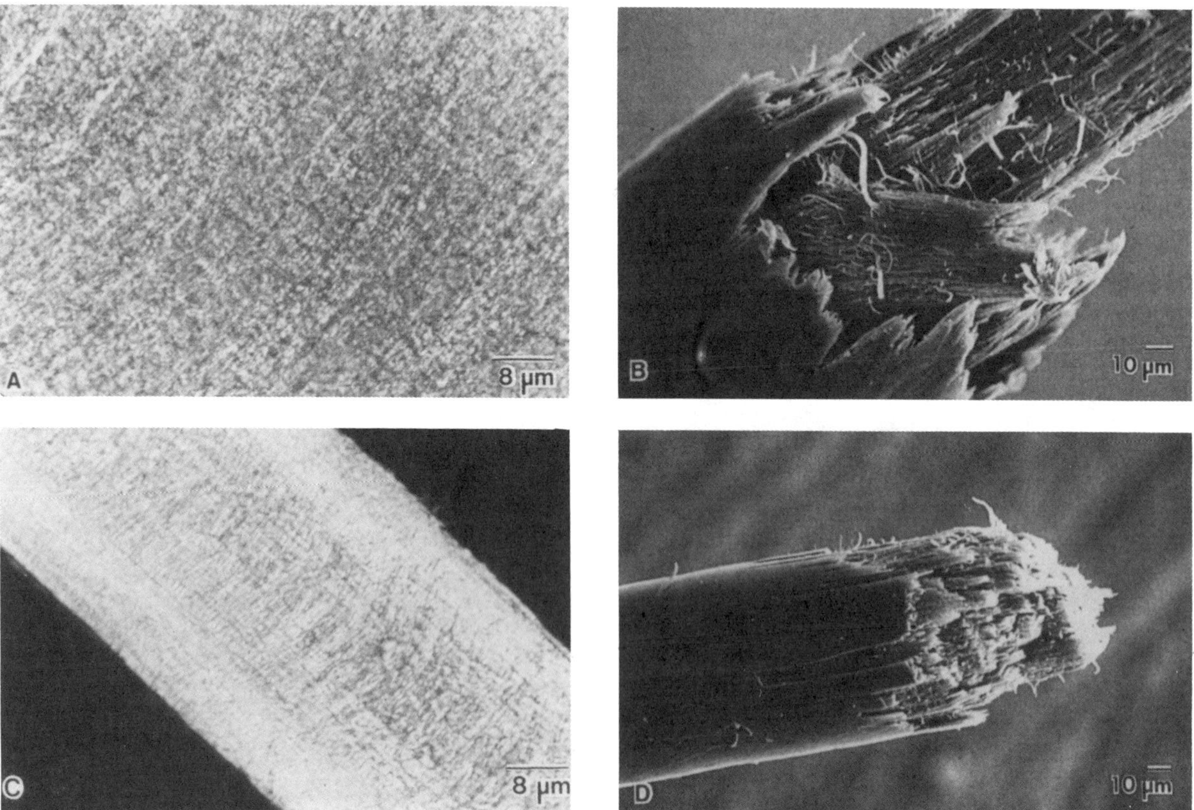

FIGURE 16 Free fall NTP strands are shown in polished sections in circularly polarized light (*A, C*), and also by SEM of fractures (*B, D*). The less oriented strand appears more uniform in domain texture (*A*) and also exhibits a coarser woody fracture (*B*). Some orientation is observed in the more highly oriented strand (*C*), and the fracture morphology is more uniform (*D*). (From Ref. 51.)

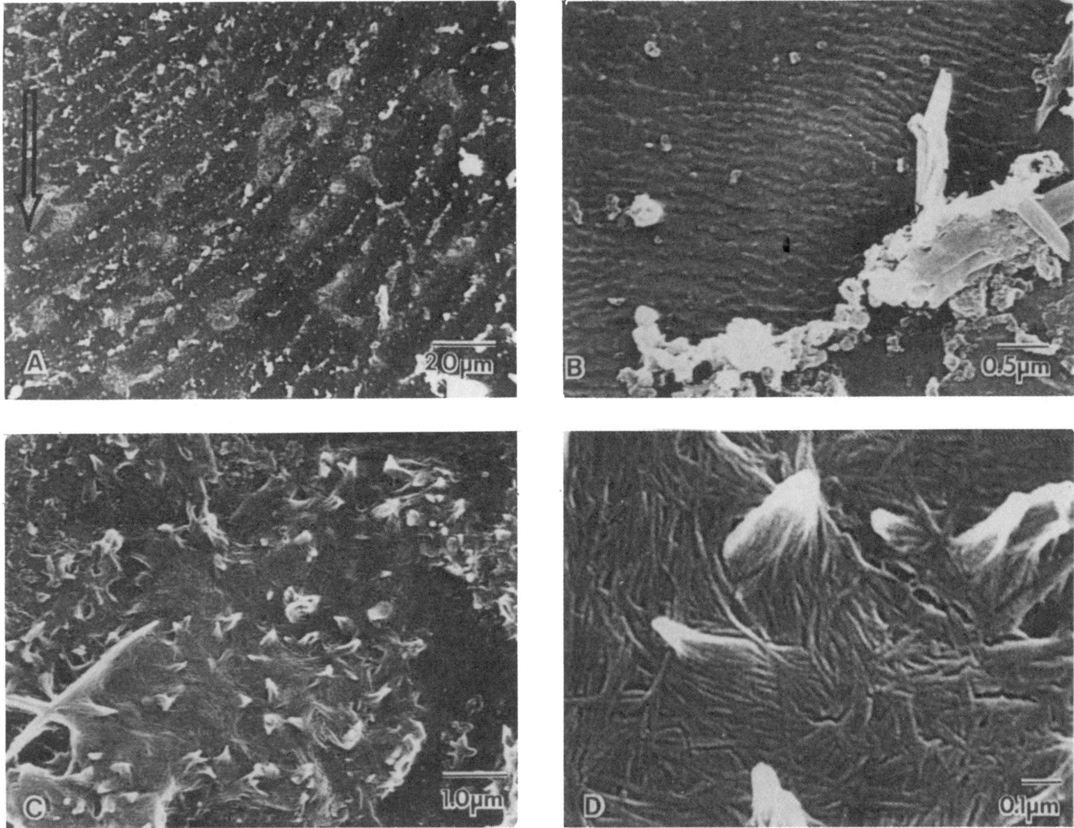

FIGURE 17 **Plasma etching with argon reveals the fine internal structure of an extrudate in these SE images. The overview (*A*) shows the flow lines of the polymer. Grainlike domains are observed in a region midway between the skin and core (*B*), with internal fine structural detail (*C*) that suggests local orientation. Microfibrillar structures are observed at higher magnification (*D*).** (From Ref. 51.)

in the "degree of crystallinity" in copolymers containing both mesogenic and nonmesogenic units; and other more specialized methods.

A simple visual observation of a sample on a hot plate may be the first indication of the ability of a polymer to melt or soften directly into a liquid-crystalline state. Such polymers will mostly show an opalescent or pearly appearance when their LC melts are showed or stirred. R. W. Lenz and J. Lin [55] observed that the "stir opalescence" brightness of copolymers containing varying amounts of mesogenic and nonmesogenic units can be taken as a semiquantitative indication of their degree of liquid-crystallinity.

Optical microscopy. The simplest characterization of liquid crystals can be done by polarized light microscopy. This is usually carried out in the initial stages of an investigation of a new polymer because thermal analyses alone do not indicate what type of phase is formed and can be misleading. The appearance of a particular texture of the melt usually depends on the mesophase structure. Therefore, this method is the direct identification method to indicate the type of mesophase present.

The texture observation of low molecular weight compounds is usually much easier than that of polymeric compounds because their lower melt viscosities lead to quicker appearance of melt structures. Polymeric materials may take minutes or hours to indicate recognizable textures, and the polymers can degrade or cross-link during that time. Some polymers show no definite texture, possibly because of their high molecular weight, and may consequently be identified as nematic phase [56,57], or the sample may not develop the equilibrium texture and may be incorrectly identified [58].

The textures exhibited by polymers are generally identical to those of low molecular weight compounds. The textures most often referred to in the literature are the threaded Schlieren textures (typical of nematics) and the focal-conic or fan-shaped textures (typical of smectics) [59]. Like their low molecular weight counterparts, cholesteric polymers also show oily streaks and parallel disclinations [60].

A more quantitative approach to microscopy involves measuring the depolarized light intensity with a photocell and plotting this quantity as a function of temperature. This approach is particularly useful in determining the clearing transition from rather broad transitions [61].

X-ray diffraction. Wide-angle X-ray diffraction (WAXD) is a useful method for identification of the struc-

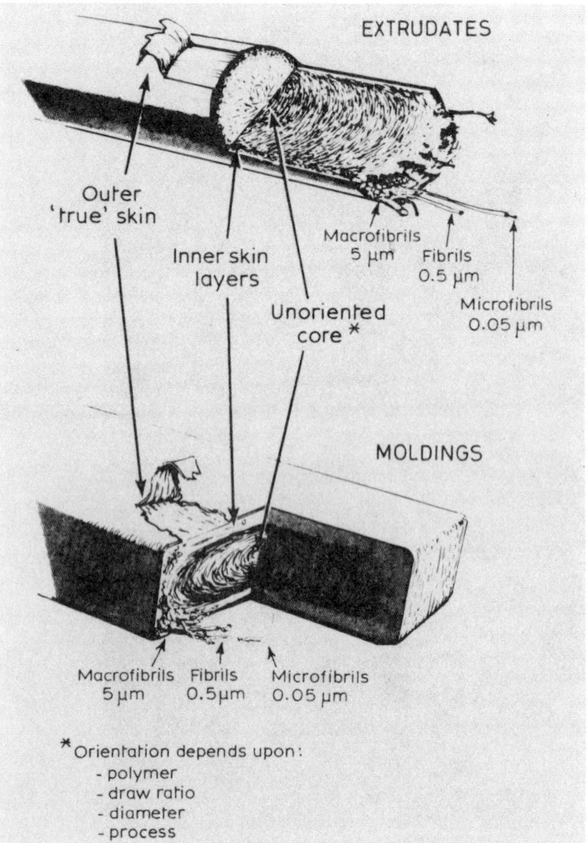

FIGURE 18 **A structural model is shown that provides a schematic of the macrostructures in moldings and extrudates—structures such as layers, bands, and skin–core textures. Process changes appear to affect the macrostructures, while the nature of the fine structures appears similar overall.** (From Ref. 52.)

ture of a thermotropic LC polymer melt because microscopy may not always produce an unambiguous determination of the type of mesophase present. Nematic structures from WAXD patterns form a diffuse ring at $4–5 \times 10^{-4}$ μm on a flat film diffractogram arising from the interchain spacings [62]. Cholesteric mesophases generally resemble nematic mesophases when observed by both small-angle X-ray diffraction (SAXD) and WAXD. However, the smectic mesophase, in contrast, produces both diffuse rings at $4–5 \times 10^{-4}$ μm and sharp rings at a distance generally equal to the repeat length of the monomer unit ($15–50 \times 10^{-4}$ μm) [59,63]. Biswas and Blackwell analyzed the three-dimensional structure of wholly aromatic main-chain LC copolymers by X-ray methods [64]. This paper described calculation of the cylindrically averaged transforms of single chains.

Thermal analysis. Thermal analysis is the most common technique for studying liquid-crystalline compounds and polymers. This technique offers the transition temperatures, heats, and entropies of transition. There is a large amount of literature concerning the thermal behavior of low molecular weight compounds, and most studies of LCPs have been performed using a DSC or DTA equipment.

The sample history can greatly affect the melting temperatures of polymeric materials. However, it is commonly agreed that the clearing of anisotropic–isotropic transition is much less subject to effects of thermal treatment and shows little undercooling. Therefore, the clearing transition is close to an equilibrium process. Then the entropy of this transition, ΔS_i, can be calculated from the clearing temperature and from the enthalpy of the transition, ΔH_i. ΔS_i values should provide some indication of the order present in the system if the isotropic state is assumed to have equal disorder in all systems [65], and the order should be related to the type of LC phase. Ober et al. [66] have determined ΔS_i and ΔH_i values for many different main-chain LCPs, particularly those with polymethylene spacers, and correlated them with the type of mesophase formed. It was found from most of those polymers that, when the homologous series was studied, an even-odd effect in the entropy of clearing was observed that was the same as the even-odd properties of both the melting and clearing transition temperatures. The interpretation of the variation in relative values of ΔS_i along such a series may be complicated by changes in the type of mesophase formed from nematic to smectic by going from an even to an odd number of methylene units [67].

Cheng [68] used DSC to carry out nonisothermal and isothermal kinetic studies of copolyesters with different 1,4-dihydroxybenzoic acid and 2,6-dihydroxynaphthoic acid compositions. Different transition behavior was observed in the nonisothermal experiments among those copolymers, and it was strongly dependent on crystallization conditions (cooling rate) and the chemical structure of their counits.

Dynamic mechanical properties. Dynamic mechanical measurement, like thermal analysis, provides extensive information concerning the characteristics of polymeric materials. Evaluations may be conducted over a wide range of temperatures with a small amount of sample. This characterization technique and thermal analysis (see mechanical properties of self-reinforced composites) are well suited for quality control of various types of polymer blends.

The dynamic mechanical spectra of polymer blends obtained as a function of temperature exhibit transition peaks similar to absorption bands in spectroscopy. In fact, such peaks represent the absorption of mechanical energy by the test materials. The proper interpretation of these spectra is used to characterize the molecular structure. The storage modulus, loss modulus, glass transition, secondary relaxations, energy absorption, etc., are evaluated as a function of temperature and frequency (time) of test. These properties may be obtained from measurements using a rheometrics mechanical spectrometer (RMS), mechanical energy resolver (MER), dynamic mechanical thermal analyzer (DMTA), dynamic mechanical analysis (DMA), or other such equipment.

The dynamic mechanical thermal analysis on LCPs and their blends can be found from Refs. 69–72.

Small-angle light scattering. The small-angle light scattering (SALS) method is applied to polymers to characterize their tertiary order or morphology at dimensions in the region of 0.05 to 1 μm. This experimental method has proven useful in the study of the spherulitic crystalline morphology [73].

LCPs apparently organize into domains of around 1 μm in size in the nematic and smectic phases. However, there is little knowledge about either the shape of these domains, the interrelationships between neighboring domains, or environmental effects (temperature and shear) on their structures. Stein and coworkers [73,74] applied the SALS method to investigate the effects of temperature on the domain morphology of thermotropic LCPs by using H_v and V_v polarization with photographic and photometric methods of analysis. It has been observed for nematic polyesters that approximately circular scattering patterns were obtained and that the scattering intensity was azimuthally independent, with H_v intensities being comparable to V_v over almost all of the temperature range of the nematic mesophase. These observations imply that the macroscopic scattering elements (domains) were approximately spherical in shape and that there was little or no correlation between director axes in neighboring domains. Similar preliminary results have been obtained for a smectic mesophase of an LCP [67].

Electron microscopy and electron diffraction. Donald and Windle [75] applied electron microscopy and diffraction to investigate an unusual band structure formed by LCPs on quenching after being oriented in the nematic state. Thomas and coworkers [76] have also applied this method to morphological investigations of oriented thin films of LCPs. For the oriented polymers, it was concluded that the LC state did not contain a domain structure, but consisted instead of a continuous phase containing various types of disclinations [76].

Order parameter measurements. Order parameter measurements are important for relating physical properties of LCPs to their molecular structure. The molecular order in LCPs is generally described in terms of two order parameters, micro-order and macro-order. The former characterizes the orientation of the repeating units within a molecular domain, whereas the latter characterizes the macroscopic alignment of the domains.

Various methods have been used to measure order parameters. These include the use of refractive index and dielectric and magnetic susceptibilities. These methods require knowledge of the corresponding molecular quantities of the polymer on which the interpretations are based [77]. Other experimental methods that are independent of such information were developed. Lenz [67] and Kothe and Berthold [78] used electron spin probes in polymers to measure liquid-crystalline order from the temperature and angular dependence of their electron spin resonance (ESR) spectra. Lenz used a cholestane probe containing a nitroxide free radical to dissolve in the same main-chain liquid crystal polyester. This probe is stable up to 180°C in the polymer melt. The orientation of the mesogenic units should be taken in the polymer backbone because of their elongated, rod-like shape. The ESR line shapes should then give valuable information about the micro- and macro-order of the polymer [79].

Deuteron nuclear magnetic resonance (DNMR). Pulsed DNMR offers unique possibilities for studying molecular order and dynamic in solid polymers and LCPs [80]. The complete orientational distribution function for individual structural elements can be determined from the DNMR analysis. By analyzing the line shapes of deuterium absorption, solid echo, and spin alignment spectra, both type and time scale of rotational motions can be determined over an extraordinarily wide range of characteristic frequencies, from around 10 MHz to 0.01 Hz. The spin lattice relaxation of the deuterons, moreover, can be exploited to extend the dynamic range to faster motions, up to 10 GHz, to detect motions that do not change the NMR spectra, and to characterize the heterogeneous nature of the molecular dynamic.

Rheology. Considerable theoretical and experimental analysis has been done on the rheology of low molecular weight thermotropic compounds [81,82]. Liquid crystals are oriented by electromagnetic fields and by mechanical stress or shear, and differ in their degree of orientation. This, in turn, affects their melt viscosity. The rheological behavior of low molecular weight liquid crystal compounds and LCPs is known to be greatly dependent on the nature and also on the texture of their mesophase. The degree of liquid crystallinity of copolymers containing both mesogenic and nonmesogenic units can be measured by the relative values of the melt viscosities as a function of composition.

The rheology of nematic LCPs has some similarities to and differences from that of low molecular weight thermotropic compounds and isotropic polymer melts and solutions. LCPs exhibit lower viscosity in the nematic state than in the isotropic state. They show anisotropic viscosity dependent upon the orientation of the flow field. The flow behavior of LCPs shows three characteristic regions of shear thinning as proposed by Onogi and Asada (Fig. 19) [83,84]. In region I, corresponding to low shear rate, they exhibit yield stress where piled domains of LCP are not strongly oriented in shear flow. The domains are stable, and their size depends on stress. They flow as entities in the shear field and can be easily oriented in the elongational flow field. Region II corresponds to a Newtonian-like plateau that involves flow of dispersed polydomains. This is followed by region III, in which LCPs exhibit shear thinning and viscoelasticity. Region III involves flow of monodomain continuous phase.

Rheological studies of two classes of thermotropic LC materials have received attention in the literature. The

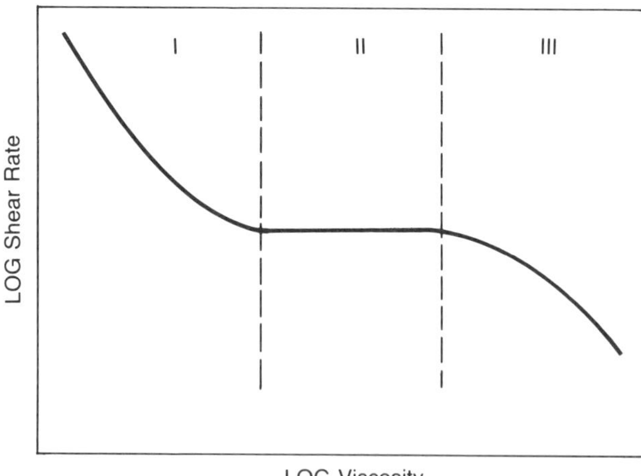

FIGURE 19 **Three regions of flow behavior.** (From Ref. 84.)

most extensive investigations have involved aromatic polyesters of varying compositions. Wissbrun [85]; Jerman and Baird [86]; Prasadarao, Pearce, and Han [87]; Baird [88]; and Cogswell [89] have presented studies of the shear viscosity function of these materials. It seems clear that the viscosity increases indefinitely at low shear stresses, and yield values are observed [85,87,89,90]. Wissbrun and Cogswell observe the shear viscosity to increase with shear rate over limited shear rate ranges for certain materials. Generally, the shear viscosity is found to decrease uniformly with increasing temperature. Jackson and Kuhfuss [91] have considered the composition dependence of the shear viscosity of hydroxybenzoic acid/polyethylene terephthalate (HBA/PET) copolymers and observed that when enough HBA is included to make the material liquid-crystalline, the viscosity shows a substantial decrease.

The second class of thermotropic LC materials that has been studied is cellulose ethers. The shear viscosity characteristics of these materials have been investigated by Shimamura, White, and Fellers [90] and more notably by Suto, White, and Fellers [92]. They observed that liquid-crystalline hydroxypropylcellulose (HPC) and ethyl cellulose (EC) exhibit yield values in shear flow. The problems associated with normal stress measurements have also been discussed in Ref. 92 for a cone plate geometry using total force measurements. The response obtained is similar to those observed in particle-filled polymer melts exhibiting yield values [93].

The presence of domains with different director orientations in nematic melts causes the lack of die swell (elastic recovery), long relaxation times, the important role of shear history and thermal history on rheological behavior, and unusual yield stress behavior [94]. The quantitative study of the rheology of LCPs is still in the early stage [95,96]. Much more research is needed before one can make any conclusions based on specific models. An understanding of the rheological properties and physical states of the LCP phase is very important, not only for theoretical interest, but also for the optimization of processing conditions to obtain maximum physical and mechanical properties.

Mechanical properties and processing. The mechanical properties, of all the properties of plastic materials, are often the most important because virtually all service conditions and the majority of end-use applications involve some degree of mechanical loading. Nevertheless, these properties are the least understood by most design engineers. The material selection for a variety of applications is quite often based on mechanical properties such as tensile strength, modulus, elongation at yield and at break, and impact strength. In practical applications, polymers are seldom subjected to a single, steady deformation without the presence of other adverse factors, such as environments and temperature. The mechanical properties are measured from tests conducted in a laboratory under standard test conditions, so the danger of selecting and specifying a material from these values is obvious. A thorough understanding of mechanical properties, tests employed to determine such properties, and the effects of adverse conditions on mechanical properties over a long period is extremely important.

Composites are a major application of thermotropic LCPs where both the adhesion of the polymer to the filler, such as glass fibers, and the orientation of the LCPs on the fibers are quite important for physical properties. In Figure 20*a*, a polarized light micrograph of a polished thin section of a glass fiber reinforced NTP shows fibers (black isotropic) in a nematic matrix. Some indication of the orientation of the polymer on the glass fiber surfaces can be observed. This is confirmed by SEM examination of a fracture surface (Fig. 20*b–d*).

Figure 21 shows a general structural model for highly oriented liquid-crystalline fibers. The model was developed for the NTP fibers. Duska [97] studied the mechanical properties of unfilled and filled LCPs. Three grades of Xydar copolyester polymers were studied, including neat resin (SRT-300), a 50% glass-filled composite (MD-5), and a 50% mineral-filled composite (FC-130). The pure resin is made from a 1 : 2 : 1 molar ratio of the following monomers: *p,p′*-biphenol, *p*-hydroxybenzoic acid, and terephthalic acid. Physical properties of these polymers were found to be functions of mold temperature, injection speed, injection pressure, part thickness, and type of filler. The measured properties were tensile strength, tensile modulus, elongation, flexural strength, and flexural modulus. The highest value of tensile strength occurred with a part made with low injection speed, low injection pressure, and low mold temperature for the unfilled resin. The observed tensile strength behavior may be explained by a consideration of the skin–core phenomenon. The major contribution to strength in the flow direction is provided by the skin layers. The thickness of these layers relative to that of the core is increased at low injection speed, low injection pressure, and low mold temperature as a result of frozen-in orientation enhanced by favorable cooling conditions. In the case of the glass- and mineral-filled compounds, the variations in tensile strength as a function of processing

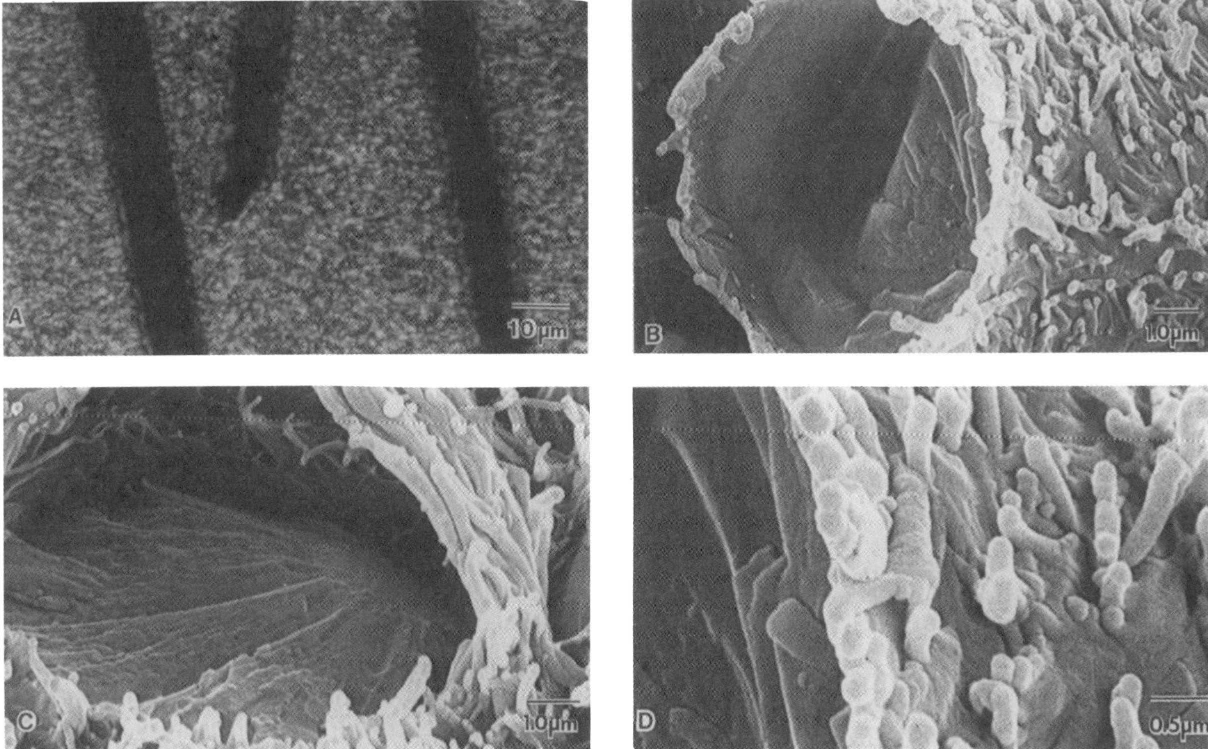

FIGURE 20 A glass fiber reinforced LCP composite is shown to have interesting morphology. (*A*) A polished thin section is shown in polarized light to exhibit a fine domain texture with some orientation of the polymer on the glass surfaces. (*B–D*) SEM fracture views show the tenacious adhesion of the LCP to the fibers. Fibrillar structures are oriented parallel to the fiber surface, and submicrometer-sized domains are observed (*D*). (From Ref. 51.)

parameters are not as pronounced as with the unfilled resin. The glass- and mineral-filled compounds are more forgiving in terms of processing parameter variations because of disruption of the orientation of LCPs in the presence of filler.

Thermotropic LCPs are thermoplastics and may be processed by conventional processing equipment [98]. Because of their low viscosity, anisotropic property with high strength and modulus, good electrical properties, and high chemical resistance, these polyesters are well suited for molding into various kinds of applications, such as electronic complicated shapes, aerospace parts, marine parts, medical devices, chemical processing machinery, and automotive parts.

LCPs are found to exhibit nearly zero die swell upon extrusion through a die, which is different from polyolefins [92]. Cellulose ethers and aromatic polyesters are the two classes of materials that have been extensively studied. The extrusion of liquid-crystalline cellulose ethers through dies has been investigated by Shimamura et al. [90]. The complete absence of die swell was reported for the system of 60/40 *p*-hydroxybenzoic acid/polybutylene terephthalate by Wissbrun [83] and by Jerman and Baird [86]. Observations using scanning electron microscope techniques revealed that extrudates show a fibrillar structure, and the wide-angle X-ray patterns showed the significant uniaxial orientation [92]. The orientation in the outer layers was seen to be greater than that in the core, as seen by the WAXS patterns of the peeled surfaces. This was confirmed by measurements of optical retardation, which indicated higher birefringence in the outer layers than in the core of the extrudates [99]. The fibrillation described above is probably due to the thermal quench stresses caused by the volume contraction during nonuniform cooling of the extrudates. The contraction of the extrudates is much greater in the transverse direction than in the flow direction. The quench forces fracture the material, leading to fibrillar characteristics because intermolecular lateral forces are small compared with those in the chain direction.

Ide and Ophir [100] investigated the effect of extrusion rates on the mechanical properties of extrudates obtained from a long capillary die. There was no variation with the extrusion rate, but the modulus showed a decrease over the shear rate ranging from 20 to 150 s^{-1} before leveling off. High strength and modulus were found from the extrusion of LCPs through a short die with little or no drawdown [101].

The most extensive studies of melt-spinning LCPs to form fibers are examinations of aromatic copolyesters in several patents and papers [102–109].

Orientation development in the melt spinning of HPC fibers was investigated by Shimamura et al. [90]. The crystalline orientation increased only slightly with the drawdown ratio and spin line stress, as opposed to the large variations in orientations found in extrudates.

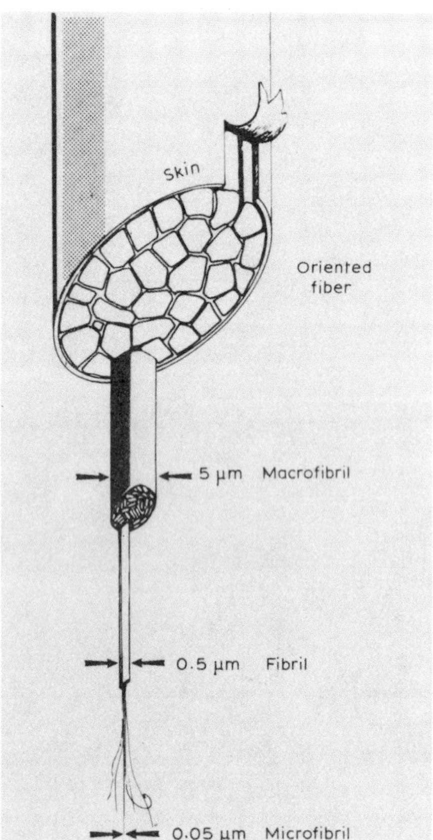

FIGURE 21 **The structures observed in uniaxially oriented LCP fibers, ribbons, and films can be summarized by this structural model. The model defines the nature of the fibrillar textures into three categories based upon size: macrofibrils, fibrils, and microfibrils. In each case the sizes of structures have been determined from complementary microscopy techniques.** (From Ref. 52.)

A complex annealing process for melt-spun aromatic copolyester fibers resulted in a substantial increase in tensile strength [106,110,111]. Solid-state polymerization is believed to be responsible to some extent for the improvement in properties. However, the extent to which the annealing steps contribute to perfecting the structure has not been properly understood.

Injection molding of polymeric LC materials has been investigated by several authors [100,112–114]. The injection-molded parts, usually end-gated tensile bars, exhibit very high modulus and tensile strength in the flow direction. Jackson et al. [113] compared the properties of glass fiber filled polyester with those of conventional polyester. Isayev and Modic [69–71] improved the low modulus and tensile strength in the transverse direction by melt-blending LCPs with thermoplastics. A comparison of LCPs and glass fiber filled thermoplastics has been done by Ide and Ophir [100]. It has been established that the skin layers of the molded specimens are more highly oriented than the core. Using ordinary injection-molding equipment, one obtains a high degree of orientation of the skin, and little if any orientation in the core. Ide and Ophir [115] discuss four different layers in injection-molded parts. Jackson and Kuhfuss [116] determined large differences in along-the-flow and across-the-flow mechanical properties of PET/*p*-HBA copolymers. High orientability of LCPs in elongational flow has been observed [117]. Thin films exhibit particularly strong anisotropy, as reported in Ref. 116. If one wants to diminish anisotropy, several avenues are open. Ide and Ophir advocate increasing the mold temperature—with obvious consequences for costs. Blending and filling are of course possible. Duska [97] reports that 50% glass filling and 50% mineral filling of Xydar (Dartco) have lowered the skin–core effect; at the same time, the processing windows became larger, while key mechanical properties are not much worse than those for neat LCP. Xydar is a copolyester based on *p,p'*-biphenol, HBA, and terephthalic acid. Another possibility is related to manipulating molding geometry. Zachariades and Economy [118,119] have developed methods based on either injection molding through a central port or cooling and simultaneously rotating one of the dies to introduce a curvilinear orientation.

Some troubleshooting phenomena in injection molding of thermoplastics that are applicable to LCP molding have been described by Oda, White, and Clark [120]. Jetting was found to be associated with low values of extrudate swell. Jetting was associated with low injection rates in the isothermal injection-molding processes with hot runners and molds. However, in injection through cold runner systems, jetting was found at high injection rates. Ide and Ophir [100], studying injection molding of LCPs, also found that the low extrudate swell results in jetting.

Since LCPs exhibit little die swell, improper gating design will cause jetting into the cavity. Gating design is very important. Gates should be large enough for the material to drag on the walls as it enters the cavity. For the gate at least as wide as its height, a gate 75–100% of the wall thickness is recommended [97]. Directing the flow of molten polymer at a core close to the gate may also prevent jetting phenomena.

In injection molding, the direction and degree of molecular and filler orientation, which control the physical properties of the molded part, are the main concern. Molded products are highly anisotropic. Therefore, the flow pattern in the cavity must be considered relative to the property requirements of the finished part. Thin sections are generally more anisotropic than thick sections. Potential weak points, such as weld lines, in any molding should be avoided. If they are unavoidable, they should be located in areas of low stress.

Vents should be located in all sections of the mold where air may become trapped by molten plastic. It has also been found desirable to incorporate vents in other areas, so that the air is not all forced through a single small opening [97].

LCPs have found applications as high modulus fibers and films with some unique properties owing to the for-

Monomers

Melt Polymerization

Polymer

Conventional Melt Spinning

As-Spun Fiber

~ 1.24 GPa Tenacity
62.1–124.1 GPa Modulus
Good Thermal, Chemical Stability

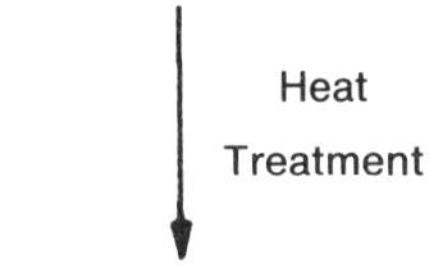

"Finished" Fiber
Up To 3.72 GPa Tenacity
62.1–124.1 GPa Modulus
Excellent Thermal-Chemical Stability

FIGURE 22 Melt spinning of LCPs. (From Ref. 32.)

mation of ordered thermotropic melts that transform easily into highly oriented extended chain structures in the solid state [32,121–128]. Thermotropic polymers are melt processable. Thick extrudates and molded parts with high strength characteristics are formed, as in fiber-reinforced thermoplastics. With lyotropic polymers, one is limited to producing one- and two-dimensional articles, that is, fibers and films. Three-dimensional parts—such as moldings and heavy extrudates—can also be made directly from thermotropic polymers.

Thermotropic LCPs may be melt-spun using conventional equipment to produce fibers with a high balance of thermal, chemical, and mechanical properties. A further improvement in these properties may be produced by heat treatment, yielding fibers with properties comparable to those of solvent spun aramids. Figure 22 shows a simple diagram of LCP fiber production, from monomer to finished fiber. LCPs are high performance polymers with an ability to be fabricated into three-dimensional shapes. One may extrude tapes, films, and heavy profile shapes by conventional melt-processing methods. Complex parts may be produced by injection molding. LCPs may also be used in composite structures as both reinforcing fibers and matrix materials. LCPs can be processed as "self-reinforced" composites (see the next section) such that the mechanics of the property distribution of LCPs and short fiber reinforced thermoplastic resins are expected to be the same.

LCP plastic articles generally exhibit a good balance of mechanical and thermal properties, low shrinkage, and high solvent resistance [32]. Properties of typical LCP tapes are summarized in Table 3 [32]. It can be observed that the mechanical properties of thin tapes are equivalent to the fiber properties. Typical injection-molded properties of LCPs are shown in Table 4. The tensile and flexural properties of the LCP moldings are comparable to those of short-fiber glass-filled thermoplastics. However, these properties of glass-filled thermoplastics are significantly lower than those of LCP extrudates as a result of the lower average molecular orientation in conventional moldings. The impact strength and heat deflection temperature of LCP extrudates are also higher than those of other resins. The mechanical property distributions of a side-gated LCP molded disc and of a neat and glass-filled conventional polymer (polybutylene terephthalate, PBT-Celanex) are shown in Figure 23 for comparison [32]. The similar response of the LCP and the glass-filled PBT should be noted. The LCP behaves as if it were a short fiber reinforced composite; that is, LCPs are "self-reinforcing" composites in molded parts.

Self-Reinforced Composites

Development of Concept

Chemical diversity has played a very essential role in the field of polymer science and technology over the last several decades. The development of numerous new polymers from a seemingly endless variety of monomers,

TABLE 3
Typical LCP Tape Properties

Width (mm)	Thickness (mm)	Tensile Strength (GPa)	Elongation at Break (%)	Modulus (GPa)
5.1	0.009	1.97	3.60	82.8
4.6	0.019	1.79	3.45	70.3
5.6	0.030	2.22	3.90	64.8

Equivalent to fiber properties—2.14–2.21 GPa tenacity, 62.1 GPa modulus.
From Ref. 32.

TABLE 4
A Comparison of Typical LCP Injection-Molded Properties with Those of Some Common Thermoplastics

	Tensile			Flexural			
Polymer	Strength (MPa)	Modulus (GPa)	Elongation (%)	Strength (MPa)	Modulus (GPa)	Notched Izod (kJ/m)	Heat Deflection (°C)
NTP	200.0	13.8	2.2	160.0	11.0	0.43	185
NTP (heat-treated)	330.3	15.9	2.6	174.5	11.0	0.75	250
Celanex 2001	56.6	—	250.0	88.3	2.07	0.05	54
Celanex 3300 (30% glass)	11.7	8.97	2.0	179.3	7.59	0.07	212
Polysulfone	70.3	—	100.0	106.2	2.76	0.05	167
Polycarbonate	62.1	—	90.0	88.3	2.07	0.75	134

From Ref. 32.

and the modification of polymers by block, graft, and random copolymerization, among other processes, highlight the pivotal position occupied by the diverse chemical approach. However, it has been realized that new chemical structures are not always needed to meet new requirements, overcome existing problems, or reduce costs. Polymer blends and composites display a broad gamut of behavior, ranging from toughened elastomers through impact-resistant plastics to fiber-reinforced thermosets and polymer-impregnated concrete. Such materials are of practical importance because their unique two-phase structure often allows for nonlinear and synergistic behavior. Polymer blends are defined as combinations of two kinds of polymers. The term *reinforcement* is used to denote the increase in rigidity and strength from dispersing inorganic fibers or particulate fillers in the polymer matrix. Since inorganic solids are intrinsically much stiffer and stronger than polymers, a mixture of the two, known as a "composite," is expected to exhibit properties that are intermediate between those of the constituents. Typical polymers that are extensively used as matrix materials include polyester, phenolic, acrylic, epoxy, silicone, alkyd, polyphenylene oxide, fluorocarbon, polyphenylene sulfide, and nylon. Most of these have been classified as engineering resins, since, according to the definition by McQuiston [129], they possess "high performance" engineering properties.

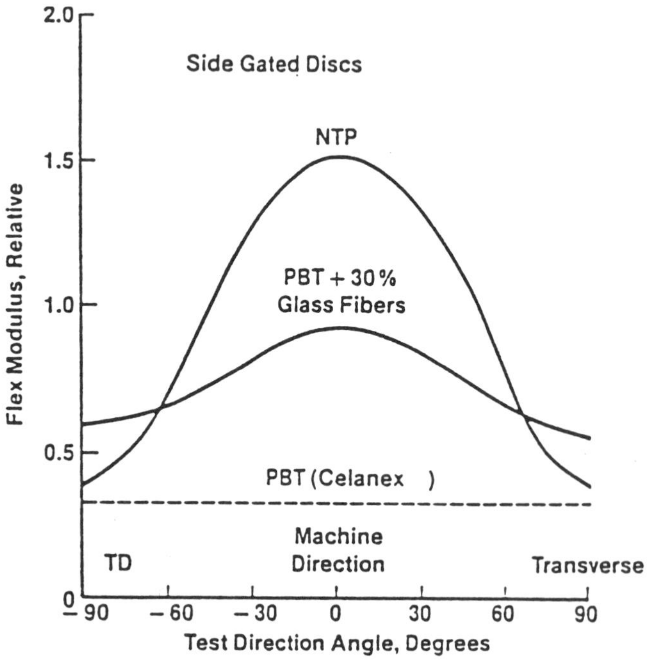

FIGURE 23 The distribution of flex modulus as a function of measurement direction in an injection-molded 0.159 cm × 10.16 cm disc. (From Ref. 32.)

The introduction of LCPs (Vectra by Celanese and Xydar by Dartco) in 1985 was the most significant advance in the area of high performance engineering thermoplastics. These have shown properties equal or superior to those of other engineering thermoplastics, and have the distinct advantage of being melt processable. The most recent additions to commercial LCPs are the E-6000 series of Mitsubishi Chemical Industries (1986), the KV series of Bayer AG, The Utrax KR series of BASF, and the Victrex SRP series of ICI, all introduced in 1987 [130] and the Granlor LCP of Granmont introduced in 1988 (Table 2).

It is well known that the physical properties of polymers, such as their mechanical, thermal, and rheological properties, can be improved by blending. Recently, the blends receiving more attention from both academic and industrial entities have been polymeric systems containing a thermotropic LCP.

LCPs are a relatively new class of polymers that combines the advantage of melt processability with outstanding mechanical properties. They are able to form products very similar to fiber-reinforced composites owing to their rigid rodlike structure and capability to form highly anisotropic crystalline structures when subjected to shear and elongational deformation above their melting point [100,131]. Thermotropic LCPs form fibers extremely readily, even in injection-molded parts. The recent major emphasis has been on utilizing the ability of such polymers to form fibrillar structures by melt-blending them with isotropic polymers to generate self-reinforced fibrous domains within the polymer matrix [69–71,132–135]. These self-reinforced composites would then behave in a manner similar to chopped glass fiber reinforced composites because of the inherent strength and stiffness of LCPs. However, processing and fabrication of chopped glass composites present some disadvantages. The most significant are the possibility of wear on the processing equipment as a result of abrasion, the increase in viscosity of the molten polymer, and difficulties in compounding [133]. LCPs show high ultimate properties along the flow direction; however, their properties along the transverse direction are usually poor [136]. Recently it has been shown that this deficiency can be substantially improved in some cases when LCPs are blended with flexible chain polymers [69,70]. This work on the polycarbonate (PC)–LCP system shows formation of the self-reinforced composite material with high modulus and high strength LCP fibers well distributed in the polymer matrix. Similar findings, though at a higher LCP concentration in the blends, have also been reported [133]. Addition of small amounts of LCP to thermoplastics also reduces their shear viscosity [69–71,132–135,137,138] and thus makes the processing easier and reduces the power consumption. The number of processing steps that are usually associated with the manufacture of chopped fiber filled composites is also reduced. These can be identified as distinct advantages.

It would be highly desirable to find an approach in which the reinforcing species is not actually present before the processing of the resin, but comes into existence during the processing. The self-reinforced formation of the reinforcing species has led to the term *self-reinforced composite* to describe materials of this type.

The premise behind this technique is that given the proclivity of such thermotropic polymers to form fibrous structures, it is likely that they would form fibrous domains or inclusions in a matrix polymer when melt-blended. Unlike the traditional chopped-glass composites, which are present before the processing of the resin, LCPs take up the reinforcing role that comes into existence during the processing (hence the term "self-reinforced"). Superficially, this appears to be similar to the "molecular composite" concept proposed by Takayanagi [139]. The domains of the reinforcing species observed as birefringent fibrous patterns by the polarized light microscopy were close to molecular levels, whereas in the case of self-reinforced composites the reinforcing domains are to be considered at a macroscopic level of observation. Hwang et al. [140–143] have also studied the concept of a molecular composite of rigid rod polymers in a flexible-coil polymer matrix.

The preparation of a blend is important from the point of view of both its properties and its economics. The objective of mixing during the preparation of blends is to bring the component materials into close contact, yielding the molecular relaxation of any nonequilibrium ingredients [144]. Mixing is generally aided by solvents and/or heat. Shearing of the mixture is additionally required.

Blending in the melt state is the most suitable in the preparation of composites containing thermotropic LCPs. It offers the advantage of introducing no foreign components (e.g., solvents) into the blend. For this reason, the simplicity and the speed of melt mixing offer economic advantages that make it the primary commercial blending method. Using proper equipment, it is possible to obtain excellent dispersion and equilibration of the components. Time, temperature, and environment for the mixing can usually be controlled. However, one may sometimes run across blending components that are difficult to mix. This difficulty may be due to large differences in the melt viscosities of the components.

Various machines can be utilized for blending, such as the Banbury mixer, twin screw extruder, or single screw extruder wih static mixer attachment. For mixing of small amounts of polymers, the Brabender mixer and Minimax machine can be used.

Rheology

There have been only a small number of investigations on blends of liquid-crystalline and flexible chain polymers [69–71,132,133,137,138,146,147] compared with the extensive work done in the polymer blend area. However, within the short span of a few years, during which LCP research gained momentum, several interesting observations have been reported in literature. ICI's patent on polymer blends combining at least one melt-processable polymer with another polymer capable of forming an anisotropic melt emphasizes the reduction in the shear viscosity of the melt-processable polymer when approximately the same processing temperature range is considered [138]. As much as a 30% reduction in the viscosity was achieved with only a 10% inclusion of liquid crystal at high shear rates. Similar observations are reported in the patent granted to Celanese for blends involving PC and wholly aromatic copolyesters [148].

The further investigations of various authors on blends of thermoplastics and liquid-crystalline wholly aromatic copolyesters also indicated significant viscosity reductions with blends of PC and nylon 6,6 with LCP [132]; LCPs with polyether sulfone (PES), polyetherimide (PEI), polyarylate, nylon, polyacetal, PBT, PC, PEEK, and polychlorotrifluoroethylene [133]; LCP with PC and PEI [69–71]; LCP with PS [134]; LCPs with PPS [136]; and LCP with polyamide (PA) [147]. Blizard and Baird [132] have reported that only a small (10 or 30%) weight fraction of LCP was required to reduce the viscosity of the thermoplastics nylon 6,6 and PC to that of the polymeric liquid crystal. It was found by Siegmann et al. [147] that in blends involving LC aromatic polyester and an amorphous polyamide, significant viscosity reductions can be obtained for blends consisting of only 5% LCP. The blend viscosity was always lower than the viscosity of the parent polymer. A typical example of such behavior is shown in Figure 24, where the shear viscosity is shown

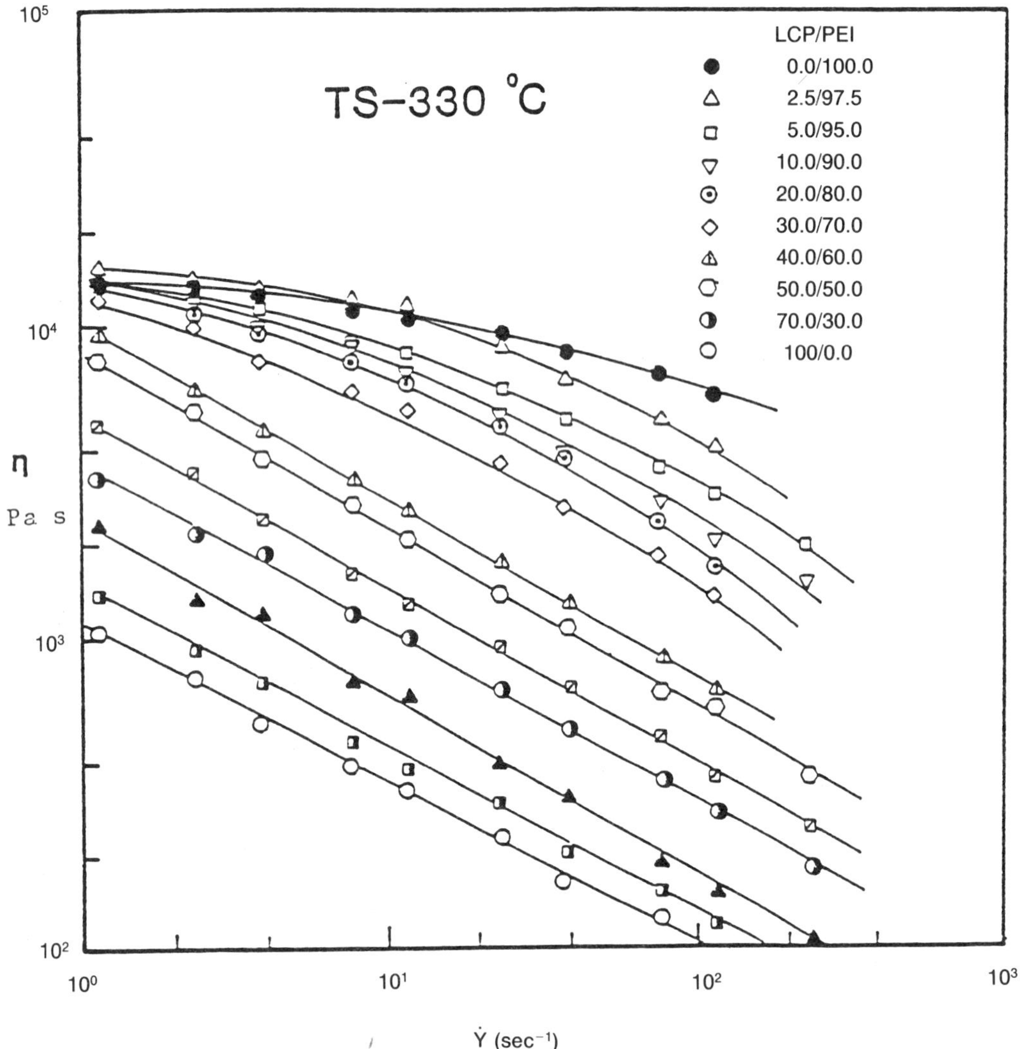

FIGURE 24 Shear viscosity η as a function of shear rate γ for PEI and LCP blends prepared at 330°C and extruded at 310°C using a die with length to diameter ratio $L/D = 28.7$.

as a function of the shear rate for PEI and LCP blends prepared by mixing in a twin screw extruder at 330°C. Some of the observations are clearly evident from this figure. First, PEI is seen to exhibit slightly non-Newtonian behavior in the range of the shear rates investigated, whereas the blends show substantial non-Newtonian behavior, with pure LCP showing the strongest shear dependence of viscosity even at low shear rates. Also, the viscosities of the blends are seen to decrease monotonically with an increasing amount of LCP in the blend, and the shear viscosity approaches that of the liquid-crystalline component at higher LCP concentrations. The viscosities of the blends are seen to lie in between those of the pure components, although at low LCP loadings (2.5%) and low shear rates the blend viscosity is seen to slightly exceed that of pure PEI. An advantage of low viscosity at high shear encountered in the molding process is that the finished shaped articles are subjected to reduced stress during fabrication so that the risk of subsequent warpage is reduced.

Mechanical Properties

Some studies of the mechanical properties of polymer blends have revealed that in the case of compatible blends, properties such as modulus and tensile strength lie above the linearly additive rule of mixtures [149–152]. Synergism, in certain cases, was reported. This was believed to be due to the volume contraction that accompanies the mixing processes in these systems [146].

Blends of thermoplastics with LCPs have been an intense field of activity for the last few years. They are usually incompatible. Despite this incompatibility, there have been a large number of reports on the improvement in mechanical properties for systems prepared by melt-blending techniques. Compatibility was achieved in blends of PBT with PET/HBA liquid-crystalline polymer in terephthalate enriched phase only [153]. Froix [148] obtained a patent for PC–wholly aromatic copolyester systems that showed improvement in properties at high LCP contents. Isayev and Modic have recently reported similar improvements in the blend properties, but at low [69,70] and high [71] LCP concentrations. Kiss [133] indicated a significant improvement in mechanical properties for blends of thermotropic LCPs and various thermoplastics at a 30% concentration of LCP. Most of the work in this area has shown that the extent of reinforcement with LCP fibers is dependent on the LCP concentration and the nature of the flexible chain polymer [69–71,132–134,136].

Isayev and coworkers [69–71] and Kiss [133] have used standard industrial processing equipment and identified two important features associated with blends involving LCPs. These features are as follows: (a) LCPs are able to reduce the power consumption associated with the processing of the neat isotropic polymer, and (b) LCPs can act as reinforcing fibers created in situ when blended with isotropic polymers; therefore, a significant improvement in mechanical properties can be achieved. In particular, Figure 25 shows Young's modulus of tensile bars of blends of PEI and LCP as a function of the LCP concentration. Tensile bars were prepared by injection molding of blends mixed using a twin screw extruder (TS) or static mixer (STM). The improvement in the modulus is evident over the whole concentration range. For the blends containing more than 50% LCP, the values exceed that of pure LCP. The reinforcement is due to fiber formation by LCP in the PEI matrix during processing. Such a fiber formation also leads to significant improvement in tensile and impact strength, as reported in [71].

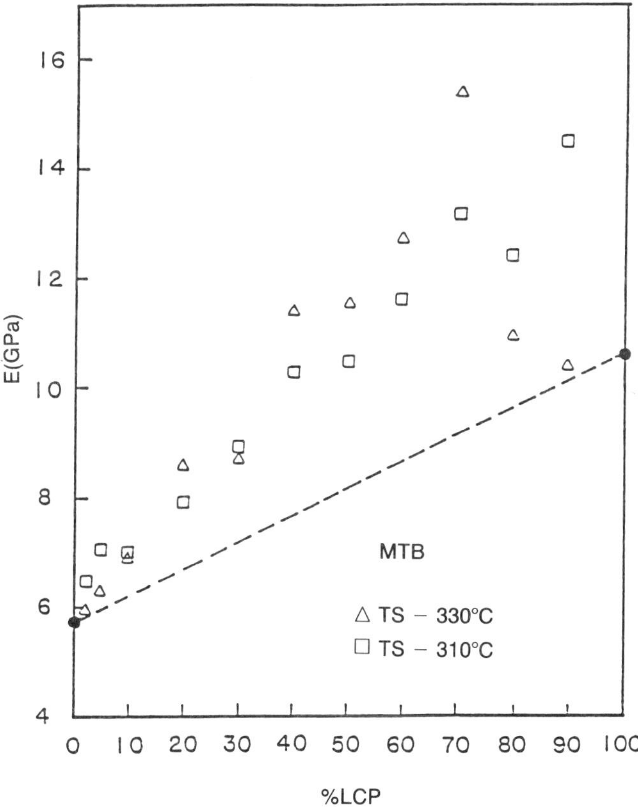

FIGURE 25 **Young's modulus *E* as a function of LCP concentration for TS blends of PEI/LCP prepared at 310°C and 330°C, and injection-molded as MTBs. The broken line is calculations according to the rule of mixtures.**

Usage

This blending technique to achieve self-reinforced composites is a new method of tailoring the property profile of engineering plastics to meet the requirements of specific applications. Usage areas or applications of self-reinforced LCPs are no different from those of the pure ones, such as in the aircraft, automotive, and marine industries.

This method has an additional advantage of not requiring, in some cases, a separate compounding step. Pellets of both polymers can simply be tumbled and the melt blending can take place inside the injection-molding machine. This is an ideal method for reducing burden when many formulations of small runs are required.

This approach is still in the early stage of development. More work needs to be done to find better systems that improve melt processing and produce better mechanical properties at the same time. The major obstacle to widespread use of this technique is probably the extremely high price that thermotropic LCPs command in the market. It is now more economical to reinforce with glass fiber; however, as the price of LCPs comes down in the near future, this technique undoubtedly will prove flexible and cost-effective enough to find niches.

Future

LCPs have been developed for highly specialized, performance-oriented applications. The following are the key market areas: (a) aircraft and aerospace, (b) chemical environments, (c) consumer and appliances, (d) electrical and electronics, (e) fiber optics, (f) military and ordnance, (g) telecommunications, and (h) transportation.

The use of conventional thermoplastics resins is predicted to grow at an 8 to 10% annual rate. However, it is expected that the growth rate of LCPs will be twice as much [154]. It is certainly more than 20% per year, and potentially much more in some market areas.

Aerospace, aircraft, military, and ordnance applications that have been already commercialized or are under development tend to involve areas where the high strength mechanical properties, high impact resistance, and excellent dimensional stability performance of LCPs are very important and long lasting, surpassing those of competing materials.

Chemical and industrial markets proliferate LCPs into devices such as pumps, metering and measuring devices, and bearing and wear applications, whereas the transportation market emphasizes automative electronic parts and under-the-hood components.

The main consumer use is now in the Tupperware area, with LCP-based Ultra 21 cookware. Appliance components, such as temperature sensors, also have some commercial significance.

There has been a dramatic expansion in the electrical/electronics and telecommunications areas, including electronic connectors, sockets for integrated circuit mounting and testing, switches, light-emitting diodes, relays, capacitors, and bobbins and coil forms for telecommunication devices. Because of their advantages in flow, precision moldability, higher productivity capacity, and high temperature performance, the new trends are toward robotic assembly, miniaturization, high temperature vapor phase, and wave reflow soldering [154].

LCPs are of particular interest for the fiber optics market. The tailorable coefficient of thermal expansion of LCPs can be mated to glass in cables, and tight-tolerance connectors and splices to join the cables can be readily produced in LCP composites. Other materials require significant, expensive machining to achieve the same precision. One of the dominant properties of LCPs is that they show nearly zero shrinkage in small injection-molded articles and extruded products. This is important for fiber-optic components, which are used over a wide range of temperatures, environments, and operating conditions.

LCP usage in the market will rapidly grow, especially in the noncookware segment, which is expected to reach 2.3–4.5 million kg in the mid-1990s. This big expansion will partially replace traditional applications of metals, thermosets, ceramics, and other high-performance thermoplastics. It is anticipated that the market in the mid-1990s will be as follows: 73% consumer/appliance, 8% industrial, 5% electrical/electronics, 5% telecommunications, 4% transportation, 3% aerospace/aircraft, and 2% other [1,2,94,154–156].

LCPs offer a variety of possible choices for today's demanding market and for the next generation of products. A wide range of applications and markets is now open. Opportunities will increase as application requirements get more stringent. The future for LCPs is virtually shining.

A. I. Isayev and T. Limtasiri

References

1. Anon., *Plast. Tech.,* Dec 1984, p. 82.
2. Anon., *Plast. Tech.,* Jan 1985, p. 11.
3. E. B. Priestley, P. J. Wojtowicz, and P. Sheng, *Introduction to Liquid Crystals,* Plenum Press, New York and London, 1975.
4. V. Krone, B. Reck, and H. Ringsdorf, *Structural Variations of Liquid Crystalline Polymers: Cross-Shape and Lateral Linked Mesogens in Main Chain and Side Group Polymers,* 16. Freiburger Arbeitstagung Fluessigkristalle, Freiburg/Br., 1986.
5. W. Brostow, *Kunststoff* 78(5), 411 (1988).
6. H. F. Kuhfuss and W. J. Jackson, Jr., US Patent 3,778,410 (1973).
7. R. W. Lenz, *Faraday Disc., 79,* 21 (1985).
8. H. R. Kricheldorf, R. Pakull, and S. Buchner, *Macromolecules, 21,* 1929 (1988).
9. W. J. Jackson, Jr., and H. F. Kuhfuss, *J. Appl. Polym. Sci., 25,* 1685 (1980).
10. K. Iimura, N. Koide, R. Ohta, and M. Takeda, *Makromol. Chem., 182,* 2569 (1981).
11. G. Wenz, *Makromol. Chem., Rapid Commun., 6,* 577 (1985).
12. O. Herrmann-Schoenherr, J. H. Wendorff, H. Ringsdorf, and P. Tschirner, *Makromol. Chem., Rapid Commun., 7,* 791 (1986).
13. H. Ringsdorf, P. Tschirner, O. Herrmann-Schoenherr, and J. H. Wendorff, *Makromol. Chem., 188,* 1431 (1987).
14. H. Finkelmann, H. Ringsdorf, and J. H. Wendorf, *Makromol. Chem., 179,* 273 (1978).
15. V. P. Shibaev and N. Plate, *Polym. Sci. USSR, 19,* 1065 (1978).
16. R. Duran, D. Guillon, P. Gramain, and A. Skoulios, *Makromol. Chem., Rapid Commun., 8,* 321 (1987).
17. A. Moeller, U. Czajka, V. Bergmann, J. Lindau, M. Arnold, and F. Kuschel, *Z. Chemie, 27,* 218 (1987).
18. V. Bergmann, K. Schwartz, A. Maedicke, J. Lindau, and F. Kuschel, *Z. Chemie, 27,* 259 (1987).
19. U. Roetz, J. Lindau, G. Reinhold, and F. Kuschel, *Z. Chemie, 27,* 293 (1987).
20. W. Kreuder and H. Ringsdorf, *Makromol. Chem., Rapid Commun., 4,* 807 (1983).
21. F. Hessel and H. Finkelmann, *Polym. Bull., 14,* 375 (1985).
22. Q.-F. Zhou, H.-M. Li, and X.-D. Feng, *Macromol., 20,* 233 (1987).
23. M. Ballauf, *Makromol. Chem., Rapid Commun., 7,* 407 (1986).
24. M. Ballauf and G. F. Schmidt, *Makromol. Chem., Rapid Commun., 8,* 93 (1987).
25. W. Weissflow and D. Demus, *Cryst. Res. Technol., 18,* K21 (1983).
26. W. Weissflow, S. Diele, and D. Demus, *Mater. Chem. Phys., 15,* 475 (1986).
27. F. Hessel, R. P. Herr, and H. Finkelmann, *Biaxial Nematic Phases—A New Type of L.C.—Side Chain Polymers. 16,* Freiburger Arbeitstagung Fluessigkristalle, Freiburg/Br., 1986.
28. B. Reck and H. Ringsdorf, *Makromol. Chem., Rapid Commun., 6,* 291 (1985).
29. H. Finkelmann, M. J. Kock, and G. Rehage, *Makromol. Chem., Rapid Commun., 2,* 317 (1981).
30. W. Gleim and H. Finkelmann, *Makromol. Chem., 188,* 1489 (1987).
31. R. Zentel and M. Benalia, *Makromol. Chem., 188,* 665 (1987).
32. G. W. Calundann, *Proceedings of The Robert A. Welch Foundation Conferences on Chemical Research, XXVI. Synthetic Polymers,* Houston, Texas, Nov. 15–17, 1982.
33. S. L. Kwolek, U.S. Patent 3,600,350 (1971).
34. S. G. Cottis, J. Economy, and B. E. Nowak, U.S. Patent 3,637,595 (1972).

35. S. G. Cottis, J. Economy, and L. C. Wohrer, U.S. Patent 3,975,487 (1976).

36. W. J. Jackson, Jr., and H. F. Kuhfuss, *J. Polym. Sci., Polym. Chem. Ed., 14,* 2043 (1976).

37. W. J. Jackson, Jr., and H. F. Kuhfuss, U.S. Patent 3,778,410 (1973).

38. F. E. McFarlane, Gordon Research Conference on Polymers, 1974.

39. P. W. Morgan, U.S. Patent 4,048,148 (1977).

40. T. E. Helminiak, F. E. Arnold, and C. L. Benner, *Amer. Chem. Soc., Polym. Prep., 16*(2), 659 (1975).

41. J. J. Kleinschuster and T. C. Pletcher (DuPont), U.S. Patent 4,066,620 (1978); R. S. Irwin (DuPont), U.S. Patents 4,232,143 and 4,232,144 (1980) and 4,245,082 (1981).

42. G. W. Calundann, U.S. Patent 4,256,624 (1981).

43. A. J. East, L. F. Charbonneau, and G. W. Calundann, U.S. Patent 4,330,457 (1982).

44. S. Garg and S. Kenig, *Prog. ACS Div. Polym. Mater. Sci. Eng., 52,* 90 (1985).

45. Y. Ide and Z. Ophir, *Polym. Eng. Sci., 23*(5), 261 (1983).

46. W. J. Jackson, Jr., and H. F. Kuhfuss, *J. Polym. Sci., Polym. Chem. Ed., 14,* 2043 (1976).

47. H. Thapar and M. Bevis, *J. Mater. Sci. Lett., 2,* 733 (1983).

48. Z. Ophir and Y. Ide, *Polym. Eng. Sci., 23*(14), 792 (1983).

49. Z. Tadmor, *J. Appl. Polym. Sci., 18,* 1753 (1974).

50. Z. Tadmor, and C. G. Gogos, *Principles of Polymer Processing,* Wiley-Interscience, New York, 1979.

51. L. C. Sawyer and D. T. Grubb, *Polymer Microscopy,* Chapman and Hall Ltd., New York, 1987.

52. L. C. Sawyer and M. Jaffe, *Proc. ACS Div. Polym. Mater. Sci. Eng., 53,* 485 (1985); and *J. Mater. Sci., 21,* 1897 (1986).

53. T. Weng, A. Hiltner, and E. Baer, *J. Mater. Sci., 21, 744* (1986).

54. E. Baer, A. Hiltner, T. Weng, L. C. Sawyer, and M. Jaffe, *Proc. ACS Div. Polym. Mater. Sci. Eng., 52,* 88 (1985).

55. R. W. Lenz and J.-I. Jin, *Macromolecules, 14,* 1405 (1981).

56. A. Blumstein, K. Sivaramakrishman, S. B. Clough, and R. B. Blumstein, *Mol. Cryst. Liq. Cryst.* (*Letters*), *49,* 255 (1979).

57. L. Strzelecki and D. van Luyen, *Europ. Polym. J., 16,* 299 (1980).

58. C. Ober, J.-I. Jin, and R. W. Lenz, *Polym. J.* (*Japan*), *14,* 9 (1982).

59. C. Ober, J.-I. Jin, and R. W. Lenz, *Makromol. Chem., Rapid Commun., 4,* 49 (1983).

60. D. Demus and L. Richter, *Textures of Liquid Crystals,* Verlag Chemie, New York, 1978.

61. G. Galli, E. Chiellini, C. Ober, and R. W. Lenz, *Makromol. Chem., 183,* 2693 (1982).

62. A. DeVries, *Mol. Cryst. Liq. Cryst., 10,* 219 (1970).

63. A. Blumstein, K. Sivaramakrishman, R. B. Blumstein, and S. B. Clough, *Polymer, 23,* 47 (1982).

64. A. Biswas and J. Blackwell, *Macromolecules, 21,* 3146 (Part 1), 3152 (Part 2), and 3158 (Part 3) (1988).

65. J. F. Johnson and R. S. Porter, Eds., *Liquid Crystals and Ordered Fluids,* Plenum Press, New York, 1974.

66. C. K. Ober, J.-I. Jin, Q. Zhou, and R. W. Lenz, *Adv. in Polym. Sci., 59,* 103 (1984).

67. R. W. Lenz, *Pure & Appl. Chem. 57*(7), 977 (1985).

68. S. Cheng, *Macromolecules, 21,* 2475 (1988).

69. A. I. Isayev and M. J. Modic, *Polym. Comp., 8,* 158 (1987).

70. A. I. Isayev and M. J. Modic, *SPE Technical Papers, 32,* 573 (1986); U.S. Patent 4,728,698 (1988).

71. A. I. Isayev and S. Swaminathan, in *Advance Composites,* ASM International, 1987, pp. 259–267; U.S. Patent 4,835,047 (1989).

72. M. R. Nobile, E. Amendola, and L. Nicolais, *Polym. Eng. Sci., 29*(4), 244 (1989).

73. R. S. Stein and J. S. Higgins, *J. Appl. Cryst., 11,* 346 (1978).

74. S. A. Jabarin and R. S. Stein, *J. Phys. Chem., 77,* 399 (1973) and *77,* 409 (1973).

75. A. M. Donald and A. W. Windle, *Polymer, 25,* 1235 (1984).

76. E. L. Thomas and B. A. Wood, University of Massachusetts, presentation at Faraday Society Meeting, Cambridge, England, 1985.

77. E. T. Samulski and D. B. Dupre, *Adv. Liq. Crystallogr., 4,* 121 (1979).

78. G. Kothe and T. E. Berthold, *Mol. Phys., 40,* 1441 (1980).

79. K. Mueller, K.-H. Wassmer, R. W. Lenz, and G. Kothe, *J. Polym. Sci., Polym. Lett. Ed., 21,* 785 (1983).

80. H. W. Spiess, *Pure & Appl. Chem., 57*(11), 1617 (1985).

81. S. Antoun, R. W. Lenz, and J.-I. Jin, *J. Polym. Sci., Polym. Chem. Ed., 19,* 1901 (1981).

82. K.-H. Wassmer, Ph.D. Thesis, University of Freiburg (1982).

83. K. F. Wissbrun, *J. Rheol., 25,* 619 (1981).

84. S. Onogi and T. Asada, in G. Astarita, G. Marrucci, and L. Nicolais, Eds., *Rheology,* Plenum, New York, 1980, Vol. 1, p. 127.

85. K. F. Wissbrun, *British Polymer J., 12,* 163 (1980).

86. R. E. Jerman and D. E. Baird, *J. Rheol., 25,* 275 (1981).

87. M. Prasadrao, E. M. Pearce, and C. D. Han, *J. Appl. Polym. Sci., 27,* 1343 (1982).

88. D. G. Baird, *Proc. IUPAC Symposium, Macromol.,* 1982, p. 753.

89. F. N. Cogswell, *Proceedings of the 10th Anniversary of the Society of Rheology, Japan,* 1983.

90. K. Shimamura, J. L. White, and J. F. Fellers, *J. Appl. Polym. Sci., 26,* 2165 (1981).

91. W. J. Jackson, Jr. and H. F. Kuhfuss, *J. Polym. Sci., Polym. Chem., 14,* 2043 (1976).

92. S. Suto, J. L. White, and J. F. Fellers, *Rheol. Acta, 21,* 62 (1982).

93. Y. Suetsugu and J. L. White, *J. Appl. Polym. Sci., 28,* 1481 (1983).

94. G. Kiss and R. S. Porter, *J. Polym. Sci., Polym. Phys. Ed., 18,* 361 (1980).

95. A. B. Metzner and G. M. Prilutski, *J. Rheol., 30,* 661 (1986).

96. R. G. Larson and D. W. Mead, *J. Rheol., 33,* 185 (1989).

97. J. J. Duska, *Plast. Eng., 42*(12), 39 (1986).

98. F. C. Jaarsma, *ANTEC, 32,* 726 (1986).

99. J. L. White, *J. Appl. Polym. Sci., 41,* 241 (1985).

100. Y. Ide and Z. Ophir, *Polym. Eng. Sci., 23,* 261 (1983).

101. Y. Ide, U.S. Patent 4,332,759 (1982); U.S. Patent 4,468,364 (1984).

102. H. F. Kuhfuss and W. J. Jackson, U.S. Patent 3,804,805 (1974).
103. S. G. Cottis, J. Economy, and J. C. Warren, U.S. Patent 3,975,407 (1976).
104. T. Pakula, M. Krysezewski, J. Grebowicz, and A. G. Galeski, *Polym. J., 6*(2), 94 (1974).
105. L. W. Kleiner, F. E. Karasz, and W. J. Mackight, *Polym. Eng. Sci., 19,* 519 (1979).
106. J. R. Schaefgen, U.S. Patent 4,118,372 (1978).
107. G. W. Calundann, U.S. Patent 4,130,545 (1978).
108. G. W. Calundann, U.S. Patent 4,067,852 (1978).
109. G. W. Calundann, U.S. Patent 4,161,470 (1979).
110. T. C. Pletcher, U.S. Patent 3,991,013 (1976).
111. J. J. Kleinschuster, U.S. Patent 3,991,014 (1976).
112. W. J. Jackson and H. F. Kuhfuss, *J. Polym. Sci., Polym. Chem., 14,* 2043 (1976).
113. W. J. Jackson, H. F. Kuhfuss, and T. F. Gray, *Proceedings of the Reinforced Plastics/Composites Institute SPI, 30th Annual Technical Conference,* 1974, Section 17, p. 1.
114. W. J. Jackson, *Br. Polym. J., 12,* 154 (1980).
115. Y. Ide and Z. Ophir, *Polym. Eng. Sci., 23,* 792 (1983).
116. W. J. Jackson, Jr., and H. F. Kuhfuss, *J. Polym. Sci., Polym. Chem. Ed., 14,* 2043 (1976).
117. S. Kenig, *Polym. Eng. Sci., 27*(12), 887 (1987).
118. A. E. Zachariades, and J. Economy, *Polym. Eng. Sci., 23,* 266 (1983).
119. J. Economy, W. Volksen, and R. H. Geiss, *Molec. Cryst. Liq. Cryst., 105,* 289 (1984).
120. K. Oda, J. L. White, and E. S. Clark, *Polym. Eng. Sci., 18,* 53 (1978).
121. P. J. Flory, "Molecular Theories of Liquid Crystals," in A. Ciferri, W. R. Krigbaum, and R. B. Meyer, Eds., *Polymer Liquid Crystals,* Academic Press, New York, 1982, Chap. 4.
122. A. Zachariades and R. S. Porter, Eds., *The Strength and Stiffness of Polymers,* Marcel Dekker, New York, 1983.
123. G. W. Gray and P. W. Winsor, *Liquid Crystals and Plastics,* Vols. 1 and 2, Wiley, Harwood, Chichester, 1974.
124. J. L. White and J. F. Fellers, in J. L. White, Ed., *Fiber Structure and Properties, Appl. Polym. Symp., 33P,* 137 (1978).
125. M. G. Dobb and J. E. McIntyre, *Properties and Applications of Liquid-Crystalline Main-Chain Polymers,* Adv. in Polym. Sci. Ser. 60/61, Springer Verlag, Berlin, 1984.
126. J. H. Wendorf, "Scattering in Liquid-Crystalline Polymer Systems," in A. Blumstein, Ed., *Liquid Crystalline Order in Polymers,* Academic Press, New York, 1978, p. 41.
127. J. F. Johnson and R. S. Porter, Eds., *Liquid Crystals and Ordered Fluids,* Plenum, New York, 1970.
128. A. Ciferri, W. R. Krigbaum, and R. B. Meyer, Eds., *Polymer Liquid Crystals,* Academic Press, New York, 1982.
129. M. McQuisten, *Plas. Eng., 57,* 18 (1980).
130. L. Charbonneau, "Synthesis and Properties of Thermotropic Liquid Crystal Polymers," lecture delivered at the 18th Akron Polymer Conference on Ordered Polymers, 1987.
131. D. J. Blundell, *Polymer 23,* 359 (1982).
132. K. G. Blizard and D. G. Baird, *Polym. Eng. Sci., 27,* 653 (1987).
133. G. Kiss, *Polym. Eng. Sci., 27,* 410 (1987).
134. R. A. Weiss, W. Huh, and L. Nicolais, *Polym. Eng. Sci., 27,* 684 (1987).
135. T. Kyu and P. Zhuang, *Polym. Comm., 29,* 99 (1988).
136. R. Ramanathan, K. Blizard, and D. Baird, *SPE Tech. Papers, 34,* 1123 (1988).
137. W. Huh, R. A. Weiss, and L. Nicolais, *Polym. Eng. Sci., 23,* 779 (1983).
138. F. N. Cogswell, B. P. Griffin, and J. B. Rosi, U.S. Patent 4,386,174 (1981); U.S. Patent 4,438,236 (1984).
139. M. Takayanagi and H. Harima, *Am. Chem. Soc., Div. Org. Coatings, Plast. Chem. Pap., 23*(2), 75 (1963).
140. W. F. Hwang, C. L. Benner, D. R. Wiff, and T. E. Helminiak, *Proc. IUPAC 28 Macromol. Symp.,* 1982, p. 827.
141. W. F. Hwang, D. R. Wiff, and C. Verschoore, *Polym. Eng. Sci., 23,* 789 (1983).
142. W. F. Hwang, D. R. Wiff, C. Verschoore, G. E. Price, T. E. Helminiak, and W. W. Adams, *Polym. Eng. Sci., 23,* 784 (1983).
143. W. W. Adams, T. E. Helminiak, A. Viswanathan, W. F. Huang, and D. R. Wiff, *Proc. IUPAC 28, Macromol. Symp.,* 1982, p. 828.
144. M. T. Shaw, "Preparation of Blends," in D. J. Walsh, J. S. Higgins, and A. Maconnachie, Eds., *Polymer Blends and Mixtures,* Martinus Nijhoff Publishers, Boston, 1985.
145. T. S. Chung, *SPE Tech. Papers, 33,* 1404 (1987).
146. E. G. Joseph, G. L. Wilkes, and D. G. Baird, *Polym. Eng. Sci., 25,* 377 (1985).
147. A. Siegmann, A. Dagan, and S. Kenig, *Polymer, 36,* 1325 (1985).
148. M. Froix, U.S. Patent 4,460,735 (1984).
149. D. Heikens, N. Hoen, W. M. Barentson, P. Piet, and H. Ladem, *J. Polym. Sci., Polym. Symp., 62,* 309 (1978).
150. E. A. Joseph, M. D. Lorenz, J. W. Barlow, and D. R. Paul, *Polymer, 23,* 112 (1982).
151. G. Kiss, A. J. Kovacs, and J. C. Wittman, *J. Appl. Polym. Sci., 26,* 2665 (1981).
152. N. Sundgren, G. Bergman, and Y. J. Shur, *J. Appl. Polym. Sci., 22,* 1255 (1978).
153. M. Kimura and R. S. Porter, *J. Polym. Sci., Polym. Phys. Ed., 22,* 1697 (1984).
154. J. R. Dole, *Chemtech,* April 1987, p. 242.
155. A. S. Wood, *Modern Plastics,* April 1985, p. 78.
156. R. W. Lenz and J. Jin, *Polymer News, 11,* 200 (1986).

Liquid Resin/Chopped Strand Process

See Manufacturing

Liquid Resin, Preform Process

See Manufacturing

Literature of Composites

Introduction

The composite industry has had tremendous growth during the 20 years since the introduction of the new strong fibers: boron, graphite, and S-glass fibers. The military and defense agencies played an important role providing a drive for higher and higher performance materials, and the aerospace industry together with materials manufacturers passed the challenge with flying colors. Along with the new materials came new applications and new methods for processing and design. The increased number of reports, meetings papers, and journal articles on the new materials, processes, and technology has contributed to the rapid growth of composites information. Composite technology is of such an interdisciplinary nature, with research results reported in publications of a wide range of subject fields (chemistry, physics, engineering, polymer science and engineering, materials science, etc.), that scientists and engineers are hard put to keep abreast of what's going on in all areas. The need for consolidation of information has resulted in a flood of conferences and symposia that center on various aspects of composite materials. Conference proceedings consequently have become a major source of composite literature today, and the number of meetings held each year all over the world continues to increase. Books published during this period have tended toward more technical and applied subjects, as reflected by many titles on mechanical properties and engineering analysis and design, and very few purely theoretical treatises. The publishing activity increased in the 1980s with the appearance of a greater number of new books, including quite a few up-to-date handbooks. New journals on composites are also being started, and the names of some older periodicals have been changed to reflect new coverage and new emphasis.

Conferences and Symposia

Conferences and symposia constitute an important source of composite literature. They provide up-to-date coverage of research going on in industrial, academic, and governmental laboratories. Papers presented in these meetings generally fall into two categories: original findings from ongoing projects and state-of-the art reviews. Most meetings are organized or sponsored by professional societies and trade associations (SAMPE, ASTM, SPI, etc.). Because of the multidisciplinary nature of composite research, however, results are widely scattered among meetings of many fields, which makes it more difficult for workers to keep informed. The problem is compounded by the fact that a significant portion of the most advanced technology is guarded in the classified or restricted files of such government agencies as the departments of defense and energy. The best-known conferences in the composites area are the ones sponsored by SAMPE (Society for the Advancement of Material and Process Engineering): The International SAMPE Symposia and the International SAMPE Technical Conference, held annually in different locations. Next is the annual Technical Conference of the Reinforced Plastics/Composites Institute, a division of the Society of the Plastics Industry. Other regularly held meetings (and publications) are listed as follows:

ACS Polymeric Materials Science and Engineering, Proceedings, covering meetings of the ACS Division of Polymeric Materials.

ACS Polymer Preprints, covering two yearly meetings of the Polymer Division of the American Chemical Society.

Advanced Composites, proceedings of the conference sponsored by the American Society for Metals and the Engineering Society of Detroit, held annually in Detroit (beginning in 1985).

American Ceramic Society (Columbus, OH), annual conference proceedings.

American Society for Composites, Proceedings of Technical Conference, held annually, (beginning in 1986), available from Technomic, Lancaster, PA.

Composites '83–'86, proceedings of the symposium sponsored by the Industrial Materials Research Institute/National Research Council of Canada, held annually.

Composite Materials: Testing and Design, proceedings of the conference sponsored by the American Society for Testing and Materials (Philadelphia), first held in 1968; eighth conference in 1987.

Conference on Fibrous Composites in Structural Design, Fourth, 1978, San Diego, CA, sponsored by NASA. Earlier conferences sponsored by Air Force Aeronautical Laboratory, Dayton, OH, held at regular intervals. The fifth and the sixth are restricted to government agencies.

International Conference on Composite Materials (ICCM), proceedings of the conference sponsored by the Metallurgical Society of AIME, Warrendale, PA, first held in 1975, then every 2–3 years in different locations; ICCM-5 in 1985.

International Conference on Composite Structures (first held in 1981, at Paisley College of Technology, Scotland), sponsored by the Institute of Mechanical Engineers and National Engineering Laboratory (U.K.). Proceedings available from Elsevier Applied Science Publishers, Barking, England.

International Conference on Fibre Reinforced Composites, University of Liverpool, first held in 1984, sponsored by Plastics and Rubber Institute, London.

Reinforced Plastics Congress, sponsored by the British Plastics Federation, Reinforced Plastics Group, London, first held in 1958.

SPE ANTEC (Society of Plastics Engineers, Annual Technical Conference) papers.

SPE NATEC (National Technical Conference) papers.

SPE RETEC (Regional Technical Conference) papers.

In addition to the regular society conferences, numerous symposia and workshops are held each year in different locations covering various aspects of composite materials. Some of the papers are published by the sponsor as proceedings as mentioned above; others not published by the official channel may be printed as a separate issue of a journal; and still others are edited and published as individual books by such commercial publishing houses as Technomic and Plenum Press. As a matter of fact, many of the composite books today are based on the results of some conference, seminar, or workshop. A few published symposium series are as follows.

ACS Advances in Chemistry Series
ACS Symposium Series
ASTM STP Series (American Society for Testing and Materials, Special Technical Publications)
Journal of Applied Polymer Science, Applied Polymer Symposia
Journal of Polymer Science, Polymer Symposia Edition
Makromolekulare Chemie, Macromolecular Symposia
MRS Symposium Proceedings Series (Materials Research Society, Pittsburgh)

In spite of all the established channels, many meeting papers never make it to the printed form. An excellent source to locate meeting papers is *Conference Papers Index* (1973–, monthly) published by Cambridge Scientific Abstracts, Bethesda, MD. This index is cumulated quarterly and annually, and its data base of more than 500,000 records is available for on-line searching. For locating published proceedings, the following sources are used:

Directory of Published Proceedings Series SEMT, InterDok Corporation, White Plains, NY, published monthly.
Index to Conference Publications Received by BLLD, British Library Lending Division, London, monthly.
Index to Scientific and Technical Proceedings and Books, Institute for Scientific Information, Philadelphia, monthly, covering more than 3000 conferences a year.

Information on forthcoming meetings is generally provided by the news journals (*Chemical and Engineering News, Polymer News, Plastics Engineering, Modern Plastics, Metals Progress, ASM News*, etc.), which carry a calendar of meetings in each issue, but the following general listings also offer good coverage:

Scientific Meetings, Special Libraries Association, New York.
World List of Future International Meetings in Science, Technology, Agriculture, and Medicine, Library of Congress. Washington, DC.
World Meetings: Outside U.S. and Canada, World Meetings Information Center, Chestnut Hill, MA.
World Meetings: U.S. and Canada, World Meetings Information Center, Chestnut Hill, MA.

Theses and Dissertations

Theses and dissertations are reports of research work undertaken by students in universities to fulfill requirements for graduate degrees. Such work, by tradition, deals only with subjects or aspects of a subject not previously investigated, and the information is considered to be original. In most universities, thesis work on composites is carried out in departments of mechanical engineering, materials science, and polymer science and engineering. Some dissertations are later reported in journals and picked up by such abstracting services as *Chemical Abstracts* and *Engineering Index;* some never appear in the open literature. To locate and retrieve reports of this type, one usually relies on the various published directories and, fortunately for composites scientists and engineers, a recent compilation by John Summerscales entitled *International Dissertations on Fiber Reinforced Polymers* (Technomic, Lancaster, PA, 1988) offers an excellent start. The following listings are all searchable on-line by subject.

Master Theses in the Pure and Applied Sciences Accepted by Colleges and Universities of the United States and Canada (Plenum Press, New York, 1957–) is an annual bibliographic index with no abstracts.
Master Abstracts [University Microfilm International (UMI), 1962–] is published quarterly with annual cumulative indexes.
American Doctoral Dissertations (UMI, 1933/34–) is an annual index to dissertations accepted by American and Canadian universities, with no abstracts.
Dissertation Abstracts International (UMI, 1938–) is a monthly abstracting publication based on dissertations submitted to UMI by more than 400 cooperating institutions in the United States, Canada, and a few European countries. Section B covers science and engineering.
Comprehensive Dissertation Index (CDI) (UMI, 1973–) is a computer data base containing an index to more than 400,000 dissertations accepted at U.S. and Canadian institutions. Considered to be the most comprehensive list of its kind, with entries dating back to 1861, CDI replaces all earlier lists by the Library of Congress and H. W. Wilson Company. Its files are kept up to date by annual supplements and are searchable on-line (DIALOG and BRS) by author, title, subject, and institution.

Theses from institutions outside the United States may be located in one of the following foreign sources.

Canadian Theses (National Library of Canada, Ottowa, 1961–) is published annually.
Index to Theses Accepted for Higher Degrees in the Universities of Great Britain and Ireland (ASLIB, London, 1953–) is an annual index to more than 7500 dissertations and theses.
Catalogue des Théses de Doctorate Soutenues Devant les Univérsitiés Françaises (1984/85–) is the annual French official bibliography of dissertations.

Jahresverzeichnes der Deutschen Hochschulschriften (VEB Verlag für Buch- und Bibliothekswesen, Leipzig).

Katalog Kandidatskikh i Doktorskikh Dissertatsii Postupivshikh v Biblioteku imeni V. I. Lenina i Gosudarstvenuyu Tsenytral'nuyu Nauchnuyu Meditsinskuyu Biblioteki (Moscow, 1958–) is the Russian-language listing of Soviet theses deposited in the V. I. Lenin Library and the State Central Library of Medicine, published quarterly.

European Dissertation Abstracts (Foundation for European University and Research Documentation, Bern, Switzerland).

Dissertations Abstracts International, Section C: European Abstracts (UMI), published quarterly.

Research work currently in progress on campuses can be monitored by such publications as the ACS *Directory of Graduate Research,* published biannually, which covers research projects in departments of chemistry, chemical engineering, biochemistry, and polymer science. Entries are arranged by discipline and then by institution, with author (investigator) index. *Scientific Research in British Universities and Colleges* is another annual listing on the British institutions published by the Department of Education and Science and the British Council. An outstanding source, unfortunately discontinued in 1981, is the *Smithsonian Science Information Exchange* (SSIE), which used to receive and process more than 100,000 research projects a year from government agencies, foundations, and universities. Some of the SSIE functions are assumed by the National Technical Information Service (NTIS) in an on-line file named *Federal Research in Progress,* which some consider to be inferior to SSIE in number of agencies covered as well as depth of indexing. The old SSIE data base is still searchable on-line via DIALOG.

Patents

Patents are the earliest records of new technology on products, processes, and uses. As progress is being achieved in all fields of human endeavor, so is the steady accumulation of patent documents each week all over the world. The primary sources of most patents are the weekly official gazettes published by individual countries, which announce the granting of legal protection to inventions (as in the United States) or the filing of unexamined patent applications (as in the quick-issue countries: Belgium, France, West Germany, the Netherlands, Japan, etc.).

It may take 3 years or more after the filing date for the U.S. Patent and Trademarks Office to grant a patent, while the quick-issue countries may publish unexamined applications 6–18 months after the filing date. Because rights of invention apply only to the country issuing the patent, patent applications are frequently filed in several countries for legal protection. Consequently this type of literature has a multinational and multilingual characteristic. The overwhelming task of procuring and keeping track of the enormous volume of multilingual records has led to the establishment of a host of commercial patent services, which provide abstracts and indexes in English for the weekly patent documents issued worldwide. In the composites area, the major patent services are

Derwent Inc.
6845 Elm Street, Suite 500
McLean, VA 22101

Chemical Abstracts Service
P.O. Box 3012
Columbus, OH 43210

IFI/Plenum Data Company
302 Swann Avenue
Alexandria, VA 22301

International Patent Documentation Center (INPADOC)
Mollwaldplatz 4
A-1040 Vienna, Austria

Derwent

Derwent Publications, Ltd, started in 1950 as a small abstract bulletin of British patents, has gradually expanded to cover 26 countries. The two main branches of Derwent are *Central Patent Index* (CPI), covering chemical patents in 12 sections, and *World Patent Abstracts* (WPA), covering nonchemical patents in 7 sections. Services offered by the two branches are quite similar, consisting of three types of publication issued at different intervals: *World Patent Index* (WPI) *Gazette, Alerting Bulletins* by country and by Derwent classification, and *Basic Abstracts Journal.*

World Patent Index Gazette is a weekly bibliographic index to all newly reported patents in the world with no abstract. It is arranged by patentee, subject (International Patent Classification, IPC), accession number, patent number, and priorities claimed, and is published 5–6 weeks after the official release.

The *Alerting Bulletin (country)* is issued one week after the *WPI Gazette,* providing short abstracts to the patents covered in the preceding week by country. Each country booklet has indexes by patentee, Derwent class, accession number (patent family), and patent number. In the classified *Alerting Bulletin* that follows a week later, the abstracts of the country *Bulletin* are rearranged by Derwent class.

Basic Abstracts Journal is published 2 weeks after the country *Bulletin* with more detailed abstracts to the basic patents entered into Derwent system. These abstracts are so full of details on the content, uses, and advantages, as well as examples and drawings, that they are regarded as the most informative patent abstracts produced today. The weekly indexes by subject (Derwent class), patentee, accession number, and patent number are cumulated quarterly and annually. The section of Derwent that concerns the composites field most is *CPI Section A: PLASDOC,* which covers patent information on plastics, polymers, and composites and is the largest of the 12 CPI units. *PLASDOC* accounts for one-third of the total CPI output, averaging about 850 basic patents a week. Its *Basic Abstracts Journal* consists of 14 separate profile booklets, which can be purchased or subscribed to individually.

Compared with other patent services, Derwent is unquestionably the leader in terms of broad coverage as well as speed of delivery. Its abstracts can be obtained just 2 months after the original publication, while *Chemical Abstracts* (CA) may take 4 months or more. Derwent also abstracts 50% more polymer-related patents than CA due to CA's indexing policy limitations, which have, however, been improved a great deal in recent years. All Derwent publications are available in printed or microfilm version and are searchable on-line through SDC (Systems Development Corporation), Questel, and DIALOG.

Chemical Abstracts

Chemical Abstracts (CA), published by Chemical Abstracts Service, a division of the American Chemical Society, is a weekly abstracting journal covering chemical publications of all types, including patents from 26 countries. Abstracts are grouped into 80 subject sections, and each weekly issue contains a number of sections with author, keyword, and patent indexes, which are cumulated semiannually and in 5-year collections. Patent entries are designated by a "P" preceding the abstract number. The patent index is arranged by country and then by patent numbers, with all equivalents listed under the basic patents, a feature made possible by the *INPADOC Patent Family* tapes since 1981. As a source of patent information, CA offers several advantages: best-known and most readily available, accurate and well-written abstracts, in-depth indexing, and well-maintained indexes and data bases. Its coverage, though sometimes uneven, is very good for the quick-issue countries. W. Germany and Japan are comparatively very advanced in science and technology, their patent papers are of tremendous interest to the industrial countries, especially the United States. The rapidly released unexamined patent applications from these two countries, normally from 6 to 12 months after filing date, are eagerly sought after by the scientific and industrial community. It takes up to three years for the U.S. Patent Office to grant patent and release the information. In the matter of speed, CA is behind Derwent by a couple of months except for the USSR, where CA is more timely because it takes abstracts directly from the Soviet abstract bulletin instead of waiting for the original documents as Derwent does. All CA files are searchable on-line via Lockheed's DIALOG, SDC, BRS (Bibliographic Retrieval Service), and STN International.

IFI/Plenum

IFI/Plenum Data Company started to offer patent service in 1952, with the publication of the *Uniterm Index to The United States Chemical Patents,* which consists of a dual dictionary-type keyword index and a volume of *U.S. Official Gazette* claims or abstracts of accessioned patents. It can be searched manually by visual comparison of columns of numbers in both volumes. This format was later converted to machine-readable files, and both printed and tape versions are available for subscription and are up-dated quarterly and annually. *IFI Comprehensive Database to the United States Chemical Patents* was created in 1972 by combining the *Uniterm Index* data base with du Pont's patent data base. It contains more than 600,000 patent references and can be searched by any one of these access points: U.S. patent number, IFI accession number, assignee, U.S. patent class, chemical compound, chemical fragments, general term.

Comprehensive Database is noted for its in-depth indexing, which averages about 50 descriptors per patent. For polymers, a special set of roles is created to index polymers in terms of monomers. The IFI/Plenum services are available on an annual subscription basis and are also searchable on-line through DIALOG, SDC, and STN International. The on-line files are known as *Claims.* Other services include the *Weekly Patent Profile* and the *Patent Intelligence and Technology Report.* The former is a weekly data base search on a special subject or assignee and is mailed to subscribers 7 days after the *Official Gazette.* The latter is a patent data analysis by subject or assignee including a six-year patent activity profile. An on-line index to the U.S. Patent Classification system including 400 classes and 90,000 subclasses, known as *Claims Class,* is also available.

INPADOC

International Patent Document Center (INPADOC) was founded in 1972 by the Austrian government in agreement with the World Intellectual Property Organization (WIPO) for the purpose of creating a central databank to store the bibliographic records of all patent documents as soon as they are issued. Each week magnetic tapes containing references to more than 16,000 patent documents from 55 patent issuing offices have been received and entered into the INPADOC system, out of which two files are currently available for on-line searching: *INPANEW* and *INPADOC. INPANEW* covers records received in the most recent 15 weeks, and *INPADOC* covers patent documents as far back as 1968, varying by country. Both files can be accessed by title, subject (International Patent Classification codes), and bibliographic data. Because of its wide geographical coverage, INPADOC offers the most complete patent family (equivalents) information in a service known as *Patent Family Service,* which can be subscribed or searched on-line through Pergamon Infoline. In addition, INPADOC has one of the largest collections of patent documents in the world: more than 6 million stored in 30,000 rolls of microfilm. Copies of patents may be obtained through IFI/Plenum (Arlington, VA), its agent for the United States and Canada.

Abstracts and Indexes

Composites literature is scattered among hundreds of publications over a number of disciplines. Access to such enormous amounts of information has been made possible by the various abstracting and indexing services in the fields. These services together provide coverage of a broad range of subjects and publications of all types (books, journals, reports, patents, conferences, dissertations, etc.). Abstracts are the best substitute for original documents that are not readily available, and they can be obtained by subscribers in a matter of weeks to a few months after the first publication date. Some services do not supply abstracts, but invariably provide bibliographic data to the original documents. In general these services have well-maintained indexes for current or retrospective searching, and their files are mostly searchable on-line through the major commercial vendors (DIALOG, SDC, BRS, STN International, Pergamon Infoline, etc.). The abstracting and indexing services used often in the composites field are:

Chemical Abstracts (1907–, weekly), Chemical Abstracts Service, Columbus, OH; on-line vendors: STN, DIALOG, SDC, BRS.

Engineering Index Monthly and Author Index (1884–, monthly), Engineering Information, Inc., New York; on-line vendors: DIALOG, BRS, STN.

International Aerospace Abstracts (1961–, semimonthly), American Institute of Aeronautics and Astronautics, New York; on-line vendors: DIALOG, ESA (European Space Agency).

Star (Scientific and Technical Aerospace Reports) (1963–, semimonthly), NASA, Washington, DC; on-line vendors: DIALOG, ESA.

RAPRA Abstracts (1923–, monthly), Rubber and Plastics Research Association of Great Britain, Shawbury, Shrewsbury, England; on-line vendors: Pergamon Orbit Infoline.

Ceramic Abstracts (1922–, bimonthly), American Ceramic Society, Columbus, OH; on-line vendors: Orbit, Pergamon Infoline.

Metals Abstracts (1968–, monthly), Institute of Metals, London, and American Society for Metals, Metals Park, OH; on-line vendors: DIALOG, SDC, STN.

Engineered Materials Abstracts (1986–, monthly), Metals Information, ASM International, Metals Park, OH.

Institute of Paper Chemistry, Abstracts Bulletin (1930–, monthly), Institute of Paper Chemistry, Appleton, WI; on-line vendors: DIALOG, SDC.

Science Citation Index (1955–, bimonthly), Institute for Scientific Information, Philadelphia; on-line vendors: DIALOG, BRS.

Of this group, *Chemical Abstracts* (CA) is by far the most comprehensive in subjects and types of literature covered. The phenomenal success of CA as a global information producer has been universally recognized, and indeed several well-known national services, including *British Abstracts* (1871–1953), *Chemisches Zentralblatt* (1930–1969), and *Bulletin Signaletique* (1940–1983), were driven out of business either because of unnecessary duplication of coverage or inability to stay solvent.

Three types of CA service are available for subscription: the regular weekly CA and the biweekly CA—Sections and CA—Selects. Abstracts are grouped into 80 subject sections, and each weekly CA contains a segment of sections with author, keyword, and patent indexes. There are semiannual cumulative and 5-year collective indexes by author, general subject, chemical substance, formula, and patent. CA—Sections contain a smaller number of abstracts according to subjects grouped (*CA: Macromolecular Sections, CA: Biochemistry Sections*, etc.) at a much lower subscription rate, but with only keyword index for locating abstracts within the issue. CA—Selects is a biweekly data base search on any of the 198 topics offered for subscription. Currently there are six composites-related topics: *CA—Selects: Carbon and Graphite Fibers, CA—Selects: Carbon Fiber Composites, CA—Selects: Ceramic Materials (j), CA—Selects: Ceramic Materials (p), CA—Selects: Epoxy Resins*, and *CA—Selects: Fiber-Reinforced Plastics.* Both CA—Selects and CA—Sections are strictly current awareness tools with no cumulative index. For retrospective searching, one must use the regular CA indexes and CA data base.

Reviews

While primary literature such as meeting papers, theses, and patents represents new knowledge, or new interpretation of old knowledge, the secondary sources—abstracts, reviews, and books—are compilations of this knowledge according to a definite plan and in a more convenient form to satisfy specific needs. Reviews belong to the second group because they repackage information gathered on a particular subject in a compact form to help scientists staying current in their respective fields of interest. Reviews may appear in a number of formats: hard-bound series, review journals, and regular journals. Following are a few examples that carry reviews on composites.

Advances in Polymer Science (1958–) Springer-Verlag, New York.

Advances in Polymer Technology (1981–) Wiley, New York.

Annual Reviews in Materials Science (1971–) Annual Reviews, Inc., Palo Alto, CA.

Developments in Composite Materials (1977–) Elsevier Applied Science, Barking, England.

Developments in GRP Technology (1983–) Elsevier Applied Science, Barking, England.

Developments in Reinforced Plastics (1980–) Elsevier Applied Science, Barking, England.

Developments in Rubber and Rubber Composites (1980–) Elsevier Applied Science, Barking, England.

Journal of Macromolecular Science C: Reviews (1967–) Marcel Dekker, New York.

Journal of Polymer Science, Macromolecular Reviews (1967–) Wiley, New York.

Polymer–Plastics Technology and Engineering (1972–) Marcel Dekker, New York.

RAPRA Review Reports (1988–) RAPRA Technology, Ltd, Shawbury, Shrewsbury, England.

Review papers on composites also appear in numerous journals. A good locating source is *CA Review Index*, which contains references to about 30,000 reviews a year and can be searched on-line through DIALOG. The *Index to Scientific Reviews* published by the Institute for Scientific Information, Philadelphia, is another source available since 1974.

Journals

Compared with publications in other branches of science, composite journals are relatively young and few in number, the oldest titles dating back to the late 1960s. The pace of publishing activity quickened a great deal during the 1980s, when many new journals were started, and older titles underwent name changes to reflect new coverage and new emphasis: *Composites Science and Technology* (formerly *Fibre Science and Technology*), *Journal of Composites Technology and Research* (formerly *Composite Technology Review*), *Mechanics of Composite Materials* (formerly *Polymer Mechanics*), and *Composites* (Paris, formerly *Plastiques Renforcés, Fibres de Verre Textile*).

The following journals are grouped according to the subjects covered. The first group contains titles devoted exclusively to composites, and the second lists journals that have good coverage on composites in many aspects of science and technology. The lists are not intended to be exhaustive, but they do include the most representative titles currently available.

Composite Journals

Advanced Composites (1986–, bimonthly) Harcourt Brace Jovanovich Publications, Cleveland, OH.

Advanced Composites Bulletin (1988–, monthly) Elsevier, New York.

Composite Materials Science (1986–, quarterly) Freund Publishing House, London.

Composite Structures (1983–, 8/year) Elsevier Applied Science, Barking, England.

Composites (1969–, 5/year) Butterworths Scientific Ltd, Guildford, England.

Composites (1960–, bimonthly) Centre de Documentation de Plastiques Renforcés et du Verre Textile, Paris.

Composites & Adhesives (1984–, bimonthly) T/C Publications, Pasadena, CA.

Composites et Nouveaux Materiaux (1984–, monthly) Société D'Editions de Lettres Françaises, Paris.

Composites Science and Technology (1969–, monthly) Elsevier Applied Science, Barking, England.

Composites Update, Center for Composite Materials, University of Delaware, Newark.

Japan Society for Composite Materials, Journal (1975–, quarterly) Tokyo.

Journal of Composite Materials (1967–, monthly) Technomic Publishing Company, Lancaster, PA.

Journal of Composites Technology and Research (1978–, quarterly) American Society for Testing and Materials, Philadelphia.

Journal of Reinforced Plastics and Composites (1982–, bimonthly) Technomic Publishing Company, Lancaster, PA.

Journal of Thermoplastic Composite Materials (1988–, quarterly) Technomic Publishing Company, Lancaster, PA.

Mechanics of Composite Materials (1965–, bimonthly) Plenum Press, New York.

Polymer Composites (1980–, bimonthly) Society of Plastics Engineers, Brookfield Center, CT.

Reinforced Plastics (1956–, monthly) McDonald Publications, London.

Report (quarterly) Communica, Ltd, Toronto, Canada.

Composite-Related Journals

Abstract Newsletter: Materials Science (weekly) U.S. National Technical Information Service, Springfield, VA.

Advanced Materials (1979–, semimonthly) Advanced Publications, Inc., Hilton Head, SC.

Advanced Materials and Processes (1985–, monthly) American Society for Metals, Metals Park, OH.

Aerospace Engineering Magazine (1981–, monthly) Society of Automotive Engineers, Warrendale, PA.

AIAA Journal (1963–, monthly) American Institute of Aeronautics and Astronautics, New York.

American Helicopter Society Journal (1956–, quarterly) New York.

Angewandte Makromolekulare Chemie (1967–, 9/year) Hüthig und Wepf Verlag, Basel, Switzerland.

International Journal of Fatigue (1979–, quarterly) Butterworths Scientific Ltd, Guildford, England.

International Journal of Polymeric Materials (1971–, 8/year) Gordon & Breach, New York.

Journal of Adhesion (1969–, 8/year) Gordon & Breach, New York.

Journal of Applied Polymer Science (1956–, 16/year) Wiley, New York.

Journal of Engineering Materials and Technology (1973–, quarterly) American Society of Mechanical Engineers, New York.

Journal of Materials Science (1966–, monthly) Chapman & Hall, London.

Journal of Materials Science Letters (1982–, monthly) Chapman & Hall, London.

Journal of Nondestructive Evaluation (1980–, quarterly) Plenum Press, New York.

Journal of Polymer Science: Polymer Chemistry Edition, Polymer Physics Edition, Polymer Letters Edition, Polymer Symposia Edition; (1946–, monthly) Wiley, New York.

Kunststoffe—German Plastics (1910–, monthly) Carl Hanser, Munich.

Kunststoffe—Plastics (1953–, monthly) Vogt-Schild AG, Solothurn, Switzerland.

Materials Engineering (1929–, monthly) Penton/IPC Publications, Cleveland, OH.

Materials Evaluation (1942–, monthly) American Society for Nondestructive Testing, Columbus, OH.

Modern Plastics (1925–, monthly) McGraw-Hill, Hightstown, NJ.

Plastics Compounding (1978–, 7/year) Resins Publications, Inc., Cleveland, OH.

Plastics Design Forum (1975–, bimonthly) Harcourt Brace Jovanovich Publications, Chatham, NJ.

Plastics Engineering (1945–, monthly) Society of Plastics Engineers, Brookfield Center, CT.

Plastics Machinery & Equipment (1972–, monthly) Harcourt Brace Jovanovich Publications, Chatham, NJ.

Plastics Technology (1955–, monthly) Bill Communications, Inc., New York.

Plastics World (1943–, monthly) Cahners Publishing Company, Boston.

Polymer (1960–, monthly) Butterworths Scientific Ltd, Guildford, England.

Polymer Engineering and Science (1961–, 22/year) Society of Plastics Engineers, Brookfield Center, CT.

Polymer Journal (1970–, monthly) Society of Polymer Science of Japan, Tokyo.

RAPRA Review Reports (1988–, quarterly) RAPRA Technology, Ltd, Shawbury, Shrewsbury, England.

SAMPE Journal (1965–, bimonthly) Society for the Advancement of Material and Process Engineering, Covina, CA.

SAMPE Quarterly (1969–, quarterly) Society for the Advancement of Material and Process Engineering, Covina, CA.

Books

This book list is compiled on the basis of availability and currency, with the majority of titles published after 1970. The subject arrangement is purely arbitrary, based mainly on what has actually been found rather than by any classification schedule, and cross-references are not provided between groups. Entries that are part of a series or a multivolume work are so indicated by a parenthetical reference to the series, including volume or item number, at the end of the reference.

One may notice from the list that some areas are well represented, and in others very little has been written at all. This is easily understandable, for the industry is still undergoing exuberant growth, which may bring in piles of new information in certain areas but nothing in those where efforts were not directed. It is hoped that this encyclopedia will consolidate some of the newly acquired technology and fill some of the existing voids.

Reference Works

Encyclopedia of Materials Science and Engineering, Michael B. Bever, Ed., Pergamon Press, Oxford, U.K., 1984, 8 vols.

Encyclopedia of Composite Materials and Components, Martin Grayson, Ed., Wiley-Interscience, New York, 1983. (Encyclopedia Reprint Series)

Encyclopedia of Polymer Science and Engineering, 2nd ed., Wiley, New York, 1984–, 15 vols. to date.

Encyclopedia/Handbook of Materials, Parts, and Finishes, Henry R. Clauser, Ed., Technomic, Lancaster, PA, 1976.

CRC Handbook of Materials Science, Charles T. Lynch, Ed., CRC Press, Boca Raton, FL, 1974–1975, 3 vols.

Modern Plastics Encyclopedia, McGraw-Hill, New York, annual.

Composites and Laminates, desktop data bank, Cordura Publications, San Diego, CA 1987.

Composites (*Engineered Materials Handbook,* vol. 1), ASM International, Metals Park, OH, 1987.

John V. Milewski and Harry S. Katz, Eds., *Handbook of Reinforcements for Plastics,* Van Nostrand Reinhold, New York, 1987.

Harry S. Katz and John V. Milewski, Eds., *Handbook of Fillers for Plastics,* Van Nostrand Reinhold, New York, 1987.

J. A. Weeton, D. M. Peters, and K. L. Thomas, Eds., *Engineers' Guide to Composite Materials,* American Society for Metals, Metals Park, OH, 1986.

George Lubin, Ed., *Handbook of Composites,* Van Nostrand Reinhold, New York, 1982.

Mel M. Schwartz, Ed., *Composite Materials Handbook,* McGraw-Hill, New York, 1984.

A. Kelly and Yu. N. Rabotnov, Eds., *handbook of Composites,* Elsevier, Amsterdam, 1983–1985, 4 vols.

G. M. Sabnis, Ed., *Handbook of Composite Construction Engineering,* Van Nostrand Reinhold, New York, 1979.

N. P. Chermisinoff and P. N. Chermisinoff, *Fiberglass-Reinforced Plastics Deskbook,* Ann Arbor Science, Ann Arbor, MI, 1978.

Robert Nicholls, *Composite Construction Materials Handbook,* Prentice-Hall, Englewood Cliffs, NJ, 1976.

Plastics Engineering Handbook, 4th ed., Van Nostrand Reinhold, New York, 1976.

G. J. Mohr et al., Eds., *SPI Handbook of Technology and Engineering of Reinforced Plastics/Composites,* 2nd ed., Van Nostrand Reinhold, New York, 1973.

Military Handbook 17, prepared by U.S. Army Materials Technology Laboratory, Watertown, MA, updated continuously.

L. R. Whittington, *Whittington's Dictionary of Plastics,* 2nd ed., Technomic, Lancaster, PA, 1978.

C. M. Bower, Ed., *Composite Materials Glossary,* 2nd ed., T/C Publications, El Segundo, CA, 1985.

Compilation of ASTM Standards and Literature References for Composite Materials, American Society for Testing and Materials, Philadelphia, 1987.

ASTM Annual Book of Standards, Vol. 15,03: Space Simulation, Aerospace Materials, High Modulus Fibers, and Composites, American Society for Testing and Materials, Philadelphia, 1987.

John Summerscales, *International Dissertation on Fibre Reinforced Polymers,* Technomic, Lancaster, PA, 1988.

A. R. Bunsell and A. Kelly, Eds., *Composite Materials: A Directory of European Research,* Butterworths, London, 1985.

Composite Materials, General

Derek Hull, *An Introduction to Composite Materials,* Cambridge University Press, New York, 1981.

Stephen W. Tsai and H. Thomas Hahn, *Introduction to Composite Materials,* Technomic, Lancaster, PA, 1980.

Lawrence J. Broutman and R. H. Krock, Eds., *Modern Composite Materials,* Addison-Wesley, Reading, MA, 1967.

Lawrence J. Broutman and R. H. Krock, Eds., *Composite Materials,* Academic Press, New York, 1974–1975, 8 vols.

K. K. Chawla, *Composite Materials,* Springer Verlag, New York, 1987.

A. C. Marshall, *Composites, Basics,* T/C Publications, El Segundo, CA, 1985.

L. Holliday, Ed., *Composite Materials,* Elsevier, Amsterdam, 1966.

NASA Langley Research Center Staff, *Tough Composite Materials: Recent Developments,* Noyes Data Corp., Park Ridge, NJ, 1985.

G. Piatti, Ed., *Advances in Composite Materials,* Elsevier Applied Science, Barking, England, 1978.

Moriya Uchida, Ed., *Advanced Composite Materials,* Kogyo Chosakai Ltd, Tokyo, 1986 (in Japanese).

Takashi Akasaka et al., *Development of Advanced Composite Materials and Their Evaluation Techniques,* CMC Ltd, Tokyo, 1985 (in Japanese).

Advanced Composites Special Topics, 2nd ed., T/C Publications, El Segundo, CA, 1983.

N. K. Hiester, *Trends in Advanced Composites,* SRI International, Menlo Park, CA, 1984 (Report 700).

A. K. Dhingra and B. R. Pipes, Eds., *New Composite Materials and Technology,* American Institute of Chemical Engineers, New York, 1982 (*AIChE Symposium Series,* vol. 78, no. 217).

J. R. Vinson and M. Taya, Eds., *Recent Advances in Composites in the United States and Japan,* Proceedings of a Symposium in 1983, American Society for Testing and Materials, Philadelphia, 1985 (ASTM STP 864).

F. D. Lemkey, H. E. Cline, and M. McLean, Eds., *In Situ Composites,* vol. 4, Proceedings of the MRS Annual Meeting, Boston, 1981, Elsevier, New York, 1983 (*MRS Symposium Proceedings,* vol. 12).

S. W. Tsai, J. D. Halpin, and N. J. Pagano, Eds., *Composite Materials Workshop,* Technomic, Lancaster, PA, 1968.

Charles Carroll-Porcznski, *Advanced Materials: Refractory Fibres, Fibrous Metals, Composites,* 2nd American ed., Chemical Publishing, New York, 1969.

B. Harris, *Engineering Composite Materials,* Brookfield Publishing, Brookfield, VT, 1986.

Mikio Morita et al., *Functional Composite Materials,* CMC Ltd, Tokyo, 1986 (in Japanese).

I. N. Frantsevich and D. M. Karpinos, *Fibrous Composites,* Coronet Books, Philadelphia, 1972.

B. W. Rosen, *Fiber Composite Materials,* American Society for Metals, Metals Park, OH, 1965.

Fiber Reinforced Composites, T/C Publications, El Segundo, CA, 1984.

G. S. Holister and C. Thomas, *Fibre Reinforced Materials,* Elsevier, New York, 1966.

John Morley, *High-Performance Fibre Composites,* Academic Press, New York, 1987.

N. G. McCrum, *A Review of the Science of Fibre Reinforced Plastics,* H. M. Stationery Office, London, 1971.

Michael R. Piggott, *Load-Bearing Fibre Composites,* Pergamon Press, Elmsford, NY, 1980.

I. M. Shologon, Ed., *Reactive Oligomers and Composite Materials Produced from Them,* Niitekhim, Moscow, 1985 (in Russian).

J. R. Vinson, Ed., *Modern Developments in Composite Materials and Structures,* American Society of Mechanical Engineers, New York, 1979.

A. A. Berlin et al., Eds., *Principles of Polymer Composites,* Springer-Verlag, New York, 1986. (Polymer-Properties and Applications, vol. 10).

J. E. Ashton and J. M. Whitney, *Theory of Laminate Plates,* Technomic, Lancaster, PA, 1970 (*Progress in Materials Science Series,* vol. 4).

Z. Hashin, *Theory of Fiber Reinforced Materials,* (NAS CR-1974) Washington, DC, Government Printing Office, 1972.

Yu. S. Lipatov, *Physical Chemistry of Filled Polymers,* trans. from Russian, Rubber and Plastics Research Association of Great Britain, Shawbury, England, 1979 (*International Polymer Science Technological Monographs,* vol. 2).

G. F. Hewitt, J. M. Delhaye, and N. Zuber, *Multiphase Science and Technology,* vol. 3, Hemisphere, New York, 1987.

International Symposium on Composite Materials and Structures, Beijing, China, 1986, proceedings edited by T. T. Loo and C. T. Sun, Technomic, Lancaster, PA, 1986.

Analysis

J. C. Halpin, *Primer on Composite Materials: Analysis,* Technomic, Lancaster, PA, 1984.

B. D. Agarwal and L. J. Broutman, *Analysis and Performance of Fiber Composites,* Wiley-Interscience, New York, 1980.

Hatsuo Ishida and Ganesh Kumar, Eds., *Molecular Characterization of Composite Interfaces,* Proceedings of the ACS Symposium on Polymer Composites: Interfaces, Seattle, 1983, Plenum Press, New York, 1985 (*Polymer Science and Technology,* Vol. 27).

L. A. Carisson and R. B. Pipes, *Experimental Characterization of Advanced Composite Materials,* Prentice-Hall, Englewood Cliffs, NJ, 1987.

A. H. Cardon and G. Verchery, Eds., *Mechanical Characterization of Loading Bearing Fibre Composite Laminates,* Proceedings of the European Mechanics Colloquium 182, Brussels, 1984, Elsevier Applied Science, Barking, England, 1985.

Fractography of Modern Engineering Materials, Composites, and Metals, American Society for Testing and Materials, Philadelphia, 1987 (STP 948).

S. K. Garg, V. Svalbonas, and G. A. Gurtman, *Analysis of Structural Composite Materials,* Dekker, New York, 1973.

James M. Whitney, *Structural Analysis of Laminated Composites,* seminar notes, Technomic, Lancaster, PA, 1987.

James M. Whitney, *Structural Analysis of Laminated Anisotropic Plates* (with software), Technomic, Lancaster, PA, 1987.

L. R. Calcote, *Analysis of Laminated Composite Structures,* Van Nostrand Reinhold, New York, 1969.

T. Kevin O'Brien, Ed., *Long-Term Behavior of Composites,* American Society for Testing and Materials, Philadelphia, 1983 (STP 813).

Composite Reliability, American Society for Testing and Materials, Philadelphia, 1975 (STP 580).

S. P. Prosen, *Composite Materials: Testing and Design,* American Society for Testing and Materials, Philadelphia, 1969 (STP 460).

Analysis of the Test Methods for High Modulus Fibers and Composites, American Society for Testing and Materials, Philadelphia, 1973 (STP 521).

C. C. Chamis, *Test Methods and Design Allowables for Fibrous Composites,* American Society for Testing and Materials, Philadelphia, 1981 (STP 734).

V. L. Blagonadezhin, G. Kh. Murzahanov, and V. P. Nikolaev, *Methods for Experimental Study of [Polymeric] Composite Materials and Structures Made of Them,* Mosk. Energ. Institute, Moscow, 1976 (in Russian).

Richard Chait and Ralph Papirno, Eds., *Compression Testing of Homogeneous Materials and Composites,* American Society for Testing and Materials, Philadelphia, 1983 (STP 808).

Yu. M. Tarnopolskii and T. Ya. Kintsis, *Static Test Methods for Composites,* trans. from Russian, 3rd ed., Van Nostrand Reinhold, New York, 1984.

S. Kessler et al., Eds., *Instrumented Impact Testing of Plastics and Composite Materials,* American Society for Testing and Materials, Philadelphia, 1987 (STP 936).

M. L. Lomax, *Acoustic Testing of Porous Materials and Composite Panels,* Rubber and Plastics Research Association of Great Britain, Shawbury, Shrewsbury, England, 1981.

Alex Vary, Ed., *Materials Analysis by Ultrasonics: Metals, Ceramics, Composites,* Noyes Data Corp., Park Ridge, NJ, 1987.

Raymond W. Meyer, *Pultrusion Technology Handbook,* Chapman & Hall, New York, 1985.

K. H. G. Ashbee, Ed., *Polymer NDE,* Proceedings of the European Workshop on Nondestructive Evaluation of Polymers and Polymer Matrix Composites, Termar do Vimeiro, Portugal, 1984, Technomic, Lancaster, PA, 1984.

Nondestructive Testing of Composite Materials, 1972–1984, T/C Publications, El Segundo, CA, 1984.

John Summerscales, Ed., *Non-Destructive Testing of Fibre-Reinforced Plastics and Composites,* vol. 1, Elsevier, New York, 1987.

R. B. Pipes, Ed., *Nondestructive Evaluation and Flaw Criticality for Composite Materials,* based on a symposium, American Society for Testing and Materials, Philadelphia, 1979 (STP 696).

Mechanical Properties

Energy Content of Reinforced Plastics Materials, International Reinforced Plastics Industry (IRPI), London, 1981.

Richard M. Christensen, *Mechanics of Composite Materials.* Wiley, New York, 1979.

G. J. Dvorak, Ed., *Mechanics of Composite Materials,* American Society of Mechanical Engineers, New York, 1983.

Z. Hashin and C. T. Herakovich, Eds., *Mechanics of Composite Materials,* Pergamon Press, Elmsford, NY, 1982.

Robert M. Jones, *Mechanics of Composite Materials.* McGraw-Hill, New York, 1975.

K. Kawata and T. Akasaka, Eds., *Composite Materials: Mechanics, Mechanical Properties and Fabrication,* based on a Japan–U.S. Conference, Elsevier Applied Science, Barking, England, 1982.

A. S. Kravchuk, V. P. Maiboroda, and Yu. D. Urzhumtsev, *Mechanics of Polymeric and Composite Materials,* Nauka, Moscow, 1985 (in Russian).

G. P. Sendeckyi, Ed., *Mechanics of Composite Materials,* Academic Press, New York, 1974 (*Composite Materials,* vol. 2).

A. Nica, *Mechanics of Aerospace Materials.* Elsevier, Amsterdam, 1981.

V. K. Tewary, *Mechanics of Fibre Composites,* Wiley-Interscience, New York, 1978.

P. P. Benham, *Mechanics of Engineering Materials,* Halsted Press, New York, 1987.

Jozsef Bodig and Benjamin A. Jayne, *Mechanics of Wood and Wood Composites,* Van Nostrand Reinhold, New York, 1982.

R. C. Laible, *Ballistic Materials and Penetration Mechanics.* Elsevier, Amsterdam, 1980 (*Methods and Phenomena,* vol. 5).

J. M. Hedgepeth, *Stress Concentrations in Filamentary Structures,* NASA TN D-882, National Aeronautics and Space Administration, Washington, DC, 1961.

James M. Whitney, I. M. Daniel, and R. B. Pipes, *Experimental Mechanics of Fiber Reinforced Composite Materials,* 2nd ed., Prentice-Hall, Englewood Cliffs, NJ, 1984.

Society for Experimental Stress Analysis, *Experimental Mechanics of Fiber Reinforced Composite Materials,* Prentice-Hall, Englewood Cliffs, NJ, 1982.

Fracture Mechanics of Composites, Proceedings of a Symposium, American Society for Testing and Materials, Philadelphia, 1976 (STP 593).

I. V. Grushetskii et al., *Fracture of Structures Produced from Composite Materials,* Zinatne, Riga, Latvia, USSR, 1986 (in Russian).

Harold Liebowitz, Ed., *Fracture,* vol. 7, *Fracture of Nonmetals and Composites,* Academic Press, New York, 1972.

S. Mindless and S. P. Shah, Ed., *Cement-Based Composites: Strain Rate Effects on Fracture,* Proceedings of an MRS symposium, Materials Research Society, Pittsburgh, 1986 (*MRS Symposium Proceedings Series,* vol. 64).

George C. Sih and E. P. Chen, Eds., *Cracks in Composite Materials,* Nijhoff, Boston, 1981.

D. Wilkins, Ed., *Effect of Defects in Composite Materials,* Proceedings of a Symposium, American Society for Testing and Materials, Philadelphia, 1984 (STP 836).

K. Reifsnider, Ed., *Damage in Composite Materials,* American Society for Testing and Materials, Philadelphia, 1982 (STP 775).

G. Dorey, *Impact and Crashworthiness of Composite Structures,* vol. 1, Elsevier Applied Science, Barking, England, 1984.

Foreign Object Impact Damage to Composites, American Society for Testing and Materials, Philadelphia, 1975 (STP 568).

James M. Fleck and Richard L. Mehan, Eds., *Failure Modes in Composites,* vol. II, Proceedings of TMA-AIME Spring Meeting, Pittsburgh, 1976, American Society for Metals, Metals Park, OH, 1976.

G. C. Sih, A. M. Studra, A. Kelly, and Yu. N. Rabotnov, Eds., *Failure Mechanics of Composites,* Elsevier, Amsterdam, 1985. (*Handbook of Composites,* vol. 3).

A. S. D. Wang et al., *Failure Analysis of Composite Laminates,* Technomic, Lancaster, PA, 1985.

Henry Brown, Ed., *Composite Repairs,* Society for the Advancement of Materials and Process Engineering, Covina, CA, 1985 (SAMPE Monograph 1).

W. A. Green and M. V. Micunovic, Eds., *Mechanical Behavior of Composites and Laminates,* Elsevier, New York, 1988.

D. W. Clegg and A. A. Collyer, Eds., *Mechanical Properties of Reinforced Thermoplastics,* Elsevier Applied Science, Barking, England, 1986.

Lawrence E. Nielsen, *Mechanical Properties of Polymers and Composites,* 2 vols., Dekker, New York, 1974.

A. K. Malmeister, S. T. Tamuz, and G. Teters, *Strength of Polymeric and Composite Materials,* 3rd ed., Zinatne, Riga, Latvia, USSR, 1980 (in Russian).

V. A. Lapitskii and A. A. Kritsuk, *Physico-Mechanical Properties of Epoxy Polymers and Glass-Fiber-Reinforced Plastics,* Naukova Dumka, Kiev, USSR, 1986 (in Russian).

F. Werren and C. B. Norris, *Mechanical Properties of Laminate Designed To Be Isotropic,* Report 1841, Forest Products Laboratory, Madison, WI, 1953.

George W. Scherer, *Relaxation in Glass and Composites,* Wiley, New York, 1986.

S. W. Tsai, H. T. Hahn, and C. T. Herakovich, Eds., *Inelastic Behavior of Composite Materials,* American Society of Mechanical Engineers, New York, 1975.

E. H. Lee, Ed., *Dynamics of Composite Materials,* Proceedings of Joint National and Western Applied Mechanics Conference, La Jolla, CA, 1972, reprint ed. from Books on Demand (UMI).

B. E. Read and G. D. Dean, *The Determination of Dynamic Properties of Polymers and Composites,* Wiley, New York, 1978.

Ramesh Talreja, *Fatigue of Composite Materials,* Technomic, Lancaster, PA, 1986.

Fatigue of Composite Materials, American Society for Testing and Materials, Philadelphia, 1975 (STP 569).

R. Evans, Ed., *Fatigue of Filamentary Composite Materials,* American Society for Testing and Materials, Philadelphia, 1977 (STP 636).

H. Thomas Hahn, Ed., *Composite Materials, Fatigue and Fracture,* American Society for Testing and Materials, Philadelphia, 1986 (STP 907).

John M. Potter, Ed., *Fatigue in Mechanically Fastened Composite and Metallic Joints,* American Society for Testing and Materials, Philadelphia, 1986 (STP 927).

K. N. Lauraitis, Ed., *Fatigue of Fibrous Composite Materials,* American Society for Testing and Materials, Philadelphia, 1981 (STP 723).

Lawrence J. Broutman, Ed., *Fracture and Fatigue,* Academic Press, New York, 1984 (*Compos. Mater.* vol. 5).

K. Friedrich and R. B. Pipes, Eds., *Friction and Wear of Polymer Composites,* Elsevier, Amsterdam, 1986 (*Composite Materials,* vol. 1).

N. P. Istomin and A. P. Semenov, *Antifriction Properties of Composite Materials Made of Fluorine-Containing Polymers,* Nauka, Moscow, USSR, 1981 (in Russian).

Adhesion in Composites

E. P. Plueddeman, Ed., *Interfaces in Polymer Matrix Composites,* Academic Press, New York, 1974 (*Composite Materials,* vol. 6).

S. S. Negmatov, *Principles of Contact Interaction of Composite Polymeric Materials with a Fibrous Material,* Fan., Tashkent, USSR, 1984 (in Russian).

First International Conference on Composite Interfaces, Cleveland, OH, 1986, *Composite Interfaces,* Proceedings edited by H. Ishida and J. L. Koenig, North-Holland, New York, 1986.

K. Kedward, Ed., *Joining of Composite Materials,* American Society for Testing and Materials, Philadelphia, 1981 (STP 749).

F. L. Matthews, Ed., *Joining Fibre-Reinforced Plastics,* Elsevier, New York, 1986.

Adhesive Bonding of Composite Materials, 1970–1985, T/C Publications, El Segundo, CA, 1985.

G. S. Koch, F. Klareich, and B. Extrum, *Adhesives for the Composite Wood Panel Industry,* Noyes Data Corp., Park Ridge, NJ, 1987.

John F. Oliver, Ed., *Adhesion in Cellulosic and Wood-Based Composites,* Plenum Press, New York, 1981 (NATO Conference Series VI—Materials Science, vol. 3).

T. Sellers, Jr., *Plywood and Adhesive Technology,* Dekker, New York, 1985.

Albert G. H. Dietz, Ed., *Composite Engineering Laminates,* MIT Press, Cambridge, MA, 1969.

D. J. Duffin, Ed., *Laminated Plastics,* Reinhold, New York, 1966.

W. S. Johnson, Ed., *Delamination and Debonding of Materials,* American Society for Testing and Materials, Philadelphia, 1985 (STP 876).

Environmental Effects

J. L. Christian, W. E. Witzell, and B. A. Stein, Eds., *Environmental Effects on Advanced Composite Materials,* American Society for Testing and Materials, Philadelphia, 1976 (STP 602).

J. R. Vinson, Ed., *Advanced Composite Materials, Environmental Effects,* American Society for Testing and Materials, Philadelphia, 1978 (STP 658).

George S. Springer, Ed., *Environmental Effects on Composite Materials,* Technomic, Lancaster, PA, 1981–1987, 3 vols.

J. C. Halpin and S. W. Tsai, *Environmental Factors in Composite Materials Design,* Air Force Materials Laboratory, Dayton, OH, 1969 (ATML TR 67-423).

A. Adsit, Ed., *Composites for Extreme Environments,* American Society for Testing and Materials, Philadelphia, 1982 (STP 768).

J. C. Halpin and N. J. Pagano, *Consequences of Environmentally Induced Dilatation in Solids,* Air Force Materials Laboratory, Dayton, OH, 1969 (AFML TR 68-395).

I. Ahmad and B. R. Noton, Eds., *Advanced Fibers and Composites for Elevated Temperatures,* Proceedings of a Conference Sponsored by The Metallurgical Society of AIME and ASM, 1979, American Society for Metals, Metals Park, OH, 1980.

Tito T. Serafini, Ed., *High Temperature Polymer Matrix Composites,* Noyes Data Corp., Park Ridge, NJ, 1987.

Gunter Hartwig and David Evans, Eds., *Nonmetallic Materials and Composites at Low Temperature,* Proceedings of the Second International Cryogenic Materials Conference, Geneva, Plenum Press, New York, 1982.

A. F. Clark, R. P. Reed, and G. Hartwig, Eds., *Nonmetallic Materials and Composites at Low Temperatures,* vol. I, Plenum Press, New York, 1979.

I. N. Cherskii, Ed., *Composite Polymeric Materials Under Low-Temperature Conditions,* Aka. Nauk SSSR, Sib. Otd., Yakutsk Fil., Yakutsk, USSR, 1983 (in Russian).

V. P. Gordienko, *Radiation Modification of Composite Materials Produced from Polyolefins,* Naukova Dumka, Kiev, USSR, 1985 (in Russian).

Design and Fabrication

Advanced Composites Design Guide, 3rd ed., prepared by L. A. Aircraft Division of Rockwell International for the Air Force Flight Dynamics Laboratory, Wright Patterson Air Force Base, Dayton, OH, 1977.

Terry L. Richardson, *Composites, A Design Guide,* Industrial Press, New York, 1987.

Stephen W. Tsai, Ed., *Composites Design,* 3rd ed., Think Composites, Dayton, OH, 1987.

Bryan R. Noton, Ed., *Composite Materials in Engineering Design,* Proceedings of the Sixth St. Louis Symposium, 1972. Reprinted ed. from Books on Demand—UMI, Ann Arbor, MI.

Fibreglass Composites Design Data, Pilkington Reinforcements Ltd, St. Helens, England, 1985.

M. A. Dorgham, Ed., *Designing with Plastics and Advanced Composites,* Proceedings of the Conference of International Association for Vehicle Design, Interscience, Ltd, Geneva, 1986.

Institution of Civil Engineers Staff, *Fiber Reinforced Materials: Design and Engineering Applications,* American Society of Civil Engineers, New York, 1977.

R. R. Shives and W. A. Willard, Eds., *Advanced Composites: Design and Applications,* Proceedings of a Symposium Held at National Bureau of Standards, Government Printing Office, Washington, DC, 1979 (NBS Special Publication 563).

D. H. Kaelble, *Computer-Aided Design of Polymers and Composites,* Dekker, New York, 1985.

J. Backlund, Ed., *Computer-Aided Engineering of Composite Materials and Structures,* Elsevier, New York, in preparation.

D. H. Kaelble, *CAD/CAM Handbook for Polymer Composite Reliability,* vol. 1, Rockwell International, Thousand Oaks, CA, 1983.

Rafaat M. Hussein, *Composite Panels/Plates: Analysis and Design,* Technomic, Lancaster, PA, 1986.

C. C. Chamis, Ed., *Structural Design and Analysis,* parts I and II, Academic Press, New York, 1975 (*Composite Materials,* vol. 7–8).

C. T. Herakovich and Y. Tarnopol'skii, Eds., *Structures and Design,* Elsevier, 1989.

M. Holmes and D. J. Just, *FRP in Structural Engineering,* Elsevier, New York, 1984.

Design, Fabrication and Mechanics of Composite Structure, Seminar Notes, Technomic, Lancaster, PA, 1984.

Engineering with Composites, vols. 1 and 2, Proceedings of Third Technology Conference of the European Chapter of SAMPE, London, 1983, Society for the Advancement of Materials and Process Engineering, Azusa, CA, 1983.

Leonard Hollaway, *Glass Reinforced Plastics in Construction: Engineering Aspects,* Halsted Press, New York, 1978.

James C. Seferis and Luigo Nicolais, Eds., *The Role of the Polymeric Matrix in the Processing and Structural Properties of Composite Materials,* Plenum Press, New York, 1983.

Eugene G. Kovach, Ed., *Properties of Wood in Relation to Its Structure,* Report of a NATO Conference, Held in Les Arces, France, 1975, NATO, Brussels, 1976.

A. Kelly, S. T. Mileiko, and Yu. N. Robotnov, Eds., *Fabrication of Composites,* Elsevier, Amsterdam, 1983 (*Handbook of Composites,* vol. 4).

M. M. Schwartz, Ed., *Fabrication of Composite Materials: Source Book,* American society for Metals, Metals Park, OH, 1985.

DoD/NASA Structural Composites Fabrication Guide, 2nd ed., Air Force Wright Patterson Aeronautics Laboratory, Dayton, OH, 1979.

Prepreg Composition, Processing and Applications, 1973–1984, T/C Publications, El Segundo, CA 1984.

J. H. Crawford, Y. Chen, and W. A. Siblay, Eds., *Defect Properties and Processing of High-Technology Nonmetallic Materials,* Proceedings of a Materials Research Society Symposium, Boston, 1983, Elsevier, Amsterdam, 1984.

Larry L. Hench and Ronald R. Ulrich, Eds., *Ultrastructures Processing of Ceramics, Glasses, and Composites,* Wiley, New York, 1984.

T. Airman and T. N. Veziroglu, Eds., *Particulate and Multiphase Processes,* Hemisphere Publishing, New York, 1987, 3 vols.

Advanced Composites Technology, T/C Publications, El Segundo, CA, 1978.

N. J. Parratt, *Fibre-Reinforced Materials Technology,* Van Nostrand Reinhold, New York, 1972.

P. K. Mallick, *Fiber-Reinforced Composite Materials, Manufacturing and Design,* Dekker, New York, 1987.

K. L. Laewenstein, *Manufacturing Technology of Continuous Glass Fibres,* 2nd ed., Elsevier, Amsterdam, 1983 (*Glass Science and Technology,* vol. 6).

John Delmonte, *Technology of Carbon and Graphite Fiber Composites,* Van Nostrand Reinhold, New York, 1981.

Composite Materials Technology, Society of Automotive Engineers, Warrendale, PA, 1986, (SP 648).

R. G. Weatherhead, *FRP Technology,* Applied Science, Publishers, Barking, England, 1980.

R. Kaiser, *Technology Assessment of Advanced Composite Materials,* Argos Associates, Winchester, MA, 1978.

Applications

Reinforced Plastics for Commercial Composites: Source Book, ASM International, Metals Park, OH, 1986.

A. Watts, Ed., *Commercial Opportunities for Advanced Composites,* American Society for Testing and Materials, Philadelphia, 1980 (STP 704).

Advanced Composite Materials Market, Frost & Sullivan, New York, 1985.

Advanced Composite Materials Market in Europe and the Far East, Frost & Sullivan, New York, 1985.

M. Salkind and G. Holister, Eds., *Applications of Composite Materials,* American Society for Testing and Materials, Philadelphia, 1973 (STP 524).

Bryan R. Noton, Ed., *Engineering Applications of Composites,* Academic Press, New York, 1974 (*Composite Materials,* vol. 3).

L. E. Kukacka, Ed., *Applications of Polymer Concrete,* American Concrete Institute, Detroit, 1981.

Raymond B. Seymour, *Polymers for Engineering Applications,* American Society for Metals, Metals Park, OH, 1987.

Reinforced Plastics in Aerospace Applications, Proceedings of a Reinforced Plastics Group Conference, Royal Aeronautical Society, London, 1973.

B. C. Hoskin and A. A. Baker, Eds., *Composite Materials for Aircraft Structures,* American Institute of Aeronautics and Astronautics, New York, 1986.

Composites: Design and Manufacture for General Aviation Aircraft, Society of Automotive Engineers, Warrendale, PA, 1985.

Second Conference on Fibrous Composites in Flight Vehicle Design, Proceedings, Air Force Flight Dynamics Laboratory, Dayton, OH, 1974.

S. V. Kulkarni et al., *Composite Materials in the Automobile Industry,* American Society of Mechanical Engineers, Fairfield, NJ, 1978.

Glass Fiber Composites in European Automobiles and Trucks, 1983 and 1984 Year Models, Owen-Corning Fiberglass Europe SA, Brussels, 1983–1984.

D. W. Wilson, Ed., *High Modulus Fiber Composites in Ground Transportation and High Volume Applications,* Papers from a Symposium Held in 1983, American Society for Testing and Materials, Philadelphia, 1985 (STP 873).

P. Lanicq, W. J. G. Bunk, and J. G. Wurm, Eds., *Advanced Materials Research and Developments for Transport Composites,* Materials Research Society, Pittsburg, 1985.

John H. Mallinson, *Corrosion-Resistant Plastic Composites in Chemical Plant,* Dekker, New York, 1988.

Composites in Bio-Medical Engineering, Proceedings of the First International Conference, Kensington, Plastics and Rubber Institute, London, 1985.

Michael Szycher, Ed., *Biocompatible Polymers, Metals, and Composites,* Technomic, Lancaster, PA, 1983.

V. V. Vasilev, *Polymeric Composites in the Mining Industry,* Nauka, Moscow, 1986 (in Russian).

S. Szpak-Szpakowski, W. Witkiewicz, and A. Zietek, *New Polymer-Based Composite Materials and Possibilities for Their Use in Shipbuilding,* Centrum Techniki Okretowe, Gdansk, Poland, 1980 (in Polish).

Mary Vance, *Aggregates—Building Materials,* Monograph, Vance Bibliographies, Monticello, IL, 1985.

E. M. Lenoe et al., Eds., *Fibrous Composites in Structural Design,* Plenum Press, New York, 1980.

J. R. Vinson, T. W. Chou, *Composite Materials and Their Use in Structures,* Elsevier, Amsterdam, 1975.

J. R. Vinson and R. L. Sierakowski, *The Behavior of Structures Composed of Composite Materials,* Kluwar Academic Publications, Norwell, MA, 1985.

Andrew C. Marshall, *Practical Sandwich Construction/Advanced Composites,* T/C Publications, El Segundo, CA, 1983.

K. Lida and A. J. McEvily, Eds., *Advanced Materials for Severe Service Applications,* Proceedings of a Japan–U.S. Joint Seminar Held in Tokyo, 1986, Elsevier, New York, 1987.

E. Scala, *Composite Materials for Combined Functions,* Hayden, Rochelle Park, NJ, 1973.

Enid K. Sichel, Ed., *Carbon Black-Polymer Composites,* Dekker, New York, 1982 (*Plastics Engineering Series,* vol. 3).

M. A. Leeds, *Electronic Properties of Composite Materials,* Plenum Press, New York, 1972 (*Handbook of Electronic Materials,* vol. 9).

Reinforced Plastics: Electrical Applications, T/C Publications, El Segundo, CA, 1984.

Fibers and Fiber Composites

M. Lewin and S. Sello, Eds., *Handbook of Fiber Science and Technology,* Dekker, New York, 1983–1984, 2 vols. (*International Fiber Science and Technology* series, vols. 1–2).

M. Lewin and J. Preston, *Handbook of Fiber Science and Technology: High Technology Fibers,* Part A, Dekker, New York, 1985 (*International Fiber Science and Technology* series, vol. 5).

M. Lewin and E. Pearce, *Handbook of Fiber Science and Technology: Fiber Chemistry,* Dekker, New York, 1985 (*International Fiber Science and Technology* series, vol. 7).

F. Happey, Ed., *Applied Fibre Science,* Academic Press, New York, 1978–1979, 3 vols.

H. S. Krassig, J. Lenz, and H. F. Mark, *Fiber Technology, from Film to Fiber.* Dekker, New York, 1984 (*International Fiber Science and Technology* series, vol. 4).

High-Tech Fibers: Key to Strong High Quality Products, Technical Insights, Fort Lee, NJ, 1985 (*Emerging Technologies* series, 16).

A. Watt and B. V. Perov, Eds., *Strong Fibers,* Elsevier, New York, 1985 (*Handbook of Composites,* vol. 1).

J. L. White, *Fiber Structure and Properties,* Wiley-Interscience, New York, 1979.

Jett C. Arthur, Ed., *Polymers for Fibers and Elastomers,* American Chemical Society, Washington, DC, 1984. (ACS Symposium Series, 260).

J. B. Donnett and R. C. Bansal, *Carbon Fibers,* Dekker, New York, 1984 (*International Fiber Science and Technology* series, vol. 3).

The Plastics and Rubber Institute, *Carbon Fibers: Technology, Uses, and Prospects,* Noyes Data Corp., Park Ridge, NJ, 1986.

Marcus Langley, *Carbon Fibres in Engineering,* McGraw-Hill, New York, 1973.

L. R. McCreight, H. W. Rauch, Sr., and W. H. Sutton, *Ceramic and Graphite Fibers and Whiskers,* Academic Press, New York, 1965.

M. S. Dresselhaus, G. Dresselhaus, J. E. Fisher, and M. J. Moran, Eds., *Intercalated Graphite*, Elsevier, Amsterdam, 1983, (*MRS Symposium Proceedings*, 20).

John G. Mohr, *Fiber Glass*, Van Nostrand Reinhold, New York, 1978.

Albert P. Levitt, *Whisker Technology*, Krieger Publishing, Melbourne, FL, 1970.

C. W. Evans, *Powdered and Particulate Rubber Technology*, Elsevier Applied Science, Barking, England, 1978.

F. S. Galasso, *High Modulus Fibers and Composites*, Gordon & Breach, New York, 1970.

W. Watt, *New Fibres and Their Composites*, Royal Society, London, 1980.

B. A. Sanders, Ed., *Short Fiber Reinforced Composite Materials*, American Society for Testing and Materials, Philadelphia, 1982 (STP 772).

L. M. Bagaasen, *Discontinuous-Fiber-Reinforced Composites*, T/C Publications, El Segundo, CA, 1985.

P. Bracke, H. Schurmans, and J. Verhoest, *Inorganic Fibers and Composite Materials*, Pergamon Press, Elmsford, NY, 1984.

R. M. Gill, *Carbon Fibres in Composite Materials*, Crane, Russak, London, 1972.

E. Fitzer, Ed., *Carbon Fibres and Their Composites*, Proceedings of the International Conference on Carbon Fibre Applications, Sao Jose das Campos, Brazil, 1983, Springer-Verlag, Berlin, 1985.

Carlos J. Hilado, Ed., *Carbon Reinforced Epoxy Systems*, Parts 1–5, Technomic, Lancaster, PA, 1974–1984 (*Materials Technology* Series, 1, 8, 9, 12, 13).

Carlos J. Hilado, Ed., *Boron Reinforced Epoxy Systems*, Technomic, Lancaster, PA, 1974 (*Materials Technology* Series, 3).

Glass Reinforced Epoxy Systems, Part 2, Technomic, Lancaster, PA, 1982. (*Materials Technology* Series, 10).

Carlos J. Hilado, Ed., *Glass Reinforced Polyester Systems*, Technomic, Lancaster, PA, 1984 (*Materials Technology* Series, 14).

H. W. Rauch, W. H. Sutton, and L. R. McCreight, *Ceramic Fibers and Fibrous Composite Materials*, Academic Press, New York, 1968 (Refractory Materials Science, vol. 3).

A. G. Evans, Ed., *Ceramic Containing Systems*, Noyes Data Corp., Park Ridge, NJ, 1986.

Richard E. Tressler, Ed., *Tailoring Multiphase and Composite Ceramics*, Plenum Press, New York, 1986 (*Materials Science Research*, vol. 20).

D. J. Hannant, *Fibre Cements and Fibre Concretes*, Wiley, New York, 1978.

Fiber Reinforced Concrete, Proceedings of a Symposium, American Concrete Institute, Detroit, 1984.

Fibro-Cement Composite, Report of Meeting, United Nations, New York, 1971 (UN 71/2B1).

Kevlar Composites, T/C Publications, El Segundo, CA, 1980.

M. Umemura, H. Hirose, et al., *Hybrid Fiber-Reinforced Composite Materials*, CMC Company, Ltd, Tokyo, 1986 (in Japanese).

N. L. Hancox, Ed., *Fibre Composite Hybrid Materials*, Elsevier Applied Science, Barking, England, 1981.

T. W. Chou and F. Ko, Eds., *Textile Structure Composites*, Elsevier, New York, 1988.

W. C. Wake and D. B. Wooten, Eds., *Textile Reinforcement of Elastomers*, Applied Science, Barking, England, 1982.

Polymer Composites

A. V. Tobalsky and H. F. Mark, Eds., *Polymer Science and Materials*, Wiley, New York, 1971.

R. B. Seymour and G. S. Kirshenbaum, Eds., *High Performance Polymers: Their Origin and Development*, Proceedings of the ACS Symposium Held in New York, 1986, Elsevier, New York, 1986.

B. Sedlacek, Ed., *Polymer Composites*, Proceedings of the 28th IUPAC Microsymposium on Macromolecules, Prague, 1985, Walter de Gruyter, Berlin, 1986.

S. A. Volfson, *New in Life, Science, and Technology*. Chemistry Series, 1: *Composite Polymeric Materials Today and Tomorrow*, Znanie, Moscow, 1982 (in Russian).

R. P. Sheldon, *Composite Polymeric Materials*, Elsevier Applied Science Publishers, Barking, England, 1982.

M. Richardson, D. Phillips, and B. Harris, Eds., *Industrial Polymeric Composite Materials*, Khimiya, Moscow, 1980 (in Russian, trans. from English).

Yu. S. Lipatov, *The Future of Polymeric Composites*, Naukova Dumka, Kiev, USSR, 1984 (in Russian).

Report of Research Briefing Panel on High-Performance Polymer Composites, Research Briefing 1984. National Academy Press, Washington, DC, 1984.

L. Salmen, A. DeRuve, J. C. Seferis, and E. B. Stark, Eds., *Composite Systems from Natural and Synthetic Polymers*, Elsevier, Amsterdam, 1986. (*Materials Science Monographs*, vol. 36).

Yu. S. Lipatov, Ed., *Composite Polymer Materials*, no. 1, Naukova Dumka, Kiev, USSR, 1979 (in Russian).

D. M. Karpinos and V. I. Oleinik, *Polymers and Composite Materials Made of Them in Industry*, Naukova Dumka, Kiev, USSR, 1981 (in Russian).

N. S. Enikolopov and N. R. Ashurov, *Polymeric Composites and Their Use in the National Economy*, Proceedings of the Second Conference, Fan, Tashkent, USSR, 1986 (in Russian).

M. O. W. Richarson, Ed., *Polymer Engineering Composites*, Elsevier Applied Science Publishers, Barking, England, 1977.

Yu. P. Belyaev, Ed., *Polymeric Composite Materials with Prolonged Storage Life and Their Use*, Leningr. Dom. Nauchno.-Tekh Propag., Leningrad, USSR, 1982 (in Russian).

G. M. Gunyaev, *Structure and Properties of Polymeric Fibrous Composites*, Khimiya, Moscow, 1981 (in Russian).

British Plastics Federation, Reinforced Plastics Group, *Guide to High Performance Plastics Composites*, London, 1980.

M. W. Gaylord, *Reinforced Plastics: Theory and Practice*, 2nd ed., Cahner, Boston, 1974.

Brian Parkyn, *Glass Reinforced Plastics*, CRC Press, Boca Raton, FL, 1970.

M. W. Ranney, *Reinforced Composites from Polyester Resins*, Noyes Data Corp., Park Ridge, NJ, 1972.

Carlos J. Hilado, *Reinforced Phenolic, Polyester, Polyimide, and Polystyrene Systems*, Technomic, Lancaster, PA, 1974.

C. B. Bucknall, *Toughened Plastics*, Elsevier Applied Science Publishers, Barking, England, 1977.

James M. Margolis, Ed., *Advanced Thermoset Composites*, Van Nostrand Reinhold, New York, 1986.

Phillips Petroleum Company, *Advances Thermoplastics Composites*, Bartlesville, OK, 1986.

Advanced Composite Thermoplastics (ACTP) Materials, Polymer Composites, Winona, MN, 1986.

N. J. Folkes, *Short Fibre Reinforced Thermoplastics*, Wiley, New York, 1982.

W. V. Titow and B. J. Lanham, *Reinforced Thermoplastics*, Halsted Press, New York, 1975.

D. Klempner and K. C. Frisch, Eds., *Polymer Alloys*, vols. I–III, Plenum Press, New York, 1973–1983. (*Polymer Science and Technology*, 10–12).

E. Tsuchida et al., Eds., *Polymer Composites, Their Function and Uses*, no. 5: *Polymer Assemblies*, Gakkai Shuppan Senta, Tokyo, 1983 (in Japanese).

J. M. Buise, D. Welsh, and R. P. Redman, Eds., *Polyurethane-Based Composite Materials*, Khimiya, Moscow, 1982 (in Russian).

Clayton A. May, Ed., *Epoxy Resins, Chemistry and Technology*, Dekker, New York, 1988.

K. Dusek, Ed., *Epoxy Resins and Composites*, 4 vols., Springer-Verlag, New York, 1985–1986, (*Advances in Polymer Sciences*, 72, 75, 78, 80).

Ernest W. Flick, *Epoxy Resins, Curing Agents, Compounds, and Modifiers, An Industrial Guide*. Noyes Data Corp., Park Ridge, NJ, 1987.

Yu. S. Lipatov, Ed., *Problems of Polymer Composite Materials*, Naukova Dumka, Kiev, USSR, 1979 (in Russian).

Paul F. Bruins, Ed., *Polyblends and Composites*, Interscience, New York, 1970 (*Applied Polymer Symposium*, 15).

John A. Manson and Leslie H. Sperling, *Polymer Blends and Composites*, Plenum Press, New York, 1976.

Chang Dae Han, *Polymer Blends and Composites in Multiphase Systems*, American Chemical Society, Washington, DC, 1984. (Advances in Chemistry Series, 206).

Donald H. Paul and Leslie H. Sperling, Eds., *Multicomponent Polymer Materials*, American Chemical Society, Washington, DC, 1986. (*Advances in Chemistry* Series, 211).

E. Martuscelli, Ed., *Polymer Blends*, vol. 1, Proceedings of the First Seminar, Capri, Italy, 1979, Plenum Press, New York, 1981.

M. Kryszewski, Ed., *Polymer Blends*, vol. 2, Proceedings of the Second Italian–Polish Joint Seminar on Multicomponent Polymeric Systems, Lodz, Poland, 1982, Plenum Press, New York, 1985.

D. J. Walsh, and J. S. Higgins, Eds., *Polymer Blends and Mixtures*, Proceedings of the NATO Advanced Institute, London, 1984, Nijhoff, Boston, 1985.

L. A. Utracki and A. P. Ploskocki, *Industrial Polymer Blends and Alloys*, Hanser International, New York, 1985.

D. R. Paul and S. Newman, Eds., *Polymer Blends*, Academic Press, New York, 1978, 2 vols.

L. H. Sperling, *Interpenetrating Polymer Networks and Related Materials*, Plenum Press, New York, 1981.

Karel Solc, Ed., *Polymer Compatibility and Incompatibility*, Proceedings of the Tenth Midland Macromolecular Meeting, 1980. Harwood Academic Press, New York, 1982. (*MMI Press Symposium Series*, vol. 2).

Metal Composites

National Materials Advisory Board, National Academy of Science, National Academy of Engineering, *Metal-Matrix Composites: Status and Prospects* (NMAB-313), Washington, DC, 1974.

J. E. Schoutens, *Introduction to Metal Matrix Composite Materials*, T/C Publications, El Segundo, CA, 1982.

Metal Matrix Composites, 1978–1983, T/C Publications, El Segundo, CA, 1983.

Carlos J. Hilado, Ed., *Carbon Composite and Metal Composite Systems*, Technomic, Lancaster, PA, 1974 (*Materials Technology* Series, 7).

Kenneth G. Kreider, Ed., *Metallic Matrix Composites*, Academic Press, New York, 1974 (*Composite Materials*, 4).

G. S. Upadhyaya, Ed., *Sintered Metal–Ceramic Composites*, Elsevier, New York, 1984 (*Materials Science Monograph*, 25).

Rutgers University, *Copper-Containing Composites*, International Copper Research Association, New York, 1970.

W. J. Renton, Ed., *Hybrid and Select Metal Matrix Composites*, American Institute of Aeronautics and Astronautics, New York, 1977.

Carlos J. Hilado, *Boron Reinforced Aluminum System*, Parts 1 and 2, Technomic, Lancaster, PA, 1974–1982. (*Materials Technology* Series, 6, 11).

M. T. Bryk, Z. T. Ilina, V. I. Chernova, et al., *Composite Metal–Polymer Materials with Dispersed Titanium*, Naukova Dumka, Kiev, USSR, 1980 (in Russian).

S. K. Bhattacharya, Ed., *Metal-Filled Polymers*, Dekker, New York, 1986 (*Plastics Engineering*, 11).

Jonathan A. Lee and D. K. Mykkanen, *Metal and Polymer Matrix Composites*, Noyes Data Corp., Park Ridge, NJ, 1987.

A. G. Terkhunov, M. I. Chernovol, and V. M. Tiunov, *Composite Metal-Containing Polymer Coatings and Materials*, Tekhnika, Kiev, USSR, 1983 (in Russian).

Julia T. Lee

Bibliography

Beck, D. P., in G. D. Shook, Ed., *Reinforced Plastics for Commercial Composites: Source Book*, American Society for Metals, Metals Park, OH, 1987, pp. 418–424.

Books in Print, 1987–1988, Bowker, New York, 1987.

Corey, J., *39th Annual Conference of Reinforced Plastics/Composites Institute*, Society of the Plastics Industry, New York, 1984. Session 7-D, pp. 1–2.

Chen, C.-C., *Scientific and Technical Information Sources*, 2nd ed., MIT Press, Cambridge, MA, 1987.

Irregular Serials and Annuals, 13th ed., Bowker, New York, 1987–1988.

Kabach, S. M., *J. Chem. Inf. Comput. Sci.*, *20*, 1–6 (1980).

Kabach, S. M., *Chemtech*, March 1980, pp. 172–177.

Lee, J. T., Ed., in *Encyclopedia of Polymer Science and Engineering*, vol. 9, 2nd ed., Wiley, New York, 1987, pp. 62–97.

Lee, Stuart M., *Dictionary of Composite Materials Technology*, Technomic Publishing Co., Lancaster, PA, 1989.

Lee, Stuart M., Ed., *Reference Book For Composites Technology* (2 Volumes), Technomic Publishing Co., Lancaster, PA, 1989.

Lotz, J. E., in M. Grayson, Ed., *Kirk-Othmer Encyclopedia of Chemical Technology*, vol. 16, 3rd ed., Wiley, New York, 1981, pp. 889–945.

Maizell, R. E., *How to Find Chemical Information*, 2nd ed., Wiley, New York, 1987.

Pebly, H. E., in G. D. Shook, Ed., *Reinforced Plastics for Commercial Composites: Source Book*, American Society for Metals, Metals Park, OH, 1987, pp. 409–417.

Subject Guide to Books in Print, 1986–1987, Bowker, New York, 1987.

Thomas, D. K., *RAPRA Rev. Rep., 1* (1), pp. 3/1–3/99 (1988).

Traceski, F. T., in *Engineered Materials Handbook*, vol. 1, *Composites*, ASM International, Metals Park, OH, 1987, pp. 40–42.

Ulrich's International Periodicals Directory, 26th ed., Bowker, New York, 1987, 2 vols.

Weeton, J. W., Peters, D. M., and Thomas, K. L., Eds., *Engineers' Guide to Composite Materials*, American Society for Metals, Metals Park, OH, 1987, pp. 9/1–9/7, 11/1–11/5, 12/1–12/2.

Westbrook, J. H., in M. B. Bever, Ed., *Encyclopedia of Materials Science and Engineering*, Pergamon Press, Elmsford, NY, 1986, pp. 527–542.

Williams, M. E., Ed., *Computer-Readable Databases*, 2 vols, American Library Association, Chicago, 1985.

Yescombe, E. R., *Plastics and Rubbers: World Sources of Information*, 2nd ed., Elsevier Applied Science Publishers, Barking, England, 1976.

Long Fiber

See Mechanical Properties

Low Observability Composites

See Stealth Low Observability Aircraft

Lubricating Composites

See Tribological Composites

Machining of Composites

Composites have become one of the most attractive groups of materials in our technological society. Their light weight and high strength characteristics make them suitable in aerospace applications that call for high strength-to-weight ratios. Their extensive use in aerospace and other industries has led to the research and development of new composites, as well as the technology for mass production to make them cost effective. In spite of the rapid advances being made in the production of composites by preshaping and preforming, secondary processes such as drilling, milling, and sawing are often used to manufacture and assemble composite components. The areas of concern in machining composites are (a) quality of the machined region or machined component using either conventional or nonconventional cutting methods and (b) rapid tool wear, especially when using conventional methods. This article describes various cutting methods used for machining composites.

Conventional Machining Methods

All traditional machining processes (turning, drilling, grinding, milling, etc.) remove material in the form of chips to produce a surface. The basic mechanics, essentially the same for all cutting processes, can be idealized by a two-dimensional model known as orthogonal cutting. The theory of metal cutting is then based on a shear process, where chip is generated ahead of the tool by shearing the material continuously along a shear plane. It is assumed that the metal is an ideal plastic body that will yield in shear on the shear plane or will be sheared in the shear zone around the cutting edge [1].

An attempt was made by Everstine and Rogers [2] to use the theory of metal cutting to investigate the machining behavior of fiber-reinforced composite materials. The analysis was restricted to plane–strain deformation of incompressible composites, reinforced by strong parallel fibers. A wedge-shaped tool was used for the cutting operation. The cutting tool was constrained to travel parallel to the surface of the workpiece, and a layer of material was removed from the surface. The analysis was based on macroscopic behavior of the composite materials and their reaction on the cutting tool. The results of this analysis led to the derivation of complete deformation and stress fields and estimates of the forces required to maintain continuous machining. The results apply to both elastic and plastic stress responses.

Quality of Machined Components in Conventional Machining

Machining is generally considered to be a finishing process, with specified dimensions, tolerances, and surface finish. The quality of a surface generated by a machining process and its characteristics are important factors when evaluating the service life of a component under dynamic loads. This section considers the quality of the machined surface offered by some conventional machining processes.

An experimental investigation was carried out by Sadat [3,4] to determine the effects of cutting speed and width of cut on the quality of the machined region of a graphite–epoxy composite material using a sawing operation. Workpieces were mounted on a holder that was bolted on a tool force dynamometer. The dynamometer itself was secured to the table of a CNC vertical spindle, Bridgeport milling machine. A schematic of cutting action is shown in Figure 1.

After each cutting test, the machined surface (edge) was examined on an optical microscope at various magnifications. The flat surfaces of the machined workpiece were bonded to the leg of a special holder and then loaded in tension, in such a manner that the direction of loading was normal to the bonded surface, which was in turn normal to the plane of the laminate. Tensile fracture load was measured this way for all the test specimens, and the fractured surfaces were then examined along the machined edge using an optical microscope. Figure 2 shows variation of interlaminar tensile strength with cutting speed. Interlaminar tensile strength increases with an increase in cutting speed.

Examination of the fractured surface in the vicinity of machined edge revealed a damaged zone where the orientations of the graphite fibers were altered. This can be seen from Figure 3, where graphite fibers are rotated in the direction of feed. The depth of the effects on zone *d*, which is the distance from the machined edge to the point of fiber rotation, was measured along the fibers that were normal to the machined edge. Figure 4 shows the effect of cutting speed on the depth of the affected zone: clearly the depth of the affected zone decreases when the cutting speed is increased. Neither the depth

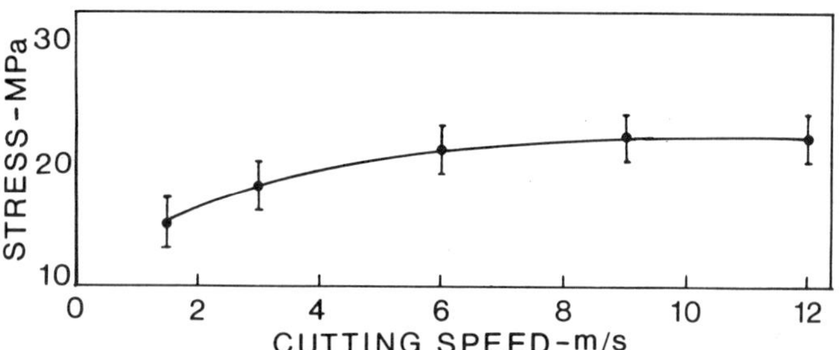

FIGURE 2 **Effect of cutting speed on interlaminar tensile strength [3,4].**

of the affected zone nor the interlaminar tensile strength was affected by changes in width of cut for the range of cutting condition selected in this investigation.

A detailed examination of the machined surface (machined edge) indicated delamination of plies and presence of cracks that eventually lead to delamination. Figure 5, a photomicrograph of a typical machined surface, shows both delamination and cracking. It was also observed that damage often occurred to the area near the back surface of the specimen where the cutting tool was exiting the workpiece.

It was suggested that in sawing graphite–epoxy composites, the tool forces that are generated during the machining operation [3,4] play an important role in controlling the quality of the machined edge (surface). Although the cutting temperature can be considered to be a controlling factor, for the cutting condition used in this investigation the maximum temperature did not exceed the softening temperature of the epoxy (182 °C).

Weisinger [5] compared the effects of various machining processes, such as shearing, abrasive cutoff, grinding, profiling, punching, and drilling on aluminum boron metal matrix composite material. None of the processes used could produce good surface finish, long tool life,

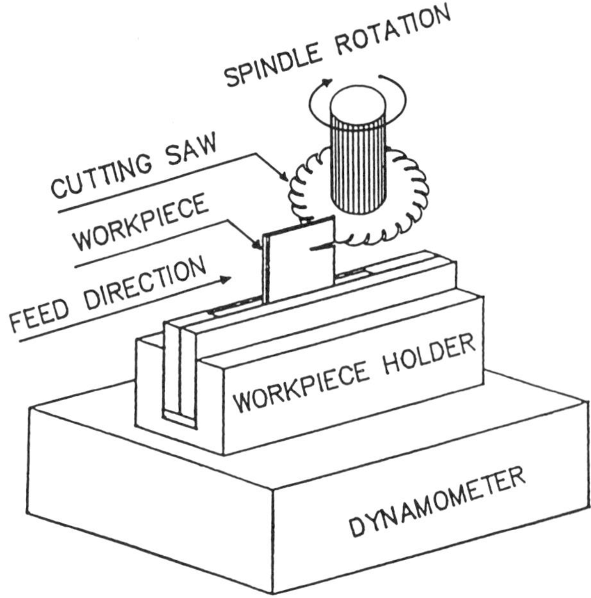

FIGURE 1 Schematic of cutting action [8].

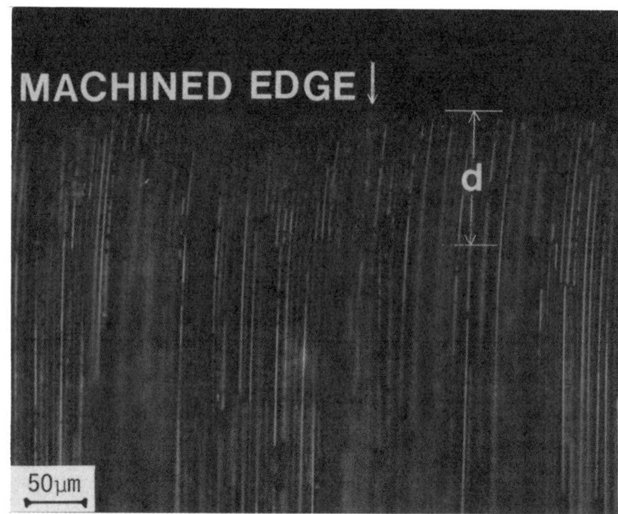

FIGURE 3 **Optical photomicrograph of the fractured surface [3,4].**

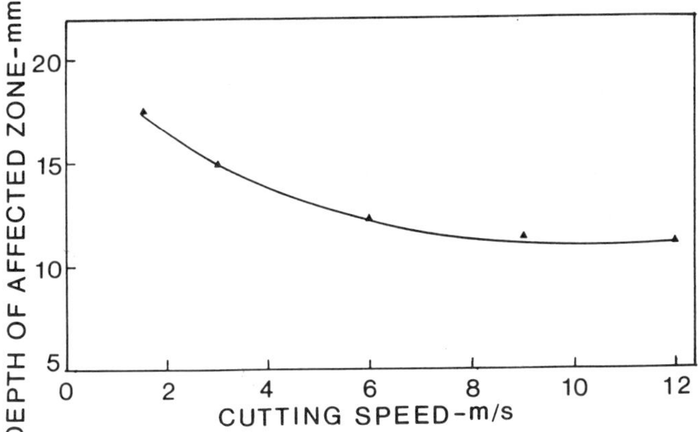

FIGURE 4 Effect of cutting speed on the depth of affected zone [3,4].

and surface features expected from high speed. The comparison was based strictly on the effect of the aforementioned processes on filament damage; hence other factors that would be considered in an overall evaluation were neglected. Weisinger ranked these processes in the following order (least damaging to most damaging): shearing, abrasive cutoff, grinding, diamond routing, punching, diamond drilling, and high speed steel drilling.

Kolesnichenko et al. [6] proposed that the most appropriate and effective method of machining Sitall-based composite material (Sitall glass is a pyroceramic type of microcrystalline glass) is grinding, because the matrix (Sitall) itself is characterized by good grindability. Their investigation showed that when Sitall was ground with an abrasive tool, the material removal was slow because of the poor heat transfer from the cutting region and the high hardness of Sitall. However, when diamond disks were used in the grinding of Sitall-based composite materials, high rates of material removal were attained with good surface finish. From the results of the experiments, optimum cutting conditions were selected for the grinding of the composites investigated using various types of grinding operation (internal and external cylindrical grinding, flat surface grinding, diamond cutting, etc.). The diamond tool was chosen for maximum material removal rate. It should be noted that because of the high mechanical strength and good wear resistance of Sitall-based composite, this material is used in sliding-bearing operations under severe conditions (high loads, high temperatures, aggressive environment). For this purpose the blanks must be machined in a way that ensures the production of good surface quality and dimensional accuracy.

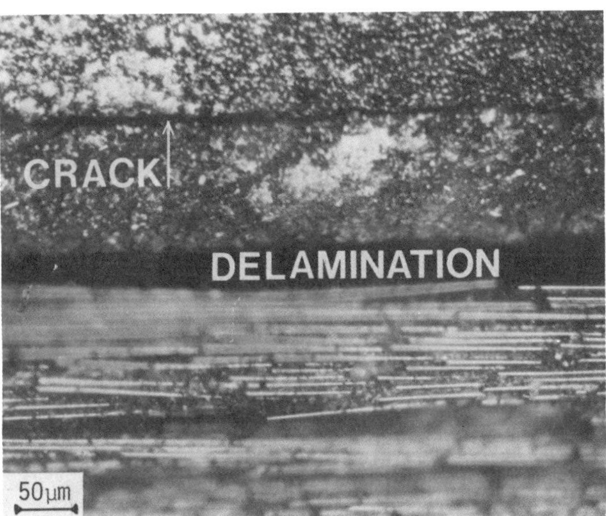

FIGURE 5 Optical photomicrograph of the machined surface [3,4].

Tool Wear in Conventional Machining

Traditional machining methods require the direct contact of the cutting tool with the workpiece during the machining operations. Because of the abrasive characteristics of most composite materials, rapid tool wear is a significant problem; hence the application of these methods is limited. Tool wear has always been a major problem in material cutting industries, because of its effect on the quality and economics of the machined components.

The effect of machining parameters on tool wear was investigated by Brian and Terry Lambert [7] in a drilling operation of boron/epoxy and boron/epoxy–titanium structure. Typical results are shown in Figures 6 and 7 where tool wear is plotted against the number of holes drilled for boron/epoxy and boron/epoxy–titanium, re-

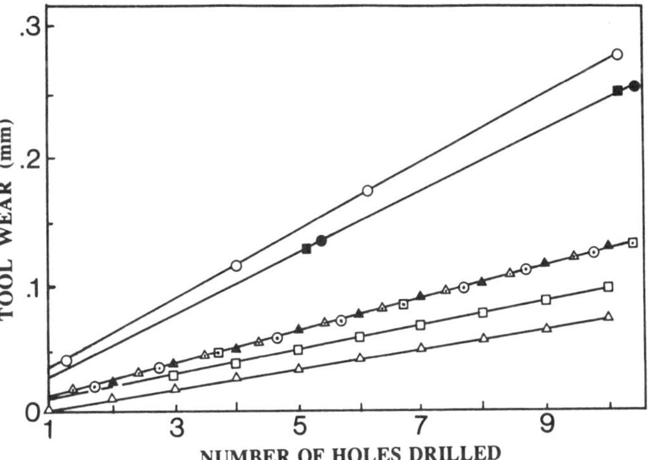

FIGURE 6 Effect of cutting speed and feed rate on tool wear when machining boron/epoxy composites: ● 0.90 m/s, 69.85 mm/min; ▲ 1.33 m/s, 69.85 mm/min; ■ 2.06 m/s, 69.85 mm/min; ○ 0.90 m/s, 41.91 mm/min; △ 1.33 m/s, 41.91 mm/min; □ 2.06 m/s, 41.91 mm/min; ⊙ 0.90 m/s, 22.35 mm/min; ◬ 1.33 m/s, 22.35 mm/min; ⊡ 2.06 m/s, 22.35 mm/min [7].

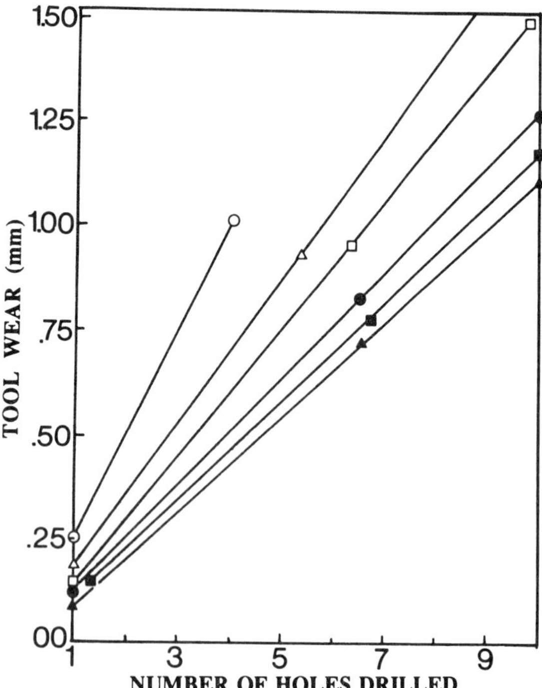

FIGURE 7 Effect of cutting speed and feed rate on tool wear when machining boron/epoxy–titanium composites: ● 0.90 m/s, 0.76 mm/min; ▲ 1.33 m/s, 0.76 mm/min; ■ 2.06 m/s, 0.76 mm/min; ○ 0.90 m/s, 1.52 mm/min; △ 1.33 m/s, 1.52 mm/min; □ 2.06 m/s, 1.52 mm/min [7].

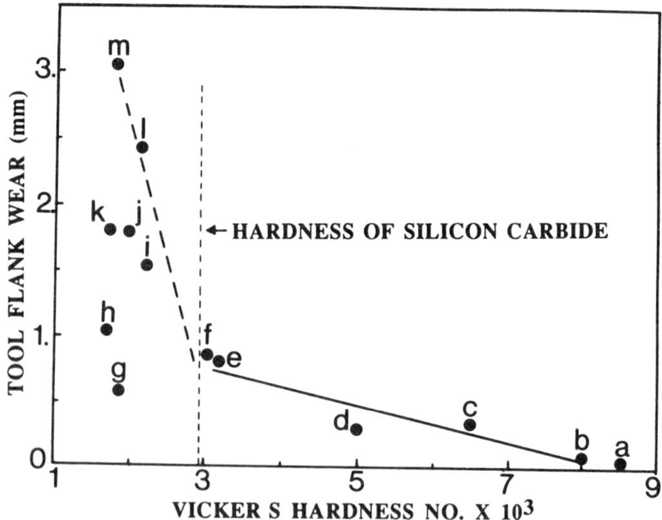

FIGURE 8 Variation of tool flank wear land with Vickers hardness of different cutting tool after 1 minute cutting time, at a cutting speed of 2.53 m/s: *a*, polycrystalline diamond (fine grain); *b*, polycrystalline diamond (coarse grain); *c*, polycrystalline cubic BN (development); *d*, polycrystalline cubic BN; *e*, hot-pressed titanium diboride–boron carbide; *f*, hot-pressed titanium diboride–titanium carbide; *g*, cemented WC–3% Co (medium grain); *h*, cemented WC–6% Co (fine grain); *i*, hot-pressed Al_2O_3–TiC; *j*, cemented WC–6% Co (medium grain), *k*, cemented WC-10% Co; *l*, Al_2O_3–ZrO_2; *m*, SiAlON [8].

spectively, using diamond-impregnated core drills. The results of the investigation show that when drilling boron/epoxy composite, the lowest tool wear occurs at a cutting speed of 1.33 m/s and a feed rate of 41.91 mm/min for the range of cutting conditions investigated (Fig. 6). Further tests were carried out to determine the effect of coolant when using core drills to machine boron/epoxy. It was found that for the range of cutting speed and feed rate used in this study, the effect of coolant on tool wear was negligible.

In a second series of tests, operation drilling was performed on a hybrid structure, consisting of alternate layers of boron/epoxy and 6Al-4V titanium alloy. The cutting tool was a 6.35 mm diameter impregnated core drill, and a water-soluble oil was the coolant. The lowest tool wear occurred at a cutting speed of 1.33 m/s (4000 rpm) and a feed rate of 0.76 mm/m (Fig. 7).

An investigation of tool wear rate when machining Al-40 vol % SiC composite was carried out by Bran and Lee [8], using a large collection of experimental and commercial cutting tools. It was found that the wear rate of tools harder than silicon carbide was inversely proportional to their hardness. Polycrystalline diamond tools showed a useful tool life, and all other cutting tools wore rapidly. The life of polycrystalline diamond cutting tools was relatively short at a moderate high cutting speed (2.5

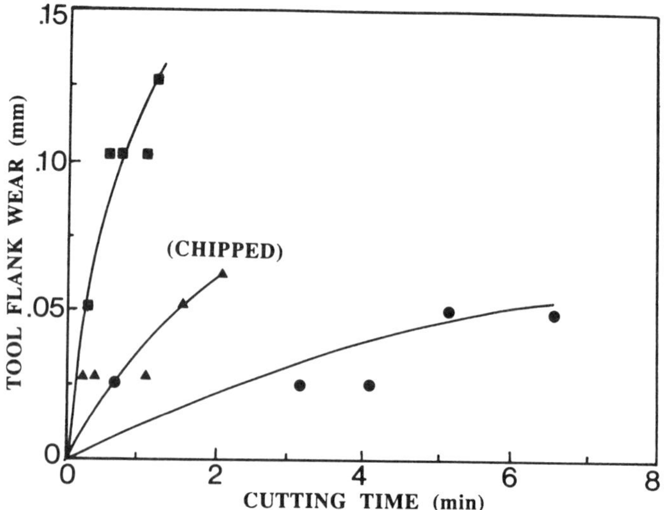

FIGURE 9 Variation of tool flank wear land of diamond cutting tools with cutting time for different cutting speeds: ●, 2.53 m/s; ▲, 6.33 m/s; ■, 14 m/s [8].

m/s); at low cutting speed, however, tool life increased significantly. The results of flank wear of various cutting tools for one minute of machining time with comparable cutting conditions are shown in Figure 8 as a function of their respective Vickers hardness. Since only diamond cutting tools showed reasonable tool life when machining Al-40 vol % SiC, the effect of cutting speed on flank wear was investigated (Fig. 9). As Figure 9 indicates, for a given machining time, an increase in cutting speed leads to an increase in flank wear. It was suggested that abrasive process was the dominant wear mechanism, and wear mechanism was also sensitive to cutting temperature [8].

Nonconventional Machining Methods

The excessive tool wear and poor quality of machined components that are associated with most conventional machining methods have made it necessary to pursue the application of other cutting methods with little or no tool contact to the workpiece. The advantage of using a noncontact tool cutting method is twofold: on the one hand, the rate of tool wear is small and easy to monitor, hence lending itself well to automation; on the other hand, surface damage due to mechanical action of the tool will be negligible because of the noncontact characteristic of the cutting tool. It should be pointed out that these methods may not be capable of inducing shape changes on a workpiece, as is possible by traditional machining methods (milling, turning, etc.); however, some nonconventional approaches have proved to be most effective for secondary processes (drilling, slitting etc.). The various types of nonconventional machining methods available for machining composite materials include:

Laser machining
Water jet and abrasive water jet machining
Electrical discharge machining
Ultrasonic machining
Electron beam machining
Electrochemical machining

Some of these methods and their applications are described next.

Laser Machining

Lasers have been used in industry for quite some time, and their application for machining or cutting is not a new concept. Laser operation is based on delivering an intense heat source that can be used to locally vaporize material and leave behind a minimal heat-affected zone. Laser processing can direct a focused beam of 0.1 mm diameter, with a power in excess of 10^8 W/cm^2 that can be used to cut various materials. There are two industrial lasers available for cutting composites [9].

1. The Nd:YAG (neodymium/yttrium–aluminum–garnet) has a wavelength of 1.06 μm with pulse rates as high as 200 pulses per second. This type of laser is effective for cutting metallic composites that do not contain an organic resin. Organic materials do not effectively absorb the 1.06 μm wavelength, hence can decompose.

2. The CO_2 laser has a wavelength of 10.6 μm and can operate with pulse lengths of the order of 10^{-4} second. It is effectively absorbed by most organic materials. This laser process has the advantage of reducing the heat-affected zone, because of the cooling that occurs between pulses.

The CO_2 laser system was used successfully for cutting fiberglass sheets [9]. A vacuum system was necessary to remove the particulates that resulted from vaporization, creating a health hazard. The heat-affected zone along the edge of the cut was narrow and the ends of the fiberglass were melted over, which prevented fraying. The cutting of Kevlar–graphite–epoxy and Kevlar–epoxy using a 1200 W CO_2 laser process was also successful. The cut edges did not exhibit signs of fraying; rather, they were smooth and required minimal secondary operation [9].

In an independent study conducted by Lawson [10], a laser of the same power level was used for cutting three reinforcing agents: Kevlar, glass, and graphite. Graphite showed poor cutting behavior because of its high dissociation temperature and thermal conductivity. Kevlar showed the best cutting behavior, followed by glass.

Water Jet and Abrasive Water Jet Machining

Water jet and abrasive water jet cutting are alternative methods of cutting used for machining composites. These systems are widely used in aerospace and other industries for cutting a wide variety of composites and other materials (reinforced polymers and plastics, most metallic and nonmetallic composites, plywood, etc.). The jet cutting nozzle can be as small as 0.13 mm in diameter with water pressure in excess of 350 MPa. The abrasive water jet cutting process that carries abrasive particles into its jet stream is suitable for cutting metallic base composites (such as boron/aluminum, SiC/Al, etc.) and ceramics. The basic components of an abrasive water jet system is shown in Figure 10 [11].

The quality of the machined surface and the volume rate of material removal is dependent on the cutting parameters [12,13]. The large number of parameters that affect the results of cutting include [13]:

1. Hydraulic parameters
    - Water jet pressure
    - Water jet orifice size
2. Abrasive parameters
    - Abrasive flow rate
    - Abrasive particle size
    - Abrasive material
3. Mixing chamber parameters
    - Length
    - Shape
    - Diameter

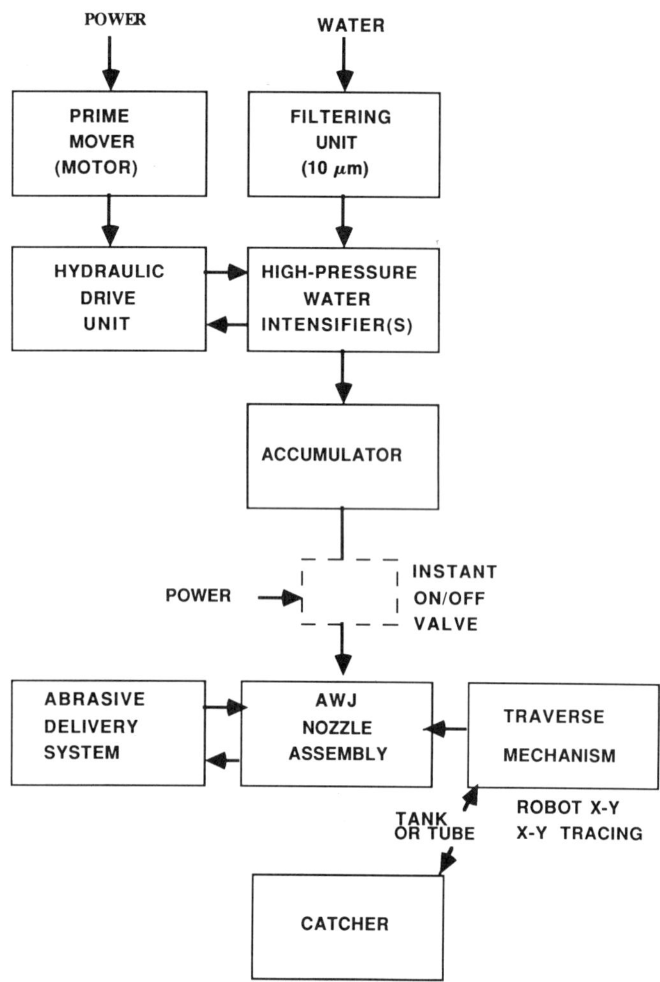

FIGURE 10 Schematic diagram of typical abrasive water jet (AWJ) system components [11].

4. Traverse parameters
   Traverse rate
   Number of passes
   Jet inclination angle
   Standoff distance
5. Type of work material

The quality of the machined edge was investigated by Brian and Terry Lambert [7] when cutting both uncured and cured graphite–epoxy composites using a water jet cutting system. Laminates were cut using a 0.754 mm diameter orifice with pressure in the range of 345–380 MPa.

When cutting uncured graphite laminates up through 24 plies (7.3 mm) with table speeds in the range of 5.08–15.24 m/min, edge conditions were acceptable. However, when the table speed was increased to 31.75 m/min and the laminate layer was increased incrementally to 48 plies (14.63 mm), the edge damage was obvious and unacceptable rough surfaces were appearing on the lower six laminates. The quality of the cut edge was acceptable when cutting up to 24 plies at a table speed of 31.75 m/min and when using a protective film at the underside, where the water jet exited the workpiece.

The cutting tests were carried out on cured graphite–epoxy laminates with thicknesses varying from 4 to 52 plies (0.56–7.26 mm), and at table speeds ranging from 0.46 to 5.08 m/min. The edge conditions were acceptable when cuts were made through 16 plies (2.24 mm) at a table speed of 1.52 m/min. However, delaminations were apparent when cutting was carried out through 52-ply material at a table speed of 0.46 m/min.

Electrical Discharge Machining (EDM)

The operation of electrical discharge machining or spark machining is based on the eroding effect of an electrical spark that is generated between an electrode and the workpiece in the presence of a dielectric fluid. The spark generated produces a localized high temperature, on the order of 12,000°C, which melts and vaporizes the material to form a small crater on the workpiece surface. This process is suitable for machining materials that possess uniform and continuous electrical conductivity. The mechanical properties of the workpiece do not affect the volume removal rate significantly. However, the volume removal rate is significantly influenced by the melting point of the workpiece. Figure 11 shows the effect of the melting point of the workpiece on the volume removal rate [14].

Electrical discharge machining is used for machining

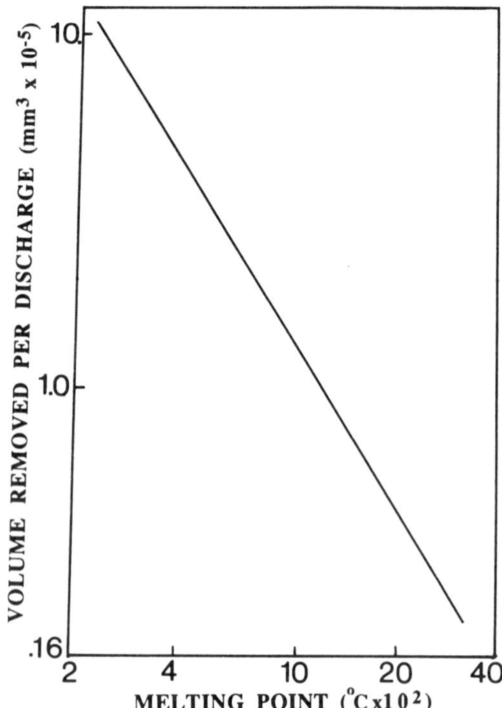

FIGURE 11 Effect of melting point of workpiece material on volume removal rate in electrical discharge machining [14].

metal matrix and other composites that exhibit good electrical conductivity. Because of the absence of directional patterns and microcracks (which result from most other cutting methods), the possibility of fatigue failures is reduced. This method can produce surface finishes of 0.25 μm and better. One of the major difficulties in EDM is high tool wear rate, (which results in inaccurate machining and adds considerably to the cost of machining. Figure 12 shows the relationship between the tool wear and melting point of the workpiece and tool [14].

Weisinger [5] investigated electrical discharge machining on boron–aluminum composites, using special carbon cutting tool (Gentrode 10) to combine drilling and countersinking in one operation, on various thicknesses of composites ranging from 0.51 to 2.18 mm. In all cases, the results were satisfactory, with no apparent damage to the filaments. The machining time for producing a hole in composite 2.18 mm thick was 6 minutes.

Today ceramics are one of the most important engineering materials in industry because of their special characteristics not found in traditional metal alloys and polymers. However, one of the drawbacks in using ceramics is their poor machinability, which is attributed to their toughness, and high hardness and abrasive characteristics. Because of their poor electrical conductivity, electrical discharge machining of ceramics is also a challenge. Kamijo et al. [15] attempted to solve this problem by increasing the electrical conductivity of the ceramic material, thus enhancing the electrical discharge machinability. Their investigation led to the development of electrical discharge machinable Si/Sub3/N/Sub4/(EDM Si/Sub 3N/Sub4/) with a conductive addition. The machinability of this ceramic is similar to that of both cemented carbide and steel, and currently existing electrical discharge machines can be used.

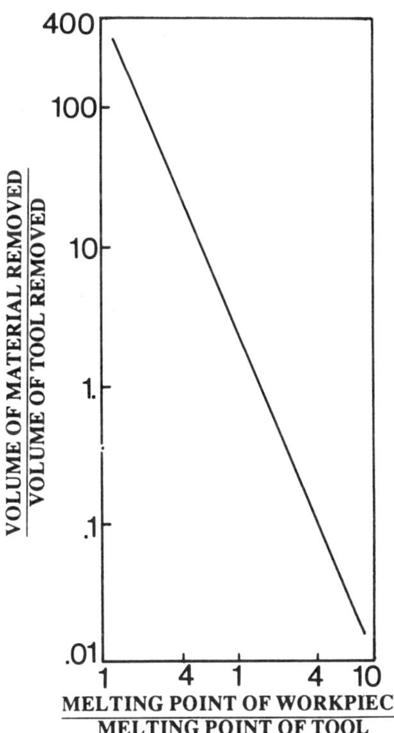

FIGURE 12 Relationship between the tool wear and melting points of the workpiece and tool material in electrical discharge machining [14].

Ultrasonic Machining

The ultrasonic machining method is based on the bombardment of the workpiece surface with abrasive particles (i.e., carried in a slurry) at high velocity. The transducer causes the attached tool to oscillate linearly at low amplitude (0.05–0.125 mm) and at high frequency (20–30 kHz). The abrasive particles used for this process are aluminum oxide, silicon carbide, boron oxide, and similar materials. Grain size can range from 100-mesh size for roughing operation to 1000-mesh size for finishing operation, and tolerances of 0.013 mm can be maintained using fine grits. This method of machining is suitable for drilling and slitting of various composites materials including ceramics.

Happe and Yeast [16] investigated the effect of machining on boron–aluminum composites in a drilling operation. This method appeared to be rapid and versatile, and tooling was inexpensive. The investigators were able to maintain high quality cuts, with good tolerances, and virtually no filament breakage. Similar results were obtained when an ultrasonic impact grinder was used for machining holes in specimens [17].

Electron Beam Machining

Electron beam machining is a thermoelectric process, generally carried out in a vacuum, in which high speed electrons impinge on the workpiece surface, and the heat generated vaporizes the material locally. This process is considered to be micromachining, but metal removal rates of approximately 0.01 mg/s can be achieved. Extremely close tolerances can be maintained using this process, with no heat-affected zone on the workpiece. This process is suitable for machining small holes and cutting and slitting of most composites.

Electrochemical Machining (ECM)

Electrochemical machining is the reverse of electroplating, where the electrolytes dissolve the products of the reaction formed on the workpiece by electrochemical action. The electrolyte is usually sodium chloride mixed with water, sodium nitrate, and other fluids that can chemically react with a specific workpiece (or attack the workpiece). For most applications the current densities range between 1.5 and 7.8 A/mm^2, and a voltage of 5–25 V is maintained. This process can be used for machining complex cavities and for slitting, drilling, and cutting most composites that exhibit continuous and uniform electrical conductivity. This process does not cause any thermal damage.

Concluding Remarks

A wide variety of both traditional and nontraditional machining methods are currently available for machining various composites. In selecting a traditional method of machining, it is important to note that because of the excessively rapid tool wear that is associated with these methods, diamond cutting tools are considered to be the most effective. This can limit the application of these methods for generating complicated shapes and features. The quality of the machined component is also an important factor in the selection of a machining method. It was shown that conventional machining can cause severe damage to the component.

Nonconventional machining methods show a promising future for machining composites. Because of advances being made in the production of versatile robots with improved accuracy, some of these methods can be fully automated, which will make them cost effective.

The demand or desire for new composites is growing rapidly. Because of the improvement in many new organic and inorganic materials and production techniques, new composites for a variety of applications are surfacing in the industrial market every year. To keep up with this advancing technology, new and improved machining techniques must advance accordingly.

Abdul B. Sadat

References

1. E. H. Lee and B. W. Shaffer, *J. Appl. Mech., 18,* 4 (1951).
2. G. C. Everstine and T. G. Rogers, *J. Compos. Mater., 5,* 94 (1971).
3. A. B. Sadat, *SAMPE Q., 19*(2), 1 (1988).
4. A. B. Sadat, *J. Compos. Tech. Res., 10,* 2 (1988).
5. M. D. Weisinger, SAE paper 700752, 2296 (1970).
6. L. F. Kolesnichenko, N. D. Nazarenko, A. I. Yuga, N. I. Vlasko, F. D. Ivashov, L. L. Sukhikh, and G. A. Sedlyar, *Sov. Powder Metall. Met. Ceram., 15*(8), 6756 (1976).
7. B. K. Lambert and T. D. Lambert, SME Paper EM77-353, (1977).
8. M. K. Bran and M. Lee, *Wear, 104*(1), 21 (1985).
9. M. D. Mello, *Proc. SPIE Int. Soc. Opt. Eng.,* 668, 288 (1986).
10. W. E. Lawson, *Composites in Manufacturing,* SME, Los Angeles, 1 (1986).
11. M. Hashish, *Proceedings of the First ASM Non-Traditional Machining Conference,* 1986.
12. M. Hashish, *ASME Trans., 106,* 8 (1984).
13. M. Hashish, *Proceedings of the 11th SME Conference on Product Research and Technology,* Carnegie-Mellon University, Pittsburgh, 1984.
14. S. Kalpakjian, *Manufacturing Processes of Engineering Material,* Addison-Wesley, Reading, MA, 1984, p. 588.
15. E. Kamijo, E. M. Honda, M. Hiyuchi, H. Takeuchi, T. Tanimka, *Sumitomo Electr. Tech. Rev., 24,* 183 (1985).
16. R. A. Happe and A. J. Yeast, *Proceedings of the 14th* SAMPE *Symp.,* 1969, pp. 75–89.
17. M. D. Weisinger, ASM paper W70-5.2 (1970).

Magnesium Metal Matrix Composites

See Metal Matrix Composites, Magnesium

Manual Lay-Up

See Manufacturing

Manufacturing

Improving the technology of manufacturing is the greatest challenge today in the field of composites. The Suppliers of Advanced Composite Materials Association (SACMA) commissioned a study in 1987 to assess the personnel needs of the composite industry. This study summarized its findings with, "The overall picture which emerges from this survey is that the advanced composites industry is composed of medium-sized fabricators with an outstanding and unmet need for processing and fabrication specialists." The study further indicates that, "The emphasis is on materials development and characterization, although 100% believe more R & D focus should be placed on processing [1]."

This emphasis on materials development and characterization has been logical because the properties can best be optimized by concentrating on materials. And the achievement of optimized properties has often been the justification for the use of composites rather than other materials. Moreover, the performance requirements of the critical parts for which composites have been used have often justified a high cost for the part, which has spurred materials development. The designer, who has traditionally focused on properties and performance, has had little incentive to worry about other factors, such as manufacturing methods, that might require compromises that would lower the performance of the part.

In some critical applications, the choice between optimized properties and manufacturing method or cost will continue to be made in favor of properties. However, as composites are considered for a wider range of applications and as cost considerations become more important, the manufacturing method will be considered, and may become an equally important factor. This is not to suggest that properties will not be important. They will be; otherwise, why choose a composite material to begin with? However, the ultimate optimization of the properties may not be required, and cost reductions from manufacturing improvements may be possible.

The complex relationships between properties and manufacturing process can best be understood through the cooperative efforts of designers and manufacturing engineers. This cooperation, which has been called simultaneous or integrated manufacturing, is needed throughout industry, but especially in the field of composites.

As the SACMA study, cited earlier indicates, the concern about manufacturing is widespread. The challenges in manufacturing are to improve the process without sacrificing the properties below the point of design, or to understand the process-property relationships so well that the process can be optimized for either properties or economics, depending on which is the dominant factor for the particular application [2]. It may even be possible, through understanding of current processes and innovative new processes, to improve the properties over those available with current manufacturing methods.

To facilitate understanding of composite manufacturing, manufacturing methods that focus on properties will briefly be introduced. Then, manufacturing concepts that focus on costs will be highlighted. The current methods of composite manufacturing will then be discussed, and some areas that might be appropriate for improvement will be suggested.

The current processes for manufacturing composites are chiefly identified or characterized by the method by which the uncured composite material is placed into or onto a mold so that it can be shaped into the final part. In the simplest and most common of these methods, the material is placed into the mold manually (manual layup). In order to maintain continuity in explaining this process, the entire manual method, from initial layup to curing, will be discussed first, with sections for each of the major process steps involved. The other composite manufacturing methods, most of them more automated than the manual layup method, will then be covered. Methods for manufacturing thermoplastic composites will also be discussed.

Because thermoset composites dominate in actual use, most of the manufacturing methods that will be discussed are best suited to making thermoset composites. (The obvious exceptions are the processes that will be discussed in the section on thermoplastic manufacturing methods.) However, even some of the traditional thermoset methods are being adapted for use with thermoplastics, and those modifications will be pointed out when the thermoplastic methods are discussed.

One of the major expenses in manufacturing, yet one of the least understood, is mold or tool making. Some aspects of this important area will be discussed.

Quality control is such an important part of manufacturing that it, too, will be considered. Because of the potential for tremendous cost savings and the overall impact on manufacturing that can be achieved from proper quality control procedures and techniques, this area will be considered in some detail.

In addition to the traditional manufacturing methods for composites, manufacturing technologies have been borrowed from many other industries, such as plastics molding, extrusion, finishing, and even resin manufacturing; metal casting, forming, and finishing; textile fiber handling, cloth pattern cutting, and layup; and engineering disciplines as diverse as bridge building and laser cutting.

In summary, many companies have recognized the need for improvements in manufacturing. Therefore, composite manufacturing is changing rapidly, and a premium is being placed on the individual who can innovate. In the field of composite manufacturing, insight and innovation are key elements to progress and success.

Manufacturing Focus on Improvement of Properties

In this section, manufacturing techniques that can be utilized to maximize properties will be discussed. These methods will, in many cases, be less than optimal from the manufacturing economics viewpoint, but will attempt to preserve or improve some particular property or properties of the final composite part. Not only is this task difficult from the manufacturing viewpoint, but the designer must also realize that the optimization of one property may compromise others. This complex interrelationship of properties should be fully discussed by the designer and the manufacturing engineer.

The logical place to begin in any attempt to improve properties is with the raw materials. Excluding the cases of synergistic behavior between materials (where the performance of two materials is enhanced by their mutual interaction), most material properties are degraded to some extent by the presence of other materials. For example, the presence of the resin matrix decreases the strength and modulus of the fibers (as we commonly measure them). The objective in making composites is, therefore, to minimize this degradation while utilizing the desired properties to the greatest extent possible. Much time and effort are expended in achieving compatibility between materials, which generally minimizes the adverse relationship. The task of manufacturing is to ensure, as far as possible, that the efforts to improve compatibility between the components are preserved and that, of course, the properties of the individual components are manifest.

The most obvious case of maintaining the properties of a component is with the fiber. Because the properties of the fiber are optimized at long fiber lengths and in specific orientations, the manufacturing method should not decrease the fiber length or disturb the orientation. These restrictions severely decrease the manufacturing options available and are the primary reason why so many of the advanced composite structures that require the ultimate in property optimization are made by hand. It is by hand layup that the fiber length and correct orientation can best be maintained in many parts. (If fiber length and orientation were not problems, some automated molding process such as injection molding could probably be used.)

Another major restriction on the manufacturing method to be chosen is that the fibers must be well wetted and thoroughly mixed with the matrix. This requirement for good mixing and wetting has given rise to the widespread use of prepreg material. (Other factors, such as convenience and environmental safety, have also contributed to the use of prepregs.) In cases where the properties are not as critical, such as with fiberglass-reinforced plastics (FRP), the use of prepreg is less common and most manufacturing is done wet. In some methods, such as filament winding, the choice of prepreg or wet is controversial or, perhaps, is based more on experience and convenience.

The presence of voids has a negative effect on the properties of the composite. Chiefly because of this problem, cures involving vacuum bagging of the part with subsequent high pressure have become standard in many applications. (The need to decrease the resin content has also dictated the use of bagging and high pressure cures.) The need to bag many layups has added significantly to the cost of the manufacturing process. Not only is the time required to do the bagging considerable, but the material costs and the costs of parts for which the bagging has failed are high. The high pressure, at the temperatures required to effect the cure, is usually obtained with an autoclave. For many parts, especially those that are very large, the need to cure the part in an autoclave has had major implications for the manufacturing method. For instance, unless the manufacturer has an extremely large (and costly) autoclave, the size of the part is often severely limited. Moreover, the manufacturing cycle is lengthened considerably because of the autoclave cycle.

The need to optimize properties has also precluded, to a large extent, the use of fillers in the resin. Fillers have been a method of reducing costs in plastics for many years, and have also been a method for imparting some specific properties, such as flame retardance and color. Because fillers usually decrease the strength of the composite, they are usually avoided. Moreover, the fillers add weight without adding appreciably to the properties most desired in the composites.

Finally, optimization of properties extends beyond the individual part to the entire structure. Therefore, manufacturing methods must allow for the integration of separate parts into a whole structure that is also optimized for properties. This requirement may mean that parts have difficult-to-manufacture shapes or special surfaces, or must be made of or joined by special materials. All of these considerations can dictate a less than optimum (for economics) manufacturing method.

Manufacturing Focus on Costs

When the focus of the manufacturing method is on economics rather than totally on product performance, several manufacturing methods become available. It should be remembered, however, that even though the economics of manufacturing will be the focus, some minimum performance specifications must be met, and those specifications are likely to be high, since composites were chosen as the materials in the first place. (By the same token, some consideration is surely given to economics in the case where the focus is on performance. What is being discussed is the ability to relax the performance specification slightly to allow for improved economics.)

The reverse of most of the points identified in the preceding section can be considered if the focus is on economics. For instance, raw material specifications can be relaxed to some extent. This is not to imply that off-spec materials can be used, but rather that the materials need not be the highest modulus or the highest strength grade available. The cost of the less stringent specification is generally lower. We may also find that the raw material suppliers may be able to use less costly manufacturing processes for the materials, even though the resulting material has slightly lower properties. The use of fillers or other cost-saving materials may be possible if lower specifications are allowed. Fillers have long been used in fiber-reinforced polyester. In fact, the material known as low-profile sheet molding compound (SMC) uses fillers to achieve the low-profile characteristic.

Fiber placement and fiber wetout can be automated if economics is the focus. In some cases the automation preserves many of the key characteristics that were present in hand layup. For instance, automated prepreg laying machines can approximate hand layup patterns if the shapes are not too complicated. Filament winding and pultrusion are automated fiber laying methods that are widely used for composite parts where certain shape restraints are met. A manufacturing technique called resin transfer molding (RTM) is now being used more widely. This technique can be highly automated and yet preserve the fiber orientation and laydown critical for good properties. It involves taking a fiber preform of the part, placing it in a mold, and then impregnating it with wet resin.

Various molding methods have been developed for thermoset materials, and several of them have good applicability to composites. Compression molding or matched-die molding has been used with FRP for many years and is now gaining support for fairly small advanced composite parts as well. The major problem with this method is that the fibers must be short enough to flow with the resin in the mold. Careful fiber placement when the mold is being filled minimizes this requirement, but the maximum fiber length is still usually limited to 10–13 cm. Another problem is that as the parts become larger, the mold and press requirements also increase dramatically. However, parts as large as sports car bodies have been successfully molded for many years using this method.

Another method for reducing costs would be injection molding. However, this method has several major drawbacks, which have not, as yet, been properly addressed. First, with only rare exceptions, the method is applicable to thermoplastic materials only. Second, the fiber length is restricted to about the size of the opening into the mold cavity, which is usually much shorter than is needed for good mechanical properties in the part. Third, composite parts are often large, and injection molding is best suited to small parts.

The need to bag, apply a vacuum, and cure in an autoclave in order to obtain the correct resin content and to reduce voids can best be met by using one of the pressure molding methods. These methods also permit very complicated shapes to be formed, at least within the limitations of fiber length that have already been discussed. The problem of complex shapes might also be addressed by developing standard composite structural components, such as I-beams, C-channels, or hat sections, that could be warehoused as separate components. These components would then be integrated into assemblies as required. If these parts were thermoplastic, they might be molded or formed before their incorporation into the finished part.

The various property and economic tradeoffs for each of the major manufacturing methods are unique. So that these tradeoffs can be better appreciated, each of the major methods will be discussed independently with a focus on the property versus economics issues.

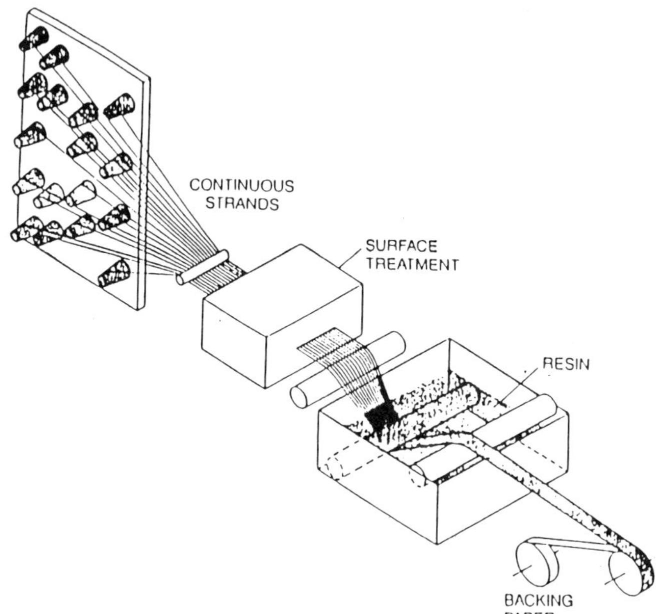

FIGURE 2 **Prepreg manufacturing.**

Manual Fiber Placement

The simplest technique, and probably the first used to make a modern composite structure, utilizes manual placement of the fibers and is called *layup, layup molding, wet layup*, or, less often, *contact liminating*. In this method, fabric or mat is saturated with liquid resin, and the layup is made by building layer upon layer to obtain the desired thickness. The impregnation of the layers is done "in process," that is, at the time the material is laid into the mold. This method is used most extensively with polyester and fiberglass, although some epoxy–fiberglass composite parts are also laid up wet (Fig. 1).

A somewhat superior product can be made, with less resin and fiber handling difficulty, by using a reinforcement that has been preimpregnated with resin and then cured slightly to increase the viscosity. This material is called *prepreg*. Normally the prepreg is made at a facility that is dedicated to its manufacture and by a method that allows careful control of the resin and fiber contents (ratio), then shipped to the site of composite manufacture (Fig. 2). Prepregs are used in applications in which the performance of the part is critical. Almost all of the resins used in composites are available in prepreg form.

In both methods, wet and prepreg, the layers are placed onto a shaped surface (the mold) by hand. Some pressure is normally applied by hand rolling, wiping with a squeegee, or sometimes vacuum bagging to remove trapped air and give better uniformity. Curing can be done at room temperature or at elevated temperatures.

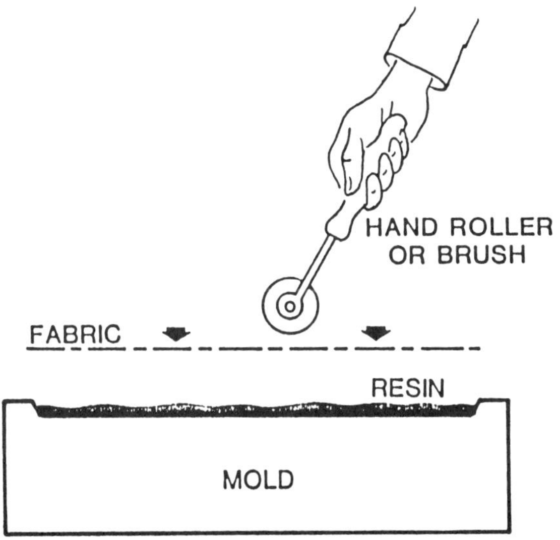

FIGURE 1 Hand layup.

Wet Layup Method

The method of laying the dry reinforcement (most often a fabric or a mat) into the mold and then applying the resin is the oldest and, perhaps, the most common means of wet layup. The wet composite is rolled by hand to evenly distribute the resin and to remove air pockets. Another layer of reinforcement is laid on top. Then more catalyzed resin is poured, brushed, or sprayed over the reinforcement. This sequence is repeated until the desired thickness is reached. The layered structure is then allowed to harden (cure) (Fig. 3).

This method is conceptually simple, does not require special handling of wet fabrics, and allows the resin to be applied only in the mold, thus helping to maintain a neat surrounding area. However, variances in resin viscosity (which are inherent in the gradual curing of precatalyzed resins) cause problems in getting good wetout (if too thick) or in having resin runoff (if too thin). Part shape can also cause difficulties in getting proper wetout.

Better uniformity is usually possible if the reinforcement is impregnated with the resin before being laid into

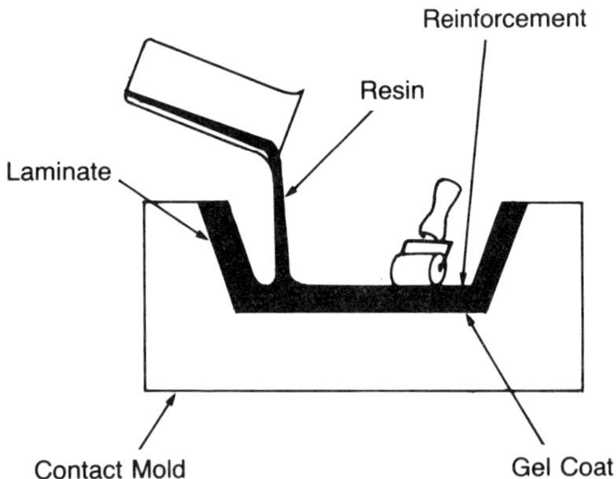

FIGURE 3 Wet layup.

the mold. The dry fabric and resin can be weighed to obtain specific fiber–resin ratios. The weighed resin can be rolled or squeegeed into the fabric more uniformly on a flat surface than on a tool because the areas of excess resin can be seen more easily. Costs are better controlled on large structures by this method of prewetting the reinforcement because the amount of excess resin required to wet out the fabric is reduced. This method also helps prevent the formation of resin-rich and resin-poor areas caused by vertical drainage. On a male mold, runoff of excess resin will flow off the layup, so that wetting the glass with excess resin while on the mold is not as critical a problem. However, this method is rarely used because of the difficulty of handling a wet sheet of reinforcement.

To prevent the composite from sticking to the mold, a *mold release* or *parting agent* is first applied to the mold. This mold release can be silicone, polyvinylalcohol (PVA), fluorocarbon, or, in some cases, a plastic film.

For many commercial applications, a layer of catalyzed resin is applied to the mold and allowed to cure to the gel (tacky) state before the reinforcement (either dry or saturated) is applied. This resin layer is called the *gel coat* and forms a protective surface layer through which fibrous reinforcements do not penetrate. Especially formulated gel-coat resins are used to improve flexibility, blister resistance, stain resistance, weatherability, and toughness.

Another approach to getting improved surface characteristics is to start by wetting out a fine-weave fabric (such as 121-style fiberglass) directly on the released mold, followed by thicker woven reinforcements. This method eliminates the resin-rich surface, which can crack and craze, especially if the structure is subjected to flexural stresses. In either approach, finer-weave fabrics are usually used near the surface of the part to prevent transfer of the weave pattern to the surface, and because air and voids can be removed more easily from these finer materials.

When mat is used instead of fabric, the mat is normally wetted on the mold to prevent unraveling and distorting during the transfer process.

Although it is not specifically required, many manufacturers using the wet layup method obtain improved parts by using a vacuum bagging method. In this method, which is described in detail in the section on vacuum bagging, a vacuum bag is placed over the layup and then sealed to the mold around the edges. A vacuum is drawn inside the bag, removing the bubbles in the part and compressing the layup to get good wetout and definition against the mold.

Parts are removed by manually pulling them from the mold. To assist in the removal, flat wooden, plastic, or metal wedges can be inserted between the part and the mold. Some manufacturers also blow low pressure air into this gap to "lift" the part from the mold, or put air ports into the sides of the mold to allow air to be introduced. Mechanical assistance is sometimes needed if the part is large.

MOLDS FOR WET LAYUP. Molds can be made of almost any material that will hold its shape and can be male (plug) or female (cavity) types (Fig. 4). The side of the part that is to be smooth and glossy should be placed against the mold. Therefore, a male mold would be used to lay up a swimming pool, since the inside of the pool must be smooth, and a female mold would be used to lay up a boat, because the outside must be smooth and glossy.

Molds made of wood, plaster, plastics, composites, and metal are common, since pressures are low and little strength is required. Molds made of composites (polyester and fiberglass) are most often used for low volume orders, since they do not hold up well to repeated use. Molds for long runs are usually made of fiberglass–epoxy or metal and would be considered permanent molds. Composite molds are especially favored because they can be formed against smooth surfaces and therefore do not require further finishing. In addition, there are no corrosion problems, as there are with some metal molds. Redesigning is usually simpler with composite tooling, since the surfaces can be recast or relaminated more easily to comply with engineering changes. The method of making a simple reinforced plastic mold is to use the model of the part as the original mold, against which the plastic is laid up (assuming that the part lends itself to serving as a model and does not have complications such as undercuts).

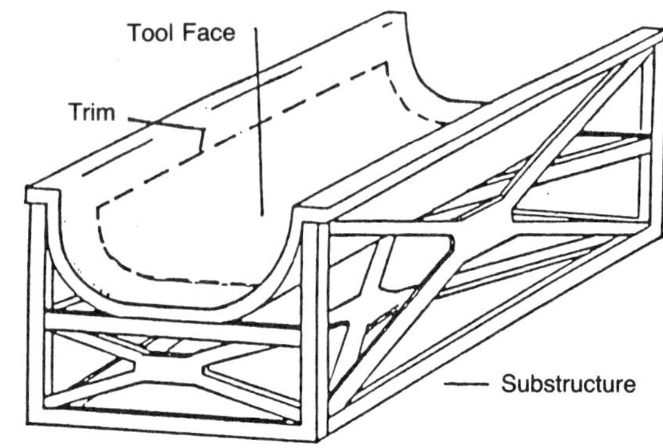

FIGURE 4 Female mold.

CURING OF WET LAYUP. Since curing is usually done at room temperature, a promoter is often added to the resin to speed up the reaction. Care must be taken to ensure that the promoter and the initiator (catalyst) are never mixed together directly. They should always be mixed into the resin in separate steps. External heating with infrared lamps or hot air blowers is sometimes used to speed the curing process.

ADVANTAGES AND DISADVANTAGES OF WET LAYUP. The advantages of wet layup are:

- Tooling can consist of any material that will hold its shape under minimal pressure.
- Tooling can be changed easily during experimental phases or to accommodate engineering redesign.
- Investment in pressure devices such as a press, autoclave, or vacuum pump is not required, although a vacuum pump is often used with wet epoxy parts and some polyester parts.
- Curing ovens are not needed.
- Semiskilled workers can be readily trained.

The limitations of wet layup are:

- Only addition-type cross-linking resins can be used, since condensation types require some form of pressure to avoid porous, poorly laminated structures.
- Product uniformity, both within a single part and from part to part, is difficult to maintain. Because of the inability to compact laminate with any pressure, the resin content is often quite high.
- Voids are common.
- Physical properties are low in comparison with those produced by other composite manufacturing methods.
- Tight-weave fabrics are difficult to saturate with high viscosity resins, resulting in low strength.
- Draining from vertical walls can be a problem, creating puddles near the base and resin-poor areas in the wall, although most resins have the appropriate viscosity to prevent this from occurring.
- High shrinkage from resin-rich areas occurs.
- There is only one finished surface.

Prepreg Method

The prepreg method can, in some regards, be viewed as an extension of the wet layup methods described above—prewetting outside the mold and then laying up the composite. In prepregging, the fibers are usually arranged in a unidirectional tape or a woven fabric. The fibers are impregnated with initiated resin, partially cured, and then rolled up for shipment. As a result of crimping and mechanical damage from the weaving process, woven prepregs are not as strong as unidirectional tapes. (Prepreg roving is also available, but it is not used in the manual layup method and will be discussed in a later section.)

In the prepreg method, the prepreg, which is supplied in rolls of convenient widths (typically 30–60 cm wide, but they can be as narrow as 8 cm and as wide as 200 cm), is normally cut to fit into the mold and laid up layer by layer until the desired thickness is achieved.

Prepregs used for manual layups are leathery and should ideally have a slight tackiness (*tack*) so that the layers will not slide over each other during layup and will stick or maintain their location in the mold. The prepreg should also be conformable to the mold, so that complex shapes can be made. This ability to conform is called *drape*. Drape and tack are often associated with each other and are dependent on the resin.

Fabric prepregs are usually used on complex contours because of their ability to drape and conform to the mold. The weave of the fabric can be a significant factor in its ability to drape. Unidirectional tape tries to follow a geodesic path on a contoured tool and is often difficult to use because it will not drape over complex contours, and will leave gaps and overlaps in severe cases.

Reactive thermoplastic resins are often stiff at room temperature and have almost no tack or drape. Therefore, they may need to be melted slightly with a hot-air blower to assist in tacking the layers together.

The resin content of the prepreg influences the tack and drape as well as the final strength of the laminate (although the laminate resin content is controlled during the resin fabrication step). In many prepreg systems, the resin content of the prepreg is higher than is desired in the finished part. This improves tack and drape, but the excess resin must be removed at some point in the manufacturing process. The removal of this excess resin assists in the removal of entrapped air and volatiles, which will flow out with the excess resin. It is essential to remove the volatiles and entrapped air, since voids within a laminate have a negative effect on the interlaminar shear strength. As a guide, the interlaminar shear strength is reduced by about 7% for each 1% of voids present up to a maximum of about 4% voids. A reasonable goal for void content in the finished laminate is 0.5% or less. The method for removing this excess resin is outlined in the section on vacuum bag assemblies.

Traditionally, the resin content of the prepreg is given as a weight percent, whereas the resin content of the finished part is given as a volume percent. The explanation for this is simply that when the prepreg is made, the resin weight percent is easy to measure and to control. However, for finished laminates, the resin volume percent is preferred, since it is related directly to the mechanical properties. Historically the trend has been towards lower resin content so that the specific strength of the composite is increased. The consistent removal of large excess amounts of resin has become a costly problem, so prepregs are now made with near-net resin contents. These prepreg materials are generally made using a hot-melt impregnation method that minimizes the volatiles remaining in the prepreg. Voids from entrapped air are minimized by debulking the material (in a vacuum or autoclave) after laying up 3 to 10 plies of material. Most laminates today are about 60% fiber volume fraction.

Because the resin has already been initiated when the

prepreg is made, prepregs have a limited shelf life before they turn into a dry and boardlike material that is difficult to use. The shelf life is usually several days to weeks at room temperature, but it can be extended by keeping the prepreg cold. The "out time," that is, the time out of the freezer, is recorded so that an estimate of the remaining useful life of the prepreg can be made. Standard tests are available to determine the percent of curing that has occurred in the prepreg and, therefore, the approximate shelf life remaining. However, many prepreg resins require high activation energies (heat) to react and cure. If parts are laid up in a mold before losing tack (from residual solvents), their life at room temperature is often very long. Upon heating, the resin will flow and cure to its full cure state, even though it appeared to be cured before it was heated.

In unidirectional prepreg, the strength in the cross-fiber direction is essentially the strength of the resin only. Therefore, to get strength in all directions, the prepreg layers are often oriented in different directions. Fabric prepregs (which may be directional too) are also laminated in different orientations to improve properties. However, as a result of crimping and mechanical damage during the weaving process, composites made of woven reinforcements are not as strong as unidirectional tapes that are cross-plied.

A ply layup pattern could be, for instance, 0° (the fiber direction), 90°, +45°, −45°, −45°, +45°, 90°, 0°. The warp direction is generally considered 0° in a fabric. Caution should be taken to ensure that the plies sequence is balanced directionally about the center so that the laminate does not twist or warp. The prepreg layup is much more precise than the wet layup method.

ADVANTAGES AND DISADVANTAGES OF PREPREG METHODS. Advantages include:

- The resin–initiator (or hardener) ratio is more accurately controlled during the premixing operations.
- Resin distribution per unit area is strictly controlled during tape manufacture. This aids greatly in providing good resin distribution in the final part.
- Health and safety problems associated with liquid resins or solvents are largely eliminated, since these are handled by the prepreg manufacturer. This manufacturer is often better prepared to deal with them, since he is accustomed to handling liquid resins in large volumes.
- The problems of poor efficiency and output can be reduced by using automated machinery for some parts.
- It provides better part definition, higher fiber content, and better consolidation than wet layup.

Disadvantages include:

- It is slow and labor-intensive compared with automated methods.
- There is a potentially high reject rate because of faulty bagging procedures.
- There is difficulty in bagging complex shapes.
- Expensive curing equipment (autoclaves) is needed.
- The inside surface is not as good as with matched-die molding.
- Long cure cycles are needed compared with matched-die molding.

Vacuum Bagging

The application of a vacuum to assist in compressing the plies together (debulking) has proven to be valuable in wet layups and necessary in prepreg layups. The vacuum provides the dual advantage of pressing the layers together and simultaneously withdrawing the excess volatiles. These volatiles could be residual solvent, low molecular weight resin components, absorbed moisture, or trapped air. The method of applying the vacuum that has been developed for composites allows the volatiles to escape freely and also permits good debulking. Further debulking and curing is occasionally provided by an autoclave, which is discussed in a later section.

VACUUM BAG ASSEMBLIES. The procedure for building up such an assembly is as follows (Fig. 5).

1. Prepare the mold by coating it with an appropriate mold release.

2. Remove prepreg materials from the freezer and bring them to room temperature before opening the protective bag to prevent contamination from water condensation.

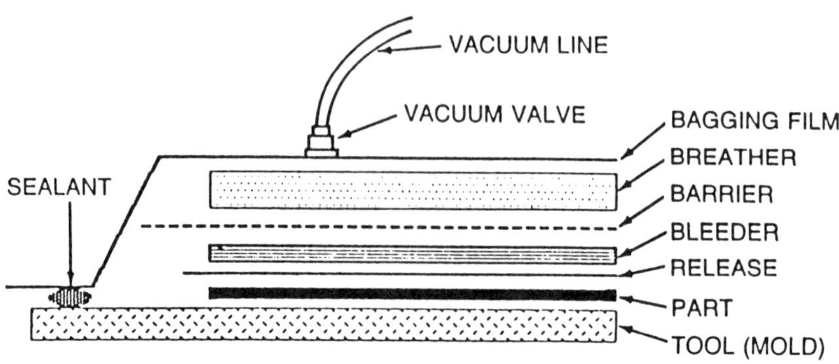

FIGURE 5 Vacuum bag assembly.

3. Build up the part by placing layers of prepreg on top of each other in the prescribed pattern and to the prescribed thickness. This buildup can either be done manually or be automated. The buildup may also involve inclusion of any inserts, ribs, or other structural members and the additional prepreg layers necessary to serve as anchoring and support for these inclusions.

4. Place the release film or peel ply material over the part. The release material is generally porous to permit excess resin to flow through, while leaving an impression on the part suitable for secondary bonding without further surface preparation.

5. The *bleeder* material is a mat that absorbs the excess resin. Common bleeder materials are polyester felt or mat, fiberglass (which may be coated with Teflon or mold release), and cotton. The important characteristic is that the bleeder material should have good absorption qualities and not compact under the pressure of an autoclave. The resin content of the final part is dependent on the ability of the bleeder material to absorb a measured amount of resin, as well as on temperature, pressure, viscosity, and amount of bleeder. The bleeder should be conformable so that it does not cause wrinkles in the assembly when it is placed under vacuum.

6. The *barrier* is a layer that limits the upward movement of the resin and prevents it from reaching or clogging the breather and vacuum lines. The barrier material must not, therefore, allow the passage of resin, but must allow air to pass. In many bagging assemblies, this material is omitted. In autoclave processing, such an omission may lead to resin-plugged vacuum lines and pumps. When the barrier is omitted in wet layup applications, resin traps are often used.

7. The *breather* material acts as a distributor for air, escaping volatiles, and gases, as well as a buffer between bag wrinkles and the part surfaces. Breather layers should be highly porous and must not collapse under vacuum, temperature, or pressure. Typical materials are fiberglass, polyester felt, and cotton, although the bleeder materials are sometimes used. For near-net or phenolic resin systems, the use of breather is sometimes omitted when the thickness of the bleeder is great and experience has shown that a separate breather system is not required.

8. Vacuum bag *sealing tape* or *sealant* is applied around the mold. This sticky material makes an airtight seal with the bagging film.

9. Thermocouples or other monitoring devices are inserted into the assembly. To ensure that the places where the thermocouple wires enter the assembly will be airtight, the outer insulation is often scraped off the thermocouple wire (especially if it has cloth insulation). The wires are then embedded in the sealant material. Thermocouples are usually placed in the outer area of a part that will be trimmed off.

10. The *vacuum bag* is then laid over the assembly and a *vacuum port* is attached through the bag, providing a fitting for attaching the vacuum hose. The bag is then pressed against the sealant to give an airtight seal all around. Vacuum bags can be made of any plastic film material that is strong enough to hold a vacuum, fit and conform to the assembly, and withstand the cure temperature without degrading. In practice, the most common material is nylon or a coextruded nylon that has been heat-stabilized.

The most common problems associated with vacuum bag processing are material quality and bag leaks. Vacuum bagging materials are often procured under vague material, construction, and performance requirements. Poor industry standards often lead to products with varying performance. For instance, uncontrolled or unregulated vacuum bag quality is the leading cause of bag failure, typified by film degradation or decomposition above 82°C. Bag leaks most often occur at the sealant–vacuum bag interface. The second most common cause of failure is handling damage to the nylon film before cure. Nylon film is hygroscopic and subject to moisture change in relation to the relative humidity of its environment. Film becomes dry and "brittle" when its moisture content falls below 2%. Dry film is susceptible to damage/cracking when excessively handled or abused (wadded).

Bridging is also a common problem in vacuum bag molding. Bridging occurs when, because of the shape of the part, the vacuum bagging materials are not pressed against all of the part surface. Some areas, therefore, are not properly pressed. (For instance, for a part with a narrow, deep channel, bagging materials may form a material bridge across the top of the channel, preventing the transfer of pressure to the bottom of the channel.) If the nylon bag itself bridges over a narrow gap, the film may be stretched beyond its limits during cure and may burst. Good bagging techniques, such as intentionally putting pleats into the bag during assembly, allow enough excess film to be present to provide bag conformity against all surfaces. Another technique is the placement of preformed rubber pads under the bag in corners and channels to fill areas where bridging is possible. This technique is particularly useful with vacuum bags made of rubber (Fig. 6).

Resin Leakage

Resin can bleed out of the laminate by either an edge bleed path or a face bleed path (Fig. 7). Edge bleeding of the laminate occurs when the resin flows in the plane of the laminate and exits along the edges. Problems can occur with edge bleeding in that the resin has a long pathway and may not reach the edge before gelation. This is overcome by judicious application of even pressure to hydraulically force a measured amount of resin from the part. In autoclave processing, edge bleeding is controlled by using a nonporous release material on the bag side of the part and/or high cure pressures (over 700 kPa), since these pressures tend to limit the alternative mode of resin escape—face bleeding.

Face bleeding is most effective in oven processing, where high pressures are not present. In this case excess resin coupled with vacuum pressure only or low autoclave pressure is used. However, most processing of composites relies on a combination of both edge and

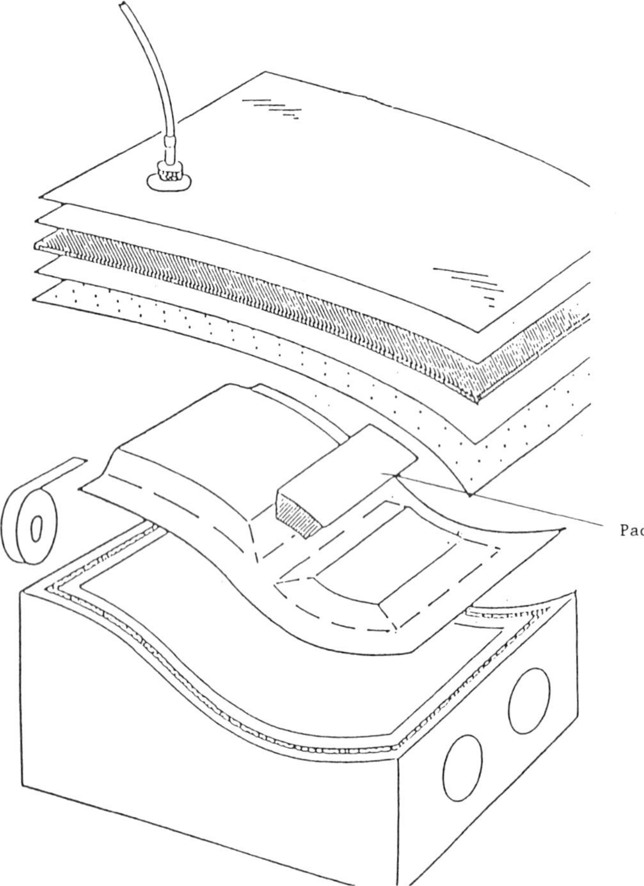

FIGURE 6 Use of rubber pad to prevent bridging.

face bleeding to control resin content and provide for unrestricted resin flow.

Curing

The prepreg method generally uses both a vacuum and an autoclave to assist in consolidating and curing the part.

The molds used in an autoclave are usually made of metal or composite and are more permanent than the typical layup mold; they must, of course, withstand the

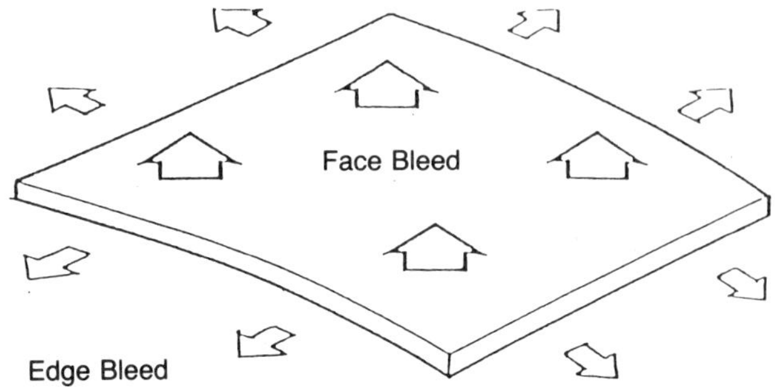

FIGURE 7 Alternative resin bleed paths.

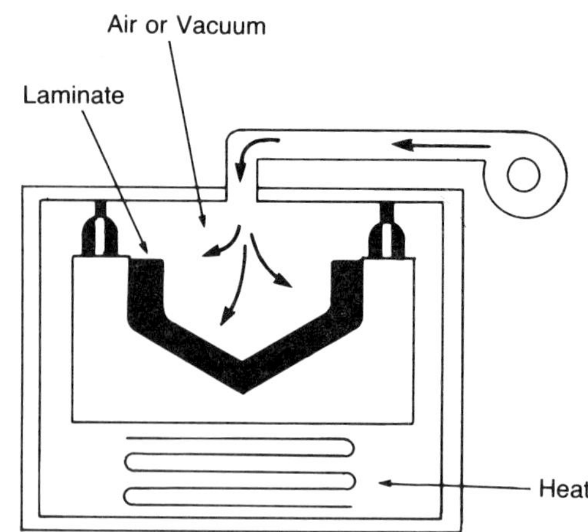

FIGURE 8 Autoclave curing.

forces of the autoclave. The autoclave provides pressure beyond that available with vacuum only and therefore gives greater compression and void elimination.

Autoclaves allow the simultaneous imposition of heat and pressure (and vacuum if a vacuum line is led directly to the part) (Fig. 8). The major difficulty with autoclaves is their high cost, which results because autoclaves are pressure vessels and must, therefore, pass stringent pressure code regulations. However, because many parts can be cured at one time in an average autoclave, labor and cure costs on a per-part basis can be quite inexpensive.

Autoclaves can be purchased as large as 8 meters in diameter and 30 meters in length. The concerns in purchasing and operating autoclaves are the uniformity of the heat flow (to avoid thermal stresses), the integrity of the closure, ensuring that the insulation is good, obtaining good and easy-to-operate controls, and ensuring that fires do not occur as a result of high temperature and pressure. (An inert gas is usually put into the autoclave to reduce the fire danger.)

Heating can be electrical, by gas, by circulating hot oil, or by steam (if the temperature requirements are not too high).

Autoclaves can be used to bond assemblies together. In this application, the principal use of the pressure is not to debulk but to ensure that the parts are kept in intimate contact.

The type, composition, and size of the tooling can affect the cure, as the heat-up rates of the tooling can vary widely. In some cases, the tooling is independently heated, which can reduce the problem of variable heating and can accelerate the cure by getting the entire assembly to cure temperature faster.

Special tooling may be required in autoclave curing to prevent bag failure, collapse of parts, or crushing of honeycomb details.

The net result is that curing in an autoclave produces a superior part than can be obtained with wet layup. Therefore, an autoclave is used extensively for making high-performance aerospace parts. For very complex parts, in fact, it is the principal method employed.

Automated Fiber Placement

Automated Fiber Laying

Some new machines and concepts have been developed to speed up the slow and labor-intensive layup process. Some of the most important of these new machines are automated tape laydown machines. These machines typically have an overhead gantry to facilitate movement of the head across large molds. The head has the capability of moving in several directions so that both simple and reasonably gentle compound contours can be followed at angles up to ±30° from vertical. The ply pattern is preprogrammed into the machine, then the head is loaded with a roll of prepreg tape (6 cm wide is typical). The machine lays the tape onto the mold in the pattern that has been programmed and does all cutting and trimming automatically.

Because the equipment uses unidirectional tape, optimum results are achieved if the machine can be programmed to follow a geodesic path for all passes. On compound contours, this is often not possible, and gaps or overlaps result. This can complicate programming and design of tape-laminated parts. To prevent overlaps and variation in thickness, small gaps are often programmed to be left in between adjacent tape paths.

The advantages in time and labor savings are obvious. Other reported advantages include more uniform parts as a result of more consistent tape laydown pressures and more precise positioning of the tape rows, nonstressing of the fibers during laydown, and the ability to easily repeat with other identical parts. Disadvantages include the need to program the machine, the inability to do some complex parts, the long production time for small parts (because of the tape-head turnaround time), and the presence of a consistent gap between adjacent paths.

Systems to program tape-laying machines by vision learning, by computer simulation, or by passing the head over the mold are now being explored, and some success has been demonstrated. These methods hold great promise for reducing the time involved in programming automated tape-laying machines.

Filament Winding

In the basic filament winding process, a continuous tape of resin-impregnated fibers is wrapped over a mandrel to form the part (Fig. 9). Successive layers are added at the same or a different winding angle until the required thickness is reached. Either the mandrel or the application head can rotate to give the fiber coverage over the mandrel, although the rotating mandrel is far more common. The head, therefore, traverses longitudinally to give the coverage. In one sense, filament winding can be thought of as a type of automated laydown machine, although filament winding normally does not use prepreg material, but instead incorporates the impregnation of the fiber tows as part of the filament winding process. This is called wet filament winding or wet winding. Dry filament winding, which is less common but is gaining acceptance as materials are developed, uses preimpregnated tow (prepreg tow) as the winding medium. We will concentrate on the most common filament winding method—wet winding with a rotating mandrel.

Parts as small as 2.5 cm in diameter and as large as 600 cm in diameter are commonly made by filament winding. The only limitations on size are those dictated by the geometries of the winding machine and the limitations in mandrel size and weight.

RESINS FOR FILAMENT WINDING. Most standard composite resins (epoxy, polyester, phenolic, some imides, silicone, and thermoplastics) can be used for filament winding provided that certain specific requirements are met. Polyesters and epoxies are the most common. As with other methods, the resin should be

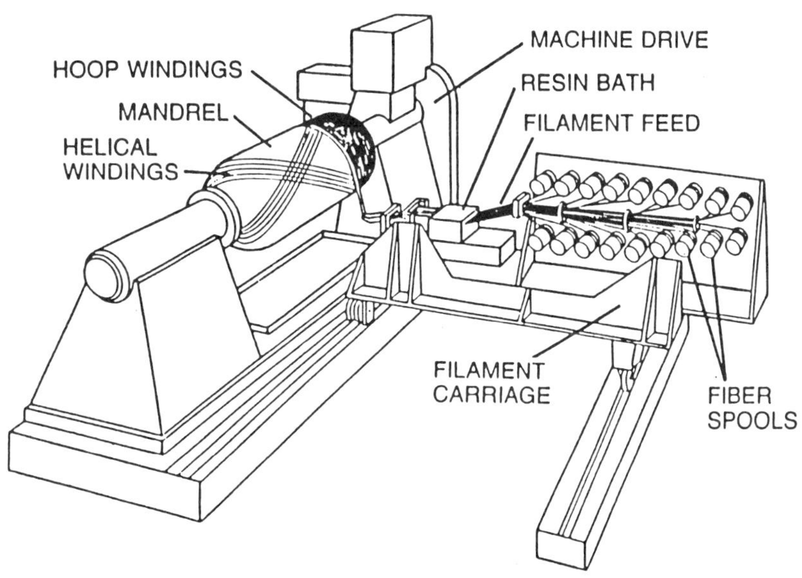

FIGURE 9 Filament winding.

low in volatile content (preferably 100% solids) so that voids within the structure can be as low as possible. The viscosity of the resin should be thick enough to coat well without dripping off the fibers or migrating onto the mandrel, yet not so thick that the fibers are poorly wetted or become fuzzy when passing through the resin bath. Resin viscosities in the 350–1500 centipoise range have been found to satisfy these requirements. Reactive and nonreactive diluents can be added to high viscosity resins, but may cause problems with physical properties (mostly due to void formation). In general, therefore, nonreactive diluents should be avoided. Solid, uncured resins can be used if a method for melting the resin without curing it prematurely can be found. A glue-gun approach in which only the tip of the resin is melted has been shown to be effective.

A pot life of several hours is necessary, and one of several days is preferred. Gelation must not occur before completion of the winding; otherwise, weakened structures could result.

Surface properties of dry resins (on prepreg tows) are critical to their successful use in filament winding. Sufficient tack is needed to prevent slipping of the bands as they are laid down if a nongeodesic path is used. (Geodesic paths are discussed in the section on the winding process.) Flow of the resin during curing is required to ensure that the layers are bonded together. The flow is a property of the B-staging of the resin. Improved tack and flow are obtained by heating the strand during winding. Despite these limitations and cautions, dry-resin filament winding has some benefits. Small shops do not need extensive resin preparation and handling facilities; close control over the consistency of the resin system is not required, as that is done by the prepreg manufacturer; and prepreg roving can be quality controlled before winding, potentially providing a more consistent part. On the other hand, prepreg tows as a raw material form are more expensive (on the order of twice the cost of wet-winding materials).

REINFORCEMENTS FOR FILAMENT WINDING. Most common continuous reinforcements for composites have been successfully used in filament winding. The most common are E-glass (for cost), S-glass (for strength), carbon (for strength with modulus), and aramids (for toughness and light weight). For best part properties, the fibers must be well collimated and laid without twisting. (The fibers wet out best and process better if they are procured from the manufacturer as untwisted fibers.)

The use of more than one type of reinforcement material may have some advantages in terms of cost and product performance. Some examples of this concept are a pressure vessel wound with graphite to give longitudinal stiffness, then overwound with fiberglass to resist the forces in the hoop direction, and a vessel wound with carbon for strength and modulus, then overwound with aramid to protect the vessel from impact damage during fabrication or use.

WINDING PROCESS. The most important part of successful filament winding is defining the relative speeds of the mandrel and the head. These two motions determine the wrapping angles and overlap and, therefore, the physical properties of the part. Two types of winding patterns are generally defined. The first is *hoop, circumferential,* or *radial* winding. In this mode, the fibers are wound perpendicular to the axis of the mandrel. This type of winding withstands hoop stresses best, as all of the fiber strength is oriented in the radial direction. During this type of winding, the lateral movement of the head is very small compared with the rotational movement of the mandrel.

The second type of winding pattern is called *helical* or *longitudinal.* This pattern withstands longitudinal forces better than hoop winding because the fibers have a force component in the axial direction. Smaller winding angles give more longitudinal strength, since the fibers have a greater longitudinal force component. A constant-angle helical path with zero slip is called the *geodesic path.* A geodesic path would be a straight-line path if the surface were converted to a flat two-dimensional pattern. Most filament-wound parts are programmed with a geodesic path.

The normal pattern is multicircuit; that is, after the first traverse, the fiber bands are not adjacent, and several circuits are required before the pattern repeats. The head normally speeds up and goes into a cross-feed motion perpendicular to the mandrel axis as it goes around the end, as the length of travel is quite large at that location. This motion also helps maintain the geodesic path. On some machines the head also rotates as it goes around the end, so that the feed is perpendicular to the surface of the vessel. Typical winding speeds are up to 100 m/min.

Polar winding is a special case of helical winding in which the fiber path passes tangent to the polar opening at one end of the vessel and tangent to the opposite side of the polar opening at the other end and forms a plane of symmetry through the part. A one-circuit path is inherent in this winding pattern, with the next path advancing the width of the tape. This is the simplest winding pattern, but it is limited to vessels with length-to-diameter ratios of 1.8 or less. This pattern is widely used to wind spherical shapes.

Tension control is also important in the manufacture of good parts by filament winding. The tension affects both resin content and void content. Typical tensions for filament winding are in the 0.2 to 7 kPa per end range. Tension is provided by eyes, drum-type brakes (sometimes magnetically controlled), scissor bars, and the drag through the resin bath.

A recent modification of the traditional winding process involves the use of multiple winding heads. By the use of two or more heads, an overlapping winding pattern (like a braid) can be achieved, which gives some improvement in product strength.

MANDRELS. For open-ended structures, the simplest mandrel is usually the best, and cylinders of cored or solid steel or aluminum are used. If a helical path is used, the domed ends of the part are cut off to remove the part from the mandrel. In some cases the design anticipates a special cut zone so that domed ends can be kept but the part can be removed from the mandrel. (The filament-wound body of the Starship airplane was of this type, with the cutting line just behind the pilot's cabin.)

Closed-end vessels require special mandrels that can be removed while keeping the ends intact. Collapsible metal (segmented) mandrels are used because they can be reused. However, these are difficult to remove if the end openings are small. Plasters have also been used, but these must be chipped out, and this process is also difficult with small end openings and, further, has the potential for damaging the composite vessel. Easily removable mandrels include sand with a water-soluble binder (generally polyvinyl alcohol, PVA), soluble salts, eutectic salts, and low melting alloys.

The inflatable mandrel has several benefits, which have led to increasing use of this method. When necessary, the mandrel can often be deflated and removed. In most applications, however, the inflatable mandrel can be left in as a liner for the composite material. Applications such as fuel tanks require this type of liner. Another benefit of the inflatable mandrel is the ability to add pressure as the winding progresses. This prevents the collapse of the inner fibers, which sometimes occurs as the outer wraps are wound onto the vessel. Increasing the internal pressure causes the inner wraps to stay in tension and not tend to buckle. The inflatable mandrel also allows the vessel to be cured with an internal pressure that is equal to the expected operating pressure of the part, thus eliminating possible stress cracking during initial pressurization.

CURING OF FILAMENT-WOUND PARTS. Autoclaves are occasionally used to cure filament-wound parts. The parts are left on the mandrel, and a mold (often in two halves) is placed over the part. The entire structure is then placed in an autoclave and subjected to a normal cure cycle.

Most applications do not require the high compaction that results from an autoclave cure, and the parts can simply be oven-cured. Care must be taken, however, that the effects of gravity are minimized during the cure, so the part is usually rotated while being cured.

Microwave and other nonoven cures have proven to be successful methods of curing filament-wound parts. These often have shorter cure cycles than either ovens or autoclaves.

ADVANTAGES AND DISADVANTAGES OF FILAMENT WINDING. Advantages include:

- It is applicable to parts of widely varying size.
- Parts with strength in several directions can be easily made.
- Filament winding has excellent material usage.
- Forming after winding and other techniques allow noncylindrical shapes to be made by filament winding.
- Flexible mandrels can be retained in the structure to serve as liners for tanks.
- Panels and fittings for reinforcement or attachment can be easily included during the winding process.
- Parts with high pressure ratings can be made.

Disadvantages or cautions include:

- Resin viscosity and pot life must be carefully chosen and monitored.
- Programming of the winding can be difficult.
- Not all shapes can reasonably be made by filament winding.
- Operational control of several key parameters (such as fiber tension) is important.

Pultrusion

In the pultrusion process, continuous reinforcement fibers are impregnated with resin and shaped by drawing through a die and are then cured (Fig. 10). This process is analogous to the extrusion of aluminum or thermoplastics (with the obvious exception that pultrusion incorporates fibers and involves thermoset resins in most cases). Pultrusion is a continuous processing method and therefore has great potential for high throughput. The major limitation of pultrusion, as with the extrusion processes, is that the cross section of the part normally must be constant, although both solid and hollow parts as well as many profiles can be made. Compliant dies that permit a change in thickness have been designed for special applications and permit some variation in cross section.

Two types of pultrusion dies are commonly used, fixed (with no movement) and floating (where one die segment floats and has pressure applied). The pressure can be applied by hydraulics, fire hoses, springs, or other methods. The use of fixed dies can generate tremendous hydraulic forces in the resin to impregnate and wet out fibers. Floating dies rarely generate more pressure in the resin than the pressure being applied to the die. Multiple dies are often used when the dies are heated and used to cure the part. Pultrusion dies can cost as much as $10,000.

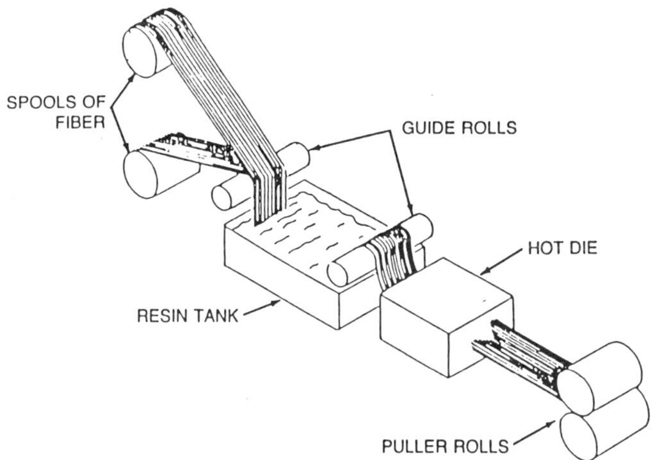

FIGURE 10 Pultrusion.

Pultrusion has a much higher material usage (95%) than layup (75%) and can therefore be considered more productive in terms of both time and material. A properly designed pultrusion die will maintain accurate resin content because of the fixed cross section. As long as the fiber volume passing through the die is held constant, as it is in normal production, excess resin will be squeezed out and will run back into the resin bath.

FIBERS AND RESIN CHOICES FOR PULTRUSION. Most of the normal continuous reinforcements for composites will work well with pultrusions. In act, the ability to use a wide variety of reinforcements and forms is an advantage of this method. Reinforcement forms used commercially include roving, mat, and fabric of fiberglass, carbon, and aramid.

The resin choices are also broad, but some limitations do exist. All resins need to cure quickly because of the continuous nature of the process and the relatively high speeds. Viscosities in the 0.5 Pa·s range are typical for pultrusion resins impregnated in-line.

Polyester is a major resin used in pultrusion, but care must be taken to ensure that the styrene content is in good control. Polyester is ideal because it shrinks slightly on curing and thus easily releases from a fixed or floating die. If the styrene content is too high, residual styrene can remain in the part and result in void areas; if it is too low, the cross-link density will not be achieved.

Epoxy has an inherent problem in pultrusion. Epoxy resins tend to stick to the die and other parts of the pultrusion equipment, causing process problems and surface defects on the parts. The modification of epoxy resins using release agents and viscosity modifiers is now beginning to be used in advanced composite manufacturing. Another method to prevent sticking utilizes a nonporous Teflon-coated cloth (Armalon) that is wrapped around the surface in contact with the die to prevent resin contamination of the die, and to reduce the friction required to pull the material through.

Imides have been used, but their high viscosity and condensation byproducts have limited their use.

Prepreg tows have been used for pultrusion, although their use is limited to special applications.

RESIN IMPREGNATION AND MATERIAL FORMING. The heart of the pultrusion process is the impregnation of the fibers with the resin and the shaping of the material into the desired shape. In most pultrusion processes, the dry fibers are pulled through a resin bath that has several mechanical rollers or other devices to assist in assuring that the fibers are well wetted. The wetted fibers are then pulled through a die or forming bushing. Alternatively, the dry fibers can be formed by passing them through a die, then passed into a cylindrical chamber made of perforated metal through which the resin can be forced to wet out the fibers. This method has the advantage that the fibers are in their final form when they are wetted. It is especially useful when the part to be made is hollow, as the fibers are formed around a mandrel and wetted in place. In either method, forming guides are often used to position the fibers for entrance into the die, especially if mats or woven or stitched materials are used.

PULTRUSION CURING. Three curing methods are traditionally associated with pultrusion. The most common of these methods is called the *tunnel oven method.* In this method, the part is usually gelled in the die and fully cured as the part travels through the oven after exiting the forming die. The length of the oven is determined by the line speed, the part dimensions, and the curing characteristics of the resin.

The second curing method is called the *split die method.* In this method, two female molds are brought up against the part as it exits the die. The line stops while the curing takes place, then continues when the curing is completed. This method is also called *pulforming.* With this system, nonuniform cross sections are possible.

The third system is called *die curing.* It involves the rapid curing of the part while it is still in the die. Curing in the die often will require use of multiple dies that start at a low temperature and increase across the dies, with a heat control zone in each die. This method is the easiest to control and achieve a steady-state process. Another method necessitates the use of microwave or rf curing so that the length of the die will not be excessive. Typical die lengths for this method are 15–60 cm. Dies are often made of ceramic or composite to allow the microwave or rf energy to reach the part. Ceramic dies are significantly more expensive and have longer fabrication lead times than metal. Composite dies do not wear very well and are less durable, requiring more frequent replacement. Microwave and rf energy are more difficult to regulate, especially if the part contains carbon fibers. The die curing system is especially useful for hollow parts that are formed on a mandrel. The mandrel extends from before the die, through the die, to just beyond the die. The parts are brought into the short curing section, and then, when the part is solidified, the part is pulled off the mandrel. Further curing can be done in a tunnel oven if necessary.

The curing process is the rate-determining step for the process. Typical speeds for pultrusion are 60–120 cm/min for parts that are 1–75 mm thick. Sizes of parts can vary from 2.5 cm to 5 m in diameter without serious limitations.

A cooling zone is generally placed before the die as a protection against premature gelation should the line stop and during start-up.

PULLERS AND CUTOFF EQUIPMENT. A large variety of pullers have been used in the pultrusion process. The most common variety is double clamp pullers, in which the part is clamped between pads and pulled forward. At the end of the pulling stroke, a second puller clamps the part and pulls it. This tandem pulling mechanism gives a continuous pulling action. Another puller has been adapted from the continuous pullers of thermoplastic extrusion processes. This puller employs double continuous belts through which the part is passed. When the belt is a cleated chain, the system is called a caterpillar puller.

Cutoffs are usually saws similar to those used for plastic extrusions such as PVC pipe. Both wet and dry saws are available. The toughness of composite reinforcements compared with that of thermoplastic resins necessitates the use of diamond or carbide saws. The blades must be frequently sharpened or changed, especially when aramid fibers are the reinforcement. The cutoff saw often travels with the part to prevent binding of the part in the saw or overheating if the part has stopped in the die or oven.

ADVANTAGES AND DISADVANTAGES OF PULTRUSION. Advantages include:

- It has a high material usage compared with layup.
- It has a high throughput rate.
- It can give high resin contents.

Some disadvantages of pultrusion are:

- Part cross sections must generally be uniform.
- Problems can arise when resin or fibers accumulate and build up at the die opening. This can increase the friction to the point that the equipment will be jammed or the fibers will break.
- When the dies are run resin-rich to account for fiber anomalies, strength is sacrificed. (Normally a good balance can be found without a significant decrease in strength.)
- Voids can be a problem if the dies are run with too much opening for the fiber volume.
- When quick curing systems are used, the mechanical properties are often sacrificed.

Matched-Die Molding

Three types of matched die molding will be discussed: preform molding, SMC molding, and BMC molding. These three methods all utilize the same type of high pressure molding equipment, but differ in the form of the material that is placed in the molds to form the part. The materials most commonly molded by this technique are fiberglass and either polyester or epoxy. The short fiber lengths generally preclude the use of this technique for high performance parts.

The equipment is a press (usually hydraulically driven) that is fitted with both male and female dies (hence the term matched-die molding). The dies are generally made of hard metal (such as tool steel) and can be highly polished and chrome-plated in order to get a fine finish. The pressures developed by the press can range up to several hundred thousand kg, which is useful for obtaining good part uniformity and compression of the voids that may develop. An alternative name for this method is *compression molding* (Fig. 11).

Compression molding can be used for both addition-type cross-linking and condensation cross-linking. When condensation polymers (such as phenolics) are molded, the condensate (usually water) must be allowed to escape to prevent gas pockets. Therefore, after the mold is closed, it is opened slightly for a few seconds to allow the gases formed by the heated condensate to escape. This process is called *degassing* or *breathing* the mold.

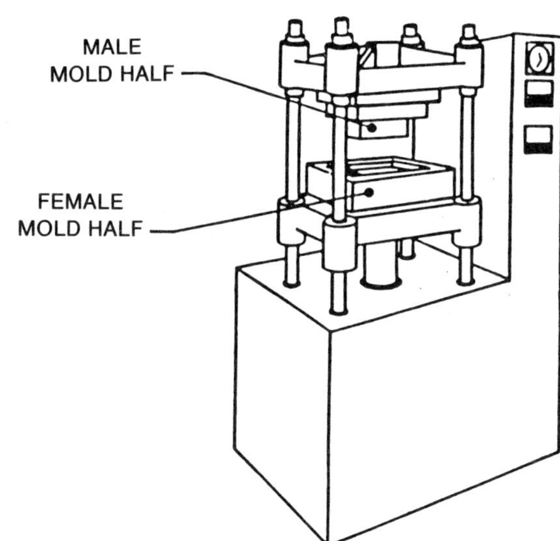

FIGURE 11 **Matched-die (compression) molding.**

PREFORM MOLDING. In preform molding, a dry mat of the reinforcing material is preformed to the approximate shape of the part and placed into the open mold. Resin is added to the preform, and the mold halves are then pressed together and heated to cure the part. During the process the resin flows, impregnating the preform, and becomes hard. The cured part is removed after the mold is opened (often with the assistance of knockout pins that are built into the mold). In a variation on the standard method, preimpregnated chopped fibers are blown onto the preform and then cured.

Because high pressures can be exerted upon the material to be molded, a higher ratio of glass to resin may be used than in layup, resulting in a stronger part (assuming that the fiber lengths are the same). The cure time in the mold depends on the temperature, the resin type, the part geometry, and the mold heating and cooling efficiency. Typically it varies from 2–20 minutes. The cure cycle can thus be very short, and a high production rate is possible.

Several methods are available for making preforms. In the *direct-fiber method,* chopped roving and a water-soluble binder to hold the fibers in place are blown against a shaped, perforated screen fixed to a turntable, which assists in getting fiber uniformity. This method is used to make preforms for very large items, such as boats. The *plenum method* employs a perforated preform screen inside a vacuum box (Fig. 12). Fibers are chopped and sprayed with binder, then sucked against the preform screen by the exhausting air. These preforms are for

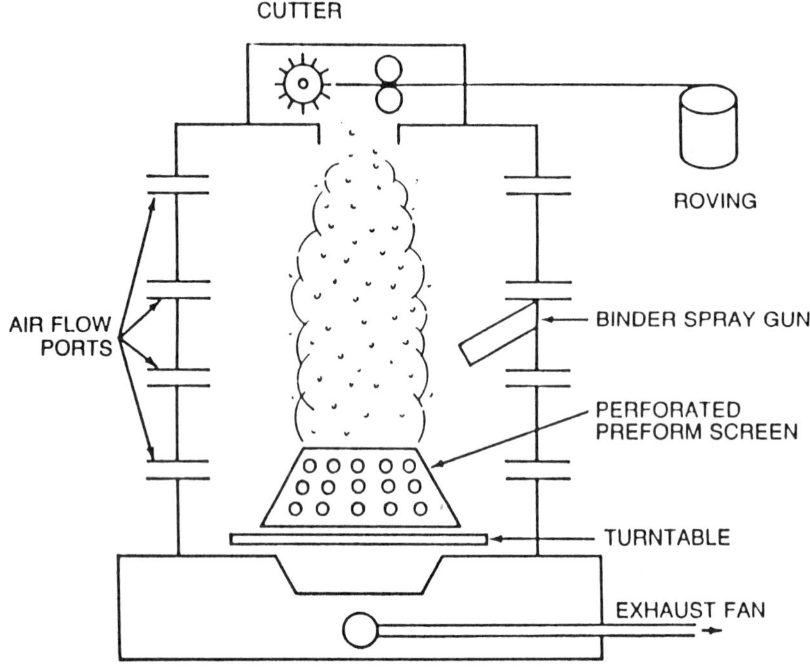

FIGURE 12 Plenum method for making preforms.

moderate-sized parts, such as chairs. Some preforms can be made simply by cutting the desired shape out of mat or cloth. When required, glass yarn can be used to stitch segments together. This is called the *cut pattern method* and is chiefly used for complex parts requiring varying thicknesses or for decorative patterns, which can be added by using special yarns.

SMC MOLDING. Sheet molding compound (SMC) is a sheet material that is made by chopping glass fibers (or swirling continuous fibers) onto a sheet of plastic film (usually polyethylene) on which a resin–initiator–filler mixture has been doctored (Fig. 13). Another film, which also has the resin mixture doctored onto it, is placed on top, and then the sandwich of resin mixture and chopped glass is passed between compaction rolls to wet the fibers and thoroughly mix the constituents. The material is then cured slightly (also called B-staging, aging, or maturing) and rolled up for shipment.

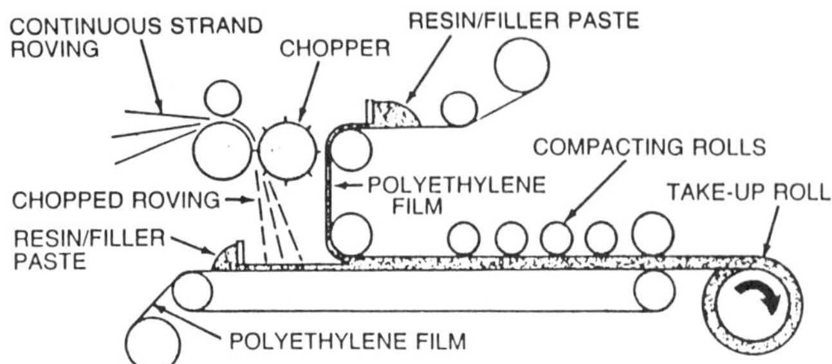

FIGURE 13 Sheet molding compound (SMC) machine.

A typical SMC incorporates about 30–50% fibers (generally 2–8 cm long), 25% resin (generally polyester), and 25–45% filler (generally clay, alumina, or calcium carbonate). Some care must be taken with SMC, as it has already had initiator added and therefore has a limited shelf life. SMC is, therefore, kept refrigerated until use. The styrene cross-linker is also quite volatile, so precautions should be taken against fire.

After aging, the SMC material has a leatherlike texture. When it is to be molded, it can be cut off the roll, the plastic film backing removed, and loaded into the compression-molding presses. Parts as large as car bodies can be made in this manner. SMC processing pressures may be as high as 5000 psi with temperatures in the range of 120–160°C and mold cycles of several minutes.

BMC MOLDING. Bulk molding compound (BMC) is a doughlike mixture of chopped fiberglass, resin, initiator, and fillers that has a composition similar to that of SMC. BMC, however, is generally mixed in bulk rather than as a sheet and is sold in a log or rope form. The resulting material is also called premix. The fiberglass content of BMC is generally 5–10% lower than that of SMC, and the fibers are generally less than 2.5 cm. Therefore, structures made from BMC are not as strong as they would be if made from SMC. The BMC material is often mixed at the molding location, but can also be purchased from commercial manufacturers who sell the material as a ready-to-mold premix.

BMC is molded by placing a weighed amount of the material (the charge) into the lower mold (Fig. 14). The molds are then closed, and pressures, temperatures, and cycles are much like those for SMC. Some special considerations must be given to the viscosity of BMC to ensure

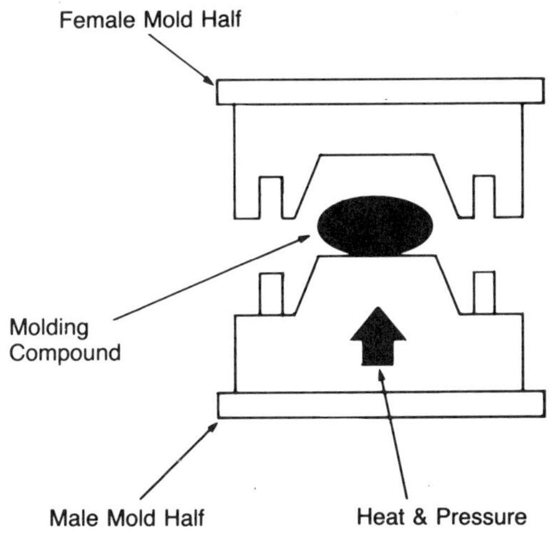

FIGURE 14 Bulk molding compound (BMC).

that all components of the mixture are evenly divided throughout the part. If the viscosity is too low, when the mold squeezes the mixture, the resin can be squeezed out of the fibers, resulting in resin-rich areas and leaving the fibers clumped. On the other hand, if the viscosity is too high, there may not be enough flow of the material to fill the mold. As BMC sits in storage, the cross-linker can evaporate or the cross-linking reaction can advance, and both effects cause changes in viscosity. Therefore, time limits are set on the use of BMC, just as on SMC.

ADVANTAGES AND DISADVANTAGES OF MATCHED-DIE MOLDING. Advantages:

- Both interior and exterior surfaces are finished.
- Complex shapes (including ribs and thin details) are possible.
- Production rate can be high.
- Labor costs are low.
- Minimum trimming of parts is needed.
- Products have good mechanical properties and close part tolerances.
- There is good consolidation of parts.

Disadvantages:

- More equipment is needed than for layup.
- Molds and tooling are costly compared with layup molds.
- Transparent products are not possible with SMC and BMC.
- Molding problems (trapped water, etc.) may cause surface imperfections, such as pitting or waviness.
- SMC and BMC have limited shelf lives.

Resin Transfer Molding

In resin transfer molding (RTM) or resin injection molding (RIM), a mold is loaded with reinforcement material, the mold is closed, and resin is then slowly injected into it (Fig. 15). The mold with the preform in it is often put under vacuum. This removes entrapped air from the reinforcement and speeds the RTM process. The reinforcement material is wetted out by the pressure of the injection.

This method has some similarities to other composite and nonreinforced plastic manufacturing methods. For example, the method of loading the reinforcement is similar to that of preform molding, but in preform molding, the resin is introduced into an open mold. Pressures are also much higher in preform molding. Transfer molding, from which this method gets its name, is a process in which a thermoset resin is injected into a closed mold. However, in traditional transfer molding, reinforcement is not incorporated. One further comparison is with reaction injection molding, a process in which two reactive resin components are mixed and then injected into a closed mold. Large, nonreinforced parts are often made by the reaction injection molding

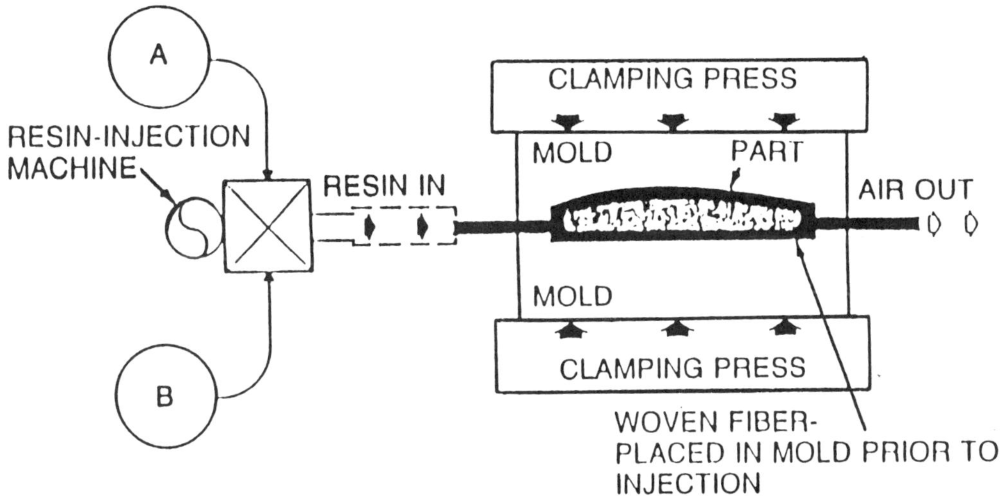

FIGURE 15 Resin transfer molding (RTM).

process. RTM is, therefore, a logical extension of several other processes to fill a specific need for improving the productivity of composite part manufacturing. RTM has been used in the automotive industry and is the process chosen for manufacturing body panels for the Fiero.

REINFORCEMENTS FOR RTM. Most standard reinforcement materials can be used, but fiberglass, carbon, and aramid are the most common. One requirement is that the reinforcement hold its shape during the injection phase. Therefore, the reinforcements are generally stitched or bonded together. Preforms are common. Reinforcement buildups in certain areas are also easily included.

Inserts of various types, such as screw receptacles or ribs, can be easily placed in the open mold at the same time as the fibrous reinforcement.

RESINS AND PUMPING EQUIPMENT. Most of the standard composite resins can be used in RTM, but the viscosity must be low enough that the fibers will be easily wetted. Typical viscosities are less than 1 Pa·s. The resin should also have about a 2 hour pot life so that injection can be slow (for wetout) without having the resin gel. Polyester and epoxy resins are the most common.

The resin pumping system can be much like the type required for spray-up molding (which is discussed elsewhere in this article). However, much simpler systems with just a ram or airline to insert the resin have also been used. Care should be taken to maintain good temperature control on the pumping/inserting mechanism to prevent premature gelation.

MOLD DESIGN FOR RTM. The design of the mold is the most critical factor in successful resin transfer molding. The mold must be constructed so that resin reaches all areas and concentrations are approximately the same throughout. This resin movement must be accomplished within the time allowed before the onset of gelation. Additionally, the resin injection process should not cause movement of the reinforcement and should be done at low pressures so that the mold will maintain its shape without massive backing methods.

Mold designers have found that RTM molds must be vented to allow the air within the mold to be pushed out by the resin. Venting is also common in compression and transfer molds. Though vents allow the escape of the air, they are too shallow to allow the resin, which is much thicker, to flow out. The mold should also have good temperature control. Even molds that are intended for room temperature–cured resins should be well insulated so that environmental conditions do not change the gel times and viscosity of the resin. Some molds are heated or designed to go into ovens to effect curing at higher temperatures.

RTM molds are usually made of composites (polyester–fiberglass), but they can be made of other materials if specific needs warrant it. Also, the low pressure requirements of RTM allow many more mold materials than would compression molding. The choice between metal molds and composites is chiefly one of volume and temperature. High volume and high temperatures dictate metal molds.

In some cases, the mold must be backed up in order to maintain its shape. This backup can be done with ribs or by simply adding mass (such as cast aluminum). The closure of the mold is also important, since the mating of the mold surfaces against a gasket is often the method of keeping resin from squirting out. Therefore, alignment pins are usually employed to ensure that the mold halves meet properly.

ADVANTAGES AND DISADVANTAGES OF RTM.

Advantages:

- Very large and complex shapes can be made efficiently and inexpensively.
- Production times are much less than with layup.
- Clamping pressure is low compared with matched metal molding.
- Surface definition is superior to that with layup.
- Inserts and special reinforcements can be added easily.
- The skill level required of the operator is low.
- Many mold materials can be used.
- Parts can be made with better reproducibility than with layup.
- The worker is not exposed to chemicals and vapors, as he or she would be with layup.

Disadvantages:

- The mold design is critical and requires great skill.
- Properties are equivalent to those with matched-die molding (assuming proper fiber wetout, etc.), but are not generally as good as with vacuum bagging, filament winding, or pultrusion.
- Control of resin uniformity is difficult. Radii and edges tend to be resin-rich.
- Reinforcement movement during resin injection is sometimes a problem.

Spray-up Methods

The spray-up method is similar to wet layup except in the method of applying the resin and reinforcement. Instead of using cloth or mat, the spray-up method uses roving that is chopped and then blown into the mold. The resin and initiator are sprayed into the mold simultaneously with the chopped fibers (Fig. 16). The mixture is hand-rolled to consolidate it in the mold, to remove air bubbles, and to ensure complete wetout of the fibers. The types of mold, application of mold release, application of gel coat, and curing are all the same as with layup.

Spray-up gives a much faster production cycle than hand layup techniques. However, the spray equipment operator must be skilled, since control of part thickness and, therefore, the physical properties of the part are largely a matter of technique.

Special equipment and materials have been developed to improve the spray-up process. Spray guns have the capability of carefully metering the resin and initiator

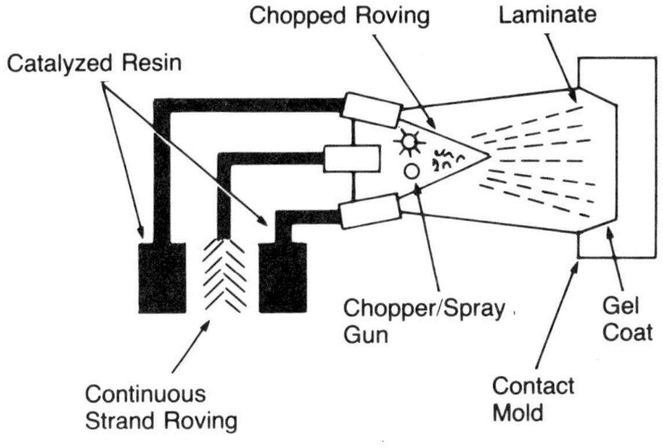

FIGURE 16 Spray-up.

with the chopping and blowing of the fiberglass (usually through two or three different openings or nozzles). Special fiberglass grades called *gun rovings* cut cleanly, wet rapidly, and conform to intricate contours (largely because of special sizings and finishes). Specially formulated resins with minimum drain-off and fast gel times are also available.

Spray-up is useful for fabricating very large structures and parts with rather complex geometries, but the only real advantage is reduced manufacturing cost compared with layup. The physical properties of the part are not as high as with layup.

Thermoplastic Processing Techniques

For critical composites (advanced composites) where the properties of the fiber are very important, the manufacturing method used for thermoplastic resins is likely to mimic one of the thermoset manufacturing methods discussed in this article, with modifications to accommodate thermoplastics, and the thermoplastic resin would be one of the high performance resins, such as PEEK or polyphenylene sulfide.

For noncritical composites, the reinforcements should be viewed as simply an upgrade of the properties that would normally be expected of the thermoplastic resin. These noncritical composites are generally made using traditional thermoplastic processing methods with only minor modifications for the reinforcement material, and the resin is likely to be one of the conventional thermoplastics, such as nylon, acetal, or polycarbonate.

Whereas thermosets must be heated to cure and turn into a solid, thermoplastics are heated to turn into a liquid and are cooled to become solid. Therefore, curing in the thermoset sense does not exist with thermoplastics.

Hand Layup of Thermoplastics

The wet layup methods are, in general, not applicable for thermoplastic resins. However, prepregs using high performance thermoplastics are available with most of the normal reinforcement fibers and in most of the fiber forms, including unidirectional and fabrics. Because the thermoplastics are solid at room temperature, thermoplastic prepregs are stiff and boardlike (no drape) and have no tack. They must be softened (slightly melted) so that the tack and drape can be improved. This is usually done by heating them with a hot air gun or some direct heating tool. Although some complex shapes may require heating of a large strip of prepreg, usually only small areas are heated to form the prepregs and tack them together.

The manufacturer of thermoplastic prepreg must take into account the higher temperatures required for thermoplastics and the much greater viscosity of the thermoplastic materials. Wetout of the fibers is certainly a problem. Increasing the temperature to lower the resin viscosity is not always possible, as in many cases decomposition begins to occur before the viscosity falls to a reasonable range. The method of lowering the viscosity that has proven to be most successful is shear thinning. Because these polymer materials are non-Newtonian, an increase in the applied shear will result in a significant drop in viscosity. Therefore fibers can be adequately impregnated by forcing the fiber and resin through a die at high temperatures. The development of thermoplastic-compatible sizes to protect the fibers during handling and to improve bonding has been another problem that is just starting to be overcome.

An advantage of thermoplastic prepreg over thermoset prepreg is that the thermoplastic prepreg has an infinite shelf life. No refrigeration or other precautions need be taken to ensure usefulness over a long period of time.

The bagging step should be essentially the same when using the thermoplastic material as with thermosets, except that the bagging materials must be able to withstand the higher temperatures and pressures that are normally used for thermoplastics. For instance, typical bags for thermoplastics are made of Kapton or Vacaloy, which can withstand 370°C. These materials are generally much more difficult to work with than their lower temperature counterparts.

The molding of thermoplastic prepregs must also account for the much higher temperatures required to obtain good flow characteristics and compaction. Most of the high performance thermoplastics melt in the 260–370°C range, compared with curing temperatures for epoxies of 120–180°C. Therefore, the mold materials and autoclave fittings and couplings must be able to withstand the higher temperatures. Many of the molds for thermoset composites will not work. Thermal expansion becomes more critical in the mold design because of the elevated temperatures. The higher viscosities of thermoplastics complicate the consolidation process, so greater pressures may be necessary.

Automated Tape Laying with Thermoplastics

Some machines have been developed for the automatic laydown of thermoplastic prepreg. The machines resemble the automatic laydown machines for thermosets, except that a special heated shoe is attached to the applicator head. As the thermoplastic prepreg tape comes between the head and the mold, the tape is heated by the shoe, which softens it and gives it the tack and drape necessary to stick to the mold or previously laid-down layers. A cooling shoe is placed just after the heated shoe to resolidify the tape and ensure that it remains in place.

A major problem with this method is that the material is "springy" and tends to raise up, causing the material in the tool to be bulky. Because of the bulk, interlaminar wrinkles may result when the final consolidation takes place.

Filament Winding of Thermoplastics

Prepreg tow is the usual material used when thermoplastics are to be filament-wound. High resin viscosity and elevated temperatures make in-line (wet) impregnation impractical. These tows are stiff, and so some preheating (melting) is usually necessary in order to get good laydown on the mandrel. Again, there are no worries about pot life with thermoplastics, but consolidation is a major problem.

Pultrusion of Thermoplastics

Several manufacturers of thermoplastic resins and independent pultruders have demonstrated the capability of pultruding thermoplastics. Shapes such as channels and rods have been tested and found to have excellent properties compared with similar thermoset parts. The resin manufacturers are not yet revealing the exact methods for manufacturing these pultruded parts.

A concept that is gaining favor in the composites industry is the manufacture of standard shapes (rods, channels, I-beams, hat structures, plates), which are then postformed to create structural members. The savings from using a high production method such as pultrusion and reducing inventories by stocking standard parts and sizes, which can then be shaped to fit specific needs, would be great.

Matched-Die Molding (Including Hydroforming or Thermoforming) of Thermoplastics

Traditional matched-die molding has been combined with a preheating step (similar to traditional thermoforming) in order to accommodate thermoplastic materials (Fig. 17). For advanced thermoplastics composites, a laminate of thermoplastic prepreg is placed in an oven capable of temperatures that will soften the thermoplastic. The thermoplastic is typically supported by a frame. When the thermoplastic sheet reaches the sag point (where the material starts to droop or soften), it is put into a matched-die mold and the mold closed. The mold is maintained at a temperature below the softening point of the thermoplastic. Therefore, the laminate will be simultaneously shaped and cooled to become a solid part. The mold can then be opened and the part removed for trimming and finishing. This method has a production cycle far shorter than a thermoset cycle (typically 5–30 min). This method obviously eliminates the need to use an autoclave, with its accompanying high costs, but may result in lower physical properties because of the shorter times involved, which do not allow proper consolidation to occur. (The very viscous thermoplastics take more time to penetrate the fiber bundles.) Some thermoplastic prepreg manufacturers are attempting to solve this problem by selling prepreg panel layups that have been preconsolidated in an autoclave.

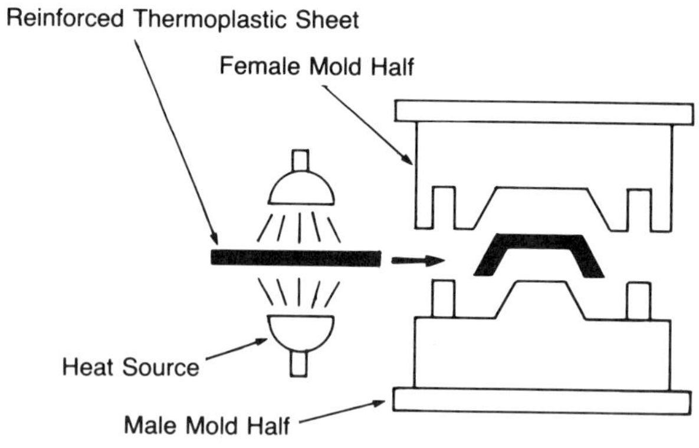

FIGURE 17 **Thermoplastic molding (externally heated).**

Some hydroforming (thermoforming) can also be done when one of the dies is flexible. The flexible die, which need not be exactly matched to the opposing die for very thin panels, will still give good consolidation and shaping of the material because of the hydrostatic pressure that is applied. (Hydrostatic pressure is that pressure that comes from pressing by a flexible or fluid material—in this case the flexible die.) For deeper contours and thick panels or laminates, the flexible die can be cast to match the metal mold. Generally the flexible die is slightly smaller than the metal to allow for the thickness of the part (Fig. 18).

Mold temperatures are typically 150–180°C. If they are cooler, the part may solidify prematurely, giving improper resin-fiber movement and part definition. Pressures vary with laminate thickness and fiber type, but would typically be 689–1723 kPa for a 1.25 cm thick carbon fiber laminate. Even though the flexible die is initially cold, degradation can occur because of the heat transferred to it by the hot laminate. High temperature rubbers (such as silicone) can be used for the entire die or as a heat barrier cap on a less expensive rubber, but even these will eventually become brittle and must be replaced.

The compressibility (hardness) of the rubber is typically 60–70 Shore A, which gives sufficient firmness to get compression while also giving good conformity to the metal mold.

To improve repeatability, a mechanical stop is usually installed to limit the travel of the press platens. When the process is controlled by pressure only, instead of by the mechanical stop, the variability of the parts has been excessive, and the resin has often been squeezed out of the prepreg.

The advantages of the flexible die in the hydroforming method are its inherent lower cost than matched metal

A) NON-MATCHED MOLDS

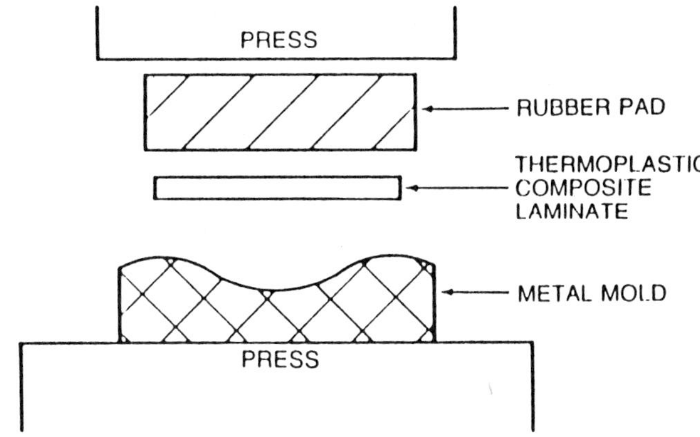

B) MATCHED MOLDS

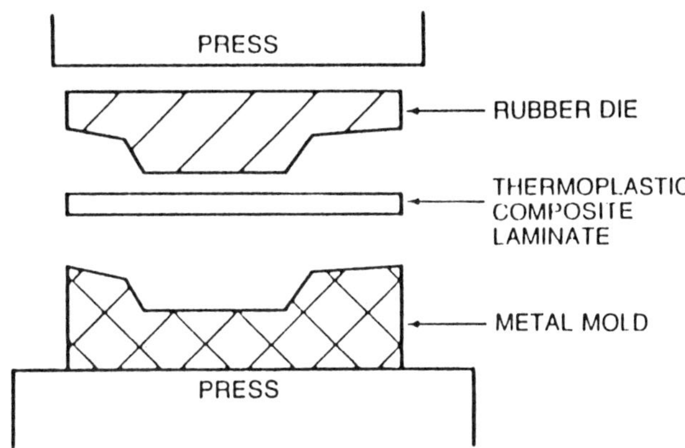

FIGURE 18 Hydroforming.

dies and the generally lower forces required, which are less likely to damage fibers.

RTM with Thermoplastics

Because of the very high melting temperatures and high viscosities of high performance thermoplastics, these resins have not been used in processes analogous to RTM.

Traditional Processes for Short-Fiber-Reinforced Thermoplastics

The most important of the traditional thermoplastic processes used to make composite materials is *injection molding*. In this process, resin pellets are melted (usually in an extruder that forms a part of the injection molding machine). The molten plastic is then injected into a closed mold. The mold is cooled so that the injected plastic solidifies, and the part is then removed (Fig. 19).

Injection molding is widely used because of the ease of automating the process, high productivity because of the short molding cycles, and the excellent detail that can be obtained with the process. Most engineering thermoplastics are readily processed by the injection molding technique.

In a typical injection molding machine, the resin enters the machine from a gravity-fed hopper. A rotating screw advances the material forward. (The screw has deep flights in the zone under the hopper, called the feed zone.) The material then moves into the melt zone, which has electrical heaters placed around the barrel and where the root of the screw gets thicker to give mechanical heating as well. The plastic is melted through the combined thermal and mechanical heating. The plastic is then conveyed by the screw into the metering zone, where the root of the screw is very thick (the flights are very shallow); the purpose is to ensure that all the material is melted and to build pressure. At the proper time in the mold cycle, the screw stops rotating and advances forward. This motion injects the plastic through the nozzle into the mold (A check valve at the

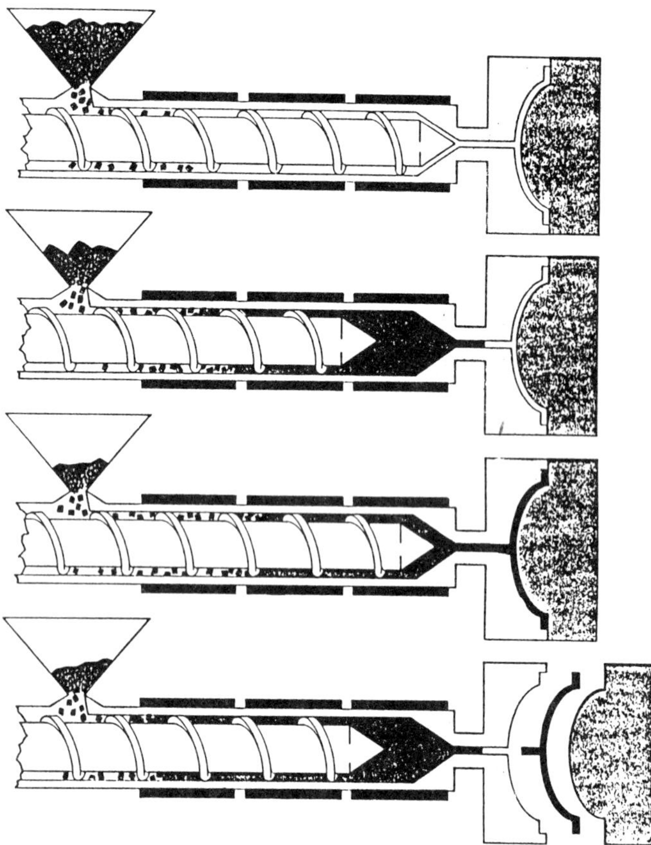

FIGURE 19 Injection molding.

end of the screw prevents the plastic from flowing backward.) The sprue and the runner system connect the nozzle with the gates (openings) of the mold cavities. In the simplest molds, the sprue and runners solidify with the part and are cut off as scrap. Some molds have hot runner systems that do not solidify, so that no trimming of the part is necessary. The mold is opened after an appropriate time for cooling of the part (often only a few seconds). Knockout pins eject the part.

The mold can be either single or multicavity; if multicavity, it can have identical cavities or the cavities can be different so that various parts can be made simultaneously. The latter is called a family mold, since several parts that go into one assembly can be molded simultaneously. Molds can also be equipped with unscrewing devices to give internally threaded parts and with sliding inserts to allow production of hollow parts. All these modifications to the simple one-cavity mold add complexity and cost to the mold, but may increase the productivity of the operation and thus reduce the cost of the final part.

The size of the injection molding machine is determined by the size of the part or parts that are to be made. Two measurements of machine size are required. The first is the *shot size*, which is simply the volume of the material to be injected (parts, runners, and sprue). Typical sizes range from 0.03–1.4 kg. The second measurement that must be known is the *clamping pressure* that will be required to hold the mold closed against the injection pressure. The area of the part and the restrictions in the sprue and runners and at the gate are major considerations in determining this pressure. Typical values are 100 to 400 MPa.

Several types of reinforcement can be used in injection molding, although particles, whiskers, and short-chopped fibers are the most common.

The use of reinforcement in injection molding must allow for several important aspects of the operation. First, the melting and injection process can damage reinforcements, greatly diminishing their positive effect. Therefore, a compromise is generally made between optimum physical properties (long reinforcements) and processing capabilities. Generally, reinforcement lengths must not be greater than the size of the gate into the mold cavity. Typical lengths for the fibers used in reinforced injection molding materials are 0.76–25.4 mm, but the longer lengths degrade. Other problems, such as flow restrictions in the runner system and excessive heating in the extruder, can arise from the use of fiber concentrations that are too high. Typical concentrations are 30–40% fiber by weight.

Improper mold design can cause serious problems in the strength of the part. If there are multiple gates into a cavity, if there are converging or diverging flows, or if the material must flow around an insert or core, two or more flows of material are created. These flows must knit together well, or the part will be weak at the knit line. The presence of reinforcements in the plastic makes the knitting problem more difficult, since the reinforcements can protrude forward of the advancing plastic front and interfere with its smooth meeting and joining with the approaching front of the other flow.

Because of the shear forces that occur in injection molding, fiber orientation is not unusual. Orientation is generally to be avoided because whatever strength is gained in one direction is lost in another. This orientation can be utilized to strengthen the part in the direction of orientation, but the process must be controlled carefully to obtain repeatable results. Recent studies have shown that thin samples and high injection speeds cause orientation. Some other studies have utilized the magnetic properties of metal reinforcements to obtain the desired orientation in a repeatable manner.

Extrusion of Reinforced Thermoplastics

Extrusion is a process for making parts of uniform cross section in high volume. The process employs an extruder similar to the one described in the section on injection molding, except that in extrusion the size of the extruder is generally much larger and the process is continuous (that is, there is no reciprocating motion), so no check value is necessary. Instead of a mold, the plastic material is forced through a die, which shapes it. The plastic is then cooled and cut to the desired length.

Extrusion of reinforced materials is not an important process for making finished parts, but is important as the method of incorporating the fibers into the resin. In this regard, extrusion can best be viewed as an excellent blending and mixing process. Fibers and reinforcements chopped to length are jointly added to the hopper and then blended by the mixing action of the extruder. Care must be taken to prevent damage of the reinforcement, but that is a function of the reinforcement concentration, reinforcement length, extruder screw dimensions with respect to the barrel, heat, and opening of the die. The plastic is extruded into a long, thin rod (like spaghetti), then chopped into pellets, which would typically be 3 to 10 mm long.

Recent new technology utilizing pultrusion technology instead of extrusion has produced injection-moldable resins with fibers 5–20 times longer and corresponding improvements in physical properties. Reinforced nylon resins made by this process are replacing magnesium, zinc, and aluminum in die castings and in automotive and appliance applications.

Thermoforming of Reinforced Thermoplastics

In thermoforming, a sheet of thermoplastic material is heated to the temperature at which the sheet begins to droop under the force of gravity, then withdrawn from the oven or other heat source and placed over a mold. A vacuum is drawn through the mold (small holes are drilled in the mold to allow this to occur), and the plastic sheet is pulled into the shape of the mold by the vacuum. Either male or female molds can be used.

A process of forming high performance thermoplastics that is similar to this (hydroforming) has already been described and holds great promise for simplifying the processing of composite sheets and, in some cases, structural shapes.

For conventional thermoplastics, thermoforming of reinforced plastics holds some promise, but currently offers only a slight improvement over nonreinforced products.

Curing

Many processing operations involve the changing of a solid material to a liquid so that it can flow to a desired shape or wet out incorporated materials. The molding of a thermoplastic resin and the casting of metals are such processes. In contrast, thermosets are typically processed at molecular weights low enough that they are already fluid at room temperature. Once the desired shaping or wetout has been accomplished, the resin is rendered solid by a chemical cross-linking reaction that is called *curing*.

For any manufacturing process, the processing parameters are, of course, chosen to obtain the optimum properties in the final part with the best possible economics. In composites, the key processing variables are time, temperature, heat-up rate, pressure, vacuum, and cool-down rate. The interrelationships among these parameters are complex, but several methods for monitoring them have been developed so that at least the major interactions can be seen. Eventually, of course, the final proof of correct curing is the combination of physical and mechanical properties of the finished part.

Cure Cycle

The usual processing procedure for thermoset curing is to increase the temperature while maintaining hydrostatic pressure (Fig. 20). As with most fluids, the viscosity of the matrix will fall with this temperature rise, thus allowing good consolidation. However, chemical cross-linking reactions will eventually overshadow this thermal thinning, and the viscosity will rise as the mobility of the resin is increasingly restricted by intermolecular cross-linking. Eventually enough cross-linking will occur to cause incipient gelation (some molecules having become large enough to span the entire specimen), and the composite will change from a liquid into a solid and will no longer permit plastic deformation. The temperature at which this phase or structural change occurs is called the *dynamic gel temperature* T_{gel}. This temperature is dependent upon the time/thermal history of the material and cannot, therefore, be fixed as a specific number for each resin. Depending on the resin, T_{gel} occurs when the matrix is 40–60% cross-linked (Fig. 21).

The interrelationships of time and temperature for a specific resin can be presented in a time-temperature-transformation (TTT) diagram. TTT diagrams show that along the gelation portion of the curve (the S-shaped

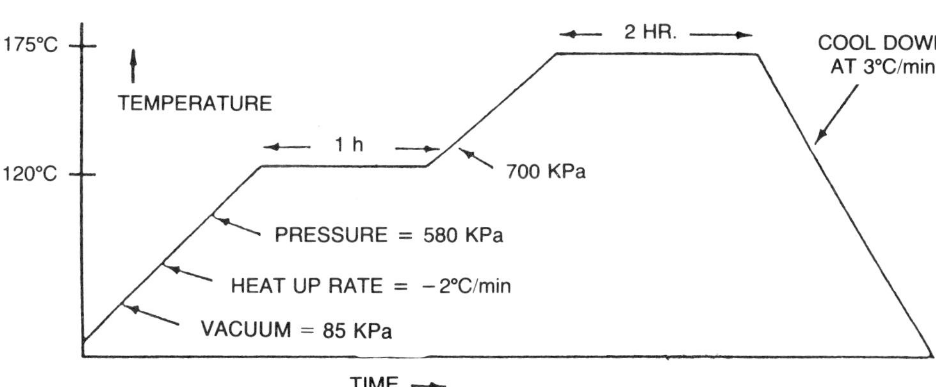

FIGURE 20 Typical epoxy cure cycle.

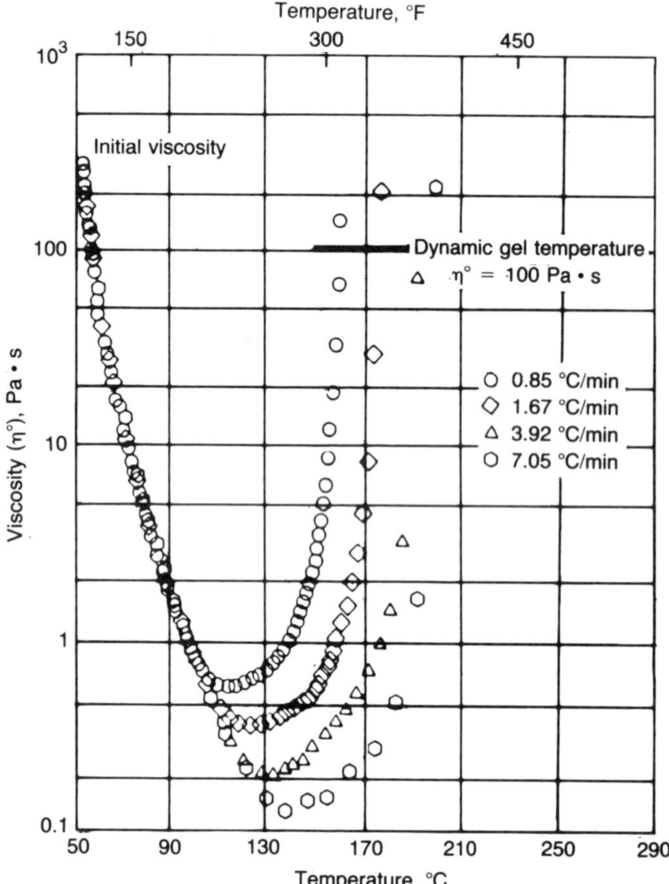

FIGURE 21 Viscosity versus temperature curve.

curve), the time for gelation decreases as the temperature increases (Fig. 22). Therefore, gelation can be achieved with either long times at moderate temperatures or moderate times at elevated temperatures.

After gelation, the composite is often given a *postcure,* which increases the cross-link density and is called *advancing the cure.* This postcure involves subjecting the part to a higher temperature for an extended period of time (often several hours). Postcures can be done while the part is still in the mold if there is no danger of part damage from differential thermal expansion between the part and the mold. Postcuring is, however, more likely to be done on a free-standing part.

A quantity that has been found to correlate with the cross-link density is the *glass transition temperature* T_g. As the cure is advanced, T_g increases as additional cross-links are formed until the entire structure is fully cross-linked. At this point the upper limiting value of T_g has been achieved.

Tooling

Tooling or molds are used to define the shape of the composite part and one or more of the surfaces, depending on the type of tooling. The requirements for the tooling are conceptually quite simple in that the tooling must provide a mechanism for giving a part the desired shape at the end of the mold cycle. In the case of layup and spray-up, this requirement is easily met by a variety of materials, since few, if any, strains are placed upon the mold during the molding cycle. Therefore, the molds for layup and spray-up can be made of composite, metal, or several other materials. The release from the composite should be assured with proper release agents, but the only other major requirement is that the mold must hold its shape.

For matched-die molding, the demands on the mold are at the opposite end of the spectrum. The molds are subjected to enormous pressure and temperature changes and must be strong enough to move a viscous molding material within the mold. These molds are generally steel and are similar to those used in compression molding.

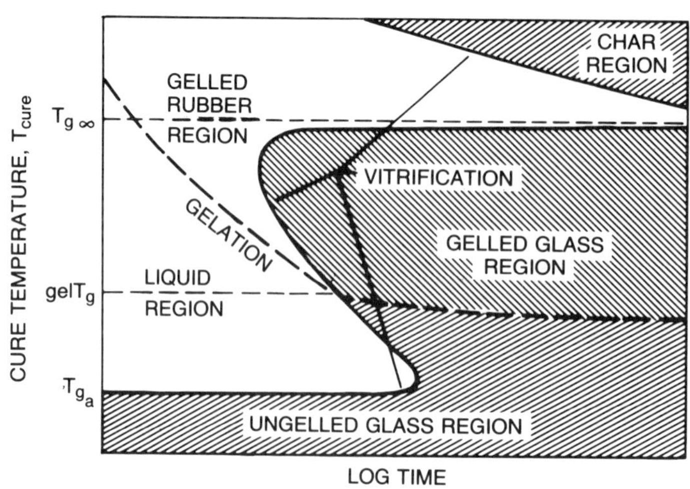

FIGURE 22 Time-temperature-transformation (TTT) diagram.

The mold requirements for vacuum bag molding, especially when an autoclave is to be used, are intermediate between the requirements for layup molds and those for matched-die molds. Typical autoclave molds must be able to withstand temperatures of 121–177°C (or up to 316°C for special applications) and the forces caused by thermal expansion. Exceptions are tools with large hollow areas that are not vented to autoclave pressures (such as a box or tube closed on both ends). A further problem with autoclave tools arises in tools that have a large number of voids, which expand and cause the tool to degrade rapidly. Therefore, strength requirements are not as severe as for matched-die molding, and the use of many materials, such as aluminum, composites, and metal-coated composites, is possible.

However, the heating process puts some special requirements on the design of the molds which must be considered. Those requirements arise because of the differences in the coefficients of expansion of the part and the mold and become much more significant as the size and complexity of the mold increases and at higher temperatures. In the curing of the part, the mold and the pliable part expand together during the initial phases of the cure heat-up. But, at a certain temperature-time relationship, the part gels; that is, it becomes hard and intractable. From that point on, the part and the mold expand at their own rates depending on their own distinctive coefficients of thermal expansion. After cure, the part and the mold cool at their own rates as well. Therefore, stresses resulting from this differential in thermal expansion are likely, especially if the part is held within the mold.

In all cases, the choice of a male or a female mold is determined by the part design and application and the ease of manufacture. For instance, a radome nose for an aircraft should be made in a female mold because the outside of the finished part must be finely finished for good aerodynamics. On the other hand, an I-beam would be made in a male mold because of the ease of manufacture. Male molds can also be used when there are serious coefficient of expansion differences between the mold and the part.

Other considerations in the choice of the mold material would include heat transfer capability, machinability, useful life, ease of repair, dimensional stability over time, and initial cost.

Post-Manufacturing Fabrication

Although composites are often molded to a near-net shape (that is, an almost finished shape), trimming, finishing, and assembly into larger structures are often required. Therefore, the trimming, finishing, and assembly of composite materials are important considerations. In some of these processes the potential for damaging or otherwise weakening the composite (through delamination, fiber fraying, drill break-through, etc.) is quite great, and care must be taken to maintain the composite's properties.

The equipment and procedures used in cutting, drilling, and machining of composites differ in some major considerations from those used for either metals or plastics. The abrasive nature of the reinforcement requires process modifications, and the type of reinforcement (aramid, carbon, glass, etc.) also makes certain modifications necessary. Composite materials can be trimmed and cut more easily with processes closer to grinding or abrasive cutting than with conventional metal cutting, which produces large chips. Cutting, drilling, and other machining processes are largely dictated by the type of reinforcement, with minor modifications to allow for the easily melted or degraded nature of the matrix. On the other hand, the processes of joining and painting are largely dependent on the matrix material, since these are chiefly surface operations.

Although composites are often drilled, milled, or cut using traditional metalworking equipment (perhaps with minor modifications in the equipment or in the cutting tools), several specialized cutting and drilling methods have proven to be especially useful in cutting and drilling composites. Water jets and lasers are two techniques that have found to be widely applicable in the composites field.

Composites are often component parts of much larger assemblies (cargo doors on an airplane, for example) and must be joined to the other parts of the assembly. In metals, this joining would most often be done with mechanical fasteners and only occasionally with adhesives. With composites, the opposite is true—most parts are adhesive-bonded, and comparatively few are joined mechanically. Therefore, the methods required to ensure good adhesion of these adhesive-bonded surfaces (and proper holding of mechanical fasteners) are critical to the proper performance of the entire assembly. The most important aspects of joining or bonding composites are the proper treatment of the bonding surfaces, the choice of a suitable adhesive (of the same resin as the composite itself), and the proper curing of the adhesive (which may be done as the composite is cured or in a separate step).

Some designs require thicknesses beyond what is reasonable with solid composite structures. Therefore, sandwich structures, in which composites serve as the skins and various lightweight materials serve as the cores, have been developed. The methods for making these sandwich materials and assembling them with other materials to form complex structures are important in today's composite world. Generally, the composites are adhesive-bonded to the inner structure of the sandwich. The inner material is often a honeycomb or rigid form material, although lightweight woods and other materials are also used (Fig. 23).

Finally, painting and coating of composites for environmental purposes or simply to improve their appearance is becoming more important as composites are used in an ever-widening assortment of products. Generally, the most important consideration in painting and coating is that the properties of the composite are not eroded. For instance, the resin should not be dissolved by the solvent in the paint. Also, the surface treatment

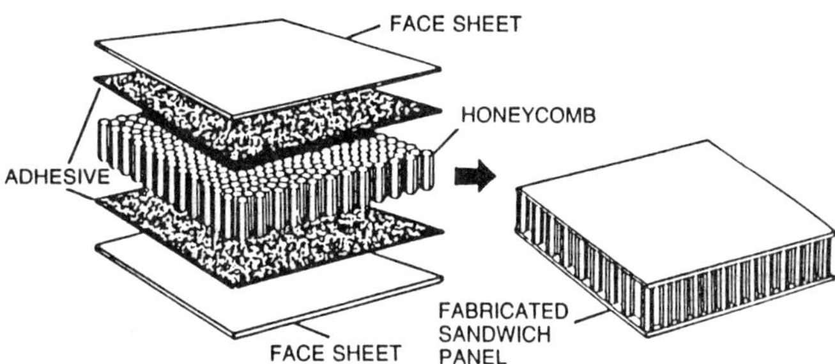

FIGURE 23 Structural sandwich construction.

that is often necessary for painting should not be detrimental to the performance of the composite.

Quality Control

The use of many composite structures in critical applications has required careful adherence to good quality control and inspection procedures. Many quality tests and specifications have been developed for certifying that composites meet the most exacting standards. In many cases these procedures require certification of the methods of manufacturing and testing as well as certification of the parts themselves. This is generally viewed as a positive step toward total statistical process control (SPC), which is the most accepted path to improved quality of products.

One difficulty in the quality area is, however, the lack of standardized procedures for some of the critical tests, especially those that relate to specific end-use performance. The establishment of these tests has been largely left to the manufacturer. Therefore, manufacturers have the responsibility of defining and implementing meaningful in-use tests so that the entire industry can have performance criteria by which their parts can be judged.

One area of quality control that needs strengthening is automatic sensing and machine adjustment or part rejection, which must be done for automated quality control. Currently, almost all quality control is done by hand after the part has been made.

Design-Manufacturing Interface

The interface between design and manufacturing has also been difficult. In the case of composite manufacturing, careful coordination of these two functions is critical. The interrelationship between properties and manufacturing methods requires that designers and manufacturing engineers coordinate closely. Furthermore, as we have already discussed, the tradeoffs between properties and economics cannot be made without the joint agreement of design and manufacturing.

The use of computers in the design of composite parts has opened up the possibility of even greater coordination between manufacturing and design. As the designer is able to pictorially represent the desired part on the computer, the manufacturing engineer will be able to use that same representation for manufacturing process design. In some cases, the tooling to make the part has been cut by numerical control machining programs generated directly by the design representation.

Contributions from Noncomposite Industries

Many of the manufacturing methods currently used for composites have had their origins in noncomposite industries. For instance, many prepreg cutting techniques are taken from the textile industry. The textile industry is also the source of many of the fiber handling and fabrication techniques used for composite fibers. The plastics industry is an obvious reference area for the handling of composite resins as well as defining some of the thermoplastic and thermoset manufacturing techniques.

In addition to these many parallels, many industries use manufacturing methods that could have great application in the composites industry. For example, tire and hose manufacturing is similar in many ways to filament winding and pultrusion. Fibers are used for reinforcement of a matrix material, and the shapes are similar in many cases. Another example would be the textile industry, where textiles are sewn, woven, or knitted into complex shapes. These techniques could serve as a basis for the making of preforms, which could lead to new methods of composites manufacturing.

Innovation and observation are the keys to improved composites manufacturing, and we should be cognizant of other industries so that applicable methods can be modified to fit the needs of the composites industry.

A. Brent Strong

[*See also* THERMOPLASTIC MANUFACTURING, DIE-LESS.]

References

1. Suppliers of Advanced Composite Materials Association (SACMA), *Personnel Needs in the Advanced Composite Materials Industry: An Assessment*, 1988.
2. A great deal of the material in this article has been extracted from the following text: A. B. Strong, *Fundamentals of Composites Manufacturing*, SME, Dearborn, MI, 1989.

Manufacturing, Economic Considerations

The cost to produce a composite part, or any part for that matter, can be subdivided into two elements: the actual cost to manufacture the part, plus the costs associated with other support activities that are necessary to run a successful manufacturing company. If to this we add the profit, all on a per unit basis, the selling price is obtained:

$$\text{Selling price} = \text{profit} + \text{manufacturing costs} + \text{support costs}$$

There are many strategies for setting the selling price, but they all essentially depend on one's competitive position. If you can offer a unique product, with high demand, you may sell it at whatever the market can bear. On the other hand, if you are new to the market, you may sell below cost (if you can afford it) to gain market share. This is called "dumping." In most cases, the selling price is positioned between these two extremes. In a very competitive market, there will be limited flexibility in setting the price. In this case, to maximize the profit, one must minimize the manufacturing and support costs. A company must make a reasonable profit over time to attract investors. The two cost components can be further subdivided as shown in Table 1.

In this article, we are primarily interested in the *manufacturing costs* to make a composite part. This information is needed in order to know which products and processes are cost-effective. Traditionally, the manufacturing costs are subdivided into direct costs (those that can be directly related to a product) and indirect costs (those that are incurred for a variety of products). By these definitions, direct costs will vary with the volume of production, whereas indirect costs will not. There are limits to this definition, but in general it is quite useful and commonly accepted. There are then only two components of direct costs: direct labor and direct materials. On the other hand, indirect costs can include a wide variety of items, including indirect labor, indirect materials, and factory and equipment costs.

In the discussion that follows, we will apply these general cost categories to a "typical" advanced composites manufacturing venture. Keep in mind that this exercise is somewhat arbitrary and that real-life circumstances may suggest slight alterations of these categories.

TABLE 1
Selling Price and Cost Components

SELLING PRICE	Profit	Profit
	Manufacturing Costs	Direct Manufacturing Costs
		Indirect Manufacturing Costs
	Support Costs	Engineering, Research and Development Costs
		General and Administrative Costs
		Marketing and Selling Costs

Direct Costs

Direct Labor

Direct labor includes all "hands-on" effort required to manufacture a specific product. For a typical composites shop this might include such activities as hand layup, tool and autoclave preparation, machinery and autoclave operation, assembly, and other activities directly associated with a product. When estimating direct labor costs, note that laborers are not 100% efficient, so some factor must be included for fatigue and sickness. Additionally, equipment is not 100% reliable, and circumstances will exist where laborers are idle during equipment downtime.

Some composites processes are quite labor intensive. This is particularly true for hand layup processes. For example, Ref. 1 shows that the ratio of direct labor cost to direct material cost for the hand layup of a graphite–epoxy airplane fuselage is about 3.5. As a comparison, Ref. 2 gives this ratio for U.S. plastics processing companies in 1985 as about 0.3. This order of magnitude difference reflects the low labor content of processes such as injection molding, which are used by the plastics processing companies, versus the high labor content of the hand layup process. Furthermore, advanced composites are often produced in small volumes, are highly inspected, and, for complicated parts or parts made with new materials, may offer low yields. All of these factors will increase direct (and indirect) labor costs.

To reduce the high labor cost content of advanced composites, there is considerable interest in implementing automated equipment. This equipment is often quite expensive, however, and so any potential savings in direct labor or elsewhere must be balanced with the capital investment. The automation of advanced composites layup is not straightforward. A close look at the cycle times for making these parts [3] shows that there are at least five important components of the cycle time for hand layup: cutting, material transfer, layup, debulking, and autoclave preparation. In some cases the debulking and autoclave preparation times can be quite long, setting a lower bound on the cycle time one can obtain from equipment that automates only steps 1, 2, and 3. Automated equipment may reduce inspection time, however, and improve yields and the overall quality and reliability of the product. The cost savings from this

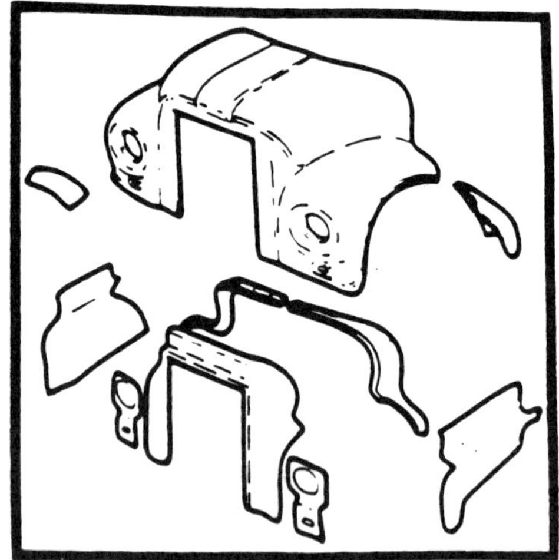

Steel Stampings (48 pieces) Matched Metal Die (10 pieces)

FIGURE 1 **Exploded views of alternative approaches to making a truck front end. The matched metal die process uses SMC.** (From Ref. 4.)

could be dramatic, particularly for those companies that are implementing 100% inspection during layup.

One important reduction in direct labor that composite materials offer involves assembly. Because of the way composite materials are manufactured—for example, by additive processes such as tape layup or filament winding, or by flow processes such as the compression molding of sheet molding compound (SMC)—it is often possible to make highly integrated composite parts of complex shapes. To make such a part from metal, in contrast may require many subparts that later have to be assembled. This point is clearly illustrated in Figure 1 [4]. The figure shows that many more parts are needed to make a steel truck front end than an SMC truck front end; hence there will be a higher assembly cost associated with the steel truck. In another paper, which compares the manufacturing costs for alternative constructions of the Airbus A300/A310 fin box, there is a clear reduction in the assembly cost for the graphite–epoxy version ("CFK-MODUL") versus the aluminum version ("AL-DIFF") [5]. This is shown in Figure 2.

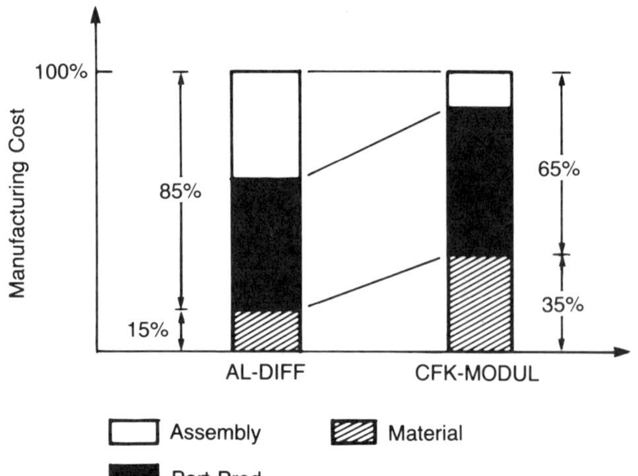

FIGURE 2 **Cost comparison for aluminum vs. graphite–epoxy fin box.** (From Ref. 5.)

Direct Materials

Direct materials include all of the components and raw materials that go into the end product. For a graphite–epoxy composite part, this would include all of the graphite–epoxy used in the part plus the scrap and waste. Examples of materials that are not strictly in this category would be bleeder and breather plies, bagging materials, raw materials purchased for tooling, and other such materials.

Some materials of this type (e.g., bleeder plies) are used only once, and therefore would vary directly with the volume of production. In this case, the material might arguably be considered as direct. Some composite materials can be quite expensive. In particular, that portion of the market that is primarily concerned with obtaining performance advantages may employ high-cost materials. The relationship between cost and performance is shown directly in Figure 3. This shows the relationship between cost per pound and modulus for several different fibers currently available. This graph illustrates that people will pay more for higher modulus materials. (Note that a graph of strength versus cost will not give a monotonic relationship!) However, it should also be kept in mind that the market size for these materials decreases rapidly as the cost and modulus go up. As the market grows, and competition intensifies, economies of scale and process improvements can be employed to reduce cost. This is illustrated in Figure 4, which shows the reduction in the cost of graphite fibers over the last 20 years, and Figure 5, which shows the increase in graphite fiber sales over the same time period [6].

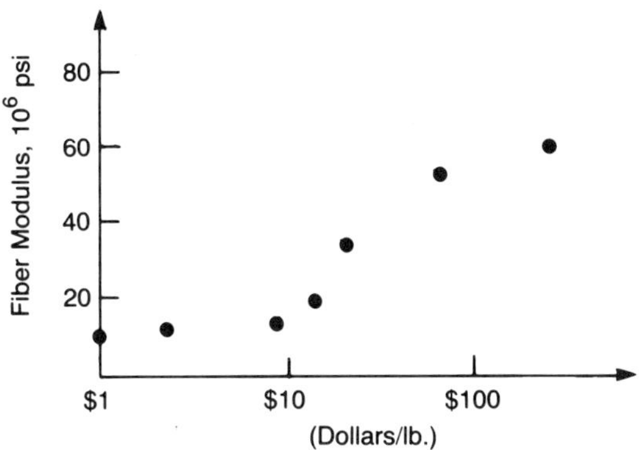

FIGURE 3 **Relationship between cost per pound and modulus for several currently available composite fibers.**

Composites manufacturers often use "preprocessed" material forms in their processing operations. These material forms usually combine the fibers and the resin in a state that makes them easier to process. Examples of preformed materials include prepregs of all types, sheet molding compound, thermoplastic–fiber pellets, and comingled and co-woven thermoplastic–fiber forms. Often processes that use these preforms are called "dry" processes, whereas processes that mix the raw resin and fibers during the manufacturing step are called "wet" processes. The value added to the preform can be substantial. For example, a prepreg may cost 2 to 4.5 times the cost of the raw materials. For this increase in cost, one usually buys convenience, a reduction in setup and processing time, and better quality control. Some processes, such as filament winding and pultrusion, may use both material forms. When deciding between the two different material forms, one must consider the production volume. In general, wet processes take a longer time to setup and to obtain a steady-state operation. Therefore, the cost advantage of wet materials may be offset by higher waste for short production runs.

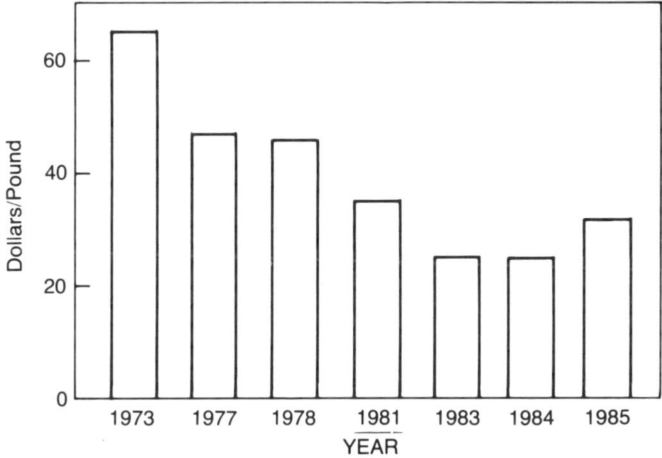

FIGURE 4 **Cost of graphite fibers.** (From Ref. 6.)

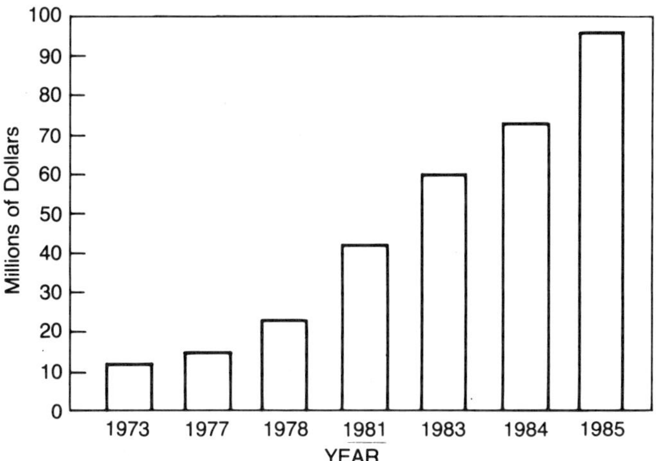

FIGURE 5 **Graphite fiber sales in the United States.** (From Ref. 6.)

Because of the very high cost of some composite materials, scrap and the possibility of recycling become important considerations. Composites manufacturers often employ nesting routines borrowed from the garment industry to get the best use of their material. In some cases, composite materials compete against other materials and processes that produce even higher amounts of scrap. This may have the effect of equalizing what may at first glance appear to be a very large direct material cost differential. A good example of this is discussed in Ref. 5. Here the cost of producing a graphite–epoxy fin box is compared with the cost of producing the same part out of aluminum. While the cost per pound of graphite–epoxy versus that of aluminum in this application is about 8:1, the high scrap rate associated with the skin milling process used to make the aluminum part reduces that ratio to about 2.5:1. This is illustrated in Figure 2.

Indirect Costs

Indirect Labor

Indirect labor includes all manufacturing personnel whose output cannot be associated with a specific product but instead is directed over a wide range of products. This includes such jobs as dispatching, shop supervision, scheduling, testing, process troubleshooting, reworking, and manufacturing engineering. For some jobs the division into direct and indirect may be somewhat arbitrary. Traditionally, batch setup and inspection are considered indirect costs. In composites manufacturing, however, tool setup and layup inspection are performed each time a part is made. In this context, and especially if these activities are restricted to a single part, they might be properly considered as direct costs. Indeed, a major goal of current cost accounting systems should be to properly allocate these costs.

Note that as more and more automation is introduced into the production system, jobs that typically fall into the indirect labor category are increasing. These include activities that are necessary to set up and program automated equipment, which often require high skill levels and, hence, high salaries. Therefore, proper allocation of these costs is an extremely important issue.

Indirect Materials

Indirect materials traditionally includes those materials required for production that do not get directly incorporated into the product. These include expendable supplies such as bleeder and breather plies, bagging materials, cork dams, material for tooling, and material for packaging.

Factory and Equipment

This category includes the expenses of running a factory, such as the cost of rent, heat, electrical utilities, water, depreciation on capital equipment, maintenance of equipment and tooling, and inventory. Some composite job shops that rely primarily on hand layup or simple automated processing aids may be relatively inexpensive to run. On the other hand, larger composite factories that employ the latest in automated machines can be quite expensive. New automated machines for laying up or cutting composites, along with their computer controls, can easily cost in excess of $1 million. Any substantial capital expenditure of this type must be properly accounted for by considering the time value of money.

Almost all composites require some kind of heating and pressurization during the forming and solidification process. The costs for the equipment to perform these operations vary with the size, capacity, temperature, and control features desired. For example, in Ref. 7, it was shown that the prices for one family of injection-molding machines varied directly with the clamping force. This is shown in Figure 6. Of course, this trend could be easily (and drastically) altered by varying the control features of the machines. This same reasoning applies to autoclaves, hot presses, and other molding equipment.

A composite part must pass through the production system in an efficient and cost-effective manner. In an ideal factory, all work stations would have similar rates and there would be no machine breakdowns. In real composites factories, this ideal is not possible to achieve. One common bottleneck that occurs in many systems is at the autoclave. While the autoclave does not necessarily have to be the slowest step, underestimates of autoclave needs and unique and/or lengthy cure requirements can greatly reduce the capacity of this step. This problem may be even more dramatic when compared with automated cutting. Many of the automated cutting machines used for composites were originally designed to cut fabrics for the garment industry, where capacity requirements are enormous compared with what is needed for composites processing. As a consequence, these machines are greatly underutilized. The rates for most other steps in the composites production system tend to fall between these extremes. Because of these problems and others, there is considerable interest in alternative factory configurations and new automated equipment or systems. These allow a more uniform flow of parts through the composites factory and more reliable scheduling. Schemes that are of considerable interest are integrally heated tools, vacuum curing with heated blankets, and cutting and "kitting." The most commonly mentioned positive effects of the automated

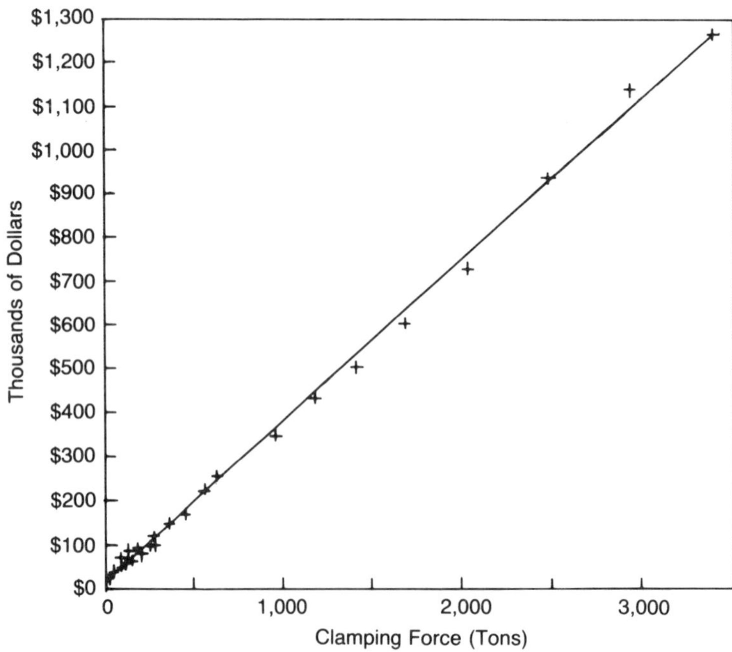

FIGURE 6 **Injection-molding machine cost vs. clamping force.** (From Ref. 7.)

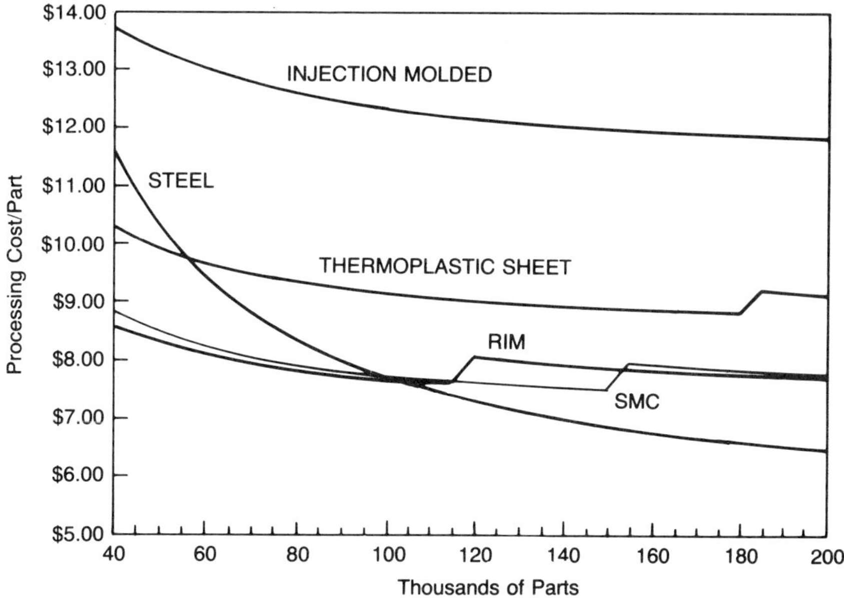

FIGURE 7 **Comparative costs for different fiber body panel fabrication technologies.** (From Ref. 8.)

systems are more efficient use of space and hence lower rent, and, for various reasons, smoother flow of parts through the factory, resulting in a reduced inventory. Given the high value added to many composite parts, a reduction in inventory can represent substantial savings.

Because indirect costs do not vary with the production volume as direct costs do, there can be a dramatic reduction in the overall cost per part as the production volume increases. This effect means that one might consider alternative manufacturing strategies depending upon the expected market size. For example, in a recent paper [8], alternative approaches were considered for the fabrication of an automobile body panel. The paper considered stamped steel sheet and four plastics fabrication technologies, including injection molding, SMC compression molding, reaction injection molding (RIM), and thermoplastic sheet stamping. The results, for the assumptions given in the paper, are shown in Figure 7. From the graph, it is apparent that the lowest cost process changes at about 100,000 parts. The steep drop in the cost for the steel part reflects a high indirect cost (capital investment in equipment), whereas the relatively constant costs for many of the plastics options reflect lower indirect costs and higher direct materials costs. Note that capital-intensive processes do not always become less expensive as the production volume increases. Once the production volume exceeds the capacity of the equipment, a whole new system and capital expense is needed to produce one more part. This was shown in Ref. 3, which compared various automated composites fabrication technologies with hand layup for a simple graphite–epoxy part. The results, which depend upon the assumptions used in the paper, are shown in Figure 8. Here, for example, the low capacity of the robotic tape layup scenario prevents it from approaching the manual production scenario at any reasonable production volume. Note that an alternative use of robots, also discussed in Ref. 3 and shown in the figure as "robotic transfer," also appears more promising.

Using Cost Information

The usual reason for assembling cost information is to make management decisions. The most reliable cost information one can get is that from a currently operating factory. This can be used, for example, to assess a company's health or the profitability of a given product or operation. In many cases, however, predictions have to be made as to what the costs for various alternatives might be in the future. Making predictions of the future always involves a certain degree of uncertainty. In fact, Ref. 9 states that even when estimating the future costs to make a product that is currently in production, the estimates can be no better than ± 5%–10%. It is important, then, to keep in mind that all cost estimating involves some degree of uncertainty. When comparing two alternatives, one should be careful not to draw conclusions when the differences between the alternatives are not too different from the expected accuracy. To account for this uncertainty, most cost estimators include some contingency amount in the estimated cost.

There are a variety of manufacturing decisions that management must make that can be aided by cost estimates. When comparing alternative options, one usually considers only the differences between the options. These are called the relevant costs. Examples of the kinds of decisions that management must make include the decision to make or buy a component, the evaluation of alternative designs for cost, and the justification of new equipment. In many of these cases, the actions required can be viewed as an investment. Management

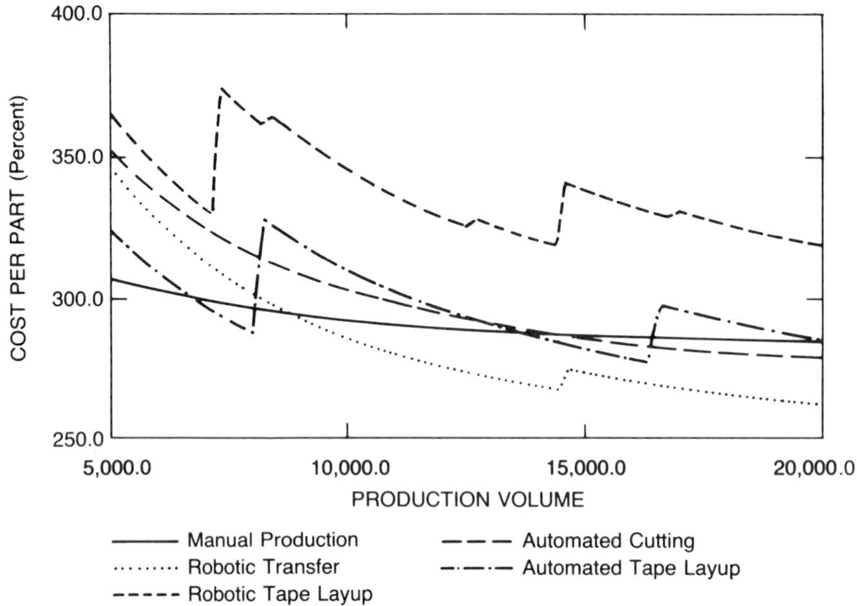

FIGURE 8 **Comparative costs for various automated and manual production techniques for the production of a graphite–epoxy part.** (From Ref. 3.)

must satisfy the owners that the proposed action will yield a reasonable return on investment. Common measures that are used to demonstrate this are given below [10].

Return on investment (ROI):

$$\text{ROI} = \frac{\text{average savings/year} - \text{cost of project/life of project}}{\text{cost}}$$

Payback period:

$$\text{Payback period (years)} = \frac{\text{cost of project}}{\text{incremental project savings/year}}$$

There are, of course, other financial measures that one can use, but these are perhaps the simplest and most frequently used. The ROI measure can be compared with the return on any other type of investment that earns interest. The payback period tells you the length of time it will take, at the estimated savings, to break even.

When using cost information to justify management decisions, one should keep in mind that while cost is extremely important, there are many other important factors. Often these factors are less definable but may yield long-term cost advantages. These include such issues as how an investment would improve one's competitive position, or how an investment would improve one's ability to satisfy customers. These kinds of arguments, for example, have been used recently to justify large capital investments in automated equipment. Such issues as the improvement of quality, the shortening of response time, and the development of a more flexible manufacturing system can all translate into a happier customer and a better competitive position. The quantification of these effects, so that management may use this in its decision making, is an area of considerable research activity. Some relevant books and articles are listed in Refs. 11–16.

Timothy G. Gutowski

References

1. H. S. Reinert, V. S Thompson, and H. R. Fenbert, *Proceedings of the 31st Int. SAMPE Symposium,* (Closed Session), 1986, pp. 58–67.
2. *Facts and Figures of the U.S. Plastics Industry,* 1986 edition, SPI.
3. S. Krolewski and T. Gutowski, *SAMPE Q., 18*(1), 42 (1986).
4. A. J. Ortega, *Proceedings, 24th Annual Tech. Conference,* SPI, 1969, pp. 1–7.
5. B. Sarh, *Proceedings of the 29th Int. SAMPE Symposium,* 1984, pp. 1477–1488.
6. J. K. Kuno, *Proceedings of the 31st Int. SAMPE Symposium,* 1986, pp. 725–737.
7. J. V. Busch, "Technical Cost Modeling of Plastics Fabrication Processes," Ph.D. Thesis, Department of Materials Science and Engineering, MIT, 1987.
8. J. V. Busch and F. R. Field III, *Proceedings of the ASM/ESD Advanced Composite Conf.,* 1987, pp. 1–9.
9. E. M. Malstrom, Ed., *Manufacturing Cost Engineering Handbook,* Marcel Dekker, Inc., New York, 1982.
10. D. T. Koenig, *Manufacturing Engineering,* Hemisphere Publishing Corporation, New York, 1987.

11. J. Meredith, *CIM Review*, Spring 1987, p. 37.
12. J. Meredith and M. Hill, *Sloan Management Review*, Summer 1987.
13. J. Meredith and N. Suresh, *International Journal of Production Research*, *24*(5), 1043 (1986).
14. G. Taguchi, *Introduction to Quality Engineering*, *Asian Productivity*, Kraus International Publications, White Plains, NY, 1986.
15. H. T. Johnson and R. S. Kaplan, *Relevance Lost, The Rise and Fall of Management Accounting*, Harvard Business School Press, Boston, 1987.
16. J. A. Hendrick, *Mechanical Engineering*, *3*(2), 64 (1989).

Manufacturing Automation, Polymer Composites

Introduction

Techniques for automating the fabrication of composite parts go back more than 30 years (filament winding and pultrusion), while other methods are still in laboratory development. The emphasis was confined almost exclusively to thermoset resin systems until very recently. Automation of thermoplastic composites parts manufacturing is being pursued aggressively but remains proprietary to the companies funding the developments. Much of the development work for the automation of thermoset composite parts has been funded by the United States Air Force, and U.S. government restrictions are currently in place to curtail the dissemination of detailed information concerning these techniques. The descriptions of some of the automated processes are thus quite general, but detailed information may be obtained from the sources cited. Addresses of several principal manufacturers are listed in Table 1. Certain process parameters have been omitted because they are so part dependent that even giving a range of values could be misleading. With the ever-increasing use of composite materials, manufacturing cost reduction is the focus of much activity for aerospace and commercial products. Automation of composite parts fabrication is regarded as very desirable for ultimately accomplishing the cost-reduction goals of the various industries.

TABLE 1
Manufacturers' Addresses

GGT Inc. CAMSCO Division
55 Gerber Road
South Windsor, Connecticut 06074

Precision Nesting Systems, Inc.
50 Spring Street
Cresskill, New Jersey 07626

GGT Inc. Aerospace Division
55 Gerber Road
South Windsor, Connecticut 06074

Gerber Scientific Instrument Company
83 Gerber Road
South Windsor, Connecticut 06074

Ingersoll-Rand Water Jet Cutting Systems
635 West 12th Street
Baxter Springs, Kansas 66713

Goldsworthy Engineering, Inc. (GEI)
23930 Madison Street
Torrance, California 90505

Prepreg Cutting

Many composite parts are fabricated by laminating layers of composite material, each layer consisting of a one-piece ply, which may be as large as 1.5 m wide by 3 m long. These plies are cut from a continuous roll of material, with standard widths measuring up to 1.5 m. Composite material wider than 30 cm is referred to as "broadgoods," a term borrowed from the garment industry. Several methods have been developed for cutting plies from broadgoods material, and these methods are described below.

Before the actual cutting takes place, it is necessary to locate the individual ply patterns onto the broadgoods to minimize the amount of unused material (which would then become scrap). The objective is to group as many parts as possible onto the smallest area of material. The process of ply pattern positioning, called ply nesting, must also take into consideration the fiber orientation of each ply (Fig. 1). Nesting efficiencies for unidirectional composite materials are on the order of 90% (i.e., leaving 10% of the material as scrap). The increases in nesting efficiencies that can be achieved depend on part shape and size. Nesting is best accomplished by a computer system because of the added complexity of fiber orientation; several systems are commercially available from GGT, CAMSCO Division, and Precision Nesting Company. The most useful nesting systems are able to electronically transfer the ply patterns directly from a computer-aided design (CAD) system into their own system. Many systems allow the "digitizing" of drawings as an input to their system. Digitizing is accomplished by manually tracing lines on a drawing with an electronic hand-held device containing several push-button commands. Automatic scanning of existing paper or Mylar drawings by an electrooptical system is another method available for transferring information into the nesting system. Two further considerations in nesting are the length of the cutting table and whether the plies must be cut in a particular order. The latter requirement is necessary only when cut plies are used immediately, as in some automated systems. If plies are to be kitted, before lamination, the order in which they are cut becomes mean-

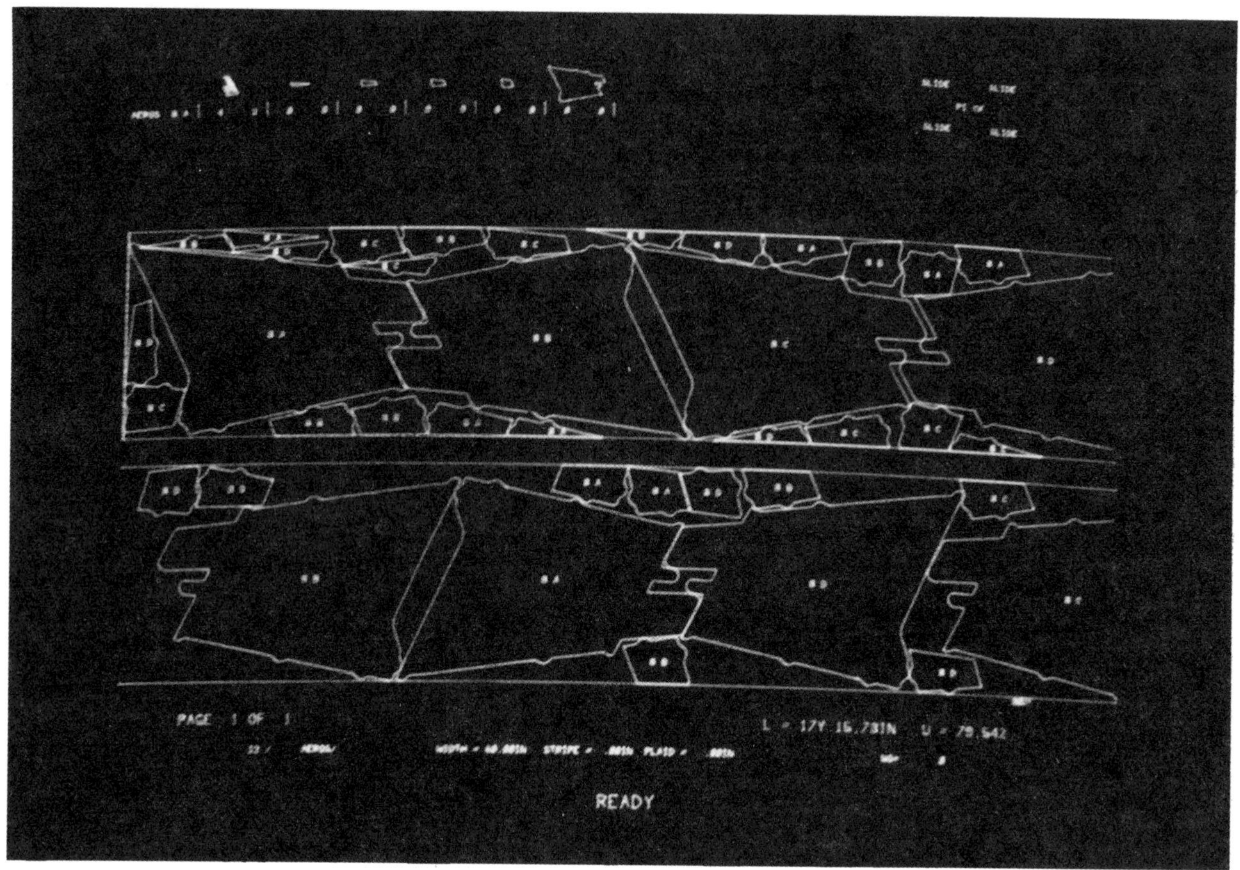

FIGURE 1 Two examples of ply nesting displayed on a CRT. (Courtesy of Gerber Garment Technology, Inc.)

ingless. As the part size becomes a significant percentage of the table size, the nesting efficiency is consequently degraded.

All the cutting technologies described here also provide computer-controlled manipulation of the cutting medium to produce the individual ply shape. Figure 2 illustrates the convention used for the cutting systems. The ideal prepreg cutting system would produce full-sized ply patterns, have clean cut edges, give 100% fiber cut along the ply periphery, cause no fraying, not alter the chemical composition of the matrix in any way, and leave no scrap material. For practical reasons, the list of ideal properties omits such characteristics as zero cut-time and zero cost. As far as automation is concerned, 100% fiber cut is mandatory. When stray fibers are still attached to both the ply and the remaining material, automated ply removal from the cutting table creates disaster as material is dragged around.

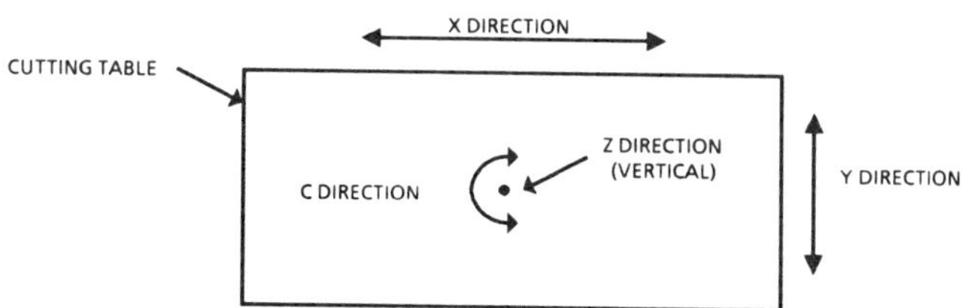

FIGURE 2 Cutting axis definition.

Reciprocating Knife

The first automated system for cutting graphite–epoxy prepreg material was developed in the late 1970s by Gerber Garment Technology, Inc. The similarity to cutting ply shapes and clothing patterns from the broadgoods materials used in these two enterprises enabled a prepreg cutting system to be developed rapidly. The material to be cut is laid onto the surface of a plastic bristle-topped table and held in place by vacuum that is sucked through the bristles from underneath. A reciprocating knife is then plunged through the material, thereby cutting it, the blade passing easily between the bristles. The effect is to have a table surface that firmly supports the material and yet allows penetration of a cutting blade without resistance. The blade can be moved in any direction through the bristles without restriction. Standard tables can be up to 12 m long with ply cutting accuracies of $\pm$ 0.13 to $\pm$ 0.5 mm. Motion in the x axis is accomplished by a bridge, which spans the table width and travels on precision metal ways. y-Axis motion is caused by a carriage riding across the bridge, and motion in the remaining axes, z (vertical) and c (rotation about z), is accomplished with a mechanism integral with the carriage. All four axes of cutter motion are controlled by a computer numerically controlled (CNC) system. The maximum linear cutting speed is 4500 cm/min.

Material can be dispensed either automatically or manually onto the cutting table from a large (660–1100 kg) roll of material. It is necessary to have a backing material between the prepreg and the plastic bristles to prevent the vacuum from drawing the fibers down between the bristles. In practice, a better quality cut is obtained if there is also a backing material on top of the prepreg, thus sandwiching the prepreg between two backing materials. A variety of backing materials may be used, including polyvinyl and paper. Once the ply patterns have been cut, it may be necessary to put some identification on them before they are removed from the table. Ply labeling has also been automated, using a system similar to that described for cutting, but instead placing a printed label onto each ply for identification or marking the plies with inkjet. Cut plies may also be removed from the cutting table automatically by vacuum pickup devices. Before ply lamination can take place, however, the backing material must be removed from both sides of the ply, and a successful method to automate this procedure has not been developed. Reciprocating knife systems also have difficulty making clean cuts in unidirectional material at shallow angles to the fiber direction. Even with these drawbacks, the reciprocating knife system is currently the most widely used system for broadgoods prepreg ply cutting.

Reciprocating Chisel

The reciprocating chisel system has many common features with the reciprocating knife system; cutting table, bridge/carriage system, computer-driven four-axis motion. The most obvious differences between the two systems are the cutting tool and the cutting table surface. The cutting surface for a chisel-based system is truly solid, made out of either metal or a resilient material. If the cutting surface is nonresilient (metal), a backing material must be used between the prepreg and the table. The chisel reciprocates rapidly up and down, cutting through the prepreg and penetrating about halfway through the backing material. When a resilient cutting surface is used, backing material is not always needed. The chisel cuts through the prepreg and merely deforms the cutting surface.

Reciprocating chisel systems require a very flat cutting surface unless a terrain-following system of some kind is used. Uniformity and consistency of the backing material thickness are additional factors to be considered. Certain proprietary systems have overcome the surface flatness problem to some extent at the expense of complicating the mechanical system. The original prototype chisel cutting system was a modified x-y graphics plotter manufactured by the Gerber Scientific Instrument (GSI) Company. An x–y plotter has several of the features required to make a ply cutting machine: smooth flat bed, bridge/carriage mechanism, computer-driven axes. GSI has now made a reciprocating chisel prepreg cutting machine one of their standard products. Plotter accuracies are usually very good, and machines based on this technology can have cutting accuracies as good as $\pm$ 0.05 mm over a 1.5 m^2 area. Chisel widths up to 3 mm have been used. The chisel width places a limit on the minimum radius that can be cut and on the speed with which a linear cut can be made. Linear cutting speed is proportional to chisel width and reciprocating speed. Typical cutting speeds range up to 7 cm/s.

Water Jet [1–3]

Water jet cutting is a method that uses a thin, high velocity stream of water as the "cutting tool." To produce the high velocity water jet, as the stream is called, pressures of 200–400 MPa are used to force the water through an orifice (0.1–0.5 mm diameter) in a sapphire or other hard material. The resultant jet of water is supersonic and produces very high noise levels. Automated prepreg cutting can be achieved with only a two-axis control system because the water stream is fairly well collimated, making it relatively insensitive to z-axis positioning. Many commercial systems do provide a z axis for added flexibility, if required. The sapphire orifice produces a fully symmetrical water stream because of its circular shape, eliminating any need for a rotational c axis. Since the cutting action is performed by the high velocity water, there is no "wear" of the "cutting tool." The sapphire orifice does require periodic replacement because of gradual water erosion.

Epoxy resins absorb moisture readily, so care must be taken with water jet cutting to minimize the amount of water contacting the resin. Water jet cutting has been used in Europe successfully; however, concerns over the moisture absorption problem have prevented widespread acceptance in the United States. Cutting requires some means of supporting the prepreg material, yet allowing the water stream to pass through. The simplest method uses a prepreg support made from a thin stainless steel, hexagonal-celled honeycomb structure, with the cell walls aligned with the direction of the water stream. The walls of the cells adequately support the material yet provide minimum resistance to the water jet. Eventually the honeycomb structure is eroded and has to be replaced.

A more complex arrangement for prepreg support uses a conveyor belt type of system with a movable slit in the surface material (Fig. 3). The four-roller system forms a unit in which the rollers are kept in the same relative position to each other, as shown. Each roller spans the full width of the support surface. The prepreg material lies on top of the support surface. As the roller unit is moved back and forth in the x direction, the slit is moved with it, following the jet of water as it traces the ply pattern on the prepreg. The effect is to leave the prepreg material stationary on the support surface while the slit is moved beneath it. In this manner the slit is always maintained directly underneath the jet stream.

To reduce the noise of the supersonic water stream, a "catcher" cup is located inside the slit mechanism. This container filled with steel ball bearings, moving in concert with the roller mechanism in the x direction, tra-

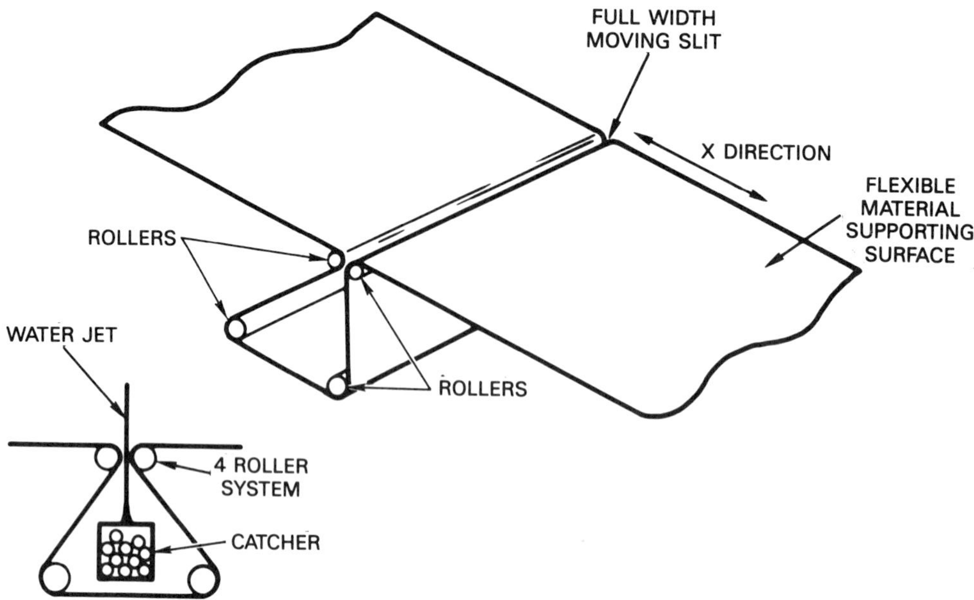

FIGURE 3 **Principle of operation of moving slit.**

verses under the slit in the *y* direction, following the carriage, such that it is always under the water stream. As the water stream impinges on the steel balls, they rotate, dissipating some of the energy of the water and containing the noise to some degree.

Ultrasonic Knife

The ultrasonic knife system provides a prepreg cutting action by ultrasonically exciting a knife blade as it traces out the ply pattern on the prepreg material. The knife blade is guided in the *x*, *y*, and *c* directions under computer control as described for the reciprocating knife. The blade is mechanically coupled to an ultrasonic transducer, which causes it to move vertically up and down about 0.127 mm at 20 kHz. Vertical blade motion is proportional to the excitation frequency and is reduced for increasing frequency. The audible noise level is just the inverse of blade motion (i.e., is reduced for increasing frequency). The ultrasonic excitation of the knife blade produces a clean cut with most materials, including paper-thin metals. The process produces some heat, which can cause the resin to flow and reseal along the cut line if the blade dwells too long in one place.

Since the vertical blade motion is limited to a maximum of 0.127 mm, it is necessary to have either a very flat cutting surface or a servo-controlled terrain following z-axis control. The system can be adjusted to cut all the way through the prepreg but only partially through the backing material. Linear cutting speeds can reach 40 m/min.

Pizza Wheel

Prepreg material can be cut with a circumferentially sharpened disc in a manner similar to that used for cutting a pizza. A sharpened metal disc spun at high speed produces a very effective prepreg cutting device. Discs may be of several diameters: the smaller the disc, the tighter the radius of cut that can be made. This method is often used to cut prepreg tape and is usually adjusted so that it does not cut all the way through the backing material. When the backing material is left in one continuous length, it can easily be wound onto a take-up reel for later disposal. Pizza wheel cutters are not generally used for broadgoods prepreg cutting because of the cutting surface flatness requirements and their limited small-radius cutting capacity. Their use has been confined mostly to prepreg tape cutting in tape-laying machines.

Laser

Conventional lasers have been used to cut prepreg materials but have been found to be generally unacceptable because of the accelerated curing of the cut edges by the laser heat energy. The "cut" edge often swells a little, producing an edge that is slightly thicker than the remaining material, preventing the edges from lying flat on top of the underlying ply.

Die Cutting [4]

Die cutting is used for automatically cutting large quantities of identical shapes. Metal dies are made to the required shape and used to stamp out the plies from the prepreg much as cookie cutters are used to cut pastry. The larger the ply shape, the more difficult this technique becomes because large dies are harder (and more expensive) to make, and greater pressure is required to cause the die to cut through the prepreg material. Indexing the prepreg material under the die automatically can produce a relatively efficient cycle time. Care must be taken in monitoring die wear, since a worn die may not cut through the prepreg completely. This method of prepreg ply cutting is not in widespread use at this time.

Prepreg Lamination

Two basic material forms use a lamination process during part fabrication: broadgoods and tape. The laminate is formed by stacking two or more layers of prepreg one on top of the other, observing the designated fiber orientations. Any pockets of trapped air between the layers must be removed before the curing process is begun. Most automated lamination techniques remove the air from between the plies during the actual prepreg material laydown.

Broadgoods Lamination

Plies cut from prepreg broadgoods material are generally in one piece and can be as large as 1.5 m × 3 m with current systems [5]. The most common method for picking up plies is by means of vacuum devices. A system may consist of vacuum cups supported by a mechanical structure and may be manipulated by a robot. When the vacuum pickup device is manipulated by a robot, it is generally referred to as an "end effector." Other ply pickup device configurations use a series of holes (through which a vacuum is drawn) distributed over a surface that may be flat or curved. The vacuum is usually turned on just before the device contacts the prepreg material. The pickup surface is sometimes resilient, to compensate for unevenness in the prepreg surface when laying down the ply.

Once a ply has been picked up, it is necessary to remove any backing materials from it before laydown. Depending on the prepreg cutting technique used, prepreg backing material removal may be automated (as in tape layers) or performed manually. If there is backing material on both sides of the prepreg, it will be necessary to remove one of the backings after ply laydown.

Automated ply pickup and laydown/lamination have been achieved with prepreg broadgoods materials [5]. Only flat laminates of unidirectional materials have been successfully automatically fabricated to date. Woven material forms may be used to fabricate gently contoured laminates automatically. Most automatic lamination methods produce flat laminates, which are then formed into their final shape by further processing.

Tape Lamination [6]

Automated tape laying has been performed with tape having widths of 25, 76, and 152 mm; the actual width used depends on the specific part size and design. The process consists of laying down strips of material alongside each other to form a wider and wider shape. Each strip, or length, of tape is referred to as a course. At the end of each course the tape must be cut to the appropriate length and shape (Fig. 4). Cutting to the correct length is relatively, easy but cutting to the exact shape at the end of a course can present problems. Straight angled cuts can be made relatively easily—for example, using a pizza wheel cutter as described previously. Cuts of complex shapes whether angles or rounded contours, are more challenging. The pizza wheel cutter angle can be varied during cutting, but this can require up to three axes of motion to ensure complete flexibility. Another method for cutting the tape ends uses a series of small chisel cutters across the width of the tape. Each cutter is independently controlled, allowing a variety of shapes to be cut; the resultant shaping is achieved through several stair-step cuts (Fig. 5). Backing material is automatically removed from the tape during laydown. If the flat laminate is of constant thickness, the ends of the courses may be trimmed after curing, eliminating the shaped end cutting problems. Fiber orientation is achieved through controlling the direction of the tape laydown. Overlaps between courses are generally disallowed for obvious reasons, which means that there is usually a small gap running the whole length between courses. These gaps are generally less than 2.5 mm, and any reduction in part strength due to them can be accounted for in the design.

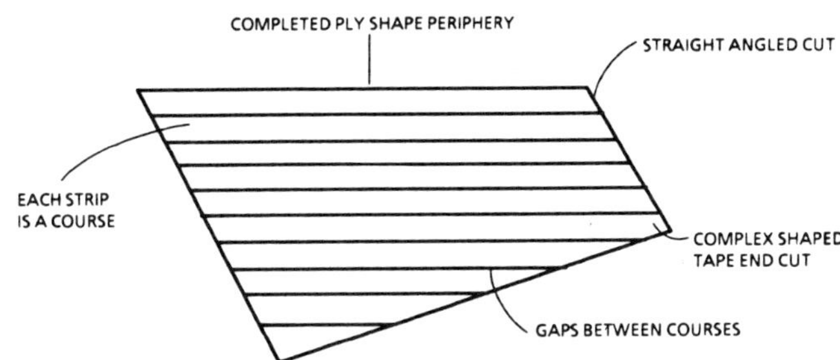

FIGURE 4 Tape layup, illustrating courses.

Typical tape-laying heads that have been developed are quite large, heavy, and expensive. In addition, they require manipulation in the x, y, z, and c axes, usually accomplished by a gantry-style robot. The robot itself must be both stiff and accurate, features that directly affect both tape laydown accuracy and robot cost. The buildup of tolerances between the mechanically sophisticated tape laydown head and the robot manipulator requires both devices to be precision designed and made. Tape laydown over gentle, simply contoured surfaces has been demonstrated, but because carbon fibers do not stretch, only flat laminates are usually fabricated.

The tape is made from unidirectional prepreg material exhibiting straight and parallel edges. There must be no material defects in the tape, since these tend to cause the tape-laying device to jam, requiring direct operator assistance. The edge and defect-free prepreg tape re-

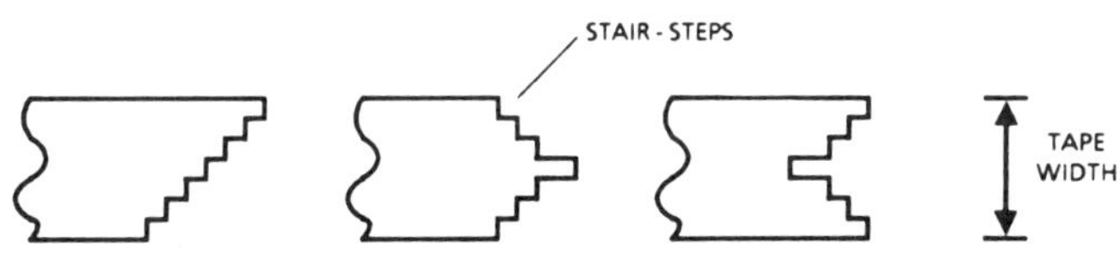

FIGURE 5 Examples of stair-step cutting.

quirements can escalate the price of the material over that for broadgoods of the same material type. Tape is carried on a spool and is backed by a carrier material that remains continuous throughout laydown. The carrier material is automatically wound onto a take-up spool for eventual disposal.

Novel forms of tape placement, in addition to the method described in the preceding paragraphs, have been devised and are commercially available. One of the alternative techniques pulls the tape out to the full length while still attached to the carrier backing material, then places the whole length down in one motion. Debulking is performed by a roller action (i.e., the air is squeezed out), and the carrier material is peeled off. Another technique uses a two-stage process [7]. The first stage cuts each tape course to length and shape, then attaches it to a continuous carrier material at a computer-controlled workstation. The second stage uses a relatively lightweight tape-laying head that contains a spool, from the first stage process, with the cut-to-length tape courses. The tape-laying head is manipulated by a gantry robot during laydown. Tape laydown is faster than the single-stage process because no tape cutting occurs during laydown.

Pultrusion [8]

Goldsworthy Engineering, Inc., of California, invented and pioneered the pultrusion process more than 30 years ago. It is one of the oldest automated methods for manufacturing composite parts. Continuous length stock can be pultruded with cross-sectional areas from a fraction of a square centimeter to thin-walled structures with outer dimensions on the order of 1 m × 0.5 m. The pultrusion process has a proven track record, and many products have been successfully manufactured using this technique.

Pultrusion is a method of fabricating a composite product of constant cross section in a continuous length. It is one of the few processes that automatically produces a continuous finished composite product from raw materials. In concept the process is quite simple: continuous resin-coated fibers are pulled through a system of heated dies to form the product, and lengths of the product are cut off as required.

In practice a pultrusion system may contain the elements shown in Figure 6. The raw material, usually glass or carbon, is contained on individual creels that typically can carry up to 12 kg of material each. The tow (bundle of fibers) is wrapped onto the creel just as string would be wound onto a bobbin. The creels are mounted in racks, and sometimes special winding wheels, so that many creels feed the pultrusion machine simultaneously. The winding wheels are able to rotate about the pull axis to provide fiber orientations other than in the pull direction. The rotation speed is synchronized to the pull speed to maintain the desired helical winding angle. As the tows leave the creel racks, they are passed through a system of guides to maintain their positions relative to each other as they pass through the resin bath. The resin can be any one of a number of thermoset materials.

As the resin-coated tows leave the resin bath, they pass through a squeeze system to remove excess resin. Varying the squeeze pressure varies the ratio of resin to fiber and is one of the parameters controlled by the system. The uncured resin–fiber combination is then pulled through a system of heated steel dies, which are sometimes chromium plated to provide a good surface finish and to reduce die friction. It is important to have the material pulled through the dies at a constant speed to produce a smooth surface finish. The linear speed can be influenced by a variety of factors, such as resin chemistry (due to age or normal batch variation), die friction (due to die expansion heating), die deformation, die wear, foreign particles, mechanism warping, and resin cure stage. The pulling mechanism must be able to overcome the variations in die friction to maintain the constant linear pulling speed.

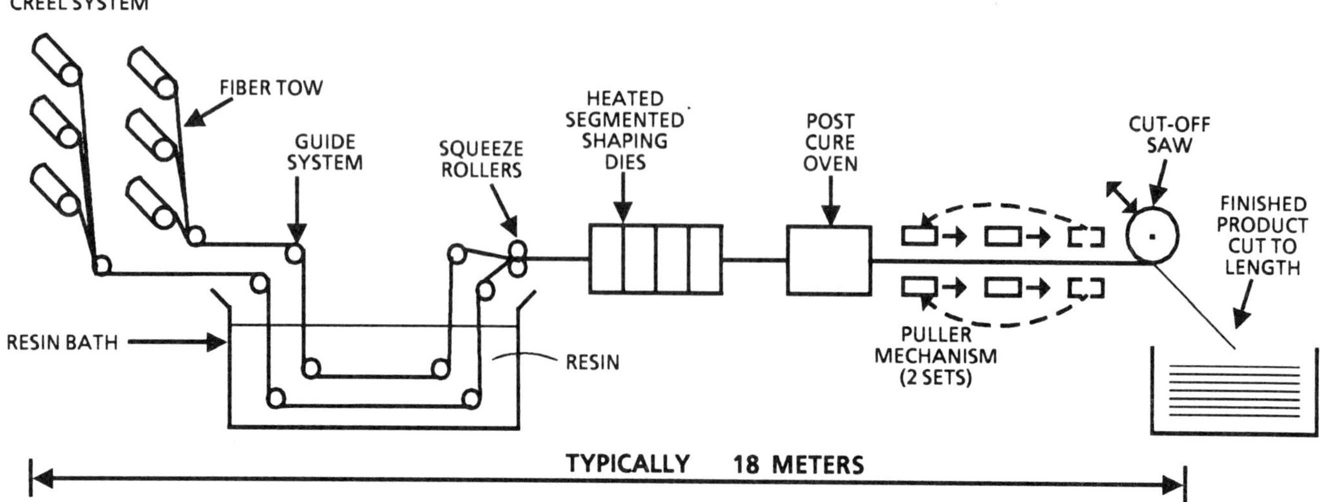

FIGURE 6 Basic pultrusion process schematic.

After emerging from the system of heated dies, the product may require further heating to complete the cure cycle; this is achieved in an oven. In addition to conventional resistance-heated ovens, the resin can be heated with radiofrequency (rf) energy or induction frequency (if) energy [9]. Radiofrequency curing is used with nonconductive fibers at a frequency of 40–70 MHz for polyester resins and 940–2450 MHz for epoxy resins. The rf energy heats the resin through the principle of dipolar molecular excitation. Sometimes it is necessary to add dipolar additives to epoxy and other resins to improve the response to dielectric excitation. The curing process occurs simultaneously throughout the profile cross section because the rf energy heats the material dielectrically, not thermally. When conductive fibers such as carbon are used, rf heating will not work and induction frequency heating is used instead. The induced current in the conductive fibers provides the heating mechanism. Die heating is most commonly accomplished by resistance heating, often with strip heaters.

Variations on the basic pultrusion heating system may include a preheater oven to dry the fibers before resin immersion (to improve fiber wetting) and rf heaters placed to preheat the resin before it enters the heated dies. In the latter case the pultruded stock is shaped after leaving the resin bath, then heated, typically to 60–74°C, before passing into the heated die set. Preheating the resin before it enters the die set can triple the pultruder output. This method of "augmented cure" was developed by Goldsworthy [9].

The pulling force is provided by an interchanging set of hydraulically powered grippers. To obtain the best surface finish on the pultruded part, a constant speed must be maintained and the change over from one gripper to the other must be carefully controlled to prevent a sudden change in pulling force. Line speeds typically range from 2.5 to 500 cm/min. Just about any shape that can be extruded with metal can be pultruded with composites: solid and tubular rods, angles, channels, I-beams, conduit, and square tubes.

Directional properties can be obtained by material selection in addition to winding tows at various helical angles. Woven roving can be used for bidirectional properties, mats and cloths for omnidirectional requirements. Additionally, combinations of materials and material forms can be used for specific applications. Although general-purpose unsaturated polyesters using styrene or vinyl monomers have been the most common matrix in the past, more recently epoxies and even thermoplastic resins are being used. Sometimes properties are enhanced by fillers, which also can be used to lower stock costs because the filler material costs less than the resin. Pigments may also be added to the resin to provide integral colors and eliminate the need to paint the finished article.

The stock may be cut off at any length by a cross cut saw, since the product is formed in a continuous length. For specific applications the stock may be left in one continuous length and used to form another structure, such as a large-diameter cylinder.

Filament Winding

The roots of filament winding can be traced back more than four decades when winding resin-coated strands directly over a rocket motor propellant was suggested [11]. The M. W. Kellogg Company developed the original equipment for winding motor cases in the United States in 1948 under contract with the Bureau of Ordnance, U.S. Navy. It was not until about 10 years later that much was done about this concept, when Thiokol Chemical Corporation conducted a feasibility study for the Army Rocket and Guided Missile Agency [12]. In 1960 Walter Kidde and Company designed and built a machine suitable for winding test specimens as part of an Air Force contract [13]. This fabrication method is ideally suited to bodies of revolution but is limited to convex shapes. This restriction is due to the necessity for maintaining tension on the filaments during winding.

The predominant type of filament winder in the industry is the simple lathe-type system. As the name implies, a mandrel is mounted between a headstock and a tailstock, and the mandrel rotation is caused by a drive mechanism in the headstock. The filaments are wound onto the mandrel as it rotates and are guided by a ring (sometimes referred to as an "eye") attached to an arm that traverses the length of the mandrel (Fig. 7). The arm can traverse in both directions, thereby creating a geodesic winding pattern onto the mandrel. The angle of the helix θ (i.e., the fiber direction) is a function of the mandrel rotational speed and the linear speed of the fiber guide arm. For a uniform, consistent part to be fabricated, the winding tension must be maintained constant.

Filaments may be preimpregnated with a variety of resin systems or wet-wound. Wet winding is accomplished by introducing the liquid resin to the dry fibers before passing them through the guide ring. Precise control of the fiber-to-resin ratio is very difficult with wet winding. When more accurate control of the resin content is required, the preimpregnated filament is preferred. Recently, experimentation with filament winding of thermoplastic (TP) resins has achieved some degree of success [14]. The thermoplastic resins have to be heated to relatively high temperatures ($\sim$ 400°C) as the filament is wound onto the mandrel. For the first layer of material, the mandrel acts as a heat sink, which can affect the cool-down rate and therefore the crystallinity of the resin. Subsequent layers are less affected by the mandrel's acting as a heat sink. The rate at which the TP material can be heated will affect the winding rate, a consideration that does not apply to thermoset resin systems. Most of the work on TP filament winding is proprietary to the corporate developer of the process because of the technical difficulties and the huge potential for cost reduction in part fabrication.

Once the winding of a thermoset resin part has been completed, it is necessary to complete the cure cycle; this is generally accomplished with the part still on the mandrel. The part must be suitably "bagged" before the curing operation can proceed. Special tooling may be assembled around the mandrel, or the mandrel may be

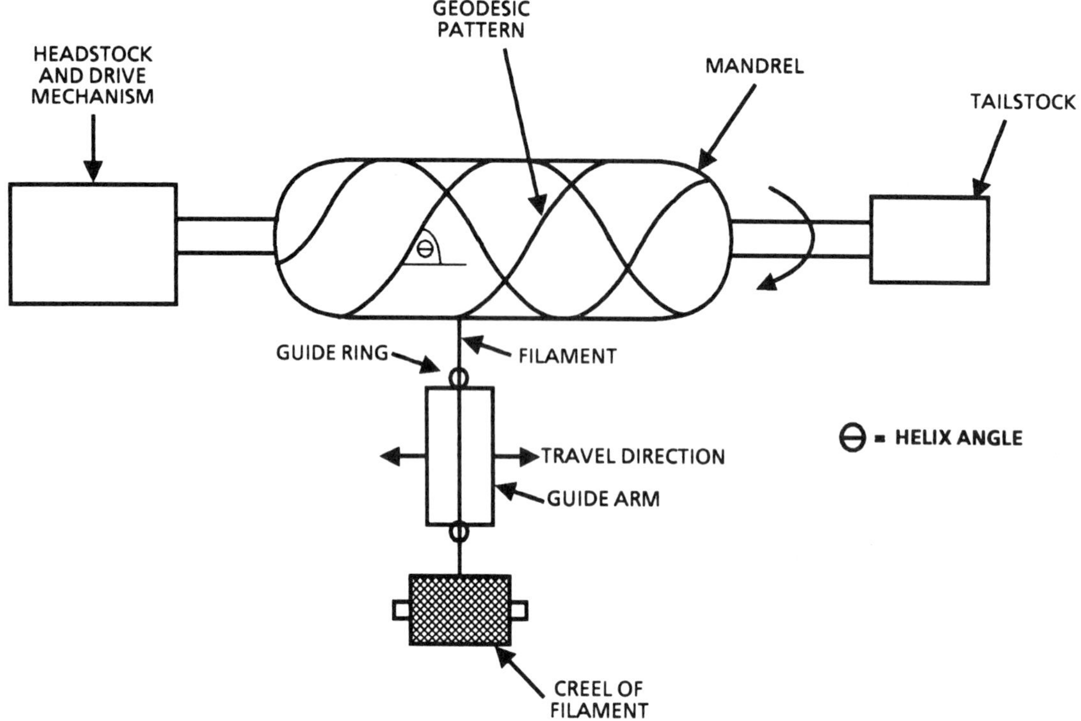

FIGURE 7 **Lathe-type filament winder.**

removed from the winding machine and placed into an autoclave. After the cure cycle, the mandrel must be removed from the part. Several techniques are used, such as dissolving the mandrel, disassembling the mandrel, or sliding the part off one end of the mandrel. It is sometimes necessary to cut the part to remove it from the mandrel.

The basic lathe-type filament-winding machine can be used to perform hoop (circumferential) winding or helical (longitudinal) winding. As the name implies, hoop winding consists of side-by-side "hoops" wound along the part length. Helical winding is accomplished by increasing the linear traverse speed in relation to the rotational mandrel speed, such that the fiber path follows a more elongated helical path. Single or multiple tows may be wound simultaneously. Winding is continued until the surface of the mandrel is completely covered and the required number of layers has been applied.

Polar Winding

Structures requiring a closed end (bottlelike structures) can be wound with a polar winding machine. Polar winding is a special case of helical winding in which the filament path can be described by the intersection of a plane passed through the part (Fig. 8). The mandrel for polar winding is supported on only one end, permitting the filaments to cover the other end completely in a continuous path.

The fiber delivery system can be stationary or movable. In a stationary system the mandrel rotates end-to-end in addition to rotating about its polar axis. The movable fiber delivery system travels around the mandrel in a polar orbit, and mandrel rotation usually is confined to simple incremental rotation.

Consideration must be given to the sagging of the unsupported end of the mandrel. When rotating the mandrel end-to-end, as in the case of a fixed fiber delivery system, care must be taken to prevent a bending or whipping action. Both forms of mandrel bending will affect winding accuracy and consistency. Mounting the mandrel in a vertical fashion eliminates the sagging problem at the expense of complicating the fiber delivery system. A further variation of polar winding is the "tumbling technique," in which the mandrel is tumbled in the longitudinal axis a few degrees as it is rotated. The fiber

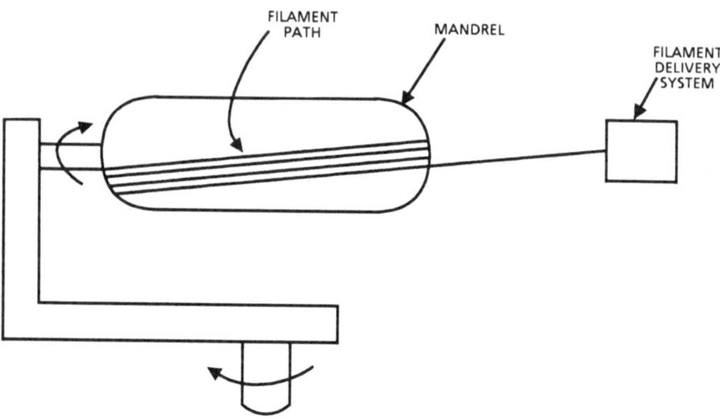

FIGURE 8 Example of polar winding.

delivery system may remain stationary for relatively small mandrel sizes during the tumbling process.

To provide more accurate fiber placement and trace a true geodesic path over the mandrel, it is often necessary to provide additional degrees of freedom to the fiber placement device. A cross-feed axis, perpendicular to the mandrel axis, aids in maintaining a geodesic path over the end dome. Adding a rotational axis to the cross-feed helps keep the fiber guide eye tangential to the point of fiber placement. Filament winding systems can use as few as two axes or as many as seven, depending on part complexity and size. Many system configurations are possible, and the best one for a specific application must be chosen carefully.

Three-Dimensional Weaving [15,16]

Three-dimensional weaving is a method of automatically weaving composite fibers into a structure such that fiber reinforcement is in three principal directions. The advantages of this method of fabrication are that the fibers reinforce all planes, eliminating delamination problems and greatly improving the mechanical properties. There are two basic types of three-dimensional construction:

1. **Cylindrical weave:** generally used for bodies of revolution to form cylinders, conical frustums, etc. Fibers are oriented in the radial, axial, and circumferential directions.
2. **Orthogonal weave:** generally used to fabricate blocks of material that can be machined to net shape after final processing. Fibers are oriented in the x, y, and z directions.

Both methods use dry fibers for weaving with a fiber volume of 45–50%. The matrix material is then added during processing by methods such as liquid resin–pitch impregnation or chemical vapor deposition. Several U.S. companies under various Air Force contracts worked on the development of three-dimensional weaving in the late 1960s and early 1970s (McDonnell-Douglas, Aerojet General, Avco, and General Electric). The techniques for automated composite three-dimensional weaving described here were developed simultaneously by Aerospatiale and Brochier S.A., financed by the Commissariat à L'Energie Atomique (CEA) of France in 1972. From that date the weaving process was refined by increasing development of automated methods. In 1977 Aerospatiale demonstrated an automated weaving machine and began manufacturing cylindrical preforms. A variety of fiber materials have been shown to be adaptable to this method of fabrication: carbon, glass, aramid, etc. The Aerospatiale technology has been licensed to Hercules, Inc., in the United States, and the Brochier S.A./CEA technology has been licensed to Avoc/Textron.

Typical applications for three-dimensional fabrication are rocket nozzles, adapter rings for rocket cases, aircraft structures, space structures, space booster tanks and shrouds, meteroid/debris shields for space structures, and fabricating blocks of material. Fiber materials can be mixed or changed during the weaving process.

CYLINDRICAL WEAVING. The Brochier S.A. (BSA) procedure [15] consists of four operations: a machined foam mandrel corresponding to the inside diameter is first prepared; then a large-diameter spool of a continuous prestiffened radial rod is provided. The radial rods are automatically cut and implanted into the foam mandrel, and finally the axial/circumferential yarns are automatically weaved/wound into the corridors formed by the radial rods.

The Aerospatiale procedure [16] consists of first assembling a series of metal discs, with holes drilled to define the initial radial and circumferential weave spacing of the preform. The discs are assembled onto metal rods that simulate the longitudinal fibers in the preform. The rod–disc assembly is then attached to a bottom vibrating plate to form a network, and the whole assembly is placed into the weaving loom and caused to rotate continuously. As the weaving takes place, a horizontal metal blade ensures compaction of the fibers. Shaping is accomplished by deflecting the metal rods during the weaving operation. Radial and circumferential weaving is first completed; then the wire rods are pushed out and replaced with fiber bundles.

ORTHOGONAL BLOCK WEAVING. In orthogonal block weaving a network of rectangular plates and longitudinal rods is first constructed. Then fibers are placed between the rods in the x and y directions in alternating layers. Finally, the rods are replaced by fibers to complete the weaving process.

Mechanical Prepreg Forming

Various proprietary techniques have been developed for automated mechanical forming of composite flat prepreg laminates [17]. These methods tend to be very part specific at this time, producing shapes such as deep corrugations with a pitch-to-depth ratio of 1:1. Robots have successfully been used to perform the forming action with specialized end-effectors and tooling.

Postcure Forming

Trimming or Routing

Automation of the trimming process was first attempted with robots, which, however, were not accurate enough to accomplish the task freehand. To overcome the robot accuracy problem, guide templates were used for the robot to follow. The robot ensured that the cutter was mechanically in contact with the template at all times, just as a human operator would use the guide. It was understood from the beginning that if the templates could be eliminated, a significant cost saving could be realized. With the manufacture of stiffer, more accurate robots, it was possible to eliminate the templates, and parts are now being trimmed in a freehand fashion [17]. Current technology allows the electronic transfer of two-dimensional part shape information from a CAD system to a robot off-line programming system. The off-line programming of the robot is then performed at a computer workstation. When completed, the program is downloaded to the robot, for the physical process of trimming to be performed. At this time minor adjustments in the robot program are usually necessary; these are done with the robot "Teach" pendant.

The robot is programmed to automatically select the correct cutter for the trimming operation. The cutters may be conventional diamond pattern ground carbide, or diamond-dusted or other conventional forms. Cutting speed ranges are a function of the material thickness and cutter design. Current accuracies are better than 0.76 mm.

Drilling [17,18]

Drilling is accomplished in a manner similar to trimming, using robots to automatically pick up the correct drill. As in the trimming process, drill guide templates were used until the improved robots became available. Guide templates have been eliminated in all but the most critical places.

Water Jet Trimming [2,3]

A system of flexible couplings routes the high pressure water jet (refer to water jet cutting earlier) to the end of the robot arm, which can then freely move to trace out any pattern downloaded to the robot controller. In many applications a garnet material is added to the jet stream after it leaves the sapphire orifice. The abrasive–water mixture is accelerated through a tungsten carbide nozzle and impacts the workpiece faster than the speed of sound. The addition of the garnet particles provides the water jet cutter with the capability to cut through metals, ceramics, composites, and sandwiches of a variety of hard and soft materials. For certain applications, water jet cutting provides additional benefits by eliminating heat-induced changes to the physical characteristics of the material.

The abrasive material helps prevent delamination when cutting composite laminates. Since there are no side loads generated from the water stream back to the manipulator, greater positional accuracies can be achieved with water jet trimming than with mechanical cutters using the same robot.

Terence F. W. Hall

References

1. R. J. Cook, "Waterjets on the Cutting Edge of Machining," *SAMPE Int. Symp., 31,* p. 1835 (1986).
2. Ingersoll-Rand Water Jet Cutting Systems, manufacturer's literature.
3. Flow Systems, Inc., manufacturer's literature.
4. "Composites Manufacturing Operations Production Integration (Automated Production of Aeropropulsion Composite Components)", Hamilton Standard, U.S. Air Force Contract F33615-78-C-5218, September 1983.
5. Final Report, "Composites Manufacturing Operations Production Integration," Northrop Aircraft Division, U.S. Air Force Contract F33615-78-C-5215, January 1986.
6. Final Report, "Composites Manufacturing Operations Production Integration," General Dynamics Fort Worth Division, U.S. Air Force Contract Number F33615-78-C-5217, December 1986.
7. Atlas II/Access Two-Phase System, Goldsworthy Engineering, Inc., Torrance, CA.
8. Anon., "Pultrusion Breakthrough Opens New Markets," *Plast. World,* March 1971. pp. 42–44.
9. Goldsworthy Engineering Inc., company brochures.
10. R. R. Roser, M. L. Skinner, K. J. Samowitz, K. L. Kemp, and B. L. Folsom, *Proceedings of the 31st International SAMPE Symposium,* 1986, pp. 810–821.
11. M. W. Kellogg Company, "Preliminary Study of Rocket Structures by Filament Winding," Research and Development Report SPD 157, Contract NORD 9999, May 5, 1978.
12. Thiokol Chemical Corporation, Redstone Division, Final Report of Work Performed on Modifications 9–13 of Contract DA-01-021-ORD-4996, June 17, 1960.
13. J. W. Kindale et al., Narmco Research and Development, Division of Telecomputing Corp., "Development of Laminating Resins, Controlled Process, and Test Methods for Glass Filament-Wound Reinforced Plastic," Quarterly Progress Reports 1–5, Contract AF33(616)-6737, October 1959–October 1960.
14. B. Latz and T. Kueterman, "Automated Fiber Placement and Consolidation of Thermoplastic Composites," Aeronca Inc., a Fleet Aerospace company, Middletown, OH., 45042; Hsin-nan Chou, McDonnell-Douglas Astronautics Company; and Phillips 66 Company, company literature.
15. P. G. Rolincik, Jr., *Proceedings of the 32nd International SAMPE Symposium,* 1987, pp. 195–207.
16. P. S. Bruno, D. O. Keith, and A. A. Vicario, Jr., *Proceedings of the 31st International SAMPE Symposium,* 1986, pp. 103–116.
17. Final Report, Composites Assembly Production Integration, Northrop Aircraft Division, U.S. Air Force Contract F33615-82-C-5012, 1988.
18. Final Report, Composites Assembly Production Integration, Grumman Aerospace Corporation, U.S. Air Force Contract F33615-82-C-5062, September 1984.

Bibliography

Composite Materials Handbook, Schwartz, Mel M., McGraw-Hill, New York, 1984.

Engineered Materials Handbook, vol. 1, *Composites,* ASM International, Metals Park, OH, 1987.

Marine Composites

See Marine Composite Processing (Supplement).

Market Evaluation*

This article discusses organic matrix advanced composites, which are generally less expensive and further along in development than metallic and ceramic matrix advanced composites; we focus on the structural area, with its less specialized applications, thus excluding the electronics market.

Early structural applications of organic matrix advanced composites included high quality leisure products such as tennis rackets and golf clubs. The major applications for these advanced composites today, however, are in the aerospace industry, which includes both military and commercial aircraft.† Advanced composites and their derivatives, which we call "engineered composites," also have potential to make significant inroads in the ordnance (i.e., nonaerospace military) and automotive industries, both of which offer large-volume applications. These derivatives are highly engineered materials designed to meet the different cost and performance requirements of nonaerospace applications. Engineered composites typically use lower cost fibers and matrix materials.

Both thermoset (e.g., epoxy and polyimide) and thermoplastic (e.g., polyetheretherketone and polyether sulfone) resins may be used in the matrices for advanced composites. Until very recently almost all applications have involved thermoset resins. A newer and faster growing type of advanced composite is based on thermoplastic resins. Advanced thermoplastic composites offer the potential of easier and faster processing than the thermoset type, as well as repairability and a wide range of performance advantages [1].

Fibers used in advanced composites include graphite, aramid (du Pont's Kevlar), and S-glass. Newer types such as oriented polyethylene (Allied-Signal's Spectra) are being evaluation for ballistic protection applications such as helmets, and liquid crystal polymer (LCP) fibers also under evaluation may be used in future composite systems.

Market Size and Growth

Advanced composites are a high value, low volume portion of the reinforced plastics business. For example, in the United States advanced composites account for less than 1% by weight (nearly 6.8 million kilograms) but approximately 5% by value (about $900 million for fabricated parts) of the 1.0 billion kilograms, $2.5 billion reinforced plastics market. Advanced composites also have a much higher annual growth rate than reinforced plastics as a group: 16–17% versus less than 3%.

* This article was adapted from an article that appeared in Arthur D. Little Decision Resources *Spectrum: Chemical Industry Overview* portfolio, September 1987.

† Throughout the article, the term "advanced composites" refers to organic matrix advanced composites unless otherwise noted.

TABLE 1
U.S. and World Markets for Advanced Composites, 1987 and 2000

	1987		2000	
	MM lb	Billion $	MM lb	Billion $
United States	15	0.9	100	5
World	25	1.5	200	10

Notes:
Volume is for intermediate product; value is for final (fabricated) product.
These market figures do not include engineered composites.
These derivatives of advanced composites have potential for significant growth (by a factor of 5-6) by the end of the century from a small base in the ordnance and automotive markets.
Source: Arthur D. Little, Inc., estimates.

The United States accounts for more than half the world market for advanced composites (Table 1). By the end of the century, U.S. demand for advanced composites should reach 45 million kilograms. However, the U.S. share of the world market is expected to decline as the commercial aerospace industry, a large end-use market, becomes increasingly international.

Aerospace applications account for 40% in volume and 60% in value of advanced composites consumed in the United States. The other large market is the recreation or leisure industry, which makes up about 30% in volume and 18% in value. The remaining markets include industrial (a fragmented market with diverse applications) and ordnance and automotive (embryonic markets with large-volume potential). The automotive market, which has received much attention in the past several years, currently accounts for less than 1% of advanced composites consumed in the United States. Most structural automotive applications of composites will involve engineered composites, selected for a combination of cost/performance benefits.

Major User–Industry Markets

Each of the market segments for advanced composites attaches different values to performance improvement, and each represents a different-sized market potential (Fig. 1). While performance is the primary concern in the aerospace industry, the need to reduce the number and cost of parts drives the ordnance and automotive markets, whose potential is much larger. Engineered composites are especially attractive for certain ordnance and automotive applications because they use lower cost materials.

Aerospace

The aerospace industry is the largest dollar-value market for advanced composites and will continue to be so

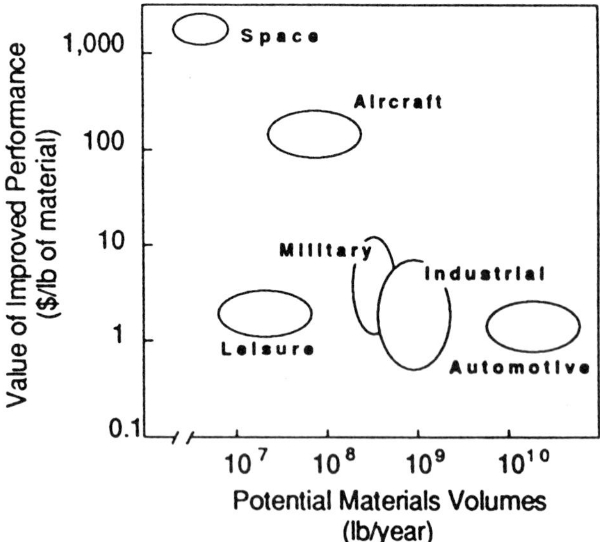

FIGURE 1 **World market potential for advanced and engineered composites as measured by value of improved performance.** (Source: Arthur D. Little, Inc., estimates.)

through the end of the century. The value of fabricated composite aircraft parts (consumed worldwide) will grow from less than $1 billion in 1986 to about $3 billion in 1995. U.S. military and commercial aircraft applications currently represent about $650 million for fabricated composite parts, which could grow to $1.5 billion–2 billion by 1995. These projections include estimates for Stealth programs (involving low visibility, low signature aircraft), which are necessarily speculative.

Substituted for aluminum in aircraft, composites offer weight reductions of 25–40%. The value of weight savings—which varies with aircraft type and mission requirements, as well as with fuel prices—is within the range of $100–1000 per 0.45 kilogram of weight saved.

The introduction of advanced composites into aircraft structure has been gradual, beginning with nonstructural applications such as radomes and hatch covers and then moving to such applications as flaps, ailerons, fairings, and nacelles. Advanced composites are now also being used for tail sections (i.e., vertical and horizontal stabilizers) and for wing skins. To date, advanced composites have achieved only 10–20% penetration of the applications for which they are technically appropriate. Continued growth is anticipated at an average annual rate of 16–17% over the next 10 years.

The military sector presently offers the greatest opportunities for advanced composites, and this circumstance will continue through the mid-1990s. Several existing fighter planes such as the AV8-B have a substantial amount of composites (up to 30% of their structural weight). At this time the major opportunities for advanced composite materials in the military sector are for the Advanced Tactical Fighter (ATF), the V-22 Osprey (vertical takeoff aircraft), and the LHX helicopter. A conservative estimate of advanced composites usage for the ATF, for example, is approximately 2268 kilograms per plane (with 150 units produced in 1995). In addition to these known opportunities, the ATA, a Navy version of the ATF, is a possibility. There are also programs involving Stealth aircraft. These programs are highly classified, and very little public information about them exists; however, substantial usage of advanced composites is envisioned.

The shift from metals to advanced composites in the commercial aerospace industry is occurring more slowly than many industry followers had expected. Development programs (including the Boeing 7J7) have experienced delays, and metals have been selected for some parts for which advanced composites were thought to have an advantage. A major stumbling block has been uncertainty about the ability to achieve cost reduction goals for final part fabrication. The major opportunity for advanced composite materials in the commercial U.S. aerospace sector is still the Boeing 7J7. The tail section of this aircraft is likely to be made of composite material. Although the wings will still be aluminum, the choice between metal and advanced composites for the wings was close, and it is likely that the next new commercial plane will have composite wings (i.e., skins, ribs, and secondary spars).

Some reasons for advanced composites' slow penetration of the overall aerospace market are as follows:

- The extreme consequences of failure and the resulting conservativeness with respect to introducing new materials and technology
- The need to develop new design technology specifically for composites
- Labor-intensive and expensive methods of fabrication.

Increasing internationalization of the commercial aircraft market and limits on the transfer of technology to outside the United States will challenge U.S. dominance of the aerospace market. With the Airbus A-320, for example, European commercial aircraft manufacturers have already gone ahead of their U.S. competitors in the use of advanced composites. At best, U.S. companies can participate in the near future by supplying parts for the Airbus. U.S. companies will find it difficult to compete in the growing European commercial aerospace industry because of the difficulties of technology transfer outside the United States.

NEW DESIGN TECHNOLOGY. Advanced composite design capability has improved enormously with experience over the past 15–20 years, and confidence in the use of composites has increased correspondingly. Thus, achievement of the projected growth for composites in aircraft applications depends largely on improved fabrication technology that will increase efficiency, reduce costs, and offer greater quality consistency.

EXPENSIVE FABRICATION. The real cost of advanced composite parts is in fabrication, which can be in the $300–$500 0.45 kg range—compared with material costs of \$15–\$50 0.45 kg and costs of preimpregnated tape or sheet product (prepreg) of \$40–\$75 0.45 kg. Fabrication requires on the average more than 10 hours of labor per pound of fabricated composite airframe component. Approximately 25% of labor is associated with inspection. Scrap is relatively low—typically 30–40%. In addition to the high expense of fabrication, part of the problem for the aerospace industry is the difficulty of accurately predicting these costs.

Advanced composite materials used in aircraft are continuously improving, and those selected for applications are changing. For example, boron–epoxy was used in the first major aircraft applications of composite material: the horizontal stabilizer of the Grumman F-14 (a Navy plane) in the late 1960s and various parts of the McDonnell Douglas F-15 (an Air Force plane) in the early 1970s. Since these applications, the properties of graphite have been improved to levels comparable to those of boron and at a much lower cost (graphite now costs much less than boron, and the improved graphite is less costly than the earlier graphite). Thus, subsequent aircraft applications (e.g., the General Dynamics F-16 and the McDonnell Douglas F-18 and AV8-B) have involved graphite fibers. The prototype LHX helicopter has an all-composite fuselage consisting of a graphite–epoxy primary structure and Kevlar–epoxy components for energy absorption and maneuverability. Sumitomo, Hoechst Celanese, and Allied-Signal have polyester LCP fibers in production or under development for use in advanced composites. Such fibers will compete with Kevlar and will have potential for use in aerospace applications.

New aircraft designs require higher skin operating temperatures for increased speed. This has led to the displacement of traditional epoxy resins by polyimides and bismaleimides for wing skins and other structural components. NASA has developed and licensed a higher performance thermoset resin (polyimide) known as PMR-15. To achieve still greater toughness and higher operating temperatures (in the 260–316°C range), thermoplastic resins such as ICI's polyetheretherketone (PEEK) and, more recently, du Pont's polyimide K-III, are being evaluated for future aircraft applications such as the ATF. These thermoplastics have the ability to maintain structural integrity at high operating temperatures.

The windows of opportunity for new materials associated with the design cycles of new and substantially redesigned aircraft are limited by time requirements and costs. Much time is required to qualify new materials, and costs are high. For example, materials are now being selected for the ATF. Prototypes were produced in 1988 and 1989, and production is scheduled to begin in 1992. The process of qualifying the composite material involved about 2000 tests over 2 years, at a cost of \$10 million. Then, flight articles will be fabricated for static test. The structural test effort, including part fabrication, will cost on the order of \$100 million.

Because of the high costs of introducing a new material, once a material has been selected and qualified for a particular application, it usually is not displaced in that application. Boron, for example, will continue to be used through the current production of the F-14 even though graphite has proved to be superior in terms of both performance and cost. A material is displaced only if it does not perform satisfactorily or if the aircraft design is revised to improve performance. For instance, aluminum was displaced by advanced composites in portions of the wing and fuselage when material and design changes to increase payload were introduced for the Harrier AV8-B (which replaced the Harrier AV8-A).

Ordnance

The ordnance market for advanced composites exists largely in the United States. This embryonic market is very small (\$20 million/year for fabricated parts), but it is expected to grow rapidly beginning in 1990, reaching \$250 million–\$500 million by 1995 and \$1 billion by the end of the century. While projections beyond the early 1990s necessarily incorporate significant technical and market uncertainties, we believe that the opportunities for advanced and engineered composites in this market will continue to increase significantly through the next decade—based on the fact that the U.S. Navy, Army, and Marines each has significant development programs for advanced composites applications.

Ordnance applications of advanced composites include mobile bridging, shelters, vehicles, weapons, ship and submarine structures, and personnel armor (Table 2). Each application has unique requirements that drive the selection of materials and the fabrication processes. Technology level and material value vary among applications.

PERSONNEL ARMOR. While most ordnance applications are in some stage of development, personnel armor is reasonably mature. Military helmets, for example, represent an early application of advanced composites. Today most Department of Defense protection helmets are made of Kevlar-based composites, which replace metal and offer substantial weight reduction. Spectra (Allied-Signal's high density polyethylene fiber) is being considered as an alternative to Kevlar for these applications; its use could result in a weight reduction.

MOBILE BRIDGING. The mobile bridging programs sponsored by the Army and Marines appear to have a high likelihood of success, with initial production on the Heavy Assault Bridge expected to begin between 1990 and 1992 and production on the Tri Arch Bridge following 2–3 years after. Both are generic bridges that would be replicated. These two major programs are likely to result in more than \$200 million of fabricated advanced composite components to be produced over 10–15 years.

The Heavy Assault Bridge, which would replace the Armored Vehicle Launch Bridge (AVLB), requires advanced composites to achieve specified weight restrictions. The major structural members of the bridge, the bottom chord and the shear webs, will be made of a graphite–epoxy composite.

The Tri Arch Bridge, which would replace the Me-

TABLE 2
Selected Ordnance Applications of Advanced and Engineered Composites

Application Areas	Applications	Materials	Likely Production Date
Personnel armor	Helmets, vests	Kevlar/polyvinyl butyral-phenolic resin system	In production
Mobile bridging	Chords, webs	Graphite/epoxy	1990–1992
	Deck, ramp	Graphite/nylon	1992+
	Tension members	Kevlar/urethane	1992+
Hardened shelters		Graphite/epoxy or Kevlar/epoxy with foam or honeycomb core	1988–1990+
Vehicles	Spall armor	Kevlar/epoxy systems	In production
	Turrets	Glass/polyester	1990+
	Integral armor, hulls	Kevlar or glass with epoxy or polyester	1992+
	Drive shafts	Graphite/epoxy	1990+
	Wheels	Glass/epoxy	1990+
Weapons	Launch tubes	Graphite/epoxy	In production
	Carriage, cradle, trails	Graphite/epoxy	1992+
	Roll bar	Glass/epoxy	1992+
Ship structures	False floors, partitions, ducts, above-deck armor	Kevlar or glass/epoxy with Nomex honeycomb core	In production
	Drive shafts	Graphite and glass hybrid/epoxy	In production
	Pumps, valves	Glass/polyester	1990
	Heat exchangers	Graphite/epoxy	1995
	Deck house	Glass/polyester	1995
Submarine structures	Control surfaces	Graphite/epoxy with foam core	1990
	Flasks, tanks, torpedo racks	Graphite/epoxy	1990+
	Stern, bow	Graphite/epoxy	1995+
	Hull	Graphite/epoxy	2000+

Source: Arthur D. Little, Inc., estimates, based on government sources.

dium Girder Bridge, represents 34 miles of bridging, for which an estimated 32 pounds of graphite per foot is required—a potential of 5 million pounds of graphite.

SHELTERS. The hardened C^3I (command, control, communications, and intelligence) system for the new multipurpose Hummer vehicle (to replace the jeep) is likely to be the first advanced composite shelter to go into production. This product, which addresses the need for compact, portable field communication centers operating on vehicles, may represent an annual market of $50 million for fabricated composites. Initial production is scheduled to begin. Initial production for other hardened shelter programs (the ISO and S280, which represent different sized shelters) is likely to begin 2–5 years later and could at least double the shelter market for advanced composites.

Nonhardened shelters (which do not house critical equipment or are not used in the front lines of battle) will likely use engineered composites.

SHIP AND SUBMARINE STRUCTURES. Marine applications of advanced composites are occurring more gradually than the ordnance applications previously discussed. Composites are already used in the bulkheads, partitions, and false flooring of the CG47 Navy ship and the sails of the NRI research submarine. Some armor, above-deck structures, and mechanical equipment (e.g., drive shafts) on surface ships, as well as ballast tanks and some internal structures in the next generation of submarines, eventually will use advanced composites. Research programs also address "further out" opportunities for composites such as the submarine hull.

Automotive

It is difficult to predict the total volume of engineered composites that will be consumed by the automotive industry because most structural applications are in relatively early stages of development (Fig. 2). Potential product liability is a significant barrier for new primary structural applications, especially wheels. In addition, because the industry is very cost sensitive and has high production rates, cost-effective fabrication and the ability to automate production will be crucial to significant market penetration for all applications [2].

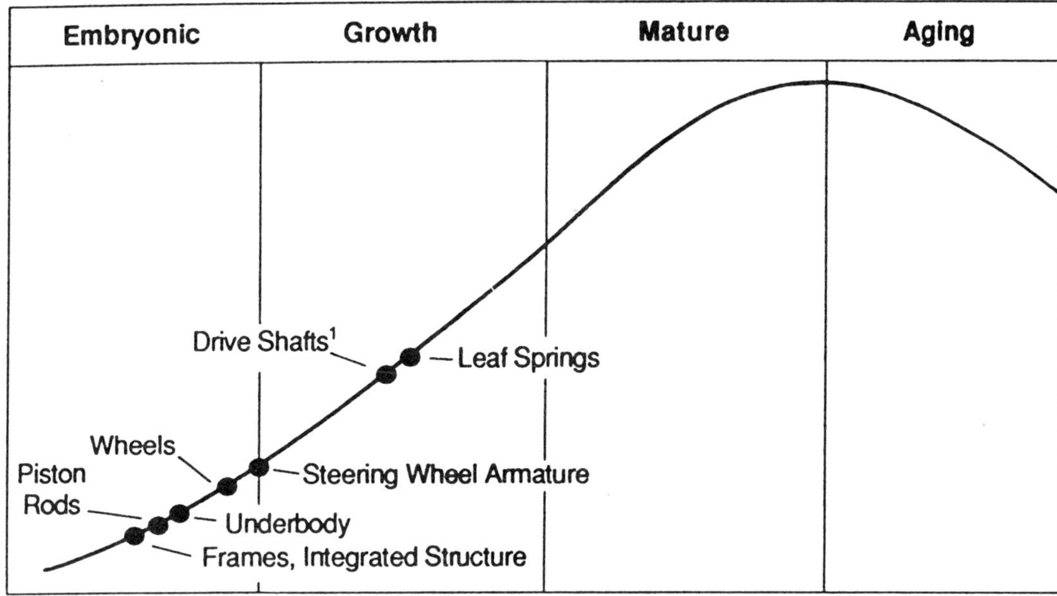

FIGURE 2 Product maturity life cycle of automotive applications of advanced and engineered composites. 1. Limited application, rear-wheel drive only. (Source: Arthur D. Little, Inc.)

LEAF SPRINGS. An application that presently represents more than 5 million pounds per year of composite materials is leaf springs. Use of some 25 million 11.3 kilograms per year in the early 1990s is projected (although this projection could slip in the mid-1990s).* Continued market growth throughout the decade is expected.

The leaf spring stands out as the most successful automotive application of engineered composites. General Motors (GM) introduced composite leaf springs in a production vehicle with the 1981 Corvette and used them in several 1987 models, including the Corvette, Seville, Toronado, Riviera, and Astrovan. Composite leaf springs are believed to be cost-competitive with steel springs because they require less space and because the transverse composite leaf spring can provide lateral stability, eliminating the need for torsion bars. In addition, composite leaf springs have significantly longer fatigue lives than the steel springs they replace. In the long term, this advantage should make them even more attractive for trucks than for cars.

Leaf springs have substantially high performance requirements; hence oriented- and continuous-fiber composites similar to those used in aerospace are being used specified. Because the fibers selected are lower in cost than graphite or aramid, leaf springs represent applications of engineered composites. For example, the materials used in the leaf springs of the 1987 GM vehicles are E-glass and epoxy resin. Vinyl ester resin is under consideration by Ford. GM's Inland Division uses filament winding followed by compression molding to fabricate the composite leaf spring. GKN also uses a two-stage fabrication process: layup followed by slicing. Alcoa/Goldsworthy has developed pulforming fabrication technology for these applications.

* Projections assume that Ford will begin to introduce composite leaf springs for light trucks and that Chrysler will purchase composite springs from GKN for its minivan.

DRIVE SHAFTS. The composite drive shaft, which was introduced in a production model in the 1985 Ford Econoline van, has been a mixed success at best. It is appropriate for rear-axle vehicles and was developed just before the move to front-wheel drive in the U.S. market. Cost did not come down sufficiently with increased production rates. For purposes of cost and weight reduction, composite drive shafts hold most promise for trucks and the larger model vans, where a single shaft can replace two steel pieces, which require an additional universal joint and mount.

Composite drive shafts have undergone an evolution in materials and design since they were introduced. The drive shaft for the 1985 Ford Econoline van consisted of a vinyl ester resin reinforced by graphite in the axial direction and glass fiber in the circumferential direction. Since then, other approaches have been tried to further reduce the high cost of reinforcement materials. The drive shaft developed for the 1987 GM C-body and K-body pickup trucks, for example, consists of an aluminum core with a graphite–epoxy prepreg tape wrap to provide added stiffness.

WHEELS. With styled wheels becoming increasingly popular in today's differentiated market, engineered composites have some potential. They offer the advantages of a cost benefit and a 20% weight savings over aluminum. A production wheel, made by Motor Wheel of fiber glass and a thermoset epoxy resin system, has been 10 years in development. It has not gone into production because of technical problems (high temperature creep) and liability concerns.

Other Markets

The leisure industry was the first to make use of aerospace materials technology. Probably because the consequences of failure are much less severe, leisure market applications for advanced composites proliferated rapidly. The market reached maturity in about a decade and now is very cost-competitive. Applications include sporting goods such as golf clubs, tennis rackets, skis, fishing rods, boat masts, and racing car chassis. Although the United States is clearly the largest market for such applications, most production has moved offshore to achieve lower fabrication costs.

The industrial market for advanced composites started with scattered applications in the 1970s and is still in an early growth stage of development. Its diverse applications today include pressure vessels, robot arms, centrifuges, and computer housings. Again, the United States is the largest market for such applications. Since no application area by itself is large enough to create a significant opportunity for a basic materials supplier, specialized fabricators pursuing niche areas will continue to be predominant.

Peter D. Hilton and Peter W. Kopf

References

1. *Spectrum: Adv. Mater. Chem. Spec.,* September 1987, pp. 1–15.
2. See also the discussion of automotive applications of composites in *Spectrum: Chem. Inf. Overview,* April 1987, pp. 1–21.

Matched Metal Die Molding

See Molding

Material Properties

See Materials Selection, Polymeric Matrix Composites

Materials Selection, Polymeric Matrix Composites

Introduction

With composite plastics, to a greater extent than with other materials, an opportunity exists to continually optimize design by focusing on material composition (Figs. 1–4), structural orientation (Fig. 5), and molding conditions (Fig. 6). Basically a composite is a matrix (plastic material) and reinforcement with usually an interface or coupling agent (Fig. 7, Table 1). There is a practical and easy approach to selecting these materials, as well as to designing and processing the materials. It is essentially no different from selecting among materials of other types—metals, aluminums, ceramics, woods, etc. [1–26].

Composites have been designed into many different products for more than a century with major new composites appearing since 1940 [4,8,17]. These materials provide a range of properties encompassing environmental conditions of all types, each with its own individ-

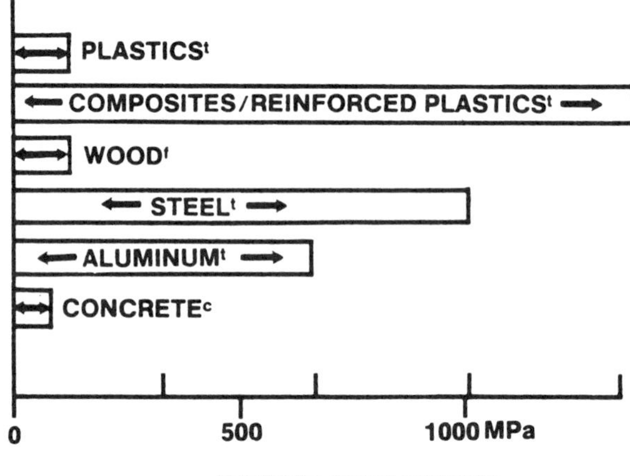

FIGURE 1 Strength.

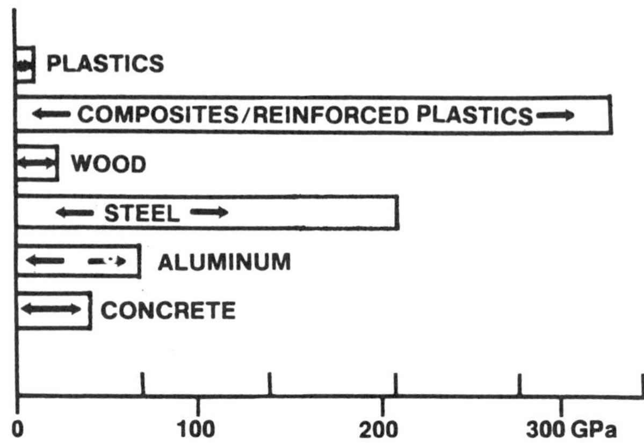

FIGURE 2 Modulus of elasticity.

Elongation	Flexural Strength	Flexural Modulus	Compressive Strength	Impact Strength IZOD	Hardness	Flammability	Specific Heat
%	MPa	GPa	MPa	kJ/m notched @ 23°C	Rockwell (except where noted)		kJ/kg·K
D638	D790	D790	D695	D256	D785	UL-94	
<1.0	179.27	11.03	165.48	0.85	Barcol 68	5V	12.56
0.4	110.32	9.65	158.59	0.43	Barcol 68	5V	12.56
1.7	310.28	13.79	220.64	1.03	Barcol 68	5V	12.56
0.5	88.26	10.89	137.90	0.23	Barcol 68	5V	12.56
0.5	87.22	9.93	—	0.15	Barcol 68	VO	12.56
1.6	689.50	34.48	310.28	2.39	M98	VO	9.63
—	206.85	11.03	206.85	1.33	Barcol 50	VO	11.73
140.0	—	0.26–0.37	—	—	S.D.* 65–75	VO	—
38.9	—	1.03	—	0.11	—	VO	—
1.3	193.06	5.17	151.69	0.69–0.80	Barcol 50	VO	12.98
1.6	317.17	15.51	186.17	1.75	Barcol 50	VO	—
3.0	193.06	7.58	117.22	0.10	M79	HB	—
3.0	199.96	7.65	165.48	0.12	R121	HB	—
1.9	241.33	5.52	182.72	0.12	M95	HB	12.56
9.0	110.32	4.14	96.53	0.11	M80	V-1	12.15
3.0	57.23	3.59	172.38	0.06	R103	HB	—
3.0	255.12	13.10	144.80	0.08	R123	V-O/5V	10.47
2.0	106.87	6.00	96.53	0.06	R107	HB	—
5.0	127.56	5.17	121.35	0.10	R107	HB	8.38–16.75
1.8	131.01	7.58	120.66	0.06	R122	HB	—
4.0	193.06	8.07	124.11	0.10	R118	HB	4.61
6.6	220.64	8.62	172.38	0.10	R120	HB	—
40.0	89.64	2.62	110.32	0.05	M78–M80	HB	14.66
30.0	108.25	2.69	89.64	0.03	R119	HB	16.75
60.0	117.22	2.90	103.43	0.04	R120, M83	V-2	12.56
110.0	93.08	2.34	86.19	0.85	M70	V-2	12.56
200.0	34.48	0.90–1.38	24.13	0.10–1.01	R50-96	HB	18.85
1.0	96.53	3.79	110.32	<0.03	R123	V-O	—
5.0	75.85	2.41–2.76	68.95	0.16–0.32	R107-115	HB	—
50.0	88.26	2.28–2.76	82.74	0.27	R115	V-1	8.38–16.75
0.5	96.53	3.79	96.53	0.02–0.02	M80-85	HB	13.82
50.0	82.74	2.28–2.76	59.30	0.04	M68-78	HB	—
50.0	96.53	2.41–3.10	75.85	0.01–0.03	M94-101	HB	14.24
22.0	—	—	448.18	—	B80	—	4.61
37.0	—	—	330.96	—	B34-52	—	4.19
40.0	—	—	551.60	—	B88	—	5.03
23.0	—	—	337.86	—	R80	—	8.79
2.5	—	—	330.96	—	Brinell 85	—	—
3.0	—	—	227.54	—	Brinell 85	—	10.47
10.0	—	—	282.70	—	Brinell 82	—	4.19

TABLE 1 (*continued*)

Material	Thermal Coefficient of Expansion	Heat Deflection (DTUL)	Thermal Conductivity	Dielectric Strength	Volume Resistivity	Relative Permittivity
	mm/m·K × 10^{-6}	°C @ 1820 kPa	W/mK	mV/m	Ohm-cm	60 Hz
	D696	D648	C177	D149	D257	D150
Glass Fiber Reinforced Thermosets	—	205 +	—	19,695.	5.7 × 10^{14}	4.40
	—	205 +	—	—	—	4.40
	16.9	205 +	—	—	—	4.40
	11.9	260	0.70	14,770.	27 × 10^{14}	4.20
	11.9	260	0.70	14,770.	—	4.20
	3.6	205 +	0.28	11,815.	> 10^{12}	—
	9.0	—	0.58	7,880.	10^{13}	4.40
	140.4	30	—	—	—	—
	95.6	—	—	—	—	—
	21.6	205 +	0.22	9,850.	—	—
	7.2	205 +	—	13,785.	10^{14}	4.20
Glass Fiber Reinforced Thermoplastics	8.5	161	—	22,845.	10^{14}	4.12
	2.7	211	0.84–1.64	19,695.	10^{13}	3.90
	3.2	255	0.22	15,755.	10^{15}	3.80
	3.2	141	0.66	19,695.	10^{16}	3.10
	4.3	132	1.21	17,330.	10^{15}	2.70
	2.0	266	0.29	14,970.	4 × 10^{15}	3.00
	3.8	99	0.20	18,315.	10^{15}	3.20
	3.6	143	0.55	19,695.	10^{16}	3.20
	3.8	102	0.40	19,300.	10^{15}	3.50
	2.5	213	1.01	14,770.	3.2 × 10^{16}	3.80
	3.1	216	0.94	20,485.	10^{16}	3.60
Unreinforced Thermoplastics	8.5	110	0.22	19,695.	10^{15}	3.70
	8.6	68–85	0.17	12,015.	4.5 × 10^{13}	4.00
	8.1	75	0.24	15,165.	1 × 10^{15}	4.00
	6.7	132	0.19	14,970.	> 10^{16}	2.96
	6.8	46–60	0.17	23,635.	> 10^{17}	2.20
	—	135	0.24	14,970.	10^{16}	—
	5.8	93–105	0.14	13,785.–19,695.	2.7 × 10^{16}	2.80
	122.4	100	0.13	15,755.	> 10^{16}	2.65
	64.8	105	0.10	20,285.	4.4 × 10^{16}	3.50
	8.1	50–85	0.15–0.24	16,545.–21,665.	4 × 10^{16}	3.10
	—	38–41	0.13	—	> 10^{16}	—
Metals	12.2	—	3.60	—	—	—
	12.1	—	5.04	—	—	—
	17.3	—	1.35	—	—	—
	25.0	—	13.25	—	—	—
	20.9	—	7.63	—	—	—
	25.2	—	6.02	—	—	—
	27.4	—	9.40	—	—	—

Arc Resistance	Water Absorption	Mold Shrinkage	Resistance to Acids, Alkalis and Organic Solvents
seconds	Percent 24 hours	in/in	
D495	D570	D955	
188	.25	—	Epoxies are highly resistant to water, alkalis, and organic solvents. Although standard epoxies have only a fair resistance to acids and oxidizing agents, they can be formulated for better resistance to these agents. They should be tested before use.
188	.10	.002	
188	.50	—	
190	.20	.001	
190	.20	.004	Polyester thermosets are available in many formulations. They can be formulated for good to excellent resistance to acids, weak alkalis, and organic solvents. However, they are not recommended for use with strong alkalis.
—	.50	.008	
80	.75	—	
—	—	—	Polyurethane is normally used for its structural properties when molded. It is highly resistant to most organic solvents. It is not recommended for use with strong acids and alkalis, steam, fuels, and ketones.
—	—	—	
—	1.30	—	
—	.50	—	
142	.29	.004	ABS is highly resistant to weak acids and alkalis and provides good resistance to most organic solvents. It is attacked by sulfuric and nitric acids and is soluble in esters, ketones, and ethylene dichloride.
120	1.30	.004	
120	.50	.002	
125	.14	.005	Acetal is highly resistant to strong alkalis. Most organic solvents do not seriously alter its properties (test before use). It is not recommended for use in strong acids.
120	.05	.003	
125	.01	.002	
80	.30	.002	Nylons, as a group, are inert to most organic solvents. They resist alkalis and salt solutions and they are attacked by strong mineral acids and oxidizing agents.
70	.24	.003	
70	.06	.002	Polyphenylene oxide provides excellent resistance to aqueous media. It is softened by aromatic hydrocarbons.
135	.06	.003	
90	.05	.003	Polycarbonates resist weak acids and alkalis, oil, and grease. They are attacked by strong acids, alkalis, organic solvents, and fuels (test before use).
129	.22	.020	
—	1.3–1.9	.005	Thermoplastic polyesters are resistant to most organic solvents, weak acids, and alkalis. They are not recommended with strong acids and alkalis or for prolonged use in hot water.
120	1.0–1.3	.008	
120	.15	.005–.007	Polypropylene provides good resistance to acids, alkalis, and organic solvents even at higher temperatures. It is, however, soluble in chlorinated hydrocarbons.
125	.03	.020	
—	< .02	.007	
—	.20–.45	.004–.009	Polyphenylene sulfide provides excellent resistance to organic solvents (below 191°C). It is unaffected by strong alkalis or aqueous organic salt solutions.
75	.07	.005–.007	
65	.20–.35	—	Polystyrene provides good resistance to alkalis, most organic solvents, weak acids, and household chemicals. It is not recommended for use with strong acids, ketones, esters, and some chlorinated hydrocarbons (test before use).
184	.08–.09	.015–.020	
—	.1–.2	.02–.025	
—	—	—	Cold rolled steels are rusted by water, oxygen, and salt solutions. Not recommended for use with acids. They do, however, have good resistance to alkalis.
—	—	—	
—	—	—	Stainless steels have poor acid resistivity (especially hydrochloride and sulfuric) as well as poor chloride solution resistivity. They do offer good resistance to alkalis and organic solvents.
—	—	—	
—	—	—	
—	—	—	Aluminum has poor acid resistance (especially hydrochloride and sulfuric) and also poor chloride salt resistance (must be chemically treated for appearance when exposed to weather).
—	—	—	
			Magnesium poorly resists acids (except hydrofluoric), however, it has good resistance to alkalis. It corrodes in the presence of salt, salt spray, or industrial atmospheres.
			Zinc offers good atmospheric resistance but is not recommended for use with strong acids, bases, and steam.

TABLE 2
Resistance of Plastic Against Chemicals

		Resistance Against[a]																
		Aromatic Solvents		Aliphatic Solvents		Chlorinated Solvents		Weak Bases and Salts		Strong Bases		Strong Acids		Strong Oxidants		Esters and Ketones		24-h Water Absorption
Plastic at °C	**Abbreviation**	**25**	**90**	**25**	**90**	**25**	**90**	**25**	**90**	**25**	**90**	**25**	**90**	**25**	**90**	**25**	**90**	**Change % by weight**
acetals		1–4	2–4	1	2	1–2	4	1–3	2–5	1–5	2–5	5	5	5	5	1	2–3	0.22–0.25
acrylics		5	5	2	3	5	5	1	3	2	5	4	4–5	5	5	5	5	0.2–0.4
acrylonitrile–butadiene–styrene	ABS	4	5	2	3–5	3–5	5	1	2–4	1	2–4	1–4	5	1–5	5	3–5	5	0.1–0.4
aramids (aromatic polyamide)		1	1	1	1	1	1	2	3	4	5	3	4	2	5	1	2	0.6
block copolymers, crystallizable		2	4	2	4	4	5	1	1	1	1	1	3	1	4	1	3	< 0.01
cellulose acetates	CA	2	3	2	3	3	4	2	3	3	5	3	5	3	5	5	5	2–7
cellulose acetate butyrates	CAB	4	5	1	3	3	4	2	4	3	5	3	5	3	5	5	5	0.9–2.0
cellulose acetate propionates	CAP	4	5	1	3	3	4	1	2	3	5	3	5	3	5	5	5	1.3–2.8
diallyl phthalates, filled	DAP	1–2	2–4	2	3	2	4	2	3	2	4	1–2	2–3	2	4	3–4	4–5	0.2–0.7
epoxies		1	2	1	2	1–2	3–4	1	1–2	1	2	2–3	3–4	4	4–5	2	3–4	0.01–0.10
ethylene–vinyl acetates	EVA	5	5	5	5	5	5	1	2	1	5	1	5	1	5	2	5	0.05–0.13
ethylene–tetrafluoroethylene copolymers	ETFE	1	1	1	1	1	1	1	1	1	1	1	1	1	1	1	1	< 0.03
fluorinated ethylene–propylenes	FEP	1	1	1	1	1	1	1	1	1	1	1	1	1	1	1	1	< 0.01
perfluoroalkoxies	PFA	1	1	1	1	1	1	1	1	1	1	1	1	1	1	1	1	< 0.03
polychlorotrifluoroethylenes	PCTFE	1	1	1	1	3	4	1	1	1	1	1	1	1	1	1	1	0.01–0.10
polytetrafluorethylenes	PTFE	1	1	1	1	1	1	1	1	1	1	1	1	1	1	1	1	0
furans		1	1	1	1	1	1	2	2	2	2	1	1	5	5	1	1	0.01–0.20
ionomers		2	4	1	4	4	4	1	4	1	4	2	4	1	5	1	4	0.1–1.4
melamines, filled		1	1	1	1	1	1	2	3	2	3	2	3	2	3	1	2	0.01–1.30
nitriles (high barrier alloys of ABS or SAN)		1	4	1	2–4	1–4	2–5	1	2–4	1	2–4	2–5	5	3–5	5	1–5	5	0.2–0.5
nylons		1	1	1	1	1	2	1	2	2	3	5	5	5	5	1	1	0.2–1.9
phenolics, filled		1	1	1	1	1	1	2	3	3	5	1	1	4	5	2	2	0.1–2.0
polyamide-imides		1	1	1	1	2	3	1	1	3	4	2	3	2	3	1	1	0.22–0.28
polyarylsulfones	PAS	4	5	2	3	4	5	1	2	2	2	1	1	2	4	3	4	1.2–1.8
polybutylenes	PB	3	5	1	5	4	5	1	2	1	3	1	3	1	4	1	3	< 0.01–0.3
polycarbonates	PC	5	5	1	1	5	5	1	5	5	5	1	1	1	1	5	5	0.15–0.35
polyesters, thermoplastic		2	5	1	3–5	3	5	1	3–4	2	5	3	4–5	2	3–5	2	3–4	0.06–0.09
polyesters, thermoset, glass-fiber-filled		1–3	3–5	2	3	2	4	2	3	3	5	2	3	2	4	3–4	4–5	0.01–2.50
polyethylenes LDPE–HDPE		4	5	4	5	4	5	1	1	1	1	1–2	1–2	1–3	3–5	2	3	0.00–0.01
UHMWPE[b]		3	4	3	4	3	4	1	1	1	1	1	1	1	1	3	4	< 0.01

TABLE 2 ***(continued)***

Plastic at °C	Abbreviation	Aromatic Solvents 25	Aromatic Solvents 90	Aliphatic Solvents 25	Aliphatic Solvents 90	Chlorinated Solvents 25	Chlorinated Solvents 90	Weak Bases and Salts 25	Weak Bases and Salts 90	Strong Bases 25	Strong Bases 90	Strong Acids 25	Strong Acids 90	Strong Oxidants 25	Strong Oxidants 90	Esters and Ketones 25	Esters and Ketones 90	24-h Water Absorption: Change % by weight
		Resistance Against[a]																
polyimides		1	1	1	1	1	1	2	3	4	5	3	4	2	5	1	1	0.3–0.4
poly(phenylene oxide)s, (modified)	PPO	4	5	2	3	4	5	1	1	1	1	1	2	1	2	2	3	0.06–0.07
poly(phenylene sulfide)s	PPS	1	2	1	1	1	2	1	1	1	1	1	1	1	2	1	1	< 0.05
poly(phenyl sulfone)s		4	4	1	1	5	5	1	1	1	1	1	1	1	1	3	4	0.5
polypropylenes	PP	2	4	2	4	2–3	4–5	1	1	1	1	1	2–3	2–3	4–5	2	4	0.01–0.03
polystyrenes	PS	4	5	4	5	5	5	1	5	1	5	4	5	4	5	4	5	0.03–0.60
polysulfones		4	4	1	1	5	5	1	1	1	1	1	1	1	1	3	4	0.2–0.3
polyurethanes	PUR	3	4	2	3	4	5	2–3	3–4	2–3	3–4	2–3	3–4	4	4	4	5	0.02–1.50
poly(vinyl chloride)s	PVC	4	5	1	5	5	5	1	5	1	5	1	5	2	5	4	5	0.04–1.00
poly(vinyl chloride)s, chlorinated	CPVC	4	4	1	2	5	5	1	2	1	2	1	2	2	3	4	5	0.04–0.45
poly(vinylidene fluoride)s	PVDF	1	1	1	1	1	1	1	1	1	2	1	2	1	2	3	5	0.04
silicones		4	4	2	3	4	5	1	2	4	5	3	4	4	5	2	4	0.1–0.2
styrene–acrylonitriles	SAN	4	5	3	4	3	5	1	3	1	3	1	3	3	4	4	5	0.20–0.35
ureas, filled		1	3	1	3	1	3	2	3	2	3	4	5	2	3	1	2	0.4–0.8
vinyl esters, glass-fiber-filled		1	3	1–2	2–4	1–2	4	1	3	1	3	1	2	2	3	3–4	4–5	0.01–2.50

[a] On a descending scale from 1 to 7.
[b] Ultrahigh molecular weight polyethylene.

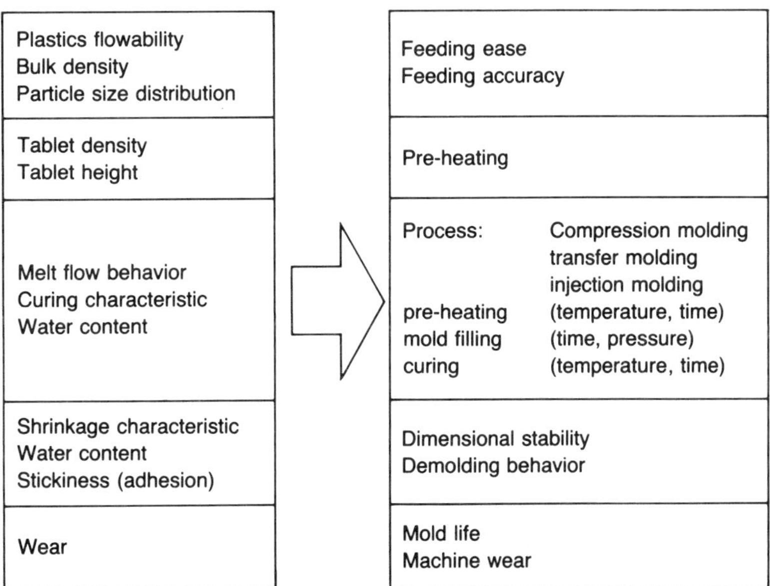

FIGURE 6 **Correlating composite's processing behavior to product performance.**

TABLE 3
Comparison of Typical Processes, Properties, and Costs Used for Composites

	Resin Transfer Molding	Injection Molding	Pultrusion	RRIM
Factor Limiting Maximum Size of Part	machine size	machine size	materials	Metering Equipment
Maximum Size to Date (m^2)	9.3	9.3	0.0058 to 4.46 × 8.92	4.6
Shape Limitations	moldable	moldable	round, rectangular	moldable
Usual Useful Production	medium	high	medium	medium-high
Volume (No. Parts/Year)	(1000–20,000)	(50,000–1,000,000)	10,000 ft.	(15,000–100,000)
Production Cycle Time	10–20 minutes	15 seconds to 15 minutes	10 to 30 minutes	1–2 minutes
Typical Glass Contest (%)	15–25	20–40	30–75	5–25
Strength Orientation	random	random	highly oriented	with flow
Strength Category	low to medium	low	high	low
Wall Thickness:				
—Minimum (mm)	0.76	0.76	1.59	2.03
—Maximum (mm)	25.4	12.7–25.4	12.7	12.7
—Tolerance (in.)	—	—	+ 0.01 ± 1.0	± .002″
—Variations	Uniform	Uniform	Uniform	Uniform
Minimum Draft:				
—to 15 cm depth	1°	1°	0–2°	1°–3°
—over 15 cm depth	1°	1° +	0–2°	3° +
Minimum Inside Radius (in.)	½ part depth	½ part depth	0.06	½ part depth
Ribs	yes	yes	no	yes
Bosses	yes	yes	no	yes
Undercuts	possible	possible	no	yes, with proper
Holes:				
—Parallel	yes	yes	no	yes
—Perpendicular	yes	yes	no	yes
—Built-in cores	yes	yes	no	yes
Metal Inserts	yes	yes	no	yes
Metal Edge Stiffners	yes	yes	no	yes
Surface Finish:				
—Number of finished Surfaces	all	2	2	2
—Quality of Surface	excellent	excellent	fair to good	excellent
—Gel-coat surface		yes	no	no
—Surfacing mat	yes	yes	no	no
—Combination with thermoplastic liner	yes	yes	no	no
Trim in Mold	no	no	no	yes
Molded-in Labels	yes	yes	no	no
Raised Numbers	yes	yes	no	yes
Translucency	yes	yes	no	no
Tool Cost	high	high	low	low-medium
Capital Equipment Cost	high	high	low	low-medium

From Ref. 6.

plastics as a binder to permit flow and compactness of the composites. Reinforcements include stainless steel wire as well as the usual types (Fig. 11). Complex metal- or glass-reinforced composites provide products of complex shapes that have no (or very little) secondary machining. Compositions include solid propellant composites that performed as structures or "containers" for an explosive.

The plastic composites used and discussed throughout the world are identified by different terms depending on past and present industry markets such as composites, reinforced plastics (RP), reinforced thermosets

Hand Lay-up Spray-up	Filament Winding	Compression: Sheet Molding Compound	Compression: Bulk Molding Compound	Preform Molding
mold size; part transport	winding machine	press rating and size	press rating and size	press rating and size
278.7	92.9	4.6	4.6	18.6
none	surface of revolution	moldable	moldable	moldable
low-medium	low-medium	high	high	high
(0–1000)	(0–1000)	(1000–1,000,000)	(1000–1,000,000)	(1000–1,000,000)
3 minutes to 24 hours	5 minutes	$1\frac{1}{2}$ to 5 minutes	$1\frac{1}{2}$ to 5 minutes	$1\frac{1}{2}$ to 5 minutes
20–35	65–90	15–35	15–35	24–45
random (usually)	highly oriented	random	random	random
medium	very high	low-medium	low-medium	medium-high
0.76	0.25	0.76	1.52	.76
38.1-up	50.8	6.35	25.4	6.35
+ 0.02	+ 0.01	+ 0.008	+ 0.005	+ 0.005
as desired	as desired	uniform desirable < 3:1	as desired	uniform desirable < 2:1
0–2°	3°	1°–3°	1°–3°	1°–3°
0–2°	3° +	3° +	3° +	3° +
0.25	0.125	0.06	0.06	0.125
yes	no	yes	yes	not recommended
yes	no	yes	yes	not recommended
avoid	no	avoid	no	no
yes	yes	yes	yes	not recommended
yes	—	undesirable	undesirable	undesirable
possible	possible	possible	possible	possible
yes	yes	yes	yes	not recommended
yes	no	no	no	yes
1	1	2	all	2
excellent	excellent	very good	excellent	very good
yes	yes	no	no	yes
yes	yes	no	no	yes
yes	yes	no	no	no
no	yes	yes	yes	yes
yes	yes	difficult	difficult	difficult
yes		yes	yes	yes
yes	yes	no	no	yes
low	low	high	high	medium
low	low	high	high	high

(RTS), reinforced thermoplastics (RTP), laminated plastics, short fiber reinforced plastics, filled reinforced plastics, reinforced engineered plastics (REP), advanced plastics composites (APC), and structural composites. They all identify or connote the combination of plastic materials and reinforcing materials. Combinations produce unique materials that can be stiff, strong, hard, elastic, rubbery, tough, rigid, opaque, electrically conductive, reduced in cost, or practically any characteristic that is desired. Unfortunately you cannot obtain all favorable properties in one composite; useful combinations are available, however (toughness, stiffness, strength, chemi-

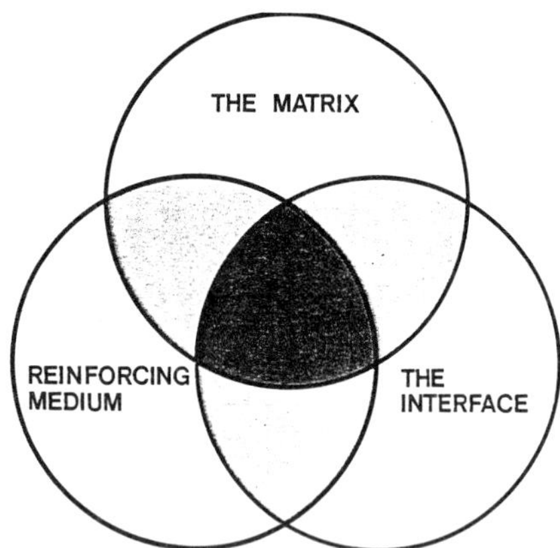

FIGURE 7 Interplay between composite constituents.

TABLE 4
Fillers and/or Additives Used in Many Composites

Natural Organic	Synthetic Organic
α-Cellulose	Aramid
Cotton	Polyacrylonitrile
Jute	Polyamide
Sisal	Polyesters
Pulp	Polyvinyl alcohol
	Rayon
Natural Inorganic	**Synthetic Inorganic**
Asbestos	Metallic (steel, aluminum, etc.)
	Boron
	Carbon, graphite
	Ceramic
	Glass

cal resistance, long-time creep resistance, smooth surface, etc.).

Reinforced or unreinforced plastics usually contain additives or fillers (Tables 4 and 5). Note that sometimes reinforcements are called additives or fillers, and also the reverse. Reinforcements usually represent fibers, filaments, whiskers, and high performance flakes. These different shapes or forms of reinforcement have provided benefits to the industry for more than a century [4,8,10,13,17,18,22].

Plastic Materials

Plastics are a family of materials, not a single material. Each type has its own distinct and special advantages, reinforced or unreinforced (Fig. 12). Most plastics fall into one of two groups: thermoplastics (TP) or thermosets (TS). Tables 6–10 provide examples of these plastics, used unreinforced or reinforced.

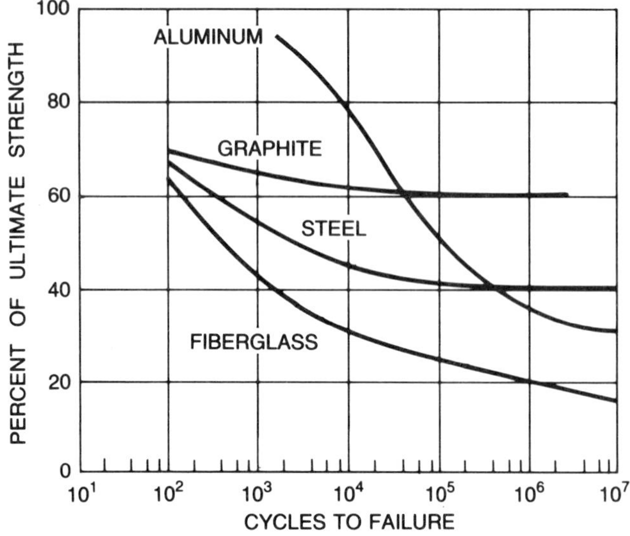

FIGURE 8 **Fatigue properties of graphite and glass fiber thermoset polyester composites, steel, and aluminum.**

The chemical composition of a plastic (also called resin, polymer, matrix, etc.) is basically organic. Very large molecules composed of chains of thousands of carbon atoms are generally connected to hydrogen atoms but also to oxygen, nitrogen, chlorine, fluorine, and sulfur [13,15,26]. The key step in the manufacture of plastics is the polymerization reaction, a chemical process in which many thousands of small monomer molecules are linked together to form large polymer molecules. The incorporation of all monomer molecules into a polymer is called addition polymerization, and the resulting material is called a thermoplastic. Condensation polymerization occurs when parts of the monomers are incorporated into the polymer and the remainder is formed as a by-product.

All plastics flow at some stage. They are soft and pliable and can be formed, usually by the application of heat, pressure, or both, into desired shapes with reinforcements. Some can be cast without pressure and some can be resoftened and rehardened by heating and cooling (thermoplastics); others cannot be softened after they have hardened (thermosets). All these plastics can be reinforced. Improvements in performance can also be gained by alloying (blending two or more plastics); an example is shown in Figure 13. Since the 1940s, thermoplastics have been added to thermoset composites to produce "smoother" surfaces. Fillers as well, as skim fiber sheets or milled fibers, can also be used to provide smooth surfaces [3,4,9,10,22].

The dividing line between thermosets and thermoplastics is not always distinct. Cross-linked thermosets are thermoplastic before cross-linking occurs; the materials may be liquid. Other chain-type polymers such as polyethylene, normally thermoplastic, can be cross-linked by high energy radiation, and they become thermosets [27–29].

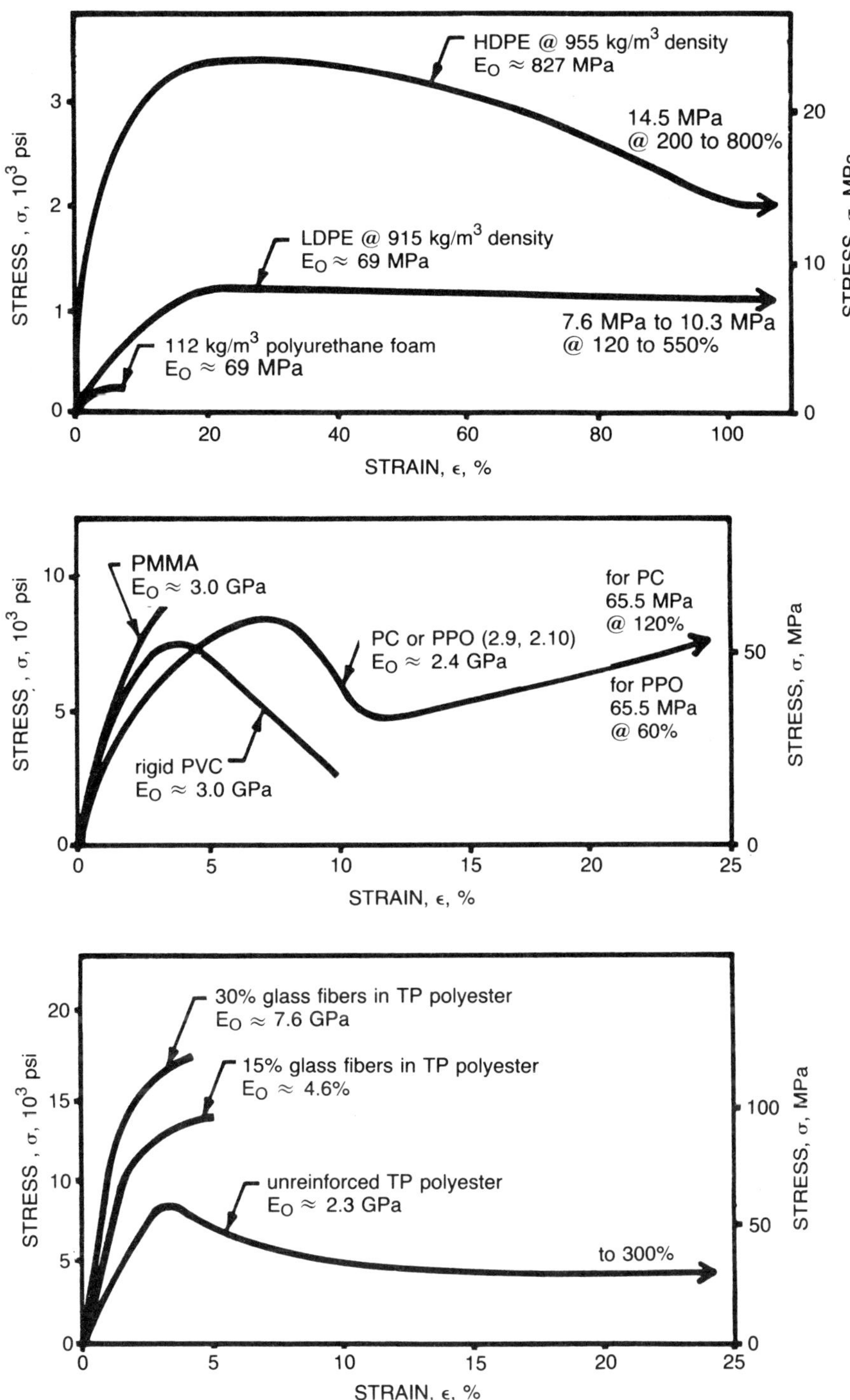

FIGURE 9 **Examples of stress–strain relationships for unreinforced and reinforced plastics.**

THERMOPLASTICS. With thermoplastics, basically no polymer changes occur during forming or processing. They are analogous to a block of ice that can be softened (i.e., turned back into a liquid), poured into any shape of a cavity, then cooled to become a solid again (Fig. 14). This cycle can be repeated, and the repetition can be accompanied by the development of heat degradation. Some plastics, when processed within controlled temperature–pressure–time cycles, can have no deterioration or an insignificant amount, while others undergo significant losses in product performances. Figures 15 and 16 show what can happen with properties of just plastic materials when they are recycled after going through a granulation process [26]. When reinforced

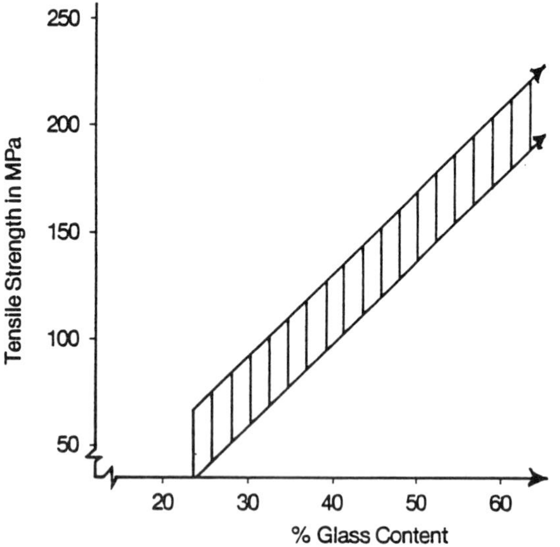

FIGURE 10 Effect on strength versus glass content for glass fiber–polyester resin composite.

plastics are granulated, fiber lengths will be shortened, and this can result in significant losses in performances (see Regrind sections in Refs. 7, 13, and 26).

Thermoplastics are characterized by their morphology (based on molecular structure), ranging from amorphous to crystalline. The degree of crystallinity has direct effects on the performance of the plastics, particularly mechanical properties. Since crystallinity can be varied widely, the structure can range from a flexible to a very stiff, high strength polymer. Since crystallinity is never complete, the term "semicrystalline" is technically the correct descriptor.

Crystalline plastics (polyethylenes, polypropylenes, nylons, acetals, etc.) have uniform and compact molecules, a structure attributed to the formation of crystals, which can be of different sizes, having definite geometric and ordered forms. The amorphous plastics such as polystyrenes, polyvinyl chlorides, ABS, and SAN are just the opposite; they lack an ordered form. Practically all thermoplastics are in the amorphous stage during heat melting. Crystalline types require higher heat loads, etc. for melting [13,26]. If the proper temperature–pressure–time cycles are not used, significant losses in performances can occur with additional cost of processing. The usual problem is that a part is molded at a faster time cycle, requiring a slight or major change in temperature and/or pressure, with the expected result of lower processing cost along with lower performances. Many molders do not recognize that with the faster cycle, the increase in heat results in an energy cost greater than the saving hoped for on the reduced time cycle [26].

To meet all the different product performance requirements both crystalline and semicrystalline materials are used. They have differences as shown in Figure 17. Crystalline types generally have different rates of shrinkage in the longitudinal and transverse directions of melt flow such as during injection molding. This characteristic can be an advantage or a disadvantage [26]. Key basic characteristics for crystalline types are sharp melting point, chemical resistance, low moisture absorption, lubricity, notch sensitivity, and relatively high shrinkage. The properties of the amorphous types include wide

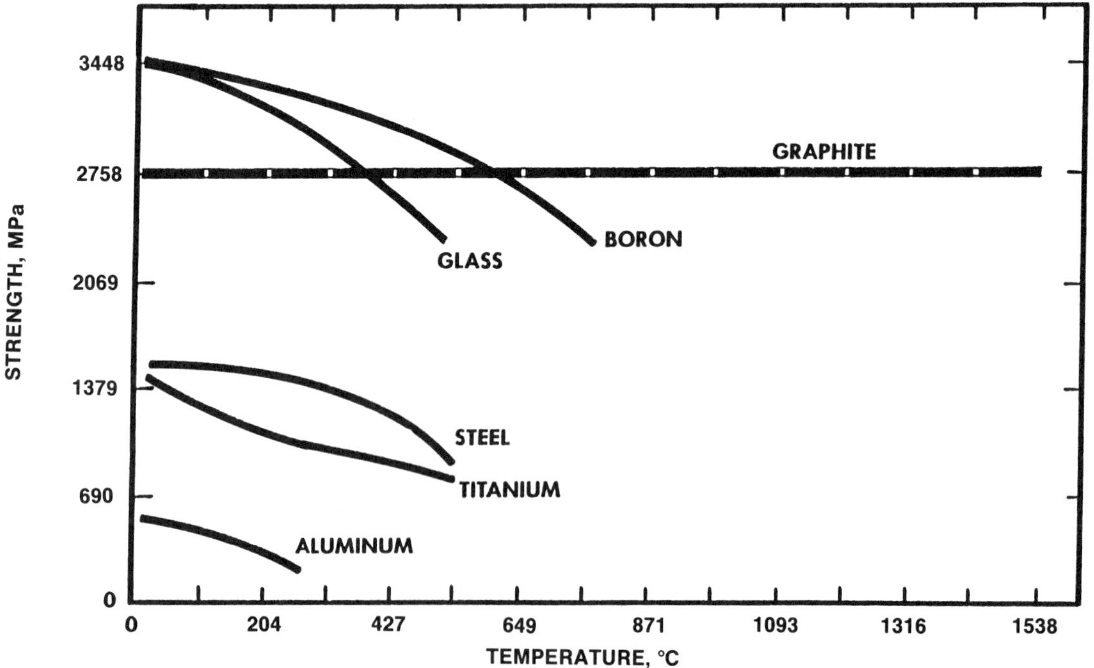

FIGURE 11 Tensile strength properties versus temperature of fibrous reinforcements used in composites.

TABLE 5
Typical Fillers Used with Thermosets

THERMOSETS	ALUMINA	CALCIUM CARBONATE	CARBON BLACK	CLAY	COTTON FLOCK	GLASS BUBBLES	GLASS FIBERS	GRAPHITE	MICA	QUARTZ	TALC	WOOLASTONITE SILICATE	WOODFLOUR
ALKYDS		•		•			•		•		•	•	•
DAP		•		•			•				•		
EPOXY	•	•	•	•		•	•			•		•	
PHENOLIC	•	•	•	•	•	•	•	•	•		•	•	•
POLYESTER		•		•		•	•					•	•
MELAMINE					•		•		•		•	•	
UREA				•	•								•
SILICONE	•			•			•			•			
URETHANE	•	•	•	•		•	•			•		•	

softening range, limited chemical resistance, moderate heat resistance, impact resistance, and low shrinkage.

It is by no means impossible or difficult to process any of these thermoplastics (or thermosets); however, there are differences in processing conditions, and to meet a composite's performance requirements, the correct analysis must be applied in designing the product shape as well as the mold. A good design permits the maximizing of material performance. Equally important is that the proper molding conditions be used. This is not a complicated procedure, but to those who assume that

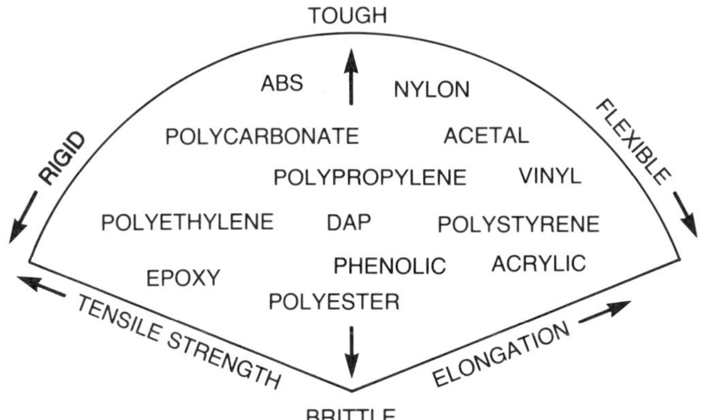

FIGURE 12 Range in mechanical properties for plastics: note that with formulation changes (via basic polymer composition, additives, fillers, reinforcements, alloying with other plastics, etc.), position of generic type plastics can move practically any place in the "pie."

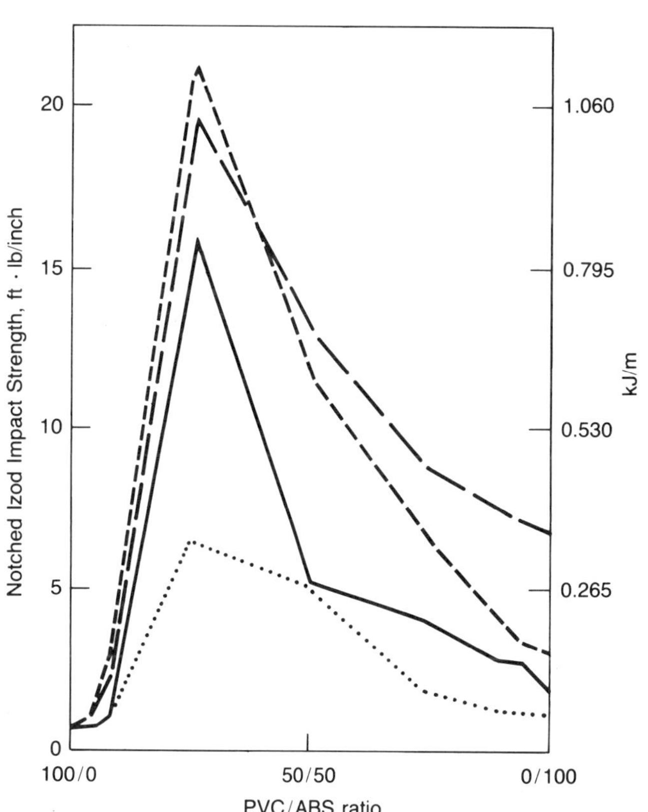

FIGURE 13 Example of how alloying affects plastic properties, resulting in synergistic benefits.

TABLE 6a
Mechanical Properties of Glass Fiber Thermoplastics Composites (ASTM Data)

Plastic	Glass Fiber Content, (wt %)	Specific Gravity, D 792	Tensile Strength (MPa), D 638	Tensile Elongation (%), D 638	Tensile Modulus (GPa), D 638	Flexural Strength (MPa), D 790	Flexural Modulus (GPa), D 790	Compressive Strength (MPa), D 695	Impact Strength, Izod Notched, (J/m), D 256
ABS	10	1.10	65	3.0	4.6	102	4.5	83	64
	20	1.22	76	2.0	5.1	107	4.9	97	59
	30	1.28	90	1.4	6.3	116	6.4	104	53
Acetal	10	1.54	72	2.4	6.6	107	6.1	69	53
	30	1.63	83	2.0	7.7	114	7.2	81	43
Nylon 6	15	1.25	104	4.0	5.9	159	5.4	97	80
	30	1.37	166	3.0	7.2	200	6.9	166	117
Nylon 6/6	13	1.23	97	4.0	6.2	173	4.5	93	53
	30	1.37	173	3.0	9.0	235	9.0	186	107
Nylon 6/12	30	1.30	135	4.0	8.3	193	7.6	138	117
Polycarbonate	10	1.26	83	9.0	5.2	110	4.1	97	107
	30	1.43	121	2.0	8.6	141	6.9	117	128
Polyester, thermoplastic	30	1.52	131	4.0	8.3	193	7.9	124	96
Polyethylene	10	1.04	36	4.0	2.5	46	2.5	35	75
	30	1.18	59	3.0	5.0	89	4.9	41	91
Polyphenylene oxide, modified	20	1.21	100	5.0	6.4	128	5.2	121	96
Polyphenylene sulfide	40	1.64	152	3.0	14.1	255	13.0	145	80
Polypropylene	10	0.98	43	4.0	2.5	54	2.4	41	43
	20	1.04	45	3.0	3.7	57	3.6	45	59
	30	1.12	47	2.0	4.4	63	4.3	47	69
Polypropylene, chemically coupled	10	0.98	50–59	4.0	3.7	72–94	3.5	43–44	64–75
	20	1.04	57–68	3.0	3.9	81–106	3.7	44–47	69–80
	30	1.12	68–83	2.0	4.6	90–131	4.6	45–48	69–91
Polystyrene									
High heat copolymer	20	1.22	90	1.2	8.3	131	7.9	110	59
High heat terpolymer	30	1.35	83	1.8	6.5	123	5.7	76	80
Polysulfone	20	1.38	97	2.5	6.0	138	5.9	124	64
	40	1.55	124	1.5	11.6	173	10.7	138	80
Polyurethane	10	1.22	33	48.0	0.7	43	0.6	35	747
PVC	20	1.58	97	3.0	0.8	145	6.9	83	80
SAN	20	1.22	100	1.8	8.6	131	7.6	121	64
	35	1.35	110	1.4	10.4	155	9.3	45	53

there is only one way of "handling" any plastic, problems are inevitable. The same situation exists with other materials (steels, glasses, concrete, etc.).

Thermoplastics have a considerable advantage over thermosets in enhancing the toughness of composites; moreover, they have unlimited shelf life and need only heat and pressure to process. However it is their potential for high volume processing, and associated low cost per part, that has expanded their use in composites. In fact, the high processing temperatures and pressures required for thermoplastic composites necessitate automated and high speed processes. Because automated processes are best able to take advantage of economies of scale, high volume production goes directly into the use of thermoplastic composite matrices.

THERMOSETS. Thermosets go through a soft plastic stage only once; then they harden irreversibly and cannot be resoftened. Because the first observed irreversible reaction requires heat, such plastics were callled "thermosetting." This term has persisted even though the reaction may occur at ordinary temperatures (no external heat applied). As an analogy for thermosets, think of an egg; once you have applied heat to "harden" it, the cycle cannot be repeated.

There are cross-linked and interlinked reactions

TABLE 6b
Thermal and Electrical Properties of Glass Fiber Thermoplastics Composites (ASTM Data)

Plastic	Glass Fiber Content (wt %)	Heat Deflection Temperature at 1.7 MPa, (°C), D 648	Coefficient of Linear Thermal Expansion (x 10^{-5} cm/cm/K), D 696	Maximum Temperatures Continuous Use (°C)	Water Absorption, 24 h (%), D 570	Volume Resistivity (Ω-cm), D 257	Dielectric Strength Dry (V/μm), D 149	Mold Shrinkage (cm/cm), D 955
ABS	10	98	4.1	77	0.3	10^{15}	17.7	0.003
	20	99	3.8	82	0.3	10^{16}	18.3	0.002
	30	100	3.1	82	0.2	10^{16}	18.9	0.002
Acetal	10	124	5.2	110	0.22	10^{14}	20.0	0.006
	30	163	4.3	127	0.2	10^{14}	18.9	0.005
Nylon 6	15	196	3.1	93	1.8	10^{15}	16.5	0.007
	30	204	2.7	110	1.3	10^{15}	16.1	0.004
Nylon 6/6	13	240	2.7	107	1.0	10^{16}	20.9	0.007
	30	252	2.3	127	0.9	10^{16}	19.7	0.004
Nylon 6/12	30	199	2.2	110	0.2	10^{13}	19.7	0.004
Polycarbonate	10	138	3.2	127	0.14	10^{15}	17.3	0.005
	30	143	2.3	132	0.12	10^{15}	18.9	0.003
Polyester, thermoplastic	30	213	2.5	121	0.06	10^{16}	23.6	0.003
Polyethylene	10	110	5.4	82	0.08	10^{16}	26.8	0.005
	30	124	3.8	93	0.06	10^{16}	24.0	0.003
Polyphenylene oxide, modified	20	143	3.6	116	0.06	10^{17}	16.5	0.003
Polyphenylene sulfide	40	266	2.0	232	0.01	10^{16}	20.0	0.002
Polypropylene	10	127	4.7	82	0.05	10^{16}	17.3	0.007
	20	132	4.3	88	0.05	10^{16}	17.3	0.006
	30	138	3.8	99	0.04	10^{16}	16.5	0.006
Polypropylene, chemically coupled	10	135–141	4.5	93	0.05	10^{16}	16.9	0.006
	20	142–146	4.1	110	0.04	10^{16}	16.7	0.006
	30	146–149	3.6	121	0.04	10^{16}	16.7	0.006
Polystyrene								
high heat copolymer	20	116	4.0	104	0.28	10^{16}	15.7	0.003
high heat terpolymer	30	149	3.6	116	0.10	10^{16}	15.7	0.002
Polysulfone	20	179	2.7	163	0.2	10^{16}	19.7	0.004
	40	185	2.3	171	0.18	10^{16}	18.9	0.002
Polyurethane	10	54	6.1	43	0.4	10^{12}	15.0	0.007
PVC	20	82	4.1	66	0.09	10^{13}	16.5	0.002
SAN	20	102	3.8	82	0.24	10^{16}	19.3	0.002
	35	104	2.9	88	0.21	10^{16}	19.7	0.001

TABLE 7
Thermoset Composites Versus Manufacturing Processes

Thermosets	Properties	Processes
Polyesters	Simplest, most versatile, economical and most widely used family of resins, having good electrical properties, good chemical resistance, especially to acids	Compression molding Filament winding Hand layup Mat molding Pressure bag molding Continuous pultrusion Injection molding Spray-up Centrifugal casting Cold molding Encapsulation
Expoxies	Excellent mechanical properties, dimensional stability, chemical resistance (especially alkalis), low water absorption, self-extinguishing (when halogenated), low shrinkage, good abrasion resistance, very good adhesion properties	Compression molding Filament winding Hand layup Continuous pultrusion Encapsulation Centrifugal casting
Phenolics	Good acid resistance, good electrical properties (except arc resistance), high heat resistance	Compression molding Continuous laminating
Silicones	Highest heat resistance, low water absorption, excellent dielectric properties, high arc resistance	Compression molding Injection molding Encapsulation
Melamines	Good heat resistance, high impact strength	Compression molding
Diallyl phthalate	Good electrical insulation, low water absorption	Compression molding

From Ref. 6.

TABLE 8
Deflection Temperature Test Data

Material	ASTM D 648 Deflection Temperature at 1820 kPa (°C)	Test Results at Deflection Temperature: Maximum Stress Load (MPa)	Test Results at Deflection Temperature: Degrees Deflection in 4 hours	Test Results at Deflection Temperature Plus 10°: Maximum Stress Load (MPa)	Test Results at Deflection Temperature Plus 10°: Degree Deflected/When Occurred
P-Sulfane	175°C	7.24	5	1.90	Failed/< 1 hour
PBT	200°C	3.62	9	1.90	Failed/< 2 hours
PET	210°C	7.24	8	3.62	Failed/< 4 hours
PEI	210°C	7.24	8	1.90	Failed/< 1 hour
PAS	215°C	1.90	Failed		
PPS	260°C	1.90	4	1.90	Failed/< 2 hours
Phenolic HR	210°C	14.48	4	14.48	7°/4 hours
Phenolic glass	270°C	14.48	1	14.48	2°/4 hours

From Ref. 20.

TABLE 9
Flexural–Temperature Test Data

Engineering Polymer	UL Temperature Index (°C)	Flexural Strength at 23°C (MPa)	Percent Retention of Flexural Strength at		
			100°C	150°C	200°C
Nylon 6/6	115	120.0	36	29	23
Nylon 6/6	120	246.8	51	41	36
PET	150	201.3	46	30	24
PET	155	259.3	51	33	26
Phenolic—GP	150	60.7	67	62	42
Phenolic—HR	160	70.3	77	64	48
Phenolic—glass	170	109.6	82	70	54

From Ref. 20.

TABLE 10
Tensile–Temperature Test Data

Engineering Polymer	UL Temperature Index	Tensile Strength at 23°C (MPa)	Percent Retention of Tensile Strength at		
			100°C	150°C	200°C
PPS	22	160.0	48.3	34.9	4.7
PES	180	157.2	85.5	57.7	13.7
Nylon 6/6	130	213.7	51.3	39.7	7.2
PET	140	131.0	38.1	21.0	
P-Sulfane	150	120.0	86.1	13.3	
PAI	200	189.6	72.8	59.1	29.9
Phenolic—GP	150	55.2	83.7	51.2	50.1
Phenolic—HR	160	58.6	71.9	57.6	52.0
Phenolic—glass	170	69.0	75.1	59.7	69.9

From Ref. 20.

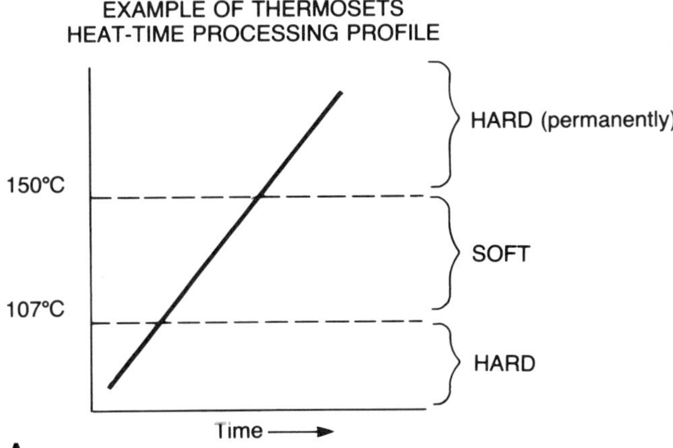

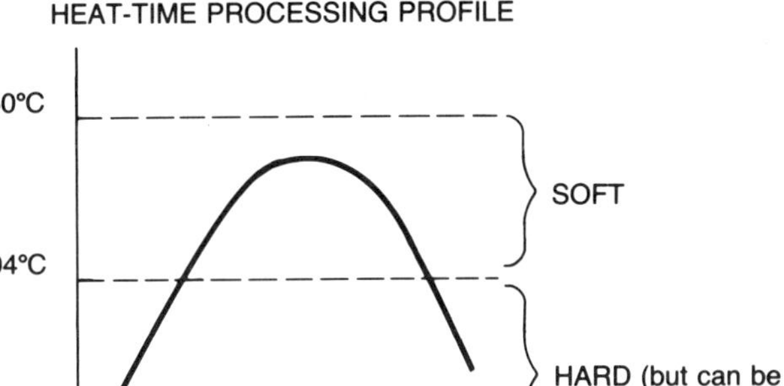

FIGURE 14 Plastics fall into one of two groups: thermosets or thermoplastics. Thermosets have reactive portions of molecules that form cross-links between the long molecules during polymerization. Therefore, once polymerized or hardened, the material cannot be softened by heating without degrading some linkages. Thermoplastics consist of long molecules, either linear or branched, having side chains or groups that are attached to other polymers. They can be repeatedly softened and hardened by heating and cooling. No chemical change generally occurs.

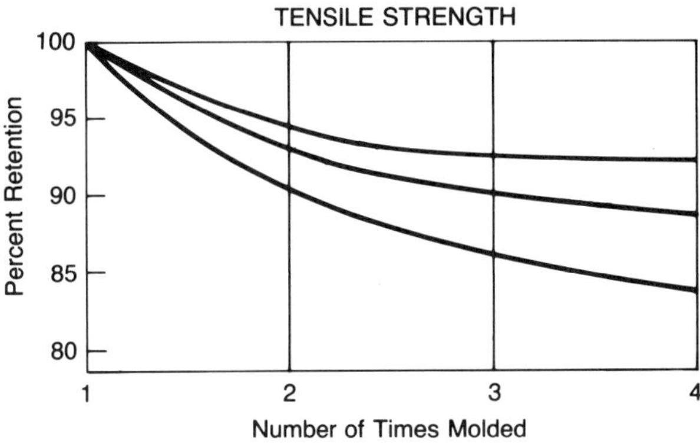

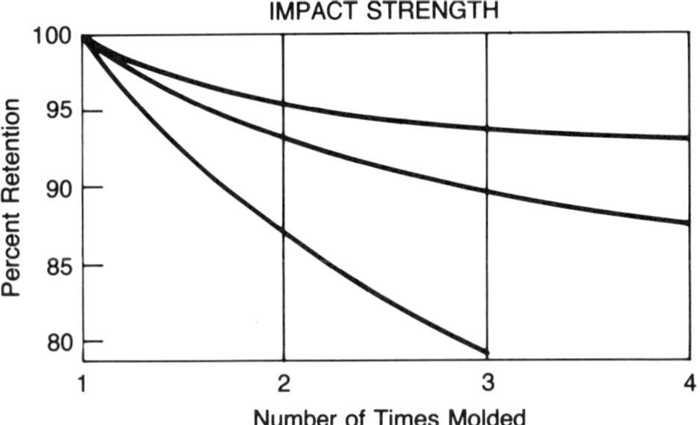

FIGURE 15 Properties of injection-molded thermoplastics used in the virgin form (first time molded) and virgin mixed with regrind. Tensile strength (top) and impact strength (bottom) of regrind and virgin thermoplastic polyester blends. In both diagrams the top curves contain 25% regrind, next at 40%, and bottom curves contain 60% regrind.

where chain molecules form in two- or three-dimensional arrays. Typical thermosets used are reviewed in Table 7. They generally are more suitable to meet the tightest tolerances and long-time use. Some thermosets have exceptional heat retention and are able to maintain performance for a long time.

The major limitations of thermoset composites are product brittleness and various problems associated with the need for plastics to be chemically reacted. The reactivity of the un-crossed-linked plastics and polymers limits their shelf life, extends the processing time, results in scrap material that may be difficult to reclaim (however, more reclaiming will occur for fillers, additives, etc.), and produces a part that usually cannot be easily repaired. (However, nonrepairability can be an advantage at times.) These limitations, by themselves, have not been sufficient to offset the economic advantage of low volume production of thermoset composites and warrant a switch to the high volume, high speed thermoplastic composites.

HEAT AND FIRE RESISTANCE. There are plastics that can provide heat resistance (Fig. 18; Tables 8–10) and plastics that have high flame resistance [4,7,15,27–30]. Generally inorganic filled reinforced thermosets provide the most heat and/or fire resistance, particularly when cost is not a criterion. Some relatively new thermoplastics offer higher than usual heat and/or fire resistance performance. These include polyphenylsulfide (PPS), polyetherimide (PEI), and liquid-crystal polymers (LCP), PEEK, etc.

Reinforcing Materials

As reviewed throughout this book, the reinforcing materials principally used are fibrous glass (at least 95% of the

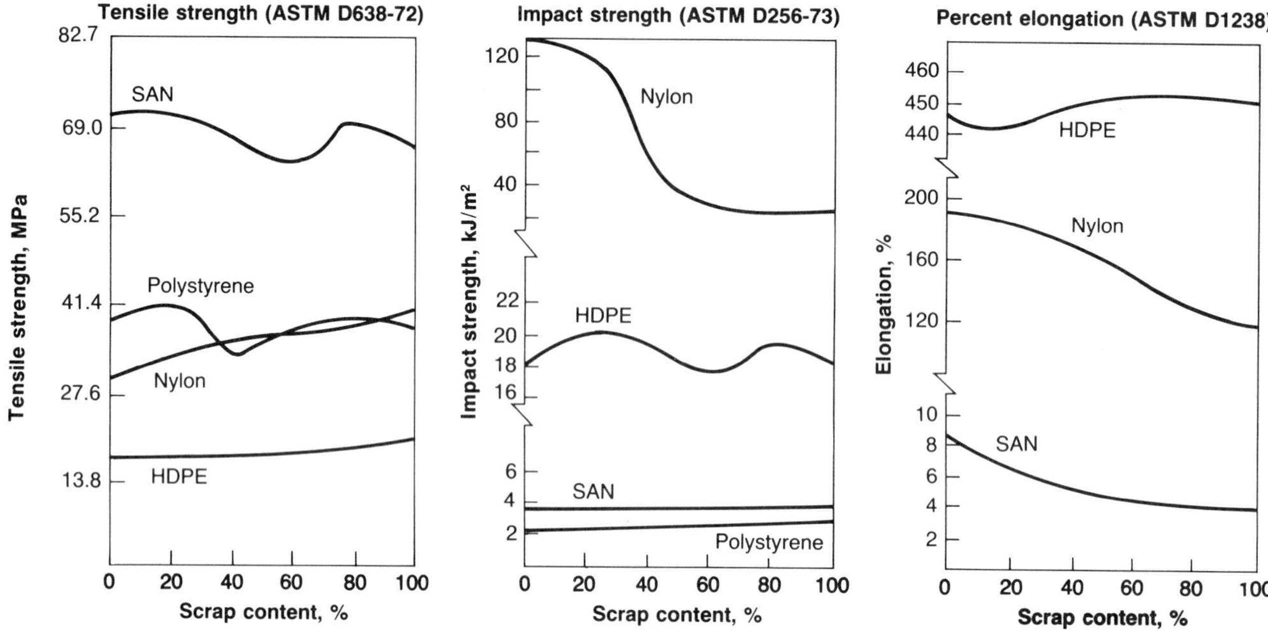

FIGURE 16 Mechanical properties using virgin and virgin–regrind mixes "once through" an injection molding machine [26].

total), graphite, carbon, and aramid (Tables 11–13; Fig. 19). Others include fibrous thermoplastics (polyester, polyamide, PE, PP, rayon, etc.), whiskers (ceramic types), cotton, wood pulp, and sisal; nonfibrous forms include powders (calcium carbonate, mica, sawdust, minerals, etc.), as well as flakes and beads (glass, ceramic, mica, etc.) (Tables 4 and 5). With flakes or similar forms, an aspect ratio (ratio of length to width) should be 10:1, to obtain the highest mechanical properties [9,10]. The fibrous reinforcements may be in the form of chopped fibers, porous mat, woven or braided fabric (Fig. 20), or continuous fibers/filaments [1,4,7–11].

Aside from the structure of the plastics, most composites contain many different reinforcements, including combinations (hybrid), with additives and/or fillers that have significant effects on properties, processibility, and cost. With the different types of reinforcement, a considerable freedom is available in selecting a particular type or combination to meet specific performance require-

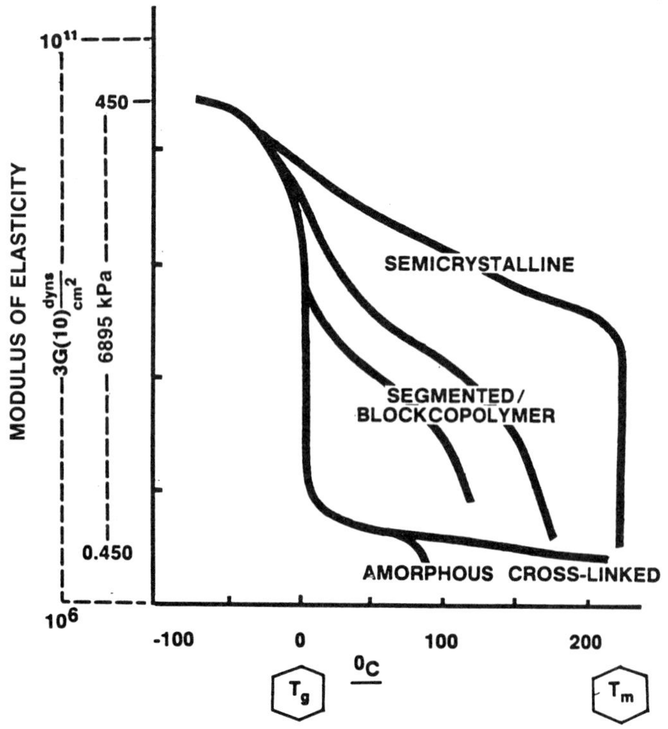

FIGURE 17 Dynamic–mechanical properties of plastics, related to viscoelasticity of plastics.

TABLE 11
Fibrous Reinforcements Used in Composites

Natural Organic	Synthetic Organic
α-Cellulose	Aramid
Cotton	Polyacrylonitrile
Jute	Polyamide
Sisal	Polyesters
Pulp	Polyvinyl alcohol
	Rayon
Natural Inorganic	**Synthetic Inorganic**
Asbestos	Metallic (steel, aluminum, etc.)
	Boron
	Carbon, graphite
	Ceramic
	Glass

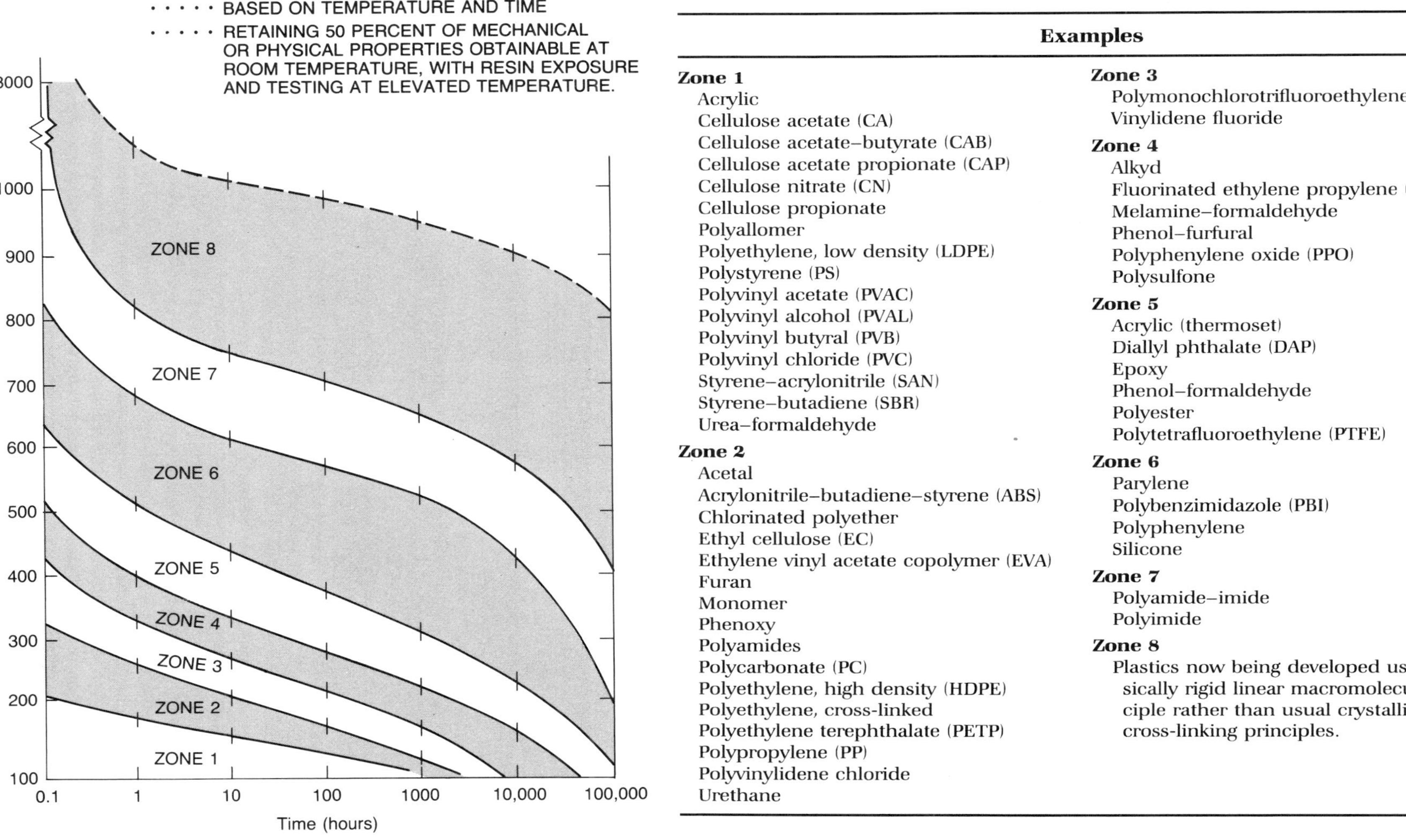

Examples

Zone 1
- Acrylic
- Cellulose acetate (CA)
- Cellulose acetate–butyrate (CAB)
- Cellulose acetate propionate (CAP)
- Cellulose nitrate (CN)
- Cellulose propionate
- Polyallomer
- Polyethylene, low density (LDPE)
- Polystyrene (PS)
- Polyvinyl acetate (PVAC)
- Polyvinyl alcohol (PVAL)
- Polyvinyl butyral (PVB)
- Polyvinyl chloride (PVC)
- Styrene–acrylonitrile (SAN)
- Styrene–butadiene (SBR)
- Urea–formaldehyde

Zone 2
- Acetal
- Acrylonitrile–butadiene–styrene (ABS)
- Chlorinated polyether
- Ethyl cellulose (EC)
- Ethylene vinyl acetate copolymer (EVA)
- Furan
- Monomer
- Phenoxy
- Polyamides
- Polycarbonate (PC)
- Polyethylene, high density (HDPE)
- Polyethylene, cross-linked
- Polyethylene terephthalate (PETP)
- Polypropylene (PP)
- Polyvinylidene chloride
- Urethane

Zone 3
- Polymonochlorotrifluoroethylene (CTFE)
- Vinylidene fluoride

Zone 4
- Alkyd
- Fluorinated ethylene propylene (FEP)
- Melamine–formaldehyde
- Phenol–furfural
- Polyphenylene oxide (PPO)
- Polysulfone

Zone 5
- Acrylic (thermoset)
- Diallyl phthalate (DAP)
- Epoxy
- Phenol–formaldehyde
- Polyester
- Polytetrafluoroethylene (PTFE)

Zone 6
- Parylene
- Polybenzimidazole (PBI)
- Polyphenylene
- Silicone

Zone 7
- Polyamide–imide
- Polyimide

Zone 8
- Plastics now being developed using intrinsically rigid linear macromolecules' principle rather than usual crystallization and cross-linking principles.

FIGURE 18 **Heat resistance properties of plastics.**

TABLE 12
Properties of the More Conventional High Performance Fiber Reinforcements

Type of Fiber Reinforcement	Specific Gravity	Density g/cm³	Tensile Strength MPa	Specific Strength (× 10⁶ cm)	Tensile Elastic Modulus (GPa)	Specific Elastic Modulus (× 10⁸ cm)
Glass						
E Monofilament,	2.54	2.547	3448	13.79	72.4	2.90
12-end roving	2.54	2.547	2565	10.26	72.4	2.90
S Monofilament,	2.48	2.491	4585	18.77	85.5	3.51
12-end roving	2.48	2.491	3792	15.67	85.5	3.51
Boron (tungsten substrate)	2.63	2.630	3103	12.04	400.	15.52
Graphite						
High strength	1.80	1.799	2758	15.62	262.	14.86
High modulus	1.94	1.938	2069	10.90	379.	19.96
Intermediate	1.74	1.744	2482	14.50	186.	10.90
Organic: aramid	1.44	1.439	2758	19.53	124.	8.79

From Ref. 3.

TABLE 13
Basic Processing Information Used For Glass Fiber Reinforced Composites: Resin Transfer Molding (RTM), Sheet Molding Compound (SMC), and Injection Molding (IM)

	RTM	SMC Compression	Injection
Process Operation			
Production requirement, annual units per press	5,000–10,000	50,000	50,000
Capital investment	Moderate	High	High
Labor cost	High	Moderate	Moderate
Skill requirements	Considerable	Very low	Lowest
Operation		Flowing, neat	Flowing, neat
Raw material inspection	Yes	Yes	Yes
Product monitoring	Visual	Visual	Visual
Finishing	Trim flash, etc.	Very little	Very little
Product			
Complexity	Very complex	Moderate	Greatest
Size	Very large parts	Big flat parts	Moderate
Tolerance	Good	Very good	Very good
Surface appearance	Gel-coated	Very good	Very good
Voids/wrinkles	Occasional	Rarely	Least
Reproducibility	Skill dependent	Very good	Excellent
Cores/inserts	Possible	Very difficult	Possible
Material Usage			
Raw material, cost	Lowest	Highest	High
Handling/applying	Skill dependent	Easy	Automatic
Waste	up to 3%	Very low	Sprues, runners
Scrap	Skill dependent	Cuts reusable	Low
Reinforcement flexibility	Yes	No	No
Mold			
Initial cost	Moderate	Very high	Very high
Cycle life	3000–4000 parts	Years	Years
Handling	Extreme care	Careful	Careful
Preparation	in factory	Special mold-making shops	
Maintenance	in factory	Special machine shops	

From Ref. 26.

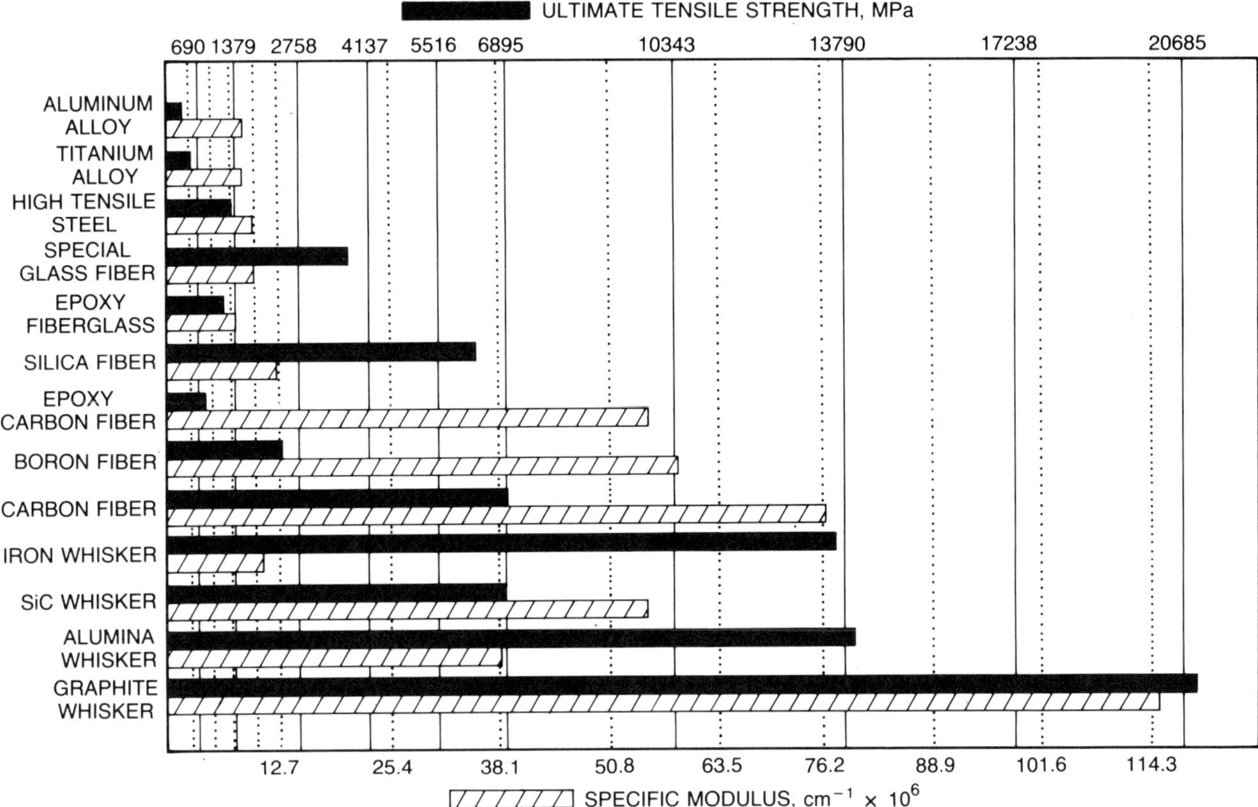

FIGURE 19 **Properties of particulate composites, includes "whisker" reinforcements.**

ments. But again, it is important to evaluate processing conditions, because damaged or reduced length of reinforcement can significantly reduce performance.

Ratio of reinforcement to plastic can be in many different proportions to provide different properties. This ratio can be identified as percent by weight or percent by volume (Figs. 10, 21, 22; Table 14). Since the density of reinforcements, additives, fillers, and plastics are usually different, it is important to specify percentages as either by weight or volume.

Additives or fillers usually are included in composites for various reasons, such as reducing cost. Calcium carbonate can provide easier flow of the composite during processing, resulting in improved properties. Addition of a few percent, by weight, of inexpensive ground limestone can increase elastic modulus and wall stiffness. These synergistic reactions are extremely beneficial, both in performance and cost, for either thermoset or thermoplastic RPs.

Additives or fillers can provide advantages but unfor-

TABLE 14
Example of Effect of Different Concentrations of Glass Fibers with a Plastic on the Properties of the Composites[a]

Property	Glass Fiber (wt %)													
	0		10		20		30		40		50		60	
Specific gravity	1.14		1.21		1.28		1.37		1.46		1.57		1.70	
Specific volume m³/mg × 10^{-10}	24.3	8.8	22.9	8.3	21.6	7.8	20.1	7.3	19.0	6.9	17.6	6.4	16.3	5.9
Tensile strength MPa	12	83	13	90	19	131.	25	172.	31	214.	32	221.	33	228.
Tensile elongation (%)	60		3.5		3.5		3.0		2.5		2.5		1.5	
Flexural strength MPa	15	103.	20	138	29	200.	34	234.	42	290.	46	317.	50	345
Flexural modulus GPa	4.0	.28	6.0	.41	9.0	.62	13	.90	16	1.10	22	1.52	28	1.93
Compressive strength MPa	4.9	33.8	13	89.6	23	158.6	27	186.2	28	193.1	29	200.0	30	206.9
Heat deflection temperature at °C @ 1820 kPa	150		470		475		485		500		500		500	
Thermal expansion mm/mmK × 10^{5}	4.5	8.1	1.6	2.9	1.4	2.5	1.3	2.3	1.2	2.2	1.0	1.8	0.9	1.6
Water absorption, 24 h (%)	1.6		1.1		0.9		0.9		0.6		0.5		0.4	
Mold shrinkage	15		6.5		5		4.0		3.5		3.0		2.0	

[a] Actual performance can be related to compound preparation, processing technique, and testing procedure.

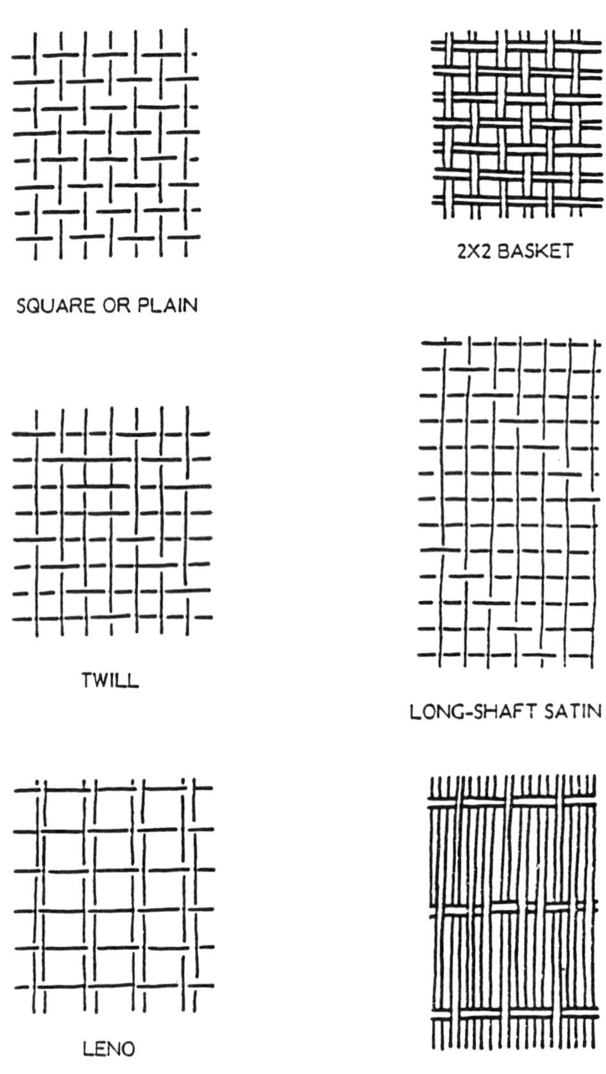

FIGURE 20 **Diagrammatic representation of fabric weaves.**

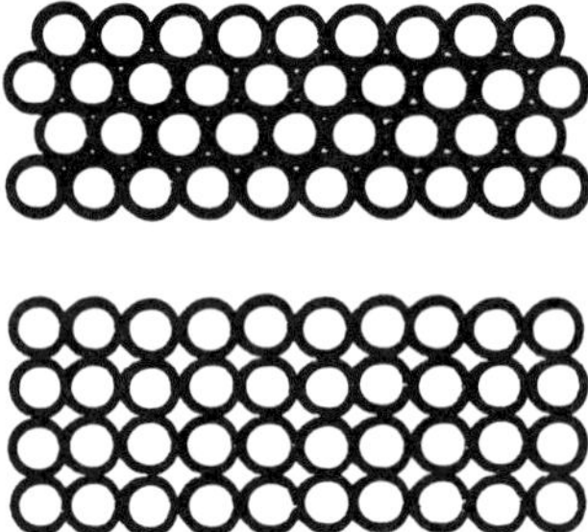

FIGURE 21 Effect of fiber packing arrangements for E-glass with epoxy plastics. *Top*: Hexagonal close packing nesting results in up to 90.8% glass by volume and 95.6% glass by weight. *Bottom*: Square packing has up to 78.5% glass by volume and 88.8% by weight.

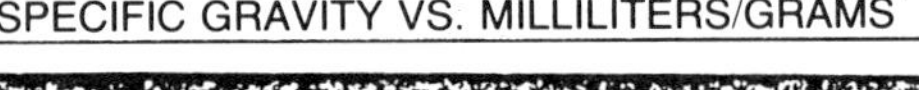

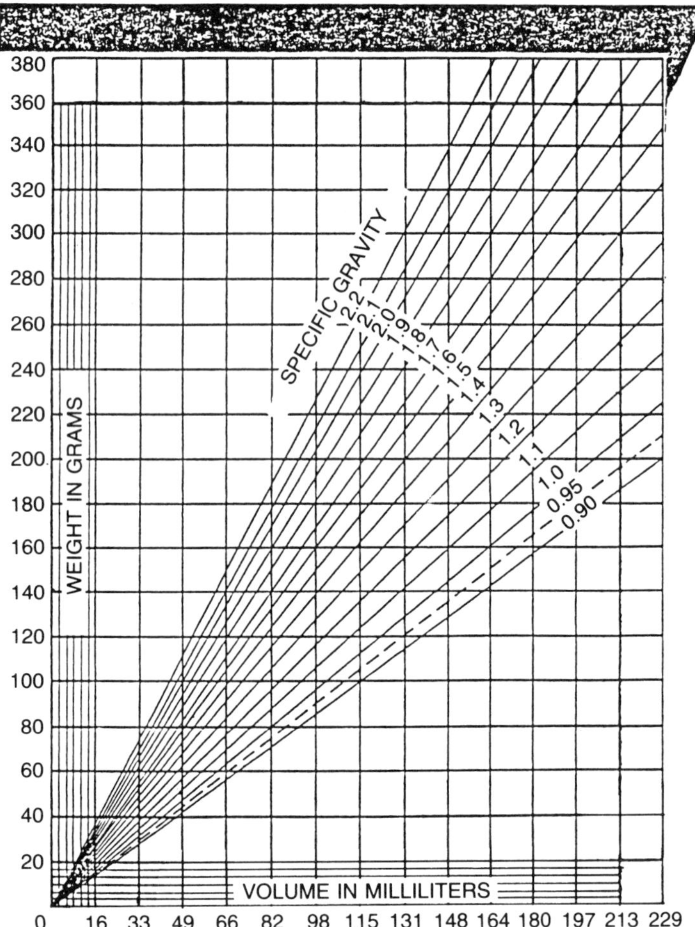

FIGURE 22 **Specific gravity versus milliliters/grams.**

tunately they could hinder melt flow or have other effects on processibility. They often contribute to frictional heating and to thermal conduction during heating and cooling cycles with screw plastication, such as in injection or compression molding [26]. The result is the speeding of the molding cycle, an effect particularly noted with iron and copper powders. Whenever it is difficult to release the part from a mold, the molding cycle is increased. This problem is commonly relieved by use of internal or external lubricants.

Additives often reduce "rubbery" melt flow. They also generally reduce mold shrinkage and often postwarpage. The presence of fillers or reinforcements can have a strong effect on rheological (flow) properties. For example, at filler loadings of 20–40%, by volume, the viscosity is increased significantly, and a sharp rise in the curve that is often observed as the shear rate is decreased suggests the presence of a mechanical yield stress. However, as usual, there are exceptions (e.g., polypropylene filled with glass beads). The beads increase viscosity quite markedly with very little effect on the flow pattern in a mold cavity. Glass beads have also been used with glass fibers (and other fibers) in practically any plastic to help disperse more evenly the reinforcing fibers.

Coupling agents that are used with certain fillers can alter the behavior of the filled composites. For example, using titanate coupling agents on the behavior of polypropylene filled with calcium carbonate (50%, by weight), there is a dramatic reduction in viscosity that effects reductions in certain mechanical properties. Particulate fillers and especially fibrous reinforcements generally increase viscosity and impede melt flow. This effect increases with fiber length and usually requires higher temperatures and/or pressures to permit satisfactory molding. Another additive that can interfere with melt flow is the type and amount used to produce cross-linking. It may be a catalyst, initiator, or hardener with themosets.

Processing

Processing composites cuts across almost all processing techniques used with all types of materials (unreinforced plastics, metals, woods, ceramics, etc.). However there are major types used as reviewed in this book and summarized in Tables 3, 6, 13, and 15–17. An example of a process that is gaining interest for use with composites is reaction injection molding (RIM) (Fig. 23) [31–36].

As a step flow process, plastics (composites) fabrication differs from most metals fabrication (Fig. 6). Although this gives the designer (and materials selector) great flexibility and provides design freedom usually not possible in metals, it requires a greater understanding to take advantage of their capabilities and interfacing material selection with processing technique that can be used to meet product performance [1,4,7,17]. Variables that influence the end results are to be recognized and understood during material selection. Some of these variables interact, hence may be advantageous or disadvantageous. The most influential molding variable is orientation of plastics [13,26] and particularly reinforcement (Fig. 24). Other important variables include plastic degradation [27–39], free volume/molecular packing and relaxation [13,26], separation of reinforcements from plastics, air or gas entrapment, weld line [26], and cooling rate of plastics.

Selecting and interrelating the optimum process with material selection encompasses a broad spectrum of possibilities. There are requirements when only one process can be used; however there are generally options that will permit one to manufacture the part to just meet material performance requirements at the lowest cost. Influencing the process to be used (and in turn material selection) are basically quantity, size, thickness, type of composite, design, and performance requirements; unfortunately, sometimes choice of process also is dependent on the equipment available. Thus it is important to interrelate all these factors, as summarized, as an example, for injection molding in Figure 25. The concept used in Figure 25 can be related to other processes. This diagram may look complicated to one who does not know the process. After a material selector starts to understand the characteristics of composites, however, Figure 25 can be understood. Recognize that variations in pro-

TABLE 15
Process Comparison of Various RTP Manufacturing Techniques

	Resin Transfer Molding	Open Molding		Cold Press Molding	Compression Molding	
		Spray-Up	Hand Lay-Up		Mat/Preform	Sheet Molding Compound
Mold construction	FRP, spray metal, cast aluminum; gasket seal, air vents, self-sealing injection port	FRP		FRP, spray metal, cast aluminum; pinch (land)	Metal; shear edge	High grade steel; shear edge
Pressure	Pressure feed pumping equipment req'd; mold halves clamped (methods range from clamp frame to pressure pod)	None		Low pressure press capable of 345 kPa* (hydraulic or pneumatic mechanical); resin dispensing equipment not req'd but recommended	Hydraulic press, normal range of 100-500 psi* 690–3, 448 kPa	Hydraulic as high as 13.8 MPa
Cure system	Room temperature				Heated; normal range of 107–163°C	Heated; normal range of 135–177°C
Resin compounding equipment	High shear type	Not needed			High shear type	
Reinforcement	Continuous strand mat, preform, woven roving	Continuous roving	Chopped strand mat, woven roving, cloth	Continuous strand mat, preform, woven roving	Continuous strand mat, preform, woven roving	Continuous roving (specific orientations for higher strength)
Part trim equipment	Yes				With optimum shear edges, minor trimming only	
Generally expected mold life (parts)	3,000	1,000		3,000	150,000+	150,000+

* Based upon projected area of part.
From Ref. 23.

TABLE 16
Plastic Processing Methods

Process	Description	Limitations
Blow	An extruded parison tube of heated thermoplastic (and even mica-reinforced type) is positioned between two halves of an open slit mold and expanded against the sides of the closed mold via air pressure. The model is opened and the part ejected. Low tool and die costs, rapid production rates, and ability to mold fairly complex hollow shapes in one piece [13].	Limited to hollow or tubular parts; wall thickness and tolerances can be difficult to control, principally used with unreinforced thermoplastics.
Calendering	Dough-consistent plastic mass is formed into a sheet of uniform thickness by passing it through and over a series of heated or cooled rolls. Calenders are also utilized to apply plastic covering to the backs of other materials. Low cost; sheet materials are virtually free of molded-in stress.	Limited to sheet materials, very thin films are not possible.
Casting	Liquid plastic, which is generally thermoset is poured into a mold without pressure, cured, and taken from the mold. Cast thermoplastic films are produced via building up the material (either in solution or hot-melt form) against a highly polished supporting surface. Low mold cost, capability to form large parts with thick cross sections, good surface finish, and convenient to low volume production.	Limited to relatively simple shapes. Most thermoplastics are not suitable for this method. Save for cast films, method becomes uneconomical at high volume production rates.
Centrifugal casting	Reinforcement is placed in mold and is rotated. Resin distributed through pipe; impregnates reinforcement through centrifugal action. Utilized for round objects, particularly pipe.	Limited to basically simple curvatures in single axis rotation. Low production rates.
Coating	Process methods vary. Both thermoplastics and thermosets widely used in coating of numerous materials. Roller coating similar to calendering process. Spread coating employs blade in front of roller to position resin on material. Coatings also applied via brushing, spraying, and dipping.	Economics dependent on close tolerance control generally.
Cold pressure molding	Similar to compression molding in that thermoset material is charged into a split mold; it differs in that it employs no heat, only pressure. Part cure takes place in an oven in a separate operation. Some thermoplastic billets and sheet materials are cold formed in a process similar to drop hammer die forming or fast cold form stamping of metals. Low cost matched tool moldings exist which utilize a rapid exotherm to cure moldings on a relatively rapid cycle. Plastics or concrete tooling can be used. With process, comes ability to form heavy or tough-to-mold materials; simple, inexpensive, and often has rapid production rate.	Compared with process such as injection molding, is limited to relatively simple shapes, and few materials can be processed in this manner.
Compression molding	Principally thermoset compound is positioned in a heated mold cavity; the mold is closed (heat and pressure are applied) and the material flows and fills the mold cavity. Heat completes polymerization and the part is ejected. The process is sometimes used for thermoplastics. Little material waste is attainable; large, bulky parts can be molded; and process is adaptable to rapid automation (racetrack techniques, etc.).	Extremely intricate parts containing udercuts, side draws, small holes, delicate inserts, etc.; very close tolerances difficult to produce.
Encapsulation	Mixed compound is poured into open molds to surround and envelop components; cure may be at room temperature with heated postcure. Generally, encapsulation includes several processes such as potting, embedding and conformal coating.	Low volume process subject to inherent limitations on materials, which can lead to product defect caused by exotherm, curing or molding conditions, low thermal conductivity, high thermal expansion, and internal stresses.
Extrusion molding	Widely used for continuous production of film, sheet, tube, and other profiles; also used in conjunction with blow molding. Thermoplastic or thermoset molding compound is fed from a hopper to a screw and barrel, where it is heated to plasticity and then forwarded, usually via a	Limited to sections of uniform cross section, principally used with unreinforced thermoplastics.

TABLE 16 (*continued*)

Process	Description	Limitations
	rotating screw, through a nozzle possessing the desired cross section configuration. Production lines require input and takeoff equipment that can be complex. Low tool cost, numerous complex profile shapes possible, very rapid production rates, can apply coatings or jacketing to core materials (such as wire).	
Filament winding	Excellent strength-to-weight here. Continuous, reinforced filaments, usually glass, in the form of roving are saturated with resin and machine wound onto mandrels having shape of desired finished part. Once winding is completed, part and mandrel are cured; mandrel can then be removed through porthole at end of wound part. High strength reinforcements can be oriented precisely in direction where strength is required. Good uniformity of resin distribution in finished part; mainly circular objects such as pressure bottles, pipes, and rocket cases [7].	Limited to shapes of positive curvature; openings and holes can reduce strength if not properly designed into molding operations.
Injection molding	Very widely used method. High automation of manufacturing is standard practice. Thermoplastic or thermoset is heated to plasticity in cylinder at controlled temperature, then forced under pressure through a nozzle into sprues, runners, gates, and cavities of mold. The resin undergoes solidification rapidly, the mold is opened, and the part ejected. Injection molding is growing in the making of glass-reinforced parts. High production runs, low labor costs, high reproducibility of complex details, and excellent surface finish [26].	High initial tool and die costs; not economically practical for small runs.
Laminating	Material, usually in form of reinforcing cloth, paper, foil, metal, wood, glass fiber, plastic, etc., preimpregnated or coated with thermoset resin (sometimes a thermoplastic) is molded under pressure greater than 6895 kPa into sheet, rod, tube, or other simple shape. Excellent dimensional stability of finished product; very economical in large production of parts.	High tool and die costs. Limited to simple shapes and cross section profiles.
Matched-die molding	A variation of the conventional compression molding, this process employs two metal molds possessing a close-fitting, telescoping area to seal in the plastic compound being molded and to allow trim of the reinforcement. The mat or preform reinforcement is positioned in the mold and the mold is closed and heated under pressures of 1034–2758 kPa. The mold is then opened and the part is removed.	Prevalent high mold and equipment costs. Parts often require extensive surface finishing.
Pultrusion	This process is similar to profile extrusion, however, it does not provide flexibility and uniformity of product control, and automation. Used for continuous production of simple shapes (rods, tubes, and angles) principally incorporating fiberglass or other reinforcement. High output possible.	Close tolerance control requires diligence. Unidirectional strength usually the rule.
Rotational molding	A predetermined amount of powdered or liquid thermoplastic or thermoset material is poured into mold; mold is closed, heated, and rotated in the axis of two planes until contents have fused to inner walls of mold; mold is then opened and part is removed. Low mold cost, large hollow parts in one piece can be produced, and molded parts are essentially isotropic in nature.	Limited to hollow parts; production rates are usually slow, principally used with thermoplastics.
Slush molding	Powdered or liquid thermoplastic material is poured into a mold to capacity; mold is closed and heated for a predetermined time to achieve a specified buildup of partially cured material on mold walls; mold is opened and unpolymerized material is poured out; semifused part is removed from mold and fully polymerized in oven. Low mold costs and economical for small production runs.	Limited to hollow parts; production rates are very slow; limited choice of materials that can be processed, principally used with thermoplastics.

(*continued*)

TABLE 16 (*continued*)

Process	Description	Limitations
Thermoforming	Heat-softened thermoplastic sheet is positioned over male or female mold; air is evacuated from between sheet and mold, forcing sheet to conform to contour of mold. Variations include vacuum snapback, plug assist, drape forming, etc. Tooling costs are generally low, large part production with thin sections possible, and often comes out as economical option for limited part production.	Limited to parts of simple configuration, high scrap, and limited number of materials from which to choose, principally used with thermoplastics.
Transfer molding	Related to compression and injection molding processes. Thermoset molding compound is fed from hopper into a transfer chamber where it is heated to plasticity; it is then fed by means of a plunger through sprues, runners, and gates into a closed mold where it cures; mold is opened and part is ejected. Good dimensional accuracy, rapid production rate; very intricate parts can be produced.	High mold cost; high material loss in sprues and runners; size of parts is somewhat limited.
Wet lay-up or contact molding	Number of layers, consisting of a mixture of reinforcement (generally glass cloth) and thermosetting resin are positioned in mold and roller contoured to mold's shape; assembly is usually oven cured without the application of pressure. In spray molding, which is a modification, resin systems and chopped fiber are sprayed simultaneously from spray gun against mold surface. Wet layup parts are sometimes cured under pressure utilizing vacuum bag, pressure bag, or autoclave and depending on method employed, this process can be called open molding, hand layup, spray-up, vacuum bag, pressure bag, or autoclave molding. Little equipment required, efficient, low cost, and suitable for low volume production of parts.	Not economical for large volume production; uniformity of resin distribution difficult to control; only one good surface; limited to simple shapes.

TABLE 17
Compatibility of Materials and Processes for Fiber Glass Composites

	Thermosets					Thermoplastics										
	Polyester	Polyester SMC	Polyester BMC	Epoxy	Polyurethane	Acetal	Nylon 6	Nylon 6/6	Polycarbonate	Polypropylene	Polyphenylene sulfide	ABS	Polyphenylene oxide	Polystyrene	Polyester PBT	Polyester PET
Injection molding	●		●	●	●	●	●	●	●	●	●	●	●	●	●	
Hand lay-up	●			●												
Spray-up	●			●												
Compression molding	●	●	●	●							●					
Preform molding	●			●												
Filament winding	●			●												●
Pultrusion	●			●												
Resin transfer molding	●													●	●	●
Reinforced reaction injection molding	●			●	●		●									

(a)

(b)

(c)

(d)

FIGURE 23 **RIM equipment.** (*a*) Piston metering RIM machine by Battenfeld, with closed-loop control of throughput and high metering accuracy. Compounding units used are for the blending of glass fiber and other reinforcements and/or fillers with liquid polyol. (*b*) The Battenfeld RIM metering machine is equipped with special tanks and agitators required by high viscosity (50 Pa·s) composites. (*c*) RIM machine mold clamp system; 2446×10^3 N press with auxiliary clamping cylinders.

cessing directly influence material selection and performance.

The growth in the use of composites in certain markets has resulted in a need for high productivity manufacturing processes to meet material performance. Currently most composites utilize thermosets. These plastics are usually well suited for impregnation into reinforcing fibers by manual means, since they are low viscosity liquids that can be handled at room temperature with little difficulty. In low production runs, a wide variety of custom shapes can be manufactured without a high degree of automation or mechanization. Manual and semiautomated production techniques used with thermosets have a strong economic advantage over fully

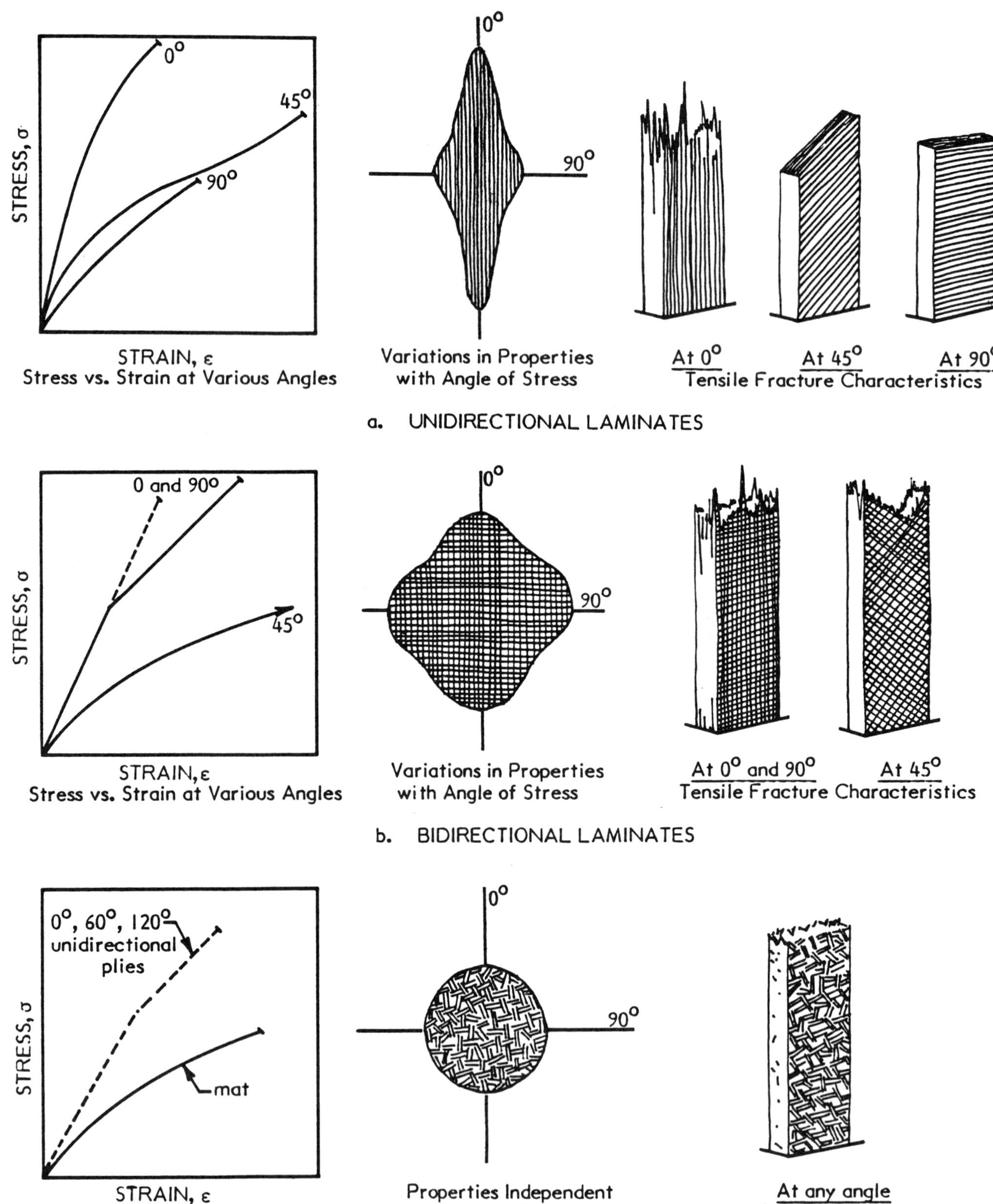

FIGURE 24 Directional properties of reinforced thermoset plastics versus different fiber orientations.

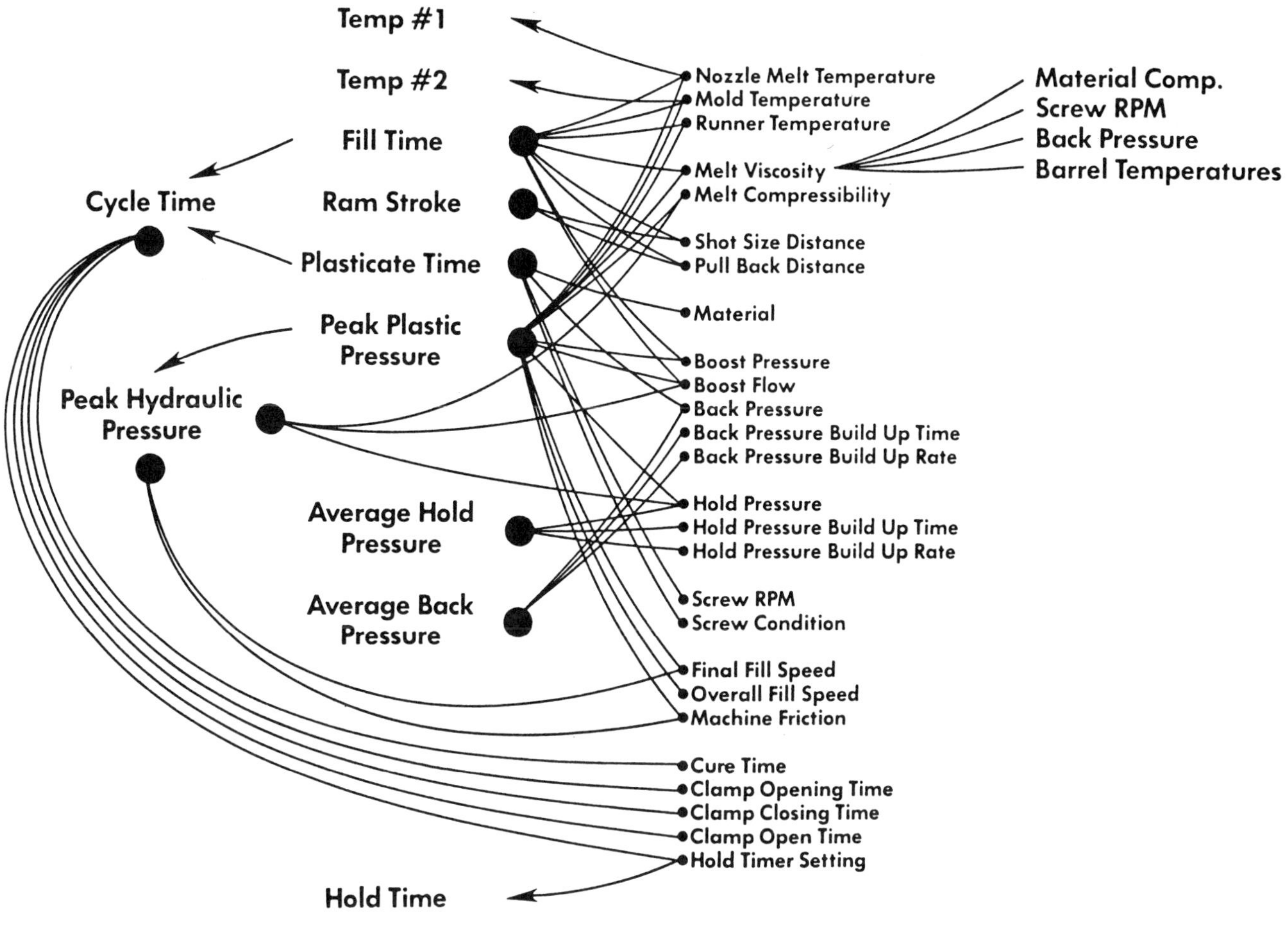

FIGURE 25 Injection molding machine controls [26].

automated thermoplastic methods of manufacturing. Recognize that certain thermoplastics and thermosets are economical only in low volume runs. With high volume runs the thermoplastics (unreinforced) are generally recognized as the major material, although for many decades reinforced thermoplastics have been used in the high volume, high speed, mass production processes.

Behavior of Composites

A representative cross section of the myriad of composites available to the material selector has been described. This section reviews their structural behaviors. Concepts are provided as background. Obviously there is no universal method to describe the behavior of all composites, just as no single reference covers the behavior of other materials (metals, etc.). Nonetheless, all composites show many similarities in behavior. The differences are frequently in terms of magnitude, not of a kind. Their stress–strain and strength behaviors vary widely but are dependent on generic type, and on the specific compositions. Different factors interact to alter behavior over a very wide range. Key parameters to consider are listed briefly [2,37,38].

Magnitude and duration of stress, strain, and temperature. At a given condition of one of these three parameters, both the magnitude and duration of the other two affect structural response and strength.

Environment. Chemicals, ultraviolet radiation, water, and other environments can have profound influence with certain composites on performance, hence may dominate the material or design selection.

Reinforcements. There is a relationship of type of reinforcement and amount used in relation to different properties, with maximum performance based on a well-defined ratio of reinforcement to plastic.

Process. The process used may dictate the structural performance of the product.

Thermoplastics, reinforced or unreinforced, can be bent, pulled, or squeezed into various shapes. But eventually, especially if you add heat, they return to their original form. This behavior, known as plastic memory, can be annoying; when properly applied, however, it offers some interesting possibilities in eliciting performance from selected material. When most materials, other than thermoplastics, are bent, stretched, or compressed, they somehow alter their molecular structure. This time/temperature-dependent change in mechanical properties results from stress relaxation and other viscoelastic phenomena. When the change is an unwanted limitation it is called creep. When it is desired it is called plastic memory.

Viscoelastic Behavior

The relationship between stress and strain, or structural response, of composites varies from viscous to elastic (Figs. 26–28). Most materials display a response that is intermediate; thus they are called viscoelastic. Type of composite, stress, strain, time, temperature, and environment all play a significant role in determining whether the response is mostly viscous, elastic, or viscoelastic.

Viscoelasticity is a complex subject, but it can be demonstrated by simple models and analogies. The terminology that follows characterizes viscoelastic behavior, including the idealized components of models.

Elastic response is represented by a Hookeian solid, as modeled in engineering terms by a linear spring. Stress is proportional to strain and independent of time; response to stress is instantaneous; there is no permanent or irrecoverable deformation; all energy used to deform the spring is stored and is fully recoverable.

Viscous response is represented in engineering terms by a Newtonian fluid, as modeled by a dashpot. Stress is proportional to strain rate, making behavior time-dependent. Recovery is nil when stress is removed. Energy to deform the dashpot is dissipated completed during the deformation process.

Creep is the time-dependent increase in strain of a viscous or viscoelastic material under sustained stress. Some of the time-dependent deformation is recoverable with time after release of stress. Creep experiments are usually performed under constant load conditions; when stresses are high, the product may "neck" and the cross section supporting the load may be reduced significantly at some point during the test. Unless otherwise indicated, the creep stress based on original cross-sectional area (engineering stress) will be used rather than the "true" creep stress, which is based on the reduced cross-sectional area, that occurs on necking. This reflects typical practice, which is to report "engineering" creep stress.

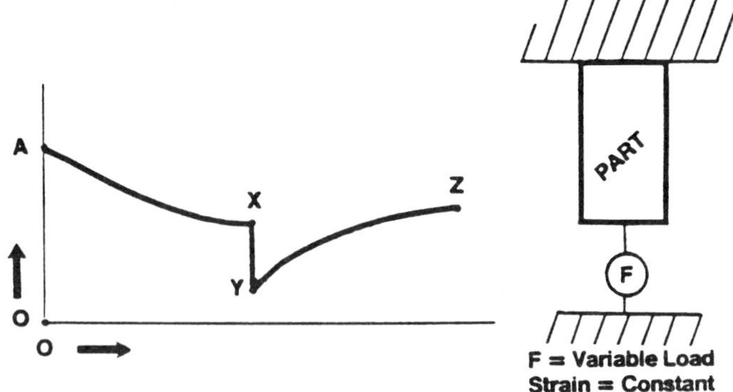

O-A: *Instantaneous loading* produces *immediate strain.*
A-X: With strain maintained gradual *elastic relaxation* occurs.
X-Y: *Instanteous deformation* occurs when load is removed.
Y-Z: *Viscoelastic deformation* gradually occurs as residual stresses are relieved. Any permanent deformation is related to type plastic, amount & rate of loading and fabricating procedure.

FIGURE 27 **Relation of strain-stress-time; stress-relaxation.**

Relaxation is the time-dependent decay in stress of a viscoelastic material under sustained strain. Some of the deformation is recoverable with time after release of the sustained strain. Unless the imposed initial strain is above the yield point, the cross section remains fairly close to the original, throughout the test. This defers from creep behavior as previously reviewed.

Recovery is the extent to which an element returns to its original configuration after release of stress or strain.

Linear viscoelastic response refers to viscoelastic response in which stress and strain are related by a single modulus, which depends only on duration of the applied stress and strain for a given temperature. This dif-

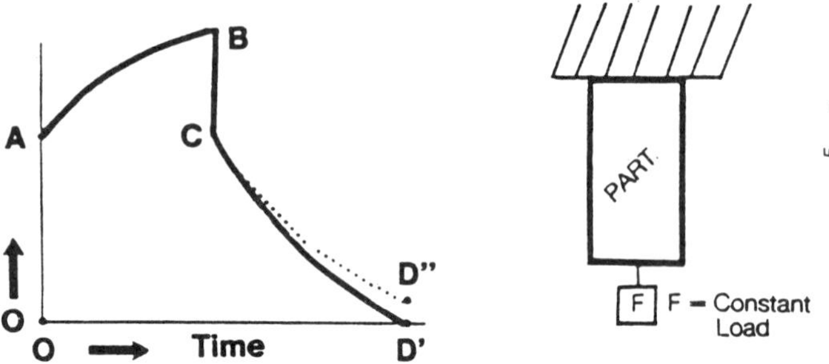

O-A: *Instantaneous loading* produces *immediate strain.*
A-B: *Viscoelastic deformation* (or creep) gradually occurs with sustained load.
B-C: Instantaneous *elastic recovery* occurs when load is removed.
C-D: Viscoelastic recovery gradually occurs; where no permanent deformation (D') o vith a permanent deformation (D"-D'). Any permanent deformation is related to type plastic, amount & rate of loading and fabricating procedure.

FIGURE 26 **Relation of stress-strain-time; creep.**

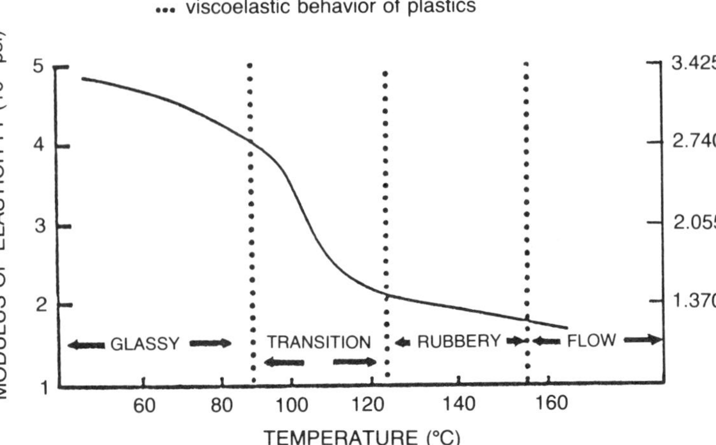

FIGURE 28 **Example of plastics modulus of elasticity versus temperature.**

fers from nonlinear viscoelastic response, in which the modulus depends on the magnitude and duration of stress or strain.

Selection Guide

Composite selection is no less challenging than design or process selection. You must first be capable of identifying finished part requirements; then you translate these requirements into physical and mechanical properties, compare materials, and make a final selection. The number and variety of materials on the market (or what you can produce if the capabilities exist) can make the task of composite selection seem overwhelming.

Failures can be avoided by properly predicting material performance requirements. Advanced computer methods as well as the usual engineering analysis are available for calculating stresses in complex structures when they exist. Early and proper comprehensive analysis permits determination of the effects of temperature, loading rate, environment, material defects, and others (required based on product performance requirements) on the performance or structural reliability of the material selected. This information, when required, can be supported by stress–strain–behavior data collected in the selection/evaluation procedure. Engineering plastics data bases can be very useful. More than 15,000 grades of plastics (including more than a thousand grades of composites) are available . . . and more become available.

If you are involved in a special area of materials, the probability exists that you can easily have your own data bank on available composites, etc. If this is not the case, or if you are not able to keep your data base up to date, a data base is available to meet your requirements. Certain plastic manufacturing plants will make available their data bank (under certain conditions), or you can purchase materials information on disk or tape. Perhaps the ultimate is an on-line computer multisupplier source of data, costs, etc. that is kept up to date [13,26,40–47].

With composites, to a greater extent than with other materials, an opportunity exists to optimize design and processing techniques by focusing on materials composition and structural orientation. The interrelationship of composite, performance, shape, processing, and cost is more important for the composites (as well as unreinforced plastics) than for other materials such as steel and wood. With these others the designer is usually limited

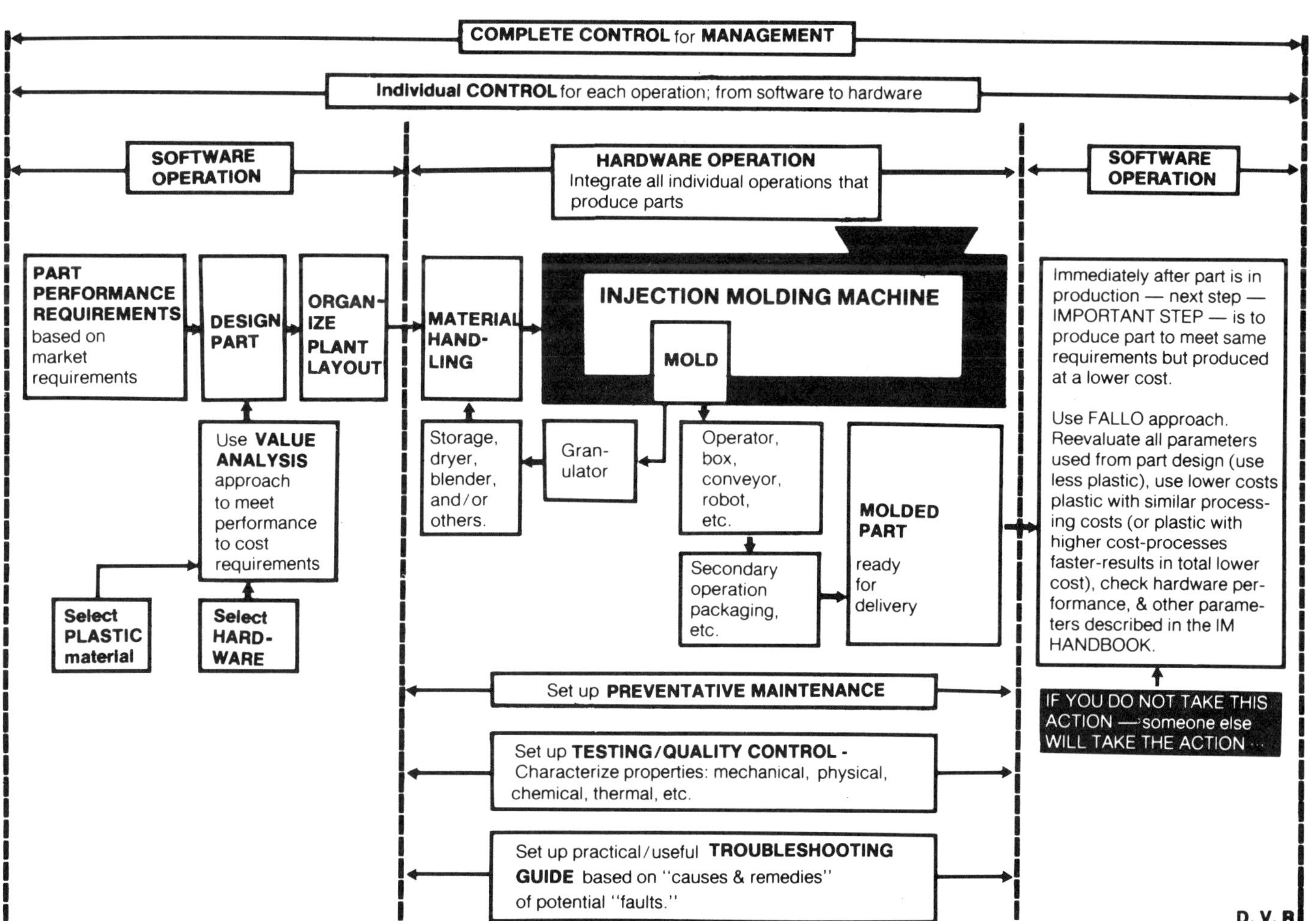

FIGURE 29 The complete molding operation: The FALLO approach (Follow *ALL* Opportunities).

to specific forms or profiles that are bent, welded, and/or assembled by more familiar procedures. Although there are more than a thousand composites, only about a hundred are used in large quantities.

There are the usual or obvious factors that influence material selection. Understanding how to obtain maximum performance at the lowest cost requires interfacing with other important operations in the *complete molding operation* to manufacture composite products. Figure 29 summarizes what should be considered to ensure a good return on investment. This block diagram pertains to injection molding unreinforced or reinforced plastics and can be used as a guide for other processes [13,26]. The FALLO (Follow *ALL* Opportunities) illustrated here approach highlights that one should "follow all opportunities" that exist with composites in a logical approach.

D. V. Rosato

References

1. *Composites Annual Conference Reprint Books,* SPE, Brookfield, CT, issued since 1944.
2. P. Richards, "Push Plastics to their Design Limits," *Plastics World 1988 Directory,* Newton, MA, pp. 381–385.
3. M. A. Dorgham and D. V. Rosato, "Designing with Plastics and Advanced Plastics Composites," *Proceedings of the International Association for Vehicle Design,* Interscience Enterprises Ltd., Geneva, Switzerland, 1986.
4. G. Lubin, *Handbook of Fiberglas and Advanced Composites,* Van Nostrand Reinhold, New York, 1982.
5. D. V. Rosato et al., *Markets for Plastics,* Van Nostrand Reinhold, New York, 1969.
6. "Fiberglas Plus Design; A Comparison of Materials & Process for Fiber Glass Composites," Owens-Corning Fiberglas Corporation, Grandville, OH, 1987.
7. D. V. Rosato and C. S. Grove, *Filament Winding,* Wiley, New York, 1964.
8. J. V. Milewski and D. V. Rosato, "History of Reinforced Plastics," National ACS Meeting, Houston, March 23–28, 1980.
9. J. V. Milewski, *Plast. Comp.,* Denver, November–December 1979.
10. D. V. Rosato, *Ind. Eng. Chem., 54*(8), New York, pp. 30–37 (1962).
11. B. Sanders, *Short-Fiber Reinforced Composite Materials,* ASTM-STP 772, Philadelphia, 1982.
12. B. Miller, *Plastics World,* Newton, MA, November 1986, pp. 33–34.
13. D. V. Rosato, in *Blow Molding Handbook: Technologies, Performance, Markets, and Economics,* Hanser, New York, 1988, Ch. 24.
14. C. D. Han et al., *Poly. Eng. Sci.,* J. Wiley & Sons, New York, *26*(6), (March 1986).
15. D. V. Rosato and R. T. Schwartz, *Environmental Effects on Polymer Materials,* Vols. I and II, Wiley, New York, 1968.
16. D. V. Rosato, "All Reinforced Plastics Military Airplane: Successful Flight Test," Technical Report, Wright-Patterson Air Force Base, Ohio, 1944.
17. A. G. H. Dietz, *Composite Engineering Laminates,* MIT Press, Cambridge, MA, 1969.
18. D. V. Rosato, "Molding Reinforced Metal, Plastic, and Metal-Plastic," Technical Report, Wright-Patterson Air Force Base, Ohio, 1944.
19. A. S. Wood, *Mod. Plast.,* New York, March 1988, pp. 42–49.
20. "Performance Temperature: Facts and Fiction," Durez Molder Newsletter, Winter 1988. (Durez, P.O. Box 535, North Tonawanda, NY 14120.)
21. D. V. Rosato, Reinforced Plastics Seminar, University of Lowell, Lowell, MA, 1986.
22. R. F. Jones, *Short Fiber Reinforced Plastics,* Methuen, New York, 1988.
23. G. Chamberland, *Des. News,* Newton, MA, January 1988, pp. 24–25.
24. S. J. Shah and R. E. Nunn, *Plast. Eng.,* Brookfield, CT, November 1987, pp. 33–36.
25. R. B. Seymour and G. S. Kirschenbaum, *High Performance Polymers,* Elsevier, New York, 1986.
26. D. V. Rosato, *Injection Molding Handbook,* Van Nostrand Reinhold, New York, 1986.
27. N. Grassie, *Development in Polymer Degradation,* Vol. 6, Elsevier Applied Science, Barking, England, 1985.
28. N. Grassie, *Development in Polymer Degradation,* Vol. 7, Elsevier Applied Science, Barking, England, 1987.
29. W. Schnabel, *Polymer Degradation, Principals and Applications,* Hanser, New York, 1981.
30. Y. T. Wu, *Mod. Plast.,* New York, March 1988, pp. 89–96.
31. D. R. Dreger, *Machine Des.,* Cleveland, OH, October 1987, pp. 92–98.
32. *Fly Fisherman,* Naples, FL, *19*(2), 5 (1988).
33. D. Nelson, *Plast. Eng.,* Brookfield, CT, November 1987, pp. 29–32.
34. G. E. Reynolds, *Plastics and Rubber: Processing and Applications,* Elsevier, New York, Vols. 7 and 8, 1987; Vols. 9 and 10, 1988.
35. B. Miller, *Plast. World,* Newton, MA, January 1988, pp. 26–31.
36. D. V. Rosato, Reinforced Plastics Design Seminar, SPI Annual Conference, Washington, DC, 1984.
37. G. Chamberland, *Des. News,* Newton, MA, January 1988, pp. 55–64.
38. W. G. Taft, *Machine Des.,* Cleveland, OH, January 1988, pp. 99–103.
39. D. V. Rosato, Filament Winding Design Parameters, Reinforced Plastics Seminar, University of Lowell, Lowell, MA, 1985.
40. E. C. Bernhardt, *Computer-Aided Engineering for Injection Molding,* Hanser, New York, 1983.
41. PLASPEC, Engineering and Marketing Data Bank, Plastics Technology, continuously updated.
42. N. E. Rouse, *Machine Des.,* Cleveland, OH, November 1987, pp. 108–112.
43. A. J. Klein, *Plast. Des. Forum,* Denver, CO, March–April 1988, pp. 39–49.
44. E. C. Bernhardt and G. Bertacchi, Plastics & Computer Inc., Montclair, NJ, 1988, personal communications.

45. D. V. Rosato, *Blow Molding Handbook,* Hanser, New York, 1989.

46. D. V. Rosato, *Plastics Processing Data Handbook,* Van Nostrand Reinhold, New York, 1990.

47. D. V. Rosato and D. P. DiMattia, *Designing with Plastics,* Van Nostrand Reinhold, New York, 1991.

Matrix-Controlled Failure Modes of Polymeric Composites

Advanced polymeric composites are composed of highly heterogenous and anisotropic plies where fibers are embedded in resin matrices. The composites are designed to derive their principal mechanical properties from the fibers. The polymeric matrices, being much weaker than the fibers, are not used for load-carrying purposes in the composites. Instead, through their relative ease of processing, the matrices impregnate and stabilize the fibers in the plies as well as the overall composite structures. For either the unidirectional or woven fabric type of ply, fibers are essentially distributed two-dimensionally in the ply. The desired composite properties are, therefore, mainly developed in the plane of the laminate. In directions lacking direct fiber reinforcement, laminate performance is largely controlled by the resin matrix. This is especially evident for the two major matrix-controlled failure modes, delamination and transverse matrix cracking.

Delamination is separation between plies as a result of out-of-plane stresses in the laminate, as shown in Figure 1*a*. Transverse matrix cracking (Fig. 1*b*), on the other hand, occurs within the plies along the fiber direction as a result of in-plane loading perpendicular to the fibers. Both failure modes can occur at relatively low load levels, well before the full load capacity of the fibers is reached. Such matrix-controlled cracks can lead to premature failure, or at least rapid property degradation of the structures. The matrix-controlled failure modes, therefore, dictate how effectively the desired fiber properties can contribute to the composite performance. One good example is the drastic compression strength reduction (up to 60–70%) caused by impact-induced matrix damage in the laminates. In this case, the failure mode is changed from gross buckling for an undamaged laminate into easy local buckling and propagation of the impact damage zone. Because the matrix plays the role of weak link in controlling laminate fracture, much of the emphasis in developing strong composites has been centered on toughening, or improving the fracture resistance, of the matrix materials. In this article, the matrix-controlled failure modes will be discussed from the viewpoints of fracture characterization, failure mechanisms, and their effects on structural performance.

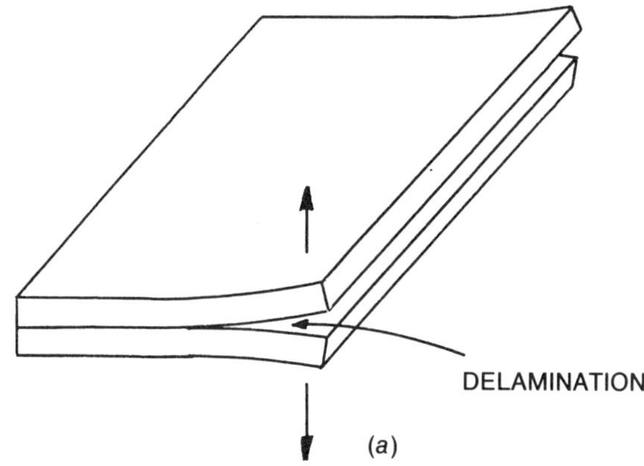

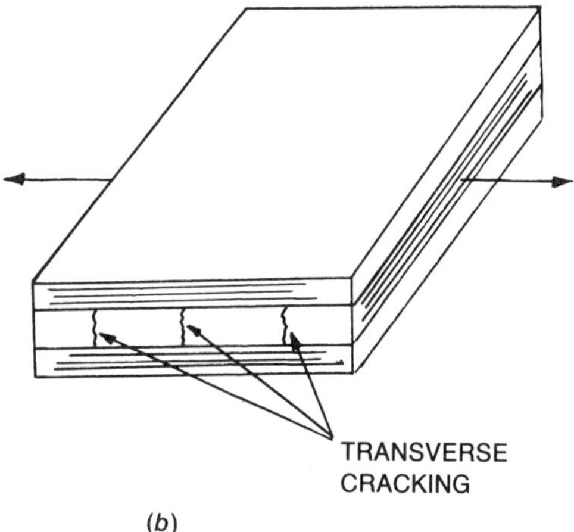

FIGURE 1 The matrix-controlled failure modes: (*a*) delamination and (*b*) transverse cracking in composites.

Fracture of Composites and Matrix-Controlled Failure Modes

Advanced composites are mainly used for high performance applications where the stiffness and strength of the materials are the critical factors to consider. Stress analysis techniques have been well established to provide guidelines for selecting laminate layups to achieve the specified stiffness. Few rules, however, exist to accurately predict composite failure. The major reason is that composites usually fail in an extremely complicated manner, where different mechanisms can be activated either sequentially or simultaneously. Direct applications of fracture mechanics to such failures by treating the composites as homogeneous materials have not always been successful. Unlike monolithic materials, such as metals and polymers, the concept of self-similar crack propagation usually does not apply to composites at the macroscopic level. Instead, the normally complex composite failures consist of three basic failure modes: fiber breakage, delamination, and transverse cracking. Their combined effects result in structural failure.

In order to predict the composite strength, the failure processes involving these basic modes need to be accurately accounted for. The matrix-dominated damage modes, being the obvious weak link in composites, are especially important from the standpoint of toughening or strengthening of composites. Conceptually, better fracture resistance of these mechanisms will ultimately lead to stronger composites. The matrix cracks, self-similar in nature, can be described and characterized by the fracture mechanics principles. Such an approach provides not only a material index and physical insight for developing resin matrices but also a data base and a manageable scheme for structural analysis.

To apply the fracture mechanics concept, matrix cracks have to be precisely defined in terms of the three basic crack tip loading modes, I, II, and III (Fig. 2), and their combinations (mixed modes). Mode I results from opening of the fracture surfaces to advance the crack. Mode II loading corresponds to in-plane shear of the crack surfaces in the direction of crack propagation. In mode III, the crack surfaces are subjected to antiplane shear, with relative movement parallel to the crack front and perpendicular to the crack growth direction. The mixed modes can be any numerical combination of crack tip stresses associated with the basic modes I, II, and III. For all the loading modes discussed, the fracture resistance can be defined in terms of the critical strain energy release rate G (e.g., G_{Ic} for mode I, G_{IIc} for mode II, G_{IIIc} for mode III, and G_c for mixed modes), with the physical meaning of fracture energy per unit crack area.

The microscopic fracture process in a bulk resin is not necessarily identical to that in the resin matrix of a composite. Matrix cracks can propagate under different loading modes (I, II, III, or mixed) along well-defined crack paths in the thin resin layers between the tightly spaced fibers. In contrast, homogeneous materials (e.g., metals and polymers) have difficulty in sustaining failures other than mode I because cracks tend to take paths that favor mode I propagation. Even for mode I matrix-controlled fractures, the fibers in the composites provide the matrix cracks with a constraint condition not observed in the mode I fracture of bulk resin. As a result, matrix cracks cannot always be simply related to the fracture behavior of the bulk resin.

In the following sections, matrix-controlled cracks of modes I and II and mixed modes are discussed separately. Mode III, delamination, which has been less studied than the other loading modes, is briefly mentioned. A section will also be included to address the rate, cycle, and temperature dependent failure behavior of matrix cracks. It should be recognized that not all the individual fracture modes are well understood. Their interactions, which are likely to occur in composite structures, are even less clear. However, some approaches did demonstrate the possibility of relating the basic failure modes to general composite failure; these will be covered in the last section.

Mode I Fracture

Mode I matrix-controlled failures, especially delamination, have been extensively studied in the past. Test methods to characterize G_{Ic} associated with matrix cracks have been well established. The most commonly used technique for mode I delamination is the double cantilever beam (DCB) test [1–11]. A large amount of G_{Ic} data from the DCB tests is also available in the literature. In contrast, mode I transverse cracking has been less investigated, and its characterization is less standardized. Many test methods have been used to measure G_{Ic} of this type of failure, such as single or double notched [12], three-point bending [13], center notched [9], double torsion [14–16], and double cantilever beam [17] techniques. Transverse cracking is physically similar to delamination except for the possible difference in resin layer thickness along the intralaminar and interlaminar crack paths, respectively, associated with the two modes. Such similarity has been confirmed experimentally [18] by comparing the G_{Ic} values of the two matrix-controlled failure modes.

Materials development to improve laminate fracture resistance often involves toughening of the resin materials. The matrix resins can be modified through resin

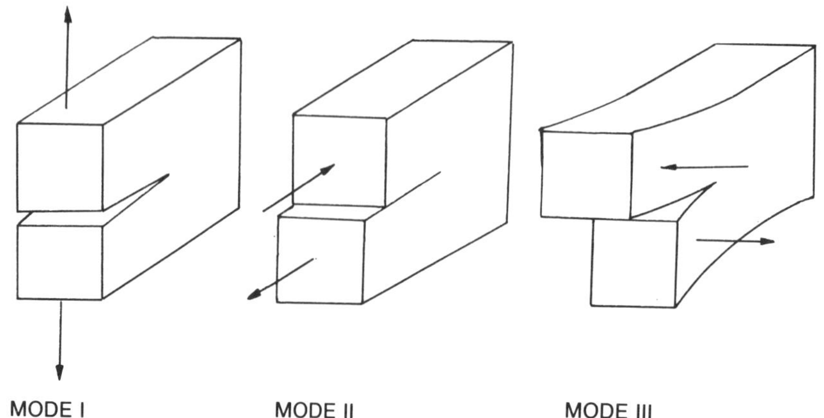

FIGURE 2 The basic crack tip loading modes: mode I (opening), mode II (shearing), and mode III (tearing).

chemistry variations and/or addition of modifiers to the base resins, all for the purposes of altering the intrinsic fracture behavior of polymers. However, the translation from resin toughness G_{Ic} to laminate G_{Ic} has been found to be neither direct nor proportional. A comprehensive comparison between resin G_{Ic} and delamination G_{Ic}, compiled by Hunston [19] for a wide range of thermoset and thermoplastic resins, can best demonstrate this point. For relatively brittle resins ($G_{Ic} < 200\ J/m^2$), delamination G_{Ic} was found to be more than two times that of the resin. For tougher resins, further increases in resin G_{Ic} increased delamination G_{Ic} by only one-third as much. A plot of delamination G_{Ic} against resin G_{Ic} is given in Figure 3 to show such a trend. Hunston noted that the toughness of brittle resins is fully transferred to composites. For tougher polymers, the partial transfer of the toughness was caused by the fibers restricting the crack tip deformation zone in the polymers. In other words, the matrix-controlled cracks are closely related to not only the resin but also the geometric constraints imposed by the tightly spaced fibers.

The crack tip deformation zone mentioned by Hunston [19] and other researchers [3] is in agreement with the fracture energy consideration and experimental observation. From the energy principles of fracture mechanics [20,21], homogeneous materials such as metals and polymers dissipate fracture energies by producing plastic flow at the crack tip. The fracture toughness of composites observed can also only be interpreted by the crack tip resin plastic deformation process [16]. Direct observations of the deformation zone at the delamination crack tip were made by Bradley et al. [22–25] with the scanning electron microscope (SEM). Such zones can extend to several fiber diameters on either side of the crack path, depending on the matrix toughness. Fiber debonding and breakage were also observed in the zones. Generally, the mode I fracture surfaces are dominated by resin fracture with varying degrees of resin deformation.

Only a few attempts have been made to establish quantitative descriptions of the mode I matrix-controlled failure processes. A model based on a hypothetical tensile bar to represent the crack tip resin deformation was proposed by Bradley et al. [22,25]. Based on the resin strain to failure, delamination G_{Ic} can be estimated. A simple calculation using this approach overestimated G_{Ic} of a graphite-reinforced composite, indicating a discrepancy of strain to failure between neat resin and the delamination crack tip.

From a somewhat different viewpoint, an in situ failure model [16,26,27] focused on the unique crack tip stress distribution in the thin resin layer between fibers. The crack tip deformation zone induced by such stresses determines the laminate G_{Ic} expressed as a function of resin G_{Ic} and several other resin variables, such as resin modulus and yield stress. This model has interpreted a number of failure phenomena [26,27], including the nonlinear translation from resin G_{Ic} to delamination G_{Ic} reported by Hunston [19]. Because resin G_{Ic} and other resin properties do not vary in the same way, their combined effect results in such a nonlinear G_{Ic} translation from resin to laminate. In general, more detailed crack tip analysis of the matrix cracks is needed before laminate G_{Ic} can be more accurately predicted from resin variables.

Mode II Fracture

Mode II fracture characterization requires pure shear loading applied at the crack tip. One test gaining increasing acceptance is the end notched flexural (ENF) method [28,29] for delamination G_{IIc} measurement where unidirectional beam specimens are under simple three-point loading. Based on similar principles, Bradley et al. [11,22–25,30] conducted mode II delamination tests using split laminate specimens. Other G_{IIc} tests include the Acran test [31] for delamination and the three rail shear [32] and thin tube [33] tests for transverse cracking.

SEM observations of mode II fracture surfaces often showed pronounced hackle marking in the resin matrices [28,30,31], as shown in Figure 4. In comparison, such a hackle pattern is either missing or much less severe on

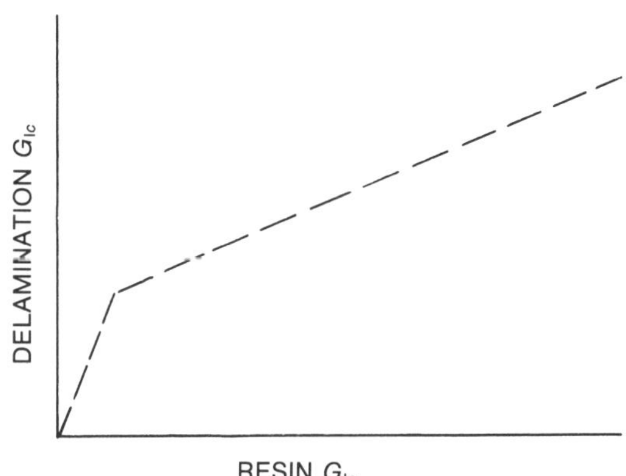

FIGURE 3 The general trend of delamination G_{Ic} as a function of resin G_{Ic} for a wide range of materials. (From Ref. 19.)

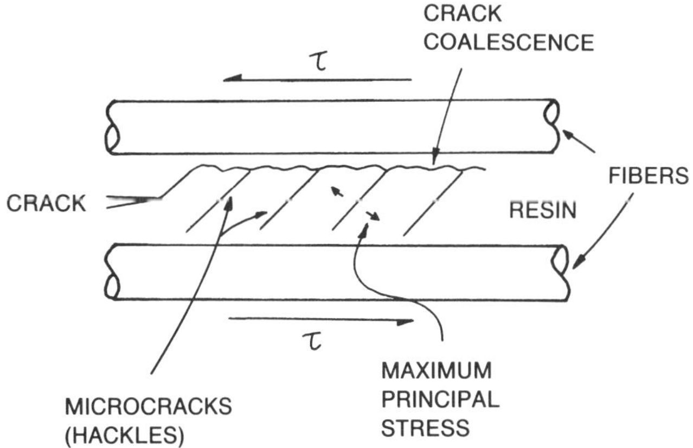

FIGURE 4 Schematic of the crack tip failure process of the mode II matrix crack.

the mode I fracture surfaces. The formation of the hackles along the mode II crack path is due to the maximum principal stress that induces microcracks at the crack tip (Fig. 4). The microcracks can grow only between the fibers. The coalescence of these microcracks results in macroscopic crack propagation [30,31]. In general, microcracking in the resin between the fibers increases the fracture surface area and thus yields a more tortuous crack path than mode I fracture. As a result, matrix cracks in composites have been observed to have higher G_{IIc} than G_{Ic}. The increase of toughness (sometimes tenfold) from mode I to mode II was especially pronounced for brittle matrices where the hackle marking was distinct. For ductile systems, microcracking was often suppressed by resin plastic deformation to give a less dramatic difference (2 to 3 times) between G_{Ic} and G_{IIc}. Such effects were also clearly observed, for example, between brittle thermoset and ductile thermoplastic composites [34].

Theoretical models to describe the mode II failure process and to predict G_{IIc} from resin properties are yet to be developed. Experimental evidence indicates that G_{IIc} is not simply related to any single resin variable. For example, loading rates had little effect on G_{IIc} measured by the ENF method over a wide range of rates [34]. Resin properties, on the other hand, vary significantly with the loading rates.

Mode III Fracture

Although mode III matrix crack growth undoubtedly exists during composite failure, such a mode of fracture has not been extensively studied. The main reason has been the difficulty in inducing a pure antiplane shear (mode III) condition in the crack tip. The only published work on mode III fracture was that of Donaldson [35,36], who developed a split cantilever beam method to measure delamination G_{IIIc}. Although slight mode I (opening) loading existed in the test, its effect was minimized. G_{IIIc} measured by this method was higher than both G_{Ic} and G_{IIc} for the materials studied. The mode III fracture surfaces also appeared to be somewhat different from the mode I or mode II ones, indicating that different mechanisms are involved. Overall, mode III fracture deserves to be further studied.

Mixed-Mode Fracture

The fracture processes in composite structures are more likely to involve matrix cracks driven by mixed loading modes than cracks driven by pure mode I, II, or III. Mixed-mode fracture can have any combination of opening (I), shear (II), and tearing (III) crack tip stresses. Test methods developed for measuring mixed-mode toughness G_c can be categorized into crack propagation type and edge delamination type.

Test methods of the former type are designed to induce combined loading to propagate cracks in specimens that already have preintroduced cracks. These tests are mostly limited to combined mode I and II fracture, because other combinations are hard to generate experimentally. Among the tests developed, the crack lap shear technique [37–39] has received considerable attention. However, other methods, such as the split laminate [22–25], the mixed-mode flexural [28], and the Acran methods [31] for delamination and the off-axis tensile method [32] for transverse cracking, have also been used.

The edge delamination approach can be attributed to the original work by O'Brien [40–45], who established the methodology and theoretical basis of the test. Laminate specimens with multidirectional plies are subjected to simple tension to induce delamination at the specimen edges. The delamination results from the unique phenomenon of out-of-plane edge stresses [46,47] as a result of in-plane loading of the laminates. Although the specimens do not have preintroduced cracks, the crack initiation and growth process can be described by fracture mechanics [40–43]. The edge delamination toughness G_c can thus be simply related to the onset strain of delamination and the laminate modulus variation due to such damages. The loading mode of fracture can, in principle, be any combination of modes I, II, and III, depending on the laminate layups.

Several experimental studies [28,39,45] have addressed mixed-mode fracture toughness as a function of mode I and mode II contributions. In most cases, mixed-mode G_c lies between the G_{Ic} and G_{IIc} values. For relatively brittle matrices with $G_{IIc} > G_{Ic}$, mixed-mode G_c increases with increasing mode II loading, as a result of the increasing amount of matrix microcracking along the macroscopic crack paths. Toughened systems, on the other hand, often lack hackle marking on both mixed-mode and mode II fracture surfaces. In general, mixed-mode G_c can be approximated as being linearly related to G_{Ic} and G_{IIc} [39,45]. Although such a linear relationship does not necessarily have a physical basis, it does provide a useful guideline for design analysis for at least mixed modes I and II.

Rate, Cycle, and Temperature Effects on Matrix Cracks

All polymeric materials are inherently viscoelastic. Their fracture and other material properties are dependent on conditions such as temperature and rate of applied load. As a result, matrix cracks in composites would, in principle, be affected by the resin viscoelastic behavior when subjected to various temperatures, loading rates, and cyclic loads. However, the dependency of matrix cracks on such conditions is not necessarily consistent with that of neat resins. This largely reflects different controlling failure mechanisms, which are yet to be clarified.

The rate sensitivity of mode I [34,48] and mode II [34,49] delamination has been studied in the past. In general, G_{Ic} and G_{IIc} of several brittle and toughened composites composed of unidirectional plies were insensitive to a large range of loading rates. Only at very high rates did such toughness values decrease with loading rate. However, G_{Ic} of a woven fabric based composite [50] steadily decreased within the entire loading rate range studied.

Limited work has been published [28,49,51] on the effect of temperature on mode I and mode II delamination. Based on the available data, G_{Ic} can increase [51] or decrease [28] with increasing temperature, depending on the materials. Mode II delamination fracture resistance also showed mixed sensitivity to temperature [49,51].

Studies on fatigue of mode I [52], mode II [29], and mixed-mode [42,53,54] delamination have been reported in the literature. In many cases, the crack growth rate per cycle can be expressed as a power law of the applied strain energy release rate. One major finding from these studies was that the threshold strain energy release rate G_{th}, below which no crack growth occurred at certain cycles (e.g., 10^6), was significantly lower than the static fracture toughness. The characterization of G_{th} is, therefore, important for considering composites for cyclic load applications.

Structural Performance as a Function of Matrix Cracks

Composite structural failures undoubtedly involve matrix-controlled crack growth of the different loading modes mentioned above. Such damage can accumulate in different ways as the laminate construction and loading conditions vary. To accurately predict laminate performance, it is necessary to account for the progressive development of matrix-dominated damage. Such a task appears to be complicated, considering the enormous variations in the laminate structures. However, the approach developed by Wang et al. [55–58] offers great promise for describing the detailed failure processes of composites.

Wang's approach basically employs the finite element analysis procedure to predict the onset of both delamination and transverse cracking. The numerical technique calculates the strain energy release rates related to the matrix cracks in the laminate. When these rates reach the matrix crack toughness values (e.g., G_{Ic} and G_{IIc}), crack growth will occur. In other words, the experimentally determined toughness values of the crack modes, discussed before, can be used for failure prediction. In principle, the type and loading mode of the crack and its location can be determined from the analysis. Wang et al. [55–58] and Law [59] successfully predicted the detailed matrix cracks observed in a series of $(+25/-25/90_n)_s$ graphite–epoxy laminates. Other cases of various laminates under different loading conditions were also investigated [60], with excellent agreement between analytical and experimental results.

The fracture energy consideration of stress analysis has several important implications. First, crack formation was clearly shown [55–60] to be affected not only by the material properties but also by the geometric constraints between plies. The simple "ply strength" consideration for composite analysis is, therefore, not accurate. Second, the methodology can, in principle, treat the matrix crack growth of all modes, thus allowing the material parameters (G_{Ic}, G_{IIc}, etc.) to be incorporated in the structural designs. Without such a detailed description of the failure processes, the toughness values can be used only for ranking and selecting materials. The fracture mechanics method, although it has great potential, can by no means solve all the design problems at the moment. For example, not all the matrix fracture modes, such as mode III and other mixed modes, are well understood and fully characterized. The interactions of different cracks in the composites still need to be better addressed in the analysis.

Shaw Ming Lee

References

1. J. M. Scott and D. C. Phillips, *J. Mater. Sci. 10,* 551 (1975).
2. F. X. de Charentenay and M. Benezeggagh, *Proceedings of ICCM-3,* Vol. 1, 1980, pp. 186–197.
3. W. D. Bascom, J. L. Bitner, R. J. Moulton, and A. R. Siebert, *Composites, 11*(1), 9 (1980).
4. D. F. Devitt, R. A. Schapery, and W. L. Bradley, *J. Comp. Mater., 14,* 270 (1980).
5. D. J. Wilkins, J. R. Eisenmann, R. A. Camin, W. S. Margolis, and R. A. Benson, "Characterizing Delamination Growth in Graphite-Epoxy," in *Damage in Composite Materials,* ASTM STP 775, American Society for Testing and Materials, Philadelphia, 1982.
6. J. M. Whitney, C. E. Browning, and W. Hoogsteden, *J. Reinf. Plast. Composites, 1*(4), 297 (1982).
7. W. D. Bascom, G. W. Bullman, D. L. Hunston, and R. M. Jensen, *Proceedings of the 29th National SAMPE Symposium,* 1984, pp. 970–978.
8. F. X. de Charentenay, J. M. Harry, Y. J. Prel, and M. L. Benzeggagh, "Characterizing the Effect of Delamination Defect by Mode I Delamination Test," in *Effects of Defects in Composite Materials,* ASTM STP 836, American Society for Testing and Materials, Philadelphia, 1984.
9. J. M. Whitney and C. E. Browning, "Material Characterization for Matrix Dominated Failure Modes," in *Effects of Defects in Composite Materials,* ASTM STP 836, American Society for Testing and Materials, Philadelphia, 1984.
10. D. L. Hunston, R. J. Moulton, N. J. Johnston, and W. D. Bascom, "Matrix Resin Effects in Composite Delamination: Mode I Fracture Aspects," in *Toughened Composites,* ASTM STP 937, American Society for Testing and Materials, Philadelphia, 1987.
11. W. M. Jordan and W. L. Bradley, "Micromechanisms of Fracture in Toughened Graphite-Epoxy Laminates," in *Toughened Composites,* ASTM STP 937, American Society for Testing and Materials, Philadelphia, 1987.
12. R. J. Sanford and F. R. Stonesifer, *J. Comp. Mater.,* (5) 241 (1971).
13. A. Parvizi, K. W. Garrett, and J. E. Bailey, *J. Mater. Sci. 13,* 195 (1978).
14. S. M. Lee, *J. Mater. Sci. Lett., 1,* 511 (1982).
15. G. D. M. DiSalvo and S. M. Lee, *SAMPE 14*(2), 14 (1983).
16. S. M. Lee, *J. Mater. Sci., 19,* 2278 (1984).
17. D. C. Phillips and G. M. Wells, *J. Mater. Sci. Lett., 1,* 321 (1982).

18. S. M. Lee, *J. Comp. Mater., 20,* 185 (1986).
19. D. L. Hunston, *Composites Technology and Review, 16*(4), 176 (1984).
20. J. F. Knott, *Fundamental of Fracture Mechanics,* Halsted Press, New York, 1973, p. 110.
21. D. Broek, *Elementary Engineering Fracture Mechanics,* Sijthoff and Noordhoff, The Hague, The Netherlands, 1978, pp. 119–120.
22. W, L. Bradley and R. N. Cohen, "Matrix Deformation and Fracture in Graphite-Reinforced Epoxies," in *Delamination and Debonding of Materials,* ASTM STP 876, American Society for Testing and Materials, Philadelphia, 1985.
23. W. M. Jordan and W. L. Bradley, "Micromechanisms of Fracture in Toughened Graphite-Epoxy Laminates," in *Toughened Composites,* ASTM STP 937, American Society for Testing and Materials, Philadelphia, 1987.
24. M. F. Hibbs, M. K. Tse, and W. L. Bradley, "Interlaminar Fracture Toughness and Real-Time Fracture Mechanism of Some Toughened Graphite/Epoxy Composites," in *Toughened Composites,* ASTM STP 937, American Society for Testing and Materials, Philadelphia, 1987.
25. E. A. Chakackery and W. L. Bradley, *Polym. Eng. and Sci., 27*(1), 33 (1987).
26. S. M. Lee, *Polym. Eng. and Sci., 27*(1), 77 (1987).
27. S. M. Lee, "Failure Mechanism of Delamination Fracture," in *Composite Materials Testing and Design,* ASTM STP 972, American Society for Testing and Materials, Philadelphia, 1988.
28. A. J. Russel and K. N. Street, "Moisture and Temperature Effects on the Mixed-Mode Delamination Fracture of Unidirectional Graphite/Epoxy, in *Delamination and Debonding of Materials,* ASTM STP 876, American Society for Testing and Materials, Philadelphia, 1985.
29. A. J. Russel and K. N. Street, "The Effect of Matrix Toughness on Delamination: Static and Fatigue Fracture Under Mode II Shear Loading of Graphite Composites," in *Toughened Composites,* ASTM STP 937, American Society for Testing and Materials, Philadelphia, 1987.
30. M. F. Hibbs and W. L. Bradley, "Correlations between Micromechanical Failure Processes and Delamination Toughness of Graphite/Epoxy Systems," in *Fractography of Modern Engineering Materials,* ASTM STP 937, American Society for Testing and Materials, Philadelphia, 1987.
31. L. Acran, M. Acran, and I. M. Daniel, "SEM Fractograph of Pure and Mixed Mode Interlaminar Fracture in Graphite/Epoxy Composites," in *Fractography of Modern Engineering Materials,* ASTM STP 948, American Society for Testing and Materials, Philadelphia, 1987.
32. S. L. Donaldson, *Composites, 16*(2), 103 (1985).
33. G. S. Giare and D. Campbell, *Engineering Fracture Mechanics, 27*(6), 683 (1987).
34. A. J. Smiley and R. B. Pipes, *Proceedings of American Society for Composites 1st Technical Conference,* 1986, pp. 434–449.
35. S. L. Donaldson, *Proceedings of ICCM-6,* Vol. 3, 1987, pp. 274–283.
36. S. L. Dondaldson, "Mode III Interlaminar Fracture Characterization of Composite Materials," M. S. Thesis, University of Dayton, 1987.
37. R. L. Rankumar and J. D. Whitcomb, "Characterization of Mode I and Mixed-Mode Delamination Growth in T300/5208 Graphite Epoxy," in *Delamination and Debonding of Materials,* ASTM STP 876, American Society for Testing and Materials, Philadelphia, 1985.
38. P. D. Mangalgiri and W. S. Johnson, Preliminary Design of Crack-Lap shear Specimen Thickness for Determination of Interlaminar Fracture Toughness" *Journal of Composites Technology and Research,* Summer 1986, p. 58.
39. W. S. Johnson and P. D. Mangalgiri, "Influence of the Resin on Interlaminar Mixed-Mode Fracture," in *Toughened Composites,* ASTM STP 937, American Society for Testing and Materials, Philadelphia, 1987.
40. T. K. O'Brien, "Characterization of Delamination Onset and Growth in a Composite Laminate," in *Damage in Composite Materials,* ASTM STP 775, American Society for Testing and Materials, Philadelphia, 1982.
41. T. K. O'Brien, N. J. Johnston, D. H. Morris, and R. A. Simonds, *SAMPE J. 18*(4), 8 (1982).
42. T. K. O'Brien, "Mixed-Mode Strain-Energy-Release Rate Effects on Edge Delamination of Composites," in *Effects of Defects in Composite Materials,* ASTM STP 836, American Society for Testing and Materials, Philadelphia, 1984.
43. T. K. O'Brien, "Interlaminar Fracture of Composites," NASA Tech. Mem. 85678, 1984.
44. T. K. O'Brien, "Analysis of Local Delaminations and Their Influence on Composite Laminate Behavior," in *Delamination and Debonding of Materials,* ASTM STP 876, American Society for Testing and Materials, Philadelphia, 1985.
45. T. K. O'Brien, N. J. Johnston, I. S. Raju, D. H. Morris, and R. A. Simonds, "Comparison of Various Configurations of Edge Delamination Test for Interlaminar Fracture Toughness," in *Toughened Composites,* ASTM STP 937, American Society for Testing and Materials, Philadelphia, 1987.
46. R. B. Pipes and N. J. Pagano *J. Comp. Mater., 4,* 538 (1970).
47. S. S. Wang and F. G. Yuan, *J. Applied Mechanics, 50,* 835 (1983).
48. I. M. Daniel, I. Shareef, and A. A. Aliyu, "Rate Effects of Delamination Fracture Toughness of Graphite/Epoxy," in *Toughened Composites,* ASTM STP 937, American Society for Testing and Materials, Philadelphia, 1987.
49. T. J. Chapman, A. J. Smiley, and R. B. Pipes, *Proceedings of ICCM-6,* Vol. 3, 1987, pp. 295–304.
50. S. Mall, G. E. Law, and M. Katouzian, *J. Comp. Mater., 21,* 569 (1987).
51. P. Davis and F. X. de Charentenay, *Proceedings of ICCM-6,* Vol. 3, 1987, pp. 284–294.
52. R. Martin, *Proceedings of the American Society for Composites, 3rd Technical Conference,* 1988, pp. 688–700.
53. A. Poursatip, "The Characterization of Edge Delamination Growth in Laminates under Fatigue Loading," in *Toughened Composites,* ASTM STP 937, American Society for Testing and Materials, Philadelphia, 1987.
54. T. K. O'Brien, *Proceedings of the American Society for Composites, 1st Technical Conference,* 1986, pp. 404–420.
55. A. S. D. Wang and F. W. Crossman, *J. Comp. Mater.* Supplementary, *14,* 71 (1980).
56. F. W. Crossman and A. S. D. Wang, *J. Comp. Mater.* Supplementary, *14,* 88 (1980).
57. A. S. D. Wang, *Proceedings of ICCM-3,* 1980, pp. 170–185.
58. F. W. Crossman and A. S. D. Wang, "The Dependence of Transverse Cracking and Delamination on Ply Thickness in

Graphite/Epoxy Laminates," in *Damage in Composite Materials*, ASTM STP 775, American Society for Testing and Materials, Philadelphia, 1982.

59. G. E. Law, "A Mixed Mode Fracture Analysis of $(\pm 25/90_n)_s$ Graphite/Epoxy Composite Laminates," in *Effects of Defects in Composite Materials*, ASTM STP 836, American Society for Testing and Materials, Philadelphia, 1984.

60. A. S. D. Wang, M. Solomiana, and R. B. Bucinell, "Delamination Crack Growth in Composite Laminates," in *Delamination and Debonding of Materials*, ASTM STP 876, American Society for Testing and Materials, Philadelphia, 1985.

Mechanical Properties

See Materials Selection, Polymer Matrix

Mechanical Testing

See Testing, Mechanical

Melt Casting

See Metal Matrix Composites, Casting

Metal Fibers

See Metal Matrix Composites, Continuous Fiber; Metal Matrix Composites, Tungsten Fiber Reinforced

Metal Matrix Composites, Aluminum

The technology and evolution of metal matrix composite systems has developed over the past 20 years, with primary support from and emphasis on aerospace and airframe requirements and, more recently, automotive and electronic applications. At present, a wide variety of composite material systems with filament and tow reinforcements is available, producing anisotropic materials and discontinuously reinforced concepts, including particulate, chopped fiber, and whisker reinforcements to create isotropic composites. Primary matrix metals being used are aluminum, magnesium, copper, titanium, and most recently "superalloy" systems.

This article will cover aluminum matrix composites primarily and is divided into an introduction background section followed by two sections on the basic types of composite material. First are the continuous graphite-reinforced metals, which approach a zero coefficient of thermal expansion (CTE) and provide high thermal conductivity and very high specific stiffness (modulus of elasticity/density) levels. This class of composites offers the material systems designer a conductive, nonoffgassing material, up to 5.2×10^{11} Pa stiffness, with a density less than that of aluminum. This type of composite is ideal for space truss structural members, thermally stable elements, electronic or heat-transfer elements, and "stable" platforms. These state-of-the-art graphite–aluminum materials are available in sheet and plate form and as simple structural elements, including tubes, at graphite loading levels of $\sim$ 50% and, via the recently developed "DWG" processes, with a thin-ply capability of 0.09 mm/layer. This thin-layer capability is important in optimizing and sizing many space structure elements (Table 1).

Boron–aluminum composites (utilizing a CVD-produced boron filament with a nominal 1.4 mm diameter and typically used in a uniaxial orientation) are presented as a subset of the continuously reinforced materials.

The second composite material section summarizes DWAl 20, a family of ceramic-particulate-reinforced (powder-metal-processed) metals with isotropic properties, low to moderate CTE, and moderate stiffness. This family of materials is easily processible by conventional forging, extruding, and shear spinning, and it utilizes existing basic metals industry facilities and equipment. This system represents a producible, low-cost opportunity to combine many of the advantages of composites with basic, conventional plant and equipment to fabricate intermediate to high property, isotropic composite items. These attributes suggest that discontinuously reinforced composites are an ideal material selection for a broad spectrum of applications. They may be used as a basic airframe material by taking advantage of enhanced mechanical characteristics and their demonstrated extrudability, or they may be used for space structures such as satellites and antenna structures, where rigidity per unit density and thermal expansion control are key design criteria.

These materials are also being evaluated for higher use temperatures, wear resistance, and the use of REVLITE composites (ceramic-particulate-reinforced aluminum) especially developed for automotive and internal combustion engine usage. DWAl 20 is available in most aluminum alloys, magnesium, copper, and, developmentally, "superalloy" matrices, using a variety of ceramic-particulate reinforcements.

TABLE 1
Properties of Carbon Fiber Types

Fiber Type	Density (g/cm^3)	Young's Modulus (GPa)	Tensile Strength (GPa)	Electric Resistivity (Ω·m)	Thermal Conductivity (W/m·K)
High strength (PAN)	1.7–1.8	230–250	2.8–4.0	12–30	7–10
Ultrahigh strength (PAN)	1.7–1.8	260–290	4.1–5.7	14–20	7–9
High modulus (PAN. mesophase pitch)	1.8–2.0	350–550	1.7–3.5	5–10	60–200
Ultrahigh modulus (mesophase pitch)	2.0–2.2	600–900	2.1–2.5	1–4	400–2500

The advantages of composite materials over conventional metals or plastics come from the ability to design a variety of mechanical and physical characteristics, with or without directionality, into a single material. This is accomplished by a series of general steps:

1. Selecting and designing the combination of reinforcements and matrix material
2. Combining the reinforcement and the matrix to achieve the desired distribution
3. Consolidation, bonding, or casting to combine the constituents
4. Secondary processing

Obviously, the details of each step will vary. Figure 1 shows schematically the most widely used processing techniques: diffusion bonding of continuous, uniaxial, boron–aluminum and graphite–aluminum composites (using preinfiltrated precursor wires), as well as powder-metallurgy-processed discontinuously reinforced composite. A wide variety of fabrication techniques, including casting, squeeze casting, plasma spraying, "Osprey," and various electro and electroless coating techniques are also applicable to producing metal matrix composites.

A simple way to consider and compare the attributes of various types of metal matrix composites is by examining the stress–strain curves for each type. Figure 2 superimposes curves for the three primary aluminum matrix composites—ceramic particulate, boron filament, and graphite tow reinforcements—on the curve for conventional 6061-T6. Although these curves are somewhat idealized, the two uniaxially reinforced continuous-fiber composites are markedly stiffer and stronger—but also show very low ductility to fracture.

The DWAl 20, ceramic particulate reinforced composite (25 vol % 6061-T6) shows a more conventional stress–strain response (exhibiting ductility), but has mechanical properties intermediate between those of the unreinforced material and those of the continuous-fiber composites. The exact positions and shapes of the composite stress–strain curves can be varied by matrix heat treatment, reinforcement volume percent (vol %), or reinforcement orientation effects. It must also be kept in

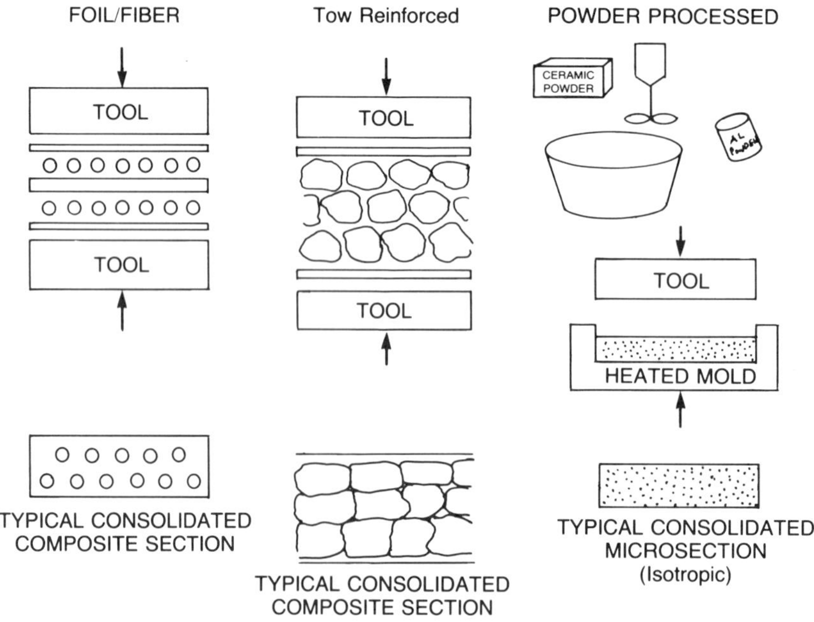

FIGURE 1 **Schematic of composite fabrication processes.**

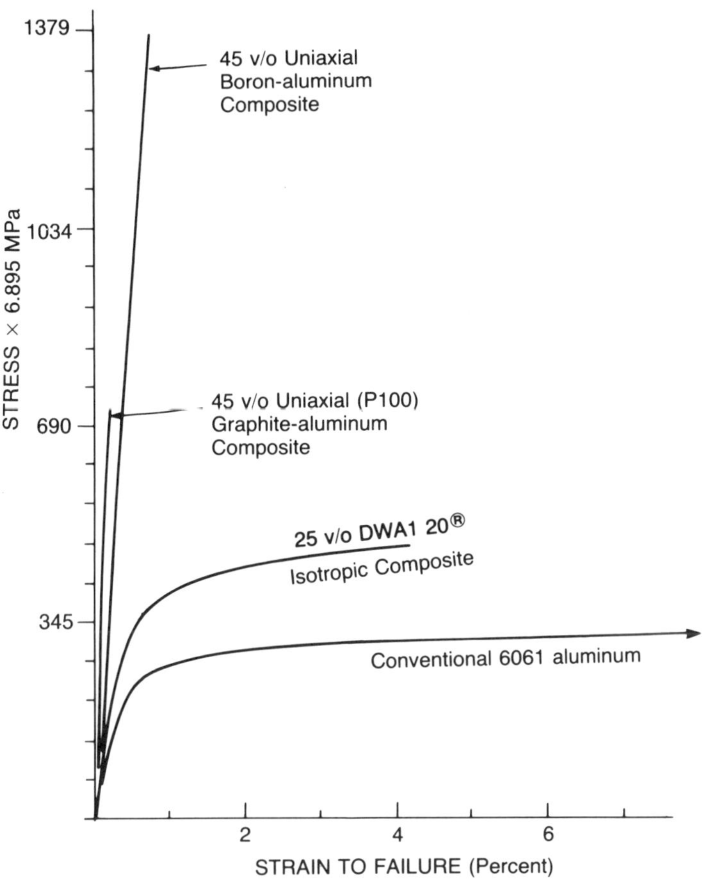

FIGURE 2 **Comparative tensile stress–strain curves.**

mind that all continuously reinforced composite material exhibits anisotropic characteristics. In Figure 2, if the transverse-direction stress–strain curve for the boron–aluminum material were to be included, it would be relegated to the extreme lower left-hand corner of the chart: strength 152 MPa, modulus ~ 145 GPa, $\varepsilon_f \sim 0.4\%$. Similarly, the graphite–aluminum composite transverse stress–strain response shows even lower mechanical characteristics.

The two continuously reinforced composite systems that have been most widely studied are the continuous-filament-reinforced composites, such as boron–aluminum or its more recent form, SCS-Al, and the graphite-aluminum system. Basic processing for these materials is diffusion bonding; Figure 3 is a schematic of the process for boron–aluminum monolayer forms that can be used for subsequent layup into more complex reinforcement patterns or shapes and rebonded. Since a wide variety of continuous reinforcements are being used, Figure 4 summarizes the basic uniaxial mechanical characteristics for the two forms of composite being discussed.

For the boron and single-filament SiC materials, high tensile strengths with very high compression strengths, particularly for the B-Al, are typical, with a modulus about the same as that of steel, but a density approximately the same as that of aluminum. The graphite-reinforced aluminum, pitch 828 GPa (120 million) modulus tow reinforcement, shows a much stiffer resulting composite, but with a less efficient compressive capability.

A review of the continuously reinforced processing concepts and materials follows.

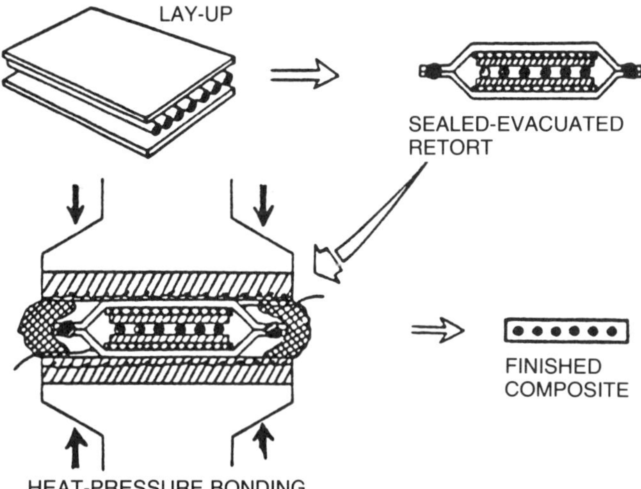

FIGURE 3 **Fabrication process for monolayer boron–aluminum composites.**

Fiber	Matrix	Density (mg/cm × 10^{-3})	Axial Modulus (GPa)	Transverse Modulus (GPa)	Axial Tensile (MPa)	Axial Compressive (MPa)	Transverse Tensile (MPa)
None	Aluminum[c]	6.1	69	69	290	241	290
P120 graphite	Aluminum	5.6	448	34	690	345	34
Boron	Aluminum	5.9	207	38	1241	2758	138
Silicon carbide[a]	Aluminum	6.6	207	38	1241	1724	103
Silicon carbide[b]	Aluminum	6.0	131		1034		69

* Fiber volume fraction = 50%
[a] Monofilament.
[b] Multifilament yarn.
[c] 6061 T6.

FIGURE 4 **Representative properties of some fiber-reinforced metal matrix composites.*** (From Ref. 1.)

Diffusion Bonding

The first successful method for secondary fabrication of graphite-reinforced MMCs was diffusion bonding. This technique starts with liquid metal infiltrated precursor wires, as shown in Figure 5. These wires are collimated and held together in a matt with a fugitive binder that thermally decomposes and leaves no residue. Layers of these matts are stacked together to make the desired number of plies and the desired orientation of fibers. Graphite—aluminum composites are normally made in the unidirectional configuration. Thin (0.089 mm) aluminum foils are placed at the bottom and the top of the wire plies to complete the "green" composite. After this green composite is placed in a retort that is heated to the proper temperature, pressure is applied for a set amount of time, then removed, and the completed composite is removed from the retort. The technique can be applied to ion-plated, sputter-coated, or electroplated fibers as well as to LMI precursors.

The diffusion-bonding technique uses the combination of time, temperature, and pressure to deform the precursor wires into intimate contact and to fully knit the wires into a composite by diffusion. This technique has been applied to graphite fiber reinforced aluminum, magnesium, copper, silver, tin, and lead matrices. If an improper combination of time, temperature, and pressure is used, incomplete consolidation or fiber damage by reaction can result. In 1984, a thin-ply graphite-reinforced aluminum composite designated DWG was introduced. It had ply thicknesses approaching 0.075 mm along with a fiber content between 50 and 60 vol %. A summary of the data derived from this work appears in Table 2. As with other forms of this composite, the modulus values are those predicted by the rule of mixtures, while the strength values are lower than those predicted by it.

FIGURE 5 **Scanning electron micrograph of precursor wire.**

A typical microsection for this composite (Fig. 6) shows that the fiber content is high and uniform. This process has been used to make sheets, plates, and structures. Among the structures produced is a 50 mm diameter tube that has a wall thickness of 0.81 mm and eight plies aligned at ± 15° with respect to the axis for expansion control. This type of structure becomes very attractive for space structures in which high stiffness and low thermal distortion are critical.

Flat sheet and plate, as well as many shapes, have been fabricated by the diffusion bonding technique. Examples of some of the shapes are shown in Figures 7 to 9. Figure 7 is a generic hat-stiffened panel made by spot-welding diffusion-bonded hat stiffeners to a sheet of graphite aluminum. Figure 8 [3,4] is a thin-walled tube 38 mm in diameter and 2 m long. Table 3 shows mechanical properties for uniaxial graphite–aluminum tubes processed from LMI precursor. Figure 9 is a high-gain antenna boom for the Hubble space telescope made with a diffusion-bonded sheet of P100 graphite fibers in 6061 aluminum. This structure is 3.6 m long with internal dimension tolerances of ± 0.15 mm along the entire length so that the tube can act as a wave guide. This structure also requires the stiffness and low-expansion characteristics of this composite.

Graphite–metal composites offer the materials designer the ability to control a variety of thermophysical characteristics as well as the basic stiffness and strength

TABLE 2
Tensile Property Summary for DWG-Produced Composite

Fiber	Elasticity (GPa)	Ultimate Tensile Strength, Longitudinal (MPa)	Ultimate Tensile Strength, Transverse (MPa)
P55	207–221	520–620	30–50
P75	276–296	620–720	30–50
P100	379–414	550–830	30–50
P120[a]	469–558	590–880	30–50

[a] Preliminary data.

response. For example, the above-mentioned Hubble space telescope boom, as well as many of the applications for thin-walled cylindrical tubes, requires very precise control of thermal distortion. This is typically measured by the CTE of the material. Since most high modulus pitch-graphite tows exhibit a very low or negative CTE when combined with a metal matrix, the resultant composite will have a very low CTE. The longitudinal CTE of composite material containing unidirectionally aligned fibers can be predicted based on the following formula:

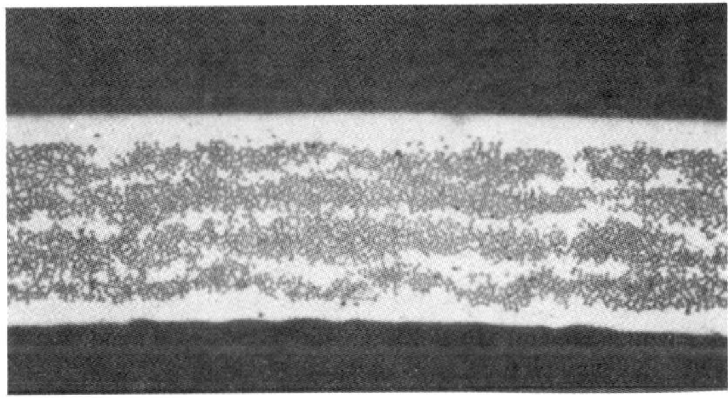

FIGURE 6 Four-layer P100 DWG-Al tube micrograph (0.38 mm thick).

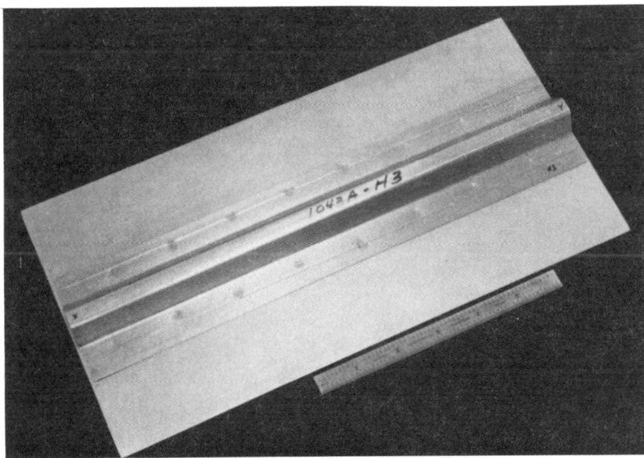

FIGURE 7 Spot-welded graphite–aluminum composite-stiffened panel (sheet and hat stiffeners made by diffusion heating).

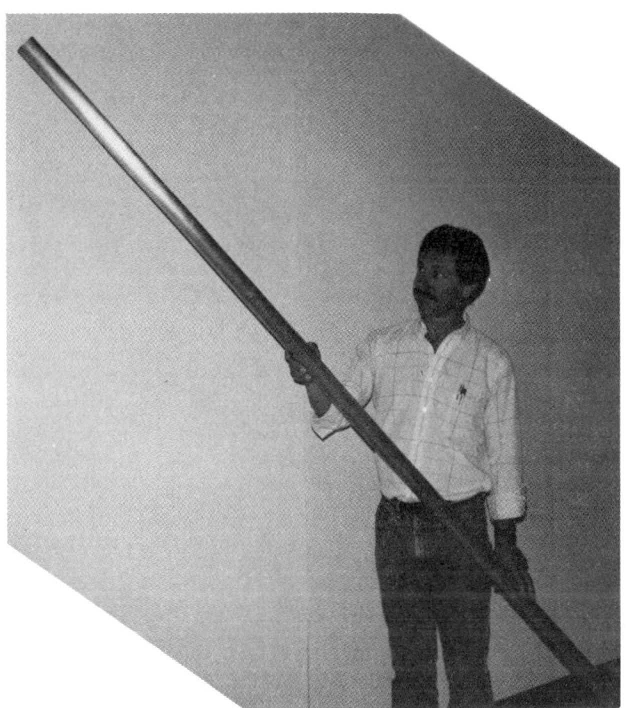

FIGURE 8 Thin-walled tube (38 mm diameter, 2 m long).

TABLE 3
Tubing Fabricated Using LMI Wire

Configuration	Ultimate Tensile Strength (MPa)	Elastic Modulus (GPa)	Graphite (vol %)
19.1 mm diameter,	720	345	45
three-ply	676	345	45
	710	331	45
Average	703 + 20	338 + 7	45
25.4 mm diameter,	720	296	40
two-ply	703	283	40
	641	310	40
Average	690 + 34	296 + 14	40

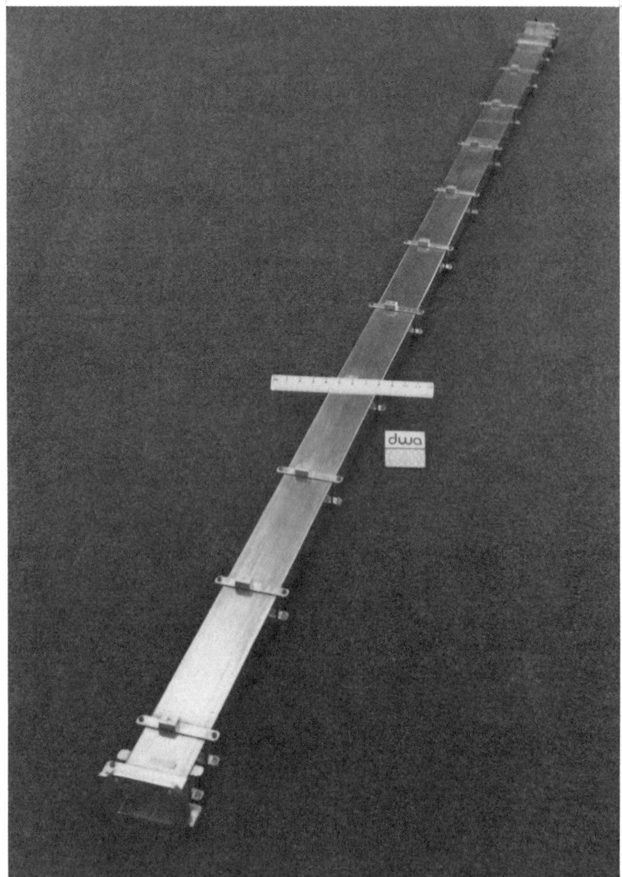

FIGURE 9 **High-gain antenna mast for Hubbell space telescope, produced with diffusion-bonded P-100 fiber-reinforced aluminum.**

$$\text{CTE} = \frac{E_m \cdot \alpha_m \cdot V_m + E_f \cdot \alpha_f \cdot V_f}{E_m \cdot V_m + E_f \cdot V_f}$$

Similarly, the pitch-graphite tow exhibits very high thermal conductivity along the tow. This enables graphite–aluminum composite materials to have thermal conductivity similar to that of pure copper, and well above that of aluminum alloy; Table 4 exhibits this effect.

Casting

Casting of continuous-graphite-fiber MMCs is an especially appropriate fabrication technology when these materials are to be used in complex-shaped parts, such as the joints that connect the load-bearing members of a truss structure or the components of an internal combustion engine. It is also excellent for producing MMC materials in thick sections or in sections containing various fiber orientations, because mechanical pressure is not applied to the material during fabrication, nor are frictional losses a factor in consolidation, nor is there a problem of damaging cross-plied fibers by forcing them to conform to the small irregularities of the underlying layers. However, casting is not as well suited to the efficient fabrication of thin, flat sheets because of the difficulty of feeding molten metal over long distances through very thin cross sections.

Several different casting processes have been successfully used with graphite-reinforced metals. These include permanent mold, split-mold plaster, and investment casting. Vacuum or pressure assist is often used in penetration of the fiber preform by the molten metal. Die casting is generally not used, because the inherent high speed of this process is not compatible with the time required for full penetration of the interfiber channels.

Graphite fibers are not normally wetted by molten metals. Because wettability is a major advantage in the casting process, a proprietary fiber coating has been used that, when applied to the fibers before casting, makes them wettable by molten magnesium alloys.

Graphite–metal composites exhibit good machinability compared with graphite–epoxy composites, which are difficult to machine. Most castings require a certain amount of finish machining in order to achieve close dimensional tolerances or provide particular surface finishes. Unlike many reinforcement materials used in MMCs, such as oxides and carbides, graphite is not an abrasive. While the fibers themselves do not act as a lu-

TABLE 4
Thermal Conductivity Measurements (Room Temperature)

Material	Direction	Conductivity (W/CM·K)	Specific Conductivity (K/ρ)
60V/0 P100/6061	Longitudinal	3.25	37.8
60V/0 P100/6061	Transverse	0.62	
60V/0 P100/6061	Z	0.42	
60V/0 P120SHK/6061	Longitudinal	4.56	53.0
40V/0 P100/6061	Longitudinal	3.08	35.0
40V/0 P100/6061	Transverse	1.04	
40V/0 P100/6061	Z	0.65	
Copper		3.80	12.7
Aluminum (6061-T6)		1.80	18.4

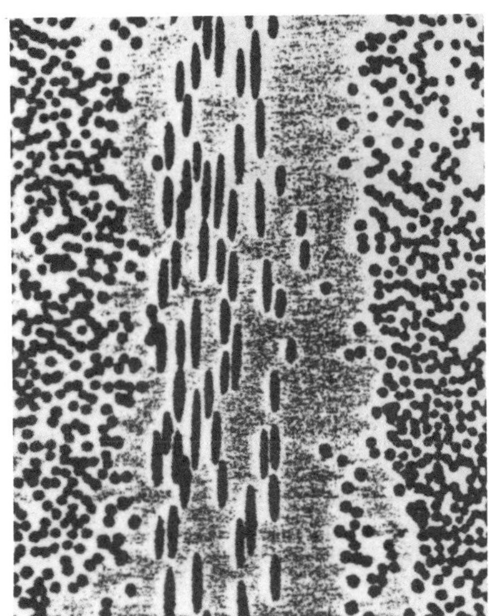

FIGURE 10 Microstructure of graphite–magnesium casting containing cross-plied fibers (magnification: 100×).

bricant, neither do they lead to excessive tool wear; and in fact they tend to act to break up chip formation. The only cautionary note on this subject is that deep tool cuts should be avoided because the rather weak fiber–matrix bond could result in delamination under high shearing forces.

As mentioned previously, casting is an extremely versatile process for producing complex shapes or shapes containing complex fiber orientations. Infiltration of the fibers by the metal is generally excellent, as shown in the microstructure in Figure 10. In general, although the development time required for a particular shape is longer for graphite fiber reinforced castings than for conventional unreinforced castings (because of the extra difficulties involved with fiber layup and metal infiltration), suitable parameter adjustment should allow any part that can be cast in conventional materials to be cast also in graphite-reinforced magnesium and, when suitable fiber coatings become available, other graphite-reinforced metals as well.

Discontinuously Reinforced Composites

The discontinuously reinforced metal composites technology represents a material system with theoretical roots going back to the 1940s and perhaps earlier. Analytical consideration showed that dispersed rigid particles in an elastic (lower modulus) continuum would result in the optimum material form for a variety of mechanical property enhancements, including increased elastic stiffness, increased wear resistance, moderate CTE, and typically decreased material ductility. Although many early processing efforts were undertaken, only nonstructural (specialized) material such as cermets resulted. During the mid-1970s, with the development of a source of less costly SiC whisker (using rice hulls as raw material) as a reinforcement, there was a renaissance of interest in discontinuously reinforced materials. In the last 8 to 10 years, by virtue of the new, high quality (and purity) ceramic particulates and metal powders now available, high mechanical property, low cost, isotropic composites with excellent ductility have resulted.

Today we have discontinuously reinforced composite materials available that (1) can be tailored and (2) can use existing facilities, equipment, and technology to produce a wide variety of "real" parts—automotive, leisure, airframe, antenna, and many thermal management control structures. The need for special technologies or procedures for incorporating these new material systems into structures is minimal, since these MMCs can be described as "special aluminums" or, better yet, "super-aluminum."

Fabrication of discontinuously reinforced composites is primarily by powder-metallurgy processes, with casting also actively being pursued. The basic criteria of uniform reinforcement distribution and bond quality determine the resultant composite properties. Figure 11 summarizes key factors affecting whisker–or chopped fiber–metal, circa 1970; many of these factors are also important for the newer particulate-reinforced metals.

The initial applications of discontinuously reinforced composites emphasized the enhanced structural characteristics, particularly stiffness and use temperature. Extruded airframe stiffeners, control surfaces, and automotive (piston) engine parts were fabricated and tested during the early 1980s. The last three years have seen a continuation of interest in structural parts, but have also seen a major new thrust to take advantage of the many desirable physical property attributes that can be designed into isotropic, lower cost particulate-reinforced composites. High thermal stability and low CTE combined with good thermal conductivity and conventional

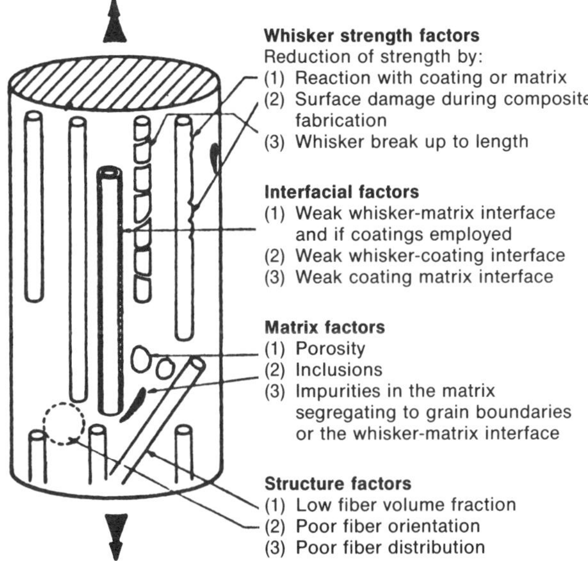

FIGURE 11 Factors affecting metal matrix composite properties.

manufacturing processes allow incorporation of these composites into many advanced electronic and heat-transfer devices.

The following sections address both areas of discontinuous composite application.

Discontinuously reinforced metal matrix composites are a class of materials that exhibit a predictable blend of the reinforcement and matrix properties.

During the 1960s, work was done on aluminum with alpha aluminum oxide (Al_2O_3) whiskers, but the cost of the whiskers was high, and the strengths achieved were lower than expected because of bonding difficulties with the alumina whiskers. These difficulties were never overcome.

During the late 1970s, composites with short, staple, polycrystalline alumina fibers, SiC whiskers made from pyrolyzed rice hulls, and SiC particulates have been investigated. The alumina fibers were first used to reinforce the ring land area of diesel pistons. These pistons were made by a squeeze-casting process that was described in an article by J. Dinwoodie et al. [6,7], shown in Figure 12. The short fibers did not increase the ultimate strength of the matrix alloy at room temperature; however, strength is retained to temperatures of approximately 300°C rather than 200°C for the base alloy (Fig. 13). The elastic modulus of the composite is substantially increased over that of the matrix at all temperatures (Fig. 14). In addition, the incorporation of the fibers decreases the coefficient of thermal expansion (Table 5) and increases the hardness of the metal composite (Table 6). This combination of properties has made the piston a success.

L. Ackerman et al. [9] reported on composites containing polycrystalline alumina fibers and SiC whiskers in an aluminum casting alloy. These composites were also made by squeeze casting. In this study, the room-tem-

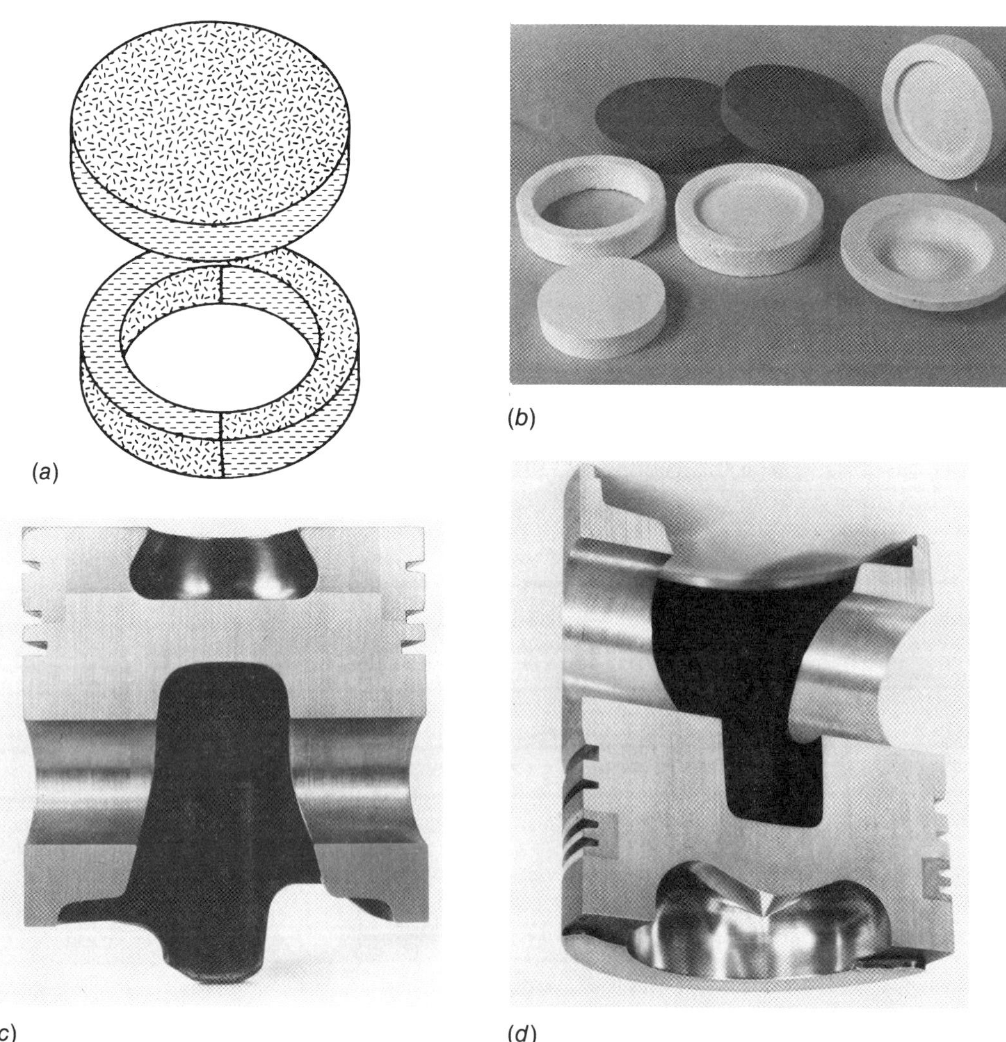

FIGURE 12 Drawing and photograph of reinforcement and photographs of reinforced piston sections. (*a*) Possible orientations in preforms. (*b*) Simple preform shapes used for piston reinforcement. (*c*) Reinforced combustion bowl. Courtesy of the AE Group. (*d*) Ring groove areas of sectioned pistons. Courtesy of Toyota Motor Corporation. (From Ref. 7.)

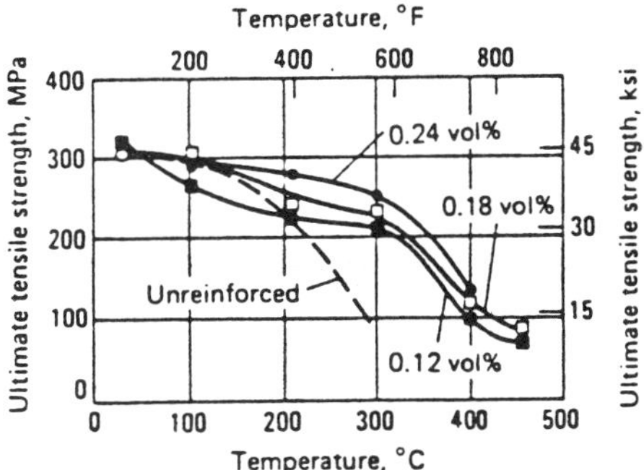

FIGURE 13 Effect of temperature on the tensile strength of Al-9Si-3Cu-based composites. (From Ref. 7.)

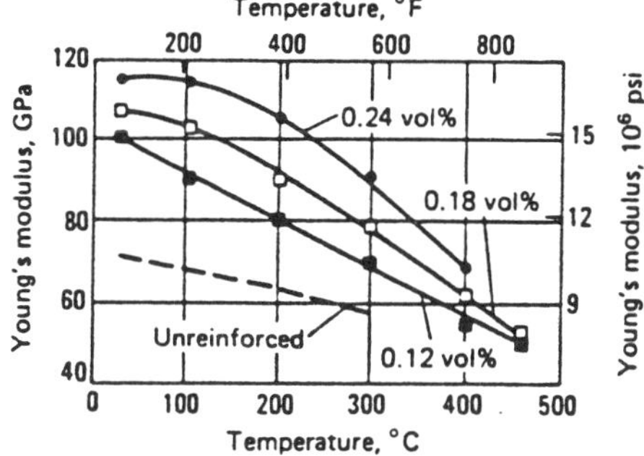

FIGURE 14 Effect of temperature on the modulus of A1-9Si-3Cu-based composites. (From Ref. 7.)

TABLE 5
Coefficient of Thermal Expansion, α

Fiber (vol %)	α (in-plane) (10^{-6}/K)	α (normal) (10^{-6}/K)
0	2.03	2.03
0.12	1.66	1.76
0.18	1.54	1.66
0.24	1.55	1.57

From Ref. 7.

TABLE 6
Hardness Values at 25°C

Fiber (vol %)	Vickers hardness No., HV10
0	131
0.12	179
0.18	190
0.24	212

(From Ref. 7.)

perature elastic modulus and yield strength were improved by the addition of both reinforcements. The SiC whiskers resulted in higher modulus values as well as higher yield strengths (Table 7). As in the previous study, strength properties at elevated temperatures were retained (Table 8). The strength of either composite containing 20 vol % reinforcement at 350°C was equal to or greater than the strength of the matrix at 250°C. The fatigue limit at 10^7 cycles was improved from 80 MPa in the matrix alloy to 109 MPa in the 20 vol % polycrystalline alumina system, and up to 131 MPa in the SiC whisker system.

During the 1980s, DWA Composite Specialties, Inc., produced and tested a variety of internal combustion engine parts using REV-LITE (particulate-reinforced aluminum, powder-metallurgy-processed composites). Testing in race cars, racing motorcycles, and small air-cooled aircraft engines has demonstrated enhanced performance and life (Fig. 15).

Composites of SiC particulates in aluminum alloys have been studied by several investigators [10–12]. The majority of the work used composites that were made by blending atomized powders with the SiC reinforcement. The choice of matrix alloys extended from 1100, that is, no alloy additions, to any of the rapidly solidified, high-strength alloys. The blend was compacted into a billet, which was vacuum-hot-pressed and hot-worked into a usable form. The composites behave in a manner similar to that of new, high-strength aluminum alloys made by the powder-metallurgy technique. The most common primary breakdown process has been extrusion. With the particulate reinforcement, extrusion through conical as well as shear face dies is acceptable practice. Composite extrusion is shown in Figure 16, while a typical extrusion cross section is shown in Figures 17 and 18 (SiCp/Al alloy refers to SiC particulate). Table 9 shows typical extruded 25 vol % DWAl 20 data. Microstructures of extruded composites are shown in Figures 19 and 20 for 20

FIGURE 15 A variety of internal combustion engine parts using REV-LITE.

TABLE 7
Tensile Data Obtained on Polycrystalline Alumina and SiC Whisker-Reinforced Aluminum Alloy

	Yield Strength (0.2%)			Ultimate Tensile Strength			Young's Modulus		
Fiber, vol %	MPa	Standard Deviation	Range of Measurement	MPa	Standard Deviation	Range of Measurement	GPa	Standard Deviation	Range of Measurement
Polycrystalline alumina									
0	210	3.8	9.5	297	1.8	3.5	71.9	4.5	13
0.05	232	4.2	10.4	282	6.5	15.1	78.4	2.3	6
0.12	251.5	14.6	38.3	273	19.6	49.6	83.0	7.8	21
0.20	282.5	11.3	25.2	312	16.0	42.3	95.2	2.7	7
SiC whisker									
0	210	3.8	9.5	297	1.8	3.5	71.9	4.5	13
0.12	266.5	4.2	10.6	359	33.6	85.6	95.3	1.6	6
0.16	264.5	0.6	1.6	374	8.0	23.0	90.0	3.7	9
0.20	298	4.0	10.2	383.6	15.2	38.8	111.0	5.0	13

TABLE 8
Yield Strength and Ultimate Tensile Strength of Aluminum Alloy Reinforced with Polycrystalline Alumina and SiC Whiskers at Different Temperatures

	350°C		300°C		250°C	
Fiber (vol %)	Yield Strength (MPa)	Ultimate Tensile Strength (MPa)	Yield Strength (MPa)	Ultimate Tensile Strength (MPa)	Yield Strength (MPa)	Ultimate Tensile Strength (MPa)
Polycrystalline alumina						
0	35	55	. . .	70	70	115
0.05	54	63	79	88	112	134
0.12	68	74	. . .	. . .	. . .	. . .
0.20	110	112	154	155	186	198
SiC whiskers						
0	35	55	. . .	70	70	115
0.12	94	124	153	180	197	226
0.16	120	147	. . .	. . .	. . .	. . .
0.20	163	184	207	235	268	284

TABLE 9
Tensile Test Data for Extruded Shapes 25 vol % SiC Particulate/6061 Aluminum Composite

Shape No.	Extrusion Area Ratio	No. Tests	Modulus (GPa)	Yield (MPa)	Ultimate (MPa)	Strain (GPa)	Strain %
2829	63.2 : 1	16	119	420	478	2931	4.25
2830	32.6 : 1	118	121	425	491	2697	3.91
2833	48.6 : 1	54	122	440	511	2676	3.88
2835	32.5 : 1	22	122	426	492	1690	3.90
2837	53.5 : 1	32	121	426	492	2690	4.87
2841	65.2 : 1	38	118	313	484	2738	3.97

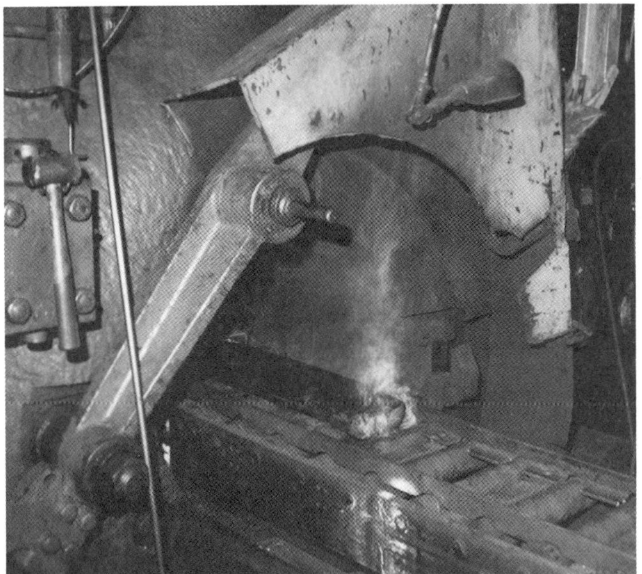

FIGURE 16 Extrusion of 360 mm diameter SiC particulate-reinforced aluminum composite.

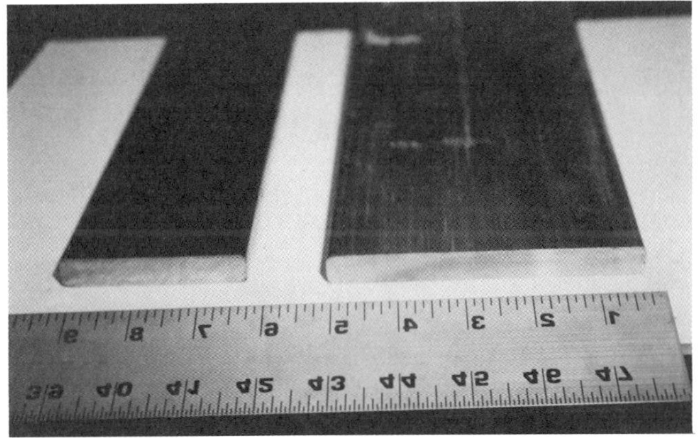

FIGURE 17 Extrusion cross section (SiCp/alloy).

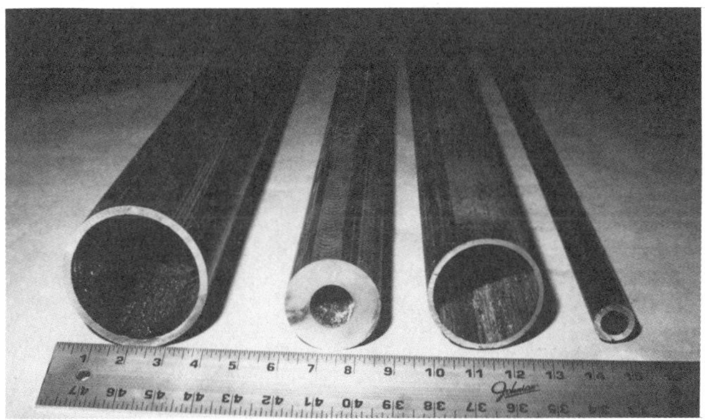

FIGURE 18 Extruded shapes of SiCp/Al alloy.

vol % SiC and 30 vol % SiC reinforcement levels in a 6061 aluminum matrix. These photomicrographs show the uniform distribution of the SiC particulates in longitudinal and transverse planes of the composites.

These SiC particulate-reinforced aluminum composites have transverse tensile properties that are within 5% of the longitudinal properties. In situations requiring multidirectional reinforcement, these composites can outperform fiber-reinforced composites. Their shear strength is greater than that of the matrix alloy. This increased shear strength is also reflected in an increase in pin-bearing strength (Table 10). These bearing-strength data are for tests conducted with the center of the pin hole 1.5 and 2 times the pin diameter from the

TABLE 10
Pin-Bearing Strengths of SiC Particulate Reinforced Aluminum

Composite	Edge Distance (Multiplied by Pin Diameter) from Edge	Bearing Yield Strength (MPa)	Bearing Ultimate Strength (MPa)
20 vol % SiC-7091 (T-6) (Ref. 12)	1.5	689.5	724.0
	2.0	1000.0	1310.0
	3.0	1000.0	1448.0
25 vol % SiC-2124	2.0	827.4	—

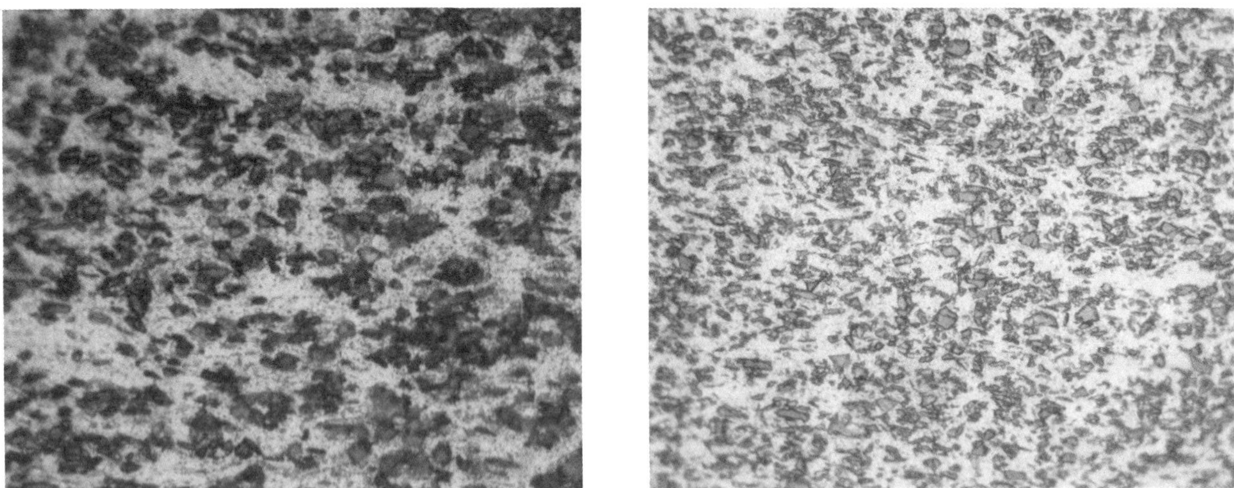

FIGURE 19 Microstructure of extruded 20 vol % SiC/6061 reinforced composite.

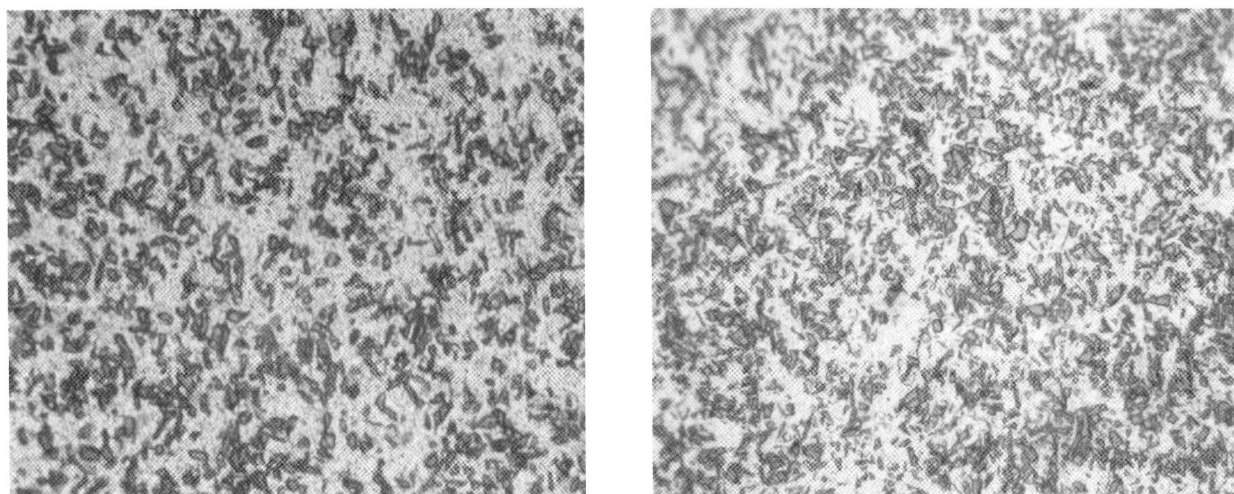

FIGURE 20 Microstructure of extruded 30 vol % SiC/6061 reinforced composite (magnification: 300×).

TABLE 11
Comparison of Material Thermal Properties and Density

	Density (g/cm^3)	Thermal *K* (W/M/°C)	Specific Thermal *K* (K/ρ)	CTE (mm/mm·K × 10^{-6})
Alumina	3.58	22–36	170–279	6.5
Conventional metal systems				
Copper	8.88	391	1220	17.5
Aluminum	2.71	221	2255	23.4
Molybdenum	10.15	146	400	5.2
Kovar	8.13	17	58	5.4
Copper/Invar/Copper	8.13	131	447	5.4
DWAl 20 systems				
40 vol % SiCp/6061 Al	2.88	128	1230	11.9
50 vol % SiCp/High K Al	2.94	200	1869	9.4
55 vol % SiCp/High K Al	2.97	205	1898	8.9

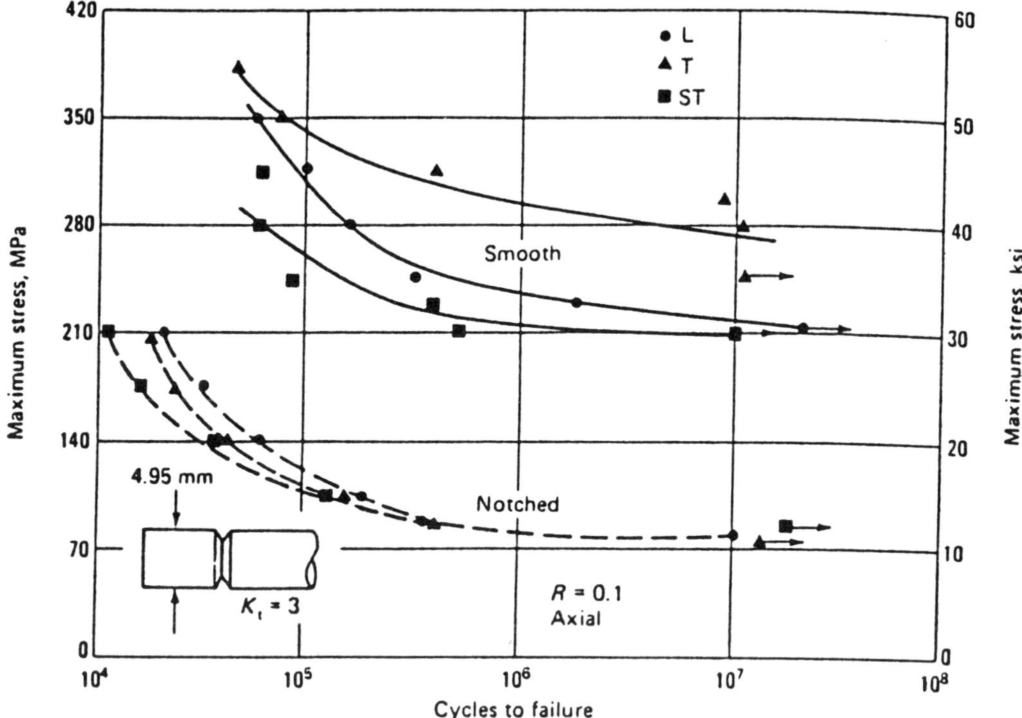

FIGURE 21 **Axial fatigue properties of smooth and notched specimens taken from plate in different orientations.**

sample edge. Typical edge distances for graphite- or glass-reinforced epoxy composites are 5 times the pin diameter. The SiC–aluminum composites are able to save material and weight by reducing the excess edge material at joints.

The fatigue properties of these composites are shown for S/N-type fatigue tests (Fig. 21). This behavior is for both smooth- and notched-type samples.

The expansion coefficient of the aluminum is decreased as SiC is added to the composite (Fig. 22), from 23×10^{-6}·K for aluminum to 11×10^{-6}·K for 40 vol % SiC composites. This expansion behavior is isotropic. A design of an advanced-composite optical system gimbal by General Electric Company [13] used 40 vol % SiC particulate-reinforced 6061 aluminum in low-expansion and joint areas, and graphite–epoxy in ultra low expansion areas. It won first place in the Materials Engineering 1985 competition (Fig. 23). Because of its low-expansion behavior, this composite is also being used in place of beryllium for guidance components, as shown in Figure 24.

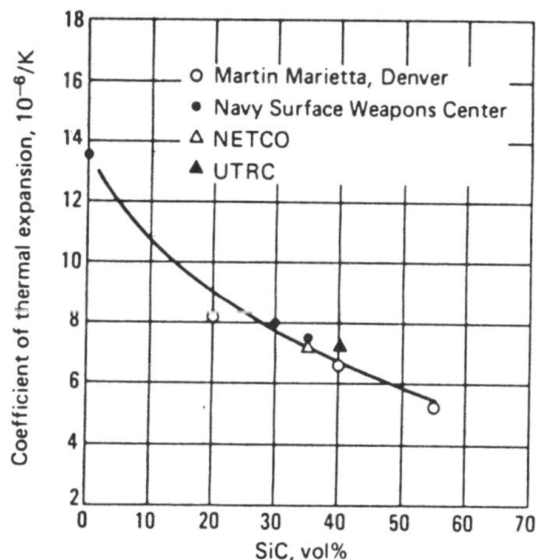

FIGURE 22 **Coefficient of thermal expansion as a function of reinforcement level, SiC–aluminum composites.**

The most recent application area for discontinuously reinforced composite expands on the General Electric gimbal application. The combinations of physical property and thermal conductivity capabilities possible with this family of isotropic-response composites suggest a variety of electronic enclosure and thermal management applications. For most electronic device systems, the basic cold plate, enclosure, and attach design require a CTE close to that of the ceramic substrate to minimize thermally induced stresses during operation. High thermal conductivity is also helpful in rapidly dissipating the heat, with low density desirable for advanced designs where weight is critical.

Table 11 shows the relationships with alumina for several conventional metals and a series of particulate-reinforced aluminum composites. The very high thermal conductivities of copper and aluminum are desirable. The aluminum also has the highest specific conductivity level; however, both metals are unacceptable for most designs because they have CTE values more than twice that of the typical electronic substrate material, alumina.

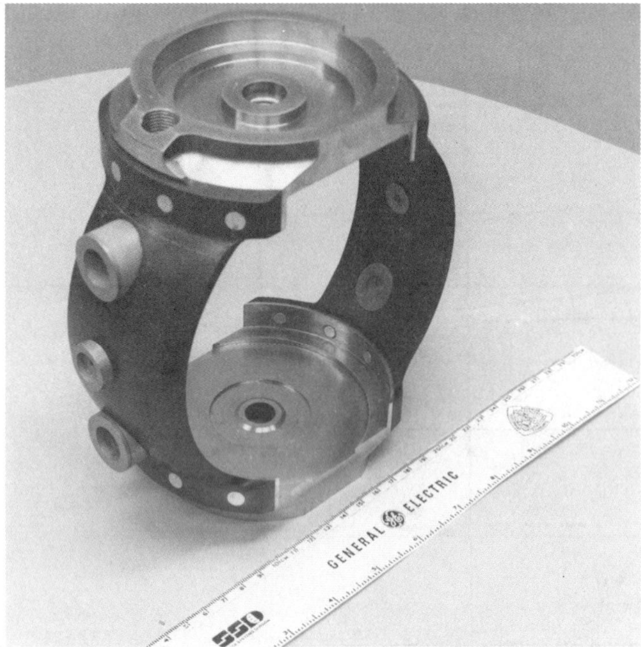

FIGURE 23 General Electric's first place: Materials Engineering Competition, 1985.

Copper–iron–copper and molybdenum create moderate thermal conductivity and CTE levels; however, a huge weight penalty must be paid. The SiC–aluminum composite materials show similar conductivity to the copper–iron–copper, but at about one-third the weight, while also exhibiting CTE levels that can be accommodated by the alumina.

MARKET STATUS

The final section of this chapter will briefly summarize the market status of the three basic reinforced aluminum composite systems presented.

TABLE 12
Yearly Production of Metal Matrix Composites

Time Frames	Graphite Metals (kg/Yr)	Boron–Aluminum Systems (kg/Yr)	Discontinuously Reinforced (kg/Yr)
1960/1970	N.A.	230	N.A.
1970/1980	140	910	230
1980/1985	360	230	230,000
Projected:			
1985/1990	910	450	>91,000

Advanced metal matrix composite material markets have been developing for the past 25 years. The primary market emphasis has come from the aerospace, airframe, and Department of Defense segments. Throughout this time period, the market demand was typically exceeded by developing fabrication capacity. Table 12 summarizes relative U.S. markets for each basic type of material. Since no detailed statistics are maintained for composite production, these levels should be treated as estimates only.

The applications considered for MMC usage are now too numerous to list here in detail; key applications of discontinuously reinforced and graphite–aluminum-type materials were presented earlier in this article. In addition, primary considerations for continuous boron filament reinforced aluminum composites are represented by the use of boron–aluminum tubular truss structure on the U.S. space shuttle (mid-1970s to today), and the use of small stiffening strips of boron–aluminum in a resin composite tennis racket (circa 1975–1980).

Recent market studies have all shown major growth potential for metal matrix composites during the 1990s. From current market estimates of $15 to $20 million per year, primarily research-based; markets up to $400 million per year by the year 2000 have been projected. This market will be primarily in automotive and aerospace (including electronics and thermal management) areas.

FIGURE 24 SiC composite used in place of beryllium for guidance components. First production use of discontinuously reinforced aluminum by DWA, primary producer of these parts.

The major markets based on material volume will be automotive and some leisure applications, while space structures, airframes, and electronic materials will utilize aluminum, magnesium, copper, titanium, superalloy, and intermetallic matrices.

J. F. Dolowy, Jr., W. C. Harrigan, Jr., and Mark R. van den Bergh

References

1. Carl H. Zweben, "Metals and Ceramic Matrix Composites," UCLA Short Course, 1987.
2. *Engineered Materials Handbook*, Vol. 1, *Composites*, ASM International, Metals Park, OH, 1987.
3. M. R. van den Bergh and J. F. Dolowy, "Aerospace and Space Application for Metal Matrix Composites," presented at AIAA Aerospace Engineering Conference, 1985.
4. B. A. Webb, J. F. Dolowy, Jr., and E. C. Supan, "Metal Matrix Materials for Navy Systems Program," Naval Sea Systems Command, Final Report (N00024-83-C-5303), September 1983.
5. Calow & Moore, *Composites*, 1971.
6. T. Donomato, K. Funatani, N. Miura, and N. Miyake, "Ceramic Fiber Reinforced Piston for High Performance Diesel Engine," Paper 830252, Society of Automotive Engineers, March 1983.
7. J. Dinwoodie, E. Moore, C. A. J. Langman, and W. R. Symes, in W. C. Harrigan, Ed., *Proceedings of the International Conference on Composite Materials V*, A.I.M.E., 1985.
8. L. Ackermann, J. Charbonnier, Desplanches, and H. Koslowski, in W. C. Harrigan, Ed., *Proceedings of the International Conference on Composite Materials V*, A.I.M.E., 1985.
9. D. L. McDaniels, *Metall. Trans. A*, *16A*, 1105 (1985).
10. R. J. Arsenault, *Materials Science Eng.*, *64*, 171 (1984).
11. C. R. Crowe, R. A. Gray, and D. F. Hasson, in W. C. Harrigan, Ed., *Proceedings of the International Conference on Composite Materials V*, A.I.M.E., 1985.
12. B. Mosdale, Air Industries, private communication, 1981.
13. 1985 Top Twenty Awards, "Materials Engineering," General Electric Company, 1985.

Metal Matrix Composites, Casting Processes

Metal matrix composites (MMCs) are engineered combinations of two or more materials (one of which is a metal) where tailored properties can be attained by systematic combination of different constituents. A variety of procedures for producing these modern MMCs, including conventional casting by foundry techniques, have become available recently. The potential advantage of preparing these composite materials by foundry techniques is near-net-shape fabrication in a simple and cost-effective manner. In addition, casting processes lend themselves to manufacture of large numbers of complex-shaped components of composites at the fast rates required by automotive and other consumer-oriented industries.

Structurally, cast metal matrix composites consist of dispersions of continuous or discontinuous fibers, whiskers, or particles in an alloy matrix, which solidifies in the restricted spaces between the reinforcing phase to form the bulk of the matrix. By carefully controlling the relative amounts and distribution of the ingredients constituting a composite, as well as by controlling the solidification conditions, MMCs can be given a tailored set of useful engineering properties that cannot be realized with conventional monolithic materials. In addition, the solidification microstructure of the matrix gets refined and modified as a result of the fibers and particles, indicating a possibility of controlling microsegregation, macrosegregation, and grain size in the matrix. This represents an opportunity to develop new matrix alloys.

From a technological standpoint of property-performance relationship, as well as from the viewpoint of processability and manufacturability, the interface between the matrix and the reinforcing phase (fiber or particle) is of central importance. Solidification processing of metal matrix composites allows tailoring of the interface between the matrix and the fiber to suit specific property-performance requirements. Considerable work on casting of MMCs, including Al–graphite, Al–silicon carbide, and Mg–graphite composites, is going on at several research organizations, including a major effort at the University of Wisconsin—Milwaukee involving the casting industry, the composites industry, and manufacturers of engines. Figure 1 shows the comparison of properties and costs of Al–SiC composites with those of other materials on the basis of data from Dural Aluminum Corporation. These types of cast composites represent a new product opportunity for foundries.

Casting Techniques

A basic requirement for casting of MMCs is initial intimate contact and intimate bonding between the ceramic phase and the molten alloy. This is achieved either by mixing the ceramic dispersoids into molten alloys, in fully or partially molten states, or by pressure infiltration of preforms of ceramic phase by molten alloys. Because of the poor wettability of most ceramics with molten metals, intimate contact between fiber and alloy can be promoted only by artificially inducing wettability or by using external forces to overcome the thermodynamic surface energy barrier and viscous drag. Mixing techniques generally used for introducing and homogeneously dispersing a discontinuous phase in a melt are [1]:

1. Addition of particles to a vigorously agitated fully or partially molten alloy.
2. Injection of discontinuous phase in the melt with the help of an injection gun.

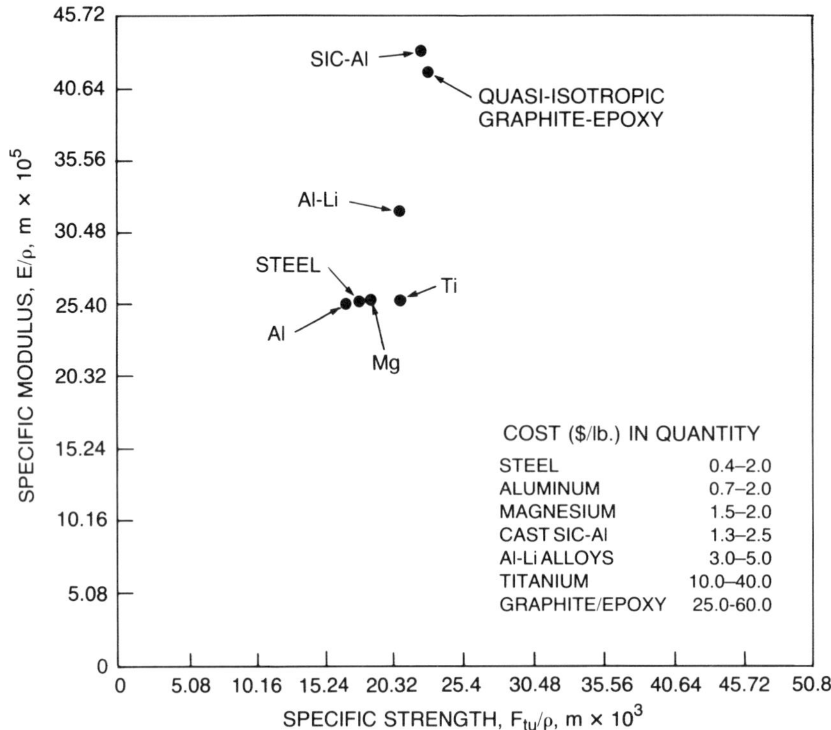

FIGURE 1 **Material, properties and cost comparison of cast SiC–Al composites.** (Courtesy D. M. Schuster.)

3. Dispersing pellets or briquettes, formed by compressing powders of base alloys and the ceramic phase, in a mildly agitated melt.
4. Addition of powder to an ultrasonically irradiated melt. The pressure gradients resulting from the cavitation phenomenon promote homogeneous mixing of ceramics in metallic melts.
5. Addition of powders to an electromagnetically stirred melt. Turbulent flow conditions achieved through electromagnetic stirring are used to obtain a uniform suspension.
6. Centrifugal dispersion of particles in a melt.

In all these techniques, external force is used to (1) transfer a nonwettable ceramic phase into a melt, and (2) create a homogeneous suspension in the melt. The melt-particle suspension thus created can be cast either by conventional foundry techniques, such as gravity or pressure die casting or centrifugal casting, or by novel techniques, such as squeeze casting (liquid forging) and spray co-deposition. The various techniques used to solidify melt-particle slurries are discussed below.

Sand and Die Castings

The slow freezing rates obtained in sand molds leads to preferential concentration of particles lighter than Al alloys (e.g., mica, graphite, porous alumina) near the top surface of sand castings and segregation of heavier particles (sand, zircon, glass, SiC, etc.) near the bottom part of castings. Depending upon the intrinsic hardness of dispersed particles, these high-particle-volume fraction surfaces serve as selectively reinforced surfaces, for instance, tailor-made lubricating or abrasion-resistant contacting surfaces for various tribological applications.

The relatively rapid freezing rates in metallic molds generally give rise to more homogeneous distribution of particles in cast matrix. Figures 2 through 4 show microstructures of permanent-mold gravity die casting of Al alloys containing dispersions of graphite, zircon, and

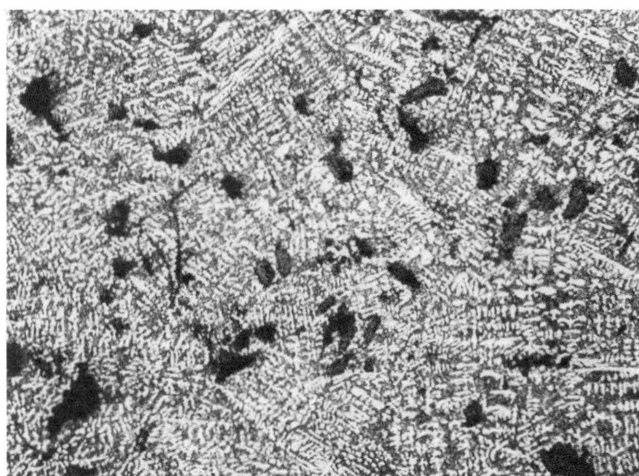

FIGURE 2 **Optical photomicrograph showing uniform distribution of graphite particles in an Al-Si alloy matrix composite solidified in a permanent mold (Magnification: ×50).**

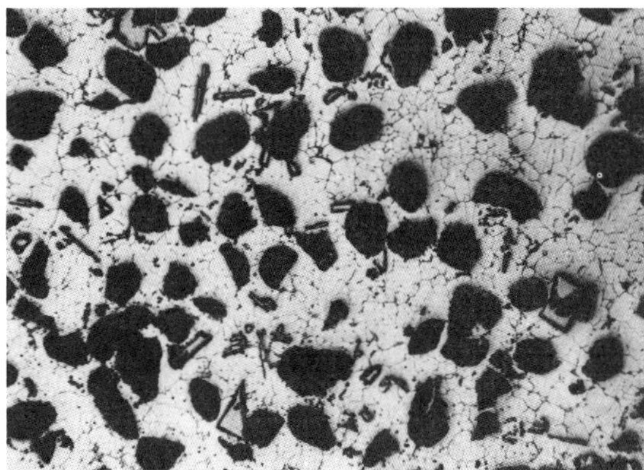

FIGURE 3 **Optical photomicrograph showing uniform distribution of zircon particles in an Al-Si alloy composite solidified in a permanent mold (Magnification: ×200).**

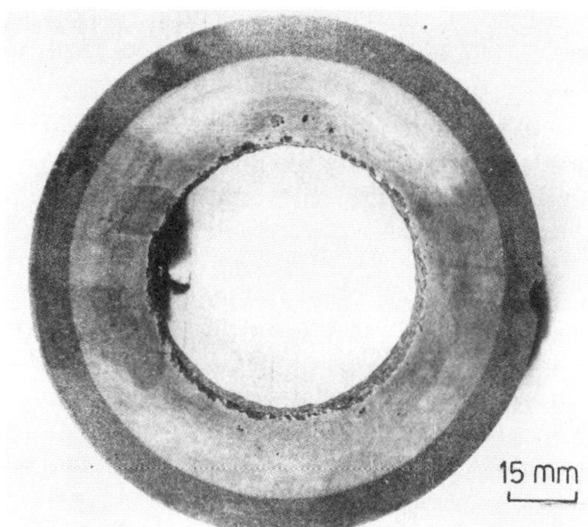

FIGURE 5 **Centrifugal casting of Al alloy–zircon particle composite showing segregation of zircon at the outer rim of the casting.**

rice husk ash particles. Further improvements in particle distribution can be achieved by use of water-cooled molds, copper chills, and other agitation techniques.

Centrifugal Castings

Solidification in rotating molds of composite melts containing particle dispersions of graphite, mica, porous alumina, zircon, and carbon microballoons exhibits two distinct zones—a particle-rich zone near the inner circumference for lower density particles and a particle-impoverished zone near the outer circumference. The outer zone is particle-rich if the particles are denser than the melt, as is the case with zircon and silicon carbide particles in aluminum. As a result of centrifugal acceleration in rotating molds, the lighter graphite, mica, and porous alumina segregate near the axis of rotation, producing high particle volume-fraction-surfaces for bearing applications. Up to 8% by weight mica and graphite, and up to 30% by weight zircon particles could be incorporated in selected zones of Al alloy castings by this technique. Figures 5 and 6 show sections of typical centrifugal castings of Al–zircon and Al–graphite composites, respectively, where the heavier zircon is seen to segregate near the outer circumference, producing hard, abrasion-resistant surfaces; the lighter graphite is seen to concentrate near the inner periphery, producing wear-resistant solid lubricated surfaces for antifriction applications like bearings or cylinder liners.

Compocasting

Particulates and discontinuous fibers of SiC, alumina, TiC, silicon nitride, graphite, mica, glass, slag, MgO, and boron carbide have been incorporated into vigorously agitated partially solidified aluminum alloy slurries by the compocasting technique. The discontinuous ceramic phase is mechanically entrapped between the

FIGURE 4 **Optical photomicrograph showing uniform distribution of rice husk ash particles in an Al-Si alloy matrix composite solidified in a permanent mold (2 cm = 500 μm).**

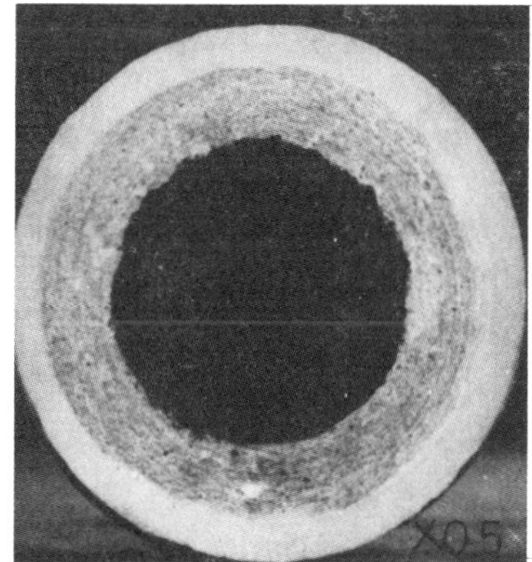

FIGURE 6 **Centrifugal casting of Al alloy–graphite particle composite showing segregation of graphite particles at the inner periphery of the casting.**

proeutectic phase present in the alloy slurry, which is held between its liquidus and solidus temperatures. This semifusion process allows near-net-shape fabrication by extrusion or forging, since deformation resistance is considerably reduced as a result of the semifused state of the composite slurry.

Pressure Die Casting

Pressure die castings of composites allows larger-sized and more intricate component shapes to be rapidly produced at relatively low pressures ($\leq$ 15 MPa) for an equivalent capital expenditure. Pressurized gas and hydraulic ram in a die-casting machine have been employed to synthesize porosity-free fiber and particle composites. It has been reported that high pressures, short infiltration paths, and columnar solidification toward the gate produced void-free composite castings. The pressure die-cast particle composites exhibit lower bulk and interfacial porosities, more uniform particle distribution, less agglomeration of particles, and occasional exfoliation/ fragmentation of soft particles (e.g., graphite in Al alloy), with melt penetrating into fine exfoliation cleavages. High concentrations (60 wt % or more) of zircon ($ZrSiO_4$) particles can be uniformly dispersed in pressure die-cast Al–Si–Mg alloys. Pressure die castings of eutectic Al–Si alloy (7 wt % graphite) and Al–(4–12%)Si–(0.5–10%)Mg–alumina particle composites showed considerable improvement in particle distribution, particle-matrix bonding, and elimination of porosities, as confirmed by SEM and ultrasonic velocity measurements.

Squeeze Casting

Squeeze casting or liquid forging of metal matrix composites is a recent development that involves unidirectional pressure infiltration of fiber preforms or powder beds in order to produce void-free, near-net-shape castings of composites [2,3]. The Saffill fiber-reinforced pistons of aluminum alloys made by Toyota have been in use in heavy diesel engines for some years now, and a considerable amount of work has been done on pressure casting of ceramic-particle- and fiber-reinforced metal matrix composites for industrial applications. The processing variables, in decreasing order of their importance, governing evolution of microstructures in squeeze-cast MMCs are:

1. Fiber-preheat temperature
2. Interfiber spacing
3. Infiltration pressure
4. Infiltration speed
5. Metal-superheat temperature

Squeeze casting of composite melts into finished shapes promotes fine equiaxed grain structure as a result of large undercoolings and rapid heat extraction. Preheated particles or fiber preforms are inserted into a metal die and infiltrated with molten metal under high pressure (70–200 MPa), followed by solidification under pressure. Alternatively, whiskers or particles may be mixed with molten metal prior to squeeze casting. Al alloy composites containing SiC and Al_2O_3 powders, S-Alumina (Saffill) fibers, and silicon nitride whiskers have been fabricated by the squeeze-casting process. These ceramic particles and fibers are poorly wetted by metallic alloys (contact angle $> 90°$), and their infiltration requires large hydrostatic pressure to overcome the capillarity pressure. Also, frictional forces arising from the viscosity of the melt tend to oppose fluid flow through interfiber channels or interparticle corridors. Additional pressure is, therefore, required to overcome the viscous friction.

Several theoretical models describing elastic deformation and fracture of fiber preforms during infiltration, as well as the solidification process and microstructure evolution in pressure-cast MMCs, have been proposed in the literature [4,5]. Mortensen et al. [5] have proposed a generalized model of heat and solute transport during solidification under pressure in MMCs. A numerical solution of this model was obtained by specifying boundary conditions and auxiliary conditions such as seepage velocity (Darcy's law), solute conservation, continuity equation for incompressible flow, and a scheil-type equation relating fraction of metal solidified to alloy composition. This model agrees well with the experimental data, although it neglects several complicating features, such as the presence of moving boundaries, finite compressibility of the preform, enthalpy of interfacial reactions, and the presence of oxide film on the moving liquid front [5].

The evolution of microstructures in unidirectionally infiltrated MMCs has been modeled based on mechanisms of dendrite arm coalescence and solid-state diffusion in constricted regions between fibers [6]. These mechanisms manifest themselves in several structural modifications, such as macro- and microsegregation, scale of the structure (dendritic to featureless), modification of eutectic, and preferential segregation of the eutectic on the fiber surface.

The properties of fiber- or particle-reinforced castings depend on the processing variables as well as the matrix-reinforcement combination chosen. Controlled experiments on process optimization have been carried out to determine the optimum range of molten metal temperature, fiber temperature, infiltration speed, and infiltration pressure [7]. Proper control of process variables eliminates preform deformation during pressurization, freeze choking, and strength variability, and promotes interfacial bonding and a fine-grained equiaxed structure, which improve the mechanical properties of reinforced casting. Casting techniques not employing pressure (e.g., mechanical stirring followed by gravity casting in permanent molds) have been extensively studied by Rohatgi and coworkers, and mechanical and tribological properties on a large number of individual composite systems such as Al_2O_3/Al, SiC/Al, and graphite/Al have been reported in the literature. (See Refs. 1, 8 and 11.)

Vacuum Infiltration Process

Several fiber-reinforced metals (FRMs) are prepared by the vacuum infiltration process [4]. In the first step, the fiber yarn is made into a handleable tape with a fugitive binder in a manner similar to producing a resin matrix composite prepreg. Fiber tapes are then laid out in the desired orientation, fiber volume fraction, and shape, and are then inserted into a suitable casting mold. The fugitive organic binder is burned away, and the mold is infiltrated with molten matrix metal. The best castings in terms of soundness (voidage, porosity, and distribution of fibers) and mechanical properties are obtained if the mold assembly is preheated and pre-evacuated before the liquid metal is introduced.

Investment Casting

In investment casting of metal matrix composites, filament winding or prepreg handling procedures developed for fiber-reinforced plastics (FRPs) are used to position and orient the proper volume fraction of continuous fibers within the casting. The layers of reinforcing fibers are stacked in the proper sequence and orientation, and the fiber preform thus produced is infiltrated either under pressure or by creating a vacuum in the permeable preform. Continuous graphite fiber reinforced Mg has been produced by this method by MCI in Cleveland in collaboration with Martin Marietta [5].

The Effect of Remelting and Degassing

There have been limited studies on remelting and degassing of MMCs, mainly on Al-graphite particle composites. Controllable loss of graphite occurs as a result of remelting and degassing aluminum–graphite particle composites with nitrogen, and graphite distribution remains reasonably uniform even after three remelts. The hot-tearing tendency of Al-graphite composite alloys is not aggravated by the presence of graphite particles. The Al–11.8 Si alloy is inherently not prone to hot tearing, and the addition of up to 4 wt % Cu-coated graphite particles (equivalent alloy composition: Al–11.8 Si–2.75 Cu) does not produce any measureable tendency to hot tear in test pieces of up to 20 cm length. These areas of remelting, degassing, and mold design for casting metal matrix composites need further study, and data similar to what are available for noncomposite conventional alloys need to be developed.

Fluidity of Composites

Discontinuous Fibers and Particles

Dispersions of discontinuous fibers and particles into metallic melts impart viscous, slurrylike characteristics to the latter, and thereby affect the flowability or fluidity. The spiral fluidity of aluminum alloys containing dispersions of graphite, alumina, and mica particles decreases linearly with increasing particle surface area of the dispersed particles per unit weight, as shown in Figure. 7. This occurs presumably because of the progressive increase in effective viscosity of a suspension with increasing particle volume fraction. However, the fluidity values of particle-filled composite melts are adequate for making gravity cast composites at low volume fractions of particles in the range investigated.

Continuous Fibers

In the case of metal matrix composites reinforced with continuous fibers made with pressure infiltration, the

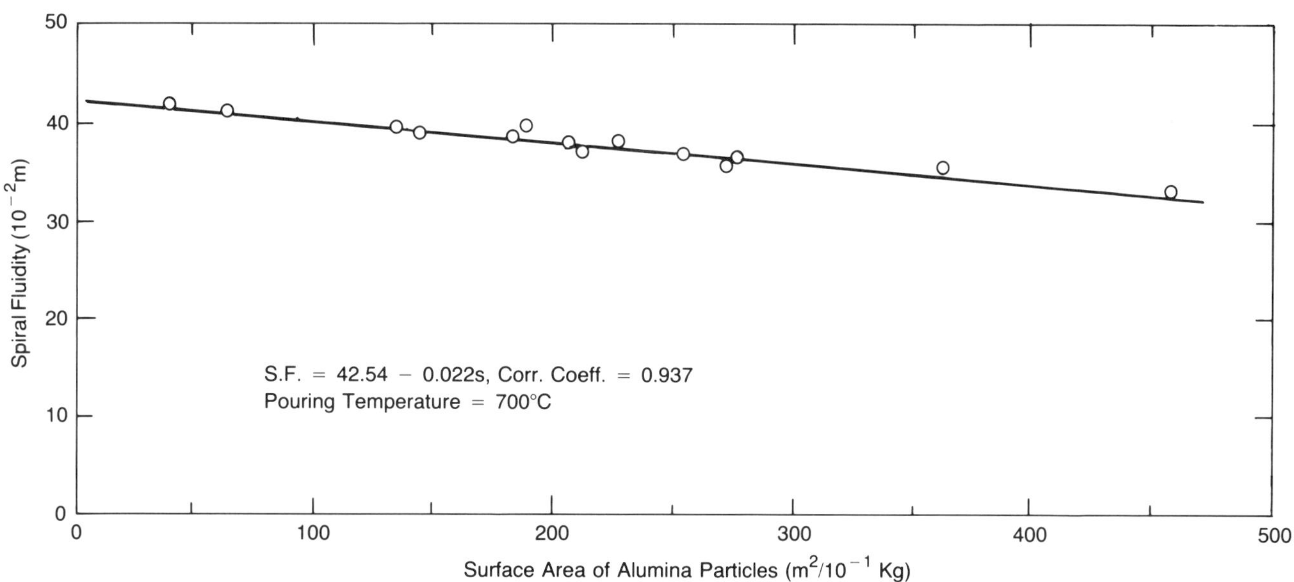

FIGURE 7 **Spiral fluidity of Al-Si alloy as a function of surface area per unit weight of pre-heat-treated alumina particles.**

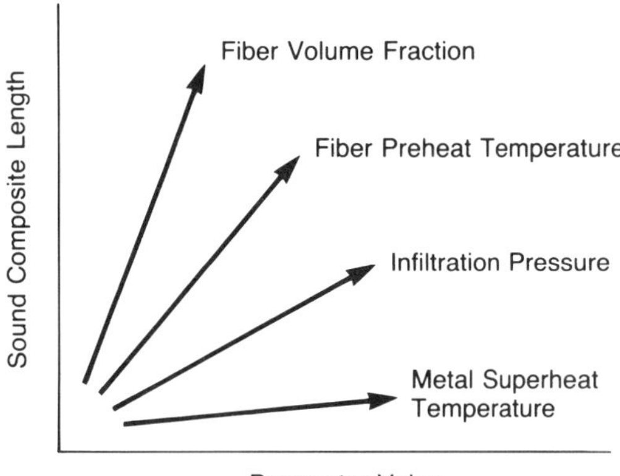

FIGURE 8 Schematic summary of the effect of processing parameters on infiltration length in fiber-reinforced metal matrix composites. (Courtesy of J. A. Cornie.)

infiltration length and preform permeability are indicators of fluidity. Experiments show a strong dependence of permeability and infiltration length on fiber volume fraction and fiber temperature, and a weaker dependence on metal temperature (Fig. 8). In the presence of metal superheat, the solid metal sheath formed initially around cold fibers (fiber temperature below the melting point of the metal) remelts, and the length of the remelted region always remains a fixed fraction of the total infiltration length for the case of constant applied pressure, no external heat extraction, and instantaneous heat transfer between metal and fiber. For the case of preheated fibers (fiber temperature above the melting point of the metal) under identical conditions, metal flow continues indefinitely unless external heat extraction causes cessation of flow to the point where the flow channel closes as a result of solidification from the external heat sink. Impurity content in the metal significantly affects infiltration length.

Microstructures

Al-Si and Al-Cu alloys have been chiefly used as matrix materials in a wide variety of cast metal matrix composites containing graphite and ceramic particles, carbon and glass microballoons, and discontinuous or continuous ceramic fibers. Microstructures of these MMCs show that primary aluminum in hypoeutectic Al-Si and Al-Cu alloys tends to avoid the discontinuous ceramic phase (SiC, alumina, graphite, mica, etc.) and nucleates in the interstices between particles or fibers unless special surface modification techniques are used to promote heterogeneous nucleation on the fiber surface. As a result, the particle is always found in the last freezing eutectic mixture. The discontinuous ceramic phase also tends to modify or refine the structure; for example, eutectic Si in Al-Si alloys gets modified, whereas primary Si is refined, when solidification occurs in the presence of a high volume fraction of ceramic phase. Figure 9 shows modification of eutectic silicon as a result of the presence of graphite particles in Al-Si alloys.

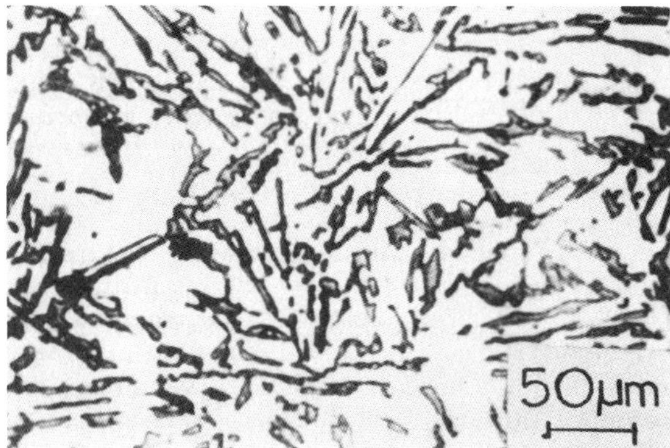

(*a*)

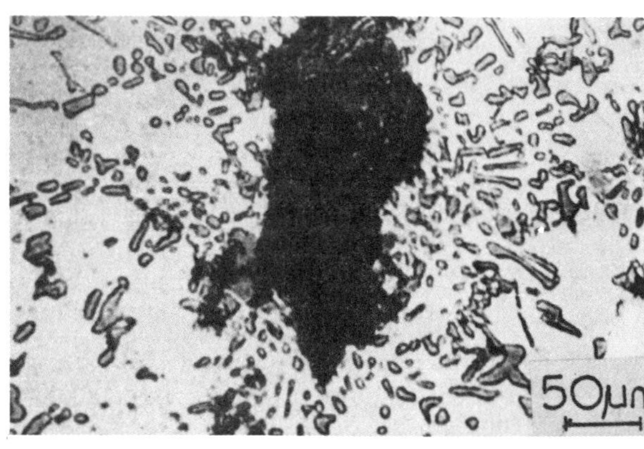

(*b*)

FIGURE 9 (*a*) Micrograph of as-cast Al-Si alloy showing needlelike eutectic silicon. (*b*) Micrograph of as-cast Al-Si alloy showing modification of eutectic silicon as a result of the presence of graphite particles.

Microstructures in fiber-reinforced MMCs can be modulated in a predetermined manner by controlling the interfiber spacing and cooling rate. If the cooling rate is sufficiently high or the fiber volume fraction sufficiently low, the matrix alloy solidifies uninfluenced by the fibers. At lower freezing rates, the dendritic structure is restricted by the fibers and microsegregation is reduced, and at sufficiently slow cooling rates the matrix can be rendered free of microsegregation.

Properties and Applications

Modern fiber-reinforced or particle-filled MMCs produced by casting techniques find a wide variety of applications (Table 1) because of the low cost of their fabrication and the specificity of achievable engineering properties. Some of these properties are high longitudinal and transverse strengths at normal and elevated temperatures; near-zero coefficients of thermal expansion; good electrical and thermal conductivities; and excellent antifriction, antiabrasion, damping, and machinability

TABLE 1
Selected Potential Applications of Cast Metal Matrix Composites

Composite	Applications	Special Features
Gr/Al	Bearings	Cheaper, lighter, self-lubricating; conserves Cu, Pb, Sn, Zn, etc.
Gr/Al, SiC-Al_2O_3/Al Fiber FP/Al	Automobile pistons, cylinder liners, piston rings, connecting rods	Reduced wear, antiseizing, cold start, lighter, conserves fuel, improved efficiency
Gr/Cu GRADIA (Hitachi)	Sliding electrical contacts	Excellent conductivity and antiseizing properties
SiC/Al	Turbocharger impellers	High temperature use
Glass or carbon bubbles in Al		Ultralight materials
Cast carbon/Mg fiber composites	Tubular composites for space structures	Zero thermal expansion, high temperature strength, good specific strength and specific stiffness
Zircon/Al SiC/Al SiO_2/Al	Cutting tools, machine shrouds, impellers	Hard, abrasion-resistant materials
Al-char Al-clay	Low cost, low energy materials	

properties. Modern FRMs like Gr/Mg can achieve a zero thermal expansion coefficient to very high temperatures and are therefore ideally suited for various structural applications in space [5]. The high temperature strength of MMCs is enhanced by reinforcements such as SiC whiskers or continuous Borsic (B fibers coated with SiC) fibers. SiC (Nicalon)/Al has excellent high temperature strength up to 500°C; above this temperature, however, debonding and decohesion between fibers and matrix cause fiber pullout and failure of the material. In the case of particle-filled MMCs, the strength and modulus are not as significantly altered (as in continuous-fiber composites), but tribological (wear, friction, galling) properties show marked improvements [6]. Soft solid lubricant particles like graphite and mica improve the antiseizing properties of Al alloys, whereas hard particles like silicon carbide, aluminum oxide, tungsten carbide, titanium carbide, zirconium oxide, silicon oxide, and boron carbide greatly improve their resistance to abrasion. Particle additions can also give rise to better damping and conductivity of the matrix alloy. For example, the damping capacity of aluminum and copper alloys is considerably enhanced when graphite powder is dispersed in them. Graphite/Al or Cu alloys have higher damping capacity than conventional vibration insulating alloys, including cast irons [7], and the damping capacity is considerably more stable at high temperatures. Sliding electrical contacts made from copper–20 graphite alloy perform better than those made from the sintered materials that are generally used, since the alloy combines excellent resistance to seizure with high electrical conductivity. Current collectors of Cu–20 graphite alloys in pantographs show higher performance than conventional powder metallurgy–produced contacts.

Selective reinforcement of metals with ceramic fibers (e.g., aluminosilicate fibers in Al) is used in automotive parts, such as diesel engines [8]. Enhanced wear resistance and higher use temperature at equivalent cost have made such selectively reinforced MMCs a potentially very useful class of modern composite materials [8,9]. Figure 10 shows a photograph of a Toyota piston selectively reinforced with ceramic fibers and produced by squeeze infiltrating a fiber assembly with molten Al alloy. Connecting rods have been made from fiber FP–Al–SiC composites by casting techniques. Figure 11 shows photographs of fan bushes, journal bearings, a cylinder block, and a piston made from cast Al–Si-graphite particle composites. In all these cases, graphite particles were stirred in molten aluminum alloys, followed by solidification of the suspensions using conventional foundry techniques. The use of graphite in automobile engine parts considerably reduces the wear of the cylinder liner as well as improving fuel efficiency and engine horsepower at equivalent cost [9–12]. The most promising application of cast aluminum–graphite particle composite alloys is for bearings, which would be cheaper

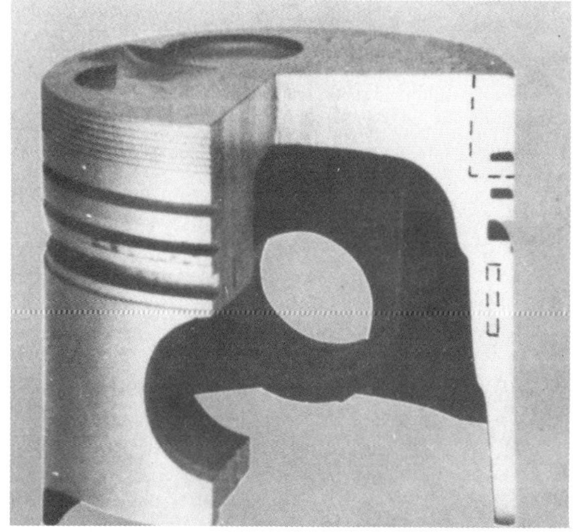

FIGURE 10 **Squeeze-cast selectively reinforced Toyota piston made from ceramic fiber reinforced aluminum.**

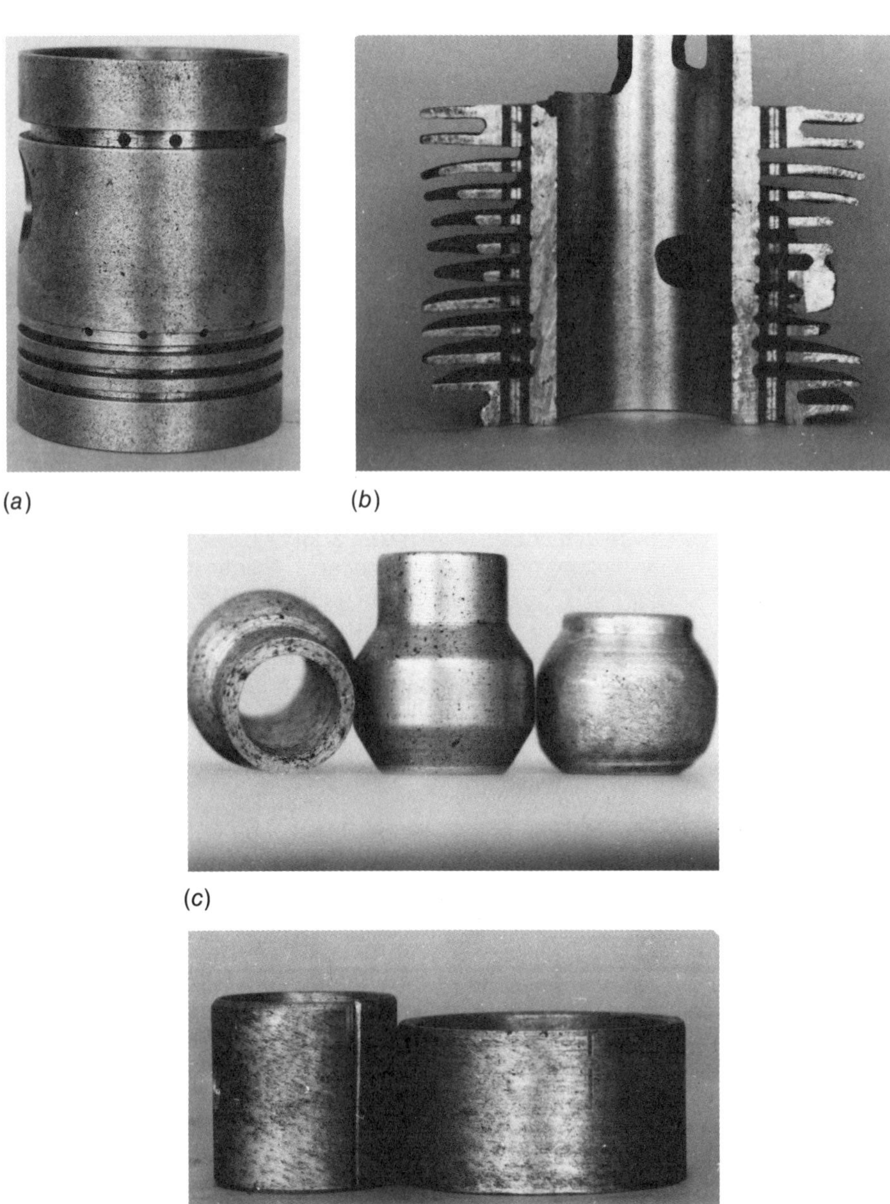

FIGURE 11 Photographs showing engineering products made from cast metal matrix composites: (*a*) graphite–aluminum piston for an automobile engine; (*b*) graphite–aluminum cylinder block; (*c*) graphite–aluminum fan bushes; (*d*) journal bearings of graphite–aluminum.

and lighter than the bearings currently being made out of Cu, Pb, Sn and Cd containing alloys, in addition to being self-lubricating. The cast aluminum–graphite fan bushings shown in Figure 11*c* experience considerably reduced wear as well as temperature rise during trial runs.

The use of cast aluminum–graphite alloy pistons in single-cylinder diesel engines with a cast iron bore reduces fuel consumption and frictional horsepower losses [11]. Because of its lower density, the use of aluminum–graphite composite in internal combustion engines reduces the overall weight of the engine. Such an engine will not seize during cold start or failure of lubricant, because of the excellent antiseizing properties of graphitic–aluminum alloys. In a Brazilian Grand Prix 42-lap race in 1975, a racing car with a cast aluminum–graphite particle composite engine liner (produced by AE Borgo) won the race even though the radiator broke and the cooling fluid was completely lost after 27 laps [9].

Dural Composite Corporation has developed a proprietary process [13,14] for producing cast MMC of Al/SiC showing acceptably uniform particle distribution, high strength, and high stiffness. Cast Al alloys reinforced with ceramic phase are being proposed for use as turbocharger impellers that run at high temperatures. Figure 12 shows an investment-cast silicon carbide/A 357 Al

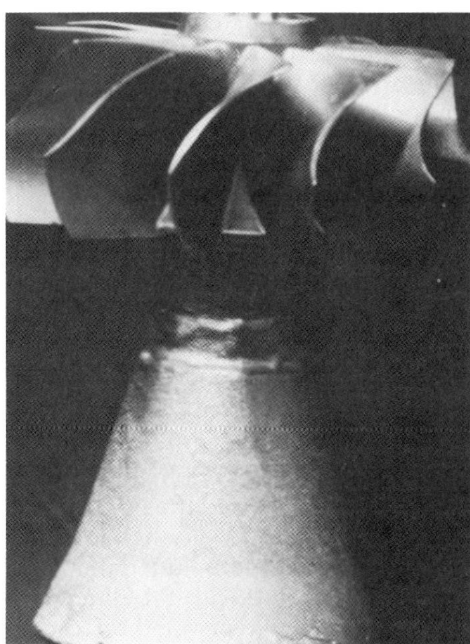

FIGURE 12 A silicon carbide particle-reinforced investment-cast turbocharger impeller. (Courtesy of D. M. Schuster, DACC.)

composite cast by Dural Co. to net shape of a prototype turbocharger impeller. Figure 13 shows cast graphite/Mg composite tubes produced at Martin Marietta for space structure applications; these tubular composites have a near-zero coefficient of expansion [15–19].

Conclusions and Recommendations

Considerable progress has been made in introducing particles and fibers into molten alloys, especially aluminum base, and casting the suspensions using conventional foundry techniques. These conventional foundry techniques are leading to the possibility of low cost mass manufacture of metal matrix composites. The advent of newer techniques, such as squeeze casting or pressure infiltration of fiber preforms by liquid metals, has provided a commercially viable production route for manufacturing porosity-free, net-shape metal matrix composite castings, thereby making possible selective reinforcement of regions in the casting where specific properties are required. In addition to the advantages of the reinforcing phase, the microstructure of the matrix alloy and the chemistry and morphology of the metal-ceramic interface can be suitably modeled and tailored by using solidification processing techniques. A variety of issues of both fundamental and engineering origin still need to be addressed for improved design and synthesis of cast MMCs; some of these include modification of the metal-ceramic interface by surface engineering; modeling of solidification morphology on reinforcing fibers; study of interactions between ceramic particulates and solidifying interfaces; elimination of porosity and freeze choking during infiltration; uniformity of distribution of reinforcements within a cast component; and reproducibility between casting and batches of castings in large-scale manufacture.

FIGURE 13 Graphite fiber–magnesium tubes produced by filament-wound vacuum casting process. (Courtesy of Martin Marietta Co.)

Pradeep Rohatgi

Acknowledgments

The author is grateful to several researchers for their critical comments and valuable suggestions on an earlier draft of this paper. He is especially thankful to James Cornie, D. M. Schuster, K. K. Chawla, and R. Asthana for their valuable suggestions and help in the preparation of the manuscript. Thanks are also due to the Word Processing Center of the University of Wisconsin—Milwaukee for their meticulous typing. The author would also like to thank several researchers at MIT and the University of Wisconsin—Milwaukee in the United States and the Regional Research Laboratories at Bhopal and Trivandrum; IITs at Kanpur and Delhi; the Indian Institute of Science at Bangalore; the Universities of Banaras, Jaipur, and Roorkee; the National Metallurgical Laboratory at Jamshedpur; the Defence Metallurgical Research Laboratory at Hyderabad; India Pistons at Madras; Bharat Heavy Electricals Ltd.; Escorts at Patiala; Hindustan Aluminum at Renukoot; and La Prenca at Bombay in India.

References

1. P. K. Rohatgi, R. Asthana, and S. Das, *International Metals Reviews, 31,* 115 (1986).
2. J. A. Cornie, Y. M. Chiang, D. R. Uhlmann, A. Mortensen, and J. M. Collins, *Ceramic Bulletin, 65,* 293 (1986).
3. A. Mortensen, M. N. Gungor, J. A. Cornie, and M. C. Flemings, *Journal of Metals, 38*(3), 30 (1986).

4. T. W. Clyne and J. F. Mason, *Metall. Trans., 18A*, 1519 (1987).
5. A. Mortensen, et al., *Proc. of World Materials Congress*, 1988, pp. 7–13, ASM International, Metals Park, Ohio.
6. A. Mortensen, et al., *Metall. Trans., 19A*, 709 (1988).
7. H. Fukunaga, *Proc. of World Materials Congress*, ASM International, Metals Park, Ohio, 1988, pp. 101–108.
8. P. K. Rohatgi, "Cast Metal Matrix Composites," in *Metals Handbook*, Vol. 15, *Castings*, 9th ed., ASM International, Metals Park, Ohio, 1988, pp. 840–854.
9. D. M. Goddard, *Metal Progress, 4*, 49 (1984).
10. M. Misra, S. P. Rawal, D. M. Goddard, and J. Jackson, "Novel Processing Techniques in Fabricating Graphite-Magnesium Composites for Space Applications," Phase I Tech. Rep., MCR–85–711, for Naval Sea Systems Command Contract No. 24–84–C–5306, 1985.
11. S. V. Prasad and P. K. Rohatgi, *J. Metals, 39*(11), 22 (1987).
12. "Hitachi Graphite Dispersed Cast Alloy-Gradia," Hitachi Chemical Co., Ibaraki, Japan, 1981.
13. T. Donomoto, K. Funatani, N. Miura, and N. Miyake, *Proceedings of the SAE Int. Congress and Exposition*, Paper No. 830252, 1983.
14. Rino Casellato, A. E. Borgo (Italy), private communication, 1988.
15. L. Bruni and P. Iguera, *Automotive Engineer, 3*, 29 (1978).
16. P. K. Rohatgi et al., *Wear, 60;* 205. (1980).
17. Jeremy W. Holt and AE Engine Components Inc., "Silgraf—A New Silicon/Graphite/Aluminum Alloy for Cylinder Liners," Copia Ricevuta 29/09/1987, 16.05 G3, p. 20.7.
18. D. M. Schuster, "Low Cost, High Performance Silicon Carbide Reinforced Aluminum Castings, Forgings, Extrusions and Rolled Sheet," Dural Aluminum Composite Corporation (DACC), San Diego, 1987.
19. D. M. Schuster, M. Skibo, and F. Yep, *J. Metals, 39*(11), 60 (1987).

Metal Matrix Composites, Continuous Fiber, Fracture

Review of Earlier Studies

An excellent and very comprehensive review of the notched behavior of composite laminates is given by Awerbuch and Madhukar [1]. The present authors do not intend to duplicate this review, but rather focus in more detail on some particular experimental studies and modeling concerning matrix plasticity and transverse crack growth in continuous fiber metal matrix composites.

It was hoped by many of the early investigators that the same fracture tests that were used for metals (e.g., ASTM E 399) could be used for metal matrix composites (MMCs), since linear elastic fracture mechanics (LEFM) for metallic specimens was reasonably well understood. As is evidenced by the findings of the following researchers, the extent and influence of the plastic zone and the constraints imposed by the much stiffer, linearly elastic fibers significantly changed the behavior to make LEFM inappropriate for most cases.

Waszczak [2] investigated the applicability of LEFM to boron/aluminum (B/Al) composites for both unidirectional and cross-plied specimens, and measured specimen elongation and crack-opening displacement (COD) during static loading. He found that load-versus-COD curves for three-point bend and for center-notched specimens were quite nonlinear. Waszczak suggested that this nonlinearity was due primarily to large-scale plasticity in the 6061 aluminum matrix. He noted that there was no significant thickness effect for the MMC specimens. Waszczak then found that the K_Q values calculated from ASTM E 399 were high compared with experimental data. He suggested the use of a pseudoplastic zone to account for matrix plasticity, as is commonly done in metals. Waszczak concluded that although there were problems associated with the direct application of LEFM to MMC fracture, the method did show some promise and that further investigation of the method was warranted.

Sun and Prewo [3] also studied the fracture toughness of boron/aluminum composites, using both unidirectional and cross-plied laminates. In their study, 0.14 mm boron fiber reinforced 6061 aluminum was used in compact tension specimens over a wide range of ratios of notch width to specimen width, a/W. The investigators showed that the fracture toughness of these composites could be characterized well, provided the crack growth was collinear with the initial premachined crack. They attempted to relate the fracture toughness of a [0/90] specimen to that of its constituent plies, but found that a simple rule-of-mixtures approach overpredicted the experimentally observed value. Sun and Prewo noted that this overprediction was expected, since the 90° plies severely constrained the fracture of the 0° plies, preventing the occurrence of any interfiber shear and subsequent crack blunting.

Boron/aluminum laminates containing discontinuities in the form of holes and slits were tested in static tension by Mar and Lin [4] and by Peters [5]. Mar and Lin noted that the length of the discontinuity, not the shape, appeared to control the stress at which fracture occurred. They used a two-parameter formula for data correlation and found that the problem could be modeled as if the discontinuity were a crack with its tip at a bimaterial interface. Peters attempted to use LEFM to analyze double-edge notched specimens but found that because of large-scale plasticity in the matrix, LEFM was in general not applicable to these MMCs, even when the crack extension was collinear with the original machined notches. He noted that LEFM does not account for such factors as different constituent strengths and stiffnesses and for fiber volume fraction.

Awerbuch and Hahn [6–8] measured crack opening displacements (COD) during static loading for both B/Al and BSiC/titanium composite specimens using a laser interferometric technique, and monitored crack tip damage (both fiber and matrix) during the tests. The load–

COD curves for B/Al specimens were found to be highly nonlinear, while those for BSiC/Ti were found to be less nonlinear. These investigators found excellent agreement between predicted (using the Whitney–Nuismer point–stress failure criterion [9]) and observed fracture strengths and load–COD curves. They found, however, that the resistance curve method commonly used in metal fracture was not applicable to MMC fracture. Longitudinal yield zones in the aluminum matrix composites were observed, providing a crack-blunting mechanism as observed by other investigators. They concluded by noting that the characterization of fracture behavior of composite laminates depends strongly on the test procedure employed and on the mode of failure; moreover, specific analytical and testing procedures must be developed for each material system.

A considerable amount of study involving B/Al composites has been performed by Reedy [10–12]. In these studies, Reedy used notched 0.14 mm boron-reinforced aluminum as in earlier studies but fabricated specimens with different matrix properties. He discovered that, in general, the amount of crack-opening displacement varied inversely with matrix shear strength. Reedy also conducted three-point bend tests with various specimen sizes to further study B/Al fracture behavior. He found that large-scale matrix plasticity prevented the definition of a fracture toughness for the material based on LEFM. Reedy measured the amount of longitudinal matrix yielding at the end of the notch using a series of strain gages, and found that the length of this yield zone could be calculated using a linear elastic–power law hardening model with reasonable accuracy.

Several NASA investigators, including Poe and Sova [13–15] and Johnson et al. [16,17], have also studied damage growth and fracture of metal matrix composites. Poe and Sova monitored fiber and matrix damage growth using radiography during static loading of specimens containing slits and again found large matrix yielding. A strain-based fracture criterion developed by Poe and Sova [14,15] was shown to give good results for a number of different laminates as long as the principal load-carrying plies were 0° plies.

Johnson showed that the damage growth could be predicted by analytically modeling (using a finite element model) the stress state in the fibers ahead of the notch. When the stress in an unbroken fiber reached a predetermined value, damage would propagate. It was discovered that first fiber failure occurred at about 50% of ultimate stress in unidirectional composites, while first fiber failure occurred very near the ultimate failure stress in multidirectional laminates. Johnson et al. also noted that while the matrix yielding in unidirectional laminates tended to reduce the stress concentration factor, the yielding actually increased the fiber stress concentration factors in multidirectional laminates because of yielding of the off-axis plies. A related investigation by Post et al. [18] studied the damage growth in the fibers and matrix using Moiré interferometry techniques. They also noted the large matrix yield zones that occur in unidirectional aluminum matrix specimens.

A detailed experimental study of the damage growth in unidirectional B/Al specimens containing notches was conducted by Jones and Goree [19,20]. The fiber damage was monitored using radiography, while the growth of damage in the aluminum matrix was studied using a brittle lacquer. It was found that the internal damage and the residual strength could be predicted with reasonable accuracy using an improved form of the shear lag model developed by Dharani et al. [21], which included longitudinal matrix damage and stable transverse crack extension. The extent of matrix yielding predicted by the model was in excellent agreement with the length of the severely cracked region in the brittle coating, as shown in Figure 1. The extent of constrained fiber damage predicted by the model also agreed well with experimentally observed values (Fig. 2).

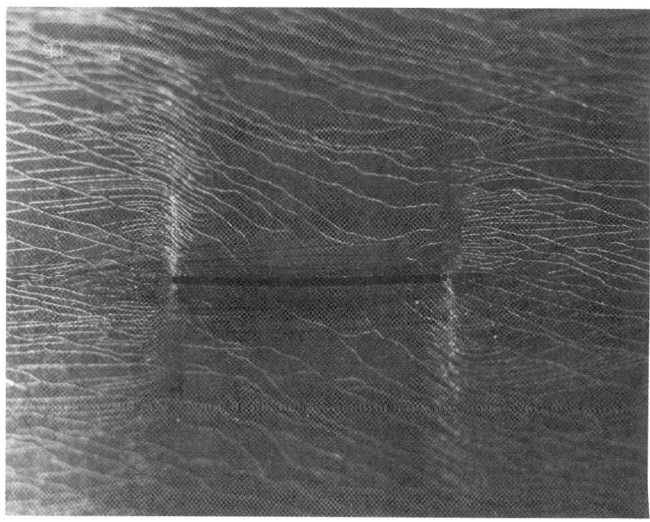

FIGURE 1 Brittle coating pattern showing severe damage region.

FIGURE 2 Typical radiograph showing broken fibers at notch tip.

It was also demonstrated that the first fiber failure occurred at roughly 50–60% of the ultimate failure load, regardless of the initial flaw size. Jones and Goree [19,20] also showed that an increase in the matrix yield strength actually made a unidirectional laminate more notch sensitive. This behavior was correctly predicted by the improved shear lag model [21]. It was suggested that this structural type of stress analysis was more appropriate for MMC than traditional continuum fracture mechanics, since these traditional methods could not account for the different types of damage growth in the multiphase composite material.

Tsangarakis et al. [22] used silicon carbide fiber reinforced aluminum composites in double-edge notched and compact tension specimen configurations. They found the fracture toughness of SiC/Al to be highly specimen dependent as in earlier studies. Kenaga et al. [23] have suggested that B/Al composites can be characterized as elastic–plastic materials. They performed unloading tests and showed that aluminum MMC exhibited plasticity in the classical sense, and described this behavior using an anisotropic elastic–plastic analysis. Their testing was, however, limited to tensile loading; the extension to more complex types of loading needs further investigation.

Other material systems have also been investigated, including the P100/6061 graphite/aluminum studied by Nardone and Strife [24,25]. These researchers performed static fracture tests on unnotched tensile coupons having various stacking sequences. Taya and Daimaru [26] performed tensile and three-point bend tests on P55/6061 graphite/epoxy to verify a model based on fracture surface energy. Finally, Shetty and Chou [27] conducted tensile and compressive tests on unidirectional FP/aluminum and tungsten/aluminum composites to study the strengthening and failure mechanisms of these materials.

Analysis

A number of parametric models have been developed in an attempt to predict the fracture of these MMCs. Among the best known are those of Whitney and Nuismer [9] and Poe and Sova [14,15]. Characteristic of these and most of the other recent models is a foundation based on either linear elastic fracture mechanics or a maximum stress (strain) failure criterion. Some more detailed stress analyses using both two- and three-dimensional finite element elasticity solutions have also been given, by Reedy [28] and by Johnson [16,17].

The models presented here, based on the authors' research over the past few years, fall between the parametric models and the more complete elasticity solutions. That is, we have attempted to retain much of the fundamental material behavior but have made some simplifying assumptions to reduce the complexity from that of a full-elasticity solution. For unidirectional MMCs, the large difference between the fiber and matrix extensional stiffness and the typically low matrix yield stress suggest use of the shear lag stress–displacement relations as used by Hedgepeth [29]. This can then be viewed as a two-phase structural model with a simplified stress or strain failure criterion.

Improved Shear Lag Model Including Matrix and Fiber Damage

Goree et al. [19–21] extended the theory of Hedgepeth to include both fiber damage and matrix yielding and splitting. Following Goree and Gross [30], the laminate is modeled as a two-dimensional region, having a single row of parallel, identical, equally spaced fibers, separated by matrix (Fig. 3). The initial damage is taken to consist of an arbitrary number of broken fibers such that all breaks lie along the x axis. Longitudinal matrix damage is introduced at the end of the initial notch as in Ref. 30. The additional notch tip transverse damage consists of an arbitrary number of broken fibers, which are constrained by the adjoining matrix and/or the unbroken fibers through the thickness. These fibers in the transverse damage zone will be referred to as constrained fibers. It is mathematically untractable to account precisely for the distribution of fiber damage occurring in the region ahead of the initial notch. The model assumes all the breaks to occur on the x axis and accounts for the loss in stiffness by assuming that these fibers will carry a

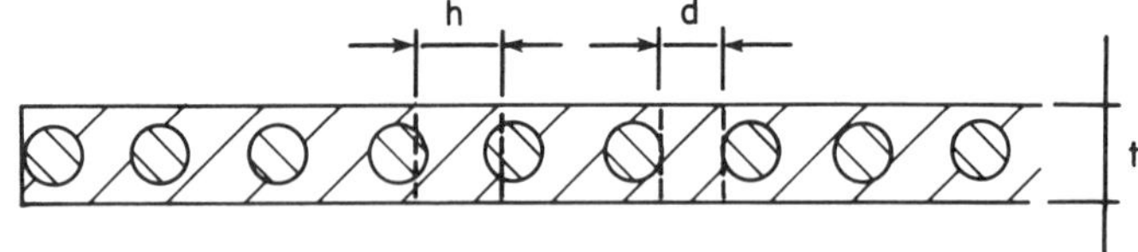

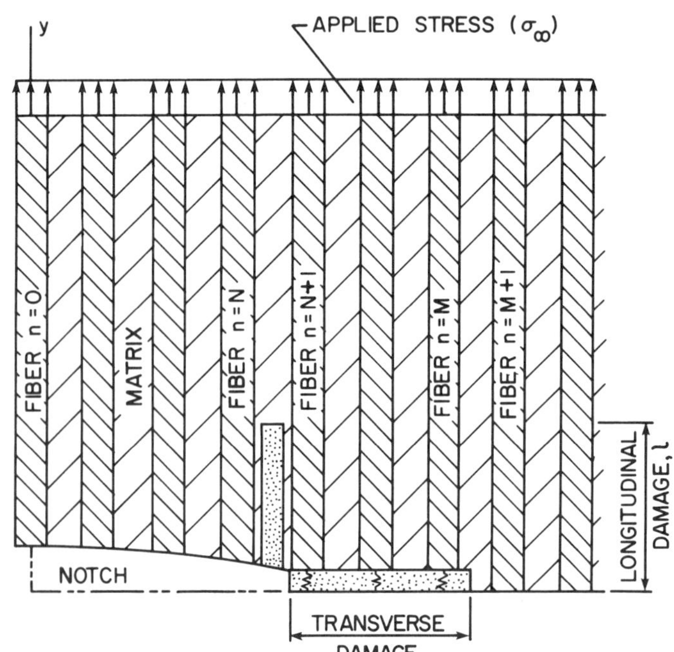

FIGURE 3 First quadrant of symmetric center-notched lamina with crack tip damage.

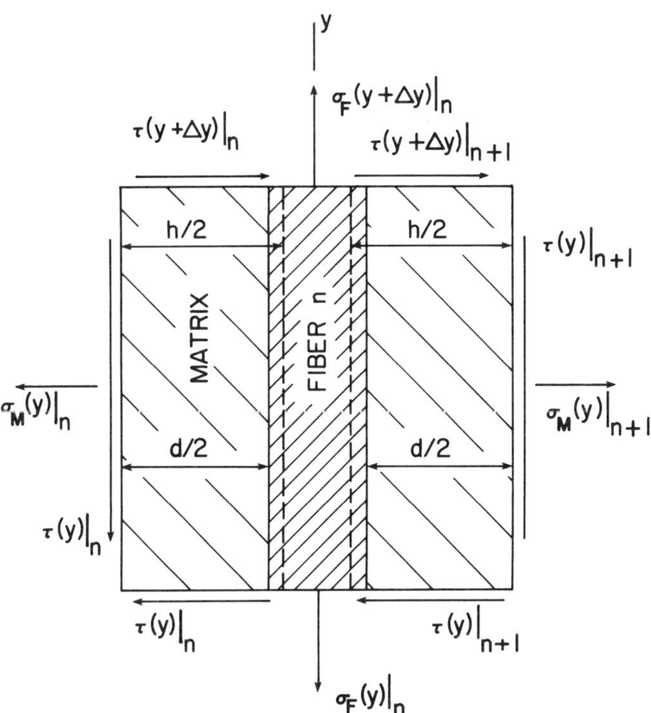

FIGURE 4 Free-body diagrams of fiber–matrix elements.

reduced load by defining a stiffness coefficient r such that

$$r = \frac{\text{stress in the constrained fiber}}{\text{stress in the first unbroken fiber}}$$

Figure 4 gives the free-body diagrams of a generic fiber n, the last broken fiber N, and the first unbroken fiber $N + 1$. Yielding between fibers, N and $N + 1$, is assumed to occur when the matrix shear strain reaches the yield strain γ_0. To account for this longitudinal damage in the form of yielding, it is assumed that the shear stress is constant (τ_0) in the yielded region, $0 \leq y \leq l$, so that

$$\tau|_{N+1} = -\tau_0 \qquad \text{for} \quad 0 \leq y \leq l \tag{1}$$

With reference to the free-body diagrams (Fig. 4), the equilibrium equations in the longitudinal and transverse directions are given by

$$\frac{A_F}{t}\frac{d\sigma_F|_n}{dy} + \tau|_{n+1} - \tau|_n - \delta_{nN}\langle y-l\rangle\{\tau|_{n+1} + \tau_0\} + \delta_{n(N+1)}\langle y-l\rangle\{\tau|_n + \tau_0\} = 0 \tag{2}$$

$$\sigma_M|_{n+1} - \sigma_M|_n + \frac{h}{2}\frac{d}{dy}\{\tau|_{n+1} + \tau|_n - \delta_{nN}\langle y-l\rangle[\tau_0 + \tau|_{n+1}] - \delta_{n(N+1)}\langle y-l\rangle[\tau_0 + \tau|_n]\} = 0 \tag{3}$$

where δ_{ij} is the Kronecker delta, A_F is the cross-sectional area of the fiber, t is the lamina thickness, and

$$\langle y-l\rangle = 1 \qquad \text{for} \quad y \leq l$$
$$\langle y-l\rangle = 0 \qquad \text{for} \quad y > l$$

The following assumptions are now made regarding the stress–displacement relations [29]:

Axial stress in fiber n

$$\sigma_F|_n = E_F\frac{dv_n}{dy} \tag{4}$$

Matrix shear stress between fiber n and $n + 1$

$$\tau|_{n+1} = \frac{G_M}{h}(v_{n+1} - v_n) \tag{5}$$

Matrix transverse stress

$$\sigma_M|_{n+1} = \frac{E_M}{h}(u_{n+1} - u_n) \tag{6}$$

where v_n and u_n are the axial and transverse displacements of fiber n, E_F and E_M are the Young's moduli of the fiber and the matrix, respectively, G_M/h is the equivalent shear stiffness of the matrix, and h is the shear transfer distance.

These assumptions simplify the equilibrium equations by removing the transverse displacement dependence from the longitudinal equilibrium equation. The fiber stress and the matrix shear stress can then be determined without solving the transverse equilibrium equation. If we consider only longitudinal equilibrium, the equation in the longitudinal direction in terms of displacement is

$$\frac{E_FA_Fh}{G_Mt}\frac{d^2v_n}{dy^2} + v_{n+1} - 2v_n + v_{n-1} - \delta_{nN}\langle y-l\rangle\left\{v_{n+1} - v_n + \frac{h}{G_M}\tau_0\right\} + \delta_{n(N+1)}\langle y-l\rangle\left\{v_n - v_{n-1} + \frac{h}{G_M}\tau_0\right\} = 0 \tag{7}$$

The equilibrium equation (Eq. 7) can be expressed in a nondimensional form by incorporating the following changes in the variables:

$$y = \psi\eta \qquad \text{and} \qquad v_n = \phi V_n \tag{8}$$

where

$$\psi = \left\{\frac{A_FE_Fh}{G_Mt}\right\}^{1/2}$$

$$\phi = \left\{\frac{A_Fh}{E_FG_Mt}\right\}^{1/2}\sigma_\infty$$

The resulting equilibrium equation in terms of the nondimensional variables is given by

$$\frac{d^2V_n}{d\eta^2} + V_{n+1} - 2V_n + V_{n-1} = \{\delta_{n(N+1)} - \delta_{nN}\}f(\eta) \quad (9)$$

where $f(\eta)$ is a new unknown function, such that

$$f(\eta) = g(\eta) - \bar{\tau}_0 = V_N - V_{N+1} - \bar{\tau}_0 \qquad \text{for} \quad \eta < \alpha \quad (10)$$

and

$$f(\eta) = 0 \qquad \text{for} \quad \eta \geq \alpha \quad (11)$$

in which

$$\bar{\tau}_0 = \frac{G_M}{h}\phi\tau_0 \qquad \text{and} \qquad \alpha = \frac{y}{\psi} \quad (12)$$

This differential-difference equation may be reduced to a differential equation by introducing the even-valued transform as

$$\bar{V}(\eta,\theta) = \frac{V_0}{2} + \sum_{n=1}^{\infty} V_n(\eta)\cos(n\theta) \quad (13)$$

from which

$$V_n(\eta) = \frac{2}{\pi}\int_0^{\pi} \bar{V}(\eta,\theta)\cos(n\theta)\,d\theta \quad (14)$$

Making use of the transformation above and the orthogonality property of the circular functions, the equilibrium equation may be written as

$$\frac{d^2\bar{V}}{d\eta^2} - \delta^2\bar{V} = -\langle\alpha - \eta\rangle D^2 f(\eta) \quad (15)$$

where

$$\delta^2 = 2[1 - \cos(\theta)] = 4\sin^2\left(\frac{\theta}{2}\right)$$

and

$$D^2 = \cos(N\theta) - \cos[(N+1)\theta]$$

The solution to the problem of vanishing stresses and displacements at infinity and uniform compression on the ends of the broken fibers will now be sought. The complete solution is obtained by adding the results corresponding to uniform axial stress and no broken fibers to the following solution. The appropriate boundary conditions are:

$$\text{as} \quad \eta \to \infty,\ V_n = 0 \qquad \text{for all fibers} \quad (16)$$

$$\text{at} \quad \eta = 0,\ \frac{dV_n}{d\eta} = \bar{\sigma}_{F|n} = -1 \qquad \text{for broken fibers} \quad (17)$$

$$\text{at} \quad \eta = 0,\ \frac{dV_n}{d\eta} = \bar{\sigma}_{F|n} = -1 + r\bar{\sigma}_{F|M+1} \qquad \text{for constrained fibers} \quad (18)$$

$$\text{at} \quad \eta = 0,\ V_n = 0 \qquad \text{for unbroken fibers} \quad (19)$$

where $\bar{\sigma}_{F|M+1}$ is the normalized stress in the first unbroken fiber at $\eta = 0$, and r is the stiffness coefficient defined earlier.

The complete solution to Eq. (15) satisfying vanishing stresses and displacements at infinity is given by

$$V_n(\eta) = \frac{2}{\pi}\int_0^{\pi} e^{-\delta n}\sum_{m=0}^{N} B_m\cos(m\theta)\cos(n\theta)\,d\theta + \frac{1}{2}\int_0^{\alpha} f(t)\{C_n(|t-\eta|) - C_n(t+\eta)\}\,dt \quad (20)$$

where

$$C_n(\xi) = \frac{2}{\pi}\int_0^{\pi}\frac{D^2}{\delta}e^{-\delta\xi}\cos(n\theta)\,d\theta$$

and the Fourier constants B_m and the function $g(\eta)$ are obtained by solving the following coupled series and integral equations:

$$\frac{2}{\pi}\int_0^{\pi}\left\{-\delta\sum_{m=0}^{N} B_m\cos(m\theta) + D^2\int_0^{\alpha} e^{-\delta t}g(t)\,dt - D^2\bar{\tau}_0\int_0^{\alpha} e^{-t}\,dt\right\}\cos(n\theta)\,d\theta = -1 \quad (21)$$

$$(n = 0, 1, \ldots, N)$$

$$\frac{2}{\pi}\int_0^{\pi}\left[-\delta\sum_{m=0}^{N} B_m\cos(m\theta) + D^2\int_0^{\alpha} e^{-\delta t}g(t)\,dt - \bar{\tau}_0 D^2\int_0^{\alpha} e^{-\delta t}\,dt\right]X\{\cos(n\theta) - r\cos[(M+1)\theta]\}\,d\theta = -1 + r \qquad (n = N+1, \ldots, M) \quad (22)$$

$$\begin{aligned} g(\eta) = {} & \frac{2}{\pi}\int_0^{\pi} e^{-\delta\eta}\sum_{m=0}^{N} B_m\cos(m\theta)\{\cos(N\theta) - \cos[(N+1)\theta]\}\,d\theta + \frac{1}{2}\int_0^{\alpha} g(t)\{C_N(|t-\eta|) \\ & - C_N(t+\eta) - C_{N+1}(|t-\eta|) + C_{N+1}(t+\eta)\}\,dt \\ & - \frac{\bar{\tau}_0}{2}\int_0^{\alpha}\{C_N(|t-\eta|) - C_N(t+\eta) - C_{N+1}(|t-\eta|) + C_{N+1}(t+\eta)\}\,dt \end{aligned} \quad (23)$$

The condition that

$$g(\alpha) = \bar{\tau}_0 \quad (24)$$

must also be satisfied.

Consistent Shear Lag Model Including Matrix and Fiber Damage

The classical shear lag model is simplified because the equilibrium equations become uncoupled as a result of the shear lag stress–displacement relations. The solution is found to give an accurate measure of the axial fiber stress and the shear stress between fibers, but a very poor measure of the transverse matrix normal stress. The model also does not have the freedom to remove the shear stresses from the crack surfaces. This motivated Sendeckyj and Jones [31–33] to develop a new shear lag model, which unlike the classical shear lag model gives correct values for matrix normal stresses as well as fiber stresses and matrix shear stresses. In addition, the traction-free boundary condition on the crack surface is satisfied. The correct matrix normal stresses can be quite important for problems such as matrix splitting and delamination, which are governed by mode I crack extension. The initial development for the model is presented below for the case of no damage other than the original notch. This model has not been extended to include matrix yielding or stable crack extension.

As above, the composite can be modeled as an elastic matrix reinforced by equally spaced, parallel, elastic fibers as shown in Figure 3. Let d and h be the width and spacing of the fibers, respectively. In this case the damage consists of broken fibers only. Using the coordinate system shown in Figure 3, a typical element of the composite can be isolated (Fig. 5). For clarity of the derivations, it is convenient to replace the element in Figure 5 with the equivalent homogeneous element by "smearing out" the properties of the fiber and matrix using micromechanics.

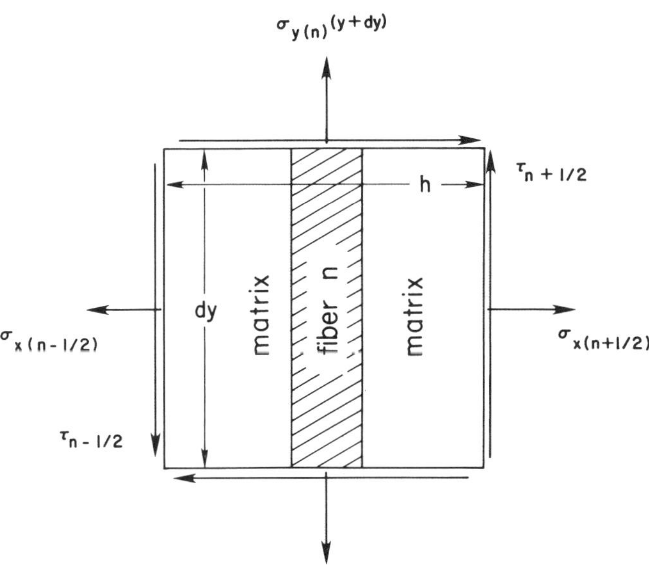

FIGURE 5 Free-body diagram of orthotropic fiber–matrix element.

As can be seen from Figure 5, and the nth typical element is bounded by the $(n + \frac{1}{2})$ and $(n - \frac{1}{2})$ midnode lines in the x direction. The equilibrium equations for the element shown are:

$$\sigma_{x(n+1/2)} - \sigma_{x(n-1/2)} + \frac{h(\tau_{n+1/2,y} + \tau_{n-1/2,y})}{2} = 0 \quad (25)$$

$$\tau_{n+1/2} - \tau_{n-1/2} + h\sigma_{y(n),y} = 0 \quad (26)$$

The constitutive relations are approximated as:

$$\begin{aligned}
\sigma_{x(n+1/2)} &= \frac{C_{11}(u_{n+1} - u_n)}{h} + C_{12}v_{n+1/2,y} \\
\sigma_{x(n-1/2)} &= \frac{C_{11}(u_n - u_{n-1})}{h} + C_{12}v_{n-1/2,y} \\
\sigma_{y(n)} &= \frac{C_{12}(u_{n+1/2} - u_{n-1/2})}{h} + C_{22}v_{n,y} \\
\tau_{n+1/2} &= C_{66}\left[u_{n+1/2,y} + \frac{v_{n+1} - v_n}{h}\right] \\
\tau_{n-1/2} &= C_{66}\left[u_{n-1/2,y} + \frac{v_n - v_{n-1}}{h}\right]
\end{aligned} \quad (27)$$

along with the following relationships between the displacements at the node and midnode:

$$\begin{aligned}
u_{n+1/2} - u_{n-1/2} &= \frac{u_{n+1} - u_{n-1}}{2} \\
v_{n+1/2} - v_{n-1/2} &= \frac{v_{n+1} - v_{n-1}}{2} \\
u_{n+1/2} + u_{n-1/2} &= 2u_n
\end{aligned} \quad (28)$$

Upon substituting Eq. (27) and Eq. (28) into the equilibrium equations (25 and 26) and rearranging, we get:

$$C_{11}(u_{n+1} - 2u_n + u_{n-1}) + \frac{h(C_{12} + C_{66})(v_{n+1,y} - v_{n-1,y})}{2} + h^2C_{66}u_{n,yy} = 0 \quad (29)$$

$$C_{66}(v_{n+1} - 2v_n + v_{n-1}) + \frac{h(C_{12} + C_{66})(u_{n+1,y} - u_{n-1,y})}{2} + h^2C_{22}v_{n,yy} = 0 \quad (30)$$

It should be noted that these equations are fully consistent with the two-dimensional equilibrium equations for an orthotropic material, and therefore should give stresses and displacements that are consistent with the theory of elasticity. For composites with the fiber stiffnesses much higher than matrix stiffnesses, the composite stiffnesses C_{ij} may be approximated as:

$$C_{22} = c_f C^f_{22}$$
$$C_{11} = \frac{C^m_{11}}{1 - c_f}$$
$$C_{66} = \frac{C^m_{66}}{1 - c_f} \quad (31)$$
$$C_{12} = c_f C^f_{12} + (1 - c_f) C^m_{12}$$

where c_f represents the fiber volume fraction, and the superscripts f and m refer to the fibers and the matrix, respectively. As in the preceding section, the equilibrium equations (29 and 30) can be written in the nondimensional form as:

$$E(U_{n+1} - 2U_n + U_{n-1}) + \frac{H(V_{n+1,\eta} - V_{n-1,\eta})}{2} + U_{n,\eta\eta} = 0 \quad (32)$$

$$(V_{n+1} - 2V_n + V_{n-1}) + \frac{H(U_{n+1,\eta} - U_{n-1,\eta})}{2} + FV_{n,\eta\eta} = 0 \quad (33)$$

where

$$\eta = y/h$$
$$U_n = u_n/h$$
$$V_n = v_n/h$$
$$F = C_{22}/C_{66} \quad (34)$$
$$E = C_{11}/C_{66}$$
$$L = C_{12}/C_{66}$$
$$H = 1 + L$$

These difference-differential equations can be reduced to ordinary differential equations by assuming that the normalized displacements U_n and V_n are the Fourier coefficients of functions $\bar{U}(\eta,\theta)$ and $\bar{V}(\eta,\theta)$, that is,

$$U_n = \frac{1}{2\pi}\int_{-\pi}^{\pi} \bar{U}(\eta,\theta)e^{in\theta}\,d\theta \quad (35)$$

$$V_n = \frac{1}{2\pi}\int_{-\pi}^{\pi} \bar{V}(\eta,\theta)e^{in\theta}\,d\theta \quad (36)$$

with the inverse formulas

$$\bar{U}(\eta,\theta) = \sum_{-\infty}^{\infty} U_n e^{-in\theta}$$
$$\bar{V}(\eta,\theta) = \sum_{-\infty}^{\infty} V_n e^{-in\theta} \quad (37)$$

Upon substituting Eqs. (35) and (36) into Eqs. (29) and (30), we obtain

$$\bar{U}_{,\eta\eta} - 2E(1 - \cos\theta)\bar{U} + iH\sin\theta\,\bar{V}_{,\eta} = 0 \quad (38)$$

$$F\bar{V}_{,\eta\eta} - 2(1 - \cos\theta)\bar{V} + iH\sin\theta\,\bar{U}_{,\eta} = 0 \quad (39)$$

The solution is then sought for Eqs. (38) and (39) satisfying the following boundary conditions:

$$\begin{aligned} U_n = V_n = 0 \text{ at } \eta = \infty &\quad \text{for all } n \\ \tau_n = 0 \text{ at } \eta = 0 &\quad \text{for all } n \\ V_n = 0 \text{ at } \eta = 0 &\quad \text{for all intact fibers} \\ \sigma_{y(n)} = -1 \text{ at } \eta = 0 &\quad \text{for all broken fibers} \end{aligned} \quad (40)$$

The solution of these ordinary differential equations (Eqs. 38 and 39), satisfying the boundary conditions (Eq. 40), is:

$$\bar{V} = A_1 e^{-D^1\eta} + A_2 e^{-D^2\eta}$$
$$\bar{U} = -i\left\{\frac{(FD_1^2 - \delta^2)A_1 e^{-D_2\eta}}{HD_1\sin\theta} + \frac{FD_2^2 - \delta^2)A_2 e^{-D_2\eta}}{HD_2\sin\theta}\right\}$$
$$D_{1,2} = \left\{\frac{(1 - \cos\theta)[\alpha \pm (\alpha^2 - 4EF)^{1/2}]}{F}\right\}^{1/2} \quad (41)$$
$$\alpha = 1 + EF - \frac{H^2(1 + \cos\theta)}{2}$$
$$A_1 = -\frac{FD_2^2 - 2(1 - \cos\theta) + H\sin^2\theta}{F(D_1^2 - D_2^2)}\sum_{m=0}^{N} a_m e^{-im\theta}$$

and

$$A_2 = -\frac{FD_1^2 - 2(1 - \cos\theta) + H\sin^2\theta}{F(D_1^2 - D_2^2)}\sum_{m=0}^{N}$$

with the Fourier constants a_m being obtained from the stress boundary condition on the broken fibers ($n = 0, \ldots, M$) given by:

$$\frac{C_{66}}{2\pi}\int_{-\pi}^{\pi}\sum_{m=0}^{N} A_m\Big[\{2[1 + (EF)^{1/2}] - [1 + L^2(EF)^{1/2}](1 + \cos\theta)\} [F(1 - \cos\theta)]^{1/2}\{2[1 + (EF)^{1/2}]^2 - H^2(1 + \cos\theta)\}^{-1/2}\Big]\cos[(n - m)\theta]\,d\theta = 1 \quad (42)$$

for $n = 0, \ldots, M$

The axial and transverse stresses along $\eta = 0$ are maximum and are given by

$$\sigma_{y(n)} = -C_{66} \sum_m a_m I_{n-m}$$

and

$$\sigma_{x(n)} = -C_{66} \sum_m a_m J_{n-m}$$

where I_{n-m} and J_{n-m} are functions [31] of the properties E, F, H, and L. Therefore, once the constants a_m are known, the displacements and stresses of any fiber n can be obtained.

Classical Shear Lag Model for a Strip of Finite Width

In this section we present a solution developed by Goree and Dharani [34,35] for a unidirectional strip of finite width, containing an arbitrary number of broken fibers to form a rectilinear notch. Some modifications to the solution cited [34,35] are made for the sake of consistency with the preceding solutions. With reference to Figure 6, the governing equations are given as:

$$\sum_{m=1}^{N-N^*} B_m \frac{2}{\pi} \int_0^{\pi} \delta \cos[(N^* + m)\theta] \cos[n\theta]\, d\theta + \frac{2}{\pi} \int_0^{\pi} Cw \cos[n\theta] \int_0^1 e^{-\delta t} g(t)\, dt\, d\theta = 1 \tag{43}$$

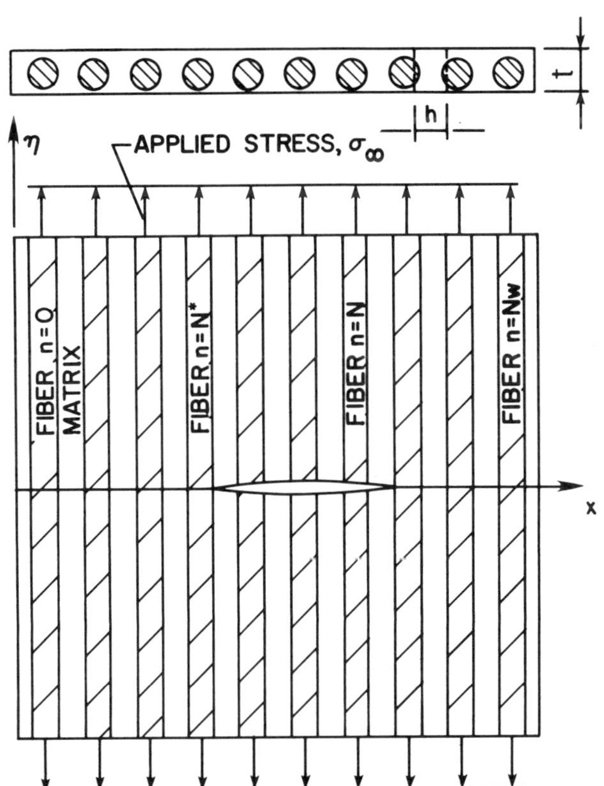

FIGURE 6 Finite width unidirectional strip.

for all broken fibers, and

$$g(\eta) = -\sum_{m=1}^{N-N^*} B_m \frac{2}{\pi} \int_0^{\pi} \cos[(N^* + m)\theta] \cos[(NW)\theta] e^{-\delta\eta} + \int_0^{\infty} \frac{2}{\pi} \int_0^{\pi} \frac{\cos[(NW)\theta]}{\delta} CwD(\delta,\eta,t)\, d\theta\, g(t)\, dt \tag{44}$$

where

$$Cw = \cos[(NW)\theta] - \cos[(NW + 1)\theta]$$

$$g(\eta) = -V_{NW}$$

$$D(\delta,\eta,t) = e^{-\delta|\eta-t|} - e^{-\delta(\eta+t)}$$

Therefore the solution of the finite width strip reduces to one series equation coupled with one linear integral equation. The location and the number of broken fibers are arbitrary except that the fiber breaks must be along the x axis. The solution obtained holds for a central notch, an edge notch, an off-center notch, or for multiple notches along the x axis.

Results and Conclusions

This section presents some typical results predicted by the analytical models and discusses the validity and limitations of these simple mechanistic models in the light of some experimental results.

Models

IMPROVED SHEAR LAG MODEL. Some interesting observations concerning the case of broken fibers with no additional longitudinal or transverse damage can be made, with particular emphasis on the behavior of the fiber stresses in front of the notch. If the total number of broken fibers is large and the notch is symmetric about the y axis with N as the last broken fiber (as treated in the analysis), then the normalized maximum fiber stress given by Hedgepeth solution [29] can be expressed [21] as:

$$\bar{\sigma}_F|_{N+1} = \left(\frac{\pi}{2}\right)^{1/2} N^{1/2} \tag{45}$$

From this expression it may be observed that in the absence of additional damage, the maximum fiber stress varies as the square root of the crack length. Failure is assumed to occur when the stress in the first unbroken fiber reaches the unnotched fiber ultimate stress. This failure criterion is similar to the "point stress" criterion of Whitney and Nuismer [9]. The authors [21] have also shown that the stress distribution at the crack tip may be approximated by a square root decay, similar to the Griffith crack in an isotropic homogeneous material, using an equivalent crack length approach as

$$\bar{\sigma}_F|_n = \frac{n}{[n^2 - (N + C_N)^2]^{1/2}} \quad \text{for} \quad n \geq N + 1 \tag{46}$$

where $(N + C_N)d$ represents the equivalent half-crack length with d representing the fiber centerline spacing. The parameter C_N is approximately equal to a constant and independent of notch length. The value of C_N is given by:

$$C_N = 1 - \frac{1}{\pi} = 0.6817 \tag{47}$$

Using this value for C_N, the stress in fibers $n > N + 1$ in front of the notch can be computed from Eq. (46) and compared with the results of classical shear lag solutions (without longitudinal and transverse damage). These two cases coincide, as shown in Figure 7, indicating that a central notch in a unidirectional laminate possesses a square root type of stress distribution with an equivalent notch length given by $(N + C_N)d$ for all N.

An analogous investigation [21] concerning the maximum fiber stress and fiber stress decay at the notch tip in the presence of longitudinal and transverse damage showed that most of the preceding observations are no longer valid: the notch tip fiber stress as determined from the present model with damage does not have a square root form, and an equivalent notch length with a constant C_N does not exist. An example of the fiber stress distribution for seven broken fibers, including yielding or transverse damage, is given in Figure 7, illustrating the reduction in stress concentration when compared with the case of broken fibers only. It is indicated later in this article, based on results presented in earlier [21], that both longitudinal yielding and transverse damage are necessary to match the experimental results for unidirectional MMC.

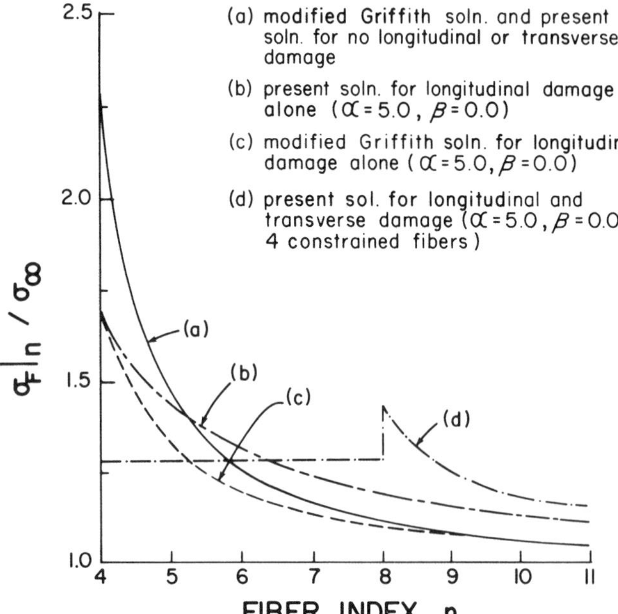

FIGURE 7 Comparison of notch tip stress distribution for seven broken fibers.

CONSISTENT SHEAR LAG MODEL. As discussed earlier, the consistent shear lag model [31], in addition to satisfying the traction-free boundary condition on the crack surface, gives a very accurate measure of normal stresses in the matrix. However, since the equilibrium equations are coupled, the incorporation of any other damage such as longitudinal matrix yield and constrained fiber breakage into the model is very difficult and has not been attempted.

SHEAR LAG MODEL FOR A STRIP OF FINITE WIDTH. Analytical results for the finite width strip with broken fibers forming a central notch and subjected to a uniform remote axial stress are presented in Table 1. The stress concentration factor in the first unbroken fiber at the notch tip is obtained for various notch widths (numbers of broken fibers). Following Hedgepeth [29], the stress concentration factors for a unidirectional infinite region are obtained and compared with those for a strip of finite width to arrive at the finite width correction factor for a given notch width. These finite width correction factors are shown in Table 1, in which $2a$ is the notch width and W is the width of the plate. Also given in Table 1 are the corresponding finite width correction factors for an isotropic strip, as given by Dharani [35]. From these results it appears that the finite width correction for the isotropic material and the unidirectional composite finite width strips are not significantly different for small to moderate aspect ratios.

Experiments

An experimental investigation dealing with the fracture behavior of unidirectional boron/aluminum composite laminates was conducted by Jones and Goree [19,20] to validate the results of the improved shear lag model. The specimens used in this investigation were all unidirectional composite panels consisting of 0.142 mm diameter boron fibers in a 6061 aluminum matrix. Some panels were heat-treated after the initial manufacture such that the aluminum matrix would possess the properties of a T6 temper instead of an annealed, as-fabricated temper. Equal numbers of one-, two-, four-, and eight-ply specimens were fabricated with a thickness of approximately 0.178 mm per ply. Center notches of various widths were introduced by electrostatic discharge machining (EDM). Specimens were tested under uniaxial tensile loading. The longitudinal and transverse damage

TABLE 1
Finite Width Correction Factors

2a/W	Unidirectional Composite	Isotropic
0.1667	1.0113	1.0204
0.25	1.0251	1.0379
0.3333	1.0458	1.0672
0.50	1.1160	1.1828
0.75	1.3928	1.5776

modes were monitored: a brittle lacquer coating was used to detect the yielding in the matrix, while X-ray techniques were used to determine the number of broken (but constrained) fibers at the crack tip.

For a given initial notch width, these experimental results [19,20] show that the dimensions of the transverse damage region, the length of the longitudinal yield region, the COD, and the stress at which the fiber breaks occurred were approximately the same for the four laminate thicknesses tested, and they agreed very well with those predicted by the improved shear lag model [21]. This similarity in behavior regardless of the specimen thickness (for up to eight plies) was observed for all initial notch widths, thus clearly indicating that a thick unidirectional panel (within the range tested) can be modeled as an equivalent monolayer.

It was also shown [19] that the mathematical model based on the classical shear lag assumption could predict accurately the load–COD curves and notched strengths for various notch widths if certain laminate properties were known. Specifically, these properties are the normalized matrix yield stress and the effective shear stiffness of the fiber–matrix region, which are calculated from an experimental load–COD curve. By using these properties, the load–COD and the notched strengths for any number of broken fibers (notch width) could then be predicted accurately. Agreement was also demonstrated [21] for the narrow specimens and short notch lengths presented by Awerbuch and Hahn [8]. The extensive results of Jones and Goree [19,20] further reinforce the validity of the improved shear lag model, incorporating the longitudinal and transverse damages. Figure 8 presents a typical load–COD curve for a four-ply unidirectional boron/aluminum composite panel for specimens having an overall width of 73.1 mm and 89 initially broken fibers, along with the curves predicted by the shear lag model. The values of material parameters used in the calculations are [21] τ_0 = 88 MPa and G_M/h = 115 TPa/m for 6061-0 aluminum matrix and τ_0 = 308 MPa and G_M/h = 172 TPa/m for 6061-T6 aluminum matrix. The agreement between the experimental and predicted results is excellent for 6061-T6 matrix through the entire loading range. However, the agreement is not as good for 6061-0 matrix at higher load levels. This deviation may be attributed to the fact that the longitudinal yielding in the case of 6061-0 is not limited to the region between the last broken fiber and the first unbroken fiber, as evidenced by the extensive brittle lacquer cracking beyond $n = N + 1$. The model does not, in the present form, account for such extensive yielding beyond the first unbroken fiber.

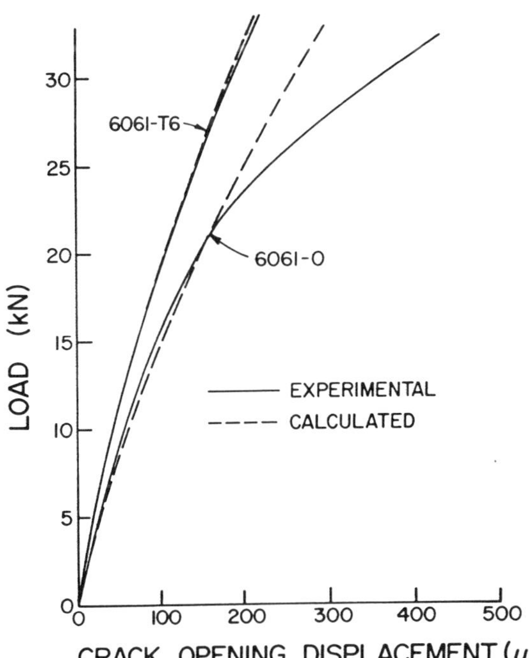

FIGURE 8 Comparison of analytical and experimental load versus crack-opening displacement curves.

The constrained fiber breaking observed through radiographs seems to be limited to about 7 and 3 fibers, respectively, for 6061-0 and 6061-T6 matrices irrespective of the notch width. These results are again in good agreement with the model predictions [21]. The extent of the transverse damage zone indicates the degree of stable notch extension; the stable notch extension is larger for the fully annealed specimens than for the heat-treated specimens.

The strength curves in Figure 9 show the applied remote stress at unstable fracture as a function of the number of broken fibers (notch width) from the experimental results as well as from various model predictions. All the models predict the right trend, but the agreement between experimental results and predictions from models, which have the square root type of crack tip behavior, is poor for short crack lengths. This is also true of the shear lag model incorporating only the longitudinal damage [30]. The model incorporating a transverse damage zone in addition to the longitudinal matrix yielding [21], however, predicts strength values in very close agreement with the experimental results and seems to account for an essential damage mode.

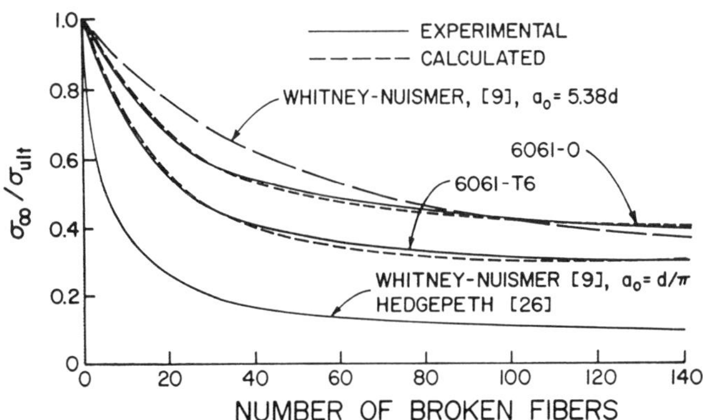

FIGURE 9 Strength curves: comparison of the predictions of various models.

It should be noted that most of the tests cited in this article deal with aluminum matrix composites. Clearly MMC will have its greatest impact on future designs in the areas where higher temperatures are encountered. This suggests the use of titanium or other high temperature materials for the composite matrix. These materials have considerably higher yield strengths and stiffnesses than aluminum, and may therefore behave in a totally different manner when damage growth is considered. The applicability of linear elastic or other standard fracture mechanics techniques as well as standard test techniques need to be studied and understood before these materials can achieve their maximum utility. The development of high temperature test techniques for composite materials has just begun and will require capabilities very different from the testing procedures available at this time.

James G. Goree, L. R. Dharani, and Walter F. Jones

References

1. J. Awerbuch and M. S. Madhukar, *J. Reinf. Plast. Compos., 4,* 1 (1985).
2. J. P. Waszczak, *J. Aircr., 13,* 770 (1976).
3. C. T. Sun and K. M. Prewo, *J. Compos. Mater. 11,* 164 (1977).
4. J. W. Mar and K. Y. Lin, *J. Compos. Mater., 11,* 405 (1977).
5. P. Peters, *J. Compos. Mater., 12,* 250 (1978).
6. J. Awerbuch and H. T. Hahn, *J. Compos. Mater., 12,* 222 (1978).
7. J. Awerbuch and H. T. Hahn, "Crack-Tip Damage and Fracture Toughness of Boron/Aluminum Composites," *J. Compos. Mater., 13,* 82–107 (April 1979).
8. J. Awerbuch and H. T. Hahn, "Fracture Behavior of Metal Matrix Composites," *Proceedings of the Society of Engineering Science, Recent Advances in Engineering Science,* 1977, pp. 343–350.
9. J. M. Whitney and R. J. Nuismer, *J. Compos. Mater., 8,* 253 (1974).
10. E. D. Reedy, Jr., *J. Mech. Phys. Solids, 28,* 265 (1980).
11. E. D. Reedy, Jr., *J. Compos. Mater. Suppl., 14,* 118 (1980).
12. E. D. Reedy, Jr., *J. Compos. Mater., 16,* 495 (1982).
13. J. A. Sova and C. C. Poe, Jr., "Tensile Stress–Strain Behavior of Boron/Aluminum Laminates," NASA TP-1117, 1978.
14. C. C. Poe, Jr. and J. A. Sova, "Fracture Toughness of Boron/Aluminum Laminates with Various Proportions of 0° and ± 45° Plies," NASA TP-1707, November 1980.
15. C. C. Poe, Jr., *Eng. Fract. Mech., 17*(2), 153 (1983).
16. W. S. Johnson, C. A. Bigelow, and Y. A. Bahei-El-Din, "Experimental and Analytical Investigation of the Fracture Processes of Boron/Aluminum Laminates Containing Notches," NASA TP-2187, July 1983.
17. W. S. Johnson and C. A. Bigelow, "Elastic–Plastic Stress Concentrations around Crack-Like Notches in Continuous Fiber Reinforced Metal Matrix Composites," NASA Technical Memorandum 89093, February 1987.
18. D. Post, R. Czarnek, D. Joh, J. Jo, and Y. Guo, "Elastic–Plastic Deformation of a Metal-Matrix Composite Coupon with a Center Slot," NASA Contractor Report 178013, November 1985.
19. J. G. Goree and W. F. Jones, "Fracture Behavior of Unidirectional Boron/Aluminum Composite Laminates," NASA Contractor Report 3753, December 1983.
20. W. F. Jones and J. G. Goree, "Fracture Behavior of Unidirectional Boron/Aluminum Composite Laminates," *Mechanics of Composite Materials—1983,* ASME AMD, vol. 58, 1983, pp. 171–178.
21. L. R. Dharani, W. F. Jones, and J. G. Goree, *Eng. Fract. Mech., 17,* 555 (1983).
22. N. Tsangarakis, B. O. Andrews, and C. Cavallaro, *J. Compos. Mater., 21,* 481 (1987).
23. D. Kenaga, J. F. Doyle, and C. T. Sun, *J. Compos. Mater., 21,* 516 (1987).
24. V. C. Nardone, and J. R. Strife, *J. Mater. Sci., 22,* 592 (1987).
25. V. C. Nardone and J. R. Strife, *J. Mater. Sci. 23,* 194 (1988).
26. M. Taya and A. Daimaru, *J. Mater. Sci., 18,* 3105 (1983).
27. H. R. Shetty and T. W. Chou, *Metall. Trans. A, 16A,* 853 (1985).
28. E. D. Reedy, Jr., *J. Compos. Mater., 18,* 595 (1984).
29. J. M. Hedgepeth, "Stress Concentrations in Filamentary Structures," NASA TN D-882, May 1961.
30. J. G. Goree and R. S. Gross, "Analysis of a Unidirectional Composite Containing Broken Fibers and Matrix Damage," *Eng. Fract. Mech., 13,* 563 (1979).
31. G. P. Sendeckyj, and W. F. Jones, "An Improved First Order Shear Lag Theory for Unidirectional Composites with Broken Fibers," *Eng. Fract. Mech.* (in press).
32. W. F. Jones, "On the Accuracy of Higher Order Shear Lag Models," Engineering Science Preprint ESP22/85046, Society of Engineering Science, October 1985.
33. W. F. Jones, "Consistent Shear Lag Modeling of Damage in Unidirectional Composite Laminates," Final Report USAF Contract F49620-85-C-0035, November 1985.
34. J. G. Goree and L. R. Dharani, "Shear Lag Analysis of a Hybrid, Unidirectional Composite with Fiber Damage," NASA Contractor Report 3682, April 1983.
35. L. R. Dharani, "Analysis of a Hybrid, Unidirectional Laminate with Damage," Ph.D. dissertation, Clemson University, August 1982.

Metal Matrix Composites, Corrosion

Metal matrix composite (MMC) materials have developed substantially over the past 25 years. Developmental efforts have involved aluminum-, copper-, magnesium-, titanium-, and lead-based MMCs, but the primary emphasis has been on aluminum-based materials. Interest in use of these composites for structural applications is due to the higher attainable strength and stiffness properties compared with materials prepared by conventional alloying.

In the past, MMC use in actual components was limited by the high cost of the composites. However, recent advances in fabrication processes and the availability of less expensive reinforcing materials have improved the potential for lower cost MMCs [1]. Successful application of MMCs in the marine environment additionally requires adequate corrosion resistance.

This article reviews the ambient temperature–corrosion characteristics of MMCs with potential application in marine environments, with major concentration on aluminum-based composites. MMC structural characteristics, design criteria, and coatings for optimum protection also are discussed.

Structural Characteristics

Metal matrix composites basically consist of a nonmetallic reinforcement incorporated into a metallic matrix. Reinforcements, characterized as either continuous or discontinuous fibers, typically make up 20 vol % or more of the composite. Reinforcements in continuous fiber (f) composites include graphite (Gr), silicon carbide (SiC), boron (B), and alumina (Al_2O_3). Fabrication techniques for these composites vary from chemical vapor deposition coating of the fibers, liquid metal infiltration, and diffusion bonding to liquid metal infiltration and direct casting to near-net shape. Discontinuous fiber composites almost exclusively consist of SiC as whisker (w) or particulate (p). These MMCs are produced using modified powder metallurgy techniques [2]. The cross sections shown in Figure 1 exemplify typical continuous and discontinuous reinforced MMCs.

Corrosion Behavior

General Corrosion of Aluminum-Based MMCs

GRAPHITE–ALUMINUM COMPOSITES. These composites exhibit accelerated corrosion in marine environments when both graphite fibers and aluminum are simultaneously exposed. Assuming that the edges of the graphite–aluminum composite are masked off to prevent exposure of the graphite and the aluminum, only the aluminum surface foils will initially be exposed to the environment. The aluminum surface foils will pit, at an average rate of 25–36 μm/year in seawater and 0.5–0.8 μm/year in the marine atmosphere (1100, 6061, and 5000 series Al). There may also be pits present with depths much greater than the average rates reported [3]. Crevice corrosion of the aluminum foils may also occur at the edges as a result of the crevice formed between the aluminum surface foil and the masking material.

Eventually the pitting and crevice corrosion processes penetrate the foils and result in exposure of the graphite–aluminum composite matrix below, at which point the corrosion rate becomes accelerated. Pfeifer [4] reported that the graphite–aluminum corrosion proceeds preferentially along foil–foil, wire–wire, and wire–foil interfaces in the composite. Severe exfoliation occurs be-

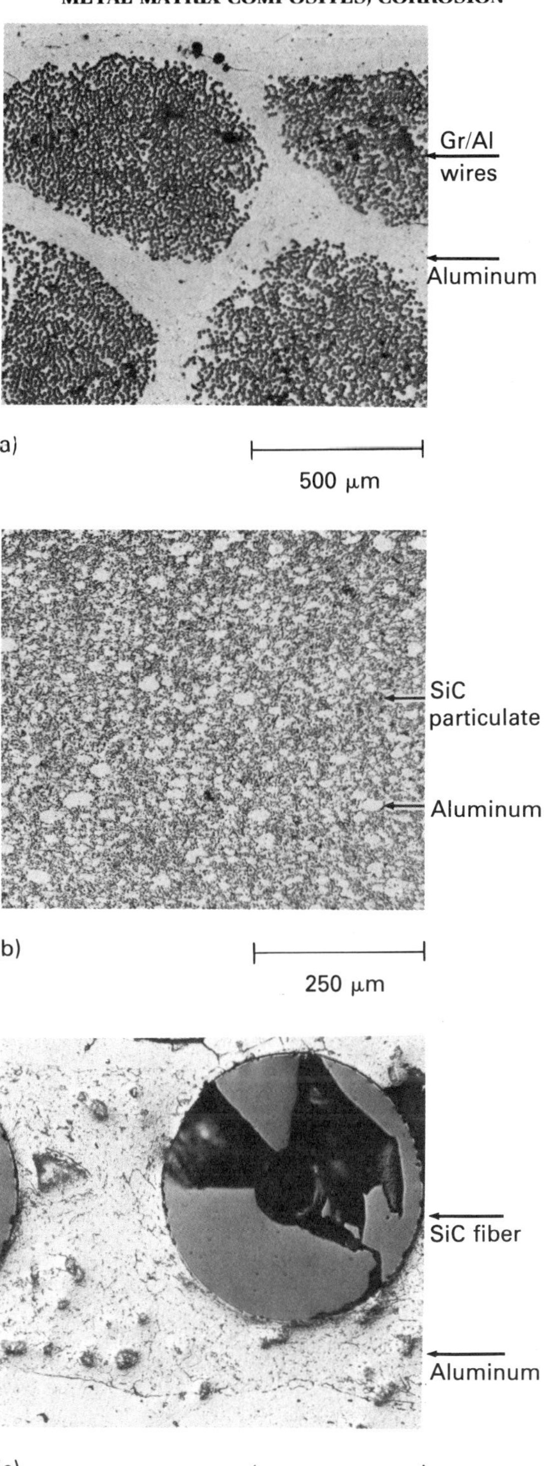

FIGURE 1 Cross sections of typical fiber-reinforced MMCs. (*a*) Continuous-fiber reinforced graphite–aluminum composite. (*b*) Discontinuous silicon carbide $_p$–aluminum composite. (*c*) Continuous-fiber silicon carbide–aluminum composite.

cause of wedging of the $Al(OH)_3$ corrosion products within the composite. The natural tendency for exfoliation of the aluminum alloy matrix will also contribute to the MMC corrosion. For instance, 6061 Al can exfoliate when intermetallic compounds (formed during fabrication) containing manganese, copper, iron, and aluminum are present [5].

Figure 2 presents an example of severe graphite–aluminum corrosion (known as catastrophic failure). This catastrophic condition can occur within 30 days in seawater after the graphite–aluminum exposure. Accelerated corrosion in the marine atmosphere and splash/spray environments is less rapid after graphite–aluminum exposure than that observed in seawater, but catastrophic failure can occur within a 6-month period [6]. The accelerated corrosion is believed to result from aluminum carbides, formed at the reinforcement–matrix interface during fabrication, altering the properties of the aluminum surface film at these localities and rendering the composite more susceptible to breakdown [4,7]. Crevice corrosion behavior of the aluminum matrix alloy may also contribute to the accelerated corrosion [5].

The aluminum surface foils alone provide reasonably good corrosion protection to the composites. Marine exposure tests of graphite–aluminum MMCs with 6061, 5056, and 1100 Al surface foils (Gr/Al edges masked) revealed no pitting penetration through the foils to expose the graphite–aluminum composite wires below through a 20-month exposure period [6]. Pitting of the foils, which

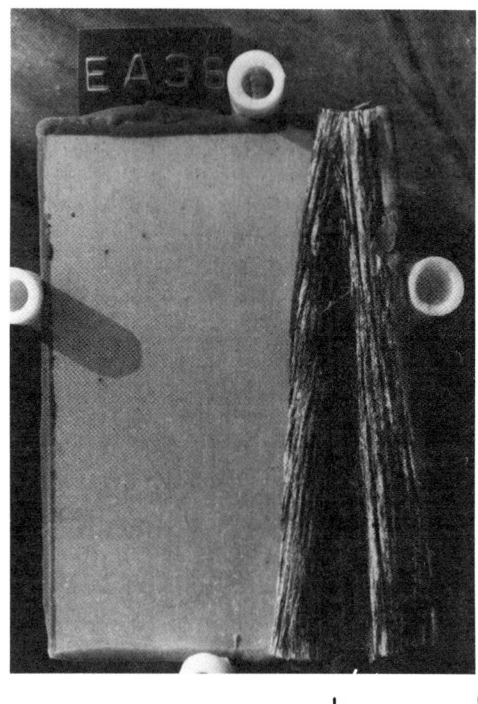

FIGURE 2 Catastrophic failure of a graphite–aluminum MMC after 6 months in a marine atmosphere.

occurred on a majority of the graphite–aluminum panels, ranked as light pitting in the splash/spray zone and marine atmosphere and localized pitting in filtered seawater. The aluminum surface foil thickness on these MMCs varied from 0.13 to 0.19 mm. Based on the average pitting rates reported above, the surface foils should protect these composites for a minimum of $3\frac{1}{2}$ years before the graphite–aluminum wires are exposed to the environment.

In summary, graphite–aluminum composites undergo severe corrosion in marine environments when the graphite and the aluminum are mutually exposed. Assuming that there is no graphite–aluminum exposure, experimental data indicate that aluminum surface foils (0.13–0.19 mm thick) will provide a minimum of 20 months' protection to the MMCs. Theoretical calculations (based on *average* rates only) suggest that the surface foils will protect the composites for a minimum of $3\frac{1}{2}$ years. This service life can be extended by increasing the thickness of the aluminum surface foils and also by applying corrosion-resistant coatings. Major emphasis should be placed on preventing exposure of the graphite and the aluminum, and frequent inspections of the composite should be made while the component part is in service.

SILICON CARBIDE–ALUMINUM. Marine corrosion of silicon carbide–aluminum composites is much less severe than that observed on graphite–aluminum MMCs. The degree of silicon carbide–aluminum corrosion resistance can be markedly affected by the MMC fabrication and processing methods. Paciej and Agarwala [8] reported poor overall corrosion resistance for SiC_w/7091-T6 Al in 3.5% NaCl because of elemental segregation, voids, inhomogeneity of reinforcement phase, unstringered PM7091 Al, partial MMC recrystallization, nonuniform plastic deformation, residual stresses, and cold-worked regions. Discontinuous silicon carbide–aluminum MMCs that are adequately processed to avoid the aforementioned problems are, however, susceptible to localized corrosion. Mild to moderate pitting has been reported on SiC whisker- and particulate-reinforced composites containing 6061 and 5000 series Al matrices, exposed for a maximum of 42 months in splash/spray and marine atmospheric environments. The degree of corrosion present on the composites is slightly greater than for unreinforced aluminum alloys (average rate of 0.5–0.8 μm/year [3]).

Silicon carbide–aluminum composites immersed in natural seawater are susceptible to significantly more severe corrosion than is typical for those used in the aforementioned environments. Silicon carbide–aluminum panels in seawater undergo pitting, both localized at the edges and distributed uniformly across the surface. The extent of pitting varies from minimal attack through 33 months' exposure to extensive corrosion as high as 249 μm/year.

Corrosion rates for silicon carbide–aluminum MMCs in seawater are also generally higher than is typical for unreinforced aluminum alloys (average rate of 25–36 μm/year [3]). This was reported by the author [6] for discontinuous silicon carbide in 6061 and 5000 series Al matrices. Contrary to these findings, Lore and Wolf [10] noted little difference in weight loss measurements between SiC/6061 Al and 6061 Al in NaCl.

Corrosion of discontinuous silicon carbide–aluminum is believed to occur at the silicon carbide–aluminum interfaces [6,9,10]. A variety of theories have been proposed to account for the concentration of corrosion at these interfaces. Paciej and Agarwala [9] contend that the corrosion concentration may result from a loss of the protective oxide film on the aluminum matrix at localized areas because of the presence of the silicon carbide or possibly from postfabrication processing. The author [6] suggests that corrosion at the interfaces is due to the crevices formed there, which are preferential sites for pitting. Evidence of the pitting concentrated at the silicon carbide–aluminum interfaces in both whisker and particulate composites is seen in Figure 3.

Electrochemical studies of discontinuous silicon carbide–aluminum MMCs containing 6061 and 5000 series aluminum matrices [7,11] have demonstrated that the presence of the silicon carbide does not increase the composite's susceptibility for pit initiation. Research on SiC/2024 Al [11] did show a more electropositive pitting potential for the composite relative to the 2024 Al; however, this difference in pitting potential may be due to the difference in microstructure between the composite matrix and the 2024 Al [2].

Continuous-fiber silicon carbide–aluminum composites also undergo localized corrosion [6]. These composites are susceptible to both crevice corrosion and pitting. Seawater ingress into the silicon carbide–aluminum composite matrix will result in crevice corrosion at the SiC fiber–Al matrix interfaces, which accelerates the corrosion rate and eventually results in delamination of the aluminum surface foils. However, the rate of silicon carbide–aluminum corrosion is much less severe than is typical for graphite–aluminum. Figure 4 contrasts the extent of corrosion evident on the silicon carbide–aluminum panels described above.

In summary, silicon carbide–aluminum MMCs are susceptible to localized corrosion in marine environments. Generally, the susceptibility for pit initiation is similar between the composites and unreinforced alloys; however, the rate of pit propagation is higher for the composites. Silicon carbide–aluminum corrosion in seawater is more severe than in the other marine environments. Corrosion-resistant coatings are recommended on these composites to enhance their service life.

BORON–ALUMINUM. A complete review of corrosion properties of boron–aluminum composites can be found in Ref. 2. This section synopsizes significant findings from that review.

Boron–aluminum MMCs experience severe corrosion in chloride environments and are significantly less corrosion resistant than unreinforced aluminum alloys. The concentration of corrosion in these composites has been found both at fiber–matrix interfaces and at the bonds between foils [12,13]. The accelerated corrosion at these sites has been attributed to imperfect bonding and fissures in the composite and emphasizes the need for eliminating fabrication flaws to reduce boron–aluminum corrosion in chloride environments [12]. Corrosion at the fiber–matrix interfaces has also been attributed to the presence of aluminum boride, formed during fabrication [13].

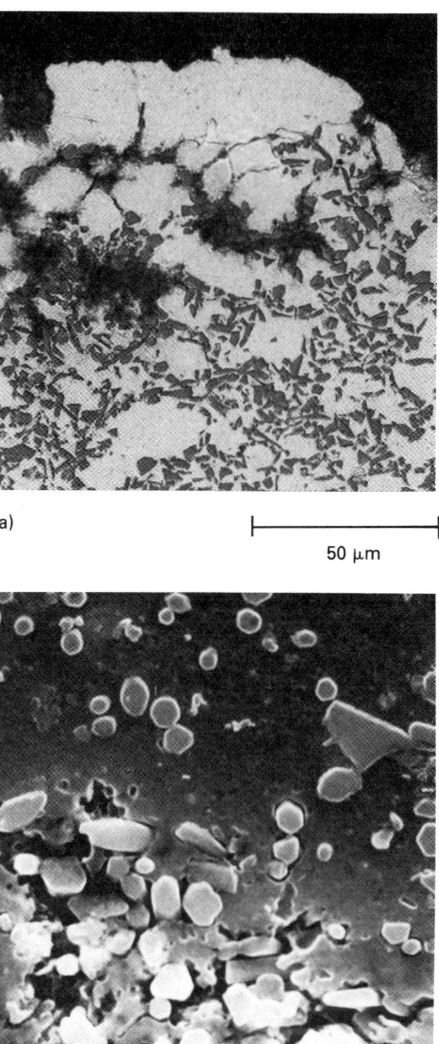

FIGURE 3 **Cross sections of discontinuous silicon carbide–aluminum MMC panels. (*a*) Silicon carbide $_p$/6061 Al MMC after a 230-day, tidal-immersion exposure. (*b*) Silicon carbide $_w$/6061 Al MMC after a 60-day filtered-seawater exposure.**

ALUMINA–ALUMINUM. A review of alumina–aluminum corrosion properties is contained in Ref. 2, and the significant findings are summarized here.

To obtain good wettability and bonding in the alumina–aluminum MMC, the aluminum matrix is alloyed to form a bonding compound between the fiber and the

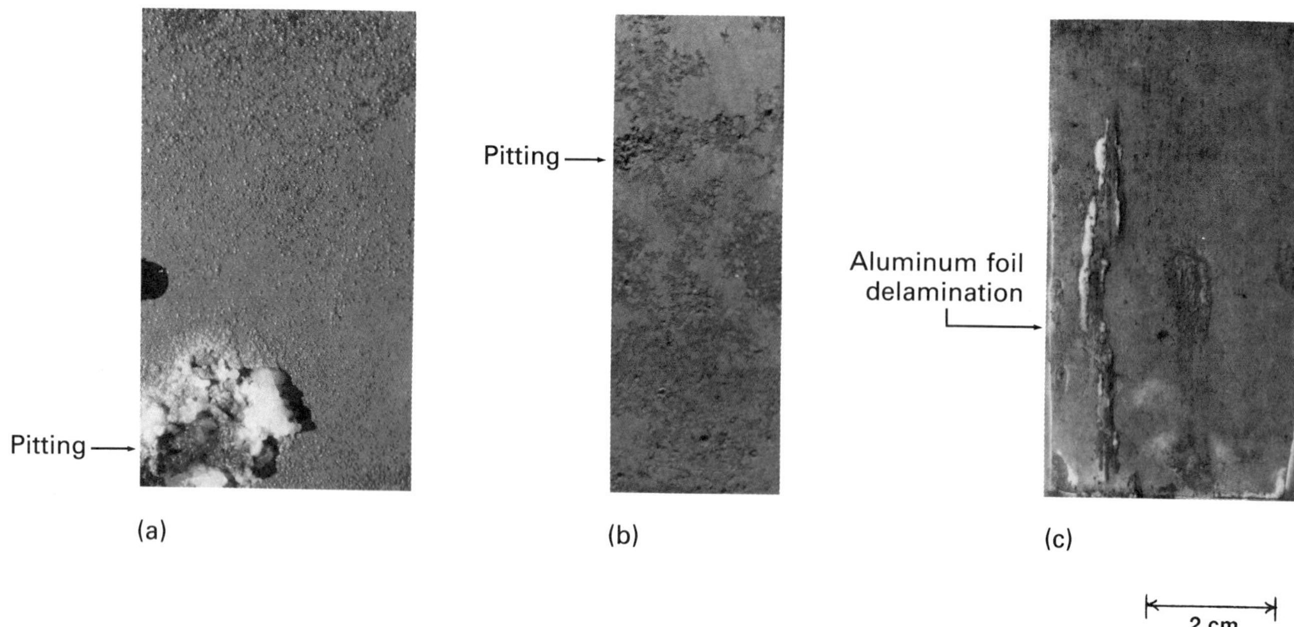

FIGURE 4 **Silicon carbide–aluminum MMC panels after exposure to filtered seawater. (*a*) Silicon carbide $_w$/6061 Al after a 4-month exposure. (*b*) Silicon carbide $_p$/6061 Al after a 24-month exposure. (*c*) Silicon carbide $_f$/6061 Al after a 33-month exposure.**

matrix. Corrosion studies of alumina–aluminum–2Li MMCs (containing a $Li_2O–5Al_2O_3$ bond layer) in NaCl solutions identified minimal attack at the fiber–matrix interfaces. The corrosion rate of the MMCs (based on weight loss measurements) was only slightly higher than 6061-T6 Al [14].

Corrosion evaluations of alumina–aluminum–2Mg MMCs in NaCl–H_2O_2 [15] identified pitting at the fiber–matrix interfaces, presumably because of the Mg_5Al_8 precipitates present. Research on Al_2O_3/6061 Al MMCs [2] also reported preferential corrosion at the fiber–matrix interface. These findings suggest that the corrosion resistance of alumina–aluminum composites is highly dependent on the bonding compound formed at the fiber–matrix interface. To date, no severe corrosion problems have been identified with alumina–aluminum–lithium composites.

Stress Corrosion Cracking of Aluminum-Based MMCs

Stress corrosion cracking (SCC) properties of graphite–aluminum were researched by Davis et al. [16] and Phillips [17]. Davis reported an initial stress-dependent corrosion mechanism for graphite–aluminum, which then shifted to a corrosion-dominated, stress-assisted, time-dependent failure as the exposure period in seawater increased. Phillips also noted a corrosion-dominated mechanism at longer exposure times but suggested that the failures were creep-related as well.

SCC testing of boron–aluminum MMCs [17] at lower stress intensities ($<80\%$ overload fracture toughness) identified no failures within the 1000-hour test limit. Delayed time failures were reported at high stress intensities for boron–aluminum MMCs evaluated in both air and seawater. It was suggested that the failures resulted from room temperature creep.

Corrosion Fatigue of Aluminum Based MMCs

The seawater and air fatigue properties of Gr/6061 Al MMCs are superior to 6061-T6 Al; however, both composites and unreinforced aluminum alloys exhibit a degradation in seawater fatigue properties as compared with the corresponding air fatigue properties [16]. Discontinuous SiC/6061 Al MMCs also retain improved fatigue properties over unreinforced 6061 Al in chloride environments (Fig. 5) [18]. It has been suggested [19] that the improved corrosion fatigue properties of silicon carbide–aluminum MMCs are due to an increased resistance to crack initiation.

Copper-Based MMCs

To date, very limited corrosion research has been conducted on copper-based MMCs. These composites offer substantial potential in marine applications because of their antifouling characteristics as well as their potentially higher strength and stiffness and lower density than that which can be attained in conventional copper alloys.

A corrosion evaluation of copper-based MMCs was performed by the author. The composites evaluated are some of the first copper MMCs ever produced and are

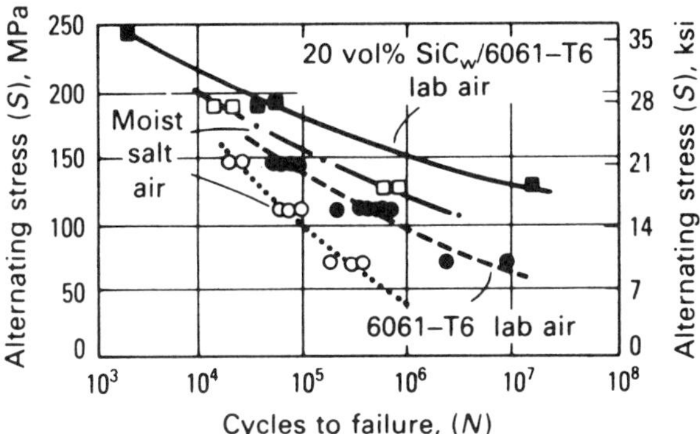

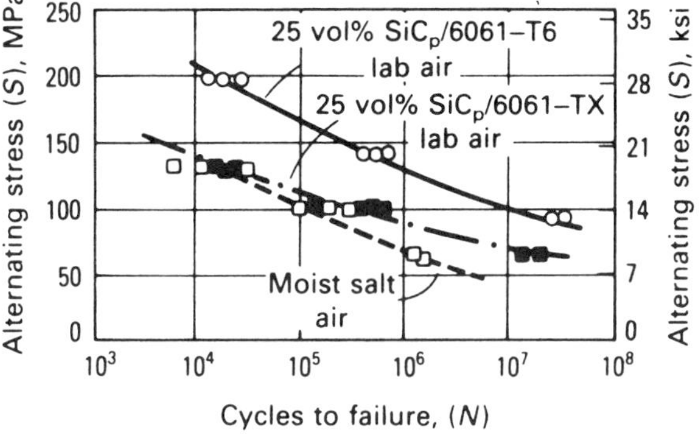

FIGURE 5 **Corrosion fatigue of discontinuous silicon carbide–aluminum in moist salt air.** (From Ref. 18.)

not optimized materials. Many of them contain porosity, inconsistent sizes of reinforcement, and an uneven reinforcement distribution. The specific composites evaluated included continuous fiber Gr/Cu, SiC/Cu-Sn, and SiC/Cu-Ti as well as discontinuous TiC_p/Cu, TiC_p/Cu-Al, Si_3N_{4w}/Cu, B_4C/Cu, SiC_w/Cu, and Al_2O_3 (chopped fiber)/Cu-Ti.

Overall, copper MMCs displayed minimal evidence of corrosive attack. Filtered seawater specimens developed a green corrosion product film on their surfaces, presumably copper hydroxide chloride, $Cu_2(OH)_3Cl$ [20]. Localized corrosion was commonly present on edges, on surfaces, and adjacent to crevice sites. Localized corrosion outside the crevice area is typical for copper alloys [3]. Localized corrosion present on the surfaces and edges was possibly due to porosity in the as-fabricated composites, which provide continuing corrosion paths during seawater exposure. Figure 6 shows the typical corrosion found on copper-based MMCs and controls in filtered seawater.

In the marine atmosphere, copper-based MMCs developed a thin, green corrosion product film on their surfaces. This film was much thinner than the film formed in seawater, and it developed more slowly. Graphite–copper MMCs showed evidence of surface foil blistering within 4 months' exposure in this environment; however, this degradation may have been due to imperfect bonding of the composite.

In general, copper-based MMCs display a marine corrosion resistance that is similar to that of copper alloys. Porosity in some of the composites could have contributed to increased corrosion. Further development efforts are required to improve the overall structure and the fiber–matrix bonding in these materials.

Magnesium-Based MMCs

Very limited research has been conducted on graphite–magnesium composites, owing to the known extreme

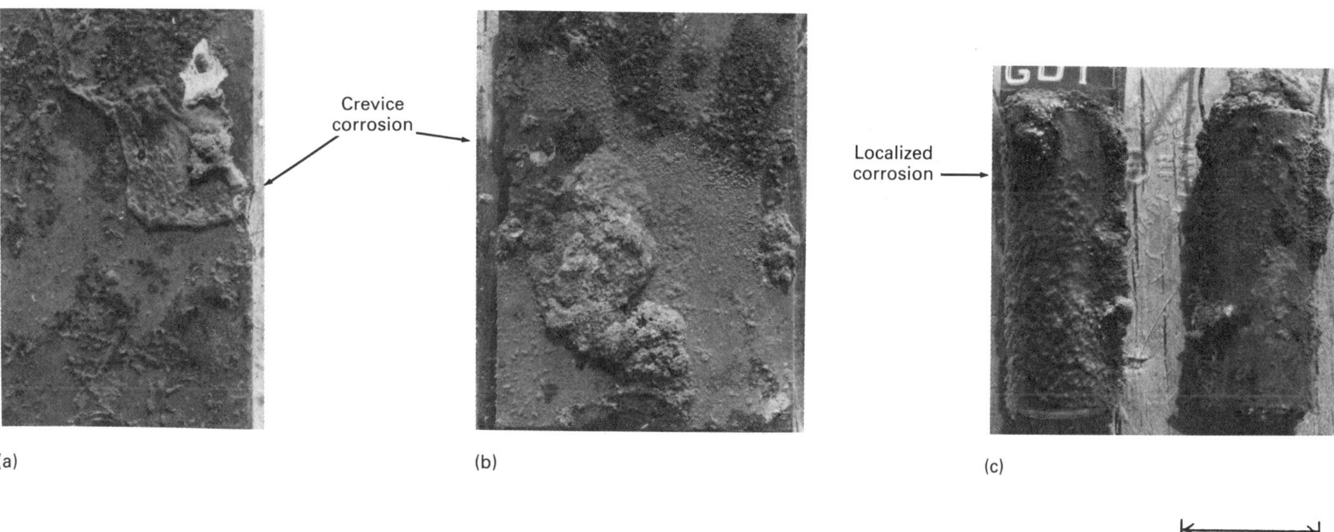

FIGURE 6 **Copper-based MMC and control specimens after exposure to filtered seawater. (*a*) Unreinforced 99.95% Cu after a 24-month exposure. (*b*) Titanium carbide $_p$/copper MMC after a 21-month exposure. (*c*) Silicon nitride $_w$/copper MMC after a 21-month exposure.**

reactivity of magnesium. Graphite–magnesium MMCs are susceptible to severe corrosion in chloride-containing solutions. Research by Trzaskoma [21] on graphite–magnesium with AZ91C wires and AZ31B foils identified severe corrosion after only 5 days' immersion. Corrosion was especially severe in areas where both the graphite fibers and the magnesium matrix were exposed, attributable to galvanic corrosion processes.

As the volume percentage of graphite is increased, the MMC corrosion rate also increases. The corrosion rate of composites containing only 10 vol % graphite (1.8 mg/cm^2 day) is still unacceptable for component applications. Thus to effectively use graphite–magnesium MMCs in chloride environments, fiber exposure to the environment must be eliminated and the corrosion resistance of the magnesium matrix must be improved. Ferrando [22] suggests the selection of low impurity magnesium alloys and the use of a fiber surface treatment to improve corrosion resistance.

Coatings for Corrosion Control

Coatings recommended in this section are applicable to aluminum-based MMCs. To date, no research has been published on corrosion protective coatings for copper- or magnesium-based MMCs. Figures 7 and 8 illustrate the corrosion evident on select coated continuous and discontinuous fiber MMCs.

Continuous Fiber Composites

A variety of coatings have been evaluated for protecting continuous fiber graphite–aluminum [4,6]. Organic coatings were identified as providing excellent graphite–aluminum corrosion protection. In service, composites protected with an organic coating must be frequently inspected, since this coating provides only barrier protection, and any coating holidays are potential sites for corrosion attack of the composite.

Noble metal coatings such as nickel and titanium, applied by chemical vapor deposition, physical vapor deposition, and electroplating methods, also provide barrier protection; however, graphite–aluminum corrosion at coating flaw sites is more accelerated for these coatings than for an organic coating. This is due to the highly unfavorable anodic (Al)/cathodic (noble coating) area ratio formed at a flaw site. Consequently, noble metal coatings are not recommended for protecting aluminum-based MMCs.

Sulfuric acid anodizing has good marine corrosion resistance for graphite–aluminum MMCs. Earlier studies

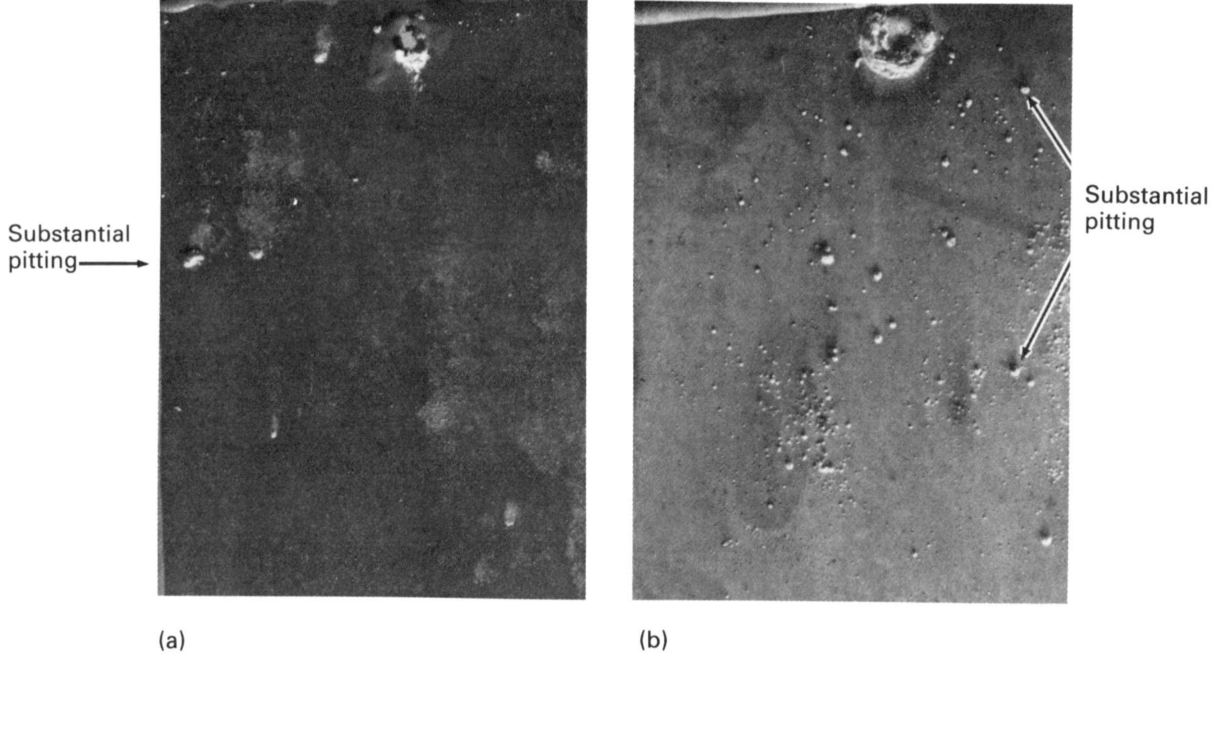

FIGURE 7 Coated continuous-fiber graphite–aluminum MMCs after 1 month in filtered seawater. (*a*) Chromate–phosphate conversion coating. (*b*) Electrodeposited aluminum–manganese coating.

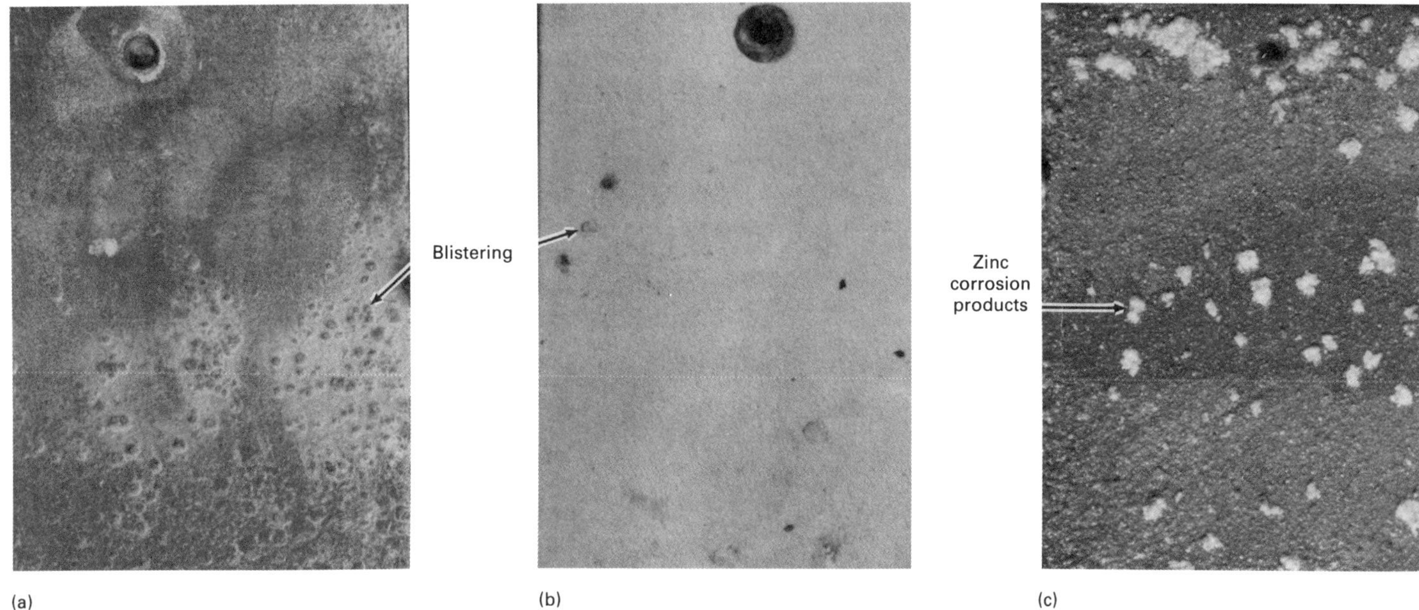

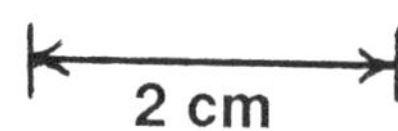

FIGURE 8 **Coated discontinuous silicon carbide $_p$/aluminum MMCs after seawater exposure. (*a*) Coated with ion vapor-deposited aluminum; 4-month exposure. (*b*) Coated with plasma-sprayed aluminum oxide; 18-month exposure. (*c*) Coated with arc-sprayed zinc; 9-month exposure.**

in filtered seawater [23] showed that the anodized layer thins as a function of exposure time; however, current results indicate that anodizing will provide at least 28 months' protection for graphite–aluminum. Chromate phosphate conversion and electrodeposited aluminum–manganese coatings on graphite–aluminum are less corrosion resistant than sulfuric acid anodizing. Graphite/aluminum MMCs with these coatings are susceptible to pitting and/or blistering within the first 3 months of marine exposure.

Aluminum arc spray coatings exhibit excellent corrosion resistance in marine environments, but this coating is not recommended for use on conventional graphite–aluminum MMC. The reduced thickness of the surface foils on these composites [0.13–0.19 mm thick] prohibits grit blast surface preparation without severe warpage occurring. Thermal spraying could be used for protecting graphite–aluminum only if the surface foils were thick enough [0.51 mm or greater] to prevent warpage.

In summary, sulfuric acid anodizing [0.03 mm thick], organic coatings [0.13 mm thick], or a combination of both are recommended for corrosion protection of continuous-fiber graphite–aluminum. These coatings will provide a minimum of 28–33 months' protection in the marine environment, assuming that there are no coating defects to expose the graphite–aluminum MMC. These coatings should also provide a similar degree of protection for continuous-fiber silicon carbide–aluminum, boron–aluminum, and alumina–aluminum.

Discontinuous Fiber Composites

Both metallic and ceramic coatings have been investigated for protecting discontinuous silicon carbide–aluminum MMCs [6]. Aluminum flame- and arc-sprayed coatings exhibit excellent silicon carbide–aluminum corrosion resistance for a minimum of 33 months. The corrosion resistance of zinc arc-sprayed coatings is below that of the aluminum thermal spray coatings. Zinc coatings protect the silicon carbide–aluminum substrate by zinc dissolution. Although zinc coatings can provide more effective cathodic protection than aluminum, their mechanism of protection restricts their useful life. Zinc-coated composites immersed in seawater exhibited pitting of the silicon carbide–aluminum substrate between 4 and 18 months' exposure, indicative of a loss of protection by the zinc coating.

Aluminum oxide plasma-sprayed coatings applied to silicon carbide–aluminum MMCs exhibited varied results in the marine environment. The performance variations were attributable to differences in the quality of coating application rather than to poor corrosion resistance of the coating. Aluminum oxide coating performance ranged from excellent condition after 33 months' exposure to coating blistering extensive enough to expose the substrate after 4 months.

Aluminum oxide coatings will provide only barrier protection to the silicon carbide–aluminum composites. Aluminum thermal-sprayed coatings, on the other hand, do have some sacrificial protection capabilities. The aluminum coatings will protect exposed silicon carbide–aluminum areas to a greater extent than a barrier coating.

Ion-vapor-deposited aluminum provides good corrosion protection to silicon carbide–aluminum MMCs in the marine atmosphere. This same coating tends to blister within 1 month of exposure in filtered seawater. The blistering is assumed to be occurring at silicon carbide particle sites, since these particles project from the surface and prevent a uniform surface for coating.

Aluminum, applied by flame or arc spraying, is recommended for silicon carbide–aluminum protection in the marine environment. A coating thickness of 0.13–0.20 mm is optimum. Organic topcoats may also be applied for added protection. Proper thermal spray and organic coating surface preparation and application procedures must be followed to avert premature failure in service.

Design for Corrosion Prevention

For long-term utilization of metal matrix composite components in service, effective coating protection must be employed. Consequently, MMC design should take into consideration ease of initial coating application as well as coating maintenance [24]. A simple component design is optimum for ensuring effective coating application. The more complicated the design, the more difficult it is to obtain an adherent, uniform coating. Difficult areas to coat, such as sharp edges and corners, overlaps, rivets, fasteners, and welds, should be eliminated if possible during design. Also, recesses or low spots should be avoided, since these areas will collect water and contribute to a lower coating corrosion resistance. Where a joint is required in the design, keep in mind that successful coating application is more easily accomplished in welds than in rivets or fasteners. For maintenance considerations, it is imperative that all areas to be coated be readily accessible.

Denise M. Aylor

[*See also* Polymer Composites, Corrosion.]

References

1. S. G. Fishman, *ASTM Standardization News* (1986).
2. M. Metzger and S. G. Fishman, *Ind. Eng. Chem. Prod. Res. Dev., 22,* 296 (1983).
3. W. K. Boyd and F. W. Fink, *Corrosion of Metals in Marine Environments,* Metals and Ceramics Information Center, Columbus, OH, 1978, pp. 44, 57–67, 85–87.
4. W. H. Pfeifer, in *Hybrid and Select Metal Matrix Composites: A State of the Art Review,* American Institute of Aeronautics and Astronautics, New York, 1977, pp. 231–252.
5. C. R. Crowe, "Localized Corrosion Currents from Gr/Al and Welded SiC/Al Metal Matrix Composites," NRL Report 5415, February 1985.
6. D. M. Aylor and P. J. Moran, NACE Corrosion '86, Preprint 202 (1986).
7. D. M. Aylor and P. J. Moran, *J. Electrochem. Soc., 132,* 1277 (1985).
8. R. C. Paciej and V. S. Agarwala, "Metallurgical Variables Influencing the Corrosion Susceptibility of a P/M Al/SiC_w Composite," Report NADFC-86071-60, October 1985.
9. R. C. Paciej and V. S. Agarwala, NACE Corrosion '87, Preprint 221 (1987).
10. K. D. Lore and J. S. Wolf, The Electrochemical Society, Abstract 154 (1981).
11. P. P. Trzaskoma, E. M. McCafferty, and C. R. Crowe, *J. Electrochem. Soc., 130,* 1804 (1983).
12. A. J. Sedriks, J. A. S. Green, and D. L. Novak, *Met. Trans., 2,* 871 (1971).
13. S. L. Pohlman, *Corrosion, 34,* 156 (1978).
14. A. R. Champion, W. H. Krueger, H. S. Hartmann, and A. K. Dhingra, in B. Noton et al., Eds., *Proceedings of the Second International Conference on Composite Materials,* The Metallurgical Society of AIME, 1978, p. 883.
15. J. Y. Yang and M. Metzger, The Electrochemical Society, Abstract 155 (1981).
16. D. A. Davis, M. G. Vassilaros, and J. P. Gudas, *Mater Performance, 21,* 38 (1982).
17. W. L. Phillips, "Sharp Notch SCC of B/Al and Gr/Al Composites," NRL Report 3616, October 1977.
18. C. R. Crowe and D. F. Hasson, Strength of Metals and Alloys, *ICSMA, 6*(2), 859 (1982).
19. S.-S. Yau, Ph.D dissertation, North Carolina State University, Raleigh, 1983.
20. J. M. Popplewell, NACE Corrosion '78, Preprint 21 (1978).
21. P. P. Trzaskoma, "The Corrosion Behavior of a Graphite Fiber/Magnesium Metal Matrix Composite in Aqueous Chloride Solution," NRL Report 5640, September 1985.
22. W. A. Ferrando, "Corrosion Resistant Mg-Based Materials for Naval Applications," NSWC TR 85-88, April 1985.
23. D. M. Aylor and R. M. Kain, *Mater. Performance, 23,* 32 (1984).
24. C. G. Munger, *Corrosion Prevention by Protective Coatings,* National Association of Corrosion Engineers, Houston, 1984, pp. 173–191.

Metal Matrix Composites, Design Methodology

Metal matrix composites (MMCs) have distinct, and in some cases overwhelming, advantages in space and aircraft applications. However, their use is harder to justify in more conventional and earthbound applications. The technology of MMCs is still relatively immature compared with that of reinforced polymer composites, so that many problems of using MMCs with or instead of conventional materials remain to be solved. In commercial applications, the motivation for using composites, particularly the reinforced polymer types, focuses on longevity and low fabrication cost instead of improved strength and stiffness, and weight reduction. To some extent manufacturing costs can be minimized by integrating structural elements and components, thereby reducing labor costs in assembly. Economically speaking, MMCs are very expensive relative to conventional materials, and on cost alone they do not appear to be competitive with unreinforced steel and aluminum. Numerous cost trend projections have been made over the years, and the main feature of these projections is a downward trend in cost based on rising production. But these trends in cost reduction have been slower to be realized than was anticipated some years ago, making it difficult to foresee how cost can decline to a level slightly above the cost of composite constituents, even when these increase with inflation. There has been much uncertainty in predicting the production volume because most MMCs are attractive in spacecraft and aircrafts, and uses of MMCs in these applications are not always clearly perceived, and because MMCs face some competition from newer materials. However, a word of caution is in order: cost comparisons of raw materials can be misleading. It is the cost of the function performed by a given component or subsystem that must be considered. Such a value analysis is quite complicated and must take into account the service life of the system, as well as many other considerations [1,2].

In some cases, MMCs offer superior performance, for example, in high temperature applications such as diesel engine pistons and connecting rods. In structures where weight, stress density, volume, life cycle costs, wear resistance, and thermal stability are overwhelming considerations, MMCs are finding numerous applications. Part of the difficulty in broadening the applications of MMCs is related to the immaturity of the technology and insufficient reliable design data, particularly for safety-critical applications. Consequently, the decision to utilize MMCs must be based not on cost considerations alone, but on technical and operational grounds. Therefore, to help designers make such decisions and assist them in their choice of materials, various design methodologies based on structural design optimization and efficiency analysis are presented in this chapter.

General Considerations in Design

In structural design there are well-established methods that have proven their utility in a wide variety of structures. Any structural design begins by determining the design requirements, which include the functional requirements as well as the economic and performance objectives. Functionality has been defined [3] as a measure of performance capabilities versus cost, weight, and geometric envelope. Therefore, to meet these requirements, the designer must choose materials, configurations, and manufacturing processes, and, unlike for conventional materials, composite structures require that the material be designed along with the structure [4]. A key point is that improvements in the design are achieved by design cycle iterations, which can be done with the use of structural indices [5] or structural efficiency analysis [4] during the terminal design phase. During this phase, improvements are also made by changing composite constituent properties and laminate configuration [4].

When considering the use of MMCs in structural applications, it is not meaningful to consider each new structural concept through every stage of analytical detail and reiteration of analysis. Such an approach is extremely tedious and may hide the implications of changes in strength, stiffness, density, wear, and other such factors upon the performance of the structure. Moreover, it would lead to the development of methodologies that become dependent on the design of the structure. Consequently, one seeks general design methods because of their (hopefully) wide applicability, and design optimization methods to simplify and reduce the number and complexity of analytical iterations that must be accomplished before the design satisfactorily meets the design requirements.

In considering structures in this manner, it is necessary to consider first the operational and cost criteria and the technical criteria that affect the design process. These criteria are:

- Operational and cost criteria
  - Weight
  - Indirect weight savings
  - Cost
  - Maintainability
  - Deployability
  - Machinability
  - Repair
  - Performance
  - Reliability
  - Life cycle cost
  - Simplicity
  - Inspection
  - Safety
- Technical criteria
  - Strength
  - Stiffness
  - Stress density
  - Critical crack length
  - Work of fracture
  - Corrosion resistance
  - Geometrical effects and structural stability
  - Weight
  - Wear
  - Repairability

These criteria are not given in order of their importance. There are interrelationships among operational and cost criteria and technical criteria. For example, in some structural elements or components the need for high stiffness may become a determining factor that overrides cost. On the other hand, the problems of routine inspection for defects, cracks, and other incipient failures may preclude access to some areas of the structure. This would necessitate critical crack lengths that are rather large for ease of visual inspection. This implies a high work of fracture, which is not necessarily obtainable with some MMCs. For example, the critical crack length for mild steel based on the Griffith theory is about 0.5 to 1.0 m, while for SiC whisker reinforced aluminum the same theory gives the critical crack length as 10 mm. The need for visual inspection may soon be replaced by more sophisticated techniques, such as nondestructive testing and inspection. Fracture mechanics for MMCs is not yet well developed, which means that high risks in cost and in safety would be involved in ignoring this aspect of material science and engineering where it is critical. These tradeoffs must be considered and evaluated during concept evaluation and the initial design analysis [5].

In this list of technical criteria there is a criterion called stress density. Stress density is defined in the following manner: quite often structural designs require forces and loads to be transmitted from one component to another through some confined space, through a joint or attachment, for example, that must occupy a small volume as a result of equipment or structural constraints. This is shown in Figure 1. Thus, a high stress density is the applied force divided by the material cross section of the force transmitter in a small volume. Obviously, from the design standpoint it may not be feasible to increase the thickness of the force transmitter material because of volume constraints. Consequently, high strength (or high stiffness) materials must be employed to meet the required stress density and hence meet the design requirements. Since stress density is a design variable, there seems to be no general way to define it as a material property.

In conventional design processes, what has been called a linear approach [3] is generally used. This involves the sequence of customer, marketing, analysis and design, manufacturing and quality control, and material suppliers, and is therefore a queued and not a simultaneous process. In composite design, such a linear approach would be ineffective because the component configurations, material specifications, and manufacturing methods must be developed simultaneously. This leads to an interdisciplinary approach [3].

General design considerations consist of analyzing qualitatively elements and substructures or components of a design to uncover potential technical and operational weaknesses resulting from the use of conventional materials, then deciding whether or not these potential deficiencies can be overcome by the use of MMCs.

A structure can be conceptually disassembled into its constituent elements so that high stresses, wear, dynamic loads, and other critical factors may be identified [5]. Figure 2 shows an exploded view of the hypothetical system shown in Figure 1. Each constituent element is

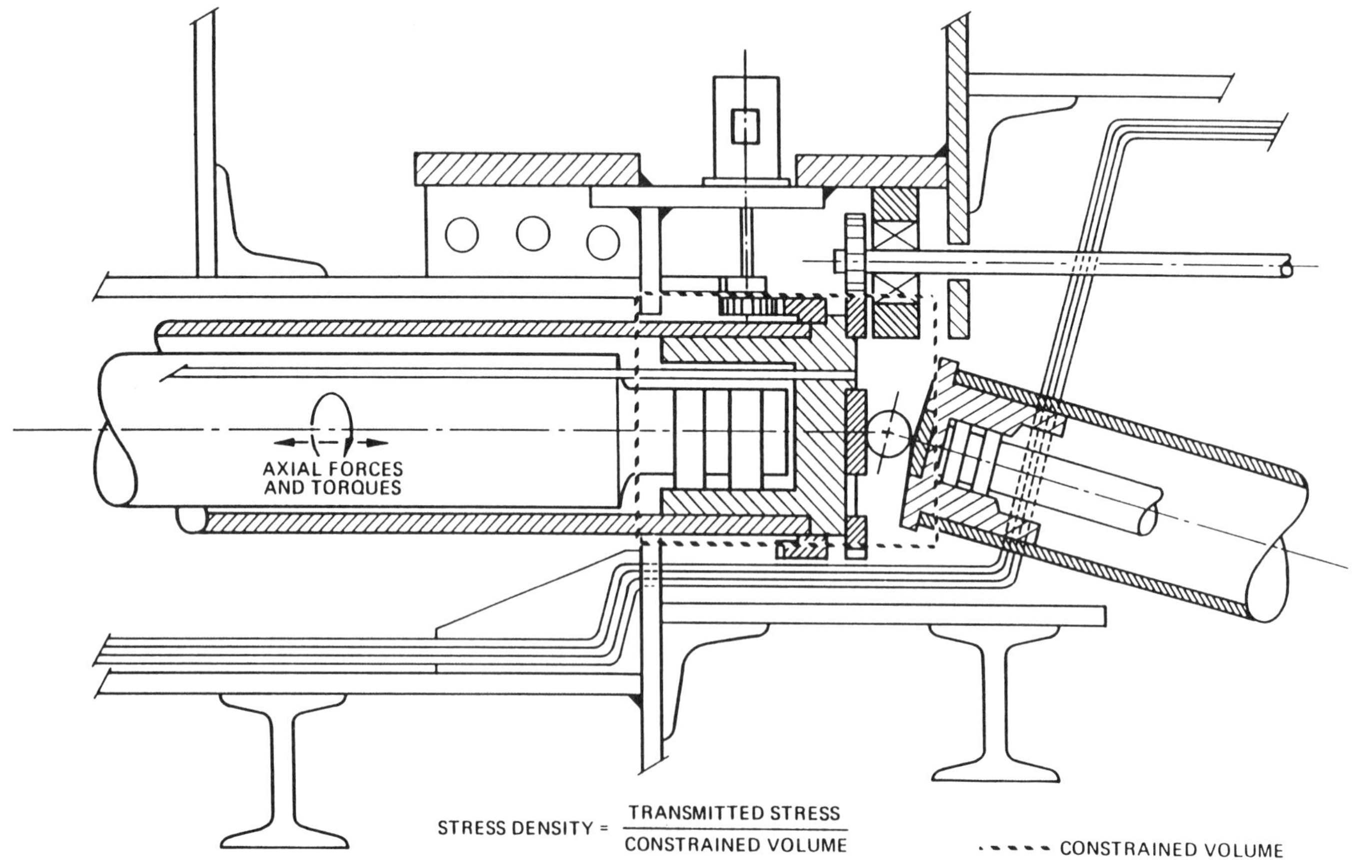

FIGURE 1 **Conceptual system used to define stress density.** (From Ref. 5.)

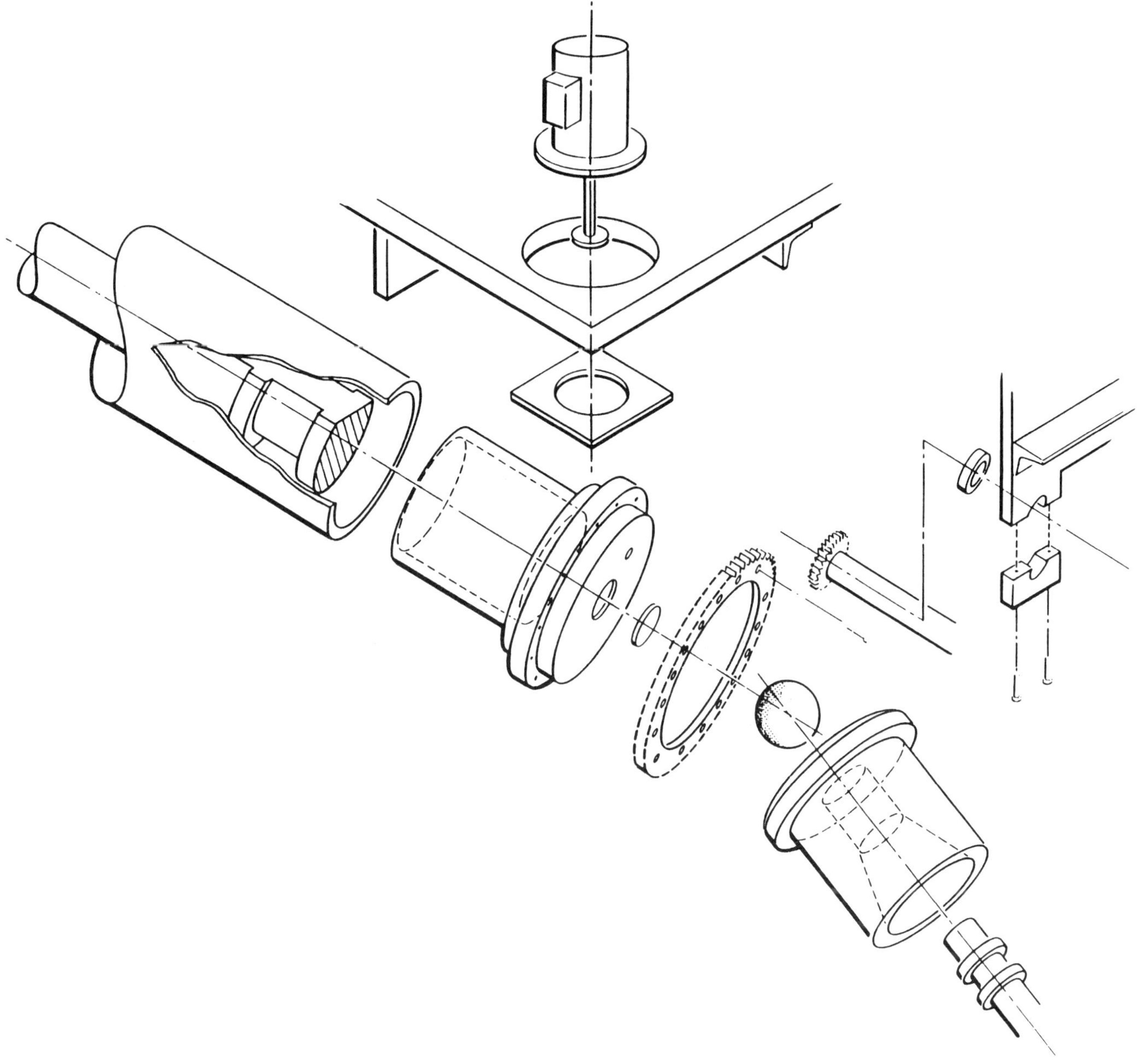

FIGURE 2 Disassembly of system shown in Figure 1. (From Ref. 5.)

shown, and free-body diagrams are prepared for each, as shown, for example, in Figure 3. These are the standard steps of stress analysis, except that all information such as allowable maximum dimensions (constraint) and limits on strain and stress rates is specified, and time-dependent forces are indicated by superposition so that critical dynamic effects may be seen. These diagrams then allow the designer to develop a complete physical and intuitive picture of the operation of the system. Members loaded in torsion or in compression need to be examined for possible loading limits set by buckling constraints of the material and design geometry. If cyclic loads exist, effects of long-term fatigue must be examined. MMCs with better high temperature strength characteristics than conventional unreinforced metals could perhaps be useful here. In the case of cyclic loads, dynamic buckling analysis may be required. If, in addition, such structural elements are impulse loaded, conventional static buckling analysis is not adequate, since most materials exhibit stress–strain behavior that is dependent on strain rates [6,7].

To assist in isolating regions of a system, subsystem, or component to determine their design and operational adequacies, a number of questions such as those shown in Table 1 should be considered. These are by no means all the possible questions that may arise during a detailed analysis of a design problem. The purpose of the enumeration given in Table 1 is to focus attention on the volume of the element or component under consideration, and reveal any requirements for thermal or other

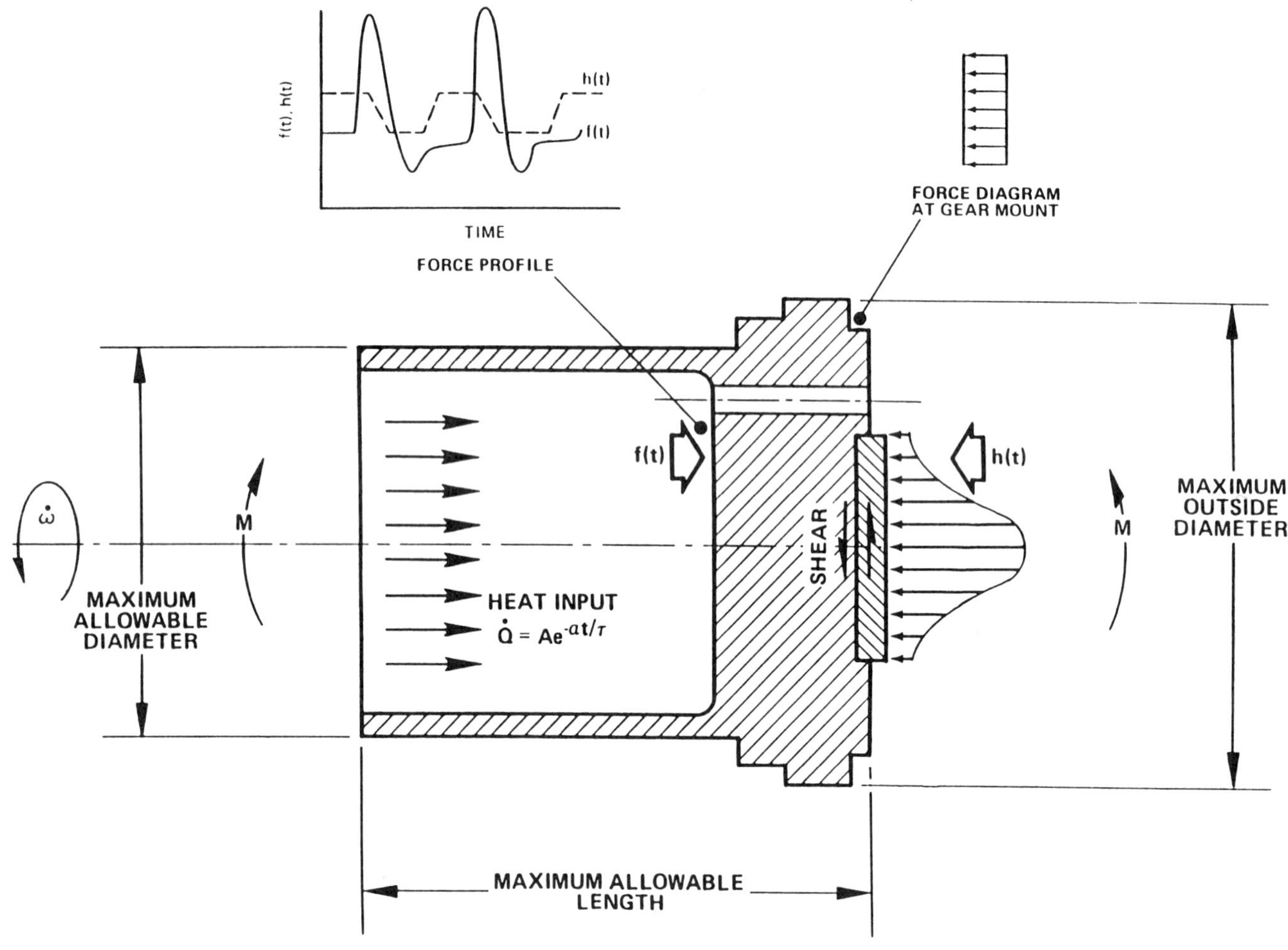

FIGURE 3 Complete description of forces, moments, dynamic loads, heat inputs, etc., on a typical component of Figure 1. (From Ref. 5.)

TABLE 1
Questions That Relate to Structural Design Adequacy

1. Are there regions, holes, bolts, weldments, webs, and so on that are more highly stressed than adjacent regions?
2. How large are the stress concentration factors?
3. Are the applied stresses anywhere likely to be larger than the material strength? (Ordinary metals tend to be strain-rate dependent in their strength; that is, an impulse-loaded steel has somewhat greater yield strength than for equal static loading.)[3]
4. Are wear and corrosion likely to result in short life? Is this lifetime lower than the required operational lifetime?
5. Are bending moments resulting in flexural rigidity values *EI* greater than the material can withstand? Would increasing *E* (or *I*) result in a safe structure? Longer life and smaller volume? Lower weight?
6. Are fatigue and toughness requirements met?
7. Are there serious impulsive loads that are likely to cause brittle fracture?
8. Are safe crack lengths detectable in routine inspection? What is an acceptable crack length?
9. Are the maintenance and repair needed between operation easily and readily performed? Are they costly?
10. Can stressed regions be increased in dimensions rather than solving the problem with high-strength, high-stiffness advanced materials?

From Ref. 5.

treatments or the need for nonconventional materials. Advanced composites may also be used to advantage with hybrid composites in which some components are cellular solids [8–11]. Examination of design considerations, (Figures 1–3) shows that all loads on a structure can be reduced to the following types:

- Tension
- Compression
- Flexure
- Torsion
- Shear

with the following failure states:

- Fracture
- Buckling
- Wear
- Corrosion
- Plastic yielding

The process of analysis is summarized in Figure 4, which shows that the cycle begins with the definition of operational and cost requirements, leading to structural performance requirements. This, in turn, leads to structural design and, through various interation cycles between analysis and material modifications, to an acceptable structure. The method just described assumes that the use of advanced composite materials is decided upon for each design exercise as these exercises progress. Thus, in its broadest sense, structural analysis indicates load magnitudes for which conventional materials are ill-suited so that new materials are indicated.

Metal Matrix Composite Equations

In this section, the equations currently used to calculate the properties of MMCs are summarized. The longitudinal elastic modulus is given to a good approximation by the rule of mixture, or:

$$E_L = E_f V_f + (1 - V_f) E_m \tag{1}$$

where E_f and E_m are the fiber and matrix elastic moduli, respectively, and V_f is the fiber volume fraction. The additional terms not shown depend on the difference in the Poisson's ratio of the matrix and fiber, usually a negligible term. The transverse elastic modulus is given approx-

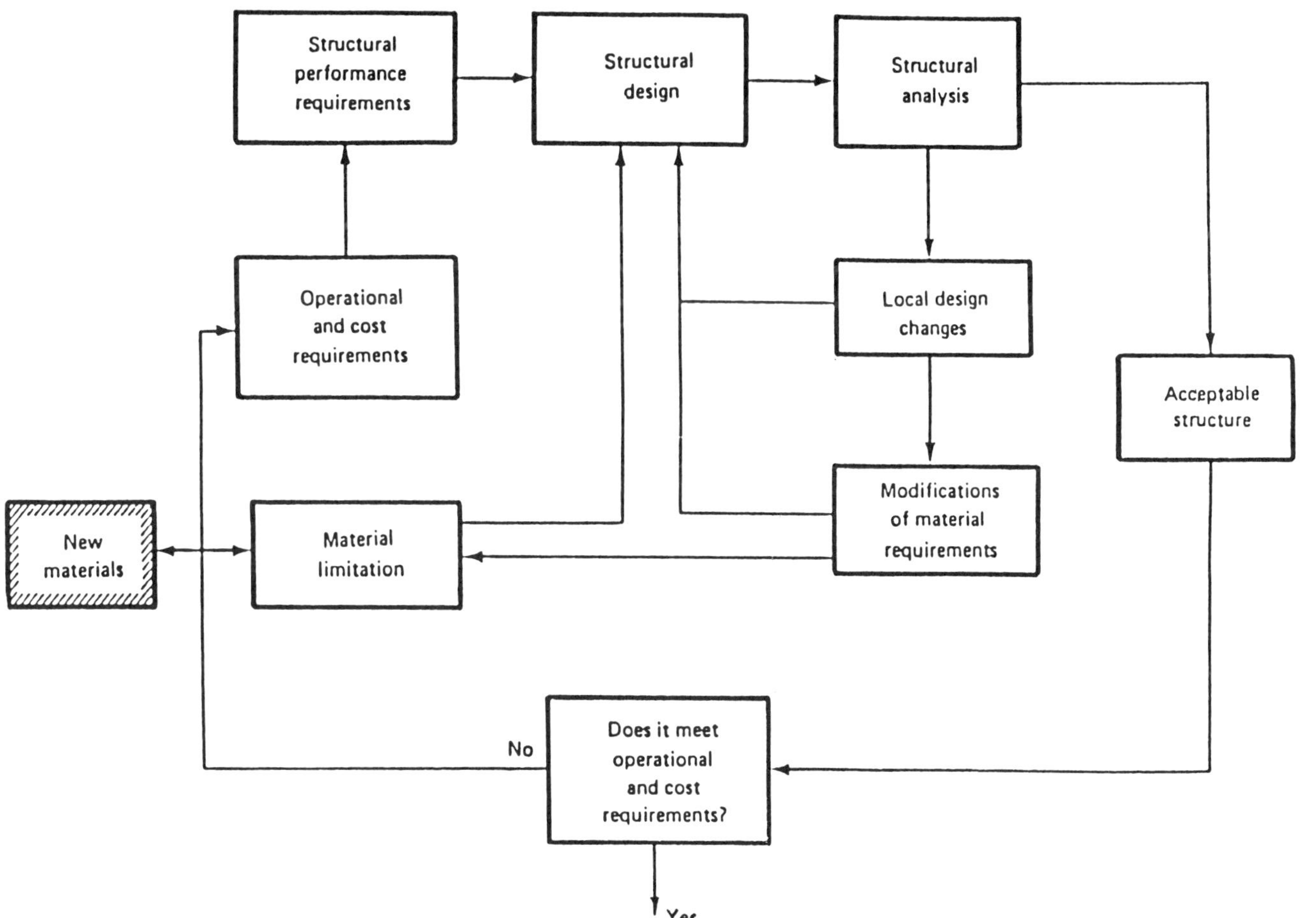

FIGURE 4 Schematic for material selection procedure. (From Ref. 5.)

imately by:

$$\frac{1}{E_T} = \frac{V_f}{E_f} + \frac{(1 - V_f)}{E_m} \tag{2}$$

Equations (1) and (2) can be written in terms of a constant K_0 relating the fiber to the matrix elastic modulus, or:

$$E_f = K_0 E_m \tag{3}$$

so that Eqs. (1) and (2) become:

$$E_L = [(K_0 - 1)\,V_f + 1]\,E_m \tag{4}$$

$$E_T = \left[\frac{1}{1 - (1 - 1/K_0)V_f}\right] E_m \tag{5}$$

These expressions are useful for parametric analysis, since K_0 is constant for any value of V_f for a given composite type.

The composite strength along the fiber direction is given by:

$$\sigma_L = \sigma_{fu}\left[\left(\frac{K_0 - 1}{K_0} - \frac{S_c}{2S}\right) V_f + \frac{1}{K_0}\right] \qquad S \geq S_c \tag{6}$$

where σ_{fu} is the ultimate tensile strength of the fiber, and S and S_c are the fiber length-to-diameter aspect ratio and critical aspect ratio, defined by:

$$S = \frac{L}{2d} \tag{7}$$

and

$$S_c = \frac{\sigma_{fu}}{2\tau_i} \tag{8}$$

where L is the fiber length and d is its diameter. Here, τ_i is the interfacial shear stress between the fiber and the matrix, a quantity that must be determined experimentally. For E-glass/epoxy, $\tau_i = 6$ MPa, while for MMCs, $\tau_i \simeq \sigma_{my}/2$, where σ_{my} is the matrix yield strength. For short-fiber-reinforced composites, $S < S_c$, the experimental values of composite yield strength can be obtained from tabulated data or it can be calculated from the following expression:

$$\sigma_L = V_f S \tau_i + (1 - V_f)\sigma_{my} \qquad S < S_c \tag{9}$$

where it has been assumed that $E_m \varepsilon_c = \sigma_{my}$, where ε_c is the composite yield strain.

Note that as the fiber becomes very long relative to S_c—that is, as $S \to \infty$—$S_c/2S \to 0$ and Eq. (6) reduces to:

$$\sigma_L = \frac{\sigma_{fu}}{K_0}[(K_0 - 1)V_f + 1] \qquad S \to \infty \tag{10}$$

which has the same form as Eq. (4).

Unidirectional fiber-reinforced composites are weak when stressed in a direction normal to the fibers. This follows from the fact that the fibers usually contribute no transverse strength and the fiber–matrix interface is usually weak [12]. Letting σ_i define the interfacial tensile strength, imperfectly bonded composites have a transverse strength determined by:

$$\sigma_T = \sigma_i V_f + (1 - V_f)\sigma_m \tag{11}$$

However, $\sigma_1 \neq 2\tau_i$. When the interface is very strong, so that $\sigma_i > \sigma_{mu}$, where σ_{mu} is the ultimate matrix tensile strength, then:

$$\sigma_T \simeq \sigma_{mu} \tag{12}$$

Equation (12) represents an upper bound. The upper bound is generally difficult to achieve because of the stress concentration created by the fiber [12]. In the case of fibers that can easily be split lengthwise during the application of a transverse stress (which is sometimes the case for boron fibers), where the transverse fiber strength is σ_{ft}, less than σ_i or σ_{mu}, then:

$$\sigma_t = V_f \sigma_{ft} + (1 - V_f)\sigma_{mu} \tag{13}$$

Equation (13) represents a lower bound on the transverse strength of unidirectional fiber-reinforced composites. Because transverse strength is usually low and the transverse modulus is relatively high, the transverse failure strain is low.

The Poisson ratio ν_{LT} is related to ν_{TL} by the usual ratio, or:

$$\begin{aligned}\nu_{LT} &= -\frac{E_L}{E_T}\nu_{TL} \\ &= [(K_0 - 1)V_f + 1][1 - (1 - 1/K_0)V_f]\nu_{TL}\end{aligned} \tag{14}$$

using Eqs. (4) and (5), so that:

$$\nu_{TL} = V_f \nu_f + (1 - V_f)\nu_m \simeq \frac{1}{3}\left(1 - \frac{1}{4}V_f\right) \tag{15}$$

for most MMCs. The right-hand side of Eq. (15) is obtained from the following considerations. For typical fiber materials, $\nu_f = 2$ for Al_2O_3, 0.21 for boron, 0.35 for carbon, and 0.19 for SiC, while for typical matrix materials, $\nu_m = 0.345$ for Al, 0.343 for Cu, 0.287 − 0.295 for steel, and 0.291 for Mg. From these data, the average values of Poisson's ratios are $\langle \nu_f \rangle = 0.24 \simeq \frac{1}{4}$ and $\langle \nu_m \rangle = 0.32 \simeq \frac{1}{3}$. This gives the numerical coefficients in Eq. (15).

Thermal expansion of composite materials can be of some concern. Useful expressions were developed by Schapery [13]. The longitudinal coefficient of thermal expansion is:

$$\alpha_L = \frac{\alpha_f V_f K_0 + \alpha_m (i - V_f)}{(K_0 - 1)V_f + 1} \tag{16}$$

using Eq. (3), and the transverse coefficient of thermal expansion is [14]:

$$\alpha_T = \alpha_{Tf}V_f + (1 - V_f)\alpha_{Tm} + \frac{V_f(1 - V_f)(\nu_f E_m - \nu_m E_f)}{V_f E_f + (1 - V_f)E_m}(\alpha_{Lf} - \alpha_{Lm}) \tag{17}$$

where the subscripts *Tf* and *Tm* stand for transverse fiber and transverse matrix, respectively, and *Lf* and *Lm* stand for longitudinal fiber and longitudinal matrix, respectively. For graphite fibers, $\alpha_{Tf} \neq \alpha_{Lf}$ [12,14], but for most other fibers, $\alpha_{Tf} \simeq \alpha_{Lf}$. Moreover, $\alpha_{Tm} = \alpha_{Lm}$. Thus, setting $\alpha_{Tf} = \alpha_{Lf} = \alpha_f$ and $\alpha_{Tm} = \alpha_{LM} = \alpha_m$, and using Eqs. (3) and (4), Eq. (17) reduces to the parametric form in K_0 and V_f, or:

$$\alpha_T = \alpha_f V_f + (1 - V_f)\alpha_m + \frac{V_f(1 - V_f)(\nu_f - \nu_m K_0)}{(K_0 - 1)V_f + 1}(\alpha_f - \alpha_m) \tag{18}$$

The electrical conductivity of MMCs is an open issue: there are numerous macroscopic models, only one microscopic model [15–18], and little experimental data. The available experimental data do not go below 80 K (liquid nitrogen temperature). Macroscopic models are based on the rule of mixtures but fail to predict correctly either the longitudinal or the transverse conductivity. Roig and Schoutens [15,16] proposed a microscopic model that predicts the longitudinal conductivity in the absence of a magnetic field down to liquid hydrogen temperature (~ 4 K), or:

$$\frac{\sigma_{CL}}{\sigma_m} = (1 - V_f)\left[1 - 1.77\left(\frac{V_f}{1 - V_f}\right)T^{-1.08}\right] \tag{19}$$

and when used at room temperature reduces to:

$$\sigma_{CL} = \sigma_m(1 - V_f) \tag{20}$$

where σ_{CL} is the longitudinal composite conductivity and σ_m is the matrix conductivity. This theory was later extended to include the effects of a magnetic field parallel to the fibers and the E-field [17,18].

Transverse electrical conductivity (or resistivity) macroscopic models fail to predict experimental data [20] by a factor as large as 2.6. An equation for predicting the transverse electrical resistivity, in the absence of a magnetic field, as a function of fiber volume fraction is [19]:

$$\rho_\perp = \rho_0 f(V_f)(1 + 2.667\, V_f) \tag{21}$$

where the function $f(V_f)$ is given by:

$$f(V_f) = \left[\left(1 - \beta - \frac{\pi}{2}\right) + \frac{2}{(1 - \beta)^{1/2}}\tan^{-1}\left(\frac{1 + \beta}{1 - \beta}\right)^{1/2}\right] \tag{22}$$

where:

$$\beta = 2\sqrt{\frac{V_f}{\pi}} \tag{23}$$

Equation (21) is valid for $0 \leq V_f \leq \pi/4$. The values given by Eqs. (20) and (21) apply to room temperature. To correct for higher or lower temperatures for aluminum, the following resistivity scaling can be used for conceptual design purposes:

$$\rho_m(T) = \rho_m(T_R)\left(\frac{295}{T}\right)^C \tag{24}$$

where $\rho_m(T)$ is the matrix resistivity at any temperature between 80 and 400 K, $\rho_m(T_R)$ is the matrix resistivity at room temperature, T is the temperature in kelvins, and C is a constant. $\rho_m(T_R)$ depends on the type of aluminum alloy. For longitudinal conductivity, $C = -1.45$, and for transverse conductivity, $C = -0.74$ [20].

The plate and shell buckling moduli are given by:

$$E_\rho = \frac{E}{1 - \nu^2} \tag{25}$$

and

$$E_s = \frac{E}{(1 - \nu^2)^{1/2}} \tag{26}$$

for unreinforced or reinforced isotropic materials,[1] where ν is Poisson's ratio and E is the elastic modulus. For unidirectional fiber-reinforced composite materials, it has been shown elsewhere [21] that these equations can be represented by:

$$E_\rho = \frac{1}{2}\left(\frac{\sqrt{E_T E_L}}{1 - \sqrt{\nu_{TL}\nu_{LT}}} + 2G_{TL}\right) \tag{27}$$

for plate buckling modulus. For shell buckling modulus, the smaller value of the following equations is used in design analysis:

$$E_{S1} = \left[\frac{2G_{TL}(E_T E_L)^{1/2}}{1 - \sqrt{\nu_{TL}\nu_{LT}}}\right]^{1/2} \tag{28}$$

or

$$E_{S2} = \left(\frac{E_T E_L}{1 - \nu_{TL}\nu_{LT}}\right)^{1/2} \tag{29}$$

where [12]:

$$\frac{1}{G_{TL}} = \frac{V_f}{G_f} + \frac{(1 - V_f)}{G_m} \tag{30}$$

Here, G_f and G_m are the fiber and matrix shear moduli, respectively. These are related to the elastic moduli by:

$$G_{f,m} = \frac{E_{f,m}}{2(1 + \nu_{f,m})} \tag{31}$$

where the subscript stands for either fiber or matrix.

[1] For example, whisker-reinforced metals may be considered to have isotropic mechanical properties even though they are quasi-isotropic.

Theory of Laminated Anisotropic Plates

Before proceeding to the subject of design optimization, we believe it necessary to present, in an abbreviated form, some results from the theory of laminated anisotropic plates. The theory of anisotropic elasticity has a long and distinguished history, and the reader is referred to the work of Lekhnitskii [22] and Hearmon [23]. The body of literature in this field is very extensive, as the reader may judge from examining the proceedings of a recent conference on the subject [24]. Recently, an excellent book by Whitney [25] was published discussing the subject extensively. The interested reader may also purchase the LAMPCAL code with this book [25], which is designed for PCs to calculate the buckling and vibration properties of anisotropic plates. In this section we present some of the important results of the theory, as these concepts will be used in later sections.

For thin plates, where the x and y coordinates are in the plane of the plate and the z axis is normal to it ($\varepsilon_z = 0$, $\gamma_{xz} = \gamma_{yz} = 0$), the strains are:

$$\varepsilon_x = \frac{\partial u}{\partial x} = \frac{\partial u_0}{\partial x} - z\frac{\partial^2 w}{\partial x^2} \tag{32}$$

$$\varepsilon_y = \frac{\partial v}{\partial y} = \frac{\partial v_0}{\partial y} - z\frac{\partial^2 w}{\partial y^2} \tag{33}$$

$$\gamma_{xy} = \frac{\partial u_0}{\partial y} + \frac{\partial v_0}{\partial x} - 2z\frac{\partial^2 w}{\partial x \partial y} \tag{34}$$

where u, v, and w are the x, y, and z components of displacements, respectively. If the second-order derivatives in Eqs. (32) to (34) represent the curvatures, then in matrix form:

$$[\varepsilon] = [\varepsilon^0] + z[k] \tag{35}$$

where the $[k]$ matrix is the curvature matrix [12]. Transforming the strains in the x, y, and z coordinates to the laminate reference axes, in order to determine the laminate constitutive relations, it then follows that:

$$[\varepsilon]_{123} = [T][\varepsilon]_{xyz} \tag{36}$$

where the subscripts refer to the coordinate axes and the $[T]$ matrix is the well-known matrix:

$$[T] = \begin{bmatrix} m^2 & n^2 & 2mn \\ n^2 & m^2 & -2mn \\ -mn & mn & (m^2 - n^2) \end{bmatrix} \tag{37}$$

with $m = \cos\theta$ and $n = \sin\theta$. An identical transformation applies to the stress matrix. The reader is referred to Schoutens [12] for a complete discussion of the stress–strain relations in tensor form as they apply to orthotropic materials. A so-called specially orthotropic lamina is one in which the principal material axes are aligned with the natural body axes for the problem. Then the stress–strain relation is:

$$[\sigma]_{123} = [Q][\varepsilon]_{123} \quad \text{or} \quad [\sigma]_{xyz} = [Q][\varepsilon]_{xyz} \tag{38}$$

By a sequence of matrix operations [12], we have:

$$[\sigma]_{xyz} = [\bar{Q}][\varepsilon]_{xyz} \tag{39}$$

where the stress matrix elements are σ_x, σ_y, and τ_{xy} and the strain matrix elements are ε_x, ε_y, and γ_{xy}. The $[Q]$ is the lamina stiffness matrix, or:

$$\begin{aligned}
\bar{Q}_{11} &= Q_{11}\cos^4\theta + 2(Q_{12} + 2Q_{66})\sin^2\theta\cos^2\theta + Q_{22}\sin^4\theta \\
\bar{Q}_{12} &= (Q_{11} + Q_{22} - 4Q_{66})\sin^2\theta\cos^2\theta + Q_{12}(\sin^4\theta + \cos^4\theta) \\
\bar{Q}_{22} &= Q_{11}\sin^4\theta + 2(Q_{12} + 2Q_{66})\sin^2\theta\cos^2\theta + Q_{22}\cos^4\theta \\
\bar{Q}_{16} &= (Q_{11} - Q_{12} - 2Q_{66})\sin\theta\cos^3\theta + (Q_{12} - Q_{22} + 2Q_{66})\sin^3\theta\cos\theta \\
\bar{Q}_{26} &= (Q_{11} - Q_{12} - 2Q_{66})\sin^3\theta\cos\theta + (Q_{12} - Q_{22} + 2Q_{66})\sin\theta\cos^3\theta \\
\bar{Q}_{66} &= (Q_{11} + Q_{22} - 2Q_{12} - 2Q_{66})\sin^2\theta\cos^2\theta + Q_{66}(\sin^4\theta + \cos^4\theta)
\end{aligned} \tag{40}$$

where, for a specially orthotropic lamina:

$$\begin{aligned}
Q_{11} &= \frac{E_{11}}{1 - \nu_{12}\nu_{21}} \\
Q_{12} &= \frac{\nu_{12}E_{22}}{1 - \nu_{12}\nu_{21}} = \frac{\nu_{21}E_{11}}{1 - \nu_{12}\nu_{21}} \\
Q_{22} &= \frac{E_{22}}{1 - \nu_{12}\nu_{21}} \\
Q_{66} &= G_{12}
\end{aligned} \tag{41}$$

where E_{11} is the modulus of elasticity along the fiber, E_{22} is the modulus of elasticity perpendicular to the fiber, G_{12} is the shear modulus, and ν_{ij} are Poisson's ratios, where $\nu_{ij}/E_{ii} = \nu_{ji}/E_{jj}$; $i, j = 1, 2, 3$ [12]. The stresses in the kth layer of a laminate, expressed in terms of the laminate midplane surface strains and curvatures, is:

$$\begin{bmatrix} \sigma_x \\ \sigma_y \\ \tau_{xy} \end{bmatrix}_k = [Q]\left\{\begin{bmatrix} \varepsilon_x^0 \\ \varepsilon_y^0 \\ \gamma_{xy}^0 \end{bmatrix} + z\begin{bmatrix} k_x \\ k_y \\ k_{xy} \end{bmatrix}\right\}_k \tag{42}$$

Because the $\bar{Q}_{ij}$ components can be different for each layer of the laminate, the stress variation through the laminate thickness is not necessarily linear even though the strain variation is linear.

The resultant forces and moments are given by [12]:

$$[N] = \sum_{k=1}^{n} [\bar{Q}]_k \left\{ \int_{z_{k-1}}^{z_k} [\varepsilon^0]\, dz + \int_{z_{k-1}}^{z_k} [k] z\, dz \right\} \tag{43}$$

$$[M] = \sum_{k=1}^{n} [\bar{Q}]_k \left\{ \int_{z_{k-1}}^{z_k} [\varepsilon^0]\, z\, dz + \int_{z_{k-1}}^{z_k} [k] z^2\, dz \right\} \tag{44}$$

Since $[\varepsilon^0]$ and $[k]$ are not functions of the z coordinates but refer to midplane values, they are removed from under the integral signs, resulting in the matrix equations:

$$[N] = [A][\varepsilon^0] + [B][k] \tag{45}$$

$$[M] = [B][\varepsilon^0] + [D][k] \tag{46}$$

where:

$$A_{ij} = \sum_{k=1}^{n} (\bar{Q}_{ij})_k (z_k - z_{k-1}) \qquad \text{Extensional stiffnesses} \tag{47}$$

$$B_{ij} = \frac{1}{2}\sum_{k=1}^{n} (\bar{Q}_{ij})_k (z_k^2 - z_{k-1}^2) \qquad \text{Coupling stiffnesses} \tag{48}$$

$$D_{ij} = \frac{1}{3}\sum_{k=1}^{n} (\bar{Q}_{ij})_k (z_k^3 - z_{k-1}^3) \qquad \text{Bending stiffnesses} \tag{49}$$

The B_{ij} components in Eq. (45) imply coupling between bending and extension of a laminate. For example, it is impossible to load in tension a laminate that has nonzero B_{ij} components without also bending or twisting it: an extensional force results not only in extensional deformations but also in twisting and or bending of the laminate. Moreover, such a laminate cannot be subjected to moments without at the same time suffering extension of the middle surface.

The equations of equilibrium for a laminated plate are [25]:

$$\frac{\partial N_x}{\partial x} + \frac{\partial N_{xy}}{\partial y} = 0 \tag{50}$$

$$\frac{\partial N_y}{\partial y} + \frac{\partial N_{xy}}{\partial x} = 0 \tag{51}$$

$$\frac{\partial^2 M_x}{\partial x^2} + 2\frac{\partial^2 M_{xy}}{\partial x\, \partial y} + \frac{\partial^2 M_y}{\partial y^2} = -9(x,y) \tag{52}$$

where N_x, N_y, and N_{xy} are the force resultants, M_x, M_y, and M_{xy} are the moment resultants, and $q(x,y)$ is the distributed transverse loading. The force resultants can be defined in terms of Airy stress functions, and substituting these functions into the relation between curvatures and deflection results in:

$$[M] = [C^*][N] + [D^*][k] \tag{53}$$

where $[C^*] = [B][A]^{-1}$ and $[D^*] = [D] - [B][A^{-1}][B]$. Now substituting the elements of these matrices into Eq. (52) gives the first governing equation, or [25]:

$$\begin{aligned} C_{12}^* \frac{\partial^4 u}{\partial x^4} &+ (2C_{62}^* - C_{16})\frac{\partial^4 u}{\partial x^3\, \partial y} + (C_{11}^* + C_{22}^* - C_{66}^*)\frac{\partial^4 u}{\partial x^2\, \partial y^2} \\ &+ (2C_{61}^* - C_{26}^*)\frac{\partial^4 u}{\partial x\, \partial y^3} + C_{21}^*\frac{\partial^4 u}{\partial y^4} - D_{11}^*\frac{\partial^4 w}{\partial x^4} \\ &- 4D_{16}^*\frac{\partial^4 w}{\partial x^3\, \partial y} - 2(D_{12}^* + 2D_{66}^*)\frac{\partial^4 w}{\partial x^2\, \partial y^2} \\ &4D_{26}^*\frac{\partial^4 w}{\partial x\, \partial y^3} - D_{22}^*\frac{\partial^4 w}{\partial y^4} - -9(x,y) \end{aligned} \tag{54}$$

This fourth-order equation has two unknowns: the Airy stress function u and the deflection w in the z direction. To find a solution, another equation must be found. A solution in terms of u and w gives all quantities of interest, since knowing u gives the stress resultants N_x, N_y, and N_{xy}, and knowing the deflection gives the curvatures. Knowing the stress resultants and the curvatures, we can invoke the constitutive equations to obtain the moment resultants and midplane strains.

The other governing equation may be found from the relationship of the midplane strains as a function of the midplane displacements, or:

$$\varepsilon_x^0 = \frac{\partial u_0}{\partial x} \qquad \varepsilon_y^0 = \frac{\partial v_0}{\partial y} \qquad \gamma_{xy}^0 = \frac{\partial u_0}{\partial y} + \frac{\partial v_0}{\partial x} \tag{55}$$

Differentiating ε_x^0 twice with respect to y and differentiating ε_y^0 twice with respect to x and adding is equal to γ_{xy}^0 differentiated once with respect to y and once with respect to x, giving:

$$\frac{\partial^2 \varepsilon_x}{\partial y^2} + \frac{\partial^2 \varepsilon_y^0}{\partial x^2} = \frac{\partial^2 \gamma_{xy}^0}{\partial x\, \partial y} = \frac{\partial^3 u_0}{\partial x\, \partial y^2} + \frac{\partial^3 v_0}{\partial x^2\, \partial y} \tag{56}$$

This is the compatibility condition for thin plates. Now the midplane strains are substituted into Eq. (56), thus resulting in the second governing equation, or:

$$\begin{aligned} A_{22}^*\frac{\partial^4 u}{\partial x^4} &- 2A_{26}^*\frac{\partial^4 u}{\partial x^3\, \partial y} + (2A_{12}^* + A_{66}^*)\frac{\partial^4 u}{\partial x^2\, \partial y^2} - 2A_{16}^*\frac{\partial^4 u}{\partial x\, \partial y^3} \\ &+ A_{11}^*\frac{\partial^4 u}{\partial y^4} - B_{21}^*\frac{\partial^4 w}{\partial x^4} - (2B_{26}^* + B_{61}^*)\frac{\partial^4 w}{\partial x^3\, \partial y} \\ &- (B_{11}^* + B_{22}^* - B_{66}^*)\frac{\partial^4 w}{\partial x^2\, \partial y^2} - (2B_{16}^* - B_{62}^*)\frac{\partial^4 w}{\partial x\, \partial y^3} \\ &- B_{12}^*\frac{\partial^4 w}{\partial y^4} = 0 \end{aligned} \tag{57}$$

and this equation has two unknowns u and w. With Eqs. (54) and (57), and with the appropriate boundary conditions, we can in theory solve for the unknowns. In practice, the solution of Eqs. (54) and (57) has proven very difficult, and few solutions have been reported [12,25]. For certain types of laminate geometries, simplifications can be obtained. If midplane symmetry is assumed, then there is no coupling between bending and stretching, so that

$$[B] = 0, [B^*] = [C^*] = 0, \text{and } [D^*] = [D] = 0,$$

and Eq. (57) reduces to:

$$A_{11}^* \frac{\partial^4 u}{\partial y^4} - 2A_{16}^* \frac{\partial^4 u}{\partial x\, \partial y^3} + (2A_{12}^* + A_{66}^*) \frac{\partial^4 u}{\partial x^2\, \partial y^2} - 2A_{26}^* \frac{\partial^4 u}{\partial x^3\, \partial y} + A_{22}^* \frac{\partial^4 u}{\partial x^4} = 0 \tag{58}$$

With the appropriate boundary conditions, Eq. (58) governs the in-plane or plane stress problem for a midplane symmetric laminated plate. It is the same equation as that governing a homogeneous (same properties through the thickness) anisotropic plate. Using the same simplifications as those used to arrive at Eq. (58) reduces Eq. (54) to:

$$D_{11} \frac{\partial^4 w}{\partial x^4} + 4D_{16} \frac{\partial^4 w}{\partial x^3\, \partial y} + 2(D_{12} + 2D_{66}) \frac{\partial^4 w}{\partial x^2\, \partial y^2} + 4D_{26} \frac{\partial^4 w}{\partial x\, \partial y^3} + D_{22} \frac{\partial^4 w}{\partial y^4} = 9(x,y) \tag{59}$$

With appropriate boundary conditions this equation governs the bending of a laminated plate with midplane symmetry. In Eqs. (58) and (59), the unsymmetric differentiations $\partial^4/\partial x\, \partial y^3$ and $\partial^4/\partial x^3\, \partial y$ render these equations intractable by any of the methods used for isotropic plates. A further simplification consists of requiring that the in-plane behavior be that of a specially orthotropic plate, or $A_{16} = A_{26} = 0$, which reduces Eq. (58) to:

$$A_{11}^* \frac{\partial^4 u}{\partial y^6} + (2A_{12}^* + A_{66}^*) \frac{\partial^4 u}{\partial x^2\, \partial y^2} + A_{22}^* \frac{\partial^4 u}{\partial x^4} = 0 \tag{60}$$

Similarly, if the bending behavior is specially orthotropic, then $D_{16} = D_{26} = 0$ and Eq. (59) becomes:

$$D_{11} \frac{\partial^4 w}{\partial x^4} + 2(D_{12} + 2D_{66}) \frac{\partial^4 w}{\partial x^2\, \partial y^2} + D_{22} \frac{\partial^4 w}{\partial y^4} = 9(x,y) \tag{61}$$

Now we present some of the solutions for simply supported laminated plates subjected to a normal distributed load.

Specially Orthotropic Plates

The governing equation is:

$$D_{11} w_{,xxxx} + 2(D_{12} + 2D_{66})\, w_{,xxyy} = D_{22} w_{,yyyy} = q(x,y) \tag{62}$$

where $w_{,xxxx} = \partial^4 w/\partial x^4$ and so on, and where $q(x,y)$ is transverse load on the plate. We have the boundary conditions for simply supported edges:

$$\begin{aligned} x &= 0, a & w &= 0 \\ y &= 0, b & w &= 0 \end{aligned} \tag{63}$$

and the moment equations are:

$$\begin{aligned} M_x &= -D_{11} w_{,xx} - D_{12} w_{,yy} = 0 \\ M_y &= -D_{12} w_{,xx} - D_{22} w_{,yy} = 0 \end{aligned} \tag{64}$$

and the solution is [25]:

$$W = \frac{a^4}{\pi^4} \sum_{m=1}^{\infty} \sum_{n=1}^{\infty} \frac{q_{mn}}{D_{mn}} \sin \frac{m\pi x}{a} \sin \frac{n\pi y}{b} \tag{65}$$

with:

$$D_{mn} = D_{11} m^4 + 2(D_{12} + 2D_{66}) \left(\frac{mna}{b}\right)^2 + D_{12} \left(\frac{na}{b}\right)^2 \tag{66}$$

If $q(x,y) = q =$ constant, then:

$$q_{mn} = \begin{cases} \dfrac{16q_0}{\pi^2 mn} & m, n \text{ odd} \\ 0 & m, n \text{ even} \end{cases} \tag{67}$$

For a simple load P distributed uniformly over a rectangle of dimensions $c < a$, $d < b$ with coordinates $\zeta < a$ and $\eta < a$, where ζ and η are the x and y coordinates at the center of the rectangle:

$$q_{mn} = \frac{16P}{\pi^2 mncd} \sin \frac{m\pi\zeta}{a} \sin \frac{n\pi\eta}{b} \sin \frac{m\pi c}{2a} \sin \frac{n\pi d}{2b} \tag{68}$$

When $c \rightarrow 0$, $d \rightarrow 0$, P becomes a concentrated load with coordinates $x = \zeta$ and $y = \eta$, and:

$$q_{mn} = \frac{4P}{ab} \sin \frac{m\pi\zeta}{a} \sin \frac{n\pi\eta}{b} \tag{69}$$

Symmetric Angle-Ply Laminate

The governing equation is:

$$\begin{aligned} D_{11} w_{,xxxx} &+ 4D_{16} w_{,xxyy} + 2(D_{12} + 2D_{66}) w_{,xxyy} \\ &= 4D_{26} w_{,xyyy} + D_{22} w_{,yyyy} = q(x,y) \end{aligned} \tag{70}$$

where $w_{,xxyy} = \partial^4 w/\partial x^2 \partial y^2$, etc. . . The boundary conditions are:

$$\begin{aligned} x &= 0, a & w &= 0 & M_x &= -D_{11} w_{,xx} - D_{12} w_{,yy} - 2D_{16} w_{,xy} = 0 \\ y &= 0, b & w &= 0 & M_y &= -D_{12} w_{,xx} - D_{22} w_{,yy} - 2D_{16} w_{,xy} = 0 \end{aligned} \tag{71}$$

A Fourier series solution allows a separation of variables but does not satisfy the boundary conditions because of the terms $4D_{26}w_{,xyyy}$ and $4D_{16}w_{,xxxy}$. These difficulties are overcome by using either the Ritz or the Galerkin method to obtain an approximate solution. The maximum deflection is at $x = a/2$ and $y = b/2$, and:

$$w_{\max} = 0.00425\frac{a^4q_0}{D_{11}} \tag{72}$$

and if D_{16} and D_{26} are neglected, the maximum deflection is:

$$w_{\max} = 0.00324\frac{a^4q_0}{D_{11}} \tag{73}$$

The error in neglecting D_{16} and D_{26} is approximately 24%. An exact solution gives:

$$w_{\max} = 0.00452\frac{a^4q_0}{D_{11}} \tag{74}$$

Antisymmetric Cross-Ply Laminates

This problem has extensional stiffness and A_{11}, A_{12}, $A_{22} = A_{11}$, and A_{66} are nonzero; the nonzero bending extension coupling stiffnesses are B_{11} and $B_{22} = -B_{11}$; and the bending stiffnesses D_{11}, D_{12}, $D_{22} = D_{11}$, and D_{66} are nonzero. The equilibrium equations are:

$$A_{11}u_{,xx} + A_{66}u_{,yy} + (A_{12} + A_{66})v_{,xy} - B_{11}w_{,xxx} = 0 \tag{75}$$

$$(A_{12} + A_{66})u_{,xy} + A_{66}v_{,xx} + A_{11}v_{,yy} + B_{11}w_{,yyy} = 0 \tag{76}$$

$$D_{11}(w_{,xxxx} + w_{,yyyy}) + 2(D_{12} + 2D_{66})w_{,xxyy} - B_{11}(u_{,xxx} - v_{,yyy}) = q \tag{77}$$

and the boundary conditions for simply supported edges are:

$$\begin{aligned}
x = 0,a: w = 0 \quad & M_x = B_{11}u_{,x} - D_{11}w_{,xx} - D_{12}w_{,yy} = 0 \\
v = 0 \quad & N_x = A_{11}u_{,x} + A_{12}v_{,y} - B_{11}w_{,xx} = 0 \\
y = 0,b: w = 0 \quad & M_y = -B_{11}v_{,y} - D_{12}w_{,xx} - D_{11}w_{,yy} = 0 \\
u = 0 \quad & N_y = A_{12}u_{,x} + A_{11}v_{,y} + B_{11}w_{,yy} = 0
\end{aligned} \tag{78}$$

and:

$$\begin{aligned}
u &= \sum_{m=1}^{\infty}\sum_{n=1}^{\infty} A_{mn}\cos\frac{m\pi x}{a}\cos\frac{n\pi y}{b} \\
v &= \sum_{m=1}^{\infty}\sum_{n=1}^{\infty} B_{mn}\sin\frac{m\pi x}{a}\sin\frac{n\pi y}{b} \\
w &= \sum_{m=1}^{\infty}\sum_{n=1}^{\infty} C_{mn}\sin\frac{m\pi x}{a}\sin\frac{n\pi y}{b}
\end{aligned} \tag{79}$$

If the load is only a single term in the Fourier series, then:

$$q = q_0 = \sin\frac{\pi x}{a}\sin\frac{\pi y}{b}$$

Substituting Eq. (79) into Eqs. (75) and (76), equating like Fourier terms, and solving the resulting simultaneous algebraic equation for the Fourier coefficients, gives:

$$A_{mn} = q_{mn}\frac{R^3b^3B_{11}m}{\pi^3D_{mn}}[A_{66}m^4 + A_{11}m^2n^2R^2 + (A_{12} + A_{66})n^4R^4]$$

$$B_{mn} = -q_{mn}\frac{R^4b^3B_{11}n}{\pi^3D_{mn}}[(A_{12} + A_{66})m^4R^4 + A_{11}m^2n^2R^2 + A_{66}n^4R^4]$$

$$C_{mn} = q_{mn}\frac{R^4b^4}{\pi^4D_{mn}}[(A_{11}m^2 + A_{66}n^2R^2)(A_{66}m^2 + A_{11}n^2R^2) - (A_{12} + A_{66})^2m^2n^2R^2]$$

where $R = a/b$ and:

$$\begin{aligned}
D_{mn} = \{&[(A_{11}m^2 + A_{66}n^2R^2)(A_{66}m^2 + A_{11}n^2R^2) \\
&- (A_{12} + A_{66})m^2n^2R^2] \\
&*[D_{11}(m^4 + n^4R^4) + 2(D_{12} + 2D_{66})m^2n^2R^2] \\
&- B_{11}^2[A_{11}m^2n^2R^2(m^2 + n^4R^4) \\
&+ 2(A_{12} + A_{66})m^4n^4R^4 + A_{66}(m^8 + n^8R^8)]\}
\end{aligned}$$

and q_{mn} are the coefficients of the Fourier series representing the load. Since u, v, and w are given by Eq. (79), these can be used to calculate the force and moment resultants. The deflection of the plate for the case of $B_{11} = 0$, we find $u = v = 0$ and:

$$w = \frac{R^4b^4}{\pi^4}\sum_{m=1}^{\infty}\sum_{n=1}^{\infty} q_{mn}\frac{\sin\frac{m\pi x}{a}\sin\frac{n\pi y}{b}}{[D_{11}(m^4 + n^4R^4) + 2(D_{12} + 2D_{66})m^2n^2R^2]} \tag{80}$$

Antisymmetric Angle-Ply Laminates

This problem has the following nonzero terms: extensional stiffness: A_{11}, A_{12}, A_{22}, and A_{66}; bending extension coupling stiffness: B_{16} and B_{26}; bending stiffness: D_{11}, D_{12}, D_{22}, and D_{66}. The equilibrium equations are:

$$A_{11}u_{,xx} + A_{66}u_{,yy} + (A_{12} + A_{66})v_{,xy} - 3B_{16}w_{,xxy} - B_{26}w_{,yyy} = 0 \tag{81}$$

$$(A_{12} + A_{66})u_{,xy} + A_{66}v_{,xx} + A_{22}v_{,yy} - B_{16}w_{,xxx} - 3B_{26}w_{,xyy} = 0 \tag{82}$$

$$D_{11}w_{,xxxx} + 2(D_{12} + 2D_{66})w_{,xxyy} + D_{22}w_{,yyyy} - B_{16}(3u_{,xxy} + v_{,xxx}) - B_{26}(u_{,yyy} + 3v_{,xyy}) = q \tag{83}$$

For hinged edges, free in the tangential direction, the boundary conditions are (25):

$$\begin{aligned} x = 0,a: w = 0; \quad & M_x = B_{16}(u_{,y} + v_{,x}) - D_{11}w_{,xx} - D_{12}w_{,yy} = 0 \\ u = 0; \quad & N_{xy} = A_{66}(u_{,y} + v_{,x}) - B_{16}w_{,xx} - B_{26}w_{,yy} = 0 \\ y = 0,b: w = 0; \quad & M_y = B_{26}(u_{,y} + v_{,x}) - D_{12}w_{,xx} - D_{22}w_{,yy} = 0 \\ v = 0; \quad & N_{xy} = 0 \end{aligned} \tag{84}$$

and:

$$\begin{aligned} u &= \sum_{m=1}^{\infty}\sum_{n=1}^{\infty} A_{mn} \sin\frac{m\pi x}{a}\cos\frac{n\pi y}{b} \\ v &= \sum_{m=1}^{\infty}\sum_{n=1}^{\infty} B_{mn} \cos\frac{m\pi x}{a}\sin\frac{n\pi y}{b} \\ w &= \sum_{m=1}^{\infty}\sum_{n=1}^{\infty} C_{mn} \sin\frac{m\pi x}{a}\sin\frac{n\pi y}{b} \end{aligned} \tag{85}$$

Substituting Eq. (85) into Eqs. (81) to (83), equating like Fourier terms, and solving the resulting algebraic equations for the Fourier coefficients, there follows:

$$A_{mn} = q_{mn}\frac{R^4b^3n}{\pi^3 D_{mn}} [(A_{66}m^2 + A_{22}n^2R^2)(3B_{16}m^2 + B_{26}n^2R^2) - m^2(A_{12} + A_{66})(B_{16}m^2 + 3B_{26}n^2R^2)]$$

$$B_{mn} = q_{mn}\frac{R^3b^3m}{\pi^3 D_{mn}} [(A_{11}m^2 + A_{66}n^2R^2)(B_{16}m^2 + B_{26}n^2R^2) - n^2R^2(A_{12} + A_{66})(3B_{16}m^2 + B_{26}n^2R^2)]$$

$$C_{mn} = q_{mn}\frac{R^4b^4}{\pi^4 D_{mn}} [(A_{11}m^2 + A_{66}n^2R^2)(A_{66}m^2 + A_{22}n^2R^2) - (A_{12} + A_{66})^2m^2n^2R^2]$$

where:

$$\begin{aligned} D_{mn} = \{&[(A_{11}m^2 + A_{66}n^2R^2)(A_{66}m^2 + A_{22}n^2R^2) - (A_{12} + A_{66})^2m^2n^2R^2] \\ &*[D_{11}m^4 + 2(D_{12} + 2D_{26})m^2n^2R^2 + D_{22}n^4R^4] \\ &+ 2m^2n^2R^2(A_{12} + A_{16}) *(3B_{16}m^2 + B_{26}n^2R^2)(B_{16}m^2 + 3B_{26}n^2R^2) \\ &- n^2R^2(A_{66}m^2 + A_{22}n^2R^2) *(3B_{16}m^2 + B_{26}n^2R^2)^2 \\ &- m^2(A_{11}m^2 + A_{66}n^2R^2)(3B_{16}m^2 + B_{26}n^2R^2)^2\} \end{aligned}$$

and q_{mn} are the coefficients of the Fourier series representation of the load. Proceeding as above, the deflection for the case $B_{16} = B_{26} = 0$, we find $u = v = 0$, and [25]:

$$w = \frac{R^4b^4}{\pi^4}\sum_{m=1}^{\infty}\sum_{n=1}^{\infty} q_{mn} \frac{\sin(m\pi x/a)\sin(n\pi y/b)}{[D_{11}m^4 + 2(D_{12} + 2D_{66})m^2n^2R^2 + D_{22}n^4R^4]}$$

which is the deflection for the homogeneous orthotropic plate.

Structural Indices

One of the most useful tools in the study of optimum structural design is the application of the principles of dimensional similarity. In structural engineering, this is accomplished with the use of structural indices. A structural index can be defined approximately as a measure of the loading intensity. This means a comparison between the magnitude of the load to be carried and the distance over which this load must be transmitted. The structural index is comparable, in structural design, to the coefficients used in aerodynamics, except that it has dimensional form [26]. To convert structural indices to dimensionless form, it is only necessary to divide them by a stress, usually the elastic modulus. One of the great advantages of this approach is that design proportions that are optimum (minimum weight) for a particular structure are also optimum for structures of any size, provided they all have the same structural index.

In determining optimum design on a weight–strength basis, the two major variables are the material and the configuration of the structure. For simple tension, the optimum configuration is a straight line, and the proportions of cross sections have no direct effect on the weight. Consequently, the weight–strength factor is determined entirely by material properties. In the case of compression, the optimum configuration is also a straight line, but the size and shape of the cross section, as well as the constraints (for example, fixed ends, lateral support, and so on), play an important role in the buckling strength of the member (Fig. 5). In this case, the material properties and configuration are interrelated. Therefore, for any structural index, it is possible to find a combination of material and configuration that will result in the lightest structure. Obviously, the problem of carrying a load in compression is more difficult and more sophisticated than that of carrying the same load in tension. Consequently, for designs for which weight is a consideration, this problem can become quite severe [26].

The use of numerical factors, dimensionless numbers, or structural indices is not new, and such factors have been developed extensively to deal with structural problems in aircraft [26] and spacecraft design. They show the consequence of choosing some parameters in preference to others and the potential effects of these choices upon the structural design. It is recognized that weight reduction, improved fracture toughness, better area

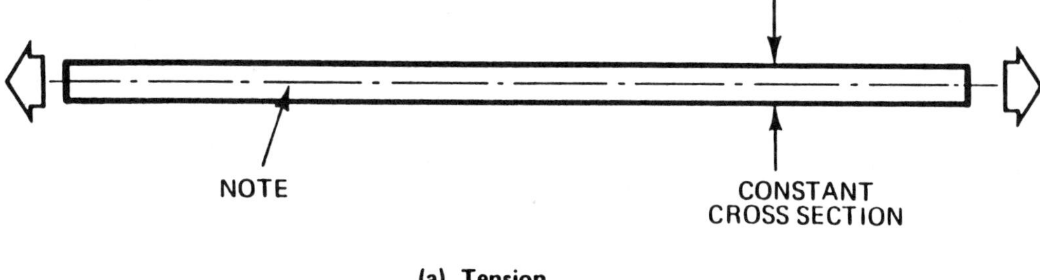

(a) Tension

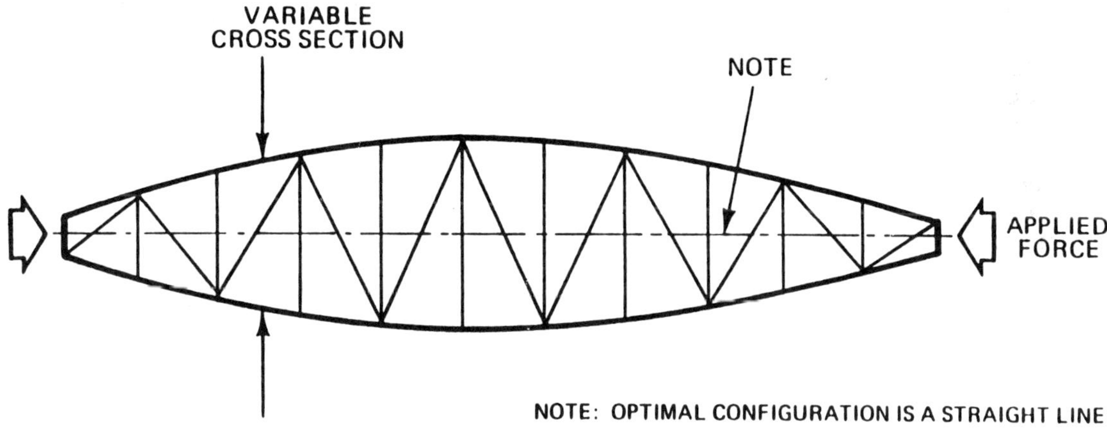

(b) Compression

FIGURE 5 Illustration of (a) tension and (b) compression load effects on structural geometry. (From Ref. 5.)

weight–stiffness or weight–strength ratios, and improved buckling properties can be understood for broad classes of designs. Consequently, various structural indices can be used to test for the effects of changing geometry or material properties to improve the design by lowering critical stresses.

The qualitative analysis presented in the first section of this chapter can now be extended to include structural indices for comparing the performance of various materials and geometries. Examination and comparison of these indices will show the areas of potentially damaging stresses so that alternative design courses can be followed, different geometries implemented, or different materials used. Iterations are thus implied in design analysis.

Weight–Strength

If weight is an important consideration in a structural design, then two important numerical factors in choosing among materials are the specific strength, σ/ρ, and the specific stiffness, E/ρ, where σ is the strength, E is the elastic modulus, and ρ is the material density. These factors can also be related to a characteristic material thickness, so that:

$$\frac{\text{Weight/unit area}}{\text{Strength}} = \left(\frac{\rho}{\sigma}\right) t \tag{86}$$

and:

$$\frac{\text{Weight/unit area}}{\text{Stiffness}} = \left(\frac{\rho}{E}\right) t \tag{87}$$

where these factors have dimensions of length. The characteristic thickness of the material can be applied to shell or plate thickness, or to any other critical dimension that supports a load. The values of σ and E are the material properties, not the stress due to the load. A material with the highest strength or stiffness efficiency would have the lowest factors given by Eqs. (86) and (87). In composite materials, it is the usual practice to compare specific strength and specific stiffness, which are σ/ρ and E/ρ, respectively.

Impact Resistance

Another parameter relevant to structure design, particularly where dynamic loads are involved, is the ratio of impact resistance to the dynamic load, or:

$$\frac{\text{Impact resistance}}{\text{Energy delivered}} = \frac{I_R}{E_d} \tag{88}$$

The impact resistance is a material property that is not very well understood for MMCs at the present time. The energy delivered is the integral of the external force over the time of its application, or:

$$E_d = \int_0^\tau F(t)\,dt \tag{89}$$

where $F(t)$ is a time-varying force. This time-varying force may be periodic and, mathematically speaking, well behaved, in which case the force may be represented by a sine wave or a Fourier series. The force may be impulsive, in which case it can be approximated by a narrow square pulse or even a step function, depending on the nature of the problem. In the case of a pulse produced by a shock, either an airblast from an explosion or stress waves in the material, the force function may be represented by an exponential pulse as shown in Figure 6. Such a force is given by [7]:

$$F(t) = AP(t) = AP_0te^{-t/t_m} \tag{90}$$

where A is the area over which the force is applied, t is time, t_m is the time to maximum pressure, and:

$$P_0 = 0.368\frac{P_{max}}{t_m} \tag{91}$$

where P_{max} is the maximum or peak pressure at time t_m. Sometimes, for analytical simplification, it is more convenient to use a triangular pulse, as shown in Figure 6, whose integral is nearly equivalent to the integral of the actual pulse. In that case:

$$E_d \simeq \int_0^\tau F(t)\,dt = \frac{1}{2}AP_{max}\tau \tag{92}$$

where τ is the maximum duration of the triangular pulse. Most airblasts are well represented by Eq. (90) when sufficiently far from the explosion source (linear region), and thus Eq. (90) can be easily integrated. The problem arises from the fact that t_m and P_{max} depend on the distance from the source. This phenomenon is known as dispersion, and similar effects occur in stress waves propagating in solids. Consequently, for conceptual design, Eq. (92) may be preferable to Eqs. (90) and (91). Using Eq. (92) in Eq. (88) gives:

$$\frac{I_R}{E_d} = \frac{2I_R}{P_{max}t} \tag{93}$$

Efficiency of Columns and Plates

Gordon [27] and Shanley [26] have shown that the efficiency of plates and columns in terms of supporting compression loads is given by:

$$e_c = C_0\left(\frac{\sqrt{E}}{\rho}\right)\left(\frac{\sqrt{F}}{L^2}\right) \tag{94}$$

for a column, and for a plate by:

$$e_f = C_1\left(\frac{\sqrt[3]{E}}{\rho}\right)\left(\frac{F^{2/3}}{L^{5/3}}\right) \tag{95}$$

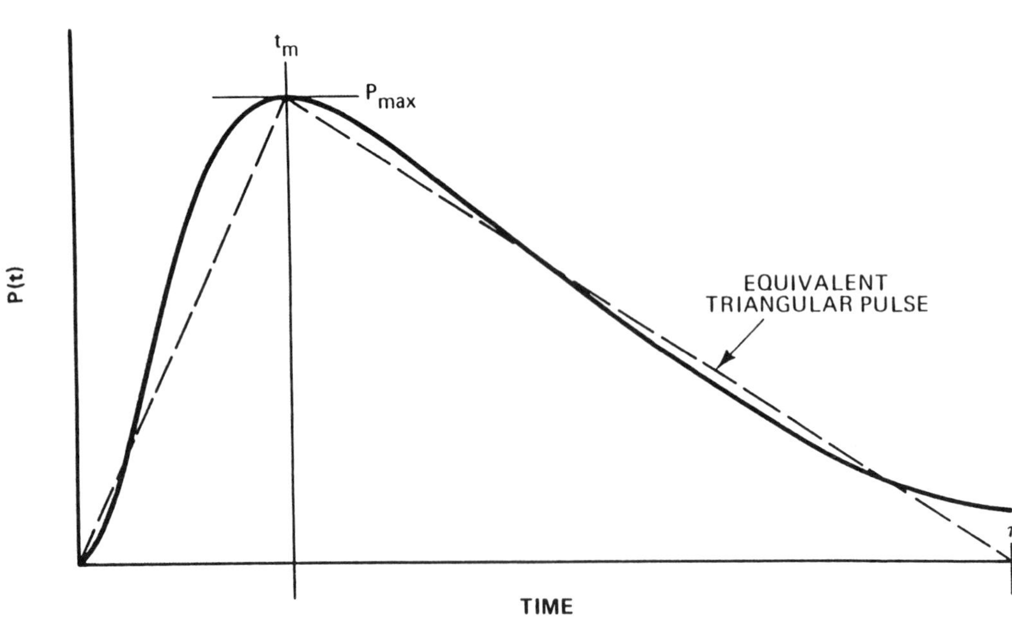

FIGURE 6 Airblast pressure pulse and triangular approximation. (From Ref. 5.)

where C_0 and C_1 are constants, E and ρ are as mentioned above, L is the length of the plate or column along the loading direction, and F is the load. Because Eqs. (94) and (95) are used for comparing materials (that is, the efficiency of column design 1 versus design 2), the constants cancel in the ratios e_{c1}/e_{c2} or e_{f1}/e_{f2}. The elastic modulus for conventional materials or composite materials given in the previous section is to be used when making numerical comparisons.

In Eqs. (94) and (95), the terms $\sqrt{E}/\rho$ and $\sqrt[3]{E}/\rho$ are called material parameters or material efficiency criteria [27]. The terms $\sqrt{E}/L^2$ and $F^{2/3}/L^{5/3}$ are called structure loading coefficients [27]. The parameters account for the weakening in compression-loaded members as a result of the Euler effect (buckling). The critical stress for buckling is always less than the material compressive yield strength because buckling is a geometric effect. These are useful indicator numbers because they point to the direction where improvements are most easily made, either geometric or material.

The structure loading coefficients indicate that weight savings are achieved with the highest value of these coefficients, that is, by using the most compact and highly loaded devices, those cases where MMCs have distinct advantages over conventional materials. However, not much can be done to raise the structure loading coefficients themselves because of fundamental limitations. The effects of low structure loading coefficients can be offset to some extent by the material efficiency criteria ($\sqrt{E}/\rho$ for a column or $\sqrt[3]{E}/\rho$ for a plate), by using material with high specific stiffness values such as composites. Some values for conventional materials and MMCs are shown in Table 2. The values in this table show that wood has the highest specific properties even though it has a low value of elastic modulus and density. But for the high specific stiffness materials, graphite–aluminum composite shows the highest values of the composites.

Flexural Rigidity

In the simple bending of beams, the elastic deformation results in a curvature of radius R. If the bending forces result in a moment M, then $MR = EI$, where EI is the flexural rigidity with I the moment of inertia about the neutral axis. A beam with great stiffness has a high value of EI, and since for a beam of height h and unit width, $I = h^3/12$, then $EI = Eh^3/12$. The weight of the beam per unit length and width is ρh. Comparing two materials, a composite and a conventional material (designated by the subscripts c and o, respectively), for the same flexural rigidity, their thicknesses will be related by:

$$E_o h_o^3 = E_c h_c^3 \tag{96}$$

and the ratio of their weight is $W_o/W_c = \rho_o h_o/\rho_c h_c$. Using Eq. (96) gives:

$$\frac{W_o}{W_c} = \frac{\rho_o/E_o^{1/3}}{\rho_c/E_c^{1/3}} \tag{97}$$

The values given in Table 2 show the advantage of using composite materials when bending stiffness should be maintained at the lowest weight penalty. Clearly, the most efficient material is the one with the highest $E^{1/3}/\rho$.

Efficiency Structural Indices for Plates and Shells

The plate and shell efficiency structural indices when these structural elements are loaded in axial compres-

TABLE 2
Stiffness–Density Parameters for a Number of Materials

Material	Density, ρ ($g \cdot cm^{-3}$)	Longitudinal Elastic Modulus, E (GPa)	E/ρ ($\times 10^5$ $cm^2 \cdot s^{-2}$)[a]	$E^{1/3}/\rho$ (Plate)	$E^{1/2}/\rho$ (Column)
Aluminum	2.70	71	26.3	1.53	3.12
Magnesium	1.74	42	24.1	2.04	3.73
Polyethylene	0.93	0.2	0.22	0.63	0.48
Steel	7.87	212	26.9	0.76	1.85
Titanium	4.51	120	26.6	1.09	2.43
Tungsten	19.3	411	21.3	0.39	1.05
Wood (Sitka spruce)	0.39	13	33.3	6.03	9.25
Zirconium	6.49	94	14.5	0.70	1.49
SiC (whisker)/Al (20 vol %)	2.7	100	37.1	1.72	3.70
Graphite/Al (50 vol %)	2.7	390	144.4	2.71	7.31
B/Al (50 vol %)	2.50	230	92.0	2.45	6.07
Borsic/Ti (45 vol %)[b]	3.68	220	59.8	1.64	4.03
Al_2O_3/Al (50 vol %)	3.60	210	58.3	1.65	4.03
SiC (cont)/Al (50 vol %)	2.93	230	78.5	2.09	5.18
SiC (cont)/Ti (35 vol %)	3.93	260	66.2	1.62	4.10
Graphite/Ni (50 vol %)	5.34	310	58.1	1.27	3.30

[a] Thus, for example, 26.3 is 26.3×10^5 $cm^2 \cdot s^{-2}$.
[b] Borsic is a trade name for boron filament coated with SiC.
From Ref. 5.

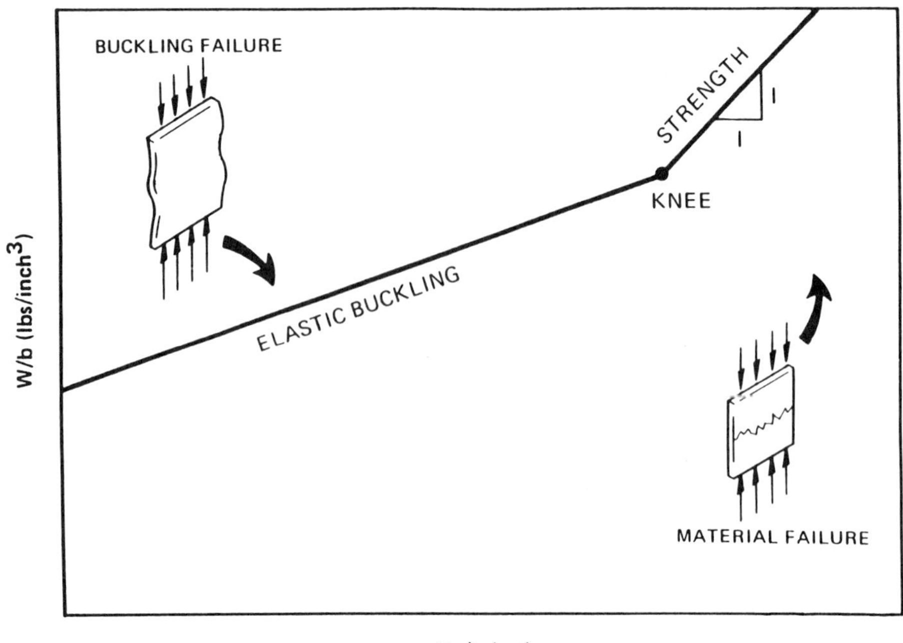

FIGURE 7 The efficiency of flat plates in axial compression. (From Ref. 5.)

sion are now discussed. The derivations are reported elsewhere [4,5,21]. A simple analysis shows that the weight index W/b or W/R versus the load index N_x/b or N_x/R for plates and shells, respectively, exhibits a knee as shown in Figures 7 and 8 [21]. Here, N_x is the axial compressive load, b is the plate width, R is the shell radius, and W is the weight. In Figures 7 and 8, the right-hand sides of these curves have slopes of 45°, representing the region in which the structure is limited by material strength. The left-hand sides of these curves correspond to the elastic stability limited region, that is, the region limited by plate or shell buckling, as is shown by the

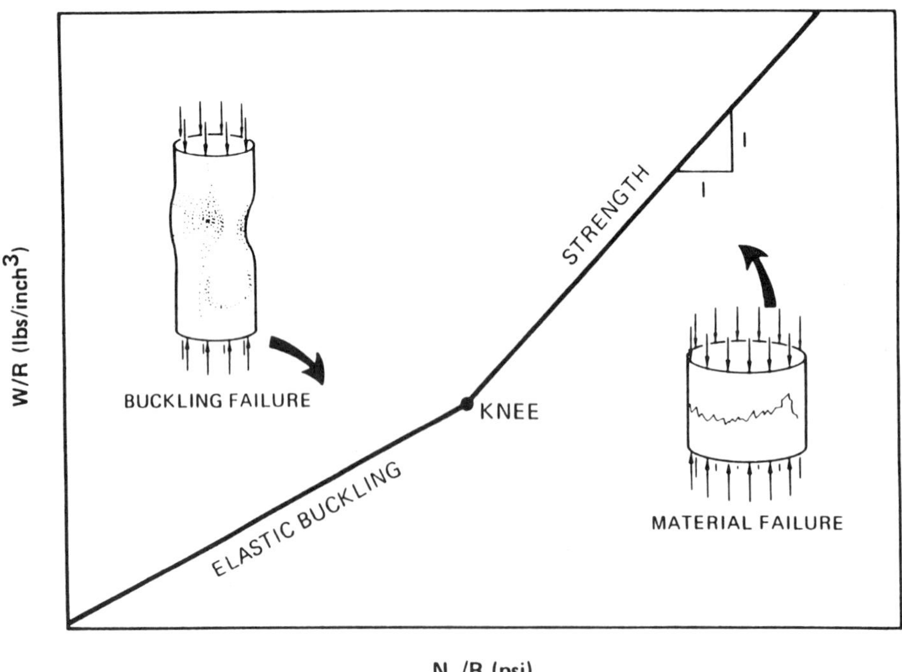

FIGURE 8 The efficiency of circular shells in axial compression. (From Ref. 5.)

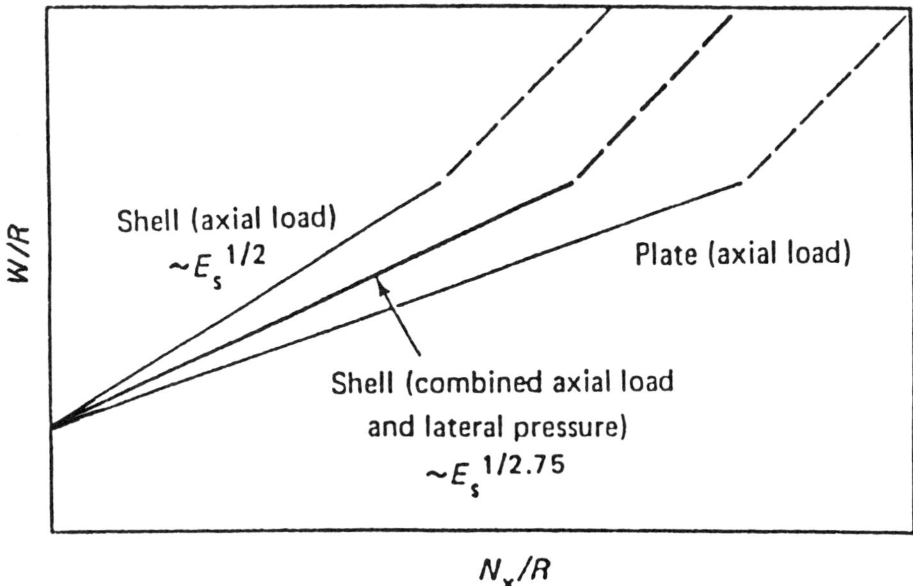

FIGURE 9 **Diagram showing the slope of shells loaded in axial compression and lateral pressure relative to axial compression alone for shells and plates.** (From Ref. 5.)

sketches in these figures. The knee is a discontinuity corresponding to the point at which failure is (hypothetically) simultaneously a material and a stability failure [21]. For shells subjected to lateral pressure and combined load, the corresponding slopes of the elastic stability region of the curve vary between those for shells and those for plates under axial load, decreasing from $E_s^{1/2}$ for shells under axial load to $E_s^{1/2.75}$ for shells under axial compression and lateral pressure, as shown in Figure 9. This result is approximate because the length of the shell must also be considered. It can be shown that below the knee [5]:

$$\frac{W}{b} = \rho\left(\frac{t}{b}\right) = \rho\left(\frac{N_x/b}{\pi^2 E_p/3}\right)^{1/2} \quad \text{(plate)} \qquad (98)$$

and:

$$\frac{W}{R} = \rho\left(\frac{t}{R}\right) = \rho\left(\frac{N_x/R}{E_s/2\sqrt{3}}\right)^{1/3} \quad \text{(shell)} \qquad (99)$$

In Eqs. (98) and (99), E_p is the plate buckling modulus given by Eq. (25) for unreinforced materials and Eq. (27) for composite materials, and E_s is the shell buckling modulus given by Eq. (26) for unreinforced materials and Eq. (28) or (29) for composite materials.

In general, the designer has no control over the values of N_x/b or N_x/R to be satisfied. If the design value of the load index is below the knee, simply increasing the strength of the material will not reduce the weight. If the design value is above the knee, simply increasing the buckling efficiency is equally useless [21]. Consequently, the knee itself is an important index in efficient design [21]. The efficiency index for a plate in axial compression loading is:

$$e_p^* = \left[\left(\frac{\sqrt[3]{E_p}}{\rho}\right)\left(\frac{\sigma_{cu}}{\rho}\right)\right]^{1/2} \quad \text{(plate)} \qquad (100)$$

and:

$$e_s^* = \left[\frac{E_s \sigma_{cu}}{\rho_3}\right]^{1/3} \quad \text{(shell)} \qquad (101)$$

where σ_{cu} is the ultimate compressive strength of the material. Equations (100) and (101) can now be used to obtain a loading intensity at which the efficiency indices are attained. For a plate, this is [5,21]:

$$\left(\frac{N_x}{b}\right)_{e_p^*} = \frac{\sigma_{cu}^{3/2}}{\sqrt{\pi^2 E_p/3}} \qquad (102)$$

and for a shell in axial compression, it is [5,21]:

$$\left(\frac{N_x}{R}\right)_{e_s^*} = 2\sqrt{3}\left(\frac{\sigma_{cu}^2}{E_s}\right) \qquad (103)$$

In the above analysis, the composite density can be obtained from the rule of mixtures, identical to Eq. (1), where ρ_c, ρ_f, and ρ_m are substituted for E_L, E_f, and E_m, respectively.

Design Data

Various types of design data can be found in Refs. 28 and 29. These data are useful for conceptual design and preliminary analyses. For final design, actual test data must be used, and scale-model tests should be performed for design verification, particularly in cases where public safety is concerned. At the present time, there are no design data available for MMCs from which commercial structures may be safely built without resorting to material or demonstration model testing. Consequently, until design handbooks are published that contain MMC design data that can be applied with confidence, designers and engineers will have to develop their own data bases in particular design instances.

Design Practices

Good structural design is a compromise among design requirements and constraints. Criteria established by design and analysis determine the degree of emphasis to be placed upon each factor considered. As discussed, the designer now has at hand an additional class of materials that are very stiff and strong in a single direction. Moreover, these materials can be prepared so that different fiber directions can be exploited, thus producing tailorable component properties. This requires that the load paths be well defined. Also, reliability is an important consideration, and a structure utilizing new materials must have a reliability at least as great as that of the structure it replaces. Moreover, considerations of the design factors discussed in this chapter, combined with experience and common sense, will result in designs that demonstrate the potential of fiber-reinforced materials.

Jacques E. Schoutens

References

1. J. E. Schoutens, *J. Metals*, June 1985, p. 43.
2. B. R. Noton, "Cost Drivers in Design and Manufacture of Composite Structures," in *Engineered Materials Handbook*, Vol. 1, *Composites*, ASM International, Metals Park, OH, 1987.
3. F. S. Principe, M. M. Monib, and D. R. Linsenmann, "Design Requirements," in *Engineered Materials Handbook*, Vol. 1, *Composites*, ASM International, Metals Park, OH, 1987.
4. B. W. Rosen and N. F. Dow, "Overview of Composite Materials Analysis and Design," in *Engineered Materials Handbook*, Vol. 1, *Composites*, ASM International, Metals Park, OH, 1987.
5. J. E. Schoutens, and D. A. Zarate, *Composites*, *17*(3), 188 (1986).
6. H. E. Lindberg and A. L. Florence, *Dynamic Pulse Buckling—Theory and Experiment*, Handbook No. DNA 6503H, SRI International, Menlo Park, CA, 1983.
7. C. C. Hudson and J. E. Schoutens, *ISA Trans.*, *20*, 13 (1982).
8. J. E. Schoutens, *Composite Structures*, *4*, 267 (1985).
9. J. E. Schoutens, *J. Mater. Sci.*, *20*, 4421 (1985).
10. J. E. Schoutens and D. I. Higa, *J. Mater. Sci.*, *21*, 1943 (1986).
11. J. E. Schoutens and D. I. Higa, *J. Mater. Sci. Lett.*, *7*, 1263 (1988).
12. J. E. Schoutens, *Reinforced Metals: Theory and Experiment*, Artech House, Norwood, MA, to be published.
13. J. Schapery, *J. Comp. Mater.*, *2*, 380 (1969).
14. M. H. Kural and B. K. Min, *J. Comp. Mater.*, *18*, 519 (1984).
15. F. S. Roig and J. E. Schoutens, *J. Mater. Sci.*, *21*, 2409 (1986).
16. F. S. Roig and J. E. Schoutens, *J. Mater. Sci.*, *21*, 2767 (1986).
17. F. S. Roig and J. E. Schoutens, *J. Mater. Sci.*, *22*, 3749 (1987).
18. F. S. Roig and J. E. Schoutens, *J. Mater. Sci.*, *22*, 4002 (1987).
19. J. E. Schoutens and F. S. Roig, *J. Mater. Sci.*, *22*, 181 (1987).
20. D. Abukay, R. V. Rao, and S. Arajs, *Fiber Sci. & Techn.*, *10*, 313 (1977).
21. N. F. Dow and E. Derby, personal communication, Materials Science Corp., Blue Bell, PA, 1984.
22. S. G. Lekhnitskii, *Theory of Elasticity of an Anisotropic Elastic Body*, Holden-Day, New York, 1963.
23. R. F. S. Hearmon, *An Introduction to Applied Anisotropic Elasticity*, Oxford University Press, Oxford, 1961.
24. T. T. Loo and C. T. Sun, Eds., *Proceedings of International Symposium on Composite Materials and Structures*, 1986.
25. J. M. Whitney, *Structural Analysis of Laminated Anisotropic Plates*, Technomic Publishing Co., Lancaster, PA, 1987.
26. F. R. Shanley, *Weight-Strength Analysis of Aircraft Structures*, 1st ed., McGraw-Hill Book Co., New York, 1952.
27. J. E. Gordon, *Structures—Or Why Things Don't Fall Down*, Decapo Paperbacks, London, 1982.
28. C. C. Chamis, "Laminated and Reinforced Metals" in Kirk-Othmer, Ed., *Encyclopedia of Chemical Technology*, 3d ed. John Wiley and Sons, New York, 1981, Chapter 13.
29. M. R. Piggott, *Load Bearing Fiber Composites*, Pergamon Press, Toronto, 1980.

Metal Matrix Composites, Discontinuously Reinforced, Microstructure and Mechanical Property Correlations

The realization that single-crystal whiskers could effectively reinforce metals to produce a material with higher strength and elastic modulus than the host matrix occurred in the 1950s. During the remainder of that decade considerable research on the growth and characterization of the whiskers and their incorporation into various alloys was conducted [1]. Interest in whisker and discontinuous-fiber reinforcements waned somewhat in the 1960s, partially as a result of the development of continuous boron, graphite, and silicon carbide filaments, which offered the potential for aerospace materials with higher and more consistent mechanical properties and, at that time, lower costs. The majority of research and development on composite systems was directed toward continuous-fiber composites until the late 1970s under Department of Defense sponsorship.

In the mid-1970s, a potentially inexpensive silicon carbide whisker material was developed [2] that generated renewed interest in discontinuous-fiber-reinforced metal matrix composites. Since that time, considerable work has been carried out on whisker-, fiber-, and particulate-reinforced composite systems. The driving force for this effort was to develop commercial interest in this class of materials by a traditionally conservative and cost-conscious industry. Mechanical property levels in these materials do not rival the high-technology continuous-fiber composites; however, the potentially low cost of this type of reinforcement and the fact that the composites can be produced by a variety of conventional metal forming techniques minimizes fabrication costs and makes these composites more amenable for industrial products.

To obtain a material with optimum properties, it is necessary to understand all the factors that control or influence those properties. Hence knowledge of how a material fails and how this, in turn, is related to the atomistic makeup of the material is necessary and crucial. Microstructure can be influenced by several factors, notably primary fabrication (type), secondary fabrication, and thermomechanical treatments. Each of these can influence orientation of grains and/or reinforcement, internal structural phases, and bonding between reinforcement and matrix, all of which are important with respect to ultimate mechanical properties.

It is the purpose of this article to demonstrate how microstructural features are related to fracture initiation and propagation, and hence to mechanical property levels, in discontinuously reinforced metal matrix composites.

Reinforcements

Discontinuous fibers and whiskers received initial emphasis in the revival of interest in this type of composite. Silicon carbide whiskers and composites therefrom produced by Advanced Composite Materials Corp. (formerly ARCO) [3] dominated most early research. Since that time, many other types of whiskers, fibers, and especially particulates have been developed for reinforcement of metals [4–12]. These are primarily ceramic-type materials with moderate to high strength and modulus, low elongation to failure, and light weight. Whisker and fiber diameters range from submicron to approximately 10 μm, with lengths up to several hundred microns. The particulate materials are basically equiaxed or flake shaped depending on their method of manufacture.

The predominant compositional basis of these materials has been SiC, Al_2O_3, B_4C, TiB_2, Si_3N_4, and oxide and aluminosilicate fibers and particles. Proprietary concerns by manufacturers prevent description of the processing of these reinforcements; however, Table 1 presents a selected list of some of the more popular ones.

Matrix Alloys

Aluminum, magnesium, and titanium alloys have been the predominant matrices utilized in the formation of

TABLE 1
Reinforcements for Composites

Manufacturer	Designation	Composition	Diameter (μm)
Whiskers			
American Matrix, Inc. Knoxville, TN	SiC, B_4C	SiC, B_4C	1–3
Tokai Carbon Co. Tokyo, Japan	Toka Whisker	SiC	0.5
Tateho Chemical Inc. Hyogo-Pre, Japan	SCW, SNW	SiC, Si_3N_4	0.1–1.6
Discontinuous Fibers			
ICI, Mond Division Cheshire, England	Saffil	Al_2O_3	3
Sohio Niagara Falls, NY	FiberFrax	50 Al_2O_3–55.02	2–3
Babcock & Wilcox Lynchburg, VA	Kaowool	Al_2O_3–SiO_2–B_2O_3	2
Powders			
American Matrix, Inc. Knoxville, TN	SiC, B_4C, ZrB_2		
Norton Worcester, MA	SiC		
Union Carbide Cleveland, OH	TiB_2		
Superior Graphite Chicago, IL	SiC		

these composite systems because of the emphasis on light weight and obtaining higher specific properties. The particular alloy selected depends upon the fabrication route for the composite system. Typically, matrix alloys that offer the best properties for that class of material have been chosen without regard to any potential alteration in properties as the result of the addition of a ceramic reinforcement. For powder metallurgy composites, the aluminum alloys 6061, 2024, 2124, and 7XXX with slight composition variations are the primary matrices used. For casting matrices, emphasis has been on the Al–Si and Al–Mg type alloys. The basic philosophy that has been used is to "add the best reinforcement to the best alloy matrix and hope for the best." This is, as will become apparent, not always the optimum approach and certainly has little scientific basis.

Fabrication Techniques

Two processing routes have dominated composite fabrication of the discontinuous-reinforced composites, both standard techniques in the metallurgical industry. These are powder metallurgy (P/M) and casting or liquid metal infiltration techniques. Both techniques are described more thoroughly elsewhere [13–14]; however, they will be discussed briefly here because the ultimate microstructure generated is strongly dependent on processing.

Powder Metallurgy

In the P/M process, matrix powders and reinforcements are mixed together to form a uniform blend of the two components. The mixture is then degassed to remove any volatile components, such as adsorbed gases and hydrated oxides. This step is crucial to subsequent consolidation [15]. An ingot is then formed by either hot pressing in vacuum or hot isostatic pressing (HIP) at an appropriate temperature. Pressing temperature is regarded as a critical step in the formation procedure, with some sources stating that exceeding the solidus temperature of the matrix alloy is necessary to obtain adequate bonding with the reinforcement [16]. This procedure is, of course, dictated by the particular alloy used. For example, if a rapidly solidified powder (RSP) that depends on metastable dispersoids for its strengthening is used as a matrix, then exceeding the solidus would obviously destroy the inherent properties; hence consolidation has to be carried out below the solidus temperature.

Secondary processing—such as extrusion, rolling, forging, and superplastic forming—is also an important fabrication step. It is only through these procedures that one can obtain usable components. Specific parameters that have been established for the metal alloys (host matrices) are not necessarily pertinent for the composite ingot. For example, flow stress of the matrix alloy is considerably altered by the addition of ceramic reinforcement; hence strain rates and temperatures for deformation are altered. The primary concern during secondary processing is to minimize damage to the reinforcement. This is particularly acute with small diameter fibrous types, as fracture would reduce the aspect ratio (length: diameter), which would affect the composite strengthening mechanism. It is inevitable that fiber damage occurs; however, it is essential to minimize this.

Liquid Metal Infiltration or Casting

Two main liquid metal processes have been used for processing composites, compocasting or rheocasting [17] and squeeze casting [18]. The intrinsic value of this type of processing is that it is much less expensive to implement than P/M and frequently net-shape components can be produced, leading to lower cost products by the elimination of secondary processing. The basic difference between the two is that in compocasting, the ceramic reinforcement is added to the molten matrix alloy and dispersed uniformly by agitation before casting the charge. In squeeze casting, a preform of the reinforcement is produced in a skeletonlike structure that is held together by sintering or a binder, and the molten metal is forced into this structure by pressure. Squeeze casting has the advantage in that the amount and position of the reinforcing phase are much more controllable, and hence more consistent products are obtained. The casting processes sound quite simple to implement; however, there are a host of variables, including wetting, reaction, viscosity, thermal effects, reinforcement distribution, and porosity, that must be controlled in order to produce a viable product [19–22].

The major disadvantage of cast composites (and cast materials in general) is that their mechanical properties are generally inferior to those of P/M products, but as has been stated, lower costs for materials and processing coupled with higher volume production make composite components using this processing technique attractive for commercial applications.

Microstructure

Powder Metallurgy

Extrusion of various shapes from P/M billets will result in orientation of the fibers along the extrusion direction (Fig. 1). Depending on the quality of blending, the fibers can be homogeneous with minimal contact, or matrix-rich areas can exist. Fiber aspect ratios are usually decreased from starting dimensions as a result of the primary and secondary fabrication procedures [23]. A typical histogram of aspect ratio for an SiC whisker–Al composite is shown in Figure 2. Most probable mean aspect ratios of 4–8 are normal for this type of system. Early samples produced by this technique contained considerable amounts of particulate or "shot" that resulted from the fiber production process. Subsequent refinement has led to whiskers of more uniform diameter and a minimum of particulate [24].

With particulate reinforcement, fracture can occur in the larger particles during fabrication such that particles with sizes of 16 μm are reduced to 5 μm, but 5 μm

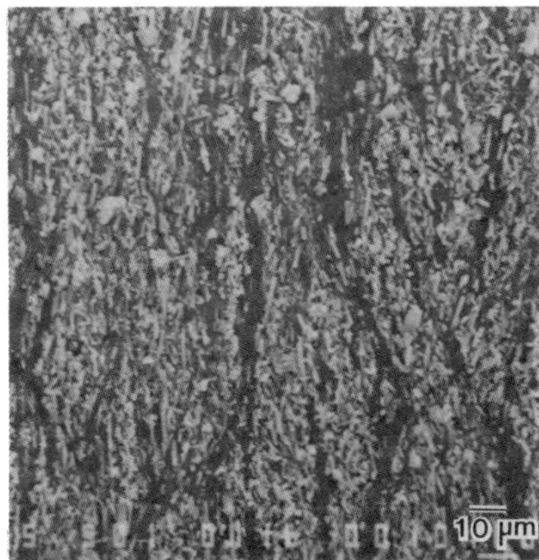

FIGURE 1 **Microstructure of a P/M extruded SiC whisker reinforced composite showing orientation.**

FIGURE 3 **Whisker cluster with no matrix in between.**

particles tend to maintain their dimension [25]. Most particles used for reinforcement tend to an oblate or nonspherical shape with aspect ratios of 2–4:1.

Clustering of the whiskers or particles is frequently observed throughout the microstructure [25,26]. This is particularly prevalent with longer whiskers or fibers and smaller diameter particles. When clustering occurs, there is a high probability of little or no matrix powder being incorporated with the cluster of reinforcement, resulting in a void in the composite and particle-particle contact (Fig. 3). Prior particle boundaries (PPB) that exist in the hot pressed billet as a result of surface oxides on the matrix powders are generally disrupted and the oxides strung out along grain boundaries after the extrusion step. Matrix grain and subgrain size in discontinuous MMCs is quite small (approximately 0.5–2 μm), which eliminates the need for normal grain refiners in the matrix alloy [27].

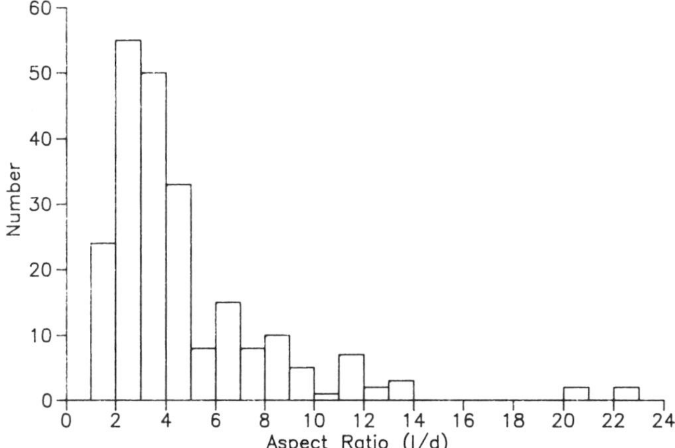

FIGURE 2 **Histogram of whisker aspect ratio (*l/d*) after primary and secondary (extrusion) processing.**

Depending on the matrix alloy used for a particular composite system, several phases can be present in the matrix continuum. Nutt and Carpenter [28] have identified seven such phases in addition to SiC and Al in the SiC whisker–2124 composite system. Large intermetallic constituent particles of $(Fe,M)_3Si_2Al_{15}$ and $FeCu_2Al_7$ were present that would not dissolve with solutioning treatment. The precipitate phase Al_2CuMg and $Al_{20}Mn_3Cu_2$ dispersoids are routinely found in the matrix of this system. In matrices with Mg as an alloying ingredient, the Mg preferentially diffuses to the particle interface, where it will react with Al_2O_3 to generate MgO along this region [28–29]. The MgO tends to form in clusters along the interfacial region. In other matrix alloys, phases characteristic of the particular system form in a similar manner.

Formation of some of these phases is indigenous to the processing parameters (such as exceeding the solidus temperature) or result from whiskers being in contact with one another or too closely spaced where preferential precipitation occurs (Fig. 4). Both of these phenomena severely influence mechanical properties, as will be pointed out later. The interface, which has special significance in composite technology because of load transfer implications, is typically nonuniform along the surface of the reinforcement [27]. Interfacial phase formation is probably a function of the surface of the reinforcement; however, in P/M composites it does not appear to be excessive. Precipitation can be enhanced by deforming the material (through secondary processes of stretching, rolling, etc.) before heat treatment. Dislocations induced by the mechanical deformation serve as nucleation sites for the precipitates to form [30].

Dislocations are a very important part of the microstructure in discontinuous-reinforced MMC. Differences in the thermal expansion coefficients of the matrix and reinforcement phases cause severe stresses to develop in the matrix during cooling from the fabrication stage [31]. These stresses are sufficiently high that plastic deformation, and hence dislocation generation by being

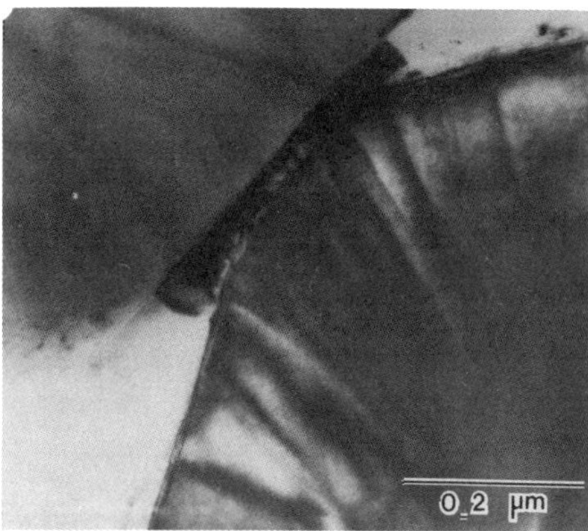

FIGURE 4 Preferential precipitation of intermetallic phase between closely spaced whiskers.

punched out at whisker ends, occurs to accommodate the stresses [30,32]. The dislocations are typically piled up at the interfacial region with densities as high as 10^{13} m^{-2} and are quite nonuniform (see Fig. 5) [32]. Postfabrication heat treatment tends to redistribute the dislocations more uniformly throughout the matrix.

Dislocations have been shown to play a dominant role in microstructural development and aging behavior [30]. Additionally, they undoubtedly play a major role in the strengthening mechanism for MMCs. Arsenault and coworkers have suggested that Orowan strengthening is the major contributor to the increase in strength of the material [32–36]. Other investigators, however, disagree and have suggested that composite strengthening mechanisms are the most dominant [30,37,38]. Evidence for composite strengthening was presented in the TEM observation of fractured whiskers in samples that were tensile tested (Fig. 6) [39]. Nardone and Prewo [37] maintain

FIGURE 5 Dislocations around reinforcing particle generated by fiber–matrix difference in thermal expansion.

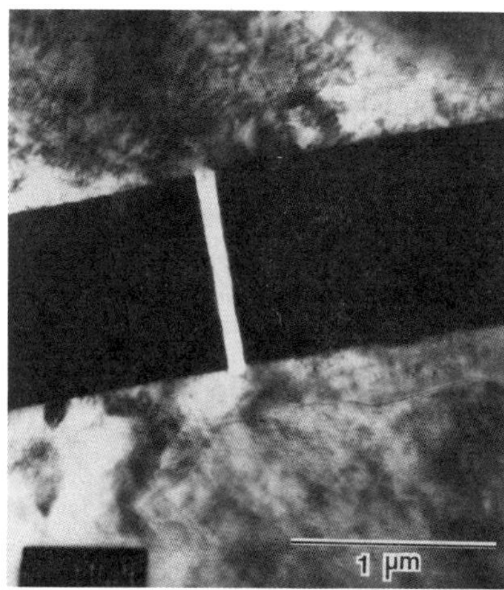

FIGURE 6 Transmission electron micrograph of a whisker fractured during tensile testing.

that composite strengthening is the major contributor to the increased strength and modulus. These authors used a modified shear lag theory to predict strengths based on composite strengthening and obtained good agreement with measured values. Contributory information to support the composite mechanism is that longitudinal strengths (that is, whiskers aligned along the test axis) are consistently 20–25% greater than transverse strengths. It is contended that if Orowan strengthening were the sole source, these values would be equal.

As shown earlier, the mean aspect ratio in SiC whisker-reinforced P/M composites is approximately 4. The modified shear lag theory predicts that the critical whisker aspect ratio of 4 is sufficient for composite strengthening. Fractured whiskers that have been observed by TEM mostly have an l/d value greater than this, with many whiskers not fractured at all [40]. The majority of whisker ends observed at the base of microvoids on fracture surfaces do, in fact, appear to be ends and not fractured halves [41,42]. Nutt and coworkers [43,44], using computer modeling, predicted high stresses at whisker ends, and in TEM studies showed that voids did preferentially nucleate at whisker ends during the fracture process (Fig. 7).

Cast MMC

Basically all the microstructural features described for P/M MMC also apply to cast materials. Primary differences are random orientation of fibrous reinforcements in the as-cast condition, much greater porosity, clumping (or clustering) of the reinforcing phase, and dendritic segregation in compocast materials [19,45–49]. Porosity is obviously enhanced because of the fact that surfaces of the added particles act as nucleation sites for gas bubbles. Clumping is due to interparticle attraction and entanglement (in the case of fibers) and is much more prevalent in smaller diameter particles and larger aspect ratio fibers [45]. Dendritic segregation is primarily a func-

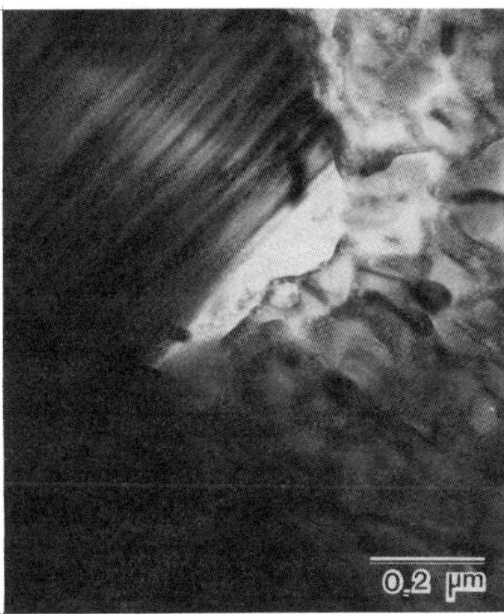

FIGURE 7 **Void formed at a whisker tip during tensile test. This is the region of greatest stress concentration.**

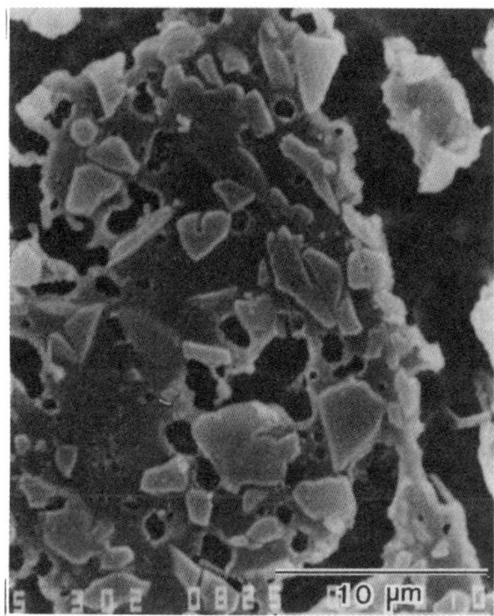

FIGURE 8 **Primary silicon surrounding reinforcing particles in a cast composite.**

tion of cooling rate of the cast material, but the degree is also dependent upon particle size and perhaps particle composition [46].

Squeeze-cast MMCs generally have much better reinforcement distribution than compocast materials. This is due to the fact that a preform is used that contains the desired volume fraction of reinforcement rigidly attached to one another so that movement is inhibited. Consequently, clumping and dendritic segregation are eliminated. Porosity is also minimized, since pressure is used to force molten metal into interfiber channels, displacing the gases. Grain size and shape can vary throughout the infiltrated preform as a result of heat flow patterns [47,50]. Secondary phases typically form at the fiber–matrix interface, since the lower freezing solute-rich regions diffuse toward the fiber ahead of the solidifying matrix. Some studies suggest that alloying species such as Mg, Li, and Cu enhance bonding between matrix and reinforcement, forming a spinel-type phase at the interface [51–54]. However, other works have not detected segregated phases or reaction products [55,56]. The role of alloying constituents and their influence on microstructure and mechanical properties is not completely clear at present.

Binder materials, such as silica, that are used to hold preforms together generally dissolve as the molten metal is forced into the preform [50]. The silicon from the compound tends to nucleate and grow as primary silicon on or around the reinforcing particles (Fig 8). This can, however, be dispersed with appropriate heat treatment.

Several different types of matrices have been used for cast MMCs. The most common have been Al–Si casting alloys and Al–Mg alloys (the Mg assists in wetting the ceramic reinforcements). Additionally, wrought alloys have been evaluated. Cook et al. [51] have shown that matrix microstructure and response to heat treatment are altered by the presence of the fibers in squeeze-cast materials. It was concluded that established alloys may not be the optimum for composite matrices but should have their composition modified to minimize deleterious microstructural features.

Secondary fabrication techniques are used with cast MMCs, but not to the extent that they have been used with P/M MMCs. This is primarily due to the desire to reduce materials cost by achieving net-shape components.

Mechanical Properties

P/M MMC

Table 2 presents representative mechanical property data for some discontinuous-reinforced MMC systems. The values vary considerably depending on fabrication technique, secondary processing, heat treatment, matrix alloy, and other factors. It can be seen that tensile strength can be almost doubled and elastic modulus is increased by 50–75% over matrix alloy values. Accompanying these impressive increases is a dramatic decrease in elongation to failure and fracture toughness. The decrease in ductility is to be expected when a brittle second phase is added to a ductile matrix (particularly when good bonding exists); however, the magnitude exceeds expectations. Reasons for this are given in the next section.

In general, longitudinal tensile strength and modulus are higher for whisker and fiber MMCs; however, these materials are more anisotrophic, having transverse values that are approximately 20–30% less than the longitudinal values and more closely equivalent to particulate data [13,33,35,57–59]. This indicates that variations in fiber orientation can influence observed properties and may be one reason for scatter in reported data.

Other mechanical property tests in which the reinforced alloy is reported to perform better than the unreinforced material are fatigue, creep, and crack growth rate [26,60,61]. In each of these studies, the whisker-reinforced material gave better properties than particulate

TABLE 2
Mechanical Properties of P/M Composites

Composite and Orientation	Volume Fraction (%)	Yield Strength (MPa)	Ultimate Tensile Strength (MPa)	Elastic Modulus (GPa)	% Elongation ε_p	Fracture Toughness ($MPa/m^{-1/2}$)	Reference
Whisker reinforced							
1100Al–SiC_w (L)	18	183	324		5.0		57
6061Al–SiC_w (L)	20	443	584	122	1.8	7.1	59
6061Al–SiC_w (T)	20	409	480	91	3.8	6.3	59
2124Al–SiC_w (L)	15	573	718	114	5.3	55	13
2124Al–SiC_w (T)	15	386	559	95	8.5	59	13
7075Al–SiC_w (L)	18	437	617		2.8		57
Particle reinforced							
1100Al–SiC_p (L)	19	110	199	95	16		57
6061Al–SiC_p (L)	25	345	410	99	4.4	15.8	59
6061Al–SiC_p (T)	25	350	409	98	4.0	14.5	59
6061Al–B_4C_p (L)	20	396	464	101	4.7		57

composites. Some studies [62] do not agree with the foregoing statement, showing superior data for the particulate-reinforced material. This indicates that much of the data in the literature are difficult to compare because the materials evaluated were made at different times or by different producers. Steady improvements in raw material and composite processing make much data outdated in a short period of time. One area where there is general agreement that particulate-reinforced MMCs outperform the whisker material is in ductility (elongation to failure) and fracture toughness.

Cast MMC

As would be expected, mechanical properties for cast MMC are considerably lower than those for their P/M counterparts. Table 3 presents data from this type of material. Squeeze-cast composites always give superior properties to compocast materials, primarily because of the lack of porosity. Secondary processing such as extrusion or rolling generates higher mechanical properties [63–66], and in some instances data similar to those for P/M composites have been obtained. Arsenault and Wu [67] have stated that there is no difference in strength between P/M and cast discontinuous-reinforced MMC if particle size, volume fraction, and distribution are the same. Any differences observed were attributed to segregation and void density in the cast material. It was also stated that dislocation arrangements and density are the same in both types of material.

One area in which cast composite properties appear to excel is elevated temperature tests. Addition of fibers or whiskers extends use temperature by 100°C over the unreinforced matrix [64], and in some cases data superior to those for P/M composites tested at 300° and 350°C are given.

TABLE 3
Mechanical Properties of Cast Composites

Composite	Volume Fraction (%)	Yield Strength (MPa)	Ultimate Tensile Strength (MPa)	Elastic Modulus (GPa)	% Elongation ε	Reference
Fiber reinforced						
Al–Al_2O_3	20	321	362	92	1.1	63
Al–Si–Al_2O_3	20	283	312	95		64
Al–Si–SiC_w	20	298	383	111		64
Al–Cu–Al_2O_3	20	335	384	86	2.2	50
Particle reinforced						
2014Al–SiC_p	10	461	484	94	6.9	50
6061Al–Al_2O_{3p}	15	317	359	87	5.7	66
2014Al–Al_2O_{3p}	15	476	504	92	2.2	66
A356–S_iC_p	20	297	317	85	0.6	66

Microstructure–Mechanical Property–Fracture Correlation

Table 4 presents a list of microstructural factors that influence mechanical properties and fracture in discontinuously reinforced MMCs. Since the P/M material is in a somewhat more advanced stage of development than the cast material, the majority of data reported in the literature are for this material. Consequently, emphasis in this section will be on P/M MMCs.

Virtually all studies have shown that there is an increase in elastic modulus with reinforcement addition. When there is good bonding between reinforcement and matrix, the modulus increases monotonically with the amount of reinforcement added. Whisker and fiber reinforcements that are aligned in the test direction generally give higher values than particulates with the same volume fraction or whiskers and fibers of random or transverse orientations [13]. However, it has also been reported that elastic modulus is isotropic and independent of type of reinforcement, being controlled solely by

TABLE 4
Microstructure–Mechanical Property–Fracture Correlations for Metal Matrix Composites[a]

Microstructure Condition	Mechanical Property Response
Addition of reinforcement	Increase in strength, modulus, fatigue life, creep properties, abrasion resistance, impact strength, high temperature strength, decrease in ductility (elongation to failure), and fracture toughness
Reinforcement type	In general, fibrous reinforcements give higher mechanical properties than particulate at equal volume fraction. Particulate reinforcement, however, gives higher elongation to failure and fracture toughness
Reinforcement orientation	Fibrous reinforcement aligned along test axis gives approximately 25% higher strength than particulate or transverse fibrous reinforcement. Fatigue and creep properties are improved in aligned fibrous composites. Ductility and fracture toughness is generally lower in the aligned material
Reinforcement distribution	Banding and/or clustering enhances crack initiation and growth, and hence lowers strength, ductility, and toughness
Particle size	No effect on modulus; strength properties decrease with particle size increase
Aspect ratio	Influences modulus, strength, fracture toughness, ductility, and fracture mechanism
Interface condition	Strong bonding increases modulus and strength but decreases ductility. Can be embrittled as a result of excessive precipitation and/or diffusion of alloying ingredients or impurities to interface
Matrix phases	Normal precipitate phases increase yield and ultimate strength. Impurity particles and preferential precipitation decrease strength, fracture toughness, fatigue, creep, and ductility
Heat treatment	Heat treatment increases mechanical properties; however, overaging minimizes the benefits. Precipitation kinetics can be altered by the addition of reinforcements; hence times and temperatures for peak aging may differ
Secondary processing	Secondary processing affects microstructure and hence mechanical properties. Extrusion aligns fibrous reinforcements but induces banding or reinforcement-free areas. Rolling homogenizes microstructure, giving higher properties, but can damage matrix–reinforcement bonds and lead to overaging. Material must be heat-treated to regain properties

[a] These comments assume well-bonded reinforcements.

the amount present [58] and independent of particle size [68]. The fact that modulus improvements are obtained in all discontinuously reinforced MMCs is proof that load transfer is occurring in the materials and that the reinforcements are contributing to composite properties through interfacial bonds.

Mechanical properties in discontinuously reinforced MMCs are strongly dependent on processing type and conditions and on heat treatment. Hot-pressed or HIPed ingots generally give lower values as a result of randomness in orientation of whisker or fiber reinforcements and the fact that there are prior particle boundaries (i.e., oxide decorated) between matrix powder particles. Extrusion aligns the whiskers or fibers, but at the expense of a reduced aspect ratio that minimizes the composite strengthening effect. The oxides around the PPBs string out, which creates better bonding between matrix powder particles.

Banding occurs in extruded MMCs whether they are fibrous or particulate-reinforced, leaving uneven reinforcement distribution and matrix-rich areas (Fig. 9). This effect is reported to generate considerable scatter in mechanical property data as a result of variations in localized stress fields. An example of a fracture surface from an extruded whisker-reinforced composite with matrix-rich areas is shown in Figure 10. The ductile-appearing knife edge is such a region.

There have been reports that hot rolling tends to homogenize reinforcement distribution, yielding a more uniform microstructure and eliminating the fracture behavior shown in Figure 10 [69–71]. Tensile data from rolled samples in these studies were generally more consistent, but more importantly, elongation to failure and fracture toughness are increased dramatically. These improvements were attributed to the increase in strain and work hardening of the matrix [70].

It has been observed that it is very important to heat-treat the composites after hot rolling to attain the enhanced properties [39,72]. Testing the material in the as-rolled condition gives lower tensile strength, which can be attributed to two factors. First, interfacial damage can occur such that full advantage cannot be realized from the reinforcements [39]. This damage (approximately 20% strength reduction in an 1100 Al–20 vol % SiC whisker composite) can be repaired by heating the composite to above 350°C for 1 hour or by subjecting the material to a normal T6 temper. Figure 11*a* and *b* depicts the fracture surface of composites tested in the as-rolled condition and in the heat-treated condition. Whisker pullout is obvious in the rolled sample, whereas the heat-treated material does not display this phenomenon. In situ transmission electron microscope studies demonstrated that voids created at the interface by rolling could be closed by the heat treatment.

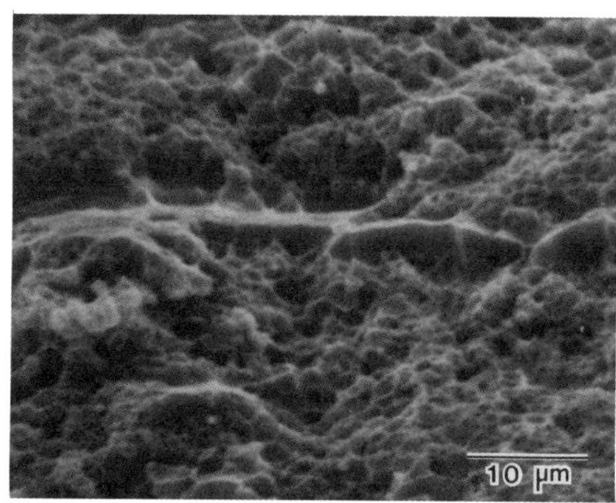

FIGURE 10 **Fracture surface of a composite with matrix-rich areas. The ductile knife edge is such an area.**

FIGURE 9 **Matrix-rich area in an extruded P/M composite.**

The second factor that leads to reduced strength in as-rolled composite sheet with a precipitation-hardened matrix (such as 2124) is accelerated aging [72]. If rolling procedures are performed at temperatures above the 250°–300°C range (which coincides with an exotherm marking a precipitation reaction), then precipitation kinetics are greatly accelerated during cool-down. This leads to overaging and formation of brittle precipitates, which severely influence the fracture characteristics of the material. Figure 12 shows the microstructure of an as-rolled sheet of 2124 Al–15 vol % SiC whisker composite that was tested in tension. The scanning electron microscope (SEM) micrograph of the area below the fracture surface was taken in the compo (backscattered) mode, which differentiates compositional species in the material. The large light particles are Al_2Cu intermetallic precipitates that have coarsened as a result of accelerated overaging from the rolling process. Figure 13 shows how these precipitates fracture prematurely during the application of a load, generating a nucleation site for composite failure and a low energy path for crack propagation.

Figure 14 is the same material that has been subjected to a T6 heat treatment. The large precipitates have been dissolved (solutionized) and do not appear in the microstructure of the composite. Additionally, any interfacial damage that occurred from the rolling has been healed

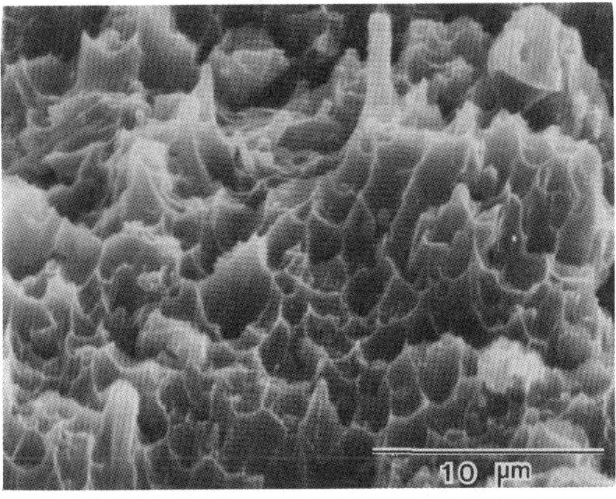

(a)

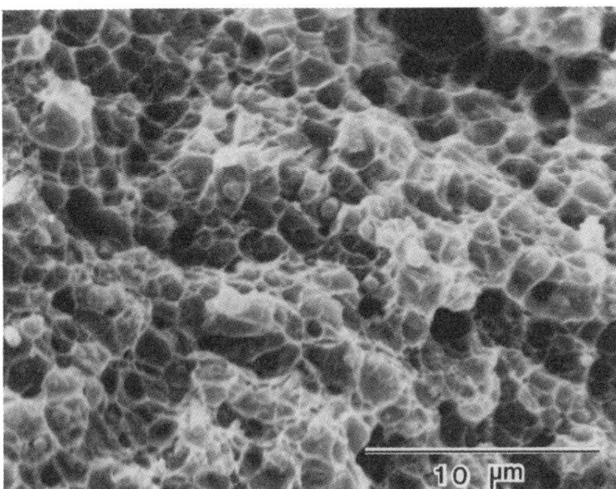

(b)

FIGURE 11 (*a*) Fracture surface of an as-rolled composite showing whisker pullout. (*b*) Fracture surface of the same composite that has been heat-treated to repair interfacial region; hence no pullout.

FIGURE 13 Scanning electron micrograph showing fracture of the intermetallic precipitates during tensile testing.

with the heat treatment. Tensile strength for the hot-rolled samples subjected to T6 was increased by 50% over the as-rolled sample [71,72]. A similar increase was noted between rolled and T6 samples in the 6061–SiC particle composite system [70].

The major contributors to premature failure and reduced ductility in discontinuous-reinforced MMC have been large, brittle impurity particles in the matrix of the composite. These particles have been identified as SiC or SiO_2 (50 μm or greater in diameter) from the whisker process [73], intermetallic constituent particles resulting from impurities in the matrix alloy (Fe, Cr, Mn rich particles 20–50 μm in diameter) [71,73], and/or divorced eu-

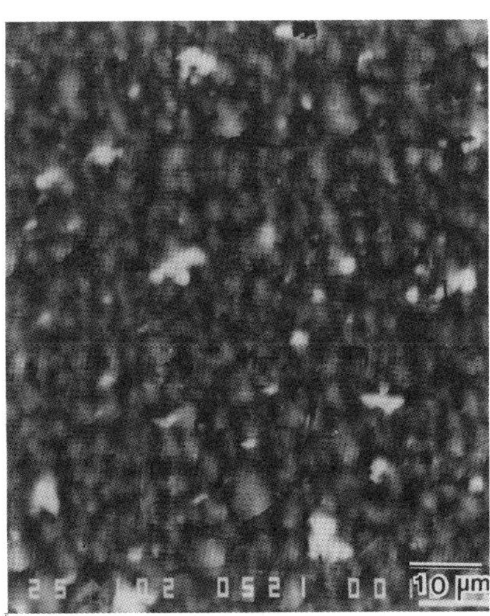

FIGURE 12 Scanning electron micrograph showing intermetallic precipitates (light areas) formed during rolling.

FIGURE 14 The same composite sample as in Figures 12 and 13 that has been subjected to T-6 heat treatment, dissolving the precipitates.

tectic intermetallic particles and dispersoids (Al_2CuMg and $Al_{20}Mn_3Cu_2$, 3–5 μm in diameter) [74]. The latter two are primarily a result of matrix alloying ingredients and exceeding the solidus temperature during processing [73]. Heat treatment will not dissolve these constituent particles. Figure 15 shows a fracture surface dominated by these large brittle particles, which offer a low energy crack path. Improved processing techniques by the whisker manufacturer has led to the removal of most of the particulate associated with whisker production as well as generating whiskers of more uniform diameter and aspect ratio.

The intermetallic constituent particles are a result of matrix composition and processing parameters (i.e., exceeding the solidus temperature during the hot-pressing stage). This type of defect, along with aluminum oxide from the matrix powder surface, can occupy 5–10 vol % of the composite. Figure 16 depicts an unreinforced 2124 aluminum sample that was processed with the same procedure as whisker composites (i.e., hot pressed above the solidus temperature and extruded). The sample was tensile tested. Intermetallic constituent particles, 3–5 μm in diameter, that have prematurely cracked can be seen. It can also be noted that voids have formed adjacent to oxide particles in the stringers. These defects obviously have a deleterious effect on mechanical properties and ductility [28,74].

It has been determined that this type of defect can be minimized by altering matrix chemistry, particularly by eliminating grain refining additives, such as manganese in 2124, and minimizing the normal impurities iron and chromium [73]. Another process alteration to eliminate the intermetallic divorced eutectic particles is to avoid fabrication above the solidus temperature [73]. The implementation of these observations in the processing stage of the materials has generated composites with considerably enhanced and more uniform ductility as well as strength, fracture toughness, fatigue, and creep properties.

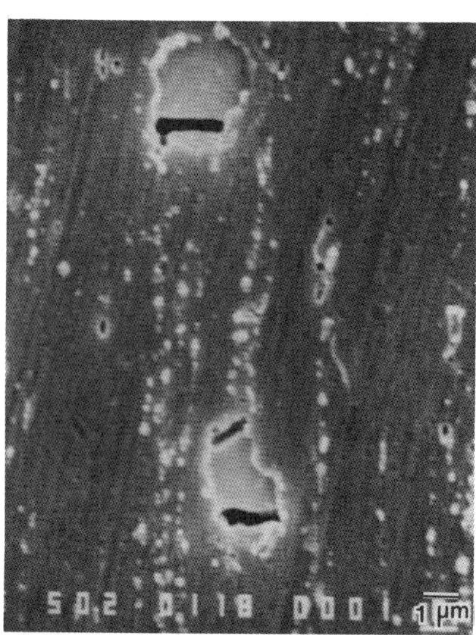

FIGURE 16 **Divorced eutectic particles formed in matrix from exceeding the solidus temperature that fracture prematurely upon application of stress.**

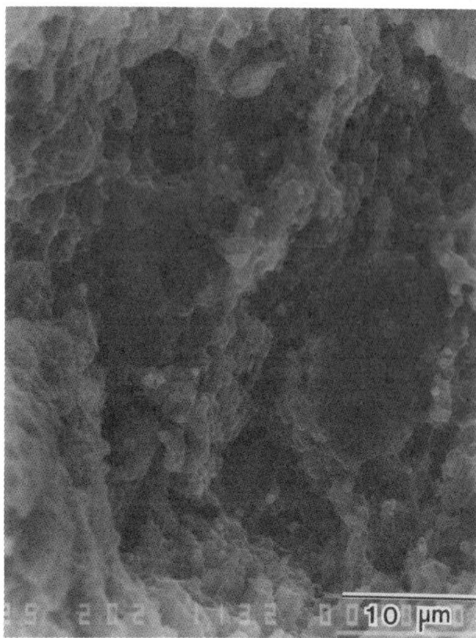

FIGURE 15 **Fracture surface of a composite sample, showing brittle particles that initiated failure.**

In addition to the deleterious nature of intermetallic phases forming as dispersoids or divorced eutectics, heterogeneous nucleation of intermetallic phases on the reinforcement surface (particularly in closely spaced whiskers or particles), along with diffusion of Mg to the interfacial region to form MgO clusters, generates a brittle and nonuniform region along the interface [28]. Precipitation kinetics in the composite matrices are altered as a result of the addition of a ceramic reinforcing phase when compared with the unreinforced alloy [30,75,76]. This has been attributed to the large difference in thermal expansion coefficient between matrix and reinforcement generating dislocations in the matrix upon cooling from elevated temperature, which, coupled with small grain size [72], present paths for high diffusivity and hence accelerated precipitation. Dislocations created by material deformation during secondary fabrication will also influence precipitation. Because of the altered precipitation characteristics, it has been observed that tempers to achieve peak aging may be different from those for the unreinforced matrix [76].

Lewandowski et al. [25] have shown that matrix temper has a strong effect on fracture toughness and ductility as well as on fracture mode in SiC particle-reinforced aluminum. In underaged materials, SiC particulate fracture was the primary mode of failure, with larger particles preferentially fracturing first. The interfacial region did not display any sign of matrix precipitation effects. The overaged samples tended to more enhanced precipitation at the reinforcement–matrix interface, which reduced interfacial strength, leading to particle–matrix decohesion and void nucleation. This was particularly prevalent in clustered regions of the reinforcing particles. High local volume fractions, and hence many closely spaced particles, tended to produce accelerated precipitation (as shown in Fig. 4) and regions that fractured or underwent particle–matrix decohesion at lower levels of deformation. This resulted in composite samples with lower fracture toughness–tensile strength combinations. The clustered regions are obvious sites for fracture initiation [25] and offer a low energy path for crack propagation [26].

Williams and Fine [26] have demonstrated that, in the absence of extrinsic defects, fatigue crack initiation is due to debonding at the tips of SiC whiskers. Additionally, crack growth appeared to be faster along the SiC–Al interface than through the matrix. This is in keeping with the analysis of Nutt and coworkers [29,43,44] and Crowe et al. [59] in that voids nucleate at the edges of whisker ends where severe stress concentrations exist and grow under increasing plastic strain. As damage accumulates, premature failure can occur, leading to low composite ductility.

All of the above indicates that interfacial strength plays a large role in the deformation and fracture process. Fiber–matrix bond strength in discontinuous-reinforced MMCs has been reported as being quite high, in the 1600–1900 MPa range [44,77], but this can be lowered by processing and thermal aging [25,39]. Precipitation of brittle intermetallics and oxide accumulation at the interface tend to weaken this region and will provide paths for the linkup of microcracks [29]. Hence, tensile strength, fatigue properties, creep, fracture toughness, and ductility will all be affected by the crack selectively seeking the brittle regions. Other studies have indicated that high temperature creep and tensile strength can be enhanced by adding lithium to the matrix alloy [65,78). This was attributed to strengthening of the interfacial region.

Virtually all fracture surfaces in tested discontinuous-reinforced MMC display a dimpled appearance, implying failure by a void initiation and growth mechanism. This implies a strong matrix contribution to crack growth and fracture. The marginal aspect ratio of the reinforcement, which minimizes its contribution to composite strengthening, can account for the matrix and/or interfacial failure mode. A review of the theoretical and experimental background for void nucleation in discontinuously reinforced MMC by Hunt et al. [79] demonstrates the importance of microstructure, interfacial bonding, and reinforcement characteristics in the fracture of these materials.

Microstructure–mechanical property–fracture correlations for cast MMCs have not been explored to as great an extent as for the P/M material. In general, many of the same phenomena described for P/M composites are controlling in cast systems. Porosity is the primary extrinsic defect, particularly in compo- and rheocast material, that causes lower mechanical properties when compared with P/M counterparts. However, nonuniform reinforcement distribution caused primarily by dendritic segregation, reinforcement clustering, large eutectic particles of alloying constituents, and excessive interfacial reaction contribute to the lower strength and ductility. Further research should minimize these factors, thereby generating properties comparable to those of P/M composites. Achieving equal properties along with the potential for a lower cost material will direct future emphasis to research and development of cast systems.

In summary, microstructural effects on properties and fracture in discontinuous-reinforced MMC appear to be divided into extrinsic and intrinsic factors. The extrinsic factors are primarily due to defects from various processing stages of the composites, namely, impurity particles accompanying the matrix and reinforcement, constituent particles and dispersoids from exceeding the solidus temperature during fabrication, coarsened precipitates from rolling, and clustering or nonuniform distribution of the reinforcements. It is possible to minimize these factors by rigid processing specifications, and, in fact, considerable progress has been made over the past five years. Critical properties such as ductility and fracture toughness have been doubled by improved processing techniques.

The intrinsic factors are more fundamental to the composite system and have not been defined completely to this point. Addition of ceramic reinforcement to a standard matrix alloy has been shown to alter many of its characteristics and properties. For example, precipitation phenomena, grain size, work hardening, plastic flow localization, and ductility are all changed as a result of reinforcement addition. Each of these can influence deformation behavior, and hence crack initiation and growth in the composite. To optimize performance of the composite, the most logical route would appear to be to alter or tailor matrix composition to eliminate many of the microstructural factors that have been shown to be deleterious to the system.

Acknowledgement

The author acknowledges the Office of Naval Research, Dr. S. G. Fishman, Program Manager, for support of this work under contract No. N00014–85-K-0179 and Mrs. Joanna Shipp for typing the manuscript.

Franklin E. Wawner

References

1. A. Levitt, *Whisker Technology*, John Wiley & Co., New York, 1970.
2. J. Lee and I. Cutler, *Ceram. Bull., 54*, 195 (1975).
3. Advanced Composite Materials Corp., data sheet, Greer, SC, 1987.
4. *SiC and B_4C*, American Matrix Inc., data sheet, Knoxville, TN, 1987.
5. *SiC Whiskers*, Tateho Chemical Industries, data sheet, Hyogo-pre, Japan, 1986.
6. *Tokawhisker*, Tokai Carbon Co., data sheet, Tokyo, Japan, 1985.
7. *Saffil*, ICI Corp., Mond Division, data sheet, Cheshire WA74QD, England, 1986.

8. *FiberFrax*, SOHIO Engineered Materials Co., data sheet, Niagara Falls, NY, 1984.
9. *Kaowool*, Babcock and Wilcox, data sheet, Lynchburg, VA, 1985.
10. *Silicon Carbide*, Norton Co., data sheet, Worcester, MA, 1987.
11. *Titanium Diboride*, Union Carbide Corp., data sheet, Cleveland, OH, 1987.
12. *Silicon Carbide*, Superior Graphite Co., data sheet, Chicago, IL, 1987.
13. J. Cook and W. Mohn, *Engineered Materials Handbook*, Vol. 1, *Composites*, ASM International, Metals Park, OH, 1987, pp. 896–902.
14. M. Toaz, *Engineered Materials Handbook*, Vol. 1, *Composites*, ASM International, Metals Park, OH, 1987, pp. 903–910.
15. P. Hood and J. Pickens, U.S. Patent No. 4,463,058 (July 1984).
16. A. Divecha, S. Fishman, and S. Karmarkar, *J. Met.*, *33*, 12 (1981).
17. C. Levi, G. Abbaschian, and R. Mehrabin, *Metall. Trans. A*, *9A*, 697 (1978).
18. T. Clyne and M. Bader, in W. Harrigan, J. Strife, and A. Dhingra, Eds., *Proceedings of the 5th International Conference on Composite Materials*, The Metallurgical Society, Warrendale, PA, 1985, pp. 755–771.
19. P. Ghosh, S. Ray, and P. Rohatgi, *Trans. Japan Inst. Met.*, *25*, 440 (1984).
20. M. Surappa and P. Rohatgi, *Met. Trans.*, *12B*, 327 (1981).
21. M. Skibo, P. Morris, and D. Lloyd, in S. Fishman and A. Dhingra, Eds., *Cast Reinforced Metal Composites*, ASM International, Metals Park, OH, 1988, pp. 257–261.
22. T. Clyne and J. Mason, *Met. Trans.*, *18A*, 1519 (1987).
23. S. Nair, J. Tien, and R. Bates, *Int. Metals Review*, *30*, 275 (1985).
24. ARCO data, 1986.
25. J. Lewandowski, C. Lin, and W. Hunt, in P. Kumar, K. Vedula, and A. Ritter, Eds., *Proceedings of Symposium on Processing and Properties for Powder Metallurgy Composites*, The Metallurgical Society, Warrendale, PA, 1987, pp. 117–137.
26. D. Williams and M. Fine, in W. Harrigan, J. Strife, and A. Dhingra, Eds., *Proceedings of the 5th International Conference on Composite Materials*, The Metallurgical Society, Warrendale, PA, 1985, pp. 639–670.
27. P. Liaw, J. Greggi, and W. Logsdon, *J. Mat. Sci.*, *22*, 1613 (1987).
28. S. Nutt and R. Carpenter, *Mat. Sci. and Eng.*, *75*, 169 (1985).
29. S. Nutt, in A. Dhingra and S. Fishman, Eds., *Proceedings of Symposium on Interfaces in Metal Matrix Composites*, The Metallurgical Society, Warrendale, PA, 1986, pp. 157–167.
30. T. Christman and S. Suresh, "Microstructural Development in an Aluminum Alloy–SiC Whisker Composite," Brown University Report No. NSF-ENG-8451092/1/87, June 1987.
31. R. Arsenault and M. Taya, *Acta Metall.*, *35*, 651 (1987).
32. M. Vogelsang, R. Arsenault, and R. Fisher, *Metall. Trans. A*, *17A*, 379 (1986).
33. R. Arsenault and R. Fisher, *Scripta Met.*, *17*, 67 (1983).
34. Y. Flom and R. Arsenault, *J. Met.*, *38*, 31 (1986).
35. R. Arsenault and S. Wu, *Mat. Sci. and Eng.*, *96*, 77 (1987).
36. M. Taya and R. Arsenault, *Scripta Met.*, *21*, 349 (1987).
37. V. Nardone and K. Prewo, *Scripta Met.*, *20*, 43 (1986).
38. V. Nardone, *Scripta Met.*, *21*, 1313 (1987).
39. C. Harris and F. Wawner, in P. Kumar, K. Vedula, and A. Ritter, Eds., *Proceedings of Symposium on Processing and Properties for Powder Metallurgy Composites*, The Metallurgical Society, Warrendale, PA, 1987, pp. 107–116.
40. F. Wawner, unpublished results, 1987.
41. R. Arsenault and C. Pande, *Scripta Met.*, *18*, 1131 (1984).
42. T. Nieh, J. Wadsworth, and D. Chellman, *Scripta Met.*, *19*, 181 (1985).
43. S. Nutt and M. Duva, *Scripta Met.*, *20*, 1055 (1986).
44. S. Nutt and A. Needleman, *Scripta Met.*, *21*, 705 (1987).
45. J. McCoy, C. Jones, and F. Wawner, *SAMPE Q.*, *19*, 37 (1988).
46. J. McCoy and F. Wawner, in S. Fishman and A. Dhingra, Eds., *Cast Reinforced Metal Composites*, ASM International, Metals Park, OH, 1988, pp. 237–242.
47. T. Clyne, in F. Matthews, N. Buskell, J. Hodgkinson, and J. Morton, Eds., *Proceedings of 6th International Conference on Composite Materials*, Elsevier Applied Science Publishers, London and New York, 1987, pp. 2.275–2.286.
48. P. Ghosh and S. Ray, *J. Mat. Sci.*, *22*, 4077 (1987).
49. F. Hosking, F. Portillo, R. Wunderlin, and R. Mehrabian, *J. Mat. Sci.*, *17*, 477 (1982).
50. S. Harris, *Mat. Sci. and Tech.*, *4*, 231 (1988).
51. C. Cook, D. Yun, and W. Hunt, in S. Fishman and A. Dhingra, Eds., *Cast Reinforced Metal Composites*, ASM International, Metals Park, OH, 1988, pp. 195–204.
52. A. Banerji and P. Rohatgi, *J. Mat. Sci.*, *17*, 335 (1982).
53. T. Kobayashik, M. Yosino, H. Iwanari, M. Niinomi, and K. Yamamoto, in S. Fishman and A. Dhingra, Eds., *Cast Reinforced Metal Composites*, ASM International, Metals Park, OH, 1988, pp. 205–210.
54. A. Das, A. Clegg, B. Zantout, and M. Yakoub, in S. Fishman and A. Dhingra, Eds., *Cast Reinforced Metal Composites*, ASM International, Metals Park, OH, 1988, pp. 139–147.
55. S. Ahmed, V. Gopinathan, and P. Ramakrishnan, in S. Fishman and A. Dhingra, Eds., *Cast Reinforced Metal Composites*, ASM International, Metals Park, OH, 1988, pp. 149–153.
56. G. Cappleman, J. Watts, and T. Clyne, *J. Mat. Sci.*, *20*, 2159 (1985).
57. T. Nieh and D. Chellman, *Scripta Met.*, *18*, 925 (1984).
58. D. McDanels, *Metall. Trans. A.*, *16A*, 1105 (1985).
59. C. Crowe, R. Gray, and D. Hasson, in W. Harrigan, J. Strife, and A. Dhingra, Eds., *Proceedings of the 5th International Conference on Composite Materials*, The Metallurgical Society, Warrendale, PA, 1985, pp. 843–866.
60. T. Nieh, *Metall. Trans. A*, *15A*, 139 (1984).
61. V. Nardone and J. Strife, *Metall. Trans. A*, *18A*, 109 (1987).
62. W. Logsdon and P. Liaw, *Eng. Fract. Mech.*, *24*, 737 (1986).
63. J. Dinwoodie, E. Moore, C. Langman, and W. Symes, in W. Harrigan, J. Strife, and A. Dhingra, Eds., *Proceedings of the 5th International Conference on Composite Materials*, The Metallurgical Society, Warrendale, PA, 1985, pp. 671–685.
64. L. Ackermann, J. Charbonnier, G. Desplanches, and H. Koslowski, in W. Harrigan, J. Strife, and A. Dhingra, Eds., *Proceedings of the 5th International Conference on Composite Materials*, The Metallurgical Society, Warrendale, PA, 1985, pp. 687–698.
65. A. Sakamoto, H. Hasegawa, and Y. Minoda, in W. Harrigan, J.

Strife, and A. Dhingra, Eds., *Proceedings of the 5th International Conference on Composite Materials,* The Metallurgical Society, Warrendale, PA, 1985, pp. 699–707.

66. *Duralcan,* Dural Aluminum Composites Corp., data sheet, San Diego, CA, 1989.
67. R. Arsenault and S. Wu, in S. Fishman and A. Dhingra, Eds., *Cast Reinforced Metal Composites,* ASM International, Metals Park, OH, 1988, pp. 231–236.
68. Y. Flom and R. Arsenault, in F. Matthews, N. Buskell, J. Hodgkinson, and J. Morton, Eds., *Proceedings of 6th International Conference on Composite Materials,* Elsevier Applied Science Publishers, London and New York, 1987, pp. 2.189–2.198.
69. T. Nieh and R. Karlak, *J. Mat. Sci. Letters, 2,* 119 (1983).
70. W. Harrigan, G. Gaebler, E. Davis, and E. Levin, in J. Hack and M. Amateau, Eds., *Proceedings of Symposium on Mechanical Behavior of Metal Matrix Composites,* The Metallurgical Society, Warrendale, PA, 1982, pp. 169–180.
71. T. Nieh, R. Rainen, and D. Chellman, in W. Harrigan, J. Strife, and A. Dhindra, Eds., *Proceedings of the 5th International Conference on Composite Materials,* The Metallurgical Society, Warrendale, PA, 1985, pp. 825–842.
72. R. Schueller and F. Wawner, *J. Comp. Mat.,* 1990. In press.
73. H. Rack, in P. Kumar, K. Vedula, and A. Ritter, Eds., *Proceedings of Symposium on Processing and Properties for Powder Metallurgy Composites,* The Metallurgical Society, Warrendale, PA, 1987, pp. 155–168.
74. F. Wawner, "Characterization of Microstructure and Its Effect on the Fracture Behavior of SiC Whisker Reinforced Aluminum Alloys," Progress Report No. 10, Lockheed-Georgia Co., 1984.
75. J. Papazian, *Met. Trans. A, 19A,* 2945 (1988).
76. H. Rack, in F. Matthews, N. Buskell, J. Hodgkinson, and J. Morton, Eds., *Proceedings of 6th International Conference on Composite Materials,* Elsevier Applied Science Publishers, London and New York, 1987, pp. 1.382–2.389.
77. Y. Flom and R. Arsenault, *Mat. Sci. and Eng., 77,* 191 (1986).
78. D. Webster, *Metall. Trans. A, 13A,* 1511 (1982).
79. W. Hunt, O. Richmond, and R. Young, in F. Matthews, N. Buskell, J. Hodgkinson, and J. Morton, Eds., *Proceedings of 6th International Conference on Composite Materials,* Elsevier Applied Science Publisher, London and New York, 1987, pp. 2.209–2.223.

Metal Matrix Composites, Infiltration

Infiltration is a priori one of the simplest methods for fabricating metal matrix composites. A preform of the reinforcing phase is typically placed within a die, into which metal is then metered. The metal then penetrates, in the molten state, the interstices between the fibers to yield a metal matrix composite. The metal will infiltrate the preform spontaneously if it wets the fibers, but most often the application of pressure of the metal is necessary, either because wetting is not favorable or because speed in fabricating the composite is desired. There are several variants of the infiltration process, depending on the shape of the composite produced (long wire, small equiaxed part, etc.), or the nature of the reinforcement–metal combination, in particular its wetting characteristics. One example is the squeeze casting process, schematically described in Figure 1 for a selectively reinforced casting. The infiltration process is suited for fabricating composites where molten metal–reinforcement chemical reactivity is not excessive, and is particularly attractive with lower melting point metals such as aluminum, magnesium, zinc, lead or tin. In all but the few cases in which very slow infiltration or very large reinforcements are used, the kinetics of the process are mainly determined by viscous drag in the metal as it flows through the narrow interstices of the reinforcement. Our present knowledge of the process is based on the fundamentals of fluid flow through porous media, on solidification science and technology, and on recent work on issues specific to the process itself. In the following, the principal equations pertaining to the process are reviewed, along with practical considerations relevant to optimization of the process.

Consider the infiltration of a composite such as the simplified composite part described in Figure 2. Molten metal is poured at temperature T_m into a cavity of initial temperature T_c that contains a preform of the fibers, preheated to a temperature T_f. The metal infiltrates the preform under a positive pressure differential $\Delta P\mu$ between the pressure in the metal at the entrance of the preform P_o and the pressure at the infiltration front. This latter pressure is equal to the pressure in the gaseous atmosphere in the noninfiltrated portion of the preform, P_a, diminished by the pressure drop at the infiltration front $\Delta P\gamma$:

$$\Delta P_\mu + \Delta P_\gamma = P_o - P_a$$

Several approaches have been adopted for calculating the capillary pressure drop at the infiltration front ΔP_γ. A simple analysis of the energy required for reversible infiltration of a porous body yields the average value of ΔP_γ [1,2]:

$$\Delta P_\gamma = (\sigma_{FL} - \sigma_{FA})S_f = -\sigma_{LA} \cos\theta \, S_f \quad \text{provided } \theta \text{ exists} \tag{1}$$

where σ_{FL} is the fiber/liquid metal interfacial energy, σ_{FA} is the fiber surface energy, σ_{LA} is the surface tension of the metal, θ is the contact angle of the metal on the reinforcement, and S_f is the specific surface area (i.e., the fiber surface area per unit volume of metal matrix) of the region of the preform that is being infiltrated at time t. For the case of a fibrous preform consisting of fibers of uniform radius r_f, S_f is given by:

$$S_f = \frac{2V_f}{r_f(1 - V_f)} \tag{2}$$

Fluid flow within the infiltrated portion of the preform is governed by Darcy's law, which correlates the pres-

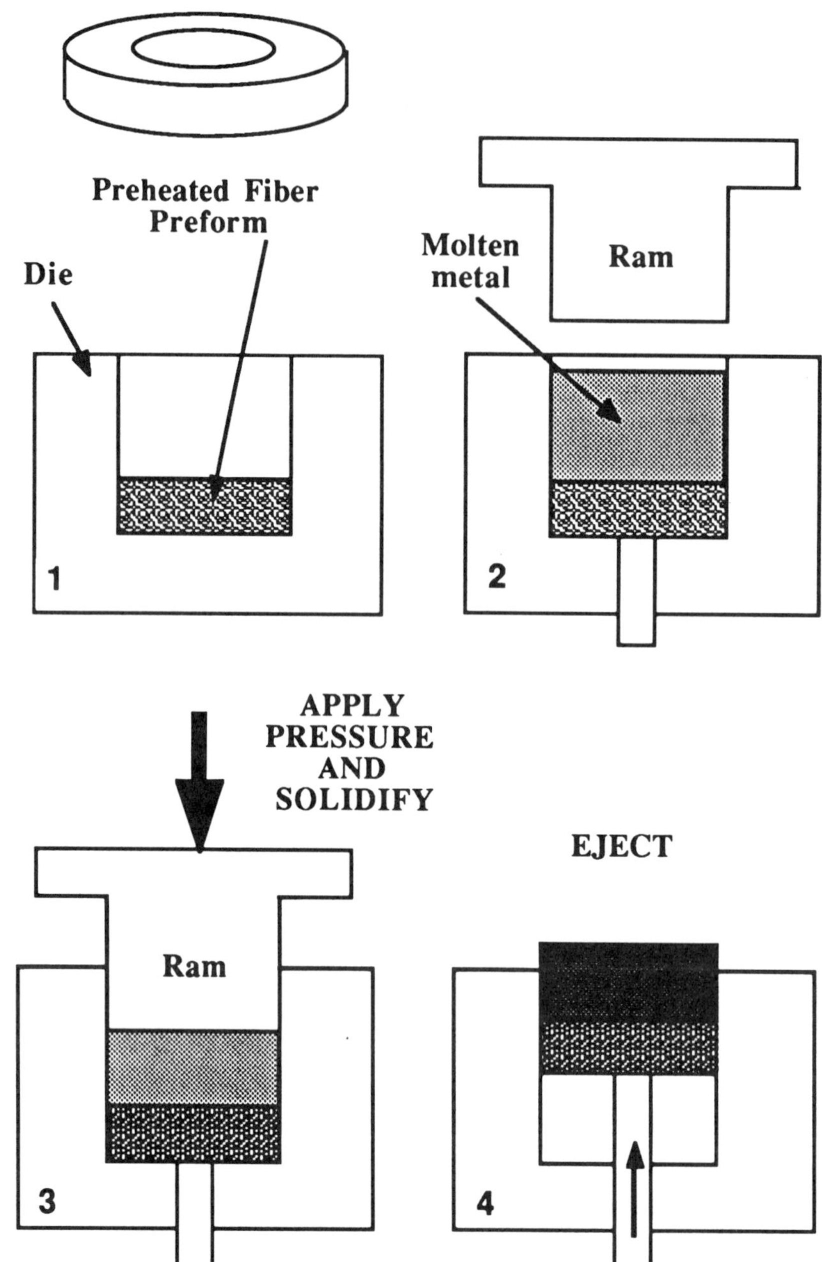

FIGURE 1 The squeeze casting process as used for manufacturing selectively reinforced automotive pistons: (1) the preheated preform is inserted into the die; (2) molten metal is poured into the die; (3) pressure is applied with a ram and the metal infiltrates the preform and solidifies; (4) the part is ejected.

sure gradient in the liquid and the superficial velocity:

$$\mathbf{v}_o = -\frac{\mathbf{K}}{\mu}[\text{grad}(P) - \rho_m \mathbf{g}] \qquad (3)$$

where

$\mathbf{v}_o$ = superficial velocity of liquid metal
μ = viscosity of liquid metal
$\mathbf{K}$ = (symmetric) permeability tensor of the preform
P = pressure in the metal
ρ_m = density of the metal
$\mathbf{g}$ = gravity field vector

provided the relevant Reynolds number:

$$R_e = \frac{2r_f \rho_m v_o}{\mu V_f}$$

is less than about 1 [3]. This will generally be the case in the infiltration of metal matrix composites because of the fine scale of the reinforcement (i.e., small r_f). Flow of the molten metal through the preform is thus governed by viscous energy dissipation and is analyzed on the scale of an elementary volume element dV that is small relative to the casting, but comprises at least several fibers so

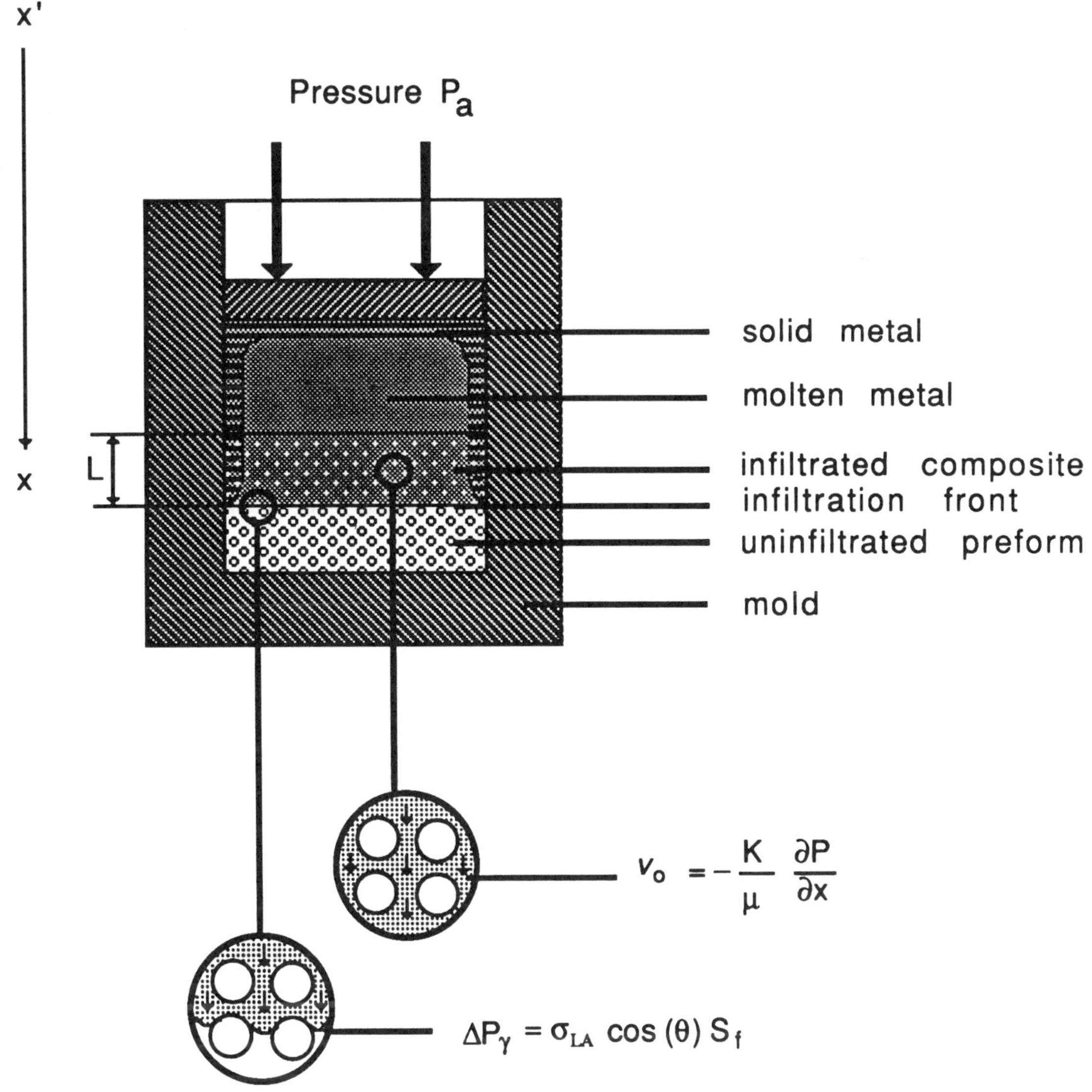

FIGURE 2 **Elementary description of the infiltration process.**

that fluid velocity and pressure can be averaged over this volume element. Fluid flow is then measured by the "superficial" velocity or "seepage" velocity $\mathbf{v}_o$, a vector whose direction is given by the average flow direction of the liquid and whose magnitude is equal to the volume of liquid flowing per unit time through a unit surface cut into the porous medium perpendicular to the average flow direction. Calculation of **K** can be based on the Blake-Kozeny approximation [4–8] or on appropriate relationships derived for the particular preform in question [9–11].

If we assume that both the liquid matrix and the preform are incompressible, the continuity equation is:

$$\text{div}\ (\mathbf{v}_o) = 0 \tag{4}$$

The pressure drop ΔP_μ is then the integral of the pressure gradient from the preform entrance to the infiltration front, and increases with increasing flow velocity of the metal, increasing viscosity of the metal and decreasing permeability of the preform. The energy required for the infiltration process is then the sum of two terms, corresponding respectively to the volume of metal displaced multiplied by the sum of the two terms ($\Delta P_\mu + \Delta P_\gamma$) of the total driving pressure gradient. The former term measures an irreversible energy expenditure as a result of viscous losses in the flowing metal, while the second term measures the minimum energetic requirement of the process (positive or negative according to wettability of the preform) resulting from capillarity.

For the simplified case of unidirectional infiltration along a principal direction of the permeability tensor, Eq. (3) is written:

$$v_{ox} = -\frac{K}{\mu}\frac{\partial P}{\partial x}$$

which is Darcy's law along direction $x'x$, where K is the permeability of the preform along that direction. This equation and the continuity equation Eq. (4) link the infiltration velocity, $dL/dt = v_o/(1 - V_f)$, with the pressure gradient in the infiltrated preform (L is the infiltrated preform length and t is time). Provided the permeability K of the preform is known everywhere and is independent of infiltration kinetics, the problem is then one of uncoupled fluid flow in that the infiltration kinetics can be derived independently to yield, for unidirectional infiltration, the position of the metal front as a function of time:

$$L^2 = \int_0^t \frac{2K\Delta P_\mu(t)}{\mu\,(1 - V_f)}\,dt \tag{5}$$

$$= \frac{2K\Delta P_\mu}{\mu(1 - V_f)}\,t \quad \text{if the applied pressure is constant}$$

We have so far ignored all heat and mass transport phenomena, and these are often important for the infiltration of metal matrix composites. First, infiltration is rarely performed within a mold held at a temperature above the melting point of the matrix, because chemical attack of the mold by the metal would be excessive and because sealing with a piston for application of pressure would be difficult. Hence, concommitant with infiltration, solidification of the matrix takes place from the mold walls, as illustrated in Figure 2. Furthermore, the fibers may initially be held at a temperature below the melting point or the liquidus of the metal matrix either deliberately in order to reduce chemical reactions between the two phases or as a result of cooling of the preform by the mold after insertion and before infiltration. Infiltration is still possible in this case because the metal can heat the fibers by local solidification at the infiltration front, to release latent heat that will raise the fibers temperature so they can coexist with liquid metal, as in Figure 3. Provided this solid metal does not choke flow of the liquid metal, infiltration will continue, albeit at a slower rate. With a highly pure metal matrix, the resulting solid metal will uniformly coat the fibers, thus protecting the latter and decreasing the apparent preform permeability. With an alloyed matrix, the problem is more complex because mass transport must be treated as well and because the metal freezes over a range of temperatures, which complicates heat flow calculations. Furthermore, the morphology of solidification of an alloy or even a metal of commercial purity differs from that of a high purity metallic matrix when the fibers are cold. The solid metal no longer takes the form of a uniform coating on the fibers because its solidification is controlled by solute diffusion, and prediction of preform permeability is therefore more difficult [12].

Heat and solute transport can also be analyzed on the scale of the same volume element dV if heat transfer is rapid enough within that element to equalize temperature and if diffusion is rapid enough to homogenize the liquid composition within dV. One can then characterize the volume element dV with a single temperature T and a single liquid solute composition, given by the concentration C_{iL} in solute i of the liquid. The heat transfer equation is then:

$$\rho_c c_c \frac{\partial T}{\partial t} + \rho_m c_m \mathbf{v}_o \cdot \text{grad}(T) = \text{div}[\mathbf{k}_c \text{grad}(T)] + (1 - V_f)\,\rho_m L \frac{\partial g_s}{\partial t} \tag{6}$$

where

$\rho_c c_c$ = volumetric heat capacity of composite,
$= \rho_f c_f V_f + \rho_m c_m (1 - V_f)$
$\rho_f c_f$ = volumetric heat capacity of fibers
$\rho_m c_m$ = volumetric heat capacity of metal
V_f = volume fraction of fibers in the composite
$\mathbf{k}_c$ = thermal conductivity tensor in the composite
T = temperature
g_s = fraction solid
L = latent heat of solidification of the metal

and solute conservation dictates:

$$\frac{\partial[(1 - V_f)\overline{C}_i]}{\partial t} = -\text{div}\,(C_{iL}\mathbf{V}_o) \tag{7}$$

where

C_{iL} = concentration of solute i in liquid metal phase
$\overline{C}_i$ = average concentration of solute i in total metal phase for each solute element, if density differences between liquid and solid metal phases are neglected. Depending on the solidification mechanism, an additional relationship correlates the average concentration with the liquid metal composition and fraction solid metal g_s.

$$\overline{C}_i = f(C_{jL}, g_s) \tag{8}$$

while the composition of the liquid and the temperature in dV are linked by the phase diagram if equilibrium is assumed at the solid/liquid metal interface:

$$C_{iL} = g(T) \tag{9}$$

In the case of a binary alloy, there will be eight unknowns, v_{ox}, v_{oy}, v_{oz}, P, T, C_L, g_s, and C at each point of the preform and one parameter, time. Equations (3), (4), and (6) to (9) yield eight relations among these unknowns. The problem is therefore defined by specification of the boundary conditions and of a relation giving the value of the permeability(ies) at each point of the infiltrated preform as a function of the nature of the fibers, including their volume fraction and the volume fraction of solid metal present g_s.

In the case of a pure metal, liquid and solid metal can only coexist at its melting point T_M if equilibrium is assumed at the solid–liquid interface. Therefore, the problem is much simplified, since there can be no heat transport where solid metal is present, and therefore g_s cannot vary continuously in the composite. Furthermore, with a pure metal, the composition is no longer an issue, and Eqs. (7) to (9) are irrelevant. As mentioned

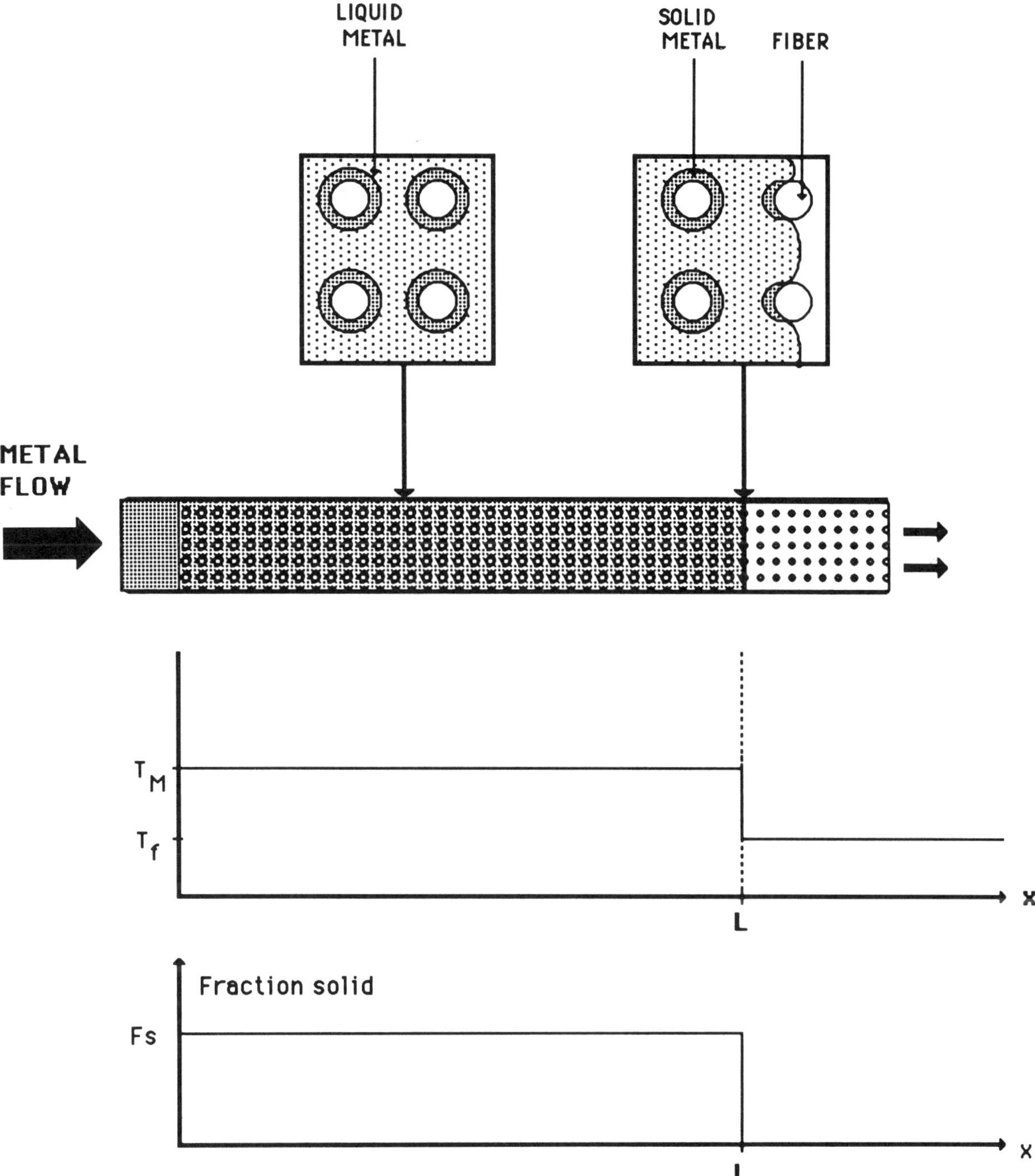

FIGURE 3 Schematic description of unidirectional infiltration of a fiber preform transverse to the fibers by a pure metal, for the case of fibers initially below the matrix melting point and no external cooling from the mold.

above, in the case of a pure unalloyed matrix, this solid metal must solidify as a sheath surrounding the fibers, since solidification is limited by heat transfer to the cold fibers. This assumption is substantiated by measured infiltration kinetics, predicted above by Eq. (3), but with a changed value of permeability in the region close to the infiltration front, which can be predicted by increasing the effective fiber diameter and volume fraction to account for a volume fraction solid g_s in the metal phase. g_s is calculated by thermal equilibration at the infiltration front:

$$g_s = \frac{\rho_f c_f V_f (T_M - T_f)}{\rho_f L (1 - V_f)} \tag{10}$$

Formation of this solid metal phase will substantially influence the kinetics of infiltration when the fibers are initially below the melting point of the metal. Low T_f will be deleterious from the point of view of infiltration kinetics, but may result in vastly improved composite properties as a result of reduced chemical reactions between the fibers and the metal [13]. Experiments with high purity aluminum have demonstrated the validity of these assumptions, and the physics of infiltration by a pure metal are presently rather well understood [7,10,11,14].

The problem of initially cold fibers with an alloyed matrix is more complex, as indicated above. Experiments have been reported in which it was shown that macrosegregation results in the composite after infiltration of alumina fiber preforms with Al–4.5 wt % Cu when the preform was initially at a temperature below the liquidus of the metal [15]. Theoretical analysis of the problem [16], allows prediction of the concentration profile in the composite.

If the preform is compressible, the fibers will move during infiltration at elevated pressures, a phenomenon first investigated by Clyne et al. [8,17]. This may result in variations of fiber volume fraction during infiltration, thus complicating prediction of infiltration kinetics and of final composite microstructure.

The effect of concommitant solidification from the mold walls must generally be taken into account as well. For example, macrosegregation observed in a composite with an alloyed matrix will be a result of the dual contributions of solidification from the fibers and of solidification from the mold walls. Both will interact with fluid flow to redistribute solute in the composite, and the macrosegregation profiles reported earlier are a result of both processes. Solidification resulting from external cooling also influences the infiltration front geometry and the kinetics of the process. With a constant applied pressure, for example, the total infiltrated length of composite is determined solely by external cooling. Also, feeding of solidification shrinkage will depend on external cooling from the mold walls. The importance of mold temperature and external cooling must not, therefore, be underestimated in interpreting occurrences of incomplete infiltration.

Other critical issues pertain to the entire casting, inasmuch as the usual foundry concern with part integrity, feeding shrinkage, and other such factors, remains justified. A first concern is that the casting be filled before the metal chokes in the riser as a result of solidification from the mold walls. The ability to predict the kinetics of both infiltration and solidification is therefore critical. Any gas present in the mold before infiltration must be evacuated, and prediction of the location of the infiltration front during infiltration is therefore important in order to vent these gases. Solidification shrinkage in an infiltrated composite is also a new problem, inasmuch as the reinforcing phase prevents massive flow of semisolid metal and also hinders flow of the liquid metal to some extent. It has been observed that hot tears often result at the boundaries of the preform where these meet unreinforced metallic regions of the casting. Fukunaga et al. were successful at preventing shrinkage voids by directional solidification toward a riser [18].

In general, there is at least one moving boundary in any infiltration problem: the infiltration front. In many cases there will also be a second moving boundary delineating the region of fully solid metal, depicted as well in Figure 2, which is formed by concomitant solidification of the whole casting as a result of external cooling at the mold wall. The anisotropic and inhomogeneous nature of many fiber preforms may further complicate the problem because the permeability of preforms consisting of parallel fibers is greater by a factor of about 2 along the fiber axis compared with that along an axis transverse to the fibers [9]. Flow of the metal will therefore be much more rapid parallel to the fibers. Furthermore, the distribution of the fibers is never regular, and there will be regions of high volume fraction fiber, and therefore opposing higher resistance to penetration by the liquid metal, because of increased capillary forces in these regions (Eqs. 1 and 2) and because of a much lower permeability K in these regions. The infiltration front will then be very irregular, as metal will first penetrate the preform along wider channels parallel to the fibers, and later complete infiltration of the preform into regions of higher local V_f in non-wetting situations.

The presence in this problem of moving boundaries, the anisotropy of the permeability tensor, and the presence of inhomogeneous packing in most fiber preforms complicate its solution immensely for the most general case. The approach in research has therefore been to utilize model systems that facilitate analysis and data acquisition. This approach has allowed (1) determination of fundamental physical phenomena underlying the problem, in particular validation or definition of the equations presented above, and (2) suggestion of possible approximations facilitating treatment in cases of practical interest.

Under various forms and denominations (squeeze casting, pressure casting, fiber tow infiltration, and so on), infiltration processes for metal matrix composite fabrication have been gaining in importance very significantly over the past decade. The kinetics of the process are to a large extent determined by the viscosity of the metal (Eq. 3) and their attractiveness is thus accentuated since the viscosity of a molten metal is much lower than that of a molten glass or polymer. Their promise in terms of economy, versatility, manufacturing speed, near-net shape capability and relative simplicity is strong, and well illustrated by the fact that the first commercially mass-produced metal matrix composite (the Toyota diesel engine piston) is manufactured by one such process, namely squeeze casting.

Andreas Mortensen

References

1. P. C. Carman, *Soil Science, 52,* 1 (1941).
2. A. Mortensen and J. A. Cornie, *Metall. Trans., 18A,* 1160 (1987).
3. S. Ergun, *Chem. Eng. Prog., 48,* 89 (1952).
4. A. E. Scheidegger, *The Physics of Flow Through Porous Media,* 3rd ed., University of Toronto Press, 1960.
5. G. H. Geiger and D. R. Poirier, *Transport Phenomena in Metallurgy,* Addison Wesley, Reading, Mass., 1973, sec. 3.4.

6. R. B. Bird, W. E. Stewart, and E. N. Lightfoot, *Transport Phenomena*, Wiley, New York, 1960, p. 196.
7. H. Fukunaga and K. Goda, *Bull. Japan. Soc. Mech. Eng., 27,* 1245 (1984).
8. T. W. Clyne, M. G. Bader, G. R. Cappleman, and P. A. Hubert, *J. Materials Science, 20,* 85 (1985).
9. G. W. Jackson and D. F. James, *Can. J. Chem. Eng., 64,* 364 (1986).
10. A. Mortensen, L. J. Masur, J. A. Cornie, and M. C. Flemings, *Metall. Trans. A, 20A,* 2535 (1989).
11. L. J. Masur, "Infiltration of Fibrous Preforms by a Pure Metal," Ph.D. thesis, MIT, 1988.
12. L. J. Masur, A. Mortensen, J. A. Cornie, and M. C. Flemings, in F. L. Matthews, N. C. R. Buskell, J. M. Hodgkinson, and J. Morton, Eds., *Proceedings of the Sixth International Conference on Composite Materials ICCM 6,* London, Elsevier, 1987, pp. 2.320–2.329.
13. H. Fukunaga and K. Goda, *J. Japan Inst. Metals, 49,* 78 (1985).
14. L. J. Masur, A. Mortensen, J. A. Cornie, and M. C. Flemings, *Metall. Trans. A., 20A,* 2549 (1989).
15. Experimental work by L. J. Masur, cited by J. A. Cornie, A. Mortensen, and M. C. Flemings, in F. L. Matthews, N. C. R. Buskell, J. M. Hodgkinson, and J. Morton, Eds., *Proceedings of the Sixth International Conference on Composite Materials ICCM 6,* London, Elsevier, 1987, pp. 2.297–2.319.
16. A. Mortensen, V. Michaud, J. A. Cornie, and M. C. Flemings, "Infiltration of Fiber Preforms by Al-4.5 wt%Cu Part I: Theory", *Metall. Trans.,* in print.
17. T. W. Clyne and J. F. Mason, *Metall. Trans., 18A,* 1519 (1987).
18. H. Fukunaga, S. Komatsu and Y. Kanoh, *Bull. Japan Soc. Mech. Eng., 26,* 1814 (1983).

Metal Matrix Composites, Magnesium

Introduction

Metal matrix composites are becoming a large and important class of materials. They offer previously unattainable levels of strength-to-weight and stiffness-to-weight ratios. Various metals such as aluminum, magnesium, and titanium have been reinforced with particulates, platelets, whiskers, short discontinuous fibers, and continuous fibers. Magnesium's light weight and natural affinity for wetting or bonding to such reinforcing materials make it one of the best choices as a matrix metal. Nonetheless, the use of magnesium as a matrix in metal matrix composites has received relatively little attention. The properties of magnesium matrix composites, although not widely publicized, are outstanding. For example, a composite of magnesium reinforced with continuous graphite fibers possesses the highest specific stiffness of any common engineering material.

The information contained in this article has been condensed from a review of the literature. Computer searches in Metadex and the U.S. patents produced a total of 235 citations. The key words used in the computer search were magnesium base alloys, matrix composites, magnesium, and composite materials. Other literature references were identified by searching the open literature, and some internally generated results are reported. Undoubtedly, some omissions have occurred, and the authors apologize in advance.

Magnesium Alloys Reinforced with Continuous Fibers

Aluminum Oxide Fibers

FIBER PROPERTIES. In recent years, Al_2O_3 fibers have been successfully used to reinforce magnesium alloys. Aluminum oxide fibers have several advantages over boron fibers and graphite fibers, such as chemical stability, electrical nonconductivity, and high temperature capability. Several Al_2O_3 fibers have become available. Tyco Laboratories has produced single-crystal continuous monofilament Al_2O_3 fibers by drawing from a melt. However, these fibers were found to lose their strength readily at high temperatures as a result of twinning. Imperial Chemical Industries produces short chopped Al_2O_3 fibers, mostly for insulation applications. Minnesota Mining and Manufacturing Company produces continuous Al_2O_3-based filaments for insulation, but these fibers have a relatively low modulus. In addition, the Sumitomo Company of Japan produces continuous Al_2O_3–SiO_2 fibers, but again these fibers have a lower modulus than 100% alpha Al_2O_3 fibers. The DuPont Company has been developing a continuous 100% alpha Al_2O_3 fiber, designated as fiber FP, and has done a considerable amount of development work on magnesium alloys + fiber FP composites [1]. In addition, DuPont has developed an Al_2O_3–ZrO_2 fiber designated PRD-166, which is claimed to have even better properties [2].

FIBER FP PROPERTIES. Fiber FP is a polycrystalline, greater than 99% pure, alpha Al_2O_3 fiber with an average grain size of 0.5 μm. Fiber FP is available in the form of continuous multifilament yarn with about 210 filaments per yarn. Each filament is approximately 20 μm in diameter, has a round cross section, and is coated with a thin layer (approximately 50 Å thick) of SiO_2. The SiO_2 coating is employed to promote wetting by molten metals. Each filament has a cobblestone surface, which is said to promote a strong metal-to-fiber bond in metal matrix composites [1]. In addition, the filaments have excellent stiffness and strength properties (see Table 1).

The fibers are very stable with respect to high temperature exposures in air [1], and fibers do not lose their strength when exposed to molten magnesium [3].

Since fiber FP is available as multifilament yarn, it is flexible and easy to handle, and it can be woven into two-dimensional fabrics. The cost of fiber FP is projected to be lower than that of boron, SiC, or graphite fibers [1,4].

TABLE 1
Properties of DuPont Alumina

E	345 to 379 GPa
UTS	1379 MPa
Elongation	0.4%
Density	3.90 g/cm^3

E = elastic modulus
UTS = ultimate tensile strength

Fiber FP is available in 454 g bobbins (approximately 2100 m of fiber). It is also available as tapes with fugitive binder, as fabrics, and as chopped staple (3.175 mm to 12.7 mm lengths) [5].

MAGNESIUM ALLOY–FIBER FP COMPOSITES: FABRICATION. DuPont has produced magnesium and magnesium alloy + fiber FP composites, with up to 70 vol % FP, by a potentially low cost liquid metal infiltration process [6–8]. First, a tape containing aligned fibers in a fugitive binder is prepared. The tapes are then laid up in the desired fashion to produce a preform. Next, the preform is loaded into a steel mold, and the fugitive binder is burned off at about 600°C. After the mold assembly is completed, the mold is preheated and evacuated with a vacuum pump. Molten magnesium or magnesium alloy is then vacuum-infiltrated into the mold at temperatures of around 700°C. The casting process is occasionally assisted by pressure in addition to vacuum. The metal is then allowed to solidify to form the composite part [4,9–11]. This process has produced excellent castings with good fiber–matrix bonding, low porosity (< 0.5 vol %), [3], and an excellent fiber distribution [7]. Castings sometimes show an extremely fine grain size [3].

DuPont has also produced magnesium alloy ZE41–30 vol % FP composite castings containing chopped fibers by undisclosed casting methods [3]. Finally, DuPont has investigated wrought FP–magnesium composites. Composite billets have been successfully extruded, forged, and rolled [12,13].

PROPERTIES OF UNIAXIALLY REINFORCED MAGNESIUM–FP COMPOSITES

Room temperature axial properties for Mg–50 vol % FP. The properties of continuous fiber FP reinforced unalloyed magnesium castings are widely reported in the literature. [3,5–7,9,11,14–18]. Ranges of axial tensile properties reported in the literature are given in Table 2.

High temperature axial properties for Mg–50 vol % FP. One of the most outstanding features of FP–Mg composites is their ability to retain axial stiffness and strength at high temperatures [3,5–7]. For instance, a Mg–50 vol % FP composite retains over 90% of its room temperature ultimate tensile strength at 316°C [6]. Table 3 shows the effect of temperature on the ultimate compressive strength of these composites [3]. No loss in the compressive modulus occurs even at 427°C.

There is virtually no loss in flexural properties until the temperature exceeds 371°C. Similarily, axial fatigue properties are only slightly affected by high temperatures [3].

Transverse properties for unidirectional Mg–50 vol % FP. The transverse tensile, compressive, and thermal expansion properties of uniaxial Mg–50 vol % FP are also widely reported in the literature [3,5–7,15,19]. The range of reported values is given in Table 4. The transverse tensile and compressive strengths are greater than those of pure magnesium, indicating an excellent fiber-to-matrix bond.

Unidirectional Mg–fiber FP composites—effects of fiber volume fraction and orientation on tensile and fatigue properties. Page et al. [16–18] have published four excellent papers on the effect of orientation on the properties of unidirectional Mg–FP composites. The tensile properties of two Mg–FP composites (35 vol % and 50 vol % FP) are given in Table 5. As the angle between the applied stress and the fiber direction (θ) is increased, both the elastic modulus and especially the ultimate tensile strength are substantially reduced. In fact, the transverse UTS is reduced to a value that is about the same

TABLE 2
Axial Properties for Magnesium–50 Vol % FP Undirectional Composites

Property	Units	Composite	Pure Mg
Tensile modulus	GPa	193–228	44.8
Ultimate tensile strength	MPa	482–572	—
Tensile strain to failure	%	0.3–0.5	—
Compressive modulus	GPa	193–228	44.8
Compressive strength	MPa	1930–2586	—
Shear modulus	GPa	40.7	16.5
Shear strength	MPa	69–83	—
Fatigue strength*	% of UTS	65–75	—
CTE	mm/mm· K × 10^{-6}	6.8	25.4

* 10^7 cycle endurance limit, tension-tension, R = 0.1.

TABLE 3
Effect of Temperature on the Modulus of Elasticity of a Mg–50 Vol % FP Composite Tested in Compression

T (°C)	UCS (MPa)	*E* (GPa)
RT	1931	207
316	1262	207
427	779	207

order of magnitude as that for unreinforced magnesium [16]. The volume fraction of fiber in the composite has a strong effect on the axial properties, but it has much less effect on the off-axis properties.

Fractography of specimens tensile-tested in the axial direction (i.e., $\theta = 0°$) showed as fracture perpendicular to the fiber direction—the fibers themselves were fractured. Many fractures were seen to initiate at the interface of process defects and the magnesium matrix, but the tensile properties seemed to be fairly insensitive to such defects. These defects include large Al_2O_3 particulate clumps and broken fiber segments lying perpendicular to the stress axis. Failure of all the off-axis 55 vol % samples and the 35 vol % samples tested at 45° and 90° occurred along the fiber axis by decohesion of the fiber–matrix interface. A small reaction zone of MgO crystals was present at the fiber–matrix interface, and failure of off-axis samples actually occurred at the Mg matrix–MgO reaction zone interface. For Mg–35 vol % FP samples tested at $\theta = 22.5°$, the fracture occurred at an angle close to 90° to the applied stress by a mixture of interfacial decohesion and fiber cracking. For off-axis samples, process defects did not play a role in fracture initiation.

Page et al. [16] also discuss the fatigue properties of Mg–35 vol % FP and Mg–55 vol % FP composites. The fatigue strength scaled roughly with the ultimate tensile strength. Therefore, the composite materials showed excellent axial fatigue strength. Off-axis fatigue properties were substantially reduced. It was also found that the volume fraction of FP has a strong influence on axial fatigue properties but a negligible influence on off-axis fatigue properties [16].

Fractography of axial fatigue specimens (i.e., $\theta = 0°$) showed fracture perpendicular to the fiber direction—the fibers themselves were fractured. Fatigue cracks initiated at the interface of process defects with the magnesium matrix, but the fatigue strength appeared to be independent of the size and number of such defects. Off-axis fatigue samples showed a different behavior: The fracture was parallel to the fiber direction. In the fatigue region, the fracture surfaces indicated a combination of interfacial (MgO–Mg interface) delamination and cracking of the Mg matrix. However, cracking of the Mg matrix

TABLE 4
Transverse Properties for Unidirectional Mg–50 Vol % FP

Property	Units	Composite	Pure Mg
Tensile modulus	GPa	90–110	45
Ultimate tensile strength	MPa	69–207	—
Tensile strain to failure	m/m, %	0.24–0.42	—
Ultimate compressive strength	MPa	241–276	—
Room temperature, CTE	mm/mm·K $\times 10^{-6}$	13.6	—
Compressive modulus	GPa	102	45

TABLE 5
Tensile Properties for Mg–35 and 55 Vol % FP Unidirectional Composites as a Function of Orientation

Vol % Al_2O_3	θ (degrees)	*E* (GPa)	UTS (MPa)	Total Strain at Failure (%)	Reference
35	0.0	149	383	0.26	16
	22.5	116	207	0.52	16
	45.0	90	108	0.45	16
	90.0	85	104	0.42	16
55	0.0	197	608	0.30	16
	22.5	155	196	0.81	16
	45.0	124	157	0.58	16
	90.0	90	67	0.24	16

was dominant. The major role of the fibers appears to be channel cracks through the matrix in a direction parallel to the fibers. In the overload region, failure was by delamination at the MgO–Mg interface in all but the 35 vol % material tested at $\theta = 22.5°$ (where fracture was by a combination of interfacial decohesion and fracture across the fibers).

In conclusion, it was found that the axial properties were strongly influenced by the fiber content, while the off-axis properties were relatively insensitive to fiber content. Off-axis loading resulted in substantial reductions in both the tensile and fatigue properties. These reductions coincided with a change in fracture morphology from flat fracture across the fibers to fracture parallel to the fiber direction. Both a weak interface and a weak matrix were found to be responsible. It was suggested that alloying to increase the matrix and fiber–matrix interface strength might improve off-axis properties.

Chan et al. discuss fracture toughness properties as a function of orientation for Mg–FP composites [20].

PROPERTIES OF UNIAXIALLY REINFORCED MAGNESIUM ALLOY–FP COMPOSITES

Axial properties for magnesium alloy–FP composites. In addition to unalloyed magnesium, magnesium alloys have also been reinforced with FP-Al_2O_3 fiber. Properties of EZ33A, ZE41A, and QH21A uniaxially reinforced with 35–55 vol % FP are given in the literature [3,4,11,17–19,21]. Generally, axial properties do not vary much with the matrix alloy, and the axial properties are very similar to those of unalloyed Mg–FP composites (Table 6). The specific properties of magnesium alloys reinforced with fiber FP are excellent [15].

Page et al. claim that ZE41–FP uniaxial composites are not as strong as Mg–FP uniaxial composites [17]. In both materials, a reaction zone consisting of MgO particles was found at the fiber–matrix interface. For the ZE41 matrix, the MgO particles in the reaction zone were larger, and the total reaction zone thickness was larger. However, the reduction in cross section of the fibers was very small, and alloying element segregation was not seen. It was suggested that the larger MgO particles in the ZE41–FP composite effectively acted as notches on the fiber surface and thereby degraded fiber strength. The degraded fiber strength resulted in lower axial properties. On the other hand, in a separate paper, Page and Leverant claim that there is little reduction of fiber strength in ZE41–FP composites [18]. In fact, properties of ZE41–35 vol % FP were slightly better than those of Mg–35 vol % FP.

TABLE 6
Axial Room Temperature Tensile Properties of Magnesium Alloys Uniaxiallly Reinforced with 55 Vol % Fiber FP

Matrix	E (GPa)	UTS (MPa)
QH21A	158	505
EZ33A	158–194	500
ZE41A	179–214	490–531
Mg	193–228	483–572

Tranverse properties for magnesium alloys–FP composites. The tranverse properties of uniaxially reinforced magnesium alloy–FP composites are strongly influenced by the matrix properties [3,4,11,17–19, 21]. For instance, it was found that the transverse UTS of as-cast undirectionally reinforced magnesium alloy–50 vol % fiber composites is directly proportional to the strength of the unreinforced matrix [3]. This is indicative of good fiber-to-matrix bonding.

Orientation effects for magnesium alloy–FP composites. Detailed information on the effects of orientation for unidirectionally reinforced ZE41–FP composites has been published [11,17,18]. Table 7 gives some tensile and fatigue properties of ZE41–35 and 55 vol % FP and Mg–35 vol % FP composites as a function of orientation [11,16–18].

The off-axis properties are greatly improved for the ZE41 matrix versus the magnesium matrix, and this has been attributed to both matrix and interface strengthening [11,18].

For ZE41–FP composite tensile tested at both $\theta = 0°$ and $\theta = 22.5°$, the fracture occurred perpendicular to the applied stress and involved flat fracture across the fibers. For $\theta = 45°$, the fracture surface exhibited a mixed-mode failure. Failure occurred both across the fibers, as in axial failure, and parallel to the fibers. The cracking parallel to the fibers occurred primarily within the matrix, but also at the MgO reaction zone–Mg alloy interface. Failure of transverse specimens did not involve fiber fracture. Instead, failure of transverse specimens involved cracking parallel to the fibers by a combination of matrix and interfacial failure. Thus, as θ was increased from 0 to 90°, a transition in the failure mode from fiber-controlled fracture to matrix–interface-controlled fracture occurred. The effect of alloying was to delay this transition to larger values for θ. The transition was delayed because alloying increased both the matrix strength and the fiber–matrix interface strength, the latter via the formation of a thicker MgO reaction zone [11,18]. The involvement of fiber fracture for $\theta = 22.5°$ and 45.0° is responsible for the improved off-axis properties.

Fatigue properties of ZE41–FP composites were also investigated [11,17,18]. The off-axis fatigue properties of ZE41–FP composites were also greatly improved relative to those of Mg–FP composites. Axial fatigue properties were less strongly affected. Again, fractography revealed similar trends with respect to the effect of the angle of misorientation on the failure mode, and the good off-axis fatigue properties can be explained by an increased involvement of fiber fracture in the failure mode.

THERMAL DAMAGE OF UNIAXIALLY REINFORCED MAGNESIUM AND MAGNESIUM ALLOY–FP COMPOSITES. In many metal matrix composite systems, a fiber–matrix reaction significantly degrades properties af-

TABLE 7
Tensile and Fatigue Properties for Magnesium and Magnesium Alloy ZE41 Unidirectionally Reinforced with Fiber FP

Matrix	Vol % FP	Angle of Misorientation θ (°)	E (GPa)	UTS (MPa)	Total Strain at Failure (%)	10^7 Cycle Fatigue Strength (MPa)	Reference
Mg	35	0.0	149	383	0.26	—	16
		22.5	116	207	0.52	—	16
		45.0	90	108	0.45	—	16
		90.0	85	104	0.42	—	16
ZE41	35	0.0	163	438	0.29	—	18
		22.5	127	301	0.63	—	18
		45.0	97	234	0.99	—	18
		90.0	83	190	0.53	—	18
ZE41	55	0.0	179–214	483–531	0.28	317	11,17
		22.5	151–165	379–414	0.34	186	11,17
		45.0	136	296	0.56	152	11
		90.0	108	228	0.36	131	11

ter long exposures to high temperatures. Thus, several papers measuring the effect of high temperature exposures on the room temperature tensile strength of Mg–, EZ33–, and QH21–FP composites have been published [19,21]. Isothermal exposures of up to 150 hours at 350°C resulted in no significant loss in the room temperature axial properties for Mg–FP, EZ33–FP, and QH21–35 and 50 vol % FP composites. Similarly, isothermal exposure for 120 hours at 450°C did not degrade the axial properties of EZ33–FP and QH21–FP composites. However, the room temperature axial tensile strength of Mg–FP composites was significantly reduced. In addition, the transverse tensile strengths of all three composite systems were reduced by thermal exposure. Thermal cycling (up to 3000 cycles) between 50°C and 250°C had no effect on the room temperature axial tensile properties of EZ33–FP composites. These rather severe tests show that Mg alloy–FP composites have considerable high temperature capabilities; however, as one author concludes, prolonged use at temperatures greater than or equal to 300°C should be avoided [19].

Thermal degradation was attributed to matrix softening, fiber–matrix debonding, and void growth at the fiber–matrix interfaces, and not to any fiber degradation via a fiber–matrix reaction [19,20). In a separate study, Mg–FP composites were fabricated, and the fiber–matrix interfaces were examined in some detail [22]. A small reaction zone consisting of $MgAl_2O_4$ spinel phase was identified. A spinel phase was also identified in Al-Mg alloys–FP composites [23]. This is in contrast to the MgO reaction zone suggested in Refs. 11 and 16–18. Nonetheless, the reaction zone was said to provide a diffusion barrier and protect the fiber at high temperatures. Mg–FP composites heated at 177°C for 100 hours showed little change in the thickness of the reaction zone.

CROSS-PLY ORIENTATIONS OF MG AND MG ALLOY–FIBER FP COMPOSITES. Several papers discuss the properties of cross-ply Mg- and Mg alloy–FP composites [1,3].

POTENTIAL APPLICATIONS FOR MG- AND MG-ALLOY–FP COMPOSITES. Potential applications for Mg–FP and Mg alloy–FP composite castings include helicopter transmission housings; automobile connecting rods, pistons, valves, and rocker arms; aircraft and missile engine components; military hardware such as ordnance, armor, and tanks; and superconductor restraints in fusion power reactors [1,3,9,15,24–28].

MG ALLOY–SUMITOMO'S ALUMINUM OXIDE FIBERS. As mentioned earlier, the Sumitomo Company of Japan produces continuous Al_2O_3–SiO_2 fibers from 0.6 to 400 μm in diameter. The alumina content is in the range of 72 to 98 wt %; the balance is SiO_2. Preferably, the fiber contains no alpha Al_2O_3 phase. Typically, the fiber has a tensile strength greater than or equal to 979 MPa and an elastic modulus of 96–296 GPa. Sumitomo's fiber is stable to prolonged exposures at 1000°C in air, and its density is 2.5 to 3.5 g·cm^{-3} [29].

In several U.S. patents, the properties of Mg alloys reinforced with Sumitomo's continuous Al_2O_3–SiO_2 fibers are described [29–33]. The particular fiber used in patent examples had the following properties:

85% Al_2O_3/15% SiO_2; no alpha Al_2O_3
UTS = 1469 MPa
E = 227 GPa
Diameter = 24 μm

The composites were undirectionally reinforced with 50 vol % continuous Al_2O_3/SiO_2 fiber, and they were fabri-

TABLE 8
Flexural Strengths of Mg Alloys Uniaxially Reinforced with Continuous Alumina–Silica Fibers

Matrix	Flexural Strength (MPa)	Reference
Mg–1.05 wt % Sn	510	30,32
Mg–1.02 wt % Cd	497	30,32
Mg–0.99 wt % Sb	621	30,32
Mg–0.08 wt % Cs	620	31
Mg–2.40 wt % Ba	710	31
Mg–2.40 wt % Bi	669	31
Mg–40 to 90 wt % Zn	> 1172	33
Pure Mg	393–448	30,31,32,33

cated by vacuum- and pressure assisted liquid metal infiltration techniques. Various elements were added to the magnesium matrix to (1) enhance wetting and (2) decrease any fiber–matrix reaction. Flexural strengths were reported (Table 8).

Carbon (Graphite) Fibers

FIBER PROPERTIES. Carbon fibers offer greater stiffness than any other fiber. In addition, carbon fibers are amendable to large-scale production, and they are available from a large number of producers [34]. The carbon or graphite fibers are produced from a variety of initial sources, the major ones being petroleum pitch and polyacrylonitrile (PAN). Carbon fibers are also available with a wide range of mechanical properties depending on the level of graphitization.

Carbon fibers generally fall into three different categories—high strength intermediate modulus fibers (HSIM), high modulus fibers (HM), and ultra-high modulus fibers (UHM). The defining characteristics are:

Category	UTS (MPa)	*E* (GPa)
HSIM	2414–3621	172–276
HM	—	345–414
UHM	—	483 and above

Generally, the higher the graphitization level, the higher the modulus. The low density of carbon fibers (on the order of 2 g·cm^{-3}) also makes them attractive as a reinforcement for metals. Finally, carbon fibers can have a negative coefficient of thermal expansion. Some typical properties of carbon fibers are given in Table 9 [35]. Carbon fibers are usually used as continuous multifilament yarn with as many as 10,000 filaments or more per yarn. Each filament is on the order of 10 μm in diameter.

TABLE 9
Properties of Graphite Fibers

	Fiber Type			
	T300	IM6	P55	P100
UTS, MPa	3103	4378	1724	2241
E, GPa	229	276	379	724
Density, g/cm^3	1.73	1.73	2.0	2.15
CTE, mm/mm K × 10^{-6}	− 0.5	N/A	− 0.9	− 1.6
K, w/m·K	8.5	N/A	100	520
K, % IACS	2%	N/A	24%	124%

FABRICATION OF MG–CARBON FIBER COMPOSITES. Their low density, excellent stiffness, and negative CTE make carbon fibers a very attractive candidate for reinforcing magnesium alloys. However, Mg–carbon fiber composite systems are not without numerous problems. One of the foremost problems involves fabricating composites. Liquid metal infiltration techniques require that the liquid metal both readily wet the carbon fibers and not degrade the fiber properties by excessive reaction. Unfortunately, magnesium and magnesium alloys do not readily wet carbon fibers, and alloying elements in magnesium alloys often attack carbon fibers. Diffusion bonding techniques to form metal matrix composites are not without their problems either. In a review article on composite fabrication, Cornie et al. claim that vacuum hot pressing/diffusion bonding cannot be easily applied to make metal matrix composites with fibers less than 20 μm in diameter unless (1) the tows have been previously infiltrated with metal to form wires, (2) the tows have been spread and coated or plated with the matrix metal to the desired volume fraction, or (3) the tows have been infiltrated with a fine powder slurry of the matrix and binder [36]. Forming metal matrix composite wires before vacuum hot pressing/diffusion bonding via liquid metal infiltration also requires that the liquid matrix metal readily wet the carbon fiber and not degrade the fiber properties. Nonetheless, because of the attractive properties of Mg–carbon fiber composites, considerable effort has been expended on overcoming the wettability and fabrication difficulties.

Generally, carbon fibers are coated before metal matrix composite manufacture in order to both promote wetting and inhibit reaction. Many coating systems have been investigated. Among the first systems investigated were metal coatings applied to the graphite fibers by ion plating methods. Metal coatings such as Ni, Ta, Ag, Al-Ag alloys, Co, Cu-Ni alloys, and Cu-Co alloys have been investigated [37–40]. However, problems were often encountered. For instance, with Cu-Ni and Cu-Co coatings, the coating degraded the fiber strength [37]. With Mg–Ni-coated graphite fiber composites prepared by squeeze casting, the thick Ni coating reacted with the Mg matrix to form excessive amounts of brittle Mg-Ni compounds and, thereby, degraded the composite strength [41].

Also, in some of the early work on carbon fiber reinforced metal matrix composites, Na and K were investigated as wetting agents [42,43]. Carbon fibers were first immersed in a bath of molten Na (liquid at high temperatures) or Na-K alloy (liquid at room temperature) to coat the fibers. The coated fibers were then immersed in a bath of Al or other matrix metal. Unfortunately, Na and K severely embrittle Mg and Mg alloys [44], and this technique is, therefore, not very amenable to Mg. Thick Ti coatings were applied to graphite fibers by evaporating Ti with an electron beam welding unit onto the fiber surfaces, and Mg–graphite fiber composites were successfully prepared [45]. Also, the Japanese have coated graphite fiber with Al metal by plasma spraying [27].

More recently, coatings containing a metal compound have proven to be excellent. Gaseous metal chlorides (i.e., Ti, Si, Zr, Ha, Va, Nb, Cr, and B chlorides) have been reacted to deposit metal boride, metal carbide, metal oxide, and metal nitride coatings on carbon fibers [46–53]. For example, in Fiber Materials, Inc.'s patented process [47,48,50,51], a yarn or tow of carbon fibers is passed through a reactive gas mixture containing $TiCl_4(g)$, $BCl_3(g)$, and Zn (vapor) to deposit a Ti_xB_y/TiC/$Ti_xB_yC_z$ coating on the carbon fiber. Postulated reactions are:

$$TiCl_4 + Zn = TiCl_2 + ZnCl_2$$
$$2BCl_3 + 3Zn = 2B + 3ZnCl_2$$
$$2TiCl_2 + XB = TiB_x + TiCl_4$$
$$2TiCl_2 + C = TiC + TiCl_4$$

The coated fiber is then immediately drawn through a molten bath of the matrix metal (i.e., Mg or Mg alloy) to form a composite wire [51]. Typical coatings are on the order of 0.01 to 2 μm thick. The Aerospace Corporation has further refined the above process by alloying the matrix metal with Ti and B; the Ti_xB_y/TiC/$Ti_xB_yC_z$ coating is thereby protected from degradation during subsequent processing (i.e., diffusion bonding) [53]. Alloying with Ti and B is more applicable to Al alloy matrices where the solubilities of Ti and B are appreciable; their solubilities in Mg are not great [44].

Very recently, thin amorphous metal oxide coatings have been deposited on carbon fibers by passing the fibers through organic solutions containing organometallic compounds and/or metal chlorides. This is followed by hydrolysis and/or pyrolysis of the fibers [54–56]. Possible coatings include Si, Ti, V, Li, Mg, Na, K, Zr, and B oxides. For example, carbon fiber can be passed through an ultrasonically agitated bath containing tetraethoxysilane and $SiCl_4$ dissolved in toluene. The fiber tow can then be passed through a steam chamber and a drying furnace to form a silica coating thereon. The postulated reactions are:

$$Si(OC_2H_5)_4 + 2H_2O = SiO_2 + 4C_2H_5OH$$
$$Si(OC_2H_5)_4 = SiO_2 + 2C_2H_5OH + 2C_2H_4$$
$$SiCl_4 + 2H_2O = SiO_2 + 4HCl$$

The coated fiber is then readily wet by molten Mg.

These methods have been found to be excellent for coating relatively low modulus carbon fibers, but they are of limited utility for coating ultra-high modulus graphite fibers. Ultra-high modulus fibers have a smoother, more chemically inert surface. The Aerospace Corporation found that by applying a thin amorphous carbon coating to the ultra-high modulus graphite fibers before coating the fibers with a metal oxide using the procedures of U.S. Patent 4,376,803, an improved fiber-to-metal-oxide bond is achieved [57]. The amorphous carbon coating is applied by pyrolizing petroleum pitch on the fiber surface. The amorphous carbon coating simulates the surface of lower modulus carbon fibers; that is, it is rough and porous. Alternatively, the ultra-high modulus graphite fiber can be electrochemically activated before applying a metal oxide coating in order to improve the fiber-to-metal-oxide bond [58]. Naerheim et al. describe a method of applying a TiO_2 or ZrO_2 coating to carbon fibers by wet chemistry methods based on $TiCl_4$ or $ZrCl_4$ [56].

Several methods have been used to make Mg– or Mg alloy–carbon fiber composites. Two methods have received attention in the past. First, coated fiber tows are pulled through a molten bath of the matrix metal to form metal matrix composite precursor wires as in U.S. Patent 4,082,864 [50]. The MMC wires are then consolidated into a composite article by vacuum hot pressing techniques or pultrusion [59–63]. Vacuum hot pressing (VHP) techniques usually involve placing the MMC precursor wires between metal foils and diffusion bonding under high pressures and temperatures [59]. In VHP, the metal matrix can be either solid or liquid. In pultrusion, precursor wires are placed between metal foils and then pulled through a heated die. The die in turn squeezes this laminate together to form a composite structure.

Second, simple, inexpensive, air-stable coatings have recently allowed near-net-shape casting of continuous carbon fiber reinforced Mg articles [35,61,62,64–68]. The coated fibers are flexible, and they can be bent, woven, and braided. They are formed into prepreg sheets, by filament winding. The prepreg sheets are placed in a mold, and any organic binder in the prepreg is burned off. Finally, the mold is infiltrated with molten Mg or Mg alloy, usually with a vacuum or pressure assist. Plaster investment casting techniques have worked best. This near-net-shape casting technique has been called filament winding vacuum casting or FWVC.

Less common fabrication techniques that do not involve MMC precursors wires or FWVC have also been used to produce Mg–carbon fiber composites. For example, powder metallurgical techniques have been used to fabricate Mg–carbon fiber composites. In U.S. Patent 3,888,661, alternate layers of AZ91 powder, chopped graphite fibers, and Ti powder were liquid phase hot-pressed between two dies to form an MMC plate [69]. The Ti powder acted as a wetting agent. Similarly, Levitt et al., fabricated Mg–graphite fiber composite plates by placing alternate layers of Mg powder and continuous aligned graphite fibers between two dies and liquid phase hot pressing to laminates [45]. No titanium powder was used, but the fibers were coated with Ti. Finally, in Naerheim et al., 15 vol % of chopped graphite fiber coated with Ti or Zr compounds was mixed with Mg powder and liquid phase hot pressed to form MMC articles [56]. Mg alloy–carbon fiber composites have also been formed by squeeze casting, using both coated and uncoated fibers [41,70].

The leading producers of Mg–carbon fiber composites are Materials Concept, Inc. (MCI) and Dolowy Webb & Associates, Composite Specialties, Inc, (DWA). MCI produces composites by vacuum hot pressing and pultrusion using MMC precursor wires [62,63]. They can produce Mg–graphite fiber composite tubing, hats, tees, cees, angles, and rods by pultrusion, both unidirectional and cross-ply [63]. Perhaps even more important, MCI has been a leading developer of FWVC techniques [35,64,66–68]. They have produced Mg–graphite fiber flat plates, T sections, irregular specialty parts, rods, round tubes, and hollow cylinders by FWVC techniques. Both thin (1.27 mm) and thick (50.8 mm) wall castings have been made. A simulated rotary engine housing weighing about 7 kg was successfully cast [68].

DWA also produces Mg–carbon fiber MMC parts: sheet, plate, and structural parts such as angles, zees, cylindrical struts, both unidirectional and cross-ply [71,72]. DWA recently developed a new ultra-thin-ply (89 μm) Mg–graphite fiber composite called DWG composite, both unidirectional and low angle cross-ply [72]. The thin ply allows the incorporation of a very high volume fraction of graphite fibers in composite parts (i.e., > 50 vol %). No CVD or ion-plated coating on the fibers is necessary. The development of high volume fraction Mg–graphite fiber composites is important in the effort to make zero CTE composites. DWG Mg–graphite composites have been made into tubular struts.

Cordec Corporation also produces Mg–graphite fiber MMC strip or sheet up to 1524 m long [73].

PROPERTIES OF MG–CARBON FIBER COMPOSITES

Microstructure. For both cast and diffusion-bonded Mg–graphite fiber composites, properly fabricated Mg–graphite fiber composites show excellent infiltration of the matrix into the fibers with no voids [74]. Little or no evidence of interfacial reaction is seen [75]. Nonetheless, even in well-fabricated composites, fracture surfaces show fiber pullouts, indicating a weak fiber-to-matrix bond.

Mechanical. The mechanical properties of carbon fiber reinforced Mg alloys are discussed in the literature [27,35,45,60,64,67,68,72,74,76–79]. Typically, composite properties of properly fabricated composites correspond well with rule-of-mixtures predictions, especially the elastic modulus in tension. Table 10 lists some representative room temperature properties of 38 vol % P100 graphite fiber reinforced AZ91 produced by diffusion-bonding techniques [74]. Table 10 also gives rule-of-mixtures predictions. Table 11 lists some properties of AZ91C–graphite fiber composite castings [79].

The longitudinal elastic modulus of unidirectional graphite fiber reinforced Mg alloys can be outstanding if ultra-high modulus fibers are used (i.e., 310 GPa for cast AZ91C–40 vol % P100 graphite fibers, Table 11). In fact, the specific modulus of certain Mg alloy–graphite fiber

TABLE 10
Room Temperature Properties of Diffusion-Bonded AZ91–38 Vol % Unidirectional P100 Graphite Fiber Composite

Property[a]	Average Value	Rule-of-Mixtures Value
E_L	313 GPa	282 GPa
E_T	26 GPa	45 GPa
E_{CL}	367 GPa	282 GPa
E_{CT}	20 GPa	45 GPa
G_{LT}	13 GPa	16 GPa
UTS_L	422 MPa	414 MPa
UTS_T	21 MPa	21 MPa
UCS_L	270 MPa	83 MPa
UCS_T	105 MPa	83 MPa
Y_{LT}	25 MPa	21 MPa
CTE_L	1.1 mm/mm K $\times 10^{-6}$	1.8 mm/mm·K $\times 10^{-6}$
CTE_T	25.7 mm/mm K $\times 10^{-6}$	25.2 mm/mm·K $\times 10^{-6}$

[a] E = elastic modulus
G = shear modulus
UTS = ultimate tensile strength
UCS = ultimate compressive strength
Y = shear strength
CTE = coefficient of thermal expansion
Subscripts: L = longitudinal
T = transverse
C = in compression

TABLE 11
Typical Properties of Graphite–AZ91 Castings Produced by Filament Winding Vacuum Casting Techniques

Fiber Type	Fiber Content/ Orientation	Casting	UTS_L (0°) (MPa)	UTS_T (90°) (MPa)	E_L (0°) (GPa)	E_T (0°) (GPa)	CTE_L (0°) (ppm/°C)
P55	40%/0°	Rod	724	ND[a]	172	ND	ND
P100	35%/0°	Rod	724	ND	248	ND	ND
P75	40%/± 16° plus 9%/90°	Hollow cylinder	448	61	179	86	1.3
P100	40%/± 16°	Hollow cylinder	559	38	228	30	− 0.072
P100	40%/0°	Plate	421	28	310	34	ND
P55	40%/0°	Plate	483[b]	21	159	21	2.3
P55	39%/0° plus 10% 90°	Plate	276	103	117	34	4.5
P55	20%/0° plus 20%/90°	Plate	241[b]	241	90	90	ND

[a] Not determined
[b] Equivalent ultimate tensile strength at 399°C

composites is higher than that of any other common engineering material, including Al alloy–graphite fiber composites [27,71,77,80]. Excellent longitudinal ultimate tensile strengths can also be achieved (i.e., > 690 MPa). Moreover, as Table 11 indicates, longitudinal properties are retained even at temperatures as high as 399°C.

Although the longitudinal modulus and ultimate tensile strength of unidirectional Mg alloy–graphite fiber composites are excellent, the transverse modulus and strength are very poor (Tables 10 and 11). This has been attributed to a low fiber-to-matrix bond strength [27,60]. One way of overcoming this problem has been to diffusion-bond unreinforced Mg alloy face sheets to Mg–graphite fiber composites [27,60]. For composites with face sheets, the transverse UTS can then be predicted by a rule-of-mixtures calculation, assuming that all of the transverse strength comes from the unreinforced face sheets [60]. Alternatively, small volume fractions of off-axis fiber reinforcement can improve transverse properties (Table 11).

Another important property of graphite fiber reinforced Mg alloys is their low coefficient of thermal expansion CTE [59,66,81–83]. Tables 10 and 11 give some thermal expansion data for Mg alloy–graphite fiber composites. Some more detailed thermal expansion data for AZ91-P55 graphite fiber composite castings are given in Table 12 [67]. The CTE of the composite materials varies from 0 to 9 mm/mm K $\times$ 10^{-6} versus 26.1 mm/mm K $\times$ 10^{-6} for unreinforced AZ91 and 12.6 for Be and many steels. Combined with good thermal conductivity, the low CTE imparts excellent dimensional stability with respect to temperature changes to Mg alloy–graphite fiber composites. A measure of dimensional stability with respect to temperature fluctuations is the thermal deformation coefficient, the transverse thermal conductivity divided by the longitudinal CTE, or K_T/CTE_L. The thermal deformation coefficient of Mg–graphite fiber composites can be better than those of any known engineering materials [27,77,80].

Theoretically, it should be possible to produce unidirectional Mg–graphite fiber composites with a zero CTE, and considerable effort has been extended to reach this goal. Table 13 gives the volume percent of graphite fiber

TABLE 12
Thermal Expansion Data for GR–Mg Castings

% Fiber in Test Direction	Temperature Range (°C)	CTE mm/mm· K $\times$ 10^{-6}	Thermal Expansion (%)
40	RT–204	2.3	0.04
	400–399	0.0	0.00
	RT–399 (av.)	1.1	0.04
30	RT–204	4.5	0.08
	400–399	0.7	0.01
	RT–399 (av.)	2.5	0.09
10	RT–204	8.8	0.15
	400–399	0.9	0.02
	RT–399 (av.)	4.7	0.17
0	RT–399 (av.)	25.9	0.97
Aluminum (calculated)	RT–204	23.4	0.41
	RT–399		0.88
Steel (calculated)	RT–204	12.6	0.22
	RT–399		0.47

From Ref. 67.

TABLE 13
Predicted Volume Percent of Graphite Fibers in Magnesium Necessary to Achieve a Zero CTE

Constituent	*E* (GPa)	CTE (mm/mm·K × 10^{-6})	Vol % Gr for Zero CTE
Mg or Mg alloy	44.8	23.4	NA
Graphite P55	379.2	− 0.9	77
Graphite P75	517.1	− 1.3	62
Graphite P100	689.5	− 1.6	48
Graphite P120	827.4	− 1.8	41

From Ref. 60.

in Mg necessary to achieve a zero CTE [77]. Data extrapolation predicts a CTE of 0.18 mm/mm·K × 10^{-6} for 48 vol % Mg–P100 graphite fiber composites, not zero as predicted by theory, but still very low. The predicted properties (extrapolated from available data) of 48 vol % P100 graphite fiber reinforced Mg are given in Table 14 [60,77,80].

Until very recently, obtaining composite materials with a volume fraction of graphite fibers high enough to attain a zero CTE was impossible [60]. This was further complicated by the need to diffusion-bond face sheets to composite panels in order to obtain adequate transverse strength. However, the development of DWG Mg–graphite fiber composites, with their ultra-thin plies and high volume percent graphite, and similar developments should enable a zero CTE to become a reality [70].

Thermal cycling. The effect of thermal cycling on a unidirectional AZ91–42 vol % P100 graphite fiber composite has been reported [82]. The longitudinal elastic modulus was not greatly affected by thermally cycling between ± 100°C. However, some interesting observations were made with respect to thermal expansion. The thermal expansion behavior of unidirectional Mg–graphite fiber composites exhibited complex behavior, including nonlinearity, history and time dependence, and hysteresis loops. During thermal cycling, the CTE was actually both positive and negative depending on the elastic and plastic state of the matrix. These effects make it difficult to control the dimensions of composite parts. Nonetheless, the total strain after 18 cycles at + 100°C was only 200 μm/m. The effective CTE was 3.2 mm/mm K × 10^{-6}. The thermal expansion behavior was adequately described by an elastic-plastic model, where the fiber was assumed to be arthotropically linearly elastic, and the matrix was assumed to be isotropic elastic-plastic. If there was a long time between the thermal cycles, isothermal creep in the matrix occurred as a result of internal stress relaxation, and the model failed.

In a separate study, Kural and Min suggest that one could decrease the effects of matrix plasticity by increasing the TYS of the matrix metal (i.e., by heat treatment) or by preconditioning the composite [83]. Preconditioning consists of subjecting the composite to a low temperature so that the matrix will remain elastic during the operational temperature range. This could be accomplished by exposing the composite to liquid nitrogen. Preconditioning was proven effective for 6061 Al–P100 graphite fiber composites.

TABLE 14
Estimated Properties for Mg–48 Vol % P100 Graphite Fibers Based on Data Extrapolation

In-plane Properties	Physical Properties
E_L, GPa = 352	P, g/cm^3 = 1.829
E_T, GPa = 20	CTE$_L$, mm/mm·K × 10^{-6} = 0.2
G_{LT}, GPa = 16	CTE$_T$, mm/mm·K × 10^{-6} = 23.4
UTS$_L$, MPa = 634	K_L, w/m K = 104
UTS$_T$, MPa = 34	K_T, w/m K = 31
	C_p, J/kg·K = 962

From Ref. 80.

Corrosion. Unfortunately, Mg–graphite fiber composites exhibit very poor corrosion performance [84–87]. In the presence of an electrolyte, graphite serves as a very efficient cathode and accelerates the anodic dissolution of magnesium. Samples exposed to salt water exhibit severe corrosion wherever graphite fibers are exposed. One natural remedy is to eliminate any exposed graphite fibers from the sample surface (i.e., by coating). Possibilities include cladding with unreinforced face sheets, applying ceramic coatings by sputtering (i.e., plating), and/or painting (applying epoxy coatings). Epoxy coatings have shown promise [85]. However, even if composites are coated and no graphite fiber surfaces are exposed, corrosion can be a serious problem if the coating is penetrated. Another possible solution to corrosion problems is to coat the graphite fibers before manufacture of the metal matrix composite to render them nonconducting.

Damping. The damping characteristics of Mg–graphite fiber composites have also received some attention [88,89]. Certain unreinforced magnesium alloys have excellent damping characteristics [44]. Moreover, as a result of the weak graphite fiber–Mg bond formed in Mg alloy–graphite fiber composites, the damping capacity for composites can by higher than that for unreinforced magnesium alloys [88,89].

POTENTIAL APPLICATIONS. Magnesium and magnesium alloy–graphite fiber composites exhibit the best combination of specific stiffness and thermal deformation resistance of any known engineering material [27,77,80,87]. This leads to a unique clear advantage of Mg–graphite fiber composites for weight-critical, stiffness-critical, and/or alignment-critical space applications [27,77,80,87]. The literature discusses potential uses for Mg–graphite fiber composites such as space platforms and structures, large deployable space antennas,

TABLE 15
Property Comparison of Mg + Graphite and Aluminum Casting Alloy A132

Material	*E* (MPa)	TYS (MPa)	UTS (MPa)	CTE (mm/mm·K × 10^{-6})	Density (g/cm³)
Cast Al alloy A132 T65 Temper[a]	73	296	324	19.1–21.1	2.74
AZ91 + 40 Vol % P55[b]	158–172	483–724	483–724	2.3	1.88

[a] Properties from ASM Handbook, Volume 8, p. 952, Ref. 161.
[b] Properties from Table 11, this paper, Ref. 79.

satellite structures, mirrors, and space optics substances [9,27,59,60,62,64–66,71,74,77,78,80,86,87,90–92]. Other materials that compete for these same space applications include Be, graphite fiber reinforced epoxy, and graphite fiber reinforced Al [27,60,74,77,80,92].

Although Be exhibits high stiffness and relatively light weight, it has a coefficient of thermal expansion of about 12.6 mm/mm·K × 10^{-6}, and it is expensive to manufacture and machine. In addition, Be demonstrates poor fracture properties, and it poses a severe health hazard [74]. Graphite fiber reinforced epoxy composites have different problems, including moisture absorption, outgassing, poor high temperature capability (which leads to charging), poor creep strength, poor microyield strength, poor resistance to microcracking, and poor resistance to laser attacks [27,60,77,80,92]. Therefore, both graphite fiber reinforced aluminum and graphite fiber reinforced magnesium composites have received serious attention.

The use of Mg–graphite fiber composite tubular spars and joints for space structures has been investigated [66,91]. The composite parts were produced by FWVC techniques. In addition, the use of Mg–graphite composites for antenna ribs has been discussed in the literature [77,80]. Most such antennas currently under construction use graphite fiber composites. Other government/military applications such as portable bridges, helicopter antennas, and missile components also use reinforced epoxy composite ribs. However, based on initial studies [80], the substitution of Mg–graphite ribs could lead to a 362% improvement in antenna performance. Other applications are cited in Refs. 27, 84, and 93.

The use of Mg–graphite fiber composite castings for rotary engine housings has also received a considerable amount of attention [35,66–68,71,91,94,95]. Rotary engines have been advocated for propeller-driven aircraft where the engine housing, rotor, and crankshaft are critical. The excellent specific stiffness and temperature resistance and the low CTE of Mg–graphite fiber composites make them quite attractive. Compared with aluminum alloy A132, an AZ91–40 vol % P55 graphite fiber composite is about 2.3 times stiffer and has only one-eighth the CTE (Table 15) [79].

MCI has been developing a Mg–graphite fiber composite rotary engine housing for propeller-driven aircraft through a NASA-funded program [35,66–68,71,91,94,95]. In the first phase of the program, an intermediate rotary engine housing was produced by FWVC techniques. A steel-enclosed plaster mold was used. The demonstration part contained 30 vol % P55 graphite fibers in a 0°/90° orientation in a matrix of AZ91C. The demonstration part weighed approximately 6.8 kg [98]. In phase II, CAD will be used to design a full-scale housing. Preliminary results indicate that the volume loading of graphite fibers required will be 40%. An important goal of phase II is to develop suitable wear- and chemical-resistant surfaces. Various possibilities are discussed by Goddard et al. [68]. Finally, the capability to supply commercial-quality castings needs to be developed, and prototypes need to be supplied. Mg alloy–graphite fiber rotary engine housings may also find use in racing cars, pleasure boats, and industrial engines and compressors [35,67,68,71].

Boron Fibers

FIBER PROPERTIES. Boron fibers were developed in 1960, and continuous filaments were marketed in 1965 [90,96]. The two major U.S. producers of B fiber are AVCO and Composite Technologies, Inc. SPNE of France and Shinku Kakin of Japan also produce B fibers. Boron fibers consist of minute crystals of B (2 and 3 nm in diameter) deposited on a tungsten or carbon filament by a reactive chemical vapor deposition process. The W filaments are typically about 13 μm in diameter. The B fibers are available as monofilaments of 100 to 200 μm in diameter. B fibers are also available with diffusion barrier coatings of SiC or B_4C. The coated fibers are primarily for use in aluminum. Table 16 lists typical properties of B fibers [90].

TABLE 16
Typical Properties of Boron Fibers

Core	Fiber	Coating	*E* (GPa)	UTS (MPa)	*P* (g/cm³)
W	B	None	400	3620	2.5
W	B	SiC	400	3103	2.7
C	B	None	358	3448	2.2
W	B	B_4C	379	3792	2.5

The B fibers possess excellent stiffness and strength properties at a low density, and accordingly, they have received a lot of attention as a reinforcement for magnesium. In fact, B fiber reinforced metals including magnesium were among the first MMCs developed [90]. Al–B fiber composite tubular struts are used on the space shuttle [90,96].

B fibers are also attractive because of their stability with respect to reaction with magnesium, even when composites are fabricated using liquid metal infiltration techniques. Most studies indicate that there is no reaction between coated or uncoated B fibers and molten magnesium [96–102]. At least one paper, nonetheless, suggests that a MgB_2 film forms at the fiber–matrix interface in composites. This paper discusses alloying additions that can be made to the magnesium matrix that theoretically reduce the fiber–matrix reaction [103]. On the other hand, the stability of uncoated B fibers with respect to molten aluminum is not good. Therefore, in magnesium alloys containing Al, a fiber–matrix reaction may occur [102].

FABRICATION OF MAGNESIUM- AND MAGNESIUM ALLOY–B FIBER COMPOSITES. Davies et al. discuss many ways in which Mg–B fiber composites can be fabricated [98]. They conclude that the only fabrication routes with a potential for low cost are plasma spraying and continuous casting for tape or wire preform and diffusion bonding and continuous casting for structural shapes. Accordingly, most Mg–B fiber composites reported in the literature have been fabricated by these methods. The Soviets have published several papers on Mg–B fiber composites fabricated by plasma spraying of tape preforms followed by diffusion bonding [97,104–107]. In the United States, AVCO and American have produced composites using plasma spraying and diffusion bonding techniques [108]. One company, General Technologies Corporation (GTC), has published several papers with respect to continuous casting of composite tape and wire preforms and structural shapes [98–101].

GTC casts Mg–B fiber composite preform tape or wire [100]. Boron fibers coming off a filament payoff rack are passed through a molten magnesium bath in a casting crucible. The casting crucible contains the matrix reservoir, a filament collimation plate, and an exit orifice. The composite tape or wire exiting the melt is then allowed to solidify and collected on a take-up system. Often, before the composite completely solidifies, a twist is imparted to the tape or wire preform [100,101]. The twist wrings out excess metal from the preform. By varying the degree of twist, the volume fraction of B fiber in the wire is controlled. For example, Mg–25 vol % B fiber wire contains one revolution in every 7.6 cm, and Mg–75 vol % B fiber wire contains one revolution in every 2.5 cm. The preform tape can be round or rectangular in cross section, and it may contain from 15 to 40 filaments. The most common preform tape consists of round spoolable wire containing 19 filaments [99]. MMC wire has been cast at a rate of up to 4572 m/h [99]. The tape or wire preform is flexible, and this simplifies handling of the brittle B filaments without breakage.

GTC has also continuously cast various structural shapes containing up to 500 filaments in Mg–B fiber composite. Selective placement of the reinforcement was also demonstrated [99]. Composite structural shapes were continuously cast using MMC wire or tape preform. Multiple composite tapes or wires are drawn through a magnesium melt in a manner similar to the B filaments in preform manufacture. In continuously cast structural shapes from preforms, very little matrix metal is added. GTC has refined the process so that composite shapes can be continuously cast with less than 0.5 vol % porosity (99). The cost of Mg–B fiber composites made by continuous casting is discussed [98–100]. The cost of composite shapes is said to be material dependent, while labor and equipment costs are secondary [98].

Mg–B fiber composites can be made by diffusion bonding and continuous casting as outlined, but only simple shapes are possible. Accordingly, other liquid metal infiltration techniques were investigated with the goal of casting complex three-dimensional shapes [102,109,110]. Banker made flat tensile bars and hoop-test rings of Mg–B fiber composite by a pressure-vacuum liquid metal infiltration process [102]. Toyota mentions squeeze casting Mg–B fiber composites in U.S. Patent 4,492,265 [110]. Also, Lawrence has cast complex Mg–B fiber composite shapes by a self-generated vacuum casting technique [109]. Other useful fabrication technology is discussed in the patent literature [31,111–115].

PROPERTIES OF MG- AND MG ALLOY–B FIBER COMPOSITES. Room temperature mechanical properties of unidirectional Mg–B fiber composites are given in Table 17. Excellent elastic moduli and ultimate tensile strengths can be obtained. Banker suggests that simple rule-of-mixtures predictions for tensile properties are not adequate, since the strength of B fibers actually varies in a statistical manner [102]. The B filament is said to be the source of fracture for all testing conditions. Failure starts by the nucleation of cleavage cracks at the W core–B interface. Plastic deformation of the matrix causes a stress concentration of the Mg matrix–B fiber interface. This in turn forms cracks at the Mg–B interface, which accelerates the cleavage cracks at the W core–B interface. Alexander suggests that failure in compression tests is by brooming at one end [100]. Alexander also discusses the importance of the twist often imparted to the fibers in continuously cast Mg–B fiber composites. The twist, giving a cablelike structure, extends the progressive filament fragmentation range of the composite, and this leads to a greater tolerance of deformation.

High temperature tensile properties have been reported [105,106,116]. Some results are given in Table 18 for a Mg–approximately 55 vol % B fiber composite produced by plasma spraying and diffusion bonding techniques. The tensile strength decreases with increasing temperature. Nonetheless, excellent high temperature strength is exhibited. As the temperature is raised, the

TABLE 17
Room Temperature Mechanical Properties of Mg–B Fiber Composites

Matrix	Vol % B_f	Ref.	*E* (MPa)	UTS (MPa)	UCS (MPa)	Fabrication Method[a]	Tensional Shear Modulus (GPa)	Tensional Shear Strength (MPa)	Flexural Modulus (GPa)	Flexural Strength (MPa)	Fatigue Strength at 10^7 Cycles
Mg	67	102	—	1455	—	LMI	—	—	—	—	—
Mg–9 Wt % Al	55	102	—	910	—	LMI	—	—	—	—	—
Mg	54	106	214	972–1172	—	PSDB	—	—	—	—	588 MPa
Mg	60	109	290–310	965	—	SGV	—	—	—	—	—
HZ32	60	109	—	1007	—	SGV	—	—	—	—	—
Mg	30	101	214	951	—	CC	—	—	—	—	—
Mg	69	101	—	—	3151	CC	—	—	—	—	—
Mg	—	99	262	1517	—	CC	50	—	—	—	—
AZ92	54	99	248	—	3275	CC	50	—	—	—	—
AZ92, EZ33, Mg	50–65	99	—	2068	—	CC	—	—	—	—	—
EZ33	50–65	99	—	2413	—	CC	—	—	—	—	—
Mg	50–65	99	—	1448	—	CC	—	—	—	—	—
Mg	50–65	99	—	1400	—	CC	—	—	—	—	—
AZ92, Mg	50–65	99	290	—	—	CC	—	148	—	—	75% of UTS
Mg	50–65	99	—	—	—	CC	—	152	—	—	—
Mg	35	100	—	—	2062	CC	—	—	—	—	—
Mg	69	100	—	—	3275	CC	—	—	—	—	—
Mg	25	100	—	—	—	CC	—	—	241	1117	—
Mg	75	100	—	—	—	CC	—	—	241	1579	—
Mg	60	110	—	1269	—	SC	—	—	—	—	—

[a] LMI = vacuum-pressure liquid metal infiltration
PSDB = plasma spraying and diffusion bonding
SGV = self-generated vacuum
CC = continuously cast
SC = squeeze cast

plastic deformation of the matrix is intensified. This increases the stress concentration at the filament–matrix interface, which favors breaking up of the brittle filaments.

The corrosion behavior of Mg–B fiber composites have also been reported [106,107]. Immersion testing has been used to study the corrosion behavior in aqueous NaCl solutions. Exposure to a tropical atmosphere was also studied. Rapid corrosion of the Mg matrix occurred wherever exposed B filaments were present. Electrochemical methods were used to determine that a galvanic cell was set up between the exposed W core and the Mg matrix. The B itself did not serve as an efficient cathode. Sealing any exposed W cores improved corrosion performance.

TABLE 18
Elevated Temperature Tensile Properties for a Mg–55 Vol % B Fiber Composite

Temperature (°C)		*E* (GPa)
20	213	972–1172
300	—	896
400	—	786
500	—	765

POTENTIAL APPLICATIONS FOR MG– AND MG ALLOY–B FIBER COMPOSITES. Applications for Mg–B fiber composites have not been widely reported in the literature. However, two applications have been mentioned [100]. First, a 19.8 m long Mg–B fiber continuously cast composite beam has been used as a structural reinforcement for a sailboat mast. Second, the potential use of Mg–B fiber composites for low visibility radio antennas has also been discussed. For portable military radios, the antennas need to have adequate stiffness and density to be self-supporting at small diameters (1.14 mm diameter). Mg–B fiber rod seems appropriate, but it currently does not have adequate ductility.

TABLE 19
Typical Properties of Nippon's Nicalon SiC Fiber

Filament diameter	10 to 15 μm
Cross section	Round
Filaments/yarn	500
Density	2.55 g/cm^3
UTS	2448–2937 MPa
E	179–193 GPa
CTE	3.1 mm/mm·K × 10^{-6}
Crystal form	beta

Silicon Carbide Fibers

Magnesium alloys have also been reinforced with other fibers, such as SiC fibers. ANCO Corporation and Nippon Carbon Company both produce continuous SiC fiber with excellent properties (Table 19) [96,117]. Nonetheless, the literature contains little published information on Mg–SiC fibers. The Japanese have published a paper on squeeze-cast Mg alloys uniaxially reinforced with 40 vol % 10 to 15 μm diameter Nippon SiC fiber [118]. The SiC fibers were arranged unidirectionally in a preform and held in place by an organic binder. The organic binder was burned off, and liquid metal was infiltrated into the preform at high pressures. Two Mg matrix alloys were studied, Mg–6 wt % Al and Mg–10 wt % Al. In both cases, a solute-rich eutectic layer formed around the SiC fibers, the thickness of which was proportional to the alloy content. This eutectic layer was said to decrease the tensile strength of the composite, but no numbers were given to substantiate the claim.

A Japanese-held U.S. patent also describes Mg alloy–SiC fiber composites [119]. A special continuous silicon carbide fiber containing 0.01 to 40 wt % free carbon was used. The free carbon was said to react with alloying elements in the Mg alloy matrix to form carbides, thereby promoting wettability. Composites were made by liquid metal infiltration and powder metallurgical techniques, and room temperature tensile properties were reported (Table 20). Excellent elastic moduli and UTS values were reported. Several other U.S. patents also mention Mg alloys reinforced with SiC fibers, although no specific examples were given [29–33,114,115].

Steel Fibers

Magnesium alloys reinforced with 0 to 50 vol % 140 μm diameter steel wire were studied [97,105,116]. The steel (nominal 13% Cr, 13% Ni, 2% Mo, 0.3% Si, 0.4% Mn, 0.05% C, balance Fe) fibers had a tensile strength of about 2827 MPa. Composites were made by hot-pressing plasma-sprayed preforms. No reaction between the Mg matrix and the steel fibers was detected [97]. Some property data was given for a unidirectional 18 vol % steel fiber reinforced Mg composite by Maksimovich et al. [105]. The room temperature ultimate tensile strength and tensile elongation were about 276 MPa and 13%, respectively. The tensile strength dropped off steadily with an increase in temperature, and the elongation increased. Fractography indicated that the steel filaments necked during tensile testing and separated from the matrix. Cracks formed at the interface and subsequently proceeded into the matrix.

Battelle Columbus Laboratories reports on Mg alloy La141A (Mg–14% Li–1% Al) unidirectionally reinforced with 0.10 mm diameter steel filaments [120]. (The steel was nominally 14% Cr, 13% Co, 5% Mo, 0.17% Co, 0.16% Mn, 0.03% N, 0.10% Ni, 0.13% Si, 0.18% V, 0.011% P, 0.012% S, balance Fe.) Composite materials were fabricated by vacuum-assisted liquid metal infiltration techniques. In this case, a definite reaction occurred at the filament–matrix interfaces. Mg alloy La141 possesses excellent specific strength; however, it shows poor creep strength even at room temperature. Thus, a study of the room temperature creep of La141 reinforced with steel fibers was undertaken. The room temperature creep of a composite material was controlled by the creep of the fibers rather than that of the matrix. Tensile properties of

TABLE 20
Properties of Unidirectionally Reinforced Mg Alloy + SiC Fiber Composites

	Fiber							
Matrix	Dia. (10^{-6}m)	% Free Carbon	Vol. % in MMC	MMC[a] Fab. Method	Fiber Coating	E (GPa)	UTS (MPa)	Failure Strain (%)
Mg, 10Al, 0.5 Mn	10	4	25	LMI	None	—	717	—
Mg, 4Zn, 2Y, 0.6 Zr	15	6	32	LMI	None	—	855	—
Mg, 1Mn, 0.25 Si, 0.1 Ca	—	5	30	PM	None	—	296	—
AZ92	10	10	25	LMI	None	—	634	—
Mg, 3Al, 1.3 Zn, 1 Mn	20	15	40	PM	Ni	158	931	1.5
Mg, 3Al, 1.3 Zn, 1 Mn	20	15	40	PM	Cu	179	965	1.2
Mg, 3Al, 1.3 Zn, 1 Mn	20	15	40	PM	Fe-Cr	165	848	1.6
Mg, 3Al, 1.3 Zn, 1 Mn	20	15	40	PM	SiO_2	158	834	1.8
Mg, 3Al, 1.3 Zn, 1 Mn	—	—	0	PM	—	45	214	—

[a] LMI = liquid metal infiltration
PM = powder metallurgy

TABLE 21
Tensile Properties of a LA141 + 32 Vol % Steel Fiber Composite and Its Constituents

Material	25°C, UTS (MPa)	200°C UTS (MPa)
As received wire	4047	3689
Exposed wire[a]	3413	3206
LA141A matrix	145	14
MMC	758	484

[a] Exposed to molten magnesium alloy LA141A.

a composite material reinforced with 32 vol % steel filaments are given in Table 21.

U.S. Patent 3,863,485 describes a process for producing chopped steel filament reinforced magnesium composites [121]. Chadwick describes a Mg–60 vol % steel wire (fiber) composite produced by squeeze casting [70].

Other Continuous Fibers

UBE Industries, Ltd. of Japan produces some unique inorganic continuous fibers composed of Si, Ti or Zr, C, and O [122]. These fibers contain one of the following:

1. An amorphous material consisting of Si, Ti or Zr, C, and O
2. An aggregate of ultrafine crystals, amorphous SiO_2, and amorphous TiO_2 or ZrO_2
3. A mixture of (1) and (2)

The ultrafine (< 500 A) crystals of (2) contain B-SiC, *k* MC, and a solid solution of B-SiC, MC, and MC1-x, where $0 < x < 1$ and M = Ti or Zr. The inorganic fibers are claimed to be very stable with respect to liquid Mg and Al alloys. Magnesium and Mg alloys reinforced with these unique fibers are described in U.S. Patent 4,146,690 [122]. For example, a Mg alloy (Mg–1.3 wt % Al–1.3 wt % Zn–1.0 wt % Mn) was reinforced with 30 vol % randomly oriented chopped fibers (1 mm length) via PM hot isostatic pressing techniques. The 13 μm diameter fibers had the following composition: 45 wt % Si, 3 wt % Ti, 25.4 wt % C, and 24.7 wt % O. The fibers had an UTS of 3034 MPa and an elastic modulus of 156 GPa. The composite material had an UTS of 490 MPa. Magnesium and magnesium alloys were also unidirectionally reinforced with continuous fibers.

Similarly, UBE Industries, Ltd. produces unique inorganic continuous fibers composed of Si, Ti or Zr, N, and O and containing one of the following:

1. An amorphous material consisting of Si, Ti or Zr, N, and O
2. An aggregate of ultrafine crystals, amorphous SiO_2, and amorphous TiO_2 or ZrO_2
3. A mixture of (1) and (2)

The ultrafine (< 500 A) crystals of (2) contain Si_2N_2O, MN, Si_3N_4, and/or MN1-x, where M = Ti or Zr and $0 < x < 1$. U.S. Patent 4,622,270 describes a magnesium composite containing elemental Mg as a matrix and 30 vol % unidirectionally aligned continuous inorganic fibers as a reinforcement [123]. The inorganic fibers contained 47.9 wt % Si, 3.0 wt % Ti, 25.6 wt % N, and 22.1 wt % O. The fiber diameter was 13 μm, the fiber UTS was 2937 MPa, and the fiber elastic modulus was 166 GPa. The Mg composite was prepared by hot rolling and diffusion bonding techniques, and it had an UTS of 699 MPa and an elastic modulus of 73 GPa. These same fibers were chopped to a length of 1 mm and used to reinforce Mg alloy (Mg–3 wt % Al–1.3 wt % Zn–1 wt % Mn) via powder metallurgy (hot isostatic pressing) techniques. The randomly oriented chopped fiber composite has an UTS of 538 MPa. This type of fiber was also said to have excellent stability with respect to molten magnesium and aluminum alloys.

Magnesium Alloys Reinforced with Discontinuous Fibers

Discontinuous Fibers: Types Used in Magnesium Composites and Properties

Magnesium alloys have also been reinforced with discontinuous alumina, alumina–silica, and mineral fibers. Alumina (Al_2O_3) and alumina-silica (Al_2O_3–SiO_2) fibers are produced on a large scale for established applications such as insulating lining for industrial furnaces. As a result, Al_2O_3 and Al_2O_3–SiO_2 fibers are relatively inexpensive. The fibers are usually produced as continuous mat. The mat can be shredded to a bulk form or milled to give very short fibers in various controlled lengths. Alumina fibers (arbitrarily defined for purposes of this paper as fibers with > 80% alumina, the balance being SiO_2) are very hard and temperature resistant. Alumina fibers are available with various crystal structures. Alpha Al_2O_3 is the hardest phase. Alumina–silica fibers are less expensive than alumina fibers; however, they are not as hard or temperature resistant. Al_2O_3–SiO_2 fibers are available with various Al_2O_3:SiO_2 ratios. Amorphous Al_2O_3–SiO_2 fibers are widely available, but they are relatively soft; they can be hardened by heat treating to produce a crystalline mullite phase [124,125]. Mineral fibers are produced, spun from molten rock, molten slag, or molten rock-slag mixtures to produce rockwool, slagwool, and mineral wool, respectively [126]. Mineral fibers are very inexpensive. Some typical fiber properties are given in Table 22 [124,126,127].

Fabrication and Mechanical Properties of Composites

Toyota has done a considerable amount of work on magnesium alloys reinforced with discontinuous Al_2O_3 fibers, Al_2O_3–SiO_2 fibers, mineral fibers, mixtures of Al_2O_3 and mineral fibers, and mixtures of Al_2O_3–SiO_2 fibers and mineral fibers [125,126,128–132]. Most of their work has been aimed at producing wear-resistant composites at a relatively low cost. Composite materials were prepared

TABLE 22
Typical Fiber Properties of Various Al_2O_3–SiO_2, Al_2O_3–SiO_2, and Mineral Fibers

Ref.	Fiber Type	Diameter (10^{-6} m)	Length (mm)	Crystal Form	Composition	*E* (GPa)	UTS (MPa)
124	Fiberfrax	2.5	0.8	Amorphous	49 Al_2O_3/51 SiO_2	103	1724
124	Fiberfrax	3	0.8	Mullite	71 Al_2O_3/28 SiO_2	152	827
127	Saffil RF	3	0.5–50	Mainly delta	96–97 Al_2O_3/3–4 SiO_2	296	2000
127	Saffil RG	3	0.5–50	Alpha & delta	96–97 Al_2O_3/3–4 SiO_2	296–324	1000–2000
126	Mineral	1–10	0.01–100	—	35–50 SiO_2/20–40 CaO/10–20 Al_2O_3/3–7 MgO/1–5 Fe_2O_3/0–10 Others	—	—

by squeeze-casting molten metal into ceramic fiber preforms. The preforms, formed by vacuum-forming operations, contained the discontinuous fibers in a random-planar orientation. The fibers were held together in the preform with an inorganic binder such as colloidal SiO_2 or Al_2O_3. When fabricating Mg alloy–amorphous Al_2O_3–SiO_2 fiber composites by squeeze casting, a reaction between the molten metal and the fibers was often a problem [125–128]. For example, in fabricating AZ91 reinforced with 7.8 vol. %, the reinforcing fibers reacted with the molten magnesium alloy matrix, and composite properties accordingly suffered [125]. However, when Al_2O_3–SiO_2 fibers of the same composition but heat treated to produce 62 wt % mullite were used, no reaction was detected [125].

In several U.S. patents held by Toyota, the authors point out the need to limit the amount of nonfibrous particulate in the ceramic preform, particularly particles larger than 150 μm in diameter [125,126,128–132]. Large nonfibrous particles were found to be detrimental with respect to composite strength and machinability [125,129]. Moreover, in lubricated metal-on-metal sliding wear tests, where the MMC was tested against an unreinforced metal, large nonfibrous particles were found to cause excessive wear of the mating unreinforced metal part. Provided that large nonfibrous particles were eliminated, excellent wear test results could be obtained. For example, AZ91 was reinforced with 7.8 vol % of discontinuous Al_2O_3–SiO_2 fiber (55 wt % Al_2O_3/45 wt % SiO_2) with 62% mullite phase. The fiber was treated to remove nonfibrous particles before MMC manufacture. The MMC part was tested against a mating part of a bearing steel in a metal-on-metal lubricated sliding wear test. The amount of MMC wear was 25 μm, and that of the mating part was negligible. In contrast, it was impossible to complete testing when using unreinforced AZ91 because of excessive wear [125]. In a separate patent, similar results were claimed for an AZ91–10 vol % mineral fiber composite [126].

In Mg–discontinuous Al_2O_3 fiber composites, Toyota suggests that the amount of alpha Al_2O_3 crystalline form should be controlled within certain limits [128]. If the fraction of alpha phase exceeds 60%, machinability becomes a problem. Also, the MMC becomes abrasive, causing excessive wear of mating parts in lubricated metal-on-metal sliding wear tests. For example, Toyota claims good results for an EZ33–Al_2O_3 fiber composite when the proportion of alpha phase within the alumina was 34% [128]. Good wear versus cast iron, fatigue strength, machinability, elastic modulus, and hardness were claimed, although no numbers were given to substantiate these claims.

Finally, Toyota found that by using mixtures of Al_2O_3–SiO_2 fibers and mineral fibers, excellent wear results could be obtained [130–132]. For example, a series of composite materials were prepared containing 8 vol % of discontinuous alumina fiber–mineral fiber hybrid mixtures in a matrix of AZ91 [130]. In these composites, the volume proportion of Al_2O_3 fibers to total fibers was varied from 0 to 100%. The composite materials were tested against a bearing steel in a lubricated metal-on-metal sliding wear test. Very low wear of both the MMC test sample and the mating part were obtained, especially when the relative volume proportion of Al_2O_3 fibers to total fibers was > 40%. Thus, it is possible to substitute a substantial amount of very inexpensive mineral fibers for the alumina fibers and still achieve good wear test results.

In addition to Toyota, Chadwick has published some information on AZ91–16 and 25 vol. % discontinuous Al_2O_3 fiber (Saffil) composites fabricated by squeeze casting [70]. The high temperature properties are improved on order of magnitude relative to those of unreinforced AZ91.

Potential Applications

Light metal alloys reinforced with discontinuous Al_2O_3 fibers, Al_2O_3–SiO_2 fibers, mineral fibers, or mixtures thereof are relatively inexpensive, and, therefore, they show considerable promise for commercial automotive applications. Peters et al. have discussed the use of discontinuous-fiber-reinforced metals for such automotive applications as pistons, connecting rods, wrist pins, intake valves, brake pads, and hydraulic cylinders and pumps [28,124]. The metal matrix composite shows excellent resistance to wear, burning or seizure, and thermally induced cracking. Similar Mg MMCs have been mentioned for pistons by several other authors [129,133,134]. Itonda holds a U.S. patent for fiber-reinforced magnesium alloy materials including discontinuous Al_2O_3–SiO_2 fiber reinforcements [133]. An example in this patent describes a piston made out of magnesium alloy (Mg–10 wt % Al–2 wt % Si–0.7 wt % Zn) reinforced with discontinuous Al_2O_3–SiO_2 fibers. The MMC piston was superior to one made out of an unreinforced magnesium alloy and an unreinforced aluminum alloy on the basis of hardness tests, CTE measurements, and wear during actual operation of the piston.

Magnesium Alloys Reinforced with Ceramic Whiskers

Whisker Properties

Ceramic whiskers such as sapphire whiskers (Al_2O_3), alpha SiC whiskers, beta SiC whiskers, Si_3N_4 whiskers, and graphite whiskers can also be used to reinforce magnesium alloys. Ceramic whiskers are elongated single crystals with excellent stiffness and strength properties. Hahn et al. discuss the use of sapphire, alpha SiC, and beta SiC whiskers [135]. Sapphire whiskers have a large fraction of side growths and a broad range of whisker diameters, and they contain some nonwhisker debris. Alpha SiC whiskers, although the most widely available of the three whiskers discussed by Hahn et al., require beneficiation to remove a substantial fraction of nonwhisker contamination debris, and nearly the entire furnace product exceeds the critical l/d ratio. Accordingly, beta SiC whiskers are claimed to be the best reinforcement. Both alpha and beta SiC whiskers are present as intertwined whisker mats, and the whiskers require separation before MMC manufacture. Si_3N_4 and graphite whiskers are discussed elsewhere [96,137,138]. With respect to reinforcing Mg alloys, alpha SiC whiskers have received the most attention. Typical alpha and beta SiC whisker properties are given in Table 23.

Fabrication of Magnesium Alloy–Alpha Silicon Carbide Whisker Composites

POWDER METALLURGY. Most magnesium alloy–alpha SiC whisker composites reported in the literature have been produced by powder-metallurgical (PM) techniques, particularly liquid phase hot pressing. First, magnesium alloy powder is combined with the SiC whiskers and some liquid vehicle to produce a slurry. The liquid vehicle is then removed, leaving an intimate mixture of the magnesium alloy powder and the SiC whiskers. The powder mixture is subsequently consolidated by the application of heat and pressure. During consolidation, the temperature is controlled so that the matrix is in the partially solid plus partially liquid state. Special procedures may be taken during consolidation to avoid whisker damage [135]. Although near-net-shape magnesium alloy–alpha SiC whisker composite parts can be fabricated by liquid phase hot pressing, usually a billet is produced. The as-pressed billet contains a network of MgO skins (from the magnesium particles), which decreases ductility and fracture toughness. Extrusion or forging of the billet is necessary to break up this network of oxide skins. Figure 1 illustrates the process.

The blending of the magnesium alloy powder and the ceramic whiskers is a critical step. As-produced alpha SiC whiskers are produced as intertwined whisker mats; thus, the SiC whiskers require deagglomeration. DWA Composite Specialties has produced whisker-reinforced MMCs by mixing the matrix powder and whiskers in a fluid organic binder [141]. The slurry is then spread into sheets, which are air dried. The organic binder is removed in a vacuum retort before consolidation via liquid phase hot pressing. Hahn et al. have dispersed magnesium alloy powder and alpha SiC whiskers by stirring in isopropyl alcohol [135]. The composite slurry is then filtered and dried before liquid phase hot pressing. Hood and Pickens have deagglomerated SiC whiskers by mixing (ball milling) in a polar solvent [142]. The deagglomerated whisker slurry is then combined in a metal powder slurry. The metal powder is also in a polar solvent. The combined ceramic whisker–metal powder slurry is quickly filtered and dried to produce a green body, and composites are made by liquid phase hot pressing. A unique way of achieving an adequate whisker distribution is discussed by Divecha [143]. Metal powder and whiskers are mixed in molten camphene. The mol-

TABLE 23
Typical Properties of SiC Whiskers

Type	Silar SC-9	Silar SC-10	Tokomax
Average diameter (μm)	0.6	0.6	0.1–0.5
Length range (μm)	10–80	10–80	50–200
Density (g/cm^3)	3.2	3.2	3.19
Whisker content (%)	80–90	70–80	—
Particle content (%)	10–20	20–30	Less than 1 wt %
Free carbon (max. wt %)	0.10	0.20	Negligible
SiO_2 content (max. wt %)	0.75	1.50	Trace
Estimated UTS (MPa)	6900	6900	2937–13,700
Estimated E (GPa)	690	690	393–683
Heat resistance in air (°C)	1760	1760	1600
Crystal type	Alpha	Alpha	Beta

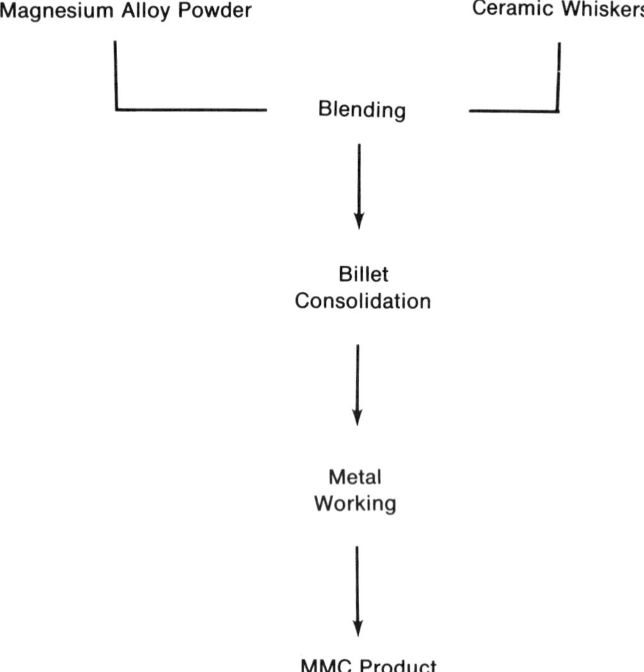

FIGURE 1 PM fabrication flow chart.

ten slurry, which exhibits a poor whisker distribution, is then frozen to form a billet. Subsequently, the billet is extruded a minimum of three times at room temperature to improve the ceramic whisker distribution. After this, the organic vehicle is removed by vacuum evaporation, and composites are made by liquid phase hot pressing.

LIQUID METALLURGY. Magnesium alloy–alpha SiC whisker composites have also been produced by casting techniques. For instance, SiC whisker preforms containing an inorganic binder have been infiltrated with molten metal by squeeze casting [136,144,145]. In making the preform, SiC whisker mats are usually broken up and mixed with an inorganic binder (i.e., colloidal SiO_2 or Al_2O_3) in some fluid medium. Mixing is done in order to deagglomerate the SiC whiskers. The preform is then formed by filtering and drying. U.S. Patent 4,548,774 describes a method for producing beta SiC whisker sponge-like cakes from ashed rice hulls that do not have to be broken up and deagglomerated [144]. The as-produced cakes can be infiltrated to give a good MMC material. However, no magnesium examples are given in the patent. Finally, using alpha SiC whisker preforms, magnesium alloy–SiC whisker composites have been fabricated by a self-generating vacuum-casting technique [146]. A SiC whisker preform is immersed in a molten magnesium alloy for 15 minutes to 2 hours. The magnesium reacts with the entrapped gas within the preform to create a vacuum. The pressure differential thus created is sufficient to cause infiltration.

WHISKER ALIGNMENT. In whisker-reinforced MMCs, it is often desirable to align the whiskers in one direction. Whiskers have been aligned in magnesium alloy–SiC whisker liquid phase hot-pressed composite billets before consolidation. This has been accomplished by extrusion of the green body and/or by magnetic means [135,143,147,148]. Magnesium alloy–SiC whisker billets previously consolidated by liquid phase hot pressing have also been extruded to achieve whisker alignment in the extrusion direction [135]. Hahn et al. claim that extruding consolidated billets to achieve whisker alignment is a superior technique compared with magnetic alignment [135].

Finally, magnesium alloy–SiC whisker composite billets prepared through squeeze casting techniques have been thixoformed to produce whisker alignment [145]. A consolidated composite billet is heated to a partially liquid–partially solid state. Since the whiskers are held together by an inorganic binder, the billet retains its shape even if the percent liquid in the metal matrix approaches 100%. The composite billet is then subjected to plastic processing (i.e., extrusion or forging). As the volume percent of liquid in the matrix is increased, the whisker alignment is improved, and less whisker damage occurs.

LEADING PRODUCERS. The leading producer of magnesium alloy–SiC whisker composites is Advanced Composites Materials Corporation of Greer, SC [138,142,149–151].

Mechanical Properties of Magnesium Alloy–Alpha Silicon Carbide Whisker Composites

Some limited room temperature mechanical property data for magnesium alloy–alpha SiC whisker composites is reported in the literature [135,136,139–140,145–147,149,152]; it is summarized in Table 24. Relative to unreinforced magnesium and magnesium alloys, excellent improvements in stiffness and strength are obtained at the expense of a significant reduction in ductility. The elastic modulus is improved by as much as 200%, and the UTS is improved by as much as 70%.

The UTS is given as a function of temperature for a Mg-3 wt % aluminum alloy reinforced with 30 vol % alpha SiC whiskers by Hahn et al. [135]. It is apparent that the UTS falls off with temperature in a manner similar to that for unreinforced AZ31; however, useful strengths are obtained at much higher temperatures. Hahn et al. found evidence of whisker pullouts at elevated temperatures. This is indicative of a loss of shear strength of the matrix and/or an insufficiently high whisker aspect ratio.

Potential Applications for Magnesium Alloy–Alpha Silicon Carbide Whisker Composites

Potential applications are not widely discussed in the literature.

TABLE 24
Mechanical Properties of Mg and Mg Alloy–Alpha SiC Whisker Composites

Material	Fabrication Method[a]	Reference	*E* (GPa)	TYS (MPa)	UTS (MPa)	Elong. (%)	CYS (MPa)	UCS (MPa)
AZ31	C & E	149	45	221	290	15	—	—
AZ31–10 vol % SiC	LPHP & E	147	—	—	393[b]	—	—	—
AZ31–20 vol % SiC	LPHP & E	147	—	—	462[b]	—	—	—
A3–10 vol % SiC	LPHP, MA	135,147	—	—	296[b]	—	—	—
A3–20 vol % SiC	LPHP, MA	135,147	103	—	324[b]	—	—	—
A3–20 vol % SiC	LPHP & E	135,147	—	—	420[b]	—	—	—
A3–40 vol % SiC	LPHP, MA	135,147	138	—	386[b]	—	—	—
Mg	C & E	152	45	96	186	16	—	—
Mg–10 vol % SiC	LPHP & E	149,151,152	70	317	358	1.5	—	—
Mg–20 vol % SiC	LPHP & E	149,151,152	100	414	448	0.9	—	—
Mg	SGV	146	—	48	69	—	48	—
Mg–10 vol % SiC	SGV	146	—	110	117	—	158	303
Mg–30 vol % SiC	SGV	146	—	—	—	—	296	441
A3	LPHP	147	—	—	248[b]	—	—	—
A3	LPHP & E	147	—	—	303[b]	—	—	—
AZ31	?	135	—	—	303[b]	—	—	—
AZ91–10 vol % SiC	SC & T-20	145	—	—	303[b]	—	—	—
AZ91–20 vol % SiC	SC & T-20	145	—	—	331[b]	—	—	—
AZ91–30 vol % SiC	SC & T-20	145	—	—	352[b]	—	—	—
AZ91–10 vol % SiC	SC & T-60	145	—	—	331[b]	—	—	—
AZ91–20 vol % SiC	SC & T-60	145	—	—	420[b]	—	—	—
AZ91–30 vol % SiC	SC & T-60	145	—	—	469[b]	—	—	—
AZ91	SC	145	—	—	276[b]	—	—	—
AZ91	SC	136	—	—	193[b]	—	—	—
AZ91–16 vol % SiC	SC	136	—	—	352[b]	—	—	—

[a] C & E = cast billet and extrude
LPHP & E = liquid phase hot press billet and extrude
LPHP, MA = liquid phase hot press billet, magnetically aligned whiskers
SGV = self-generated vacuum casting
LPHP = liquid phase hot-pressed billet
SC & T-20 = squeeze-cast billet and thixoform part with matrix of 20% liquid
SC & T-60 = squeeze-cast billet and thixoform part with matrix of 60% liquid
SC = squeeze cast

[b] Estimated from graph

Magnesium Alloys Reinforced with Ceramic Particles

Introduction

Magnesium alloys have also been reinforced with ceramic particles. Ceramic particles are produced on a large scale, and, thus, they are widely available and relatively inexpensive. Silicon carbide, B_4C, and Al_2O_3 particles have probably received the most attention [71,149–151]. Magnesium alloy–ceramic particle composites have been fabricated in a variety of fashions, including both powder-metallurgical (PM) and liquid-metallurgical (LM) techniques. Magnesium alloy–11 particle composites have also been produced by PM techniques [153].

Powder Metallurgically Produced Composites

FABRICATION. The PM fabrication of magnesium alloy–ceramic particle composites is similar to that of magnesium alloy–alpha SiC whisker composites. First, the magnesium alloy powder is blended with the ceramic particulate. The powder mixture is then consolidated, often by liquid phase vacuum hot pressing techniques [151,154]. Although near-net-shape parts can be fabricated by consolidation techniques such as liquid phase vacuum hot pressing, usually a composite billet is produced. The billet usually contains a network of MgO skins, which are deleterious with respect to certain mechanical properties (i.e., ductility and fracture properties). Extrusion, forging, or rolling the billet can be used to break up the network of oxide skins and improve the ductility and fracture properties [154]. Figure 2 illustrates this process.

The initial step in the PM manufacture of ceramic particle reinforced magnesium alloys involves the blending of the ceramic particles with the magnesium alloy powder. This step is critical in achieving a good ceramic particle distribution in the final composite. Although very little information concerning the blending opera-

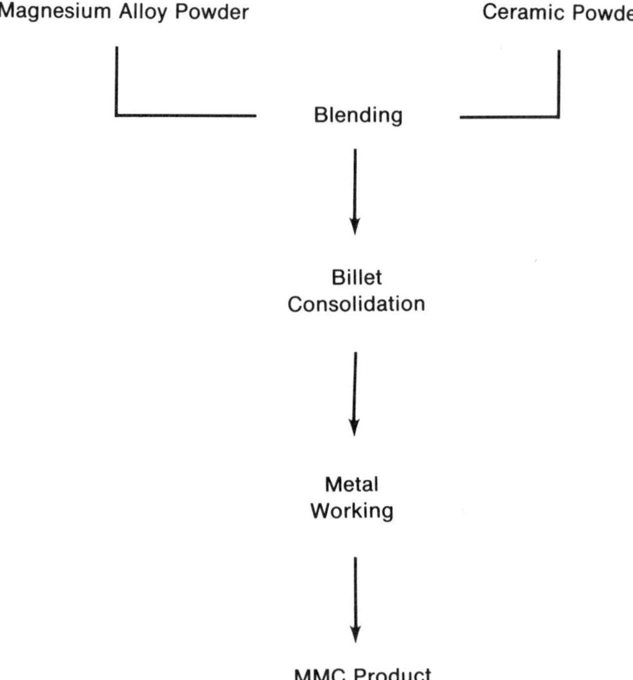

FIGURE 2 PM fabrication flow chart.

tion is available in the literature, it is possible that the same techniques that were employed for magnesium alloy–ceramic whisker composite might be used [135,141–143]. In fact, U.S. Patent 4,259,112 also claims applicability to ceramic particle reinforced metal matrix composites [141].

A data packet was received from Advanced Composite Materials Corporation (ACMC) with respect to their proprietary blending operations [155]. Two blending operations were mentioned, dry ball mill mixing and impact dispersion. No details were given.

U.S. Patents 4,619,699 and 4,647,304 discuss cryogenic ball milling as a means of conducting the blending operation [156,157]. The low temperatures are claimed to produce a fine grain size in the metal powder and the final composite.

Some limited information on the billet consolidation process used by ACMC to produce ZK60–SiC particulate composite billets was received from Battelle [158]. A two-step liquid phase vacuum hot pressing technique was said to be employed. The first step is 14 hours in duration at a temperature of 575°C. The second step is 4 hours in duration at a temperature of 600°C. The consolidated billets were then scalped and homogenized for 35 hours at 300°C plus 8 hours at 400°C before extrusion.

Some limited information on the extrusion process used by ACMC to produce ZK60–20 vol % SiC particulate composite extrusions was also received from Battelle [154,158]. Billets of 20 cm diameter were extruded into either 5.4 cm or 5.7 cm diameter round rods. A conical die was used. The billets were lubricated with graphite, and they were extruded at about 0.9 $m \cdot min^{-1}$. The billet temperature ranged from 204 to 357°C. The die and container were kept at a temperature approximately 38°C lower than that of the billet. Extrusion was said to be difficult.

Finally, some limited information on forged ZK60–10 and 20 vol % SiC particulate composites is given by Smith et al. [154]. Extruded round rods (5.7 cm diameter) produced by ACMC were used as the starting material. Forgings of an unspecified shape were produced by conventional closed die forging using a process similar to aluminum forging. The die and metal temperatures were equal and ranged from 232 to 316°C.

DWA COMPOSITE SPECIALTIES, INC. Dolowy Webb and Associates Composite Specialties, Inc. (DWA) and ACMC are two of the leading PM producers of magnesium alloy–ceramic particle composites [71,150–152,154,155,158]. DWA produces both Mg–B_4C particulate and Mg–SiC particulate composites. Some limited property data for extruded Mg–B_4C particulate appear in Table 25 [159]. These composites are claimed to be more or less isotropic. The strength and stiffness of the composites are greater than those of most aluminum alloys, while maintaining the low density of magnesium. The composites also have wear properties approaching those of steel and expansion properties near those of titanium [71]. The properties of magnesium alloy–SiC particulate composites are claimed to be similar to those of magnesium alloy–B_4C particulate composites [160].

ADVANCED COMPOSITE MATERIALS CORPORATION. ACMC also produces both magnesium alloy–B_4C particulate composites and magnesium alloy–SiC particulate composites. Limited room temperature longitudinal tensile properties for composite extrusions are given in Table 26. The data were taken from Refs. 152, 154, and 155. Again, excellent stiffness and strength properties are achieved. In fact, the specific modulus and UTS of ZK60-T5–20 vol % (X3) SiC particulate are superior to those of the strongest extruded aluminum alloy, 7075-T6 (Table 27) [154,161]. ACMC claims that SiC particulate is a more effective reinforcement than B_4C particulate [162]. They employ very fine SiC particles; for example, their X3 SiC particulate is 3 to 4 μm in size [154].

TABLE 25
Mechanical Properties of Mg Alloy–B_4C Particle Composite Extrusions (Longitudinal Tensile Properties)

Material	*E* (GPa)	TYS (MPa)	UTS (MPa)	Strain %
AZ61—typical	45	228	317	17.0
AZ61–15 vol % B_4C	61	229	300	2.0
AZ61–20 vol % B_4C	65	254	284	1.0
AZ61–25 vol % B_4C	78	258	331	1.3
ZK60-T5—typical	45	303	365	11.0
ZK60-T6–15 vol % B_4C	73	388	429	1.1
ZK60-T6–20 vol % B_4C	78	393	447	1.0
ZK60-T6–25 vol % B_4C	86	418	456	0.8
AZ90-T6–25 vol % B_4C	90	460	477	1.1

TABLE 26
Mechanical Properties of PM Composite Extrusions

Blending Technique[a]	Matrix	Reinforcement Type	Reinforcement Size	Reinforcement Vol %	Heat Treat	Reference	*E* (GPa)	TYS (MPa)	UTS (MPa)	Elong. %
—	Mg[b]	—	—	0.0	F	152	45	96	186	16.0
—	Mg	B_4C	—	10.0	F	152	56	260	292	1.2
—	Mg	B_4C	—	20.0	F	152	59	273	334	0.8
—	AZ31[b]	—	—	0.0	F	152	45	221	290	15.0
—	AZ31[c]	—	—	0.0	F	152	54	345	427	1.4
—	AZ31	SiC	—	20.0	F	152	102	465	482	0.9
DBM	Mg	SiC	X3	20.0	F	155	72	386	472	4.2
DBM	Mg	SiC	X4	20.0	F	155	70	368	456	5.8
MPD	Mg	SiC	X4	17.5	F	155	68	409	477	3.7
MPD	Mg	SiC	X4	17.5	F	155	69	381	450	5.0
MPD	Mg	SiC	X1	15.0	F	155	62	422	476	2.0
—	ZK60	SiC	X3	15.0	T5	154	78	398	488	3.2
—	ZK60	SiC	X3	20.0	T5	154	84	428	528	2.6

[a] DBM = dry ball milled
MPD = impact dispersion
[b] Cast billet
[c] PM billet

Smith et al. give some additional data for composite extrusions produced by ACMC: ZK60 reinforced with 15 and 20 vol % SiC particulate (X3 specification) in the T5 temper [154]. The CYS of these materials was found to be approximately equal to the TYS, and this has been attributed to a very fine grain size in the composite extrusions (i.e., 1 to 10 μm). (Generally, for unreinforced magnesium alloy extrusions, the CYS is significantly lower than the TYS [44].) Smith et al. found that the fracture surfaces exhibited very small dimples and a large amount of MgO inclusions. The authors suggest that during the fracture process, small MgO particles created voids that initiated small dimples and caused the low fracture toughness.

Composite extrusions were also forged, and the forged properties were measured [154]. Forging often led to slight increases in TYS, UTS, and tensile failure strain. Fracture properties as measured by short notch bend energy tests were significantly improved via forging (i.e., by as much as 50%). Presumably, forging was effective with respect to further breaking up and dispersing any MgO skins.

MG–B PARTICLE COMPOSITES. Huseby et al. have reported limited data on a Mg–25 vol % B particulate composite produced by PM techniques [153]. Small diameter extruded rods of both the composite and an unreinforced magnesium control were tested. The Young's modulus of the composite was measured as a function of temperature, and it was found to vary only from 78 GPa at 25°C to 74 GPa at 500°C. The material shows an excellent retention of stiffness at high temperatures. The composite had a room temperature CYS of 237 MPa versus 178 MPa for the unreinforced magnesium control. The composite exhibited equiaxed small grains and only slight crystallographic texture, indicating that the composite extrusion probably had approximately isotropic mechanical properties.

Liquid Metallurgically Produced Composites

INTRODUCTION. As mentioned previously, magnesium alloy–ceramic particle composites have also been fabricated by liquid-metallurgical (LM) techniques, in which the ceramic particles are added to liquid metal before solidification. The success of LM fabrication techniques hinges on the ability of the molten magnesium to wet the ceramic particles. Unfortunately, most ceramic particles are not readily wet by molten metals [163,164].

TABLE 27
Extruded Mg Composite Versus Al Alloy A7075-T6

Material	Heat Treatment	*E/p* (m)	TYS/*p* (m)	UTS/*p* (m)	Reference
ZK60–20 vol % X3 SiC	T5	3.86×10^6	1.97×10^4	2.43×10^4	154
7075 Al	T6	2.64×10^6	1.84×10^4	2.09×10^4	161

Although several techniques to overcome wettability problems are available [165], only a few are reported in the literature for magnesium–ceramic particle composites.

THIXOMIXING. Several investigators have promoted the wetting of ceramic particles in magnesium alloy–ceramic particle composites by rapidly stirring the ceramic particles into a partially solidified melt of the magnesium alloy (i.e., thixomixing) [166–173].

The original patents were issued to MIT [168–170]. The work was later elaborated upon by Battelle [167], the Rheocast Corporation [171], and The Dow Chemical Company [173]. The primary solid metal particles within the partially solidified melt mechanically interact with the ceramic particles, preventing them from either floating or sinking. The molten magnesium alloy and the ceramic particles gradually interact, and the particles become wet by the metal. Once all the particles have been introduced into the melt, the material may be (1) cast into a near-net-shape part, (2) cast into a billet and later thixoformed [174], or (3) heated to a totally liquid state and then cast into a near-net-shape part or billet.

Magnesium alloys reinforced with $CaO{\cdot}SiO_2$ particulate, CaO particulate, SiC particulate, Al_2O_3 particulate, and MgO particulate have been made using thixomixing techniques [167,171].

WETTING AGENTS. Amax, Inc. has taken a different approach to making composites by LM fabrication techniques. Amax was recently granted a patent on the use of Li as a wetting agent [175]. The addition of 0.2 to 0.7 wt % Li to magnesium alloys is claimed to enhance the wetting of SiC and TiC particles and discontinuous fibers. The ceramic is simply stirred into the molten Li-containing magnesium alloy. An AZ91–14.7 wt % (+ 325, − 200 mesh) SiC particulate composite is cited as an example. The Mg alloy contained 0.58 wt % Li. Composite material allowed to solidify in the melting crucible exhibited a 25% increase in hardness relative to an unreinforced AZ91 control sample.

THE DOW CHEMICAL COMPANY. The Dow Chemical Company has also produced magnesium alloy–ceramic particle composites using proprietary LM fabrication techniques. A wide variety of magnesium alloys have been reinforced with SiC and Al_2O_3 particulates [176–179]. The molten composite material is usually cast into either ingot or billet form. Composite ingot has been remelted, and composite parts have been produced from such melts by conventional foundry techniques (i.e., permanent mold casting, sand casting, and high pressure die casting). Cast billets of composite material have also been extruded using conventional techniques, and extruded stock has been successfully forged.

Dow has studied two types of composites: magnesium alloys reinforced with low volume fractions (0–5 vol %) of Al_2O_3 particulate, and magnesium alloys reinforced with high volume fractions (16–30 vol %) of SiC particulate. Since magnesium alloy–Al_2O_3 composites exhibit excellent abrasive wear resistance, the first group of materials has been called "wear-resistant magnesium" [177–179]. Die-cast test panels were tested on a Taber wear test machine, and the results are plotted in Figure 3 [179]. The tabor wear index (TWI) is proportional to the volume loss of the material during the test; therefore, the lower the TWI, the better the abrasive wear resistance. Figure 4 indicates that the addition of as little as 1 vol % Al_2O_3 to die-cast AZ91 results in significant improvements. Room temperature tensile properties are not largely affected by small Al_2O_3 additions (Fig. 4) [179].

Die-cast magnesium alloys reinforced with large volume fractions of SiC particulate show significant room temperature tensile property improvements. Table 28 gives the room temperature tensile properties for four die-cast materials [180]. Separately die-cast tensile bars of all the materials were cast under the same conditions. The addition of 20 vol % SiC particulate to die-cast AZ91 results in an elastic modulus improvement of 50%, a TYS improvement of 53%, and an UTS improvement of 29%. The tensile strain to failure was reduced; nonetheless, the composite material still exhibited some plasticity. The die-cast tensile bars of Z6–20 vol % SiC particulate composite were given a T6 heat treatment (750°F for 4 hours + water quench + 167°F for 24 hours + 300°F for 48 hours). Although die-cast materials are not generally heat treated because of the risk of porosity and blistering, the short solution heat treatment did not cause any blistering. The specific TYS and UTS of the Z6–SiC composite are superior to those of die-cast A380 aluminum alloy.

The high temperature strength properties of magnesium die castings are also improved by a reinforcement with SiC particulate. For AZ91–SiC composites, short-

TABLE 28
Room Temperature Properties of Die-Cast Alloys and Composites

Material	Heat Treat-ment	E (GPa)	TYS (MPa)	UTS (MPa)	Strain at Failure (%)		P (g/cm^3)
					Plastic	Total	
AZ91	F	37	155	196	—	3.0	1.80
AZ91–20 vol % SiC	F	56	237	252	0.4	0.9	2.08
Z6–20 vol % Sic	T6	61	287	319	0.9	1.4	2.10
380 Al	F	71	175	279	—	3.9	2.71

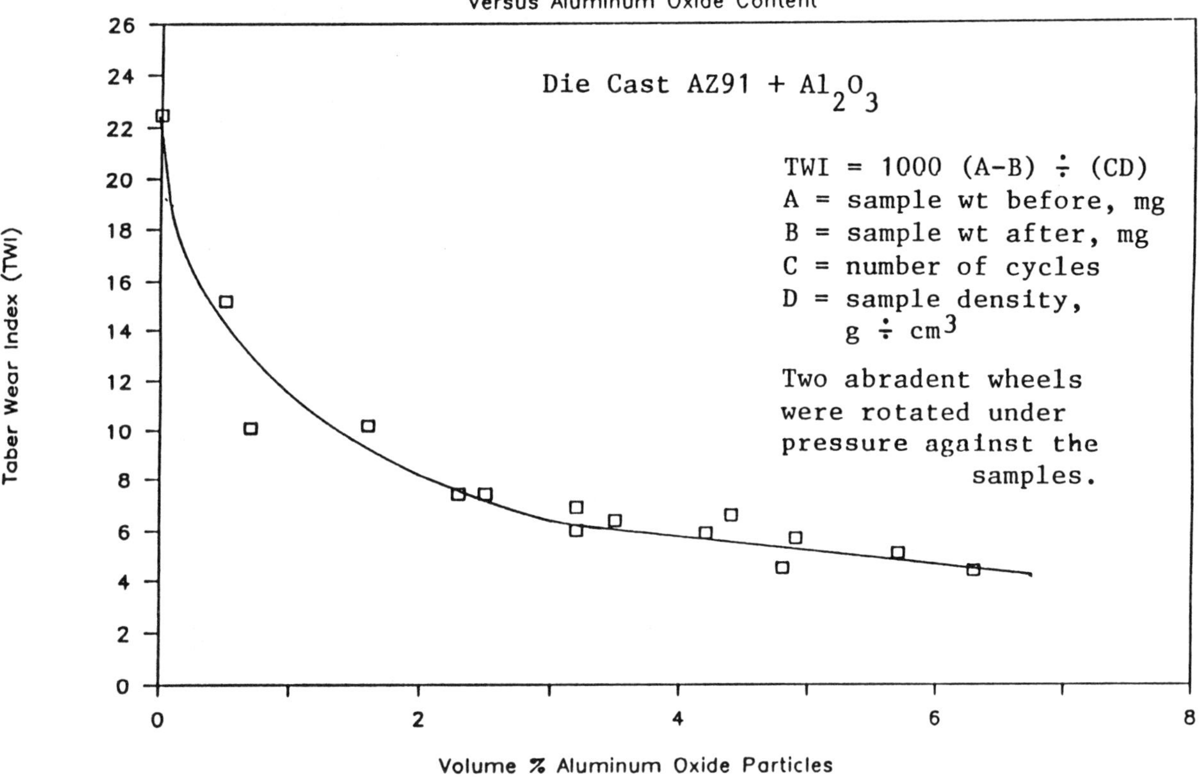

FIGURE 3 Abrasion resistance.

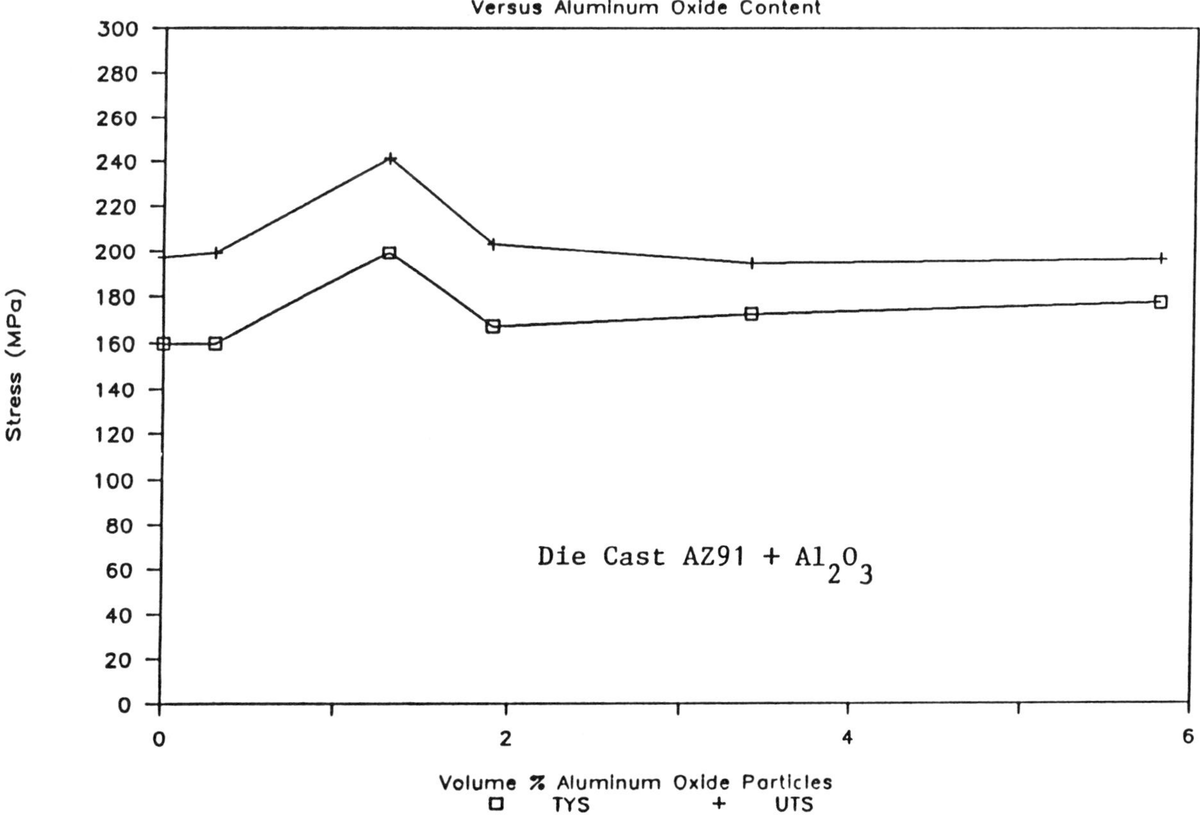

FIGURE 4 Tensile properties.

TABLE 29
Results of 100 Hour Creep Tests for Die-Cast AZ91 and AZ91 Reinforced with 25 Vol % SiC Particulate

Material	Stress Needed to Produce a Given Amount of Total Extension (MPa)	
	0.2% Total Extension	0.5% Total Extension
AZ91	5.5	15
AZ91–25 Vol % SiC	11.0	27

term high temperature properties are improved [178]. In addition, the creep strength of die-cast AZ91–25 vol % SiC particulate composite at 350°F is roughly double that of unreinforced AZ91 (Table 29) [180].

Extruded magnesium alloys reinforced with SiC particulate have also been studied by Dow. Previously reported property results are summarized in Table 30 [154,178,181]. Again significant improvements are achieved relative to unreinforced magnesium alloys, especially in elastic modulus.

However, as stated by Smith et al., improved uniformity in the distribution of the SiC particulate within the composite extrusions was needed [154]. Smith et al. studied the fracture properties of Dow's LM-produced ZK60–SiC and Z6–SiC composite extrusions. The fracture properties were found to be superior to similar composites produced by PM techniques [154].

Recent process improvements by Dow have led to the development of magnesium alloy–SiC particle composite extrusions with improved room temperature properties (Table 31) [180–182]. The improved properties are at least partly due to an improved SiC particle distribution (Fig. 5). Excellent stiffness and strength properties are attained in the Z6–SiC and ZE62–SiC composite extrusions. In fact, the specific properties compare favorably with those of extruded aluminum alloy 7075-T6 (Table 32) [161,180]. The CYSs of the composite materials in Tables 30 and 31 are much lower than the TYSs. This is due to a relatively coarse grain size (i.e., 10 to 20 μm) [154].

Sand and permanent-mold-cast magnesium alloy–SiC composites have been investigated. Table 33 lists the room temperature tensile properties of four magnesium materials [180–181]. The magnesium alloy–SiC particle composite materials show significant improvements in elastic modulus and TYS relative to the unreinforced magnesium alloys.

TOYOTA. Some very interesting work appears in two U.S. patents issued to Toyota [183,184]. In U.S. Patent 4,508,682, a metal matrix is reinforced with a unique rein-

TABLE 30
Previously Reported Room Temperature Properties of Extruded Mg Alloys and Composites

Material	Heat Treatment	E (GPa)	TYS (MPa)	UTS (MPa)	Elongation (%)	CYS (GPa)	Reference
AZ31	F	44	155	281	25.0	—	178
AZ31–15 vol % SiC	F	60	204	299	5.4	—	178
AZ61	F	48	205	307	20.3	131	178
AZ61–20 vol % SiC	F	79	256	323	2.1	186	178
ZK60	T5	45	283	352	14.0	200	181
ZK60–17.5 vol % SiC	T6	72	320	381	1.9	236	154
Z6–17.5 vol % SiC	T6	73	331	375	1.4	234	154

TABLE 31
Recent Room Temperature Properties of Extruded Mg Alloys and Composites

Material	Heat Treatment	E (GPa)	TYS (MPa)	UTS (MPa)	Elongation (%)	CYS (GPa)	Reference
AZ80	T5	45	262	345	6	214	181
AZ80–21 vol % SiC	T6	70	334	391	1.3	257	182
ZK60	T5	45	240	352	14	207	181
Z6–20 vol % SiC	T6	72	392	448	1.9	292	182
Z6–20 vol % SiC	T6[a]	75	418	465	1.9	—	180
ZE62–20 vol % SiC	T6[a]	75	428	465	1.3	313	182

[a] Improved T6

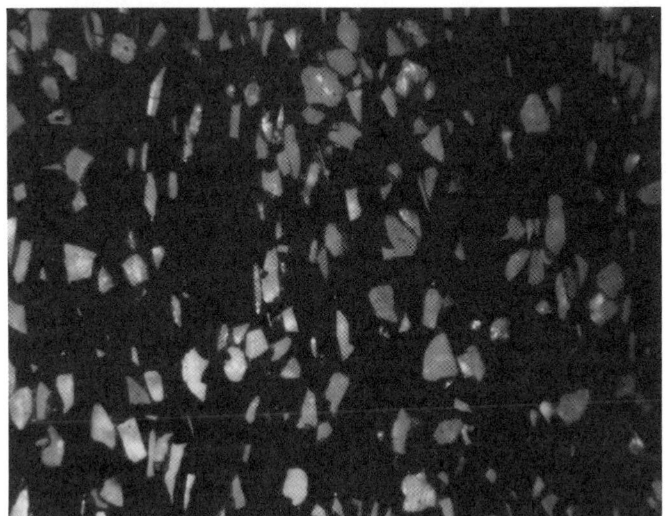

FIGURE 5 **Extruded Z6 + 20 vol. % 1000 Grit SiC, longitudinal section, as-extruded, phosphopicral etch.**

forcing particle [183]. The reinforcing particle is actually a composite particle that contains a metal core and a ceramic surface. The reinforcing particles are added to a molten metal matrix bath by a gas stream. Two magnesium examples are given. In the first example, a magnesium matrix is reinforced with 4 vol % of composite particles containing a magnesium core and a MgO surface. The reinforcing particles are about 0.03 μm in diameter. The MgO coating is approximately 0.0050 μm thick. In the second example, an unspecified magnesium alloy matrix is reinforced with 7 vol % of composite particles (7 μm in diameter) containing a Si core and a SiC surface.

The average thickness of the SiC layer is 0.1 μm. Both composites exhibited excellent wear resistance. The later composite had an UTS of 296 to 324 MPa versus 234 MPa for an unreinforced magnesium control.

In U.S. Patent 4,533,413, a metal matrix is also reinforced with unique composite reinforcing particles [184]. This time the composite reinforcing particles contain ultrafine ceramic particles dispersed within a larger metal matrix particle. Again, the reinforcement particles are gas-injected into a molten bath of the matrix metal. Two magnesium examples are cited. In the first example, magnesium alloy MC2F is reinforced with 4 vol % of composite particles containing Si_3N_4 embedded in metallic Si. In the second example, Mg alloy MC2F is reinforced with 7 vol % of composite particles containing AlN embedded in metallic aluminum. The UTS of both of the composites was in the range of 331 to 372 MPa versus 234 MPa for an unreinforced magnesium control. Both composites exhibited excellent wear resistance.

Potential Applications

Shook and Green discuss some potential applications for magnesium alloys reinforced with low volume fractions of Al_2O_3 particulate [179]. Various prototype automotive parts are studied, including an oil pump cover, a camshaft cover, a cog tooth timing belt pulley, and a poly-V-pulley. Composite material was studied based on its im-

TABLE 32
Specific Tensile Property Comparison Mg Composite Extrusion versus Al

Material	Heat Treatment	E/p (10^6 m)	TYS/p (10^4 m)	UTS/p (10^4 m)	Reference
Z6–20 vol % SiC	T6[a]	1.43	7.98	8.89	180
A7075 Al	T6	1.04	7.23	8.21	161

[a] Improved T6

TABLE 33
Room Temperature Tensile Properties for Sand and Permanent-Mold-Cast Composites

Material	Casting Method[a]	Heat Treatment	E (GPa)	TYS (MPa)	UTS (MPa)	Elongation (%)	Reference
AZ91	SC	T6	45	124	223	5.8	181
AZ91–15 vol % SiC	SC	T6	58	208	263	2.8	180
ZK61	SC	T6	45	179	276	5.0	181
Z6–20 vol % SiC	PMC	T6	72	280	286	0.3	180

[a] SC = sand cast
PMC = permanent mold cast

proved wear properties. A composite oil pump cover plate has been installed on an automobile, and it now has in excess of 160,900 km of successful road testing [185].

Other potential uses of magnesium alloy–ceramic particle composites are mentioned in the literature. Wada et al. suggest that magnesium–ceramic particle composites may find uses in pulleys, sheaves, chain enclosures, bearing surfaces, connecting rods, and pistons [175]. Albertson et al. suggest applications such as brake pads and clutch facings [172]. In several places, the use of Mg–B_4C particulate composites is mentioned for the replacement of aluminum in munitions components where specific stiffness is critical [92,186]. Finally, Mg–B_4C composites are suggested for replacement of Be in dimensionally critical applications such as joints in space structures and missile guidance system components, where isotropic properties are required [151].

Magnesium–ceramic particle composites may find uses in the same types of applications that Al–SiC particle composites are currently being considered for, that is, pistons, connecting rods, cylinders, wrist pins, push rods, rocker arms, and cam followers for air-cooled engines [28,37,71,187].

Summary

A review of the literature indicates that magnesium alloys have been reinforced with a wide variety of ceramic materials, including such continuous fibers as alpha Al_2O_3 fibers, Al_2O_3–SiO_2 fibers, carbon (graphite) fibers, boron fibers, SiC fibers, steel fibers, and continuous fibers composed of Si, Ti or Zr, C or N, and O. Continuous-fiber-reinforced composites have been fabricated by both liquid metal infiltration and diffusion bonding techniques. Liquid metal infiltration techniques look most promising, and some day complex composite castings of magnesium alloys reinforced with continuous fibers may compete on a cost basis with conventional machined metal mill products. However, currently the high price of continuous fibers has driven up the cost of composites.

Magnesium alloys reinforced with continuous fibers have shown excellent room temperature stiffness and strength properties. Elastic moduli up to about 345 GPa and UTSs of from 482 to 2410 MPa have been reported. Moreover, axial properties of unidirectional composites are retained to very high temperatures, i.e., on the order of 320°C or more. Potential applications such as helicopter transmission housings and automobile connecting rods look promising for magnesium alloys reinforced with DuPont's fiber FP. Moreover, graphite fiber reinforced magnesium composites possess the best combination of specific stiffness and thermal deformation resistance of any known engineering materials.

Therefore, such composites have become extremely attractive candidates for dimensionally critical space applications. The use of magnesium–graphite fiber composite castings for rotary engine housings on propeller-driven aircraft has also recently received considerable attention. Magnesium alloys reinforced with B fibers have been continuously cast to produce structural shapes.

Magnesium alloys have also been reinforced with discontinuous alumina fibers, alumina–silica fibers, mineral fibers, and mixtures thereof. The fibers are produced on a large scale for established applications and are, accordingly, relatively inexpensive. Composite parts are made by infiltrating preforms of chopped fibers with molten metal using squeeze casting techniques. Such fiber-reinforced magnesium composites possess excellent wear properties. Moreover, these composites have the potential to be low cost. As a result, magnesium alloys–discontinuous fiber composites show considerable commercial potential for automobile applications, such as pistons.

Alpha SiC whiskers have also been used to reinforce magnesium alloys. Composite materials have been made both by PM techniques and by infiltrating molten metal into whisker preforms. Generally, composite billets are made. Composite parts can then be fabricated by conventional metal working techniques such as extrusion and forging. The properties of composite materials reported in the literature are attractive. For example, the room temperature tensile properties of an extruded Mg–20 vol % SiC whisker composite are E = 100 GPa, TYS = 417 MPa, UTS = 448 MPa, and failure strain = 0.9%. However, SiC whiskers are expensive, and the whiskers are often broken up during extrusion or forging.

Finally, magnesium alloys have been reinforced with widely available and relatively inexpensive ceramic particles, such as SiC, B_4C, and Al_2O_3. Composite materials have been made by both PM and LM techniques. When PM techniques are used, composite billets are usually made first, and composite parts are subsequently made via conventional metalworking techniques. When LM techniques are used, composite parts can also be made by conventional foundry techniques. Composite extrusions possess excellent properties; the specific stiffness and strength for some composite extrusions are superior to those of aluminum alloy 7075-T6. Composite castings have also exhibited good room temperature tensile properties and improved abrasive wear resistance. Magnesium alloy–ceramic particle composites have the potential for low cost, especially when they are produced by LM techniques. Therefore, magnesium–ceramic particle composites are attractive for commercial automotive applications. In addition, they are attractive for applications requiring good specific stiffness and isotropic properties, such as joints in space structures and munitions components.

B. A. Mikucki and S. O. Shook

References

1. A. K. Dhingra, Inc., *Proceedings of the 16 National Symposium on Polymers in the Service of Man*, American Chemical Society, 1980, pp. 88–123.

2. Hans Lilholt, *Compos. Sci. and Tech., 30,* 2 (1987).
3. A. K. Dhingra and W. H. Krueger, *Proceedings of the 36th World Conference on Magnesium,* 1979, pp. 1–28.
4. A. R. Champion, W. H. Krueger, H. S. Hartmann, and A. K. Dhingra, *Proceedings of the Second International Conference on Composite Materials,* 1978, pp. 883–904.
5. "Fiber FP Alumina Fiber, Reinforcement Fiber for Advanced Composites," company literature, E. I. du Pont de Nemours & Co., Inc., Textile Fibers Department, Advanced Materials Research and Development, Experimental Station, Wilmington, DE.
6. H. J. Nusbaum and H. S. Hartmann, *Proceedings, Fourth Metal Matrix Composites Technology Conference,* 1981, paper 31/1–16.
7. "Fiber FP Reinforced Composites," company literature, E. I. du Pont de Nemours & Co., Inc., Textile Fibers Department, Pioneering Research Laboratory, Experimental Station, Wilmington, DE, 1978.
8. A. K. Dhingra, *Proceedings, Composite Materials: Mechanics, Mechanical Properties and Fabrications,* Japan Society for Composite Materials, 1981, pp. 239–244.
9. J. W. Lenski, *Proceedings, A Collection of Technical Papers on Structures and Materials,* AIAA/ASME/ASCE/AHS, 1979, pp. 333–340.
10. A. K. Dhingra, *Phil. Trans. R. Soc. Lond. A, 294,* 559 (1980).
11. J. Nunes, E. S. C. Chin, J. M. Slepetz, and N. Trangarakis, *Proceedings, Fifth International Conference on Composite Materials: 1CCM-V,* The Metallurgical Society, 1985, pp. 723–745.
12. A. K. Dhingra, A. R. Champion, and W. H. Krueger, *Proceedings of the First Metal Matrix Composite Workshop,* Institute for Defense Analysis, 1975, Paper No. C-501.
13. A. K. Dhingra, U.S. Patent 4,036,599 (August 15, 1975).
14. R. M. Lamothe and J. Nunes, in R. Chait and R. Papirno, Eds., *Compression Testing of Homogeneous Materials and Composites,* ASTM STP 808, American Society for Testing and Materials, Philadelphia, PA, 1983, pp. 241–253.
15. W. H. Krueger and A. K. Dhingra, *Proceedings of the American Institute of Chemical Engineers* Summer National Meeting, 1981.
16. J. E. Hack, R. A. Page, and G. R. Leverant, *Metall. Trans. A, 15A,* 1389 (1984).
17. R. A. Page, J. E. Hack, R. Sherman, and G. R. Leverant, *Metall. Trans. A, 15A,* 1397 (1984).
18. R. A. Page and G. R. Leverant, *Proceedings, Fifth International Conference on Composite Materials: 1CCM-V,* The Metallurgical Society, 1985, pp. 867–886.
19. R. T. Bhatt, "Modulus, Strength and Thermal Exposure Studies of FP-Al_2O_3/Aluminum and FP-Al_2O_3/Magnesium Composites," NASA Technical Memorandum 82868, NASA Lewis Research Center and AVRADCOM Research and Technology Laboratories, Cleveland, OH.
20. K. S. Chan, J. E. Hack, and R. A. Page, *Metall. Trans. A, 15A,* 756 (1984).
21. R. T. Bhatt, in J. E. Hack and M. Amateau, Eds., *Mechanical Behavior of Metal Matrix Composites, Proceedings of the 111th Annual Meeting of the American Institute of Mining, Metallurgical, and Petroleum Engineers,* The Metallurgical Society of AIMME, 1983, pp. 51–64.
22. S. C. Chin and M. H. Richman, "An Investigation of Interfacial Reactions in Metal Matrix Composites," Contract No. DAAG46-80-C-0027; AMMRC TR 82-14, 1982.
23. A. Munitz, M. Metzger, and R. Mehrabian, *Metall. Trans. A, 10A,* 1491 (1979).
24. J. W. Lenski, Jr., "Helicopter Transmission Vibrations and Noise Reduction Program, Volume III," Boeing-Vertol Company, Contract No. DAAJ02-74-C-0040; Report No. D 210-11442-1, 1979.
25. F. Folgar, J. E. Widrig, and J. W. Hunt, *Proceedings of the SAE International Congress and Exposition,* 1987, Technical Paper No. 870406.
26. F. Folgar, J. E. Widrig, and J. W. Hunt, *Proceedings of the 32nd International SAMPE Symposium,* 1987, pp. 1559–1568.
27. L. Rubin, *Proceedings of the 24th Int. SAMPE Symposium,* Vol. 2, 1979, pp. 1236–1249.
28. "Metals in Power Train and Chassis," *Adv. Mater. Process. Inc. Metal Progress,* May 5, 1987, p. 40.
29. H. Okamoto and K. Yamatsuta, U.S. Patent 4,515,866 (May 7, 1985).
30. Osaka Kohji Yamatsuta and Shiga Ken-ich Nishio, U.S. Patent 4,526,841 (April 17, 1984).
31. Osaka Kohji Yamatsuta and Shiga Ken-ichi Nishio, U.S. Patent 4,489,138 (July 23, 1981).
32. Osaka Kohji Yamatsuta and Shiga Ken-ichi Nishio, U.S. Patent 4,547,435 (Oct. 15, 1985).
33. Osaka Kohji Yamatsuta and Shiga Ken-ichi Nishio, U.S. Patent 4,465,741 (Aug. 14, 1984).
34. Nelvin K. Hiester, "Trends in Advanced Composites," SRI International, Business Intelligence Program, Report No. 700, Menlo Park, CA, 1984.
35. D. Goddard, W. Whitman, R. Pumphrey, and C. M. Lee, *Proceedings of the AIAA/ASME/SAE/ASEE 22nd Joint Propulsion Conference,* 1986, Paper No. AIAA-86-1559.
36. J. A. Cornie, Y.-M. Chiang, D. R. Ohlmann, A. Mortensen, and J. M. Collins, *Ceramic Bulletin, 65*(2), 293 (1986).
37. W. Harrigan, "Fifth International Conference on Composite Materials, Conference Preview," *J. Metals, 37*(6), 57 (1985).
38. F. P. La Iacona, U.S. Patent 3,894,677 (July 15, 1975).
39. R. V. Sara, U.S. Patent 3,473,900 (Feb. 21, 1967).
40. R. V. Sara, U.S. Patent 3,571,901 (June 13, 1969).
41. I. W. Hall, *Metallography, 20*(2), 237 (1987).
42. R. T. Pepper and E. G. Kendall, U.S. Patent 3,770,488 (April 6, 1971).
43. A. P. Levitt and H. E. Band, U.S. Patent 4,157,409 (June 5, 1979).
44. E. F. Emley, *Principles of Magnesium Technology,* Pergamon Press, New York, 1966, pp. 229, 239, 305, 499, and 509.
45. A. P. Levitt, E. Di Cesare, and S. M. Wolf, *Metall. Trans., 3,* 2455 (1972).
46. A. W. Denham and B. A. W. Redfern, U.S. Patent 3,827,129 (Aug. 6, 1974).
47. W. L. Lachman, R. A. Penty, and F. J. Apul, U.S. Patent 3,860,443 (Jan. 14, 1975).
48. F. P. La Iacona, U.S. Patent 3,894,677 (July 15, 1975).
49. R. T. Pepper and T. A. Zack, U.S. Patent 4,072,516 (Feb. 7, 1978).
50. E. G. Kendall and R. T. Pepper, U.S. Patent 4,082,864 (April 4, 1978).

51. E. G. Kendall and R. T. Pepper, U.S. Patent 4,145,471 (March 20, 1979).

52. F. S. Galasso, and R. D. Veltri, U.S. Patent 4,214,037 (July 22, 1980).

53. W. C. Harrigan, Jr., R. H. Flowers, and S. P. Huson, U.S. Patent 4,223,075 (Sept. 16, 1980).

54. H. A. Katzman, U.S. Patent 4,376,803 (March 15, 1983).

55. T. Donomoto, A. Tanaka, M. Okada, A. Kitamura, and T. Kyono, U.S. Patent 4,419,389 (Dec. 6, 1983).

56. Y. Naerheim and M. W. Kendig, *Proceedings of the 18th Gathering Momentum*, The International Technical Conference 1986, pp. 898–908.

57. H. A. Katzman, U.S. Patent 4,376,804 (March 15, 1983).

58. Herbert Heisslec, Eckart Hitsch, and Wolfgang Scheer, U.S. Patent 4,050,997 (Sept. 27, 1977).

59. M. H. Kural and A. M. Ellison, September/October 1980, p. 20.

60. B. J. Maclean and M. S. Misra, in *Proceedings, Mechanical Behavior of Metal/Matrix Composites*, The Metallurgical Society/AIME, 1983, pp. 195–212.

61. S. P. Rawal, L. F. Allard, and M. J. Misra, *Proceedings of the Conference on Interfaces in Metal-Matrix Composites*, The Metallurgical Society/AIME, 1986, pp. 211–225.

62. Dr. David B. Goddard, Materials Concepts, Inc., 666 N. Hague Avenue, Columbus, OH, private communications, Oct. 22, 1986.

63. "FMI and MCI Announce Capacity Increases," *DOD Metal Matrix Composites Information Analysis Center, Current Highlights, 4*(2), 6 (1984).

64. D. M. Goddard, *Metal Progress*, April 1984, p. 49.

65. G. Clark, S. Paprock, W. Meyerer, and D. Kizer, *Proceedings of the 11th Natl. SAMPE Technical Conference*, 1979, pp. 341–343.

66. "Graphite/Magnesium Casting R&D," *DOD Metal Matrix Composites Information Analysis Center, Current Highlights, 7*(1), (1987).

67. D. M. Goddard, *AFS Transactions, 86*(123), 667.

68. D. M. Goddard, W. R. Whitman, and R. L. Pumphrey, "Graphite/Magnesium Composites for Advanced Lightweight Rotary Engines," SAE International Congress and Exposition, 1986, Paper No. 860564.

69. A. P. Levitt and E. Di Cesare, U.S. Patent 3,888,661 (April 16, 1974).

70. G. A. Chadwick, *Proceedings of the Conference on Magnesium Technology*, The Institute of Metals, London, 1987, pp. 75–82.

71. K. Marsden, *J. Metals, 37*(6), 59 (1985).

72. "DWA, Where Concepts Become Reality, All Metal-Matrix Composites," company literature, DWA Composite Specialties, Inc., 21119 Superior Street, Chatworth, CA, 1985.

73. R. J. Weimer, "Unique Advantages of VOM MMC Precursors," company literature, Cordec Corp., 8270-B Cinder Bed Road, Lorton, VA, 1985.

74. R. F. Monks and P. A. Seamon, *Proceedings of the Conference on Advanced Composites: The Latest Developments*, ASM International, Metals Park, OH, 1986, pp. 257–262.

75. A. Mortensen, J. A. Cornie, and M. C. Flemmings, *J. Metals, 40*(2), 12 (1988).

76. R. J. Boan, *J. Reinforced Plastics and Composites, 3*, 63 (1984).

77. W. D. Wade and A. M. Ellison, in reference 27, pp. 1265–1275.

78. C. A. Ginty and C. C. Chamis, *Proceedings of the 29th National SAMPE Symposium and Exhibition*, 1984, pp. 979–993.

79. D. M. Goddard, Table of Property Data, Materials Concepts, Inc., Columbus, Ohio, 1988.

80. H. H. Armstrong, in reference 27, pp. 1250–1264.

81. E. G. Wolff, E. G. Kendall, and W. C. Riley, *Proceedings of the Conference on Advances in Composite Materials*, Vol. 2, 1980, pp. 1140–1152.

82. E. G. Wolff, D. K. Min, and M. H. Kural, *J. Materials Science, 20*, 1141 (1985).

83. M. H. Kural and B. K. Min, *J. Composite Material, 18*(6), 519 (1984).

84. M. H. Richman, A. P. Levitt, and E. S. Di Cesare, *Metallography, 6*(6), 497 (1973).

85. F. Mansfeld and S. L. Jeanjaquet, *Corrosion Science, 26*(9), 727 (1986).

86. P. P. Trzaskoma, *Corrosion, 42*(10), 609 (1986).

87. W. A. Ferrando, "Corrosion Resistant Magnesium Based Materials for Naval Applications," Report No. NSWC TR 85-88, Naval Surface Weapons Center, 10901 New Hampshire Avenue, Silver Spring, MD, 1985.

88. S. G. Fishman, *Proceedings of the Conference, Role of Interfaces on Material Damping*, The American Society for Metals, 1985, pp. 33–41.

89. E. F. Crawley and M. C. Van Schoor, *J. Composite Materials, 21*, 553 (1987).

90. W. K. Kinner, *Materials Enineering, 91*(1), 64 (1980).

91. "Graphite-Magnesium Gains Attention as Composite," *Metal Working News*, July 6, 1987, p. 24.

92. "Ceramics and Composites," *Metal Progress*, January 1985, p. 46.

93. C. Zweben, "Metal Matrix Composites, History, Overview, and Applications," Paper No. 3441, Metal Matrix Composites Information Analysis Center, 816 State Street, PO Drawer QQ, Santa Barbara, CA.

94. D. C. Brown, Ed., "Magnesium Composites for Rotary Engine Parts," *Magnesium Monthly Review, 15*(7), 3 (1986).

95. D. C. Brown, Ed., "Graphite-Magnesium Composites," *Magnesium Monthly Review, 16*(6), 7 (1987).

96. "Functional Inorganic Fibers," Report No. DIRASS E6-6, DIA Research Institute Inc., Tokyo DIA Building, 28-25, Shinkawa 1-Chome, Chuo-Ku, Tokyo 104, Japan, 1984.

97. D. M. Karpinos, V. Kh. Kadyrov, A. I. Gordienko, and V. P. Dzeganovskii, *Soviet Powder Met. Ceram., 17*(12), 927 (1978).

98. L. G. Davies, W. M. Powers, and R. G. Shaver, *SAMPE Q., 3*(1), 32 (1971).

99. R. G. Shaver, *Composite Materials in Engineering Design*, 1973, pp. 232–241.

100. J. A. Alexander, "Fabrication of Rod Composite Forms in the Mg-B System," Paper from Metal-Matrix Composites

Clearinghouse for Federal Scientific and Technical Information, Springfield, VA, 1969, pp. 68–73.

101. "Continuous Casting Could Be Composite Cost Cutter," *Iron Age, 22,* 96 (1969).

102. J. G. Banker, *SAMPE Q., 5*(2), 39 (1974).

103. K. I. Partnoi, G. B. Stroganov, V. I. Bogdanov, A. V. Mikhailov, and D. L. Fuks, *Met. Sci. and Heat Treat., 22*(5–6), 361 (1980).

104. A. F. Voitenke, V. K. Samatin, and A. I. Gordienko, *Sov. Powder Metall. Met. Ceram., 21*(9), 715 (1982).

105. G. G. Maksimovich, E. N. Lyutii, O. I. Eliseeva, V. Kh. Kadirov, and A. I. Gordienko, *Met. Sci. and Heat Treat., 21*(11–12), 830 (1979).

106. A. F. Stroganova and M. A. Timonova, *Met. Sci. and Heat Treat., 20*(9–10), 834 (1978).

107. M. A. Timonova, G. I. Spiryakina, V. F. Stroganova, and L. A. Zolotareva, *Metalloved, Term, Obrab. Met., 11,* 33 (1980).

108. R. Fisher, American, Inc., 8948 Fulbright Avenue, Chatsworth, CA, private communication, 1983.

109. G. D. Lawrence, *AFS Transactions, 80,* 283 (1972).

110. T. Donomoto and A. Tanaka, U.S. Patent 4,492,265 (Jan. 8, 1985).

111. M. Basche, R. Fanti, and S. F. Gulasso, U.S. Patent 3,556,836 (Feb. 24, 1967).

112. R. G. Schierding and T. L. Tolbert, U.S. Patent 3,676,916 (Jan. 2, 1970).

113. R. C. Tucker, Jr., U.S. Patent 3,840,350 (June 2, 1971).

114. J. H. Lemelson, U.S. Patent 3,889,348 (June 17, 1975).

115. H. Okamoto and K. Nishio, U.S. Patent 4,338,132 (July 6, 1982).

116. D. M. Karpinos, V. Kh. Kadyrov, A. I. Gordienko, V. P. Moroz, and P. A. Verkovodov, *Sov. Powder Metall. Met. Ceram., 19*(2), 141 (1980).

117. "Nicalon Silicon Carbide Continous Fiber," company literature, Nippon Carbon Company, Ltd., Japan, 1984, pp. 1–7.

118. H. Fukunga and K. Goda, *Bull. Jpn. Soc. Mech. Eng., 28*(235), 1 (1985).

119. S. Yajima, J. Hayashi, M. Omori, and H. Kayano, U.S. Patent 4,134,759 (Jan. 16, 1979).

120. B. A. Wilcox and A. H. Clauer, *Trans. Met. Soc. AIME, 245,* 935 (1969).

121. R. G. Schierding and T. L. Tolbert, U.S. Patent 4,614,690 (Sept. 30, 1986).

122. T. Yamamura, M. Tokuse, and Y. Waku, U.S. Patent 4,614,690 (Sept. 30, 1986).

123. T. Yamamura, M. Tokuse, and T. Furushima, U.S. Patent 4,622,270 (Nov. 11, 1986).

124. B. Peters, *Automotive Engineering, 94*(12), 62 (1986).

125. T. Dohnomoto, M. Kubo, and H. Kito, U.S. Patent 4,590,132 (May 20, 1986).

126. M. Kubo, T. Dohnomoto, A. Tanaka, and Y. Tatematsu, U.S. Patent 4,615,733 (Oct. 7, 1986).

127. "Saffil Alumina Fiber: Cost-Effective Reinforcement for Metal Matrix Composites," company literature form no. GC/12174/1Ed/152/584, Imperial Chemical Industries PLC, Mond Division, The Heath, Runcorn, Cheshire WA7 4QD, England, 1984.

128. T. Donomoto, M. Koyama, J. Miyake, and Y. Furva, U.S. Patent 4,457,979 (July 3, 1984).

129. T. Donomoto, M. Koyama, Y. Ruwa, N. Miura, and T. Sakakibara, U.S. Patent 4,576,863 (March 18, 1986).

130. T. Dohnomoto and M. Kubo, U.S. Patent 4,595,638 (June 17, 1986).

131. T. Dohnomoto, U.S. Patent 4,601,956 (July 22, 1986).

132. T. Dohnomoto and M. Kubo, U.S. Patent 4,664,704 (May 12, 1987).

133. K. Ban, T. Arai, and A. Daimaru, U.S. Patent 4,279,289 (July 21, 1981).

134. M. E. Toaz and M. D. Smalc, U.S. Patent 4,587,177 (May 6, 1986).

135. H. Hahn, A. P. Divecha, P. J. Lare, and R. A. Hermann, "Some Recent Advances in Preparation of Whisker Reinforced Metal Composites," in *Materials Technology and Interamerican Approach,* ASME, New York, 1968, pp. 231–235.

136. T. Donomoto, Y. Tatematsu, and A. Tanaka, U.S. Patent 4,530,875 (July 23, 1985).

137. F. H. Simpson, U.S. Patent 4,388,255 (June 14, 1983).

138. A. P. Divecha, U.S. Patent 4,060,412 (Nov. 29, 1977).

139. "Silar Silicon Carbide Whiskers," company literature, ARCO Metals Company, Silag Operations, Rt. 6, Box A, Greer, SC, 1984.

140. "Tokamax Silicon Carbide Whiskers," company literature, Tokai Carbon Co., Ltd., Aoyama Bldg. 2-3, Kita-Aoyama 1 Chome, Minato-Ku, Tokyo, Japan, 1983.

141. J. F. Dolowy, Jr., B. A. Webb, and E. C. Supan, U.S. Patent 4,259,112 (March 31, 1981).

142. P. E. Hood and J. O. Pickens, U.S. Patent 4,463,058 (July 31, 1984).

143. A. P. Divecha, U.S. Patent 4,060,412 (Nov. 29, 1977).

144. M. Akiyama, J. Yamada, and M. Takahata, U.S. Patent 4,548,774 (Oct. 22, 1985).

145. K. Funatani, T. Donomoto, A. Tanaka, and Y. Tatematsu, U.S. Patent 4,548,253 (Oct. 22, 1985).

146. G. D. Lawrence, U.S. Patent 3,529,655 (Sept. 22, 1970).

147. "Light Metal Composites. Review of Current Progress," *Light Metal Age, 28*(1–2), 10 (1970).

148. A. P. Divecha, F. Ordway, Jr., R. A. Herman, O. B. Van Blaricon, and H. Hahn, U.S. Patent 3,668,748 (June 13, 1972).

149. D. C. Brown, Ed., *Magnesium Monthly Review, 13*(6), 6 (1984).

150. D. C. Brown, Ed., *Magnesium Monthly Review, 15*(7), 3 (1986).

151. H. J. Rack and P. W. Niskanen, *Light Metal Age, 42*(1–2), 9 (1984).

152. P. A. Roth, data sheets, ARCO Chemical Company, Silag Operation, Route 6, Box A, Greer, SC, 1986.

153. I. C. Huseby, J. S. Shyne, and O. D. Sherby, *International Journal of Powder Metallurgy, 9*(2), 91 (1973).

154. M. T. Smith, E. M. Patton, and C. A. Lavendar, "Characterization and Processing of Silicon-Carbide Reinforced Magnesium Composites," preprint of paper prepared for SAMPE conference, 1987.

155. P. A. Roth, data packet, Advanced Composite Materials Corporation, Route 6, Box A, Old Buncombe Road, Greer, SC, 1988.

156. R. Petkovic-Luton and J. Vallone, U.S. Patent 4,619,699 (Oct. 28, 1986).
157. R. Petkovic-Luton and J. Vallone, U.S. Patent 4,647,304 (March 3, 1987).
158. C. A. Lavender, Battelle Pacific Northwest Laboratories, Battelle Blvd., Richard, WA, private communications, Sept. 9, 1987 and Jan. 22, 1987.
159. W. C. Harrigan, Jr., data sheets, DWA Composite Specialties, Inc., 2119 Superior St., Chatsworth, CA, 1985.
160. W. C. Harrigan, Jr., DWA Composite Specialties, Inc., 2119 Superior St., Chatsworth, CA, private communication, Jan. 16, 1986.
161. "Properties of Aluminum and Aluminum Alloys," in Taylor Lyman, Ed., *ASM Handbook,* 8th ed., Vol. 1, American Society for Metals, Metals Park, OH, 1961, pp. 948–952.
162. P. E. Hood, Advanced Composite Materials Corporation, Route 6, Box A, Old Buncombe Road, Greer, SC, private communication, Jan. 16, 1986.
163. M. Humenik, Jr. and W. D. Kingery, *J. American Ceramic Society, 37,* 18 (1952).
164. A. Bondi, *Chemical Reviews, 52,* 417 (1953).
165. P. Rohatgi, *Modern Casting,* April 1988, p. 47.
166. J. A. Cornie, Metal Matrix Composites Laboratory, Materials Processing Center, Massachusetts Institute of Technology, 77 Mass. Avenue, Room 4-415, Cambridge, MA, private communications, May 1985 and January 1987.
167. F. C. Bennett, P. C. Burrier, and S. L. Couling, "Naval Processes for Casting Partially Solid Magnesium and Magnesium Alloys—Sixth Interim Progress Report to the Dow Chemical Company," Battelle Columbus Laboratories, 505 King Avenue, Columbus, OH, Aug. 4, 1977.
168. R. Mehrabian and M. C. Flemings, U.S. Patent 3,936,298 (Feb. 3, 1976).
169. M. C. Flemings, R. Mehrabian, and D. B. Spencer, U.S. Patent 3,948,650 (April 6, 1976).
170. R. Mehrabian and M. C. Flemings, U.S. Patent 3,951,651 (April 20, 1986).
171. F. C. Bennett and S. L. Couling, U.S. Patent 4,174,214 (Nov. 13, 1979).
172. C. E. Albertson, J. A. Horwath, and D. W. Lahua, U.S. Patent 4,409,298 (Oct. 11, 1983).
173. E. K. Keith, U.S. Patent 4,432,936 (Feb. 21, 1984).
174. M. C. Flemings, R. G. Rick, and K. P. Young, *Materials Science and Engineering, 25,* 103 (1976).
175. T. Wada, G. T. Eldis, and D. L. Albright, U.S. Patent 4,657,065 (April 14, 1987).
176. E. K. Keith, W. E. Mercer, II, and C. R. Dick, U.S. Patent 4,605,438 (Aug. 12, 1986).
177. S. O. Shook, *Proceedings of the 42nd Annual World Magnesium Conference,* 1985, pp. 19–25.
178. B. A. Mikucki, S. O. Shook, W. E. Mercer, II, and W. G. Green, *Proceedings of the 43rd Annual World Magnesium Conference,* 1986, pp. 13–23.
179. S. O. Shook and W. G. Green, "Improving Magnesium Wear Resistance—A Composite Approach," Society of Automotive Engineers 1985, Technical Paper Series—850421.
180. B. A. Mikucki, W. G. Green, and W. E. Mercer, II, The Dow Chemical Company, Freeport, TX, previously unpublished property data, 1987.
181. "Magnesium Mill Products and Alloys," The Dow Chemical Company, Midland, MI, Form No. 141-481-83, 1982, pp. 16–19.
182. M. T. Smith and C. A. Lavendar, Battelle Pacific Northwest Laboratories, Richland, WA, testing of Dow produced materials, 1987.
183. H. Miura, H. Satou, T. Natsume, and H. Katagiri, U.S. Patent 4,508,682 (April 2, 1985).
184. H. Miura, H. Jato, T. Natsume, and H. Katagiri, U.S. Patent 4,533,413 (Aug. 6, 1985).
185. G. MacFarlane, General Motors Technical Center, Engineering Bldg./C1 Advanced Engine, 30200 Mound Road, Warren, MI, private communication, December 1987.
186. D. C. Brown, Ed., *Magnesium Monthly Review, 14*(2), 4 (1985).
187. "Composite Engine Components Upgrade Performance," *Magnesium Monthly Review, 16*(2), 16 (1987).

Metal Matrix Composites, Rapid Solidification Processing

Continuous-length fiber-reinforced metal matrix composites are a family of materials that are beginning to gain acceptance in the aerospace industry, where increased performance and not necessarily economics can dictate the use of new materials. These composites combine the high strength and stiffness of modern fibers with the environmental resistance and toughness of a metallic matrix in order to create a component with better overall material performance than is possible with present monolithic metals. Gains in terms of higher strength-to-weight ratios, combined with high temperature corrosion and oxidation resistance, are expected with the use of continuous-fiber composites. These types of advanced materials have significant potential for use in aerospace propulsion and structural applications [1–3] and advanced space power systems [4–5].

More than just an experimental curiosity, rapid solidification processes (RSP) are coming into their own as a viable materials fabrication alternative, as is evident from the numerous conferences and topical reviews recently dedicated to this subject [6–22]. As is apparent from these references and Table 1, rapid solidification encompasses a broad spectrum of technologies, from powder-producing atomization techniques to the plasma spray process. Therefore, as interest in fiber-reinforced metal matrix composite materials increases, so does the application of rapid solidification technologies to the processing of continuous-length fiber composites. Whether used to create novel microstructures (e.g., by melt spinning) or simply because of the convenience of the fabrication technique (such as the thermal spray processes), rapid solidification techniques have become an integral part of the overall fabrication scheme for metal matrix composites.

TABLE 1
Rapid Solidification Techniques

Process Name	Description	Product/Dimensions	Cooling Rate (K/s)	References
Atomization Methods				
Fluid atomization (normally twin jets) -Gas (Ar, N_2)	High pressure fluid impacting a continuous stream of liquid metal.	For gas: Low O_2 contamination -Spherical and smooth powder 50/100 μm diameter Supersonic G.A.: 10–50 μm	Gas: 10^2–10^3 SSGA: up to 10^6	45–53 122–127
	Ultrasonic G.A.: Disintegration occurs by high intensity pulsed waves.	Ultrasonic G.A.: <30 μm	USGA: 10^5	128–130
-Water		For water: High O_2 contamination -Irregular particulates 75/200 μm diameter	Water: 10^2–10^4	45, 122–124
Vacuum atomization (soluble gas)	Molten metal supersaturated with gas under pressure is suddenly exposed to vacuum; gas expands, causing liquid to atomize.	-Spherical powder 40/150 μm diameter	10^4–10^5	53,133
Drum splat quenching	Gas-atomized droplets directed to impinge on a rotating drum.	Flakes 50/100 μm thick 1–3 mm diameter	10^4–10^5	134
Rotating electrode process (REP)	Alloy in electrode form is rotated (~ 250 rps) while it is melted by arc plasma/beam. Molten metal is ejected centrifugally and solidifies in an inert gas filled chamber.	-Spherical/smooth powder 150/200 μm diameter	10^2	135
Rapid solidification rate (RSR) (centrifugal atomization)	Molten metal is ejected into a rotating water-cooled cup/disk, resulting in fine droplets cooled by high flow helium gas.	Spherical powder <100 μm diameter	10^5	136–138
Twin roll atomization	Mechanical atomization process where stream of molten metal is directed to high speed contra rotating rolls.	Variable powder/flake 200 μm thick	10^5–10^6	134–141
Laser-melting/spin atomization	Focused (CO_2) laser beam is used to melt the top of a rotating rod. Droplets are expelled by the centrifugal force and cooled by inert gas.	Spherical particles 40 μm diameter	> 10^4	129
Electrohydrodynamic atomization (EHDA)	Electric field is applied to the surface of liquid metal and causes a droplet to be emitted.	Droplets > 0.01 to 150 μm diameter	10^7	142
Spark erosion technique	Repetitive spark discharge between two electrodes immersed in dielectric fluid. Spark vaporizes a small amount of metal that immediately freezes.	> 0.5 μm diameter Contamination from dielectric fluid.	10^5–10^6	143
Electron-beam melting combined with splat quenching (EBSQ)	A focused electron beam melts the bottom tip; molten drops fall onto the surface of a rotating disk.	Flakes 50 mm diameter 50 μm thick	10^6	129,144

(continued)

TABLE 1 (*continued*)

Process Name	Description	Product/Dimensions	Cooling Rate (K/s)	References
Chill Methods				
Melt spinning (CBMS)	Molten metal is expelled out onto a rotating wheel (flat or notched).	Ribbons 25/50 μm thick	10^5–10^7	145–152
Crucible melt extraction (CME)	Molten metal solidifies on the edge of a water-cooled disk and flies off.	Filaments or fibers 20–100 μm thick	10^5–10^6	153,154
Melt drag (overflow)	Overflow of a molten metal from a reservoir onto a chill rotating surface.	Filaments, foils, particulates	10^4–10^6	155–157
Pendant drop (PDME)	Filament is extracted from molten end of a rod suspended just above the rotating wheel.	Filaments, fibers	10^5–10^6	158
Rapid spinning cup (RSC)	Stream of molten metal is ejected onto a thick layer of rotating liquid located in the interior wall of a spinning cup.	Spherical to irregular powder/flakes 50 mm diameter 50 μm thick	10^6	134 159,160
Piston and anvil twin pistons	Droplet of molten metal is impacted by piston(s).	Splat 5–300 μm thick	10^4–10^6	41,161
Plasma spray deposition	Molten metal is propelled onto a substrate by a hot ionized gas emanating from the plasma torch. If deposited layers are kept very thin, rapid solidification is possible. Potential for near net shape. Coherent deposit.	Porous layer > 50 μm thick	10^3–10^6	102 162,163
Arc spray	Electrically opposed charged wires of the alloy to be sprayed are fed together to produce a controlled arc. The molten metal is atomized and by a stream of gas is projected onto a substrate.	Porous film >50 μm thick	10^2–10^5	111,112

While methods for processing metallic materials have been established for many years, the fabrication of fiber-reinforced metal matrix composites is still primarily in the laboratory stage of development. Special problems associated with the fabrication of continuous-fiber composites include the control of fiber spacing, complete densification of the matrix while maintaining matrix purity, and optimal bonding of the matrix to the fiber (good chemical bonding without excessive reaction). It is our intention in this article to demonstrate the viability of metal matrix composite fabrication and the solution to many of these problems through the use of rapid solidification processing techniques.

At this stage in composite technology, very little can be said about large-scale production processes for fabrication of continuous-fiber composites, though a number of techniques have been developed on a laboratory or pilot plant scale. Some of the more common techniques include unidirectional solidification of eutectic or constitutionally appropriate alloys, liquid metal infiltration, conventional powder-metallurgy processes, hot pressure bonding of a suitable matrix phase around fiber bundles or mats via the powder cloth or fiber/foil technique, or consolidation of matrix precoated fibers (e.g., by thermal spraying or plating).

The first of these processes, unidirectional solidification or the generation of in situ composites, has a number of naturally imposed restrictions that limit its usefulness as a manufacturing procedure. The types of material that can be fabricated into composites by this technique are generally limited to eutectic or other constitutionally appropriate alloys. The fiber volume fraction in most of these systems is low, being controlled by the thermodynamic phase relationships of the system, and cannot be varied significantly [23,24]. Finally, off-axis properties such as transverse strength and ductility are generally poor [25–27], and there is no convenient method for modifying off-axis properties through angle-ply procedures.

With liquid infiltration techniques, proper wetting between fibers and matrix without significant degradation of the fibers is important and yet difficult to achieve [28,29]. The relative instability of available advanced filaments and their reactive nature in most metal melts has minimized the practicality of liquid-state composite fabrication. Furthermore, fiber spacing is extremely difficult if not impossible to control. This technique may be useful for producing small volumes of discontinuous-fiber composite materials using metals with low melting points, such as magnesium and aluminum alloys. It would, however, be impractical to infiltrate presently available fibers with refractory metals or high melting point intermetallic compounds because of severe chemical attack of the fibers [29].

Presently, components under consideration for aerospace applications, such as B/Al composite panels [30] or W/superalloy turbine blades [31–33], are produced by diffusion bonding or hot pressure consolidation of prefabricated monotapes. Monotape production is one area where rapid solidification techniques can have a great impact on the fabrication of metal matrix composites. Thermal spray processes such as arc spray and plasma spray techniques inherently involve rapid solidification. These techniques can be directly utilized to produce composite monotapes, which can then be cut, molded, and bonded into appropriate composite structures. Monotapes or simple composite plates can also be fabricated from rapidly solidified constituents by the powder cloth or foil/fiber technique. Whether an RS material is used as a composite matrix phase to take advantage of the benefits gained from having a rapidly solidified microstructure, or because a rapid solidification process is the most convenient process available for matrix processing, RS processes are becoming increasingly important to composite fabrication.

Detailed composite fabrication process parameters are almost universally considered proprietary or restricted information. Therefore, the authors' intent in this article is to describe general rapid solidification processes and their advantages, followed by a discussion of the application of rapid solidification processing to composite monotape and component fabrication.

Rapid Solidification Processes

Over a quarter of a century ago, a new frontier for the processing of materials was established by the efforts of Paul Duwez [34–38] and others [39–42]. While Duwez was not the first to investigate rapid quench rate techniques [43–44], it is his efforts that are recognized as beginning the revolution in rapid solidification technology. During one of his initial experiments, Duwez melted a small amount of metal in the bottom of a tube. By sudden gas pressurization the molten metal was ejected onto a copper plate, resulting in a "splat" of frozen metal with a solidification rate on the order of 10^6 K/s. Since then, thousands of technical papers have been published in the area of rapid solidification technology (RST), and new commercial applications are emerging every day [45–53].

Today the term *rapid solidification*, which is defined as a "rapid" quenching from the liquid state, covers a broad range of material processes. Traditionally, for a process to be considered in the rapid solidification regime, its cooling rate would have to have been on the order of 10^4 K/s or greater. Now the concept of rapid solidification is more general, and processes with lower cooling rates, such as some powder-producing techniques, have been included under the RST definition.

Rapid solidification technology, therefore, incorporates a large number of different processes. These processes can be classified either as a function of their resulting products (e.g., powder-producing techniques), as a function of the resulting microstructure (grain and particle size), or as a function of cooling rate (generally measured by secondary dendrite arm spacing). A useful comparison of secondary dendrite arm spacing versus solidification rate for several different alloys is shown in Figure 1 [54–56]. Superimposed on the figure are some common processing techniques, indicating the range in homogenization that can be achieved for different processes.

For solidification rates in excess of about 10^5 K/s (as observed during melt spinning), the resulting microstructure becomes cellular rather than dendritic. Figure 2, a transmission electron bright-field image, shows the cellular structure formed in a melt-spun NiAl alloy containing 0.5 at % W. A fine dispersion of tungsten particles helps delineate the cell and grain boundaries.

Common products obtained by rapid solidification techniques include powders, flakes, ribbons, wires, and foils. An essential factor associated with almost all current rapid solidification techniques is that the as-solidified product is very small (μm size) in at least one dimension. In other words, the surface-to-volume ratio for the product is very large. This is essential in obtaining very high cooling or solidification rates.

An important objective of all rapid solidification processes is to produce a solid phase of uniform composition with no micro- or macro-scale segregation of alloying elements [57]. Other effects that may result from rapid solidification rates are an increase in the solid solubility of alloying elements, a significant decrease in grain size, and possibly formation of unusual metastable crystalline phases or production of noncrystalline (amorphous) phases. For most applications, however, rapidly solidified alloys must be consolidated into bulk forms, either monolithic or composite. The consolidation processes and parameters (especially time and temperature) are very critical, since many of the microstructures exhibited by RS materials are metastable. Furthermore, many RS materials are intended for use at elevated temperature; therefore, prolonged exposure at elevated temperature during both consolidation and service could destroy any benefits that accrued from the

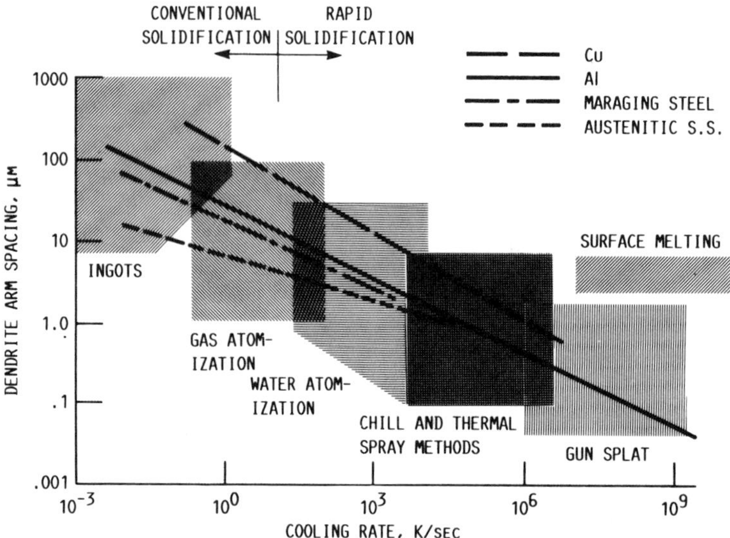

FIGURE 1 **Dependence of secondary dendrite arm spacing on cooling rate for several types of alloys.** (From Refs. 54–56.)

microstructural modifications introduced by RS processing. The addition of insoluble elements or those that will form fine precipitates can result in both pinning of the grain boundaries, controlling grain coarsening during the consolidation process, and increased strength of the alloy [58–60]. Thus increased homogeneity, grain refinement, and strengthening effects remain the primary advantages of RS materials in high temperature structural applications. The main techniques available for rapid solidification processing are summarized in Table 1, and the more common techniques are also shown schematically in Figure 3. Numerous reviews describing these techniques have been published [24,45–53,61,62]; therefore, only a brief description of the processes most relevant to composite fabrication is included here. Details on the numerous processing parameters that can affect final product quality, such as melt composition, melt superheat, chamber atmosphere, crucible composition, crucible configuration, and substrate condition, composition, and temperature, can be found in the references listed in Table 1. For simplicity, the different processes have been divided into two main categories: (1) atomization methods and (2) chill methods. A third general category for rapid quenching is surface techniques, which includes such processes as laser surface melting. This category is not included in the following discussion, since it is not relevant to composite fabrication.

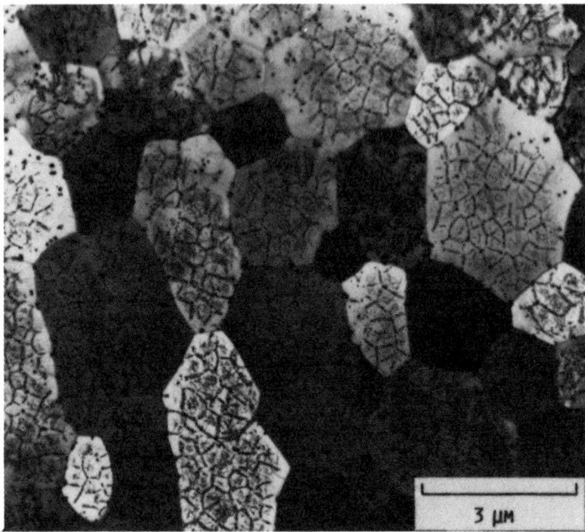

FIGURE 2 **TEM bright field image of the cellular structure of a melt-spun NiAl + 0.5 at % W alloy.**

Elemental and prealloyed powders used in typical powder-metallurgy processes are commonly produced by atomization techniques. Several examples of the morphology and microstructure of powders produced by atomization methods are shown in Figure 4. The common feature of all atomization processes is that they are based on the disruption of a molten metal stream by external forces, such as another fluid (gas or water), mechanical disturbances, or electric fields.

Figure 3*a* is a schematic representation of a two-fluid atomization process. This process consists of a high pressure fluid impacting a continuous stream of liquid metal, resulting in a broad distribution of powder particle sizes. Figure $4a^1$ and $4a^2$ shows examples of the morphology and microstructure of NiCrAlY powder produced by argon gas atomization. Ultrasonic and supersonic atomization techniques can greatly reduce the range of particle size distribution and also increase the cooling rate. This is accomplished by use of high velocity gas pulses that break the molten metal stream into very fine droplets ($< 30\ \mu$m). The fine liquid droplets then solidify convectively at high rates, resulting in a narrow size range of fine powder.

Rotary methods are also very common atomization-type processes. The rotating electrode process (REP) is one such method for producing high quality, spherical

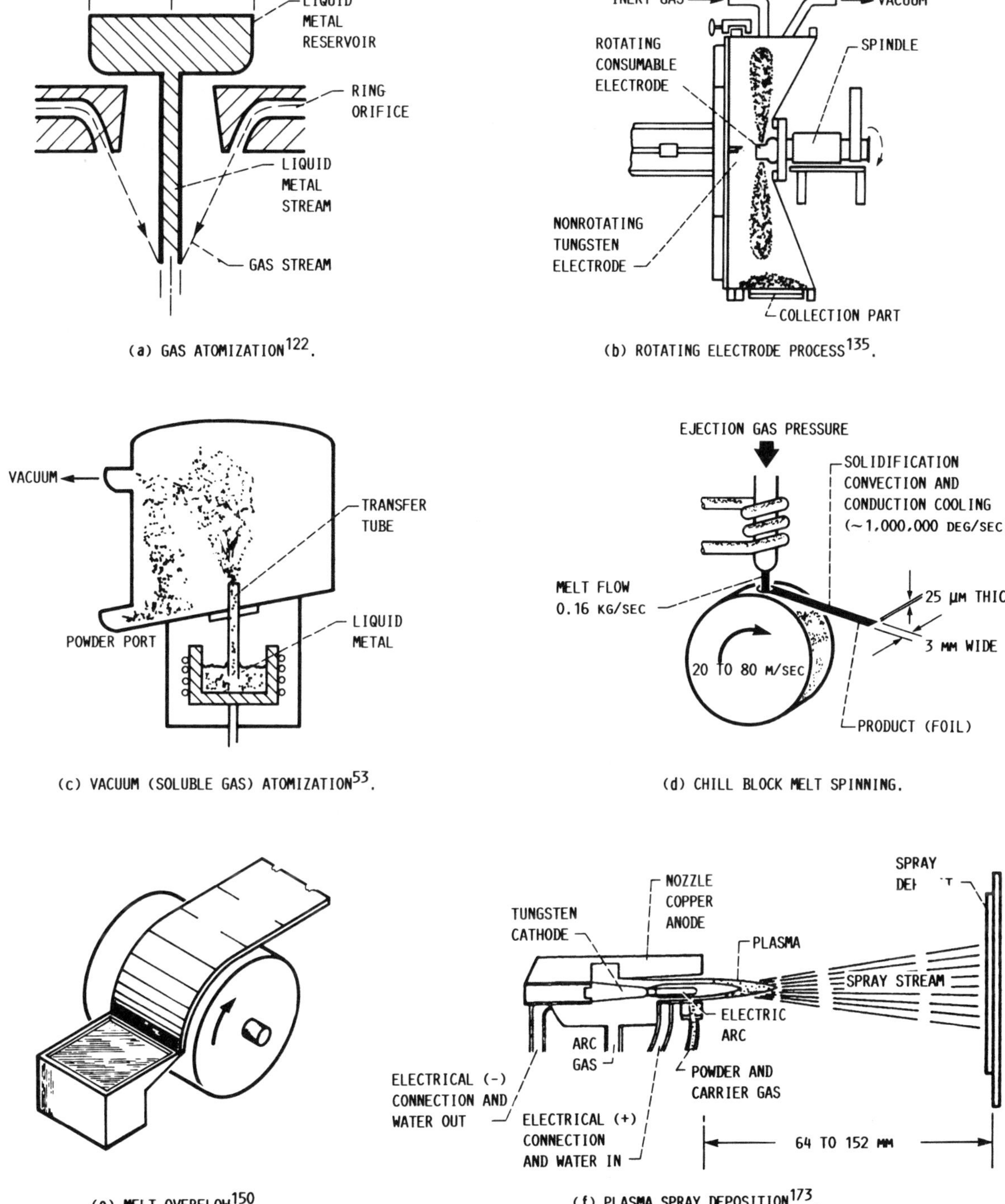

FIGURE 3 **Schematic representation of RS powder and ribbon producing techniques.**

powders. This process, illustrated in Figure 3*b*, is based on the melting of an alloy electrode rotated at a relatively high rate of speed. The molten alloy droplets are then ejected centrifugally and solidify in an inert environment. Figure 4b^1 and 4b^2 gives examples of the extremely spherical nature of Ti–24Al–11Nb (at %) powders produced by this technique. One of the biggest advantages to this technique is that the molten metal never comes in contact with a mold material, thus eliminating the chance for reaction between the metal and any crucible material. The disadvantage to this technique is the additional processing step necessary to produce the prealloyed electrodes used in this process.

The chill methods, based on the work of Duwez, obtain their high cooling rates by heat extraction through a substrate. Solidification rates in excess of 10^6 K/s are readily achievable with these systems. The chill techniques normally make use of a moving substrate so that

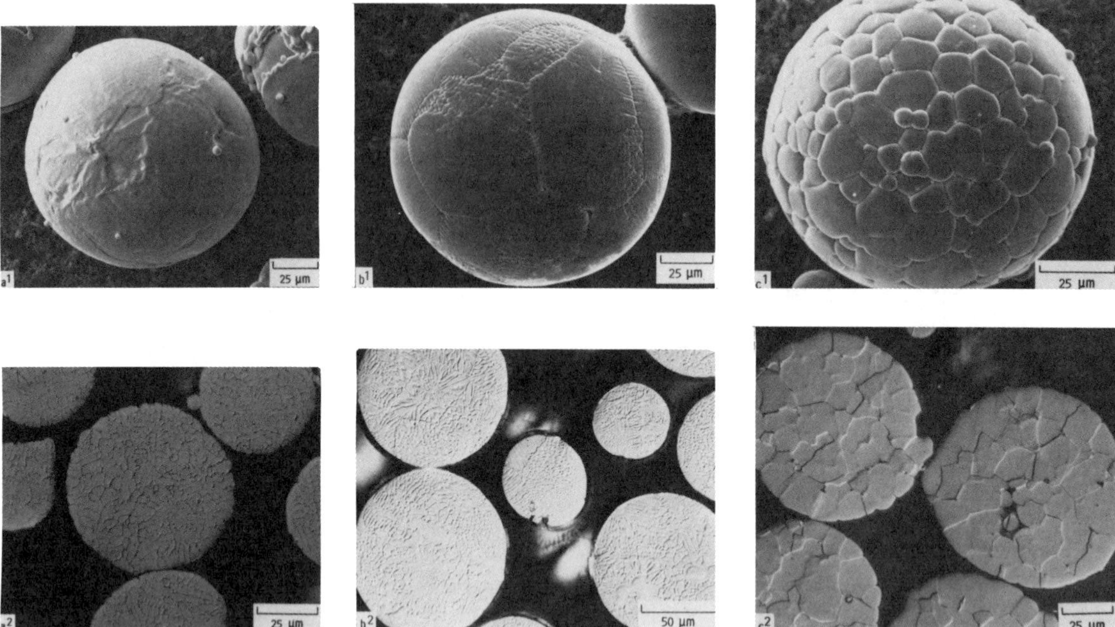

FIGURE 4 **Examples of the morphology[1] and microstructure[2] of powders prepared by various atomization techniques: (*a*) NiCrAlY powders produced by argon atomization; (*b*) Ti–24Al–11Nb powders produced by the rotating electrode process; (*c*) NiAl powders produced by vacuum atomization.**

localized heating of the substrate and the deposit is minimized. The most common technique is the chill block melt spinning (CBMS) method and its variations, free-fall melt spinning (FFMS) and planar flow melt spinning (PFMS). Figure 3*d* shows a schematic setup of the CBMS process. During the CBMS process, the molten metal is ejected from the crucible in which it was melted, through a nozzle, and onto a rotating wheel, with solidification occurring during contact with the substrate. The wheel can be either flat, producing ribbon, or notched, resulting in flake production. The product then flies off into a catcher device before one complete revolution of the wheel. This process is presently being used as a research tool as well as on a pilot plant scale production basis to produce RS material. Melt overflow (Fig. 3*e*) and melt drag processes avoid the use of ceramic crucibles with built-in orifices by making use of water-cooled Cu hearths, thereby minimizing contamination of the melt. With the elimination of ceramic molds, it is possible to more cleanly melt reactive metals such as titanium and silicon-based alloys. Both of these modified chill block methods are being used quite successfully on a commercial basis. With the increased demand for high melting point refractory metal and intermetallic compounds for aerospace applications, new melting processes and crucible innovations are being combined with the melt spinning techniques. For example, arc melting, plasma melting, and levitation processes are just a few of the innovations being implemented with the melt spinning process.

The main drawback in using rapidly solidified products is the need to safely handle and consolidate the RS materials into fully dense components. Several general techniques are used for consolidation, such as hot pressing, hot isostatic pressing, forging, extrusion, and dynamic and explosive compaction processes [16]. However, near-net-shape materials can also be formed by thermal spray processes. With these processes, a continuous and controlled deposition of atomized product is sprayed onto a substrate to build up multiple layers of rapidly solidified material. Several techniques have been developed for thermal spraying with near-net-shape products in mind. They include spray atomization techniques such as the Osprey process [46,62], plasma spraying [42], and the arc spray process [63]. The latter two processes are already used on a regular basis for the fabrication of composite monotapes [63,64]. Fabrication of composite components and monotapes by the thermal spray processes as well as from RS products such as flake, ribbon, and powder will be described in the following sections.

Use of Rapidly Solidified Constituents in Monotape or Composite Fabrication: Powder Cloth and Foil/Fiber Techniques

Any useful composite fabrication technique must produce a component that meets minimum prescribed design parameters and yet still be versatile enough to create parts of complex shape. For example, a common requirement in many components designated for composite reinforcement is the incorporation of hollow cooling or weight reduction passages [65–67]. The entire fabrication scheme must also be cost-effective and reproducible.

The powder cloth and fiber/foil techniques come as close as any process to meeting these requirements and are the most cost-effective [68] and readily adaptable techniques for commercial exploitation [69]. These processes are generally known as hot pressure bonding techniques because the individual composite components, matrix and fiber, are formed into composite monotapes or panels by a static pressure consolidation process carried out at elevated temperature. The general concept is shown schematically in Figure 5. From this process, simple sheets or plates can be formed and used as honeycomb facing or creep formed into cylindrical or complex blade shapes [70]. This process is also an important source of monolayer filament tapes or monotapes, which are used to form complex composite shapes by laminate layering and pressing techniques [31, 65–67].

Monotape fabrication by the powder cloth technique simply consists of sandwiching a fiber mat between two metal powder filled "cloths" and consolidating. The entire processing sequence from matrix alloy selection to finished composite monotape is shown schematically in Figure 6. The powder cloth is formed by combining the prealloyed matrix powder (produced by any of the techniques described previously) with a suitable organic binder, usually Teflon powder [71,72], and blending the mixture with the aid of a high purity stoddard solution [72]. Once the slurry of powder, binder, and stoddard solution is appropriately mixed, the majority of the stoddard solution is evaporated off by the application of low heat. The remaining doughlike mixture is then rolled into a pliable powder cloth. During the blending and rolling process, the polymeric binder forms an interlocking network, holding the powder particles together in a clothlike sheet. A typical powder cloth, an example of which is shown in Figure 7, generally contains 4–15% organic binder by weight.

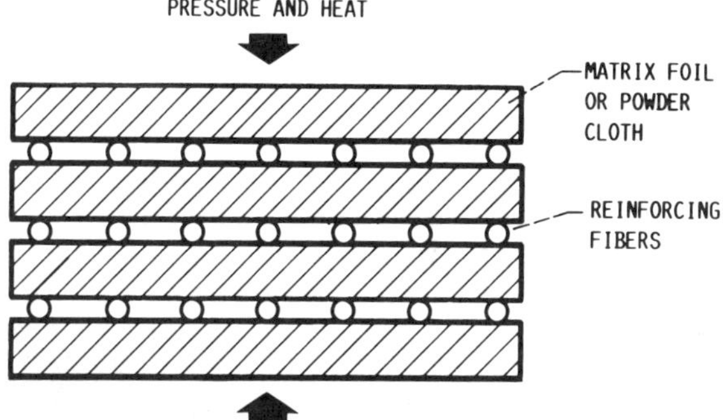

FIGURE 5 **Hot pressure bonding technique for composite fabrication.**

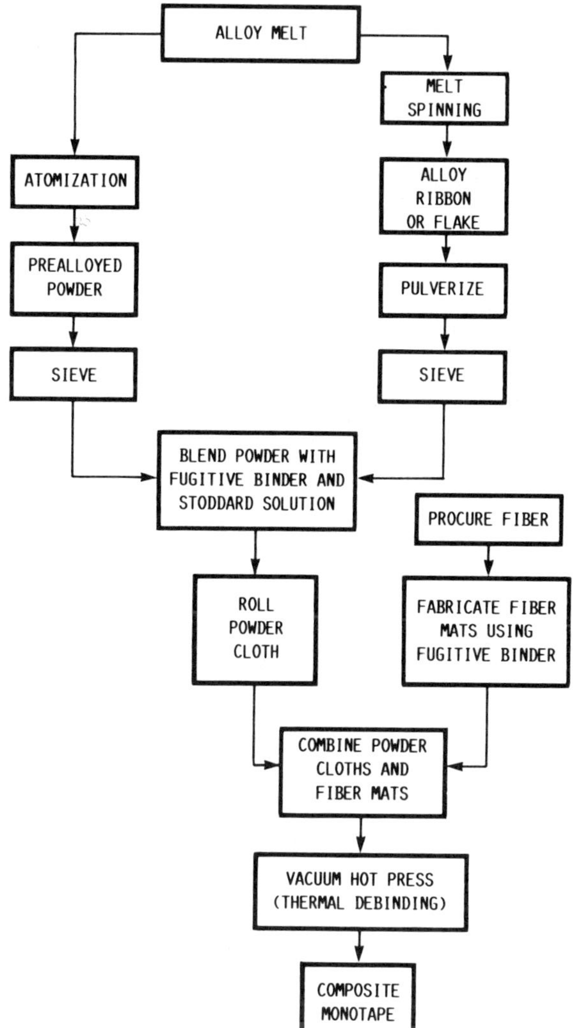

FIGURE 6 **Composite monotape fabrication from rapidly solidified constituents.**

The SiC fiber mat shown in Figure 7 is also held together by a fugitive binder. The fibers are first wound onto a lathe-mounted drum or madrel using a translating wire guide assembly to provide a desired uniform spacing of the fibers. Then an organic binder dissolved in a suitable solvent (for example, polystyrene dissolved in toluene [71,73,74]) is sprayed or painted over the fibers. The organic binder can also be applied directly to the filaments during the winding procedure by passing the fibers through the binder solution just before they are wound onto the mandrel [74]. As the solvent evaporates, a polymer film is left, encasing the fibers. The fiber mat is then removed from the drum and cut into segments that fit into the hot press.

A monotape is formed by sandwiching the fiber mat between powder cloths and consolidating. Several layers of alternating powder cloths and fiber mats can also be laid up at the same time to form a composite panel. The binder removal or thermal debinding step can take place in the hot press, under a vacuum or a low pressure inert

FIGURE 7 **Examples of an actual powder cloth and fiber mat used in the hot pressure bonding process of composite fabrication.**

gas atmosphere, usually without leaving a residue [71,73]. A light retaining pressure is applied to the stack of powder cloths and fiber mats when they are introduced into the hot press to hold everything in place during the binder burnout stage before the final pressing temperature and pressure are reached [73]. Therefore, both binder removal and partial or complete consolidation of the monotape or composite panel can occur in a single step [73,75]. A pictorial review of the steps involved in the powder cloth technique for composite fabrication is shown in Figure 8. More complex components (other than simple plates) can be formed by stacking cut monotapes in desired orientations and then hot pressure bonding them together, as illustrated in Figure 9 [76]. This process is described in more detail in a later section.

The powder cloth process is one of the most versatile methods for producing both fiber composite finished products and "building block" materials in the form of composite monotapes. Almost all metals and ceramics can be produced in powder form, so that the choice of matrix material for this process is limitless. The powder cloth technique as described above or with slight variations has been used in the past on W/superalloy [77,78], aluminide intermetallic matrix [72,75], W–FeCrAlY [65,68,71], and UO_2–W composites [79–82]. These and other composite systems made by the powder cloth technique are listed in Table 2. Examples of more recent composite systems fabricated by the powder cloth technique are shown in Figure 10. It should be noted that the SiC/Ti–24Al–11Nb composite in Figure 10 was fabricated from the same REP powders as shown in Figure 4*b*, and the W–NiAl composite was produced from the vacuum atomized powder shown in Figure 4*c*.

The minimum requirement for judging the viability of a composite fabrication technique for a system with continuous aligned fibers should be the attainment of a rule of mixtures (ROM) strength. If attained, this would indicate that sufficient bonding between the matrix and the fiber has occurred to permit load transmittal between the matrix and fiber phases. No loss of strength (compared with ROM) would also mean that bonding occurred without severe reaction between the fiber and matrix phases, that the fibers were not damaged during

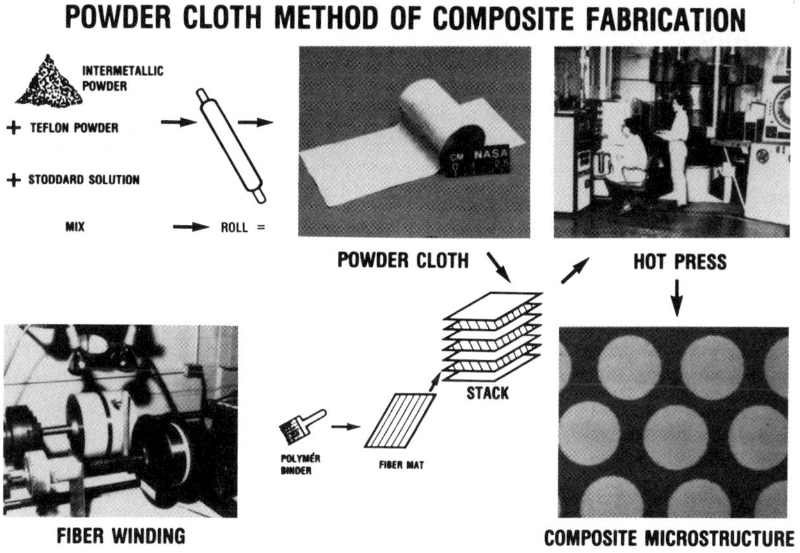

POWDER CLOTH METHOD OF COMPOSITE FABRICATION

INTERMETALLIC POWDER
+ TEFLON POWDER
+ STODDARD SOLUTION
MIX → ROLL =
POWDER CLOTH
HOT PRESS
STACK
POLYMER BINDER
FIBER MAT
FIBER WINDING
COMPOSITE MICROSTRUCTURE

FIGURE 8 Powder cloth method of composite fabrication.

the hot pressing procedure, and that the matrix was not embrittled during processing. In other words, if a rule of mixtures strength is obtained, then the composite processing technique is capable of producing a composite part without significantly degrading the strength of any of the individual components of the composite (fiber or matrix).

Viability of the powder cloth technique for producing intermetallic matrix composites has been demonstrated by Brindley [83] by attainment of near rule of mixtures strengths in the SiC/Ti–24Al–11Nb system when low oxygen starting materials were used. These results are shown in Figure 11. Near rule of mixtures strengths have been reported in other composite systems fabricated by the powder cloth process as well [71].

Almost all composite systems fabricated to date by the powder cloth technique, including those illustrated in Figure 10, have made use of this technique because it was convenient. However, the fine grain sizes that can be achieved by RS processing can also be extremely beneficial for properties. Schulson and Barker [84] have shown that fine grain sizes are preferable for low temperature ductility in intermetallics, while Whittenberger [85] has shown that a fine grain size can even increase the elevated temperature strength of nickel aluminides. With this in mind, a fine-grained NiAl alloy was produced by melt spinning for use as a composite matrix phase, this is presently under investigation at NASA Lewis Research Center. In preliminary studies a composite produced with this fine-grained NiAl alloy has demonstrated improved thermal cycling resistance over NiAl composites made from conventional powders. An example of a NiAl-based composite made from melt-spun ribbon is shown in Figure 12. Subμm-size tungsten particles are responsible for pinning the grain boundaries and retention of the fine grain size in this alloy even after high temperature consolidation [58,86].

The foil/fiber process for forming composites or monotapes is very similar to the powder cloth technique just discussed. Foil/filament arrays are formed by winding fibers with a predetermined filament spacing onto a large-diameter mandrel, forming a fiber mat (Fig. 7), or by winding fibers directly onto the surface of the matrix foil, which is placed over the mandrel or drum. A fugitive binder (e.g., polystyrene or an acrylic [71,73,74,87]) is then used to hold the filaments in place before the consolidation process. This time, however, the matrix is in the form of a dense foil instead of a powder cloth. With the application of heat and pressure, the binder phase is removed by volatilization, and the matrix is forced to flow between the filaments until opposite surfaces meet and all remaining voids are filled [87]. Diffusion bonding will occur if the foil surfaces are clean and oxide-free. Optimum fabrication parameters (time, temperature,

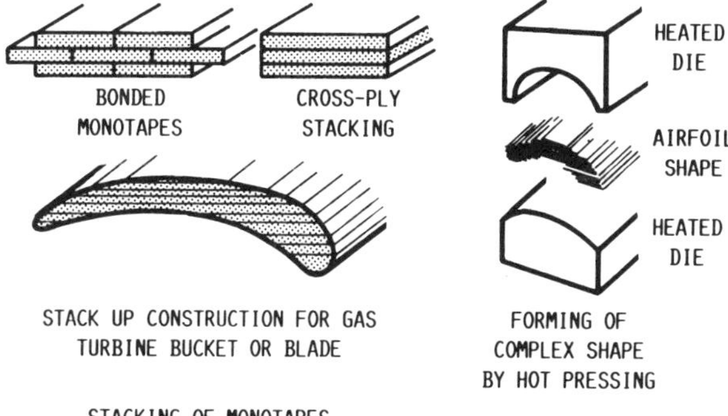

FIGURE 9 **Fabrication of complex-shaped composite parts by diffusion bonding of stacked monotapes.**

TABLE 2
Fabrication of Fiber-Reinforced Composite Systems

Process	Composite System Fiber/Matrix[a]	Product Form	References
Powder cloth	W (1.5% ThO_2)/FeCrAlY	JT9D-7F turbine/blade plates	31,65,120
Powder cloth	SiC/Ti–24Al–11Nb	Plates	72,83,160
Powder cloth	W/NiAl, SiC/NiAl	Plates	75
Powder cloth	SiC,B,W,Al_2O_3/Fe–40Al	Plates	75,166
Powder cloth	W/Superalloy	Plates	77,78
		Turbine blades and vanes	119,167
Powder cloth	UO_2/W	Honeycomb structural panels	79–82
Powder cloth	HfC–SiC/Superalloy	Plates	101
Foil/fiber (RS foil by PDME)	Borsic,SiC,B_4C/Ti–6Al–4V, Ti–11.5Mo–6Zr–4.5Sn, Ti–15v–3Cr–3Sn–3Al	Plates	99
Foil/fiber (wrought foil)	W/Superalloy	Plates	77,78
Foil/fiber (WF)	B/Al	Plates	89–92,98,109
		Structural tubes	168
		Compressor fan blades	169,170
Foil/fiber (WF)	Borsic/Al	Plates	93–95,109
Foil/fiber (WF)	Be/Al	Plates	96
Foil/fiber (WF)	Stainless steel/Al	Plates	97
Foil/fiber (WF)	SiC/Al	Plates	98
Foil/fiber (WF)	Borsic/Ti–6Al–4V Borsic/Ti–11.5Mo–6Zr–4.5Sn	Plates	109
Combination: foil/fiber and powder cloth	Borsic/Ti–6Al–4V	LP1 fan blade	74,171
Combination: foil/fiber and plasma spray	Borsic/Ti (foil) + Al (powder)	Plates	108
Plasma spray	B/Ti–6Al–4V	Ring structure	74
Plasma spray	Borsic and B/Al (Al Matrix Alloys: 2024, 6061, 5052, SAP, 713, 1145)	Monotapes, plates	106–109
Plasma spray	Borsic/Ti–6Al–4V Borsic/Ti–11.5Mo–6Zr–4.4Sn	Plates	109
Plasma spray	W/W	Rocket nozzle inserts	110
Plasma spray	SiC/Al	Missile body casings	172
		Structural panel Z-stiffeners	
Arc spray	W/Cu	Combustion liner	3
Arc spray	W/Nb or Nb-1Zr	Plates, tubes	4,5,174
Arc spray	W/FeCrAlY	Plates	31,33,66,111,112,175,176
Arc spray	W/Incoloy 907 W/316 SS W/Waspalloy	Plates	33,175,176
Arc spray	W/Kanthal	Tubes	177

[a] All compositions in at %.

pressure), determined primarily by trial and error [73,88], are dependent upon matrix composition, filament type, and filament spacing. Foil thickness and filament spacing dictate the volume fraction of reinforcement.

Rapid solidification is not the only method, or even the most common, for forming thin foils of metals. Before the advent of RSP, thin foils primarily of Al and Al alloys were made by conventional rolling techniques for use in composite fabrication by the foil/fiber process [89–98]. The benefit of RS-processed foils lies in improvements in mechanical properties as a result of microstructural refinement, as well as the economic advantages gained from producing foils by RS techniques compared with wrought methods [99]. Furthermore, some alloys, including many of the high temperature alloys of interest as matrix materials, cannot be mechanically worked into foils thin enough for use in composite fabrication. On the other hand, almost any alloy can be melted and cast into thin foils by modified melt spinning or melt overflow techniques.

Presently, RS titanium alloy foils are the primary material of interest for further fabrication into continuous-

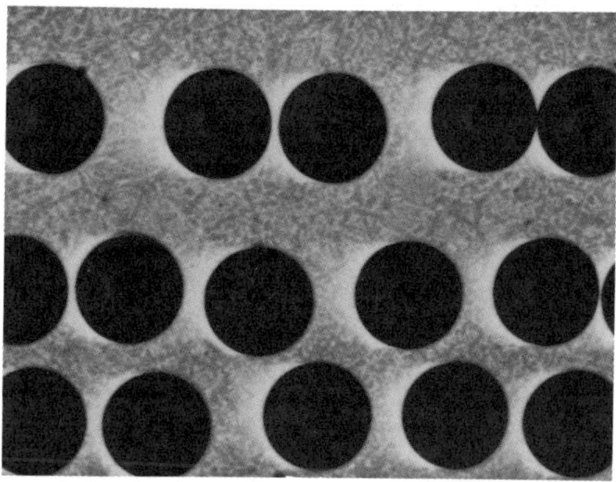

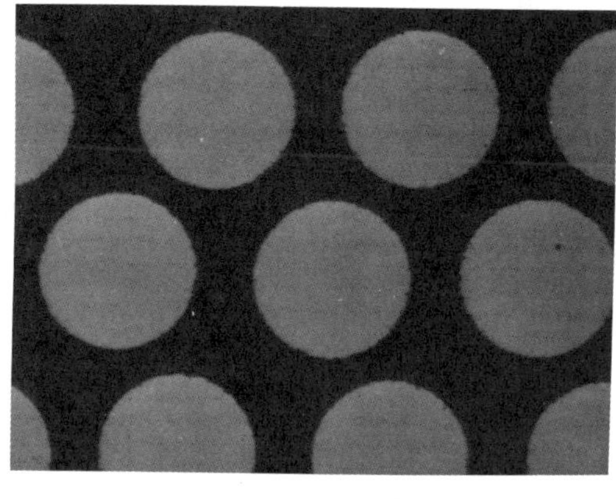

0.10 MM

W/NiAl

FIGURE 10 Intermetallic matrix composites produced by the powder cloth technique.

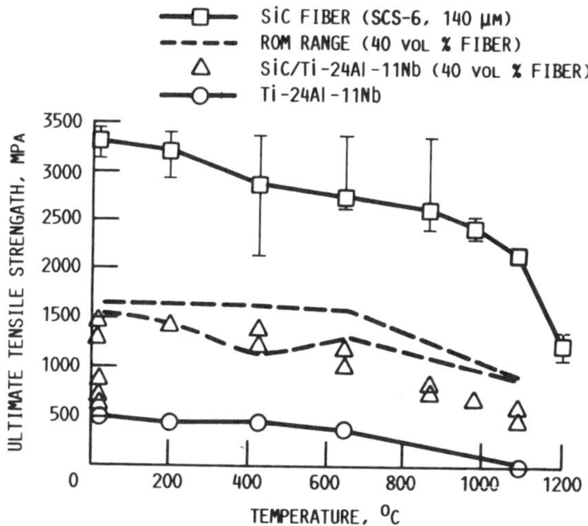

FIGURE 11 Tensile strength versus temperature for SiC/Ti–24Al–11Nb composite indicating close agreement to rule of mixtures strengths for low oxygen containing composites.

fiber composites using the foil/fiber process [99,100]. The major reason for this is that two major problems encountered in titanium metal matrix composite fabrication can be solved by use of rapid solidification technology. The first problem is the relatively high cost of wrought titanium alloy foil, which requires numerous cycles of vacuum annealing and cold rolling using a low reduction ratio in order to prevent cracking. In contrast, RS technology can be economically advantageous, as thin sheet can be produced directly from the melt [99]. The second major problem is the magnitude of the matrix–fiber reaction zone that typically appears during processing, resulting in severe degradation of composite properties. Higher loads at lower temperatures cannot be used to correct for this problem because the maximum pressing load is controlled by the diametral compression strength of the fiber. However, plastic flow occurs at lower temperatures in fine-grained RS foils than in wrought foils. This results in less reaction zone growth [99] and less chance of damaging ceramic fibers during the pressing procedure because of the lower loads and temperatures necessary for consolidation.

Composite plates suitable for mechanical testing or as structural panels can be fabricated in a single step by

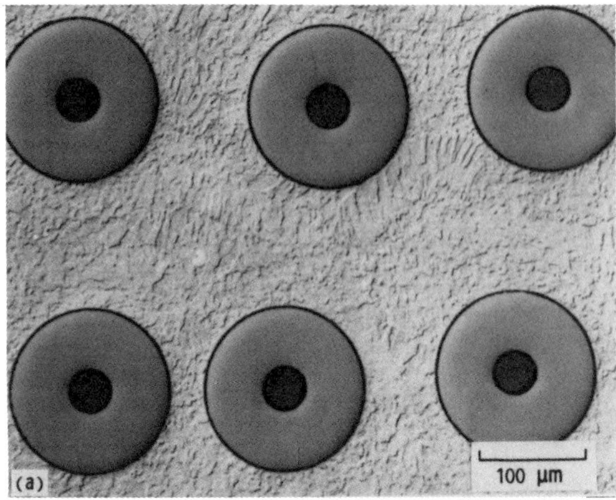

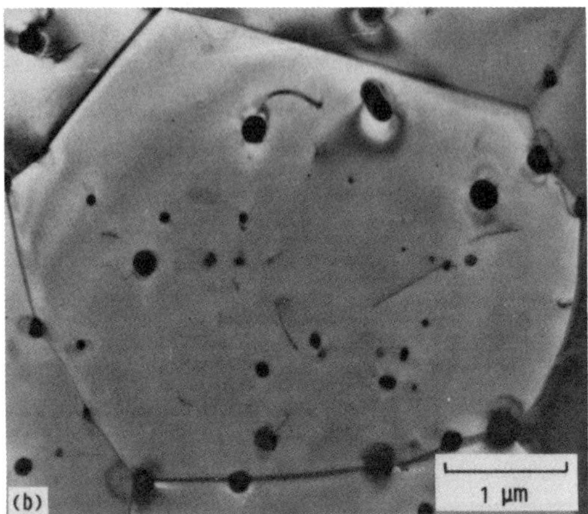

FIGURE 12 (*a*) SiC/NiAl + 0.5 at % W composite made from pulverized melt-spun ribbon by the powder cloth technique and (*b*) TEM image of the NiAl + 0.5 at % W melt-spun ribbon after annealing at 1300°C for 1 hour, showing grain boundaries pinned by W particles.

laying up alternate layers of foil sheets and fiber mats and subsequently consolidating. RS foils can even be cast directly onto the fibers by a melt overflow technique, which would also lend itself to a semicontinuous type of process. Other minor variations on the powder cloth and fiber/foil techniques exist [e.g., 65,74,101,108], but the basic principles are similar to what has been described.

Thermal Spray Techniques for Monotape Fabrication

The use of thermal spray processes to fabricate composite monotapes is a natural extension of forming structural components [102] and coatings [103,104] by plasma spray deposition. These processes are considered rapid solidification techniques because cooling rates, even though varying from about 10^3–10^6 K/s, fall well within the rapid solidification regime. This wide variation in cooling rate is primarily due to differences in distance between the spray gun and the substrate and differences in substrate temperature. In fact, one of the first techniques used to study the rapid solidification of materials was plasma spraying [42], and it is still used today as a tool to study rapidly quenched metals [105].

Kreider [106–109] was the first person to optimize a thermal spray technique for continuous-fiber composite fabrication and did so primarily for aluminum matrix composites using a plasma spray process. Today there are basically two types of thermal spray processes used for composite monotape fabrication: the plasma spray process [106–110], shown in Figure 3*f*, and a more recent technique developed by Westfall [111,112] known as the arc spray process, which is described in Figure 13. These and the previously discussed hot pressure bonding techniques for composite fabrication are compared in Table 3.

The primary differences between the two thermal spray techniques are the design of the spray gun and the type of feedstock necessary for the spraying process. For a plasma spray system, fine spherical powder is used as the feed material for spraying. The powder to be deposited is injected into a plasma stream, usually within the throat of the gun, as illustrated in Figure 3*f*. The plasma

TABLE 3
General Comparison of Monotape Fabrication Techniques

	Powder Cloth	Foil/Fiber	Arc Spray	Plasma Spray
Oxygen contamination	Low to intermediate	Low	Low	High
Practical maximum fiber loading	40–50 vol %	40–50 vol %	> 50 vol %	> 50 vol %
Starting cond. of matrix	Powder	Foil or sheet	Wire	Powder
Limiting size of monotape	Hot press die (cm^2)	Hot press die (cm^2)	Fiber wound drum (m^2)	Fiber wound drum (m^2)
Min. equipment requirements	Vacuum hot press	Vacuum hot press	Arc spray facility + HIP or hot press	Plasma spray facility + HIP or hot press

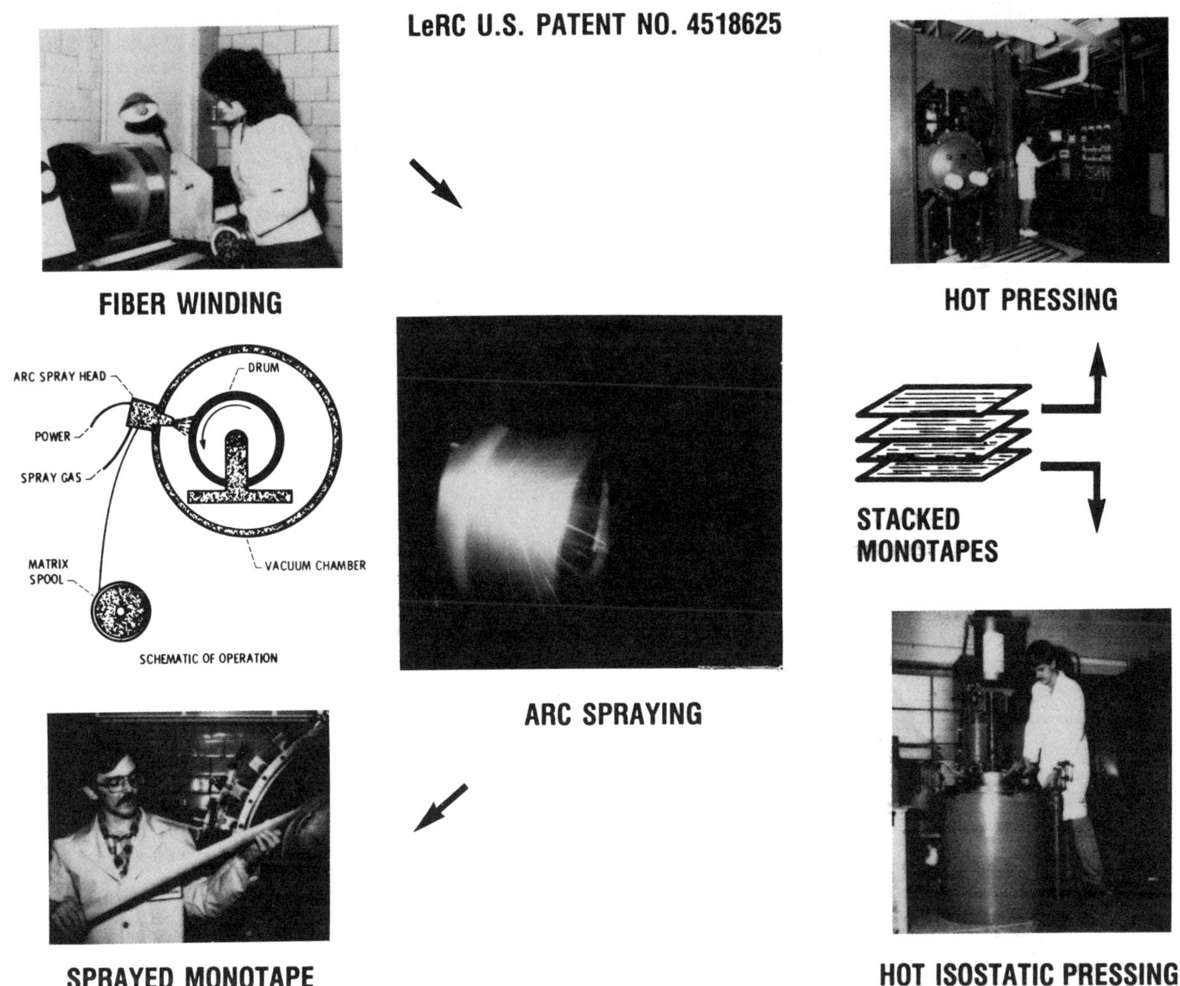

FIGURE 13 Arc spray method of composite fabrication.

is generally at 10,000 K and traveling on the order of Mach III [102]. Therefore, if the powder is too coarse in size, it will not melt before it reaches the substrate, and if it is too fine in size, it will vaporize in the plasma stream. General practice is to use powder screened to a particular mesh size, for example, 400 mesh or an average diameter of about 37 μm [102].

For the arc spray technique, the feed material for the matrix phase is in the form of continuous wire, generally 0.16–0.32 cm in diameter. During this process, two wire feeds of opposite charge are forced through the arc spray gun at a controlled rate. This causes an electric arc to be struck between the wires, melting the wire tips. The resulting molten metal droplets are then sprayed onto the substrate by a stream of argon gas that passes through the gun and directly past the arc [4]. An advantage of the arc spray process over plasma spraying is that the as-deposited matrix material has a much lower oxygen content because the starting material (wire rather than powder) will almost always be cleaner [113]. As-consolidated oxygen levels for arc sprayed composites is generally on the order of several hundred ppm [4,113], while for plasma-sprayed materials the oxygen level is an order of magnitude higher [113]. The primary disadvantage of the arc spray process is that it procures continuous strands of prealloyed wire of brittle alloys which is not always convenient or even possible. Clad bimetallic and hollow core elemental powder filled composite wires, however, can potentially be used for creating most desired matrix compositions. For example, W–nickel aluminide monotapes have been successfully fabricated by the arc spray process using a composite wire formed from elemental constituents as feed stock for the spray gun [114].

A flow diagram of the general processing scheme used for monotape fabrication by thermal spray processes is shown in Figure 14. A protective foil wrapped drum or mandrel is covered with a predetermined spaced array of fibers (similar to forming a fiber mat), which are then sprayed with a layer of matrix material. The mandrel is usually rotated and traversed in front of a stationary spray gun to ensure spraying of an even layer of matrix alloy. This sprayed matrix material mechanically bonds

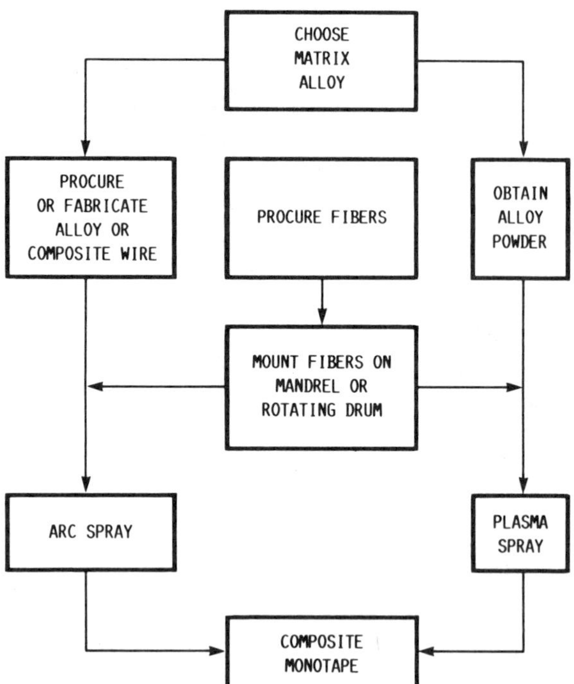

FIGURE 14 Monotape fabrication by thermal spray processes.

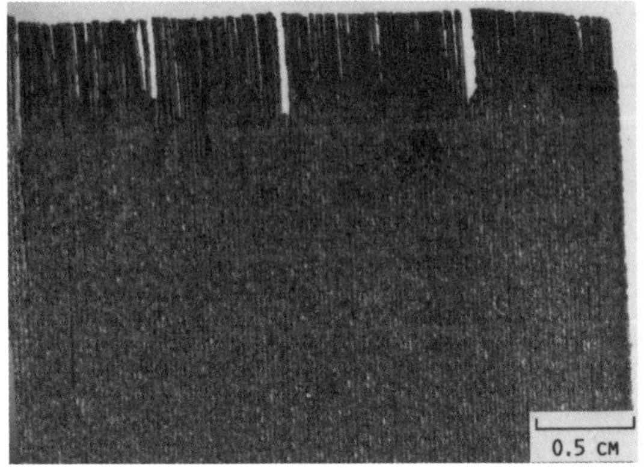

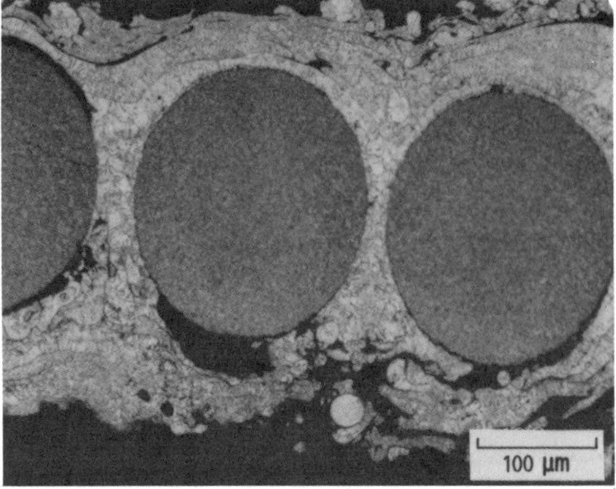

FIGURE 15 W/Nb monotape prepared by the arc spray process.

the wires in place by filling the interstices between and even to a limited extent behind the fibers. The monotape can then be removed from the mandrel, or it can be overwrapped with another layer of fibers and resprayed, forming a unidirectional composite panel. A typical example of a thermally sprayed W–Nb monotape and the resulting RS microstructure can be observed in Figure 15. The as-sprayed monotapes contain 10–30% porosity depending on spraying conditions and drum speed [70,112]. Hot consolidation processes are then used to fully consolidate the monotape or composite component.

Component Fabrication from Composite Monotapes

Thermal spray processes and the hot pressure bonding techniques are a convenient source of prefabricated components (monotapes), which can be easily processed into composite parts with angle-ply structures. Although the potential for high purity product exists, monotapes are currently being produced with a range of interstitial impurity levels depending upon fabrication method. Residual oxygen levels resulting from the plasma spray process, about 2000–5000 ppm depending on the alloy being deposited [113,115], can be an order of magnitude greater than those resulting from any of the other techniques [4,72,75,113]. This can be a problem in some materials, since high oxygen contents can have a devastating effect on matrix alloys that are prone to oxygen embrittlement, such as titanium-based alloys [113,115] and intermetallics [118]. The powder cloth, foil/fiber, and arc spray processes can all produce monotapes with oxygen levels in the hundreds of ppm range [4,72,75,113]. While the arc spray process has the lowest oxygen pickup of the thermal spray techniques, the foil/fiber process is potentially the cleanest of all the techniques, especially if grooved foils are used to hold the fibers in place instead of organic binders. While all the binder can be theoretically burned off in the powder cloth and foil/fiber processes, there is a tendency with these processes for a slight pickup in C (20–500 ppm) [72,75] as a result of the presence of trapped organic binder. Carbon pickup is not a problem associated with any of the thermal spray processes.

An advantage of fabrication of monotapes by thermal spray processes is the elimination of the initial hot press and binder removal step that is necessary with the hot pressure bonding procedures. This can result in savings of both fabrication time and cost [112]. Another advantage of the thermal spray processes is that the matrix material can be sprayed onto relatively complex shapes. Shapes of revolution can be produced simply by winding, and more complex shapes can be produced by winding fiber onto flexible frames and then deforming or orienting the frames [106]. Experimental rocket nozzles of tungsten wire and tungsten matrix have been produced by this process [110], demonstrating the versatility inherent in the thermal spray techniques.

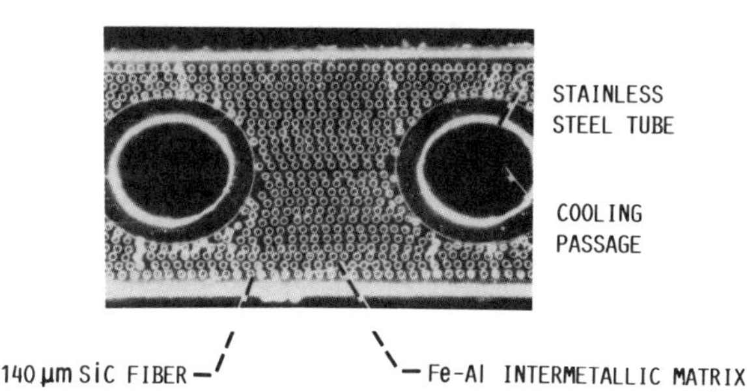

FIGURE 16 Composite panel with cooling passages prepared by the powder cloth technique.

Hot consolidation of preforms or stacks of monotapes produced by any of the techniques previously discussed has been extensively used to form plate- or sheet-type composite components (for examples see Table 2). Even when structures are restricted to this simple geometry, there is considerable room for flexibility. Figure 16 is an example of the cross section of a structural panel made from a SiC/Fe–40Al (at %) composite containing stainless steel cooling passages. This particular panel was fabricated by the powder cloth technique to demonstrate the versatility of the process.

The use of monotapes to form more complex components has been demonstrated by the fabrication of composite turbine blades [31,65,119,120]. Figure 9 is an illustration of a simplified version of this process. However, even this process is capable of producing solid blade shapes that are close to final dimensions and in need of only limited touch-up machining [31,65,119]. Intricate demonstration components such as a composite turbine blade containing internal cooling passages and trailing-edge cooling slots have also been fabricated [31]. Furthermore, an economic assessment of the processing routine [121] indicates that the manufacturing cost of a W fiber reinforced turbine blade should be competitive with current blade production costs using a directional solidification technique.

All the composite fabrication techniques discussed have the advantage of maintaining accurately controlled fiber volume fractions and fiber distribution. This is because the spacing between fibers in any particular layer is extremely uniform and can be fixed to whatever distance is deemed appropriate. Furthermore, the thickness or total volume of the matrix phase applied can also be accurately controlled. This is a significant advantage over conventional powder-metallurgy techniques such as extrusion and molten metal infiltration techniques, in which fiber placement and spacing cannot be controlled. In addition, a structural component can even be produced with a gradient of fiber reinforcement through the thickness of the part. This can be easily accomplished in all the techniques discussed by simply varying the spacing between fibers in the different layers.

Conclusions

In this paper we have attempted to point out the advantages and versatility of composite fabrication techniques that in one form or another involve the use of rapid solidification technology. The potential for new composite processing schemes incorporating state-of-the-art RSP is nearly limitless as are variations in the microstructure which can be achieved through new processing techniques. Therefore, it is to the advantage of the processing engineer to make the best use of RS technologies in order to optimize not only the composite processing procedure but composite properties as well.

Acknowledgments

The authors are indebted to all individuals at NASA LeRC who have contributed to this paper. Special thanks to P. Brindley for allowing us to publish her results on the SiC/Ti–24Al–11Nb system, D. Petrasek for helpful discussions on composite research and for pointing out key references, T. Leonhardt for the metallography performed on all samples, and A. D. Tenteris for reviewing the original manuscript.

R. D. Noebe and I. E. Locci

References

1. D. L. McDanels and J. R. Stephens, "High Temperature Engine Materials Technology—Intermetallic and Metal Matrix Composites," NASA TM 100844, 1988.
2. J. R. Stephens and M. V. Nathal, "Status and Prognosis for Alternative Engine Materials," in S. Reichman et al., Eds., *Superalloys 1988*, The Metallurgical Society, Warrendale, PA, 1988, p. 183.
3. J. R. Stephens, "High Temperature Metal Matrix Composites for Future Aerospace Systems," NASA TM 100212, 1987.
4. L. J. Westfall, D. W. Petrasek, D. L. McDanels, and T. L. Grobstein, "Preliminary Feasibility Studies of Tungsten/Niobium Composites for Advanced Power Systems Applications," NASA TM 87248, 1986.
5. J. R. Stephens, D. W. Petrasek, and R. H. Titran, "Refractory Metal Alloys and Composites for Space Power Systems," NASA TM 100946, 1988.
6. P. W. Lee and J. H. Moll, Eds., *Proceedings of 2nd ASM International Conference on Rapidly Solidified Materials*, ASM International, 1988.
7. F. H. Froes and S. J. Savage, Eds., *Proceedings of a 7-Session Symposium on Enhanced Properties in Structural Metals via Rapid Solidification*, 1986, ASM International, 1987.
8. S. M. L. Sastry and B. A. MacDonald, Eds., *Proceedings of the Conference on Mechanical Behavior of Rapidly Solidified Materials*, 1985, The Metallurgical Society, 1986.

9. F. H. Froes and D. Eylon, Eds., *Proceedings of Symposium on Titanium Rapid Solidification Technology*, The Metallurgical Society, 1986.
10. R. Mehrabian, B. H. Kear, and M. Cohen, Eds., *Proceedings of 1st International Conference on Rapid Solidification Processing*, 1977, Claitor's Publishing Division, Baton Rouge, LA, 1978.
11. R. Mehrabian, M. Cohen, and B. H. Kear, Eds., *Proceedings of 2nd International Conference on Rapid Solidification Processing*, 1979, Claitor's Publishing Division, Baton Rouge, LA, 1980.
12. N. J. Grant and B. C. Giessen, Eds., *Proceedings of 2nd International Conference on Rapidly Solidified Metals*, 1975, M.I.T. Press, Cambridge, MA, 1976.
13. P. W. Lee and R. S. Carbonara, *Proceedings of an International Conference on Rapidly Solidified Materials*, American Society for Metals, 1985.
14. B. Cantor, Ed., *Proceedings of 3rd International Conference on Rapidly Quenched Metals*, The Metals Society, 1978.
15. T. Masumoto and K. Suzuki, Eds., *Proceedings of 4th International Conference on Rapidly Quenched Metals*, 1981, The Japan Institute of Metals, 1982.
16. S. Steeb and H. Warlimont, *Proceedings of 5th International Conference on Rapidly Quenched Metals*, Elsevier North-Holland Physics Pub., Amsterdam, 1985.
17. "Rapidly Quenched Metals VI," in *Proceedings of 6th International Conference on Rapidly Quenched Metals*, 1987, *Mat. Sci. and Eng.*, *97–99*, 1988.
18. "Amorphous and Metastable Microcrystalline Rapidly Solidified Alloys: Status and Potential," Publication NMAB-358, National Academy of Sciences, Washington, DC, 1980.
19. H. Jones and C. Suryanarayana, *J. Mat. Sci.*, *8*, 705 (1973).
20. B. H. Kear and B. C. Giessen, Eds., *Proceedings of Materials Research Society Meeting, Symposium F*, 1981, Elsevier North-Holland, New York, 1982.
21. B. H. Kear and B. C. Giessen, Eds., *Proceedings of Materials Research Society Meeting, Symposium F*, 1983, Elsevier North-Holland, New York, 1984.
22. B. C. Giessen, D. E. Polk, and A. I. Taub, Eds., *Proceedings of Materials Research Society Meeting, Symposium F*, 1985, Elsevier North-Holland, New York, 1986.
23. M. R. Jackson and J. L. Walter, in F. D. Lemkey, H. E. Cline, and M. McLean, Eds., *In Situ Composites IV*, North-Holland, New York, 1982, p. 103.
24. M. R. Jackson, *Met. Trans.*, *8A*, 905 (1977).
25. N. S. Stoloff, "Mechanical Behavior of High Temperature Eutectic Composites," in J. L. Walter, M. F. Gigliotti, B. F. Oliver, and H. Bibring, Eds., *Conference on In Situ Composites—III*, Ginn Custom Pub., Lexington, MA, 1979, p. 357.
26. M. McLean, "Mechanical Behavior of Eutectic Superalloys," in F. D. Lemkey, H. E. Cline, and M. McLean, Eds., *In Situ Composites IV*, North-Holland, New York, 1982, p. 1.
27. M. F. X. Gigliotti, M. R. Jackson, S. W. Yang, M. F. Henry, and D. A. Woodford, "Transverse Mechanical Behavior of High-Strength Monocarbide Eutectics," in F. D. Lemkey, H. E. Cline, and M. McLean, Eds., *In Situ Composites IV*, North-Holland, New York, 1982, p. 21.
28. D. C. Phillips, "High Temperature Fibre Composites," in F. L. Matthews, N. C. R. Buskell, J. M. Hodgkinson, and J. Morton, Eds., *ICCM & ECCM*, Elsevier Applied Science Pub., New York, 1987, p. 2.1.
29. S. Nourbakhsh, F. L. Liang, and H. Margolin, *Adv. Materials & Manufacturing Processes*, *3*, 57 (1988).
30. M. Miller and A. Robertson, "Boron/Aluminum and Borsic/Aluminum Metal-Matrix Analysis, Design, Application and Fabrication," in W. J. Renton, Ed., *Hybrid and Select Metal Matrix Composites: A State of the Art Review*, American Institute of Aeronautics and Astronautics, 1977, p. 99.
31. D. W. Petrasek, E. A. Winsa, L. J. Westfall, and R. A. Signorelli, "Tungsten Fiber-Reinforced FeCrAlY—A First Generation Composite Turbine Blade Material," NASA TM 79094, 1979.
32. W. T. Chandler, "Materials for Advanced Rocket Engine Turbopump Turbine Blades," NASA CR-174729, 1983.
33. D. W. Petrasek and J. R. Stephens, "Fiber Reinforced Superalloys for Rocket Engines," NASA TM 100880, 1988.
34. P. Duwez, W. Klement, and R. H. Willens, *Nature*, *187*, 869 (1960).
35. P. Duwez and R. H. Willens, *Trans. AIME*, *227*, 362 (1963).
36. P. Duwez, R. H. Willens, and W. Klement, *J. Appl. Phys.*, *31*, 1136 (1960).
37. P. Duwez, R. H. Willens, and W. Klement, *J. Appl. Phys.*, *31*, 1500 (1960).
38. P. Duwez, *Trans. ASM*, *60*, 607 (1967).
39. G. Falkenhagen and W. Hofmann, *Z. Mefallkde*, *43*, S.69 (1952).
40. I. S. Miroshnichenko and I. V. Salli, *Industrial Laboratory*, *25*, 1463 (1959).
41. P. Pietrohowsky, *Rev. Sci. Instr.*, *34*, 445 (1963).
42. M. Moss, D. L. Smith, and R. A. Lefever, *Appl. Phys. Letters*, *5*, 120 (1964).
43. R. S. Carbonara, T. A. Gaspar, R. V. Raman, R. E. Maringer, and J. L. McCall, "State-of-the-Art Review of Rapid Solidification Technology," MCIC Report 81-45, October, 1981.
44. J. E. Flinn, *Rapid Solidification Technology for Reduced Consumption of Strategic Materials*, Noyes Publications, Park Ridge, NJ, 1985.
45. A. Lawley, *J. Met.*, *33*, 13 (1981).
46. A. Lawley, *J. Met.*, *38*, 15 (1986).
47. N. J. Grant, *J. Met.*, *35*, 20 (1983).
48. S. J. Savage and F. H. Froes, *J. Met.*, *36*, 20 (1984).
49. F. H. Froes and R. Carbonara, *J. Met.*, *40*, 20 (1988).
50. J. V. Wood, *Materials and Design*, *4*, 673 (1983).
51. R. S. Carbonara, "Rapidly Solidified Powders," in *ASM Metals Handbook, Desk Edition*, American Society for Metals, Metals Park, OH, 1985, pp. 25.22–25.24.
52. R. W. Cahn, *Ann. Rev. Mater. Sci.*, *12*, 51 (1982).
53. *Metals Handbook*, Vol. 7, *Powder Metallurgy*, 9th Ed., ASM International, Metals Park, OH, 1988.
54. N. J. Grant, "A Review of Various Atomization Processes," in R. Mehrabian et al., Eds., *Rapid Solidification Processing: Principles and Technologies*, Claitor's Publishing Division, Baton Rouge, LA, 1978, p. 230.
55. H. Matyja, B. C. Geissen, and N. J. Grant, *J. Inst. Metals*, *96*, 30 (1968).
56. V. K. Sarin and N. J. Grant, *Met. Trans.*, *3*, 875 (1972).
57. V. Laxmanan, "On Producing an Alloy of Uniform Composition During Rapid Solidification Processing," in F. H. Froes and S. J. Savage, Eds., *Processing of Structural Metals by Rapid Solidification*, ASM International, Metals Park, OH, 1987, p. 41.

58. I. E. Locci, T. A. Bloom, and M. G. Hebsur, "Rapid Solidification Research at the NASA Lewis Research Center," in P. W. Lee and J. Moll, Eds., *Rapidly Solidified Materials: Properties and Processing*, ASM International, Metals Park, OH, 1988, p. 207.

59. P. R. Rios, *Acta Metall., 35*, 2805 (1987).

60. C. H. Worner and A. Cabo, *Acta Metall., 35*, 2801 (1987).

61. H. Jones, "Some Developments in Techniques for Rapid Solidification," in F. H. Froes and S. J. Savage, Eds., *Processing of Structural Metals by Rapid Solidification*, ASM International, Metals Park, OH, 1987, p. 77.

62. D. L. Erich, *Int. J. Powder Metal., 23*, 45 (1987).

63. L. J. Westfall, "Composite Monolayer Fabrication by an Arc-Spray Process" in D. L. Houck, Ed., *Thermal Spray: Advances in Coatings Technology*, ASM International, Metals Park, OH, 1988, p. 417.

64. K. G. Kreider, L. Dardi, and K. Prewo, "Metal Matrix Composite Technology," AFML-TR-71-204, 1971.

65. P. Melnyk and J. N. Fleck, "Tungsten Wire/FeCrAlY Matrix Turbine Blade Fabrication Study," NASA CR-159788, 1979.

66. J. R. Lewis, "Design Overview of Fiber-Reinforced Superalloy Composites for the Space Shuttle Main Engine," NASA CR-168185, 1983.

67. J. R. Lewis, "Design Overview of Fiber-Reinforced Superalloy Composites for the Space Shuttle Main Engine," presented at the Advanced High Pressure Oxygen/Hydrogen Conference, 1984.

68. D. W. Petrasek, E. A. Winsa, L. J. Westfall, and R. A. Signorelli, "Tungsten Fiber Reinforced FeCrAlY—A First Generation Composite Turbine Blade," in I. Ahmad and B. R. Noton, Eds., *Advanced Fibers and Composites for Elevated Temperatures*, The Metallurgical Society of AIME, Warrendale, PA, 1980, p. 136.

69. L. W. Davis, "How Metal Matrix Composites Are Made," in *Fiber-Stengthened Metallic Composites, ASTM STP 427*, Am. Soc. Testing Mats., Philadelphia, 1967, p. 69.

70. J. A. Alexander, *Metals Eng. Quarterly, 10*, 22 (1970).

71. I. Ahmad, D. N. Hill, J. Barranco, R. Warenchak, and W. Heffernan, "Reinforcement of FeCrAlY with Silicone Carbide (Carbon Core) Filament," in I. Ahmad and B. R. Noton, Eds., *Advanced Fibers and Composites for Elevated Temperatures*, The Metallurgical Society of AIME, Warrendale, PA, 1980, p. 156.

72. P. K. Brindley, P. A. Bartolotta, and S. J. Klima, "Investigation of a SiC/Ti–24–Al–11Nb Composite," NASA TM 100956, 1988.

73. I. J. Toth, W. D. Brentnall, and G. D. Menke, "A Survey of Aluminum Matrix Composites," in J. Weeton and E. Scala, Eds., *Composites: State of the Art*, The Metallurgical Society of AIME, New York, 1974, p. 139.

74. G. R. Sippel and M. Herman, "Fiber Reinforced Titanium Alloy Composites," in J. Weeton and E. Scala, Eds., *Composites: State of the Art*, The Metallurgical Society of AIME, New York, 1974, p. 211.

75. J. W. Pickens, G. K. Watson, R. D. Noebe, P. K. Brindley, and S. L. Draper, "Fabrication of Intermetallic Matrix Composites by the Powder Cloth Process," NASA TM 102060, 1989.

76. J. W. Weeton, *Mach. Des., 20*, 142 (1969).

77. W. D. Brentnall and I. J. Toth, "Fabrication of Tungsten Wire Reinforced Nickel-Base Alloy Composites," NASA CR-134664, 1974.

78. W. D. Brentnall and D. J. Moracz, "Tungsten Wire-Nickel Base Alloy Development," NASA CR-135021, 1976.

79. D. D. Johnson, V. G. Leak, and R. E. Pettman, "Research and Development on Tungsten, Uranium Dioxide, Fuel-Element Configurations," NASA CR-54218, 1964.

80. D. D. Johnson and R. E. Pettman, "Fabrication of Tungsten-UO_2 Composites by a Powder Metallurgy Process," NASA CR-54342, 1965.

81. D. D. Johnson, R. E. Pettman, and D. E. Jackson, "Fabrication of Tungsten-UO_2 Composites by a Powder Metallurgy Process," NASA CR-54424, 1965.

82. D. D. Johnson and R. E. Pettman, "Fabrication of Tungsten-UO_2 Composites by a Powder Metallurgy Process," NASA CR-54766, 1965.

83. P. K. Brindley, S. L. Draper, M. V. Nathal, and J. I. Eldridge, in P. K. Liaw and M. N. Gungor, Eds., "Fundamental Relationships Between Microstructure & Mechanical Properties of Metal-Matrix Composites," The Minerals, Metals & Materials Society, Warrendale, PA, 1990, p. 387.

84. E. M. Schulson and D. R. Barker, *Scripta Met., 17*, 519 (1983).

85. J. D. Whittenberger, *J. Mater. Sci., 23*, 235 (1988).

86. I. E. Locci, R. D. Noebe, J. A. Moser, D. S. Lee, and M. V. Nathal, "Microstructure, Properties, and Processing of Melt Spun NiAl Alloys," in C. C. Koch et al., Eds., *High Temperature Ordered Intermetallic Alloys, III*, MRS Symposium Proceedings, Vol. 133, 1989, p. 639.

87. I. J. Toth, W. D. Brentnall, and G. D. Menke, *J. Met., 24*, 19 (1972).

88. S. T. Mileiko, "Fabrication of Metal-Matrix Composites," in A. Kelly and S. T. Mileiko, Eds., *Fabrication of Composites*, Elsevier Science Pub. B. V., Amsterdam, 1986, p. 221.

89. A. Toy, G. D. Atteridge, and D. I. Sinizer, "Development and Evaluation of the Diffusion Bonding Process," AFML-TR-66-350, 1966.

90. A. L. Cunningham and J. A. Alexander, "The Fabrication, Evaluation and Mechanical Properties of Aluminum Matrix Composites," in Society of Aerospace, Material and Process Engineers, *Advances in Structural Composites*, Western Periodicals Co., North Hollywood, CA, 1967, Paper AC-15.

91. E. G. Wolff and R. J. Hill, "Research on Boron Filament/Metal Matrix Composite Materials," AFML-TR-67-140, 1967.

92. W. F. Stuhrke, "The Mechanical Behavior of Aluminum-Boron Composite Materials," in *Metal Matrix Composites*, ASTM STP 438, Amer. Soc. Testing Mater., Philadelphia, 1968, p. 108.

93. K. C. Antony and W. H. Chang, *Trans. ASM, 61*, 550 (1968).

94. I. J. Toth, "Creep and Fatigue Behavior of Unidirectional and Cross Plied Composites," in *Composite Materials: Testing and Design, ASTM STP 460*, Amer. Soc. Testing Mater., Philadelphia, 1969, p. 256.

95. J. H. Young and R. G. Carlson, "Advanced Composite Material Structural Hardware Development and Testing Program," AFML-TR-70-140, 1970.

96. A. Toy, *J. Mater., 3*, 43 (1968).

97. P. W. Jackson and D. Cratchley, *J. Mech. and Phys. of Solids, 14*, 49 (1966).

98. J. A. Alexander, A. L. Cunningham, and K. C. Chaung, "Investigation to Produce Metal Matrix Composites with

High Modulus, Low Density Continuous Filament Reinforcements," AFML-TR-67-391, 1968.

99. D. Eylon, C. M. Cooke, and F. H. Froes, "Production of Metal Matrix Composites from Rapidly Solidified Titanium Alloy Foils," in F. H. Froes and D. Eylon, Eds., *Titanium Rapid Solidification Technology*, The Metallurgical Society, Inc., Warrendale, PA, 1986, p. 311.

100. T. Gasper, L. E. Hackman, S. W. Scott, D. J. Chronister, and W. A. T. Clark, "Direct Cast Titanium Alloy Strip by Melt Overflow," in F H. Froes and S. J. Savage, Eds., *Processing of Structural Metals By Rapid Solidification*, ASM International, Metals Park, OH, 1987, p. 247.

101. J. A. Cornie, W. R. Lovic, and A. T. Male, "Effect of Interface Reactions on Properties of HfC-Coated SiC Fiber/Superalloy Matrix Composites," in J. A. Cornie and F. W. Crossman, Eds., *Failure Modes in Composites IV*, The Metallurgical Society of AIME, Warrendale, PA, 1979, p. 236.

102. M. R. Jackson, J. R. Rairden, J. S. Smith, and R. W. Smith, *J. Met., 33*, 23 (1981).

103. J. Fairbanks, Ed., *Proceedings of the 1987 Coatings for Advanced Heat Engines Workshop*, CONF-870762, U.S. Dept. of Energy, Washington, DC, 1987.

104. D. L. Houck, Ed., 1987, *Thermal Spray: Advances in Coatings Technology*, Proceedings of the National Thermal Spray Conference, ASM International, Metals Park, OH, 1988.

105. M. R. Jackson, P. A. Siemers, J. R. Rairden, R. L. Mehan, and A. M. Ritter, "Composite Structures Produced by Low Pressure Plasma Deposition," in P. Kumar et al., Eds., *Processing and Properties for Powder Metallurgy Composites*, The Metallurgical Society, Inc., Warrendale, PA, 1988, p. 45.

106. K. G. Kreider and G. R. Leverant, "Boron Fiber Metal Matrix Composites by Plasma Spraying," AFML-TR-66-219, 1966.

107. K. G. Kreider and V. M. Patarini, *Met. Trans., 1*, 3431 (1970).

108. K. G. Kreider, R. D. Schille, E. M. Breinan, and M. Marciano, "Plasma Sprayed Metal Matrix Fiber Reinforced Composites," AFML-TR-68-119, 1968.

109. K. G. Kreider, L. Dardi, and K. Prewo, "Metal Matrix Composite Technology," AFML-TR-71-204, 1971.

110. T. A. Greening, "Advanced Wire-Wound Tungsten Nozzles," AFRPL-TR-67-181, 1967.

111. L. J. Westfall, "Arc Spray Fabrication of Metal Matrix Composite Monotape," U.S. Patent 4,518,625 (1985).

112. L. J. Westfall, "Tungsten Fiber Reinforced Superalloy Composite Monolayer Fabrication by an Arc-Spray Process," NASA TM 86917, 1985.

113. E. Erturk and H. D. Steffens, "Low Pressure Arc Spraying in Comparison with Low Pressure Plasma Spraying," in *Advances in Thermal Spraying*, Proc. of the 11th Int. Thermal Spray Conf., Pergamon Press, New York, 1987, p. 83.

114. G. K. Watson and J. W. Pickens, 2nd Annual HITEMP Review-1989, NASA CP-10039, 1989, p. 50-1.

115. P. K. Brindley, NASA Lewis Research Center, Cleveland, personal communication, 1988.

116. G. Welsch and W. Bink, *Met. Trans., 13A*, 889 (1982).

117. H. Conrad, *Prog. Mater. Sci., 26*, 123 (1981).

118. N. S. Stoloff and R. G. Davies, *Prog. Mater. Sci., 13*, 3 (1966).

119. P. J. Mazzei, G. Vandrunen, and M. J. Hakim, "Powder Fabrication of Fiber Reinforced Superalloy Turbine Blades," in *Advanced Fabrication Techniques in Powder Metallurgy and Their Economic Implications*," AGARD-CP-200, AGARD, Paris, 1976.

120. D. W. Petrasek, R. A. Signorelli, T. Caulfield, and J. K. Tien, "Fiber Reinforced Superalloys," NASA TM 89865, 1987.

121. C. F. Barth, D. W. Blake, and T. S. Stetson, "Cost Analysis of Advanced Turbine Blade Manufacturing Processes," NASA CR-135203, 1977.

122. A. Lawley, *Ann. Rev. Mater. Sci., 8*, 49 (1978).

123. A Lawley, *Int. J. Powder Met., 13*, 169 (1977).

124. J. K. Beddon, *The Production of Metal Powders*, Heyden and Sous, London, 1978.

125. J. B. See and G. H. Johnson, *Powder Tech., 21*, 119 (1978).

126. C. F. Dixon, *Can. Met. Quart., 12*, 309 (1973).

127. P. U. Gummeson, *Powder Met., 15*, 67 (1972).

128. V. Anared, A. J. Kaufman, and N. J. Grant, "Rapid Solidification of Modified 7075 Aluminum Alloy by Ultrasonic Gas Atomization," in R. Mehrabian, B. H. Kear, and M. Cohen, Eds., *Rapid Solidification Processing: Principles and Technologies II*," Claitor's Publishing Division, Baton Rouge, LA, 1980, p. 273.

129. S. M. L. Sastry, T. C. Peng, P. J. Meschter, and J. E. O'Neal, *J. Met., 35*, 21 (1983).

130. G. Rai, E. Lavernia, and N. J. Grant, *J. Met., 37*, 22 (1985).

131. S. Smoll and T. J. Bruce, *Int. J. Powder Met., 4*, 7 (1973).

132. J. J. Dunkley, *Wire Industry, 45*, 365 (1978).

133. J. M. Wentzell, *J. Vac. Sci. Technol., 11*, 1035 (1974).

134. R. V. Raman, A. N. Patel, and R. S. Carbonara, *Progress in Powder Metal., 38*, 99 (1982).

135. G. Friedman, "Production of Titanium Powder by the Rotating Electrode Process," in *Advanced Fabrication Techniques in Powder Metallurgy and Their Economic Implications*, AGARD Conf. Proc. No. 200, 1976, Paper P.1.

136. R. E. Anderson, A. R. Cox, T. D. Tillman, and E. C. Van Reuth, "Use of RSR Alloys for High Performance Turbine Foils," in R. Mehrabian, B. H. Kear, and M. Cohen, Eds., *Rapid Solidification Processing: Principles and Technologies II*," Claitor's Publishing Division, Baton Rouge, LA, 1980, p. 273.

137. A. R. Cox, J. B. Moore, and E. C. Van Reuth, "Rapidly Solidified Powders, Their Production, Properties and Potential Applications," in *Advanced Fabrication Processes*, AGARD Conf. Proc. No. 256, 1978, Paper 12.1.

138. P. R. Holiday, A. R. Cox, and R. J. Peterson, "Rapid Solidification Effects on Alloy Structures," in R. Mehrabian, B. H. Kear, and M. Cohen, Eds., *Rapid Solidification Processing: Principles and Technologies*, Claitor's Publishing Division, Baton Rouge, LA, 1978, p. 246.

139. Y. V. Murty and R. P. I. Adler, *J. Mater. Sci., 17*, 1945 (1982).

140. A. R. E. Singer, A. D. Roche, and L. Day, *Powder Metal., 23*, 81 (1980).

141. H. Ishii, M. Naka, and T. Masumoto, "Amorphous Metallic Powder Prepared by Roller Atomization," in T. Masumoto and K. Suzuki, Eds., *Rapidly Quenched Metals. IV*, The Japan Institute of Metals, Sendai, Japan, 1982, p. 33.

142. J. Perel, J. F. Mahoney, B. E. Kaleusher, and R. Mehrabian, "Electrohydrodynamic Techniques in Metals Processing," in J. J. Burke, R. Mehrabian, and V. Weiss, Eds., *Advances in Metal Processing*, Plenum, New York, 1981, p. 79.

143. S. F. Cogan, J. E. Rockwell, F. H. Cocks, and M. L. Shepard, *J. Phys. E: Sci. Instrum., 11*, 174 (1978).

144. T. C. Perg, S. M. L. Sastry, and J. E. O'Neal, "Rapid Solidification Processing of Titanium Alloys," in R. Mehrabian, B. H. Kear, and M. Cohen, Eds., *Rapid Solidification Processing: Principles and Technologies III*, Claitor's Publishing Division, Baton Rouge, LA, 1978, p. 129.

145. R. B. Pond, U.S. Patent 2,879,566 (1959).

146. R. B. Pond and R. Maddin, *Trans. AIME., 245*, 2475 (1969).

147. R. E. Maringer, U.S. Patent 4,215,084 (1980).

148. R. W. Jech, T. J. Moore, T. K. Glasgow, and N. W. Orth, *J. Met., 36*, 41 (1984).

149. P. G. Boswell and G. A. Chadwick, *J. Phys. E: Sci. Instrum., 9*, 523 (1976).

150. H. H. Liebermann, "Gas Boundary Layer Effects in Processing Glassy Alloy Ribbons," in B. Cantor, Ed., *Rapidly Quenched Metals III, Vol. I*, The Metals Society, London, 1978, p. 34.

151. R. B. Pond, R. E. Maringer, and C. E. Mobley, "High Rate Continuous Casting of Metallic Fibers and Filaments," in *New Trends in Materials Fabrication*, ASM Seminar Series, Metals Park, OH, 1976, p. 128.

152. M. C. Narashimham, U.S. Patent 4,221,257 (1980).

153. R. E. Maringer and C. E. Mobley, *J. Vac. Sci. Tech., 11*, 1067 (1974).

154. R. E. Maringer and C. E. Mobley, "Advances in Melt Extraction," in B. Cantor, Ed., *Rapidly Quenched Metals III, Vol. I*, The Metals Society, London, 1978, p. 49.

155. L. E. Hackman and T. Gaspar, *Industrial Heating, 36* (1986).

156. I. G. Butler, W. Kurz, J. Gillot, and B. Lus, *Fiber Science and Technology, 5*, 243 (1972).

157. J. Hubert, F. Mollard, and B. Lux, *Z. Metallk, 64*, 835 (1973).

158. R. E. Maringer, C. E. Mobley, and E. W. Collings, "An Experimental Method for the Casting of Rapidly Quenched Filaments and Fibers," in N. J. Grant and B. C. Giessen, Eds., *Rapidly Quenched Metals*, M.I.T. Press, Cambridge, MA, 1976, p. 29.

159. I. Ohnaka, T. Fukusako, T. Ohmichi, T. Masumoto, A. Inoue, and M. Hagiwara, "Production of Amorphous Filaments by In-Rotating-Liquid Spinning," in T. Masumoto and K. Suzuki, Eds., *Rapidly Quenched Metals IV*, The Japan Institute of Metals, Sendai, Japan, 1982, p. 31.

160. T. Masumoto, I. Ohuaka, A. Inoue, and M. Hagiwara, *Scripta Met., 15*, 293 (1981).

161. R. W. Cahn, K. D. Krishnanand, M. Laridjani, M. Greenholtz, and R. Hill *Mater. Sci. Eng., 23*, 83 (1976).

162. D. Apelian, M. Paliwal, R. W. Smith, and W. F. Schilling, *Intern. Met. Rev., 28*, 271 (1983).

163. R. W. Cahn, "Rapid Solidification by Plasma Spraying," in R. Mehrabian, B. H. Kear, and M. Cohen,Eds., *Rapid Solidification Processing: Principles and Technologies*, Claitor's Publishing Division, Baton Rouge, LA, 1978, p. 129.

164. J. N. Fleck, "Fabrication of Tungsten Wire/FeCrAlY-Matrix Composite Specimens," TRW-ER-8076, TRW Inc., Cleveland, OH, 1979.

165. P. K. Brindley, "SiC Reinforced Aluminide Composites," in N. S. Stoloff et al., Eds., *High Temperature Ordered Intermetallic Alloys, II*, Materials Research Society, 1987, p. 419.

166. S. L. Draper, D. J. Gaydosh, A. K. Misra, and M. V. Nathal, "Compatibility of Fe40Al with Various Fibers," Presented at 13th Annual Conf. on Composite Materials and Structures, NASA CP-3054, 1990, p. 191.

167. P. J. Mazzei, G. Van Drunen, and N. Chung, "High Temperature Artificial Composite-Alloys for Industrial Gas Turbines," in J. A. Cornie and F. W. Crossman, Eds., *Failure Modes in Composites IV*, TMS-AIME, Warrendale, PA, 1978, p. 219.

168. J. D. Forest, "Boron/Aluminum Tube Constructions for Advanced Vehicle Applications," General Dynamics, Convair Div., March 1975.

169. J. W. Brantley and R. G. Stabrylla, "Fabrication of J79 Boron/Aluminum Compressor Blades," NASA CR-159566, 1979.

170. C. T. Salemme and S. A. Yokel, "Design of Impact Resistant Boron/Aluminum Large Fan Blades," NASA CR-135417, 1978.

171. E. C. Stevens and D. K. Hanink, "Titanium Composite Fan Blades," Final Report AFML-TR-70-180, 1970.

172. J. A. McElman, "Continuous Silicon Carbide Fiber MMCs," in *Vol. 1, Engineered Materials Handbook, Composites*, ASM International, Metals Park, OH, 1987, p. 858.

173. *Thermal Spraying: Practice, Theory and Applications*, American Welding Society, Miami, 1985.

174. D. W. Petrasek and R. H. Titran, "Creep Behavior of Tungsten/Niobium and Tungsten/Niobium-1 Percent Zirconium Composites," NASA TM 100804, 1988.

175. J. L. Yuen, *Proceedings of the Advanced High Pressure Oxygen/Hydrogen Conference*, 1984.

176. Y. L. Yuen, "Screening Evaluation of Candidate Fiber-Reinforced Superalloys for Space Shuttle Main Engine Turbopump Blade Application," Part I: NASA CR-175085; Part II: NASA CR-180838, 1987.

177. S. M. Arnold, "A Qualitative Examination of the Fabrication Process for a Tubular Metal Matrix Composite Test Specimen," NASA TM 100271, 1988.

Metal Matrix Composites, Tungsten Fiber Reinforced Copper

Fiber-reinforced metal matrix composites offer a wide range of properties to meet specific design and application requirements. They combine the strength and modulus of a fiber with the ductility and oxidation resistance of a matrix (Fig. 1). Most of the current emphasis on metal matrix composites is on low density, high modulus fibers such as graphite, silicon carbide, or boron fibers reinforcing matrices of aluminum, magnesium, titanium, or intermetallics. Most of the applications being considered for metal matrix composites are in a relatively low temperature range. Spacecraft applications include antenna and structural applications where the temperatures range from 366 K in the sun to 166 K in shadow conditions. Airframe applications range up to 478 K for supersonic flight regimes. For applications such as these, the high room temperature strength/density and stiffness/density of composites offer major design advantages.

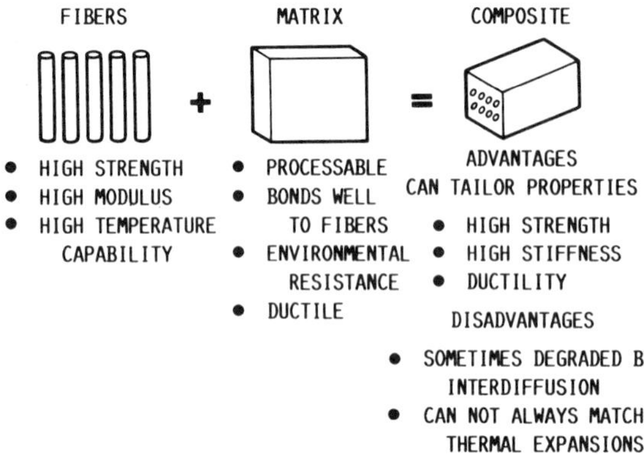

FIGURE 1 **Advantages and disadvantages of combining a fiber and a matrix into a composite.**

However, the propulsion system environment in aircraft or spacecraft presents a much more severe set of operating conditions. Increased engine efficiency and reduced fuel burn are the prime goals sought by engine designers. These goals place significant challenges on engine materials since they can be achieved only through the use of lower weight materials, higher creep-strength/density materials, higher temperature operation, and increased rotational speed with its accompanying increased stress on rotating components [1].

The main emphasis of metal matrix composite research at NASA Lewis has traditionally been focused on materials for aircraft engine applications. The majority of these applied research programs have been aimed at two mains areas of engine components: creep-resistant tungsten fiber reinforced superalloy matrix composites for the high temperature turbine section of the engine [2], and bird-strike-resistant boron fiber reinforced aluminum matrix composites for the fan section [3].

From its start, the NASA Lewis efforts on metal matrix composites have focused on improvement of high temperature properties. The first step in the development of composites for elevated temperature service was to develop a metal matrix composite model system, to analyze the behavior of the model composite, and to generate a data base to allow prediction of properties for other composite systems. To serve as a model system for the analysis of the behavior of metal matrix composites, the tungsten fiber reinforced copper (W/Cu) composite system was chosen. The Lewis Research Center used these results to publish the first systematic, in-depth analysis of the behavior of metal matrix composites in 1959 [4] and the first mass-reader publication on metal matrix composites in 1960 [5].

Tungsten fiber reinforced copper composites were ideal as a model system for analyzing the behavior of metal matrix composites because the two components, tungsten wire and copper matrix, were mutually insoluble in each other, readily available, low cost, and easily fabricated into composites by liquid infiltration. The availability of materials allowed fabrication of large numbers of test specimens covering a range of fiber contents and fiber diameters. The mutual insolubility of the components allowed a detailed analysis of the stress–strain behavior to determine the contributions of each component to the properties of the composite.

A series of research programs were conducted to investigate the stress–strain behavior of W/Cu composites; the effect of fiber content on the strength, modulus, and electrical conductivity of the composites; and the effect of alloying elements on the behavior of the tungsten wire and the composite. Later programs investigated the stress-rupture, creep, and impact behavior of the composites at elevated temperatures. The results of this research were used to select candidate fibers and matrices for the development of usable tungsten fiber reinforced superalloys for high temperature aircraft and rocket engine turbine applications. It is the purpose of this article to summarize the W/Cu composite efforts conducted at NASA Lewis, to describe some of the results obtained, and to describe more recent work using W/Cu composites as high strength, high thermal conductivity composite materials for high heat flux, elevated temperature applications.

Materials and Fabrication

The basic W/Cu model composite work used commercially drawn tungsten filament wire (General Electric type 218CS). This wire was selected for study because of its high tensile strength, its high recrystallization temperature, its availability in a wide range of diameters, and its relative ease of handling.

High purity OFHC copper was selected as the primary matrix material for these composites. This choice was based upon copper's melting point (which is below the temperature at which the properties of tungsten wire are seriously damaged by recrystallization), copper's insolubility in tungsten, and the ability of molten copper to wet tungsten. In addition, copper alloys were also used as matrix materials. Selected alloying additions were added to the pure copper matrix to determine the effect of alloying elements on the reinforcement behavior of the tungsten wire and on the properties of the composites.

Composites were fabricated by liquid phase infiltration. Continuous, unidirectional tungsten fibers were packed in ceramic tubes to the desired fiber content. A slug of copper was placed over the fiber bundle and the entire assembly was placed in a furnace and heated to 1478 K for 1 hour in either a vacuum or a hydrogen atmosphere. The molten copper flowed over the tungsten wires by gravity and capillarity and fully infiltrated the fiber bundle to form a sound, fully dense composite. After infiltration, the composite rods were removed from the ceramic tubes. Some rods were centerless ground into test specimens, while others had threaded grips brazed onto their ends to make threaded round test specimens.

This method of fabrication allowed production of large numbers of fully dense, pore-free composites, with accurately aligned unidirectional fiber orientation, to be used for analysis of the behavior of W/Cu composites. All testing was done in the longitudinal direction, and all properties were determined in the direction parallel to the fiber axis.

Stress–Strain Behavior of Tungsten Fiber Reinforced Copper Composites

Analysis of the stress–strain curve of metal matrix composites is the key to understanding the behavior and predicting the mechanical properties of these composites. The W/Cu composite system is an ideal model for evaluating the stress–strain behavior of composites [6,7]. Tungsten and copper are mutually insoluble and have no interfacial reaction. Both the fiber and the matrix undergo plastic deformation at failure, and the properties of each component are very reproducible and consistent.

A set of stress–strain curves of W/Cu composites is presented in Figure 2. These curves show that there are four stages of deformation in a ductile-fiber/ductile-matrix composite such as W/Cu. These stages are shown schematically in Figure 3 for the fiber, matrix, and composite. The stress on the composite at any point on the stress–strain curve can be represented by a volume percent weighted rule of mixtures relation connecting the stress on the fiber and on the matrix at that strain:

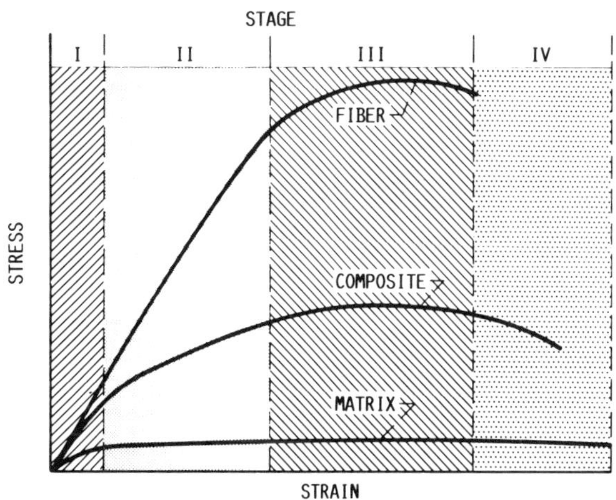

FIGURE 3 Schematic representation of four stages of stress–strain behavior of metal matrix composites. (After Refs. 6 and 7.)

$$\sigma_c^* = \sigma_f^* V_f + \sigma_m^* V_m \tag{1}$$

where σ is stress, V is volume fraction, and the subscripts c, f, and m refer to the composite, fiber, and matrix, respectively. The superscript * refers to the stress on each component in the condition in which it exists within the composite at that strain.

In stage I behavior, both the fiber and the matrix undergo elastic strain. Since each component is straining elastically, the composite also exhibits elastic behavior. The modulus of elasticity, E, can be predicted using a rule of mixtures relation connecting the moduli of the fiber and the matrix:

$$E_c = E_f V_f + E_m V_m \tag{2}$$

This linear relation between the moduli of the fiber and the matrix is shown by a plot of the dynamic modulus of elasticity data from W/Cu composites over a range of fiber contents (Fig. 4). The line shown on the figure represents the rule of mixtures prediction connecting the modulus of the copper matrix and the tungsten reinforcing fiber.

In stage II behavior, the fibers continue to strain elastically, while the matrix has passed into plastic strain. For W/Cu composites, the fibers continue to strain elastically up to about 0.4% strain, while for copper, the proportional limit is at about 0.04% strain. This strain transition gives rise to a secondary modulus E^1, which can be predicted by:

$$E_c^1 = E_f V_f + \frac{d\sigma_m}{d\varepsilon} V_m \tag{3}$$

where $d\sigma_m/d\varepsilon$ is the slope of the stress–strain curve of the plastically deforming matrix. A plot of the secondary modulus of elasticity of W/Cu composites, measured

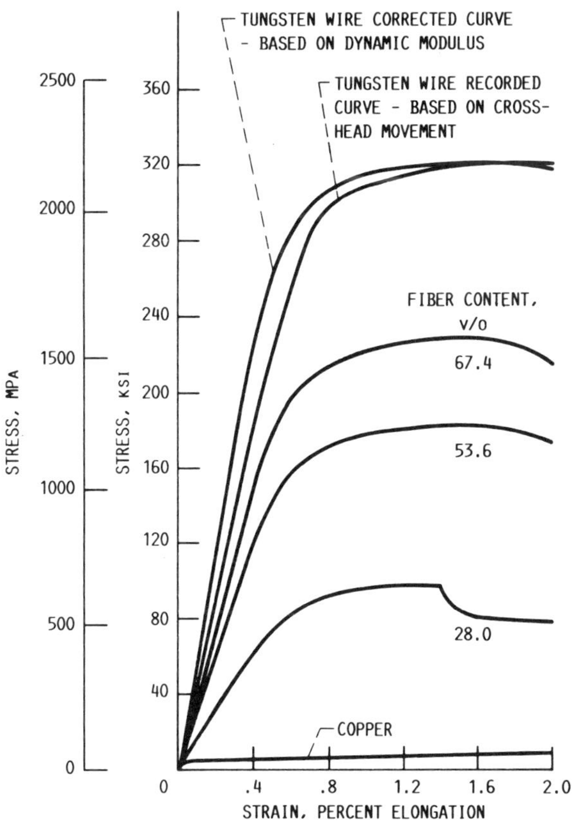

FIGURE 2 Stress–strain curves for tungsten wire, copper, and tungsten fiber reinforced copper composites. (From Refs. 6 and 7.)

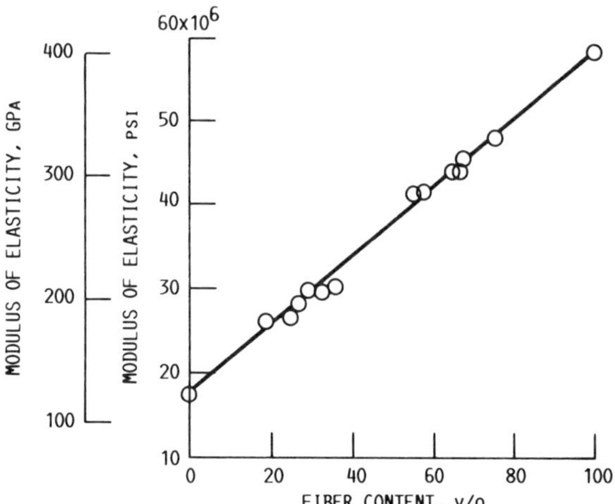

FIGURE 4 **Dynamic modulus of elasticity of tungsten, copper, and tungsten fiber reinforced copper composites.** (From Refs. 6 and 7.)

from stress–strain curves, is presented in Figure 5 over a range of fiber contents. The line shown on the curve represents a least-squares fit of the data, and the end points fall at the near-zero slope of the stress–strain curve of copper and at the initial modulus of the still elastic tungsten fiber.

In stage III behavior, both the fiber and the matrix are straining plastically. Again, the stress on the composite is equal to a volume percent weighted rule of mixtures relation with the stress on the fiber and on the matrix at the same equivalent strain. The yield strength of the composite can be predicted by:

$$\sigma_{y_c} = \sigma_{y_f} V_f + \sigma'_m V_m \tag{4}$$

where σ'_m is the stress on the matrix at the strain at which the yield stress of the fiber is measured. For most matrices, stress increases resulting from work hardening are not significant over this strain range, compared with the differences in strength between the fiber and the matrix. Therefore the value of the yield strength of the matrix, σ_{y_m}, could be substituted for the value of σ'_m in Eq. (4) if actual stress-strain data are not available. Yield strength data for W/Cu composites are shown in Figure 6. The yield strength of the composites was calculated using a 0.2% offset, based upon the secondary modulus for convenience of measurement. Because of the flatness of the copper stress–strain curve in the plastic region, the differences in yield strength, based upon offset from the initial or secondary modulus, should not be significant. The line shown represents a least-squares fit of the data obtained. Extrapolation of this curve to the end points shows good agreement with the anticipated yield strengths of the copper matrix and the tungsten reinforcing fiber.

Stage III behavior continues until the ultimate strength of the fiber is reached, which coincides with the strain at which the ultimate tensile strength of the composite is also reached. The ultimate tensile strength of the W/Cu composites can be predicted using the equation:

$$\sigma_c = \sigma_f V_f + \sigma^*_m V_m \tag{5}$$

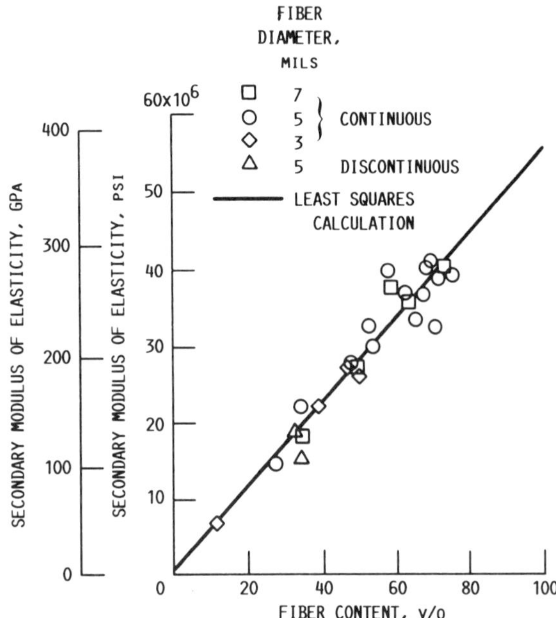

FIGURE 5 **Secondary modulus of elasticity of tungsten fiber reinforced copper composites.** (From Refs. 6 and 7.)

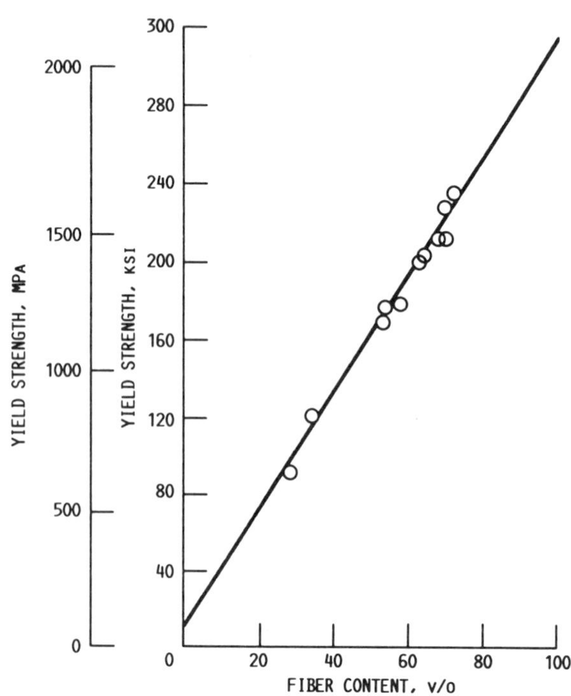

FIGURE 6 **Yield strength (based on secondary modulus) of tungsten fiber reinforced copper composites.** (From Refs. 6 and 7.)

where σ_m^* is the stress on the matrix at the strain at which the fiber reaches its ultimate tensile strength. Ultimate tensile strengths of W/Cu composites are presented in Figure 7 over a range of fiber contents. The line shown on the figure represents the rule of mixtures prediction from Eq. (5). The experimental data for W/Cu composites show excellent agreement with the predicted ultimate strength line.

During stage IV behavior, the fibers reach their ultimate tensile strength and start to fail. Initially, the fibers start to break at random locations. Eventually, fiber breaks align in a failure plane and the remaining fibers in that cross section fail and the load drops rapidly. The composite is held together by ligatures of ductile, unreinforced matrix that continue to strain until they also fail. The stress on the composite drops as these ligatures break.

The effect of fiber content on failure strain is shown over a range of fiber contents in Figure 8. At low fiber contents, matrix ligatures are of sufficient size to continue to strain, with the composite reaching a strain to failure of up to 10% and the remaining copper forming a localized point of unreinforced copper before the last ligature failed. At higher fiber contents, the composite strain to failure dropped to about 4–5%. Since the fibers had broken at about 2–3% strain, the weakest fibers, with the lowest failure strains, will break first, and the composite will continue to strain at nearly full load until the main fracture plane is established. After the fibers at the fracture edge have broken, additional strain on the composite is occurring from the unreinforced copper ligatures only, at very low stresses.

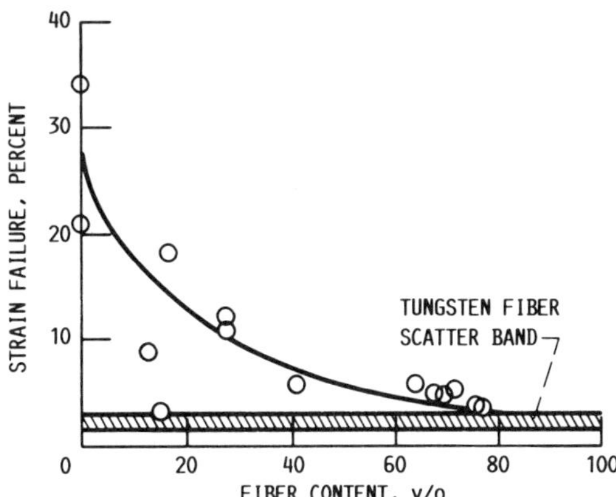

FIGURE 8 Failure strain of tungsten fiber reinforced copper composites. (From Refs. 6 and 7.)

The rule of mixtures relation was also observed for tensile test results for W/Cu composites at elevated temperatures. The ultimate tensile strengths of W/Cu composites are plotted as a function of fiber content on Figure 9 for a series of temperatures up to 1255 K [8]. Results of tensile tests on pure copper were reported for temperatures up to 1089 K and on type 218CS tungsten wire up to 922 K in the same refererence.

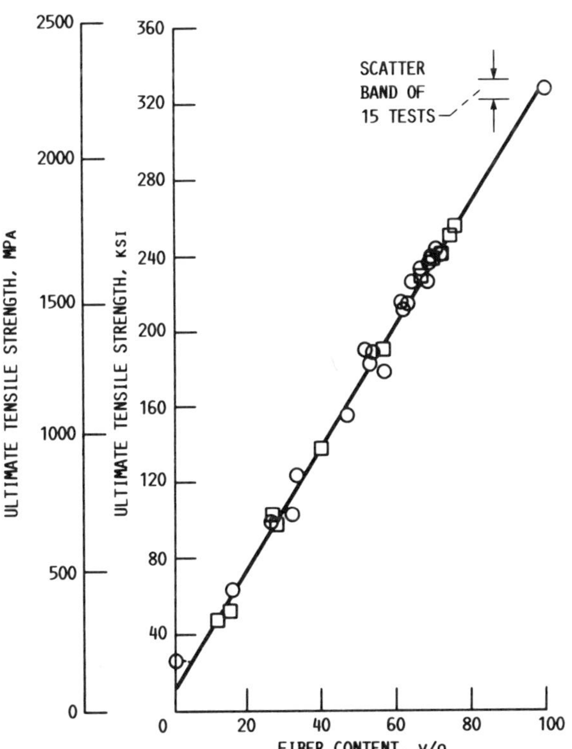

FIGURE 7 Ultimate tensile strength of tungsten fiber reinforced copper composites. (From Refs. 6 and 7.)

Properties of Composites Reinforced With Discontinuous Fibers

The composites discussed previously were all reinforced with continuous unidirectional tungsten fibers. Other

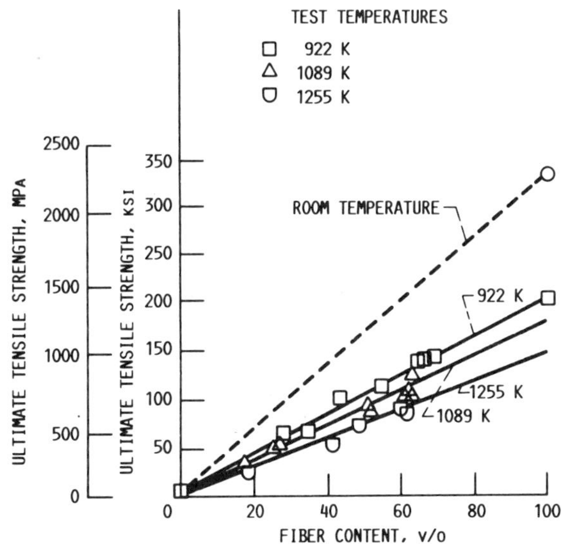

FIGURE 9 Ultimate tensile strength of tungsten fiber reinforced copper composites at various temperatures. (After Ref. 8.)

composites were also fabricated with uniaxially oriented, discontinuous fiber reinforcement and tested in the direction parallel to the fibers. Tungsten wire of 0.127 mm diameter was chopped to lengths of 0.975 mm and infiltrated with copper. Results of tensile tests on these composites, with a reinforcing fiber length-to-diameter aspect ratio of 75, are shown in Figure 10 for a range of fiber contents. The ultimate tensile strengths of the composites showed good agreement with the rule of mixtures prediction line used for composites reinforced with continuous fibers.

Although the stress–strain behavior and tensile strength relations observed for discontinuous-fiber-reinforced composites are similar to those observed for composites with continuous reinforcement, the mechanism by which such composites are strengthened is different. In a composite reinforced with continuous fibers, the load is carried by the fibers along their full length. When random fiber breaks appear, part of the load is carried around the break by interfacial shear transfer through the interface and the matrix. In a composite reinforced with discontinuous fibers, however, all the load must be carried from fiber to fiber by interfacial shear transfer through the matrix.

The stress distribution along the length of a discontinuous fiber is shown schematically in Figure 11 [9,10]. The ends of the fibers, called the *ineffective length*, can carry a very high shear stress but only a low tensile stress. Thus, short fibers will not contribute their full tensile

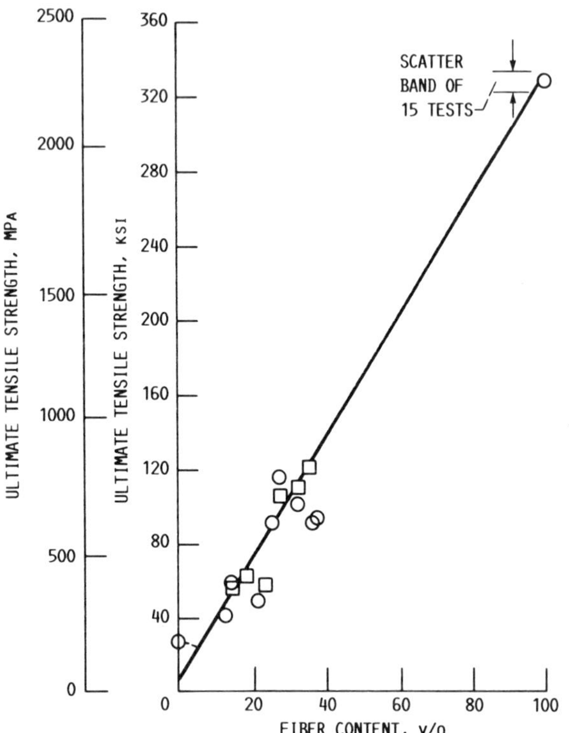

FIGURE 10 **Ultimate tensile strength of copper matrix composites reinforced with discontinuous tungsten wires with an aspect ratio of 75.** (From Refs. 6 and 7.)

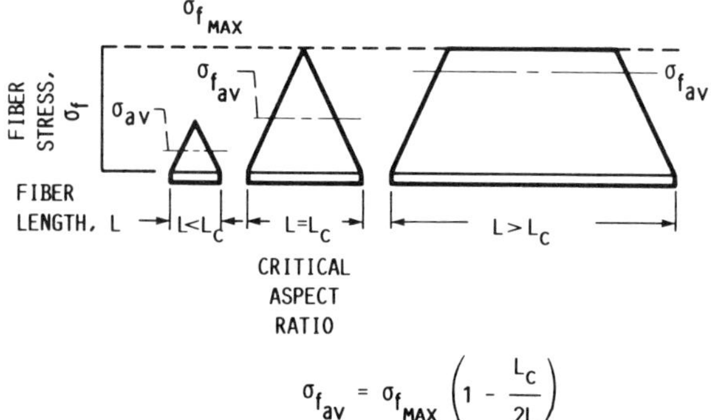

FIGURE 11 **Tensile stress gradients in fibers of various lengths.** (From Refs. 9 and 10.)

strength to a composite. At longer lengths, the shear stress and tensile stress will balance and the fiber can support the full tensile load. The length at which a portion of the fiber starts to carry a stress equal to its full tensile strength is called the critical length L_c, it may be calculated by:

$$L_c = \sigma_f \frac{D_f}{2\tau} \tag{6}$$

where σ_f is the tensile strength of the fiber, D_f is the fiber diameter, and τ is the shear strength of the matrix. As the fiber length increases, a greater portion of the fiber will be able to carry its full tensile stress. For a continuous fiber, the ineffective length at the ends of the fiber becomes insignificant and the fiber can support its full tensile strength. For a discontinuous fiber, the ineffective length at the ends of the fibers cannot carry a full tensile load and the average tensile stress that can be carried is the ratio of the effective length $L - L_c$ to the total fiber length L as given by:

$$\overline{\sigma}_f = \sigma_f\left(1 - \frac{L_c}{2L}\right) \tag{7}$$

where $\overline{\sigma}_f$ is the average fiber stress, σ_f is the ultimate tensile strength of the fiber, L is the fiber length, and L_c is the critical length of the fiber. The tensile strength of a discontinuous-fiber-reinforced composite depends on the tensile strength of the fiber, the length of the reinforcing fiber, and the shear strength of the matrix or the fiber–matrix interface, whichever is less. This calculation of the average fiber strength for discontinuous-fiber reinforcement modifies the composite strength prediction of Eq. (5) to:

$$\sigma_c = \sigma_f \left(1 - \frac{L_c}{2L}\right) V_f + \sigma_m^{*} V_m \tag{8}$$

Thus, for longer fibers, the ultimate tensile strength of discontinuous-fiber composites will approach that of continuous-fiber composites. As the aspect ratio becomes shorter, the composite strength is lower, but the composite will still fail in tension. At fiber lengths below

L_c, the composite will fail by shear pullout of the fiber from the matrix, at a much lower stress than the continuous-fiber composites.

The effect of fiber aspect ratio on the tensile strength of copper matrix composites reinforced with discontinuous fibers becomes more important at elevated temperatures because of the reduced shear strength of the copper matrix. The effect of several low fiber aspect ratios was reported in Ref. 10 for discontinuous-fiber-reinforced W/Cu composites tested at 873 K. The deviations from continuous-fiber composite behavior increased significantly with decreasing fiber aspect ratio (Fig. 12).

Results reported in Ref. 9 show that W/Cu composites reinforced with discontinuous fibers of higher aspect ratios can be used to produce composites with high strengths at elevated temperatures. Composites with a discontinuous-fiber aspect ratio of 200 approached the strength values predicted for continuous-fiber-reinforced composites at 755 K, while composites reinforced with fibers with an aspect ratio of 100 had lower strengths. Similar trends were observed for W/Cu composites tested at 1089 K, except that the deviation from continuous-fiber behavior was greater (Fig. 13). At fiber aspect ratios of 200, the W/Cu composites failed in tension, but at lower strengths than the continuous-fiber composites. At fiber aspect ratios of 100, the W/Cu composites appeared to be approaching the critical aspect ratio. Some composites failed in tension, while others started to fail by shear pullout of the fiber at lower stresses.

A detailed study was conducted to determine the critical aspect ratio for W/Cu composites at various temperatures [11]. A pullout specimen was used, in which a hole was drilled into tungsten buttons of various thicknesses. A 0.254 mm diameter tungsten fiber was placed in the hole, and copper was infiltrated between the wire and the walls of the hole. Tensile tests were conducted by pulling on the bare end of the wire and on the button. The fiber shear length was determined by the thickness of the tungsten button. For longer shear lengths, failure occurred by tensile failure of the tungsten wire (Fig. 14). At shorter shear lengths, the wire remained intact and pulled out of the matrix in a shear pullout failure. The critical aspect ratio was experimentally measured by determining the aspect ratio at which failure underwent a transition from shear pullout to fiber tensile failure. The experimentally determined critical aspect ratio increased with increasing temperature (Fig. 15). The rise was fairly minor up to 755 K, but increased rapidly above this temperature.

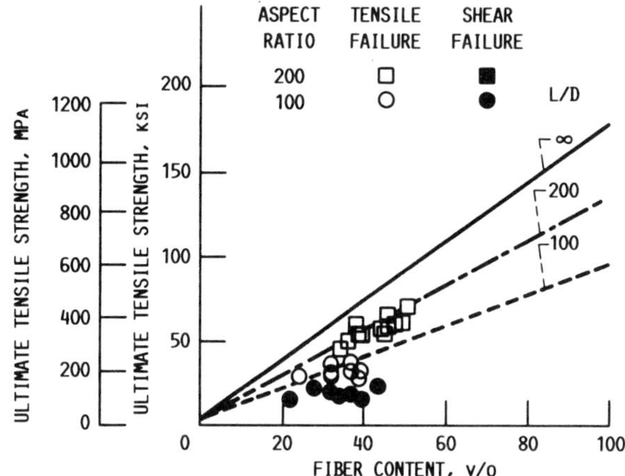

FIGURE 13 **Effect of aspect ratio on ultimate tensile strength of copper matrix composites reinforced with discontinuous tungsten wire, tested at 1089 K.** (From Ref. 9.)

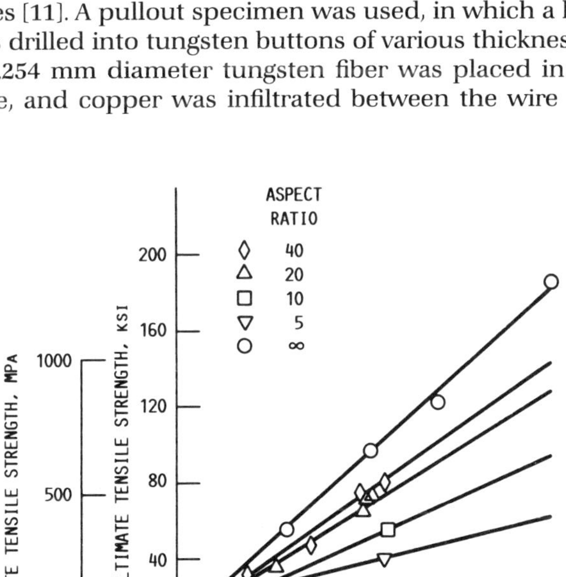

FIGURE 12 **Effect of aspect ratio on ultimate tensile strength of tungsten fiber reinforced copper composites reinforced with discontinuous tungsten wire tested at 523 K.** (From Ref. 10.)

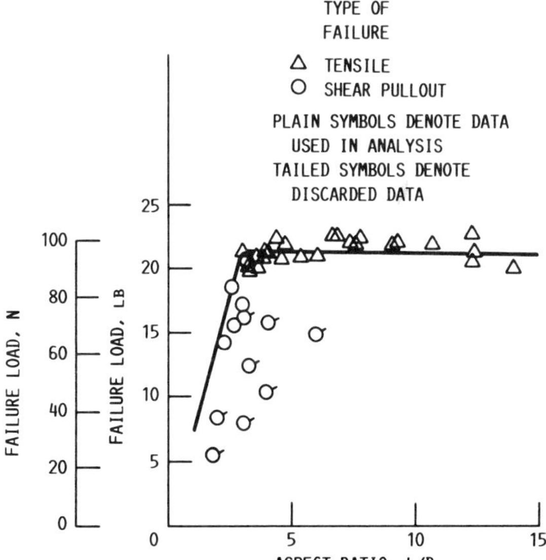

FIGURE 14 **Effect of aspect ratio on failure load and failure mode of tungsten wire–copper matrix pullout specimens.** (From Ref. 11.)

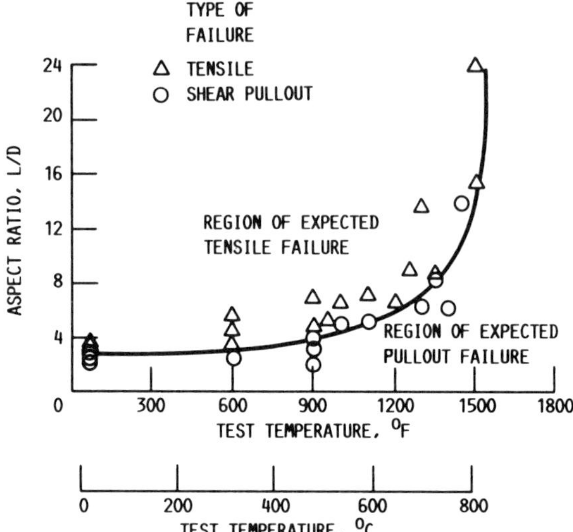

FIGURE 15 **Effect of test temperature on critical aspect ratio for tungsten wire–copper matrix pullout shear specimens.** (From Ref. 11.)

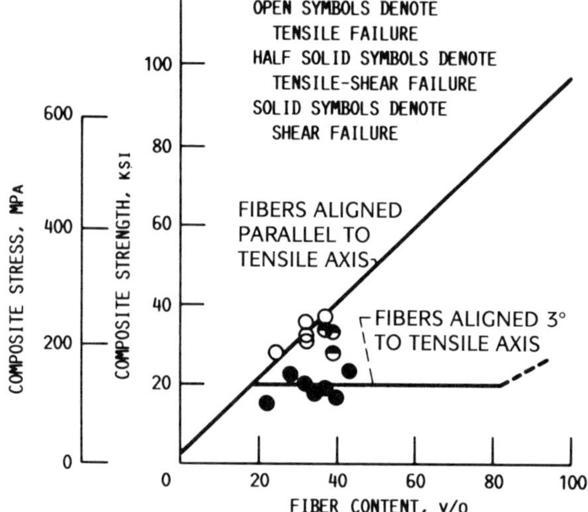

FIGURE 17 **Composite strength as a function of fiber content and orientation for copper matrix composites reinforced with discontinuous tungsten fibers with an aspect ratio of 100 at 1089 K.** (From Ref. 9.)

The fracture behavior of unidirectional discontinuous-fiber W/Cu composites is also influenced by the orientation of the fibers. There are three potential fracture modes in fiber-reinforced composites: tensile failure of the fiber, shear failure at the fiber–matrix interface, and tensile failure of the matrix (Fig. 16). A composite will fail at the lowest strength condition predicted from these three potential failure modes at a given fiber orientation [9]. At very low angles from uniaxiality, a composite will fail at a high stress by tensile failure of the fiber. With increasing misalignment, the fracture mode changes to shear failure at the fiber–matrix interface, with an accompanying large decrease in strength. At a fiber misalignment of about 65°, the fracture mode changes to tensile failure in the matrix.

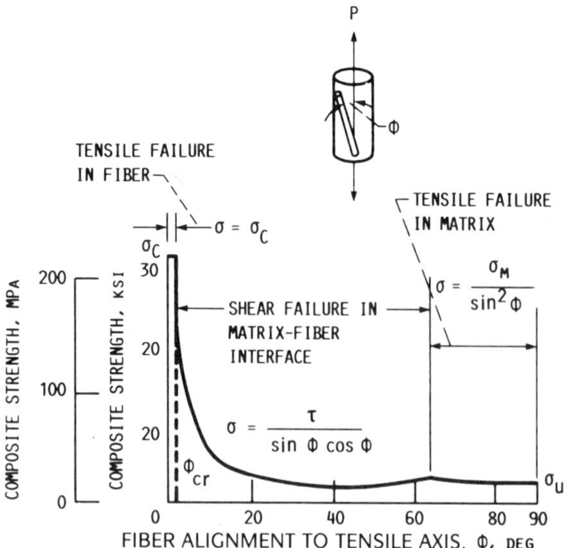

FIGURE 16 **Effect of fiber alignment on strength of discontinuous tungsten fiber reinforced copper composites.** (From Ref. 9.)

These calculations indicate the importance of maintaining axiality in discontinuous-fiber-reinforced composites. As the temperature increases, the matrix shear strength drops rapidly, thus reducing the allowable misalignment of fibers for effective composite reinforcement. At a temperature of 1089 K, composite strength starts to drop at a misalignment of about $\frac{1}{2}$°. With misalignments of only 3°, shear pullout failure predominates (Fig. 17) and the composite strength has dropped by one-half [9]. Thus, for discontinuous-fiber-reinforced composites, maintenance of axiality is very important, much more so than with continuous-fiber reinforcement.

Effect of Alloying Additions on Properties of Tungsten Fiber Reinforced Copper Matrix Composites

The previously described model studies of W/Cu composites used a pure copper matrix, which is insoluble in tungsten. The effect of reactive matrices on composite properties was studied using copper binary alloy matrices containing elements with varying solubility in tungsten [12,13]. The composites were tested at room temperature, and a microstructural analysis was made to determine the types of reactions occurring at the fiber–matrix interface. The tensile strengths of the tungsten fiber–copper alloy matrix composites studied were reduced to some degree when alloying with the tungsten fibers occurred. Several of the composite systems studied, however, showed very little reduction in tensile strength relative to composites made with pure copper matrix.

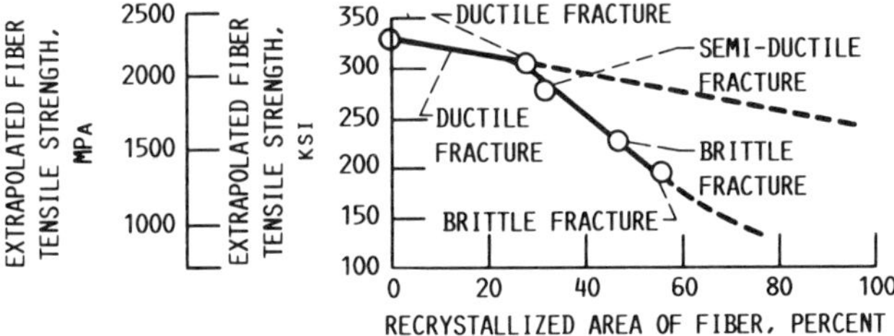

FIGURE 18 **Effect of recrystallized area on extrapolated tensile strength of fiber for tungsten fiber reinforced copper–5% cobalt alloy matrix composites.** (From Refs. 12 and 13.)

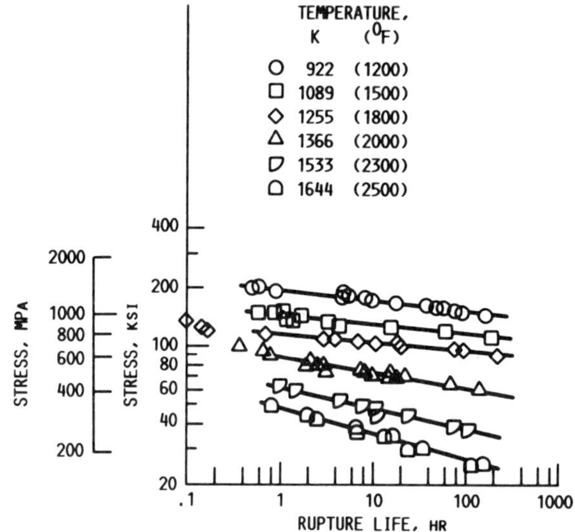

FIGURE 19 **Effect of stress on rupture life for as-drawn 0.127 mm diameter tungsten wire at various temperatures.** (From Ref. 15.)

Three types of reactions were observed to occur at the tungsten fiber–copper alloy matrix interface: A diffusion-penetration reaction accompanied by a recrystallization of the grains at the periphery of the tungsten wire, formation of a two-phase zone, and a solid solution reaction without subsequent recrystallization. The first type of reaction was observed with alloying additions of cobalt, aluminum, and nickel, and caused significant degradation of composite strength properties. The second type of reaction was observed with alloying additions of titanium and zirconium. The third type of reaction was observed with additions of niobium and chromium. These last two types of reactions did not seriously affect the properties of the composites studied [12,13].

A correlation between the tensile strength and ductility of the composite was observed. In general, those materials that had the best strengths also had the best ductilities. The strength and ductility behavior was also correlated with depth of penetration measurements [12,13]. The greater the penetration of the alloying element into the tungsten wires, the lower the tensile strength and ductility of the composite (Fig. 18). Property degradation of the composite was greater than that predicted by the simple rule of mixtures relation based on the volume fractions of the reaction zone and the still intact fiber. This fact, along with the observed correlations between the ductility and tensile strength of the fibers in the composite, suggested that damage was due to a notch-embrittling effect. It was also found that the diffusion-triggered penetration-recrystallization reaction at the fiber–matrix interface could be prevented by combining the damaging alloying element with one that did not cause this type of reaction and was compatible with the fiber.

Stress-Rupture and Creep Properties of Tungsten Fiber Reinforced Copper Composites

The initial analyses of W/Cu composites focused on the short-time tensile properties of composites. However, in high temperature aircraft engine applications, the long-term creep behavior and stress-rupture properties are of greater importance. A series of programs were conducted to determine the stress-rupture properties of tungsten wire and to relate the wire properties to the stress-rupture properties and creep behavior of W/Cu composites with an analysis similar to that used for stress–strain and tensile behavior prediction.

The initial phases of this program consisted of the determination of the stress-rupture properties of 0.127 mm diameter tungsten wire over a temperature range from 922 to 1644 K. The tests were conducted in a specially designed wire stress-rupture testing apparatus described in Ref. 14. Results of these tests are shown in Figure 19. Tungsten wire retained its fibrous microstructure and exhibited ductile fracture behavior at temperatures up to 1255 K. At a test temperature of 1533 K, the tungsten wire undergoes a transition in behavior. At short rupture times, the microstructure of the wire retains its heavily worked condition, while at longer times and higher temperatures, recrystallization starts to remove the strong, ductile fibrous structure and replace it with more equiaxed recrystallized grains, causing reduced strength and ductility in the fiber.

Composites were fabricated using type 218CS tungsten wire and a pure copper matrix. Stress-rupture and creep tests were conducted at temperatures of 922 K and 1089 K. A creep curve of typical W/Cu composites at each test temperature is shown in Figure 20. Analysis of the creep behavior of the composite, combined with the creep behavior of the copper matrix, allowed calculation of the creep behavior of the tungsten wire, which could not be measured conveniently [16].

Analysis of the creep curves of a number of W/Cu specimens indicated that there are seven stages of creep behavior in a ductile fiber–ductile matrix metal matrix composite such as W/Cu (Fig. 21). The first stage consisted of initial elastic loading, analogous to a high temperature tensile test, where both the tungsten fibers and the copper matrix carry a portion of the load elastically.

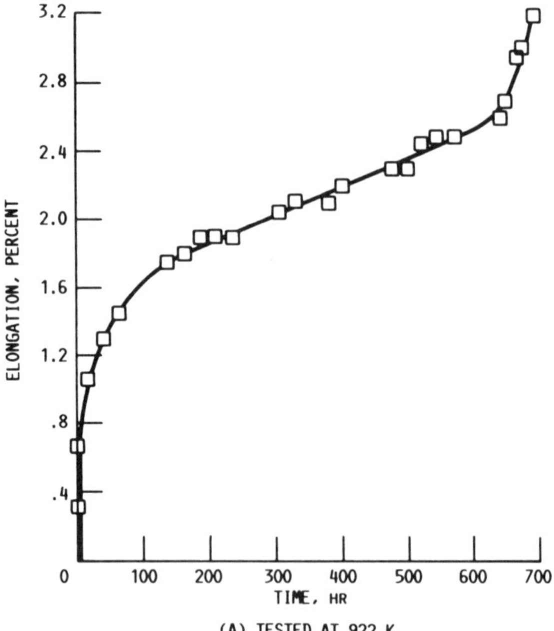

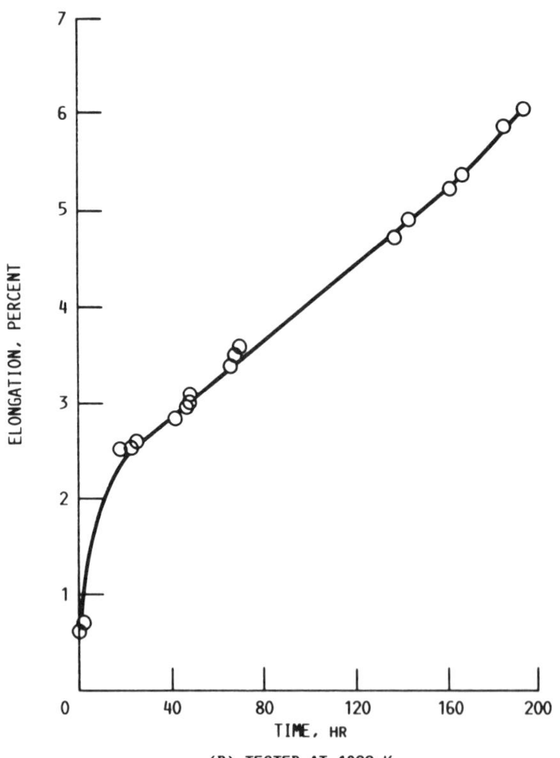

FIGURE 20 **Typical creep curves of tungsten fiber reinforced copper composites.** (From Ref. 16.)

The load carried by each component can be predicted using the stage I analysis of stress–strain behavior at the stress level used in the creep test.

After a fiber–matrix stress equilibrium is reached during initial loading, the composite starts to elongate with time by first-stage creep. As the creep rate decreases, there is a continuing rebalancing of stresses between the fiber and the matrix. If the entry into second-stage steady-state creep occurs at different times for the two components, then the stronger component, the fiber, forces the weaker matrix to adopt its deformation behavior after a short transition period. This change in the normal behavior of the components gives rise to a quasi-first-stage creep governed by the creep characteristics of the fiber.

Eventually a balance is reached between work hardening and recovery, and a constant creep rate is attained. With the attainment of second-stage creep, with its constant creep rate, there is an additional rebalancing of stresses between the components. This stress equilibrium is maintained for a fairly high fraction of test life.

After a constant second-stage creep rate is established, the stress on the composite that causes a given creep rate may be calculated from a log stress–log creep rate curve. If the equilibrium of forces is considered, the stress distribution on the fiber and matrix can be predicted by:

$$\sigma_{C_{\dot{\varepsilon}_i}} = \sigma_{f_{\dot{\varepsilon}_i}} V_f + \sigma_{m_{\dot{\varepsilon}_i}} V_m \qquad (9)$$

where $\sigma_{C_{\dot{\varepsilon}_i}}$ is the stress on the composite to give a creep rate of $\dot{\varepsilon}_i$, and $\sigma_{f_{\dot{\varepsilon}_i}}$ and $\sigma_{m_{\dot{\varepsilon}_i}}$ are the stresses on the fiber and on the matrix to give the same creep rate. Since it is difficult to experimentally measure creep rates for fibers, this equation allows fiber creep rates at various stresses and temperatures to be calculated when the creep rate of the composite is measured and the stress to cause that creep rate is known for the matrix.

The creep rates of W/Cu composites were measured experimentally and a log fiber content–log creep rate relation was plotted for a series of composite stresses. An example of these plots is shown in Figure 22. The fiber contents required for a given creep rate were plotted in Figure 23 for each composite stress to give a rule of mixtures relation of composite stress to fiber content for a series of creep rates. Extrapolation of these data in the low fiber content region showed that the stresses for a given creep rate are in good agreement with the creep rates for unreinforced copper at these temperatures.

After a period of second-stage creep, most materials pass into third-stage creep. Again, the onset of third-stage creep of the fiber and the matrix may occur at different times. Materials such as copper show a considerable amount of third-stage creep prior to failure. From the data for W/Cu composites in this investigation, it appears that the tungsten wire remained in second-stage creep for a much longer time than the copper matrix. Since the fiber and the matrix are bonded together in the composite, the stronger, rate-controlling component, the fiber, forces the more easily deformed matrix to remain in second-stage creep. This process gives rise to quasi-second-stage creep behavior in the composite. This behavior causes the stress on the matrix to be reduced to a value that will enable it to remain in second-stage creep. The lowering of the stress on the matrix is compensated by an increase in the stress on the fiber. Quasi-second-stage creep would be expected to con-

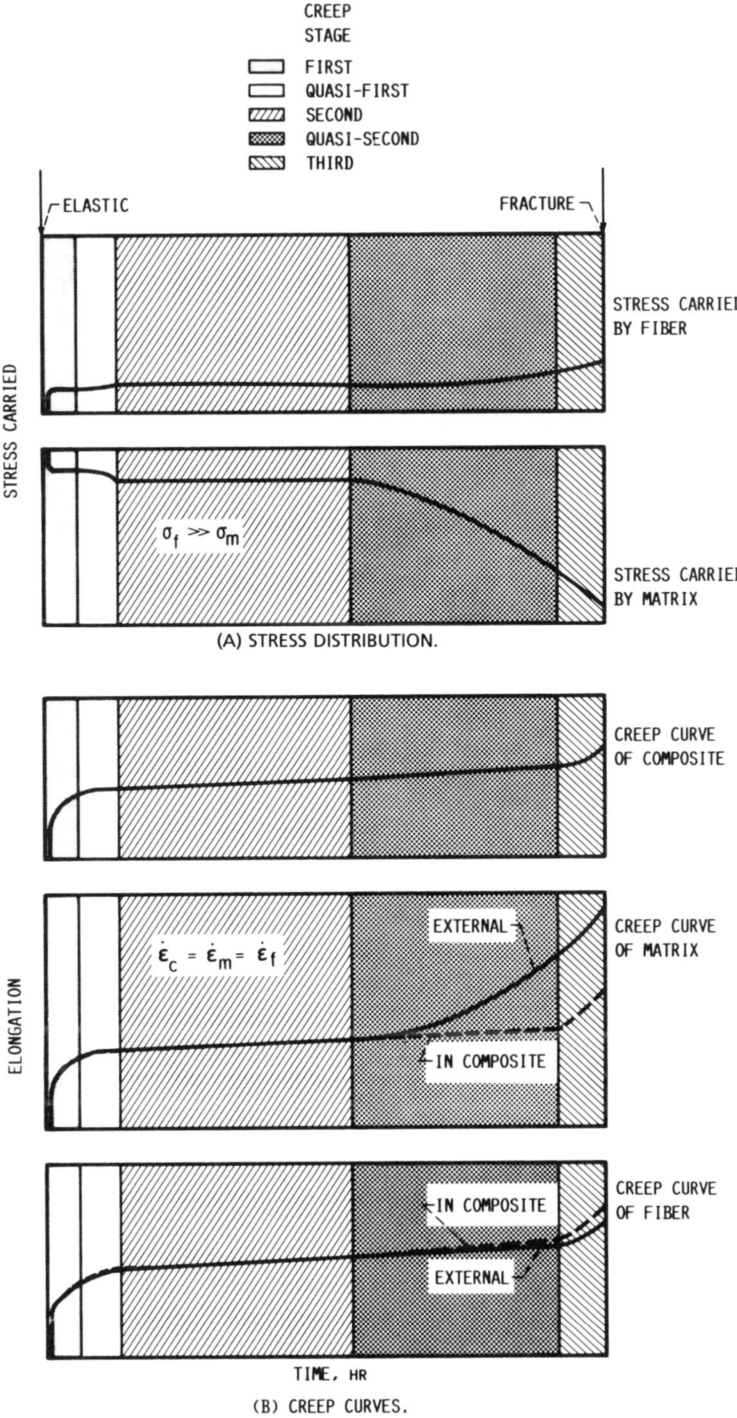

FIGURE 21 Schematic of creep curves and stress distribution curves showing seven stages of creep behavior in tungsten fiber reinforced copper composites. (From Ref. 16.)

tinue until the rate-controlling fiber enters third-stage creep, at which time a new stress distribution would be set up between the fiber and the matrix.

Third-stage creep continues until the fibers start to fail and the initiation of composite fracture begins. The individual fibers in a composite have a scatter band of rupture times at a given stress. The fibers within the composite would be expected to have a scatter band similar to that of fibers tested externally. As the first fibers start to fail at random locations, the remaining fibers in the composite must support a stress slightly higher than that originally encountered when all the fibers were intact. With successive fiber fractures, the actual stress on the remaining fibers would continue to

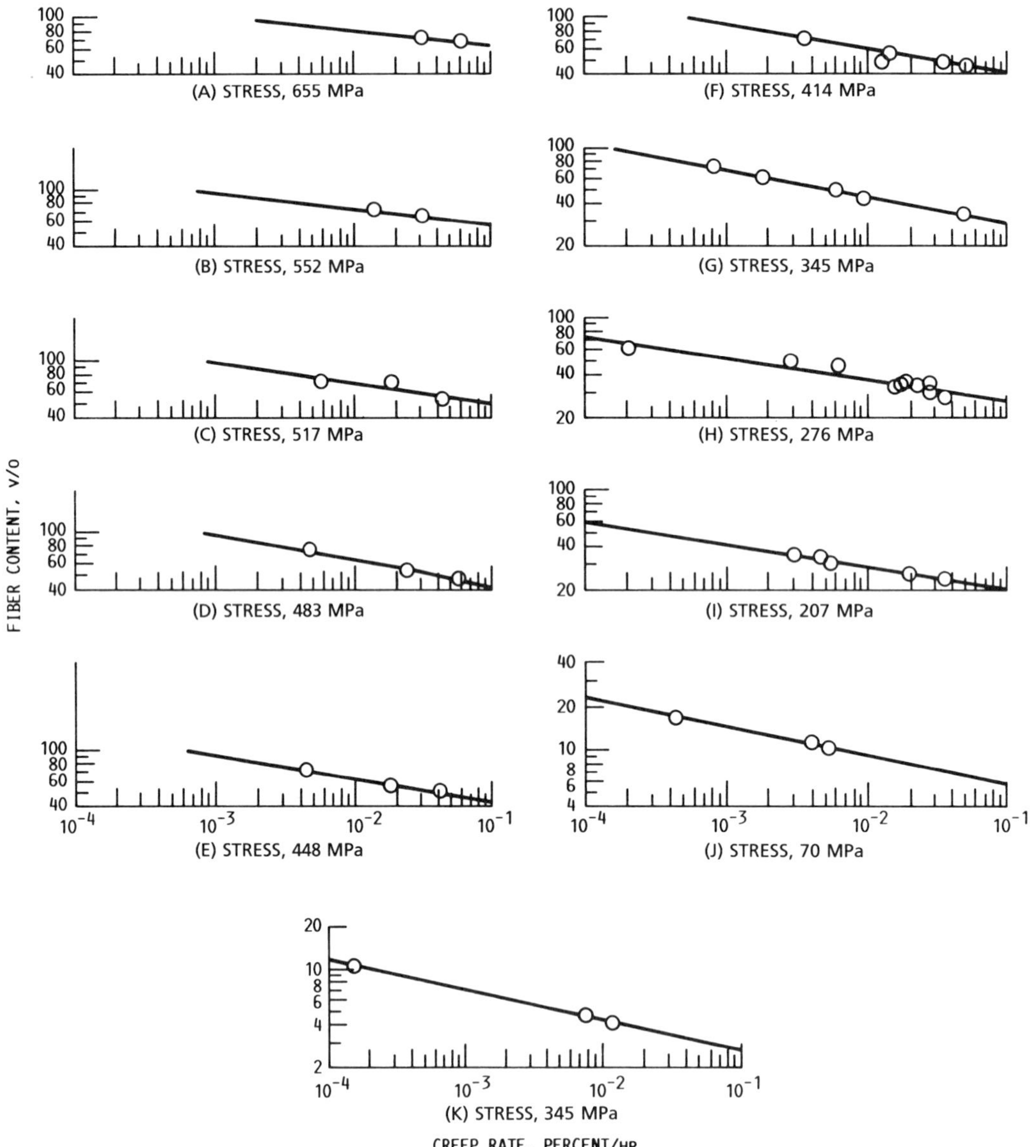

FIGURE 22 Fiber content as a function of creep rate at various stresses for tungsten fiber reinforced copper composites tested at 1089 K. (From Ref. 16.)

increase. Eventually, the stress exceeds the strength of the remaining fibers, and fracture of the composite occurs. The actual time to rupture would depend on the number of fibers present and on the rupture-time scatter band of the fibers.

Results of stress-rupture tests indicated that the rupture time of a composite may be predicted by a rule of mixtures relation similar to that used to predict composite creep rates. For a given composite stress, the log of the rupture time is a function of the log of the fiber content (Fig. 24). The fiber contents required for given rupture times were determined for a number of composite stresses and plotted on Figure 25 over a range of fiber contents. This plot shows that a linear rule of mixtures relation exists between the composite stress and the fiber content for a given rupture time. The stress to cause rupture of a composite at a given time can be predicted from the properties of the fiber and matrix by:

$$\sigma_{c_t} = \sigma_{f_t} V_f + \sigma_{m_t} V_m \tag{10}$$

where the stress on a composite that will cause rupture in time t can be calculated from the stress-rupture properties of the fiber and the matrix. Handbook values or experimental data can be used directly in Eq. (10) for prediction of the properties of composites.

The fracture behavior observed from the stress-rupture tests of W/Cu composites indicated that the stress redistributions postulated during quasi-second-stage and third-stage creep were valid. Unreinforced copper had a brittle stress-rupture failure with a very small reduction in area and a great deal of intercrystalline crack-

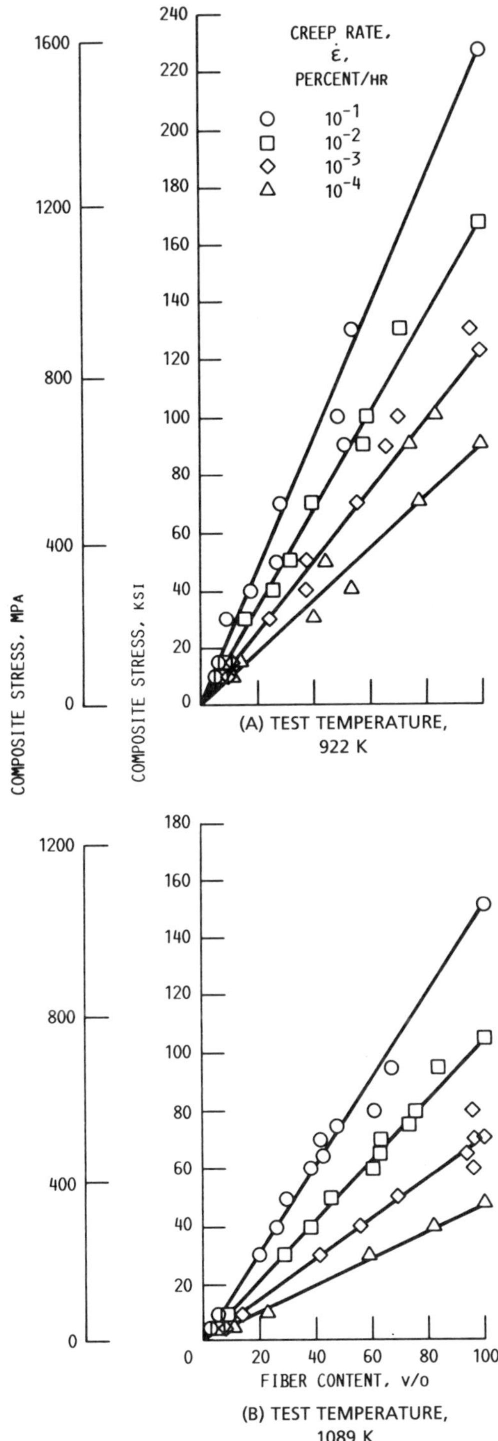

FIGURE 23 **Stress for a given creep rate as a function of fiber content for tungsten fiber reinforced copper composites.** (From Ref. 16.)

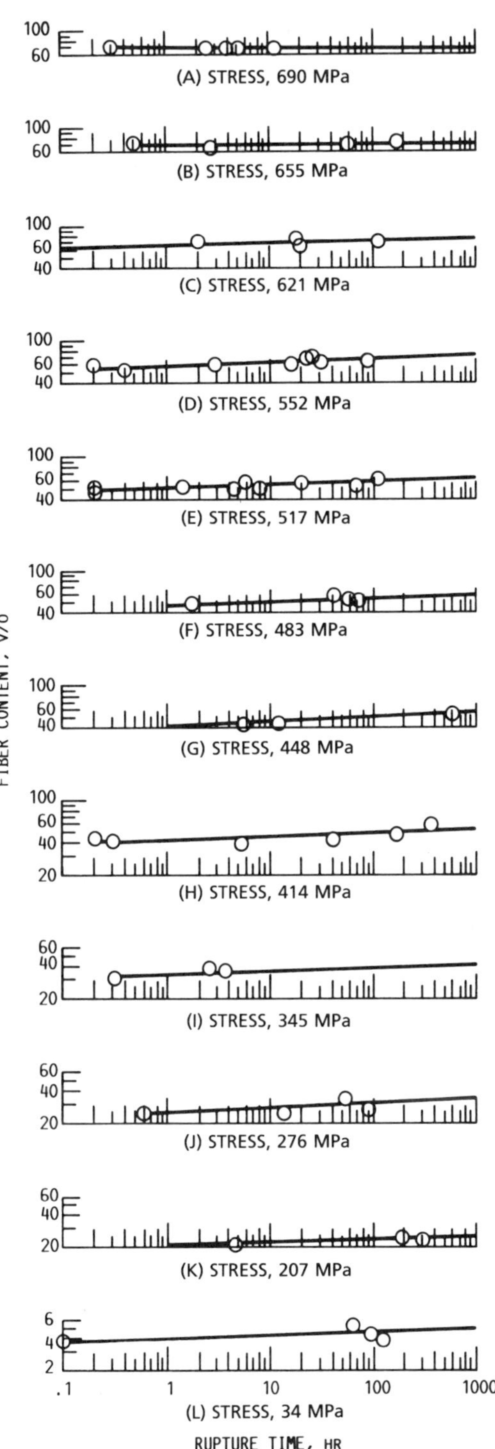

FIGURE 24 **Fiber content as a function of rupture time at various stresses for tungsten fiber reinforced copper composites tested at 1089 K.** (From Ref. 16.)

ing at the fracture edge. This brittle-type behavior is typical of materials tested at a very high fraction of their melting points.

The incorporation of a small amount of reinforcing fibers (about 10 vol %) changed this brittle-type behavior to a very ductile failure in which the copper almost formed a point at fracture. Since there was no intercrystalline stress-rupture cracking in the matrix, it appears that the matrix of the composite failed in second-stage creep and did not undergo a large amount of the third-stage creep normally found in copper. The majority of

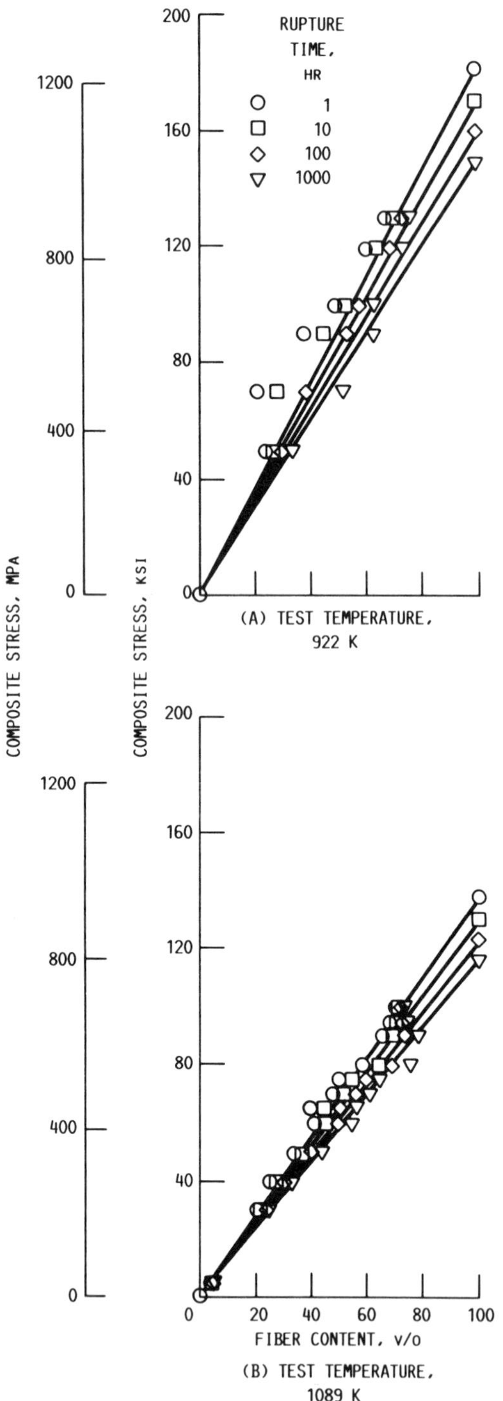

FIGURE 25 **Stress to cause rupture in 1, 10, 100, and 1000 hours as a function of fiber content for tungsten fiber reinforced copper composites.** (From Ref. 16.)

the strain probably occurred almost instantaneously after fracture of the fibers.

Composites with higher fiber contents had a less ductile fracture. Although the fibers had about the same reduction in area as when tested individually (about 80%), the composites had an apparent ductility of about 20% reduction in area. The fiber–matrix bond had been destroyed at the fracture during necking of the fibers, and the fibers continued to neck in their normal behavior until they failed. The copper between the fibers would then start into third-stage creep for a short time until the fibers broke, at which time the entire load would be imposed upon the matrix and the remaining matrix would fail instantaneously.

Although the W/Cu composites tested in this study were chosen primarily as a mutually insoluble model system, the results showed that the composites themselves were very strong. The 100 hour rupture strength and the 100 hour rupture strength/density ratio at 1089 K of two superalloys commonly used in this temperature range are compared with those of W/Cu composites in Figure 26. The plot shows that the 100 hour rupture strength of the composites compares favorably with that of the superalloys at intermediate fiber contents. At higher fiber contents (>60 vol %), the properties of the W/Cu composites were superior to those of the superalloys. In spite of the high density of the tungsten wire, the 100 hour rupture strength/density ratio of the W/Cu composites also compares favorably with that of the superalloys. These results are even more remarkable considering that the 100 hour rupture strength of unreinforced copper at 1089 K is 2.6 MPa, while a 50 vol % W/Cu composite has a 100 hour rupture strength of about 483 MPa.

The stress-rupture and creep results obtained in this study of the W/Cu model system served as the basis for methods of predicting potential metal matrix composite behavior at high temperatures and demonstrated the feasibility of using tungsten fiber reinforced composites for this type of application. Further work on investigating other fibers, such as W–l-2ThO_2, W–3Re, W–0.3Hf–0.04C, and W–4Re–0.4Hf–0.02C alloys, has demonstrated that these have much better high temperature tensile and creep properties than the unalloyed type 218 tungsten wire [17]. These fibers were prime candidates used for the development of tungsten fiber reinforced superalloy composites [2].

In addition to the stress-rupture and creep tests conducted on continuous-fiber W/Cu composites, tests were also conducted to determine the effect of critical aspect ratio on the stress-rupture life of dicontinuous-fiber composites [18]. Button-head pullout specimens with 0.254 mm diameter tungsten wire infiltrated with copper were tested at 922 K and 1089 K for nominal shear pullout stress-rupture times of 1, 10, and 100 hours. These tests used the same type of pullout specimen mentioned earlier and reported in Ref. 11. A summary of the results is shown in Figure 27. The observed critical aspect ratios were greater than in short-time tensile tests, but were within the same order of magnitude at the test temperatures used. Failure times were controlled by the shear properties of the copper matrix at fiber embedment lengths less than the critical aspect ratio, and by the tensile properties of the tungsten fiber at greater lengths than the critical aspect ratio.

These results indicate that the critical aspect ratio at 1089 K has only to be lengthened from about 15 for short-

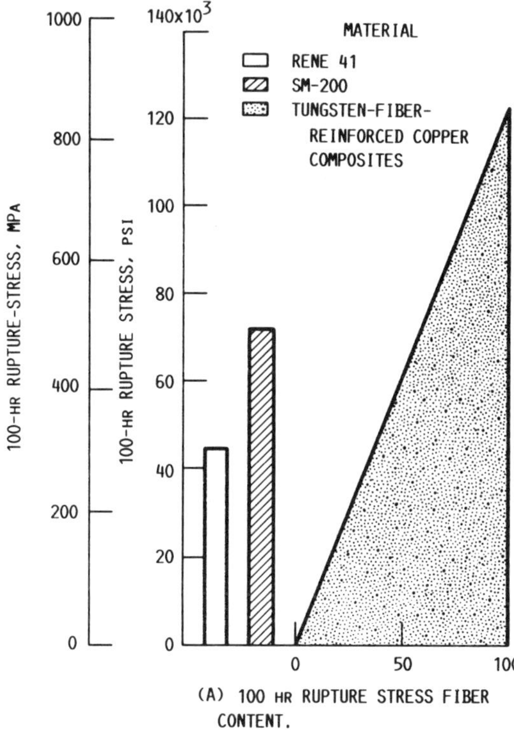

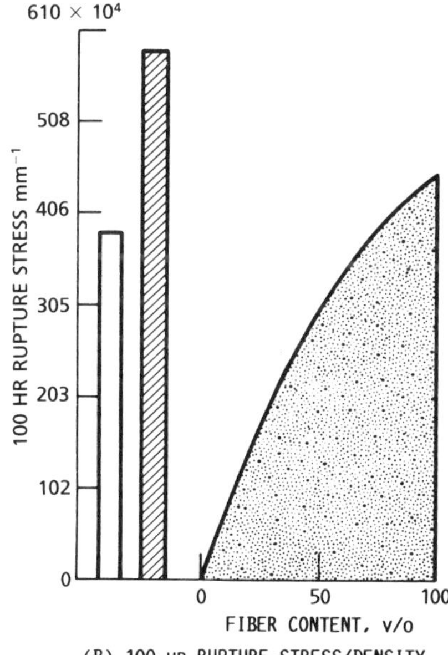

FIGURE 26 **Comparison of 100 hour rupture-stress and 100 hour rupture-stress/density ratio for tungsten fiber reinforced copper composites and superalloys at 1089 K.** (From Ref. 16.)

time tensile tests to about 30 for a 1000 hour life in stress-rupture service. The most important significance of the results reported in Ref. 18 deals with the joining of metal matrix composites for high temperature service. Extrapolation of the results to a higher strength 0.38 mm diameter W–2ThO$_2$ wire in a nickel alloy matrix indicates that

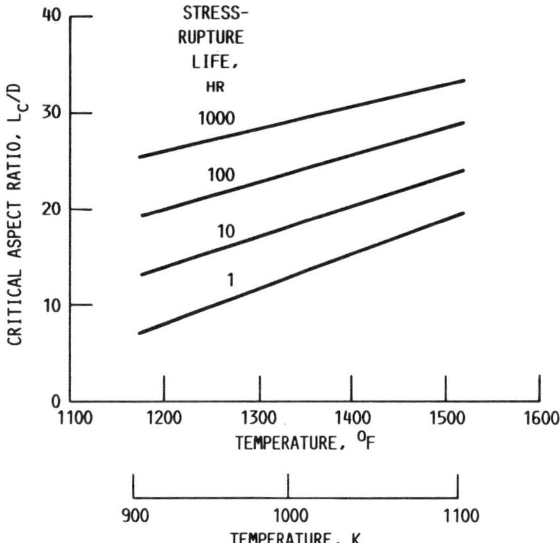

FIGURE 27 **Calculated critical aspect ratios of tungsten wire/copper pullout specimens in stress rupture at 1089 K.** (From Ref. 18.)

a fiber length of only 1.14 cm would be required to achieve the critical length for 1000 hour stress-rupture life at 1366 K. At the critical length, the reinforcement would only be 50% efficient, but increasing the fiber length to 11.5 cm would increase the efficiency to 95%. This means that composite panels could be joined with a simple scarf joint with an overlap of only 5.75 cm for a 95% efficient joint for 1000 hour stress-rupture life at 1366 K. The same efficiency could be obtained for short-time tensile applications at 1366 K with an overlap of 2.28 cm.

Shear stress-rupture pullout tests were also used to investigate the effect of alloy matrix composition on the critical aspect ratio [19]. Tungsten fiber pullout specimens were fabricated using copper alloys containing up to 10 wt % nickel. Nickel is soluble in tungsten, and nickel additions to copper matrices caused recrystallization of the tungsten fiber and reduction in strength and ductility of the composites. However, alloy additions of nickel would also increase the shear strength of the copper matrix, and reaction with the fiber should tend to increase the strength of the fiber–matrix interface. The results of these tests indicated that nickel additions to the copper matrix reduced the critical aspect ratio both in short-time tensile tests and in long-time shear stress-rupture tests (Fig. 28). This reduction in critical aspect ratio was caused by two factors: the weakening of the strength of the fiber and the increase in the shear strength of the matrix and the fiber–matrix interface.

Impact Behavior of Tungsten Fiber Reinforced Copper Composites

Materials for advanced gas turbine blades and vanes must have high stress-rupture strength, oxidation resis-

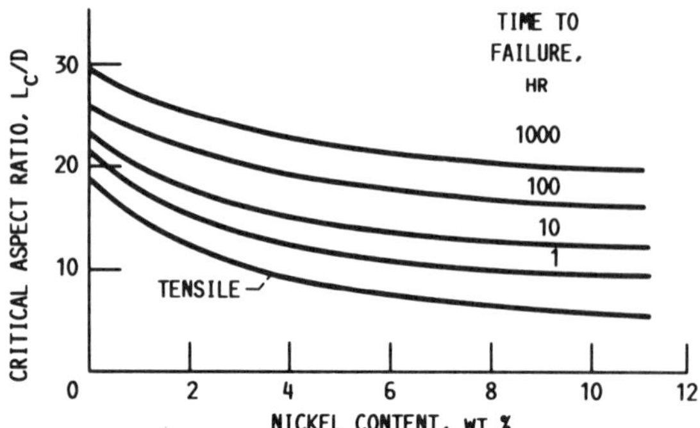

FIGURE 28 **Critical aspect ratio in tension and stress-rupture for tungsten wire–copper-nickel alloy pullout specimens tested at 1089 K.** (From Ref. 19.)

tance, and impact damage resistance. As part of the W/Cu composite model system program, pendulum impact tests were conducted on notched and unnotched miniature Izod specimens from room temperature to 810 K using a modified Bell Telephone Laboratory type miniature Izod testing machine [20]. Results of room temperature impact tests are shown in Figure 29. Unnotched specimens with less than 39 vol % fiber bent in a ductile manner and were forced out of the testing machine without breaking. At fiber contents higher than 39 vol %, the W/Cu composites broke on impact, with the impact strength decreasing with increasing fiber content. In tests at 420 K and 810 K, all specimens bent in a ductile manner while absorbing an impact value in excess of 12.88 J before being forced out of the testing machine. Notched miniature Izod specimens were also tested; they fractured at an impact strength below that of the unnotched specimens.

Subsequent studies were conducted to determine the effect of changing fiber or matrix toughness on the impact strength of W/Cu composites [21]. A set of specimens were fabricated at an infiltration temperature of

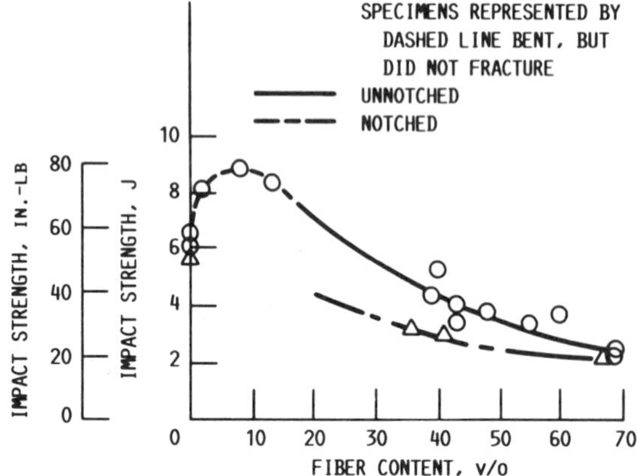

FIGURE 29 **Effect of fiber content on room temperature impact strength of tungsten fiber reinforced copper composites.** (From Ref. 20.)

1700 K, which caused partial recrystallization of the tungsten fibers. Partial recrystallization of the wires reduced the tensile strength and ductility of the composites. This thermal treatment reduced the impact strength of the composites at room temperature. However, the impact strength approached that of the unrecrystallized fiber composites at 810 K.

The impact behavior of the W/Cu composite is controlled by the relative toughness and ductility of the fiber and the matrix and by the volume fraction of each component. Because of their mutual insolubility, there is no fiber–matrix reaction, and the bond between the fiber and the matrix is primarily mechanical in nature. The impact strength of the composite is primarily a measure of energy absorbed during elastic-plastic deformation of the fibers and the matrix. The ductile-to-brittle transition temperature (DBTT) of as-drawn tungsten wires is around room temperature. W/Cu composites fabricated at 1478 K also had a DBTT around room temperature, while the partially recrystallized fiber W/Cu composites had a DBTT somewhat above room temperature. The unalloyed copper matrix is ductile at all testing temperatures, while the tungsten fibers are ductile above their DBTT. Both types of W/Cu composites were above the DBTT at elevated temperatures and showed ductile impact behavior [21].

Impact specimens with Cu–7Ni or Cu–10Ni alloy matrices showed a more brittle behavior [21]. Unreinforced copper–nickel alloys had lower impact strengths than unalloyed copper at room temperature. Nickel alloy additions to copper also cause recrystallization in the fiber and reduce the strength and ductility of the composite. The amount of property degradation is a function of the extent of the diffusion-triggered recrystallization in the fiber (Fig. 18). The impact test results indicated that the greater the depth of penetration of the diffusion-triggered penetration-recrystallization zone, the lower the impact strength and ductility of the copper–nickel alloy matrix composites (Fig. 30). At small reaction depths, the composites were brittle at room temperature, but were ductile at elevated temperatures. With larger reaction depths, the Cu–Ni matrix composites exhibited brittle behavior throughout their entire testing range up to 811 K and did not exhibit a transition to ductile behavior at a DBTT.

Electrical Conductivity of Tungsten Fiber Reinforced Copper Composites

Copper and tungsten are two of the best electrical conductive materials available. As part of the analysis of W/Cu composites program, composites covering a wide range of fiber contents were fabricated and tested to determine the effect of fiber content on room temperature electrical resistivity and conductivity [22].

Results of electrical resistivity and its reciprocal, conductivity, are plotted in Figure 31 as a function of fiber content. These results showed that the electrical resistivity was a hyperbolic function of fiber content and the electrical conductivity was a linear rule of mixtures type

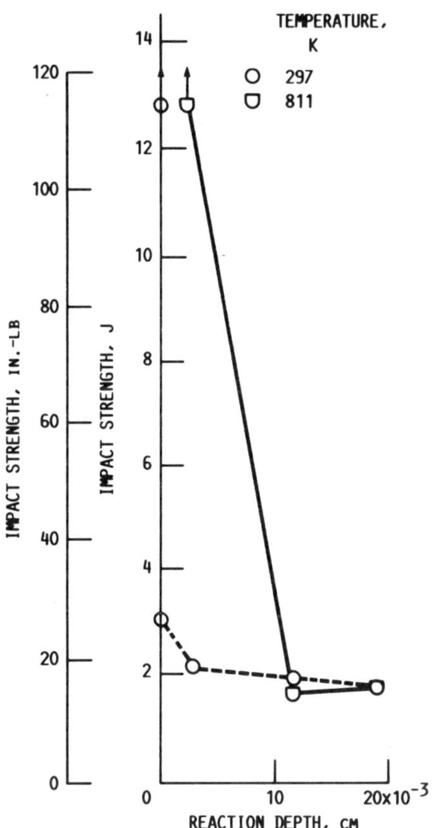

FIGURE 30 **Effect of reaction depth on impact strength of unnotched tungsten fiber reinforced copper-nickel alloy matrix composites (approx. 55 vol % tungsten fibers).** (After Ref. 21.)

function of fiber content. The electrical conductivity of the composites can be expressed as:

$$K_c = K_f V_f + K_m V_m \tag{11}$$

where K is the electrical conductivity and V is the volume fraction of fiber or matrix.

Because of the high strength and relatively high electrical conductivity of W/Cu composites, they have potential as practical high strength electrical conductors. Using a W/Cu composite electrical conductor as a structural member could produce significant potential weight savings in spacecraft applications. Since an application as a high-strength electrical conductor calls for a material that is a compromise between strength and conductivity, a comparison was made on the basis of the ratio of ultimate tensile strength to resistivity. The composite values are plotted over a range of fiber contents in Figure 32 and are compared to current conductors and high-strength electrical cables. The strength/resistivity ratio for W/Cu composites increased rapidly above a fiber content of about 10 vol %, reached a maximum at about 70 vol %, and then fell off to that of the tungsten wire. The ratio of strength to resistivity for W/Cu composites in the 50–75 vol % fiber range was about three to seven times that of the other conductors.

For applications in which density is also important, the materials were also compared on the basis of ultimate tensile strength/density to resistivity in Figure 33. The ratio increased with increasing fiber content to reach a maximum at about 50 vol % and then dropped to the value of tungsten wire. The plot shows that for W/Cu

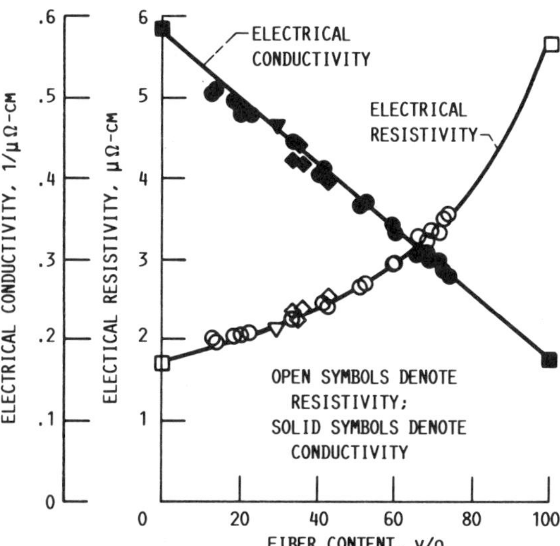

FIGURE 31 **Electrical resistivity and conductivity of tungsten fiber reinforced copper composites.** (From Ref. 22.)

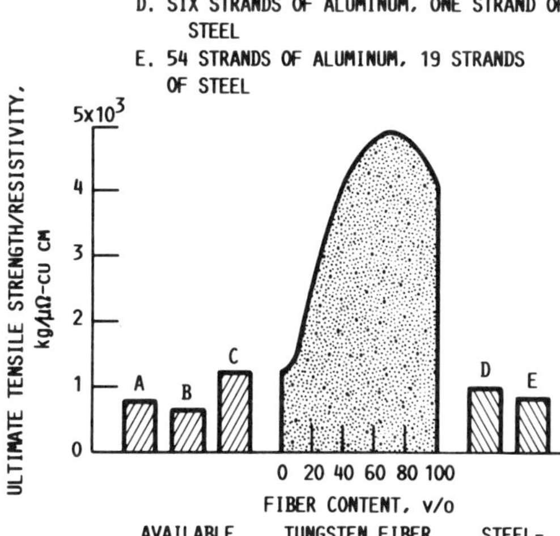

FIGURE 32 **Comparison of ratio of ultimate tensile strength to resistivity for tungsten fiber reinforced copper composites with that for other electrical conductors.** (From Ref. 22.)

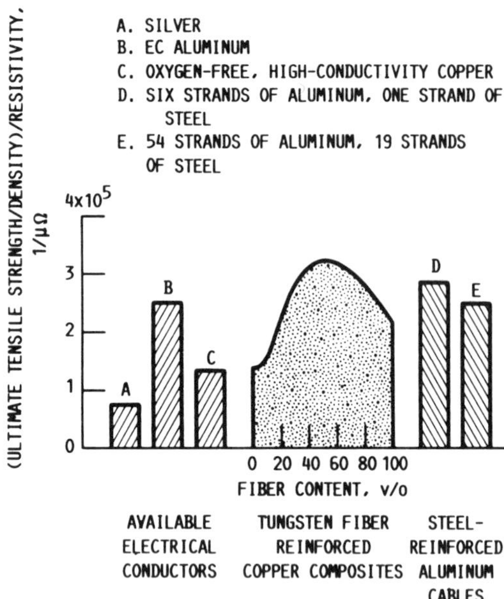

FIGURE 33 **Comparison of ratio of ultimate tensile strength/density to resistivity for tungsten fiber reinforced copper composites with that for other electrical conductors.** (From Ref. 22.)

composites with a fiber content of 50 to 60 vol %, the W/Cu composites are about 10–15% better than the best high strength electrical cable available, about 30% better than aluminum, and more than twice as good as copper.

These plots show that W/Cu composites have good potential as high strength, high electrical conductivity materials. A fiber content of 55–60 vol % would probably be the best compromise between electrical conductivity, strength, and strength/density. A further potential advantage is that because of their very high tensile and creep strengths at elevated temperatures, W/Cu composites could also be used at much higher temperatures than current electrical conductors. This would allow the composites to carry increased current amperages without structural damage, since conductor overheating would be a far less severe problem than with standard conductors.

Thermal Conductivity of Tungsten Fiber Reinforced Copper Composites

Because of the high temperature strength and good electrical conductivity shown by W/Cu composites, they also could have a good potential as high strength, high temperature thermal conductivity materials, since thermal conductivity is usually proportional to electrical conductivity. Efforts are currently underway to apply W/Cu composites to rocket thrust chamber liners to take advantage of their high temperature strength and thermal conductivity properties.

Advanced rocket engines, such as the space shuttle main engine, must be capable of repeated use. As a consequence, the combustion chamber walls undergo numerous thermal cycles with significant thermal strain. Currently these high pressure thrust chambers are life-limited as a result of the plastic strain levels encountered in the hot-gas-side wall during each thermal cycle. This high strain level is caused by the large hot-gas-side wall to outer-surface wall temperature difference that exists during the burn portion of the cycle. The large plastic strains from these repeated thermal cycles appear to cause thinning of the cooling passage wall at the centerline until the wall thins to a point where it can no longer sustain the high pressure load and finally fails in fatigue, as shown in Figure 34.

Based upon the high temperature properties of W/Cu composites, a preliminary fabrication demonstration and testing program was conducted to determine the feasibility of using W/Cu composites as rocket thrust chamber combustion liner materials. Materials for the composites were chosen on the basis of strength and thermal conductivity of W/Cu composites using 0.2 mm diameter 3D (W–3Re) fiber and a Cu–0.15Zr (Amzirc) matrix, which has about the same thermal conductivity as OFHC copper. Fiber contents were selected from design

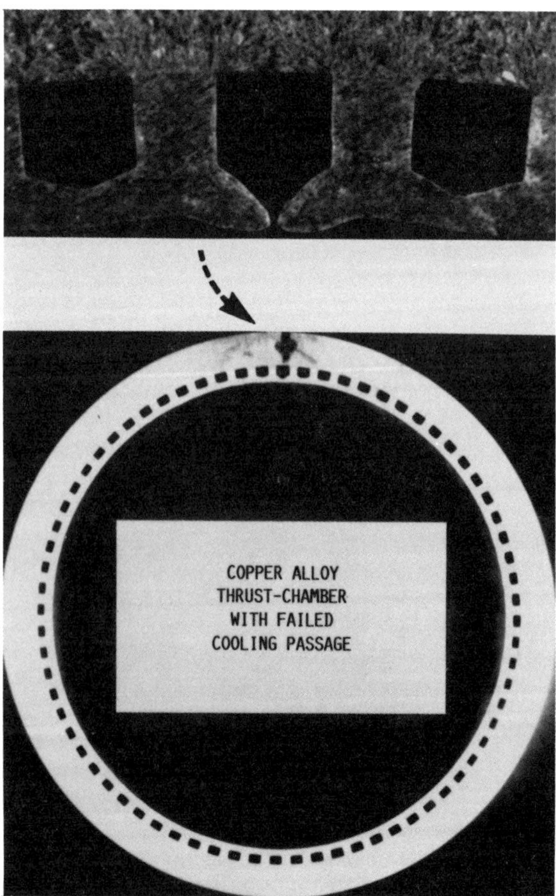

FIGURE 34 **Rocket thrust chamber with enlargement of fatigue failure in cooling passage wall.** (From Ref. 23.)

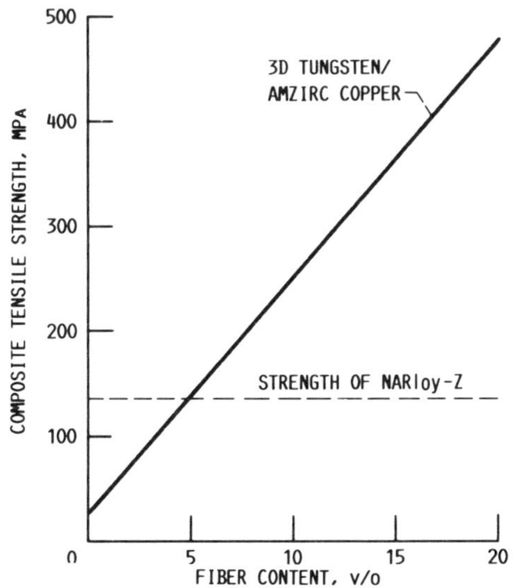

FIGURE 35 **Effect of fiber content on projected tensile strength of tungsten fiber reinforced copper alloy matrix composites at 866 K.** (From Ref. 23.)

projections of the strength and thermal conductivity of W/Cu composites for a minimum-weight configuration (Figs. 35 and 36). The goal was to get a strength improvement over the current liner material, NARloy–Z (Cu–3Ag–0.5Zr), with a minimal loss in thermal conductivity.

A 10 vol % fiber content was chosen for the composite to meet the strength, conductivity, and weight requirements. The composite tubes were 8.5 times stronger than similar unreinforced copper tubes in internal pressurization tests at 1228 K. In addition, the W/Cu composite tubes were thermal cycled from 339 K to 866 K and exhibited no thermal distortion after 2000 cycles. Several cylindrical combustion liner test specimens were fabricated for future evaluation in rocket engine test firings

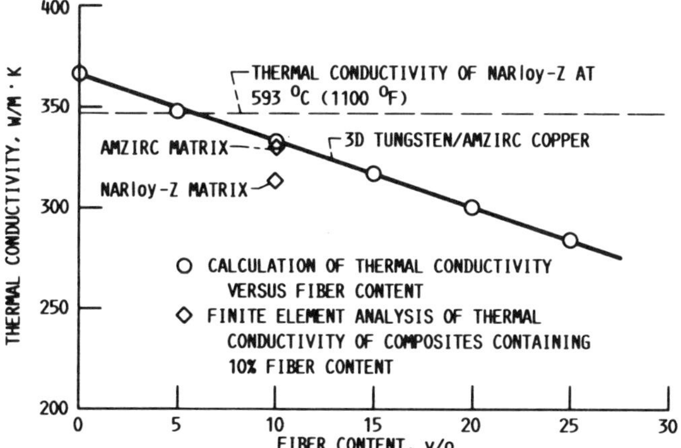

FIGURE 36 **Effect of fiber content on projected thermal conductivity of tungsten fiber reinforced copper alloy matrix composites at 866 K.** (From Ref. 23.)

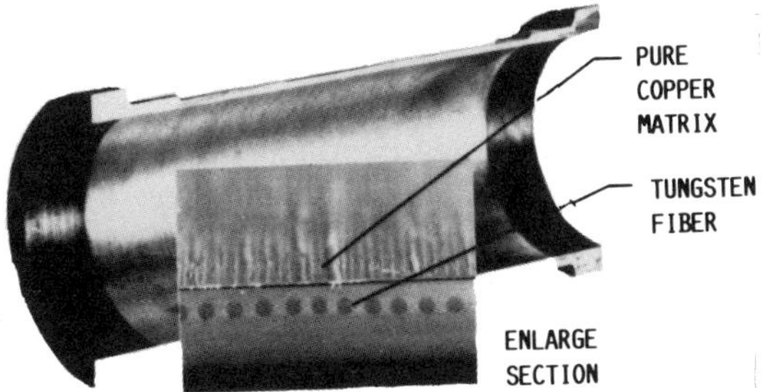

FIGURE 37 **Tungsten fiber reinforced copper cylindrical combustion liner test specimen.** (From Ref. 23.)

(Fig. 37). The final phase of the program involved making a combustion liner fabrication demonstration in which a one-third-size hourglass-shaped composite combustion liner was successfully fabricated.

Concluding Remarks

NASA Lewis Research Center has conducted a series of research programs to evaluate the properties of tungsten fiber reinforced copper matrix composites. The W/Cu composite system was chosen as a model system because of the mutual insolubility of tungsten and copper, which allowed the properties of the composites to be characterized and analyzed. Most of the effort was aimed at using pure copper matrices, but some copper alloys were also studied to give insight into diffusion/reaction kinetics. In the course of these studies, stress-strain behavior, tensile and creep behavior, and impact and conductivity behavior were analyzed.

The choice of the W/Cu composite system and the use of liquid infiltration fabrication allowed composites to be made over a wide range of fiber contents in this series of programs. This allowed a number of specimens to be made and tested so that properties could be analyzed in depth. Because of the mutual insolubility of the tungsten fibers and the copper matrix, the effect of the properties of each component could be evaluated independently. Because the tungsten fibers and the copper matrix form a strong bond at the fiber–matrix interface, the components of the composite always strain equally, with the actual strain being a stress–strain-related balance between the properties of the two components. Analysis of these stress–strain relations allowed the rule of mixtures predictions to be developed.

At the time of the initial W/Cu studies, metal matrix composites technology was in its infancy. The high strength W/Cu composites tested in these studies generated a lot of interest, and the rule of mixtures analyses first developed in these studies gave the initial impetus for the development of metal matrix composites. Since then, the rule of mixtures relations developed in these studies have been universally adopted as the criteria for measuring composite efficiency. While most of this work was aimed at analyzing W/Cu composites as a model system, the properties have sufficient potential that these composites are starting to be considered for applications where high strength, high conductivity materials are needed for specialized applications at elevated temperatures.

David L. McDanels

References

1. D. L. McDanels, T. T. Serafini, and J. A. DiCarlo, *J. Mater. Energy Systems, 8*(1), 80 (1986). (Also, NASA TM-87132.)
2. D. W. Petrasek and R. A. Signorelli, in M. M. Schwartz, Ed., *Fabrication of Composite Materials Source Book,* American Society for Metals, Metals Park, OH, 1985, pp. 14–81. (Also, NASA TM-82590.)
3. D. L. McDanels and R. A. Signorelli, "Improvement of High-Velocity Ballistic Impact Behavior of Boron/Aluminum Composites," NASA TM-83683, 1984.
4. R. W. Jech, D. L. McDanels, and J. W. Weeton, in *Composite Materials and Composite Structures,* Syracuse University Press, Syracuse, NY, 1959, pp. 116–139.
5. D. L. McDanels, R. W. Jech, and J. W. Weeton, *Met. Prog., 78*(6), 118 (1960).
6. D. L. McDanels, R. W. Jech, and J. W. Weeton, "Stress-Strain Behavior of Tungsten-Fiber-Reinforced Copper Composites," NASA TN D-1881, 1963.
7. D. L. McDanels, R. W. Jech, and J. W. Weeton, *Trans. Met. Soc. AIME, 233*(4), 636 (1965).
8. D. W. Petrasek, *Trans. Met. Soc. AIME, 236*(6), 887 (1966). (Also, NASA TN D-3073.)
9. D. W. Petrasek, R. A. Signorelli, and J. W. Weeton, in *Fiber Strengthened Metallic Composites,* ASTM STP-427, American Society for Testing and Materials, Philadelphia, 1967, pp. 149–175. (Also, NASA TN D-3886.)
10. A. Kelly and W. R. Tyson, in V. F. Zackay, Ed., *High-Strength Materials,* John Wiley & Sons, New York, 1965, pp. 578–602.
11. R. W. Jech and R. A. Signorelli, "The Effect of Interfiber Distance and Temperature on the Critical Aspect Ratio in Composites," NASA TM X-52347, 1967. (Also, NASA TN D-4548.)
12. D. W. Petrasek and J. W. Weeton, "Alloying Effect on Tensile Properties and Micro-Structure of Tungsten-Fiber-Reinforced Composites," NASA TN D-1568, 1963.
13. D. W. Petrasek and J. W. Weeton, *Trans. Met. Soc. AIME, 230*(5), 977 (1964).
14. R. W. Jech, R. H. Springborn, and D. L. McDanels, *Rev. Sci. Instrum., 35*(3), 314 (1964).
15. D. L. McDanels and R. A. Signorelli, *Met. Eng. Q., 6*(3), 51 (1966). (Also, NASA TN D-3467.)
16. D. L. McDanels, R. A. Signorelli, and J. W. Weeton, in *Fiber Strengthened Metallic Composites,* ASTM STP-427, American Society for Testing and Materials, Philadelphia, 1967, pp. 124–148. (Also, NASA TN D-4173.)
17. D. W. Petrasek, "High-Temperature Strength of Refractory-Metal Wires and Consideration for Composite Applications," NASA TN D-6881, 1972.
18. R. W. Jech, "Influence of Fiber Aspect Ratio on the Stress-Rupture Life of Discontinuous Fiber Composites," NASA TN D-5735, 1970.
19. R. W. Jech, "Critical Aspect Ratio for Tungsten Fibers in Copper-Nickel Matrix Composites," NASA TM X-3311, 1975.
20. E. A. Winsa and D. W. Petrasek, in *Composite Materials: Testing and Design,* ASTM STP-497, American Society for Testing and Materials, Philadelphia, 1972, pp. 350–362. (Also, NASA TM X-67810.)
21. E. A. Winsa and D. W. Petrasek, "Factors Affecting Miniature Izod Impact Strength of Tungsten-Fiber-Metal Matrix Composites," NASA TN D-7393, 1973.
22. D. L. McDanels, *Trans. ASM, 59,* 994 (1966). (Also, NASA TN D-3590.)
23. L. J. Westfall and D. W. Petrasek, "Fabrication and Preliminary Evaluation of Tungsten Fiber Reinforced Copper Composite Combustion Chamber Liners," NASA TM-100845, 1988.

Metal Matrix Composites, Tungsten Fiber Reinforced Superalloys*

Introduction

The key to the development of more efficient heat engines lies in the use of higher operating temperatures, which in turn relies on the development of higher temperature materials. Heat engine efficiency is proportional to maximum cycle operating temperature. Increases in these operating temperatures have been accomplished through improvements in the properties of nickel- and cobalt-based superalloys as well as by the incorporation of sophisticated schemes for the air cooling of turbine components. The gap between the highest temperature at which superalloys can be used for long-time service [870–950°C] and the gas inlet temperatures for which engines are being designed [1300–1600°C] is progressively widening. Thus, there is an increasing need to employ more air cooling in the turbine section. However, as increasing amounts of compressor discharge air are diverted to cool hot-section components, the efficiency of a gas turbine engine drops. This, coupled with the fact that nickel- and cobalt-based superalloys have been developed nearly to their technological limits, indicates that new types of material are necessary to further raise engine efficiency. The challenge for materials engineering is to develop a new generation of high temperature materials that are qualified for long-life application at material operating temperatures higher than 1000°C.

The need for improved materials at elevated temperatures has stimulated research in many areas, including efforts to develop fiber-reinforced superalloy matrix composites. A number of fibers have been studied for such use, including ceramic whiskers of submicrometer diameter, continuous length ceramic filaments, boron

* Reprint of a NASA publication. NASA Technical Memorandum 82590, January 1981.

filaments, carbon filaments, and refractory metal alloy wires. Attempts to attain high temperature strength with superalloy matrix composites using ceramic whiskers, continuous length ceramic filaments, boron filaments, or carbon filaments as the reinforcing fiber have been unsuccessful to date. Boron and carbon react and are virtually dissolved when incorporated into superalloy materials. Attempts to develop coatings to prevent such reactions have been unsuccessful. Ceramic whiskers and continuous length ceramic filaments also have suffered from degraded fiber strength because of fiber–matrix reaction. The fiber–matrix reaction causes fiber surface flaws, which act as local stress concentrations. These stress concentrations become nucleation sites for premature failure. The thermal expansion mismatch between ceramic fibers and typical candidate matrix alloys can induce severe stresses in composite materials. The combination of fiber surface roughening from reaction of the fiber with the matrix combined with thermally induced stress can cause catastrophic fracture of composite test specimens even during fabrication. Because of these factors, relatively low strength values have been achieved with such nonmetallic fibers compared with those theoretically possible. Given the high use temperature potential of such composites, however, research efforts to overcome the problems should and will continue.

The theoretical specific strength potential of refractory alloy fiber reinforced superalloys is less than that of ceramic fiber reinforced superalloys. However, the more ductile metal fiber systems are more tolerant of fiber–matrix reactions and thermal expansion mismatches. Also, the superalloy matrices can protect high strength refractory metal fibers from environmental attack. In laboratory tests, refractory fiber reinforced superalloy composites have demonstrated stress–rupture strengths significantly above those of the strongest superalloys. Tungsten fiber reinforced superalloy composites, in particular, are potentially useful as turbine blade materials because of their many desirable properties, including good stress–rupture and creep resistance, oxidation resistance, ductility, impact damage resistance, and microstructural stability. The potential of tungsten fiber reinforced superalloys (TFRS) has been recognized and has stimulated research to develop this material for use in heat engines.

This status review of TFRS research emphasizes the promising property data developed to date, the status of TFRS composite airfoil fabrication technology, and the areas requiring more attention to assure their applicability to hot-section components of aircraft gas turbine engines. We review refractory metal fiber and matrix alloy development first, then discuss fabrication techniques for TFRS and property results of importance to their use at high temperatures. Specific criteria for employing TFRS as a turbine blade material are presented next. From these criteria emerged a first-generation TFRS material, tungsten alloy fiber/FeCrAlY, which is currently under evaluation. The final section discusses property, design, fabrication, and fabrication cost data for this system.

Fiber Development

Refractory metal wires have received a great deal of attention as fiber reinforcement materials in spite of their poor oxidation resistance and high density. When used to reinforce a ductile and oxidation-resistant matrix, they are protected from oxidation and their specific strength is much higher than that of superalloys at elevated temperatures. The majority of the studies conducted on refractory wire–superalloy composites have used tungsten or molybdenum wire, available as lamp filament or thermocouple wire, as the reinforcement material. These refractory alloys were not designed for use in composites nor for optimum mechanical properties in the temperature range of interest for thermal engine application, namely 1000–1200°C. Lamp filament wire such as 218CS tungsten was most extensively used in early studies. The stress–rupture properties of 218CS tungsten wire were superior to those of rod and bulk forms of tungsten and showed promise for use as reinforcement of superalloys. The need for stronger wire was recognized, and high strength tungsten, tantalum, molybdenum, and niobium alloys for which rod and/or sheet fabrication procedures had already been developed were included in a wire fabrication and test program [1–4]. Table 1 gives the chemical compositions of these alloys. The approach just described precluded development of new alloys specifically designed for strength at the intended composite use temperatures. Table 2 summarizes the stress–rupture and tensile properties determined for the wires developed and compares them with commercially available wire (218CS, W-1 ThO_2, and W-3Re).

Excellent progress was made in providing wires with increased strength compared to the strongest wires available previously. The ultimate tensile strengths obtained for the wires at 1093 and 1204°C are shown in histogram form in Figure 1. Tungsten alloy wires were fabricated having tensile strengths 2.5 times that obtained for 218CS tungsten wire. The strongest wire fabricated, W-Re-Hf-C, had a tensile strength of 2165 MN/m^2 at 1093°C, which is more than 6 times as strong as the strongest nickel- or cobalt-based superalloy. The ultimate tensile strength values obtained for the tungsten alloy wires were much higher than those obtained for molybdenum, tantalum, or niobium alloy wire. When density is taken into account, the tungsten alloy wires show a decrease in advantage compared to tantalum, niobium, or molybdenum wire (Fig. 2). Still, the high strength tungsten alloy wires as well as molybdenum wires offered the most promise.

The elevated stress–rupture strength of such wire is more significant than the tensile strength, since the intended use of the material is for long-time applications. Figure 3 gives the 100-hour rupture strength at 1093 and 1204°C for the various wire materials and compares it with superalloys. The rupture strength of tungsten alloy fibers was increased by a factor of 3 at 1093°C from about 434 MN/m^2 for 218CS tungsten to 1413 MN/m^2 for W-Re-Hf-C wire. The tungsten alloy wire was superior in stress–rupture strength to the other refractory wire ma-

TABLE 1
Chemical Composition of Wire Materials

Material	Weight Percent of Component									
	W	Ta	Mo	Nb	Re	Ti	Zr	Hf	ThO_2	C
Tungsten alloys										
218CS	99.9	—	—	—	—	—	—	—	—	—
W-1ThO_2	bal	—	—	—	—	—	—	—	0.95	—
W-2ThO_2		—	—	—	—	—	—	—	1.6	—
W-3Re		—	—	—	2.79	—	—	—	—	—
W-5Re-2ThO_2		—	—	—	4.89	—	—	—	1.78	—
W-24Re-2ThO_2		—	—	—	22.54	—	—	—	1.7	—
W-Hf-C		—	—	—	—	—	—	0.37	—	0.030
W-Re-Hf-C		—	—	—	4.1	—	—	.38	—	.021
Tantalum alloys										
ASTAR 811C	8.2	bal	—	—	—	1.13	—	.91	—	.027
Molybdenum alloys										
TZM	—	—	bal	—	—	.45	0.085	—	—	.031
TZC	—	—	bal	—	—	1.18	.27	—	—	.12
Niobium alloys										
FS85	10.44	27.95	—	bal	—	—	.85	—	—	.031
AS30	20	—	—	bal	—	—	1	—	—	—
B88	28.3	—	—	bal	—	—	—	1.94	—	.58

From Refs. 3 and 4.

terials except for a tantalum alloy, ASTAR 811C, which was stronger than most of the tungsten alloy materials at 1093°C. The strongest tungsten alloy wire, W-Re-Hf-C, was more than 16 times as strong as superalloys at 1093°C. Figure 4 gives the ratio of the 100-hour rupture strength to density for refractory metal wires and superalloys. Again the stronger tungsten wire materials are superior to the other refractory metal wires. When den-

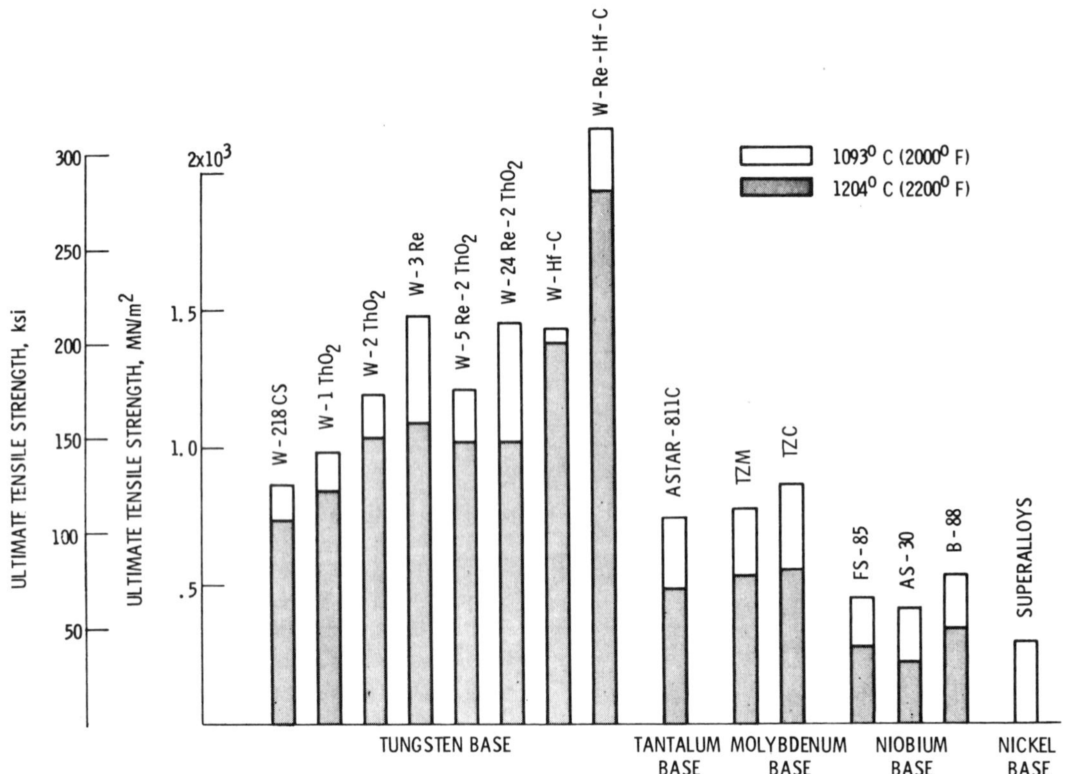

FIGURE 1 Ultimate tensile strength for refractory metal wires and superalloys [3,4].

TABLE 2
Representative Properties of Refractory-Alloy Wires

Alloys	Density, gm/cm^3	Wire Diameter, mm	Ultimate Tensile Strength MN/m^2	Stress for 100-hr Rupture MN/m^2	Stress/density for 100-hr Rupture, cm × 10^3
A. 1093°C Data					
Tungsten alloys					
218CS	19.1	.20	869	434	234
W-1ThO_2	19.1	.20	979	531	282
W-2ThO_2	18.9	.38	1193	655	356
W-3Re	19.4	.20	1475	476	249
W-5Re-2ThO_2	19.1	.20	1213	483	254
W-24Re-2ThO_2	19.4	.20	1455	345	183
W-Hf-C	19.4	.38	1427	1110	584
W-Re-Hf-C	19.4	.38	2165	1413	744
Tantalum alloys					
ASTAR 811C	16.9	.51	745	579	351
Molybednum alloys					
TZM	10.0	.38	779	290	295
TZC	10.0	.13	862	262	267
Niobium alloys					
FS85	10.5	.13	455	303	295
AS30	9.7	.13	421	214	224
B88	10.2	.51	531	331	328
B. 1204°C Data					
Tungsten alloys					
218CS	19.1	.20	745	317	170
W-1ThO_2	19.1	.20	841	372	198
W-2ThO_2	18.9	.38	1034	483	257
W-3Re	19.4	.20	1082	317	168
W-5Re-2ThO_2	19.1	.20	1020	303	160
W-24Re-2ThO_2	19.4	.20	1014	193	102
W-Hf-C	19.4	.38	1386	765	404
W-Re-Hf-C	19.4	.38	1937	910	480
Tantalum alloys					
ASTAR 811C	16.9	.51	490	262	157
Molybdenum alloys					
TZM	10.0	.20	531	131	135
TZC	10.0	.13	545	124	127
Niobium alloys					
FS85	10.5	.13	276	159	155
AS30	9.7	.13	228	—	—
B88	10.2	.51	345	193	190

From Refs. 3 and 4.

sity is taken into account, the strongest tungsten wire material, W-Re-Hf-C, is more than 7 times as strong as the strongest superalloys at 1093°C and more than 13 times as strong at 1204°C.

The processing schedules used to fabricate the newer high strength wires were not optimized to provide maximum strength at 1093 and 1204°C. Much more work is needed to maximize their properties. Considerable opportunity exists to develop wire processing schedules tailored for fiber–matrix composite use. The eventual application of TFRS composites will justify the added effort to further improve wire properties.

Matrix Alloy Development

The matrix is the exposed component of fiber-reinforced composites and therefore must be able to withstand high temperatures and an environment that can result in catastrophic oxidation and hot corrosion. The primary function of the matrix is to bind the fibers into a useful body and to protect the fibers from oxidation and hot corrosion. The matrix must be relatively ductile compared with the fibers to facilitate load transfer from the matrix to the fiber. It also must be capable of evenly redistributing local stress concentrations and resisting

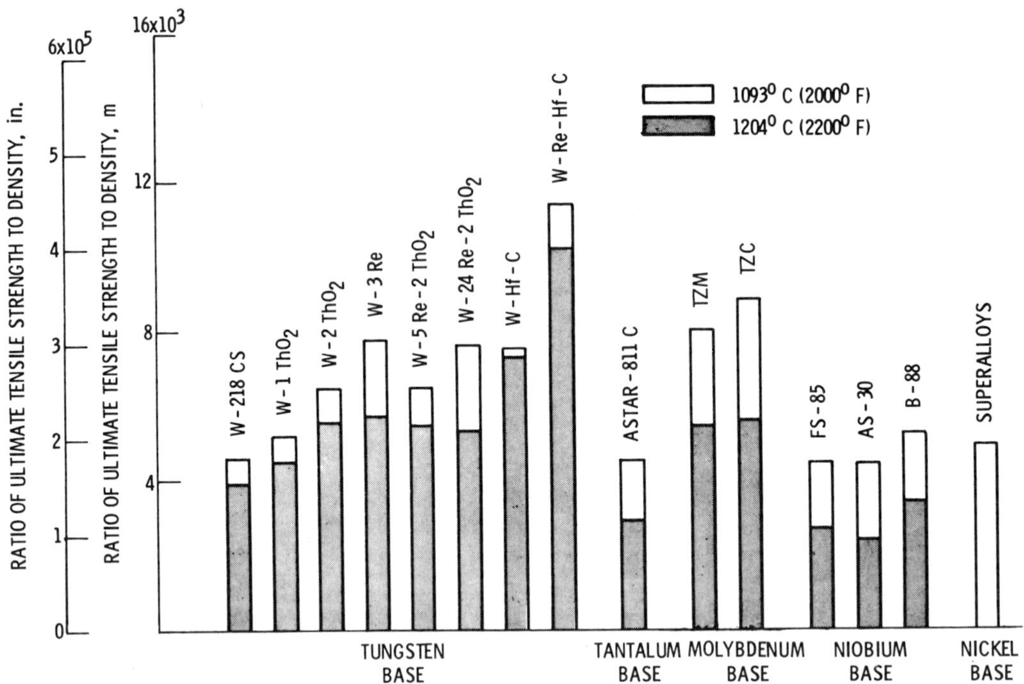

FIGURE 2 Ratio of ultimate tensile strength to density for refractory metal wires and superalloys [3,4].

abrasion and impact damage from foreign objects. The matrix and reinforcing fiber must be able to coexist without mutually induced disintegration, which can result from chemical interactions that can degrade both the fiber and matrix properties. The most important factor in the initial selection of matrix composition is the ability of the matrix to form a good bond with the fiber without the occurrence of excessive reaction, which could degrade the fibers properties.

For high temperature use, nickel-, cobalt-, and iron-based superalloys are preferred as the matrices for refractory metal fiber composites because they have demonstrated strength and ductility at elevated temperatures as well as good oxidation and hot corrosion resistance.

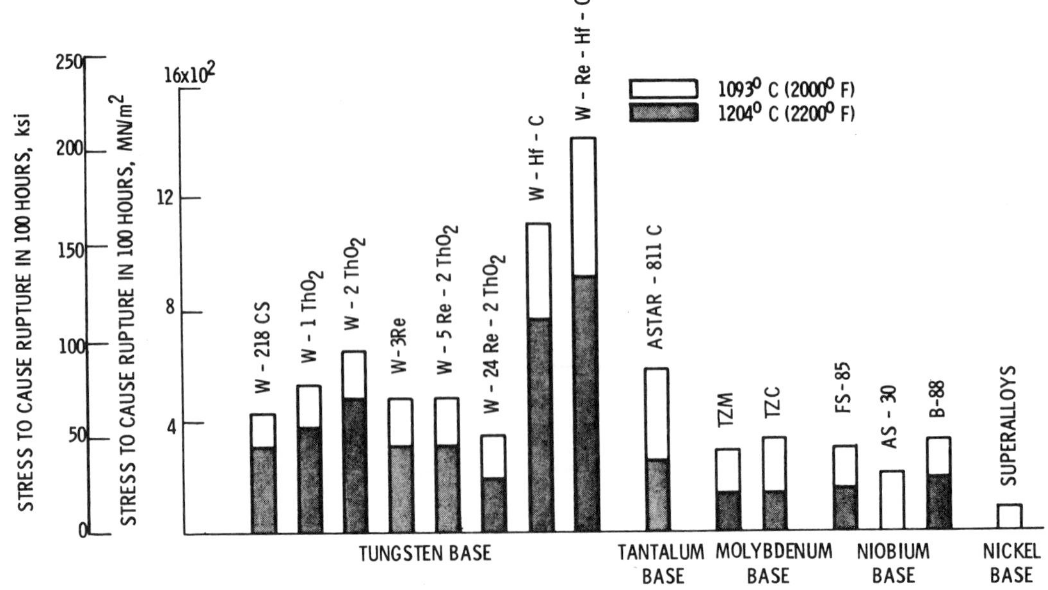

FIGURE 3 Stress to cause rupture in 100 hours for refractory metal wires and superalloys [3,4].

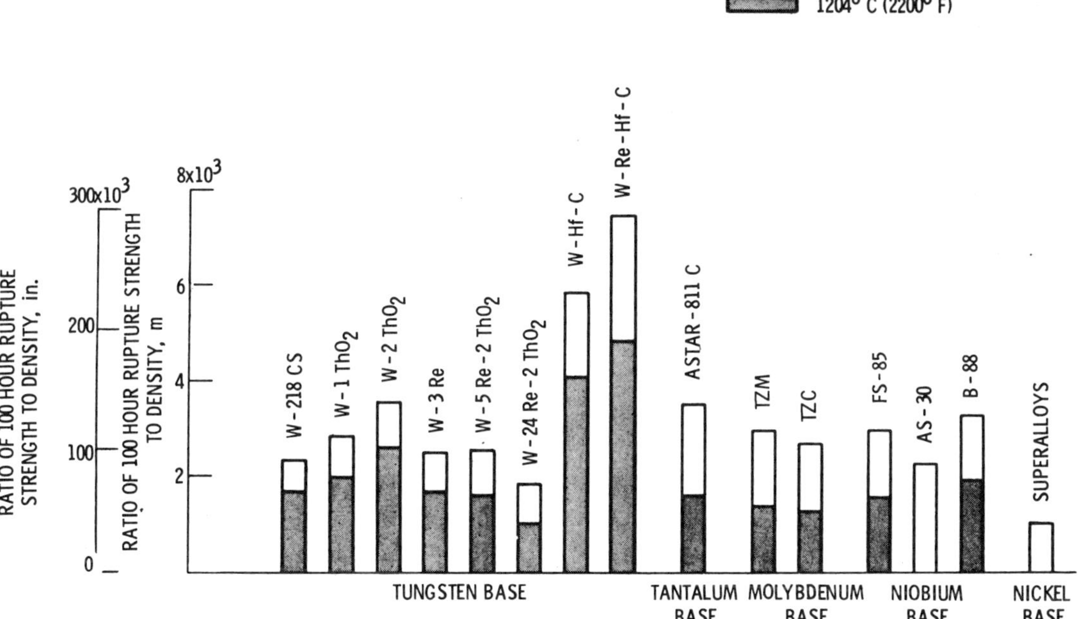

FIGURE 4 **Ratio of 100-hour rupture strength to density for refractory metal wires and superalloys [3,4].**

A large proportion of the research effort conducted on refractory fiber composites has been on fiber–matrix compatibility, with an effort being made to develop structurally stable composites by choosing a matrix composition that does not severely degrade the properties of the reinforcing fiber. One of the first systematic examinations to determine the effect of alloying reactions on the strength and microstructure of refractory metal fiber composites was reported in Ref. 5, where copper-based binary alloys were used as a matrix for tungsten fiber composites. The effects of alloying element additions to copper on the strength and microstructure of tungsten fiber composites were compared with mutually insoluble pure copper matrix composites exposed under the same conditions. The alloying elements studied were aluminum, chromium, cobalt, niobium, nickel, zirconium, and titanium. Data obtained for solute elements in this system can be related to the expected behavior of the same elements in superalloys. These effects served as the basis for modifying superalloy matrix composition to control the fiber–matrix reaction. Three types of fiber–matrix reaction were found to occur: (a) a diffusion–penetration reaction accompanied by a recrystallization of a peripheral zone of the tungsten fiber, (b) precipitation of a second phase with no accompanying recrystallization, and (c) a solid solution reaction with no accompanying recrystallization in the fiber. Peripheral recrystallization was caused by diffusion of cobalt, aluminum, or nickel into the tungsten wire. Compound formation occurred with titanium and zirconium. Chromium and niobium in copper formed a solid solution with tungsten with no accompanying recrystallization of the tungsten fiber. The greatest damage to composite properties occurred with the penetration–recrystallization reaction, while the two-phase and solid solution reactions caused relatively little damage. Figure 5 shows recrystallization of tungsten fibers in a matrix of copper + 10% nickel.

A number of studies have been conducted on nickel-induced recrystallization of tungsten fibers [6–8]. Recrystallization could be induced at low temperature by the presence of solid nickel on the surface of the tungsten wire. Once initiated, nickel-induced recrystallization required a continued source of nickel for propagation of the recrystallization front. Palladium, aluminum, manganese, platinum, and iron also greatly lowered the recrystallization temperature of tungsten [6]. Based on such findings, superalloy matrix compositions were developed that caused limited reaction with the fiber and minimal fiber property loss on work reported by Petrosek et al. [9] These superalloys contained high weight percentages of refractory metals to reduce diffusion penetration of nickel into tungsten. Additions of titanium and aluminum to the matrix were also made to form intermetallic compounds, which would further reduce the diffusion of nickel into tungsten. A typical alloy that was developed was Ni-25W-15Cr-2Al-2Ti. The fiber stress to cause rupture in 100 hours at 1090°C was reduced only 10% in the composite compared with the equivalent fiber rupture strength tested in a vacuum outside a composite.

The problems of obtaining structure-stable composite materials from the nickel–tungsten and nickel–molybdenum systems were further examined by the Soviets [10]. These experiments showed that in reinforced metal composite materials in which the matrix and fiber form

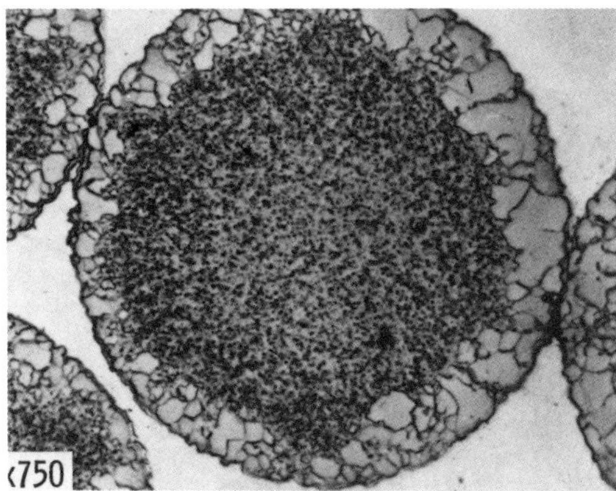

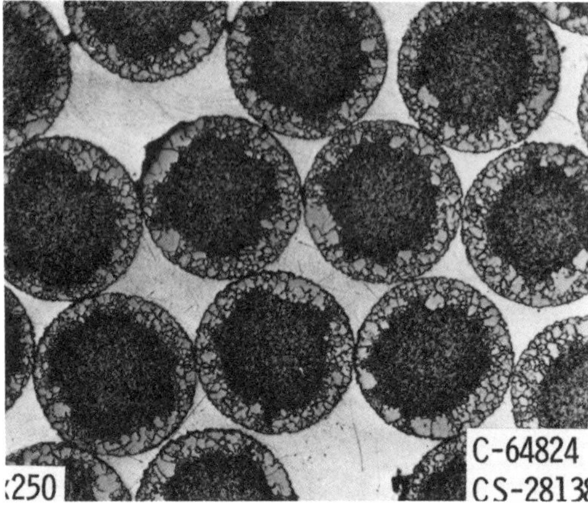

FIGURE 5 **Recrystallization of tungsten fibers in a copper plus 10% nickel matrix [9].**

restricted solid solutions in the absence of intermetallic compounds, fiber dissolution can be minimized by alloying the matrix with the fiber material up to a concentration that is close to the solubility limit. However, when the matrix and the fiber react to form intermetallic compounds, matrix saturation is not effective in controlling fiber attack by dissolution.

The effect of the composition of nickel and cobalt alloys on the structural stability of composite materials reinforced with tungsten fibers was determined by Klypin et al. [11]. The rate of interaction between the fiber and the matrix was determined from the extent of recrystallization of the fiber, the formation of intermediate phases at the interface, the solution of the fiber in the matrix, and the formation of diffusional porosity. In composites having a nickel matrix, a zone of accelerated recrystallization of the fiber was observed which increased with annealing time. The recrystallization rate of the fiber was highest with a matrix of pure nickel and somewhat lower with a nickel–chromium matrix. Alloying of nickel–chromium with iron, titanium, aluminum, and especially tungsten and niobium, slowed down the recrystallization process considerably. The stability of the fibrous structure of the tungsten fiber was highest in a matrix of cobalt or cobalt alloys. The accelerated recrystallization of tungsten fibers in a matrix of nickel–cobalt was due to the presence of nickel. Unfortunately, intensive solution of tungsten fibers occurred in a matrix of cobalt alloys and especially in pure cobalt.

Reaction of tungsten fibers with cobalt-based matrices was determined by Mirotvorskii et al. [12]. Fiber-reinforced binary and multicomponent cobalt-based alloys were found to have a strong propensity to intermetallic compound formation at the matrix–fiber interface. Four cobalt-based alloys were identified that did not react with tungsten fibers after exposure for 1 hour at 1300°C.

Work reported by Soviet investigators [13] on the reaction of tungsten fibers with iron-based matrices concluded that certain such alloys can be successfully employed as matrices for composite materials reinforced with fibers of tungsten and its alloys. The matrix composition can be chosen so that it forms no intermetallic compounds with the tungsten fibers and the fibrous structure of the drawn tungsten wire is not affected by heating to 1300°C. In most of the binary alloys investigated no recrystallization of the tungsten fibers occurred during annealing for 1 hour at 1300°C. Dissolution of the fibers in the matrices was detected only rarely. However, in the majority of the matrices studied, layers of intermetallic compounds formed at the matrix–fiber interface. In the case of binary iron matrices alloyed with 5 and 10% aluminum, no reactions were observed. After annealing for 1 hour at 1300°C no intermetallic compounds were formed in the binary iron matrices alloyed with 5–10% aluminum, 10% silicon, or 25% titanium, molybdenum, or nickel. The presence of nickel in multicomponent iron-based alloy matrix reinforced tungsten fiber composites accelerates the recrystallization of the fibers, while increasing the amount of chromium and zirconium inhibits recrystallization. Complex alloying therefore offers a means of suppressing fiber recrystallization and the formation of intermetallic compounds at the interface between tungsten fibers and the iron-based matrix. Studies conducted on the reaction between tungsten fibers and a matrix of Fe-24Cr-5Al-1Y [14] showed that only 10% of a 0.38 mm diameter fiber was reacted after exposure for 1000 hours at 1090°C.

Soviet investigators [15] further studied the reaction of tungsten fibers with binary alloys of iron, nickel, and cobalt alloyed with 5, 10, or 25% of one of the following elements: aluminum, copper, silicon, zirconium, niobium, molybdenum, or tungsten. The composite samples were annealed at 1200–1400°C for 1 hour. The interaction between fiber and matrix was investigated by metallographic analysis and by change in hardness. The results of this study for samples annealed at 1200 and 1300°C are summarized in Table 3, which shows the number of compositions investigated for each matrix system and the relative number of cases, based on a percentage, that resulted in the following reactions with the tungsten fiber: recrystallization, formation of an intermetallic compound, a diffusional penetration into the fiber, and no detectable recrystallization or reaction with the fiber. At 1200 and 1300°C contact between the tungsten fiber and the nickel binary alloys almost always induced recrystallization of the fiber. Nickel binary alloys containing 25% aluminum and 25% copper did not

TABLE 3
Comparison of Fiber-Matrix Reactions for Various Matrix Materials

Annealing Temperature °C	Matrix	No. Compositions Investigated	Relative No. of Cases, Percent: Recrystallization	Intermetallic Compound	Diffusion Penetration	No. Recrystallization	No. Rection
1200	Ni-base	27	93	55	—	7	4
	Co-base	29	10	83	12	90	10
	Fe-base	30	3	30	—	97	70
1300	Ni-base	27	96	63	—	4	4
	Co-base	19	21	84	—	79	10
	Fe-base	30	20	80	3	80	13

From Ref. 15.

cause recrystallization of the fibers. As Table 3 indicates, the interaction of the tungsten fibers with binary alloys of cobalt generally did not lead to recrystallization of the fibers. At all annealing temperatures, however, a layer of intermetallic compounds was formed at the fiber–matrix interface. This process was suppressed at 1200 and 1300°C only by alloying of the cobalt matrix with aluminum. Alloying of cobalt with nickel induced recrystallization of the fibers in addition to the formation of intermetallic compounds. The tungsten fiber was not recrystallized in most of the iron binary alloys studied. There was no interaction of any kind at 1200°C with alloys of iron and aluminum, silicon, titanium (10 and 25%), zirconium, niobium, chromium (10 and 25%), molybdenum, and tungsten. An intermetallic layer was formed in most of the iron alloys at the fiber–matrix interface at 1300°C; there was no interaction at this temperature only in the case of the matrix with 5–25% aluminum or 25% titanium.

The results of the preceding investigations show that selective alloying additions to nickel-, cobalt-, and iron-based matrix materials offer a means of suppressing fiber recrystallization and the formation of intermetallic compounds at the fiber–matrix interface. A number of matrix compositions have been identified, particularly for iron-based alloys, in which no detectable reaction occurs with tungsten fibers after short-time exposures at temperatures up to 1200°C.

The selection of matrix composition based on compatibility calls for a compromise between two opposing requirements. A strongly bonded interface must be formed between the matrix and the fiber to allow efficient stress transfer and to maintain continuity during heating–cooling cycles, and at the same time a destructive reaction must be prevented while at the operating temperature. The first requirement implies the initiation of a chemical reaction, while the second requirement involves the prevention of chemical reaction. A continuing reaction at the operating temperature would be acceptable if the rate of reaction were slow enough to give an adequate service life. Studies conducted on the rate of reaction between tungsten fibers and nickel-, cobalt-, and iron-based alloy matrix materials indicate that compositions can be obtained that satisfy these requirements.

Most of the matrix compositions investigated that resulted in minimum reaction with the fibers involved the formation of intermetallic compounds, which served to reduce interdiffusion. Thus, intermetallics might be sought as a naturally occurring diffusion barrier. Use of a suitable protective barrier between fiber and matrix offers the possibility of a wider range of composition selection for composites for high temperature application. However, as pointed out by Morris and Burwood-Smith [16], the introduction of a second interface and a deposited coating, whose possible breakdown in service at high temperatures would cause a catastrophic decrease in strength, is not an attractive proposition to engine manufacturers and operators. Signorelli [17] reported that although diffusion barrier coatings on reinforcing wire are a potentially effective way to achieve control of fiber–matrix interaction, techniques attempted to date have not resulted in reproducible, successful barrier coatings for refractory alloy wire. Optimism continues, however, that such natural or deposited coatings are possible and will offer increases in both strength and use temperature. Tradeoffs in compound composition and ductility offer a fruitful area for continued studies.

Sufficient evidence has been accumulated on matrix–fiber reactions to project that matrix alloy compositions with chemical compatibility and adequate bonding can be obtained with iron-, nickel-, and cobalt-based alloy systems. However, as discussed later, several additional requirements must be satisfied in selecting a matrix alloy; these include ductility, thermal expansion compatibility, creep strength, strain hardening rate, oxidation and corrosion resistance, and ease of fabrication. The basis for selecting matrix compositions to fill the host of requirements is in an early state of development. The approach taken in the NASA–Lewis program has been to select alloys with chemical compatibility to control fiber–matrix reactions and with ductility and low strain hardening to resist cracking from thermally induced cyclic stresses. Other requirements have been given consideration but were subordinated to those two needs.

Variation of matrix composition within the composite

geometry has also been used to better satisfy the variation in requirements within a component. For example, a higher creep strength material with less thermal strain and lower oxidation resistance at moderate temperatures can be used in the base portion of a blade because it is a better match to fill the requirements in this portion of the component.

Matrix alloy selection is a complex activity, and much more experience is needed before guidelines can be established. Matrix–fiber reaction is one requirement that can be met, but the challenge is to satisfy simultaneously the many additional needs.

Composite Fabrication

The fabrication of matrix and fibers into a composite with useful properties is one of the most difficult tasks in developing refractory wire reinforced superalloys. Fabrication methods for refractory wire–superalloy composites must be considered to be in the laboratory phase of development. Production techniques for fabrication of large numbers of specimens for extensive property characterizations are not yet available.

Fabrication methods can be classified as either solid phase or liquid phase depending on the condition of the matrix phase during its penetration into a fibrous bundle. In liquid phase methods the molten matrix is cast using investment casting techniques so that the matrix infiltrates the bundle of fibers in the form of parallel stacks or mats. The molten metal must wet the fibers, form a chemical bond, and yet be controlled to prevent the degradation of the fibers by dissolution, reaction, or recrystallization. This control is difficult at very high temperatures, since liquid metal temperatures of 1500°C are involved and contact time at temperature must be less than 1 minute [18]. Larger fiber diameters have been used to increase the size of the unreacted wire core, and matrix alloy compositions have been selected to reduce solute diffusion into the fibers.

Liquid phase methods are particularly suited to the preparation of fiber-reinforced superalloys for gas turbine application because casting is the basic fabrication technique presently used for turbine blades and vanes. The technology developed for casting cooled blades via shell molds containing silica inserts could be applied directly to the fabrication of a reinforced blade [16]. Experience has shown, however, that the use of fibers having a diameter less than 0.075 cm in composites prepared by liquid metal infiltration leads to displacement of the wires. This is explained by the lack of rigidity of the unsupported span of the wire, permitting surface tension to pull the fibers into a tight bundle near the midspan [16]. Thus, control of alignment and spatial distribution of fibers at low volume fractions is difficult.

A similar technique was attempted [19] to cast tungsten fiber reinforced cobalt-based alloy test specimens. It was reported, however, that the matrix composition inside the fiber bundle became richer in tungsten than the matrix surrounding the fiber bundle, and there appeared to be separation of the filament bundle from the matrix. An alternate means of using liquid state technology is to coat the wire with the matrix by passing it as a single strand or small bundles rapidly through a molten bath of matrix. The coated wires could then be diffusion bonded in closed dies to form the composite component.

Solid phase processing requires diffusion, which is time and temperature dependent. Solid phase processing temperatures are much lower than liquid phase processing temperatures; diffusion rates are much lower, and reaction with the fiber can be less severe. The prerequisite for solid state processing is that the matrix be in either sheet, foil, or powder form. Hot pressing or cold pressing followed by sintering is used for consolidation of the matrix and fiber into a composite component.

Use of matrix materials in the form of sheet or foil involves placing the reinforcing fibers between layers of the matrix sheet or foil, which are then pressed together. They may be hot pressed or alternately cold pressed followed by diffusion bonding. An example of this type of processing is reported by Karpinos et al. [20]. One of the most promising methods of manufacture of composite sheet materials is that of vaccuum hot rolling, which gives high productivity and enables large sheets to be manufactured.

Processing parameters for the manufacture of composite sheet material by vacuum hot rolling have been reported [21]. Several layers of tungsten fibers were placed in alternate layers between matrix alloy sheets; the pack was wrapped in nickel foil, covered with thin mica sheets, and placed in a carbon steel sheath, which was sealed by gas welding. The resultant pack was evacuated, heated, and rolled. On the basis of the results obtained, the authors claim that it is possible to select rolling conditions ensuring maximum high temperature strength in the resultant material, adequate adhesion between fibers and matrix, freedom from porosity, and strong welding between matrix sheets in the composite. A disadvantage of the use of thin sheet or foil is that most high strength superalloys are not available in this form, although the problem stems from economic considerations rather than technical limitations.

The powder metallurgy approach is one of the most versatile methods for producing refractory fiber–superalloy composites and has yielded some excellent results. Almost all alloy metals can be produced in powder form. However, the large surface area of the fine powders is easily contaminated and introduces impurities that must be removed. High capital cost equipment is necessary to apply pressure and temperature in an inert atmosphere. Most powder fabrication techniques limit fiber content to 40–50 vol %. Despite these disadvantages, powder processing has been used to achieve control of matrix–fiber reactions and has resulted in excellent composite properties.

Slip casting of metal alloy powders around bundles of fibers followed by sintering and hot pressing was developed for the solid state fabrication of refractory fiber–superalloy composites [9]. The slip-cast slurries consisted of a mixture of powders and an organic gel in water. Slurries were converted in a vibratory mold to a solid "green" composite, which was sintered using a heating schedule designed to drive off the organic binder material and sinter the powders and fiber together without oxidizing either. The sintered composite was then isostatically hot pressed to full density. This method is capable of achieving good matrix consolidation and bonding between fiber and matrix without excursions into the liquid metal region, which would greatly increase fiber–matrix reactions. Although the slipcasting + sintering + hot pressing technique has demonstrated excellent success for uniaxially reinforced specimens, it is not regarded as an ideal method for component fabrication because most applications require some cross-ply fiber orientation, which is not easily accomplished with slipcasting. Control of fiber distribution was also found to be a problem.

Brentnall and Toth [22] developed a fabrication procedure using solid phase processing in which fiber distribution, alignment, and fiber–matrix reaction could be accurately controlled. Matrix alloy powders were blended with a small quantity of organic binder (Teflon) and warm rolled into high density sheets. During rolling, the Teflon formed an interlocking network of fibers that held the powder particles together. Fiber mats were made by winding the fibers on a drum and then spraying them with a binder. The fiber array was cut from the drum and flattened to form a fiber mat. Precollimated fibers in mat form were sandwiched between layers of matrix powder sheet, and the matrix was densified and extruded between fibers by hot pressing. Fiber–matrix and matrix–matrix metallurgical bonding was achieved while preserving uniform fiber distribution and eliminating any voids. This procedure resulted in the fabrication of a single layer of fibers contained in the matrix material, which was termed a monotape. Monotapes can be cut into any shape desired with any orientation of fiber desired and subsequently stacked up and hot pressed into a desired component. This approach appears to be the most viable method currently available for the fabrication of tungsten fiber reinforced superalloy composite components for use in heat engines.

Composite Properties

The principal reason for most of the work on refractory fiber–superalloy composites has been to produce a material capable of operation as highly stressed components such as turbine blades in advanced aircraft and industrial gas turbine engines at temperatures of 1100–1200°C or higher. Such an increase in temperature above the current limit of about 1010°C for superalloys would permit higher turbine inlet temperatures and markedly decreased cooling, thus improving engine performance and efficiency. An increase in blade temperature of 50°C over current limits would be a significant improvement [18]. A review of gas turbine blade material property requirements [23–26] indicates creep resistance, stress-rupture strength, low cycle fatigue, thermal fatigue resistance, impact strength, and oxidation resistance as properties of primary concern for turbine blade applications. The section that follows reviews the results obtained for

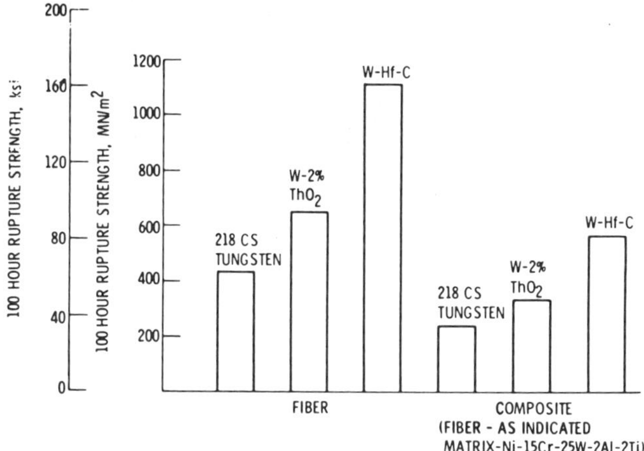

FIGURE 6 **Comparison of 100-hour rupture strength at 1093°C for fibers and 70 vol % fiber composites.**

refractory fiber–superalloy composites for these property areas.

Stress–Rupture Strength

At temperatures of 1100°C and above, a superalloy matrix contributes very little to the rupture strength of the composite compared with the contribution of the refractory fibers. Fiber stress-rupture strength, volume fraction of fiber, and the degree of fiber–matrix reaction all control the stress-rupture strength of the composite. Figure 6 compares the 100-hour rupture strength at 1093°C for various fibers and composites containing 70 vol % of these fibers, from NASA studies [9,27,28]. The matrix composition (Ni-15Cr-25W-2Al-2Ti) was the same for all the composites, and as indicated in the hostogram, the stronger the fiber, the greater the stress-rupture strength of the composite. The effect of fiber content on the stress-rupture strength of a composite is shown in Figure 7 [29]. Stress-rupture strength increases linearly as the fiber content increases.

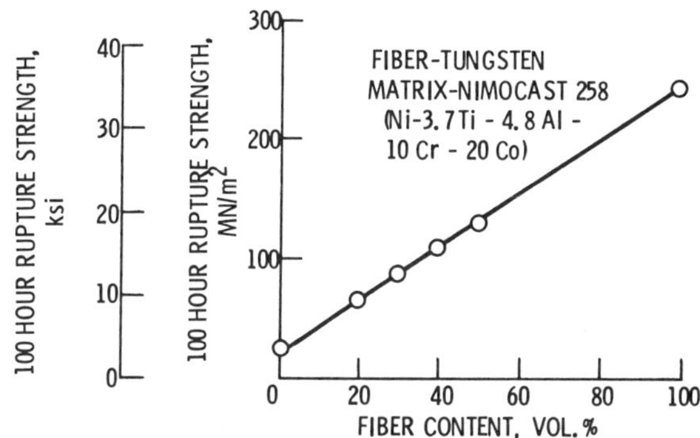

FIGURE 7 **Effect of fiber content on 100-hour composite rupture strength at 1100°C [29].**

Table 4 and Figure 8 offer comparisons of the 100-hour rupture strength at 1093°C (2000°F) for some of the composite systems that have been investigated [9,14,19,27–32]. Where possible, comparisons were made for composites containing 40 vol % fiber. It should be noted that higher values would be obtained for these composite systems when the fiber content was increased. Also shown in Figure 8 are the values for the 100-hour rupture strength for unreinforced alloys and for the strongest commercially available superalloys. The 100-hour stress-rupture strength of all the alloys investigated was substantially increased by the addition of tungsten fibers. All the 40 vol % fiber composites had a 100-hour rupture strength greater than that for the strongest commercially available superalloys. The W-Hf-C fiber composite system is the strongest composite system obtained to date. A 40 vol % W-Hf-C fiber content superalloy composite is more than 3.5 times as strong in rupture for 100 hours at 1100°C as the strongest commercially available superalloys. The composite containing a larger amount of fiber reinforcement (56 vol % W-1% ThO_2 wire) in FeCrAlY also had an impressive stress-rupture strength, more than 2.5 times that for the strongest commercially available superalloys.

The density of these composite materials is greater than that of superalloys, and this factor must be taken into consideration. The stresses in turbine blades, for example, are a result of centrifugal loading; therefore, the density of the material is important. A comparison of the specific strength properties of composites and superalloys is therefore significant. Figure 9 compares the ratio of the 1100°C 100-hour rupture strength to density for composites and superalloys. Even when density is taken into account, the stronger composites are much superior to the strongest commercially available superalloys. The composite containing 40 vol % W-Hf-C wire is almost 2.5 times as strong as the strongest superalloys.

The comparison of stress-rupture strength between composites and superalloys is even more favorable for the composite when long application times are involved. Figure 10 plots stress to rupture versus time to rupture for three different fiber compositions, each having the

TABLE 4
Rupture Strengths and Compositions for Composites and Superalloys

A. 100 hr Rupture Strength at 1100°C for Composites and Superalloys

Ref.	Alloy	Wire	Wire Diameter mm	Vol. %	Density gm/cc	100 hr Rupture Strength MN/m²	Stress-Density for 100 hr Rupture m
30	ZhS6	VRN tungsten	0.3–0.5	40	12.5	138	1125
29	EPD-16	—	—	—	8.3	51	635
29		tungsten	0.25	40	12.7	131	1040
31	Nimocast 713C	—	—	—	8.0	48	613
31		tungsten	1.27	20	10.3	93	927
19	MARM322E	—	—	—	—	48	—
19		W-2% ThO_2	0.08	40	—	207	—
9	Ni,Cr,W,Al,Ti	—	—	—	9.15	23	254
9		218CS (tungsten)	0.38	40	13.3	138	1058
27		W-2% ThO_2	0.38	40	13.0	193	1513
28		W-Hf-C	0.38	40	13.3	324	2491
14	FeCrAlY	W-1% ThO_2	0.38	56	12.5	831 hr rupture strength- 242 MN/m² (35 ksi)	1957
32	FeCrAlY	W-Hf-C	0.38	35	11.3	242	2147

B. Nominal Composition of Matrix Alloys (Weight %)

Alloy	Composition
ZhS6	Ni-12.5Cr-4.8Mo-7W-2.5Ti-5Al
EPD-16	Ni-6Al-6Cr-2Mo-11W-1.5Nb
Nimocast 713C	Ni-12.5Cr-6Al-1Ti-4Mo-2Nb-2.5Fe
MARM322E	Co-21.5Cr-25W-1ONi-0.8Ti-3.5Ta
Ni,Cr,W,Ti,Al	Ni-15Cr-25W-2Ti-2Al
FeCrAlY	Fe-24Cr-5Al-1Y

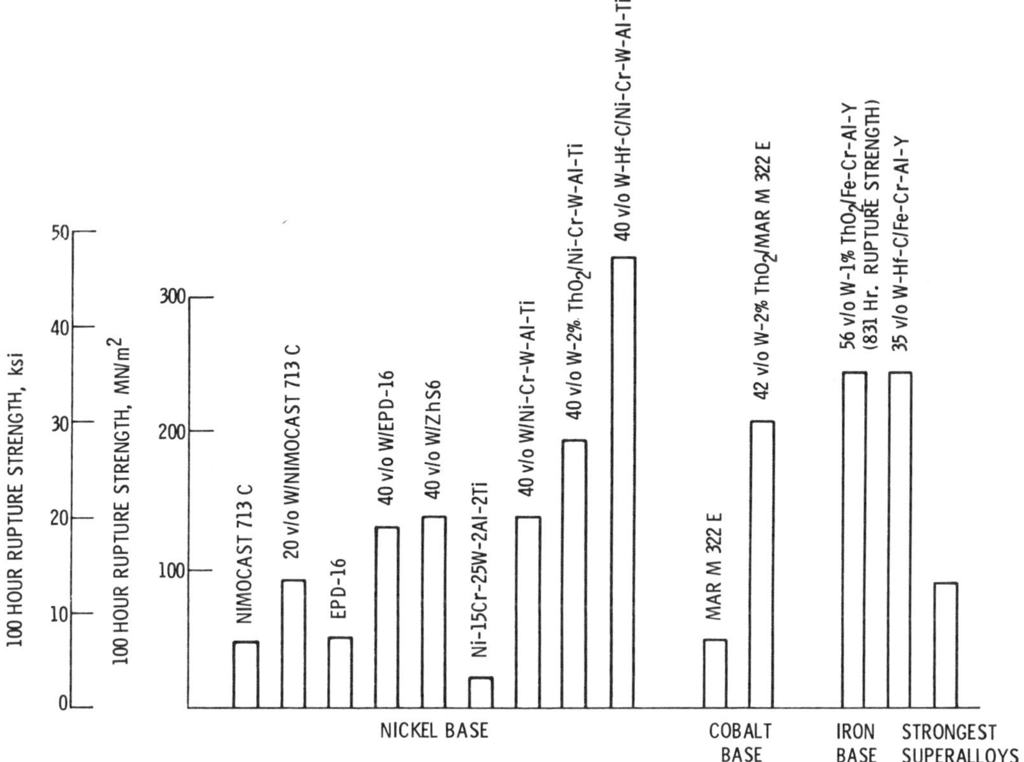

FIGURE 8 **Comparison of 100-hour rupture strength at 1093°C for composites and superalloys.**

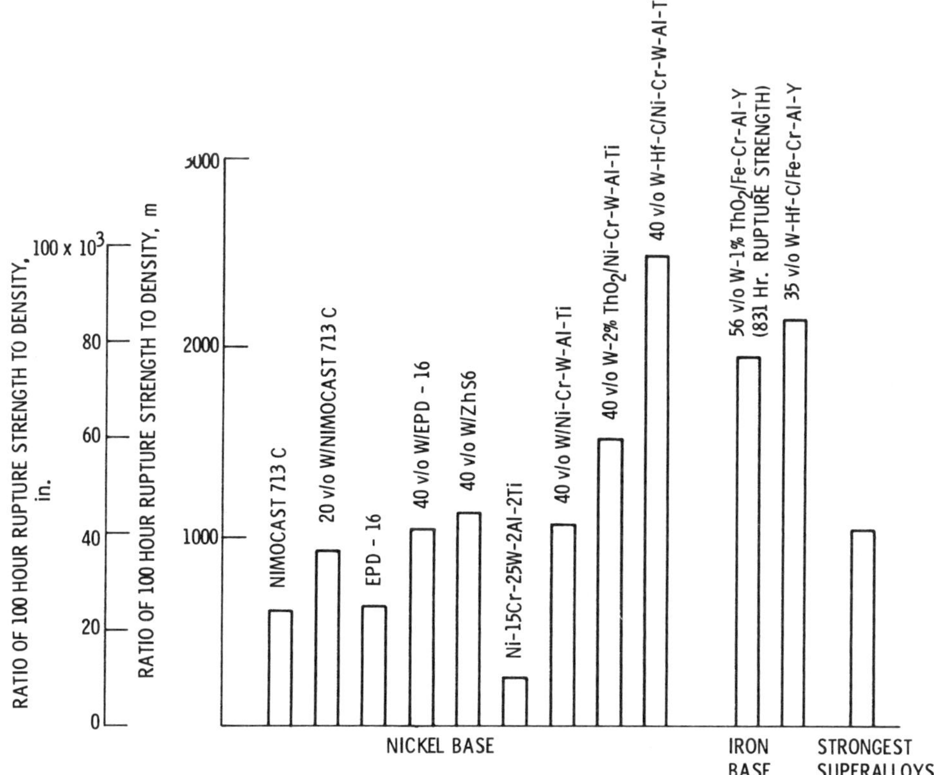

FIGURE 9 **Comparison of the ratio of 100-hour rupture strength to density for composites and superalloys at 1093°C.**

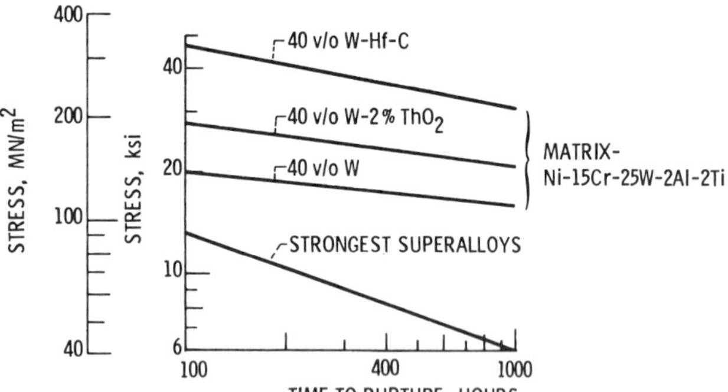

FIGURE 10 **Stress versus time to rupture comparison for composites and superalloys at 1093°C.**

same matrix material, versus the strongest superalloys. All the fiber composite systems are stronger relative to superalloys for rupture in 1000 hours than for rupture in 100 hours at 1093°C. The ratio of stress (to cause rupture) to density versus time to rupture is plotted in Figure 11. The specific stress-rupture strength advantage for the composite also increases with time to rupture. The 40 vol % tungsten fiber composite, for example, has about the same specific (density-corrected) strength for rupture in 100 hours as superalloys but is almost twice as strong as superalloys for rupture in 1000 hours. For currently required blade lives of 5000–10,000 hours, this advantage becomes even greater.

Figure 12 compares the range of values for the 100-hour rupture strength for tungsten fiber reinforced superalloy composites tested at 1093°C (2000°F) with the range for the stronger cast superalloys as a function of temperature. The strongest TFRS composite has the same rupture strength at 1093°C as the strongest superalloy at 915°C. This represents a material use temperature advantage for the composite of 145°C versus the strongest superalloy. Figure 13 shows the density-corrected values for rupture in 100 hours as a function of temperature. When density is taken into consideration, the composite has a material use temperature advantage of 110°C over the strongest superalloys.

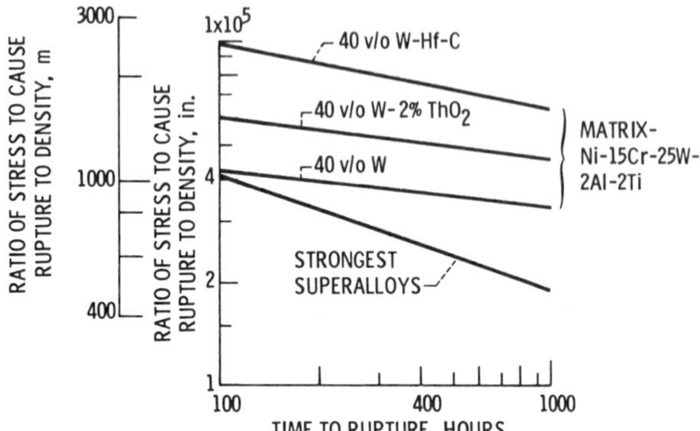

FIGURE 11 **Sress (to cause rupture) to density ratio for composites and superalloys at 1093°C.**

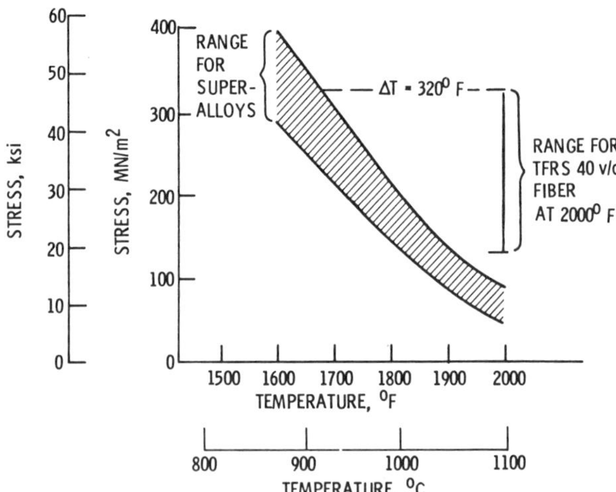

FIGURE 12 **Comparison of 100-hour rupture strength for TFRS and superalloys.**

Creep Resistance

The creep-rupture properties of Nimocast 713C reinforced with tungsten or tungsten-5% rhenium wire were evaluated and compared with the data determined for vacuum-cast Nimocast 713C by Morris and Burwood-Smith [31]. Typical composite creep curves are shown in Figure 14, together with a comparative curve for the unreinforced matrix. The creep curves for both materials exhibit the three characteristic stages of creep associated with conventional materials. Essentially, reinforcement reduces the second-stage minimum creep rate markedly for a given applied stress; this effect is due to the presence of the more creep-resistant fibers. The reduction in minimum creep rates observed on reinforcing Nimocast 713C suggests that the stronger, more creep-resistant component, the fiber, controls the creep behavior. The lack of evidence of creep deformation in the matrix of the

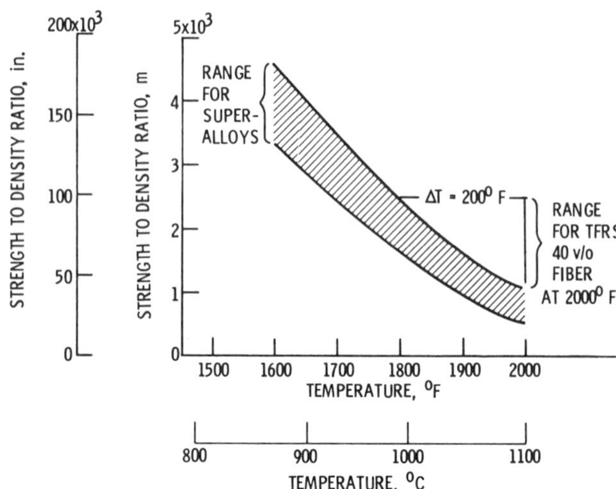

FIGURE 13 **Comparison of the ratio of 100-hour rupture strength to density for TFRS and superalloys.**

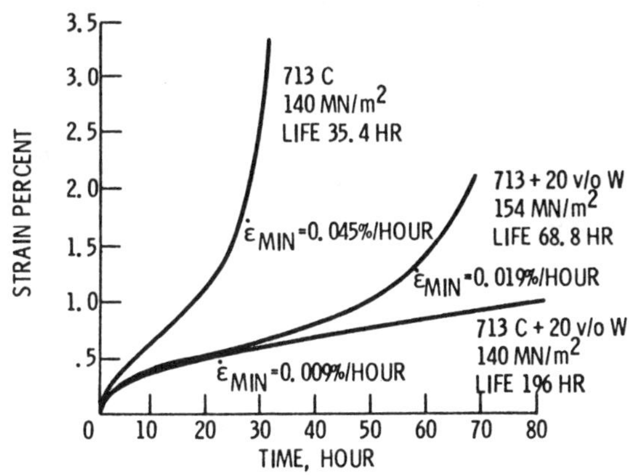

FIGURE 14 **Comparison of typical creep behavior of Nimocast 713C with and without tungsten reinforcement at 1100°C [31].**

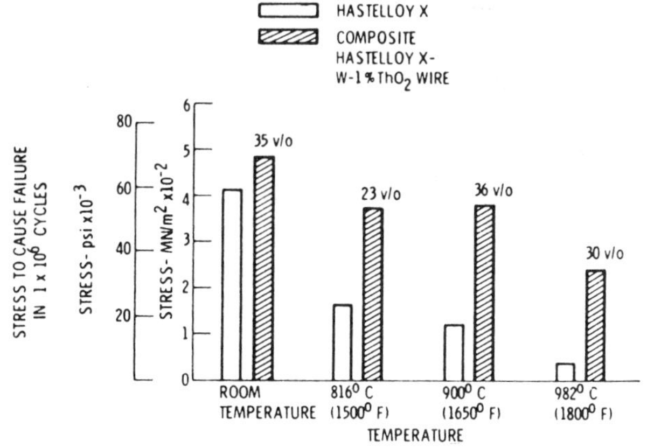

FIGURE 15 **Stress for failure in 1×10^6 cycles for Hastelloy X and composite [33].**

composite, except at the matrix–fiber interface adjacent to the fracture surface, also suggested that the behavior is controlled by the reinforcement. Similar results were obtained with W-1% ThO_2 reinforced Hastelloy X composites [33], W-1% ThO_2 reinforced FeCrAlY composites [14], and tungsten–nickel composites [34].

Fatigue

High temperature materials in gas turbines are subject to cyclic stresses and strains. These can lead to the development of cracks and failures, which conventionally are discussed in three separate groupings, depending on the magnitude and cause of the stresses:

High cycle fatigue
Low cycle fatigue
Thermal fatigue

HIGH CYCLE FATIGUE. High cycle fatigue failures can result from cyclic mechanical loads that lead to small-scale yielding (e.g., applied stresses that cause only localized plastic deformation while the rest of the structure is loaded in the elastic range). High cycle fatigue is often simulated in the laboratory by testing specimens under cyclic loads, the results being presented as an *S–N* curve in which the applied cyclic stress amplitude (*S*) is plotted as a function of the number of cycles to failure (*N*). Typically, the number of cycles to failure is greater than 10^5. Rarely are tests carried beyond 5×10^8 cycles, while turbine blades in service may experience 10^{12} or more cycles [35].

High cycle fatigue tests have been conducted on W-1 % ThO_2/Hastelloy X composite specimens [33]. Fatigue tests were performed at 10–15 Hz using direct stress, tension–tension, axially loaded specimens. The specimens were cycled from a minimum stress to a selected maximum stress and back to the minimum stress. The load ratio *R* ranged from 0.1 to 0.7 for unreinforced Hastelloy X and from 0.1 to 0.2 for the composite. The composite load ratios represented a more severe test. The stress ratio *A* ranged from 0.16 to 0.82 for unreinforced Hastelloy X and from 0.52 to 0.91 for the composite. Test temperatures were room temperature, 816, 900, and 980°C. Composite specimens were unidirectionally reinforced and contained 23, 30, and 37 vol % fibers. The stress to cause failure in 1×10^6 cycles versus temperature is plotted in Figure 15. Unreinforced Hastelloy X data are given for comparison. The composites were stronger at all temperatures, ranging from 1.2 times as strong at room temperature to 4 times as strong at 980°C. The ratio of fatigue strength to ultimate tensile strength for the same materials is plotted in Figure 16. For all test temperatures, the ratio for the composite was higher than that for the Hastelloy X, indicating that high cycle fatigue resistance of the composite is controlled by the fiber.

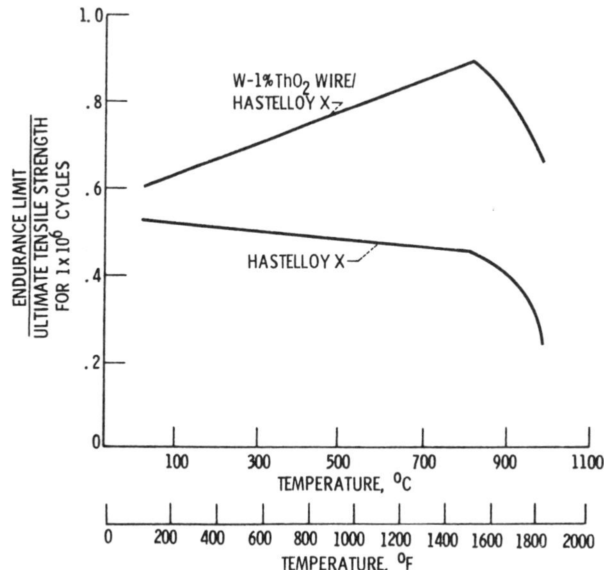

FIGURE 16 **Ratio of endurance limit to ultimate tensile strength for Hastelloy X and composite tested in axial tension–tension [33].**

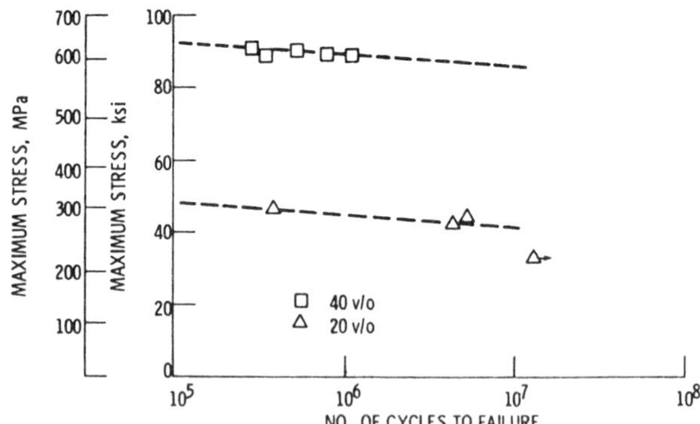

FIGURE 17 Stress as a function of number of cycles to failure for W-1%ThO_2/FeCrAlY composites tested at 760°C [36].

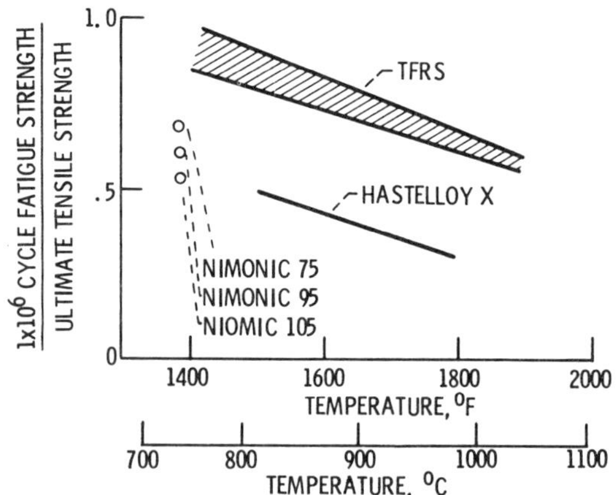

FIGURE 19 High cycle fatigue strength ratio comparison for TFRS and superalloys.

The high cycle fatigue strength for W-1% ThO_2/FeCrAlY composites was determined at 760 and 1038°C by Fleck [36]. Fatigue tests were performed at 30 Hz using direct tension–tension, axially loaded specimens. The specimens were cycled from a minimum stress to a maximum stress and back to the minimum stress. The load ratio *R* was 0.3 and the stress ratio *A* was 0.5. Composite specimens containing 20, 35, and 40 vol % fibers were tested. Figure 17 is a plot of maximum stress versus the number of cycles to failure for specimens tested at 760°C. The maximum stress versus number of cycles to failure for specimens tested at 1038°C is plotted in Figure 18.

The results again indicate that fatigue is controlled by the fiber. The composite containing 40 vol % fiber had a 1×10^6 cycle fatigue strength to ultimate tensile strength ratio of 0.9, at 760°C. Similar ratios for some superalloys range from 0.52 to 0.67 at the same test temperature [37]. Thus the composites' response to high cycle fatigue appears to be superior to that of superalloys. Figure 19 is a plot of the 1×10^6 cycle fatigue strength to ultimate tensile strength ratio for some superalloys and the range of values obtained for TFRS composites showing the advantage for the composite. Push–pull and reverse bend fatigue strength data were determined for a tungsten–superalloy composite by Dean [29]. The fatigue strength measured in push–pull tests at 20, 300, and 500°C was substantially increased by the introduction of 40 vol % tungsten wires. With cantilever specimens tested in reverse bending, a significant increase in fatigue strength also resulted from the incorporation of tungsten wires.

LOW CYCLE FATIGUE. Low cycle fatigue is characterized by high cyclic loads that lead to failure after a relatively small number of load cycles, usually less than 10^4–10^5. Fracture under low cycle fatigue conditions is generally associated with large-scale yielding.

Limited work has been reported on the low cycle fatigue behavior of refractory fiber–superalloy composites. The low cycle fatigue behavior for tungsten fiber reinforced nickel was determined at room temperature by Nilsen and Sovik [38]. Specimens containing tungsten fibers of 11–25 vol %, 500 μm diameter or 20–28 vol %, 100 μm diameter were fabricated by a liquid metal infiltration process and tested in fatigue. Specimens containing 8 or 10 vol % of 300 μm diameter tungsten fibers were fabricated by a powder and subsequent forging process and also tested in fatigue. Fatigue tests were performed at about 150 Hz using direct stress, tension–tension, axially loaded specimens. Figure 20 shows the ratio of the maximum stress for fatigue failure to ultimate tensile strength for the range of cycles investigated. The observed fatigue ratios for the composite specimens were much higher than some superalloys referenced by the author. The fatigue ratio reported for Nimocast 713C

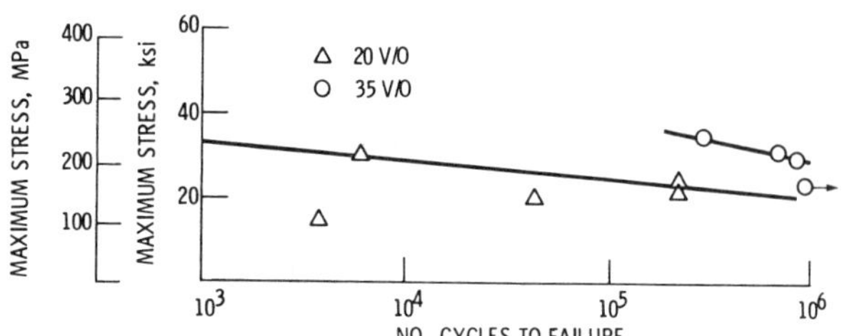

FIGURE 18 Stress as a function of number of cycles to failure for W-1%ThO_2/FeCrAlY composites tested at 1038°C [36].

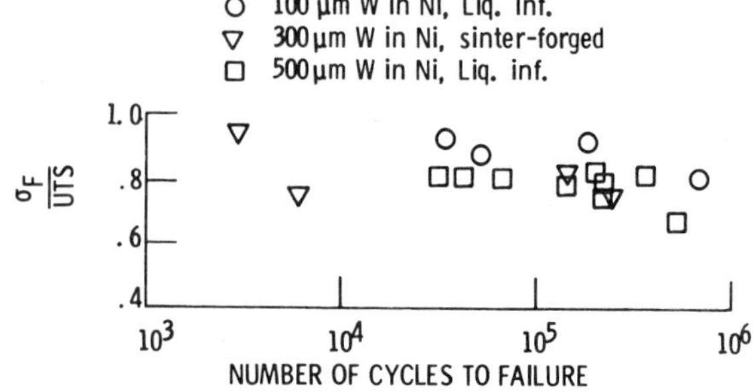

FIGURE 20 Ratio of fatigue strength (σ_F) to ultimate tensile strength (UTS) for Ni/W continuously reinforced composites [38].

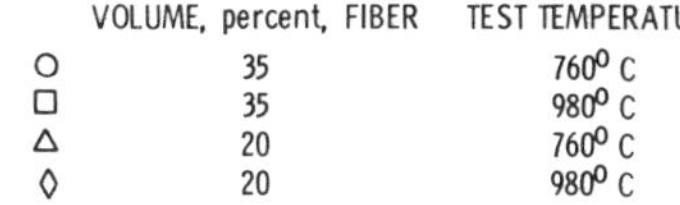

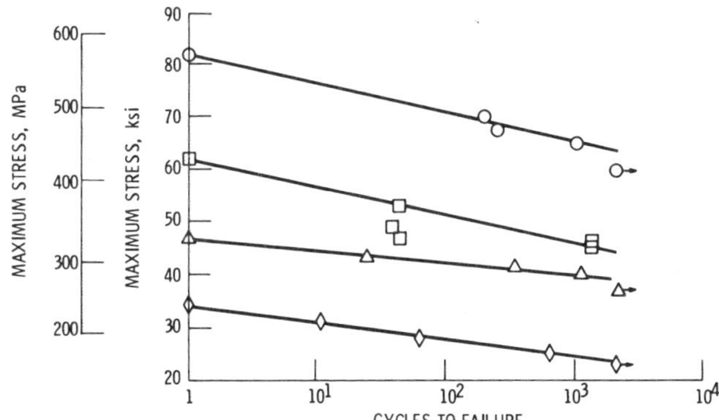

FIGURE 21 **Maximum stress versus cycles to failure for W-1%ThO_2/FeCrAlY composites [36].**

for 10^8 cycles was 0.24; for Incoloy 901 and Udimet 700 for 10^7 cycles the respective ratios were 0.14 and 0.17. As shown in Figure 20 the fatigue ratio obtained for the composites was greater than 0.65 at 10^6 cycles.

Low cycle fatigue tests were conducted at 760 and 980°C on 20 and 35 vol % W-1% ThO_2 fiber reinforced FeCrAlY composites [36]. Fatigue tests were performed at 0.65 Hz and a stress cycle of from 5.5 MN/m² to a maximum stress. The load ratio R was approximately 0.01 and the stress ratio A was 1. The low cycle fatigue results (Fig. 21) indicate that the fiber controls low cycle fatigue strength, as was the case for high cycle fatigue behavior. The 35 vol % fiber content specimens had much higher values of fatigue strength versus cycles to failure than did the 20 vol % fiber specimens. The ratio of fatigue strength to ultimate tensile strength versus cycles to failure is plotted in Figure 22. Very high values were obtained at both 760 and 980°C, indicating that the composite has a high resistance to low cycle fatigue in this temperature range.

THERMAL FATIGUE. Thermal fatigue failures are caused by the repeated application of stress that is thermal in origin. Rapid changes in the temperature of the environment can cause transient temperature gradients in turbine engine components. Such temperature gradients give rise to thermal stresses and strains. Thermal fatigue failure is the cracking of materials caused by repeated rapid temperature changes.

Superimposed on stresses generated by temperature gradients, in the case of the composite, are internal stresses caused by the difference in expansion coefficients between the fibers and matrix. The mean coefficient of thermal expansion from room temperature to 1100°C for superalloys ranges from 15.8 to 19.3 × 10^6/K and is approximately 5 × 10^{-6}/K for tungsten. Because of the large difference in expansion coefficients between the fiber and matrix and the resulting strains, thermal fatigue is believed to be the most serious limitation on composite usefulness.

A number of investigators have developed analytical methods to calculate the dependence of composite deformation on cyclic, geometric and constituent deformation parameters [39–41]. The results of these calculations illustrate the possible effects of several variables on deformation damage parameters. Because of the difference in expansion coefficients, the matrix is strained in tension upon cooling and in compression upon heating while the fiber is strained in compression upon cooling and tension upon heating.

Figure 23 shows the stress on the matrix as a result of heating and cooling. During cooling (A–B) the matrix stress increases continuously in tension because the matrix contracts faster than the fiber. After reaching the lowest temperature, point B, and upon reheating, the matrix is strained elastically in compression up to point C and the stress falls linearly. At point D the temperature and stress are high enough to allow matrix creep, and yielding of the matrix occurs. Because the temperature continually increases above point D, creep becomes

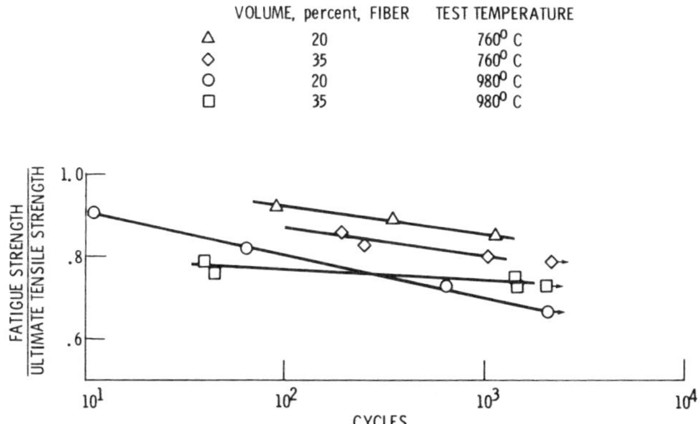

FIGURE 22 **Ratio of fatigue strength to ultimate tensile strength versus cycles to failure for W-1%ThO_2/FeCrAlY composites [36].**

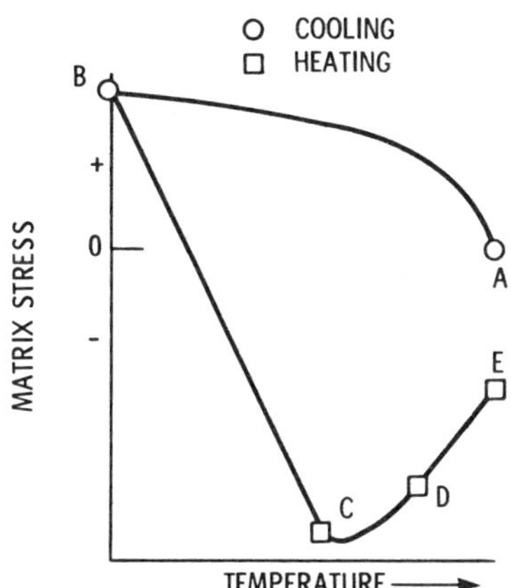

FIGURE 23 **Matrix stress as a function of heating and cooling.**

much easier and the stress on the matrix continuously decreases until the original maximum temperature is achieved, point *E*. If, upon reaching point *E*, the temperature were held constant, the matrix stress would fall further because of matrix creep with time. The magnitude of the stress on the matrix depends on the volume fraction fiber present in the composite. The larger the volume fraction fiber present in the composite, the higher the stress on the matrix and the lower the temperature at which the matrix would yield upon heating. Yielding of the matrix could also occur in tension during cooling. Work reported by Garmong [39] indicates that the hysteresis loop of matrix stress versus temperature caused by plastic deformation of the matrix stabilizes after a few cycles, so that a steady-state plastic compression–tension fatigue results when no external stress is present. Total cyclic plastic strain increases by a law of the form:

$$\text{total strain} = \text{strain per cycle} \times \text{number of cycles}$$

The ability of the matrix to accommodate plastic strain thus controls the number of cycles to failure for the composite if plastic deformation of the matrix governs the failure mode of the composite in fatigue.

A similar type of plot could be constructed for the stress on the fiber as a function of temperature. The stress on the fiber during cooling would be compressive, and upon heating the stress would be tension. At low volume fraction fiber contents, the compressive stress on the fiber during cooling could be large enough to cause plastic flow of the fiber as indicated by Baranov and Yakovleva [40]. The stress on the matrix and fiber also depends on the maximum cycling temperature, the heating–cooling rate, and the creep rate. Strain-induced damage increases markedly with increases in the maximum cycling temperature. Decreasing the heating–cooling rate or increasing the creep rate has the same effect. A decrease in heating–cooling rate increases the range of plastic strain per cycle, simultaneously decreasing the maximum matrix stress. Three types of cycling damage have been noted to date: plastic flow of the fiber in compression, matrix fracturing, and fiber–matrix interface debonding.

A number of studies have been conducted on the response of tungsten fiber–superalloy composites to thermal cycling. Table 5 compares the data obtained for several composite systems. Cylindrical specimens of 40% W/Nimocast 258 were cycled between room temperature and 1100°C in a fluidized bed to obtain rapid heating and cooling [29]. Metallographic examination after 400 cycles revealed no apparent damage at the fiber–matrix interface. Presumably no matrix or fiber cracks were observed, or they would have been reported as such. Cylindrical specimens of 13% W/Nimocast 713C were cycled in a fluidized bed in the temperature ranges shown in Table 5 [16]. Cracking occurred after relatively few cycles except for the specimens cycled from 20 to 600°C. The cracking was not extensive for the 550–1050°C cycle tests, but was extensive in the 20–1050°C cycle tests. The bond between the fiber and matrix was reported to be severely degraded by thermal cycling to 1050°C.

Thermal cycle tests were conducted on specimens of reinforced sheet material having a matrix of EI435 (Nichrome) and volume fiber contents of 14, 24, or 35% [42]. The specimens were heated in an electric resistance furnace for 2.5 minutes up to a temperature of 1100°C, followed by a water quench to room temperature. The number of cycles for debonding between the fiber and matrix to occur was determined as a function of fiber content. As shown in Table 5, the number of cycles for debonding to occur decreased with increasing fiber content.

The tests also indicated that one of the inherent requirements of the fiber, from the viewpoint of obtaining the best thermal fatigue response, is a strict uniformity of spacing of the fibers throughout the body of the material. Thermal fatigue failure occurred first where fibers were in contact with one another. Concurrently, where fibers were congested, cracks formed in the matrix along surfaces paralleling the fiber axis.

Tests were also conducted on reinforced EI435 sheet material [43]. The fiber contents investigated were 15 or 32%. The specimens were heated by passage of an electric current. Specimens were heated and cooled in 30 seconds in the temperature ranges shown in Table 5. Irreversible deformation occurred after cycling for all the 15 vol % fiber content specimens but not for the 32 vol % fiber content specimens. During the initial stages of cycling, warpage and bending were observed. A length decrease was observed during the entire test. With an increase of the number of cycles, the length and rate of the dimensional change diminished. After 1000 cycles from 600 to 1100°C, the length of the specimen decreased 20%.

Heat treatment had a considerable effect on dimensional instability of the composite. As a result of annealing the specimens, their propensity for deformation during thermocycling decreased. After annealing at 1100°C for 4 hours, specimens cycled from 570 to 1000°C for 1000 cycles did not change their shape. Annealing reduced the yield strength of the matrix. The level of stresses arising in the fibers as a consequence of the difference in expansion coefficients was determined by the resistance to plastic deformation of the matrix. With a decrease in the yield strength of the matrix, the level of stress on the fiber decreased and the fibers did not plastically deform. Cracks at the interface were observed for specimens containing 15 vol % fiber, while cracks in the matrix between the fibers were observed for specimens containing 32 vol % fibers.

Several different nickel-base composite systems were thermally cycled by Brentnall and Moracz [44]. Specimens containing 35 or 50 vol % fiber were heated by passage of an electric current. The specimens were heated to 1093°C in 1 minute and cooled to room temperature in 4 minutes. All the 35 vol % fiber content specimens were warped after 100 cycles, while the 50 vol % fiber content specimens were not. Two of the nickel-based composite materials containing 35 vol % fiber experienced a decrease in specimen length after 100 cy-

TABLE 5
Thermal Cycling Data for Tungsten/Superalloy Composites

Ref.	Composite Material	Heat Source	Cycle	No. of Cycles	Remarks
29	40 v/o W/Nimocast 258	Fluidized bed	RT-1100°C	400	No apparent damage to interface
16	13 v/o W/Nimocast 713C	Fluidized bed	20°–600°C 550°–1050°C 20°–1050°C	200 2–12 2–25	No cracks Cracks at interface Cracks at interface
42	W/EI435 (14, 24, and 35 v/o)	Electric resistance furance	RT-1100°C 2.5 min to temp. Water quench	100	No. of cycles for fiber-matrix debonding 14 v/o–90 to 100 cycles 24 v/o–60 to 70 cycles 35 v/o–35 to 50 cycles
43	W/EI435 (15 and 32 v/o)	Self resistance	30 sec to heat and cool 480°–700°C 500°–800°C 530°–900°C 570°–1000°C 600°–1100°C	1000	All 15 v/o specimens warped and had a specimen length decrease. Cracks at interface. 35 v/o specimens did not deform externally but matrix cracks between fibers observed.
44	W/NiWCrAlTi (35 and 50 v/o)	Self resistance	1 min to heat 4 min to cool	100	35 v/o warpage and shrinkage 50 v/o no damage
	W/NiCrAlY (35 and 50 v/o)		RT-1093°C	100	35 v/o warpage 50 v/o no damage
	W/2IDA (35 and 50 v/o)			100	35 v/o warpage and shrinkage 50 v/o no damage
	50 v/o W/NiCrAlY		427°–1093°C	1000	Internal microcracking
45	30 v/o W-1%ThO_2/FeCrAlY	Passage of electric current	1 min to heat 4 min to cool RT-1204°C	1000	No damage Surface roughening but no cracking

cles. The most ductile matrix materials (NiCrAlY) showed the least amount of damage after 100 cycles. The 50 vol % fiber content specimen containing a NiCrAlY matrix was cycled from 427 to 1093°C for 1000 cycles and experienced internal microcracking in the matrix between the fibers.

Specimens containing 30 vol % W-1% ThO_2 fibers in a matrix of FeCrAlY were exposed to 1000 cycles from room temperature to 1204°C [45]. The specimens were heated up to 1204°C in 1 minute and cooled to room temperature in 4 minutes. Surface roughening occurred, as shown in Figure 24, but there was no matrix or fiber cracking after the 1000 cycle exposure.

As indicated in Table 5, a composite system has been identified, W-1% ThO_2/FeCrAlY, that can be thermally cycled through a large number of cycles without any apparent damage. Except for the 40% W/Nimocast 258 composite system, which withstood 400 cycles without any apparent damage, all the other systems investigated evidenced some type of damage. These systems would be limited to applications in which the component would be exposed to very few thermal cycles. Only a limited number of systems have been investigated to date, and a need exists to identify other thermal fatigue resistant systems. The results obtained indicate that a ductile matrix that can relieve thermally induced strains by plastic deformation is required for composite thermal fatigue resistance.

Impact Strength

Composite materials must be capable of resisting impact failure from foreign objects or from failed components that may pass through the engine if they are to be considered for use as a blade or vane.

Factors affecting the impact strength of tungsten fiber metal matrix composites were investigated by Winsa and Petrasek [46]. Miniature Izod and standard Charpy impact strength data were obtained for a tungsten fiber reinforced nickel-based alloy (Ni-25W-15Cr-2Al-2Ti). It has been found that composite properties as measured by the miniature Izod impact test correlate closely with composite properties as measured by various ballistic impact tests, and it was concluded that the miniature Izod test is a reasonable screening test for candidate turbine blade and vane materials [47]. The Izod impact

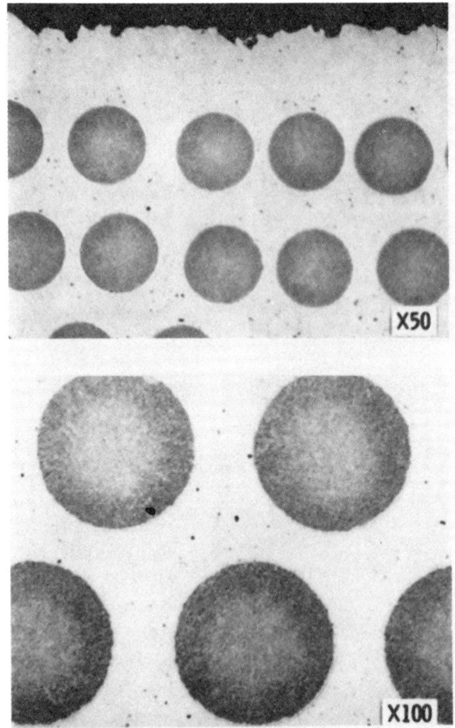

FIGURE 24 Photomicrographs of thermally cycled tungsten wire reinforced FeCrAlY composite [45]. (Photomicrographs courtesy of Irving Machlin.)

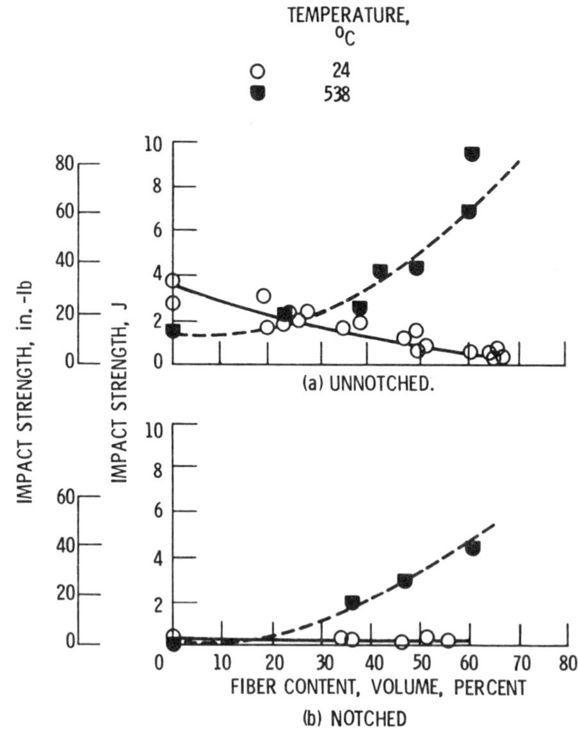

FIGURE 25 Impact strength of unnotched and notched HIP tungsten/superalloy as a function of fiber content [53].

strength of unnotched and notched specimens as a function of fiber content is plotted in Figure 25 for two test temperatures. Impact strength decreased with increasing fiber content at the lower temperature, but increased with increasing fiber content at the higher temperature.

Figure 26 plots impact strength as a function of temperatures and fiber content. There is a sharp increase in impact strength for the 60 vol % unnotched specimen at 260°C, the ductile–brittle transition temperature (DBTT) for the fiber. In general, unnotched composites had higher impact strength than the matrix at temperatures above the DBTT of the fiber and lower impact strength than the matrix below the DBTT of the fiber. The matrix's contribution to impact strength for the composite is most significant at low temperatures, while the fiber controls higher temperature impact strength [> 260°C].

The effect of fiber content on notch sensitivity was also determined. The ratio of the composite's notched impact strength per unit area to its unnotched impact strength per unit area is plotted as a function of fiber content in Figure 27. The notch sensitivity of the composite decreased with increasing fiber content both above and below the DBTT of the fiber. Heat treatment or hot rolling improved the room temperature impact strength of the composite. Heat treatment increased the impact strength of the notched unreinforced matrix by

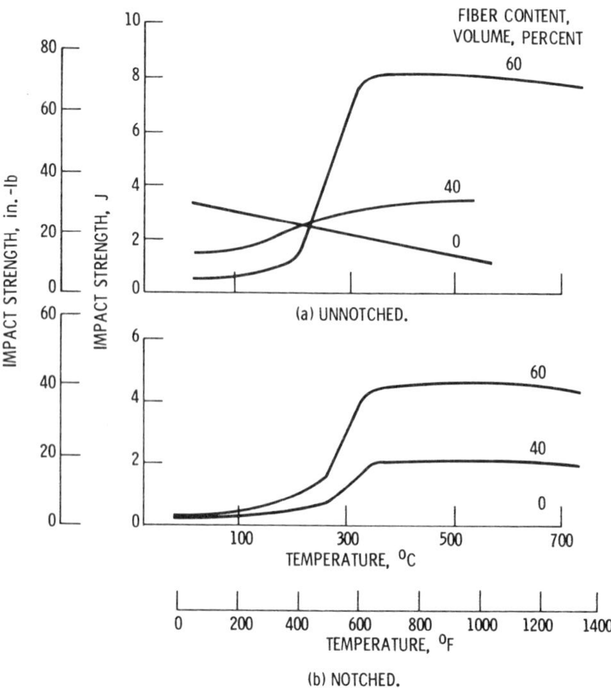

FIGURE 26 Impact strength of unnotched and notched tungsten/superalloy as function of temperature and various fiber contents [53].

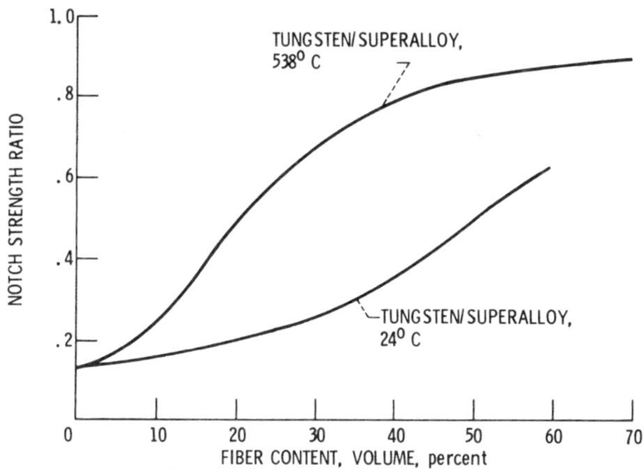

FIGURE 27 **Notch strength ratio as function of fiber content for tungsten/metal composites tested at 24 and 538°C [53].**

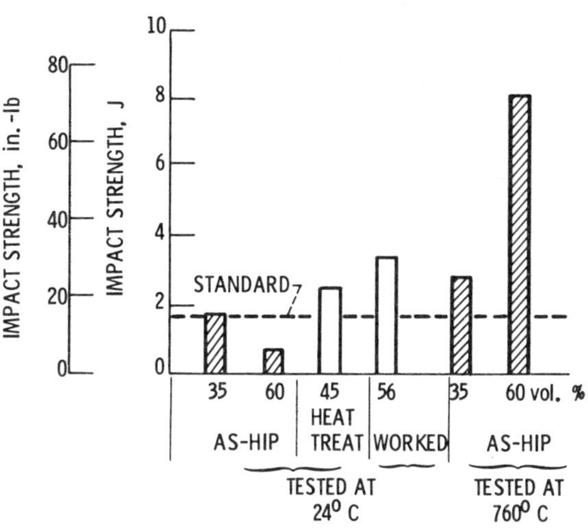

FIGURE 28 **Miniature Izod impact strengths of unnotched tungsten/nickel-base superalloys compared to minimum impact strength standard used to screen potential turbine blade and vane materials [53].**

almost 4 times and nearly doubled the impact strength of a 45 vol % fiber content composite. Round rolling increased the impact strength of a 56 vol % fiber content composite by nearly 4 times. The improved impact strength for the composite was related to improved matrix impact strength.

An additional objective of work conducted by Winsa and Petrasek [46] was to determine whether the potential impact resistance of tungsten fiber–superalloy composites was sufficient to warrant their consideration as turbine blade or vane materials. Alloys with miniature Izod impact values less than 1.7 J have been successfully run as turbine blades [48,49]. Based on this, the value of 1.7 J (15 in.-lb) was taken as the minimum value for Izod impact strength to indicate whether a material is worth further evaluation for possible turbine blade use. Figure 28 compares the tungsten fiber–superalloy impact strength values with the minimum standard. At room temperature as fabricated, composites containing fiber contents greater than 35% did not meet the minimum requirement. Heat treatment and hot working, however, improved the impact strength so that high fiber content composites met the minimum requirement. At 760°C, the higher fiber content as-fabricated composites have impact strengths distinctly above the minimum requirement. High Charpy impact strength values were obtained at 1093°C, 37.3 J for a 60 vol % fiber content specimen, implying that most of this strength is maintained to at least 1093°C. The impact strength potential for tungsten fiber–superalloy composites thus appears to be adequate for turbine blade and vane applications.

Results of a further comparison of the room temperature and 760°C miniature Izod impact strength for some superalloys and other composite systems are plotted in Figure 29. Values for the Inconel-713C, 25% W/Nichrome, and WI-52 were obtained from Ref. 26, while data for the 25% W-1% ThO_2/Hastelloy X are from Ref. 47. Inconel-713C and Guy alloy represent past turbine blade materials, while the WI-52 alloy is representative of an older vane material. The room temperature impact strength values obtained for the composites all exceed the minimum standard, which is equal to the value for Guy alloy, as well as that for the vane material WI-52. At 760°C the 25% W/Nichrome composite bent but did not fracture when impacted at 127 J at a velocity of 54 cm/s. At 760°C the 25% W-1% ThO_2/Hastelloy X composites bent and cracked, but the crack did not extend or completely propagate through the cross section of the specimen. The impact strength of the composites were thus much higher than that obtained for Inconel 713C or WI-52 at 760°C and should be adequate for turbine blade or vane applications.

Oxidation and Corrosion

The gaseous environment in the gas turbine engine is highly oxidizing, with oxygen partial pressures of the order of 2–4 atm [50]. However, this environment also contains significant amounts of combustion product impurities including sulfur from the fuel and alkali salts ingested with the intake air. Under these conditions, an accelerated oxidation may be encountered, sometimes, but not always, accompanied by the formation of sulfides within the alloy: this is commonly referred to as hot corrosion.

The basic design of the composite material assumes that the superalloy matrix will provide oxidation resistance, including protection of the tungsten fibers. Superalloys that are used for hot section engine components are oxidation resistant for material operating temperatures up to about 980°C. Above a material temperature of 980°C it is necessary to coat or clad the material to provide the required oxidation resistance. Claddings that are used for superalloy oxidation protection such as NiCrAlY and FeCrAlY are oxidation resistant to temperatures above 1090°C. These materials may also be

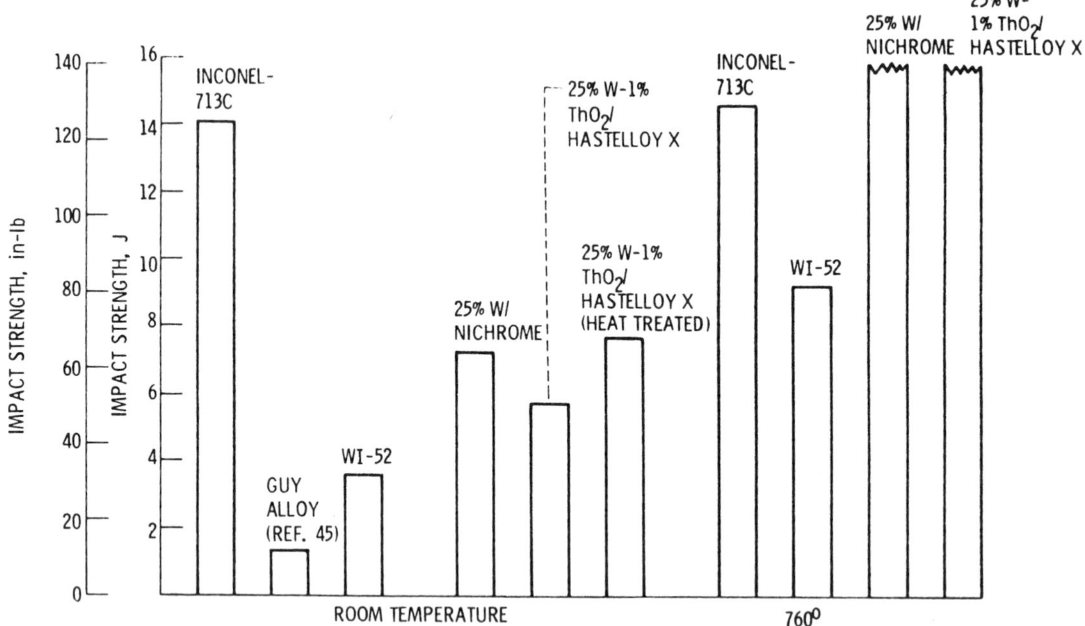

FIGURE 29 **Miniature Izod impact strength for superalloys and composites.**

considered to be the matrix for composites, so that the composite does not have to be coated or clad for high temperature oxidation resistance.

In preliminary oxidation studies on a nickel-based superalloy reinforced with W-1% ThO_2 fibers and clad with Inconel [27], the specimens were exposed in air at 1100°C for times up to 300 hours. Figure 30 shows a transverse section of a clad composite specimen exposed for 50 hours at 1093°C. The Inconel cladding was oxidized, and a coherent oxide scale formed on the Inconel. Oxidation had not progressed to the composite, and the surface fibers were not affected by the oxidation of the cladding. Composite specimens of W-1% ThO_2/FeCrAlY having completely matrix-protected fibers were exposed to static air at 1038, 1093, and 1149°C for up to 1000 hours. The weight change in 1000 hours was 0.3 mg/cm^2 for 1038°C and 1.26 mg/cm^2 for 1149°C. These values are in agreement with values obtained for the matrix material without any reinforcement [51]. Oxidation did not progress to the surface fibers.

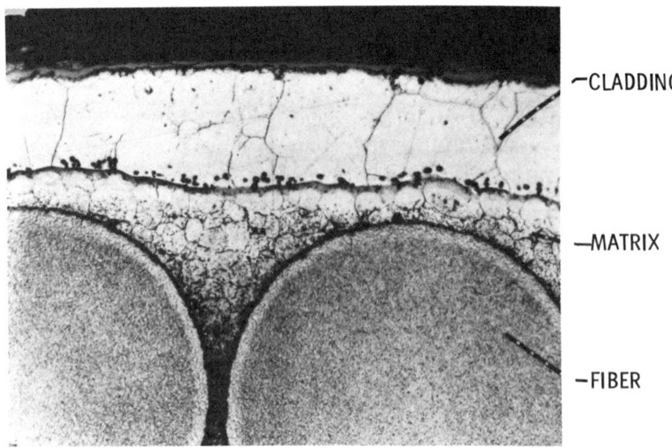

FIGURE 30 **Transverse section of oxidized refractory fiber–nickel alloy composite [27].**

The oxidation and corrosion resistance of composite materials having exposed fibers is also an important consideration. Although the fibers in the composite need not be designed to be exposed to the engine environment, an understanding of their high temperature oxidation and corrosion behavior is desirable in the event of a coating, cladding, or matrix failure during service (e.g., from impact due to foreign objects passing through the engine). Figure 31 illustrates the principal paths for oxidation and corrosion of exposed fibers. Oxidation proceeding perpendicular to the fibers (through the blade or vane thickness) would destroy the exposed fibers, but intervening matrix would prevent oxidation of subsequent layers. Thus only a partial loss of strength would result. Oxidation parallel to the fibers (along the blade or vane span) potentially is more severe, since all the exposed fibers in the cross section potentially could be oxidized along their entire length. However, studies conducted to evaluate oxidation along fibers showed only limited oxidation penetration along the fibers [45,50,52]. Brentnall and Moraez [45] found that after 10 hours of exposure to static air at 1200°C, the fibers in W-1% ThO_2/FeCrAlY were oxidized to a depth of only 2.5 mm. After 100 hours of exposure at 1100°C, the fibers in a W/Nimocast 258 composite oxidized to a depth of 1.3 mm in static air and to 2.5 mm in a low velocity simulated engine exhaust gas stream moving at 1.8 m/s [52]. Measure-

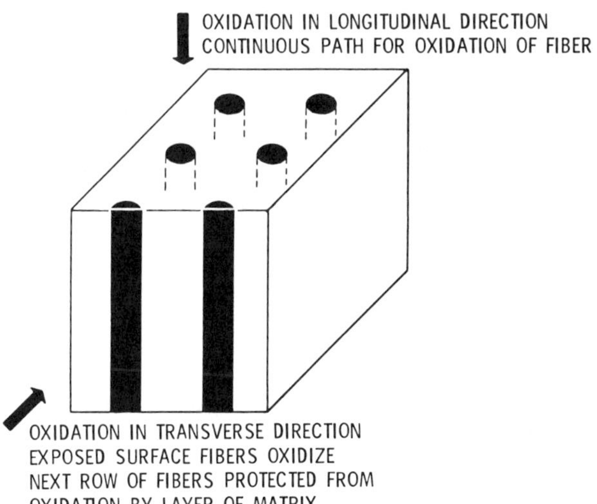

FIGURE 31 **Principal paths for exposed fiber oxidation.**

ments of the depth of fiber oxidation for end-exposed tungsten fibers in oxidized tungsten fiber reinforced Ni-20% Cr specimens obtained from photographs presented by Ed Dahshan et al. [50] indicate similar values. However, these investigators observed considerable distortion and degradation of the matrix surrounding the oxidized tungsten fiber. Typical weight gain–time curves for the oxidation of 40 vol % tungsten fiber content nickel–20% chromium material tested at 900 and 1000°C in 1 atm oxygen are shown in Figure 32 [50]. While tungsten can form volatile oxides, the large weight gains reflected in these data indicate the possible formation of complex matrix-tungsten oxides.

The hot corrosion behavior of tungsten fiber reinforced Ni–20% Cr composite specimens was also examined under the following exposure conditions: (a) sulfidation in H_2–10% H_2S, (b) presulfidation in H_2–10% H_2S followed by oxidation in oxygen, and (c) oxidation in 1 atm oxygen after precoating with Na_2SO_4 [50]. During sulfidation, only the matrix formed sulfides; the fibers remained unaffected. Consequently, presulfidation, although having a dramatic effect on the oxidation of the matrix, did not damage the fibers. The presence of sodium sulfate was also not critical. Thus, hot corrosion conditions were not harmful to the tungsten-reinforced composites studied, and catastrophic loss of the exposed tungsten fibers did not occur upon exposure to a high temperature oxidizing environment.

Thermal Conductivity

High thermal conductivity is desirable in a turbine blade material to reduce temperature gradients; this, in turn, results in reduced thermally induced strains, which can cause cracks or distortion. In addition, higher thermal conductivity can reduce coolant flow requirements in some impingement-cooled blades, leading to greater engine efficiency or durability [53]. High conductivity for some turbine blade applications could result in unacceptable higher disc temperatures. However, proper design could alleviate the problem [54]. The conductivity of tungsten fiber–superalloy composites is markedly superior to that of superalloys. The thermal conductivity of tungsten is much higher than that for superalloys, and the more tungsten added to a composite, the greater the conductivity. Thermal conductivity of the composite is greatest in the direction of the fiber axes, since there is a continuous path for conduction along the tungsten fibers. Conduction perpendicular to the fiber axes is lower because the heat cannot find a continuous path through tungsten. The thermal conductivity of some tungsten fiber–superalloy composites was determined as a function of temperature [53]. Figure 33 shows the thermal conductivity values obtained for a composite containing 65 vol % fibers in a nickel-based alloy and tested in the

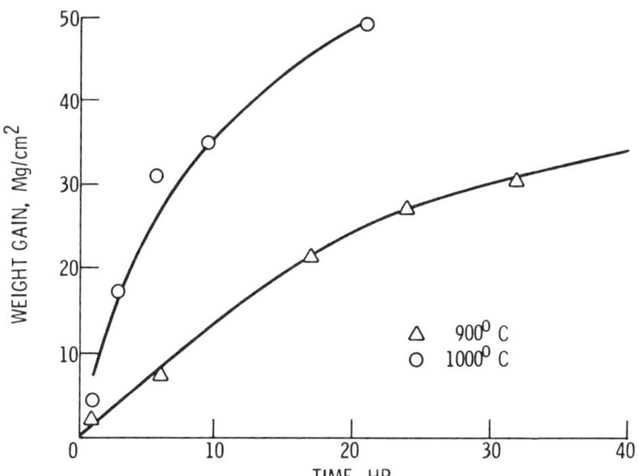

FIGURE 32 **Weight gain–time curves for the oxidation of W-reinforced Ni-20Cr at 900 and 1000°C containing end-exposed fibers [50].**

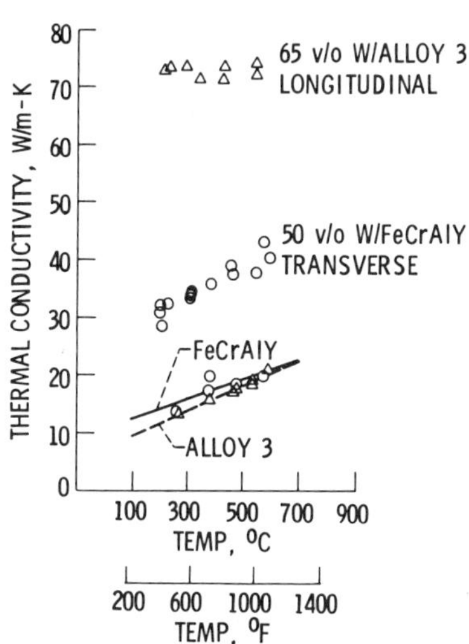

FIGURE 33 **Measured thermal conductivity of TFRS composites and matrix alloys [53].**

direction of the fiber axis and for a composite containing 50 vol % fiber and tested in the direction perpendicular to the fiber axis. The longitudinal thermal conductivity is seen to be much higher than the transverse conductivity. Also shown are values for the matrix materials. The composites have much higher values for thermal conductivity over the entire temperature range. The results [53] indicate that at blade conditions [30–60 vol % fiber content and 730–1130°C, the transverse (through-the-wall) conductivity of the composite will be 35–50 W/m·K. The longitudinal (spanwise) conductivity will be about 45–65 W/m·K. Typical superalloys have conductivities of about 25 W/m·K at 1030°C.

Composite Turbine Blade Material Requirements

Relatively small increases in turbine component use temperature, such as 6°C, are beneficial and result in cost savings in aircraft engine operation. Numerous modifications in engines, including material changes, have been made to further such gains. However, the introduction of a fiber-reinforced superalloy component, such as a tungsten fiber reinforced superalloy turbine blade, would represent a large material change. Such a change would require a larger potential gain to justify the considerable effort necessary to make the application possible.

It has been assumed in the program at NASA–Lewis that a composition selected as a serious turbine component material candidate must achieve at least a 50°C use temperature advantage over the best currently used superalloys, and the goal of our program is to permit a 100–150°C use temperature increase.

The properties of the composite are controlled by the fiber and matrix properties. Therefore, the properties needed in the composite dictate the fiber and matrix materials that may be used. Hence, selecting the composite involves determining the best compromise in the combination of fiber and matrix alloys used. Specific composite fiber and matrix requirements are as follows.

Fiber Property Requirements

Creep–rupture and mechanical fatigue strength must be adequate to permit at least a 50°C composite blade metal temperature advantage over current superalloys.

Toughness must be adequate at operating temperatures to ensure the needed foreign object damage (FOD) resistance.

Fiber costs must be low, so that blade fabrication costs can be kept acceptably low.

Matrix Property Requirements

Compatibility with the fibers is required at fabrication and operating temperatures so that fiber strength is not excessively degraded by interdiffusion.

Mechanical and thermal fatigue resistance is needed at operating temperatures. Thermal fatigue damage in the matrix can be initiated by the thermal expansion coefficient mismatch between fiber and matrix as well as by large temperature gradients.

Oxidation and hot corrosion resistance are needed in the matrix for use at temperatures up to 1100°C. Not only is this a severe temperature in general, but at these levels the fibers lack oxidation resistance and must be protected by the matrix.

Density must be low, to help offset the high density of the tungsten fibers. This is a particularly important consideration in aircraft engines, where weight must be minimized because high blade densities can lead to high disc weights.

Toughness and ductility must be high at low temperatures because the matrix imparts impact damage resistance to the composite at low temperatures.

Shear creep strength must be adequate to allow fiber angle plying for airfoil chordwise strength and torsional strength.

First-Generation Composite Turbine Blade Material Selection

Tungsten-fiber/FeCrAlY was identified as a promising first-generation turbine blade material because of the excellent combination of complementary properties possible with this combination of fiber and matrix [55]. The matrix provides a high melting point, as well as low density, excellent oxidation and hot corrosion resistance, limited fiber–matrix interdiffusion at proposed blade temperatures, and excellent ductility, to aid in thermal fatigue resistance. The fiber provides high stress–rupture, creep, fatigue, and impact strength, along with high thermal conductivity. Properties reported for this material indicate that it has adequate properties for turbine blade use and could permit turbine blade operating temperatures more than 50°C greater than those of current directionally solidified (DS) superalloy blades.

Composite Component Fabrication

When a material has demonstrated adequate properties for application as a turbine blade material, designers next consider whether complex shapes such as hollow turbine blades can be designed and fabricated from the candidate material at reasonable cost. The composite fabrication techniques selected must result in a composite whose properties meet those required for application of the composite. The processes must be capable of producing component shapes to required dimensions, incorporating both uniaxial and off-axis fiber positioning, providing uniform matrix cladding to prevent fiber oxidation, and providing for cooling or weight-reduction passages, if necessary. The combined fabrication techniques must also be cost effective and reproducible.

Investment casting has been considered for fabrication of composite blades. However, there are two obstacles to overcome. A way must be found to hold the fibers in positive "angle plied" alignment during infiltration by the molten matrix. Also, small-diameter fibers, which must be used in hollow blades because of a space limitation, are subject to damage caused by fiber–matrix interdiffusion and to displacement during casting. A W-2% ThO_2/MAR M322 composite JT9D turbine blade was fabricated by a casting technique reported more than 10 years ago [19]. Separation of the fiber bundle from the matrix occurred in the concave areas of the blade. Test specimens fabricated by the same process revealed that outer fibers are exposed to higher temperatures than inner fibers. Furthermore, the matrix composition became richer in tungsten, which indicated dissolution of the fibers. Given the current state of the art, investment casting would appear to be more suitable for fabrication of uniaxially reinforced solid blades in which large fibers can be used.

Diffusion bonding of monolayer composite plies is currently the most promising, cost-effective method of fabrication for a hollow blade. The composite plies consist of aligned tungsten fibers sandwiched between layers of matrix material. This approach has the capability for accurate fiber distribution and alignment; moreover, it limits fiber–matrix interdiffusion during fabrication. This approach also is capable of producing blade shapes that are close to final dimensions; hence only root machining and touchup grinding is needed. Solid boron–aluminum and boron–titanium fan and compressor blades have been fabricated using this approach. A solid W-2% ThO_2/conventionally cast MAR-M200 prototype airfoil for potential application in an advanced industrial gas turbine engine was also fabricated [56], using a similar approach.

The feasibility of fabricating a composite hollow turbine blade was successfully demonstrated in work reported by Melnyk and Fleck [57]. A JT9D-7F first–stage, convection-cooled blade was selected as the model from which a W-1% ThO_2/FeCrAlY composite blade was designed. The major purpose of the fabrication effort was not only to demonstrate the feasibility of fabricating a hollow blade but to also show that design requirements could be met in the fabricated blade. The design features incorporated into the fabricated blade are indicated in Figures 34 and 35. The external airfoil was identical to

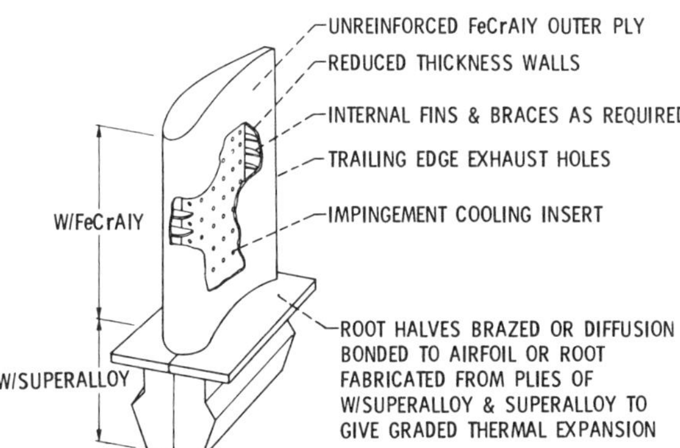

FIGURE 34 **Schematic of FRS JT9D blade designed for fabrication feasibility experiments.**

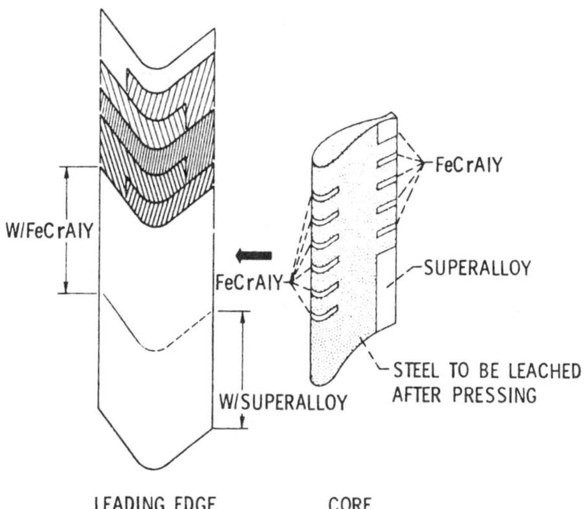

FIGURE 35 **Schematic of airfoil ply configuration in JT9D blade.**

that of the current MAR-M200 (DS), JT9D-7F blade. However, the airfoil walls were designed to be thinner, to reduce the composite blade weight to within 10% of the current blade weight and to allow for more efficient air cooling. An impingement cooling insert was added to improve cooling efficiency. The walls of the airfoil were built up of composite plies containing W-1% ThO_2 fibers. The inner and outer plies of the wall consisted of unreinforced FeCrAlY for oxidation resistance. The composite airfoil extends down into the blade root. The matrix used in the airfoil above the root was FeCrAlY for oxidation and thermal fatigue resistance. The matrix used in the airfoil within the root was an alloy optimized for shear strength and thermal fatigue resistance. The root requires high strength superalloy or a composite built up of plies graded for differing thermal expansion to minimize thermal fatigue problems at the airfoil–root interface. The blade was designed to have a potential 50°C use temperature advantage over the current MAR-M-200 (DS) blade.

Figure 36 shows the fabrication sequence used to produce the hollow composite turbine blade. A tungsten fiber mat was sandwiched between powder sheets of FeCrAlY and hot pressed to form a monotape. The monotape was cut into the plies necessary to arrive at the final blade dimension, and the plies were stacked around a steel core. Root inserts and outserts could also be stacked around the assembly or could be attached in a secondary fabrication step. The entire assembly was then placed in a refractory metal die, heated, and pressed to arrive at the proper airfoil contour. After pressing, the steel core was removed by leaching out with an acid. A tip cap was welded on to the end of the airfoil, and an impingement cooling insert was placed in the leached out cavity and brazed to the root of the blade.

Figure 37 shows the as-fabricated W-1% ThO_2/FeCrAlY composite hollow JT9D-7F airfoil containing a bonded-on end cap and trailing-edge coolant slots. Figure 38

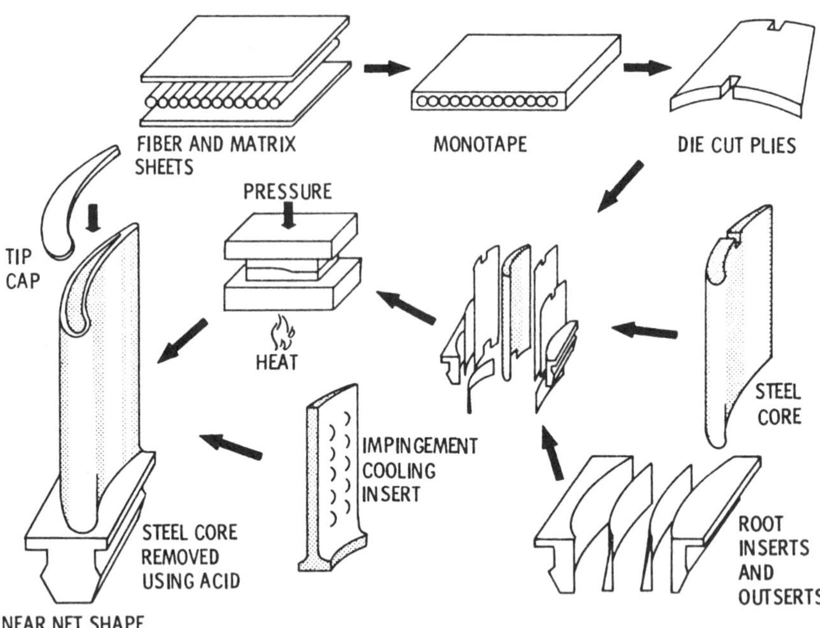

FIGURE 36 **TFRS blade fabrication process.**

shows the composite airfoil, which was brazed to a high strength superalloy arc root. A cross section of the composite airfoil appears in Figure 39. Excellent fiber alignment and fiber distribution were obtained, and the fibers were fully protected by a layer of FeCrAlY in the interior and exterior of the airfoil.

Successful fabrication of a hollow composite airfoil has demonstrated that this material can be fabricated into the complex design shapes for hot turbine section components. While vanes and combustion liners have not been fabricated, these components are less complex than the blade and can be considered for future programs.

The fabrication process sequence used to produce a hollow composite blade was used in a fabrication cost study [58]. Fabrication costs were estimated for high technology turbine blades fabricated using three different materials. The same turbine blade configuration, a first-stage JT9D-7F blade was used for each material. Directionally solidified eutectic (DSE), an oxide dispersion strengthened superalloy (ODSS), and W-1% ThO_2/FeCrAlY blade manufacturing costs were compared with the cost of producing the same blade from a DS superalloy, the current blade material. The relative costs are shown in Figure 40. The study indicates that W/FeCrAlY

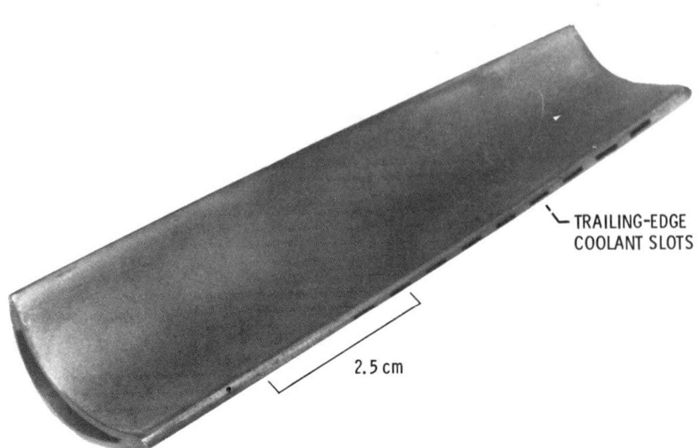

FIGURE 37 **Composite hollow airfoil.**

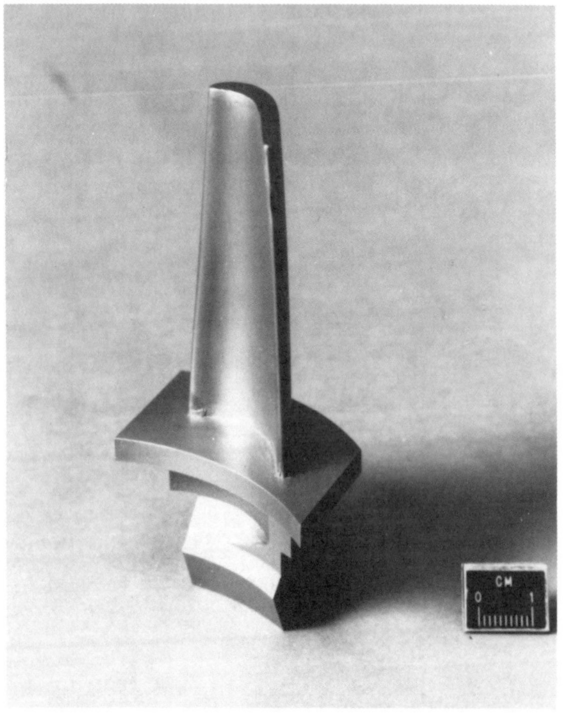

FIGURE 38 **Hollow composite blade.**

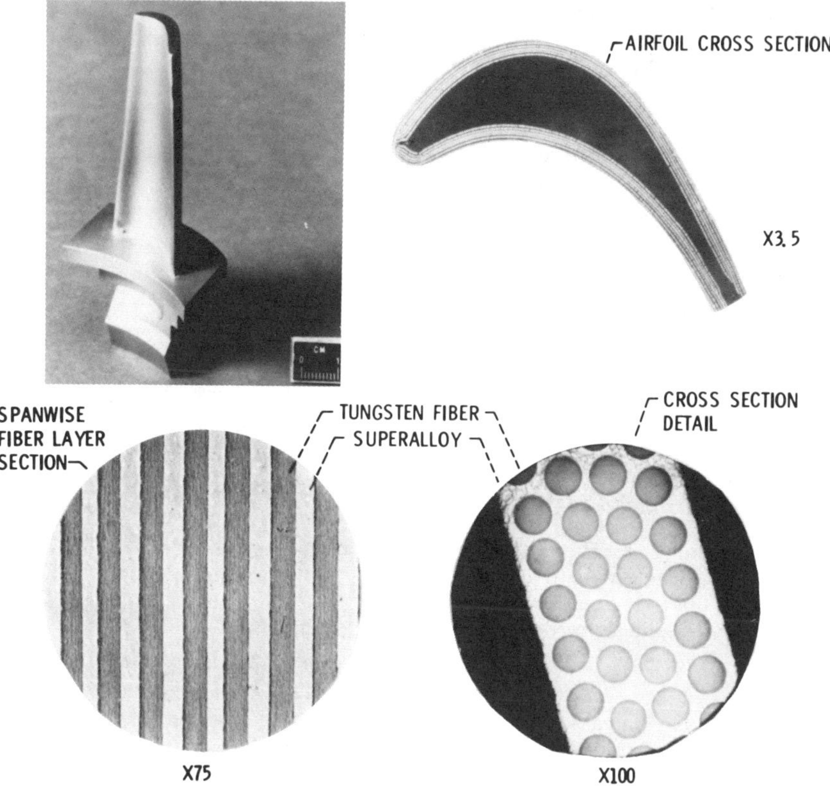

FIGURE 39 Tungsten fiber–superalloy composite blade.

manufacturing costs should be competitive with costs of manufacturing this blade by directional solidification of a superalloy—provided the projected manufacturing yields can be realized in actual commercial production of blades.

Concluding Remarks

Exploratory development and material property screening have indicated that tungsten fiber reinforced superalloy composites have considerable potential for application as advanced high temperature turbine engine component materials. A first-generation TFRS composite has been selected to serve as a demonstration system to evaluate the merit and problems of this family of composites for turbine engine applications. Based on the data obtained from the development and property screening, thoriated tungsten wire in a FeCrAlY matrix has potential to become a viable candidate for application.

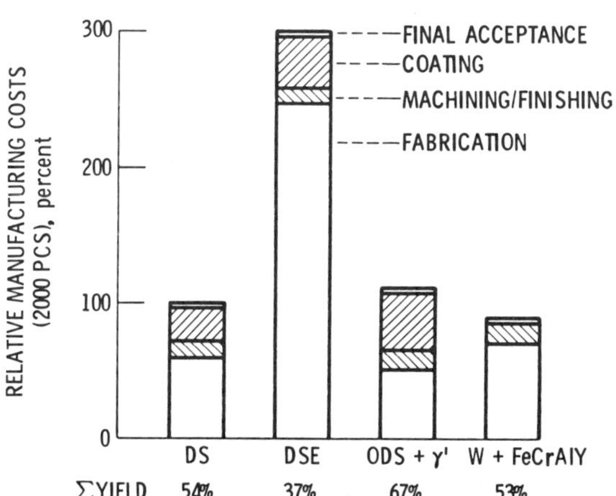

FIGURE 40 Cost analysis for JT9D first-stage turbine blade [58].

The choice of commercially available thoriated tungsten wire was made despite the appreciable difference in properties obtainable when stronger wires such as W-Hf-C were used because of current cost and availability. The manufacturing technology effort to make the stronger wire available can more readily be justified after TFRS has become better accepted for application. Test parameters can be evaluated equally well with the less expensive, readily available wire.

The choice of FeCrAlY as a matrix is based on the ductility, low strain hardening, and oxidation and corrosion resistance of the alloy; and good chemical compatibility with tungsten fibers. The composite can be made to resist oxidation and corrosion by a thin layer on the external and internal surfaces of a component. The ductility and low strain hardening rate of the matrix are vital to resist the plastic deformation that occurs from thermally induced strains from thermal gradient and thermal expansion mismatch between the fiber and matrix. FeCrAlY is not the ideal matrix because its thermal expansion coefficient is much higher than the fibers and its shear strength is low at elevated temperatures. As in the case of the thoriated tungsten wire, however, it is a reasonable choice as a first-generation system to indicate that TFRS is a viable material for further improvement.

Use of tungsten/FeCrAlY composite material could permit engine turbine blade temperatures more than 50°C higher than those possible using conventional superalloys. Moreover, blade fabrication studies have demonstrated the feasibility of producing hollow W/FeCrAlY turbine blades at a cost competitive with DS superalloy blade costs. Still, a great deal of work remains to be done on this material to aid in its transition from laboratory feasibility to rig testing of prototype hardware and then on to manufacturing technology and detailed design.

Donald W. Petrasek and Robert A. Signorelli

References

Publishers's Note. References for this article are as up to date as possible in view of the fact that (author's statement): "The work that the manuscript addresses was classified shortly after publication and an updating would be impossible for an international publication."

1. L. H. Amra, L. F. Chamberlin, F. R. Adams, J. G. Tavernelli, and G. J. Polanka, "Development of Fabrication Process for Metallic Fibers of Refractory Metal Alloys," NASA CR-72654, 1970.
2. G. W. King, "Development of Wire-Drawing Processes for Refractory-Metal Fibers," NASA CR-120925, 1972.
3. D. W. Petrasek and R. A. Signorelli, "Stress-Rupture and Tensile Properties of Refractory-Metal Wires at 2000 and 2200°F (1093 and 204°C)," NASA TN D-5139, 1969.
4. D. W. Petrasek, "High-Temperature Strength of Refractory-Metal Wires and Consideration for Composite Applications," NASA TN D-6881, 1972.
5. D. W. Petrasek and J. W. Weeton, "Alloying Effect on Tensile Properties and Micro-Structure of Tungsten-Fiber-Reinforced Composites," NASA TN D-1568, 1963.
6. T. Montebano, J. Brett, L. Castleman, and L. Seigle, *Trans. Metall. Soc. AIME, 242,* 1973 (1968).
7. J. Hoffman, S. Hofmann, and L. Tillmann, *Z. Metallkd., 65,* 721 (1974).
8. H. Gruenling and G. Hofer, *Z. Werkstofftech., 5*(2), 69 (1974).
9. D. W. Petrasek, R. A. Signorelli, and J. W. Weeton, "Refractory-Metal–Fiber–Nickel-Base-Alloy Composites for Use at High Temperatures," NASA TN D-4787, 1968.
10. D. M. Karpinos, L. I. Tuchinskii, L. R. Vishnyakov, L. N. Pereselentseva, and L. N. Klimenko, *Fiz. Khim. Obrab. Mater., 6,* 107 (1972).
11. B. A. Klypin, A. M. Maslov, and S. B. Maslenkov, *Metalloved. Term. Obrab. Met., 5,* 6 (1977); trans. *Met. Sci. Heat Treat. Met., 19*(5–6), 343 (1977).
12. V. S. Mirotvorskii and A. A. Ol'shevskii, *Poroshk. Metall. 7*(187), 57 (1978); trans. *Sov. Powder Metall. Met Ceram., 17*(7), 536 (1978).
13. V. S. Mirotvorskii and A. A. Ol'shevskii, *Poroshk. Metall., 7*(163), 46 (1976); trans. *Sov. Powder Metall. Met. Ceram., 15*(7), 534 (1976).
14. W. D. Brentnall, "Metal Matrix Composites for High Temperature Turbine Blades," TRW-ER-7790-F, TRW, Inc., Cleveland, 1976, and NADC-76225-30, Naval Air Development Center, Warminster, PA, 1976.
15. V. S. Mirotvorskii and A. A. Ol'shevskii, *Metalloved. Term. Obrab. Met., 11,* 12 (1979); trans. *Met Sci. Heat Treat. Met., 21*(11), 826 (1980).
16. A. W. H. Morris and A. Burwood-Smith, "Fiber Strengthened Nickel-Base Alloy," *High Temperature Turbines,* AGARD-CP-73-71, January 1971.
17. R. A. Signorelli, "Review of Status and Potential of Tungsten Wire: Superalloy Composites for Advanced Gas Turbine Engine Blades," NASA TM X-2599, 1972.
18. R. J. E. Glenny, *Proc. R. Soc., Ser. A, 319,* 33 (1970).
19. I. Ahmad and J. M. Barranco, *SAMPE Q., 8,* 38–49 (1977).
20. D. M. Karpinos, L. I. Tuchinskii, L. R. Vishnyakov, and V. Ya. Fefer, *Poroshk. Metall., 6*(150), 20 (1975); trans. *Sov. Powder Metall. Met. Ceram., 14*(6), 447 (1975).
21. V. P. Severdenko, A. S. Matusevich, and A. E. Piskarev, *Porshk. Metall., 6*(138), 51 (1974); trans. *Sov. Powder Metall. Met. Ceram., 13*(6), 476 (1974).
22. W. D. Brentnall and I. J. Toth, "Fabrication of Tungsten Wire Reinforced Nickel-Base Alloy Composites," NASA CR-134664, 1974.
23. P. R. Sahm, "Eutectic and Artificial Composite Superalloys," in P. R. Sahm, Ed., *Proceedings of the Third Symposium on High-Temperature Materials in Gas Turbines,* Elsevier, Amsterdam, 1974, pp. 73–114.
24. R. J. E. Glenny and B. E. Hopkins, *Phil. Trans. R. Soc. (London), Ser. A, 282*(1307), 105–118 (1976).
25. W. Endres, "Design Principles of Gas Turbines," in P. R. Sahm, Ed., *Proceedings of the Third Symposium on High-Temperature Materials in Gas Turbines,* Elsevier, Amsterdam, 1974, pp. 1–14.
26. A. R. Stetson, B. Ohnysty, R. J. Akins, and W. A. Compton, "Evaluation of Composite Materials for Gas Turbine Engines," AFML-TR-66-156, Part 1, Air Force Materials Laboratory, Wright-Patterson AFB, June 28, 1966.
27. D. W. Petrasek and R. A. Signorelli, "Preliminary Evaluation of Tungsten-Alloy–Fiber–Nickel-Base-Alloy Composites for Turbojet Engine Applications," NASA TN D-5575, 1970.
28. D. W. Petrasek and R. A. Singorelli, "Stress-Rupture Strength and Microstructural Stability of Tungsten–Hafnium–Carbon-Wire Reinforced Superalloy Composites," NASA TN D-7773, 1974.
29. A. V. Dean, *J. Inst. Met., 95,* 79 (1967).
30. V. M. Chubarov, Yu V. Levinskii, S. E. Salibekov, A. F. Trefilov, L. V. Grachev, E. M. Rodin, M. Kh. Levinskaya, and L. V. Dvoichenkova, *Probl. Prochn., 3*(7), 100–104 (1971); trans. *Strength Mater., 3*(7), 856 (1972).
31. A. W. H. Morris and A. Burwood-Smith, *Fibre Sci. Technol., 3*(1), 53 (1970).
32. G. I. Friedman and J. N. Fleck, "Tungsten Wire-Reinforced Superalloys for 1093°C (2000°F) Turbine Blade Applications," NASA CR-159720, 1979.
33. R. H. Baskey, "Fiber-Reinforced Metallic Composite Materials," AFML-TR-67-196, Air Force Materials Laboratory, Wright Patterson AFB, 1967.
34. A. Kannappan and H. F. Fischmeister, "High Temperature Stability of Tungsten Fiber Reinforced Nickel Composites," in M. Tilli, Ed., *Proceedings of the Fourth Nordic Symposium on High Temperature Materials Phenomena,* Vol. II, *Physical Metallurgy,* Helsinki University of Technology, Esbo, 1975, pp. 85–98.

35. M. O. Speidel, "Fatigue Crack Growth at High Temperatures," in P. R. Sahm, Ed., *Proceedings of the Third Symposium on High-Temperature Materials in Gas Turbines*, Elsevier, Amsterdam, 1974, pp. 207–255.

36. J. N. Fleck, "Fabrication of Tungsten-Wire/FeCrAlY-Matrix Composites Specimens," TRW Report ER-8076, TRW, Inc., Cleveland, 1979.

37. C. T. Sims and W. C. Hagel, *The Superalloys*, Wiley-Interscience, New York, 1972.

38. N. Nilsen and J. H. Sovik, "Fatigue of Tungsten Fiber Reinforced Nickel," in *Practical Metallic Composites; Proceedings of the Spring Meeting*, Institution of Metallurgists, London, 1974, pp. B51–B54.

39. G. Garmong, *Metall. Trans.*, *5*, 2183 (1974).

40. A. A. Baranov and E. V. Yakovleva, "Deformation of a Composite Material During Thermocycling, II," *Probl. Prochn.*, *8*, 50–53 (1975); trans. *Strength Mater.*, *7*(8), 966 (1976).

41. V. V. Gaiduk, A. S. Lavrenko, and Yu V. Sukhanov, *Probl. Prochn.*, *9*, 108 (1972).

42. G. I. Dudnik, F. P. Banas, and B. V. Aleksandrov, *Probl. Prochn.*, *5*, 99 (1973); trans. *Strength Mater.*, *5*(1), 106 (1973).

43. F. P. Banas, A. A. Baranov, and E. V. Yakovleva, *Probl. Prochn.*, *6*, 82 (1975); trans. *Strength Mater.*, *7*(6), 744 (1976).

44. W. D. Brentnall and D. J. Moracz, "Tungsten Wire-Nickel Base Alloy Composite Development," TRW ER-7849, TRW, Inc., Cleveland; NASA CR-135021, 1976.

45. W. D. Brentnall and D. J. Moracz, "Tungsten Wire-Nickel Base Alloy Composite Development," TRW ER-7849, TRW, Inc., Cleveland; NASA CR-135021, 1976.

46. E. A. Winsa and D. W. Petrasek, "Factors Affecting Miniature Izod Impact Strength of Tungsten-Fiber-Metal-Matrix Composites," NASA TN D-7393, 1973.

47. B. Ohnysty and A. R. Stetson, "Evaluation of Composite Materials for Gas Turbine Engines," AFML-TR-66-156, Part 11, December 1967.

48. R. A. Signorelli, J. R. Johnston, and J. W. Weeton, "Preliminary Investigation of Guy Alloy as a Turbojet-Engine Bucket Material for Use at 1650°F," NASA RM E56119, 1956.

49. W. J. Waters, R. A. Signorelli, and J. R. Johnston, "Performance of Two Boron-Modified S-816 Alloys in a Turbojet Engine Operated at 1650°F," NASA Memo 3-3-59E, 1959.

50. M. E. El-Dahshan, D. P. Whittle, and J. Stringer, *Oxid. Met.*, *9*, 45 (1975).

51. C. S. Wukusick, "The Physical Metallurgy and Oxidation Behavior of Fe-Cr-Al-Y Alloys," GEMP-414, General Electric Company, Cincinnati, OH, 1966.

52. A. V. Dean, "The Reinforcement of Nickel-Base Alloys with High Strength Tungsten Wires," NGTE-R-266, National Gas Turbine Establishment, Pyestock, England, 1965.

53. E. A. Winsa, L. J. Westfall, and D. W. Petrasek, "Predicted Inlet Gas Temperatures for Tungsten Fiber Reinforced Superalloy Turbine Blades," NASA TM-73842, 1978.

54. H. J. Gladden, "Air Cooling of Disk of a Solid Integrally Cast Turbine Rotor for an Automotive Gas Turbine," NASA TM X-3471, 1977.

55. D. W. Petrasek, E. A. Winsa, L. J. Westfall, and R. A. Signorelli, "Tungsten Fiber Reinforced FeCrAlY–A First Generation Composite Turbine Blade Material," NASA TM-79094, 1979.

56. P. J. Mazzei, G. Vandrunen, and M. J. Hakim, "Powder Fabrication of Fibre-Reinforced Superalloy Turbine Blades," in *AGARD Conference Proceedings 200* (Advanced Fabrication Techniques in Powder Metallurgy and Their Economic Implications), Advisory Group for Aerospace Research and Development, Paris, 1976, pp. SC 7.1–SC 7.16.

57. P. Melnyk and J. N. Fleck, "Tungsten Wire/FeCrAlY Matrix Turbine Blade Fabrication Study," NASA CR-159788, 1979.

58. C. F. Barth, D. W. Blake, and T. S. Stelson, "Cost Analysis of Advanced Turbine Blade Manufacturing Processes," NASA CR-135203, 1977.

Microballoons

See Fillers, Microspheres

Micromechanical Design

See Composite, Micromechanics

Micromechanisms of Delamination in Composite Materials

In this article we will consider the various micromechanisms that result in the failure of composite materials by delamination. Until the introduction of ply stitching in the mid-1980s, continuous-fiber composite materials were made with all the fiber reinforcement exclusively in the planes of the various plies in the laminate. Such composite laminates are quite strong when loaded in-plane but are relatively weak when loaded out-of-plane, the relatively weak composite matrix strength being the determining factor. Ideally, it would be desirable to design structures that do not experience significant out-of-plane loading. In practice this is not possible.

Out-of-plane loads are developed in composite materials for a variety of reasons, several of which are illustrated in Figure 1. The principal concern in aerospace design today is delamination driven by compressive stresses, which are essentially unavoidable as a result of bending moments, which are usually present in almost all structures. Compressive loads give a driving force for propagation of incipient delamination flaws introduced during manufacturing and/or delamination damage introduced by impact in service. Design strains in airplanes today are in fact driven by the post-impact compression strength, resulting in a maximum allowable strain of 0.004. If the resistance to delamination can be

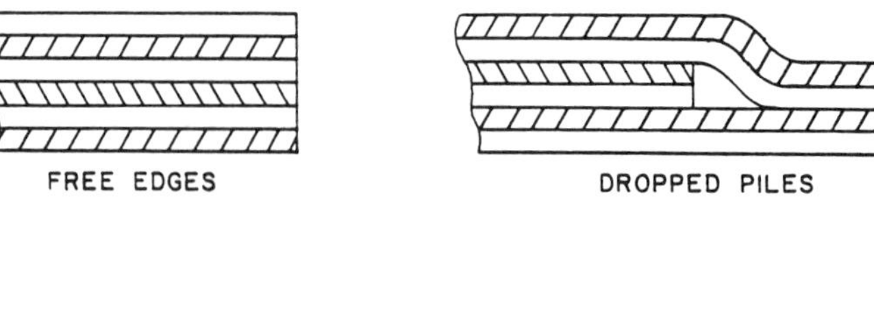

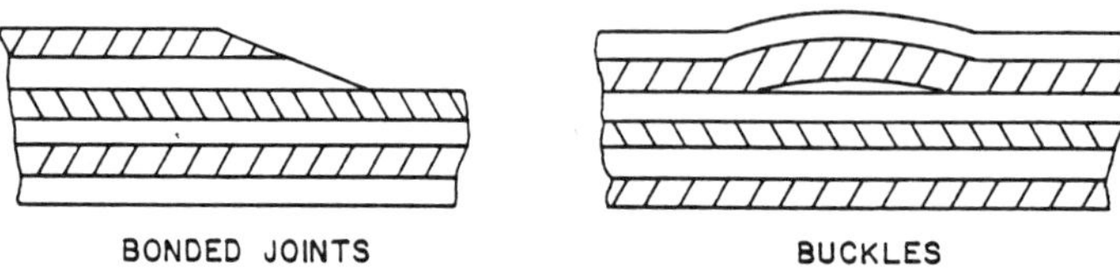

FIGURE 1 **Schematic drawings of four conditions that result in out-of-plane stresses when the primary loading is in-plane.**

substantially improved by either tougher resins or ply stitching, the design will then be driven by fatigue, with an allowable design strain of approximately 0.006.

This discussion makes it very clear that inadequate resistance to delamination results in a very significant performance penalty for polymeric composite materials. This has been recognized since the late 1970s. The simplest approach is to improve the fracture toughness of the resins used for matrices. However, the initial attempts to improve the resistance to delamination fracture toughness by simply improving the resin delamination toughness have met with disappointing results, as shown in summary by Hunston et al. [1] in Figure 2. It is

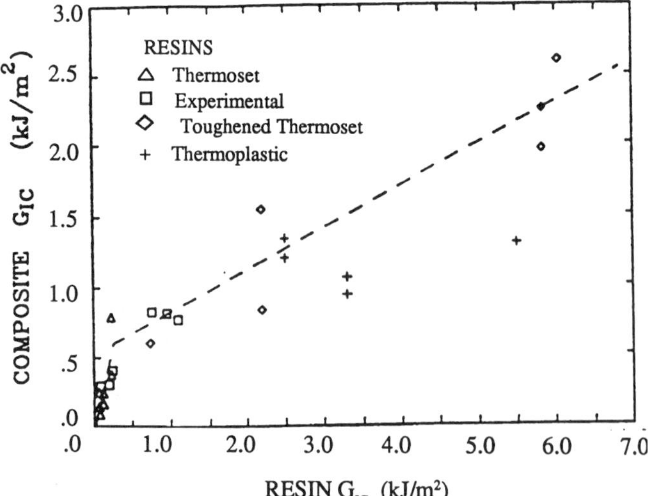

FIGURE 2 **Composite interlaminar strain energy release (for mode I loading) for steady crack growth, presented as a function of the neat resin, mode I energy release rate, G^m_{Ic}.** (From Ref. 1.)

clear from the results presented in this figure that increases in resin fracture toughness from 70 J/m² to 500 J/m² as measured by G_{1c} (to be explained presently) resulted in proportionate increases in the delamination fracture toughness. However, increasing the resin toughness above a value of approximately 500 J/m² gives a much smaller incremental increase in the delamination fracture toughness.

The considerable amount of research performed during the past 10 years to better understand the micromechanisms of delamination and the delamination fracture process have been motivated by the desire to develop composites systems that give a better translation of resin fracture toughness into delamination fracture toughness. Understanding the delamination process in detail is essential if this goal is to be realized. This article summarizes what we have learned during the past 10 years about the micromechanisms that result in delamination crack growth in composites.

Loading Modes for Fracture

Since we will be trying to relate the various fracture processes to their effect on the macroscopic delamination fracture toughness, before getting into the details it would be useful to briefly describe the three modes of delamination fracture and the simple approach used to measure the resistance to delamination crack growth. The three modes of fracture are presented in Figure 3. Whether for a composite material or an isotropic solid, the stresses that produce crack growth can be subdivided into out-of-plane (mode I), in-plane shear (mode II), and out-of-plane shear or tearing (mode III). Most composite delamination is the result of either mode I, mode II, or mixed modes I and II. Thus, most of the work done to date has focused on delamination for these conditions.

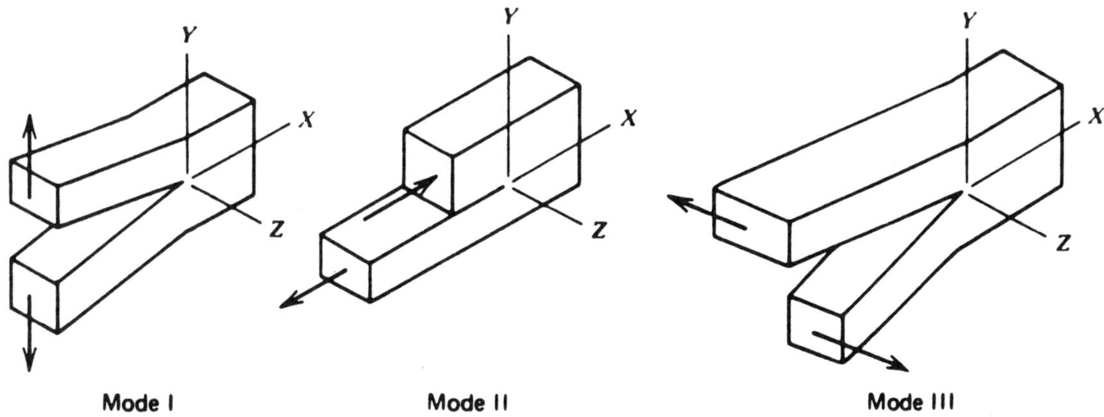

FIGURE 3 Schematic drawings illustrating the various loading modes used in fracture toughness characterization of composite materials using a fracture mechanics approach.

Critical Energy Release—The Measure of Fracture Toughness or Resistance to Crack Growth

The quantity used to characterize resistance to delamination crack growth is the critical energy release rate G_{1c} or G_{2c} for mode I or mode II delamination, respectively. This quantity may be defined by a simple application of the first law of thermodynamics to crack growth under conditions of constant displacement, as shown in Figure 4. A specimen of any geometry containing a crack of length a is loaded, giving the load-displacement curve shown in Figure 4 and a stored elastic energy given as the area under the load-displacement curve. If the crack grows to a new length of $a + da$, the specimen becomes more compliant (less stiff) and the load required to produce the observed displacement is reduced accordingly. The elastic energy stored is given by the area under $0P_1'$ in Figure 4. The area between $0P_1$ and $0P_1'$ represents the

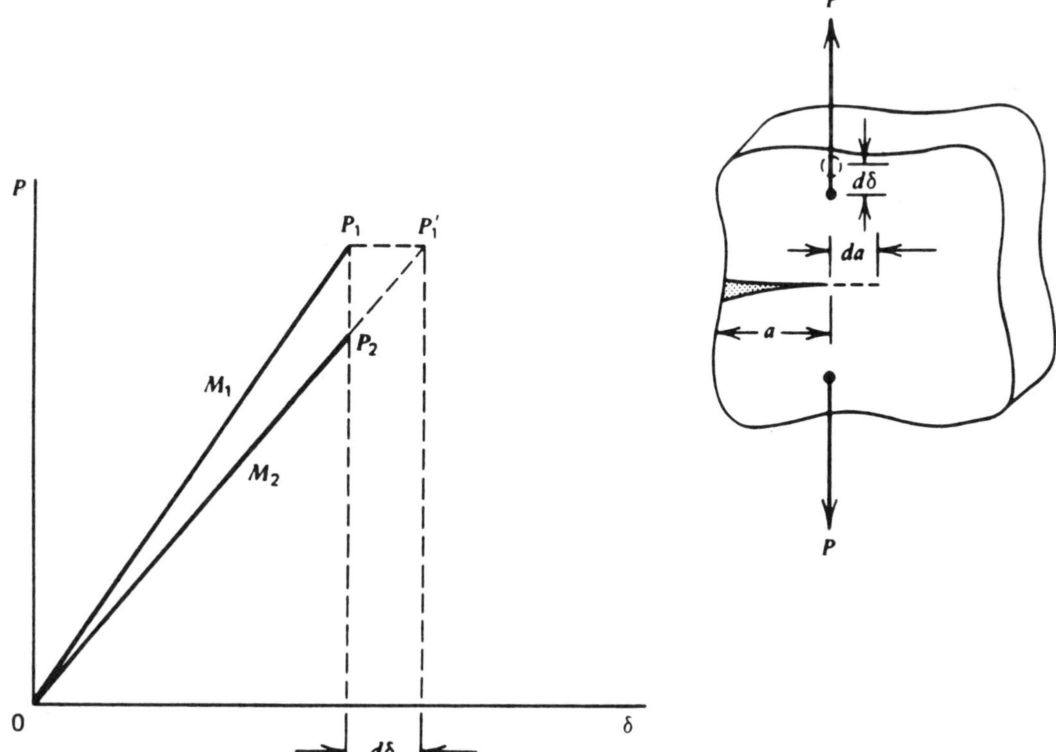

FIGURE 4 Schematic drawing of the load displacement measured in two cracked test specimens with slightly different sizes of cracks.

elastic energy released ($-\ dU$) to grow the crack da in length or $B\ da$ in area, where B is the width of the crack. It is the reduction in the stored elastic energy per unit area of increase in crack area that we call the critical energy release rate, $G_{1c} = -\ dU/B\ da$. The units are energy per unit area, usually joules/meter^2.

Standard specimen geometries used to measure G_{1c} and G_{2c} for delamination in composite materials are shown in Figure 5 along with specimens for measuring the resistance to mixed-mode delamination. The D-30 committee of the American Society for Testing and Materials (ASTM) currently has a subcommittee that is performing a round robin test program to standardize the measurement of G for the various loading conditions.

In the remainder of this article, we will first review the various micromechanisms for mode I delamination of composites with brittle and ductile matrices, including delamination of multidirection as well as unidirectional laminates. Then we will review the micromechanisms responsible for mode II delamination of composite materials with brittle and ductile matrices, after which mixed-mode delamination will be considered.

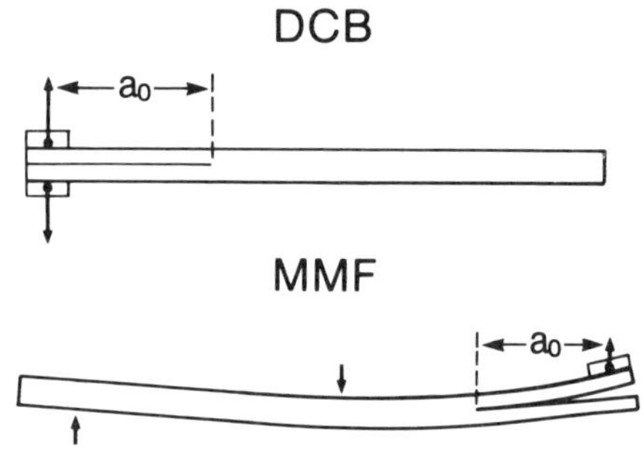

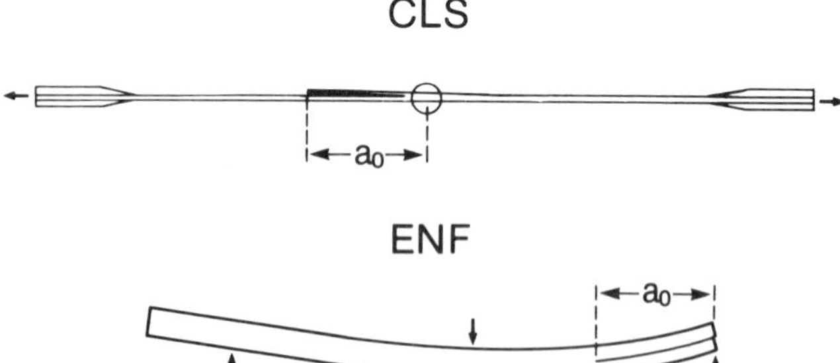

FIGURE 5 **Schematic drawings of four standard test specimens that are used to determine the resistance to delamination for various percentages of mode I and mode II energy release rate. The double cantilevered beam specimen (DCB) gives 100% mode I; the end notched flexure specimen (ENF) gives 100% mode II; the MMF specimen give 57% mode I; the crack lap shear specimen (CLS) varies in the percentage of mode I and mode II, depending on the relative thickness of the two specimens that are lapped.**

Mode I Delamination of Composites

In Situ Observation of Fracture in the SEM

The most direct approach to obtaining an understanding of the interlaminar fracture process is to observe the process directly in real time in the scanning electron microscope (SEM). For six years, our research group at Texas A&M University has been pioneering the development of techniques that not only allow observation of crack tip fracture process in the SEM but also permit measurement of the strain field that develops around a growing crack during neat resin cracking as well as composite delamination [2–8].

The edge of the specimen to be observed in the SEM must be carefully polished, as shown in Figure 6. After polishing, the surface is sputter-coated with a 0.015 μm gold-palladium film to avoid charging in the SEM. After deformation of the specimen, the coating will have very fine microcracks, which are generally visible at magnifications of 3000× or greater. The very fine microcracking observed in the higher magnification photographs to be shown presently should therefore be interpreted as indicative of resin deformation in excess of 3% strain (the strain level required to crack the gold-palladium coating, as determined experimentally). The extent of the fine microcracking gives an indication of the size of the region of nonlinear deformation.

The very fine microcracking that develops in the coating precedes the formation of real voids or microcracks in the bulk matrix (resin). As the coating microcrack density reaches a critical value corresponding to a threshold value of matrix strain, real void formation and/or resin microcracking begins. While it may appear that the coating microcracks are responsible for the much larger resin microcracks that form subsequently, this is not usually the case. The coating microcracks simply correlate with increasing matrix deformation, which is ultimately responsible for the observed matrix damage.

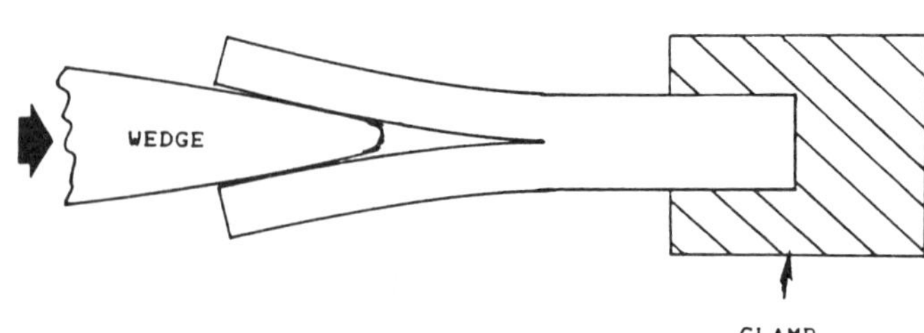

FIGURE 6 **Schematic drawing of a composite specimen being tested in the scanning electron microscope. In the drawing, the fibers run the length of the specimen, and therefore parallel to the surface being polished and viewed. The surface being viewed is the free edge (as opposed to the top or bottom of the laminate), carefully polished to a final finish of 0.1 μm using standard metallographic techniques.**

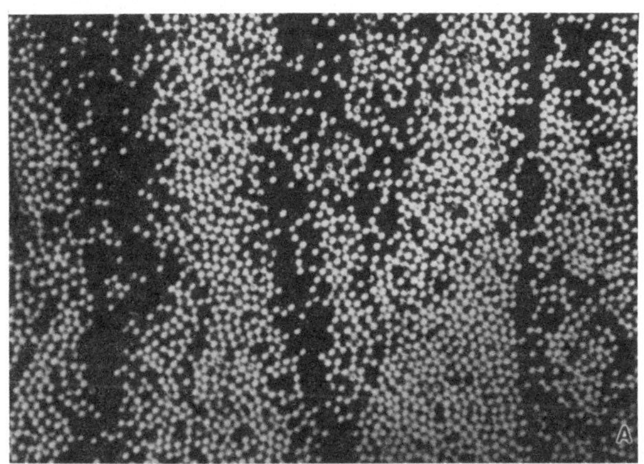

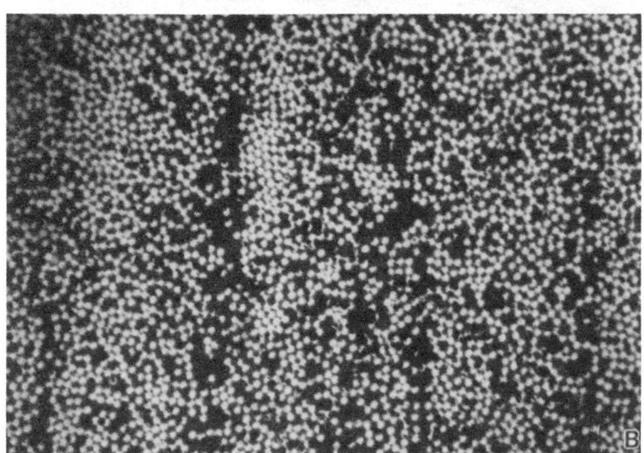

FIGURE 7 Cross section of a composite, showing the fiber volume fraction within respective plies and the resin-rich region between plies. Note that this surface is perpendicular to the one seen in Figure 6. (*a*) (top) AS4/3501-6; (*b*) (bottom) an experimental resin with AS4 fibers. 200×.

FIGURE 8 In situ observation of mode I delamination of AS4/3502 in the SEM. The appearance of fiber termination is misleading and results from the orientation of some fibers not being perpendicular to the polishing plane during specimen preparation. Note that crack propagation appears to be by interfacial debonding. The fine cracks in the matrix (probably due to cracking of the gold-palladium coating) indicate some matrix deformation in conjunction with crack propagation. 1000×.

Microstructure of Polymeric Composite Materials

The microstructure of a polymeric composite material is shown in Figure 7. The cut to prepare the specimen has been taken perpendicular to the fiber direction in a unidirectional composite. The various plies are clearly seen in this composite as a result of the rather thick resin-rich region between plies. Some thermosetting composites may typically have a much thinner (sometimes difficult to distinguish) resin-rich region between plies. In both thermosetting and thermoplastic composites, the thickness of the resin-rich region between plies is dependent on the conditions of processing. The thickness of this region will also have some effect on various mechanical properties of the composite.

Delamination can occur through the center of the resin-rich region between plies or at the edge of one of the adjacent plies. It will be seen presently that delamination usually occurs through some combination of resin cracking and interfacial failure at the boundary of resin-rich region between plies and one of the adjoining plies.

Micromechanisms of Mode I Delamination of Composites

Two factors appear to dominate the mode I delamination fracture behavior in composites, the resin ductility and the interfacial bonding strength. Figures 8 through 12 show in situ observations of fracture in five different systems in order of increasing neat resin toughness.

Mode I delamination of AS4/3502 is seen to occur with little if any crack tip deformation/damage zone development (Fig. 8). Furthermore, cracking at the fiber–resin interface is common. This system has a mode I delamination toughness G_{1c} value of only 175 J/m^2, which is quite low but expected in view of the 70 J/m^2 toughness of the 3502 neat resin.

Mode I delamination of a graphite/BP907 composite is seen in Figure 9. The neat resin BP907 has a toughness of 300 J/m^2 and a composite mode I delamination fracture toughness of 450 J/m^2. Some matrix deformation and microcracking is evident around the crack tip. Again, the fracture process seems to be dominated by interfacial failure, preventing the extraction of the full resin toughness. It is only the fiber bridging behind the crack tip from fiber debonding and the greater fracture surface area developed during composite delamination compared with neat resin failure that allows it to have a higher G_{1c} value.

The delamination of a composite with a somewhat tougher matrix resin, Hexcel F155, which has rubber par-

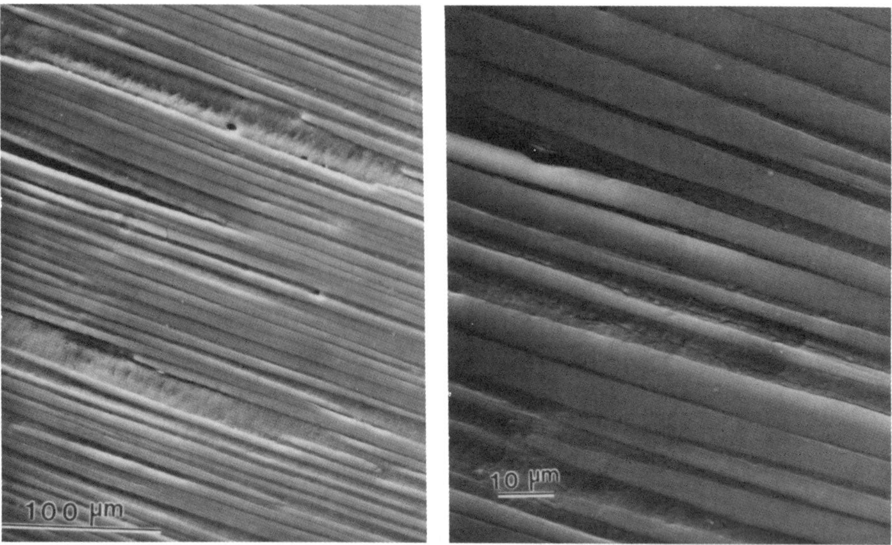

FIGURE 9 In situ observation of mode I delamination of T300/BP907 in the SEM.

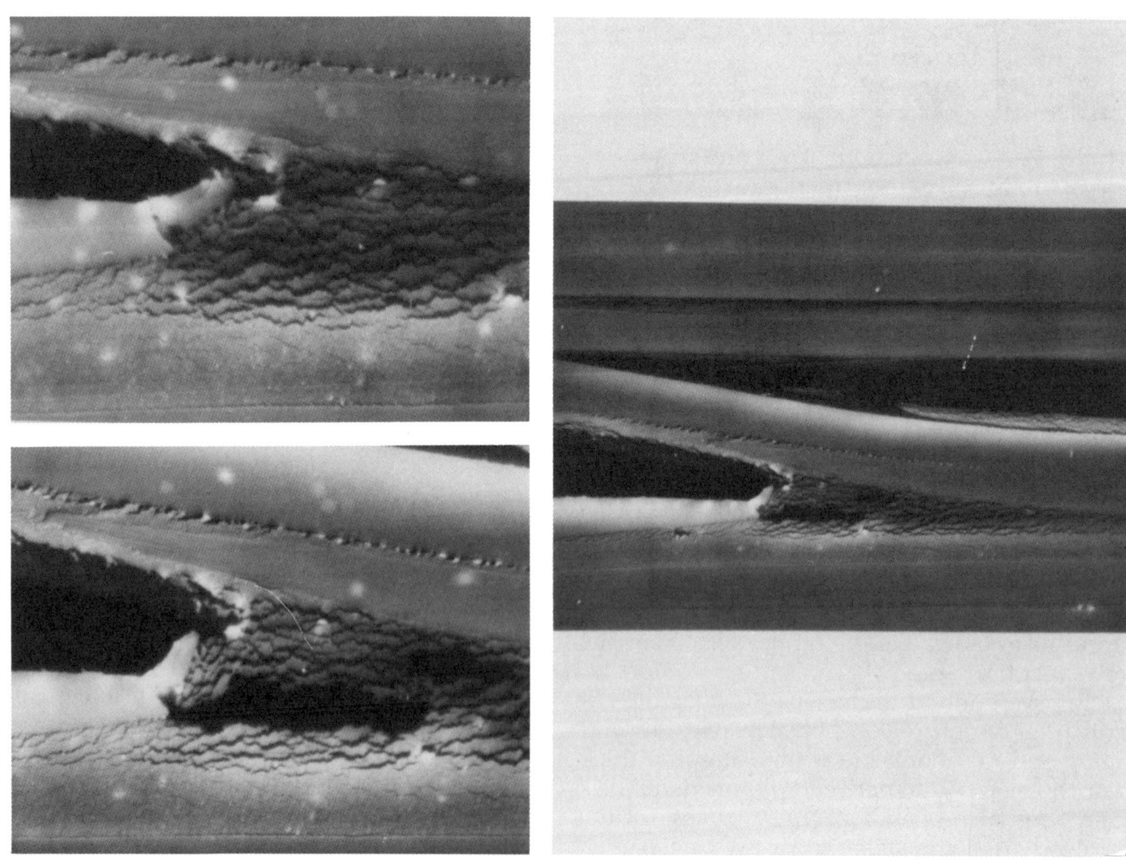

FIGURE 10 (Top) In situ delamination of AS4/3502 showing debonding with very little resin damage (at 1000×). (Bottom, left) In situ delamination of F155 showing extensive microcracking around crack tip (at 3900×). (Bottom, right) In situ delamination of F155 showing coalescing of microcracks to form macroscopic crack growth (at 3000×).

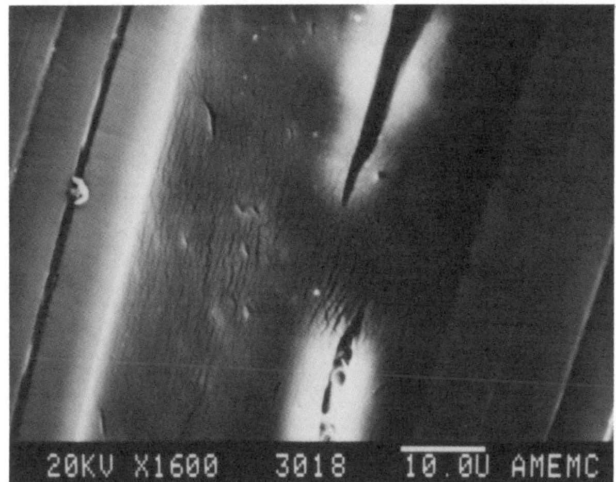

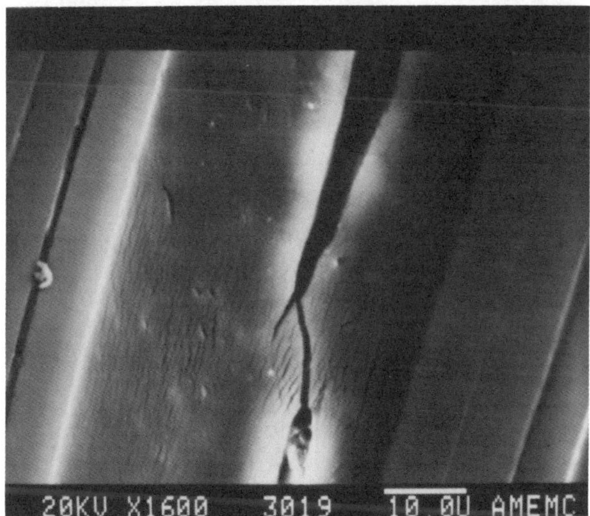

FIGURE 11 In situ observation of mode I delamination of AS4/polycarbonate in the SEM. The very fine cracking seen in the resin-rich region between plies is probably cracking in the gold-palladium film that results from deformation of the resin. The more substantial cracks are resin cracks that eventually develop from the resin deformation. Note crack development ahead of the crack tip with crack growth by subsequent coalescence. 1600×.

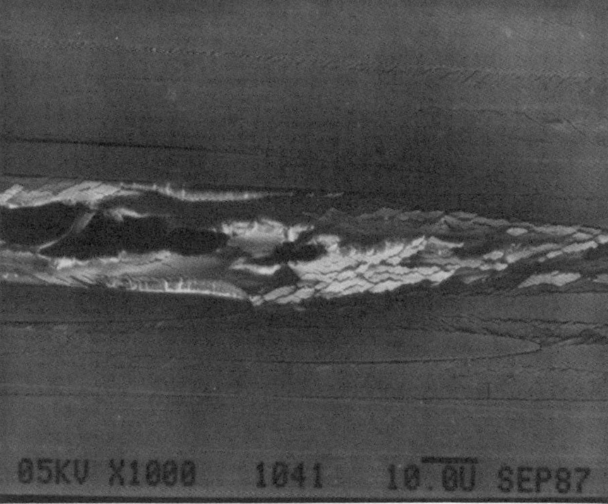

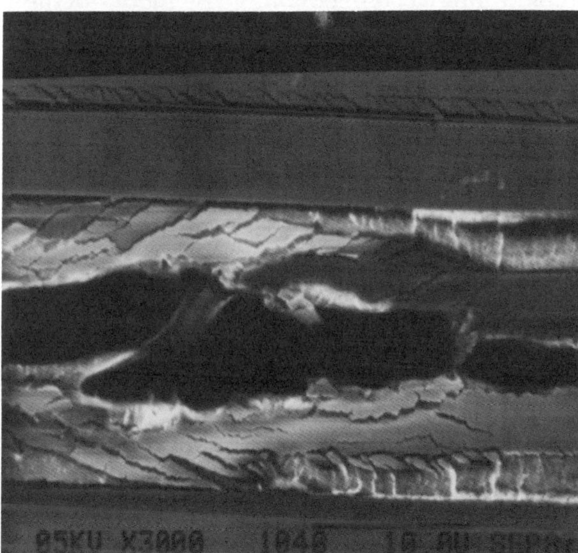

FIGURE 12 F185 mode I delamination.

ticle additions, is seen in Figure 10. The G_{1c} for the neat resin is 730 J/m^2, whereas the mode I delamination G_{1c} is 550 J/m^2. Failure in this system results from a combination of deformation-induced matrix cracking and interfacial debonding. Some deformation is apparent outside the resin-rich region between plies, although the majority of the deformation is concentrated in between the plies in the resin-rich region. The very fine microcracks that appear first are in the gold-palladium coating and occur when a critical strain is reached, as previously noted. They are soon followed, however, by resin microcracks, which coalesce to give crack advance when this process is not preempted by interfacial failure.

The last two composite material systems to be reviewed are much tougher than the three previously presented, and this will be evident from the size and intensity of the deformation zone that precedes the growing delamination crack. In Figure 11, delamination of AS4 graphite in a matrix of polycarbonate is seen. The polycarbonate has a neat resin G_{1c} value of 8100 J/m^2, with a composite mode I delamination G_{1c} value of 1600 J/m^2. Failure is again seen to be a combination of deformation-induced matrix cracking and interfacial failure. In these in situ fracture observations, the very fine microcracks seen are (as before) strain-induced cracking in the coating, whereas the coarser cracks are bulk matrix cracks, which seem to form at local heterogeneities before matrix failure. In general in this system, when matrix microcracking occurs before fiber debonding, the matrix deformation appears to be confined to the resin-rich region between plies, as indicated by the distribution of coating microcracks.

Matrix deformation and damage in Hexcel T6T/145/F185 is seen in Figure 12 to extend several fiber diameters above and below the plane of delamination. In this composite made with a ductile, rubber-toughened epoxy, the extensive deformation gives significant load redistribution ahead of the crack tip, resulting in a mode I delamination fracture toughness of 2000 J/m^2. However, this is significantly less than the fracture toughness of the neat resin, which exceeds 6000 J/m^2. While the total height of the deformation/damage zone around the crack tip in the composite is only slightly less than in the neat resin, obviously the presence of fibers that act as rigid filler significantly reduces the total deformation and thus the load redistribution that can occur, giving a lower fracture toughness. Furthermore, the fracture process in this composite is sometimes "short-circuited" by interfacial failure.

In summary, the fracture process in mode I delamination of composites is seen to be a combination of resin failure and interfacial debonding. The more ductile the resin and the larger the deformation zone ahead of the crack tip, the larger is the observed delamination fracture toughness.

Comparison of Size, Shape, and Intensity of the Deformation/Damage Zone for Mode I Delamination to Neat Resin Crack Growth

Corleto et al. [9] have used a linear, orthotropic finite element analysis to compare the stress field ahead of the crack tip in a double cantilevered beam specimen with orthotropic elastic properties to that in one with isotropic properties. The results are shown in Figure 13, where it is seen that in an orthotropic split laminate specimen the stresses decay more slowly ahead of the crack tip and the stress field tends to be more elongated than circular, compared with that in a specimen with

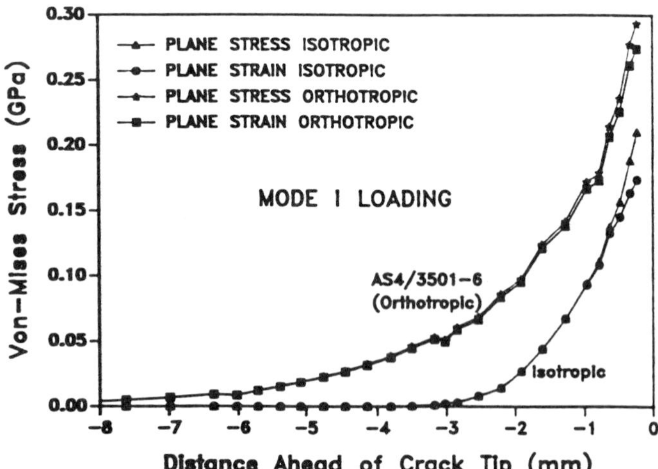

FIGURE 13 Finite element results calculating the effective stress (sometimes called the von Mises stress) as a function distance ahead of the crack tip for a double cantilevered beam specimen of an orthotropic solid (unidirectional AS4/3502) and an isotropic solid.

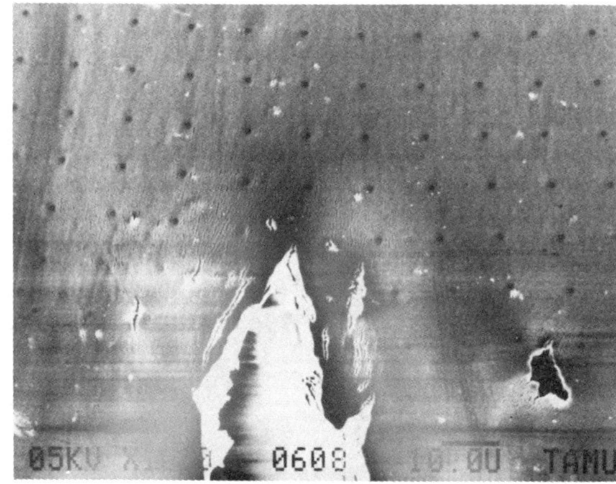

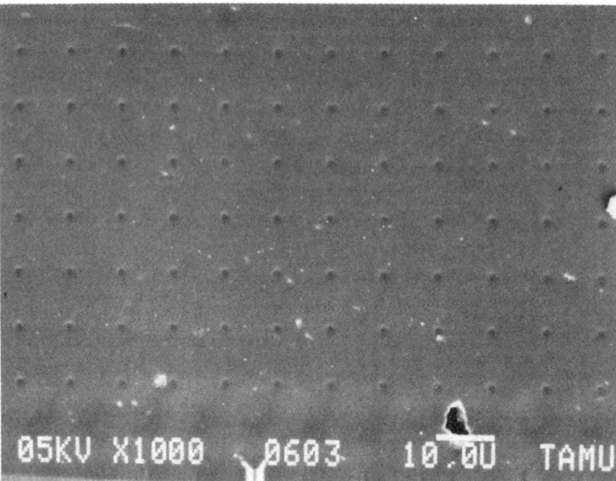

FIGURE 14 In situ observation of crack growth in Hexcel F185 (a moderately cross-linked, rubber-toughened epoxy) in the SEM. A dot map has been placed on the surface by burning small holes in the gold-palladium film. The distortion of this dot map may be used to measure the local displacements and the associated strain field. 1000×.

isotropic behavior. Similar predictions have been made by Wang et al. [10] and Crews et al. [11] for an adhesive bond.

Recently a technique has been developed at Texas A&M that allows the direct measurement of strain fields around a growing crack. This technique makes use of a very fine, regular array of dots (Fig. 14) to measure the crack tip strain field directly. The results of mode I loading of neat resin and mode I delamination of a composite utilizing a similar resin are presented in Figure 15, where it is seen that the shape of the respective strain fields is similar to that predicted in Figure 13.

An important point to note from these results is that the crack tip strain to failure is quite large (~ 15%) compared with the strain to failure measured in a tensile test (~ 1–2% for this system). Similar discrepancies have been seen in a variety of resins. Tensile elongation to failure is obviously limited by incipient flaws and/or strain localization before failure, whereas the growing crack samples such a small volume of material that such

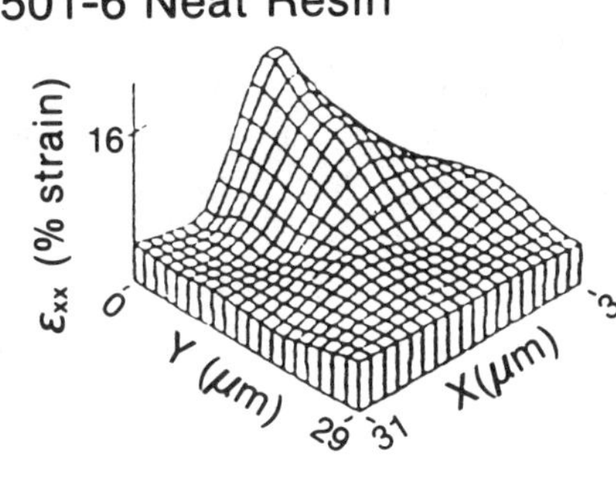

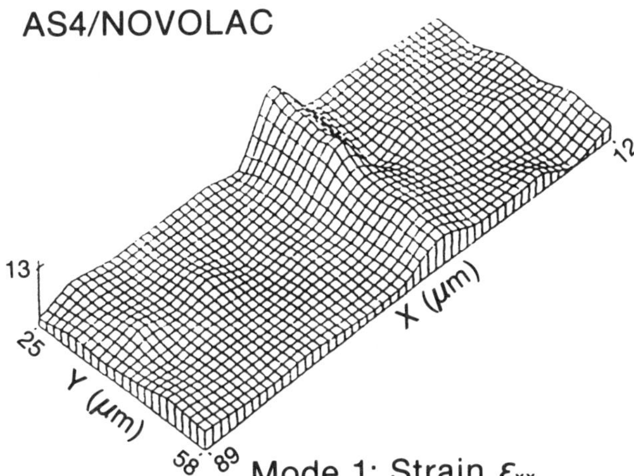

FIGURE 15 **The strain field as measured for neat resin 3501-6 and for a composite made from AS4 fibers and novolac resin, which is also a brittle epoxy.**

effects are minimized. Furthermore, the ratio of the crack tip strain to failure to the resin tensile elongation typically decreases from a factor of approximately 10 (or higher depending on incipient flaw size and brittleness of resin) to values of 3 or less in more ductile systems, as one would expect.

Heterogeneity of Resin and Interfacial Bonding

Direct observations of mode I delamination of composites in the SEM usually indicate a fracture process that is discontinuous and locally unstable, even when the macroscopic testing conditions indicate stability. This behavior is the consequence of resin and interfacial heterogeneity. The development of damage ahead of the crack tip clearly indicates material heterogeneity. For conditions of plane strain, the effect of constraint can cause the stress to rise as one moves away from the crack tip, giving damage ahead of the crack tip for a homogeneous material. However, our surface observations are all for plane stress conditions. Thus, damage ahead of the crack tip must be the result of material heterogeneity. Such behavior is seen in Figures 8 through 12, which indicate resin and interfacial heterogeneity. It is for this reason that the fracture surface after mode I delamination shows a combination of resin cracking and interfacial failure. Thus, along a given fiber, the crack may move alternately through the resin and at the interface.

Mechanism for Mode I Delamination Crack Growth

Recently, Chakachery [8] has proposed a qualitative model for mode I delamination crack growth. Her model, pictured schematically in Figure 16, suggests that the crack—which must still pass through the resin, even when a portion of the local fracture is via interfacial failure—grows from the adjacent fibers toward a line equidistant from and parallel to the two adjacent debonding fibers. This is because the local stresses along the crack tip would be highest adjacent to the debonded fibers, as indicated in Figure 16. This accounts for the fracture surface artifacts that are often found on the fracture surface after mode I delamination, as seen in Figure 17.

Delamination of Multiaxial Layups

While almost all delamination studies have been conducted with unidirectional layups (except for the edge delamination test), most composite structures utilize

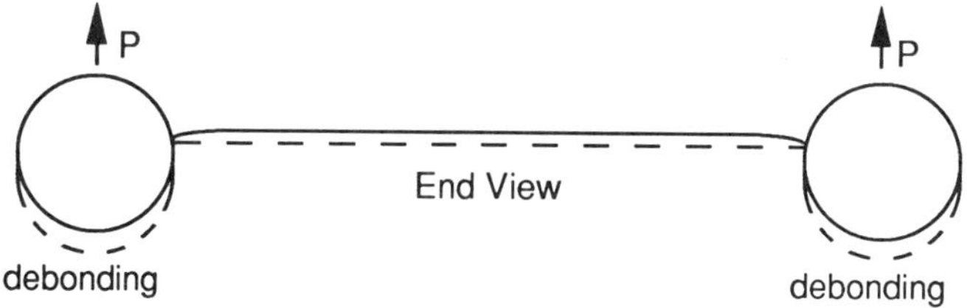

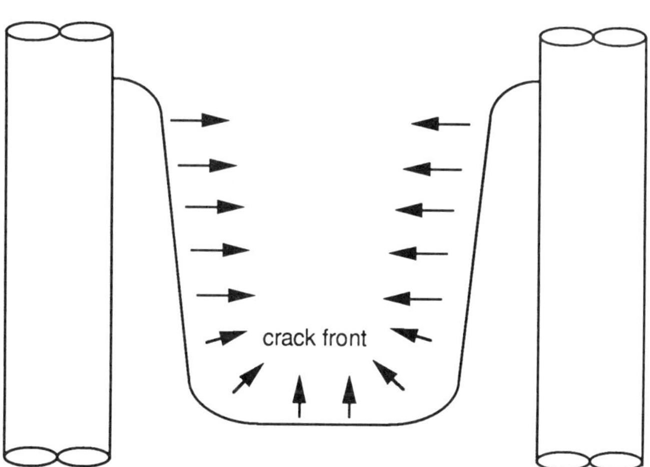

FIGURE 16 **Schematic drawing showing crack growth during mode I delamination of a composite, where crack advance is greater adjacent to fibers and slower between fibers as a result of stress concentrating effect of the fiber.**

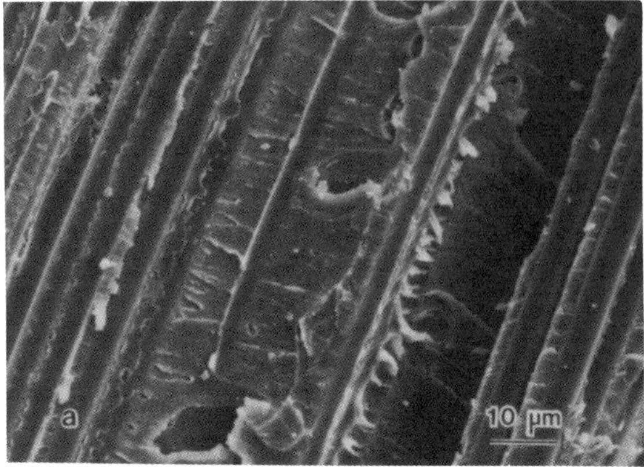

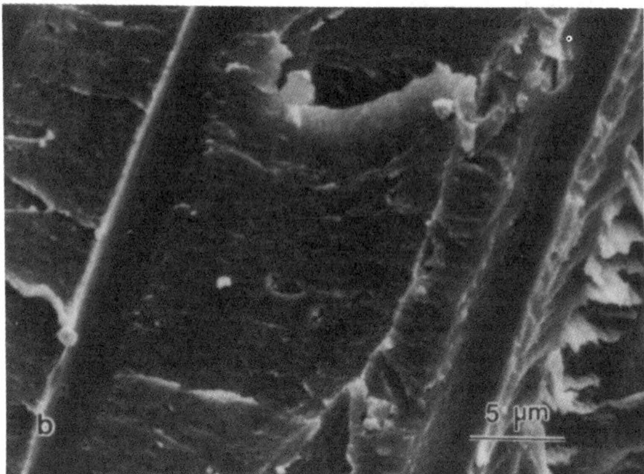

FIGURE 17 **Mode I delamination with general evidence of resin crack growth nearly perpendicular to the fibers, as suggested by the physical model proposed in Figure 16.**

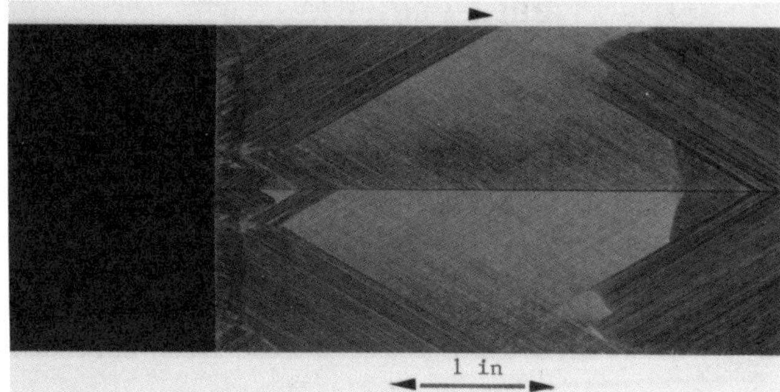

FIGURE 18 **Two fractured surfaces for mode I delamination of a double cantilevered beam specimen in which 45° plies were at the midplane. Note that closing the specimen would have the fibers on either side of the fracture running parallel, which indicates that crack growth went from an initial location (due to a Teflon insert) between 45° plies to a new location propagating within a particular ply.**

multidirectional layups. Recent work done at Texas A&M University by Jordan [6] and Goetz [12] has attempted to explore what perturbation in the fracture process results when multiaxial composites delaminate.

In the simplest perturbation, unidirectional plies are placed on either side of the Teflon crack starts, but other orientations of plies are used in making the laminate. The primary result of such a perturbation is that considerable damage and the associated energy dissipation occurs at locations other than the crack tip. Unless the conventional G_{1c} approach is replaced by a J_{1c} approach, the estimate of fracture work (initial energy release rate) will be artificially high.

When ply orientations other than 0 are used on either side of the Teflon crack starter, greater perturbation results. Typically the crack will not propagate very far in the initial region between plies, but tends to move several fiber diameters into one of the plies, after which it propagates in a similar fashion. This behavior is clearly indicated in the delaminated specimen shown in Figure 18, where the fibers on each side of the delamination crack run in the same direction when the specimen is closed on itself. A micrograph showing crack propagation in a DCB specimen that is delaminating from a Teflon starter crack bounded by $\pm$ 45° plies is seen in Figure 19. It is clear that the crack is propagating through a ply rather than between plies.

It is probably true that cracks in individual composites would also prefer to propagate in-ply rather than between plies, but find it more difficult to move out of their original plane of growth than does a delamination crack bounded by $\pm$ 45° plies. It should also be noted that crack growth from Teflon bounded by $\pm$ 45° plies occurs under mixed-mode conditions, even though the macroscopic loading may be nominally mode I. This is due to the interlaminar shear stress, which adds to the mode I loading applied to the specimen.

Fractography and Mode I Delamination

The difference in mode I delamination of composites made with ductile and with brittle resins is sometimes

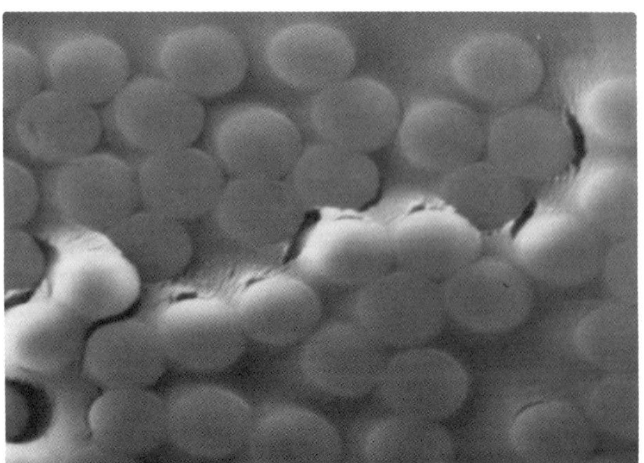

FIGURE 19 **In situ observations of mode I delamination in AS4/3502. Note that the initial crack was between two 45° plies, but crack growth has now moved into one of the plies, consistent with the observations in Figure 18.**

less dramatic than one might expect. A brittle resin will usually have an almost mirror-smooth fracture surface. A ductile resin will often have a relatively smooth (but not mirror-smooth) fracture surface as well, even though considerable crack tip stretching precedes the crack tip rupture. To the extent that the fracture process is continuous, no significant surface roughening results. Thus, direct observations of crack growth in the SEM (Figs. 8–12) or post-mortem sectioning perpendicular to the fracture plane and then thinning, as is done by Yee et al. [13], is necessary to determine the degree of deformation/damage around the crack tip, as the post-mortem fracture surface alone gives an ambiguous indication. By contrast, ductile fracture in metals produces a dimpled fracture surface that is distinctively different from that of brittle fracture, which has the characteristic river pattern associated with cleavage.

Micromechanisms of Mode II and Mixed Mode I/Mode II

A significant difference in the mode I and mode II delamination fracture processes is observed for composite materials made with brittle and ductile matrices. Figure 20 contrasts mode I and mode II delamination of AS4/3501–6. It should be emphasized that the mode I failure in Figure 20 is observed at 1500×, whereas the mode II failure is seen at only 600×. Mode I failure occurs with the development of a small deformation zone ahead of the crack tip, evidenced by the film microcracking in Figure 20. On the other hand, mode II delamination crack growth begins with the formation of a series of sigmoidal-shaped matrix microcracks ahead of the crack tip with an orientation of approximately 45° to the fiber direction (which for mode II loading is the principal normal stress plane). The visible damage extends over a distance greater than 150 μm ahead of the crack tip.

Figure 21 shows even more clearly the development of sigmoidal-shaped microcracks during delamination crack growth of a composite (AS4/3501-6) with a brittle matrix. Again crack advance is seen to be the result of coalescence of the microcracks. This distinction between the mode I and mode II delamination fracture processes for composites made with brittle resins is clearly seen in post-mortem fractography.

Figure 22 shows the fracture surface of AS4/3501-6 delaminated with mode I loading (top) and mode II loading (bottom). The very discontinuous crack growth by microcrack coalescence leaves the mode II delamination fracture surface with a ratcheted appearance. The uneven topography of the mode I fracture surface also suggests

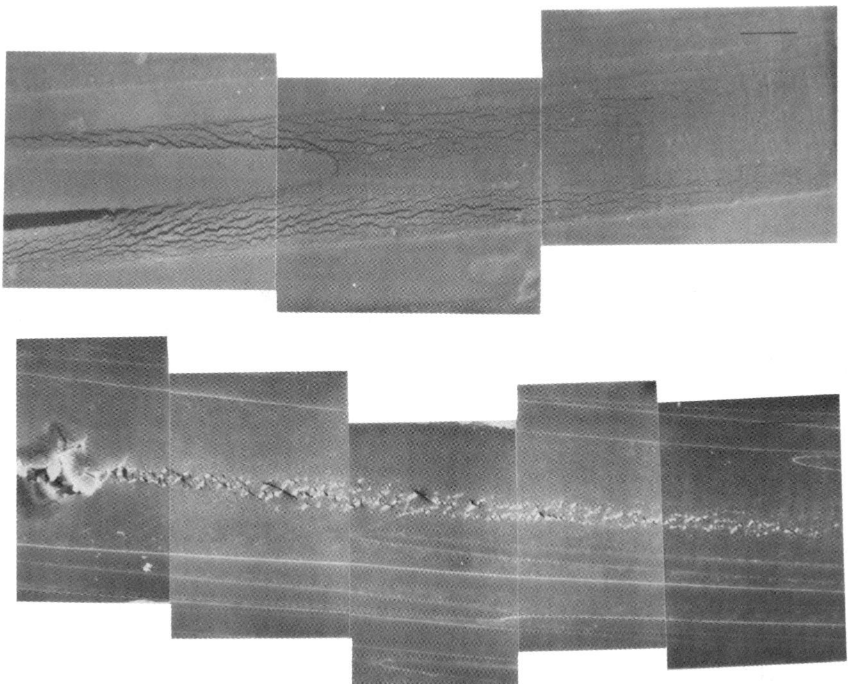

FIGURE 20 In situ observations of mode I and mode II delamination of AS4/3501-6. Note that the crack growth for mode I (top, 1500×) is primarily by debonding, with a very localized deformation field as indicated by the film microcracking. By contrast, mode II delamination gives a much more extensive damage zone ahead of the crack tip, with the eventual development of microcracks at 45° to the orientation of the primary crack (far left). Note that the magnification for mode II delamination (bottom) is only 600× and the white spots are local charging due to film rupture resulting from resin deformation.

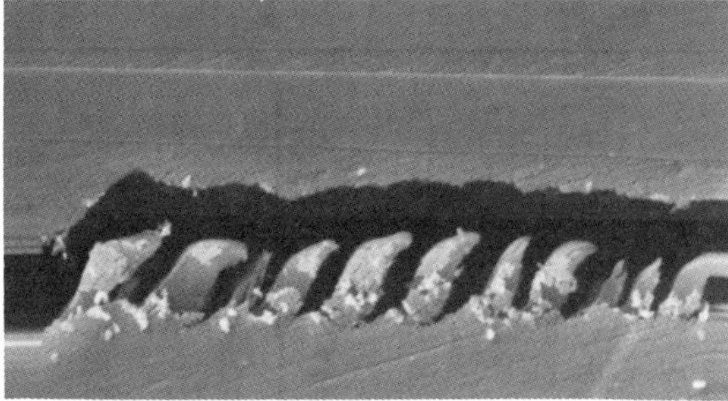

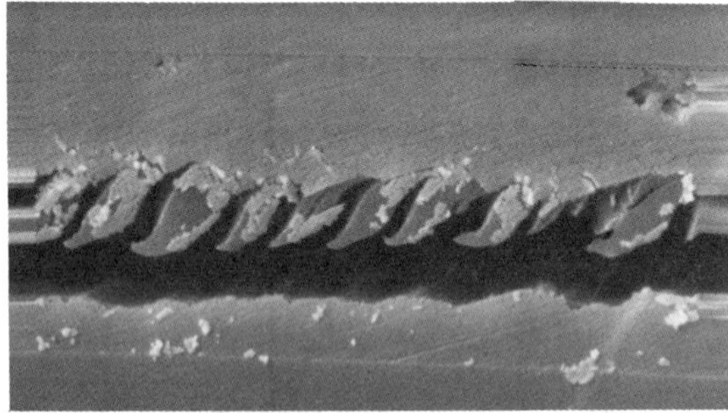

FIGURE 21 In situ observations of mode II delamination of AS4/3501-6 in the SEM. The coalescence of the sigmoidal-shaped microcracks ahead of the crack tip results in a "zipper" appearance on one fracture surface and a scalloped appearance on the opposite fracture surface. 2000×.

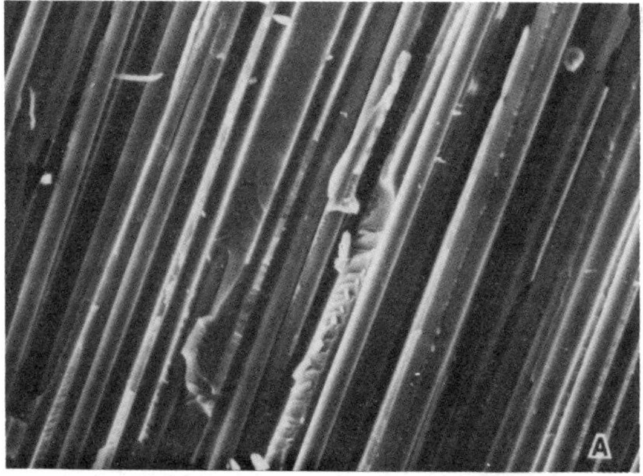

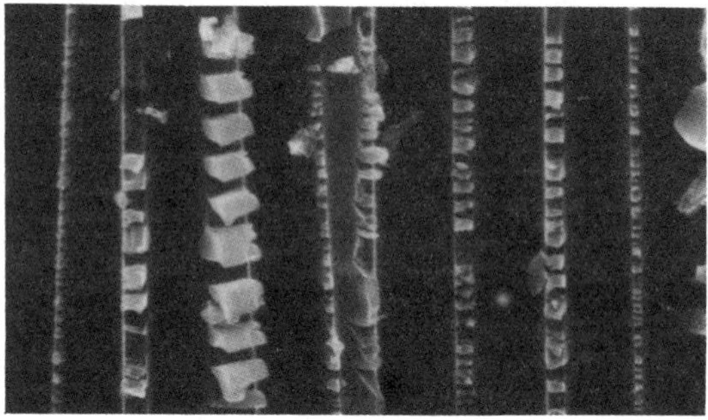

FIGURE 22 Post-mortem observations in the SEM of mode I delamination (top, 500×) and mode II delamination (bottom, 1500×) in AS4/3502. The significant differences in the two failure modes seen in Figure 20 are reflected here in the very unhackled (top) and very hackled (bottom) fracture surface morphologies.

damage development ahead of the crack tip, with crack extension occurring by coalescence of the damage and the crack tip. However, the damage development in this case is all on planes parallel to the crack plane. Thus, the fracture surface is much more flat in appearance.

For mode II delamination in a more brittle system, such as AS4/3502, many "zipper" looking facets, usually called hackles, are formed, as seen in Figure 23. This is in contrast to the behavior of this same composite when delamination occurs under mode I conditions. Depending on whether microcrack coalescence takes place at the top or the bottom of the resin-rich region between plies, a fracture surface may have hackles, scallops, or both, as seen in Figure 23. A detailed discussion of the hackle formation during mode II delamination has been presented by Hibbs and Bradley [14]. The number of hackles and their orientation, or angle with respect to the ply plane, increases as the percentage of shear loading (mode II) increases, as may be seen by comparing Figure 23 to Figure 24. As the percentage of mode II loading is increased, the angle of the plane of action of the principal normal stress increases monotonically from being parallel to the ply plane in mode I to 45° to the ply plane for pure mode II loading. The matrix fracture in composites made with brittle resins should occur in the principal normal stress plane. Therefore, the hackles are apparently the result of microcrack nucleation in advance of the crack tip and subsequent propagation in a brittle fashion through the resin-rich region on planes of principal normal stress, as shown schematically in Figure 25. As these microcracks encounter the fibers of the plies that bound the resin-rich region, they are stopped. The coalescence of these microcracks at the ply boundaries constitutes growth of the primary delamination crack.

While the concept that mode II or mixed-mode loading of a relatively brittle resin composite gives microcracking on the principal normal stress plane and that subsequent coalescence gives rise to hackle formation seems qualitatively correct, there must be more to the explanation. The post-failure fractograph for the pure mode II delamination (Fig. 22) clearly indicates a final hackle orientation much greater than the 45° (from the plane of delamination) that one would predict based on a simple stress analysis (see Fig. 25). Rotation due to shear loading of the hackles just before microcrack coalescence (Fig. 25) could alter the orientation. This rotation would not only orient the hackles more vertically but also separate the hackles near the boundary where coalescence occurs (Fig. 25). The opening of the microcracks gives the impression of material lost, which might correspond to matching hackles on the opposite surface. However, the mating fractured surfaces have dish-shaped or scalloped regions that correspond to places where the microcracks coalesce on that surface (Fig. 23).

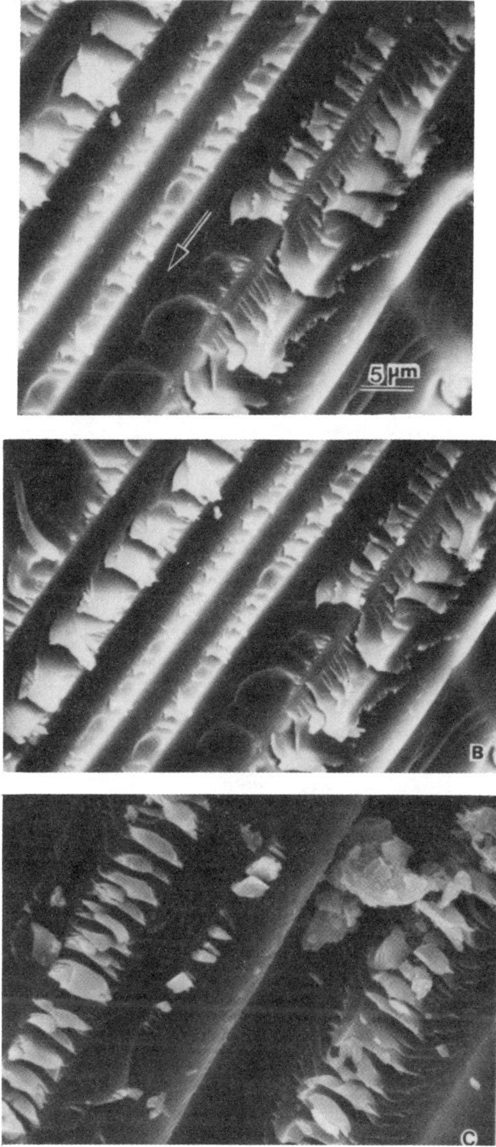

FIGURE 23 Post-mortem fractography of delamination of AS4/3501-6, with 43% of the total energy release rate being mode II. Note the combination of hackles and scallops, depending on whether the coalescence of the sigmoidal-shaped microcracks occurs on the top or bottom ply in the DCB specimen. 1500×.

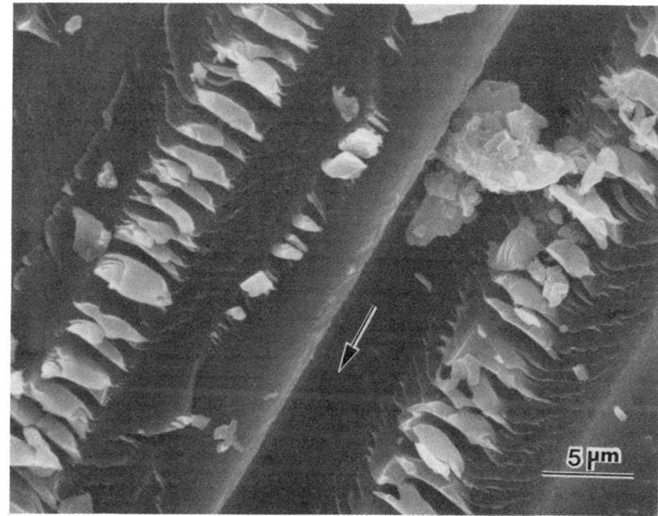

FIGURE 24 Post-mortem observation of delamination of AS4 fracture surface after mode II experimental epoxy (top, 500×) and 100% mode II (bottom, 3000×). Note that the hackle orientation increases to a more nearly vertical orientation as the percentage of mode II in the total energy release rate increases to 100% (compare Figures 23 and 24). Mode I delamination occurs with little or no hackle formation.

The mode II in situ results for AS4/3501-6 (Fig. 21) clearly show that the direction in which the hackles point depends on whether the coalescence of the microcracks occurs at the upper or lower boundary of the resin-rich region between plies. In all the mixed-mode and mode II fractured surfaces observed, the number of hackles on the upper surface appears to be the same as the number of hackles on the lower fractured surface. Therefore, no preference as to whether coalescence takes place at the top or lower boundary is indicated. Once microcrack coalescence begins on either the upper or lower boundary of the resin-rich region, it will continue on that same boundary until some obstruction is encountered.

The formation of hackles is a very energy dissipative process. Thus, the increasing density of hackles noted as one increases the percentage of mode II loading should be reflected in an increasing critical energy release rate

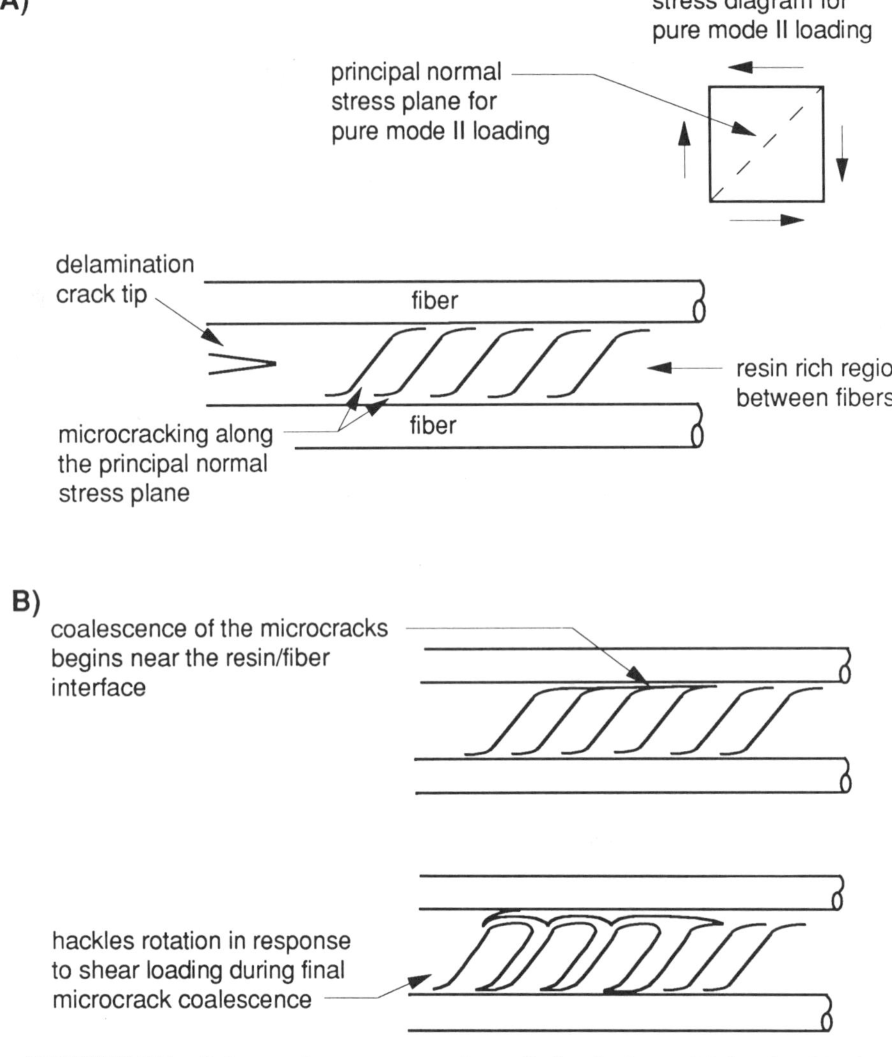

FIGURE 25 **Schematic representation of physical model for hackle formation during mode II delamination of composites made with a brittle resin. Compare this figure with Figures 20 through 24.**

G_c for composite delamination. This is observed, with G_{2c} values usually found to be 3 to 10 times G_{1c} values for composites made with brittle resins.

Contrasting Mode II Delamination in Composites with Ductile and Brittle Resins

The change in the mode II fracture process as one goes from composites made with brittle resins to composites made with ductile resins is dramatic, as seen in Figures 26 and 27 (to be compared with Figs. 22 through 24). As the neat resin toughness increases, the density and distinctive appearance of hackles is gradually replaced by a more featureless flat surface, indicating a ductile shear failure process. In systems with sufficient ductility to have mode II delamination without hackle formation, the G_{2c} value is similar to G_{1c}, as one might expect.

In comparing mode I to mode II delamination in general (for both brittle and ductile resins), one additional distinction is noted. Mode I delamination is much more likely than mode II delamination to give interfacial failure and fiber bridging behind the crack tip. This implies that the peel stress (normal stress at the resin–fiber interface) associated with mode I loading is much more likely to give interfacial failure than is the interfacial shear stress that develops during mode II loading.

Size, Shape, and Intensity of Deformation/Damage Zone for Mode II Loading

A comparison of the decay of the stress field ahead of the crack tip for mode II delamination and mode I delamination using a split laminate has been made by Corleto, Bradley, and Henriksen [9]; this is shown in Figure 28. The mode II delamination is seen to have a much more gradual stress field decay than does mode I delamination. This explains the previously observed much longer damage zone ahead of the crack tip for mode II loading than for mode I. Strain field mapping in the SEM (as

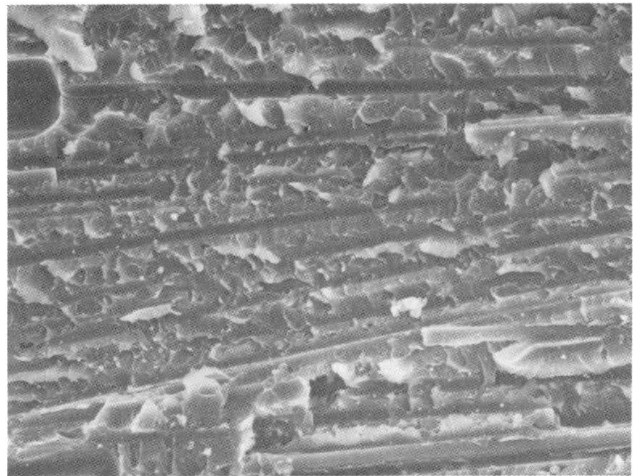

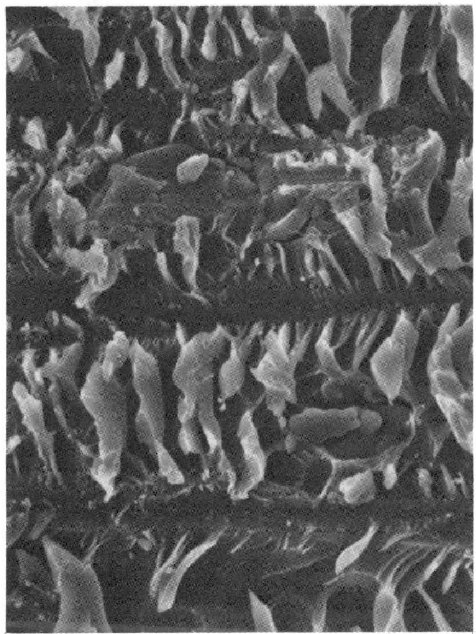

FIGURE 26 **Post-mortem fractography of mode I (top, 250×) and mode II (bottom, 1000×) delamination of T6T145/F155. Note the more ductile appearance of the fracture surface for both mode I and mode II delamination.**

previously described) indicates a very long, narrow deformation/damage zone, as seen in Figure 29. It should be noted that where microcracking is present, the strain field mapping gives an artificially high indication of the actual resin strain, combining as it does displacements due to resin deformation with displacements due to microcracking. From a fracture mechanics point of view, both resin deformation and microcracking give local softening at the crack tip and load redistribution away from the crack tip.

Summary

The fracture process is seen to depend on both the constitutive behavior of the resin (ductile or brittle) and the mode of loading (mode I or mode II). For composites with a brittle resin matrix, the fracture process for mode

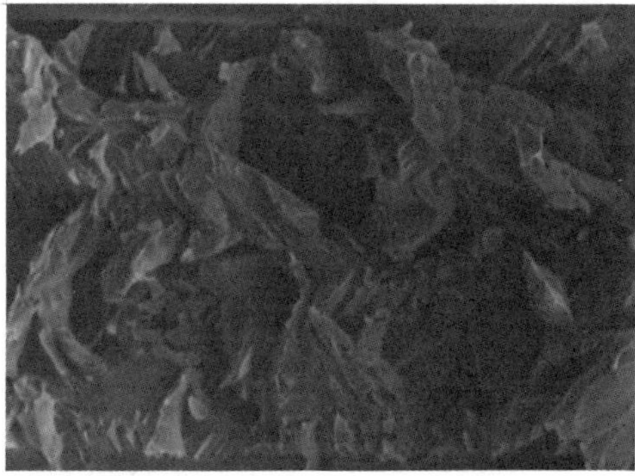

FIGURE 27 **Post-mortem fractography of a DCB specimen of T6T145/F185 that failed under mode II loading. Note the absence of any clearly formed hackles as a result of the very ductile nature of the resin.**

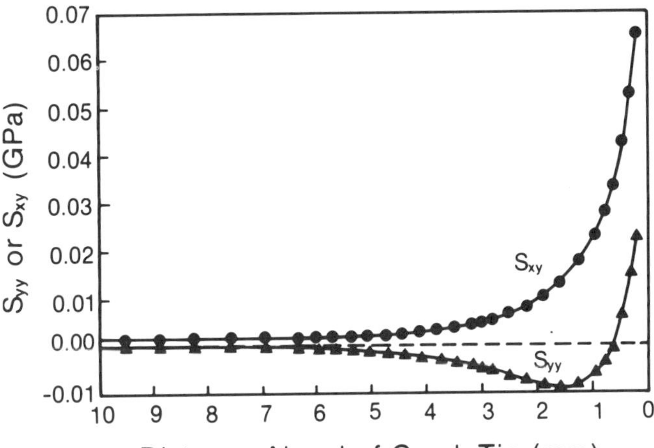

FIGURE 28 **Finite element analysis indicating the stress field ahead of a crack tip for mode I and mode II delamination in a DCB specimen. Note that the stress field decay is much slower for mode II loading (S_{xy}) than for mode I loading (S_{yy}).**

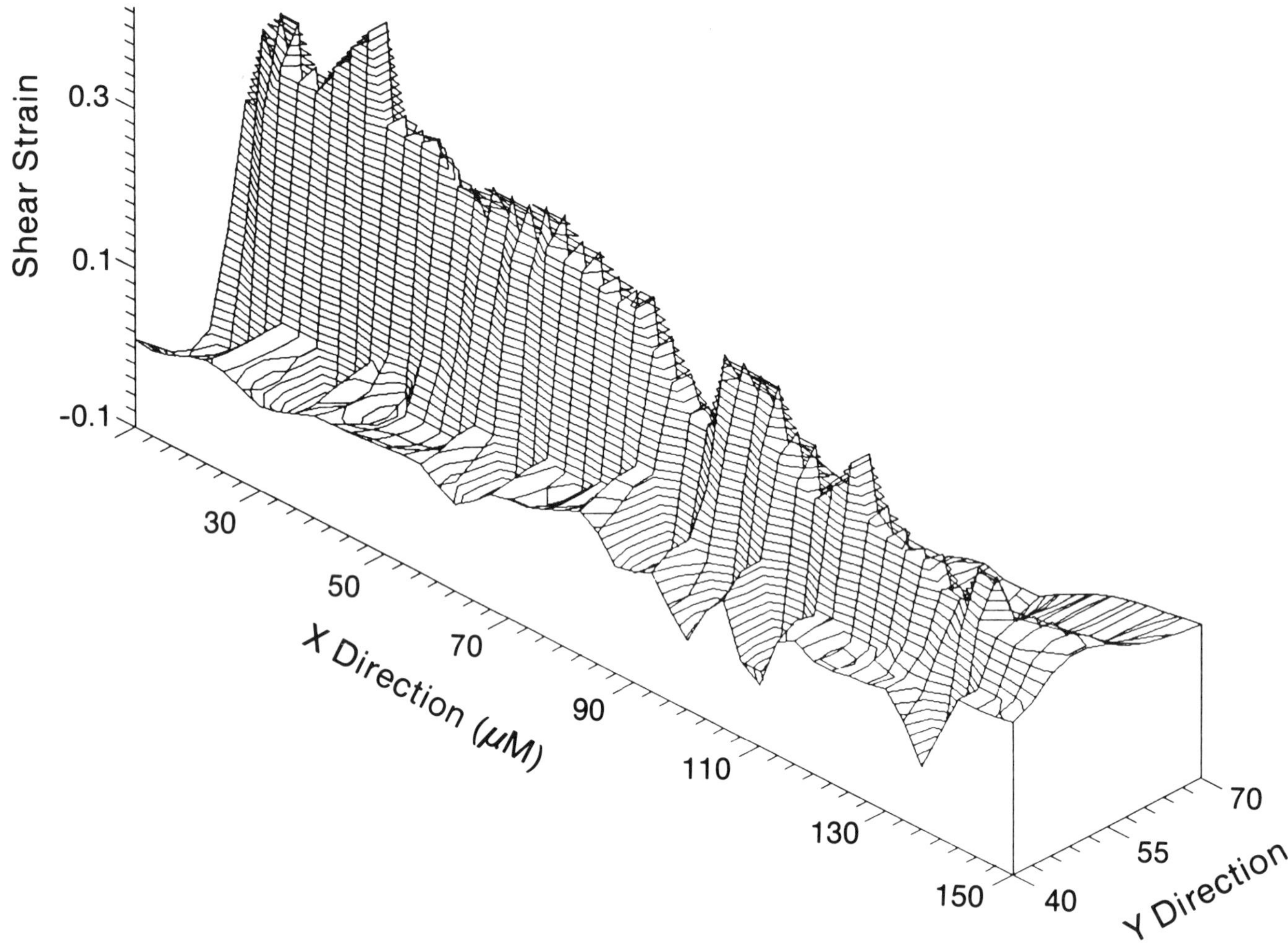

FIGURE 29 Strain field map showing the strain field ahead of a crack tip for mode II delamination in AS4/3502. Note that the strain field calculated from a distorted dot map ignores cracking. Thus, the "strain" indicated includes real resin strain and apparent strain via microcracking.

I loading occurs within a very localized deformation/damage zone around the crack tip and usually involves some combination of interfacial failure and matrix cracking. For mode II loading of brittle composites, a long, narrow damage zone of 45° microcracks develops ahead of the crack tip. The coalescence of these microcracks gives the distinctive hackled surface that is characteristic of mode II delamination of composites with brittle resins.

In both mode I and mode II delamination of composites made with very ductile resins, there is concurrent development of a quite large deformation/damage zone ahead of the crack tip. The fracture surfaces will be relatively similar, each showing considerable resin flow during the fracture process.

Walter L. Bradley

[*See also* FRACTOGRAPHY.]

References

1. D. Hunston and R. Dehl, *Proceedings of the Society of Manufacturing Engineers,* Paper EM87-355, 1987.
2. W. M. Jordan and W. L. Bradley, "Micromechanisms of Fracture Toughness in Graphite-Epoxy Laminates," in N. J. Johnston, Ed., *Toughened Composites,* ASTM STP 937, American Society for Testing and Materials, Philadelphia, 1987, pp. 95–114.
3. W. L. Bradley and R. N. Cohen, *Proceedings of the 4th International Conference on Mechanical Behavior of Materials,* 1983.
4. M. F. Hibbs, M. K. Tse, and W. L. Bradley, "Interlaminar Fracture Toughness and Real-Time Fracture Mechanisms of Some Toughened Graphite/Epoxy Composites," in N. J. Johnston, Ed., *Toughened Composites,* ASTM 937, American Society for Testing and Materials, Philadelphia, 1987, pp. 115–130.
5. E. A. Chakachery and W. L. Bradley, *Polym. Eng. and Sci.,* 27(1), 33 (1987).

6. W. M. Jordan, "The Effect of Resin Toughness on the Delamination Fracture Behavior of Graphite/Epoxy Composites," Ph.D. dissertation, Texas A&M University, College Station, Texas, 1985.
7. M. F. Hibbs, "The Effect of Resin Toughness on the Delamination Behavior of Composites," Ph.D. dissertation, Texas A&M University, College Station, Texas, 1988.
8. E. A. Chakachery, "Deformation Mechanisms in Polymeric Matrices and Carbon Fibre Reinforced Composites," Ph.D. dissertation, Texas A&M University, College Station, Texas, 1988.
9. C. Corleto, W. Bradley, and M. Henriksen, *ICCM & ECCM Proceedings of Sixth International Conference on Composite Materials,* Elsevier Applied Science, New York, 1987, pp. 3.378–3.388.
10. S. S. Wang, J. F. Mandell, and F. J. McGarry, *Int. J. of Fract., 14,* 39 (1978).
11. J. H. Crews, Jr., K. N. Shivakumar, and I. S. Raju, "Factors Influencing Elastic Stresses in Double Cantilever Beam Specimens," in W. S. Johnson, Ed., *Adhesively Bonded Joints: Testing, Analysis and Design,* ASTM STP 981, American Society for Testing and Materials, Philadelphia, 1985.
12. D. Goetz, "Determination of the Mode I Delamination Fracture Toughness of Multidirectional Composite Laminates," Ph.D. dissertation, Texas A&M University, College Station, Texas, 1988.
13. A. F. Yee, R. A. Pearson, and H. J. Sue, *Proceedings of Seventh International Conference on Fracture,* Pergamon Press, New York, 1989, vol. 4, pp. 2739–2746.
14. M. Hibbs and W. Bradley, "Correlations between Micromechanical Failure Processes and the Delamination Toughness of Graphite/Epoxy Systems," in J. E. Masters and J. J. Au, Eds., *Fractography of Modern Engineering Materials: Composites and Metals,* ASTM STP 948, American Society for Testing and Materials, Philadelphia, 1987, pp. 68–97.

Microspheres

See Fillers, Microspheres

Microwave Composites

Plastic dielectric materials are widely used in microwave engineering because of their obvious electrical properties. In the majority of applications, conventional polymers perform satisfactorily, but when the application involves structural or environmental demands, plastics composites play a significant role.

It is appropriate at this stage to give some basic definitions applicable to the microwave composites field.

The Microwave Frequency Band

Frequencies lying between 1 gigahertz (GHz) and 300 GHz are considered to be the microwave domain of the electromagnetic spectrum (Fig. 1), with most applications falling in the 1 to 20 GHz bracket. The corresponding free-space wavelength (λ) range is 300 mm (1 GHz) to 1 mm (300 GHz). A number of frequency-band letter designations are freely used; these are shown, together with some typical applications, in Figure 2.

Microwave Composites

A microwave composite is a mix of two or more materials, at least one of which, in this case, is either a thermoplastic or a thermosetting resin and the other is a reinforcement, brought together to satisfy a microwave requirement (e.g., an antenna) or a microwave requirement that arises from environmental/aerodynamic demands (e.g., a radome). In other areas the main property of interest would usually be structural. In the microwave area, structural properties will also be of great importance, particularly in a hostile environment. However, specific mixes will also be required to achieve precise electromagnetic (later shortened to "electrical," as is the usual custom and practice) properties. Thus "composite" here is taken to mean any combination of plastic and reinforcing materials that achieves a desired electrical performance in situations where structural performance is also important. Despite this definition, including purely structural composites, used in order to define a reflecting surface for microwave energy, in this chapter is

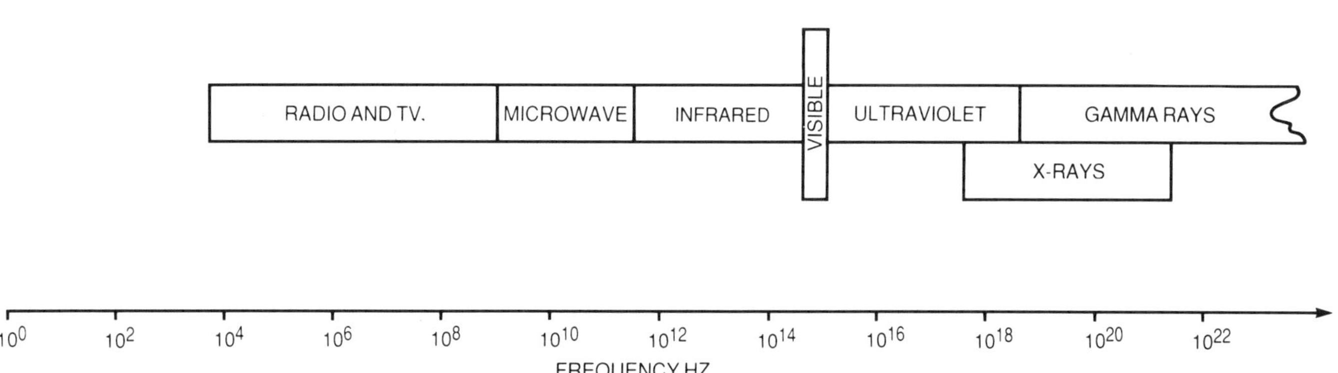

FIGURE 1 **The electromagnetic spectrum.**

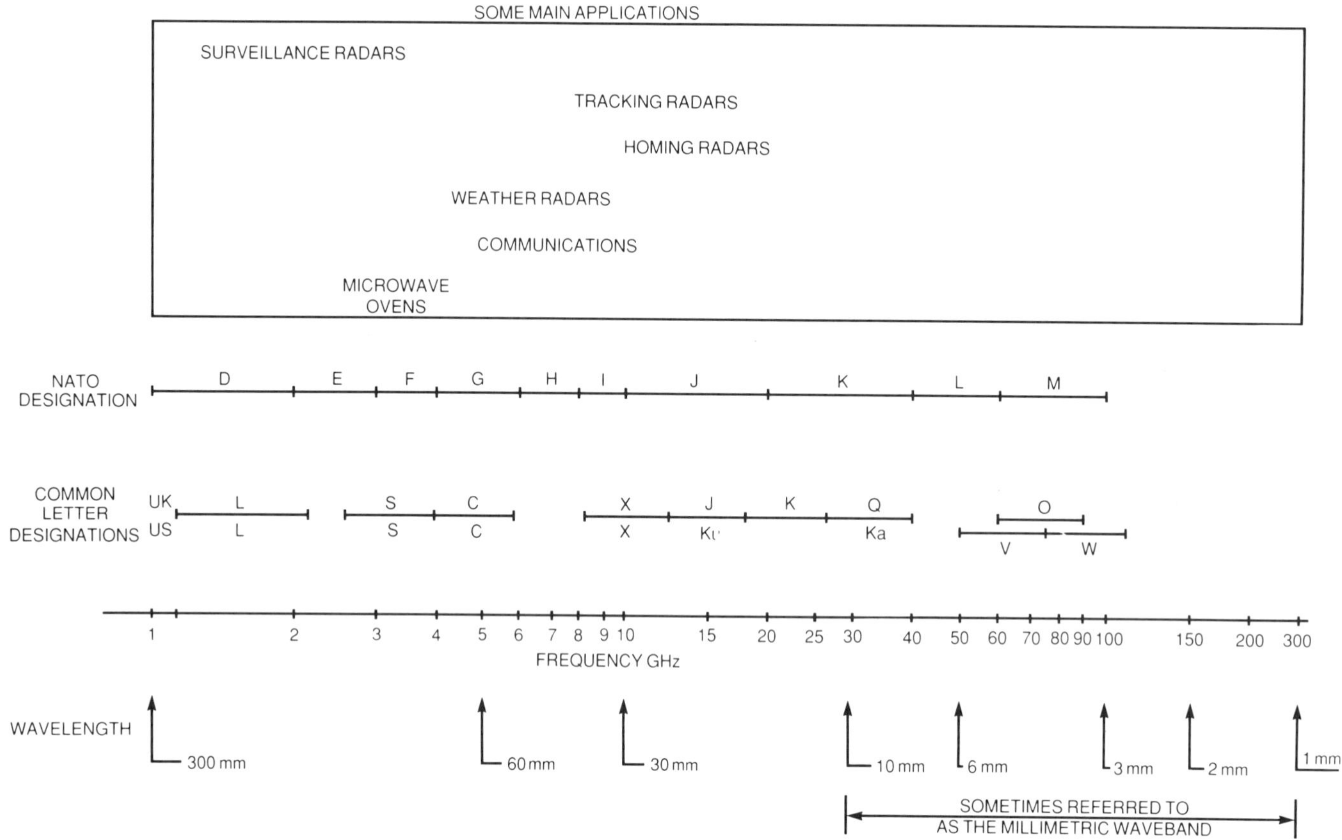

FIGURE 2 **The microwave frequency band.**

questionable. They are, nevertheless, given brief mention because of special structural problems that arise from shape, accuracy, and mechanical requirements, such as might arise in satellite antenna deployment. As will be seen, however, the emphasis is on those applications where the composite is used as a dielectric and thus contributes to the processing of the microwave signal.

Microwave Properties

Properties such as dielectric constant and loss tangent* [13,16] are of prime importance to a microwave transparency such as a radome, whereas superior structural stiffness would be required for a composite reflector antenna.

With regard to transparencies, the parochial interests of the microwave designer would favor a low dielectric constant and low loss. Thus polythene or PTFE would be a good choice. However, structural and environmental demands divert the choice away from such electrical excellence. Under these conditions, composites provide a better compromise between the mechanical and electrical performance demands. Even then, there is a wide

---

* Relative dielectric constant ε and loss tangent $\tan\delta$ fully describe the properties of a composite or any other dielectric material for microwave purposes. The full complex dielectric constant for the material is $\varepsilon' + j\varepsilon''$. The real part of this, normally referred to as the relative dielectric constant when divided by the properties of free space, controls the velocity of propagation through the material; thus:

$$\nu_{\mathrm{D}} = \frac{c}{\sqrt{\varepsilon}}$$

where c is the velocity in free space. From this we can deduce wavelength in the material as follows:

$$\text{Wavelength} \times \text{frequency} = \text{velocity}$$

hence:

$$\nu_{\mathrm{D}} = \lambda_{\mathrm{D}} \times f \quad \text{and} \quad c = \lambda \times f$$

thus:

$$\lambda_{\mathrm{D}} = \frac{\lambda}{\sqrt{\varepsilon}}$$

where λ_{D} is the wavelength in the material, λ is the wavelength in free space, and f is the frequency. The loss tangent δ is a measure of the dissipation of energy in the material:

$$\tan\delta = \frac{\varepsilon''}{\varepsilon'}$$

and the attenuation due to dissipation is approximately $(27.3 \times \sqrt{\varepsilon} \times \tan\delta)/\lambda$ dB/m (λ in meters).

range of reinforcements and matrix materials that allow a balance to be struck between these demands and the third important characteristic, which is cost.

As a word of caution, while the thermal environment can generally be accepted as due to external effects (ambient, solar, or aerodynamic), in some high power applications self-heating can predominate. In these cases, low loss tangent materials with high temperature capabilities are of prime importance.

Microwave Application Categories

Microwave devices are used in a wide range of applications from underwater to space, and their operational requirements are extremely wide and demanding. The application of composites to microwave devices can broadly be divided into two categories:

1. Those in which the composite component acts as a straightforward structure, such as a large waveguide horn or a reflector for an antenna.
2. Those in which the composite component acts as a protective window for a microwave source, as in the case of a radome.

In the case of waveguide horns, plastics composites would be used to reduce overall weight, rather than for any improvement in microwave performance that their use would bring (e.g., to reduce payload weight for a satellite or to reduce scanning inertia for a microwave antenna). Antenna reflectors would use plastic composites as the medium on which a reflecting surface is formed. The reason for doing this could be to reduce weight while retaining stiffness or to ease the making of a very accurate three-dimensional shape. Such structures can benefit from the use of carbon fiber reinforcement, which can, at low frequencies (below I-band), also serve as the reflecting surface as a result of its relatively high electrical conductivity. Examples of this approach might be found among satellite antennae. It should be pointed out that carbon fiber, being a reflector of microwave energy, cannot be used as a reinforcement for a microwave window (radome).

Where the composite structure supports a reflecting surface, this would be created by either spraying after manufacture with a metallic finish, spraying the mold with a metal reflecting surface before manufacture, or incorporating a metallic foil or mesh during manufacture. The reflecting surface thus formed would need to be accurate to within about $\pm\ \lambda/32$, which, when the diameter of the antenna may be a number of meters, may be no easy task (Fig. 3).

Microwave windows or radomes can, despite their apparent simplicity in shape and construction, be very complex components. They must allow transmission of the microwave signals with minimum loss and beam pointing error, known in the microwave business as aberration or boresight error. Radomes can range from large, very thin skin constructions (usually supported on a framework) for long-range ground tracking radars to fairly small conical-type components used in high speed missiles, where any loss or aberration can give significant performance penalties.

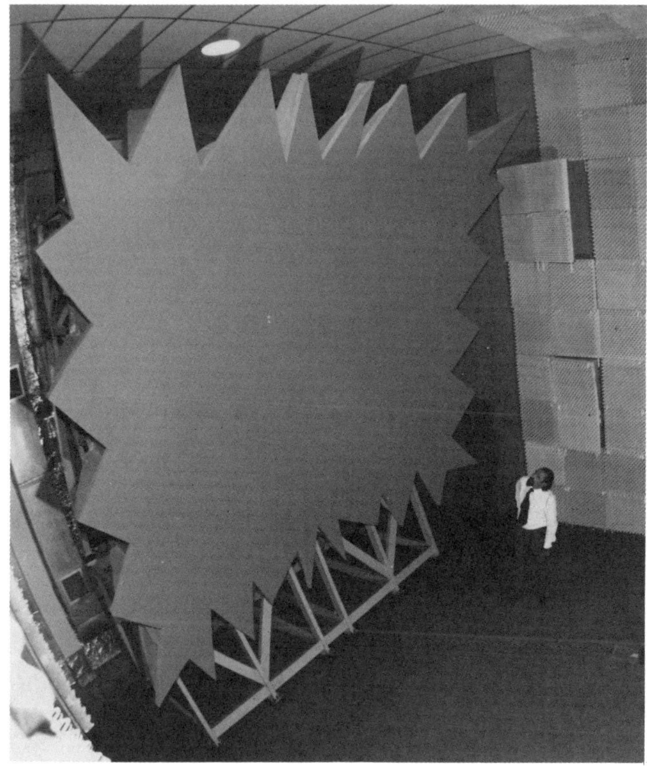

FIGURE 3 Compact site reflector.

In addition to these major categories, general applications for composites can be found in the field of shaped containers for electronic equipment that has to be screened against external electromagnetic effects and in printed circuit boards (e.g., microstrip), where materials such as PTFE reinforced by quartz might be used.

Antennae

There is a very wide range of microwave antenna types [2,3,9], of which only one basic type needs to be considered as a main use of composites. This is the reflector antenna, of which the simplest type is the paraboloid (a car headlamp is often used as a handy example), although it might appear in a number of forms covered by such terms as offset paraboloid, cassegrainian, newtonian, or gregorian (from telescope terminology). More complex versions might also use polarization processing, as in a twist-reflecting Cassegrain type [10] "or some polarization selective satellite antennas [35] or by incorporating metallic elements for dichroic reflection purposes" [35]. Figure 4 illustrates the various forms. There is, therefore, the possibility of composites being used for the pure reflector structure [11] or as a functional part of the microwave signal processing. It is the former use that, with the advent of high modulus fibers, has seen the greatest expansion, generally in the interests of reducing weight while retaining high stiffness. The latter,

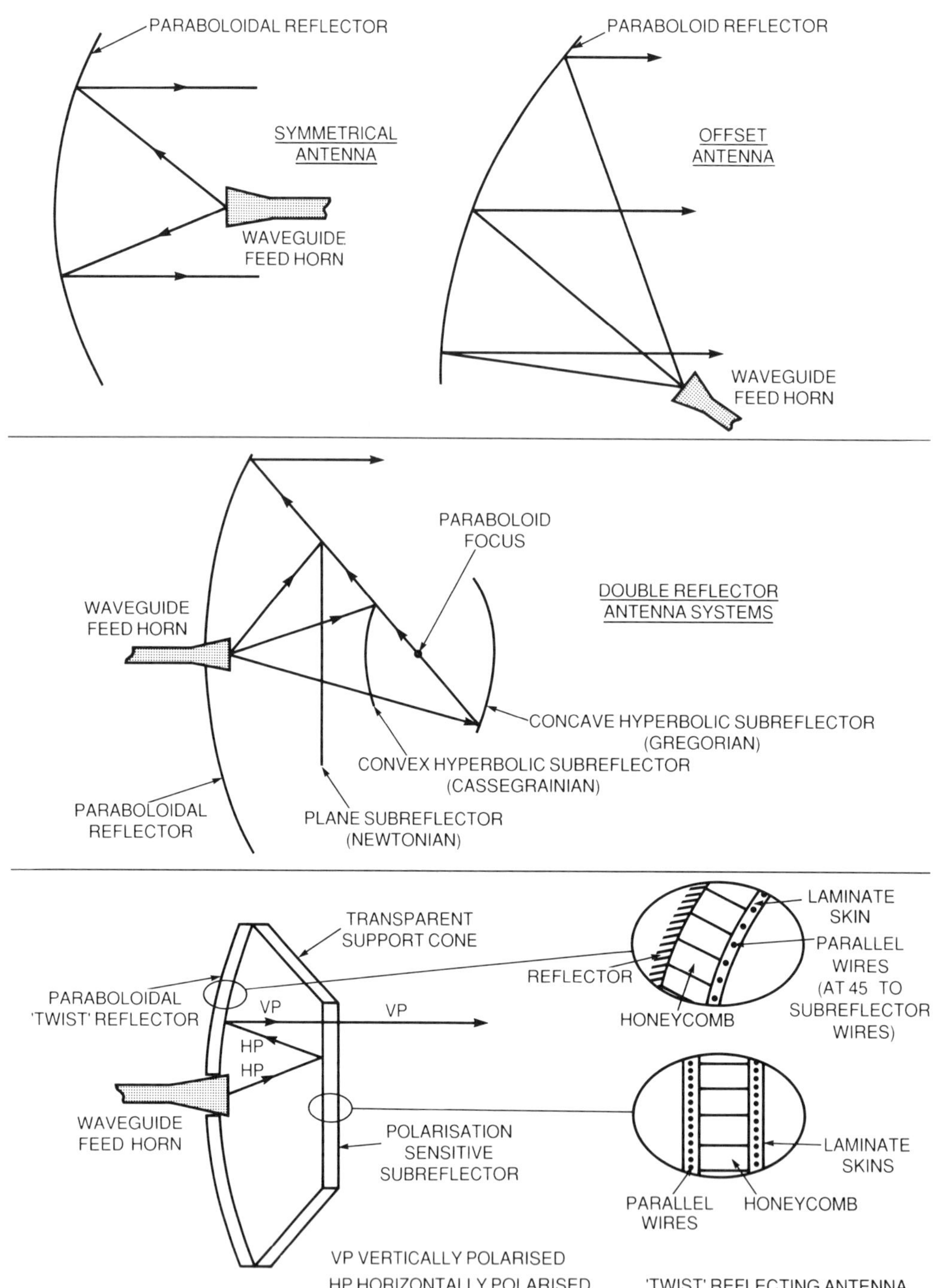

FIGURE 4 **Reflector antenna forms.**

of course, fits more correctly under the title "microwave composites."

In general, unless some good technical justification exists, simple paraboloidal reflectors have continued to be made by spin forming or to be fabricated from stretch-formed metal on the grounds of low cost. Exceptions are beginning to appear in connection with domestic DBS receiver systems. Technical justifications are mainly as follows:

1. Low weight	To reduce payload weight of a satellite. To reduce scanning inertia for a tracking antenna system. To reduce gravity distortion effects on large antennas.

2. Noncircularly symmetric profiles	Offset paraboloids. Nonparaboloidal fan beam reflector surfaces.
3. Functional part of microwave signal processing	Where part of the antenna structure must be dielectric, it may be advantageous to make the whole, including pure structure, from composite materials (e.g., twist-reflecting Cassegrain).

Simple Reflector Antennae

Composite antenna reflectors can range from a few centimeters in diameter, usually contained and protected by a radome, up to about 5 m, possibly exposed to the environment. While larger antenna reflectors are in use, they tend to be of metallic construction. Plastic composites are generally limited to one-piece constructions rather than assemblies, and here constraints arise from mold size and transportation to the site.

Antenna reflectors can be either static or moving, that is, oscillating or rotating to achieve the scanning mode required by the radar system. Small antennae tend to be made of short-fiber-reinforced polymers, whereas filament reinforcement, such as woven cloths, is used for larger versions.

Small reflectors are usually of solid construction (Fig. 5), which can be maintained to larger sizes with the use of stiffening ribs. When weight becomes excessive, sandwiches are introduced, with honeycomb or foam cores. Small reflectors can be simply attached to support structures, but for larger cases (Fig. 6) the distribution of loads from attachment points into a sandwich construction can become quite sophisticated and demanding, in terms of composite engineering know-how. Additionally, environmental conditions can constrain and dictate the materials used, which can have a considerable effect upon manufacturing costs. A typical case would be that of an antenna fitted to a satellite (Fig. 7). These antennae are subjected to the rigors of the launch vibration and g loads and then must operate with a large thermal differential between the front and rear faces. Antennae of this type have tended to make use of carbon fiber and epoxy or polyimide resin systems, which give stiffness, low thermal distortion, and, of course, light weight.

Undoubtedly it is in the satellite application area

FIGURE 6 **Honeycomb sandwich antenna (2.4 meters).**

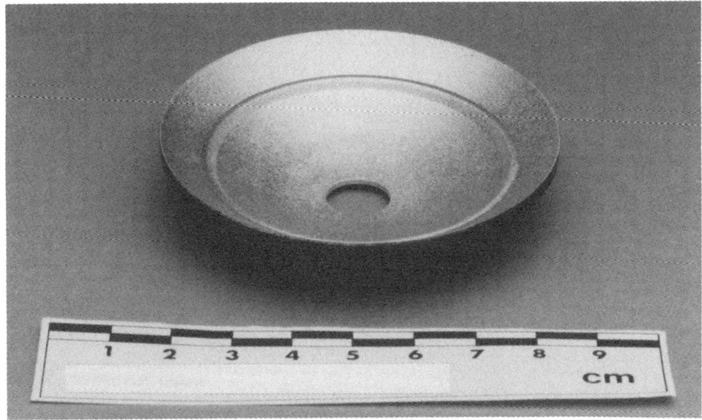

FIGURE 5 **Small 70 mm seeker reflector.**

FIGURE 7 **Carbon fiber satellite antennas.**

where major advances in structural composite technology have been and will continue to be made [35]. Moves toward higher frequencies (20–30 GHz) for communication links will place heavy demands upon reflector surface accuracy, as will larger sizes. In addition to large one-piece structures, such as the Ford Aerospace 3.7 m graphite reflector for Voyager, much ingenious work is taking place, particularly in Europe, with regard to the deployment of very large antennae, which currently encompass sizes up to some 6 m diameter.

Twist-Reflecting Antennae

A disadvantage of the simple Cassegrain system can be the obscuration of microwave energy by the subreflector. The twist-reflecting design [10] overcomes this and is particularly useful for small scanning antennae, where the Cassegrain layout is adopted so that the position of microwave source can be near the center of rotation, thus giving a low inertia solution. The principle of operation was shown in Figure 4, and a section through a typical composite construction is shown in Figure 8.

The reflecting surface of the antenna is a metal-sprayed fiber-reinforced molding. A wire grid, enclosed in a reinforced skin, is mounted normally 0.375 wavelength in front of the reflecting surface by the use of an accurately ground honeycomb separator. This forms the main reflector.

At a distance away from this, a secondary (sub) reflector is positioned, which in this case is of sandwich construction with parallel wire gratings embedded into the two skins; these are aligned at 45° to the wire grid on the main reflector. Because the signal has to pass through the secondary reflector support, a conical honeycomb sandwich transparency is used in a number of missile applications. Resin transfer molding was used for manufacture of these devices in the late 1950s.

Wire gratings have the widest uses, such as in polarization-sensitive satellite antennae, where, for one large case, laser etching has been used to define the conductive tracks. Other metallic elements (crosses, circles, swastikas, etc.) have been used to create frequency-selective (dichroic) reflecting surfaces.

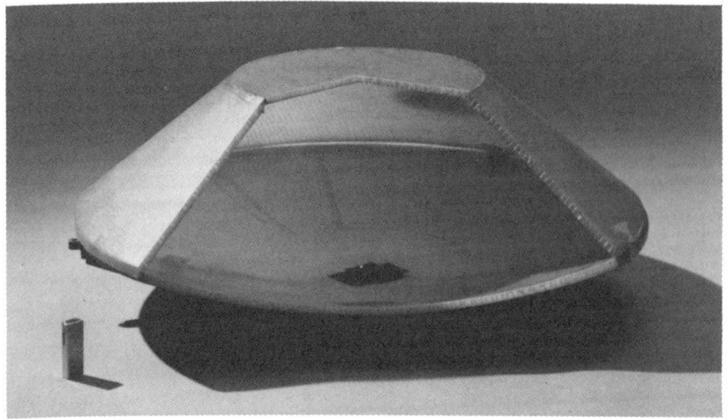

FIGURE 8 Twist reflecting Cassegrain antenna.

Radomes

The radome field [1,22,29] is almost wholly within the scope of plastics composites because of the environment to which radomes are subjected. They are, after all, there to protect the radar and to provide an aerodynamic fairing. Exceptions to the use of composites are inflatable fabrics, which have been used in the past on ground installations and on specific helicopter applications; ceramics, which are used when very high speed missiles are fitted with a microwave homing capability; and unreinforced thermoplastics where size and environment combine to permit their use.

In addition to being by far the greatest composite application area, the radome field is also the most environmentally demanding. It is this latter aspect that encouraged the enthusiastic adoption of composites as they emerged for general industrial use. Prior to that time, requirements had been satisfied, if at all, by the use of early thermoplastics, or even plywood where the environment was more severe (circa World War II). Designers were pleased to find materials becoming available (initially polyester–glass combinations) that were tough, were water-resistant, had good electrical properties, and could be manufactured relatively easily and accurately to the shapes and wall thicknesses required. This, of course, was not the end of the story, and escalating performance requirements (i.e., high speeds and hence high temperatures and low electrical loss or beam pattern distortion) have in their turn forced the use of more exotic and expensive resin systems and reinforcements. However, in most cases these requirements have not been a driver for material development, since, in the broader scheme of things, radomes use only a small quantity of composites. Nevertheless, and perhaps fortuitously, the availability of advanced composite materials has more or less kept pace with requirements.

Radomes can be subdivided into application types and detailed wall constructional forms according to the nature of the performance required [1]. While certain biases exist, all the wall constructions available can in principle be used in any of the applications. Table 1 lists the main application areas and general constructions involved.

A commonly held view is that a radome, being plastic, is simply transparent to microwave energy. This is very far from the truth. In fact, a radome will always, and sometimes seriously, degrade the performance of the protected antenna and can be classed as "a necessary evil." The degradation can be a simple loss of signal strength or may additionally involve beam pointing error, thus reducing direction-finding accuracy. Also, the radome may increase the amount of pickup of energy away from the main beam direction. A simple signal loss example will serve to illustrate that transparency in the sense implied above does not apply. Let us consider microwave energy incident upon a dielectric interface at normal incidence. If, typically, the relative dielectric constant of the material is 4, then the power lost (reflected) at the interface will amount to 11%. This is for one interface. A radome wall has at least two interfaces, and what

TABLE 1
Applications

Application	Main Radome-Using Area				Main Wall Types (Fig. 3)
	Ground	Ship	Aircraft	Missile	
Satellite communications	*	*			B, D
Communications links	*				B
Electronic surveillance			*		D
Early warning radar	*				B
Surveillance radar		*	*		B, C, D
Tracking radar		*	*		C, D
Homing radar				*	C
Weather radar			*		D
Navigation radar			*		C, D

Common Constructions

Common Constructions	Main Radome-Using Area
Geodesic space frame	Ground
One-piece domed structure	Ground, Ship
One-piece aerodynamic	Aircraft, Missile

happens as a result of the reflection from the second depends upon its time delay relative to the first (Fig. 9*a*). As a result of the sinusoidal nature of the microwave signal, we can have constructive or destructive interference, or something in between. If interference is constructive, then the total loss will rise to 36% (note that this is more than twice the simple reflection loss as a result of multiple internal reflections). If interference is destructive and a lossless material is used ($\tan \delta = 0$), the loss will be zero, since the reflections cancel. In fact, the major loss for a radome arises from interface reflections rather than from losses to heat from molecular motion in the material. Reflection losses are more difficult to handle as angle of incidence increases; hence a pointed radome on a high speed missile or aircraft is a more difficult case than a hemispherical radome.

It follows from the above that the microwave designer's task is one of arranging interfaces so that over the frequency band(s) of interest, reflection losses are minimized [1,5]. From Figure 9*a* we can see immediately that useful designs could result for small electrical thickness, or around 0.5λ. In fact, this latter value, chosen for appropriate incidence angles, has proved to be the mainstay of high performance aircraft nose radome design. This is often referred to as a half-wave solid laminate radome. There are, however, many ways of arranging for interface reflection cancellation and hence loss reduction. For completeness, the main forms, together with their essential transmission characteristics, are shown in Figures 9*b* to *f*.

[Note: Transmission efficiency (%) = 100 − reflection power (%) for a lossless ($\tan \delta = 0$) material.]

The forms of transmission curves shown illustrate typical exploitations of the build characteristics, but it will be appreciated that by a subtle choice of dielectric constants and thicknesses, a very wide range of properties can be engineered. These would generally be classed as single or multiple narrowband, broad bandpass, or low-pass. Beyond these basic types, multilayer (greater than 5) builds [5] can be exploited for improved broadband response, and additional elements such as wire gratings can prove useful [4].

For simplicity, the above examples, which give some insight into radome microwave technology, have been based upon transmission efficiency. While this is important, it is not always the overriding feature. For instance, where missile homing is concerned, aberration or boresight error considerations might predominate, and a different range of design techniques will be adopted. These might include precise control of a subtly varied radome wall thickness [6]. Equally, the choice of a particular build might not be determined only by specific microwave concerns. Thus, the majority of civil aircraft radomes are 'A' sandwiches mainly because of their low weight. Equally, where structural superiority or rain erosion conditions predominate, such as in the military field, the solid half-wave structure is favored.

This has been a brief and general introduction to the subject of radomes. Recommended further reading, in addition to the references given, is the range of "electro-

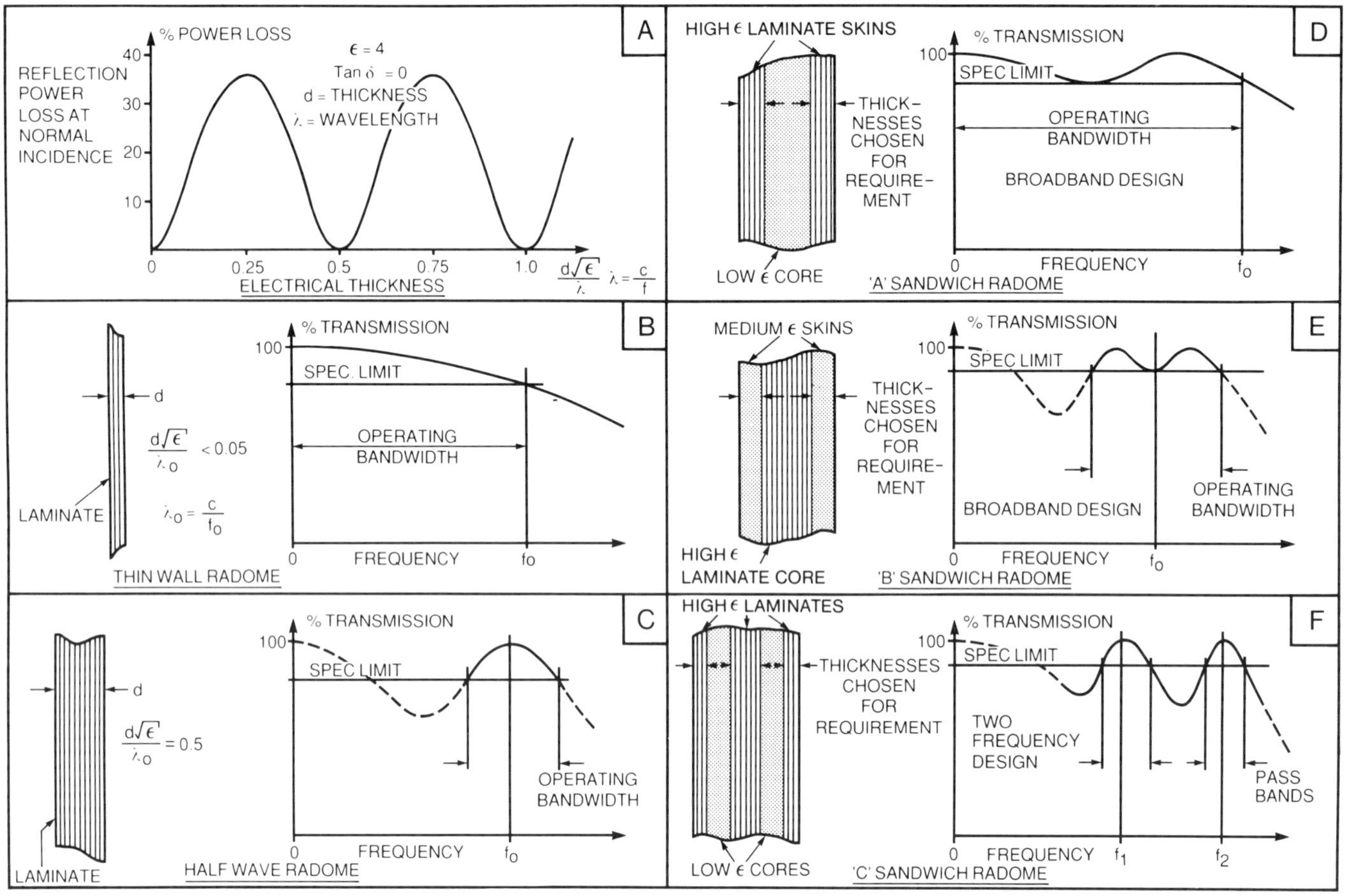

FIGURE 9 **Radome builds and frequency responses.**

magnetic window" symposium proceedings. (Most of these symposiums were held at the Georgia Institute of Technology.)

Underwater Radomes

This is a very specialized application in which a radar is, typically, mounted on a submarine periscope. Size is therefore usually small, and the radome must withstand high pressures when submerged and must drain rapidly when lifted above the sea surface. These radomes are usually of the solid wall type and may be a number of half-wavelengths thick. Very high glass content is also a feature.

Ship-Mounted Radomes

These can vary quite considerably in size and are usually of sandwich construction. Requirements vary considerably in the interests of both microwave performance and reduced top weight. For instance, the radome may just cover the antenna aperture, or it may cover the entire radar, thus allowing the equipment to be serviceable under adverse conditions. A typical one-piece ship-mounted radome is shown in Figure 10. No flanged joints are allowed over the transparency surface; therefore the woven glass fabric skins are scarfed to give constant thicknesses. These radomes can use foams or paper, Nomex or glass honeycomb, as the core. The choice varies according to the application and manufacturer or customer preference. The size can vary from about 1 to 5 m in diameter, and the shape can be spherical or

FIGURE 10 **Typical spherical ship-borne radome.**

FIGURE 11 Space frame radome (33.5 meters). (Courtesy Lightweight Structures Ltd., U.K.)

FIGURE 12 Concorde SST aircraft radome on microwave site.

straight-sided with a hemispherical top (Fig. 11). Environmental conditions are severe, and the selection of materials must take into account blast, greensea loads, and perhaps corrosive funnel gases.

Ground-Based Radomes

These can be similar to ship-borne radomes, but may be larger, and one-piece construction is often not practical. Very large radomes normally use thin composite skins supported by a metallic spaceframe structure [28]. Composite framework ribs have been used, but have shown little, if any, advantage over metal in the microwave band. However, more recently developed composite ribs, of interest for low frequency radar applications [34] and improved low loss designs in the higher bands, are increasing in use (Fig. 12). Again, environmental conditions can be severe; locations are often remote and subject to extremes of high or low temperature. Snow or ice removal techniques and differential thermal expansion characteristics of the panels and the supporting framework can be major concerns. Ease of erection on the site and over the preinstalled radar is another important design consideration.

Aircraft Radomes

This category [15,21,22,29] covers a wide range of types, all of which are to a greater or lesser extent subjected to additional flight environment factors such as temperature, rain erosion, and lightning and bird strikes.

Civil aircraft usually have a nose radome and other smaller radomes covering glide slope antennas and other such equipment. The latter tend to be of injection-molded thermoplastics, sometimes glass-reinforced, and are not particularly critical. Nose radomes range from thin skin construction on small aeroplanes to, in the majority of cases, sandwich construction, with most using Nomex honeycomb as the core material. Civil aircraft nose radars are usually for cloud and collision purposes. The main requirement is that the radome minimize signal loss, with aberration not being very critical.

One notable exception in the civil aircraft field is the radome for the supersonic Concorde aircraft [31], which is of solid half-wave construction because of the flight environment. Flight conditions for the Concorde nose radome involve a cyclic temperature range of −50–+190°C, which, together with supersonic aerodynamic requirements, dictates a slender shape. The design and construction are more typical of those for a high speed military aircraft (Fig. 13). Military aircraft radome microwave requirements are, however, in a completely different category from those for civil aircraft, with a greater variety of types and functions. A military strike aircraft (such as the Panavia Tornado [15], Fig. 14) may be fitted with a nose radome covering weapons-aiming and terrain-following radars. Other smaller radomes will be fitted on the skin of the airframe, or perhaps on a pod, for a

FIGURE 13 Panavia Tornado IDS aircraft.

FIGURE 14 Application of rain erosion coating of an aircraft radome by robot.

variety of purposes, including electronic countermeasures.

Nose radomes have the most demanding specification. Generally they are conical or ogival in shape, with structures capable of withstanding aerodynamic loads at high temperature. They must withstand the effects of rain erosion and of lightning and bird strikes. While radomes for very weight-sensitive applications have been made of sandwich construction, the majority are solid half-wavelength with the glass content varying between 55 and 85%, depending upon the loads imposed and the technique of manufacture.

Wide-ranging performance requirements, with detailed constraints on loss, aberration, and antenna radiation pattern distortion (e.g., side lobe degradation), and sometimes in more than one frequency band, lead to very sophisticated designs, all of which are made more difficult by the pointed shape.

Probably the largest high performance aircraft radome is the one fitted to the Boeing AWACS aircraft. This is fitted on a pylon on top of the aircraft and measures some 9 m in diameter and 2 m in depth. It is in two parts and is a five-layer honeycomb sandwich. It is also somewhat unique in that the skins are made of an S-glass–epoxy composite that has to withstand unusually high heat generated from the radar within. This is in addition to all the usual problems of lightning strike, static discharge, rain erosion, and so on.

While most civil aircraft radomes can have a constant wall thickness, many military aircraft require that the wall thickness vary in order to optimize performance. This leads to significant problems in manufacture, particularly when the accuracy of the wall thickness is extremely critical, for, as will be seen later in the manufacturing section, this cannot always be achieved by molding alone.

Rain erosion protection, which is normally in the form of an elastomeric coating [26], also has a significant influence on wall thickness control. It is not necessarily good enough to hand-spray an elastomer onto the outer surface. The coating is also a dielectric (usually different from that of the main body) and must be accounted for as part of the microwave design with the same order of thickness control as for the laminate. The introduction of spray application robotics has considerably eased this problem, providing, as it does, a high degree of repeatability. For instance, a deposit that is systematically thicker at the nose of the radome than at the base can prove acceptable if all the radomes are the same. This variation can then be taken into account in the basic design.

Radome matrices can range from polyester resin (usually of the low loss type) through epoxies to polyimides. For very high temperatures, polyimides of the condensation type would be preferred, but problems associated with void content, which cannot be tolerated, dictate the use of addition-type materials. Reinforcements are most commonly E-glass fiber with some D-glass and, in critical areas, quartz.

Missile Radomes

Requirements for missile radomes vary considerably depending on the environment and application. They can be subsonic or supersonic, but both can be equally critical in design and function.

Subsonic missiles are usually larger, typical examples being the Sea Skua, Sea Eagle (Figs. 15 and 16), and Exocet. The Sea Skua is helicopter-carried, and speeds are such as to allow an ellipsoidal sandwich structure. The Exocet is a long-range weapon, and the carrier aircraft may have a supersonic capability. It is more prone to rain erosion problems. This type would be a solid half-wave construction and pointed in shape.

Supersonic anti-aircraft radar homing missiles can be ground-, ship-, or air-launched. Ground-launched-mis-

FIGURE 15 Sea Skua missile.

FIGURE 16 Sea Eagle missile.

sile radomes (Fig. 17) are usually fairly large and may suffer long-term exposure to the elements. Air-to-air missiles carried as wing or underbody stores are exposed to the aircraft environment. In either case, supersonic radomes usually have a high fineness ratio (i.e., ratio of

FIGURE 17 Thunderbird ground-to-air missile.

length to base diameter), sometimes as great as 3 to 1. This considerably aggravates the design problems associated with aberration. Commonly a rugged solid half-wave construction is used.

In the case of air-carried missiles, where a number of flights may take place before firing, additional effects of sand and rain erosion can be experienced; this could dictate the use of a ceramic material even though these suffer from thermal and mechanical shock problems. This is particularly so when operational speeds are in excess of about M3.5. The incorporation of fibers into the ceramic matrix (a ceramic composite) can improve shock resistance, and no doubt these composites will be used for radomes in the future, provided that the problems of ensuring a uniform distribution of fibers (thus giving a homogeneous dielectric constant) can be overcome.

Other Applications

Electrically Screened Containers

For mainly economic reasons, designers have been using composite materials to make containers for electronic and microwave apparatus. Generally these are of solid laminates and for large quantities would be press-molded. For smaller quantities, autoclave manufacture would be common. While these are not strictly microwave composites, they are mentioned here because they sometimes have some form of screening to protect the contained equipment from external microwave or RF emissions.

Screening can be achieved by:

1. Metal, sprayed on either the inside or the outside of the component.
2. Metallic foil, mesh, or metallized fabric within the laminate.

These are well-established procedures, but great care must be taken to ensure that the electrical screen is complete, that is, that there are no nonconducting gaps within the component.

Waveguide and Feedhorns

A waveguide is a means of transmitting microwave energy from one point to another, normally via a hollow rectangular-cross-section metallic tube. The cross-sectional tolerances and avoidance of discontinuities within a waveguide are usually of great importance. Conventionally, such a waveguide would be completely of metal, either drawn, cast, and/or machined. In some applications, such as satellites, the weight of long waveguide runs can be excessive. Fortunately, because of a phenomenon known as skin effect [12], the actual thickness of the conducting wall required to pass current to support the energy propagation is small, usually considerably less than 0.025 mm. If such a thin conducting tube can be stiffened with a lightweight material, a preferable solution can be obtained.

Such waveguides have been constructed by first plating an accurate, bright-finished mandrel with a high conductivity metal (such as copper or silver). Reinforcement, perhaps filament-wound with a high modulus fiber, is then applied, and finally the mandrel is removed. Clearly, with this type of process, cross sections other than rectangular, and of varying dimensions for microwave processing purposes, can be produced. Thus feedhorns (i.e., the illuminating radiators for reflector antennae), including complicated corrugated types, and filter sections can be made in this fashion.

Manufacturing Techniques

While manufacturing techniques are discussed in detail elsewhere, it is important to discuss those that are specific to the production of microwave composites.

Structures are manufactured by press, autoclave, and/or vacuum bag molding. Radomes of the sandwich type would almost certainly be manufactured by an autoclave technique. Half-wave high performance radome manufacture is normally by filament winding or resin transfer molding. It is important to note that quality demands for radomes are very much higher than those for standard products. Radomes need to be largely void-free and to have tightly controlled limits on the wall thickness.

The filament winding process produces a radome body on a male mold positioned under a machine that is capable of winding fibers radially and axially. The orientation of these fibers is very critical, and because of bunching of the fibers towards the nose of the radome, machining operations have to be carried out at various stages of construction (Fig. 18).

FIGURE 18 U.S. aircraft radome being filament-wound. (Courtesy Brunswick Corporation.)

When winding and machining are almost complete, the radome electrical thickness is measured by a microwave interferometer (which also monitors variations in dielectric constant). By measuring the whole surface and storing the data, computer-controlled final machining optimizes the thickness to meet the electrical requirement (Fig. 19).

The resin transfer mold process utilizes a matched-mold technique; that is, male and female molds are machined within practical tolerances to obtain a wall thickness that meets the electrical requirement. The manufacturing process consists of laying up a precise amount of glass reinforcement (in either knitted shape, woven shape, broad goods, or a combination of these) on the male mold half (Fig. 20), then assembling the female to the male mold, followed by resin injection. Injection can be by vacuum, pressure, or a combination of both, at room temperature or elevated temperature depending on the characteristics of the matrix resin.

This process ensures a very constant and accurate reinforcement-to-resin ratio, which ensures an even dielectric constant throughout the component (Fig. 21). In some cases a radome produced by this process would be sufficiently accurate to meet specifications. In more critical cases, it is necessary to fine-tune the radome wall thickness. The principle here is similar to that of the

FIGURE 19 Radome grinder and one horn interferometer. (Courtesy Brunswick Corporation.)

FIGURE 20 Layup of woven shapes on male mold half.

FIGURE 21 Molded radome following removal of the female mold half.

interferometer, with the electrical thickness of the radome wall being measured at many stations over the surface. The information stored then controls an automatically sprayed varnish, which is applied to the inside surface of the radome, thus building up the thickness to the required value (Fig. 22).

Conventional sandwich radomes are made using premachined-thickness honeycomb for the cores. High performance types require greater accuracy, which is provided by machining the cores after they are bonded to the inner skin. This is done using special cutting equipment, as seen in Fig 23.

A relatively new, interesting process for manufacturing solid half-wavelength military radomes has been developed in France. In this process, preimpregnated woven glass shapes are placed on a male mold; when the layup is complete, it is covered by a pressure bag and the assembly is cured in an autoclave. Following this, the molding is removed from the mold and machined to size on a contour grinding machine. Originally the loading of the shapes was done manually, but now a special machine has been designed to do this hitherto difficult task.

Material Properties and Environmental Performance

The structural characteristics of composites are covered in detail elsewhere. The concern here is, in broad terms, to indicate the electrical properties of composites and how particular combinations can be used to the specific benefit of a design. Also briefly examined are those aspects of environmental performance, such as temperature, moisture absorption, lightning strike, and rain erosion, that are critical to some microwave applications. From this it follows that the prime concern is with those materials that are used in radome applications.

FIGURE 22 Radome electrical wall thickness measurement and optimization equipment.

FIGURE 23 **Machining of honeycomb core on a sandwich radome following molding to the inner skin to give good build dimensions.**

The broad classification for composites is in terms of the following:

1. Matrix — The basic thermoplastic or thermosetting resin system.
2. Reinforcement — The various fibers in different forms.
3. Construction — The material format used in the resulting composite (e.g., laminate honeycomb).

Tables 2 and 3 list a selection of materials under the first two headings and give some essential parameters [13]. The short-term operating temperature is essentially that which would arise from missile applications, whereas the long-term figure is for aircraft. Clearly the precise limit that should be applied is also critically dependent on size and loads. The reader should consult the references given for more detail and for structural properties [13,16,32,33].

The main matrix materials for radome construction are thermosetting polyesters and, predominantly, epoxy resins. In general, these cover an adequate range of microwave, structural, and environmental performance capabilities. Modified polyesters offer improved electrical characteristics and, like epoxy systems, an improvement in high temperature performance. A further improvement in both respects would be realized with a polyimide system [24].

The use of thermoplastics [32,33] has been limited as a result of the good performance of thermosetting systems, but there has been a steady improvement that more recently has produced materials such as polyethersulphone (PES), polyetherimide (PEI), and polyetheretherketone (PEEK). These materials—and no doubt there will be further improvements in the near future—offer good temperature and microwave performance and, most importantly, good rain erosion properties. This latter aspect, however, can be degraded when the material is reinforced with fibers.

The electrical characteristics of the final composite depend on the construction chosen. Where this is a honeycomb, as might be used as a core for an "A" sandwich, the aim is to have a material with the lowest possible dielectric constant. In general, this is in the region of 1.1 and is not open to significant modification, nor, for that matter, is modification usually required. Low density foam cores (e.g., polyurethane) will have a similar dielectric constant, but syntactic foams, which are a composite of a thermoplastic or thermosetting resin matrix and so-called microspheres, generally have a higher value—above 1.4—and increased density with greater strength. Foams allow more flexibility in the choice of dielectric constant than honeycombs, but again the requirement is generally for this to be as low as possible consistent with adequate mechanical properties.

Laminates, on the other hand, whether they stand alone or are incorporated in a sandwich build, are critically dependent upon a particular dielectric constant. Within reasonable limits, this can be adjusted by varying the ratio of reinforcement to matrix material, but more often than not this ratio is fixed according to the build/structural requirements, and the thickness is chosen to satisfy the resulting dielectric constant. A simple expression that allows the dielectric constant of a mixture to be determined with a fair degree of accuracy is as follows:

$$\log \varepsilon = \frac{V_1}{V_T} \log \varepsilon_1 + \frac{V_2}{V_T} \log \varepsilon_2$$

where

ε = the resultant dielectric constant
ε_1 = the dielectric constant of material 1
ε_2 = the dielectric constant of material 2
$\frac{V_1}{V_T}$ = the fractional volume of material 1
$\frac{V_2}{V_T}$ = the fractional volume of material 2

and

$$V_T = V_1 + V_2$$

The design degree of freedom offered by this relationship might be exploited to the full when, say, a "B" sandwich is being used for a multiple- frequency requirement [23]. Here the dielectric constants of various layers of the build (skin, core, skin) might be defined by the choice of reinforcement material alone, thus allowing a common matrix resin to be used. Thus the skins might use quartz and the core E-glass, giving a low–high–low dielectric constant combination.

TABLE 2
Matrix Materials

Type	ε (10 GHz, 20°C)	tan δ (10 GHz, 20°C)	Approx. $\frac{d\varepsilon}{dT}$	Max Operating Temperature (°C) Short-Term	Long-Term	Remarks
Polyester	2.7	0.013	1.1×10^{-3}	180	120	Low cost
Modified polyester	2.8	0.005	Negligible	280	200	Low loss but expensive
Epoxy	2.8	0.012	6×10^{-4}	290	220	Good all-round performance
Polyimide	2.7	0.005	Negligible	500	340	Excellent high temperature performance
PES	3.4	0.013	4×10^{-4}	200	150	Poor hydrocarbon and UV resistance
PEEK	3.2	0.003	Negligible	220	120	Long-term temperature moving toward 150°C with new developments. Excellent rain erosion performance in unfilled form

The relationship might be used in a different way where homogeneity is of extreme importance. This is particularly true at very high frequencies, such as in the millimeter-wave band, where relatively few reinforcement layers can be incorporated in the naturally reduced thickness wall, or where reinforcement bunching can occur. In both cases unacceptable local variations in the effective dielectric constant can be experienced. This can be overcome, consistent with adequate structural performance, by ensuring that ε_1 and ε_2 are the same or nearly so. Hence, rather than glass, the choice might be quartz, Aramid, or, in some instances, polyester fiber.

The above expression can be extended to more than two materials in the form

$$\log \varepsilon = \sum_1^n \frac{V_n}{V_T} \log \varepsilon_n$$

where $V_T = \sum_1^n V_n$. Thus a means exists for establishing the percentage of a third (or more) high dielectric constant filler, such a titanium dioxide (TiO_2), for the purpose of adjusting the frequency response of a given laminate thickness.

Naturally, even though the main loss from a radome across a frequency band is due to reflection, low tan δ materials are preferred. In general, tan δ should not be greater than about 0.015 for a laminate, and in some cases figures well below 0.01 would be required. As an indication, a radome such as that for the Concorde will lose about 13% of the microwave power (at I-band) to heat, on both transmission and reception.

A rough guide to the main characteristics of various matrix and reinforcement combinations is given in Table 4. This is by no means exhaustive and is included only as an indication of the normal thinking process used at the commencement of a new design. While polyester–glass can be considered as a cost-effective solution for environmentally undemanding cases, such as for most civil aircraft, the tendency is to use epoxy–glass as a prepreg, which because of its commercial availability gives a comparable unit cost. Epoxy–glass is also commonly found in military aircraft radomes. The use of polyimide might be indicated for a high speed missile, although this has been rare in the past because of void generation during processing, which degrades moisture absorption properties.

Moisture absorption, or lack of it, is an important feature in the choice of resin system and reinforcement. Since pure water has a dielectric constant in the region

TABLE 3
Typical Reinforcements

	ε (10 GHz, 20°C)	Tan δ (10 GHz, 20°C)	Remarks
E-glass	6.1	0.004	Low cost, good all-around performance
D-glass	4.0	0.003	For more demanding electrical specifications
Quartz	3.8	0.0002	Excellent low loss
Aramid	3.8	0.01	Good structurally, but hygroscopic

TABLE 4
Comparison Chart for Some Composite Combinations

Construction	Approximate Ranking		
	Cost	Temperature	Microwave Performance
Polyester–glass	Low	Low	Low
Modified low loss polyester–glass	High	Medium	High
Epoxy–glass	Medium	Medium	Medium
Epoxy–quartz	High	Medium	Medium
Polyimide–quartz	High	High	High
PES–glass	Medium	Low	Low
PEEK–glass	High	Medium	High

of 80 at microwave frequencies, it will be appreciated that small quantities can have a marked effect on frequency response, quite apart from loss tangent implications. Moisture absorption can occur in a number of ways, such as:

1. The hygroscopic nature of the resin or reinforcement.
2. Void generation during the manufacturing process.
3. Inadequate wetting of the reinforcement fibers by the resin.
4. Any machining operation that exposes the fibers (see the section on manufacturing techniques).

In this last case, a thin varnish or a rain erosion coating can improve resistance, but this is not totally successful, especially in the long term.

The rain erosion properties of materials and builds are a further constraint on choice. Much has been written on this subject [8,25], but the mechanisms involved are complex and, while conceptually understood, remain difficult to predict. In fact, the use of the word "erosion" is somewhat misleading in that it implies a "wearing through." The action of raindrop impact is more to search out surface defects and to initiate local delamination with an accelerating effect, or, by repeated hammering, to cause internal damage, such as at the interface glue line between the skin and core of a sandwich construction. Foam-core forms are particularly susceptible.

Rain erosion performance, then, is very dependent upon the precise nature of the design and the quality of the particular product. Apart from high product integrity, two main forms of erosion protection exist. The first, used in the Concorde, is to ensure a high surface smoothness, thus reducing the tendency for small defect damage initiation. In this case a pointed radome is seen to be a considerable advantage, since there is a high degree of correlation between damage and the normal incidence component of velocity. This is also true to some extent of the second method, in which an energy-absorbing elastomeric layer (e.g., polyurethane) is sprayed, or in some cases bonded as a sheet or boot, onto the radome surface [26]. This is by far the most common method for both civil and military aircraft. Its efficiency is very dependent upon an adequate coating thickness, usually in the region of 0.3–0.4 mm. This has to be accounted for in the electrical design, which means that laminate structure will be reduced for a given frequency of operation. This can be of great importance in the "A" sandwich, where at I band a total skin (laminate plus erosion coating) might not be much more than 1 mm thick, or for a millimeter-wave-band radome, which in half-wave form can have a similar thickness.

A final electrical feature of a laminate that has a significant bearing on environmental performance is resistance to high voltage puncture. High voltages can arise as a result of static charge buildup or direct lightning strike

FIGURE 24 **Civil aircraft radome following a severe lightning strike.**

[14]. Static discharge is normally more important for communication system interference but can lead to pinhole puncture of a build. Prevention usually takes the form of applying a very high resistance (megohms per square) coating on the radome exterior.

Lightning strike is potentially much more catastrophic (Fig. 24), the possible consequence being severe delamination or, occasionally, the total loss of the radome. Various forms of protection exist [14], with external metallic divertors being preferred. The object is to encourage lightning attachment to one such divertor, with subsequent conduction of the extremely high discharge current straight to the airframe. The resistance of the laminate to puncture is an important property for the success of this diversion, and hence to the number of such devices that must be placed on the surface. It must be noted that the conductors adversely affect the microwave performance of the assembly [27].

Resistance depends on resin and reinforcement type, on the quality of the component, and on the thickness of the build. Thus a high quality half-wave construction, such as for the Tornado, can survive with few divertors, whereas a sandwich form may need eight or more. In general, good quality resin-transfer-molded epoxy constructions produce the best puncture resistance.

Test Methods

In view of the wide range of constraints applied to microwave composites, it is not surprising that an equally wide range of additional and unusual tests need to be carried out. Some of these (e.g., the measurement of radiation patterns) are common to both radomes and antennae, but because of environmental impositions, such as rain erosion and lightning strikes, additional tests are very much a feature of radomes.

Conventional structural tests will, of course, be needed, both for basic design data and for overall component qualification. Some of these do, however, become quite sophisticated when one considers the range of aerodynamic loads and temperatures that might be simulated in a radome case [1]. This section will concentrate on some aspects of testing that will be unfamiliar to the majority of composites designers, who nevertheless will be very much aware of the general high level of process control needed to ensure component integrity and, particularly in this case, precise laminate thicknesses or reflector profiles.

Tests may be grouped as follows:

1. Design data	Basic microwave properties (e.g., ε, tan δ, and variation with temperature) Rain erosion characterization
2. Quality control	Electrical wall thickness measurement and correction
3. Qualification testing	Radiation patterns Rain erosion Bird strike Lightning strike
4. Production testing	Radiation patterns

Design Data

Determination of basic microwave properties for design purposes is normally carried out using small samples placed in a waveguide circuit. The effect of the sample on the transmission properties of the circuit can be used to determine ε and tan δ [16]. Properties do not vary much throughout the microwave frequency band, but it is normal to test close to the operational frequency. If values at elevated temperatures, are required, they can often be determined by heating the sample holder.

In some instances information is required about a complete wall construction for which waveguide testing is inappropriate. This would be true of a sandwich construction, where additional complications, such as glue line thickness, have an influence beyond simple knowledge of ε and tan δ, or in the determination of honeycomb properties. In these cases, sample panels, say 0.5 m square, would be examined for transmission amplitude and phase for various incidence angles and polarization conditions. The panel might be illuminated by a simple radiating horn or might be included in a more sophisticated quasi-optical setup. A version of the Michelson interferometer has been used for such work.

Rain erosion performance characterization has been and continues to be a most difficult subject. Methods have ranged from so-called whirling arms [8], where a sample mounted on the end of a counterbalanced arm is rotated at high speed in an artificial rain field, through rocket sledge equivalents [17] to actual flight testing of a multiplicity of samples mounted on a nose cone [19].

Despite obvious concerns regarding the application of these simulation results to the real case, the controlled conditions available with the whirling arm and the long test durations that can be used leave this as a favored method. In general, operating speeds are subsonic, but supersonic equipment has been made. More recently, and following many interesting diversions, including shotgun tests, a modified sand blast machine using plastic pellets has provided a good simulation of rain effects [18] when calibrated against whirling arm results. This is an ideal tool for establishing the relative merits of materials and builds.

In connection with lightning strike protection, the measurement of the resistance of builds to electrical puncture (electrical strength) and surface tracking characteristics (important in determining divertor effectiveness) has been carried out using large panels. The high voltage facility can in this case be fairly modest (~300 kV) [14].

Quality Control

An aspect of quality control for very high performance radomes is ensuring a precisely controlled electrical wall thickness [15]. This might be constant, tapered, or varied according to some relatively complicated profile. A preferred method of achieving this is to measure the path length (phase) of a microwave signal passed through the radome wall. This is automatically carried out for every few square centimeters (typically 4 at J band) over the required transparency area, and the data are recorded. The data gathered are then compared with the requirement, and local corrective action is taken, either adding (spray varnish) to or subtracting (grinding) wall material from the inner wall of the radome. An exception to this is with the filament-winding process, where machining to an average reading takes place on the outer surface.

Qualification

Qualification for antennae consists of measurement of radiation patterns using a conventional test range. Here transmission from a remote point is received by the test antenna (or vice versa), and the signal level is recorded as it is moved through a series of arcs or scans that adequately describe the solid angle of interest. Factors such as gain (energy-collecting capability), beam width, side lobe levels, and beam pointing accuracy, relative to some mechanical reference, would be analyzed from the recorded data. For large antennae, a conventional range can become inconveniently long. Compact sites [20] (Fig. 25) that simulate infinite range by test beam collimation can be of some use, but, in the main, aperture field sampling followed by mathematical long-range radiation pattern prediction is used.

Radomes will also, in conjunction with the specified antenna, be tested for radiation patterns, with the emphasis on the level of degradation (Fig. 26). In general, sizes are such as to allow conventional test ranges, but in some instances compact sites are used. Test variables such as frequency, scan conditions, and other factors can multiply to cause extremely long qualification times of many months.

It is not normal to qualify complete radome structures against rain erosion because of the obvious difficulties. This qualification relies on sample tests, usually on whirling-arm-type equipment. Bird strike qualification, on the other hand, is meaningless without a complete structure. Here certain bird weight rules and velocities have been laid down, and a suitable carcass is fired at the radome with an appropriate velocity and trajectory; the radome is then visually examined and, if necessary, structurally tested.

When lightning strike qualification is called for, it will be simulated in a very high voltage facility (~2 mV) and will aim to prove the divertor system. Separate high current test to demonstrate the carrying capacity of the divertors and the integrity of their attachment may be carried out on a complete structure or on suitable sample panels.

Production Test

Production antennae of the simple reflecting type can normally be passed on the basis of dimensional inspection. Twist-reflecting types will normally be tested, be-

FIGURE 25 Overall view of a microwave compact site.

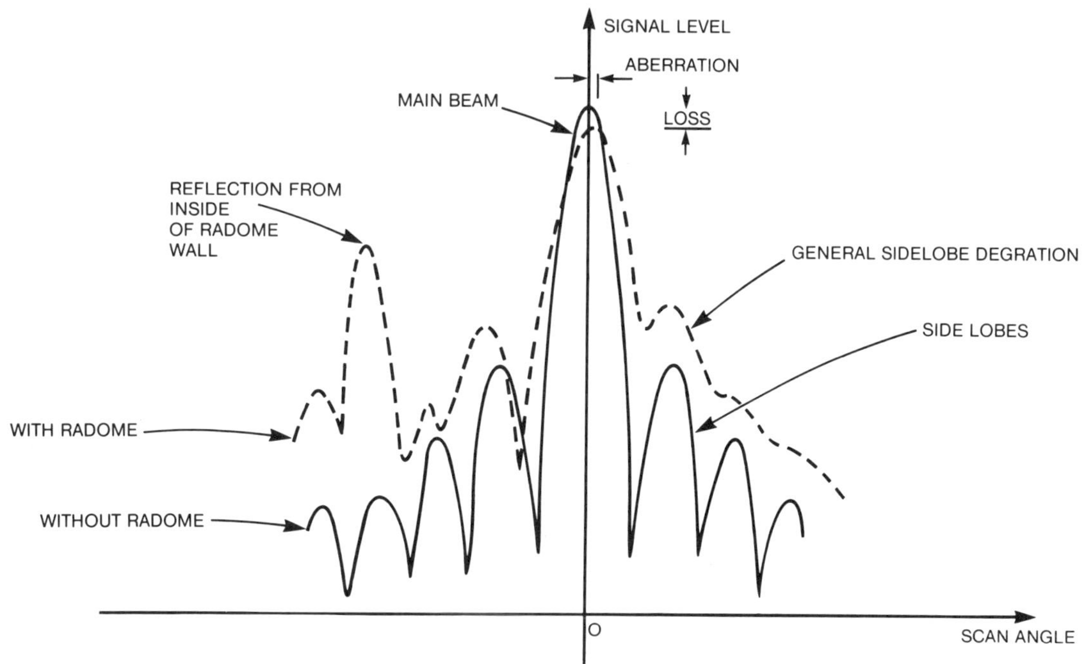

FIGURE 26 Typical antenna radiation pattern with and without radome.

cause of build variability in the transparencies and other electrical functions.

Radome production testing is normally limited to radiation pattern measurement. The extent of such testing is very dependent upon manufacturing process quality control and the tightness of the requirement. Thus, at one extreme, a consistent product from a resin transfer process allied to a relatively relaxed specification could remove the need for production test. At the other extreme, a low cost but more variable vacuum bag or autoclaved product, which most likely will be associated with a sandwich construction where quality control techniques are limited, might require 100% testing for all but the easiest of specifications. Equally, 100% testing can be necessitated by a very tight specification even when the product is subjected to the best available quality control (e.g., electrical wall thickness correction). In between these extremes, cases can be found where batch sampling can be used.

Whatever the level of testing required, every effort is made to limit the range of frequencies, scan angles, and functions recorded in order to control this potentially very time-consuming activity. The advent of fully automated test equipment and indoor compact sites, which are weather-independent, has greatly eased this problem.

Acknowledgments

Acknowledgment of assistance is given to colleagues in British Aerospace and permission to publish from the directors of British Aerospace Dynamics Division.

Additional photographs have been kindly provided by the Brunswick Corporation and Lightweight Structures, Ltd.

David A. Conti and Geoffrey F. Meades

References

1. J. D. Walton, Jr., Ed., *Radome Engineering Handbook: Design and Principles,* Marcel Dekker, New York, 1970.
2. R. C. Hansen, Ed., *Microwave Scanning Antennas,* Vols. 1, 2, and 3, Academic Press, New York, 1966.
3. A. W. Rudge, K. Milne, A. D. Oliver, and P. Knight, *Handbook of Antenna Design,* Vols. 1 and 2, Peter Peregrinus, London, 1983.
4. D. A. Conti, *Proceedings of the 10th Symposium on Electromagnetic Windows,* 1970.
5. J. R. Richmond, "Calculation of Transmission and Surface Wave Data for Plane Multilayers and Inhomogeneous Plane Layers," OSU Laboratory Report 1751-2, Contract AF33 (615) - 1081, October 1963.
6. D. T. Paris, *IEEE Trans. A & P, AP-18* and (1) (1970).
7. S. A. Morefield and J. R. Rogers, *Proceedings of 10th Symposium on Electromagnetic Windows,* 1970.
8. *Proceedings of the 2nd Meersburg Conference on Rain Erosion and Allied Phenomena,* 1967.
9. H. Jasik, Ed., *Antenna Engineering Handbook,* McGraw-Hill, New York, 1961.

10. Hannon, *IRE Trans. A & P,* March 1961.
11. R. A. Rawlinson, *Proceedings of the RPG Symposium on Reinforced Plastics in Electrical and Electronic Applications,* 1976.
12. R. A. Waldron, Theory of Guided Electromagnetic Waves, Van Nostrand Reinhold, London, 1970, p. 111.
13. "Radome Materials," AGARD Advisory Report No. 75.
14. R. H. J. Cary and D. A. Conti, *Proceedings of the Lightning and Static Electricity Conference,* 1975.
15. H. Bertram, P. Bini, and W. Kendall, *Proceedings of the Electromagnetic Window Symposium,* 1975.
16. A. R. Von Hipple, Ed., *Dielectric Materials and Applications,* MIT Press, Cambridge, MA, 1954.
17. K. N. Letson et al., *Proceedings of the 14th Electromagnetic Window Symposium,* 1974.
18. A. Campbell, *Proceedings of the 6th International Conference on Erosion by Liquid & Solid Impact,* 1983.
19. A. A. Fyall and R. B. King, *Proceedings of the Electromagnetic Window Symposium,* 1975.
20. *Compact Antenna Ranges Seminar,* 1980.
21. D. A. Conti, *IEE Proceedings, 128F*(7) (1981).
22. P. C. Wilcockson, *Radome Design and Their Properties—A Review,* IEE Conference Publication 77, 1971.
23. R. H. J. Cary, *Design of Multiband Radomes,* IEE Conference Publication 77, 1971.
24. R. G. Purcell, *Proceedings of the 13th Symposium on Electromagnetic Windows,* 1976.
25. A. A. Fyall, "Guide to Rain Erosion," RAE Technical Memo MAT 266, RAE Farnborough, January 1977.
26. G. F. Schmitt, "Polyurethane Coatings of Subsonic Radome Rain Erosion Protection," *Proceedings of the 10th Symposium on Electromagnetic Windows,* 1970.
27. R. H. J. Cary and D. A. Conti, *Proceedings of the 12th Symposium on Electromagnetic Windows,* 1972.
28. A. Kay, *IEE Trans., AP-13* (1965).
29. Crone et al., *IEE Proceedings, 128F*(7) (1981).
30. *Radome Advanced Design,* AGARD Advisory Report No. 53.
31. G. F. Meades and T. S. Powis, *Proceedings of the Electromagnetic Window Symposium,* 1967.
32. C. K. Hall, *Continuous Fibre Reinforced Thermoplastics in New Generation Radomes,* Fibre Reinforced Composites, Liverpool University, 1984.
33. C. K. Hall, *Proceedings of Military Microwave Conference,* 1984.
34. A. J. Bellworthy and J. G. A. Croll, *Space Structures, 1* (1985).
35. R. A. Stonier, *SAMPE J., 25*(1), 7 (1989).

Military Specifications

See Military Application of Advanced Composites; Specifications and Standards

Modeling

See Process Modeling

Moisture Absorption

See Environmental Resistance; Environmental Stress Cracking; Polymer Composites, Corrosion

Mold Fabrications

When fabricating a mold or tool to form a simple layup, a structural-resin transfer molded (S-RTM) advanced composite part or assembly, or a thermoplastic composite material, the best materials and processes to use are the ones that result in the production of the best quality finished shape or contour at the lowest cost.

Models, masters, patterns, molds, and tools are required prerequisites for any molded part or assembly. Composite forming molds and tools of the best quality can be produced from a mixed array of both metallic and nonmetallic materials. The most advantageous mold or tool should also provide dimensional stability during use, be easy to fabricate, maintain, and repair; and be convenient to handle and store.

Besides low cost to produce and lowest cost of operation, the factors of thermal mass and thermal uniformity in the heat-up rate play an important role in the choice of mold material and mold design. Aluminum has always been a basic material for use in molding laminated fiberglass-reinforced composites and assemblies. When graphite-reinforced structural part materials appeared in the aircraft and aerospace industries, new requirements developed. The old standby, steel, with its mass and high cost to process, fit hardly any application other than simple contours and loose tolerance applications.

Fiberglass-reinforced, low temperature laminating materials serve well for fabricating checking and inspection tools, trim fixtures, prototype and intermediate molds, and low temperature curing molds with limited production rates.

The use of reinforced, high-temperature thermoset composite materials, which was just gaining acceptance in mold and tool fabrication, took a downward turn when the U.S. government declared certain constituents harmful to the health and safety of users. It was at this point that industry started not only to look at nonmetallic mold and tool materials, but to reexamine the materials used to form metallic molds and tools as well.

New machining methods were developed to form contoured electroformed or plated nickel tools because this material demonstrated a low coefficient of thermal expansion (CTE). The need for low CTE materials directed mold and tool researchers to examine other materials, such as castable ceramics, reinforced cements, monolithic bulk graphite, special formulated mass-cast resin compounds, graphite-reinforced low temperature precure epoxy prepreg materials, and high temperature structural foams.

The introduction of reinforced thermoplastic composites broadened the requirements necessary for low CTE mold and tool materials by increasing the molding temperatures of the part materials from an average of 177°C to 399°C and higher. This narrowed the mold material choices.

The art of making of advanced composite molds is divided into a number of sequential units, and these are presented in the following pages of this article.

Mold Design

Since statics deals with forces and reactions on rigid bodies, and the strength of materials expands the science of statics to included internal reactions, these two subjects are fundamental tools needed for proper mold design.

To complement an understanding of strength of materials, stress–strain relationships and diagrams are required. These stress-strain relationships provide the equations necessary to arrive at a safe stress strength.

Stress–strain diagrams will characterize a material's strength properties. Thus, by combining statics, strength of materials, the stress–strain diagrams, and stress–strain relationships, proper loading conditions and safe stress can be calculated.

Statics studies the forces acting upon bodies at rest. A force consists of two components, magnitude and direction. If a force acts upon a mold, it is positive and the reaction of the mold on that force is negative. The elastic limit is the stress limit beyond which permanent deformation occurs within a material. This permanent deformation of set is known as inelastic action and is typical of metallic members. Composites have an elastic limit, but beyond the elastic limit, permanent deformation occurs and catastrophic failure is imminent. Yield point, on the other hand is the point at which the material displays an increase in strain without an increase in stress. Metals exhibit this property and composites do not, owing to the fact that composites cannot experience high strain without catastrophic failure. Ultimate stress is the highest value of stress observed on the stress–strain curve. In general, metallic members can be strained past the ultimate stress point, but composites cannot. Rupture strength is the point at which failure occurs.

To avoid failure in a mold or tool design, the maximum value of shear or normal stress should never be used. The analysis of static loading on tools may be accomplished through the use of the theory of elasticity. The design strength of the mold is dependent upon the magnitude of the applied load and the duration of its application. Furthermore, increased temperatures decrease the strength of composite or plastic molds and tools. Since large safety factors are always incorporated into a mold or tool design, critical loading should never occur. In summary, designing the structural surfaces, members, and substructure within a tool is basically a bookkeeping operation, in which inserting the different variables yields an equation/solution, to which the appropriate safety factors may be applied.

Since most composites are cured and shaped at elevated temperatures, as high as 400°C and higher, a tool designer must know the effects associated with these temperatures. The three modes of heat transfer—conduction, convention, and radiation—should be understood. It becomes, apparent, then, that the designer must consider statics, the strength of materials, and thermal effects in order to create a tool that is reliable throughout its service life.

Tool heat-up rate should be uniform. The three parameters that govern the heat-up rate are temperature, specific heat, and material. (See Table 1.)

The CTE will be a major consideration when sizing a mold or tool. For instance, a composite part has a smaller coefficient than a metallic tool, and the tool design must accommodate this difference. A tool design that has CTE provisions incorporated into it is considered to be "thermally shrunk."

Mold Engineering and Materials

In a word, rapidity and economy make the utilization of plastics practical in mold making today, provided there is a careful evaluation of the end use when a plastics material is selected for a specific application.

TABLE 1
Thermal Conductivity and Expansion Properties

Property	Water	Fiberglass–Epoxy	Graphite	Soft Steel	Aluminum	Nickel	Alumina 96% αAl_2O_3
Thermal Conductivity (w/cm^2°C)	20–200°C 0.00059	0.00168–0.00448	1.7588	0.4732	50°C 1.2144–2.1776	0–100°C 0.5863	0.3518
Thermal Expansion (mm/mm·K)		11–35 × 10^{-6}	50°C 7.9 × 10^{-6}	16°C 11.9 × 10^{-6}	200–300°C 25.5 × 10^{-6}	25–100°C 13.7 × 10^{-6}	25–300°C 6.4 × 10^{-6}

Further development and growth of tooling with plastics must be based upon sound engineering standards, mold design, and proper but innovative materials.

Considerations

A number of considerations should be taken into account before proceeding with the engineering of a particular mold or tool. First, some knowledge of the part design, as well as the manufacturing methods to be used, should be obtained. Second, the number of parts or assemblies that will be made from the mold, and the intricacy of each part, should be studied.

When a part with fine details and close tolerances is to be produced, designers almost universally go to a nonmetallic mold or tool that is made from the same material as the structural part. In fact, the prototype mold might well be the production mold, because of mold engineering constraints.

Further help is provided by new techniques for recording designs. Once produced manually, new designs are now automatically transferred to computer memory. Existing designs can be duplicated (digitized) for the record, while new designs can be created with computer assistance (CAD) and presented in three-dimensional form (CAM), and computers can assist with the transformation to machining (N/C) and inspection.

Third, cost and fabrication feasibility must be considered. Some of the more important details to be examined are:

- The choice of a metal or nonmetal material for the mold or tool
- The temperature at which the mold is to be used
- Part and mold tolerances
- The size of the mold or tool
- Undercuts, hardware, and inserts (if any)
- The mold wall thickness
- Tapering and draft angles
- Inside and outside radii of the tool or mold
- Special reinforcements, such as ribs or fillers,
- Warpage due to substructure support
- Parting and trim lines
- Holes, rails, and attachments
- The surface finish
- Post-mold making handling and storage

The least expensive and most accurate materials for making a lightweight mold or tool with complex contours and their comparative costs are shown in Table 2.

A final consideration in a good mold or tool design should be the rate at which the material will heat up. Table 3 shows the general rates at which some mold and tool materials conduct heat.

Mold Life

One of the most controversial topics discussed over the past years in all industries has been the expected life of nonmetallic high temperature oven and autoclave molds.

TABLE 2
Cost of Mold or Tool Surface Area

Material	Cost/0.093 m² (1988 $)
Mass-cast ceramic	$50
Integrally stiffened graphite epoxy prepreg	$200–$300
Electroformed nickel (8.9 mm thick shell)	$300

The original materials used to make molds were formulated for use at much lower temperatures than were required to cure the materials used for molded parts. As a result, designers of composite molds and tools have pursued various other metallic mold and tool design approaches for making production molds.

As advances have occurred in the formulation of composite part materials, it is only natural that advanced composite mold materials should follow. Therefore, nonmetallic molds and tools can now safely be compared with metallic ones, and because of the unique advantages of nonmetallic molds in maintenance and repair, they promise to be less expensive production mold materials in the future.

With new advances have come new considerations. Since the CTE can be matched almost exactly to that of the nonmetallic part or assembly, material accuracy must be guaranteed in the production of a mold. This is defined more precisely in Table 4.

The number of cycles for each mold in Table 4 is shown to the point at which repairs to the surface or substructure will have to be made. The number of cycles shown does not indicate the total life of the mold or tool, but rather indicates the number of uses before replacement or major repair might occur. The numbers shown are culled from reports of various industry manufacturers of aircraft and aerospace parts, and are for design use only, as they are average or approximate.

TABLE 3
General Mold Material Heat-up Rates

Fast	Impregnated monolithic graphite block (N/C machined)
	Graphite fabric reinforced epoxy
	Mass cast materials/conductive fillers
	Aluminum
	Nickel
Slow	Steel
	Mass cast ceramic*

* It is important to note that the slow thermal conductivity of mass-cast ceramic materials make these mold materials efficient and less expensive to operate. The molding surface adjacent to the part material being formed, shaped, or cured is what must be heated, not the entire mold or tool mass. This makes integral heating an excellent design tooling aid.

TABLE 4
Fabricated Mold, Tool, and Bonding Fixture Life

Type of Mold	Autoclave Cycles at 149°C Before Repair Is Needed	
	Molds/Tools	Bonding Fixtures
Sheet cast or machined steel	700	1000
Sheet cast or machined aluminum	300	500
Epoxy, fiberglass, or graphite fabric Reinforced room temperature wet layup (vacuum bagged)	30	50
Room temperature/high temperature wet layup (vacuum bagged)	50	80
High temperature wet layup (vacuum bagged)	60	90
High temperature prepreg, wet surface coat (vacuum bagged)*	100	150
High temperature prepreg, film surface coat*	200	300

* Fabricated in an autoclave at 586–690 kpa.

Extending the Life on Nonmetallic Molds and Tools

Good storage and handling practices are considered vital to the maintenance of the molding surfaces. Designs that include laminate mold or tool edges that are recessed or flush with the egg crate or backup structure edge (Fig. 1) eliminate structural failure along laminated edges during handling by eliminating delamination of plies and leaks in the mold laminate.

Rolled mold or tool edges, although time-consuming and expensive to produce, also act as "bumpers" on tools in use or storage (Fig. 2). The edges, furthermore, serve as a protective surface along the periphery of the tool and add to the stiffening of the laminate. The operators who mold the parts sometimes create shortcuts to increase their daily productivity. This often results in the careless practice of using the mold or tool surface as a table on which to trim part patterns. To eliminate scoring or cutting of mold or tool surfaces, "slip sheets" are sometimes used between the fabric/prepreg and the tool surface during the finish cutting of the ply edges. Some tool makers have also been known to use fillers that are highly abrasion resistant, in the surface coat along the tool edges, in order to reduce scoring of the surface by operators' cutting knives (Fig. 3).

Most mold making materials are not formulated for outdoor use. Therefore, it is recommended that molds and tools not be stored in direct exposure to sunlight and weather. If this is absolutely necessary, they should be stored with the molding surfaces inverted, upright and away from direct exposure to sunlight. Covers will also help to reduce the damage from exposure to outdoor weather conditions.

Although the effects of aging as a result of improper storage can affect the surface of a tool, refurbishment and selective replacement of molding surfaces can return the molding surface to like-new condition.

Automated Design and Manufacturing

Traditionally, skilled mold and tool designers have been required to prepare the daily designs, and craftsmen to

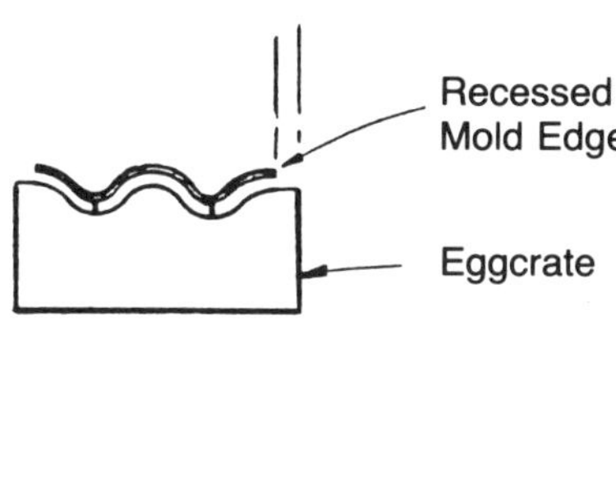

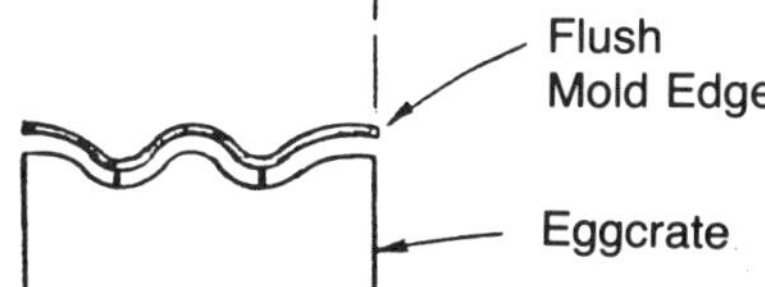

FIGURE 1 Recessed and flush mold and tool edges.

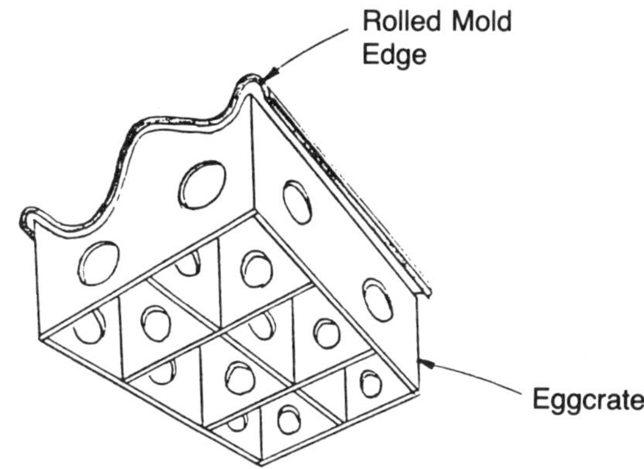

FIGURE 2 Rolled mold or tool edge.

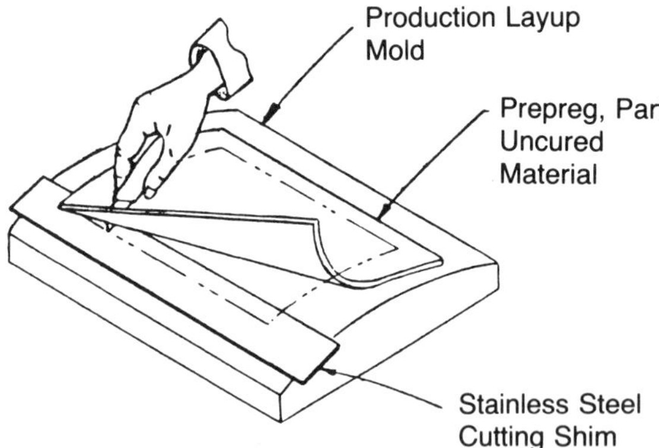

FIGURE 3 Pattern cutting aid to protect tool surface.

manually produce simple or complex mold and tool cavities and contours. Until recently, however, some demands for precision in both design and production have been impossible to achieve manually. This has sometimes resulted in conceptual shapes rather than exact contours, since only the general positions of part and assembly components could be made. In addition, conventional hand-constructed plaster models and masters, although state-of-the-art, have resulted in problems of dimensional control and accuracy, the imposition of extreme lead and turnaround times, and limitations on the number of design changes.

The introduction of automated design does not alter the basic approaches and systems required to complete a design. In fact, many of these basic approaches, whether done manually or automatically, overlap and must be applied in all cases.

A number of automated engineering design systems are available to support complex and complicated structural mold and part configuration evolution. Some of these include digitizing, which is the semiautomatic drafting and design interface to a number of automated programmable design, drafting, and machine systems. The input to automated systems is digital, and more accurate digitizers translate point and line locations to a microprocessor. Others include the CADAM* (CAD-CAM) computer-aided design and manufacturing system; multipurpose, high level function design and drafting packages such as CAD; and interactive numerical control (N/C) computer systems such as CAM, which is designed to direct a machine tool to produce a mold or tool contour automatically through an interactive numerical control computer system. CATIA† (computer graphics aided three-dimensional interactive application) is another interactive application system used for direct construction of three-dimensional objects, with provisions for viewing, analyzing, manipulating and modifying these entities. This type of system automatically produces the necessary machining data and instructions to drive N/C equipment of various kinds.

* CADAM is the registered trademark of Lockheed California Co.
† CATIA is the registered trademark of Dassault Systems, Paris, France.

Mold Materials

With the advent of plastic mold materials, certain casting and reinforcement processes were also developed.

Laminates resemble a multilayered sandwich with each of the laminations bonded together with various resin systems. The laminate is formed of multiple plies of a number of reinforcement types and configurations.

Reinforcements and fillers are added to and used with plastics to control viscosity, weight, and thermal properties; to reduce shrinkage, exothermic heat, and cost; and to increase pot life, strength, and wear resistance. They are also used to add color and texture, Fillers are controlled by the plastic formulators to meet specific requirements.

The weight of plastics can be adjusted by the use of various types of fillers. The heavyweight fillers used include metal powder and particles, barite, sand, and tar. Lightweight plastics such as foams, cast resins, and paste resins can be poured or applied over or into contours to make lightweight cores or space fillers.

The designing of tools made of fiberglass and graphite-reinforced epoxy involves the same basic considerations as those made of metals and other familiar materials. With glass- and graphite-reinforced epoxy, however, the designer has a wider choice of control factors. This choice is available in part because of the directional control of fiber strength, and partly because several reinforcing media with different characteristics can be incorporated into the individual tool. Ply orientation and stacking of plies (balanced layup) also affect strength control.

The highest strength properties are found in laminates employing thin glass or graphite cloths, at least on the outer stress surface. The overall strength decreases as the fabric thickness increases. The diminishing effect of increased thickness is greater in compression. Impact strength, however, varies directly with the thickness (i.e., thicker fabrics have high impact strengths).

FIBROUS REINFORCEMENTS. Reinforcements, aligned in various manners within a matrix, yield a composite that exhibits relatively uniform, if low, mechanical strength in all directions.

A general rule is this: Reinforcement loadings should contain no more than 38% or less than 20% resin for maximum mechanical strength. Resin, it must be remembered, contributes little to overall mechanical strength.

Composites reinforced with strands aligned parallel to each other have their maximum mechanical strength and stiffness in the direction of strand alignment.

When half of the strands are laid at right angles to the rest, the mechanical strength at either angle is less than that of the parallel alignment. Also, as the distribution of the strands varies between the 0° and 90° angles, the mechanical strength varies accordingly.

This can be offset by rotating alternate layers of reinforcement or varying the yarn distribution between the 0° and 90° directions. A balanced fabric with equal yarn distribution between the warp and filling directions will have comparable (but not necessarily equal) laminate properties in those directions.

Composites reinforced with unidirectional fabrics have their maximum mechanical strength in the direction corresponding to the greatest concentration of yarn. Since this is true, it would seem logical to align the reinforcement in a random manner within the matrix (as in a chopped-strand mat), since this would yield a composite that exhibits relatively uniform mechanical strength in all directions. Ironically, this does not happen. Instead, it leads to a composite of relatively low mechanical strength in all directions.

The type of weave selected for a tool is important, since the reinforcing material is an integral part of a laminate. Here are some types of woven fabrics (see Fig. 4).

RESIN SYSTEMS. The resin system materials used to fabricate plastic tools are, in most cases, specially formulated epoxy resin systems. These systems employ various reinforcements, fillers, thixotropic agents, and wetting agents and/or diluents to provide specific physical properties and/or handling characteristics.

Epoxy. Among all the resin systems, epoxy resins tend to rank high as matrix materials for many fiber composites, especially in tooling applications.

The reasons for this are manifold. First, epoxy resins and curing agents are available in a wide variety of viscosities and cure and use temperature ranges, and thus they can be formulated to give a broad range of properties after curing and to meet a diverse spectrum of processing requirements. Second, epoxy resins adhere well to a wide variety of reinforcing agents, fillers, and substrates. Third, because the chemical reaction between epoxy resins and a curing agent does not release any volatiles or water, the shrinkage after cure is generally lower than that of phenolic or polyester resins.

Applications. The above-mentioned qualities make epoxy resins particularly usable in various composites and structural molds and parts, in adhesives, in moldmaking and tooling compounds, and as potting and encapsulating compounds.

Since no volatiles are given off during the cure of epoxy resins, materials can be processed into void-free cured products. Furthermore, since epoxies cure with considerably less shrinkage than is encountered in vinyl polymerizations, reduced stress is seen in the cured product.

Elastomeric materials, as sheets and castable compounds, have been adapted for use in the formation of composites by taking advantage of their CTE, flexibility, and (in some instances) built-in release qualities.

Recent discoveries have even adapted low temperature elastomers for use as molding intensifiers, flexible caul plates, and pressure pads. Such compounds as latex, neoprene, ethylene propylene, polyacrylic, and fluoroelastomers have been adapted, modified, and reinforced for mold making and for tooling use.

A major advantage of the newer silicones is their use as sheeting for permanent reusable vacuum bags. These materials can be laid up and stacked to form conformal bags, thereby reducing or eliminating the bridging found in conventional film bagging.

Soft and rubbery, polysulfides have the single disadvantage of exhibiting shrinkage when exposed to prolonged temperatures over 82°C.

On the other hand, they are quickly cured at room temperature with no shrinkage, give excellent reproducibility of detail, and have a short-time resistance to very

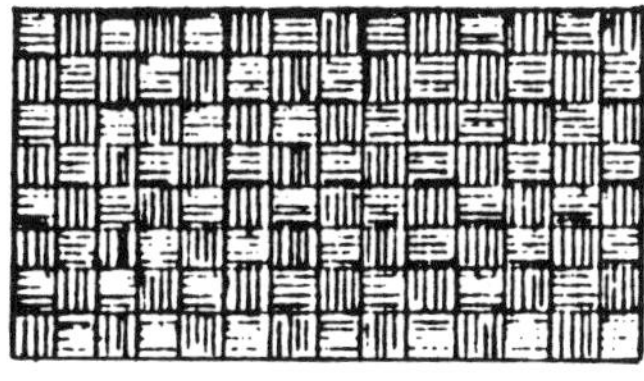

Plain

Crow Foot

Basket

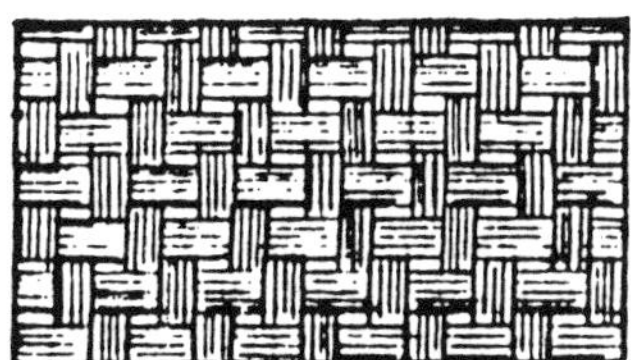

Satin Weave

FIGURE 4 Types of woven fabrics.

high temperatures—for example, they can withstand a temperature of 2760°C for up to 1 minute.

Polysulfides have the added advantage of being cheaper than either silicones or polyurethanes.

The flexibility of epoxy resin systems is varied in two ways: (1) by adding a flexibilizer to a resin–hardener mix, or (2) by the simple use of a flexible hardener.

When epoxies are at the upper hardness ranges, they are tough and strong, but highly flexible epoxies tend to cold-flow. In a general sense, these epoxies are relatively slow curing, and are prone to crack or shatter under sharp impact.

PATTERN AND MODEL MATERIALS. The materials used in the formation of master patterns and models include some of those previously mentioned. More frequently, materials such as those described in the following sections will be found in state-of-the-art patterns and models.

High quality wood pattern lumber, though sometimes difficult to obtain and expensive to purchase, is nevertheless an important material to the patternmaker. As expensive as high quality lumber is, wood is an ideal material, because it carves readily and little labor is required to eliminate defects that could adversely affect a pattern.

On the other hand, judicious buying must be practiced, since inferior, unworkable lumber is easily obtainable—and in this case, the money saved in purchasing will result in higher expenses later in the pattern-making process.

Recent developments in stabilizing wood through autoclave impregnation of the cell using thermoset materials such as polyurethanes have improved the physical characteristics of wood, but synthetic materials are still proving to be better than wood for patterns.

Wood is composed of approximately 60% cellulose and 28% lignin, plus an assortment of minor quantities of other materials. The cells of wood are similar to those of straw in minute form, and cellulose forms the framework of the cell wall. Lignin has two functions: It is the cementing material that binds the cellulose together, and it is mixed with cellulose in the cell wall proper. Color, order, and natural resistance to decay are credited to approximately 2% of the components other than cellulose or lignin.

Plaster of paris has been used for pattern, model, and mold making for thousands of years, and until a few years ago was thought of as the only usable material. Today, as a result of research and consequent development, a wide variety of specialized, scientifically formulated gypsum cements are available. The properties of these cements are designed to meet the specific patternmaking needs of the aerospace, autoclave, foundry, and plastics industries, among others.

Pattern making with gypsum cements can be done using three basic methods: pouring, screeding, and splash casting. Combined, these methods offer the advantages of accuracy, dimensional stability, economy, and adaptability.

The basic material of any gypsum is first finely ground and then calcined to form a uniform product. Controlling this calcination, as well as further processing, makes it possible to obtain superstrength plasters, termed "hard plasters" or "alphagypsum cements." As a result of new technology in model and master making, the need for new model and master pattern materials has become apparent. To match coefficients of thermal expansion, master model and mold materials have been reinforced with matching reinforcements, such as graphite and glass fibers. Slab materials of various forms have been developed, based upon machinability, cost, and availability of raw materials. Laminated paper reinforcements for high temperature slab stock, mass-cast lightweight master material, and even solid graphite block has been adapted for this use.

These materials have been used as N/C machining masters. N/C milling machines are specialized to produce repetitive contours after careful programming of the operations to be performed. The coded instructions are usually composed largely of numbers on N/C tapes, which can be stored in compact locations.

SYNTACTIC FOAMS. Syntactic foams for mass cast-in-place net molds and machining masters are advanced state-of-the-art pattern materials (see Fig. 5). Classically, foams have been produced by generating gas cells within a material. Desired densities have been obtained by controlling the sizes and number of gas cells. But syntactic foam is a different material entirely, produced by introducing hollow microspheres into a resin matrix that acts after mixing. The term *syntactic* implies an orderly arrangement of spherical filler particles. Syntacric foam is a composite material in which small preformed bubbles of glass, plastic, or ceramic are mixed with a binder, usually a thermosetting resin, and castable ceramics.

These foams are a bit heavier and a great deal stronger than gas-blown foams. In fact, they can withstand 5 to 10 times greater hydrostatic pressure than any type of plastic foam.

The eye can detect the difference between these types of foam: cells in gas-blown foam are usually large and easily discernible, whereas cells in syntactic foam/are microscopic, giving the material a homogeneous appearance.

It is well to note that the CTE of syntactic foams is low, as a result of the reduced amounts of matrix material and the structural independence of spherical fillers against the expanding matrix material.

Mechanical properties of syntactic foams vary with the types of microspheres and resin systems used.

The specific gravity of syntactic foams ranges from 0.2 to 1.5, depending upon the type and size of the microspheres, as well as the type of resin and curing agents used. Where applications require high mechanical values, additional modification may sacrifice lower density values.

Ceramic microspheres can maintain structural integrity up to a temperature of 3000°F.

Mold and Tool Substructures

Honeycomb panels are used to form substructures on laminate tools and other style molds. Stiffeners are used in conjunction with fiberglass–epoxy laminates and ep-

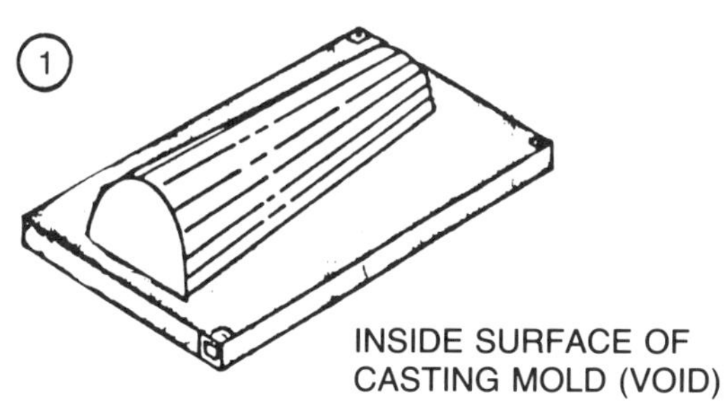

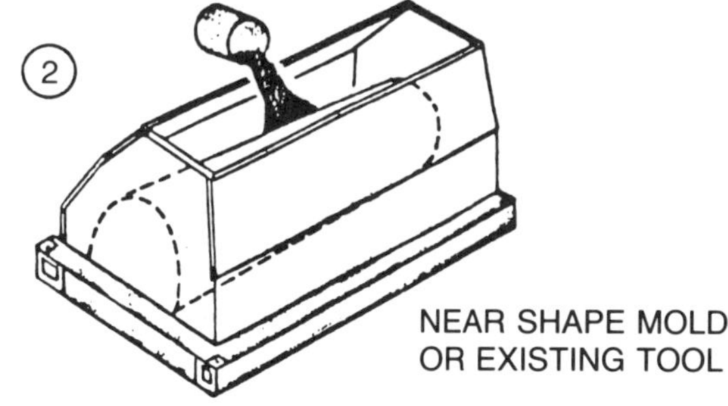

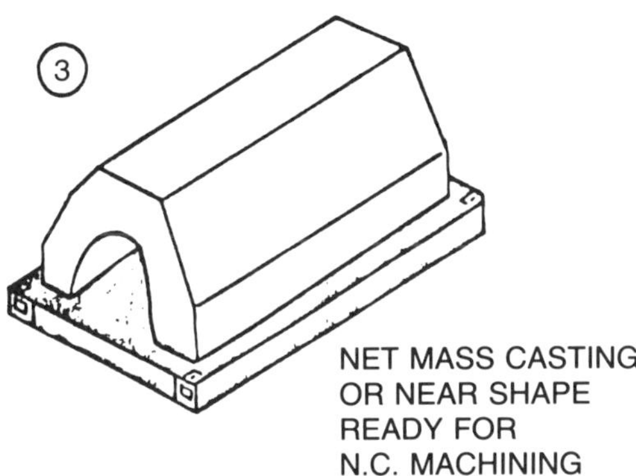

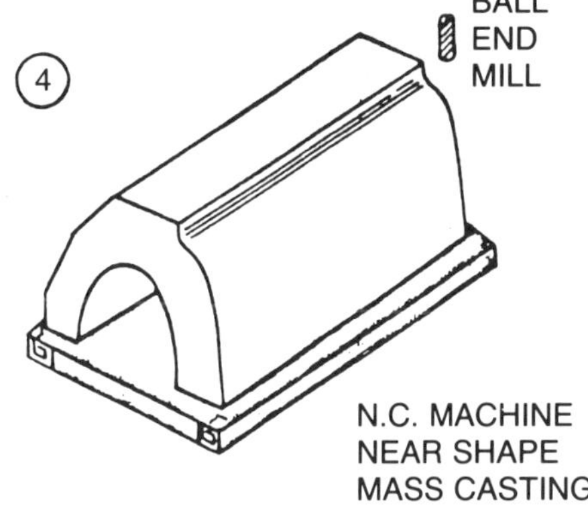

FIGURE 5 Syntactic foam casting sequence.

oxy casting material for structural support purposes on molds and tools.

Used for eggcrate back-up boards, these stiffeners are made from honeycomb flat stock consisting of three layers of fiberglass or graphite cloth for the skin and a perforated aluminum honeycomb as the core material.

The panels may be either purchased from outside manufacturers who use both prepreg or wet layup skins or fabricated in-house if the presses are available. There are three basic kinds of aluminum honeycomb.

1. Fiberglass–epoxy faced with an aluminum honeycomb core.
2. Graphite–epoxy faced with an aluminum honeycomb core.
3. Aluminum faced with an aluminum honeycomb core.

In addition to honeycomb reinforcements for back-up substructures on molds and tools, recent advances have shown that solid laminates of 6.4 mm thick fiberglass or graphite fabric reinforced boards made a more positive structural support for the tool.

In conclusion, although all of the choices demonstrated seem to indicate that in most cases one should use an eggcrate back-up structure, the recent trend is towards elimination of honeycomb and substructure totally. Integrally molded stiffeners, formed in the last stages of mold or tool layup, will be the main reason for the future elimination of the back-up structure (Fig. 6).

Preparation of Master Models, Patterns, and Master Molds

Plaster Master Models

Masters, models, and patterns are the names used to identify the totality of the shape, form, contour, and configuration of an object that must be created or duplicated. The methods used today to form these masters, models, or patterns can be classified into two general

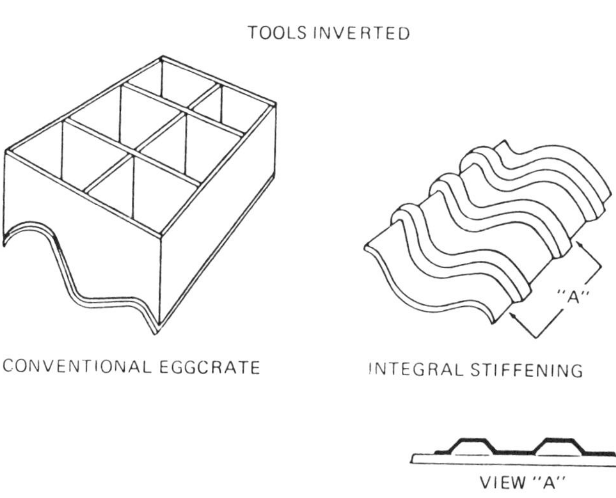

FIGURE 6 Elimination of eggcrate structure.

categories: (a) manual design and fabrication, and (b) computer-assisted design and manufacturing.

In either case, the master model maker, whether that person is a hands-on craftsman or a computer designer, must conceive the same overall shape or form in his or her mind as the structural part designer who created that shape or form. This configuration could be the master pattern necessary to form a mold copy, or it could be the actual mold contour itself, which is used directly to form the finished part.

Since it may be difficult to produce the actual mold contour directly using the first method (manual design and fabrication), a model or pattern is usually required. Since the second method (computer-assisted design and manufacturing) provides automated means for designing and producing a mold contour directly, models and patterns are superfluous.

Master patterns are sometimes prepared from lofting templates, which are flat sheet metal sections of the finished contour that have been shaped around the outer edge either by N/C machining or by hand.

Although some shops still form their templates by hand cutting and finishing of the edges, most templates today are fashioned by automation.

The automated method, similar to that used to automatically cut fabric part patterns, can also be used to lay out the lofting base so that the lofting template can be fit accurately in the base slots. Manual templates to transfer location data to the base can also be used.

If automated equipment or the base template is not available, the template base can be "laid out" by scribing the lines directly into the base surface.

The holes for the attachment of wire screens and spacers can be drilled automatically or manually, with the templates placed in stacks. Holes should be spaced approximately 6 in. apart and approximately 25.4–38.1 mm from the template edge. Good judgment with respect to their location, number, and placement will result in a secure and rigid support structure. Figure 7 suggests the form of the master pattern frame assembly.

The lofting template structure can easily be inspected while it is open and before the application of the wire screen. If inspection dictates changes, adjustments can be made at this point.

Once these adjustments have been made, the entire structure can be "tacked" in place, piece to piece, fastener to pieces, using a quick-setting, thixotropic, polyester resin system paste material.

The wire screen, usually about 6.35 mm square mesh size, is now cut to fit just inside the templates and outside of the threaded support rods, using 12.7 mm thick blocks, cut to suit. The reason for this is one of economy: more than one space can be filled at the same time only if the contour is mild or near flat.

After the application of splining material is set, dry, or cured, the reference and trim lines are scribed in the required locations. Hole patterns and any other information are transferred to the finished surface at this time. If the pattern has been made of plaster, the finished surface should be coated with several coats of sanding sealer or lacquer sealer.

The above procedures are suggested as a guide. There are numerous other methods that can be followed that will result in the same final master pattern finish.

Most of the layup molds used for prototypes or small orders are made from the plaster masters described previously. Copies of these masters, called plastic-faced plasters (PFPs), must be made to provide a contour that will withstand a temperature high enough to cure the mold laminate resin system being made from the surface. These plastic-faced plasters, which are castings of plaster over the shape to be produced, are usually made from a plaster "splash," which is a copy of the master model (Fig. 8).

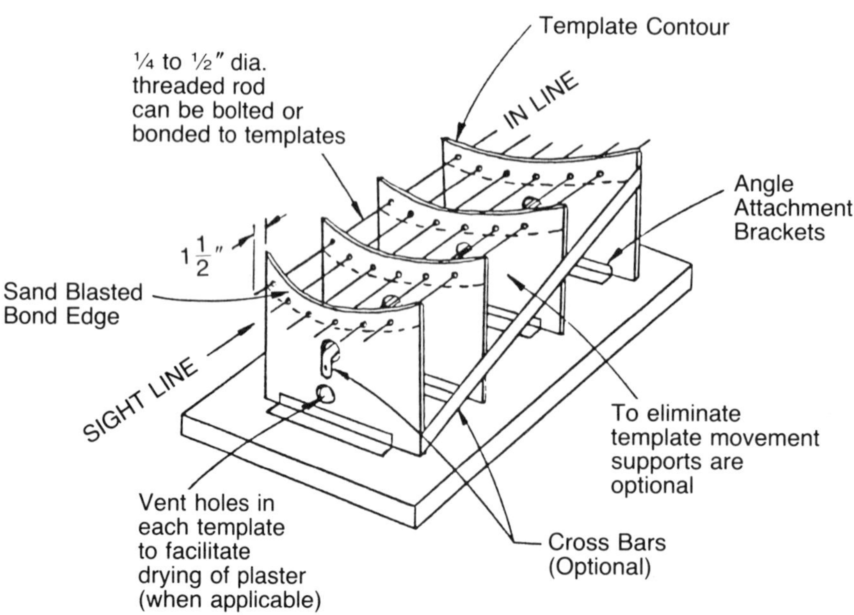

FIGURE 7 Master pattern frame assembly.

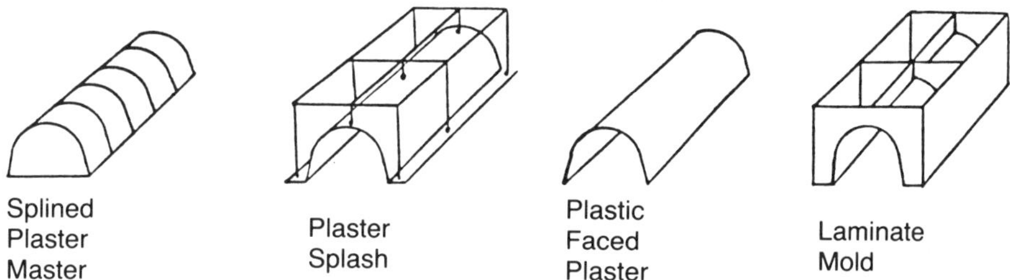

FIGURE 8 Master pattern and mold fabrication sequence.

Master Models

The N/C master, on the other hand, eliminates a number of steps, so that one can go directly to the laminate mold. An N/C-machined master may be used to produce a room temperature (RT)/high temperature (HT) or 93°C precured, 177°C free-stand post-cured laminate mold (Fig. 9). The concept of N/C-machined master models is, however, an innovation in model making, where master dimension data is available.

Undercuts, when necessary, can be formed by using wash-away plaster on plaster masters and molds. N/C-machined master model undercuts can be formed by machining a breakaway section that will fall off with the removal of the laminate mold. The breakaway section can be pinned to the master with fragile plastic or wood pins (Fig. 9).

N/C-machined masters have other advantages besides the one-step procedure that leads to the finished laminate mold.

Plastic materials can be cut with no lubricant, and this is the preferred procedure. Properly formulated plastic master slab materials produce a curl rather than waste dust when machined.

As stated earlier, the innovation in N/C-machined master models involves the availability of dimension data. When the coordinate data do not exist, CAD and CAM can be used to change a design from manual plaster to automated machining and to check or evaluate the design as it develops and evolves. Automated digitizing, the automatic recording of coordinate contour points, is one such method of accomplishing this task.

CAD/CAM not only allows for ease of design, it increases accuracy by expediting a multiplicity of functions. Designs are proven in the computer, so time spent in prototyping and development of the master is reduced. Besides developing the loft data in the computer, the loft template data can be used to create a three-dimensional wire frame skeleton of the contoured master model. Even the machining path route of the cutter can be viewed, monitored, and readjusted before the production of the master. Tooling holes, substructure patterns, trim lines, surface data, and any other mold or tool information are automatically transferred to the surface of the master model. In the final analysis, improved use of materials and reduced waste become immediately apparent in the planning of the precast slab model stock layout.

Another major advantage of N/C-machined master models is the fact that an electroformed plating mandrel can be formed in one step. The one-step forming of electroformed plating mandrels is directly associated with the machining of the master contour. Machining the contour of the master model is the same operation that is performed to produce the nickel electroform mandrel master for the production of a nickel-plated tool. To facilitate simple master model preparation, a mass-cast, near-shape master model can be prepared more quickly than a bonded one. Simple plywood near-shape molds are used to form the mass casting (Fig. 10).

Mass-Cast Master Models

A mass-cast master model can even be eliminated in some instances if the mass-casting material is properly formulated for use as a forming tool, such as in metal stamping, a stretch-draw forming die, etc., so that the contour that is then machined becomes the finished tool surface in one step.

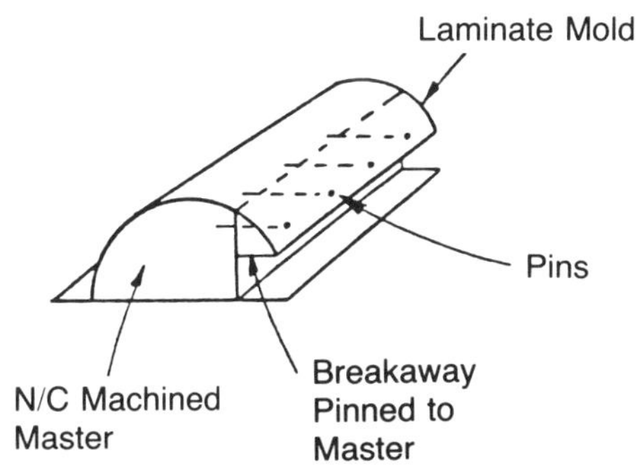

FIGURE 9 Breakaway master pattern.

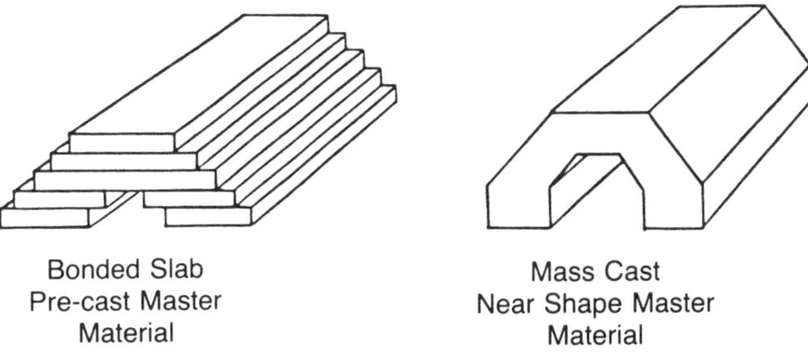

FIGURE 10 Near-shape master pattern.

The choice of the precast slab or mass-casting resin system should be a careful one. A system must be chosen that:

- Has low shrinkage
- Can operate as a master model at 93°C
- Is not abrasive and is easily machined, drilled, tapped, cut, or sanded
- Has good compressive strength (48–62 MPa is acceptable)
- Has low water absorption
- Has a hardness rating of approximately Durometer Shore "D" 65–75
- Can be easily bonded or coated with finishes
- Has a density of approximately 640 kg/m^3
- Has a tensile and flexural strength of 28 MPa or more
- Has a thermal coefficient of expansion of less than 20×10^{-6} mm/mm·K

Model Substructures

Various methods are used to support precast or mass-cast master models. Important considerations are that the plastic master should be supported by a very stable frame, and that a copy of the master (another casting, layup, etc.) should be made immediately after the finished machining and surfacing finish operations have been completed.

Steel frames, plastic frames of fiberglass–epoxy tube, fiberglass, and graphite–epoxy skins with aluminum honeycomb, as well as various combinations of the foregoing, are used to support the precast slab or mass-cast master.

There is a difference of opinion regarding the construction techniques to be used when bonding together precast slab stock. The present bonding style appears to have evolved from the formerly prevalent use of wood as master model material. The impregnated wood was classically precut and bonded, as shown in Figure 11. The new, suggested method is shown on the right.

When bonding precast slab material to itself and to the base materials, it is important to try to choose a material that has a thermal coefficient that matches that of the slab material being bonded.

At the time of bonding, precast slab materials sometimes require the squaring of bonding sides, sanding of bond surfaces, solvent cleaning of surfaces to be bonded, and embedding of mechanical fasteners between layers.

In addition, the use of surface fillers or fairing compounds to smooth and finish machined surfaces is sometimes necessary to provide an improved surface.

Treatments and Releases

MASTER MODEL AND MOLD TREATMENTS AND RELEASES. Master models, as well as molds and tools, must be treated and sometimes coated with various sealants and releases to provide the necessary functional surface upon which laminates, castings, and duplications can be made. Besides assuring positive release of the master pattern and copy, a properly chosen release also eliminates mold cleaning and produces parts ready to finish.

In addition, various coatings, films, releases, or barriers act as separators between the master patterns or mold and the part. The reasons for the release to occur is that there has been an interruption or lessening of the surface polarity.

Releases come in many forms: solid, liquid, wax, and aerosol compositions containing silanes, silicones, plastic films, carnauba, and paste waxes.

Whenever practical, releases should be chosen that, stay on the master model, mold, or tool forming the part. Releases that stay on the mold will not either contaminate the part internally as a laminate or casting or impair finishing of the part surface. A positive release of a mold, tool, or part can be affected by applying more than one release agent over another, forming a shear plane, such as sealer coating release, 2–5 wax releases.

General-purpose waxes, petroleum jelly mixtures, and stearates are used for lower temperature master pattern release applications. The general-purpose releases, some of which contain fluoroethylene propylene (FEP), should be used below 177°C.

A good grade of quick-drying lacquer is suggested for treating various wood, plaster, or porous master or model surfaces. Automobile-type primers work well, and are also supplied in clear form when required. At least

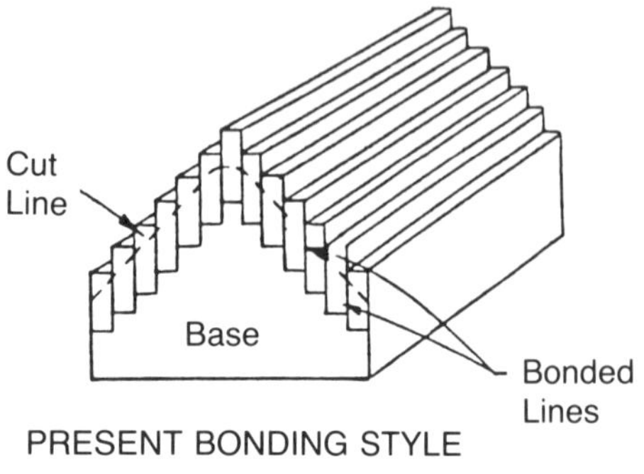

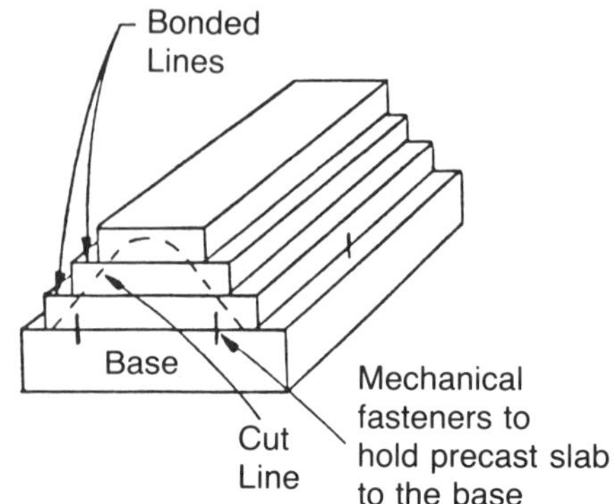

FIGURE 11 Bonding precast machinable plastic model stock material.

two coats are required before the appropriate release is applied. The first coat is usually thinned, so that it penetrates into the surface. In the case of gypsum or plaster, this first coat may be applied soon after the material has set. In the case of wood, the lacquer is meant to seal in and/or out moisture so that the wood grain will not swell and rise above the contour surface. Solid treatments like sheet wax or sheet putty are used to compensate for the thickness of the laminate or part materials. The sheet material should be applied to the thickness of the part being formed. The sheet material will then compensate for the space needed in the mold to contain the part material during the forming process. Sheet materials are applied over the sealed master.

Mold release is applied to the master pattern on the exposed surfaces, around the sheet wax, only after the wax is applied.

In the use of metal master patterns and molds, fusible PTFE and mixed dispersions or blends have been used. Permanent releases applied to the mold surface are selected only in special applications, where the coating thickness will not interfere with the overall dimensions of the contour or shape being duplicated.

The external release, preferred by almost all mold and part fabricators, can also be a curable coating that may be applied by painting, brushing, or dipping. As with many other coatings, the safety of the user and those around the user is important. Government regulations affect the users of mass quantities of treatments, coatings, and releases containing solvents, so the solvent's or solid's content, number of applications, amount to be applied, and other requirements relating to the evaporation of solvent carriers should be considered when choosing the release. Harmful solvents in the hands of the operator should be avoided at all costs.

It should be noted that routine and proper maintenance of the mold surface is critical to the life of the mold, as well as assuring that the parts being molded will not hang up and bond to the mold surface, thereby destroying the mold and part.

Cleaning of the mold surface after the application of the release is a primary maintenance function. General-purpose detergents and other commercial and industrial cleaners work well on metallic mold surfaces. Again, the release to be removed will govern what cleaner will need to be used to remove the release, or to prepare the mold for releasing.

Whether the mold is metallic or nonmetallic, cleaning can also be accomplished by using various abrasives, such as plastic, glass, and natural blasting media. In fact, the surface of a mold or tool can be cleaned by blasting in an open shop area, using equipment that collects the blasting media while it is dispensing it (Fig. 12).

Cleaned molds, as well as new molds, sometimes require sealers in addition to releases. The sealers must also contain solvents that do not affect mold surfaces, since the sealer must sometimes fill microscopic voids in the mold surface and might also require an oven back if the sealer is a coating that cures. Entrapment of an undesirable solvent and a subsequent oven bake might affect the mold surface.

Many of the composite reinforcements and fillers in resin systems abrade the mold surface in certain areas, removing the mold releases. These select areas may be touched up while the mold or tool is in service, if the proper release is chosen. Hot molds require the proper choice of solvent carrier type, and a heated mold is sometimes required for special process cycles. An external release that is applied to a sealed mold and cures as a permanent coating to the mold surface is the best release choice.

Permanent release coatings do not transfer or leave transferable residues on a fabricated mold. For part forming, a permanent release applied to a properly prepared mold surface will provide many trouble-free releases before having to be reapplied.

Now, with the advent of high temperature part materials such as the polyimides, bismaleimides, and advanced thermoplastic resins and blends, temperatures of mold use in many instances reach 399°C. The thermoset molds and metal molds must be coated with nontransferable releases as well, so that the fabricated part will not be contaminated or adhere to the mold.

Because parts bonded to a mold must be physically removed, the original plastic mold may sometimes emerge extensively marked, gouged, or damaged. It is,

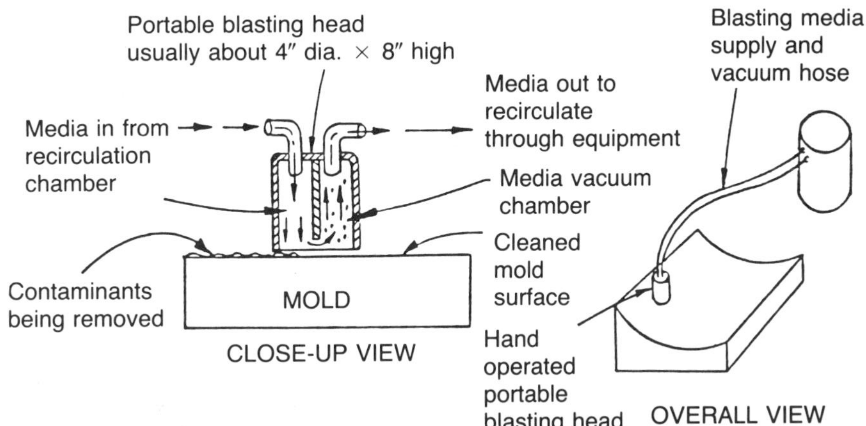

FIGURE 12 Portable vacuum sand or grit blaster.

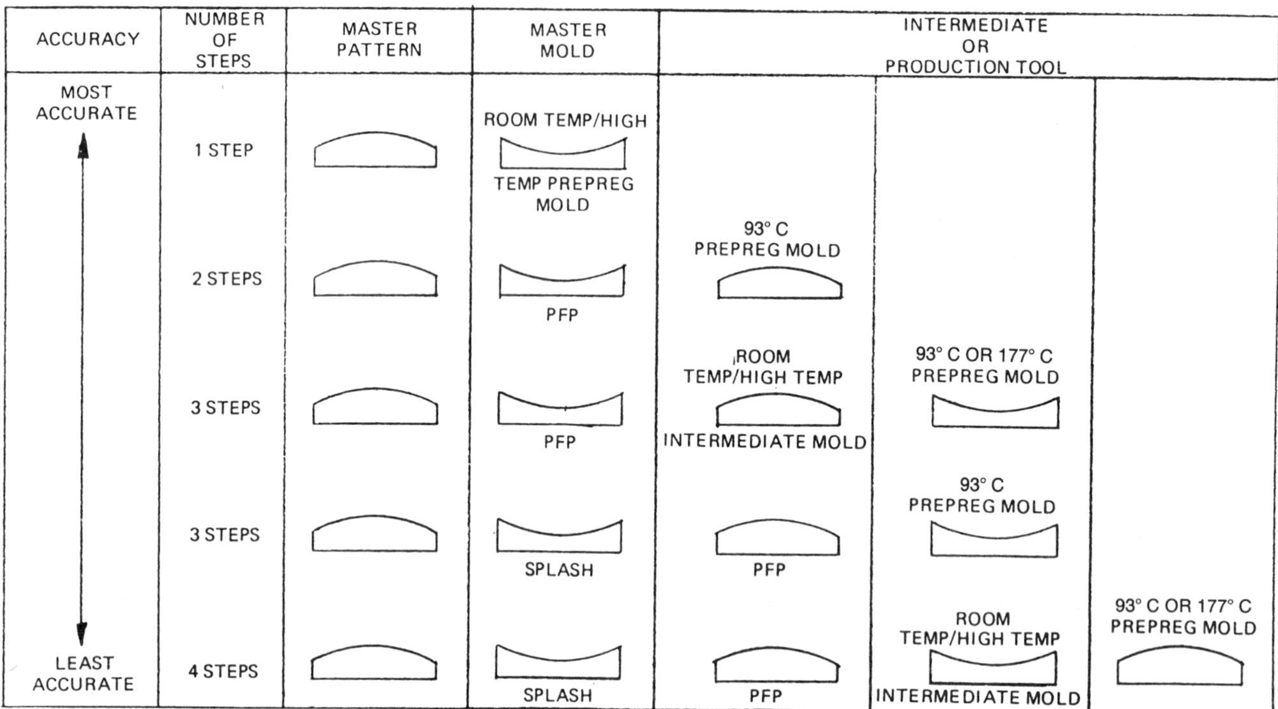

FIGURE 13 **Suggested dimensional tolerances using various molding methods.**

therefore, imperative that the proper type of mold release be selected, and that a reliable manufacturer be chosen to supply the required treatment, sealer, or release.

Intermediate Molds and Tools

In some cases, the intermediate mold must be expended during the production of the nonmetallic production mold. It is for this reason that less expensive materials and processes are used to produce these copies. Plastic-faced plaster (PFP) master molds with the front face of plastic (epoxy), reinforced or nonreinforced with fabric, is one form of inexpensive mold. This mold copy is similar to the plaster splash, associated with plaster masters.

Master molds are formed through the application of a number of materials and processes. The choice is one that involves cost in almost every case, although sometimes cost is overruled by the need for extreme accuracy or some other technical design consideration.

As of this date, the basic master mold is still plastic-faced plaster (PFP). Soon, however, the N/C-machined master model, a mass-cast mold or laminate, when temperature resistance allows, will become the basic master mold. The cost of producing master molds includes costs of both metallic and nonmetallic materials. In the formation of graphite–epoxy laminate parts and bonded assemblies, the master mold can, in some cases, be used to form the finished part.

A comparison of the weight of the different choice of materials is a logical and easy way of relating weight, density, cost, and other factors. For instance, steel, the material used for many composite molding applications, approaches 8010 kg/m^3. Thus, not only is steel heavy and difficult to machine, its cost could be prohibitive. Aluminum, although light and easy to machine, expands nearly 31 times more than cast ceramic. Also, the cost of aluminum and steel ranges from 3 to 5 times that of ceramic.

Economics thus leads us to choose only a select few of the many materials that can be used to form composite parts. The currently most economically feasible materials for forming composite parts are:

- Room temperature and 93°C cure mold and tool prepreg, which in some cases is fabricated directly from the master model.
- Nickel-plated, electroformed, or deposited over laminate mold forms. (The most economical form of this is done over N/C machined model stock.)
- Machined or cast ceramics and cementitious compounds.
- Carbon/carbon (monolithic bulk graphite).

Wet layup epoxy systems reinforced with fiberglass and graphite fabric, metals, and other master mold materials are not listed because economic factors dictate that these materials are feasible only in select cases and should be chosen for reasons other than economic ones.

The reason that the electrodeposited nickel material approach is economical is that the new master model materials available allow for the formation of plated contours directly from the machined master. Mass-cast masters also yield near-shape master molds, and the machined master model or form can be used to form a mold or tool using a room temperature or 93°C cure prepreg. The same master can then be used to form the nickel electroform tool.

Using fiberglass and, when required, graphite fabric prepregs is economically advantageous because the 93°C cure systems can be formed over PFP surfaces and directly over the machined master model when the master materials can withstand the temperature and the CTE of the model material is acceptable.

The least expensive approach to forming the master model and/or mold is to form a laminate using a room temperature curing, free-standing post-cure prepreg. This type of material (the least expensive laminate material) does not require autoclave pressures to process, and cures at room temperature.

Master models can actually be formed from any material, even wood. The CTE is not important, since the master model is never exposed to elevated temperatures. An intermediate master mold is not required, and the mold laminate can be formed directly over the master model.

Yet another inexpensive method of forming a master mold is by backing up a room temperature laminate with "pour-in-place" quick-setting polyurethane or a ceramic syntactic foam. After the two- to four-ply laminate over the surface coat is set, a two-part castable chemical or syntactic foam is poured or a froth foam system is sprayed over the rear of the laminate to stiffen the master mold surface. Sometimes the foam is covered with a few more plies of fiberglass-reinforced layup material to form a sandwich foam–laminate structure. This not only reduces the cost to produce the master mold, it also reduces the weight of the mold when very large molds are required.

Another form of inexpensive master mold can be built by producing a composite of a paint-on surface coat backed up with a premixed, frozen syntactic foam material that is supplied in sheet form, about 12.7–19.00 mm thick. The frozen sheet of foam coating is laid directly against the surface coat and either vacuum bagged in place or set in place by contact molding methods. This prefrozen foam then cures at room temperature.

This type of syntactic foam is also available in prepackaged, preweighed plastic pouches. It can be easily and quickly mixed together by kneading the resin system together inside the package. The package is then cut open and the nonflowing paste-type material is spread over the paint-on surface coat.

Syntactic foams are also used as casting materials to form master molds. Master molds as intermediate mold forms can be easily transported long distances. This is especially helpful when the mold is duplicated as a PFP and shipped elsewhere.

More accurate tolerances can be maintained in a nonmetallic mold if the systems illustrated in Figure 13 are considered when fabricating a master mold.

Fabricating Nonmetallic and Metallic Molds and Tools

Mass-Cast and Monolithic Molds and Tools

Probably the most important mass-casting material available today is castable ceramic. This type of material possesses a CTE of about 0.4×10^{-6} mm/mm·K.

Castable ceramic has a multitude of advantages over other mass-casting materials. It is prepared simply by mixing the material with water, it cures at room temperature, and when cast, it has insignificant shrinkage. This versatile material can be cast around heaters, and because the material can be thermally shocked in excess of 1093°C without damage, the heaters and/or cooling coils can be located near the molding surface of the tool (Fig. 14).

Inexpensive mixing equipment can be purchased that will handle up to drum quantities of material (Fig. 15). The cost per cubic meter is considerably less than that for some of the mass-casting epoxy systems being produced today.

There are applications for cementitious compounds that combine organic binders, organic and metallic fillers, and other compound additives. One such compound, unique to this class of formulations, uses a graded aggregate of stainless steel particles embedded in a silica-modified portland cement matrix.

The CTEs of materials in this classification approach those of steel or nickel electroform tools. Male and female molds, usually in the form of laminate layups, are used to case the cementitious compound.

Graphite, carbon/carbon, monolithic graphite, and other similar descriptive names classify this mold and tool material as graphitized carbon (carbon/graphite). In this process, carbon material is placed in an electric furnace with granular coke tamped around it. Through a series of subsequent steps, the baked carbon material is converted to a graphite at temperatures of 1427–1649°C. The CTE range of this material is $5.4\text{–}7.2 \times 10^{-6}$ mm/mm·K.

Differences in the process will produce various forms of the finished base product that will vary in density, pore size, grain size, and other internal and surface qualities. It is therefore suggested that when choosing this type of material as a mold-making substance, careful

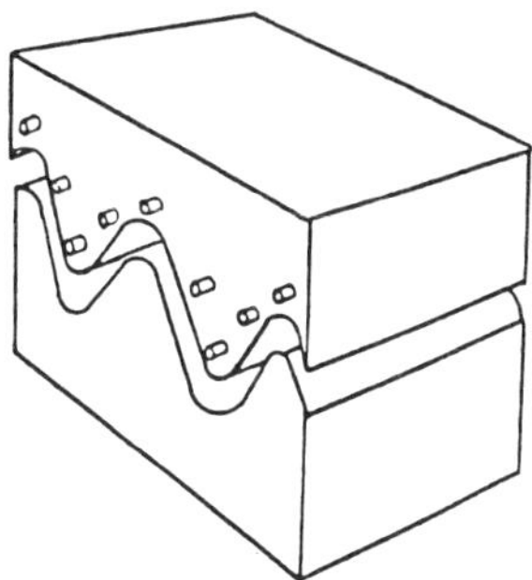

FIGURE 14 Ceramic tool with heaters cast in place.

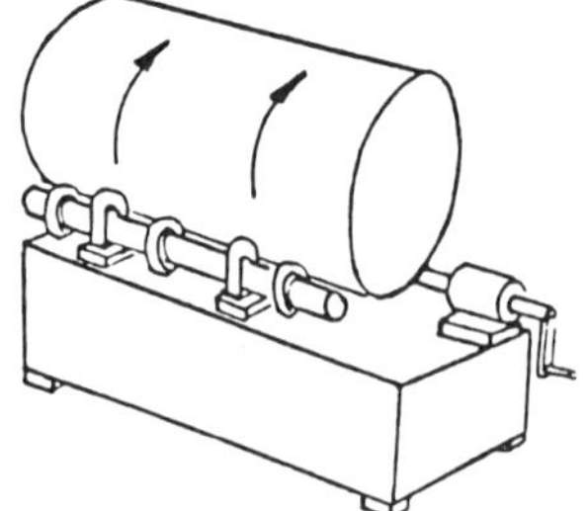

FIGURE 15 Roller mixer.

consideration should be given to all the base characteristics of the material of choice.

Unlike high temperature ceramics, high temperature ungraphitized and graphitized carbon cannot be cast into complicated shapes, only simple blocks. It can, however, be machined. However, ungraphitized carbon is so hard that diamond wheels or equivalent cutters must be used to process it. Still, formulations have been developed that allow machining to be used to form molds and tools to close tolerances. Sharp male corners and edges will have to be chamfered, since the carbon porosity will cause chips and breaks that are larger than the pores present in the material.

This material can be bonded, threaded, doweled, and grooved to be attached to larger shapes so that various size mold forms can be produced (Fig. 16).

Precast and mass-cast semirigid compounds are used at temperatures from room temperature ambient to moderate levels of up to 121°C.

The compounds are formulated using polyurethane or epoxy modified formulations.

Low shrinkage when mass casting leads the list of requirements; thus, new formulations addressing this are constantly becoming available. Blends of epoxies and urethanes, epoxy systems by themselves, and urethanes alone are the base materials usually chosen for producing the mold forms. Prototype and low-production molds are formed directly from the master pattern.

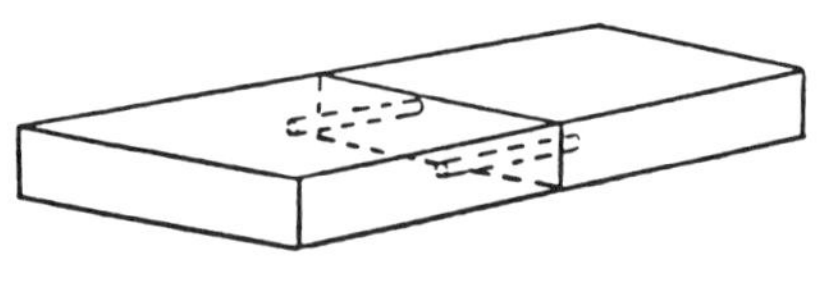

DOWELED MOLD

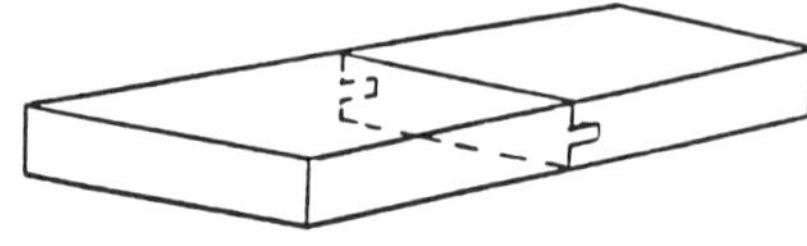

BONDED MOLD

FIGURE 16 Doweled and bonded carbon mold.

Reaction injection molds (RIM), resin transfer molds (RTM), compression molding forms, autoclave mold forms and bonding fixtures, thermoforming molds, and plating mandrels are just some of the mold forms that can be produced with this material and process technique.

The thermal coefficient of expansion is another important characteristic to consider when mass casting the mold form. This characteristic, however, is not important if the mold is used at room temperature.

Mass-casting compounds are easy to mix, and are usually formulated as one-to-one ratio compounds (resin to curing agent). Casting can be poured to any size, thickness, durometer (hardness), density, or impact resistance.

Another advantage of the mass-cast mold is that alterations and modifications are easily accomplished by sandblasting the original surface and pouring a near-shape cast surface over the original one.

The advantages also include the capability for short lead times, because mass-cast molds are produced in one step.

Expanded sheet metal, reinforcing rods, wire mesh, and other forms of reinforcement are sometimes required to secure the cast mold internally and maintain contours during shipment, use, and storage.

Other nonmetallic molds, tools, and bonding fixtures are made from fibrous reinforced laminates. These laminates can be fabricated using the wet layup or prepreg methods.

Wet Layup and Prepreg Molds

Wet layup materials are recommended only where necessary, or where the mold maker is forced by some requirement to use these materials and process techniques. Room temperature precure, high temperature post-cure prepregs exist that replace the labor-intensive, inconsistent, uncontrollable, hand wet layup mold-making process. Intermediate molds and tools, which produce the final production prepreg laminate mold or tool, can be fabricated by the wet layup process if the mold form is made from a master model that cannot be subjected to heat, or if the intermediate mold is to be expended after completion of the final production mold or tool.

It is obvious, though, that room temperature cure prepregs, supplied and stored, assure simplicity and quality.

It must be emphasized that the wet layup process is a process of last or special resort. The most accepted method of producing a laminated mold or tool is the prepreg process. Wet layup composite molds and tools have been limited in use by a number of factors:

- High resin contents
- Improper mixing of resin systems
- Short service life
- Inconsistent and nonuniform fabrication results

- Unacceptable service and cure temperatures
- High labor costs
- Toxicity problems during fabrication

Prepregs of both fiberglass and graphite fabrics offer choices of resin formulations to suit all manufacturing methods. These formulations allow the use of wood, foam, plaster, machinable plastic, and metal as the master model material.

When choosing the prepreg system, look for high hot flexural strength and thermal stability, the longest work life, but the shortest cure time at the lowest temperature, the best coefficient of thermal expansion, the highest use temperature, good drapability, and handling characteristics that allow low vacuum bag pressures to be used to cure the prepreg.

It is advisable to have the prepreg packaged in precut rectangular sheets instead of rolls, since the material can then be used immediately upon thawing and will not require cutting, which uses up the work life of the material. Furthermore, room temperature cure prepregs are very soft, flexible, and "wet," so handing is more difficult than with the 93°C and 149–177°C systems.

Room temperature prepregs are customarily laid up directly over a properly prepared master of wood, plaster, plastic, or metal, vacuum bagged, and left at room temperature under vacuum for 2–3 days. Dry plies of veil mat and fabric are usually applied in back of the surface of gel coat and laminate stack to assure void-free surfaces and laminates, and no-bleed resin layups are recommended, as the resin content of the prepreg is usually lower than 93°C cure systems.

The ideal material for a high temperature use prepreg mold or tool is a 93° cure prepreg. It offers a longer working time (up to 30 days at room temperature), higher thermal and physical properties, and much easier handling and transportation characteristics than room temperature prepreg. Furthermore, this type of prepreg can be applied as a surface or gel coat.

In fact, the 93°C cure system will usually provide the best combination of properties when compared with the room temperature and 149–177°C cure systems.

Surface coats or gel coats are sometimes required to provide an extremely smooth surface to the front of the tool laminate. Because these coatings, in some instances, are painted on, it is necessary to choose certain materials and take certain application precautions:

- Choose a flexible, not rigid, paint-on surface coat
- Do not apply too thick a coating (0.3–0.4 mm is ideal)
- Choose a room temperature set and heat post-cure system
- Do not "puddle" resin in female corners

Fiberglass epoxy prepregs, cured into a laminate, have approximately a $10–14 \times 10^{-6}$ CTE. Graphite epoxy prepregs, on the other hand, have about a $3–4 \times 10^{-6}$ CTE.

It is always a good idea to use a liquid gel or surface coat when "vacuum bagging only" forming a mold or tool. The low pressures will not compress the laminate sufficiently to affect vacuum integrity, so a coated mold or tool surface is required.

Paint-on surface coats are also used to form spray-metal-faced RIM, RTM, and SMC tools. A gel coat layer is painted on in back of the sprayed metal, thereby bonding the metal to the laminate during the co-cure of gel coat and laminate resin system.

Film gel or surface coats are available in various forms: frozen films of epoxy reinforced with lightweight fabrics (0.08–0.16 kg/m^2), veil mats, or unreinforced but supported nylon or Dacron scrims. It is also not unusual to apply a fiberglass epoxy prepreg (0.05–0.08 mm) as the surface coat of a fiberglass or even graphite prepreg laminate mold.

The film gel coats usually eliminate mark-through of the laminate prepreg.

The fabric used in the wet layup process, as well as in all other lamination patterns, should be cut into rectangles, with the largest leg of the rectangle never more than 61 cm long.

Mold and Tool Substructures

Unlike cast or ceramic-reinforced ceramic, laminated molds and tools require reinforcement or strengthening of the laminate face from the rear of the mold or tool. A framework of some type is either formed during the face laminate fabrication (integral stiffening) or added to the rear of the tool after the laminate is formed. Most mold and tool designers choose the same materials for the frame as for the front mold or tool laminate, since distortion may occur if dissimilar materials are arbitrarily used.

Some of these materials include steel and aluminum sheet, plastic or metal angles, or tubes, or a combination of these. Others include solid fiberglass or graphite fabric reinforced epoxy sheet materials (usually 6.35 mm thick), square and round tube formed from braided material, and pultruded or woven fabric-reinforced tubes.

As mentioned above, the eggcrate is formed of various materials, even wood. Usually, the wood is laminated over with a few plies of reinforced polyester or an epoxy–resin system to seal the surface and form a strong "sandwich" panel.

Whatever the material used to form the backup structure, contour boards must be cut, slit, and preassembled. The contour boards or headers may be N/C machined or cut manually. When they are manually cut, polyester film tracings, cardboard, or even sheet metal templates may be used.

The outside edges of the tool laminate face should be either flush with the outside edge of the backup structure or recessed (Fig. 17). This will protect the face laminate edges of the mold during movement, use, and storage.

A good procedure to follow in establishing the eggcrate backup structure opening size is an equal spacing of 41–46 cm. The eggcrate is laminated together, using corner "tie-in" coverage, 100% internal and 50% edge contact, when attached to the face laminate. The "tie-ins" are 15 cm wide strips of reinforcement fabric folded

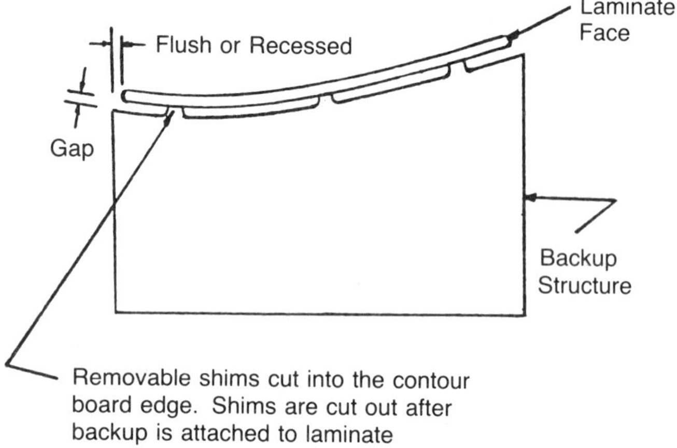

FIGURE 17 Recessed laminate tool face.

in half, impregnated, laminated to the eggcrate surface, cured, and post-cured before attachment to the face laminate.

Before bonding the substructure to the face laminate, it is wise to assure intimate contact of the face laminate to the master pattern or model. This is accomplished by peripheral bagging. This process is accomplished by simply applying a vacuum bag around the periphery of the back of the face laminate and pulling a vacuum. Atmospheric pressure will then press the laminate against the master pattern or model. Then, the preassembled, post-cured, and thermally stabilized eggcrate is attached to the back of the face laminate, as illustrated in Figure 18.

Nonmetallic tubular frames and backup structure. An eggcrate substructure is only one form of support available for a laminate mold or tool. A lightweight (and preferred) method is the use of tubular material or preformed columns. In this process, the tubes are cut, attached to each other, and formed into a frame. This frame or structure can be used to form an assembly fixture or size gauge, or attached to the rear of a laminate mold as a support frame (Fig. 19).

Integral Stiffening

Integral stiffening is still another form of mold and tool reinforcement. This is the least expensive form of reinforcement for laminate molds and tools. The reason for this is that the substructure is formed at the same time as the face laminate. Elimination of the substructure can save as much as two-thirds of the cost of a mold or tool.

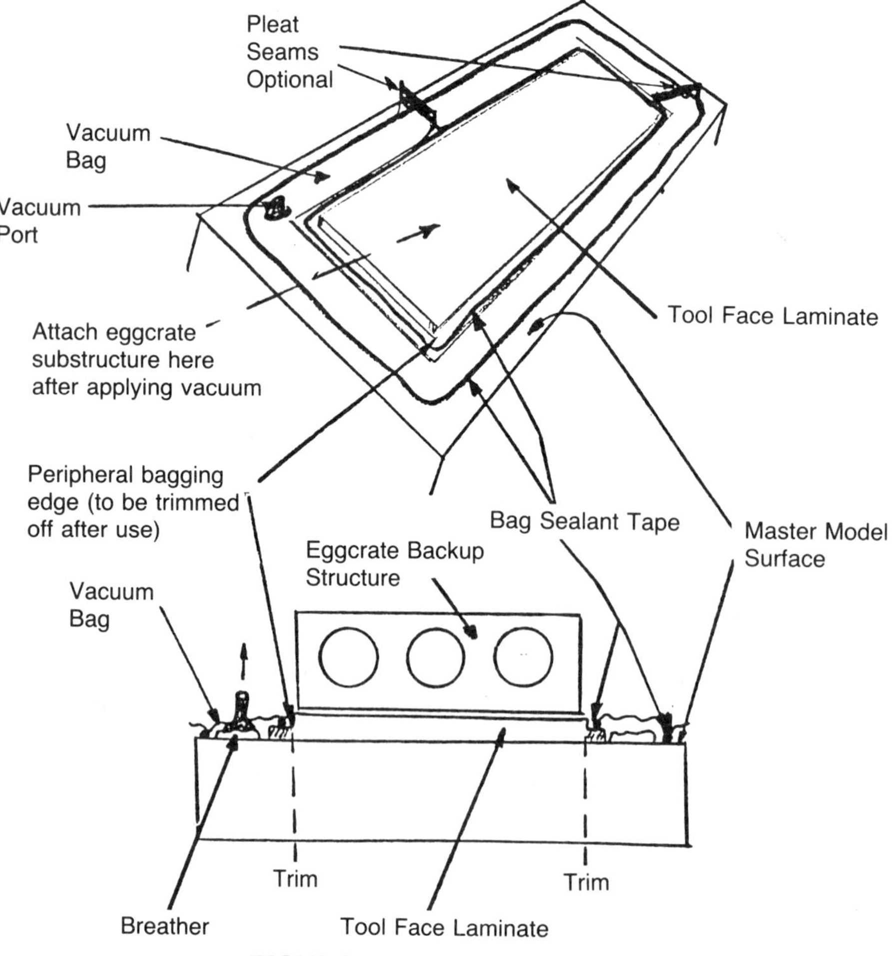

FIGURE 18 Peripheral bagging.

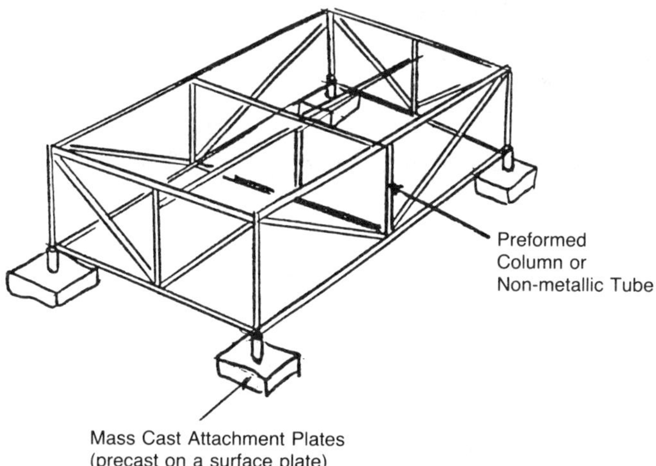

FIGURE 19 **Tubular mold or tool backup support frame.**

Figure 20 shows molded-in stiffener using a simple technique whereby high temperature foam is used to support the laminating material (wet layup or prepreg) during cure.

Integral stiffening has been used in structural part manufacture for many years. Its use in mold and tool manufacture can create substantial cost savings (Fig. 21).

A final form of integral stiffening is a technique in which stiffeners are formed by designing the mold or tool to have a built-in brace. A rolled tool edge is a good example. This rolled edge or, as it is sometimes called, extended edge also allows for a part bagging surface when a vacuum-bagged laminate part is required. The rolled edge will also provide a base upon which the mold or tool can sit or rest (Fig. 22).

Metal-Faced, Laminated, Nonmetallic Molds

With the innovation of metallic faces applied and co-cured to the face of the laminate, fiberglass- and graphite-reinforced prepreg molds, which are state-of-the-art in the aircraft and aerospace industries, have also gained application in the automotive, marine, and appliance sectors.

In this process, a spray metal layer (like a gel coat) forms the face of a laminate mold or tool. The spray metal is co-cured to the reinforced epoxy laminate and becomes the surface upon which part material is formed. Embedded heating and cooling coils maintain the temperature of the molding face so that parts made using vacuum bag, autoclave, RIM (reaction injection molding), RTM (resin transfer molding), rotational molding, blow molding, and other plastic-forming processes can be produced.

The process has evolved over the past ten years and has a solid six to eight years of proven application background in the mold- and tool-making area. Many thousands of these tools are presently in use producing a variety of commercial and military products. Metal faces eliminate many problems that are encountered when wet and film epoxies are used as the surface coat on lamination or bonding tools. Before the evolution of this process, the flame spray process was considered. This method produced excessive heat, which built up in the master model and generated excessive heat in the wire and over the wire being sprayed. The flame blowing onto

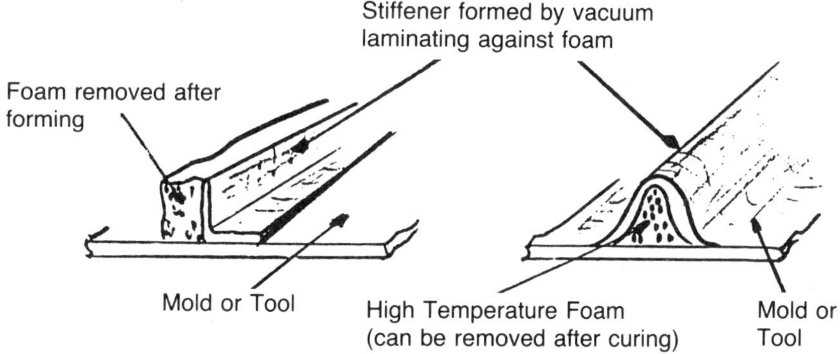

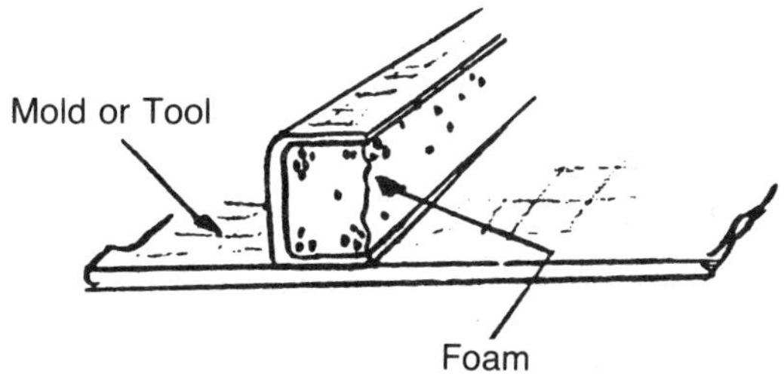

FIGURE 20 **Integral stiffener.**

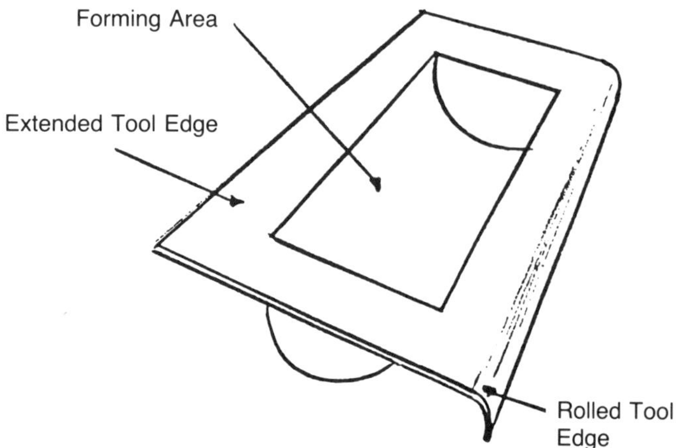

FIGURE 21 Integrally stiffened tool.

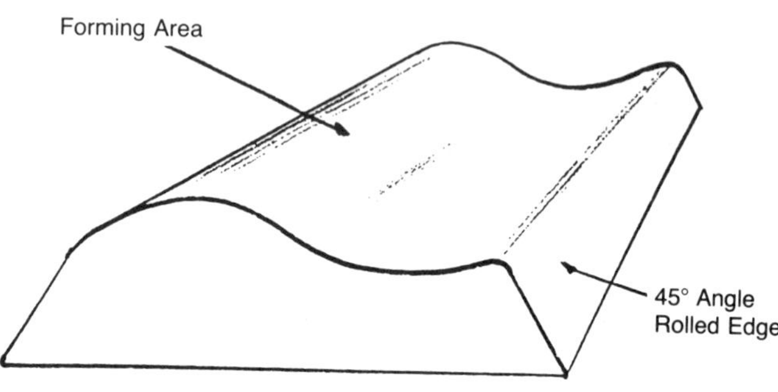

FIGURE 22 Rolled edge tool.

the pattern surface damaged both it and the coating being sprayed.

Metallic Molds

There are many methods of fabricating metallic molds for use as tools in manufacturing advanced thermoplastic and thermoset composite materials. These metallic molds are used as autoclave molds, bond fixtures, and tools and as compression, RIM, and RTM process aids.

The choice of a material and fabrication process is dependent upon the size of the tool required, complexity of contours, tool reliability and the number of finished parts in the production run.

Another important factor that determines mold fabrication choice is cost feasibility. A reasonable mold fabrication decision can only be made after careful consideration is given to all of the factors stated above.

Methods of forming metal molds include roll forming, brake forming, and stretch/draw forming.

Although the fabrication of formed aluminum and steel molds and tools is identical, the characteristics of the tool material determine the properties of the finished mold or tool. For this reason, it is important to comprehend the properties of the material from which the mold or tool will be created. The advantages and disadvantages of steel and aluminum as mold materials are as follows:

Steel

Advantages
- Dimensionally stable with temperature
- Reasonable CTE
- Readily adaptable to combination-type fixtures
- Can be brake- or roll-formed to shape

Disadvantages
- Slow machining
- High tool weight and mass
- Slow heat-up rate
- Oxidation-prone if left untreated—should be nickel-plated

Aluminum

Advantages
- Dimensionally stable with temperature
- Readily adaptable to combination-type fixtures
- Can be brake- or roll-formed to shape
- Easily machined
- Low tool weight and mass
- Good heat-up rate
- Less costly per pound in comparison to steel

Disadvantages
- High CTE
- Less scratch- and dent-resistant than steel
- Less durable than steel

Formed metallic tools have three major components: formed outside mold line (OML) skin, contour boards, and a base frame (Fig. 23).

The base frame is the main supporting structure for the entire tool. Usually, it is welded together.

The contour boards are welded to the base frame. Since they are the backup structure for the OML skin, their curvature must match the back of the OML skin at a given station.

It will soon be apparent that each metallic mold-making material and process has its own special purpose, but that electroformed nickel is by far the most universally chosen method for detailed and complicated molds.

Electroforming is an electroplating process in which pure hard nickel irons are deposited on either a conductive or a conductive-coated master. The electroplating process is actually a nickel ion transfer from a solution of sulfamate nickel electrolytes to a deposit of hard nickel particles on the conductive master.

Since the plating is an ion transfer, the process occurs very slowly, at a rate of 0.003–0.025 mm/h until a thickness of 6.35 mm or more is obtained. Although the plating rate is very slow, the plating process occurs 24 hours a day without interruption. Hence, for a typical plated thickness of 6.35 mm, the in-tank plating time required can range between 4 and 11 days. Therefore, ample lead time should be provided for any electroplating process.

Although the ion transfer during the electroplating process is provided by electric energy, little or no heat is

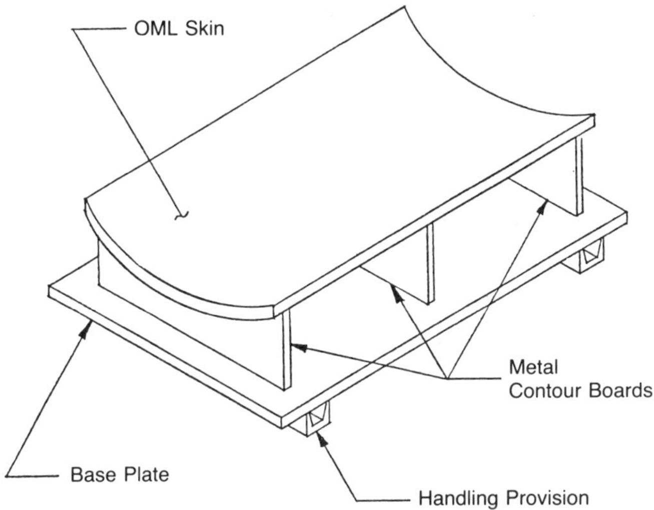

FIGURE 23 **Metallic formed tool.**

generated. The lack of heat generation enables the master to be fabricated from materials with low thermal stability, such as mass-cast and machinable plastic, wax, mixed forms of rubber, and various other forms of thermoset and thermoplastic materials.

Furthermore, unlike metal castings, electroformed nickel does not shrink, has zero porosity, and creates extremely accurate reproductions. Thus, it may be deduced that the electroplated surface can have a mirror finish if and only if the master surface has a mirror finish. Exact reproduction of contour and surface finish and the ability to fabricate a master from materials of low thermal stability and cost are the major advantages of the electroforming process.

The outstanding major disadvantages are the lengthy time needed to create a tool surface of substantial thickness, and the fact that if a large mold or tool is required, the master pattern must be transported to the plater. Large plating tanks exist at only very few locations in various countries.

Design and Production of Mold Accessories, Fittings, Support Tools, and Fixtures

Vacuum Bagging for Molds and Tools

There are a number of ways in which a laminated mold or tool (and even a part) can be formed when a laminate construction approach is used. Component parts of a laminate mold, tool, or part assembly can also be attached together using various forms of adhesive, cured under pressure. Several methods are used to apply this pressure. These include mechanical or hydraulic pressures, autoclaves, pneumatic or mechanical clamps, compression chambers or cavities, and various kinds of solids or liquid pressure.

To take advantage of atmospheric pressure for forming laminates, laminated parts, or bonded assemblies, one can apply the vacuum bag theory. If pressure must be applied over a large surface and a mechanical or hydraulic press, clamp, or other device is not practical, a vacuum bag installed over the entire surface can provide the necessary pressure.

Here is the method of bag installation:

1. A flexible sealant is installed around the master mold or tool periphery.
2. A material (breather) is provided beneath the bag and over the surface to be pressed, so that air can circulate and travel to the vacuum exit or port.
3. The bag material is installed by pressing or sealing it to the tape sealant around the mold periphery.
4. The vacuum exit or port is installed and connected, and the air is drawn from beneath the bag. Atmospheric pressure then squeezes whatever has been sealed beneath the bag at whatever pressure has been chosen, up to the capacity of the vacuum pump and/or the quality of the edge seal. The pressures usually vary from approximately 69–90 kPa unless the bagged assembly is inserted into an autoclave. Figure 24 illustrates a typical vacuum bag installation.

Vacuum bags allow for the production of large, high quality, lower cost composite parts that cannot be fabricated by hand pressure alone.

The trend in production vacuum bagging systems seems to be away from conventional disposable vacuum bag and accessory materials and techniques and toward high temperature, reusable vacuum bagging materials and systems. In order to fully appreciate the reasons for this trend, an examination of conventional materials and techniques is necessary.

The trend in conventional disposable materials, bagging films, release fabrics, peel plies and films, vacuum bag sealant tapes, and other materials in the disposable

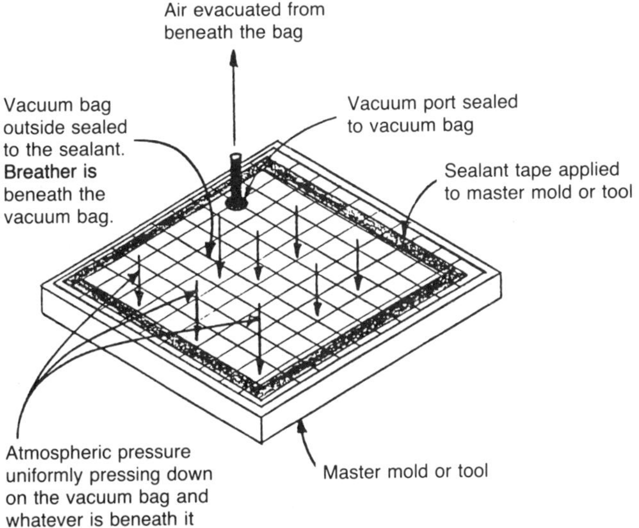

FIGURE 24 **Typical vacuum bag installation.**

category is toward the development and production of higher temperature materials (316°C or more). One of the leading factors contributing to this trend is the recent development of castable mold-making materials that can be used to 1649°C and part-making materials requiring process forming temperatures in excess of 371°C (advanced thermoplastic composites).

Usually, the vacuum bag is used to apply pressure to the surfaces of a composite part laminate or structure, either before curing a thermoset laminate layup or for forming and shaping an advanced thermoplastic composite laminate part material. Foils of stainless steel, high-temperature films, and other diaphragm materials are presently used to fabricate vacuum bags for forming these materials.

For forming a composite mold laminate or tool, the conventional disposable vacuum bag materials and system are used to compact the ply pieces of a laminate composite mold during layup, when necessary. These bag materials are also used to form the vacuum bag surface pressure required during the initial or final cure of the tool laminate.

Room temperature mold and tool laminate materials can be vacuum bagged using inexpensive bagging films, such as polyvinyl alcohol (PVA) and other semielastomeric films.

Oven cure vacuum-bagged mold and tool laminates require the use of disposable films or reusable elastomers that can withstand the oven processing temperatures.

Skin or surface bagging can take place on the surface of a master pattern or laminate mold copy if the surface being duplicated has vacuum integrity. The method is demonstrated by Figures 25 and 26.

Envelope bagging of a mold, tool, or part laminate is a positive way of assuring that no vacuum leaks can occur in the contour surface upon which the laminate layup has been applied. As with skin or surface bagging, the bag edge can be double sealed, to assure that no bag edge leaks can occur.

Figure 27 demonstrates the optional double sealing of a bag edge.

The envelope, as its name implies, is an edge-sealed pouch-type container, formed by folding and doubling over a preformed film tube of a desired diameter or a piece of bag material to form a bag that is sealed along the edges. In either case, the bag should be substantially larger than the laminate being formed, to avoid bag bridging at high points on the contour.

The laminate mold, tool, or part is completely wrapped in a bleeder–peel ply–breather stack to allow air circulation around the laminate and to protect the bag from sharp mold or tool edges. The laminate is inserted into the bag, and the vacuum port or valve is inserted into the bag surface in proper position. Finally, the bag is carefully sealed.

The master model or pattern can also be placed on a plate. Mold or tool laminate material is applied to the model surface, and a vacuum bag is then placed over it. Finally, the form is bagged down to the plate. In this approach, the master model or pattern must be able to withstand near-atmospheric pressure (approximately 69–83 kPa. A quick calculation can determine whether this is possible, so that the master pattern or model will not be damaged (Fig. 28).

After sealing the bag, draw a slight vacuum, so that the wrinkles and bridges can be adjusted out of the bag. Start from the bag edges and shift the bag around towards the laminate layup.

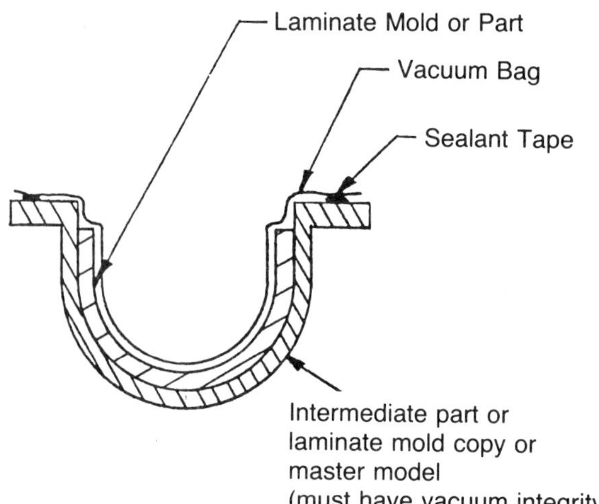

FIGURE 26 **Vacuum bag on the surface of a laminate mold copy.**

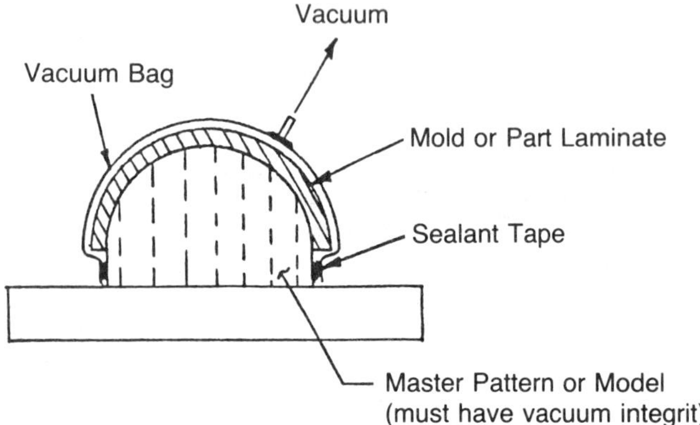

FIGURE 25 **Vacuum bag on the surface of a master pattern or model.**

REUSABLE VACUUM BAGS AND BAG-SIDE TOOLS. Reusable vacuum bag systems are production part forming devices. The subject is presented as an explication of the theory behind their functionality.

The materials used in the formation of the reusable bag can range from simple elastomeric materials such as latex rubber to exotic blends or formulations of specially catalyzed silicone or fluoroelastomers.

Conventional silicone rubber bagging materials—the most widely used materials—are prepared through chemical reaction, using peroxide catalysts. There are proprietary catalyst systems that have just recently been developed that produce an elastomeric bag rubber that

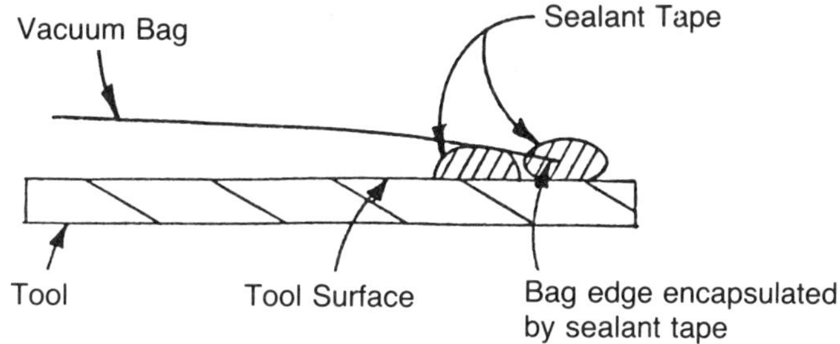

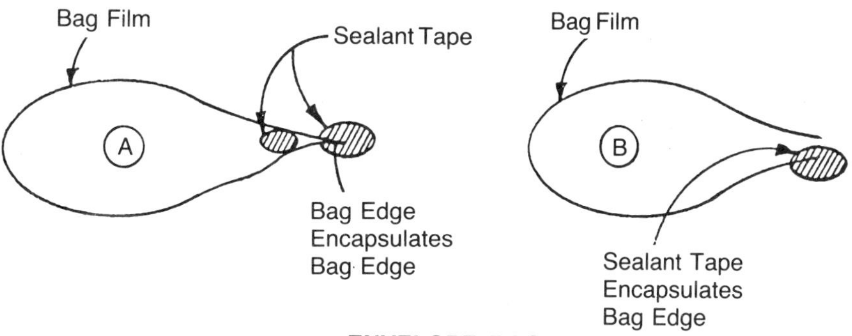

FIGURE 27 **Optional double sealed bag edges.**

is highly reversion-resistant. This system involves material chemistry and manufacturing processes that only a very limited number of manufacturers have mastered. The process involves special compounding, blending, and calendaring of these special elastomer formulations, yielding bag materials capable of operating at temperatures of 232°C and higher.

Using this new material manufacturing process, silicone bag materials have been manufactured that will not transfer silicone byproducts, oils, or other residuals and reaction byproducts during use. These materials can be supplied uncured for layup like prepreg mold laminate materials, reinforced with various fabrics made from fiberglass, Kevlar, and other materials, or cured and supplied post-processed by heat stabilizing the cured sheet

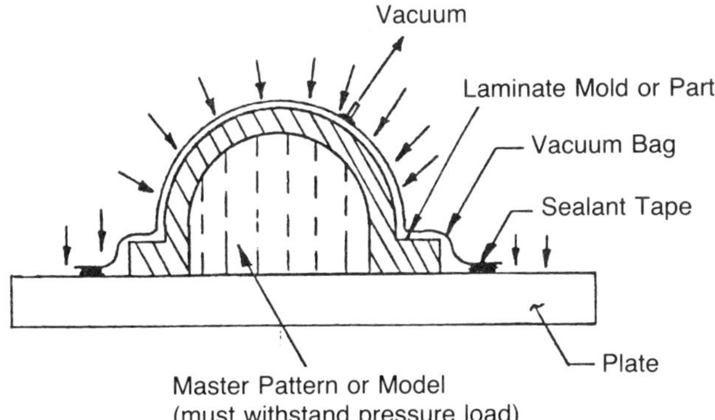

FIGURE 28 **Envelope bagging a mold or tool pattern down to a plate.**

materials to temperatures of 232°C. This post-cure step assures that the materials (when used as vacuum bag materials) will not shrink, weaken, or take a compression set.

Reusable bag systems eliminate costly labor operations (such as hand-installing sealant tape around mold peripheries), reduce film bag labor installation costs, and eliminate mold and/or laminate part loss resulting from failure of a bag seal or pleat, or of the disposable bag material itself.

The savings attainable with reusable vacuum bags can be as much as 50% over conventional materials and processes. The production-use life expectancy of an elastomeric bag made from state-of-the-art materials can be in the 500–1000 part range before maintenance must be performed on the bag surfaces. In the formation of composite laminates or bonded assemblies, the reusable vacuum bag system will pay for itself quickly if the system is used to automatically compact, densify, consolidate, and debulk the composite layup or assembly.

The bags are formed from flat sheets bonded to frames or custom bonded to patterns shaped to a size near that of the finished contour. With the new laminate part-making procedures of co-curing structural stiffeners, inserts, laminate materials, and other inclusions all at one time, the reusable, preformed, or shaped bag has unmatched advantages.

Unlike the situation posed by the disposable film bag, reusable bags allow the mold, tool, or part maker to install various types of hardware into the bag surface when required. Thermocouples, vacuum ports, stiffeners, metal support frames, hoist brackets, and internal heaters are just some of the added accessories that can be installed in or attached to a reusable bag, which may not be possible when a conventional disposable film bag is used.

Edge seal or closure systems, made as part of the bag or tool, become a reusable permanent part of the vacuum bag. For this reason, and the fact that reusable seal materials withstand cure cycle temperatures better, disposable sealant tape cannot compete and is not a good substitute. Disposable sealant tape softens upon use. Usually, leaks occur in the seal or the conventional bag film. Conventional disposable bag materials do not conform as well as reusable elastomers. Leaks can occur in spots, and failure can ruin the part, the mold, and sometimes the autoclave interior.

When using reusable bags, through-the-bag vacuum port or valve locations must be reinforced, and not used as bag lifting supports. Bag supports or lift locations can be laminated into the bag surface. Bag surfaces must be inspected and questionably thin areas patched with uncured or room temperature vulcanizing (RTV) rubber materials.

There is no limitation on the size that a reusable bag can be fabricated to, since seams are easily and smoothly formed to any contour or shape. Contoured bags can even include intensifiers, pressure pads or blocks, rails, and other inclusions that have been molded and placed beneath the bag surface.

Elastomeric, unreinforced reusable bags make better quality parts because the bag material has an elongation factor up to 700%, and as such can form to the required contour and provide intensified pressure in radius corners. This cannot be attained with conventional bag films.

Another important feature of reusable bag systems is that, if they are carefully designed, they can be stored by rolling them up and racking them in tubes for easy handling. Reusable bags should not be stored until they have been solvent-cleaned and dried. They should be stored away from sharp objects and tool sections.

Other forms of reusable vacuum bags include cast bags and sprayed elastomers that are applied over a release-coated object or actual part used to simulate the part thickness during bag fabrication. Reusable vacuum bags of this kind must have adequate space beneath them to allow for the expansion of the mold and part surfaces, and to allow for insertion of peel ply, bleeder, and breather layers, caul plates, pads, and other necessary bag accessories.

Permanent reusable elastomeric rubber bag fabrication techniques. There are a number of manufacturing methods available for forming reusable vacuum bags. The following methods are the most common:

- Uncured sheet rubbers: reinforced and unreinforced silicone
- Cured and post-cured sheet rubber: reinforced and unreinforced silicone
- Spray-on: silicone elastomer, latex elastomer, polysulfide elastomer
- Paint-on: silicone elastomer, latex elastomer, polysulfide elastomer
- Cast: silicone RTV elastomer, latex elastomer, polysulfide elastomer

Of all the elastomers used in the formation of elastomeric molds and bag-side tools, silicones are customarily recognized as the best of the flexible materials. The reasons are that, first, they possess qualities that lend themselves easily to use as a flexible material, including adjustable hardness, temperature, chemical resistance, and a predictable CTE.

Second, the use of silicones as materials for elastomeric tools also provides for easy molding of areas of a part that might have a slight undercut or zero draft.

Third, when molding from the surface of a silicone elastomer, the use of a parting agent is usually not required.

Finally, tool repair and replacement is easy. The operator simply cuts away and replaces sections and portions as required.

BAG-SIDE TOOLS. Caul pads as elastomeric bag-side tools increase the possibilities for composite laminate part molding and bonding far beyond the scope of metallic or nonmetallic rigid tooling. But sometimes a particular application does not allow for the use of silicones; therefore, the use of a nonsilicone is required.

Nonsilicone materials for bag-side tools are available processed from butyl, acrylate, and slightly higher temperature materials such as fluoroelastomers. Butyl, the lowest temperature-of-use material of the group, continues to shrink under repeated use, so the acrylates have proven to be the material of choice.

Acrylate elastomers can be applied, processed, and cured in an unreinforced or reinforced state. Reinforcement is available in various forms. Wire screen, fiberglass, or graphite–epoxy prepreg, as well as some dry fabric reinforcement materials, have been used. In all cases, the reinforcement is selectively placed so that the bag-side tool can stretch over and into radius corners as required. This is illustrated in Figure 29.

To obtain reproducible structural integrity of finished parts, a vacuum is applied after laminate layup and pressure is applied during the cure. The maximum theoretical vacuum pressure is atmospheric pressure, and the maximum obtainable vacuum pressure is approximately 76–83 kPa. During the autoclave cure cycle, the vacuum pressure is enhanced by autoclave pressures of 172–1379 kPa. Voids or leaks in the mold surface, or around fasteners or fittings, reduce the maximum obtainable vacuum pressure and should be avoided.

There are two major types of vacuum ports. The first is securely attached to the mold form. The second requires less tooling cost and is a part of the vacuum bag.

Although the configuration of the individual tool should be considered in determining the number of vacuum ports and their locations, tool surface area is the best guide for vacuum port placement. Two vacuum ports are required for a tool with a surface area of 2.8 m^2 or less, one of which should be utilized for a vacuum

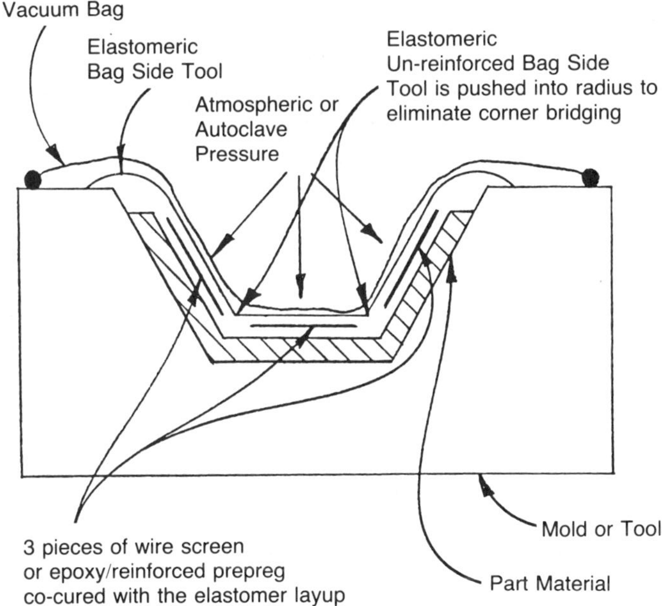

FIGURE 29 **Reinforcement and layup of an elastomeric bag-side tool.**

gauge. Four ports are required for a tool of 2.8–9.3 m^2, and six are required for a tool of 9.3–18.6 m^2.

Fittings, Support Tools, and Fixtures

For metallic molds and tools, the securely attached vacuum port has three variations for three different circumstances. Electroformed metal molds must have brass vacuum bosses bonded and electroformed into the mold. These are then drilled and tapped for vacuum lines. Thin-skinned metallic molds have vacuum bosses welded into place. For thick metal molds, vacuum ports may be drilled directly into the mold surface.

Resin traps may be added to a mold or tool design. The resin trap's function is to provide a reservoir for collection of excess resin to prevent vacuum line fouling. Excessive resin bleed can be expected when a mold has radical contours or high vertical walls. Resin traps should be incorporated into mold designs when excessive resin bleed is anticipated.

Additional accessories include locating, routing, and trim tools, ply locating templates, routing jigs, and drill fixtures.

Some locating templates are fabricated from flexible polyester film, semirigid polycarbonate sheet, and rigid metal plate.

Drill fixtures, or jigs, allow for consistent positioning of drilled holes in finished parts, molds, or tools. The drill jig is usually a laminated epoxy tool into which drill bushings have been laminated during fabrication, or potted after fabrication. The locations of these bushings are determined by coordinated positions on a mandrel or master model.

All of the conventional trim tools have their own applications, but they can also be replaced with new, more accurate, and technologically advanced trim tools. Among these are diamond wire, the water-jet router, and the laser cutter.

The diamond wire has a heat-treated high tensile strength core with a copper-plated diamond-impregnated periphery. The wire is capable of cutting the mold or part. The feed rate of the mold, part, or water jet and the orifice diameter are the two controllable factors governing the quality of the cut.

There are two types of water-jet routers. The first is a hand-held router that is used on stationary molds or parts. The second is a stationary router that necessitates the mold or part being fed into the water jet. Both types produce dust-free cuts.

Lasers can be used to cut extremely accurate molds or parts—within ±0.005 mm. These close tolerances are enhanced by the high degree of reproducibility, the absence of dust development during cutting, and the low consumption of energy. Lasers coupled with computers allow molds or parts of any shape to be trimmed or cut while leaving a smooth, polished edge. Laser cutters can be attached to five-axis machine heads and the cutting pattern programmed to provide automated trimming or cutting of laminates up to 15 cm or even greater thicknesses.

Facilities, Industrial Engineering, Mold and Tool Handling, Safety, Inspection and Quality Control

Priority facility requirements will be covered in this section. Some of these include:

- Individual materials, molds, and tools
- Transporters of contour-forming materials and mold forms
- Storage of these materials in holding areas (indoor and outdoor)
- Final production use areas such as curing ovens, autoclave systems, and inspection equipment locations
- Worker safety

While industrial engineering considerations focus on plant layout and production areas in general, a focus on the mold and tool fabrication storage and use areas will suffice to assure efficient planning in the handling of materials, molds, and tools.

It is important to note that the handling, moving, and storage of completed molds and tools accounts for a part of the total cost of industrial engineering and facility requirements, and that material and mold/tool work flow during fabrication, as well as use, will greatly affect the end product cost.

Because material and mold/tool work flow during fabrication and use greatly affect the end product cost, mold and tool shops should be separated from the production facility.

The mold and tool design, mechanical engineering, and N/C programming should start the flow, with automated design and fabrication teams located near each other so that an interface with mold and tool making can effectively take place. Generally speaking, a 2-hour travel time is the maximum reasonable distance of separation for the facilities.

Transportation and Storage of Molds and Tools

In moving molds and tools through the fabrication stages, and subsequently through production, careful planning should allow the production operators to handle the contour-forming surfaces, bonding fixtures, tooling, and accessories easily and efficiently. In some instances, it will become necessary to perform the fabrication of a mold or tool in a production area because of restrictions of material layup, size, weight, and worker skills. If this need should arise, the area used for performing the work should be planned and isolated well in advance, so that regular production is neither interrupted nor distracted by the fabrication operations.

Support equipment such as forklifts, transportation dollies, and overhead cranes should always be chosen with flexibility in mind, given the distinct and varied needs of materials and design. Handling is an important element in facility design, since it does not add value to the parts being produced, and must therefore be minimized or performed at the lowest cost for the available equipment and labor, without sacrificing efficiency.

Standardization of equipment, whether major or support, is extremely important. In order to standardize this equipment, an analysis must be done of how the mold or tool is handled by the operator at each work station, as well as the equipment used to perform the production operation, to determine the specifications of the transportation or handling device.

The storage and holding areas are directly associated with the layout of the total material and work flow. When the layout is already in existence, the distance for each move seems to be already established. However, redesign of existing areas to more firmly link material and work flow is suggested to generate cost savings. Indoor and outdoor storage that is remote from material, mold, and tool handling is obviously unwise and antithetical to certain fundamental prerequisites already mentioned.

Automation in storage areas as well as in the manufacturing area can be accomplished by operator-assisted or totally wire-guided vehicle and work platforms, as shown in Figure 30.

Figure 31 depicts a typical operator work station showing the exhausting of the heavier-than-air fumes and reactants of a thermoset resin system. Notice that the exhaust is down and away from the operator.

The matter of waste disposal has become extremely complicated. Nothing should be discarded, whether it be a resin, filler, curing agent, unique reinforcement, sol-

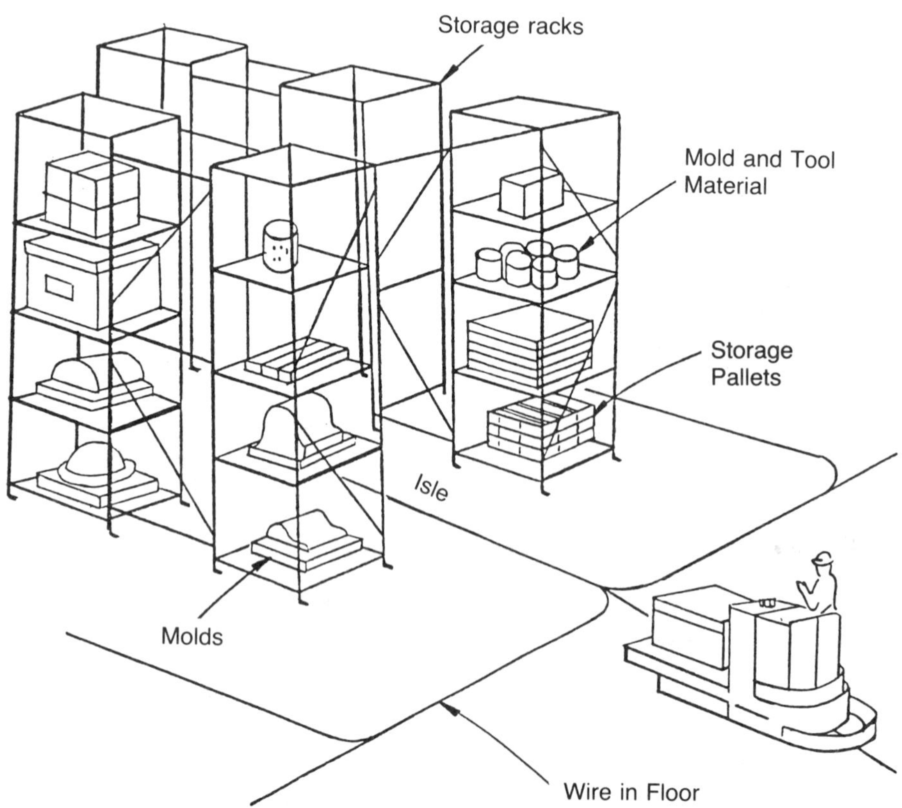

FIGURE 30 "Operator aboard" wire-guided system.

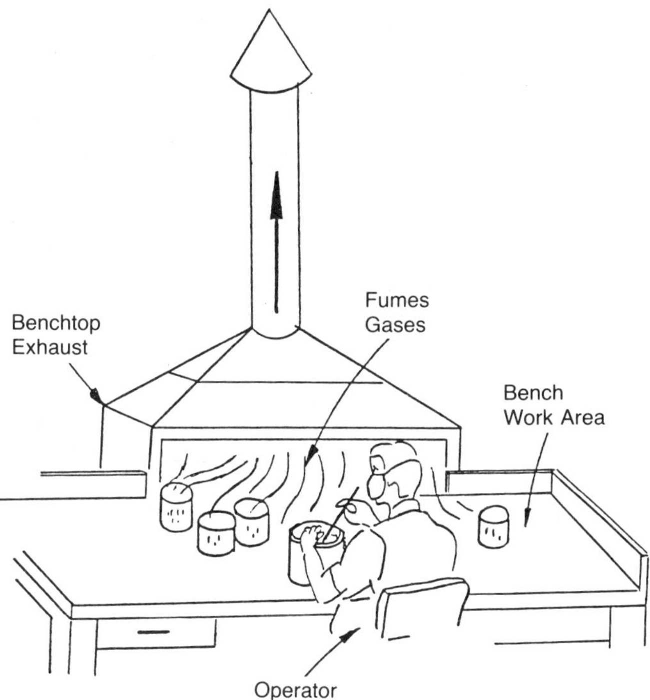

FIGURE 31 Bench top exhaust.

vent, chemical, or other substance, until it is identified and the disposal procedure documented in writing and approved by the required authority.

Inspection and Quality Control

Much emphasis has been placed upon the complicated automated methods developed for the inspection of composite parts and assemblies. But until recently there has been very little emphasis on the area of mold and tool inspection and quality control.

Quality control and the inspection of advanced composite molds and tools should start with the originator of the structural design, as well as the mold and tool designer.

The demands that are placed upon the mold or tool maker are based upon requirements to produce a part or assembly, using a special material or process. It is for this reason that there are sometimes only two methods, or perhaps even only one, for producing the complete part or assembly. Therefore, the molding or forming method might dictate the mold-making method and thereby define the inspection criteria.

Understanding what happens to the mold and tool materials during use is a good way to understand what the inspection and quality control requirements must be.

A basic example is the choice of dissimilar materials within a mold or tool, and the expectation that these materials will stay dimensionally stable during thermal cure cycles. The coefficient of thermal expansion (CTE) is only one governing factor directly linked to mold and tool thermal stability.

Some fairly exotic equipment designed to inspect materials, parts, and assemblies has been developed in recent years. This equipment can also be used to check the materials, or the mold or tool, that will form the part or assembly.

Some of these nondestructive inspection (NDI), nondestructive testing (NDT), and destructive testing (DT) methods of inspection are:

- Real-time microfocus computer-aided integrated thermography
- Shearography-double-exposure holography, vacuum assisted
- Radiographic methods
- Optical systems and equipment (fiberoptics/video)
- Electrical measuring
- Sound or sonics in air, such as acoustic emission (airscan)
- Sound in other transportation media, such as water (seascan)
- Mechanical testing
- Thermal techniques, such as survey, infrared, and thermographic reading
- Diffuse light, lasers, and liquids as penetrants
- Physical testing through various tensile, compressive, flexural, and other related methods
- Computer testing

Besides these, probably the single most critical nondestructive visual inspection process is "process inspection" itself. Thus, it is the best method that can be used to assure that there are no foreign inclusions, such as release materials, hardware, or even hand tools, included in a mass-cast tool or mold laminate.

For metallic molds and tools, monitoring of the process will always reveal a defect, if one is present when the mold or tool is being fabricated. In-process inspection of trim tools, drill fixtures, routing tools, mold forms, bonding tools, and others can also be performed using templates and checking tools.

The fabrication process for laminate molds and tools should be monitored to assure that the age and shelf life of the materials used are within allowable storage limits. To assure that the orientation of the precut plies (warp and fill direction) of a laminate is correct, the information should be noted in the design drawing or the manufacturing procedure. The gaps between ply ends, the weights of fabric materials, compaction, debulking, and final vacuum bagging are usually planned by the mold and tool designer and process engineer.

Additional inspection and quality control processes include:

- Dielectric monitoring of cure
- Loss factor measuring of viscosities
- Thermographics

And in the area of destructive or physical testing there are methods to measure the following:

- The ultimate tensile strength and modulus of elasticity of the reinforcement

- The specific gravity
- The weight per unit length of yarn or material
- The finish on the reinforcement fiber
- The resin content, as a percentage by weight
- Flow, as a percentage by weight
- Gel time in minutes
- Tack time in minutes
- The impregnated fabric or material weight
- Visual uniformity of fabric or material appearance (for example, defects in excess of 5–10% of total roll length could be cause for rejection of material roll)
- Alignment of warp and fill of impregnated fabric so as to be perpendicular to each other, with the warp parallel to roll edges
- The roll widths, size, and roll cure configuration defined by the user
- Storage life as per manufacturer's data
- Physical and mechanical properties of the cured laminate: ply thickness and weight of reinforcement to resin content; void content; and longitudinal flexural strength, modulus, and horizontal shear

In summation, quality and control, which are usually considered at the end of materials and process manufacturing, should also be considered at the very beginning of a structural, conceptual, engineering design, even before the mold and tool design process is chosen.

John J. Morena

Molding

Most natural and synthetic materials cannot be used in the form in which they are originally found or made. For instance, iron ore must be refined and then shaped, sand must be heated and formed into useful glass objects, trees must be trimmed of limbs and then sawed, and chemicals must be combined and then processed into a useful shape. At the most basic level of investigation, three major methods for changing the form of materials are used. These methods are material removal, material joining, and material reshaping. The last of these we will call molding. Therefore, molding is the reshaping of a material without removing material or joining new material. This reshaping is generally accomplished by causing the material to deform (flow) into a preset shape, which is the mold.

Molding is the most important shaping process for composite materials, and for synthetic materials in general. In almost every application, the basic shape of the composite part is produced by molding. However, because composites consist of more than one material, by definition, the molding process can be quite complicated. This is especially true for fiber-reinforced composites because the fibers are straight, stiff materials that inherently resist forming into new shapes. It is, therefore, the job of the matrix material to take the new shape and to restrain the fibers in that new shape.

The matrix material determines the moldability of the composite because the fiber contribution is often quite constant—always resisting molding. These matrix materials can be classed by their ease or method of molding. For instance, the polymeric matrices (also called plastic and resin matrices) are far easier to mold than are the metal and ceramic matrices. This ease of moldability has led to the dominant position of the polymeric matrices in advanced composites. These polymeric materials are easy to mold because they either are liquid at room temperatures or can be melted at relatively low temperatures (generally at less than 250°C), whereas metals are molded at temperatures often reaching 700°C, and ceramics at 1000°C or even higher. Therefore, the study of molding of composites is almost exclusively the study of reinforced polymeric matrices.

Polymer matrices can be conveniently divided into two categories, thermosets and thermoplastics. The thermosets, which are far more common in composites, are characterized by being cross-linked in their final, cured state. However, they are placed into or onto the mold in their uncured or partially cured state and can range in consistency from a liquid to a tacky solid. The mold and the material are then heated to effect the cure. After curing, the material becomes cross-linked and set in the shape of the mold. Because of the cross-linking that occurs, further molding of these cured materials is not possible.

Thermoplastic polymers are less commonly used in composites and are characterized by their lack of cross-linking. Without the cross-linking to give the "locked in" rigidity to the structure, these materials can be repeatedly heated and reshaped by molding. The most common methods for molding thermoplastics involve forcing the melted resin (they are usually solids at room temperature) into the mold and then cooling the mold and the part to solidify the resin in the desired shape. Because this cooling process takes far less time than the corresponding curing step for thermosets, and with the introduction of thermoplastics that can be used at relatively high temperatures (up to 300°C) and that have, in some cases, properties superior to those of thermosets, the use of thermoplastics in composites is growing rapidly.

Because of the major differences between thermosets and thermoplastics, the molding methods are also usually separated into two groups—those appropriate for thermosets, and those suited to thermoplastics. The thermoset group includes nonpressure curing (as done with wet layup), vacuum bag curing, autoclave curing, matched-die molding (including compression molding and transfer molding), resin transfer molding (RTM) and reaction injection molding (RIM), and pultrusion. The thermoplastic methods include injection molding, extrusion, blow molding, rotomolding, hydroforming, diaphragm forming, thermoforming, continuous roll forming, and casting. Some of the methods used for thermosets and some of those used for thermoplastics are not well suited for use with fiber reinforcements but are acceptable for whisker and particle reinforcement, and these cases will be pointed out. The molding of metal and ceramic matrix composites will also be discussed briefly.

The study of molding is, therefore, a study of the materials being molded and how the material dictates the method used to mold it. The opposite point of view is also true; the method of molding affects the material, and these changes should also be examined.

In general, this article will first discuss the methods of fiber placement that are common to the most common composite molding methods, then the thermoset molding methods will be examined in order of increasing complexity and pressure, and then the thermoplastic processes will be examined. Molding of metal and ceramic matrices is then considered. Finally, some discussion will be given to tooling (molds and dies), curing, and how to select the appropriate molding process.

Fiber Placement Methods for Nonpressurized and Autoclave Molding

Several fiber placement methods are common to many of the molding techniques; they are, therefore, considered together in this section. The method of fiber placement in the mold is important in determining the physical properties of the part and has some importance in determining the molding method. Where one type of molding is preferred over another for a particular method of fiber placement, that will be noted. Fiber placement also has a significant influence on the method of resin incorporation, and that will also be noted as appropriate.

Wet Layup

The most common method for placing the fibers into the mold is called *layup, layup molding, wet layup,* or, less often, *contact laminating.* In this method, fabric or mat is saturated with liquid resin, and the layup is made by building layer upon layer to obtain the desired thickness. The impregnation of the layers is done "in process," that is, at the time the material is laid into the mold. This method is used most extensively with polyester and fiberglass, although some epoxy–fiberglass composite parts are also laid up wet. The layers are usually placed onto a shaped surface (the mold) by hand. Some pressure is normally applied by hand rolling, wiping with a squeegee, or sometimes vacuum bagging to remove trapped air and to give better uniformity. Curing can be done at room temperature or at elevated temperatures.

The method of laying the dry reinforcement (most often a fabric or a mat) into the mold and then applying the resin is the oldest and, perhaps, the most common means of wet layup. The wet composite is rolled by hand to distribute the resin evenly and to remove air pockets. Another layer of reinforcement is laid on top. Then, more catalyzed resin is poured, brushed, or sprayed over the reinforcement. This sequence is repeated until the desired thickness is reached. The layered structure is then allowed to harden (cure). This method is conceptually simple, does not require special handling of wet fabrics, and allows the resin to be applied only in the mold, thus helping to maintain a neat surrounding area. However, variances in resin viscosity (which are inherent in the gradual curing of precatalyzed resins) cause problems in getting good wetout (if the resin is too thick) or in having resin runoff (if it is too thin). Part shape can also cause difficulties in getting proper wetout.

Better uniformity is usually possible if the reinforcement is impregnated with the resin before being laid into the mold. The dry fabric and resin can be weighed to obtain specific fiber–resin ratios. The weighed resin can be rolled or squeegeed into the fabric more uniformly on a flat surface than on a tool because the areas of excess resin can be seen more easily. Costs are better controlled on large structures by this method of prewetting the reinforcement because the amount of excess resin required to wet out the fabric is reduced. This method also helps prevent the formation of resin-rich and resin-poor areas as a result of vertical drainage. On a male mold, runoff of excess resin will flow off the layup, so that wetting the glass with excess resin while on the mold is not as critical a problem. However, this method is rarely used because of the difficulty of handling a wet sheet of reinforcement.

For many commercial applications, a layer of catalyzed resin is applied to the mold and allowed to cure to the gel (tacky) state before the reinforcement (either dry or saturated) is applied. This resin layer, called the gel coat, forms a protective surface layer through which fibrous reinforcements do not penetrate. Especially formulated gel-coat resins are used to improve flexibility, blister resistance, stain resistance, weatherability, and toughness.

Another approach to getting improved surface characteristics is to start by wetting out a fine-weave fabric (such as 121 style fiberglass) directly on the released mold, followed by thicker woven reinforcements. This method eliminates the resin-rich surface, which can crack and craze, especially if the structure is subjected to flexural stresses. In either approach, finer-weave fabrics are usually used near the surface of the part to prevent transfer of the weave pattern to the surface, and because air and voids can be removed more easily from these finer materials. When mat is used instead of fabric, the mat is normally wetted on the mold to prevent unraveling and distorting during the transfer process.

Prepreg Layup

A somewhat superior product can be made, with less resin and fiber handling difficulty, by using a reinforcement that has been preimpregnated with resin and then cured slightly to increase the viscosity. This material is called *prepreg.* Normally the prepreg is made at a facility that is dedicated to the manufacture of prepreg and by a method that allows careful control of the resin and fiber contents (ratio), then shipped to the site of composite manufacture. Prepregs are used in applications in which the performance of the part is critical. Almost all of the normal resins are available in prepreg form. The prepreg method can, in some regards, be viewed as an extension of the wet layup methods described above—prewetting outside the mold and then laying up the composite. In prepregging, the fibers are usually arranged in a unidirectional tape or a woven fabric. The fibers are impregnated with initiated resin, partially cured, and then rolled up for shipment. As a result of the crimping and mechanical damage of the weaving process, woven prepregs are not as strong as unidirectional tapes. (Prepreg roving is also available, but it is not used in the manual layup method and will be discussed in a later section.) In the prepreg method, the prepreg, which is supplied in rolls of convenient widths (typically 30–60 cm, but sometimes as narrow as 6 cm on as wide as 200 cm), is normally cut to fit into the mold and laid up layer by layer until the desired thickness is achieved. Prepregs used for manual layups are leathery and should ideally have a slight tackiness (*tack*) so that the layers will not slide over each other during layup, and will stick or maintain their location in the mold. The prepreg should also be conformable to the mold so that complex shapes can be made. This ability to conform is called *drape*. Drape and tack are often associated with each other and are dependent on the resin as well as the fabric. Fabric prepregs are usually used on complex contours because of their ability to drape and conform to the mold. The weave of the fabric can be a significant factor in the ability to drape, with satin weaves generally having better drape than plain weaves. Unidirectional tape tries to follow a geodesic path on a contoured tool and is often difficult to use because it will not drape over complex contours, and will leave gaps and overlaps in severe cases.

Prepregs made from reactive thermoplastic resins are often stiff at room temperature and have almost no tack or drape. Therefore, they may be melted slightly with a hot-air blower or a soldering iron to assist in tacking the layers together.

The resin content of the prepreg influences its tack and drape as well as the final strength of the laminate (although the resin content of the laminate is controlled during the resin fabrication step). In many prepreg systems, the resin content of the prepreg is higher than that desired in the finished part. This assists in tack and drape but requires that the excess resin be removed at some point in the manufacturing process. The removal of this excess resin assists in the removal of entrapped air and volatiles, which will flow out with the excess resin. It is essential to remove the volatiles and entrapped air, since voids within a laminate have a severe effect on the interlaminar shear strength. As a guide, the interlaminar shear strength is reduced by about 7% for each 1% of voids present up to a maximum of about 4% voids. A reasonable goal for void content in the finished laminate is 0.5% or less. The method for removing this excess resin is outlined in the section on vacuum bag assemblies.

Traditionally the resin content of the prepreg is given as a weight percent, whereas the resin content of the finished part is given as a volume percent. The explanation for this is simply that when the prepreg is made, the resin weight percent is easy to measure and to control. However, for finished laminates, the resin volume percent is preferred, since it is related directly to the mechanical properties. Historically the trend has been toward lower resin content so that the specific strength of the composite is increased. The consistent removal of large excesses of resin has become a costly problem, so prepregs are now made with "near-net" resin contents. These prepreg materials are generally made by a hot melt impregnation method that minimizes the volatiles remaining in the prepreg. Voids from entrapped air are minimized by debulking the material (vacuum or autoclave) after laminating 3 to 10 plies of material. Most laminates today are about 60% fiber volume fraction.

Because the resin has already been initiated when the prepreg is made, prepregs have a limited shelf life before the prepreg turns into a dry and boardy material that is difficult to use. The shelf life is usually several days to weeks at room temperature, but can be extended by keeping the prepreg cold. The "out time"—that is, the time out of the freezer—is recorded so that an estimate of the remaining useful life of the prepreg can be made. Standard tests are available to determine the percent of curing that has occurred in the prepreg and, therefore, the approximate shelf life remaining. However, many prepreg resins require high activation energies (heat) to react and cure. If parts are laid up in a mold before they lose tack (from residual solvents), their life at room temperature is often very long. Upon heating, the resin will flow and cure to its full cure state, even though it appeared to be cured before heating.

In unidirectional prepreg, the strength in the cross-fiber direction is essentially the strength of the resin only. Therefore, to get strength in all directions, the prepreg layers are often oriented in different directions. Fabric prepregs (which may be directional too) are also laminated in different orientations to improve properties. However, as a result of crimping and mechanical damage of the weaving process, composites made of woven reinforcements are not as strong as unidirectional tapes that are cross-plied. A ply layup pattern could be, for instance, 0° (the fiber direction), 90°, + 45°, − 45°, − 45°, + 45°, 90°, 0°. The warp (machine) direction is generally considered 0° in a fabric. Caution should be taken to ensure that the sequence of plies is balanced directionally about the center so that the laminate does not twist. The prepreg layup method is much more precise than the wet layup method.

Automated Fiber Laying Machines

Some new machines and concepts have been developed to speed the slow and labor-intensive layup process. Some of the most important of these new machines are automated tape laydown machines [1]. These machines typically have an overhead gantry to facilitate movement of the head across large molds. The head has the capability of moving in several directions so that both simple and reasonably gentle compound contours can be followed at angles up to ±30° from vertical. The ply pattern is preprogrammed into the machine, then the head is loaded with a roll of prepreg tape (5 cm wide is typical). The machine lays the tape onto the mold in the pattern that has been programmed and does all cutting and trimming automatically. Because the equipment uses unidirectional tape, optimum results are achieved if the machine can be programmed to follow a geodesic path for all passes. On compound contours, this is often not possible, and gaps or overlaps result. This can complicate programming and design of tape-laminated parts. To prevent overlaps and variation in thickness, small gaps are often programmed to be left in between adjacent tape paths. The advantages in time and labor savings are obvious. Other reported advantages include more uniform parts as a result of more consistent tape laydown pressures and more precise positioning of the tape rows, nonstressing of the fibers during laydown, and the ability to easily repeat with other identical parts. Disadvantages include the need to program the machine, the inability to do some complex parts, the long production time for small parts (because of the tape head turnaround time), and the presence of a consistent gap between adjacent paths. Systems to program tape-laying machines by vision learning, by computer simulation, or by passing the head over the mold are now being explored, and some success has been demonstrated. These methods hold great promise in reducing the time involved in programming automated tape-laying machines.

Filament Winding

An even more automated method of laying down fibers is *filament winding*. The basic filament winding process wraps a continuous tape of resin-impregnated fibers over a mandrel to form the part. Successive layers are added at the same or a different winding angle until the required thickness is reached. Either the mandrel or the application head can rotate to give the fiber coverage over the mandrel, although the rotating mandrel is far more common. The head, therefore, traverses longitudinally to give the coverage. In one sense, filament winding can be thought of as a type of automated laydown machine, although filament winding normally does not use prepreg material but, rather, incorporates the impregnation of the fiber tows as part of the filament winding process. This is called *wet filament winding* or *wet winding*. Dry filament winding, which is less common but is gaining acceptance as materials are developed, uses preimpregnated tow (prepreg tow) as the winding medium. Parts as small as 2 cm diameter and as large as 7 m diameter are commonly made by filament winding [2]. The only limitations on size are those dictated by the geometries of the winding machine and the limitations in mandrel size and weight. Most standard composite resins (epoxy, polyester, phenolic, some imides, silicone, and thermoplastics) can be used for filament winding provided that certain specific requirements are met. Polyesters and epoxies are the most common. As with other methods, the resin should be low in volatile content (preferably 100% solids) so that voids within the structure can be as low as possible. The viscosity of the resin should be in the range where it is thick enough to coat well without dripping off the fibers or migrating onto the mandrel, yet not so thick that the fibers are poorly wetted or become fuzzy when passing through the resin bath. Resin viscosities in the 0.35–1.5 Pa·s range have been found to satisfy these requirements. Reactive and nonreactive diluents can be added to high viscosity resins, but may cause problems with physical properties (mostly as a result of void formation). In general, nonreactive diluents should therefore be avoided. Solid, uncured resins can be used if a method for melting the resin without curing it prematurely can be found. A glue-gun approach where only the tip of the resin is melted has been shown to be effective. A pot life of several hours is necessary, and one of several days is preferred. Gelation must not occur before completion of the winding; otherwise, weakened structures could result. Surface properties of dry resins (on prepreg tows) are critical to their successful use in filament winding. Sufficient tack is needed to prevent slipping of the bands as they are laid down if a nongeodesic path is used. (Geodesic paths are discussed in the section on the winding process.) Flow of the resin during curing is required to ensure that the layers are bonded together. The flow is a property of the B-staging of the resin. Improved tack and flow are obtained by heating the strand during winding. Despite these limitations and cautions, dry resin filament winding has some benefits. Small shops do not need the extensive resin preparation and handling facilities; close control over the consistency of the resin system is not required, as that is done by the prepreg manufacturer; and prepreg roving can be quality-controlled before winding, potentially providing a more consistent part. On the other hand, prepreg tows as a raw material form are more expensive (on the order of twice the cost of wet winding materials).

Most common continuous reinforcements for composites have been successfully used in filament winding. The most common are E-glass (for cost), S-glass (for strength), carbon (for strength with modulus), and aramids (for toughness and light weight). For best part properties, the fibers must be well collimated and laid without twisting. (The fibers wet out best and process better if they are procured as untwisted fibers from the manufacturer.) The use of more than one type of reinforcement material may have some advantages in terms of cost and product performance. Some examples of this concept are a pressure vessel wound with graphite to give longitudinal stiffness, then overwound with fiberglass to resist the forces in the hoop direction; and a vessel wound with carbon for strength and modulus, then overwound with aramid to protect the vessel from impact damage during fabrication or use. A recent modification of the traditional winding process involves the use of multiple winding heads. By the use of two or more heads, an overlapping winding pattern (like a braid) can be achieved, which gives some improvement in product strength. For open-ended structures, the simplest mandrel is usually the best, and cylinders of cored or solid steel or aluminum are used. If a helical path is used, the domed ends of the part are cut off to remove the part from the mandrel. In some cases the design anticipates a special cut zone so that domed ends can be kept but the part can be removed from the mandrel. Closed-end vessels require special mandrels that can be removed while keeping the ends intact. Collapsible metal (segmented) mandrels are used because they can be reused. However, these are difficult to remove if the end openings are small. Plasters have also been used, but these require chipping out, and that process is also difficult with small end openings and further has the potential for damaging the composite vessel. Easily removable mandrels include sand with a water-soluble binder (generally polyvinyl alcohol, PVA), soluble salts, eutectic salts, rigid foams, and low melting alloys. The use of an inflatable mandrel has several benefits, which have led to increased use of this method. When necessary, the mandrel can often be deflated and removed. In most applications, however, the inflatable mandrel can be left in as a liner for the composite material. Applications such as fuel tanks require this type of liner. Another benefit of the inflatable mandrel is the ability to add pressure as the winding progresses. This prevents the collapse of the inner fibers, which sometimes occurs as the outer wraps are wound onto the vessel. By increasing the internal pressure during winding, the inner wraps stay in tension and do not tend to buckle. The inflatable mandrel also allows the vessel to be cured with an internal pressure that is equal to the expected operating pressure of the part, thus eliminating possible stress cracking during initial pressurization.

Advantages and Disadvantages of Wet Layup Fiber Placement

The advantages are:

- Semiskilled workers can be readily trained.
- Very large parts can be laid up conveniently.
- Machine investment costs are low.
- Thickness, reinforcement type, resin type, and other variables can be changed easily.

The limitations are:

- Draining from vertical walls can be a problem, creating puddles near the base and resin-poor areas in the wall, although most resins have the appropriate viscosity to prevent this from occurring.
- The process is slow and labor-intensive compared with automated methods.

Advantages and Disadvantages of Molding Using Prepreg

Advantages include:

- The resin/initiator (or hardener) ratio is more accurately controlled during the premixing operations.
- Resin distribution per unit area is strictly controlled during tape manufacture. This aids greatly in providing good resin distribution in the final part.
- Health and safety problems associated with liquid resins or solvents are largely eliminated, since these are handled by the prepreg manufacturer, who is often better prepared to deal with them since he is accustomed to handling liquid resins in large volumes.
- The problems of poor efficiency and output can be reduced by using automated machinery for some parts.
- It provides better part definition, higher fiber content, and better consolidation than wet layup.

Disadvantages include:

- Costs are much more than those of wet layup.
- Prepregs have a limited shelf life.

Advantages and Disadvantages of Filament Winding

Advantages include:

- The process is applicable to parts of widely varying size.
- Parts with strength in several directions can be easily made.
- Filament winding has excellent material usage.
- Forming after winding and other techniques allow noncylindrical shapes to be made by filament winding.
- Flexible mandrels can be retained in the structure to serve as liners for tanks.
- Panels and fittings for reinforcement or attachment can be easily included during the winding process.
- Parts with high pressure ratings can be made.

Disadvantages or cautions include:

- Resin viscosity and pot life must be carefully chosen and monitored.
- Programmming of the winding can be difficult.
- Not all shapes can reasonably be made by filament winding.
- Operational control of several key parameters (such as fiber tension) is important.

Nonpressure Molding

The simplest molding technique, and probably the first used for a modern composite structure, is nonpressure molding, that is, molding without external pressure being applied. The absence of pressure means that the molds can be made of almost any material that will hold its shape. The molds can be either male (plug) or female (cavity). The side of the part that is to be smooth and glossy should be placed against the mold. Therefore, a male mold would be used for a swimming pool, since the inside of the pool must be smooth, and a female mold would be used for a boat because the outside must be smooth and glossy.

Molds made of wood, plaster, plastics, composites, and metal are common, since pressures are low and little strength is required. Molds made of composites (polyester and fiberglass) are most often used for low volume orders, since they do not hold up well for repeated use. Molds for long runs are usually made of fiberglass—epoxy or metal and would be considered permanent molds. Composite molds are especially favored because they can be formed against smooth surfaces and therefore do not require further finishing. In addition, there are no corrosion problems, as there are with some metal molds. Redesigning is usually simpler with composite tooling than with metal, since the surfaces of composites can more easily be recast or relaminated to comply with engineering changes. A simple reinforced plastic mold is made by using the model of the part as the original mold against which the plastic is laid up (assuming that the part lends itself to serving as a model and does not have complications such as undercuts).

Curing of Wet Layup

Since curing is usually done at room temperature, a promoter is often added to the resin to speed up the reaction. Caution must be taken to ensure that the promoter and the initiator (catalyst) are never mixed together directly. They should always be mixed into the resin in separate steps. External heating with infrared lamps or hot-air blowers is sometimes used to speed the curing process.

Mold Release

To prevent the composite from sticking to the mold, a *mold release* or *parting agent* is first applied to the mold. This mold release can be silicone, polyvinylalcohol (PVA), fluorocarbon, or, in some cases, a plastic film. Parts are removed by manually pulling them from the mold. To assist in the removal, flat wooden, plastic, or metal wedges can be inserted between the part and the mold. Some manufacturers also blow low pressure air into this gap to "lift" the part from the mold, or put air ports into the sides of the mold to allow air to be introduced. Mechanical assistance is sometimes needed if the part is large. Most applications do not require the high compaction that results from an autoclave cure, and the parts can simply be oven cured. Care must be taken, however, that the effects of gravity are minimized during the cure, so the part is usually rotated while being cured. Microwave and other nonoven cures have proven to be successful methods of curing composite parts. These often have shorter cure cycles than ovens.

Advantages and Disadvantages of Nonpressure Molding

The advantages include:

- Tooling can consist of any material that will hold its shape under minimal pressure.
- Tooling can be changed easily during experimental phases or to accommodate engineering redesign.
- Investment in pressure devices such as a press, autoclave, or vacuum pump is not required, although a vacuum pump is often used with wet epoxy parts and some polyester parts.
- Curing ovens are often not needed.

Disadvantages include:

- Only addition-type cross-linking resins can be used, since condensation types require some form of pressure to avoid porous, poorly laminated structures.
- Product uniformity, both within a single part and from part to part, is difficult to maintain. Because of inability to compact the laminate with any pressure, the resin content is often quite high.
- Voids are common.
- Physical properties are low in comparison with those of parts produced by other composite manufacturing methods.
- Tight-weave fabrics are difficult to saturate with high viscosity resins, resulting in low strength.
- There is high shrinkage from resin-rich areas.
- Parts have only one finished surface.

Vacuum Pressure Molding

The application of a vacuum to assist in compressing the plies together (debulking or consolidation) has proven to be valuable in wet layups and necessary in prepreg layups. Vacuum bagging is often used in conjunction with autoclave molding, although it can be done separately and is, therefore, considered separately. Most of the features will, however, apply to vacuum bagging when done with autoclaving.

The vacuum provides the dual advantage of pressing the layers together and simultaneously withdrawing the excess volatiles. These volatiles could be residual solvent, low molecular weight resin components, or trapped air. The method of applying the vacuum that has been developed for composites allows the volatiles to escape freely and also permits good debulking. Further debulking and curing is occasionally provided by an autoclave, which is discussed in a later section.

Vacuum Bag Assemblies

A vacuum bag assembly is illustrated in Figure 1 [3]. The procedure for building up this assembly is as follows:

1. Prepare the mold by coating it with an appropriate mold release.

2. Remove prepreg materials from the freezer and bring them to room temperature before opening the protective bag to prevent contamination from water condensation.

3. Build up the part by placing layers of prepreg on top of each other in the prescribed pattern and to the

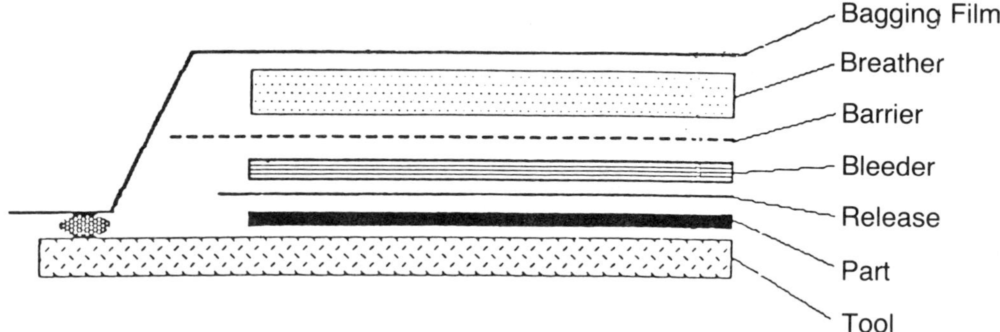

FIGURE 1 Vacuum bag assembly.

prescribed thickness. This buildup can either be done manually or be automated. This may also include placement of any inserts, ribs, or other structural members and the necessary additional prepreg layers to serve as anchoring and support for these inclusions.

4. Place the release film or peel ply material over the part. The release material is generally porous to permit excess resin to flow through, while leaving an impression on the part suitable for secondary bonding without further surface preparation.

5. The bleeder material is a mat that absorbs the excess resin. Common bleeder materials are polyester felt or mat, fiberglass (which may be coated with Teflon or mold release), and cotton. The important characteristic of the bleeder material is that it have good absorption qualities and not compact under the pressure of an autoclave. The resin content of the final part is dependent on the ability of the bleeder material to absorb a measured amount of resin as well as on temperature, pressure, viscosity, and amount of bleeder. The bleeder should be conformable so that it does not cause wrinkles in the assembly when it is placed under vacuum.

6. The barrier is a layer that controls the upward movement of the resin and prevents resin from reaching or clogging the breather and vacuum lines. The barrier material must limit the passage of resin and must allow air to pass. A film with holes punched in it is often used as a barrier. In some bagging assemblies this material is omitted. In autoclave processing this omission may lead to resin plugged vacuum lines and pumps. When the barrier is omitted in wet layup applications, resin traps are often used.

7. The breather material acts as a distributor for air, escaping volatiles, and gases, as well as a buffer between bag wrinkles and the part surfaces. Breather layers should be highly porous and must not collapse under vacuum, temperature, or pressure. Typical materials are fiberglass, polyester felt, and cotton, although the bleeder materials are sometimes used. For near-net or phenolic resin systems, the use of breather is sometimes omitted when the thickness of the bleeder is great and experience has shown that a separate breather system is not required.

8. Vacuum bag sealing tape or sealant is applied around the mold. This sticky material makes an air-tight seal with the bagging film.

9. Thermocouples or other monitoring devices are inserted into the assembly. To ensure that the places where the thermocouple wires enter the assembly will be air-tight, the outer insulation is often scraped off the thermocouple wire (especially if it has cloth insulation). Some thermocouples have plastic insulation that is specifically designed to be air-tight in these applications. The wires are then embedded in the sealant material. Thermocouples are usually placed in the outer area of a part that will be trimmed off.

10. The vacuum bag is then laid over the assembly and a vacuum port is attached through the bag or through the mold base to provide a fitting for attaching the vacuum hose. The bag is then pressed against the sealant to give an air-tight seal all around. Vacuum bags can be made of any plastic film material that is strong enough to hold a vacuum, fit and conform to the assembly, and withstand the cure temperature without degrading. In practice, the most common material is nylon or a coextruded nylon that has been heat stabilized.

The most common problems associated with vacuum bag processing are material quality and bag leaks. Vacuum bagging materials are often procured under vague material, construction, and performance requirements. Poor industry standards often lead to products with varying performance. For instance, uncontrolled or unregulated vacuum bag quality is the leading cause of bag failure, typified by film degradation or decomposition above 80°C. Bag leaks most often occur at the sealant–vacuum bag interface. The second most common cause of bag failure is handling damage to the nylon film before cure. Nylon film is hygroscopic and subject to moisture change in relation to the relative humidity of its environment. The film becomes dry and brittle when its moisture content falls below 2%. Dry film is susceptible to damage or cracking when it is excessively handled or abused (wadded).

Bridging is also a common problem in vacuum bag molding. Bridging occurs when the shape of the part does not allow the vacuum bagging materials to press against all of the part surface. These areas are not, therefore, properly pressed. (For instance, in a part with a narrow, deep channel, bagging materials may form a material bridge across the top of the channel, preventing the transfer of pressure to the bottom of the channel.) If the nylon bag itself bridges over a narrow gap, the film may be stretched beyond its limits during cure and burst. Good bagging techniques, such as pleats intentionally put into the bag during assembly, allow enough excess film to be present for bag conformity against all surfaces. Another technique is the placement of preformed rubber pads under the bag in corners and channels to fill areas where bridging is possible. This technique is particularly useful with vacuum bags made of rubber. Metal sheets can also be used when the part shape is not complex. The metal and rubber sheets are sometimes known as **caul plates.**

Resin Leakage

Resin can bleed out of the laminate by either an edge bleed path or a face bleed path. Edge bleeding of the laminate occurs when the resin flows in the plane of the laminate and exits along the edges. Problems can occur with edge bleeding in that the resin has a long pathway and may not reach the edge before gelation. This is overcome by judicious application of even pressure to hydraulically force a measured amount of resin from the part. In autoclave processing, edge bleeding is controlled by using a nonporous release material on the bag side of the part and/or high cure pressures (690 kPa), since these pressures tend to limit the alternative mode of resin escape—face bleeding.

Face bleeding is most effective in oven processing where high pressures are not present. In this case, excess resin coupled with vacuum pressure only or low autoclave pressure is used. However, most processing of composites relies on a combination of both edge and face bleeding to control resin content and provide for unrestricted resin flow.

Advantages and Disadvantages of Vacuum Bag Molding

Advantages include:

- Volatiles can be removed easily.
- Void content is lower than with wet layup.

Disadvantages include:

- Bagging of complex shapes is difficult.
- The inside surface is not as good as with matched-die molding.
- The cure cycles are long compared with those for matched-die molding.

Autoclave Molding

One of the major disadvantages and limitations of the nonpressurized cure is the high void content in the parts. In many applications, especially for aerospace, the void content must be below 2%, and in some cases below 1%, and these low void contents cannot be obtained without pressurizing the part during the cure. The most common method of pressurizing large parts, especially those in which only one side is exposed or needs to be finished, is in an autoclave. Autoclaves are large pressure vessels into which the entire part can be placed. The autoclave then allows pressure and heat to be applied to the entire part and mold system. Hence, the pressure pushes the part against the mold as the part is cured. Parts made by wet layup are usually not cured in an autoclave because these parts rarely have the high performance standards that would require the time and expense of an autoclave cure. On the other hand, layups involving prepregs generally use both a vacuum and an autoclave to assist in consolidating and curing the part.

The molds used in an autoclave are usually made of metal or composite. They are more permanent than the typical layup mold and must, of course, withstand the forces of the autoclave. The autoclave provides pressure beyond that available with vacuum only and therefore gives greater compression and void elimination.

Autoclaves allow the simultaneous imposition of heat and pressure (and vacuum if a vacuum line is led directly to the part). The major difficulty with autoclaves is the high capitalization cost, which results because autoclaves are pressure vessels and must, therefore, pass stringent pressure code regulations. However, because many parts can be cured at one time in an average autoclave, labor and cure costs on a per part basis can be quite low.

Autoclaves can be purchased as large as 10 m in diameter and 30 m in length. The concerns in purchasing and operating autoclaves are the uniformity of the heat flow (to avoid thermal stresses), the integrity of the closure, ensuring that the insulation is good, obtaining good and easy-to-operate controls, and ensuring that fires do not occur as a result of high temperature and pressure. (An inert gas is usually put into the autoclave to reduce the fire danger.) Heating can be electrical, by gas, by circulating hot oil, or by steam (if the temperature requirements are not too high). Autoclaves can be used to bond assemblies together. In this application the principal use of the pressure is not to debulk but to ensure that the parts are kept in intimate contact.

The type, composition, and size of the tooling can affect the cure, as the heatup rates of the tooling can vary widely. In some cases, the tooling is independently heated, which can reduce the problem of variable heating and can accelerate the cure by getting the entire assembly to cure temperature faster.

Special tooling may be required in autoclave curing to prevent bag failure, collapse of parts, or crushing of honeycomb details. The net result is that curing in an autoclave produces a superior part to what can be obtained with wet layup. Therefore, autoclaves are used extensively for making high-performance aerospace parts. For very complex parts, in fact, this is the principal method employed.

Advantages and Disadvantages of Autoclave Molding

Advantages include:

- Void contents can be very low.
- Part definition can be very precise, although often not as good as with matched-die molding.
- Large parts can be made if the autoclave is large enough.
- Vacuum bagging can be done in conjunction with autoclave molding.

Disadvantages include:

- Costs of autoclaves are high.
- Cure cycles are often very long.
- Very thick parts may require multiple cure cycles.

Matched-Die Molding

Three types of matched-die or compression molding will be discussed. These are preform molding, SMC molding, and BMC molding. These three methods all utilize the same type of high pressure molding equipment, but differ in the form of the material that is placed in the molds to form the part. The materials most commonly molded by this technique are fiberglass reinforcement and either polyester or epoxy as the matrix. The short fiber lengths that are characteristic of matched-die molding generally preclude the use of this technique for high-performance parts.

The equipment is a press (usually hydraulically driven) that is fitted with both male and female dies (hence the term matched-die molding). The dies are generally made of hard metal (such as tool steel), and can be highly polished and chrome plated in order to get a fine finish. The pressures developed by the press can range up to several megapascals, which is useful for obtaining good part uniformity and compression of the voids that may develop. An alternative name for this method is *compression molding*.

Compression molding can be used for both addition-type cross-linking and condensation cross-linking. When condensation polymers (such as phenolics) are molded, the condensate (usually water) must be allowed to escape to prevent gas pockets. Therefore, after the mold is closed, it is opened slightly for a few seconds to allow the gases formed by the heated condensate to escape. This process is called *degassing* or *breathing* the mold.

Preform Molding

In preform molding, a dry mat of the reinforcing material is preformed to the approximate shape of the part and placed into the open mold. Resin is added to the preform, and the mold halves are then pressed together and heated to cure the part. During the process the resin flows, impregnating the preform, and becomes hard. The cured part is removed after the mold is opened (often with the assistance of knockout pins that are built into the mold). In a variation on the standard method, preimpregnated chopped fibers are blown onto the preform and then cured.

Because high pressures can be exerted on the material to be molded, a higher ratio of glass to resin may be used than in layup, resulting in a stronger part (assuming that the fiber lengths are the same). The cure time in the mold depends on the temperature, the resin type, the part geometry, and the mold heating and cooling efficiency. Typically it varies from 2 minutes to 20 minutes. The cure cycle can thus be very short, and a high production rate is possible.

Several methods are available for making preforms. In the *direct-fiber method,* chopped roving and a water-soluble binder to hold the fibers in place are blown against a shaped, perforated screen fixed to a turntable which assists in getting fiber uniformity. This method is used to make preforms of very large items, such as boats. The *plenum method* (illustrated in Fig. 2) employs a perforated preform screen that is inside a vacuum box. Fibers are chopped and sprayed with binder, then sucked against the preform screen by the exhausting air. These preforms are for moderate-sized parts, such as chairs. Some preforms can be made simply by cutting the desired shape out of mat or cloth. When required, glass yarn can be used to stitch segments together. This is called the *cut pattern method* and is chiefly used for complex parts requiring varying thicknesses or decorative patterns, which can be added by using special yarns.

SMC Molding

Sheet molding compound (SMC) is a sheet material that is made by chopping glass fibers (or swirling continuous fibers) onto a sheet of plastic film (usually polyethylene)

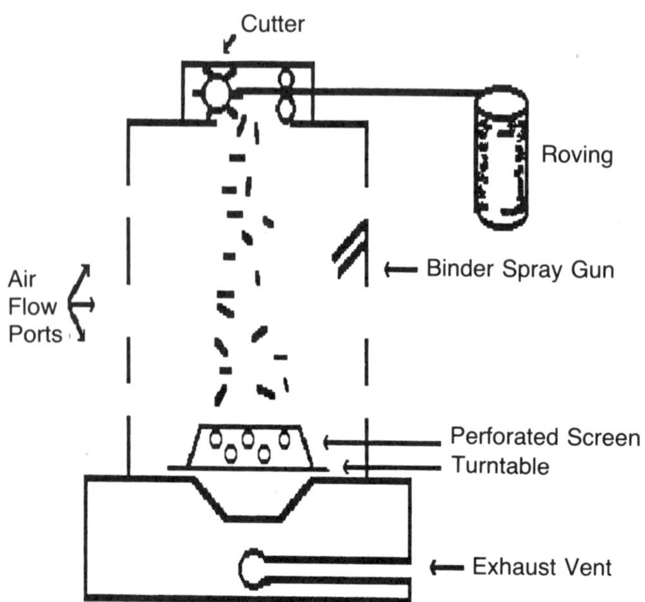

FIGURE 2 **Plenum method for making preforms.**

on which a resin/initiator/filler mixture has been doctored, as shown in Figure 3. Another film that also has the resin mixture doctored onto it is placed on top, and then the sandwich of resin mixture and chopped glass is passed between compaction rolls to wet the fibers and thoroughly mix the constituents. The material is then cured slightly (also called B-staging, aging, or maturing) and rolled up for shipment.

A typical SMC incorporates about 30–50% fibers (generally 2–5 cm long), 25% resin (generally polyester), and 25–45% filler (generally clay, alumina, or calcium carbonate). Some care must be taken with SMC, as it has already had initiator added and therefore has a limited shelf life. SMC is, therefore, kept refrigerated until use. The styrene cross-linker is also quite volatile, so caution should be taken with flames, etc.

After aging, the SMC material has a leatherlike texture. When it is to be molded, it is cut off the roll, the plastic film backing is removed, and the SMC material is loaded into the compression-molding presses. A normal technique is to build up certain areas within the mold by laying in additional strips of SMC. Parts as large as car bodies can be made in this manner. SMC processing pressures may be as high as 34 MPa with temperatures in the range of 120–200°C and mold cycles of several minutes.

BMC Molding

Bulk molding compound (BMC) is a doughlike mixture of chopped fiberglass, resin, initiator, and fillers that has a composition similar to that of SMC. BMC, however, is generally mixed in bulk rather than as a sheet and is sold in a log or rope form. The resulting material is also called premix. The fiberglass content of BMC is generally 5–10% lower than that of SMC, and the fibers are generally less than 3 cm. Therefore, structures made from BMC are not as strong as they would be if they were made from SMC. The BMC material is often mixed at the molding location, but it can also be purchased from commercial manufacturers who sell the material as a ready-to-mold premix.

BMC is molded by placing a weighed amount of the material (the charge) into the lower mold. The molds are then closed, with pressures, temperatures, and cycles much like those for SMC. Some special consideration must be given to the viscosity of BMC to ensure that all components of the mixture are evenly divided throughout the part. If the viscosity is too low, the resin can be squeezed out of the fibers when the mold squeezes the mixture, resulting in resin-rich areas and leaving the fibers clumped, On the other hand, if the viscosity is too high, there may not be enough flow of the material to fill the mold. As BMC sits in storage, the cross-linker can evaporate or the cross-linking reaction can advance, both effects causing changes in viscosity. Therefore, time limits are set on the use of BMC just as on SMC.

Advantages and Disadvantages of Matched-Die Molding

Advantages include:

- Both interior and exterior surfaces are finished.
- Complex shapes (including ribs and thin details) are possible.
- Production rate can be high.
- Labor costs are low.
- Minimum trimming of parts is needed.
- Products have good mechanical properties and close part tolerances.
- There is good consolidation of parts.

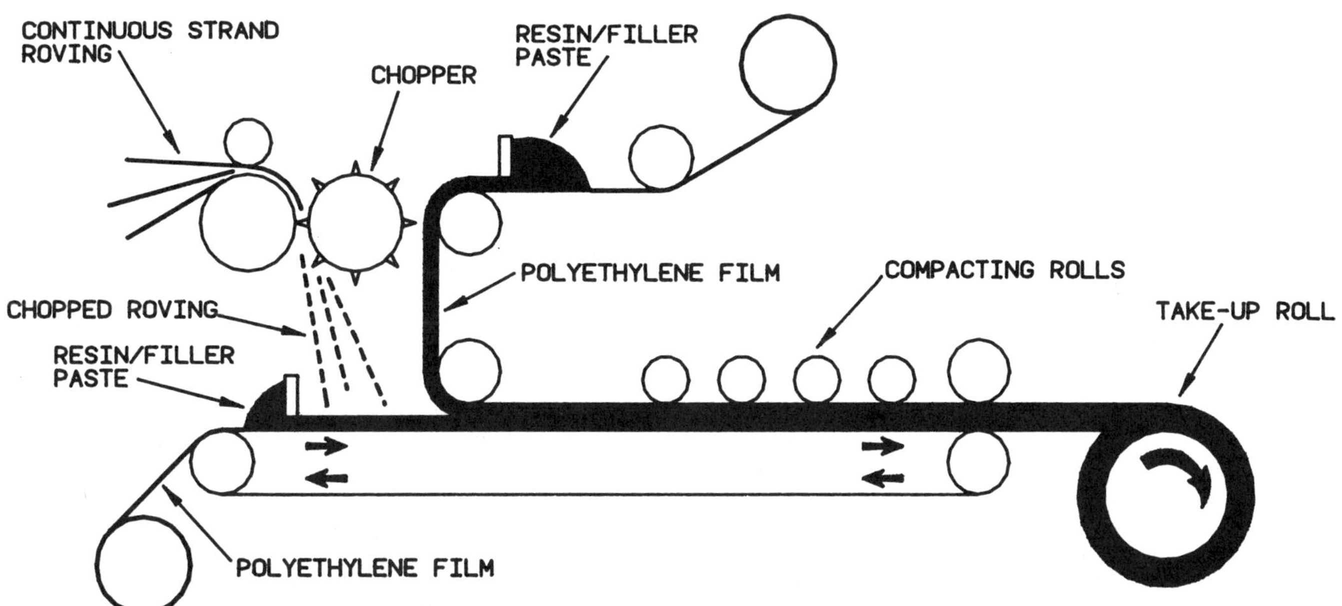

FIGURE 3 Sheet molding compound (SMC) machine.

Disadvantages include:

- More equipment is needed than for layup.
- Molds and tooling are costly compared with layup molds.
- Transparent products are not possible with SMC and BMC.
- Molding problems (trapped water, etc.) may cause surface imperfections, such as pitting or waviness.
- SMC and BMC have limited shelf lives.

Resin Transfer Molding and Resin Injection Molding

In resin transfer molding (RTM) a mold is loaded with reinforcement material (often as a preform), the mold is closed, and resin is then slowly injected into it. The mold is often put under vacuum to remove entrapped air from the reinforcement and speed the injection process. The reinforcement material is wetted out by the pressure of the injection and by the wicking action of the fibers.

This method has some similarities to other composite and nonreinforced plastic manufacturing methods. For example, the method of loading the reinforcement is similar to that in preform molding, but in preform molding, the resin is introduced into an open mold. Pressures are also much higher in preform molding. Traditional transfer molding, from which this method gets its name, is a process in which a thermoset resin is injected into a closed mold. However, in traditional transfer molding, reinforcement is not incorporated. One further comparison is with reaction injection molding (RIM), a process in which two reactive resin components are mixed and then injected into a closed mold. Large, nonreinforced parts are often made by the RIM process. RIM is, therefore, a logical extension of several other processes to fill a specific need for improving the productivity of making large parts. RIM has been used in the automotive industry and is the process chosen for manufacturing body panels for the Fiero.

Reinforcements for RTM

Most standard reinforcement materials can be used, but fiberglass, carbon, and aramid are the most common. One requirement on the form of the reinforcement is that it hold its shape during the injection phase. Therefore, the reinforcements are generally stitched or bonded together. Preforms are common. Reinforcement buildups in certain areas are also easily included.

Inserts of various types, such as screw receptacles or ribs, can easily be placed in the open mold at the same time as the fibrous reinforcement. Fiber volume contents of up to 60% are now being achieved with RTM, thus opening the possibility of using RTM for high performance parts. (Fiber concentrations above 50% usually require special and often proprietary techniques and are, therefore, not generally accomplished.)

Resins and Pumping Equipment

Most of the standard composite resins can be used in RTM, but the viscosity must be low enough that the fibers will be easily wetted. Typical viscosities are less than 1 Pa·s. The resin should also have about a 2 hour pot life so that injection can be slow (for wetout) without having the resin gel. Polyester and epoxy resins are the most common.

The resin pumping system can be much like the type required for spray-up molding (which is discussed elsewhere in this article). However, much simpler systems with just a ram or airline to insert the resin have also been used. Care should be taken to maintain good temperature control on the pumping/inserting mechanism to prevent premature gelation.

Mold Design for RTM

The design of the mold is the most critical factor in successful resin transfer molding [4]. The mold must be constructed so that resin reaches all areas and concentrations are approximately the same throughout. This resin movement must be accomplished within the time allowed before the onset of gelation. Additionally, the resin injection process should not cause movement of the reinforcement and should be done at low pressures so that the mold will maintain its shape without the need for massive backing methods.

Mold designers have found that RTM molds must be vented to allow the air within the mold to be pushed out by the resin. Venting is also common in compression and transfer molds. Although vents allow the escape of the air, they are too shallow to allow the resin, which is much thicker, to flow out. The mold should also have good temperature control. Even molds that are intended for room temperature cured resins should be well insulated so that the environmental conditions do not change the gel times and viscosity of the resin. Some molds are heated or designed to go into ovens to effect curing at higher temperatures.

RTM molds are usually made of composites (polyester/fiberglass) or metal, but they can be made of other materials if specific needs warrant it. The generally low pressure requirements of RTM allow many more mold materials to be used than for compression molding, although the techniques for obtaining high fiber contents often include additional pressure and, therefore, dictate a robust mold (often metal). The choice between metal and composite molds is chiefly one of production volume, temperature, and this special need for high pressures at high fiber volumes.

In some cases, the mold must be backed up in order to maintain its shape. This backup can be done with ribs or by simply adding mass (such as cast aluminum). The closure of the mold is also important, since the mating of the mold surfaces against a gasket is often the method of keeping resin from squirting out. Therefore, alignment pins are usually employed to ensure that the mold halves meet properly.

Uses for RTM Parts

The manufacture of large parts is the primary use for RTM. The reason for this is that the production volume and efficiency are higher for RTM than for layup, and the cost of the molding equipment is lower than for SMC. Typical parts would include machine cabinets, solar collectors, snowmobiles, bathtubs and shower enclosures, airplane wing ribs, hatch covers, car bodies, and bus shelters.

Advantages and Disadvantages of RTM

Advantages include:

- Very large and complex shapes can be made efficiently and inexpensively.
- Production times are much less than with layup.
- Clamping pressure is low compared with matched-die molding.
- Surface definition is superior to that with layup.
- Inserts and special reinforcements can be added easily.
- The skill level required of the operator is low.
- Many mold materials can be used.
- Parts can be made with better reproducibility than with layup.
- The worker is not exposed to chemicals and vapors as he would be with layup.

Disadvantages include:

- The mold design is critical and requires great skill.
- Properties are equivalent to those with matched-die molding (assuming proper fiber wetout, etc.) but are not generally as good as with vacuum bagging, filament winding, or pultrusion.
- The control of resin uniformity is difficult. Radii and edges tend to be resin-rich.
- Reinforcement movement during resin injection is sometimes a problem.

Pultrusion

In the pultrusion process, continuous reinforcement fibers are impregnated with resin, shaped by drawing through a die, and then cured [5]. This process is analogous to the extrusion of aluminum or thermoplastics (with the obvious exception that pultrusion incorporates fibers and deals with thermoset resins in most cases). Pultrusion is a continuous processing method and therefore has great potential for high throughput. Its major limitation, as with the extrusion processes, is that the cross section of the part normally must be constant, although both solid and hollow parts as well as many profiles can be made. Compliant dies that permit a change in thickness have been designed for special applications; these permit some variation in cross section. Two types of pultrusion dies are commonly used, fixed (with no movement) and floating (where one die segment floats and has pressure applied). The pressure can be applied by hydraulics, fire hoses, springs, or other means. The use of fixed dies can generate tremendous hydraulic forces in the resin to impregnate and wet out fibers. Floating dies rarely generate more pressure in the resin than the pressure being applied to the die. Multiple dies are often used when the dies are heated and used to cure the part. Pultrusion dies can cost as much as $10,000.

Pultrusion has a much higher material usage (95%) than layup (75%) and can therefore be considered more productive in both time and material. A properly designed pultrusion die will maintain accurate resin content because of the fixed cross section. As long as the fiber volume passing through the die is held constant, as it is in normal production, excess resin will be squeezed out and run back into the resin bath.

Fiber and Resin Choices for Pultrusion

Most of the normal continuous reinforcements for composites will work well with pultrusions. In fact, the ability to use a wide variety of reinforcements and forms is an advantage of this method. Reinforcement forms used commercially include roving, mat, and fabric of fiberglass, carbon, and aramid.

The resin choices are also broad, but some limitations do exist. All resins need to cure quickly because of the continuous nature of the process and the relatively high speeds. Viscosities in the 0.5 Pa·s range are typical for pultrusion resins impregnated in-line.

Polyester is a major resin used in pultrusion, but care must be taken to ensure that the styrene content is in good control. Polyester is ideal because it shrinks slightly on curing and thus easily releases from a fixed or floating die. If the styrene content is too high, residual styrene can remain in the part and result in void areas; if it is too low, the cross-link density will not be achieved.

Epoxy has an inherent problem in pultrusion. Epoxy resins tend to stick to the die and other parts of the pultrusion equipment, causing process problems and surface defects on the parts. Modified epoxy resins using release agents and viscosity modifiers are now beginning to be used in advanced composite manufacturing. Another method to prevent sticking utilizes a nonporous Teflon-coated cloth (Armalon) which is wrapped around the surface in contact with the die to prevent resin contamination of the die and to reduce the friction required to pull the material through.

Imides have been used, but their high viscosity and condensation byproducts have limited their use.

Prepreg tows have been used for pultrusion, although their use is limited to special applications.

Resin Impregnation and Material Forming

The heart of the pultrusion process is the impregnation of the fibers with the resin and the shaping of the material into the desired shape. In most pultrusion processes, the dry fibers are pulled through a resin bath, which has several mechanical rollers or other devices to assist in assuring that the fibers are well wetted. The wetted fibers are then pulled through a die or forming bushing. Alternatively, the dry fibers can be formed by passing them through a die, then passed into a cylindrical chamber made of perforated metal through which the resin can be forced to wet out the fibers. This method has the advantage that the fibers are in the final form when they are wetted. It is especially useful when the part to be made is hollow; the fibers are formed around a mandrel and wetted in place. In either method, forming guides are often used to position the fibers for entrance into the die, especially if mats or woven or stitched materials are used.

Pultrusion Curing

Three curing methods are traditionally associated with pultrusion. The most common of these methods is called the *tunnel oven method.* In this method, the part is usually gelled in the die and fully cured as the part travels through the oven upon exiting the forming die. The length of the oven is determined by the line speed, the part dimensions, and the curing characteristics of the resin.

The second curing method is called the *split die method.* In this method, two female molds are brought up against the part as it exits the die. The line stops while the curing takes place, then continues when the curing is completed. This method is also called *pulforming*. With this system, nonuniform cross sections are possible.

The third system is called *die curing*. It involves the rapid curing of the part while it is still in the die. Curing in the die often will require the use of multiple dies that start at a low temperature; the temperature increases across the dies with a heat control zone in each die. This method is the easiest to control and achieve a steady-state process. Another method necessitates the use of microwave or RF curing so that the length of the die will not be excessive. Typical die lengths for this method are 15–60 cm. Dies are often made of ceramic or composite to allow the microwave or RF energy to reach the part. Ceramic dies are significantly more expensive and have longer fabrication lead times than metal. Composite dies do not wear very well and are less durable, requiring more frequent replacement. Microwave and RF energy are more difficult to regulate, especially if the part contains carbon fibers. The die curing system is especially useful for hollow parts that are formed on a mandrel. The mandrel extends from before the die, through the die, and then just beyond the die. The parts are brought into the short curing section, then, when the part is solidified, the part is pulled off the mandrel. Further curing can be done in a tunnel oven if necessary.

The curing process is the rate-determining step for the process. Typical speeds for pultrusion are 0.5 to 2 m/min for parts that are 1–75 mm thick. Sizes of parts can vary from 2 cm to 5 m in diameter without serious limitations.

A cooling zone is generally placed before the die as a protection against premature gelation should the line stop and during start-up.

Pullers and Cutoff Equipment

A large variety of pullers have been used in the pultrusion process. The most common varieties are double clamp pullers in which the part is clamped between pads and pulled forward. At the end of the pulling stroke, a second puller clamps the part and pulls it. This tandem pulling mechanism gives a continuous pulling action. Another puller has been adopted from the continuous pullers of thermoplastic extrusion processes. This puller employs double continuous belts through which the part is passed. When the belt is a cleated chain, the system is called a caterpillar puller.

Cutoffs are usually saws similar to those used for plastic extrusions such as PVC pipe. Both wet and dry saws are available. The toughness of composite reinforcements compared with thermoplastic resins requires the use of diamond or carbide saws. The blades must be sharpened or changed frequently, especially when aramid fibers are the reinforcement. The cutoff saw often travels with the part to prevent binding of the part in the saw or overheating if the part has stopped in the die or oven.

Properties of Pultruded Parts

The physical properties of pultruded parts are generally good compared with those of similar parts made by other methods. However, parts made from all unidirectional material have very poor transverse strength. Specially woven or stitched materials that are pulled through the pultrusion operation by the 0° fibers found in these materials will often be used. This gives better overall structural integrity. The ability to control fiber tension and orientation throughout the process is especially advantageous.

A technique long used in the manufacture of rubber hose has proved to be very successful in giving radial strength to pultruded parts. In this method, the formed but not cured pultruded part is overwrapped by a layer of hoop-oriented fibers, much like filament winding on the pultruded mandrel. In a variation of this method, fabric or mat is formed simultaneously with the fibers to give an overwrap with radial components.

Uses of Pultruded Parts

Making pipe is perhaps the most obvious application for pultrusion. However, the largest market for pultruded parts is translucent building panels, which are made from fiberglass mat and polyester resin. Another important application for polyester fiberglass pultrusions is in electrical insulators.

Several structural applications have used pultruded parts (again mostly fiberglass and polyester). Examples are the supports and panels for truck trailers, door supports for automobiles, and ladder rails.

Pultruded parts are also used for pressure tanks; there the ends are separately attached. Aircraft flooring is another application that illustrates the ability of pultrusion to make a wide part.

High strength applications include pultruded rods, stringers and stiffeners in aircraft, and automobile leaf springs.

Advantages and Disadvantages of Pultrusion

Advantages include:

- Material usage is high compared with layup.
- The throughput rate is high.
- The process can give high resin contents.

Some disadvantages of pultrusion are:

- Part cross sections must generally be uniform.
- Problems can arise when resin or fibers accumulate and build up at the die opening. This can increase the friction to the point that the equipment will be jammed or the fibers will break.
- When the dies are run resin-rich to account for fiber anomalies, strength is sacrificed. (Normally a good balance can be found without a significant decrease is strength.)
- Voids can be a problem if the dies are run with too much opening for the fiber volume.
- When quick curing systems are used, the mechanical properties are often sacrificed.

Spray-up Methods

The spray-up method is similar to wet layup except in the method of applying the resin and reinforcement. Instead of using cloth or mat, the spray-up method uses roving that is chopped and then blown into the mold. The resin and initiator are simultaneously sprayed into the mold with the chopped fibers. The mixture is hand-rolled to consolidate it in the mold, to remove air bubbles, and to ensure complete wetout of the fibers. The types of mold, application of mold release, application of gel coat, and curing are all the same as with layup.

Spray-up gives a much faster production cycle than hand layup techniques. However, the spray equipment operator must be skilled, since control of part thickness and, therefore, the physical properties of the part are largely a matter of technique.

Special equipment and materials have been developed to improve the spray-up process. Spray guns have the capability of carefully metering the resin and initiator with the chopping and blowing of the fiberglass (usually through two or three different openings or nozzles). Special fiberglass grades called *gun rovings* cut cleanly, wet rapidly, and conform to intricate contours (largely because of special sizings and finishes). Specially formulated resins with minimum drain-off and fast gel times are also available.

Spray-up is useful for fabricating very large structures and parts with rather complex geometries, but the only real advantage is reduced manufacturing cost compared with layup. The physical properties of the part are not as high as with layup.

Thermoplastic Processing Techniques

In critical-use composites (advanced composites), where the properties of the fiber are very important, the manufacturing method used for thermoplastic resins is likely to mimic one of the thermoset manufacturing methods (autoclave molding or compression molding), with modifications to accommodate thermoplastics. The thermoplastic resin for these applications would be one of the high performance resins such as PEEK or polyphenylene sulfide.

In the case of noncritical composites, the reinforcements should be viewed more as an upgrade of the properties that would normally be expected of the thermoplastic resin. These noncritical composites are generally made using traditional thermoplastic processing methods with only minor modifications for the reinforcement material, and the resin is likely to be one of the conventional thermoplastics, such as nylon, acetal, or polycarbonate.

Whereas thermosets must be heated to cure and turn into a solid, thermoplastics are heated to turn into a liquid and cooled to become solid. Therefore, curing in the thermoset sense does not exist with thermoplastics.

Hand Layup of Thermoplastics and Autoclave Molding

The wet layup methods are, in general, not applicable for thermoplastic resins. However, thermoplastic prepregs using high performance thermoplastics are available with most of the normal reinforcement fibers and in most of the fiber forms, including unidirectional fibers and fabrics. Because thermoplastics are solid at room temperature, the thermoplastic prepregs are stiff and boardy (with no drape) and have no tack. They must be softened (slightly melted) so that the tack and drape can be improved. This softening is usually done by heating them with a hot-air gun or some direct heating tool. Although for some complex shapes a large strip of prepreg may need to be heated, usually only small areas are heated to form the prepregs and tack them together. To correct or improve the problem of boardiness, thermoplastic prepregs are also available in comingled and cowoven forms. In these forms, the thermoplastic matrix material is extruded into fibers that are mingled with the reinforcement fibers. This mingling can be laid together in a tow, woven together, or even braided. Because of the flexibility of the fibers, these comingled and cowoven materials are not boardy. Another method used to eliminate the boardiness of thermoplastic prepregs is to coat the reinforcement fibers with powdered thermoplastic resin. This form has the obvious problems of nonuniformity and loss of resin (unless it is heated slightly to tack the resin onto the fibers), but has proved to be acceptable for many applications.

An advantage of thermoplastic prepreg over thermoset prepreg is that the thermoplastic prepreg has an infinite shelf life. No refrigeration or other precautions need be taken to ensure usefulness over a long period of time.

The bagging step should be essentially the same when using the thermoplastic material as with thermosets, except that the bagging materials must be able to withstand the higher temperatures and pressures that are normally used for thermoplastics. For instance, typical bag materials for thermoplastics are Kapton or Vacaloy, which can withstand 350°C. These materials are generally much more difficult to work with than their lower temperature counterparts.

The molding of thermoplastic prepregs must also account for the much higher temperatures required to obtain good flow characteristics and compaction. Most of the high performance thermoplastics soften in the 200–350°C range, compared with curing temperatures for epoxies of 100–175°C. Therefore, the mold materials and autoclave fittings and couplings must be able to withstand the higher temperatures. Many of the molds for thermoset composites would not work. Thermal expansion becomes more critical in the mold design because of the elevated temperatures. The higher viscosities of thermoplastics complicate the consolidation process, so greater pressures may be necessary.

Automated Tape Laying with Thermoplastics

Some machines have been developed for the automatic laydown of thermoplastic prepreg [6]. The machines resemble the automatic laydown machines for thermosets except that a special heated shoe is attached to the applicator head. As the thermoplastic prepreg tape comes between the head and the mold, the tape is heated by the shoe, which softens it and gives it the tack and drape necessary to stick to the mold or the previously laid down layers. A cooling shoe is placed just after the heated shoe to resolidify the tape and ensure that it remains in place.

A major problem with this method is that the material is "springy" and tends to raise up, causing the material in the tool to be bulky. Because of the bulk, interlaminar wrinkles may result when the final consolidation takes place.

Filament Winding of Thermoplastics

Prepreg tow is the usual material used when thermoplastics are to be filament-wound. High resin viscosity and elevated temperatures make in-line (wet) impregnation impractical. These tows are stiff, and so some preheating (melting) is usually necessary in order to get good laydown on the mandrel. Again, there are no worries about pot life with thermoplastics, but consolidation is a major problem.

Pultrusion of Thermoplastics

Several manufacturers of thermoplastic resins and independent pultruders have demonstrated the capability of pultruding thermoplastics. Shapes such as channels and rods have been tested and found to have excellent properties compared with similar thermoset parts. The resin manufacturers are not yet revealing the exact methods for manufacturing these pultruded parts.

A concept that is gaining favor in the composites industry is manufacturing standard shapes (rods, channels, I-beams, hat structures, plates), then postforming them to create structural members. The savings would be great from using a high production method such as pultrusion and reducing inventories by stocking standard parts and sizes, which can then be shaped to fit specific needs. A difficulty with this method is the maintenance of physical properties as the reinforcements move during the reshaping operation.

Matched-Die Molding (Including Hydroforming, Diaphragm Forming, and Thermoforming) of Thermoplastics

Traditional matched-die molding has been combined with a preheating step (similar to traditional thermoforming) in order to accommodate thermoplastic materials. As applied to advanced thermoplastics composites, a laminate of thermoplastic prepreg is placed in an oven capable of temperatures that will soften the thermoplastic. The thermoplastic is typically supported by a frame. When the thermoplastic sheet reaches the sag point (where the material starts to droop or soften), it is put in a matched-die mold and the mold is closed. The mold is maintained at a temperature below the softening point of the thermoplastic. Therefore, the laminate is simultaneously shaped and cooled to become a solid part. The mold can then be opened and the part removed for trimming and finishing. This method has a production cycle far shorter than a thermoset cycle, typically 5–30 minutes. This method obviously eliminates the need to use an autoclave, with its accompanying high costs, but may result in lower physical properties because the shorter times involved do not allow proper consolidation to occur. (The very viscous thermoplastics take more time to penetrate the fiber bundles.) Some thermoplastic prepreg manufacturers are attempting to solve this problem by selling prepreg panel layups that have been pre-consolidated in an autoclave.

Some hydroforming (thermoforming) can also be done when one of the dies is flexible. The flexible die, which need not be exactly matched to the opposing die for very thin panels, will still give good consolidation and shaping of the material because of the hydrostatic pressure (usually about 700 kPa) that is applied. (Hydrostatic pressure is the pressure that comes from pressing by a flexible or fluid material—in this case the flexible die.) For deeper contours and thick panels or laminates, the flexible die can be cast to match the metal mold. (Generally the flexible die is slightly smaller than the metal to allow for the thickness of the part.) Both the nonmatched and matched molding methods are illustrated in Figure 4. Mold temperatures are typically 250–300°C. If the mold is cooler, the part may solidify prematurely, giving improper resin–fiber movement and part definition. Pressures vary according to laminate thickness and fiber type but would typically be 700–2000 kPa for a 1 cm thick carbon fiber laminate. Even though the flexible die is initially cold, degradation can occur because of the heat transferred to it by the hot laminate. High temperature rubbers (such as silicone) can be used for the entire die or as a heat barrier cap on a less expensive rubber, but even these will eventually become brittle and must be replaced. The compressibility (hardness) of the rubber is typically 60–70 Shore A, which gives sufficient firmness to get compression while also giving good conformity to the metal mold. To improve repeatability, a mechanical stop is usually installed to limit the travel of the press platens. When the process is controlled by pressure only, instead of the mechanical stop, the variability of the parts has been excessive, with the resin often squeezed out of the prepreg.

The advantages of the flexible die in the hydroforming method are its inherent lower cost than matched metal dies and the generally lower forces required, which are less likely to damage fibers.

Another method of forming composites is diaphragm forming [7]. In this process, a composite sheet is placed between sheets of the diaphragm material, and the assembly is held in a steel frame and then evacuated. The assembly is then placed in a heating chamber and formed onto a mold in a manner similar to hydrostatic forming. The advantage of this method is that the composite is restrained between the diaphragm plates, and therefore there are less likely to be major forming problems resulting from extraneous fiber movement, as there might be with the softer materials used in the hydroforming mold.

Still another molding process for thermoplastic composites is continuous roll forming. This process also involves preheating the thermoplastic composite part and then shaping it. (The process is analogous to roll forming in the shaping of steel.) In roll forming, the composite part moves through a set of rollers, which shape it. This method is most applicable to standard shapes with uniform cross sections, such as I-beams and C-channels. Fiber movement during this shaping method has not been extensively examined, although the method holds promise for stocking standard shapes and postforming.

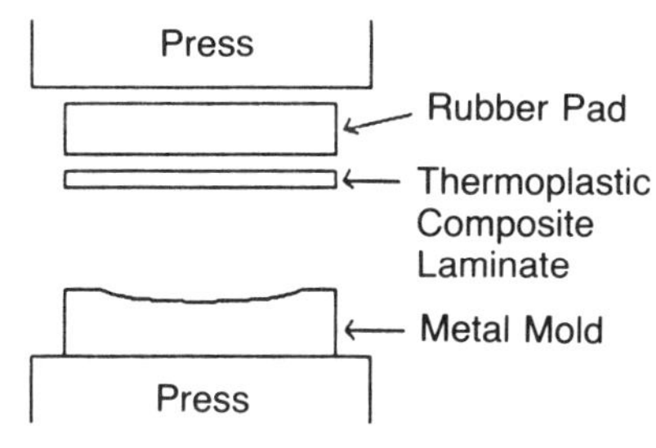

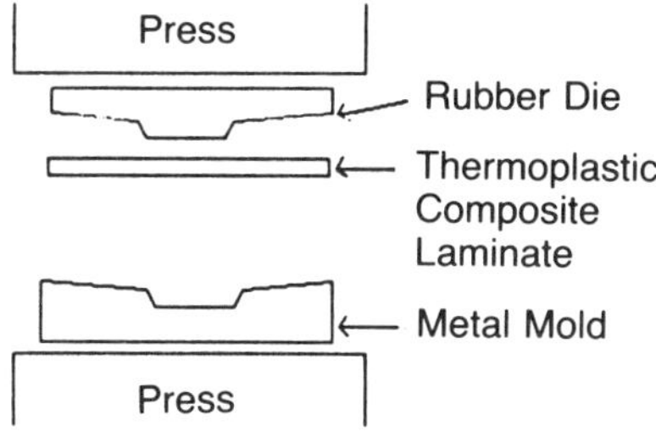

FIGURE 4 **Hydroforming of thermoplastic composites.**

RTM with Thermoplastics

Because of the very high melting temperatures and the high viscosities of high performance thermoplastics, these resins have not been used in processes analogous to RTM.

Injection Molding of Reinforced Thermoplastics

The most important of the traditional thermoplastic processes used to make composite materials is *injection molding*. In this process, resin pellets are melted (usually in an extruder that forms a part of the injection-molding machine). The molten plastic is then injected into a closed mold. The mold is cooled so that the injected plastic solidifies, and the part is then removed.

Injection molding is widely used throughout the plastics industry because of the ease of automating the process, the high productivity as a result of the short molding cycles, and the excellent detail that can be obtained with the process. Most engineering thermoplastics are readily processed by the injection-molding technique.

In a typical injection-molding machine, the resin enters the machine from a gravity-fed hopper that is mounted on top of the machine. This hopper feeds the material into one end of the barrel. A rotating screw advances the material forward. The screw has deep flights in the zone under the hopper, and this is called the feed zone. The material then moves into the melt zone, which has electrical heaters placed around the barrel and where the root of the screw gets thicker to give mechanical heating as well. The plastic is melted through the combined thermal and mechanical heating. The plastic is then conveyed by the screw into the metering zone, where the root of the screw is very thick (the flights are very shallow) to ensure that all the material is melted and to build pressure. At the proper time in the mold cycle, the screw stops rotating and advances forward. This motion injects the plastic through the nozzle into the mold. (A check valve at the end of the screw prevents the plastic from flowing backward.) The sprue and the runner system connect the nozzle with the gates (openings) of the mold cavities. In the simplest molds, the sprue and runners solidify with the part and are cut off as scrap. Some molds have hot runner systems that do not solidify, so that no trimming of the part is necessary. The mold is opened after an appropriate time for cooling of the part (often only a few seconds). Knockout pins eject the part.

The mold can be either single- or multicavity; if multicavity, the cavities can be identical or they can be different to make various parts simultaneously. The latter case is called a *family mold,* since several parts that go into one assembly can be molded simultaneously. Molds can also be equipped with unscrewing devices to give internally threaded parts and with sliding inserts to allow production of hollow parts. All these modifications to the simple one-cavity mold add complexity and cost to the mold, but may add productivity to the operation and thus reduce the cost of the final part.

The size of the injection-molding machine is determined by the size of the part or parts that are to be made. Two measurements of machine size are required. The first is the *shot size,* which is simply the volume of the material to be injected (parts, runners, and sprue). Typical sizes range from 0.03 to 1.4 kg. The second measurement that must be known is the *clamping pressure* that will be required to hold the mold closed against the injection pressure. The area of the part and the restrictions in the sprue and runners and at the gate are major considerations in determining this pressure. Typical values are 1 to 150 MPa.

Several types of reinforcements can be used in injection molding, although particles, whiskers, and short-chopped fibers are the most common.

The use of reinforcements in injection molding must allow for several important aspects of the operation. First, the screw action in melting and the injection process can damage reinforcements and therefore greatly diminish their positive effect. Therefore, a compromise is generally made between optimum physical properties (long reinforcements) and processing capabilities (short fibers). A further problem is encountered in the sprue, runner, and gate system. The fibers must be able to pass through these areas easily. Since the gate is the most restrictive part of the sprue and runner system, reinforcement lengths are generally not greater than the size of the gate into the mold cavity. Typical lengths for the fibers used in reinforced injection-molding materials are 0.7–10 mm, but the longer lengths may degrade (that is, be damaged in passing through the gate). A recent method of manufacturing injection-molding pellets has, however, greatly improved the processability of long-fiber injection molding [8]. Fiber degradation is also dependent on the injection pressure, the injection speed, and the polymer type and viscosity. Other problems, such as flow restrictions in the runner system and excessive heating in the extruder, could arise from the use of fiber concentrations that are too high. Typical concentrations are 30–40% fiber by weight. Machine wear in the barrel, on the screw, and in the mold are constant problems that arise from the use of reinforcements.

Improper mold design can cause serious problems in the strength of the part. If there are multiple gates into a cavity, if there are converging or diverging flows, or if the material must flow around an insert or core, two or more flows of material are created. These flows must knit together well or the part will be weak at the knit line. The presence of reinforcements in the plastic makes the knitting problem more difficult, since the reinforcements can protrude in front of the advancing plastic front and interfere with its smooth meeting and joining with the approaching front of the other flow.

Because of the shear forces that occur in injection molding, fiber orientation is not unusual. Orientation is generally to be avoided because whatever strength is gained in one direction is lost in another. This orientation can be used to strengthen the part in the direction of orientation, but the process must be controlled carefully to obtain repeatable results. Recent studies have shown that thin samples and high injection speeds cause orientation. Some other studies have utilized the magnetic properties of metal reinforcements to obtain the desired orientation in a repeatable manner [9].

Several part properties are enhanced by the addition of reinforcements. Table 1 indicates the differences in several properties when 13%, 33%, and 43% by weight short fibers are added to nylon. The tensile strength, flexural modulus, shear strength, and Izod impact increase, in some cases quite dramatically. The elongation drops, as does the coefficient of linear expansion.

Extrusion of Reinforced Thermoplastics

Extrusion is a process for making parts of uniform cross section in high volume. The process employs an extruder similar to the one described in the section on injection molding, except that in extrusion the size of the extruder is generally much larger and the process is continuous (that is, there is no reciprocating motion), so no check value is necessary. Instead of a mold, the plastic

TABLE 1
Effects of Fiber Reinforcement in Nylon

	Glass Content			
	0%	**13%**	**33%**	**43%**
Tensile strength (MPa)	82.8	120.8	186.3	207.0
Elongation (%)	60	3	3	2
Flexural modulus (GPa)	2.8	4.8	9.0	11.0
Shear strength (MPa)	66.2	75.9	86.2	93.1
Izod impact (J/cm)	0.54	0.48	1.19	1.34
Coefficient of linear expansion (mm/mm·K × 10^{-5})	7.2	2.7	2.3	2.2

Adapted from DuPont, *Design Handbook for DuPont Engineering Plastics,* E62618, pp. 23–24.

material is forced through a die, which shapes it. The plastic is then cooled and cut to the desired length.

Extrusion of reinforced materials is not an important process for making finished parts, but is important as the method of incorporating the fibers into the resin. In this regard, extrusion can best be viewed as an excellent blending and mixing process. Fibers and reinforcements chopped to length are jointly added to the hopper and then blended by the mixing action of the extruder. Care must be taken to prevent damage of the reinforcement, but that is a function of the reinforcement concentration, reinforcement length, extruder screw dimensions with respect to the barrel, heat, and opening of the die. The plastic is extruded into a long, thin rod (like spaghetti) and then chopped into pellets, which would typically be 1–5 mm long.

Recent new technology utilizing pultrusion technology instead of extrusion has produced injection-moldable resins with 5–20 times longer fiber lengths and corresponding improvements in physical properties. Reinforced nylon resins made by this process are replacing magnesium, zinc, and aluminum in die castings and in automotive and appliance applications.

Thermoforming or Vacuum Forming of Reinforced Thermoplastics

In thermoforming, a sheet of thermoplastic material is heated to the temperature at which the sheet begins to droop under the force of gravity, then withdrawn from the oven or other heat source and placed over a mold. A vacuum is drawn through the mold (small holes are drilled in the mold to allow this to occur), and the plastic sheet is pulled into the shape of the mold by the vacuum. Either male or female molds can be used.

A process of forming high performance thermoplastics similar to this (hydroforming) has already been described and holds great promise for simplifying the processing of composite sheets and, in some cases, structural shapes.

For conventional thermoplastics, thermoforming of reinforced plastics holds some promise, but currently offers only a slight improvement over nonreinforced technology.

Molding of Metal Matrix Composites

Metal matrix composites (MMCs) consist of high performance reinforcements in a metallic (aluminum, titanium, magnesium, copper, etc.) matrix. The reinforcements can be in the form of particles, whiskers, or fibers. The first production use of MMC was as components of the space shuttle (boron/aluminum tubes, 1974). MMCs have also been used or investigated for use as piston ring inserts, pistons, connecting rods, impellers, brake calipers, sway bars, critical suspension components, and tennis rackets.

A major problem with the incorporation of a reinforcement into metals is the inherent nonwettability of most fibers. Various approaches are being pursued to overcome this problem, such as coating SiC fibers with carbon or coating aluminum with lithium ceramic. Alternatively, high pressure is used to assist in wetting.

Material Characteristics

In order to demonstrate the improvements that can be achieved with metal matrix composites, some important mechanical properties are presented in Table 2. From the discontinuous systems, improvements in modulus and strength over those of the nonreinforced metal are now nearing 100%. Not only are the mechanical properties increased, but significant improvements in wear resistance, use temperature, thermal stability (coefficient of thermal expansion), and fatigue resistance have been noted. Generally, there is a loss in ductility and lower fracture toughness in the reinforced materials.

The continuous-fiber material properties show outstanding stiffness and strength. Good thermoelectrical and specific stiffness and strength are also characteristic. The fracture toughness and fatigue resistance are good.

When ceramic reinforcements are used, operational temperature limits of MMCs are increased over those of

TABLE 2
Properties of Selected Metal Matrix Composites

Reinforcement	Type	Matrix	Density (kg/m³)	Modulus (GPa)	Strength (MPa)
30% SiC	Particulate	6061 Al	2912	12.1	552
30% Sic	Particulate	7090 Al	2996	12.8	773
20% Sic	Whisker	2124 Al	2885	12.4	745
Gr (P100)	Fiber	AZ91C Mg	1997	36.0	800
Gr (P100)	Fiber	6061 Al	2469	33.5	828
SiC	Fiber	Cu	6380	20.7	1035
SiC	Fiber	Al	2857	20.7	1463
SiC	Fiber	Ti	3883	21.4	1656

(Reprinted courtesy of the Society of Manufacturing Engineers. Copyright 1986. Proceedings of the SME Conference, Technical Paper EM86-106-6, SME, 1986, "MMC Overview—1985" by David L. Hunn.)

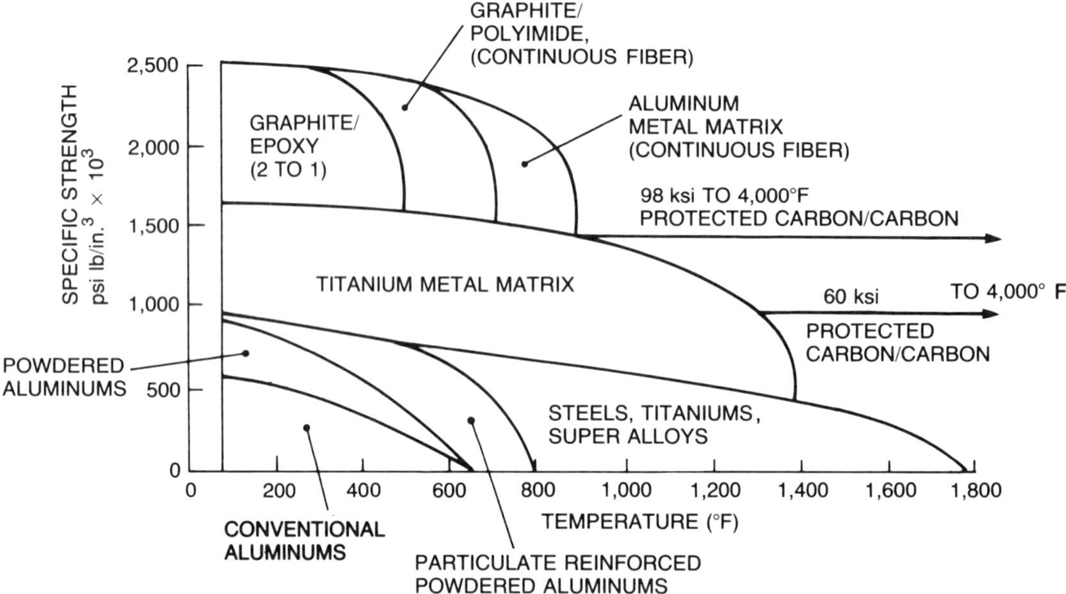

FIGURE 5 **Elevated temperature performance of a number of advanced materials.**

nonreinforced materials. This is illustrated in Figure 5, where the specific strength of a number of advanced materials is shown as a function of temperature. This improved temperature limit has strategic military materials importance, since MMCs can be substituted for conventional high alloy materials that are designed to operate at elevated temperatures.

The incorporation of a second phase (the reinforcement) into a metal significantly affects the propagation of pressure waves through the material by acting as sites for scattering and attenuation. Measurements on boron fiber reinforced aluminum have shown a fivefold increase in dynamic damping capacity.

The high thermal conductivity of a metal matrix combined with the low thermal expansion of the composite allows the design of high performance heat sinks for electronics applications. Carbon reinforced copper is of particular interest for this application.

Discontinuous Metal Matrix Composites

Particulate-reinforced metal matrix composites are produced by mixing small granular ceramic particles with the matrix material. Whisker-reinforced MMCs are formed by growing single-crystal silicon carbide whiskers and then mixing them with the matrix material. These two forms of MMCs are called discontinuously reinforced because the stiff constituent of the composite material is not a continuous fiber. Because the fibers are discontinuous and typically randomly oriented, the composite has nearly similar properties in all directions (called isotropic)—a characteristic very similar to conventional metals. These materials can be processed utilizing traditional manufacturing methods—extruding, forging, rolling, shear spinning, machining, and welding.

Discontinuously reinforced MMCs are available in a variety of forms, such as rolled sheet, extrusions, or forgings. These materials almost always begin with what is known as the powder-metallurgy process. Refined powder of the correct matrix alloy is blended with the particulate or whisker reinforcing agent. Currently, the most widely used discontinuous MMC reinforcement material is silicon carbide. The particles are only a few micrometers in size; whiskers have a more distinct length–diameter ratio. This blended material is then compacted into a billet shape and refined under heat and pressure.

Continuous-Reinforced Metal Matrix Composites

Fiber-reinforced MMCs are produced by orienting long, continuous fibers and then diffusion-bonding metallic foils, casting, hot molding, or plasma spraying the matrix around the fibers. Individual layers can be formed and then stacked in any ply orientation to give the dictated mechanical properties.

Molding of Ceramic Matrix Composites

If a material is useful for engineering applications, such as a matrix in a composite, but it is not resinous (generally organic) and is not metallic, then it is usually ceramic. This is a way of defining ceramics as what they are not—neither organic nor metallic. What they are is more difficult to define. Generally, however, we can describe ceramics as solid materials that have both positive and negative ions and typically exhibit ionic bonding, although some ceramics, such as SiC, have covalent bonding. The solid material may be either crystalline or vitreous (glasslike) or mixed. It is crystalline if the ions are arranged in an orderly array. It is vitreous if its ions are arranged with some regularity at close range but do not display long-range order.

A high degree of chemical and thermal stability is characteristic of the oxides, carbides, nitrides, borides, and other such compounds that form the basis for ceramic materials. The nonoxides can be stable to over 1000°C in the absence of oxygen, but will not tolerate long-term high temperature exposure in oxidizing atmospheres. The stability of ceramic compounds is the result of the tightness of the crystal structure, the unique ionic bonding, and the high stability arising from the relatively small, highly charged ions that are closely packed.

Uses for Ceramic Matrix Composites

Ceramics are generally cast from slurries or pressed into shape with an organic binder and then fired (or sintered or cured) at very high temperatures. The resulting materials are brittle and very sensitive to failures from defects. Under applied external loads, these cracks or flaws grow unstable in the ceramic, leading to brittle, catastrophic material failure. Since the incidence and size of these flaws are difficult to control during conventional ceramic processing and also during operation in aggressive environments, ceramics typically display a wide variation in strength properties. Ceramics are not, therefore, often used for structural parts if other materials are available that will perform at the desired temperatures. The inclusion of a reinforcement will enhance the structural capability of composites, but processing difficulties have limited the reinforcements to very short fibers only (whiskers), and so the improvements are not sufficient to allow for widespread structural use.

Ceramics' very high heat resistance is the principal reason for their use. Of additional importance is their good corrosion resistance. They are, therefore, often used with metal reinforcements or high temperature ceramic fibers (such as SiC) and serve as both the strain transfer medium and the environmental (heat) barrier.

Some of the applications include rocket nose cones, rocket nozzles, ram-jet chambers, turbine components, and possibly entire motors. Other uses for ceramics hinge on their acceptability for use in the 1000–2000°C range. Carbon matrix composites are also competing in this range, and the relative merits of each matrix material are being closely examined. Both materials have been reported to be able to withstand temperatures of 3000°C for short periods (1–2 minutes) without melting, decomposing, or being broken by thermal shock. Their high temperature uses include the skin on the space plane and other similar high performance aircraft or spacecraft parts.

While the high temperature performance potential of ceramic matrix composites remains the main area of interest, their dimensional stability may lead to their adoption in several aerospace applications. Their dimensional stability is the best of the composite materials. Another use is in electrical applications because ceramic composites have low dielectric constants and are generally more transparent to radar, microwaves, and other telecommunication frequencies than other composites.

Curing of Thermosets

Many processing operations involve the changing of a solid material into a liquid state so that it can flow to a desired shape or wet out incorporated materials. The molding of a thermoplastic resin and the casting of metals are such processes. In contrast, thermosets are typically processed at molecular weights low enough that they are already fluid at room temperature. Once the desired shaping or wetout has been accomplished, the resin is rendered solid by a chemical cross-linking reaction called *curing*.

For any manufacturing process, the processing parameters are, of course, chosen to obtain the optimum properties in the final part. In composites, the key processing variables are time, temperature, heat-up rate, pressure, vacuum, and cool-down rate. The interrelationships between these parameters are complex, but several methods for monitoring them have been developed so that at least the major interactions can be seen. Eventually, of course, the final proof of correct curing is the combination of physical and mechanical properties of the finished part.

Cure Cycle

The usual processing procedure for thermoset curing is to increase the temperature while maintaining hydrostatic pressure. As with most fluids, the viscosity of the matrix will fall with this temperature rise and thus allow good consolidation. However, chemical cross-linking reactions will eventually overshadow this thermal thinning, and the viscosity will rise as the mobility of the resin is increasingly restricted by intermolecular cross-linking. Eventually enough cross-linking will occur to cause incipient gelation (some molecules having become large enough to span the entire specimen), and the composite will change from a liquid into a solid and will no longer permit plastic deformation. The temperature at which this phase or structural change occurs is called the *dynamic gel temperature* T_{gel}. This temperature is dependent upon the time/thermal history of the material and cannot, therefore, be fixed as a specific number for each resin. Depending on the resin, T_{gel} occurs when the matrix is 40–60% cross-linked.

The interrelationships of time and temperature for a specific resin can be presented in a time-temperature-transformation (TTT) diagram such as Figure 6. The TTT diagram shows that when moving along the gelation portion of the curve (the S-shaped curve), the time for gelation decreases as the temperature increases. Therefore, gelation can be achieved with either long times at moderate temperatures or moderate times at elevated temperatures.

After gelation, the composite is often given a postcure that increases the cross-link density; this is called *advancing the cure*. This postcure is done by subjecting the

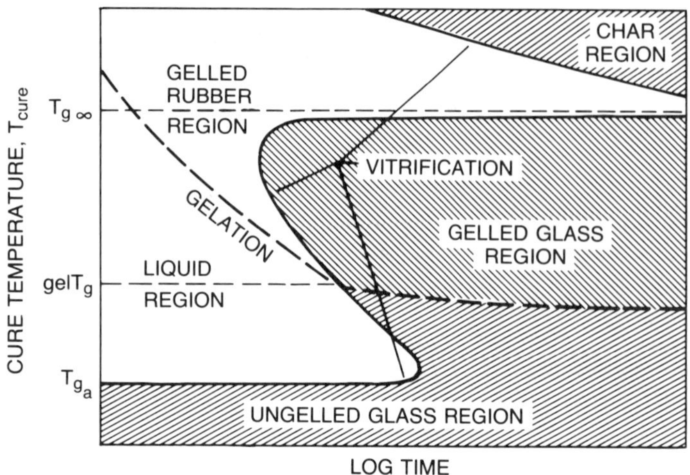

FIGURE 6 **The time-temperature-transformation diagram for isothermal cure.**

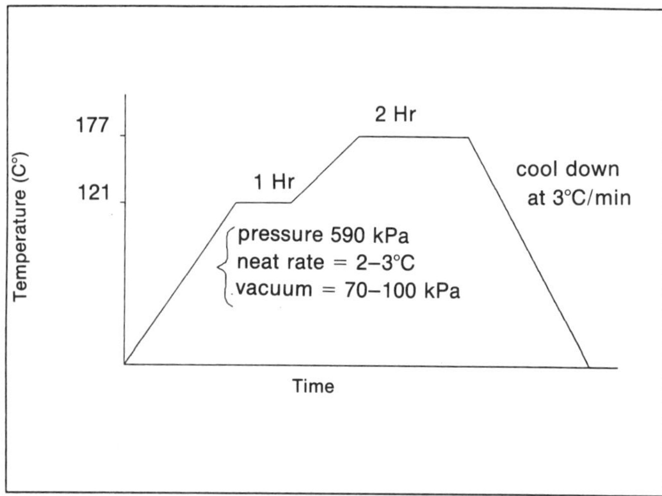

FIGURE 7 Typical epoxy cure system.

part to a higher temperature for an extended period of time (often several hours). Postcures can be done while the part is still in the mold if there is no danger of part damage from differential thermal expansion between the part and the mold. Postcuring is, however, more likely to be done on a free-standing part.

A quantity that has been found to correlate with the cross-link density is the *glass transition temperature* T_g. As the cure is advanced, T_g increases as additional cross-links are formed until the entire structure is fully cross-linked. At this point the upper limiting value of T_g has been achieved.

CURING POLYESTER RESINS. Many of the polyester composites are cured at room temperature, and so specific temperature control is not required. The curing reaction is *exothermic*, so the temperature of the system increases, causing a small decrease in viscosity for a short period and then the characteristic increase in viscosity with cross-linking.

Typical cure cycles for polyesters are several hours (such as overnight). If the parts are heated, such as with an air dryer, the reaction speeds up. Ovens are also used to increase the reaction speed and to effect a postcure.

CURING EPOXY RESINS. The vast majority of epoxy composite parts are cured at elevated temperatures in autoclaves, presses, or ovens with the simultaneous application of pressure. In ovens, the pressure is supplied by a vacuum bag, whereas in autoclaves and presses, the pressure is supplied by the equipment, although a vacuum bag is needed as a sealed membrane to allow pressure to act on the part. Two major types of epoxies are identified by their normal cure temperature, either 120°C or 175°C cures. The higher temperature cure resins find their principal use in high performance aircraft.

Epoxy cross-linking is exothermic, but because of the application of external heat, the effect of the exotherm is not observed in most cure cycles.

A typical epoxy cure cycle is illustrated in Figure 7. The vacuum bagged layup is placed in the autoclave. A vacuum of 70–100 kPa is applied, and then the autoclave pressure of 590 kPa is applied. The temperature is raised at a rate of 2–3°C/min up to 121°C. The temperature, pressure, and vacuum are held for 60–70 minutes. The pressure is then raised to 690 kPa and the vacuum bag can be released to atmospheric pressure. The temperature is then raised at a rate of 2–3°C/min to 350 and is held for 120 ± 10 minutes. The entire assembly is then cooled at a rate of 3°C/min. This stepped cure cycle is especially useful for thick parts or for those that have a high exotherm.

Another typical cure cycle is to simply ramp the temperature to 175°C, hold for two hours, and then ramp down.

This cure sequence is followed to allow for the changes in viscosity and the forces created in the cure. The vacuum is applied to remove any volatiles, and the pressure is applied to effect consolidation. The temperature is raised to a level that is below the gelation point, but when it is held at the cure temperature (in this case 120°C), the resin will gel. Up until gelation, the part has been pliable and has expanded with the mold. After gelation, the part has become rigid but has not yet developed any strength. Therefore, the temperature is held at the cure temperature until the part has developed strength through additional cross-linking. However, the rate of cross-linking is rather slow at the cure temperature, and so when the part has developed sufficient strength, the temperature is raised further to advance the cure. Stresses are induced in the part because of thermal expansion mismatching between the part and the mold, but proper design and the additional strength of the part should prevent part damage. After holding at the higher temperature for some time, the part and mold are cooled. Again, some stresses could be induced as a result of thermal expansion mismatches, but these are, hopefully, minor.

The heat-up rate can effect the cure. Figure 8 shows a study in which the viscosity is plotted against the tem-

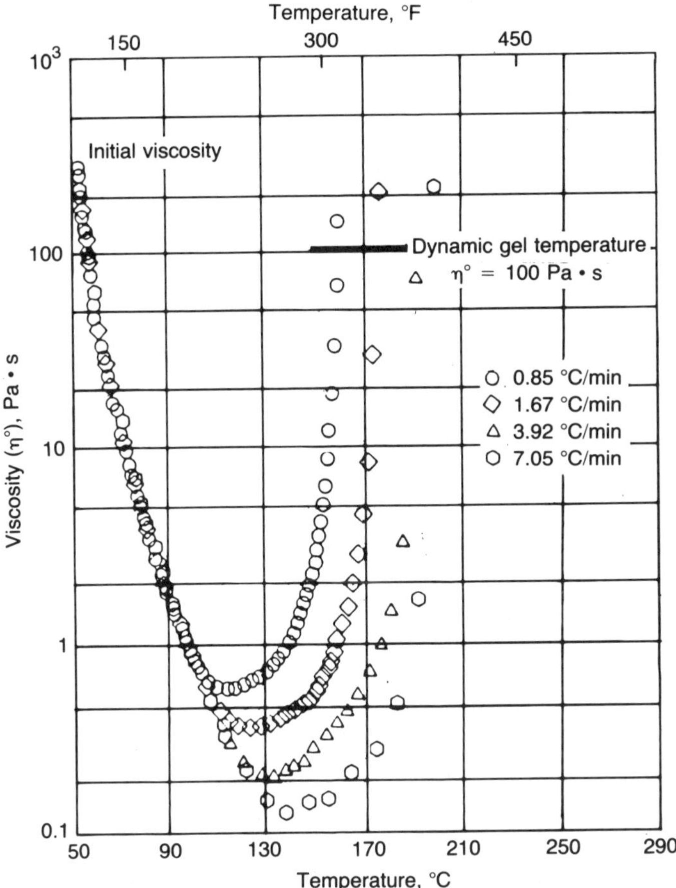

FIGURE 8 **Dynamic mechanical spectroscopy scans at different heating rates.** (Reprinted with permission from *Engineering Materials Handbook,* Vol. 1, *Composites,* ASM International, Metals Park, OH, 1987, p. 654).

perature. The data show that faster heat-up rates result in lower overall viscosity and subsequent reaching of the gelation point (when the viscosity is so high that the part is intractable) at higher temperatures.

The temperature history, which is the accumulated temperature and time profile, is obviously very important in the cure of the resins. The temperature history determines the kinetics (reaction rates), viscoelasticity, morphology (shape), phase, cross-link density, glass transition temperature, and structure of the composite.

Polyimide Curing

Typical polyimide cures for BMI and PMR-15 are much longer than those for epoxies, ranging up to 6 hours, and the temperatures are also much higher, up to 315°C.

Many polyimides are press-cured, which shortens the cure time by over 2 hours. However, polyimides are quite fragile before the postcure, and can be easily damaged just in trying to remove them from the mold. Postcures of up to 7 hours are always recommended. The advantage, therefore, is not in the total time required to cure the part, but in reducing the time inside the mold.

Cure Monitoring Equipment

The simplest and most commonly used cure sensing devices are thermocouples. These are used chiefly as means of controlling the temperature profile of the cure, and give little direct information concerning the cure itself. They are, however, critical to determining the uniformity and rate of heat-up and the time–temperature relationships during cure.

The direct measure of cure is the sensing of changes in viscosity. Rheometers (instruments that measure deformation and flow as a consequence of stress, strain, or time) can be used to monitor the viscosity of the system during curing. Rheometers can be used to define optimum heat-up rates to give good wetting and reasonable cure times. The sudden increase in viscosity that indicates gelation can be readily detected using the rheometer.

The use of differential scanning calorimetry (DSC) is an effective method of determining the heat that is absorbed or emitted during the cure and provides data that augment the rheometric data. By using DSC, the rate of chemical reaction and the resulting T_g can be detected (these are seen as two peaks in the scan, each representing a change in the heat content of the part through phase changes or chemical reactions). By running samples with various temperature profiles, the extent of cure can be approximated by following the increase of T_g with increasing heat history.

The DSC can also be an effective method for simulating a cure and thus determining the optimum method of heating to obtain the most wetout and the best physical properties with the minimum cure cycle.

A method for monitoring cure is dielectric sensing or dielectrometry. This method depends on the changes in the dielectric properties of the part that occur during the cure. When an electric force is applied across a dielectric (nonconducting) material, the molecules (dipoles) within that material attempt to align in the direction of the force. Their ability to align is influenced by such factors as the viscosity of the material, its temperature, and the volume of the molecule. A change in any of these factors brings about a measurable change in the material's dielectric behavior. During cure, as the material changes from a viscous liquid to a cross-linked solid, there is an increase in molecular weight and a decrease in ion mobility (conductivity), and hence a hindrance of dipole motion. The decrease in ion motion and dipole motion are major contributors to the change in the resin's dielectric properties.

Dielectric monitoring measures the permittivity (dielectric constant, e') and loss factor (e'') of the material. Permittivity is a function of the strength, amount, and mobility of dipoles in the resin. It usually starts out at a high constant value in the early stage of the cure because of the initial high mobility of the dipoles in the liquid resin. As the material hardens and the dipole motion is restricted, the permittivity decreases to a lower constant value. The loss factor is a function of both dipole motion and ion motion. As the material vitrifies during the cure, dipole movement causes a peak in the loss factor. The loss factor is directly proportional to ion concentration and mobility. Newly developed microprobes sense the changes in dielectric constant and permittivity and give a direct readout of the cure of the part.

Other methods for sensing cures have not been as widely developed as those just discussed. However, the use of vibrations [mechanical impedance analysis (MIA)] and of light reflectance or transmission (including infrared) is also being investigated.

Tooling

Tooling or molds are used to define the shape of the composite part and one or more of the surfaces, depending on the type of tooling. The requirements for the tooling are conceptually quite simple in that the tooling must provide a mechanism to give a part the desired shape at the end of the mold cycle. In the case of layup and spray-up, this requirement is easily met by a variety of materials, since few, if any, strains are placed upon the mold during the molding cycle. Therefore, the molds for layup and spray-up can be made of composite, metal, or several other materials. The release from the composite should be assured with proper release agents, but the only other major requirement is that the mold must hold its shape.

For matched-die molding, the demands on the mold are at the opposite end of the spectrum. The molds are subjected to enormous pressure and temperature changes and must be strong enough to move a viscous molding material within the mold. These molds are generally steel and are similar to those used in compression molding.

The mold requirements for vacuum bag molding, especially when an autoclave is to be used, are intermediate between the requirements for layup molds and those for matched-die molds. Typical autoclave molds must be able to withstand temperatures of 121–177°C, or up to 315°C for special applications, and the forces caused by thermal expansion. Exceptions are tools with large hollow areas that are not vented to autoclave pressures (such as a box or tube closed on both ends). A further problem with autoclave tools arises in tools that have a large number of voids that expand and cause the tool to degrade rapidly. Therefore, strength requirements are not as severe as those for matched-die molding, and the use of many materials, such as aluminum, composites, and metal-coated composites, is possible.

However, the heating process puts some special requirements on the design of the molds which must be considered. Those requirements arise because of the differences in the coefficients of expansion of the part and the mold and become much more significant as the size and complexity of the mold increase and at higher temperatures. In the curing of the part, the mold and the pliable part expand together during the initial phases of the cure heat-up. But, at a certain temperature–time relationship, the part gels; that is, it becomes hard and intractable. From that point on, the part and the mold expand at their own rate depending on their own distinctive coefficients of thermal expansion. After cure, the part and the mold cool at their own rates as well. Therefore, stresses resulting from this differential in thermal expansion are likely, especially if the part is held within the mold. The special considerations for handling thermal expansion will be discussed in each of the following sections on the various types of mold materials used for autoclave/vacuum bag molding.

In all cases, the choice of whether to use a male or female mold is determined by the part design and application and the ease of manufacture. For instance, a radome nose for an aircraft should be made in a female mold because the outside of the finished part must be finely finished for good aerodynamics. On the other hand, an I-beam would be made in a male mold because of the ease of manufacture. Male molds can also be used when there are serious coefficient of expansion differences between the mold and the part.

Other considerations in the choice of the mold material would include heat transfer capability, machinability, useful life, ease of repair, dimensional stability over time, and initial cost.

Autoclave Tools Made of Metal

Aluminum is the most common metal used for autoclave molds for composite materials when parts are small or simple in contours such that thermal expansion problems are not critical. Aluminum is much lighter and cheaper to machine than steel and has better conductivity. The face of the mold is machined or bent to coincide with the shape of the part. This face is reinforced by headers, eggcrate, or other structural members to ensure that the stresses of handling, weight, and expansion can be withstood without affecting part dimensions. Some care should be taken to ensure that the face material and the backing are similar alloys to reduce any mismatch of thermal expansion of these two mold parts.

The biggest problem with metal tooling is the large difference in thermal expansion coefficient between metals and composites. This difference is typically 3–7 times more expansion for the metal. Steel and Inconel have lower coefficients of thermal expansion than most metals and are often used when high temperature metal tools are required. The method of correcting for this expansion difference is to build the mold to a smaller size than the part. The thermal correction can be calculated as follows:

$$\text{Thermal correction} = \text{engineering dimension} \times (\text{CTE}_p - \text{CTE}_t) \times (T_{gel} - T_{room}) \quad (1)$$

where CTE_p is the coefficient of thermal expansion of the part, CTE_t is the coefficient of thermal expansion of the tool, T_{gel} is the gel temperature, and T_{room} is room temperature.

Caul plates are metal or composite plates that are used to give a better surface to the side opposite the mold and to restrict the movement of the part during molding, but are secondary tools to the mold itself. Caul plates are laid on the back of the laminate and follow the general contour of the desired part. These plates transmit the pressure, since they are inside the bag, and impart a smooth finish to the part. Also, a caul can be used to restrict the movement of a part. As the part expands, it meets the caul plate, and further expansion is restricted. Resin leakage, if there is any, is routed sideways.

Autoclave Tools Made of Composite Materials

The greatest advantage of tools made of composites is that the coefficients of thermal expansion of the mold and the part can be made to closely coincide. To get this matching of expansion coefficients, the resin and fibers and fiber orientation in the part and the mold should be similar. Several fiber orientations can all have the same CTE and are, therefore, equivalent in thermal expansion. For instance, the following orientations are all equivalent in CTE: 0, 90; ± 45; 0, 90, ± 45; 0_2, ± 30, ± 60, 90_2.

In practice, the exact matching of CTE is not done, since the differences between normal tools and most parts are minimal relative to the differences with other tooling materials. If matching is achieved, that obviously solves, or greatly reduces, the problem of stresses induced by expansion mismatches. This advantage, and some of the others, are most significant for large or highly complex parts. Another advantage of tooling made of composite materials is the saving in weight. A typical composite tool would be about one-third the weight of a comparable aluminum mold. The cost of composite tooling is also generally lower than that of metal tooling, as much as 50% of the cost of aluminum tools depending on complexity.

Some disadvantages, however, include the smaller number of cycles that can be run using composite parts, although high temperature resins and experience are now increasing the life expectancy of carbon composite tooling. Both high temperature epoxies and polyimides (such as Avimid N) are successfully extending the life of composite tools (in some cases to over 500 cycles). The retention of the surface finish is also a problem that is being improved. This improvement is coming from the use of high temperature capability prepreg materials or, more recently, from the use of a fine surface layer of carbon mat (veil). The tools are cured in an autoclave at 690 kPa to eliminate voids. The prepreg materials provide more uniform laminates, thus significantly reducing resin-rich and resin-poor areas.

The most common method for making composite tools (see Fig. 9) is to start with a model from which a plaster master is taken, then make a transfer tool (usually a plastic-faced plaster), and finally make the finished composite tool, which is laid up on the transfer tool and then autoclave cured. This composite tool is then reinforced as required and can be used to make molded parts.

Autoclave Tooling Made of Metal-Coated Composite Material

Many of the advantages of both metal tooling materials and composite tooling materials can be combined by coating a composite mold with a metal. The metal gives long life, good mold surfaces, faster heat-up times, ease of repair, and better temperature capabilities, while the composite gives lighter weight, lower cost, and ease of making large molds. The critical factor of the coefficient of thermal expansion is greater than for an all-composite mold, but less than metal and can be highly reproducible. The choice of metal and composite must be made carefully to get the best match of thermal expansion.

These molds are made by two methods. In one, the composite tool is fabricated first and then plated with metal through electrolytic or chemical plating. (This plating method is also used by itself to make nickel tools, which are quite common. The pure nickel, which is about 0.32 cm thick, is subsequently reinforced with substructure.) The metal coating is typically 0.00046–0.001 cm thick. In the other metal coating method, the transfer tool, which is discussed in the section on composite tools, is first coated with metal by either plating or thermal spraying. Then the composite tool is fabricated over this metalized surface. The metal then becomes the face of the composite tool. Metal thicknesses produced by this process are typically 0.15–0.32 cm. The chief draw-

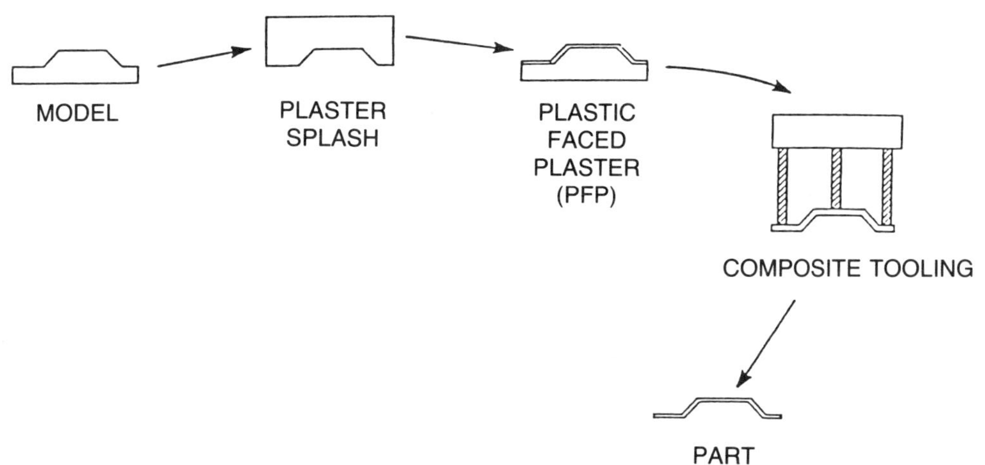

FIGURE 9 Composite tool fabrication.

back of metal-coated tooling is surface cracking after extended service.

Tools Made of Elastomeric Materials

These molds are not completely composed of elastomers, but use elastomers in the key areas where special pressure needs to be exerted. Figure 10 shows the two most common uses for elastomeric tooling. In Figure 10*a*, the hollow composite part is placed in a rigid box, which effectively constrains any outward movement of the part. Then an elastomeric material is placed inside the hollow part. As the assembly is heated, the elastomer expands and presses out against the part to give the required pressure for consolidation. One important consideration in using this method is the need to know the coefficient of thermal expansion of the elastomer and to carefully control the volume of elastomeric material; otherwise very high pressures could be generated (over 7 MPa), which could damage the part and the tool. This method is called *trapped rubber molding*.

The second method for using elastomeric tooling is illustrated in Figure 10*b*. In this method, the elastomer acts like a pressure pad to give compression in places where the bagging material may be limited in movement in order to better transfer the hydrostatic pressure of the autoclave to the desired location. These elastomeric tooling aids are sometimes called *pressure intensifiers*.

The major disadvantages of elastomeric molding are the heavy, bulky tooling that is often required to contain the pressure in trapped rubber molding, the decreased thermal conductivity of the rubber, limited durability if the rubber comes in contact with the resin, and heat aging of the rubber.

As already described in the section on hydrostatic forming of thermoplastic composites, elastomeric tools can also be used in molding thermoplastic laminates. One die is normally metal and the other is elastomeric, which may be flat (if the part contour is small and simple) or cast to mate with the metal.

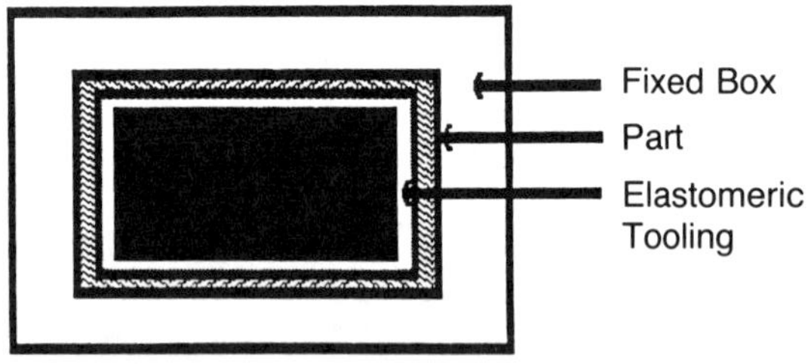

(a) Trapped Rubber Molding

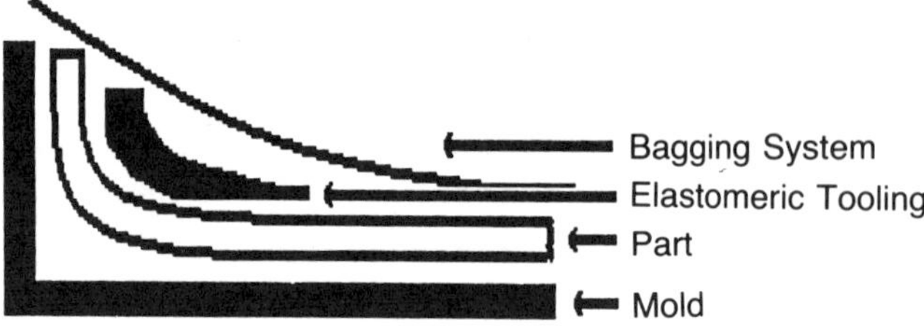

(b) Pressure Intensifiers

FIGURE 10 Use of elastomeric tooling. (*a*) Trapped rubber molding; (*b*) pressure intensifiers.

Matched-Die Molds

These molds are not used for autoclave molding but are used for compression molding of SMC/BMC or preforms, as was discussed in the section on matched-die manufacturing. The molds required for this process are metal molds that can withstand high pressures, and so they are most often made of steel.

Various types of molds are shown in Figure 11. The first type of mold is the flash type. In this mold, the amount of material charged into the mold is generally slightly more than the mold capacity. The molds do not meet at the top and thus allow the excess material to be squeezed out as a thin *flash* on the part. This flash is removed after the part is removed from the mold. The second type of mold is similar to the first except that the flash is vertical. The third type of mold, the landed plunger type, produces no flash. The male and female molds mate precisely (landed), and the resulting product does not need to be trimmed. The charge of material in this last type of mold needs to be carefully controlled, or short shots (inadequate material that will not fill the mold) or excessively thick or poorly defined parts can result.

All of these mold types may be either cold or integrally heated depending on the requirements of the particular process. Also, they generally require some provision for the escape of air. That can be done by providing an air vent or simply by opening the mold slightly for a short period just after the initial closing of the mold. Knockout pins are common in almost all matched-die molds to facilitate the removal of the part.

How to Choose a Molding Process

The most important consideration in choosing the type of molding process is the performance requirements of the part. If the specifications are critical, then a pressure process should be used to reduce the void content. Most aerospace applications today utilize some form of pressurized molding process.

The second most important criterion for choosing the molding process is the resin. If the resin is thermoset, one set of processes is available for molding, whereas if the resin is thermoplastic, another set is appropriate. The choice of resin is also based on performance criteria, although several resins may qualify under the performance requirements. Other resin-related considerations that may affect the molding method are temperature for processing, flow and viscosity, resin brittleness, and the presence of fillers. Other factors that may dictate resin choice are certification and familiarity, that is, which resin is approved for the particular application and what is the on-site familiarity with the particular resin.

Another consideration, again related to specification and performance, is the type of reinforcement, the

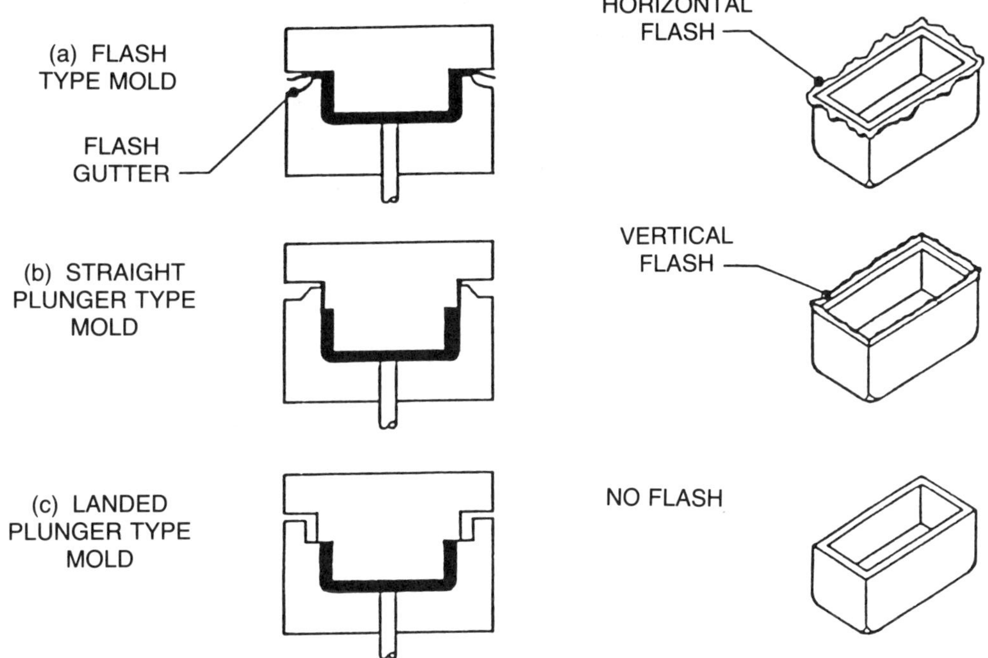

FIGURE 11 **Matched-die typical molds and parts produced.**

method of placing the reinforcement into the mold, and the performance requirements of the reinforcement after processing. Some fiber placement methods naturally lead to a particular molding method; for instance, random fibers lead to either matched-die molding or injection molding, depending on whether the resin is thermoset or thermoplastic. Some of the processes cause greater fiber degradation than others, and so the final properties must be tested to ensure that the reinforcements have not been excessively degraded.

Finally, and of increasing importance, is the economics of the molding method. Some methods, such as nonpressurized molding, are low in cost, but are limited when a large number of parts are to be made. On the other hand, matched-die molding and autoclave molding are more capital intensive, but are appropriate if a large number of parts are to be made. Another issue related to the number of parts to be made is the complexity of the part. High complexity favors hand layup and either nonpressurized molding or autoclave molding. Simple parts can be automated more readily, and that leads to the capital-intensive, nonmanual methods.

Therefore, the overall consideration of which method to choose depends upon performance, economics, and experience. In many cases, if not all, a balance between performance and economics must be made.

A. Brent Strong

References

1. L. Schnell, *Proceedings of the SME Conference, 1985, Technical Paper* EM85-105.
2. S. T. Peters and W. D. Humphrey, "Filament Winding," in *Engineered Materials Handbook*, Vol. 1, *Composites*, ASM International, Metals Park, OH, 1987, pp. 503–518.
3. R. E. Sanders and S. Taha, "Bonding Cure Considerations," in *Engineered Materials Handbook*, Vol. 1, *Composites*, ASM International, Metals Park, OH, 1987, pp. 702–705.
4. N. E. Michaels, J. Laven, and J. Bauer, *Proceedings of the SME Conference, 1985, Technical Paper* EM85-103.
5. C. M. Ma, K. Y. Lee, Y. D. Lee, and J. S. Hwang, *Proceedings of the 31st International SAMPE Symposium*, 1986, pp. 43–57.
6. J. M. Gaynor, "Automated Thermoplastic Composite Processing," Cincinnati Milacron Publication No. SP-156.
7. A. J. Smiley and R. B. Pipes, "Diaphragm Forming of Carbon Fiber Reinforced Thermoplastic Composite Materials," CCM Report 88-11, 1988.
8. J. E. Travis, D. A. Cianelli, and C. R. Gore, *Machine Design*, Feb. 12, 1987.
9. J. Theberge, B. Arkles, and D. Knabb, *Proceedings of the SME Conference, 1977, Technical Paper* EM77-177.

Molding, Ceramic Injection

Ceramic, which is made of earth, shaped with water, dried in air, and made durable by fire, represents the ideal combination of the four elements that the ancients believed compose the world. The most abundant and accessible material on the earth's crust is clay. Early humans were in constant touch with it. It was probably after seeing that their footprints remained in the clay after a heavy rain that they first tried shaping it with their hands.

Potters have certainly existed at least as long as civilization. Much of our knowledge of prehistory derives from the kind of clay and stone objects shown in Figure 1, which were fashioned and fired before there were written records. Ceramic objects first appeared between 15,000 and 10,000 B.C. with the dawn of the Neolithic age. Nomads were aware of ceramic, but did not make and use ceramic objects because of their fragility. The appearance of ceramic and its development marks an important stage in the progress of mankind.

No individual or race can claim credit for originating the shaping of ceramic. Crude forms of pottery for daily use were apparently developed in many places throughout early civilization. Relics still being discovered in ancient tombs in the Orient and Near East as well as in Egypt suggest that each group originated its early work independently. Figure 2 traces the development of mankind, materials, and tools [1].

The introduction of the potter's wheel, probably the first machine, marked the beginning of the mechanization of ceramic forming. The earliest known use of a potter's wheel was in the Mesopotamian town of Worka around 5000 B.C. Wheels were also used during the Indus Valley civilization from 3250 to 3000 B.C., and the Mayans of Central America used them for making ceramic toys. At the same time, the Chinese were making exquisite porcelain objects. In Europe, it was the Romans of the first and second centuries A.D. who developed pottery to a very sophisticated level and spread the skill of making high quality ceramics throughout their widespread empire, which included England, France, Spain, and parts of Germany.

But it is only in this century that shaping ceramics for

FIGURE 1 **An example of an early highly developed clay object.**

industrial purposes has become an important business. As early as 1929 patents were granted for ceramic molding [2,3]. As the manufacture of automobiles and thus of internal combustion engines increased, there was a need for large numbers of spark plugs, which had fairly complex shapes. New shape-forming techniques had to be developed to accommodate the demand. By 1937, the AC Spark Plug Division of General Motors was producing spark plugs in large quantities using injection molding [4]. A cross section of the machine used is shown in Figure 3.

Through the 1940s and 1950s, development efforts, particularly concerned with binder selection, rheology, and dimensional accuracy, continued in Germany, England, Japan, and the United States [5]. Spark plug insulators, refractory products, and some porcelain were emphasized. The injection molding of various alumina shapes was described in the literature, but the technology was highly proprietary [6].

During the 1950s and 1960s, fairly extensive but extremely proprietary developmental work in injection molding was actively pursued both in the United States and in Europe. At this time the man-made fiber and textile industries were growing rapidly. Higher speeds plus the mechanization of fiber production caused excessive wear in the metallic thread guides that were used exclusively in this industry. Ceramic thread guides became a natural choice. Injection molding is ideally suited for the manufacture of thread guides, which are small and come in a variety of complex shapes. (Some are shown in Fig. 4.) For similar reasons, ceramic nozzles have been gradually introduced into the welding industry. As can be seen in Figure 5, these nozzles have a complex geometry with internal threads. They too are being successfully fabricated by injection molding. Other injection-molded wear parts, also shown in Figure 5, include bushings and wire drawing dies.

The introduction of jet propulsion systems in the aerospace industry created a new interest in ceramics. The fabrication of complex jet engine components required a radical change in foundry practice. Casting the superalloys used to make these jet engine parts demands high temperatures and complex ceramic cores (Fig. 6). The manufacture of these cores has provided a new opportunity for injection molding. By the early 1960s, the spark plug manufacturers began to shift to an alternative forming method known as dry bag pressing, while injection molding became established as a routine forming technique for manufacturing cores [7]. A number of electronic parts are also being produced by injection molding. Despite these examples of successful applications, the market for injection-molded products has remained both relatively small and fragmented.

In 1970, the Advanced Research Projects Agency of the U.S. Department of Defense initiated a program to demonstrate the use of ceramics in heat engines. The production of dense silicon nitride and dense silicone carbide components for use in the hot flow path of vehicular gas turbine engines and some diesel engines is being extensively investigated. Furthermore, these complex-shaped components must meet strict dimensional

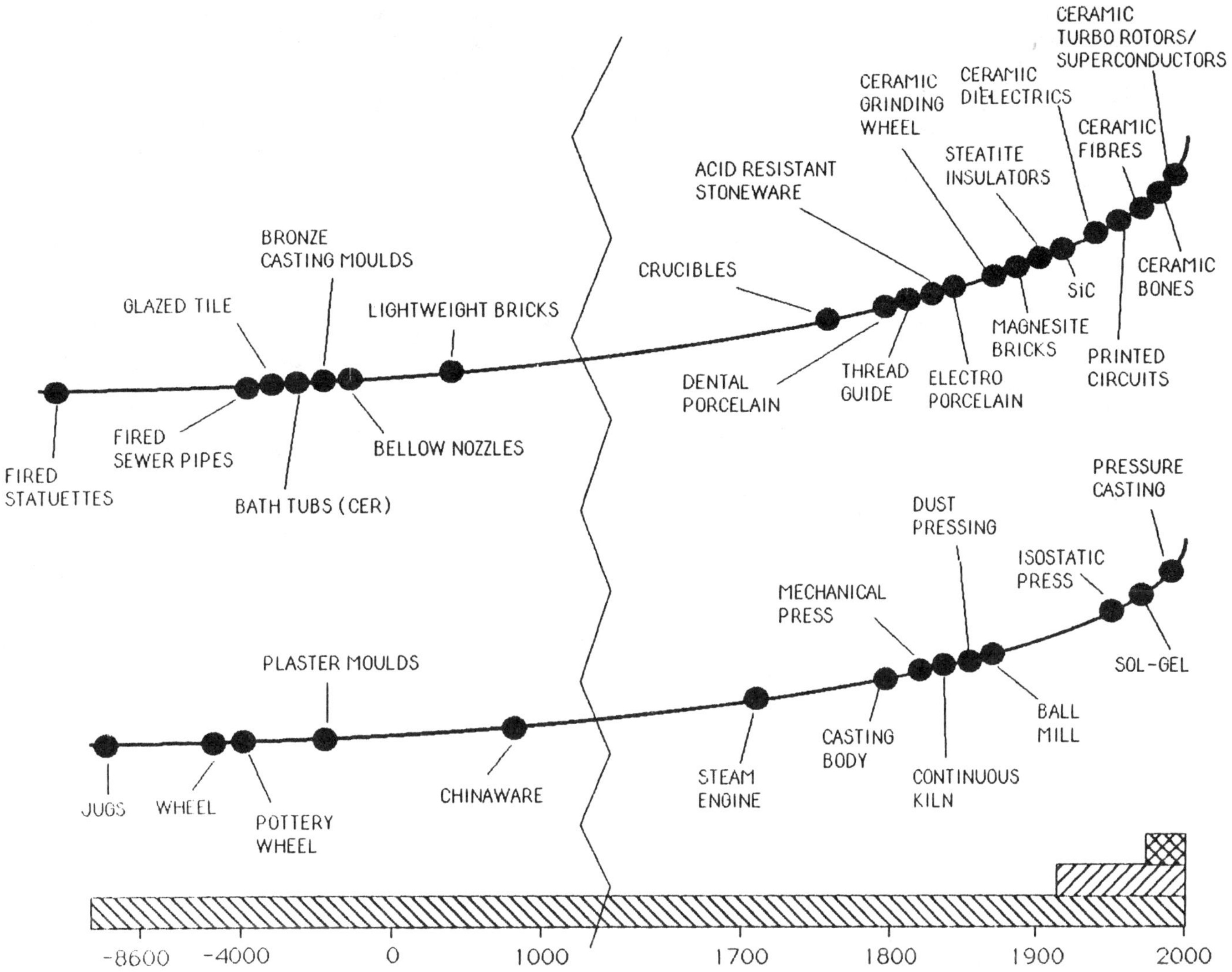

FIGURE 2 Landmarks in the history of the technology of ceramics. (Courtesy of Interceram.)

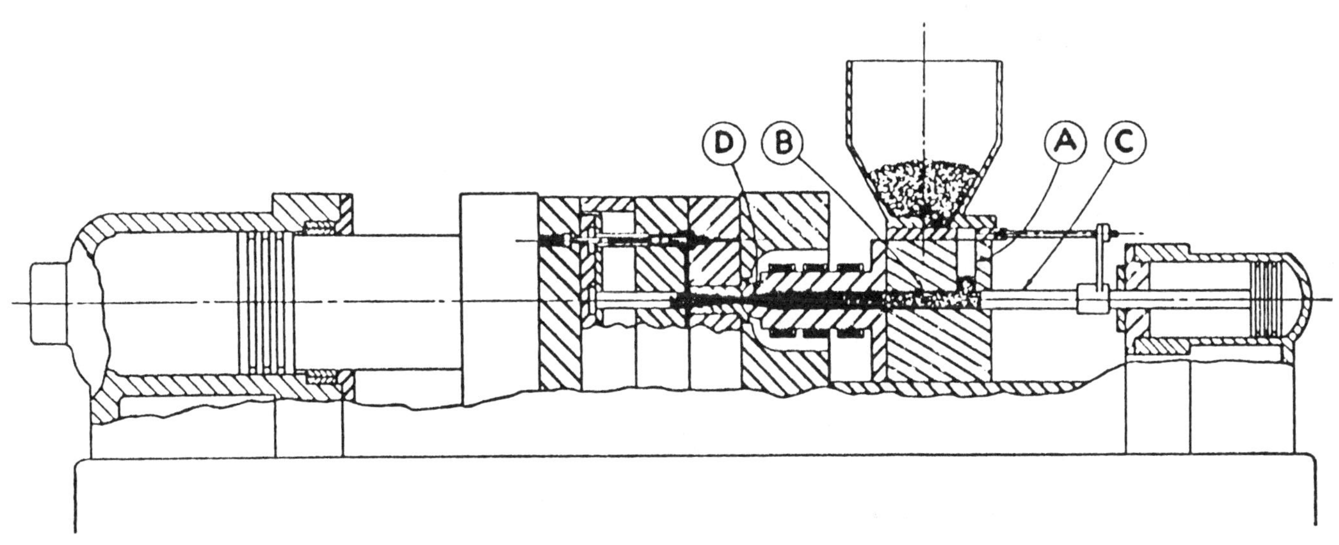

FIGURE 3 Cross section of the injection molding machine used for early versions of the spark plug.

FIGURE 4 Injection-molded thread guides.

FIGURE 5 Nozzles and assorted complex shapes. (Courtesy of Diamonite Products.)

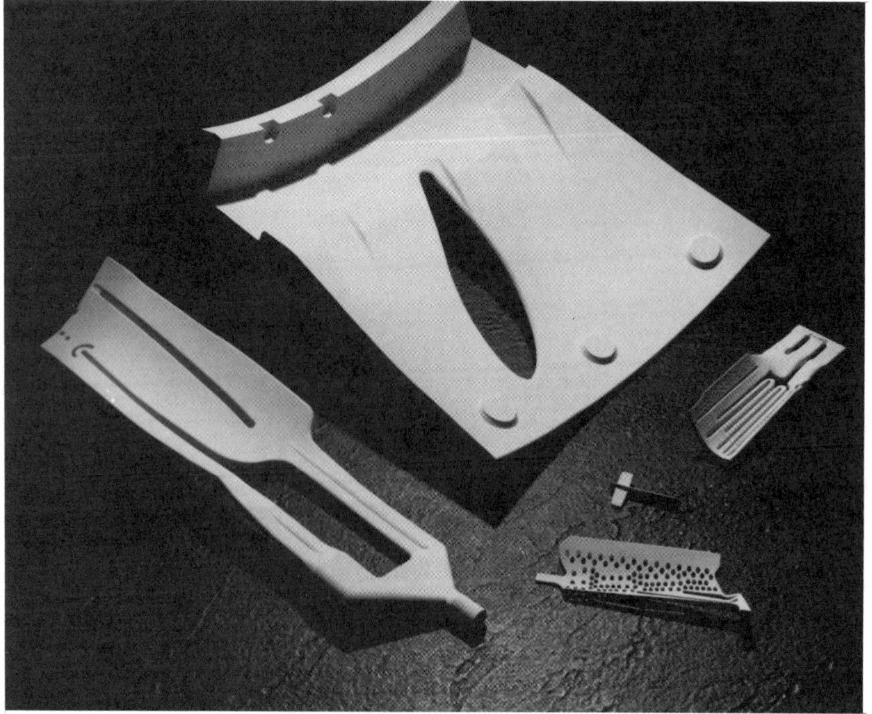

FIGURE 6 Ceramic cores used for the investment casting of alloys. (Courtesy of Howmet Turbine Components Corp.)

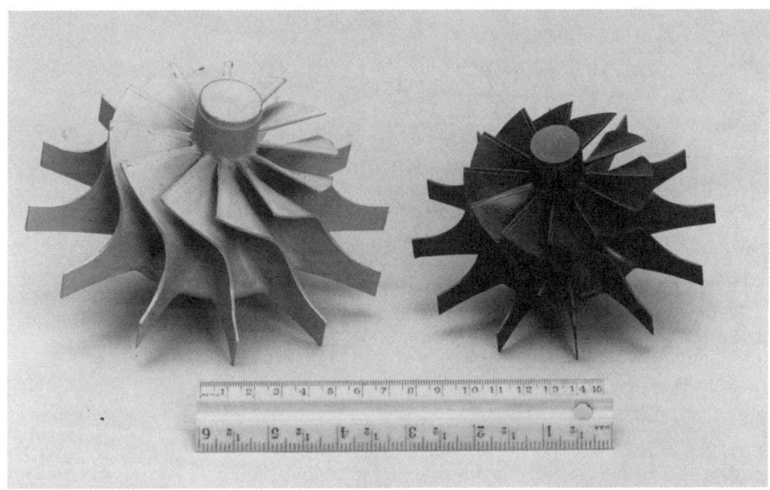

FIGURE 7 **Silicon carbide radial rotors.** (Courtesy of Carborundum/BP Corp.)

tolerances, be highly reliable, and be produced cost competitively in quantity. These requirements have renewed interest in injection molding. The heat-engine-related components most suited for injection molding include:

- Radial rotors (Fig. 7)
- Outer backplate and vanes (Fig. 8)
- Turbocharging rotor (Fig. 9)
- Combustion prechamber insert (Fig. 10)

Ceramic injection molding consists essentially of the steps shown in Figure 11. The acceptance of this process as a routine production method, however, has been extremely slow. Consequently, those currently engaged in ceramic injection molding tend to be either small companies or small divisions of large companies [8]. Undoubtedly, the age of ceramics will depend on the ability to produce complex shapes in large quantities, reproducibly and economically. The injection molding process will therefore play a key role.

Tailoring the Powder

Almost every ceramic material has been injection-molded. The objective of injection molding is to produce complex shapes with close dimensional accuracy in the as-fired state. It is, therefore, essential to use a powder that can be closely packed to minimize the amount of binder needed while maintaining the necessary rheology of the injection molding mix. From an analysis of the influence of particle size distribution on the packing characteristics and filler in an injection molding mix, it

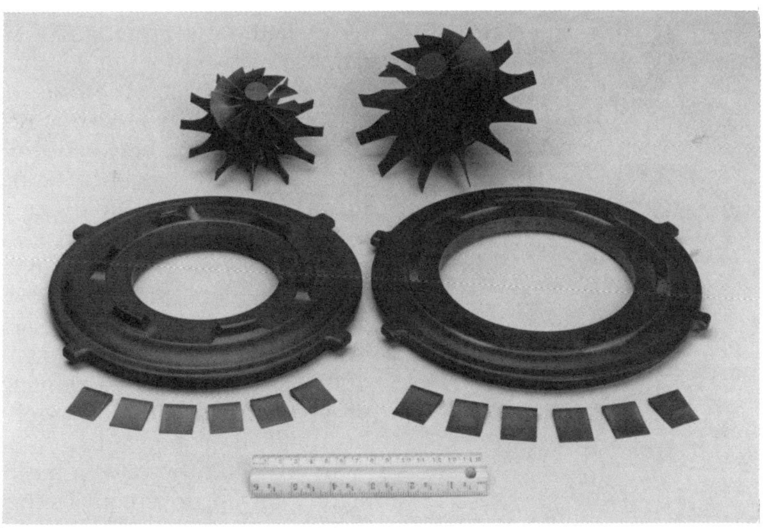

FIGURE 8 **Silicon carbide outer backplate and vanes.** (Courtesy of Carborundum/BP Corp.)

FIGURE 9 Turbocharging rotors.

FIGURE 10 Combustion prechamber inserts.

was concluded that one intermediate milling time can yield a powder with good flow behavior during molding [9].

Work on the injection molding of silicon for reaction bonding has demonstrated the importance of particle size distribution [10]. It was shown that for a constant composition containing 73.5 vol % solids, only silicon powder that was ball-milled for 140 hours can be adequately molded [10]. The particle size distribution for this powder ranges from 30 μm to less than 1 μm. It was also observed that a sample that was ball-milled for 48 hours, although it contained essentially the same particle size distribution, had a much shorter flow distance. This suggests that something other than a physical property such as particle size is responsible for the marked difference in flow between the two powders.

In general, it has been found that, irrespective of the mean size, as the particle size distribution becomes narrower, the powder becomes less moldable [11]. Alternatively, extremely high filler loading and thus high green density, which indicates maximum packing, has been achieved using a very broad size distribution [12]. It has also been shown that bimodal distribution (the presence of two different size fractions in a powder) provides improved packing volume as well as good flow behavior in highly filled suspensions.

In studying the injection molding of silicon nitride, a significant improvement in the dry compaction density (from 1.4 g/cm^3 to 1.85 g/cm^3) has been shown with powder that was ball-milled for 24 hours [13]. This improvement was explained by the breakup of the fibrous silicon

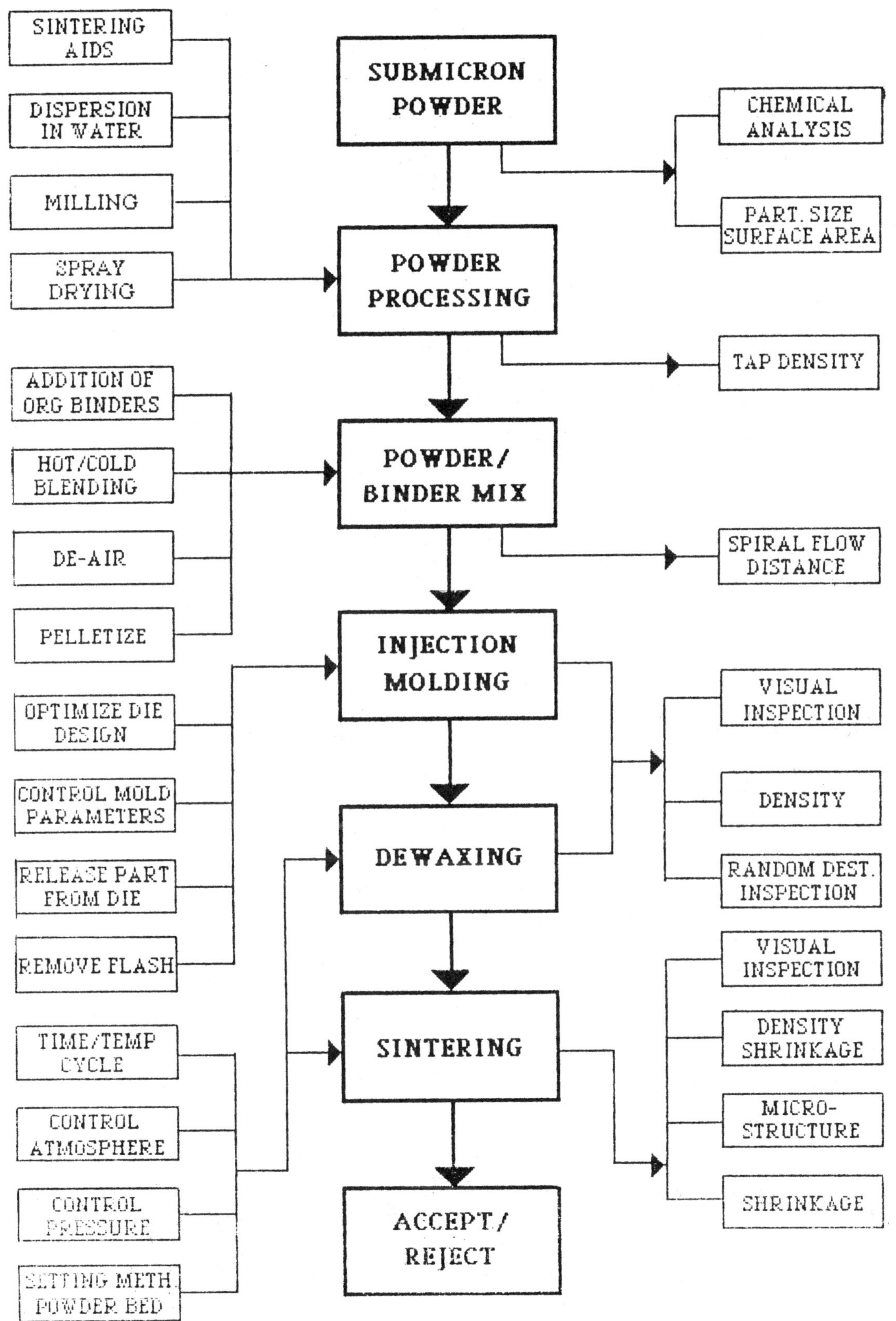

FIGURE 11 Injection molding process flow sheet.

nitride needles, followed by a reduction in their average aspect ratio (the ratio of the length of a fiber to its diameter).

Although this discussion has emphasized the importance of the packing characteristics of the ceramic powders in the final mix formulations, few guidelines have been developed. Furthermore, the trend in the ceramics industry is toward using highly reactive powders (mean particle size less than 1 μm) to achieve better end properties and to reduce the sintering temperature and time. Such powders tend to agglomerate, which can have a deleterious effect on particle packing. In recent years, however, silicon nitride, silicon carbide, and alumina, all with particle sizes of less than 1 μm and specific surface areas between 5 m/g and 20 m/g, have been successfully injection-molded [14–18].

The ability to injection-mold whisker-reinforced ceramic composites would be a major accomplishment. But thus far efforts to do so have not been very successful for two main reasons: During the injection molding of these materials, the whiskers become oriented, which is not desirable for achieving optimum mechanical properties in the composite; and excessive damage to the whiskers during the process makes it difficult to maintain the aspect ratio required for maximum packing density in the part. Injection molding of ceramic composites reinforced with a range of particle sizes has not been found to markedly improve toughness and reliability over those of injection-molded monolithic parts.

Designing a Binder System

The amount of binder required markedly affects the properties of the part being molded during subsequent processing. For optimal results, the binder content should be kept to a minimum. In addition to reducing costs and burnout time, a low binder content aids in controlling dimensional variation during binder burnout, in reducing shrinkage during sintering, and in attaining a high sintered density.

In a study of a variety of thermoplastic binder systems (ceramic powder–polymer suspensions), it was concluded that such systems should possess three flow characteristics: a flow index smaller than 1 (power law),* fluidity greater than 10 Pa·s at a shear rate of about 100 s^{-1}, and a relatively low dependence of the viscosity on temperature at the same shear rate. Furthermore, shrinkages in the binder system resulting from phase changes or thermal contraction should be kept to a minimum [19].

* The power law relates the shear stress τ and the shear rate $\dot{\gamma}$ for pseudoplastic materials in the following way:

$$\tau = \tau^0 \left| \frac{\dot{\gamma}}{\dot{\gamma}_0} \right|^n$$

where the exponent n, defined as the flow behavior of a pseudoplastic fluid, is a number greater or smaller than 1.

Almost all the natural and synthetic organic compounds that are commercially available have been tried as binders and are covered by patents. Low molecular weight polyethylene (5000–20,000), a particular grade (unspecified) of polystyrene, and, more recently, paraffin and microcrystalline waxes with small additions of other organic modifiers have been used successfully. The object is to achieve mixes that do not separate, flow freely during molding, mix uniformly, can be readily removed from the mold without sticking, and burn out from the molded shape without causing it to crack or rupture.

A binder system that will meet the above criteria has been sought since 1932. In general, wax-based formulations (including one containing a liquid epoxy as a plasticizer) have received wide acceptance [20–24]. Useful thermoplastic formulations include atactic polypropylene with or without polyethylene or ethylene–vinyl acetate but containing a phthalate plasticizer and a lubricant such as stearic acid or paraffin wax [25,26]; polystyrene ethylene–acrylic acid copolymers plus polysulphone for molding silicon carbide [27–29]; various elastomeric formulations containing rubber-related block polymers mixed with a paraffin wax plasticizer, a processing aid such as stearic acid, and an antioxidant [26]; and ethylene–vinyl acetate copolymer with various combinations of fatty acids plus a coupling agent containing an amine functional group [30,31].

Other advantageous binder materials include polybutyl methacrylate in combination with certain additives, which improves the adhesion of the binder to the powder surface, reduces the binder removal time, and minimizes the formation of internal defects during removal [14,32–34]; a polyacetal binder that volatilizes evenly and completely during binder removal without creating any internal defects [35]; and polyalkylene carbonates that burn out rapidly at temperatures as low as 300°C [36].

Thermosetting resins have also been used as binders, particularly with silicon carbide powders [37–41]. Their major advantage is the availability of residual carbon after pyrolysis and the relative ease of binder removal by thermal processes. However, unlike thermoplastics, thermosets cannot be recycled.

The major advantage of water-soluble binders is the reduction in debinding time. Water-soluble polymers such as the cellulose ethers are used for injection molding metal powders [42,43]. Combined with polyacrylamides, they produce macro-defect-free cement [44]. The most recently developed water-soluble binder formulation is a mixture of ceramic powder with a gel-forming material like agar, plus water. The water, which constitutes most of the binder, is removed first by controlled humidity drying. The small amount of the organic binder remaining is then removed by thermal processing during sintering. The dewaxing time is thus reduced to several hours from several days [45].

Capillary rheometry is a widely used way to investigate the flow behavior of a ceramic powder–binder formulation. It provides a practical and direct measure of the moldability of the mix. In this method, the apparent viscosity of the mixture is measured as a function of shear rate, temperature, or time. To date, little attention has been given to predicting the flow behavior from theoretical models [46]. Despite the extensive and growing number of binder formulations, ceramic injection molding remains more of an art than a science. There is still insufficient knowledge about the chemical compatibility of different binder components, about interfacial chemistry, and about the stability of selected ceramic powder–binder compositions at elevated temperatures and under various shear conditions.

Preparing the Mix

Mixing a large fraction of ceramic powder with a much smaller fraction of organic binder can create a moldability problem. The flow characteristics of the mix, its subsequent moldability, and the quality of the final parts all depend on the homogeneity of the mix. The preparation of the mix is therefore a key factor. The mixing process is a complex phenomenon, and its importance in developing ceramic injection molding has not been fully appreciated.

Two methods of mixing the binder with the ceramic powder have been developed [47–50]. In one, the ceramic powder and the organic binder are kneaded in a hot mixer to a homogeneous consistency at an elevated temperature. A twin-roll mill, which is well suited for processing of high viscosity materials, can also be used as a mixer. It consists of two counterrotating differential-speed rollers with an adjustable nip, and imposes intense shear stresses on the material as it passes through the nip. Because there is little transverse mixing, an operator is needed to displace the strip of material transversely as the material passes through the nip. A twin-roll mill is used to produce moldable cement paste with a very high content of solids [44]. Two types of extruder can also be used for mixing: a single-screw or a twin-screw extruder. The latter, however, achieves a greater degree of dispersion [19].

In the second mixing method, the organic binder is dissolved in a suitable solvent and the resulting solution is mixed with the ceramic powder in a ball mill or a blade-type mixer. The solvent is then removed by heating the mix for a sufficient time above the boiling point of the solvent. These two mixing method are the most widely used.

In general, indices that can be used to quantify the extent of mixing are based on the standard deviation or the variance of the composition of spot samples taken at intervals from the mixture. However, this mathematical method cannot be used for controlling the process, but only for checking its results afterwards. Unfortunately, evaluation of the quality of the mixing process is still largely based on subjective observations.

The Molding Process

The machines presently used for ceramic injection molding are a direct transfer from plastic injection molding, usually without even any modification. An injection molding machine was also suggested for making spark plug insulators [51]. Many different types of machines are available (Fig. 12). Injection molding machines, which can be broadly classified as either plunger types or screw types, are characterized according to their shot capacity, plasticizing capacity, rate of injection, injection pressure, and mold locking force (clamp pressure). They range in size from those with a shot capacity of a few grams to machines that can deliver several kilograms. Some contain added features, such as the capability to preplasticize the ceramic powder mix. Both automatic and semiautomatic machines are available, as well as fixed or rotary machines.

Shot capacity is the maximum weight of material that can be injected per shot. It usually is expressed in terms of grams of polystyrene or cellulose acetate. This rating depends both on the volume of the cylinder swept during one stroke of the plunger and on the volumetric capacity of the feed mechanism. Thus its determination requires correcting for the bulk density and specific gravity of the mix. A preferable way to express shot capacity is in terms of the volume of material that can be injected by the plunger into a mold at a specific pressure.

One factor affecting the output of a screw-type molding machine is its plasticizing capacity. This is usually defined as the weight per hour (expressed in kilograms per hour) of material that can be heated to the molding temperature. Many plunger machines are fitted with a torpedo to improve the uniformity of the temperature in the mix and with preplasticizing units to increase the plasticizing capacity. The torpedo, which is in the heated cylinder, forces the viscous mix into closer contact with the cylinder walls. Its fins act to conduct heat from the hot cylinder walls to the torpedo. In some machines, the torpedo has a separate heating unit.

The injection rate is also an important factor in determining the output of a machine. It is usually expressed as the volume of material discharged through the nozzle per second during a normal injection cycle, and depends on the pressure, the temperature, the material used, and the smallest aperture (or gate) in the flow line. Thus, with low viscosity materials and fairly large gates, the speed of the injection plunger may be the limiting factor.

The injection pressure of an injection molding machine (the pressure exerted by the face of the plunger) can vary over a wide range. However, because of pressure losses in the system, the pressure in the mold cavities is much lower than the plunger pressure. Nevertheless, the actual force exerted on the inside surface of the mold is large even if the projected area of the part (the area of the molded part parallel to the clamps) is large. Therefore, the clamp pressure is an important factor in determining the maximum projected area that can be molded on a given machine. In machines as large as 1379 MPa or more, the molds are normally closed by large, direct-acting hydraulic rams. In smaller machines (less than 1379 MPa, a toggle-type action is preferable. Up until about 15 years ago, plunger-type machines were the most widely used. However, there has been a gradual shift to screw-type machines because plunger types have

FIGURE 12 Typical injection molding machines.

been unavailable in the large sizes (over 1379 MPa) now in demand.

A number of innovative machines have been designed. One, for example, is a multiple injection molding unit mounted on a tablelike framework with an individual hydraulic cylinder and material-transfer cylinder for each molding unit. The material-transfer cylinders are all connected to a manifold, which in turn is connected to a pressurizing unit. This unit includes a hopper and a heater, plus a pressure-transmitting device for pressurizing a plastic mix to a high hydrostatic pressure. The pressurized plastic mix is forced through the manifold into all the material-transfer cylinders, then from each material-transfer cylinder into the respective molding unit. A toggle-type clamping mechanism holds each individual mold in position [52].

An unusual machine in which a refrigerated ceramic slurry is injected rapidly into a mold at high pressure before the ceramic slurry sets and becomes rigid was designed for making ceramic parts for electronic applications. Controlling the pH accelerates the setting of the ceramic slurry in the mold [53]. And a machine that operates in a vacuum removes internal flaws caused by entrapped air [54].

A screw-type machine is now being marketed that is specifically designed to mold highly abrasive ceramics. It uses screws made of an unidentified corrosion- and abrasion-resistant material. The barrels are lined with a nickel-base iron–chromium boride composite [36]. To reduce the internal shrinkage in molded parts, a machine has been designed, shown in Figure 13, that applies oscillating pressure to the ceramic mixture in the mold [55,56]. The pressure on the material is caused to fluctuate over a wide range, which results in an extension of the heat flow in the sprue. This prolongs the solidification of the material in this section of the mold. The oscillating pressure is applied by a high pressure valve assembly that can be maintained at high temperatures, mounted between the nozzle and the mold.

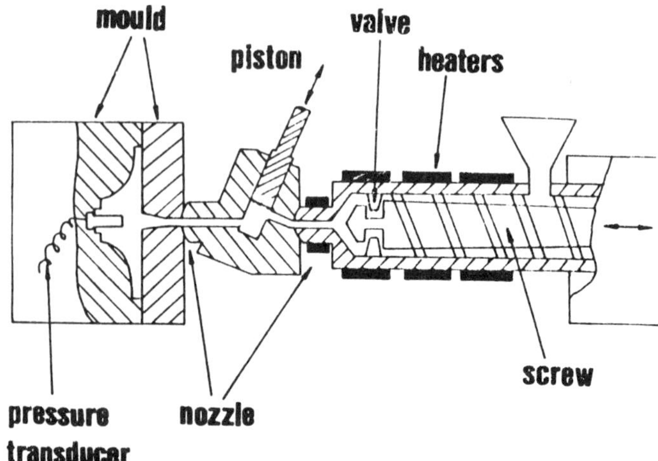

FIGURE 13 Newly designed injection molding machine to reduce internal shrinkage. (Courtesy of Edirisinghe and Evans.)

Machines for making electronic parts that operate at pressures as low as 0.2–0.6 MPa have also been built [24,57]. Operating at low pressures and low temperatures is claimed both to extend the life of the machine and to provide greater operator safety. One design (Fig. 14) consists of a tank connected with a feeder pipe, both with independent temperature control systems. A planetary mixer for blending the ceramic powder and binder mixture is located in the tank, which is connected to a vacuum pump to deair the mix during preparation. The die is clamped to the opening of the feeder pipe with an air cylinder. During molding, air pressure is applied to the surface of the ceramic mix in the tank. The molding cycle is automatically controlled.

FIGURE 14 **Low-pressure injection molding machine.** (Courtesy of Peltsman Corp.)

Designing the Mold

The efficiency of the mold design and the quality of the mold construction are both measures of the success of an injection molding operation. Before a mold is designed, however, initial consideration should be given to the design of the part. The determination of whether a part should be manufactured using injection molding should be based on factors such as its size and weight, the thickness of its sections, anticipated shrinkage, tolerances required, the draft (how much it tapers), the presence of threads, radii dimensions, and the number and size of holes. After a detailed review of all the information available about a part, the mold design should take into consideration factors such as the sprue design, the runners, the number and location of gates, the parting line (where the two halves of a mold meet), the location of vents, the mode of ejection, and the way the mold is heated and cooled. Very little has been written on the important subject of mold design [49,58–61].

Proper selection of a molding machine and an optimum mold design are essential elements of ceramic injection molding. However, ultimate success strongly depends on achieving a proper balance among the key operating parameters of pressure, temperature, and time cycles. To develop a generalizable formula for optimum processing conditions requires a more fundamental understanding of the rheology of mixes, of fluid dynamics during molding, and of the limitations on the geometry of parts that can be successfully injection-molded.

Removing the Binder

The next critical step in ceramic injection molding is the removal of the organic binder from the molded parts before the sintering step. Binder systems are generally composed of several organic ingredients with markedly different melt viscosity and decomposition ranges. Therefore, the ingredients are removed as slowly as feasible in order to create capillary passages gradually within the part with increasing temperature and time. This allows the major binder ingredient to escape at higher temperatures without causing the part to crack and fail.

Thermal degradation in air has been the principal method used for binder removal. In this process, the heat applied to the part increases the vapor pressure of the plasticizer and at the same time softens the major binder component (such as polyethylene). The part must be heated slowly to the maximum temperature to avoid vaporizing the plasticizer faster than the vapor can be dissipated through the part's pores. Otherwise, excessive vapor pressure within the part will cause cracking and blistering, or distort the softened part if it is not supported. The debinding process can take from 70 to 100 hours.

A number of approaches have been devised to improve upon the conventional thermal process for removing the binder. By controlling the rate of extraction to a preset level of weight loss in a vacuum, 13 wt % of binder can be removed from a 20 mm thick silicon nitride cylinder in 35 hours. This is close to a 50% reduction in cycle time [21]. The binder has also been successfully removed from molded parts by heat treating in a pure oxygen atmosphere with a flow rate of 1 liter per gram per hour [62]. In another method, the molded parts are packed into a material that can absorb the binder when it is converted to a liquid. At least 45% of the binder can be wicked out in less than 8 hours without causing any imperfections. A fine-grained binder-absorbing material (such as superground alumina) produces the best results [13,63].

Another method for reducing binder removal time is to add an organic compound that sublimes (such as naphtha) to the total binder composition (these compounds have a vapor pressure of 0.1 mmHg at room temperature and a boiling point no higher than 250°C). The binder can then be sublimed from the molded part within a relatively short time, from 16 to 24 hours, with minimal shrinkage. The process is carried out in a partial vacuum, by moderately heating the parts, or by a combination of both [64–66]. Vacuum distillation and various solvent extraction techniques have also been found to considerably reduce debinding time [58,67–69].

Controlled heat and gas pressure (for example, argon, helium, or nitrogen) can also be used for removing the binder. The gas pressure within the chamber is maintained above the binder's vapor pressure. The binder is condensed in a cooled chamber connected through a value with the heated high pressure chamber. The rate of binder removal from the molded part can be regulated by controlling the cooling rate within the cooled chamber or by controlling the rate at which the binder vapor can flow in the atmosphere from the part to the cooled chamber [69]. In a novel method, the binder is extracted from the molded part using a supercritical fluid* such as carbon dioxide or a chlorofluorocarbon as the solvent [70,71]. Although various approaches to developing a reliable cost-effective method for removing the binder from an injection-molded part have been investigated, no optimum solution has as yet emerged.

A major problem in ceramic injection molding is meeting dimensional specifications. Tolerances of 0.005 cm are now common, and some applications require tolerances of less than 0.0013 cm. A linear relationship has been shown between the variation in shrinkage and the binder content [49]. A number of suggestions have been made for improving dimensional accuracy [9]:

- Careful preparation of the powder to assure a closer packing of particles.
- Maintaining a higher filler loading to enhance the green density of the part.
- Accurate weighing and mixing of the binder ingredients with the ceramic powder.
- Careful control of the debinding and sintering schedule.

*A supercritical fluid is an extremely dense gas that remains a fluid above its critical temperature regardless of the applied pressure.

Acceptance of a new technology rests primarily on the quality of the product achieved by applying the technique cost-effectively. For the new and promising technology of ceramic injection molding to become economically viable, all the technical variables discussed above will need to be brought under control.

A basic understanding of such key areas of ceramic injection molding as the rheology of molding mixes, mixing techniques, molding, and mold design has been developed over the past 10 years. Based on the information generated, it is now possible to commercially produce such automotive parts as a turbocharger rotor and prechamber combustion inserts, as well as ceramic dental braces, agricultural spray nozzles, and ceramic scissors, among other products.

The current market for injection-molded ceramic parts in the United States is less than $10 million per year. But it is expected to grow at a rate of 8–10% over the next 10 years as ceramics become increasingly used in heat engines and other applications that require parts that can withstand high temperatures and corrosive environments. The ultimate widespread adoption of this technology, however, will require solving such problems as the development of organic binder systems for molding sections with thicknesses greater than 1.27 cm and holding dimensional tolerances within ± 0.001% and 0.3%.

Beebhas C. Mutsuddy and Renée G. Ford

References

1. H. Reh, *Interceram, 38*(2), 25 (1989).
2. W. J. Miller, U.K. Patent 283151 (1929).
3. W. A. Triggs, U.K. Patent 400281 (1933).
4. K. Schwartzwalder, U.S. Patent 2,122,960 (1938).
5. H. W. Hennicke and K. Neuenfeld, *Ber. Deut. Keram. Ges., 45*, 469 (1968).
6. R. Bahn and H. Blechschmidt, *Silikat Tech., 10*, 442 (1959).
7. R. L. Randolph, *Materials in Design Engineering, 54*(2), 10 (1961).
8. M. J. Edirisinghe and J. R. G. Evans, *Int. J. High Technology Ceramics*, 2($\frac{1}{4}$), 1, 249 (1986).
9. M. A. Strivens, *Bulletin of the American Ceramic Society, 42*(1), 13 (1963).
10. J. A. Mangels, "Development of Injection Molded Reaction Bonded Silicon Nitride," in Burke, Lenoe, and Katz, Eds., *Ceramics for High Performance Applications , Vol. II*, Brook Hill Publishing Company, Chestnut Hill, MA, 1978, pp. 113–131.
11. J. A. Mangels, "Development of Moldable, High Density Reaction Bonded Silicon Nitride," NASA-Lewis Research Center, Contract No. DEN 3-20, May 1978, August 1978, and January–April 1979.
12. W. J. Corbett and T. B. Peter, "Injection Molded Ceramics," *Proceedings of the CeramTec '88 Conference*, ASM International, Metals Park, OH, 1988.
13. D. L. Mann, "Injection Molding of Sinterable Silicon-Base, Non-oxide Ceramics," Technical Report AFML-TR-78-200, Final Report for Period September 1977 to October 1978.

14. T. Tanaka et al., *Yogyo-Kyokai-Shi, 93*(a), 96 (1985).
15. S. Kamiya et al., *Proceedings of the Society of Automotive Eng.*, 1985, 850523.
16. S. Wada and Y. Oyama, *Proceedings of the 2nd International Conference on Materials and Components for Engines*, 1986.
17. G. Bandyopadhyay and K. W. French, *Proceedings of the 31st International Gas Turbine Conference*, 1986.
18. J. Heinrich, E. Backer, and M. Bohmer, *Journal of the American Ceramic Society, 71*(1), c-28 (1988).
19. M. J. Edirisinghe, and J. R. G. Evans, *Materials Science and Engineering, A109*, 17 (1989).
20. R. E. Wiech, European Patent Applications 8110211.2, 81100210.4, and 81100209.6 (1981).
21. A. Johnsson et al., "Rate-Controlled Thermal Extraction of Organic Binders from Injection Molded Bodies," Mangels and Messing, Eds., in *Advances in Ceramics*, Vol. 9, *Forming of Ceramics*, The American Ceramic Society, 1983, pp. 241–245.
22. K. W. French, J. T. Neil, and L. L. Turnbaugh, U.S. Patent 4,456,713 (1984).
23. K. W. French et al., U.S. Patents 4,708,838 and 4,704,242 (1987).
24. I. Peltsman and M. Peltsman, *Interceram, 4*, 56 (1984).
25. K. Saito, T. Tanaka, and T. Hibino, U.S. Patent 4,000,110 (1976).
26. R. A. Pett et al., U.S. Patent 4,265,794 (1981).
27. R. W. Ohsorg, U.S. Patents 4,144,207 and 4,233,256 (1979 and 1980).
28. R. H. Smoak and R. S. Storm, "Interactive Development of Injection Molded Sintered Alpha Silicon Carbide Turbine Material," American Society of Mechanical Engineers Publication No. 79-GT-77, 1979.
29. T. Sugano, *Proceedings of the 1st Symposium on Research and Development of Basic Technology for Future Industry, Fine Ceramics Project*, Japan Technology Association, 1983, pp. 67–84.
30. B. C. Mutsuddy, *Proceedings of the British Ceramic Society, 33*, 117 (1983).
31. B. C. Mutsuddy, "Formulation of Injection Molding Binder Systems," *Chemical Engineering Communications*, 1988, pp. 1–17.
32. E. R. W. May, U.S. Patent 3,819,786 (1974).
33. I. A. Crossley et al., U.S. Patent 3,882,210 (1975).
34. M. S. Thomas and J. R. G. Evans, *British Ceramics Transactions Journal, 87*, 22 (1988).
35. G. Farrow and A. B. Conciatori, U.S. Patent 4,624,812 (1986).
36. M. Inoue, Y. Kihara, and Y. Arakida, *Interceram, 2*, 53 (1989).
37. T. J. Whalen, J. E. Noakes, and L. L. Terner, "Progress on Injection Molded Reaction-Bonded Silicon Carbide," in Burke, Lenoe, and Katz, Eds., *Ceramics for High Performance Applications*, Vol. II, Brook Hill Publishing Co., Chestnut Hill, MA, 1978, pp. 179–192.
38. T. J. Whalen and C. F. Johnson, *American Ceramic Society Bulletin, 60*, 216 (1981).
39. P. A. Willermet, R. A. Pett, and T. J. Whalen, *Bulletin of the American Ceramic Society, 57*, 744 (1978).
40. D. R. Fitchmun, U.S. Patent 3,947,550 (1976).
41. P. Wirth et al., German Patent 27 50 095 (1978).
42. R. D. Rivers, U.S. Patent 4,113,480 (1978).
43. N. Sarkar and G. K. Greminger, *American Ceramic Society Bulletin, 62*, 1280 (1983).
44. J. D. Birchall, A. J. Howard, and K. Kendall, European Patent 0021682 (1981).
45. A. J. Fanelli and R. D. Silvers, U.S. Patent 4,734,237 (1988).
46. B. C. Mutsuddy et al., "Computer Modeling of Fluid Flow," in Mangels and Messing, Eds., *Advances in Ceramics*, vol. 9, The American Ceramic Society, 1984, pp. 259–264.
47. Glen H. Howatt, U.S. Patent 2,434,271 (1948).
48. E. Wainer, U.S. Patents 2,593,507 and 2,593,943 (1952).
49. H. D. Taylor, *American Ceramic Society Bulletin, 45*(9), 768 (1966).
50. A Waugh, U.S. Patent 3,549,736 (1970).
51. R. W. Ehlers, U.S. Patent 2,446,872 (1948).
52. J. G. Hagerborg, U.S. Patent 3,049,757 (1962).
53. E. J. Mellen, U.S. Patent 3,222,435 (1965).
54. Werner Huther, U.S. Patent 4,412,804 (1983).
55. Demag, Kunststofftechnik, Federal Republic of Germany, UK Patent 1553924 (1979).
56. M. J. Edirisinghe and J. R. G. Evans, *Materials & Design, 8*(5), 284; *9*(2), 85 (1987).
57. P. O. Gribovsky, *Hot Molding of Ceramic Parts*, G.E.I. Moscow-Leningrad, 1961, p. 400.
58. M. A. Strivens, U.K. Patent 808,583 (1957).
59. R. T. Gilbert and I. B. Pfau, U.S. Patent 4,083,903 (1978).
60. C. L. Quackenbush, K. French, and J. T. Neil, *Ceramic Engineering Science Proceedings, 3*, 20 (1982).
61. V. V. Stanciu, "Tooling for Ceramic Injection Molding," in J. A. Mangels and G. L. Messing, Eds., *Advances in Ceramics*, vol. 9, The American Ceramic Society, 1984, pp. 239–240.
62. R. Gilissen and A. Smolders, "Binder Removal from Injection Molded Ceramic Bodies," in P. Vincenzini, Ed., *High Tech Ceramics*, Elsevier Science Publishers, Amsterdam, 1987.
63. K. Watanabe, Japanese Patent 100973 (1982).
64. E. A. Bush, U.S. Patent 3,346,680 (1967).
65. E. R. Herrmann, U.S. Patent 3,330,892 (1967).
66. R. A. Horton, U.S. Patent 3,769,044 (1973).
67. M. A. Strivens, U.S. Patent 2,939,199 (1960).
68. J. M. Adee, U.S. Patent 4,225,345 (1980).
69. R. E. Wiech, U.S. Patent 4,305,756 (1981).
70. N. Nakajima, S. Yasuhara, and M. Ishihara, U.S. Patent 4,731,208 (1988).
71. H. Nishio et al., *Proceedings of the 1st Japan International SAMPE Symposium*, 1989, pp. 1504–1508.
72. B. C. Mutsuddy and R. G. Ford, *Ceramic Injection Molding*, Van Nostrand Reinhold, New York, in press.
73. B. C. Mutsuddy, "Oxidative Removal of Organic Binders from Injection Molded Ceramics," in Stuart Hampshire, Ed., *Non-oxide Technical and Engineering Ceramics*, Elsevier Applied Science, 1986, pp. 397–408.
74. R. E. Wiech, U.K. Patent 1516079 (1978).
75. Winkelbauer et al., U.S. Patent 4,627,945, 1986.

Molding, Compression

The compression molding process for fabricating plastics dates back to the nineteenth century. The process and its associated equipment, and the materials being molded, have continually improved. Today, "the compression molding process is presently the most technologically developed and versatile way to incorporate either continuous or random chopped fibers into a structural composite" [1].

Compression molding permits the fabrication of complex shapes using high strength materials with excellent dimensional reproducibility at relatively low cost. Both thermoplastic and thermoset materials can be compression-molded, but the vast majority of compression molding is done using thermoset materials [2]. Although a number of fabrication processes are related to compression molding, this discussion will define compression molding as a process in which compound is added to the open mold cavity, then the mold is closed and heat and pressure are applied to form the material into the cavity shape. This process is shown in simplified form in Figure 1.

Compression molding of composite parts offers a number of advantages [3,4]. Combining this process with the appropriate material will provide:

- High volume production. Molding cycle times of less than 2 minutes can be achieved when molding some very large parts.
- Excellent part-to-part reproducibility. For example, one part molded from high glass content epoxy SMC maintains ± 0.08 mm over a 35 cm length where the material thickness varies from 0.6 to 5 cm, while another 7 cm diameter part molded around a steel insert maintains concentricity within ± 0.05 mm.
- Low labor cost.
- Minimum scrap.
- Excellent design flexibility.
- In common with most composites, compression molding allows part consolidation and weight reduction, and can provide significant cost savings in the right application.

Compression Molding Equipment

The two essential items required for compression molding are a press and the mold. Presses represent the largest single item of capital investment for most projects. Most compression molding presses are vertical-acting, with either the top or the bottom platen(s) movable. Major improvements in the capabilities of presses have been made recently. These have allowed the molding of components that require a much higher level of dimensional and property control at faster cycles, and thus more cost-effectively. A major improvement is in the control of platen parallelism. This is particularly important in molding larger parts, where relatively small differences in parallelism can result in major unplanned differences in part thickness. For molding these larger parts, it is also important to minimize platen deflection.

Compression molds for production are almost always matched metal dies constructed of tool steels. The mold cavities require a high polish and may be chrome-plated or given other surface treatments to increase hardness and improve lubricity. Softer materials may be used for prototype molds and small-quantity runs. Molds made from aluminum, low melting metal alloys such as Kirksite, ceramic, and even plastic can be useful. Extra care must be used to minimize scoring and other mold damage when molding high fiber content materials in such molds. Release of the molded part from a scored cavity becomes more and more difficult until acceptable parts are no longer obtained or the mold must be reworked.

Other equipment that can be useful, depending on the specific situation, includes preheaters and preformers. Preheating the material offers a number of advantages. Faster molding cycles can be achieved, and the amount of entrapped air and volatiles can be reduced. Preheating is most efficiently performed using radio-frequency (RF) energy. Since most composites are electrical insulators, RF energy heats them quickly and uniformly. RF heating should not be used with carbon fiber composites. Forced-air ovens can also be used for preheating, but they are more difficult to control since the molding compound must be spread out to ensure uniform heating.

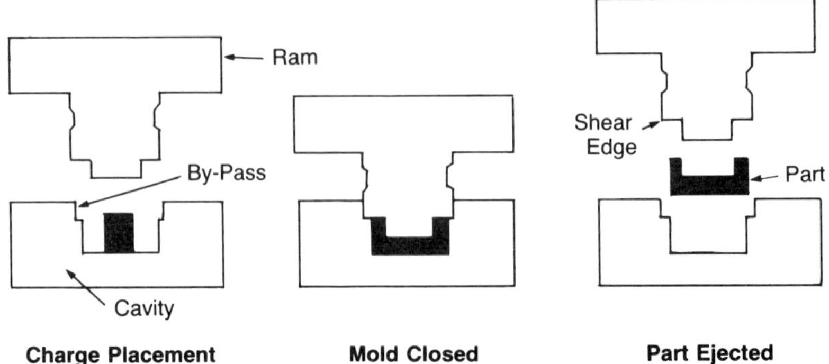

FIGURE 1 Compression molding.

Preforming can be helpful when molding materials with a high bulk factor, such as phenolic and epoxy chopped-fiber BMC (bulk molding compound). Preforming may require a separate mold and press. By densifying and reducing the volume of material to be loaded into the mold cavity, air removal is improved, preheating is easier, and overall material handling is improved, particularly for larger parts. Precompaction of SMC mold charges has recently been reported to improve part quality and speed molding cycles.

Design

The first concern in the design of parts for compression molding is the ability to get the part out of the mold. This elementary consideration can initially be overlooked when a more complex part is being considered. It is essential to consult both the mold builder and the molder early in the design of a part in order to ensure that the part is functional and can be manufactured. Undercuts, ribs, bosses, threads, inserts, holes, pins, and other such features can all be molded into the part under certain conditions. These require more complex molds and, therefore, have higher initial costs, so consideration must be given to whether a particular design feature is feasible, functional, and cost-effective.

When designing a composite part for compression molding, an early question should be, "Can several parts be consolidated into one molded part?" If so, this can usually offer both cost and performance benefits.

There are a number of publications offering design suggestions for plastics [5,6,7]. Designing for compression molding with high fiber content composites follows most of the same general principles used in designing any molded part, but with some additional considerations. Since the mechanical strength of any fiber-reinforced part will be primarily contributed by the fibers, the designer must be concerned about fiber orientation. This requires an understanding of both the design of the mold and the likely molding material and procedures. The properties of compression-molded composites are always anisotropic. There will always be some material flow during the molding process, and this will result in some degree of fiber orientation. Flaws such as knit lines may also occur. Consideration of these potential effects during the design phase can prevent future problems, and it may be possible to take advantage of fiber orientation to enhance properties. When molding a material containing continuous unidirectional fibers, it must be kept in mind that these fibers will exhibit little, if any, flow, and kinks in the fibers should be avoided.

Parts should be designed to eliminate sharp corners whenever possible, since the fibers will flow much more easily around a generous radius. The use of a draft of 1–3° on vertical walls is recommended. It is possible to mold walls with no draft, but special care must be taken to insure that there are no subtle undercuts that can cause the part to hang up in the mold. Since high fiber content materials exhibit very low shrinkage, these effects can sometimes be difficult to identify. Ribs, bosses, variable cross sections, threads, holes in sidewalls, inserts, and other design features can be utilized, depending on the material being molded. Hollow cavities can be molded using low melting metal cores.

Molds used for production-scale compression molding of high fiber content composites are almost always matched metal dies made from high quality tool steel [8]. The preferred design is fully positive in order to develop maximum properties in the molded part. Flash or semi-positive molds can cause problems when the fibers get on the land area; the fibers will act as shims and prevent the mold from closing fully. Any flash should be provided for in the direction of mold closing. Fibers, particularly glass fibers, can be quite abrasive, so mold surfaces are usually chrome-plated or given other surface treatments, such as nitriding. Areas of the mold that will be subject to unusual wear should be constructed as inserts to allow easy replacement. Most production molds will have built-in heating using electricity, hot oil, or steam. Care must be taken to ensure that uniform temperatures are achieved in all areas of the mold.

One of the most critical areas for compression molds is the bypass or telescope and the shear edge (Fig. 1). The height of the bypass must be sufficient to allow the unmolded charge to fit into the cavity. If the bypass is too short, the compound will be forced out onto the land areas of the mold, resulting in poor mold fill and/or wrong dimensions. The shear edge defines the parting line of the mold and is essential in developing back pressure in the material to provide good densification.

During the molding process, shrinkage will occur. This shrinkage is conventionally reported as a percentage based on a room temperature mold and a room temperature part. For high fiber content thermoset plastics, this nominal shrinkage is typically 0.0–0.2%, which is usually reported as 0.000–0.002 mm/mm. This nominal shrinkage of thermosets is actually a combination of two effects; one is the volumetric shrinkage (or expansion) of the chemical cure reaction, and the second is the change in dimensions that occurs during cooling of the part from molding temperature to room temperature. This dimensional change of the material is measured by the coefficient of thermal expansion (CTE), which may be different from or similar to the CTE of the mold material. Thus, the nominal shrinkage of the material may be different from the actual shrinkage during the molding cycle, and problems can be encountered in ejecting parts from the mold, since actually a hot part is ejected from a hot mold. A larger shrinkage effect can occur with high fiber content materials because of fiber orientation. Shrinkage along the fiber direction will be less than that perpendicular to the fibers.

In designing molds for parts larger than 1 m, consideration must also be given to possible deflection of the metal used in the mold and of the press platens. When prototype molds are being built from softer materials, extra-generous draft should be used. Shear edges frequently cannot be as sharp, and it may not be possible to obtain fine detail consistently.

Materials

The primary forms of polymeric composites that are compression-molded include sheet molding compound (SMC), bulk molding compound (BMC), and preforms or prepregs. Each of these material types is available commercially using various resin matrices and fiber reinforcements. The basic properties and characteristics of these materials are discussed in detail elsewhere. These characteristics control the specific compression molding process requirements [9].

BMC was one of the earliest forms of compression-molded composite material. It may be available as a gunk or soft log. In this form, there is typically a low fiber content (~20%) along with a particulate mineral filler. Physical properties are relatively low. Another form of BMC uses chopped resin-coated fibers; it contains only the fiber reinforcement and no fillers [10]. Preform molding involves placing a fiber preform into the mold, then adding the liquid resin into the open mold. This is done most frequently with glass fiber preforms and polyester resins. Prepregs can also be compression-molded in simple shapes. In order to utilize the advantages of compression molding, it is desirable to accelerate the reactivity and modify the flow compared with properties of conventional prepregs. SMC is the most versatile composite material for compression molding. A variety of fibers and matrices are available, it can mold thick and thin sections, and maximum fiber integrity is retained. Figure 2 shows the physical forms of the BMC and SMC materials.

The Compression Molding Process

The compression molding process is probably the simplest composite fabrication process in concept. Specific details depend on the part geometry, the material being molded, the mold design, and the preferences of the molder.

Commercial thermoset composites use conditions similar to those noted below:

Temperature: 135–160°C
Pressure: 3.5–15 MPa
Cycle time: 1–6 minutes

Conditions must be optimized for the particular part and the molding compound. For example, higher temperatures will provide faster cure, but too high a temperature can cause gelation of the compound before flow is complete and the mold cavity is filled. Excessive molding pressure can lead to distortion of the part.

Placement of the charge of compound in the mold is critical. If a material with continuous unidirectional fibers is used, then these fibers must be placed directly in the desired location, since little fiber flow will occur. For random-fiber materials, considerable experimentation may be necessary. Because these materials will flow, care must be taken to eliminate trapped air and knit lines in the molded part [11]. Before building the first mold charge, the operator should visualize how the material will flow as the mold closes. To minimize air entrapment, the material must flow outward toward the part-

FIGURE 2 Typical compression-moldable composite materials. Left, SMC; right, BMC.

ing line or toward other venting. Flow around inserts frequently results in knit lines on the opposite side of the insert. After the initial parts are molded, the results are analyzed and adjustments made in the charge to overcome any problems. A number of techniques are available for analyzing molded parts to detect hidden flaws. These include x-ray and other NDT methods, and destructive tests. For initial molding analysis, destructive tests are desirable in order to visually confirm flaws in critical areas. High temperature burnoff of the polymer matrix allows inspection of fiber orientation and is particularly valuable in determining proper charge placement.

Once acceptable parts are being molded, the molding cycle can be optimized. This typically involves decreasing the mold cure time and increasing the mold temperature to determine the fastest cycle that will comfortably produce acceptable parts.

Post-molding operations usually include removal of flash. For very high fiber content materials, this flash can be quite strong, and techniques such as die trimming, water-jet, and other similar methods may be desirable in a production mode. Post-curing may be necessary for some materials, particularly in applications where the use temperature is at or above the molding temperature, or where maximum chemical resistance or dimensional stability is needed.

When molding some materials—for example, polyesters and vinyl esters—surface porosity frequently occurs. In-mold coating is now being utilized for molded automotive exterior parts to overcome this problem and produce a high quality surface. In one version, the mold is opened slightly just before completion of the cure cycle, and a liquid coating system is injected into the mold. The mold is then reclosed, and the part and coating are cured together. Another version involves electrostatically coating the mold with a powder just before introduction of the mold charge, then curing as usual.

Comparison with Other Processes

A number of other processes for fabricating composites are also available. In comparison with compression molding (CM) of production parts, these other processes have advantages and disadvantages, as summarized below. Most parts can be made using any one of several processes and materials, so each application must be considered individually to determine what is best.

Injection Molding

Faster cycles than CM can mean lower piece cost and possibly lower capital investment. CM offers better control of fiber orientation, less fiber breakage, and less tendency for the part to warp. Larger parts are being molded commercially with CM, and molds are usually less expensive.

Transfer Molding

This process is typically used when inserts are required. It also develops minimal flash. The use of material is less efficient than in CM, and physical strength is usually less. Most high fiber content materials are difficult to transfer-mold.

Prepreg/Layup

Higher strength and stiffness than with compression molding can usually be obtained. Tooling is frequently less expensive. CM is much faster, can produce more complex shapes, and usually provides better dimensional control.

Resin Transfer or Liquid Resin Molding

More control of reinforcement placement and type can be achieved. Tooling is less expensive. Thin sections, edges, and other such areas may be resin-rich. Low density cores can be included. These processes offer cost advantages when relatively few parts are to be made. CM offers faster molding cycles, and more complex small parts can be molded.

Reinforced Reaction Injection Molding (RRIM)

This offers similar advantages, plus much faster cure speed, but physical properties are usually much less.

Combinations of processes can offer real benefits in the right situation, for example, compression molding the core of a component and using this as a mandrel for filament winding to give better hoop strength.

Economics

The costs of producing any part should involve all elements required for its production: material cost, labor, equipment cost, secondary operations, and so on. Comparing the expected costs of compression molding a part that is already being made in another material can be especially difficult, since the full costs of producing the existing part are not always known. The worst cost and performance situation for any composite part is when the composite part must be made "just like the existing part." An attempt should always be made, whenever possible, to adapt the design to maximize composite strengths and minimize molding problems. A major element of designing for composites is parts consolidation. Can several individual components be combined to eliminate later assembly? Can parts be molded to finished dimensions? Can threads be molded in? Can painting be eliminated? In comparing the costs of metal and composite parts, the elimination of these secondary operations frequently shifts the cost advantage in favor of composites.

The easiest way to determine the cost of a compression-molded composite part is to request a quotation from a custom molder. The molder will already have presses, and will be experienced in determining his costs. In most situations, the customer will pay for the mold and any associated special tooling. The more information that is provided to the custom molder, the better his quotation will be.

If doing compression molding in-house is desired, approximate piece costs can be determined by considering the following:

1. Material costs—part weight; material usage efficiency factor; cost per unit weight. In compression molding, usage efficiency is usually 95% or greater. Most molding compounds are sold on a weight basis.
2. Molding cost—molding cycles per hour multiplied by parts per cycle; press cost per hour divided by the number of parts produced per hour. Press cost typically includes the amortized cost of the press, plus one operator. These costs are determined by each individual molder, but they typically range from $30 per hour for a smaller press to well over $100 per hour for larger presses.
3. Secondary operations—flash trimming; painting; supplemental machining.
4. Mold cost—amortized over the useful life of the mold (or other guidelines) by dividing by the number of parts to be molded in that time frame.
5. Overhead and cost of money.

Prototypes

The cost of prototypes of compression-molded parts can be particularly vexing because of the need for a mold. This mold cost can be 60% or more of the cost of the prototype program. If the prototype program does not require a part representative of production quality, it may be possible to machine a portion or all of the details from a larger piece of molded material. However, a machined part will not exhibit the same fiber orientation as a molded part, and its strength properties are typically as much as 30% less than those of a molded part. For most prototype programs, a mold must be built. An informed decision must be made among the molder, mold builder, and customer as to whether a mold made with "soft" materials will provide parts of satisfactory quality, or whether hard tooling is required. Soft tooling is generally less expensive and can usually be built more quickly than steel molds.

Applications

Some of the best understanding of the capabilities of any process for fabricating composites comes from seeing where that process is actually being utilized. Compression molding is being used commercially to fabricate a wide variety of composite materials into high performance components. Selected examples illustrate the diversity of sizes, shapes, and functions that are being compression-molded.

The aerospace industry has long been the pioneer in developing applications for newer advanced composite materials. Recent trends toward greater cost reduction have resulted in increased use of the compression molding process because of its ability to provide lower cost, highly reproducible components with the normal lighter weight and corrosion resistance of most polymer matrix composites. Since most materials for compression mold-

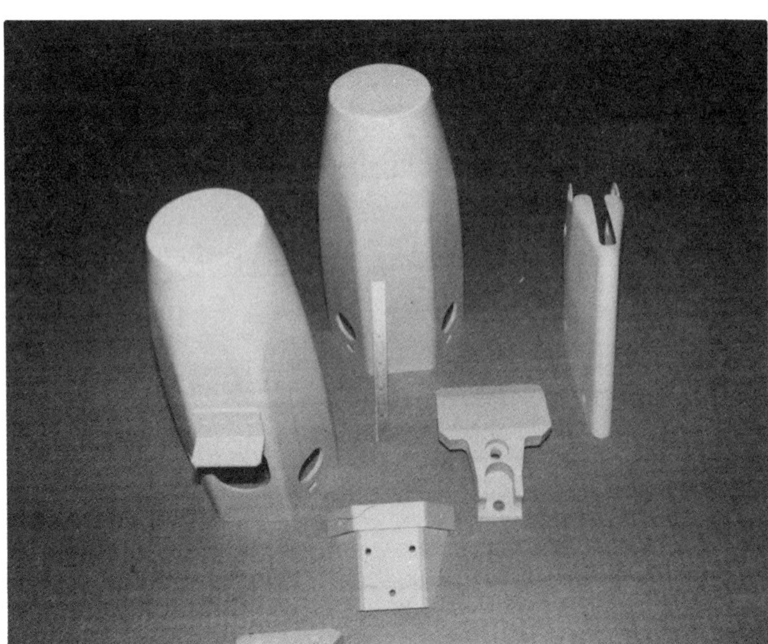

FIGURE 3 Aerospace components compression-molded of epoxy SMC.

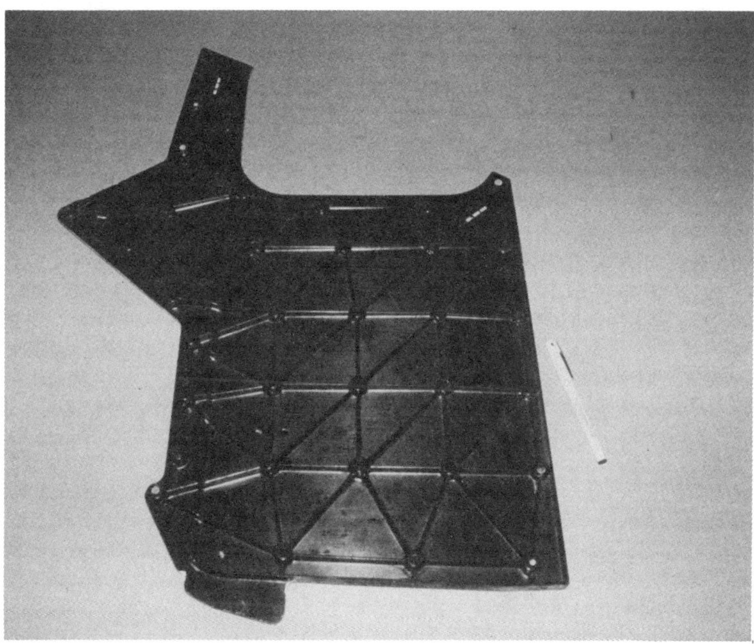

FIGURE 4 Compression-molded fairing. (Courtesy of Boeing.)

ing do not contain a high content of continuous, directional fibers, they are not currently suitable for primary structural use. They are, however, candidates for some secondary structural and nonstructural uses (Fig. 3). In components where high modulus is needed, the required stiffness can be achieved by designing ribs, flanges, and other such features, into the part.

In commercial aircraft, a number of covers, fairings, panels, and housings are compression-molded of epoxy SMC to replace aluminum (Fig. 4). Aircraft interior panels have been molded of low fiber content polyester SMC. Because of new FAA flame and burn requirements, phenolic SMC and some thermoplastic materials with much superior heat release and smoke generation properties are being evaluated for future aircraft.

Missiles have been made using exhaust nozzles compression-molded of a fiber-reinforced phenolic because of its heat resistance and ablative properties. Wings, fairings, control surfaces, bulkheads, nose cones, and other components are being compression-molded of high fiber content epoxies that meet the requirements of MIL P 46069. These molded components meet strict performance requirements and are lighter in weight and much lower in cost than the metal components they replace.

The automotive industry began using compression-molded body panels of polyester SMC containing about 25% glass fiber in the 1950s, and similar technology is now used on many vehicles. One of the first truly structural applications for compression-molded composites was a leaf spring in 1979. This was molded with about 70% continuous unidirectional glass fiber in an epoxy matrix. Tests showed that such springs (Fig. 5) could outperform conventional steel springs by a factor of 4. One of the newest automotive uses is for a composite wheel (Fig. 6). One version that is compression-molded of a proprietary matrix SMC, reinforced with both random and continuous glass, was commercially introduced in early 1989.

Limited production quantities of compression-molded pickup truck boxes are scheduled for introduc-

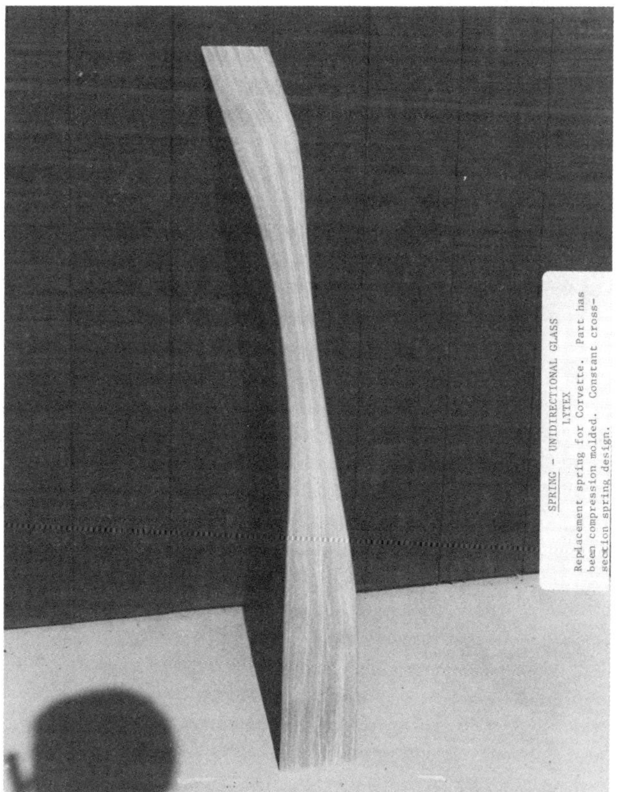

FIGURE 5 Automotive spring.

FIGURE 6 Automobile wheel. (Courtesy of Motor Wheel Corporation.)

tion in the early 1990s [12]. These are molded of vinyl ester SMC, and tests show improved toughness, impact resistance, and overall improved performance compared with steel. Compression molding offers a proven manufacturing method for large quantity requirements. Large automotive components are also compression-molded from thermoplastic composites. Figure 7 shows a load floor component for a station wagon that is compression-molded from a polypropylene sheet reinforced with 40% glass mat. The sheet molding material is also shown.

A new design for a pickup truck tailgate combines a pultruded shell of polyester–glass with bonded-in end caps. These caps are compression-molded of vinyl ester–glass SMC and include integrally molded hinges and other hardware. Several automatic transmission components are compression- or injection-molded from phenolic materials, and compression-molded epoxy SMC is used in another more highly stressed component.

Compression-molded components using composite

FIGURE 7 Load floor component for station wagon. (Courtesy of Butler Polymet, subsidiary of North American Rockwell.)

materials are used in many other industries. Several office copiers utilize compression-molded end caps on the copier fusor roll. These must maintain accurate dimensional stability at temperatures up to 225°C and are molded from epoxy SMC and bismaleimide (BMI) BMC materials. Embossing dies and counter dies are compression-molded from phenolic/rubber mats and from epoxy SMC. Figure 8 shows a centrifugal pump whose major elements are all compression-molded of glass-reinforced epoxy or vinyl ester SMC.

Glass fiber reinforced materials are by far the major composite materials that are compression-molded. Limited compression molding is being done with carbon fiber reinforced BMI and epoxy prepregs and with carbon fiber–epoxy BMCs and SMCs. Some compression molding of aramid fiber reinforced prepreg-style materials is also being done. Other organic fiber reinforced thermoset materials are not appropriate for compression molding because the required curing temperatures degrade the properties of currently available fibers.

Health and Safety Considerations

Since compression molding uses high pressures and elevated temperatures, care must be taken to minimize accidental worker exposure. All recently manufactured presses contain interlocks to prevent the press from closing before workers are clear. Some mold shops also require that presses be blocked open during any time when a worker will be inside the press closure area. Automated press loaders can reduce the chances of worker exposure to accidental closure, as well as to heated molds.

Since most of the materials used for compression molding are fully compounded, there is little problem with direct chemical exposure. One exception is when liquid polyester resins are used with fiber preforms. In this case, care must be taken to minimize exposure both to the liquid resin and to fumes, typically styrene monomer. Some fumes may be given off during the heat cure cycle. These will be most obvious when the press is opened for part removal. Positive ventilation should be used around all presses and all areas where uncured materials are handled to ensure that any fumes are pulled away from the operators.

Handling materials containing fiberglass can cause problems to certain individuals. This can occur in loading material into the mold and in flash removal or machining of the molded part. Carbon particles from carbon fiber reinforced materials can cause problems with electrical equipment.

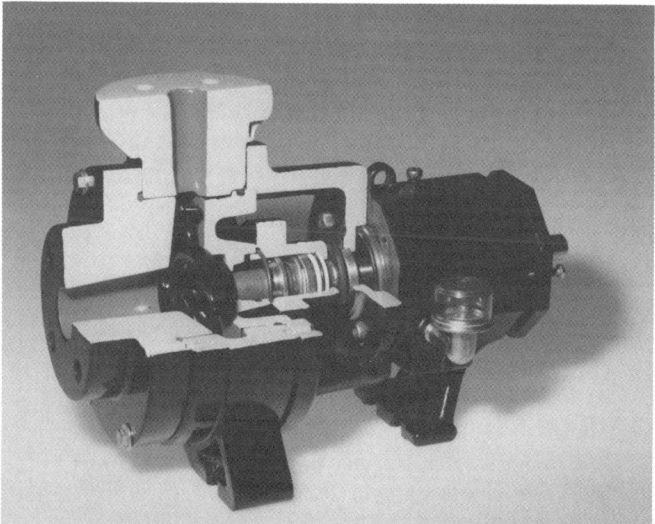

FIGURE 8 **Centrifugal pump.** (Courtesy of Warren Pumps.)

Conclusions

Compression molding is a well-established process for fabricating composites. It is particularly well adapted for those applications requiring low cost, accurate dimensional control, and good part-to-part reproducibility and can be used to fabricate large numbers of parts in a wide range of sizes and shapes. Numerous polymeric composite materials, both thermoset and thermoplastic, can be readily compression-molded, and there is broad-based knowledge and manufacturing capability.

William I. Childs

References

1. C. F. Johnson, in T. Reinhart, Ed., *Engineered Materials Handbook,* Vol. 1, *Composites,* ASM International, Metals Park, OH, 1987, p. 559.
2. S. Witzler, Ed., *Advanced Composites, 3*(5), 49 (1988).
3. R. M. Riddell, "Tailoring the Properties of Compression Molded Plastics," *Machine Design,* July 10, 1980 (no page).
4. J. J. McCluskey and F. W. Doherty, in T. Reinhart, Ed., *Engineered Materials Handbook,* vol. 1, *Composites,* ASM International, Metals Park, OH, 1987, p. 157.
5. *A Fiber Glass Design Manual,* Owens Corning Fiberglass, Toledo, OH, 1986.
6. S. Tsai, Ed., *Composites Design,* Think Composites, Dayton, OH, 1987.
7. T. Reinhart, Ed., *Engineered Materials Handbook,* vol. 1, *Composites,* ASM International, Metals Park, OH, 1987, pp. 417ff.
8. J. Kluz, *Moldmaking & Die Cast Dies for Metal Working Trainees,* National Tool, Die & Precision Machining Association, Washington, DC, 1978, chap. II, Compression Molds.
9. R. M. Riddell, "Tailoring the Properties of Compression Molded Plastics," *Machine Design,* July 10, 1980 (no page).
10. W. G. Colclough, Jr., and D. P. Dalenberg in T. Reinhart, Ed., *Engineered Materials Handbook,* vol. 1, *Composites,* ASM International, Metals Park, OH, 1987, p. 161.
11. R. B. White and R. S. Jackson, "Problems with Premix Moldings," *Modern Plastics,* March, May, July, September 1959.
12. P. Miskech and J. J. Phipps, *Proceedings of the SPI Composites Institute, 44th Annual Conference,* 1989, Paper 18-C.

Molding, Metal Injection

This presentation covers the process of metal powder injection molding and its application to forming complex-shaped high performance components. The metal powder injection molding approach allows net shaping of a variety of materials, including metals, ceramics, cermets, intermetallics, and composites. In this process, a high concentration of powder is mixed with a heated thermoplastic binder to form a low viscosity slurry. This composite slurry is shaped using conventional injection molding practices. Subsequently, the thermoplastic is removed from the compact and the powder is sintered to near full density. The process allows net-shape fabrication of complicated structures and is being applied to a variety of materials, including metal matrix composites.

Basic Concepts

Low cost, intricate plastic parts are often formed by injection molding. The resulting thermoplastic polymer components are structurally inferior in comparison with other engineering materials like steel. In spite of the property differences, plastics formed by injection molding are used because of the desirable combination of low cost and high complexity. In its simplest form, injection molding involves heating a plastic to a temperature at which flow is possible, then forcing the plastic into a shaped cavity, where it is cooled before ejection. The process has similarities to metal casting, but results in greater precision and much finer surface detail.

In turn, injection molding of polymers filled with powders is a recognized route for improving the polymer strength. Filled polymers are routinely used for applications requiring electrical conductivity. The filled polymer is a composite incorporating the fabricability of the polymer and the mechanical, thermal, magnetic, or electrical properties of the filler. The chief uses are in electromagnetic shields, antistatic surfaces, and applications where the thermal conductivity of the polymer must be increased [1,2]. As an example of the latter, thick sections of ultrahigh molecular weight polymers require slow heating during forming, but incorporation of a metal powder increases the thermal conductivity and increases productivity. The fabrication of electrically conductive polymers by injection-molding mixtures of thermoplastics and metal powders provides a basis for understanding the evolution of the current technology, which is aimed at fabrication of metallic parts where the polymer is a transient phase selected for easy forming.

A recent development has been to maximize the content of solid particles and to remove the binder during sintering. As a consequence, a new powder forming process has evolved, providing shape complexity, low cost forming, and high performance properties [3]. The key steps in this process are outlined in Figure 1. This new shaping process, termed powder injection molding, is begun by mixing selected powders and binders. The mixture is granulated and injection-molded into the desired shape. The polymer imparts viscous flow characteristics to the mixture to aid in forming, die filling, and uniform packing. After molding, the binder is removed and the remaining powder structure is sintered. The product may then be further densified, heat-treated, or machined to complete the fabrication process. The sintered compact has the desirable complex shape and high precision of plastic injection molding, but the materials are capable of performance levels unattainable with filled polymers.

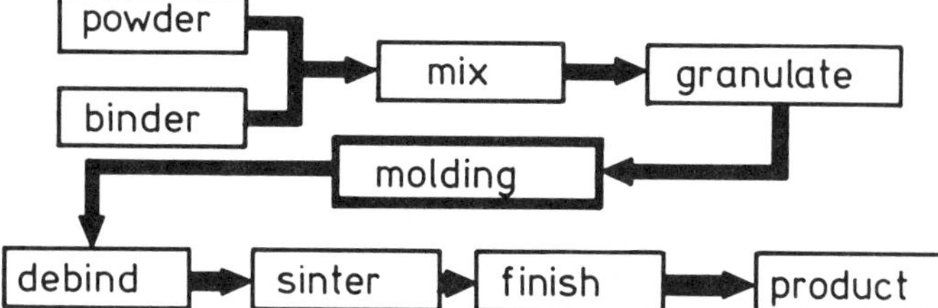

FIGURE 1 **A flow chart identifying the key steps in the powder injection molding process.**

The process of powder injection molding has been in incubation since the late 1920s. Early applications were in the forming of ceramic spark plugs, with more recent emphasis on metallic, composite, and cermet materials with complex shapes for high performance applications [4–6]. This interest accelerated as design engineers and parts fabricators realized the property and shape limitations of conventional powder processing routes. In the past 10 years major progress has been made in forming heat engine components, military hardware, aerospace components, and automotive components by powder injection molding [3,7–13]. Thus, in the recent past powder injection molding has moved into the forefront of advanced materials manufacturing.

Process Outline

As noted in Figure 1, the steps involved in forming a component by powder injection molding include the following: (1) selecting and tailoring a powder for the process, (2) mixing the powder with a suitable binder, (3) producing homogeneous granular pellets of mixed powder and binder, (4) forming the part by injection molding in a closed die, (5) processing the formed part to remove the binder (termed *debinding*), (6) densifying the compact by high temperature sintering, and (7) post-sintering processing as appropriate, including heat treatment or further densification. Figure 2 illustrates the key steps with special reference to the fabrication of aligned fiber composites.

Powders

Table 1 provides a summary of the desired properties for the powder used in powder injection molding. Small particles are used to aid sintering densification. These powders can be fabricated by a variety of techniques that

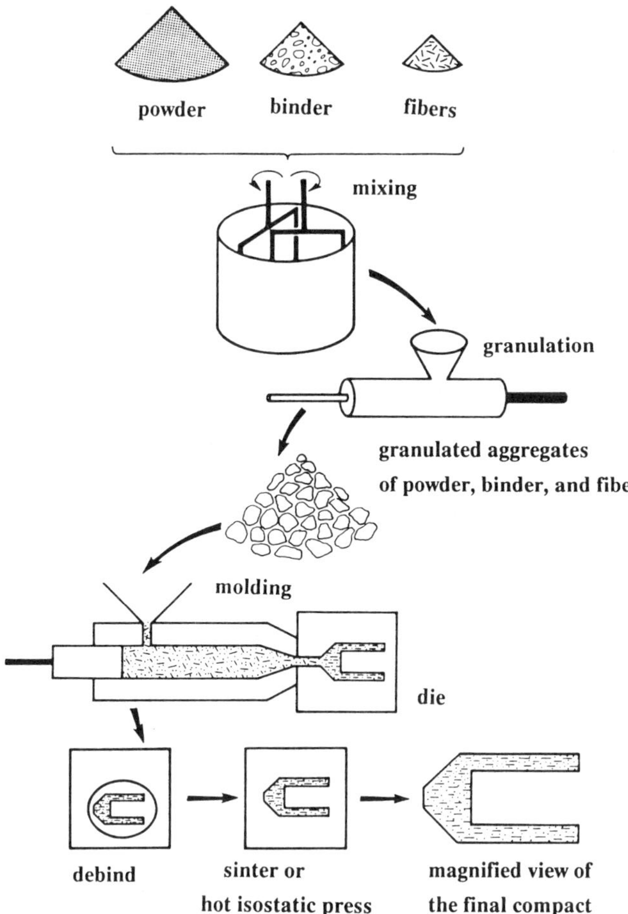

FIGURE 2 A schematic diagram of the processing steps and equipment involved in powder injection molding.

involve milling, chemical precipitation, oxide reduction, water atomization, gas atomization, or centrifugal atomization [14]. Additionally, novel metal powder production techniques are being developed to supply the needed small particles for injection molding. These approaches include plasma atomization and atomization of melts saturated with soluble gases. There is a fundamental difference between production of elemental powders, which tend to involve chemical processes such as precipitation or oxide reduction, and of prealloyed powders, which require atomization techniques. For the elemental powders, the small particle sizes are generated using close temperature and process concentration control to minimize particle growth after nucleation of the desired phase. The most popular approach is carbonyl decomposition, where a volatile metal carbonyl molecule is decomposed to form a spherical metal powder. Alternatively, with atomization approaches, the required small particles are generated using high energy delivery systems (such as high pressure gas jets) in the disintegration of the alloy melts. Details on the production of metal powders are provided in Refs. 14, 15, and 16.

Carbonyl iron and gas atomized stainless steel powders with spherical shapes and mean particle sizes of 4 to 15 μm are typical. Some progress has been made using particles as large as 100 μm, but compact shape distortion is a greater problem. In selecting a powder for the process, the goal is to attain a high packing density with the fewest processing problems. The shrinkage in sintering to full density depends on the particle packing density after binder removal. The lower the initial packing density, the greater the sintering shrinkage needed to attain the density level associated with high performance materials. Figure 3 shows a plot of the linear shrinkage versus the initial fractional powder packing density (solids content) for final fractional densities of 0.95 and 1.00. A key point is that less shrinkage occurs with high initial packing densities. This contributes to easier process and dimensional control. A low interparticle friction and spherical particle shape prove most useful for metal powder injection molding. Mixtures of differing particle sizes increase the packing density, especially when the particles are quite different in size.

Table 2 is a summary of some powders used in powder injection molding processing. Provided in this table are the relevant powder characteristics, such as particle size, particle shape, and packing density. These exam-

TABLE 1
Optimal Powder Characteristics

Tailored particle size distribution for a high packing density and low cost (mixture of lower cost large particles and higher cost small particles)
No agglomeration
Predominantly spherical (or equiaxed) particle shape
Sufficient interparticle friction to avoid distortion after binder removal; probably an angle of repose over 45°
Small mean particle size for rapid sintering, below 20 μm
Dense particles free of internal voids
Clean particle surface for predictable interaction with the binder
Minimized explosion and toxic hazards

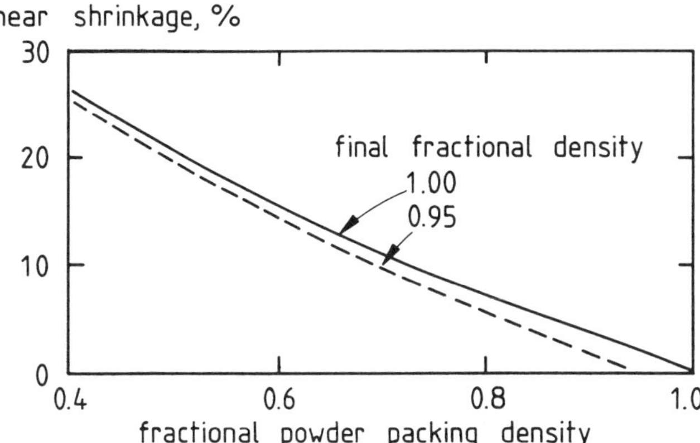

FIGURE 3 A plot of the linear shrinkage versus the initial powder packing density showing the amount of uniform shrinkage needed to attain final compact fractional densities of 0.95 or 1.00.

TABLE 2
Characteristics of Example Powders

Powder 1: 316L stainless steel
Fabrication approach = water atomization
Particle shape = rounded, irregular
Mean particle size = 6–8 μm
Pycnometer density = 7.8 g/cm^3
Tap density = 43%

Powder 2: 316L stainless steel
Fabrication approach = gas atomization
Particle shape = spherical
Mean particle size = 16 μm
Pycnometer density = 7.8 g/cm^3
Tap density = 63%

Powder 3: iron
Fabrication approach = carbonyl decomposition
Particle shape = spherical, slight agglomeration
Mean particle size = 4–6 μm
Pycnometer density = 7.6 g/cm^3
Tap density = 56%

Powder 4: tungsten heavy alloy (93% W, 7% Ni, 3% Fe)
Fabrication approach = reduced tungsten oxide mixed with carbonyl nickel and iron
Particle shape = cubic, polygonal tungsten
Mean particle size = 3 μm
Pycnometer density = 17.8 g/cm^3
Tap density = 37%

ples were selected to show a range of particle types. The first powder is a 316L stainless steel produced by high pressure water atomization. The particles are rounded, but somewhat irregular in shape, giving a tap density equal to 43% of the pycnometer density. In contrast, the second powder is also a 316L stainless steel, but in this case the particles are spherical. The spherical particle shape gives a packing density equal to 63% of the pycnometer density. Approximately 90% of the particles on a mass basis are below 35 μm in size for most powders used in powder injection molding, as shown in the particle size distributions given in Figure 4. The third powder is a low alloy steel formed by the carbonyl decomposition route, giving a sphericity index of 0.82 because of agglomeration. The angle of repose for this powder is 76°, indicative of a high interparticle friction. The near-spherical particle shape gives a tap density of 56% of the pycnometer density, while the apparent density is 35%. Solid contents as high as 68% can be formed in a well-mixed powder injection molding feedstock from this powder. The last powder is a mixture of reduced tungsten oxide and carbonyl nickel and iron, termed a W–Ni–Fe heavy alloy. The cubic particle shape (sphericity index of 0.47) for the tungsten constituent degrades the packing, and consequently considerable shrinkage is encountered in sintering.

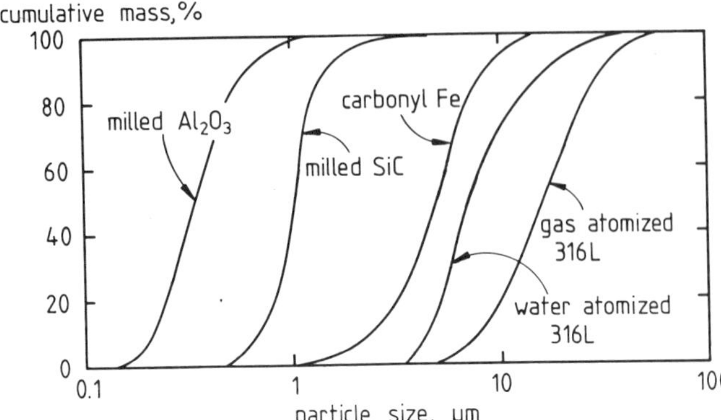

FIGURE 4 **The cumulative particle size distributions for some powders used in powder injection molding.**

Binders

Often the binder is a thermoplastic polymeric material, but water and various inorganic substances have been used successfully. A typical binder is composed of waxes and plastics, such as paraffin and polypropylene, with appropriate wetting agents to provide binder adhesion to the powder. The amount of binder ranges from 15 to 50 vol % of the mixture, depending on the powder packing characteristics. Since monosized spheres exhibit a random packing density of 64% of theoretical, a powder injection molding mixture using spheres will have over 36% by volume of binder. It is desirable to attain a high packing density of particles in the binder system while maintaining a low mixture viscosity; this dictates that there be sufficient binder to fill all interparticle voids and to lubricate particle sliding during molding. Best molding success is attained with mixture viscosities below 100 Pa·s. This viscosity is dependent on the inherent binder viscosity, mixture temperature, shear rate in molding, packing density of the powder, polar dispersants, and concentration of powder in the binder [17,18]. The loading of solid particles in the feedstock is an important molding consideration [19]. This parameter determines the amount of binder as well as the mixture viscosity and subsequent compact shrinkage in sintering. The amount of binder added to a powder should slightly exceed the void space between particles.

The binder is a temporary vehicle for homogeneously packing the powder into the desired shape and then holding the particles in that shape until the beginning of sintering. Accordingly, the binder cannot dictate the final composition, yet it has a major influence on the processing success. Binder compositions and debinding techniques are the main differences between the various powder injection molding processes.

There is no universal binder; the binder selected depends on the specific situation. Most binders contain a major phase that dictates the basic properties and several modifiers that are added to suit the particular application. Table 3 lists the attributes desired for an ideal binder. These attributes are subdivided into categories: flow characteristics, interaction with the powder, debinding characteristics, and manufacturing attributes. A primary requirement of the binder is that it allow flow and packing of the particles into the die cavity. It is mandatory that the binder wet the powder surface to aid in

TABLE 3
Desired Binder Attributes

Flow characteristics
Pure binder viscosity below 10 Pa·s at the molding temperature
Low viscosity change with temperature during molding
Strong and rigid after cooling
Small molecule to fit between particles
Powder interaction
Low contact angle and good adhesion with powder
Capillary attraction of particles
Chemically passive with respect to powder
Debinding
Multiple components with differing characteristics
Noncorrosive, nontoxic decomposition product
Low ash content, low metallic content
Decomposition temperature above molding and mixing temperatures
Manufacturing
Inexpensive and available
Safe and environmentally acceptable
Long shelf life, nonhydroscopic without volatile components
Not degraded by cyclic heating (reusable)
Lubricity
High strength and stiffness
High thermal conductivity
Low thermal expansion coefficient
Soluble in common solvents
Short chain length, no orientation

mixing and molding. Various chemicals are available to modify wetting behavior, and these are widely employed. Furthermore, it is important that the binder-powder mixture satisfy various rheological criteria that allow mixing and molding without defects. There is a viscosity range that proves successful in maintaining a dispersed powder without interfering with the molding process. This dictates a low molecular weight binder, which is typically a wax. Besides imparting a low viscosity at a high solids content, the binder must also inhibit separation or agglomeration of the solids. Generally, pure waxes prove inadequate in this respect, and the binder must include polymeric components.

A large viscosity increase is needed on cooling to hold the compact shape. If the viscosity is too low, then the binder and powder will separate during molding. Alternatively, too high a viscosity hinders mixing and molding. The shorter molecular chain length thermoplastic polymers generally satisfy the need for good forming. Also, shorter molecules minimize chain orientation in the molded component, which is important for attaining isotropic properties. Additionally, the powder and binder must be chemically passive with respect to each other. The powder should not polymerize or degrade the binder, and the binder should not corrode the powder.

The binder must be designed to minimize the debinding time and defects. Generally, a binder system that consists of several components that are not chemically intersoluble is beneficial. The multiple components allow for progressive extraction in debinding. This is beneficial because one component is removed to partially open pores. The remaining binder holds the particles in place and retains the compact shape during this first stage of debinding. Subsequently, the remaining binder vaporizes through the open pores without generating an internal vapor pressure that might cause compact failure. This sequencing of debinding stages allows for more rapid debinding than with a single-component binder.

Waxes are most frequently used in binders. These include paraffin, beeswax, and carnauba wax, as well as several waxlike short polymer chains of polyethylene or polypropylene. The waxes have low melting temperatures, good wetting, short molecular chain lengths, and low viscosities, and decompose with smaller volume changes than other polymers. Besides the major thermoplastic components, the binder will contain additives for lubrication, viscosity control, wetting, and debinding [20–23]. The most effective additives contain anchor groups that are strongly attached to the powder surface. The balance of the additive molecule should be soluble in the binder. In this respect, stearic acid or derivatives involving stearates are most important at low concentrations. These are low cost polar molecule additions that are widely used in powder processing. Table 4 provides a few example binder systems to illustrate the types of composition in routine use. The first six binders are thermoplastic systems, while the last (binder 7) is a thermosetting system, which has limited use since runners and sprues can not be recycled. Thermosetting binders offer an advantage in shape retention, but generally prove difficult to control since the viscosity changes continuously with time and temperature.

Powder–Binder Composition

It is necessary to decide on the composition of the powder–binder mixture for the feedstock. Too little binder results in a high viscosity that is difficult to mold. Also, a deficiency of binder creates voids in the mixture. During

TABLE 4
Example Binder Formulations

Binder 1
70% paraffin wax, 20% microcrystalline wax, 10% methyl ethyl ketone
Binder 2
67% polypropylene, 22% microcrystalline wax, 11% stearic acid
Binder 3
33% paraffin wax, 33% polyethylene, 33% beeswax, 1% stearic acid
Binder 4
69% paraffin wax, 20% polypropylene, 10% carnauba wax, 1% stearic acid
Binder 5
45% polystyrene, 45% vegetable oil, 5% polyethylene, 5% stearic acid
Binder 6
25% polypropylene, 75% peanut oil
Binder 7
65% epoxy resin, 25% wax, 10% butyl stearate

debinding, these voids cause cracking as a result of internal vapor buildup and lead to compact defects. Alternatively, an excess of binder is wasteful and slows subsequent processing; this will result in greater dimensional shrinkage during sintering. During molding, the excess binder will separate from the powder, leading to inhomogeneities in the molded compact and possible dimensional control problems. Furthermore, a binder excess will cause compact slumping, since the particles will settle or migrate during debinding. In the ideal case, the particles are in point contact with no voids in the binder. It is necessary that the binder fill all of the void space between the particles while maintaining a reasonably low viscosity [24,25]. In Figure 5, the theoretical and experimental density are plotted versus composition for powder injection molding feedstock mixtures. Any departure of the actual density from the theoretical density is a first indication of voids in the mixture. The point where this first occurs is the critical loading Φ_c. Simple experiments allow determination of this loading, and predictions of the critical loading are possible based on knowledge of the powder packing characteristics. In practice, the optimal loading for powder injection molding processing contains more binder than the critical content. As examples of common powder-binder compositions, the 316L stainless steel powders listed as powders 1 and 2 in Table 2 can be injection-molded at solids loadings of 45 and 55%, respectively. In contrast, powder 3 can be injection-molded at 64% solid if stearic acid is incorporated in the binder. Finally, the tungsten powder (number 4) forms an injection-moldable feedstock with a maximum of 49% powder by volume.

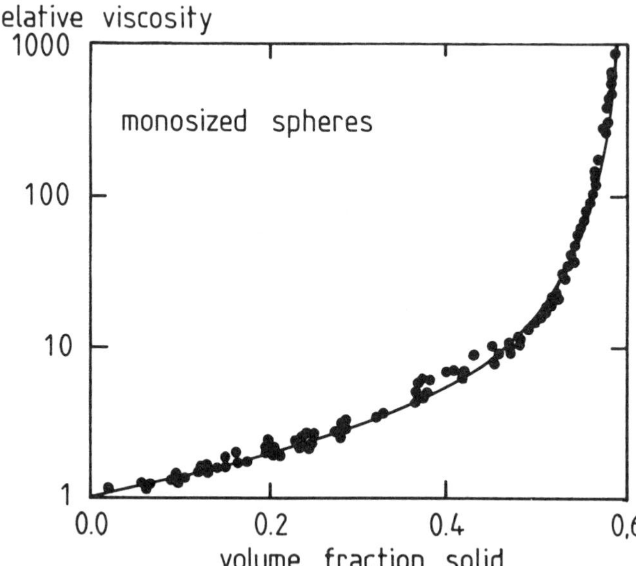

FIGURE 6 The relative viscosity on a logarithmic scale versus volume fraction of monosized spheres for a wide range of sphere sizes.

Viscosity

Mold filling is dependent on viscous flow of the feedstock into the die cavity. The most important property is the viscosity, which is a material parameter that relates the shear stress to the shear strain rate. A high viscosity makes molding more difficult. The viscosity of a powder injection molding mixture depends on temperature, shear rate, binder chemistry, powder interfacial chemistry, and solids loading. As a first estimate, the binder mixture should have a viscosity at the molding temperature below approximately 10 Pa·s. However, the viscosity is much higher when powder is added to the binder. Figure 6 shows an example of this behavior using monosized glass spheres [26]. In Figure 6 the viscosity of the mixture is divided by the viscosity of the pure binder to give the relative viscosity. There is a limiting solids loading where the relative viscosity becomes essentially infinite and the mixture is too stiff to be considered viscous. For monosized spheres, this limiting loading corresponds to the dense random packing condition with 63.7% solid.

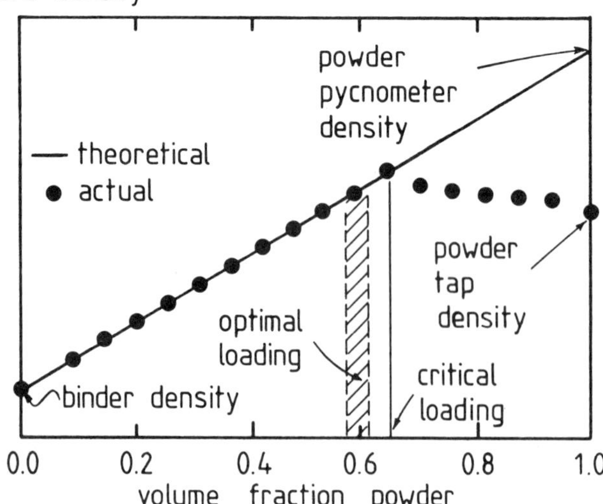

FIGURE 5 The density of powder injection molding feedstock mixtures as a function of the composition, showing the determination of the critical and optimal compositions.

The most appropriate model for the change in relative viscosity η_r (mixture viscosity divided by the binder viscosity) with relative solids loading Φ_r (solids loading divided by the critical loading) is as follows [27–29]:

$$\eta_r = A(1 - \Phi_r)^{-n} \tag{1}$$

Generally, the exponent n is 2.0. The coefficient A contains factors such as the shear rate sensitivity and particle size effects. This equation gives a relative viscosity of unity as the solids loading approaches zero. Additionally, it approaches infinite viscosity as the solids loading approaches the maximum value. Figure 7 illustrates the application of Eq. (1) to the viscosity results for several systems.

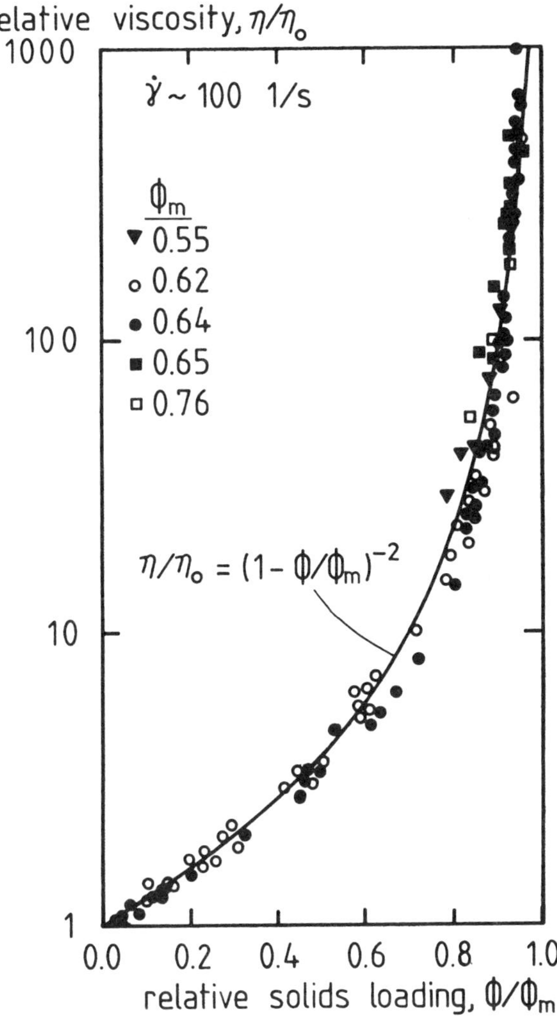

FIGURE 7 A plot of the normalized viscosity results from several studies, showing general agreement with the form predicted by Eq. (1).

Molding

The equipment used in shaping the compact is the same as that used for polymer injection molding. A typical molding machine is sketched in Figure 8. It consists of a hydraulically clamped die that is filled through a gate from a pressurized and heated barrel. A motor-driven reciprocating screw is employed to maintain mixture homogeneity and to generate the pressure needed to fill the die. The feedstock enters as granules from the loading hopper. Proper control of the molding operation is attained using a closed-loop feedback system. A schematic cross section through the die area is given in Figure 9, with the sequence of molding steps outlined in Figure 10. The first steps are die clamping and filling. A high pressure is attained rapidly and maintained until the compact becomes rigid. After sufficient cooling in the die, the compact is ejected and the cycle repeated. Figure 11 is a sketch of the pressure versus time during the actual molding cycle (approximately 20 seconds). The initial pressure rise causes mold filling in a very short time. The maximum pressure (usually less than 15 MPa) is held until the gate has chilled. Example forming cycles are given in Table 5 for alumina, silicon nitride, and steel components.

The injection molding step involves concurrent heating and pressurization of the feedstock. Because of the high thermal conductivity of powder injection molding mixtures, the molding process requires careful attention to several variables. Much progress in computer modeling and control is occurring to minimize defects that might arise in molding [30].

Debinding

After molding, the binder is removed from the compact by a process termed *debinding*. As illustrated in Figure 12, there are several options for debinding based on thermal, solvent, and capillary extraction approaches [31,32]. A basic comparison of some industrial debinding cycles

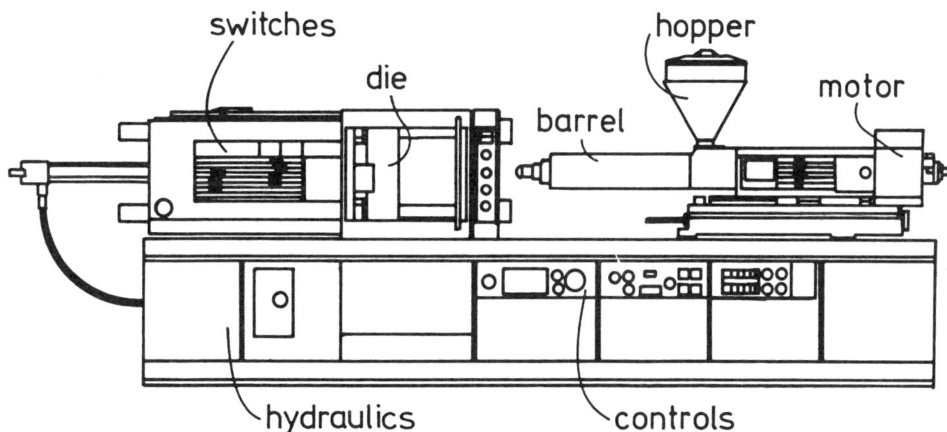

FIGURE 8 A sketch of an injection molding machine as used for powder injection molding, showing the key components.

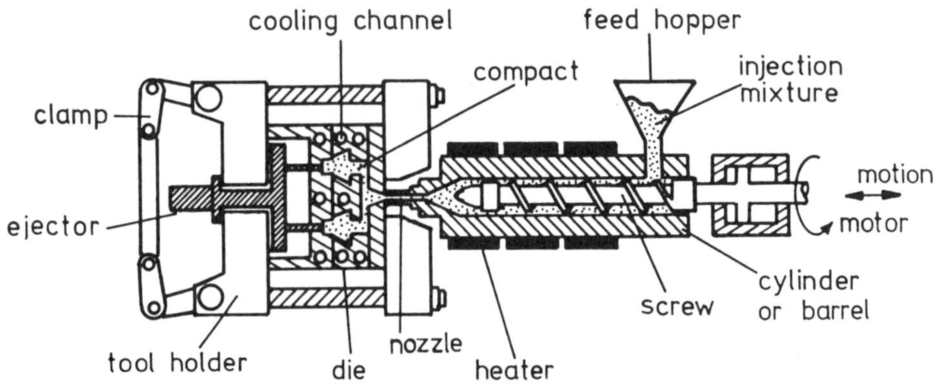

FIGURE 9 A cross-sectional view of the die and barrel areas during molding.

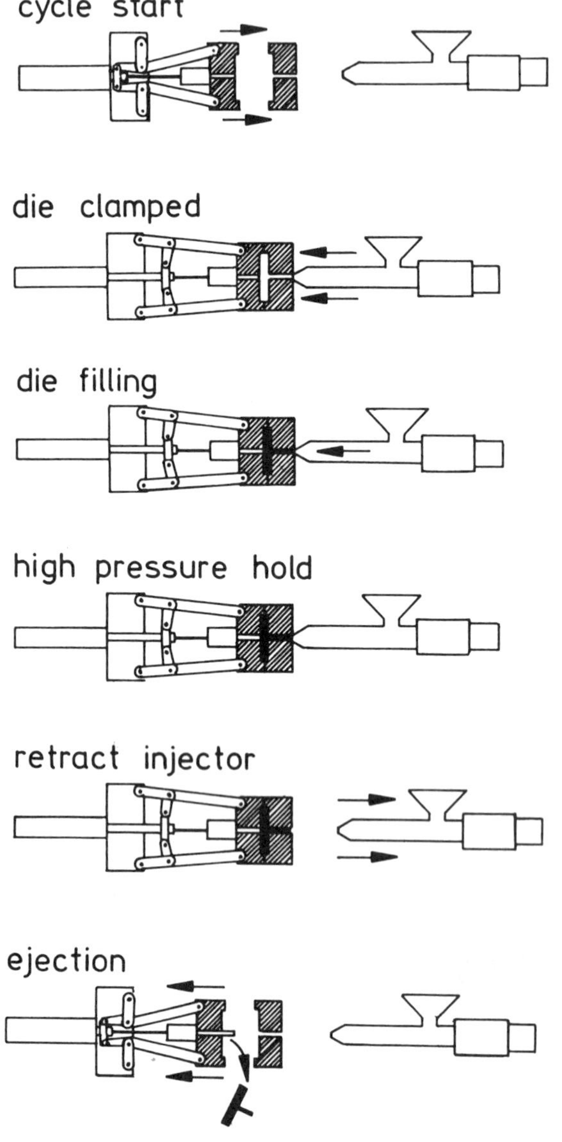

FIGURE 10 The molding sequence, showing the key points in the molding cycle.

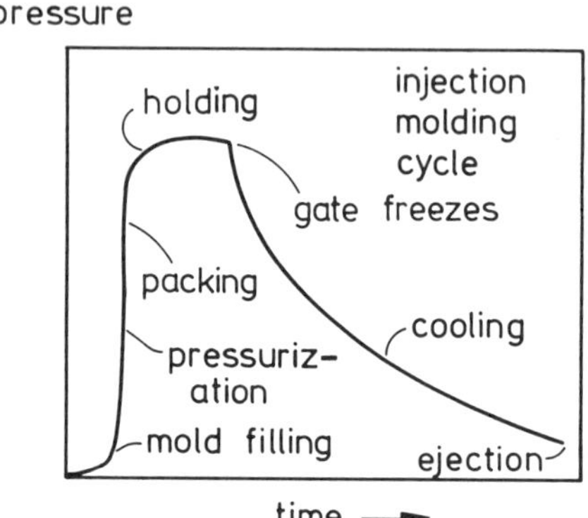

FIGURE 11 The pressure versus time during the powder injection molding cycle; a typical peak pressure might be 15 MPa and the cycle time might be 20 seconds.

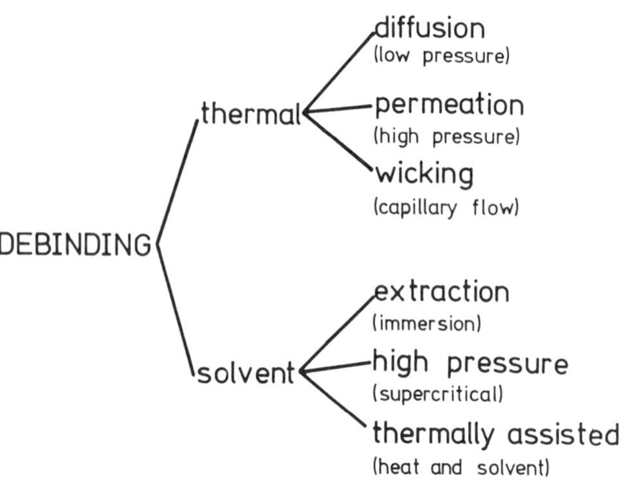

FIGURE 12 The options available for debinding powder injection molding components include both solvent and thermal decomposition steps.

TABLE 5
Example Molding Conditions

Component	Cutter	Rotor	Trigger
Material	Al_2O_3	Si_3N_4	Fe–2% Ni
Weight (g)	100	70	40
Cylinder temperature (°C)	80–110	100–140	130–180
Nozzle temperature (°C)	110	145	175
Melt temperature (°C)	125	158	185
Mold temperature (°C)	40	35	40
Screw diameter (mm)	35	40	20
Screw speed (min^{-1})	65	50	35
Injection pressure (MPa)	13	14	20
Packing pressure (MPa)	5	7	8
Filling time (s)	0.3	0.4	0.6
Packing time (s)	15	12	3
Cooling time (s)	10	40	18
Cycle time (s)	40	60	37

is given in Table 6. For metals, thermal debinding is used most frequently, where the compact is heated slowly to 600°C in air to decompose the binder and oxidize the particles. This oxidation process provides some handling strength to the compact after the binder is removed. Several defects can arise in the debinding process, especially if the heating rate is more than a few kelvins per hour. A multiple-component binder helps overcome such problems by leaving one component in the compact to hold the particles in place while a lower stability component is removed by evaporation. Alternatively, use of capillary wicking and solvent extraction processes decrease the debinding time, but these are less widely applied as yet. These also rely on multiple-component binder systems to progressively open the pore structure while holding the particles in place. Solvent extraction is a rapid technique for the first step of debinding. It has the advantage of speed and good compact shape retention. Example solvents include acetone, ethylene dichloride, and carbon tetrachloride.

TABLE 6
Example Debinding Cycles

Oxidation
- Applicable to wax-based binder
- Heat 5 K/h to 150°C in air
- Isothermal hold up to 20 hours
- Heat to 600°C for 5 hours

Wicking
- Applicable to thermoplastic binders
- Embed compact in submicrometer alumina powder
- Heat in hydrogen atmosphere in 1 K/min to 100°C
- Isothermal hold for 2 hours
- Heat 5 K/min to 500°C
- Isothermal hold for 2 hours

Supercritical extraction
- Applicable to thermoplastic binders
- Heat in Freon vapor at 10 K/min to 600°C
- Maintain pressure of 10 MPa for 6 hours

Vacuum sublimation
- Applicable to water-based binders
- Cool compact to − 20°C
- Isothermal evacuation for 8 hours
- Heat in hydrogen to 600°C for 2 hours

Solvent extraction
- Applicable to wax–polymer–oil binders
- Immerse compact in ethylene dichloride at 50°C for 6 hours
- Heat in reducing atmosphere to 600°C for 4 hours

Drying
- Applicable to gelation binders
- Heat in air to 60°C for 20 hours
- Heat to 110°C for 4 hours
- Heat in reducing atmosphere to 600°C for 4 hours

Sintering

At the end of debinding, some binder may remain in the pores to provide handling strength. The next step is *sintering*, which can be incorporated directly into the debinding cycle. Sintering provides strong interparticle bonds and removes the void space by densification. Isotropic powder packing allows for predictable and uniform shrinkage, so tooling is oversized as appropriate for the final compact dimensions and powder packing density. Computer techniques are being developed to handle tool design to account for the sintering shrinkage [33]. Also, additives are used to enhance the sintering behavior. As an example, grain growth control during the sintering of alumina is achieved by the addition of 0.1% magnesia. Likewise, vanadium carbide is added to the tungsten carbide–cobalt cemented carbide to control sintered grain size. In stainless steels, silicon forms a liquid phase that aids densification. Examples of sintering cycles and additives for some sintered injection-molded materials are given in Table 7. The furnaces used to process the compacts are both batch and continuous types with atmospheres of hydrogen, nitrogen, argon, or mixtures of these. In some instances vacuum is used with the batch furnaces. Figure 13 provides a sketch of the basic layouts of both types of furnaces. After sintering, the compact has good strength and microstructural homogeneity, with isotropic properties superior to those available with many other processing routes.

TABLE 7
Example Sintering Schedules

Material	Sintering Aids	Particle Size (μm)	Atmo-sphere*	Dew Point (°C)	Heating Rate (K/min)	Hold Temperature (°C)	Maximum Temperature (°C)	Time (min)
Al_2O_3	MgO	0.3	H	—	10	—	1600	240
316L	—	15	V, H	−40	5–10	1000	1360	90
SiC	B, C	0.2	A, V	—	10–50	1700	2100	60
Si_3N_4	Y_2O_3	0.5	N	−40	10	—	1750	120
Steel	—	4	A, H	−20	15	870	1250	60
W–Ni–Fe	—	3	H	10	5–10	1000	1500	30
WC–Co	VC	0.5	V	—	2–10	700	1400	60

* V = vacuum, H = hydrogen, N = nitrogen, A = argon

Basic Attributes

Powder injection molding is capable of producing a wide range of components from powders. A main attraction is the economical production of complex parts from high performance engineering materials. The Venn diagram shown in Figure 14 reinforces this concept. The three basic considerations are shown as overlapping circles. The intersection of these three circles is an attractive area for the application of powder injection molding.

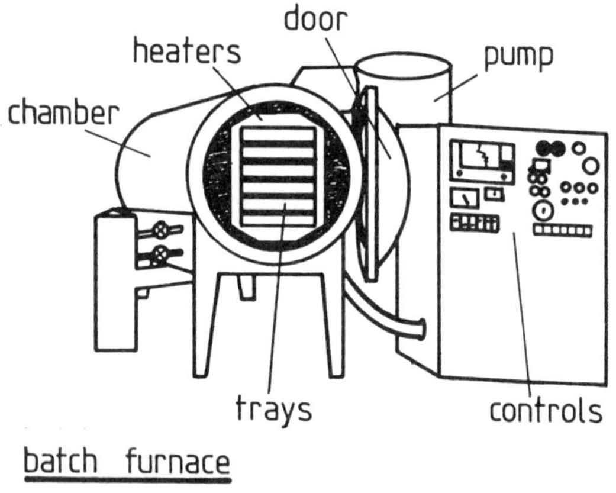

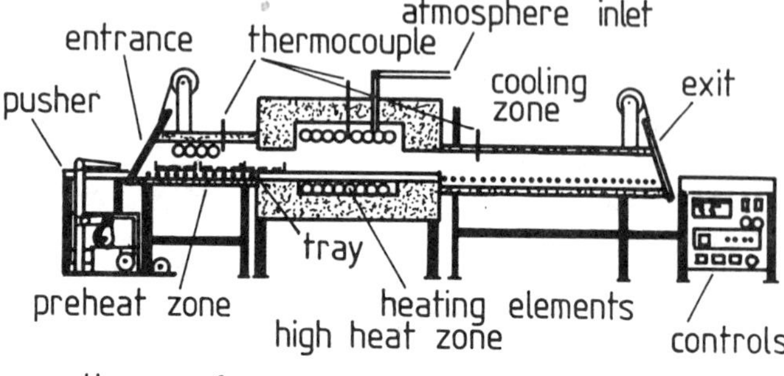

FIGURE 13 Sketches of typical batch and continuous sintering furnaces.

Because of the high final density, the powder injection molding products are suitable for high performance levels. Examples of these attributes for powder injection molding are illustrated in Figure 15, which provides photographs of a few compacts formed by powder injection molding.

Besides the primary advantages of shape complexity, low cost, and high performance, several secondary attributes are worth notice. These include near-net-shape geometries with concomitant material and processing savings through the recycling of wastes. The optimum part and mold design may be modeled by methods already established for thermoplastics [34]. Also, capital equipment costs are relatively low in contrast with other shaping technologies. Finally, the process has a high level of automation, yet is applicable to short production runs.

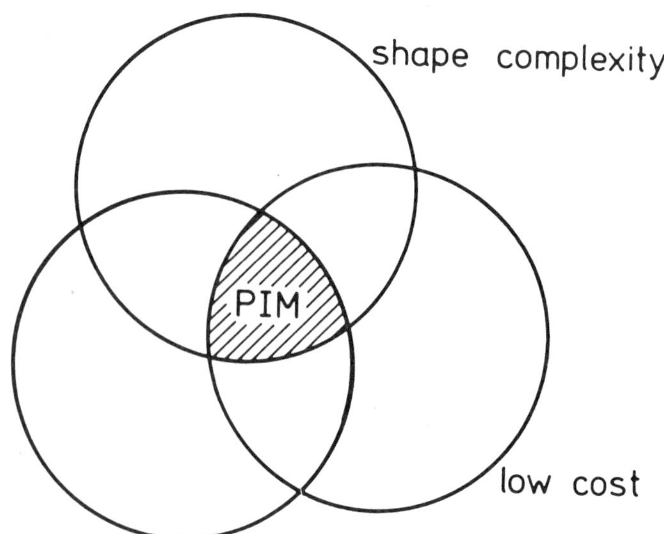

FIGURE 14 This Venn diagram identifies the optimal application of powder injection molding with the combination of concerns over processing costs, performance levels, and component shape complexity.

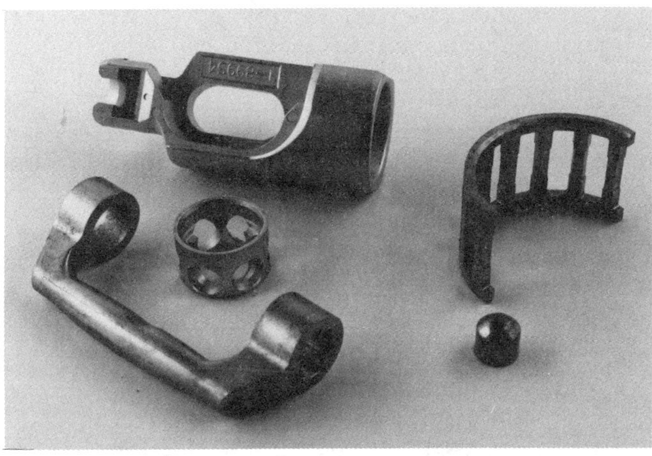

FIGURE 15 Some examples of the complex shapes being formed by the metal powder injection molding process. (Courtesy of Brunswick Corp. and Multimaterials Molding Corp.)

The high final density and good properties make powder injection molding a natural choice for the fabrication of high performance components. Table 8 provides a summary of a few systems and the measured mechanical properties. These properties are competitive with those produced by alternative fabrication approaches. The density and property advantages result from the uniform powder packing attained in the powder injection molding process. For comparison, consider the pressure gradients in die compaction [14,35]. The wall friction between the powder and the die induces pressure gradients during compaction that lead to a nonuniform powder density. In die compaction of metal powders, the friction between the powder and the die wall causes the applied pressure to decrease with depth in the powder bed. There are many intrinsic powder and processing factors that affect this behavior, including the powder size, shape, lubricant, oxide coating, die wall surface finish, and material attributes like the work hardening coefficient. However, in all cases the die wall friction causes pressure gradients, which in turn give green density gradients. Dimensional change (typically shrinkage) during sintering depends on the powder packing density; low density regions shrink more than high density regions. Accordingly, die-compacted powders exhibit nonuniform shrinkage, which creates problems with dimensional control. For this reason, die-compacted metal powders are often sintered at lower temperatures where densification is minimized, necessarily giving inferior sintered properties.

The porosity of the final powder injection molding product is the primary determinant of its properties; a lower porosity leads to improvements in such attributes as strength, toughness, ductility, conductivity, and magnetic response [36,37]. The hydrostatic forming character of powder injection molding minimizes density gradi-

TABLE 8
Example Mechanical Properties for Powder-Injection-Molded Materials

Material	Fractional Density (%)	Yield Strength (MPa)	Failure Elongation (%)
Alumina	99	510	—
304L stainless	96	185	35
316L stainless	98	205	68
17-4 PH stainless	96	900	—
Silicon carbide	95	300	—
Silicon nitride	98	740	—
Fe	95	100	28
Fe 2% Ni steel	93	185	33
Fe 2% Ni steel HT	93	1090	4
Fe 8% Ni 1% C steel	96	970	5
Fe 50% Ni	96	170	21
Fe 2% Cu steel	95	335	23
Fe 3% Si	99	345	25
90% W heavy alloy	99	600	32
97% W heavy alloy	99	710	8
WC–10% Co	99	1140	—

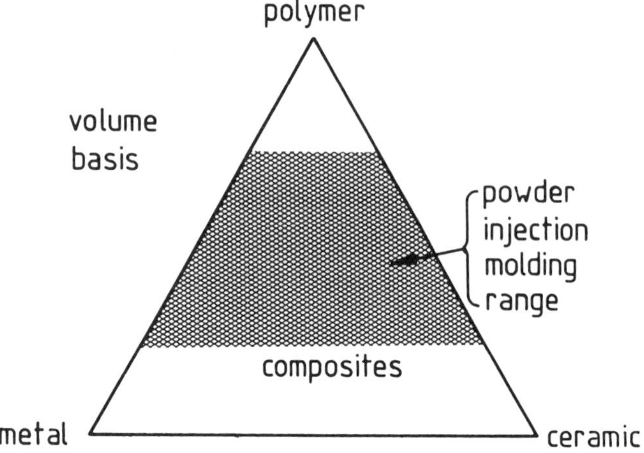

FIGURE 16 A ternary diagram composed of the three main structural material groups—metals, ceramics, and polymers. The materials involved in powder injection molding are by definition composite materials.

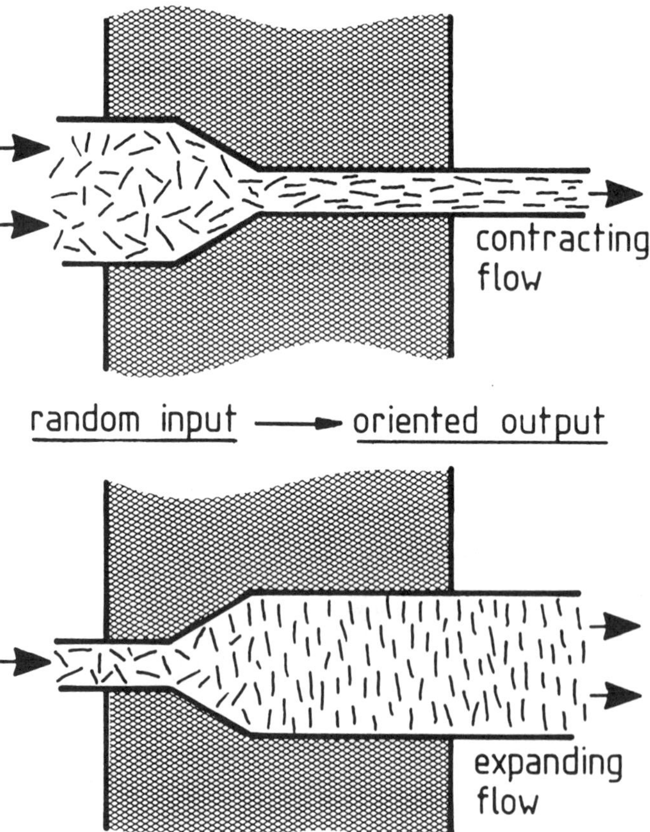

FIGURE 17 A contrast in the fiber orientation with molding gate design for composites formed by powder injection molding techniques.

ents, which allows more predictable sintering densification and higher properties. Generally the sintered density exceeds 93% of theoretical, as compared with the 85% density attained in die compaction. Furthermore, the pores in powder injection molding compacts are not interconnected and are less detrimental to properties. Consequently, powder injection molding compacts have very attractive properties [7–12,38–43].

Powder injection molding is applicable to a wide range of materials, including steels, nickel, most ceramics, cemented carbides, tungsten heavy alloys, niobium, silicon nitride, silicon carbide, superalloys, aluminide intermetallics, cobalt alloys, and fiber-reinforced metals and ceramics. One area of current research is the fabrication of composite materials by powder injection molding techniques. As Figure 16 illustrates, powder injection molding applicability falls naturally into that range termed composites. From this simple ternary diagram, with polymer, metal, and ceramic as the terminal points, it is evident that injection molding can be used to form a variety of composites. This includes metal matrix, ceramic matrix, and polymer matrix materials with dispersed fibers or particles of ceramic or metal. Figure 17 shows how the flow pattern in molding can be adjusted to place fibers into specific orientations in the compacts. This capability opens a new area of tailored composite microstructures.

One important attribute is the application of powder injection molding to the emerging high performance materials. These materials have a strong dependence on processing. Although several routes are available for producing the new materials, powder injection molding approaches are attractive because of the combination of technology, cost, and feasibility [44]. Another advantage of powder injection molding is the co-molding of different materials, that is, forming part of a component from one material and then another portion from a second material. Such components can be joined together in the green condition [45]. This option has merit for forming corrosion barriers, wear surfaces, and electrical interconnections in ceramics.

For the producer of powder-injection-molded parts, some important attributes are related to manufacturing ease, including process control, flexibility, and automation. Small production runs are possible, with as few as 5000 parts being economical. This flexibility fits well with the emerging direction in manufacturing, especially for the advanced ceramics [46]. There is always a desire to make rapid shifts in production with minimum expense. In this regard, powder injection molding is attractive, since much of the processing equipment remains invariant while the powder composition and tooling change. The process is an alternative to investment casting and die casting for metals, and to slip casting and cold isostatic compaction for ceramics, with shape being less of a cost factor in manufacturing.

As with all technologies, the bottom line is the economics. Here, powder injection molding can be defended as cost-advantageous for the more complex shapes and specialty materials. One advantage is powder injection molding's reduced number of production steps as compared with other techniques [39,47]. This is partly due to the elimination of secondary operations like grinding, drilling, and boring that typically are required for precision components. Also, since the feedstock material can be recycled, material use is nearly 100%. This is especially important for costly raw materials, such as refractory metals, specialty ceramics, and precious metals [12,38,48,49].

Process Limitations

The previous section outlined several attributes of powder injection molding. Generally, it is viable for all shapes that can be formed by plastic injection molding. Still, it is not cost-competitive with traditional die compaction for shapes with relatively simple or axial-symmetric geometries. Also, the availability of suitable powders limits the current materials selection. Other limitations can be traced to equipment size and sophistication. Large components require larger molding and sintering devices, which prove more difficult to control. Accordingly, powder injection molding is best applied to complex, small shapes [50].

The formation of thick cross sections remains a problem with powder injection molding because binder removal times scale with the square of the section thickness [31]. The upper limit on section thickness is approximately 100 mm with a total component volume below 100 cm^3. Section thicknesses as low as 1 mm have been formed by powder injection molding. Dimensional tolerances are typically in the range from 0.1 to 0.5% of the section thickness. For better dimensional control, machining or coining is required after sintering, which adds to the cost. Also, density gradients can arise during mold filling as a result of uneven flow velocities. This is especially a problem when the feedstock sees a series of thickness variations during mold filling. These density gradients cause subsequent component warpage during sintering. Thus, rapid changes in section thickness should be avoided, and if possible the variation in thickness should be held within a factor of 2. Microstructural flaws induced during molding limit the mechanical properties of the final product [10,51–53].

Growth Patterns

Early experience with powder injection molding established high production yields with good tolerance. As the engineering community better appreciates these attributes, powder injection molding will continue to grow. Although the current industry is relatively small, the anticipated growth is quite impressive [6]. Recent reports suggest growth rates as high as 50% per year, or a doubling of the sales approximately every 1.75 years. For this reason, many new ventures and development efforts are underway. Although knowledge concerning powder injection molding is restricted to a relatively small number of companies, there are similarities among established powder injection molding processes. These similarities allow careful study of the underlying science and should lead to a wider understanding of powder injection molding.

Many of the markets for powder injection molding products have been developed. They include medical and dental devices, office equipment and instrumentation parts, high temperature aerospace parts, printed circuit substrates, metalworking tools, electrical and electronic materials, military hardware, aircraft components, household appliance parts, machining tools, computer peripheral components, camera parts, firearm components, and high temperature ceramic turbines [3,6].

One of the critical issues facing those involved with powder injection molding is the lack of adequately trained personnel. Indeed, all appearances indicate that the growth rate will be dependent on the ability to engineer specific systems and solutions for the many identified applications for powder injection molding. This translates into a need for engineers who can work comfortably with this new technology. Likewise, there is a need for more research on the process basics associated with molding, powder characteristics, binder formulation, automation, and sintering.

Acknowledgment

Much of the information in this article is based on the current research at Rensselaer being performed as part of the Advanced Powder Processing program in the Center for Manufacturing Productivity and Technology Transfer.

Randall M. German

References

1. S. K. Bhattacharya, Ed., *Metal-Filled Polymers,* Marcel Dekker, New York, 1986.
2. I. L. Kamel, in E. Klar, Ed., *Metals Handbook,* Vol. 7, 9th ed., American Society for Metals, Metals Park, OH, 1984, pp. 606–613.
3. R. M. German, *Powder Injection Molding,* Metal Powder Industries Federation, Princeton, NJ, 1990.
4. K. Schwartzwalder, *Ceramic Bull., 28,* 459 (1949).
5. B. Haworth and P. J. James, *Metal Powder Rept., 41,* 146 (1986).
6. L. F. Pease, *Inter. J. Powder Met., 22,* 177 (1986).
7. T. J. Whalen and C. F. Johnson, *Ceramic Bull., 60,* 216 (1981).
8. P. A. Willermet, R. A. Pett, and T. J. Whalen, *Ceramic Bull., 57,* 744 (1978).

9. W. Engle, E. Lange, and N. Muller, in J. J. Burke, E. N. Lenoe, and R. N. Katz, Eds., *Ceramics for High Performance Applications—II*, Brook Hill Publishing Co., Chestnut Hill, MA, 1978, pp. 527–538.
10. C. F. Johnson and T. G. Mohr, in J. J. Burke, E. N. Lenoe, and R. N. Katz, Eds., *Ceramics for High Performance Applications—II*, Brook Hill Publishing Co., Chestnut Hill, MA, 1978, pp. 193–205.
11. B. C. Mutsuddy and D. K. Shetty, *Technical Aspects of Critical Materials Use by the Steel Industry*, vol. IIB, Report NBSIR 83-2679-2, National Bureau of Standards, Washington, DC, 1983, pp. 43.1–43.28.
12. R. S. Storm, R. W. Ohnsorg, and F. J. Frechette, *J. Eng. Power, 104*, 601 (1982).
13. P. K. Johnson, *Inter. J. Powder Met. Powder Tech., 15*, 323 (1979).
14. R. M. German, *Powder Metallurgy Science*, Metal Powder Industries Federation, Princeton, NJ, 1984.
15. F. V. Lenel, *Powder Metallurgy Principles and Applications*, Metal Powder Industries Federation, Princeton, NJ, 1980.
16. E. Klar, Ed., *Metals Handbook*, Vol. 7, 9th ed., American Society for Metals, Metals Park, OH, 1984.
17. R. L. Mackey and B. R. Patterson, *Prog. Powder Met., 43*, 843 (1987).
18. J. A. Mangels and W. Trela, in J. A. Mangels and G. L. Messing, Eds., *Forming of Ceramics*, American Ceramic Society, Columbus, OH, 1984, pp. 220–233.
19. J. Warren and R. M. German, in P. U. Gummeson and D. A. Gustafson, Eds., *Modern Developments in Powder Metallurgy*, Vol. 18, Metal Powder Industries Federation, Princeton, NJ, 1988, pp. 391–402.
20. M. Kimoto and S. Uchida, *J. Japan Soc. Powder Powder Met., 34*, 369 (1987).
21. G. Y. Onoda, in G. Y. Onoda and L. L. Hench, Eds., *Ceramic Processing before Firing*, John Wiley and Sons, New York, 1978, pp. 235–251.
22. B. Haworth and P. J. James, *Metal Powder Rept., 41*, 146 (1986).
23. B. Mutsuddy, *Indust. Res. Develop.*, July 1983, pp. 76–80.
24. J. A. Mangels, in J. J. Burke, E. N. Lenoe, and R. N. Katz, Eds., *Ceramics for High Performance Applications—II*, Brook Hill Publishing Co., Chestnut Hill, MA, 1978, pp. 113–130.
25. R. E. Wiech, Jr., U.S. Patent 4,445,936, (May 1, 1984).
26. R. J. Farris, *Trans. Soc. Rheol., 12*, 281 (1968).
27. T. B. Lewis and L. E. Nielsen, *Trans. Soc. Rheol., 12*, 421 (1968).
28. A. B. Metzner, *J. Rheol., 29*, 739 (1985).
29. D. I. Lee, *Trans. Soc. Rheol., 13*, 273 (1969).
30. D. Lee, K. F. Hens, B. O. Rhee, and C. M. Sierra, in P. U. Gummeson and D. A. Gustafson, Eds., *Modern Developmens in Powder Metallurgy*, Vol. 18, Metal Powder Industries Federation, Princeton, NJ, 1988, pp. 417–429.
31. R. M. German, *Inter. J. Powder Met., 22*, 237 (1987).
32. M. Kimoto and S. Uchida, *J. Japan Soc. Powder Powder Met., 34*, 369 (1987).
33. C. M. Sierra and D. Lee, *Powder Met. Inter., 20*, 28 (1988).
34. E. C. Bernhardt, Ed., *Computer Aided Engineering for Injection Molding*, Hanser Publ., Munich, 1983.
35. R. A. Thompson, *Ceramic Bull., 60*, 237 (1981).
36. R. Haynes, *Rev. Deform. Behav. Mater., 3*, 1 (1981).
37. G. F. Bocchini, *Rev. Powder Met. Phys. Ceram., 2*, 313 (1985).
38. E. Lang and N. Muller, *Powder Met. Inter., 18*, 416 (1986).
39. R. Billiet, *Proceedings P/M-82*, Associazione Italiana die Metallurgia, 1982, pp. 603–610.
40. P. C. Chen and C. K. Lim, *Prog. Powder Met., 39*, 153 (1983).
41. T. S. Wei and R. M. German, *Inter. J. Powder Met., 24*, 327 (1988).
42. G. D. Schnittgrund, *SAMPE Quart.*, July 1981.
43. E. Lange and N. Muller, *Powder Met. Inter., 18*, 416 (1986).
44. A. Bose and R. M. German, in P. U. Gummeson and D. A. Gustafson, Eds., *Modern Developments in Powder Metallurgy*, Vol. 18, Metal Powder Industries Federation, Princeton, NJ, 1988, pp. 299–314.
45. A. R. Erickson and R. E. Wiech, Jr., in E. Klar, Ed., *Metals Handbook*, Vol. 7, 9th ed., American Society for Metals, Metals Park, OH, 1984, pp. 495–500.
46. T. L. Francis, *Powder Met. Inter., 17*, 185 (1985).
47. R. J. Waikar and B. R. Patterson, in W. A. Kaysser and W. J. Huppmann, Eds., *Horizons of Powder Metallurgy*, Part II, Verlag Schmid, Freiburg, Federal Republic of Germany, 1986, pp. 661–665.
48. R. Billiet, *Prog. Powder Met., 38*, 45 (1982).
49. R. Billiet, *Inter. J. Powder Met. Powder Tech., 2*, 119 (1985).
50. J. D. Destefani, *Adv. Mater. Proc., 134*(3), 71 (1988).
51. D. W. Richerson, J. R. Smyth, and K. H. Styhr, *Ceramic Eng. Sci. Proc., 4*, 841 (1983).
52. J. A. Mangels, in J. J. Burke, E. N. Lenoe, and R. N. Katz, Eds., *Ceramics for High Performance Applications—II*, Brook Hill Publishing Co., Chestnut Hill, MA, 1978, pp. 113–130.
53. A. Nagel, G. Wingefeld, D. Agranov, and G. Petzow, in W. A. Kaysser and W. J. Huppmann, Eds., *Horizons of Powder Metallurgy*, Part II, Verlag Schmid, Freiburg, Federal Republic of Germany, 1986, pp. 636–640.

Molding, Polymer Injection

For efficient and economic production of discrete parts with any material, it is always desirable to obtain the final product using one process alone, without the need for secondary operations. This is so-called net-shape manufacturing, and injection molding is perhaps the best net-shape manufacturing process for materials that can be handled by this process. For polymers and some of their composites, injection molding can mass-produce high precision parts with complex geometries at very low cost. In engineering applications, some injection-molded plastic parts of intricate shape with tight tolerance can be snap fit together for easy assembly, while others can achieve extremely high quality to meet the optical functional requirements for such parts as optical lenses and information-storage disks.

Since the oil crisis in the early 1970s, the use of polymers as an engineering material has been increasing rapidly in order to improve the energy efficiency of both the processes themselves and the final products. Plastics today still constitute one of the world's fastest growing industries. Of all the polymer-processing methods, injection molding and extrusion each account for approximately one-third by weight of all the materials processed [1]. Injection molding achieves its prominent position because of the following major advantages [2].

- It provides high volume and high production rates.
- It has a low labor cost and is amenable to automation.
- Little or no finishing is required.
- Many different surface finishes, patterns, and colors can be produced.
- Scrap can be reused in most applications.
- Small parts can be produced to extremely high precision.
- Fillers can be added for better mechanical and physical properties or decreased material cost.
- Metal inserts can be molded into the parts.

In addition, polymeric materials and their composites offer various desirable properties, including high strength-to-weight ratio, corrosion resistance, and clarity, among others.

On the other hand, the injection-molding process is not suitable for low volume production because the costs for the machine and its auxiliary equipment and tooling are high. Moreover, the lack of fundamental understanding from a scientific viewpoint has made the operation (from mold design to process control) unpredictable. The complex properties of polymeric materials make the situation even more difficult than for most other materials. As a result, for a long time injection molding has been essentially an art rather than a science. Mold design has depended heavily on the designer's prior experience with particular types of products or materials; setting up processing parameters has required extensive trial-and-error experiments to achieve an acceptable product. Further, the industry has been dominated by small custom molders, who operate over 50% of the injection molding machines [3]. Little research effort had in fact been focused on the subject until the early 1970s.

In order to make rational decisions based on scientific principles in the injection molding operation, one has to understand and be able to analyze mathematically the flow and solidification of the polymer melt in the mold throughout the entire molding operation, the so-called molding dynamics, as it is called on p. 2 of Ref. 3. As reviewed in chap. 2 of Ref. 3, the first serious attempt at simulating the mold-filling stage was attributed to Ballman et al. [4] in 1959; they modeled the filling of a long, thin, rectangular cavity with polystyrene, but in doing so they employed some rather severe simplifying assumptions. More detailed simulation models were subsequently undertaken by many investigators, such as Kamal and Kenig [5] in 1972 and Tadmor et al. [6] in 1974. Considerable effort was also devoted during this period to experimentally investigating such properties as molecular orientation of injection-molded parts by Wales et al. [7] in 1972, and Menges and Wubken [8] in 1973. A more organized and well-focused research effort covering all aspects of injection molding was launched in 1974 as the Cornell Injection Molding Program (CIMP) at Cornell University. Its publications include 14 volumes of annual Progress Reports since 1975, which consist of its own original work as well as literature surveys of others. In addition to many publications on molding practice [9,10,11], a few recent books consider computer applications [12,13] as well as fundamental aspects [14] of injection molding. It appears, then, that the research effort during the past two decades has resulted in a reasonable basis for scientifically attacking injection molding problems.

In this chapter, after a brief description of the process and machine, the emphasis will be placed on the most recent developments of the technology from a fundamental viewpoint as well as future trends. In particular, a theoretical basis for simulating molding dynamics and for developing computer applications and process-control strategies will be presented.

The Injection Molding Process and Machines

Fundamentally, injection molding is a thermal-mechanical process that first converts a polymer resin from a solid pellet form into a highly viscous fluid under so-called plastication through an extruder barrel, which is a major component of injection molding machines. The resin absorbs thermal energy primarily from the heating bands around the barrel, but also from heat generated by shearing of the material by the screw rotation (in the case of the reciprocating-screw-type machine), as shown in Figure 1. As the injection cycle begins, the polymer melt accumulated in front of the screw in the injection chamber is injected through the nozzle into the mold, driven by the hydraulic pressure in a cylinder behind the other end of the screw.

The polymer melt enters the mold, passes through the sprue, runner, and gate, and then fills the cavity that gives the shape to the final part. In the meantime, the

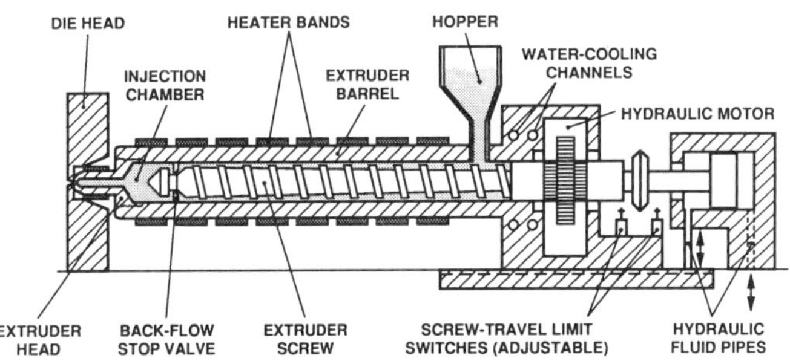

FIGURE 1 **Schematic drawing of the injection end of a reciprocating-screw machine.**

melt is subject to instant cooling when it contacts the cooled mold wall surface. Typically, the polymer forms a thin boundary layer of solidified polymer attached to the wall as the rest of the hot melt in the core region continues to flow and fill the cavity. The interface between the frozen layer and the flowing melt often exhibits the highest stress levels in the final molded part. At the end of fill, it is usually desirable to increase the hydraulic pressure substantially in order to "pack" more material into the cavity to compensate for material shrinkage resulting from continuous cooling until the gate freezes. Therefore, if a pressure sensor is inserted in the cavity, it will typically see a pressure history such as that shown in Figure 2, where three fairly distinct phases are seen, namely the filling, packing, and cooling (or holding) stages [15].

To perform the sequential melting/injection/solidification processes, two basic types of injection molding machines have been used in the industry. A majority of the machines are of the reciprocating-screw type, which was developed by W. H. Willert, with a U.S. patent issued in 1956. The other is the plunger type (in place of the screw), patented, also in the United States, by John Hyatt in 1872. In operation, the mold, which consists of at least two separate plates to form the cavity, is placed between two platens. One platen is fixed in place with respect to the machine bed, while the other can move along a set of tie bars to open or close the mold. During injection, the platens are clamped together to keep the mold closed by means of either an in-line hydraulic cylinder or a mechanical toggle mechanism [16]. The capacity of a machine is typically specified by two major parameters: the shot size and the clamp force. The shot size is defined as the maximum volume of polymer melt that can be injected during each stroke of the machine, whereas the clamp force refers to the maximum force (usually given in tons) that the machine can deliver in order to prevent the mold plates from separating ("flashing"). A large variety of machines originating from these two basic types have been developed to meet specific requirements; see Refs. 1, 2, and 11, for example.

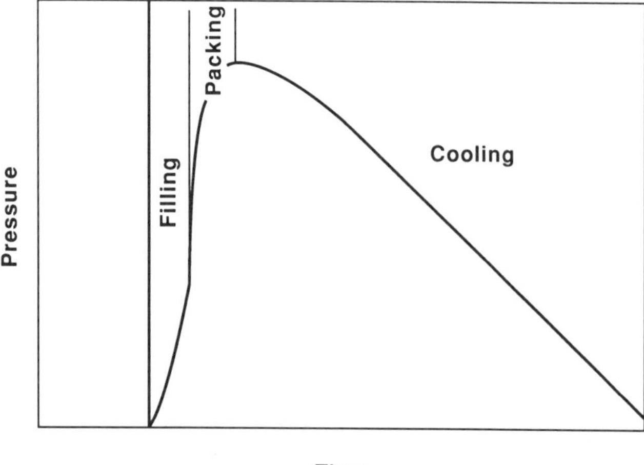

FIGURE 2 Schematic pressure history in mold cavity during injection molding in three stages: filling, packing, and cooling.

Molds and Mold Design

An injection mold is an assembly of mechanical components into which the polymer melt is injected and solidified to form the desired part. Figure 3 shows a typical commercial standard mold base in which the cavity gates and runners (material-feeding system) are machined off the A and/or B plates. In these plates there is a set of coolant passages, located relatively close to the cavity wall, that remove heat from the polymer melt in the mold. Each end of the mold assembly is clamped onto the fixed and moving platens of the machine, respectively, in order to close and open the mold at the parting line during the molding operation. Depending on the geometric complexity of the part, a mold can be as simple as a two-plate construction, as shown in Figure 3, or as complicated as a stand-alone machine with numerous in-and-out slides driven by powered actuators under automated sequence control. Such a mold can be very expensive and requires a long time to build and debug. The quality of the resulting part and its cost of manufacture are strongly influenced by mold design, construction, and excellence of workmanship [17].

As with the design of any mechanical device, the engineer has to make a series of rational decisions with the hope of attaining a near-optimal product. In the design of a mold with a single cavity, the crucial and important design decisions would be the type and location of the gate (entrance to the cavity), or sometimes the number of gates and their relative locations. These decisions may have irreversible effects on the quality and/or productivity of the final product. If a multicavity mold is contemplated in order to attain higher productivity, the design of the runner system becomes equally important.

Unfortunately, because of the large variety of part configurations and materials employed and the lack of fundamental understanding of the process and physical laws, rational decisions cannot be made readily. As a result, mold design has been handled in an ad hoc man-

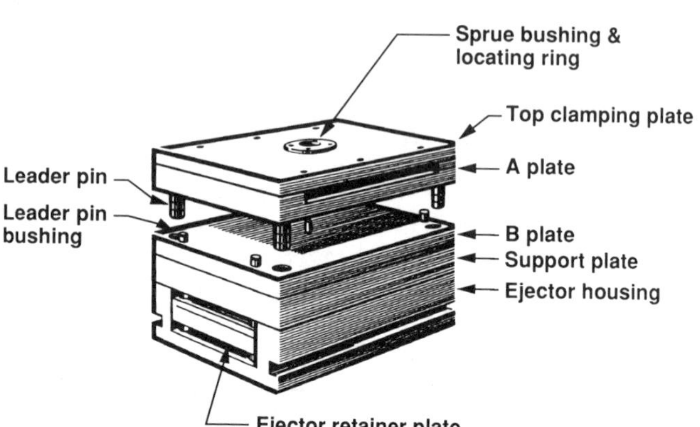

FIGURE 3 Exploded view of standard mold base.

ner for a long time, relying heavily on prior specific experiences and rules of thumb. Numerous books [1,2,9,10] and articles in trade journals have helped practicing engineers a great deal in the past. In addition, as a result of intensive research efforts, more scientific and quantitative solutions to these problems have become available in recent years. These latter advances will be briefly presented in the following sections.

Theoretical Basis for Injection Molding

In order to predict how the polymer melt will behave during the molding process, it is necessary to represent this behavior (or molding dynamics) by a set of governing mathematical equations based on physical laws. Among the earlier serious attempts made in this regard, Kamal and Kenig [5] studied one-dimensional (center-gated disk) cavity filling both analytically and experimentally. Lord and Williams [18] continued the mold filling study by also incorporating the flow through the runner. During the same period, attempts were made by a number of investigators [6,19–24] to study theoretically the filling of thin cavities of relatively simple planar geometry based on a generalized Hele-Shaw flow formulation [25]. Extending from the same modeling scheme, Hieber and Shen [26] presented a detailed formulation for simulating the mold filling of thin cavities of arbitrary planar geometry with possible inserts in the cavity. A hybrid finite element–finite difference numerical implementation was developed to compute pressure and temperature fields and keep track of the melt-front advancement in the cavity.

By making use of the control-volume concept, the scheme in [26] was extended by Wang et al. [27] to analyze the melt flow through a runner/gate/cavity system in three-dimensional space. This advance opened up a great opportunity for developing commercial computer codes, which have become a viable tool for the injection molding industry. More recent effort has extended the simulation to the post-filling stages [28,29], including the packing and cooling phases, as shown in Figure 2. For proper design of the mold cooling system aimed at minimizing the cycle time and achieving uniform cooling temperature on the cavity wall, analytical methods based on three-dimensional heat transfer have also been developed in recent years [30,31].

As described in Ref. 32, one pervasive characteristic of injection moldings is that they tend to be thin. In addition, because of the large viscosity of molten polymers, the effective Reynolds number is small. Accordingly, the filling of such parts corresponds to creeping, shear-dominated flow, which might be treated in terms of classical Hele-Shaw [25] flow as generalized to a non-Newtonian, nonisothermal, nonsteady situation. Implicit in such an approach is the assumption that memory effects (viscoelasticity) are not important; this assumption might seem surprising for such polymeric materials, but its validity has been corroborated by detailed experimental comparisons, which indicate that an inelastic modeling can quite adequately predict the shape of the advancing melt front in the cavity as well as the attendant pressure field; in fact, such a modeling can even describe the cavity pressure development well into the post-filling stage, as will be demonstrated in a later section.

In particular, if one considers the case of a thin cavity of arbitrary planar (x, y) geometry, the resulting governing equation for the pressure field can be written as (see [26] for further details):

$$\frac{\partial}{\partial x}\left(S\frac{\partial p}{\partial x}\right) + \frac{\partial}{\partial y}\left(S\frac{\partial p}{\partial y}\right) = 0 \tag{1}$$

where

$$S \equiv \int_0^h \frac{z^2}{\eta}\,dz \tag{2}$$

is a measure of the fluidity, with h denoting the cavity half-gap thickness in the z direction and $\eta(\dot{\gamma}, T, p)$ denoting the shear viscosity of the molten polymer as a function of shear rate, temperature, and pressure. For example, in terms of the power-law model:

$$\eta = m(T, p)\dot{\gamma}^{n-1} \tag{3}$$

where n is the power-law index, and:

$$m(T, p) = A\exp\left(\frac{T_a}{T}\right)\exp(bp) \tag{4}$$

where A, T_a, and b are constants. It follows that:

$$S = \frac{1}{\Lambda}\left(\frac{\Lambda}{m_0 e^{bp}}\right)^{1/n}\mathbf{S} \tag{5}$$

where

$$\Lambda \equiv |\nabla p| = \sqrt{\left(\frac{\partial p}{\partial x}\right)^2 + \left(\frac{\partial p}{\partial y}\right)^2} \tag{6}$$

and

$$\mathbf{S} \equiv \int_0^h z\left(\frac{z}{g(T)}\right)^{1/n} dz \tag{7}$$

with

$$m(T, p) = m_0 g(T)e^{bp} \tag{8}$$

and

$$g(T) \equiv \exp\left(\frac{T_a}{T} - \frac{T_a}{T_0}\right) \tag{9}$$

where T_0 denotes the injection melt temperature. Accordingly, with a representative value for the power-law index of $n \approx \frac{1}{4}$, it follows from Eq. (5) that S will be proportional to $|\nabla p|^3$, indicating that Eq. (1) is a highly nonlinear governing equation for pressure.

In addition, because of the temperature dependence

of η, as reflected by $g(T)$ in Eq. (8), it is necessary to determine the temperature field, which is modeled based upon the following energy balance:

$$\rho C_p \left(\frac{\partial T}{\partial t} + u\frac{\partial T}{\partial x} + v\frac{\partial T}{\partial y} \right) = k\frac{\partial^2 T}{\partial z^2} + \eta\dot{\gamma}^2 \tag{10}$$

that is, thermal conduction is maintained in the thin gapwise direction, and viscous heating is also included, although transverse convection is omitted.

As described in Ref. 26, a finite element–finite difference procedure has been developed for solving this problem, using triangular elements with quadratic shape functions for pressure and linear shape functions for temperature. The resulting pressure equation is obtained by applying a Galerkin procedure to Eq. (1), with the resulting nonlinear equation being solved by successive underrelaxation. On the other hand, Eq. (10) is solved basically in terms of a finite difference representation, but with the $\partial T/\partial x$ and $\partial T/dy$ terms being evaluated by means of the linear shape functions.

Whereas the numerical formulation developed in Ref. 26 involves a predictor-corrector scheme for advancing the melt front, with additional elements being added each new time step, the method developed in Ref. 27 employs a fixed finite element grid and uses a control-volume approach that makes the melt-front advancement amenable to an automated algorithm. This later development has been extended to thin three-dimensional parts, treated as a union of thin two-dimensional components, which resulted in a powerful commercial code, C-FLOW, with sophisticated computer graphics [33].

As an illustration, Figure 4 shows the melt flow front advancement as a function of time during the filling of a mold cavity for an automobile instrument panel. The cavity is fed through 17 gates via a complex runner system. Figure 5*a* indicates the pressure distribution at the instant just before the cavity is filled, while Fig. 5*b* shows the temperature distribution exactly at the time of fill. With the pressure distribution and the area perpendicular to the applied force known, the computer program calculates the required clamp force. Much other useful information can be obtained, such as weld-line locations (where two flow fronts merge) and shear stress and shear rate distributions.

Once the cavity fills, additional material is "packed" into it in order to compensate for the increasing polymer density arising from substantial cooling during the post-filling stage. That is, additional mass (typically $\lesssim$ 10% of the total mass) is required during the post-filling stage if the polymer is to continue to occupy the entire cavity volume as the polymer density increases. Accordingly, an essential ingredient of the flow dynamics during the post-filling stage is the compressibility of the polymer.

In particular, if we extend the inelastic modeling of

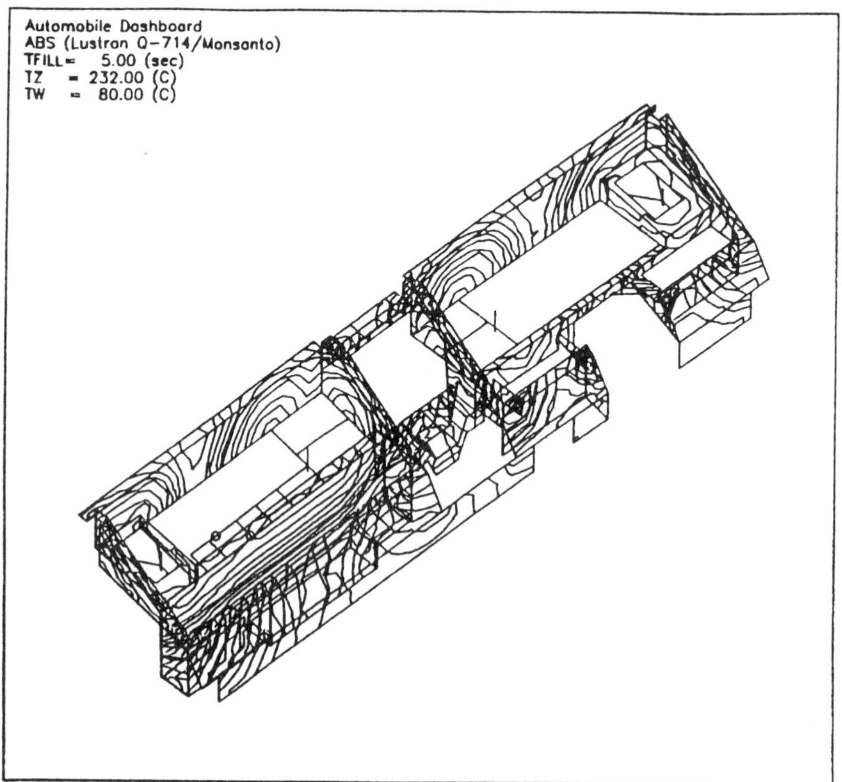

FIGURE 4 Predicted melt-front advancement in the mold cavity for an automobile dashboard.

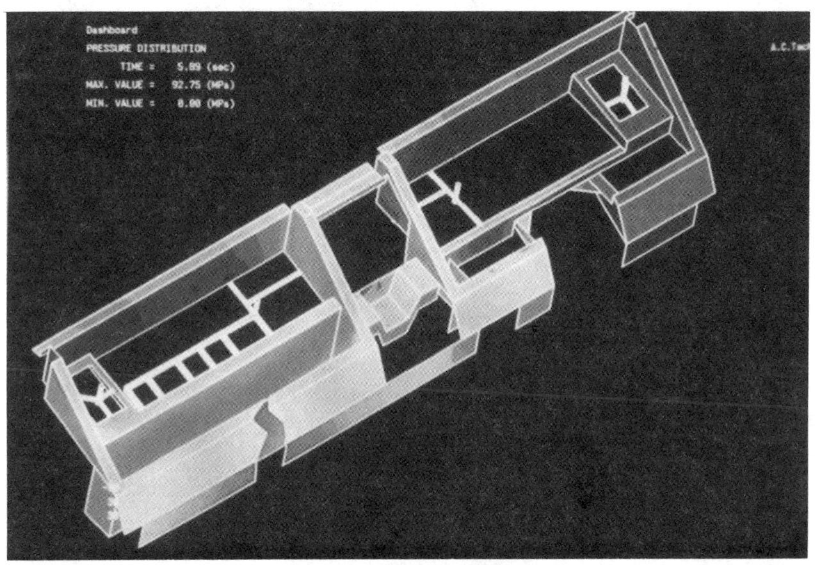

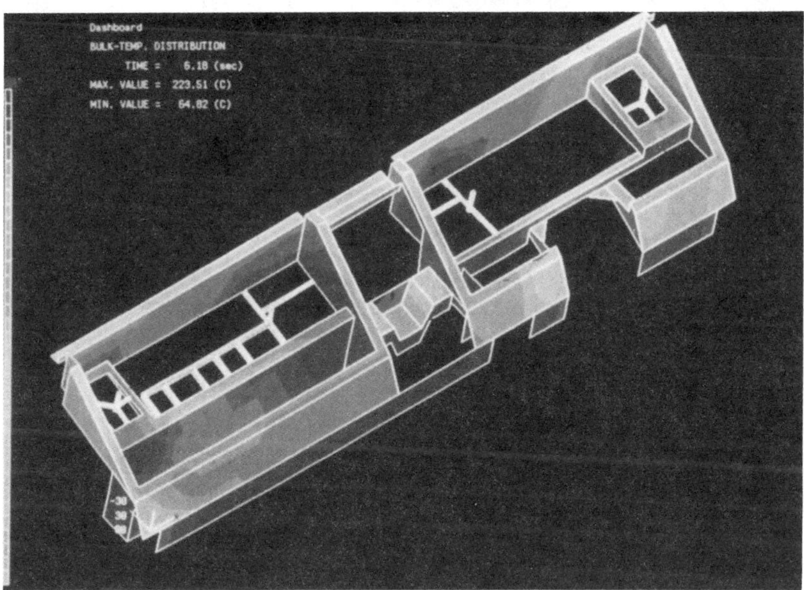

FIGURE 5 **Color-shaded plots of (*a*) pressure distribution and (*b*) temperature distribution in an automobile dashboard at instant of fill.** (From Advanced CAE Technology, Inc., Ithaca, NY.)

the previous section by incorporating compressibility, making the reasonable assumption that variations in density are small relative to the density level itself (an underlying approximation in the field of acoustics), then the major change concerns Eq. (1), which now becomes [35]:

$$h\frac{\partial \ln \bar{\rho}}{\partial t} = \frac{\partial}{\partial x}\left(S\frac{\partial p}{\partial x}\right) + \frac{\partial}{\partial y}\left(S\frac{\partial p}{\partial y}\right) \tag{11}$$

where $\bar{\rho}$ denotes the gapwise-averaged density. Accordingly, it becomes necessary to introduce an equation of state $\rho(T, p)$, which, for amorphous polymers, might be based upon the Spencer-Gilmore [36] functional form:

$$(p + \hat{p})\left(\frac{1}{\rho} - \frac{1}{\hat{\rho}}\right) = \hat{R}T \tag{12}$$

where $(\hat{p}, \hat{\rho}, \hat{R})$ are constants for the given material. In this case, Eq. (11) can then be expressed as [35]:

$$G(x, y, t)\frac{\partial p}{\partial t} - \frac{\partial}{\partial x}\left(S\frac{\partial p}{\partial x}\right) - \frac{\partial}{\partial y}\left(S\frac{\partial p}{\partial y}\right) = -F(x, y, t) \tag{13}$$

where

$$G(x, y, t) \equiv \left(\frac{1}{p + \hat{p}}\right)\int_0^h \left(1 - \frac{\rho}{\hat{\rho}}\right) dz \tag{14}$$

and

$$F(x, y, t) \equiv \int_0^h \left(1 - \frac{\rho}{\hat{\rho}}\right)\left(-\frac{\partial \ln T}{\partial t}\right) dz \tag{15}$$

with S still being defined by Eq. (2) and T still being governed by Eq. (10). Accordingly, the elliptic partial differential equation (PDE), Eq. (1), now becomes a parabolic PDE, Eq. (13), with a forcing term F due to cooling.

As would be expected, the terms involving G and F are of smaller order during the filling stage, so that Eq. (13) reduces to Eq. (1) in leading approximation. However, at the instant of fill, when the inlet pressure is abruptly changed (to the preset packing level) and the $p = 0$ (gage pressure) boundary condition along the melt front gets switched to an impermeability condition (i.e., $\partial p/\partial n = 0$ as the front reaches the outer boundary of the cavity), the term involving G first becomes important. In fact, the typical diffusion time for this first phase of the post-filling stage is on the order of 0.1 second or less, so that the pressure field rapidly approaches a uniform value corresponding to the imposed packing pressure, with the temperature remaining essentially unchanged during this short period. Thereafter, the F term in Eq. (13) becomes important as substantial cooling eventually occurs (characterized by a cooling time h^2/α of $\approx$ 10 seconds, since the half-gap thickness is typically $\approx$ 0.1 cm and the thermal diffusivity of polymers is $\approx 10^{-3}$ cm^2/s). During this latter phase, F is the driving term; in particular, the increased density resulting from decreasing T requires additional flow into the cavity, which, from the dynamic force balance, requires a pressure gradient to balance the viscous resistance. Accordingly, during this cooling phase, the pressures in the cavity begin to fall. Hence, whereas the cavity pressure traces would be characterized by a monotonic approach to the imposed holding pressure under isothermal conditions, in the actual nonisothermal situation the pressure traces are characterized by a peak value followed by decay during the cooling-dominated phase (as indicated schematically in Fig. 2).

As documented in Ref. 35, the use of the power-law modeling of viscosity in Eq. (3) is appropriate for both the filling and post-filling stages for polymers such as polystyrene and polypropylene, which shear thin at relatively low shear-stress levels ($\approx 2 \times 10^4$ N/m^2). However, for polymers such as polymethylmethacrylate and polycarbonate, which shear thin at much higher stress levels of $\approx 1 \times 10^5$ N/m^2 and 4×10^5 N/m^2, respectively, it becomes appropriate to represent the shear thinning behavior in terms of a Cross model [37], which exhibits both Newtonian behavior at low shear-stress levels and power-law behavior at high stress levels with a continuous transition between these two asymptotic limits. Concerning the temperature and pressure dependence of the viscosity, it is noted that the representation in Eq. (4) is appropriate during the filling stage but can become quite inaccurate for the post-filling stage, particularly for materials whose glass transition temperature T_g lies above the mold wall temperature. In such cases, the WLF [38] functional form is appropriate [35], with the pressure dependence incorporated through the pressure dependence of T_g.

As with the modeling of cavity filling, this two-dimensional formulation can be extended to handle three-dimensional thin parts. One of the features such a program can offer is calculation of the volumetric-shrinkage distribution in a molded part. Figure 6 shows such a distribution for an injection-molded bezel with two gates as indicated. The volumetric shrinkage in this case has a maximum value of 3.22% on the side of the bezel far from the gate, whereas the minimum value of 1.83% occurs near the gate. If two more gates are added, as shown in Figure 7, the maximum shrinkage value is significantly reduced to 2.7%, while the minimum value shows little change. Such a simulation program can be a very powerful tool to enable mold designers and molding engineers to make rational decisions to improve their overall performance.

CAE Technologies for Injection Molding

As the theory and computer simulation of molding dynamics have gradually become accepted by practicing engineers to help them solve real-world problems, these techniques have evolved into a new sector of the high-technology industry, namely CAE (computer-aided engineering) for injection molding. The first publication on this specific topic was a book sponsored by the Society of Plastics Engineers, which appeared in 1983 with the exact title of *Computer Aided Engineering for Injection Molding* [12]. Even more recently, SPE sponsored a series of three books on CAE for polymer processing, published by Hanser Publishers; they are *Fundamentals of Computer Modeling for Polymer Processing, Applications of CAE in Injection Molding,* and *Computer Models for Extrusion and Other Processes.*

In the second [13] of these three books, the editor, L. T. Manzione, pointed out in his introduction that, "The complexity of the injection-molding process makes successful production start-up on the first attempt a rare occurrence. The combination of a viscoelastic molding material, an irregular mold geometry, steep temperature gradients and other criteria such as appearance and dimensional tolerances make mold design a difficult task even for recognized experts. Computer aids for injection molding have been developed and have been accepted by the processing industry to improve productivity, part quality and performance and shorten start-up times." This development has resulted from the recent proliferation of powerful computer hardware and software advances, sophisticated computational algorithms, and the rapid growth of CAD (computer-aided design) and AI (artificial intelligence) techniques. This particular book [13] was aimed at introducing existing CAE packages that were commercially available at its time of publication, 1987. Its eight contributed chapters cover several aspects of the molding process, including cooling-system design and expert systems. Most presentations in the book are

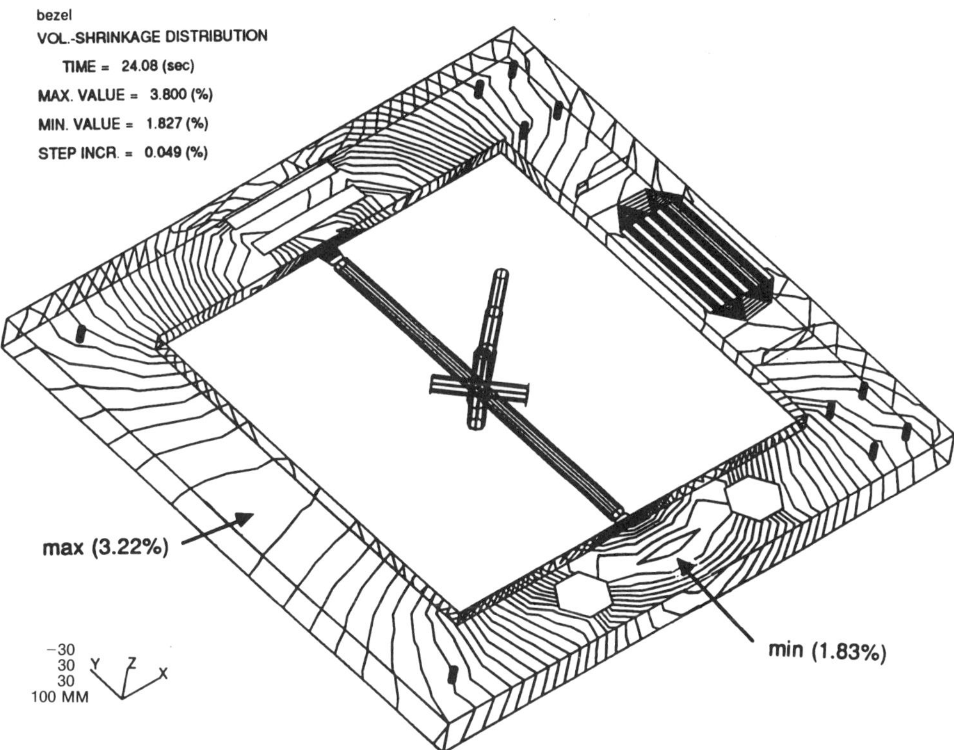

FIGURE 6 Volumetric shrinkage distribution of a molded bezel with two gates. (From Advanced CAE Technology, Inc., Ithaca, NY.)

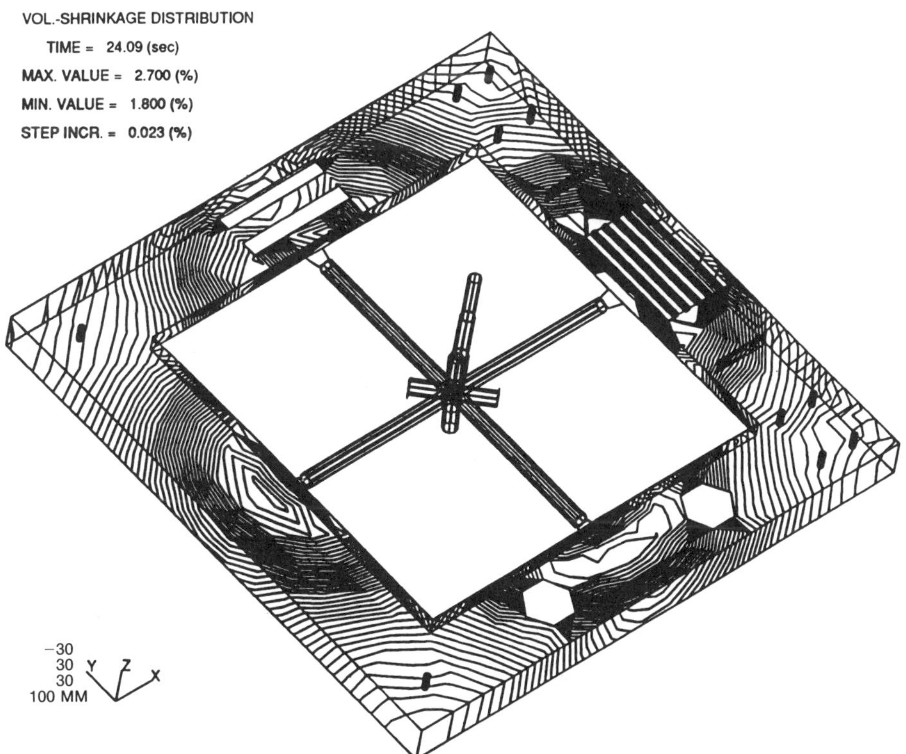

FIGURE 7 Same as Figure 6 but with four gates.

descriptive, with little detail on mathematical formulation and computational methods. Simulation packages primarily deal with mold filling. However, recent commercial releases offer simulation of the post-filling stage also (e.g., C-PACK from Advanced CAE Technology, Inc.), and attempts are being made to predict shrinkage and warpage (e.g., the prerelease of SWIS by Moldflow PTY, Ltd). Considerable attention is currently being focused on dealing with post-ejection problems, such as prediction of the properties and quality of the final molded part. For an extensive review of the related issue of residual stress effects in injection-molded products, the reader is referred to Ref. 39.

Application of Artificial Intelligence Techniques

Although great strides are being made in analyzing injection molding on the basis of physical laws and numerical treatment, both the process and the material properties are still too complex to be mastered analytically. Many variables are involved in the numerous aspects of the process, and each can affect the quality of the final product. This situation makes it logical to apply some of the existing AI techniques, such as an "expert system" to help solve complex problems based on heuristic knowledge. In particular, two chapters in Ref. 12 are dedicated to this approach.

A typical expert system consists of four major components [40], as shown in Figure 8. The user interface in the figure comprises a collection of capabilities: knowledge acquisition, debugging and experimenting with the knowledge base, running test cases (perhaps systematically, from a library), generating summaries of conclusions, explaining the reasoning that led to a conclusion (or to a question by the system), and evaluating system performance (including sensitivity of an answer to particular data items, present or absent).

The main computational (or inference) engine is in the center of the diagram, containing search guidance and inference components. It searches the knowledge base for applicable knowledge and makes inferences on the basis of current problem data. The search guidance component selects the portion of the knowledge base that it is most important to apply at any point in the problem-solving session. It may use general knowledge-base considerations, or it may make use of user-specified strategic rules (sometimes called meta-knowledge). The inference component evaluates individual rules and interconnections among concepts in the knowledge base in order to add to the working memory, which is a store of the current problem data, such as answers to questions about the problem and results of diagnostic tests.

A number of attempts have been made and prototype expert systems have been developed to cope with injection-molding problems. As with many other expert-system applications, success has been very limited, and they are not widely used. However, progress is being made in this area. For example, an intelligent resin-selection system has been developed in a recent study [41] in which almost 7000 injection-molding-grade resins are involved, each with 60 associated properties in 5 categories, as shown in Figure 9. The database is divided hierarchically into three levels: the generic family, the manufacturer, and the manufacturer's designated trade name. In operation, the user can specify multiple criteria for selection with an assignable weighting factor to each cri-

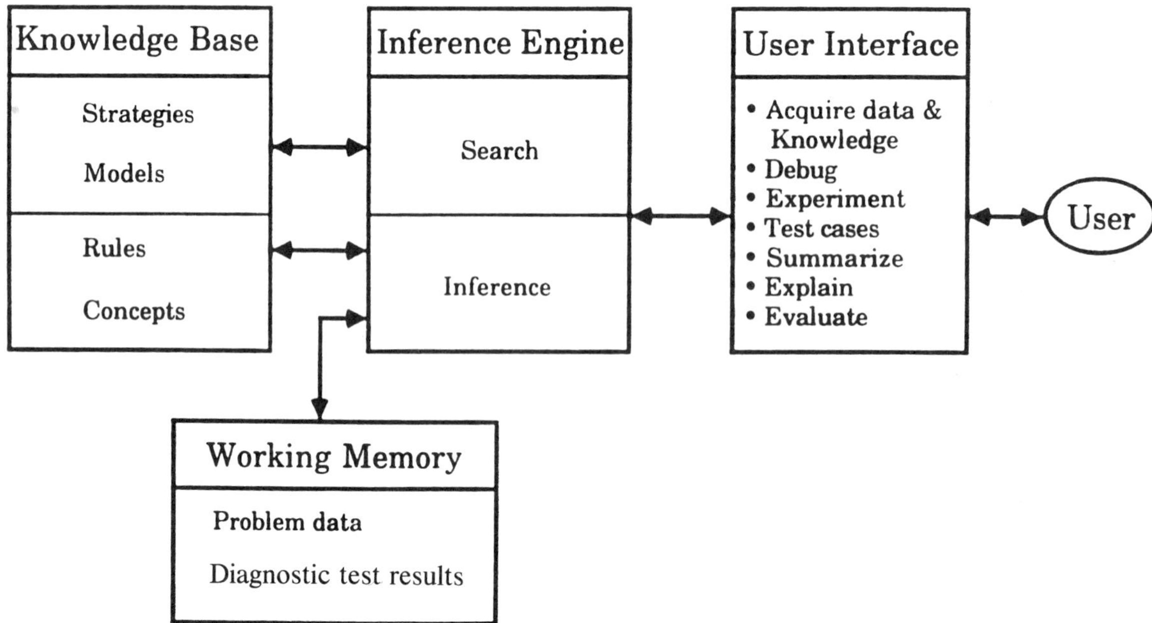

FIGURE 8 Components of an expert system. (From Ref. 40.)

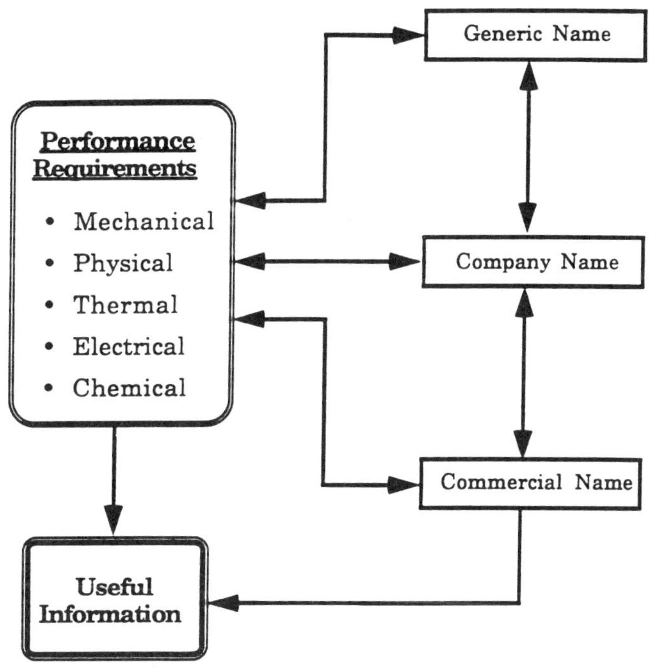

FIGURE 9 Structure of an intelligent resin-selector system. (From Ref. 41.)

terion based on its relative importance for meeting the design requirements. The system will automatically highlight all qualified resins and recommend the best choice based on the "fuzzy" algorithm [42]. The user can make comparison at any of the three levels based on a specific property. The results are ranked and some displayed in color graphics, including viscosity curves (when available).

A more recent example of applying AI techniques is to incorporate the results generated by numerical simulation with expert knowledge to interpret these results, resulting in an integrated system [43]. In particular, the system in Ref. 43 is able to predict automatically the weld-line location and possible air entrapments in the mold. It first synthesizes deterministic knowledge of melt-front position and velocity field, generated by a thoroughly verified flow-simulation program developed by the Cornell Injection Molding Program [27], into useful information such as split, merge, and evolvement of melt-front advancement. This information can then be integrated with heuristic knowledge extracted from the logical thinking and experience of mold designers and mold engineers to automatically interpret the formation and position of weld lines and air traps.

The concept and logic used to interpret the weld-line and air-trap position are rather straightforward for simple geometries. For molded parts with complex and irregular geometries, however, the task becomes extremely challenging. As an illustration of the robustness of the system developed in Ref. 43, Figure 10 shows the interpreted weld lines for an automobile dashboard. Such a capability is very important in order to fully realize the power of CAE technology as applied to injection molding. This development also illustrates that artificial intelligence and deterministic knowledge can be integrated to solve a very complex engineering problem.

Process-Control Methodologies

With the advent of engineering plastic resins, the injection-molding process is today being used for the production of many important parts, originally made of metals, in electric appliances, automotive components, and the aerospace industry where precision and quality are especially important. As a result, process-control systems, which are mainly used for increasing the repeatability and productivity of production equipment, have become the heart of today's injection molding machines. In particular, the quality requirements for precision injection-molded parts have created new challenges for the process-control field.

As noted in previous sections, the injection molding process is characterized by the successive stages of plastication, filling, and post-filling (also known as packing and cooling). Polymer resins are melted uniformly in a heated barrel by a rotating screw under a back pressure during the plastication stage, then molten polymer is quickly squeezed into the cavity of a steel mold whose temperature is controlled with coolant (usually water) passing through appropriately designed cooling channels until the cavity is completely filled. The process then instantaneously switches to the post-filling stage based upon the ram-position or hydraulic-pressure settings on a process controller. During the post-filling stage, high pressure is applied to pack more molten polymer into the cavity in order to compensate for the thermal contraction of the polymer in the cavity as it cools down rapidly into the solid state. The entire cycle is ended when the part is sufficiently cool to be ejected.

Evidently, many specific features, such as barrel temperature, coolant temperature, injection speed, hydraulic packing pressure, and shot size, have to be controlled in the injection molding process. Today, there are three commonly used control methods for injection molding machines, namely, open-loop control, incremental closed-loop control, and direct closed-loop control [44]. Open-loop control, namely, control without measurement, can be used for less important features, such as mold-open speed and cycle time, where measurement is not necessary. On the other hand, incremental closed-loop control, which adjusts the command periodically based on measurement, is applied when accuracy and measurement are important but instantaneous response is not required. In general, barrel and coolant temperatures are usually controlled by this method, which sometimes is known as PID (proportional-integral-differential) control. Finally, direct closed-loop control, which means that the function is directly measured and adjusted by controlling relevant actuators, is used widely on the hydraulic system for injection-speed, switchover, and packing pressure control.

In the past few years, a new generation of process control has evolved, providing total machine repeatabil-

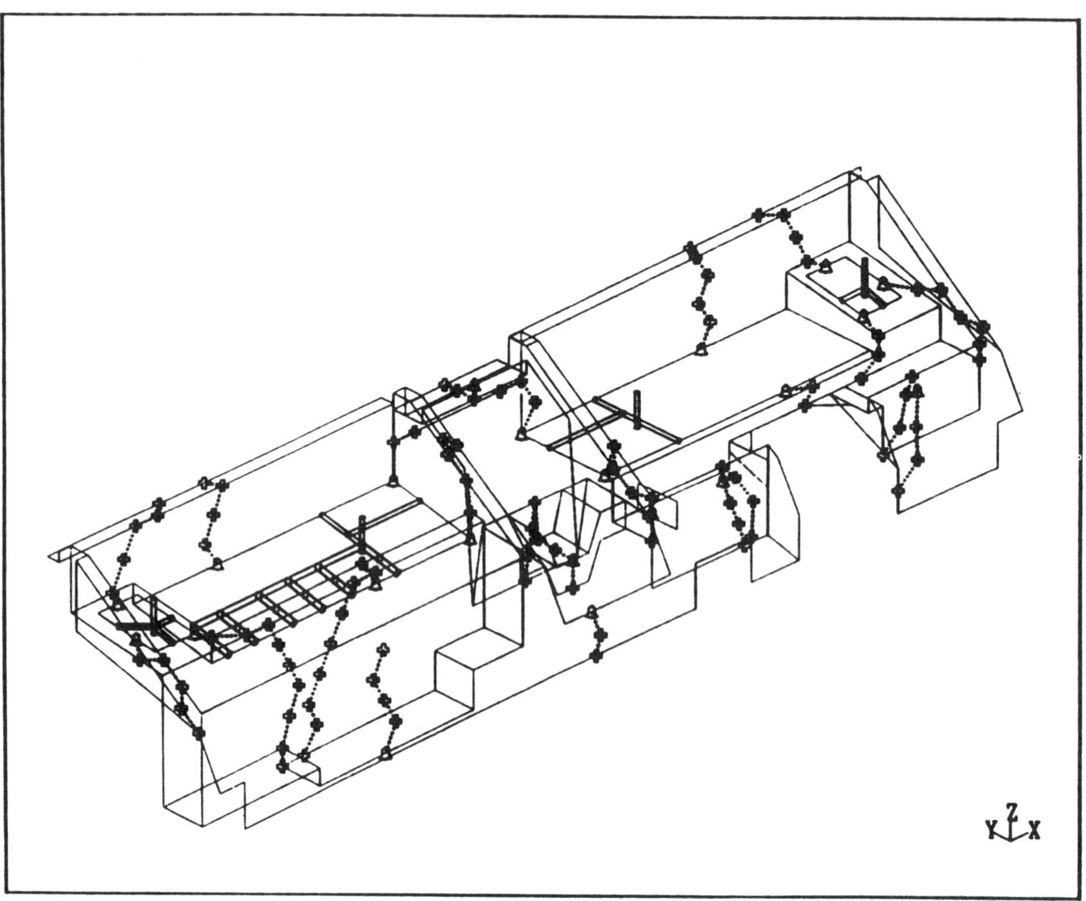

FIGURE 10 Interpreted weld-line positions for an automobile dashboard. (From Ref. 43.)

ity and foolproof flexible automation based upon the introduction of powerful microprocessors [45]. Unlike conventional closed-loop machine controls, higher-level controls, such as statistical process control [46] or p-v-T control, are used to interpret signals received from the machine. Decisions are then made by means of complex software without human intervention. Basically, self-learning or intelligent controls have been incorporated to give final adjustments based upon either statistical methods or semiempirical knowledge.

Transfer-function and frequency-response techniques for modeling the dynamics of the injection molding machines have been studied in an attempt to achieve effective process-control strategies [47–51]. Injection molding machines with variable injection-speed-profile controls have also been introduced to improve product quality. In addition, the guidelines for switchover from the filling to the post-filling stage based upon injection-time, ram-position, or hydraulic-pressure settings have been studied and used to optimize the mechanical properties of molded parts [52–54]. This has led to the advent of cavity-pressure switchover using pressure sensors mounted in the mold assembly, which has proven to be the best guideline for switchover.

It might also be noted that several recent investigations have used experimental methods to set up an empirical model for predicting the shrinkage or mechanical properties of injection-molded parts [55–58]. In addition, an adaptive process control based on an empirical model has been successfully used [59] to control part thickness and part weight. Further, statistical process control methods have been introduced by major machine manufacturers [60–62]. Although the idea of using the thermodynamic properties of polymers to describe the injection molding process was introduced initially by Spencer and Gilmore [63] in their pioneering work dating back to the early 1950s, optimizing the processing parameters based upon the p-v-T diagram of the polymer resins was only first suggested by Menges and coworkers at IKV [64,65] in the late 1970s. A modified Spencer-Gilmore equation of state was used by IKV to optimize the packing-pressure profile in order to minimize the cycle-to-cycle shrinkage variations in molded parts. More recently, discussions and assessments concerning p-v-T control or optimization have also been reported [66–69].

It might be noted that the impact of technology on process control, as far as machinery is concerned, has been the centralization of controls. Process controllers based on computers for injection molding machines can communicate with central monitoring units, where most of the processing information is decided. In fact, machines are now capable of thinking and learning with little help from operators. Computer-integrated manufacturing will be feasible in the near future with the help of new sensors, powerful microchips, and knowledge of materials. A program based on artificial intelligence such as that discussed in a previous section will eventually be able to transfer processing parameters from CAD/CAM molders and monitor the entire process in order to avoid producing even a single bad part.

K. K. Wang

References

1. D. V. Rosato and D. V. Rosato, *Injection Molding Handbook,* Van Nostrand Reinhold, New York, 1986.
2. I. I. Rubin, *Injection Molding—Theory and Practice,* John Wiley & Sons, New York, 1972.
3. K. K. Wang et al., "Computer-Aided Injection Molding System," Progress Report No. 1, Cornell Injection Molding Program (CIMP), Cornell University, January 1975.
4. R. L. Ballman, H. L. Toor, and L. Shusman, *Modern Plastics, 37*(1), 105 & *37*(2), 115 (1959).
5. M. R. Kamal and S. Kenig. *Polym. Eng. Sci., 12,* 294, 302 (1972).
6. Z. Tadmor, E. Broyer, and C. Gutfinger, *Polym. Eng. Sci., 14,* 660 (1974).
7. J. L. S. Wales, J. Van Leeuwen, and R. Van der Vijgh, *Polym. Eng. Sci., 12,* 358 (1972).
8. G. Menges and G. Wubken, *Proceedings of SPE ANTEC,* 1973, p. 519.
9. J. B. Dyne, *Injection Molds and Molding, A Practical Manual,* Van Nostrand Reinhold, New York, 1979.
10. K. Stoeckhert, Ed., *Injection Molds, 102 Proven Designs,* Hanser Publishers, Munich, 1983.
11. F. Johannaber, *Injection Molding Machines, A User's Guide,* 2nd ed., Hanser Publishers, Munich, 1985.
12. E. C. Bernhardt, Ed., *Computer Aided Engineering for Injection Molding,* Hanser Publishers, Munich, 1983.
13. L. T. Manzione, Ed., *Applications of Computer Aided Engineering in Injection Molding,* Hanser Publishers, Munich, 1987.
14. A. I. Isayev, Ed., *Injection and Compression Molding Fundamentals,* Marcel Dekker, New York, 1987.
15. K. K. Wang et al., "Computer-Aided Design and Fabrication of Molds and Computer Control of Injection Molding," Progress Report No. 11, Cornell Injection Molding Program (CIMP), Cornell University, April 1985, pp. 197–207.
16. Reference 1, pp. 39–55.
17. Reference 2, p. 86.
18. H. A. Lord and G. Williams, *Polym. Eng. Sci., 15,* 553, 569 (1975).
19. S. Richardson, *J, Fluid Mech., 56,* 609 (1972).
20. E. Broyer, Z. Tadmor, and C. Gutfinger, *Israel J. Technology, 11,* 189 (1973).
21. E. Broyer, C. Gutfinger, and Z. Tadmor, *Trans. Soc. Rheol., 19,* 423 (1975).
22. J. L. White, *Polym. Eng. Sci., 15,* 44 (1975).
23. M. R. Kamal, Y. Kuo, and P. H. Doan, *Polym. Eng. Sci., 15,* 863 (1975).
24. Y. Kuo and M. R. Kamal, *AIChE J., 22,* 661 (1976).
25. H. Schlichting, *Boundary-Layer Theory,* 6th ed., McGraw-Hill, New York, 1968.
26. C. A. Hieber and S. F. Shen, *J. Non-Newtonian Fluid Mech., 7,* 1 (1980).
27. V. W. Wang, C. A. Hieber, and K. K. Wang, *J. Polym. Eng., 7,* 21 (1986).
28. V. W. Wang and C. A. Hieber, *Proceedings of SPE ANTEC,* 1988, p. 290.
29. K. K. Wang et al., "Integration of CAD/CAM for Injection-Molded Plastic Parts," Progress Report No. 14, Cornell Injection Molding Program (CIMP), Cornell University, September 1988, pp. 155–174.
30. T. E. Burton and M. Rezayat, in Ref. 13, chap. 9.
31. K. Himasekhar, J. Lottey, and K. K. Wang, *Proc. of Supercomputers-Emerging Applications in Manufacturing,* 1989, pp. 3.1–3.11.
32. K. K. Wang and C. A. Hieber, in T. G. Gutowski, Ed., *The Manufacturing Science of Composites,* ASME Book No. H00415, 1988. pp. 87–94.
33. V. W. Wang, C. A. Hieber, and K. K. Wang, in Ref. 13, chap. 7.
34. K. K. Wang et al., "Integration of CAD/CAM for Injection-Molded Plastic Parts," Progress Report No. 12, Cornell Injection Molding Program (CIMP), Cornell University, July 1986.
35. C. A. Hieber, in Ref. 14, chap. 1.
36. R. S. Spencer and G. D. Gilmore, *J. Appl. Phys., 20,* 502 (1949).
37. M. M. Cross, *J. Colloid Sci., 20,* 417 (1965); *J. Appl. Polym. Sci., 13,* 765 (1969); *Rheol. Acta, 18,* 609 (1979).
38. M. L. Williams, R. F. Landel, and J. D. Ferry, *J. Am. Chem. Soc., 77,* 3701 (1955).
39. A. I. Isayev, in Ref. 14, chap. 3.
40. C. L. Dym, Ed., *Applications of Knowledge-Based Systems to Engineering Analysis and Design,* ASME, New York, 1985.
41. W. R. Jong and K. K. Wang, *Proceedings of SPE ANTEC,* 1989, p. 367.
42. G. Salton, E. A. Fox, and H. Wu, *Commun. ACM, 26,* 1022 (1983).
43. W. R. Jong, *Investigation of Intelligent Systems for Injection Molding of Plastics,* Ph.D. thesis, Sibley School of Mechanical and Aerospace Engineering, Cornell University, 1990.
44. C. Kirkland, *Plastics Technology, 40,* 65 (1984).
45. J. A. Sneller, *Modern Plastics, 62,* 42 (1985).
46. M. J. Sercer, I. J. Catic, and J. B. Zoric. *Proceedings of SPE ANTEC,* 1983, p. 672.
47. M. R. Kamal, W. I. Patterson, Dib Abu Fara, and A. Haber, *Polym. Eng. Sci., 24,* 686 (1984).
48. Dib Abu Fara, M. R. Kamal, and W. I. Patterson, *Polym. Eng. Sci., 25,* 714 (1985).
49. M. R. Kamal, W. I. Patterson, N. Conley, Dib Abu Fara, and G. Lohfinx, *Proceedings of SPE ANTEC,* 1986, p. 189.
50. Dib Abu Fara. W. I. Patterson, and M. R. Kamal, *Proceedings of SPE ANTEC,* 1987, p. 221.
51. I. O. Pandelidis and A. R. Agrawal, *Polym. Eng. Sci., 28,* 147 (1988).
52. L. W. Fritch, *Proceedings of SPE ANTEC,* 1987, p. 218.
53. R. A. Malloy, S. J. Chen, and S. A. Orroth, *Proceedings of SPE ANTEC,* 1987, p. 225.
54. W. Friesenbichler, W. Knappe, and R. Rabitsch, *Intern. Polym. Process., 3,* 191 (1988).

55. B. Sanschagrin, *Polym. Eng. Sci.*, *23*, 431 (1983).
56. J. J. DeLuca and S. P. Petrie, *Polym. Eng. Sci.*, *25*, 19 (1985).
57. P. Girard, G. Salloum, L. P. Herbert, and B. Sanschagrin, *Proceedings of SPE ANTEC*, 1986, p. 151.
58. R. C. Ricketson and K. K. Wang. *Proceedings of SPE ANTEC*, 1986, p. 145.
59. L. S. Shah, *Adaptive On-line Process Control of Injection Molding Using an Empirical Model*, M.S. thesis, Sibley School of Mechanical and Aerospace Engineering, Cornell University, 1987.
60. K. R. Kreisher, *Modern Plastics*, *66*, 11 (1989).
61. R. J. Groleau, *Proceedings of SPE ANTEC*, 1989, p. 1704.
62. K. B. Schultz and D. P. Werle, *Plastics Today*, *1*, 8 (1989).
63. R. S. Spencer and G. D. Gilmore, *Modern Plastics*, *27*, 143 (1950).
64. H. O. Hellmeyer, *Ein Beitrag zur Automatisierung des Spritzgiessprozesses*, Ph.D. thesis, I.K.V., RWTH Aachen, Federal Republic of Germany, 1977.
65. W. Bonqardt and G. Menges, *Proceedings of SPE ANTEC*, 1980, p. 141.
66. J. Rothe, *Kunststoffe*, *76*, 307 (1986).
67. A. Bockenheimer, *Kunststoffberater*, *32*, 26 (1987).
68. A. Smith, *Plastics Rubber Weekly*, *12*, 22 (1988).
69. Editorial, *Modern Plastics*, *66*, 138 (1989).

Molding, Producibility for Polymer–Fiber Composites

The vast horizon of polymer–fiber composites lies wide open to the innovative explorer. Using liquid plastics and pliable textiles, products with specific strength and stiffness greater than steel can be molded. Furthermore, by altering the ingredients, different functions can be served. For example, by combining resilient plastics with a wavy fiber pattern, the composite can move from elastic to tensioned as load is placed on the structure—much like biological structures. Conversely, a rigid matrix–oriented fiber composite will exhibit high tension under initial loading.

While the designer can select and manipulate ingredients, creating the ideal construct, it is the processor that produces the actual hardware. The quality of a composite part, therefore, is a function of both *structural* and *production* design. Both design issues must be analyzed simultaneously. It accomplishes no purpose to design a structure that cannot be produced or to produce a part that does not have the required structure.

This article is divided into two sections: Selecting the Composite Molding Process and Selecting Molding Tools for Composites. The issues discussed herein must be addressed before a composite design is finalized. The task of improving the producibility of low quantity, but high quality composites is a big challenge.

Selecting the Composite Molding Process

Composites processing is a system of production whereby unstable raw materials are molded into finished products of high stability. Both the material and the product are created simultaneously during product manufacture.

Composite Constituents

A composite is composed of three elements: *structure* (reinforcement), *binder* (matrix), and *interface*. Typically, fibers of glass, aramid, and graphite provide the structure. Plastics resins serve as fiber binders. Good adhesion of resin to fiber is made possible by treating the fiber surface (interface), but when fiber sizing (primer) is used, the adhesion promoting element is called the *interphase*.

Fiber Style vs. Strength

A fiber-dominated composite is superior in tensile strength to a polymer-dominated composite (see Fig. 1). Continuous fiber has higher strength than short fiber, and directionally oriented fiber significantly outperforms a random-fiber composite.

Short-Fiber Composites

For short-fiber composites, fibers with lengths of up to 1.5 mm mils can be integrated into resin stock as powders, granules, or liquids. These then become feedstocks for injection, transfer, extrusion, compression, and rotational molding or for liquid castings, all of which are very efficient production methods. However, while short fibers improve producibility properties, these formulations are limited to secondary and cosmetic structures requiring only light physical loadings.

Somewhat longer but still discontinuous fibers, 8–10 cm in length, are useful for more demanding functions. Automotive and appliance manufacturers and the building trades use these composites for semistructural applications. Small missile fins and other moderately loaded

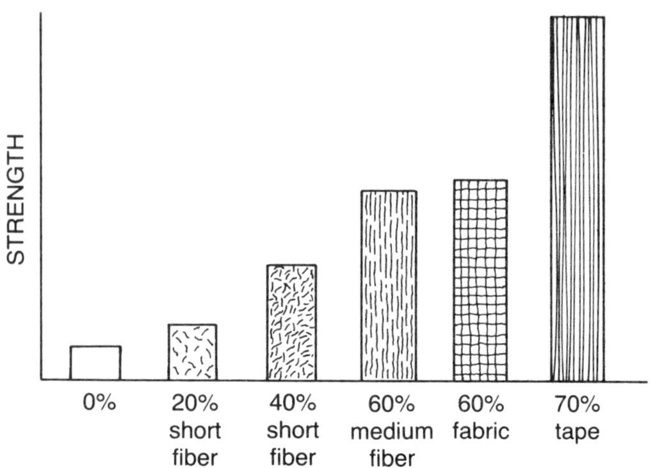

FIGURE 1 Fiber style vs. strength.

aircraft structures use these medium-length-fiber composites, even with random orientation. These formulations are still moldable by injection, transfer, extrusion, and compression molding.

With further development, medium-length but oriented-fiber composites may supplant continuous- and oriented-fiber products in many applications. For molding applications of severe contour and complexity, the medium-length oriented fiber will theoretically equal the strength of a product that, at best, is molded poorly from a continuous oriented fiber.

To repeat, short and medium-length fibers lend themselves to producibility. Short-fiber–resin formulations can be "squirted" through a nozzle into a closed mold with a cycle time of less than a minute. The parts produced by these processes can be molded net and require little or no trimming. However, higher performance can be obtained from continuous fiber, and because these are more difficult to process, the remainder of this chapter will focus on continuous-fiber composites.

Composites Producibility

Structural engineers are most interested in composites that are formed from:

1. Continuous fiber
2. Directionally oriented fiber
3. Maximum fiber loadings

These highly engineered composites are more difficult to produce than the randomly oriented short-fiber types. Unfortunately, as engineering demands increase, producibility decreases (see Fig. 2).

Fiber Drapeability

Textile fibers, in addition to their outstanding specific strength, are very supple. They comply easily to contoured shapes. This compliant attribute, *drapeability*, is important to product manufacture. Any conceivable shape can be easily hand formed by wrapping the desired number of fabric plies.

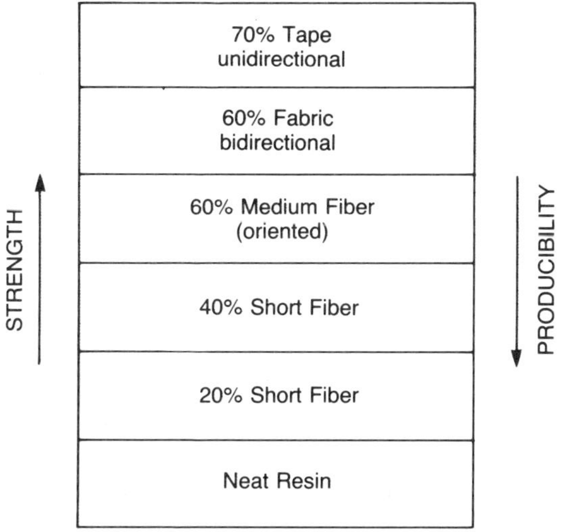

FIGURE 2 Strength vs. producibility.

Fiber Patterns

The basic element of a textile good is the *filament*. Typically the diameter of the filament is only 0.0025 to 0.1524 mm. These are plied, several thousand filaments at a time, to make a *yarn*. A yarn is used for *filament winding*, for *braiding*, for making *unidirectional tape*, or for making *short- and long-fiber mat*. Yarn is also used to weave *fabric broadgoods*. Developmental work has begun in using yarn to make three-dimensional *preforms*. These are shaped to match a specified composite product. Figure 3 shows the common textile configurations.

Plastics Matrix

Plastics fall into one of three chemical reactive categories: (1) cross-linking monomers, (2) linear condensation monomers, and (3) reformable, heat-fusible thermoplastics. The monomers in the first two categories are low to high viscosity liquids before heat reaction, but those in the third category are solid at room temperature. If a thermoplastic is heated, it too will become a liquid; however, it will be very viscous. Each of these resins must be processed differently when manufacturing composite products. Some are heated to effect a cure (hardening), while others are heated to soften them. In the latter case, the polymer will harden when cooled.

Integrating Fiber and Resin

To maximize producibility, the ideal composite material (fiber and matrix mix) would have excellent drapeability and upon being formed could be immediately hardened into a fixed and permanent shape.

The fiber has excellent drapeability. Resins, under prescribed conditions, have good plasticity. Unfortunately, when the two are combined, the result is less drapeability and less plasticity. Therefore, for ease of formability, the dry fiber should first be formed into the desired shape. Second, the resin should be integrated into the fiber, preferably as a low viscosity liquid. Third, the fiber should be permanently trapped by hardening the resin.

The process, however, is not that simple. Other factors must be considered.

1. Dry fiber, while drapeable, will not stay in a desired alignment in anything but the most simple configurations.

2. It is difficult to resin-infuse a thick fiber stack-up. Many of the desired resins will not easily penetrate into the fiber stack because they are too viscous. With thermoplastic resins, elevated temperatures are required to make them liquid. Operations in excess of 425°C appear to be unmanageable.

3. The ratio of fiber volume to resin content must be carefully prescribed and uniform to obtain optimum properties. This is a difficult task when infusing resin into a fiber bundle.

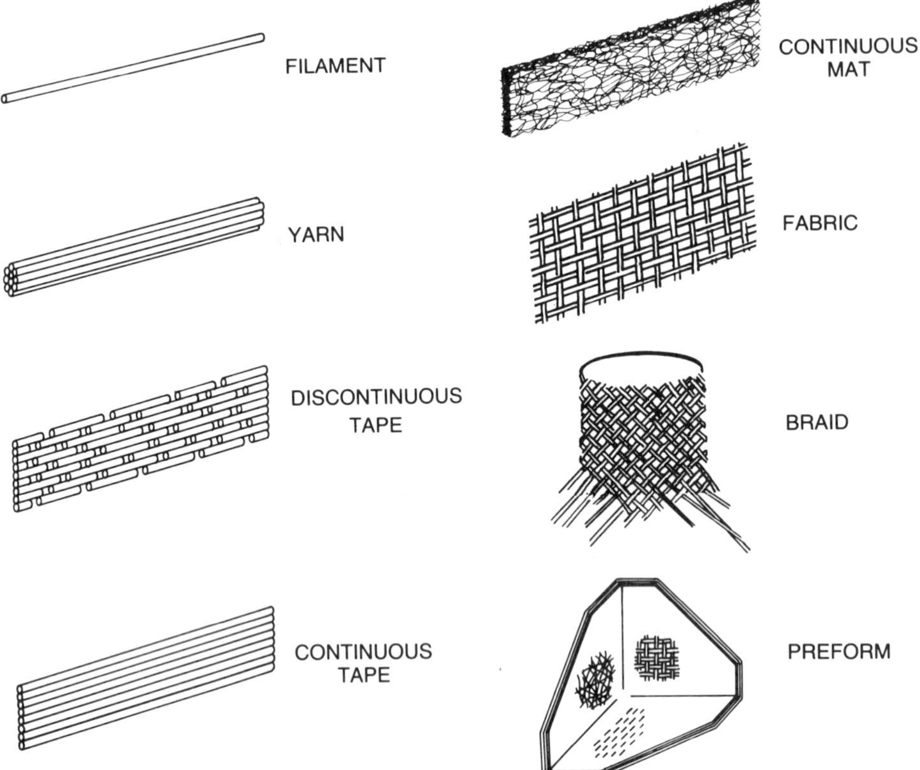

FIGURE 3 **Fiber configurations.**

4. Assuming the satisfactory accomplishment of the first three items, one must still be concerned with the phenomenon of fiber wash. Will the resin infusion process distort the desired fiber alignment?

Prepregs (Preimpregnated Fiber)

Even with such stringent requirements, the aforementioned system is a desirable one. Technological progress is being made. The most common resin integration procedure, however, is to combine the resin into collimated tape or fabric broadgoods before draping the textile into shape. These materials are called *prepregs.* This practice has brought to the trade several unique phenomena.

1. The need to store the prepreg, except for thermoplastics, at $< -18°C$ temperature. This will retard the cure of the already catalyzed resin.

2. The establishment of *tack.* This characteristic enables the material to cling to the mold wall and to itself during its placement into the mold.

3. The need for *backing film.* A film, typically of polyethylene or waxy paper, must be incorporated on at least one prepreg fiber face. This allows the prepreg to be rolled into a cylinder for storage without the material adhering tenaciously to the previous layer. Without this film, the material would be very difficult to unroll. The attached backing film also facilitates prepreg handling.

4. A concept called *out time.* Since the prepreg includes a catalyzed resin, slow curing of the resin will begin even at room temperature. Therefore, the prepreg materials should be laid up and cured, typically, in a time frame of less than 300 hours after removal from the freezer.

5. Precise *fiber–resin* ratio. Desired fiber loadings typically range from 60 to 70%. This is easily held with prepreg technology.

6. Infused resin helps to maintain *fiber alignment* and *cohesiveness.* Collimated tape and other fiber bundles tend to fray and/or separate while in the dry condition. The inclusion of a tacky rein reduces this problem.

7. Prevention of *slippage* between plies. When consolidating a set of prepreg layers into a mold, it is desirable for the materials to slip into or bottom out fully to the mold face. The desirable attribute of slippage is drastically reduced because of resin tack.

8. Use of *boardy and tackless* prepregs. For some mechanized processes such as compression molding and thermoforming, it is desirable to use a somewhat stiff and nontacky material. This is accomplished by advancing the thermosetting prepreg well into its early stages of cure or by using a thermoplastic prepreg.

There are many fiber–resin options available (see Fig. 4). Structure and process designers are advised to use the more fundamental composite materials where feasible. Why pay for an extensively processed raw material when a less processed material is sufficient?

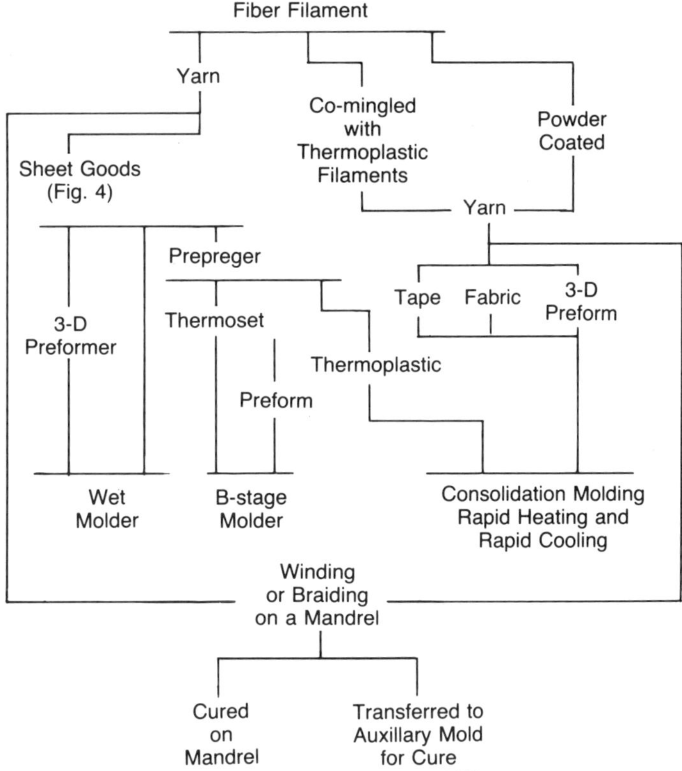

FIGURE 4 **Schematic from fiber to product.**

Building the Laminate from Laminae

The goods displayed in Figure 5 are relatively thin sheets of 0.0013–0.0038 mm thickness, not thick enough to make up a composite part. By placing ply upon ply, however, virtually any composite product can be produced. Each ply is referred to as a *lamina,* and the resultant stack is called a *laminate.* The ply stack-up shown in Figure 5 represents a lay-up of overall and partial *reinforcement plies.*

By utilizing oriented laminae, the structural designer can place fiber in a desired direction to maximize strength or stiffness. In the directions in which physical loadings are minimal, very little fiber is used. Hence, considerable weight saving can be realized. Fiber is used only where it is needed. The designer can choose the laminae (collimated tape, woven fabric, or nonwoven mat), determine whether the laminae are to be full-size or partial plies, select the fiber direction for each lamina, and finally choose the number of laminae required to complete the laminate.

Processing Considerations

MOLD-CONTROLLED SURFACE(S). A majority of long-fiber composites are mold-controlled on only one surface. The opposite face is established by the bleeder/breather pack and the bagging system. In general, the mold-controlled surface has the best finish and the highest dimensional accuracy. Where necessary, both faces may be mold-controlled by using a male-female mold set. Since a two-piece mold is considerably more expensive than a single-surface mold, single-piece molds are used wherever possible, especially for low volume production.

QUANTITY TO BE PRODUCED. When quantity requirements are small, low cost tooling and equipment is necessary. Most operations will require a predominance of human craftsmanship with relatively expensive materials in preference to mechanization and automation. While low capital expenditures are required to get pro-

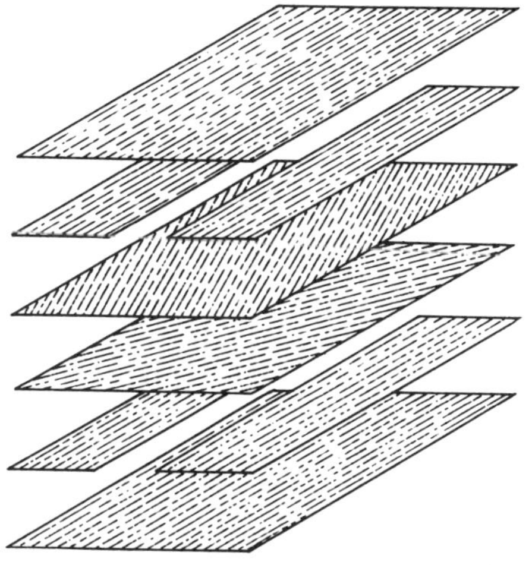

FIGURE 5 **Representative ply stack-up.**

duction started, the recurring labor and material costs for each successive part will be high.

With high quantity production, a less expensive material and decreased labor would reduce the production cost even through the initial cost of the tooling and equipment is higher.

SELECTING THE PROCESSING METHOD. The manufacturing engineer must choose a processing method or a combination of methods based upon the characteristics of the composite part to be produced. Choosing the process involves careful thought. The following case study illustrates the concept of process selection.

Case study problem. To manufacture the composite radome shown in Figure 6, several decisions must be made. The designer must choose from among the many materials and processes available. Only 50 radome parts have been ordered. This is an aircraft part that must be radar transparent. Its size is 91 cm diameter by 112 cm in length. The radome is to be placed on a low performance aircraft. To determine what manufacturing system is to be used, the following questions must be answered:

1. Which surface(s) need to be mold-controlled?
2. Which systems are cost-effective for a 50-piece production run?
3. What choices of process are available?
4. Which material will best satisfy the radar transparency criterion?
5. What is the best material configuration for the process and product?

The determination of the manufacturing method is as follows:

1. The external surface is in the aerodynamic air stream and must be smooth as molded or sanded smooth after molding. The internal surface has no precision requirement other than fit-up at the open end. This can be established with a metallic mold insert. Considering these conditions, a female, single-piece mold appears appropriate.

2. Since the radome will be used on low performance aircraft, a thermosetting epoxy matrix will be sufficient. The high temperature and more expensive thermoplastics will not be required.

3. The low quantity production does not warrant the development of a dry fiber preform. Nor does the product require two mold-controlled surfaces, as provided by resin transfer molding or compression molding.

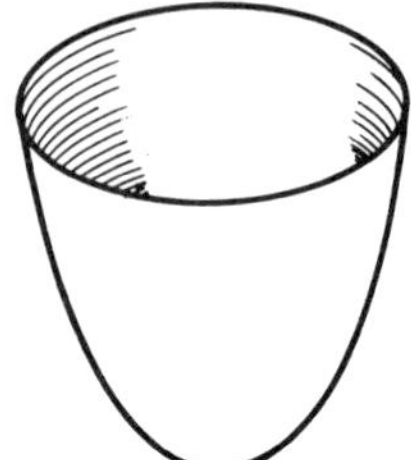

FIGURE 6 Radome.

4. The depth of draw is beyond the capability of a flat pattern stack-up. This part will need to be laid up ply by ply.

5. This aircraft part must resist water ingestion. The laminate must be dense (free from voids) and must have very good distribution of resin content. This excludes wet layup.

6. With the use of a thermosetting resin, thermoforming and diaphragm forming are not appropriate.

7. The pultrusion process produces parts with constant cross section and high aspect ratio. The radome does not fit this description; hence, pultrusion is an inappropriate process.

8. Filament winding, with on-the-mandrel curing, is a possibility, especially if two products are wound simultaneously on a siamese mandrel (see Fig. 7). These could be separated into two parts after cure. However, the mold-controlled surface is internal, and in this situation the external surface finish is the higher requirement. Nonetheless, the shape could be wound using the filament winding process as previously described. Before cure, the two fiber preforms could be separated and placed into a female mold for cure. In this arrangement, the filament winder is merely used for mechanized placement of the fiber into a preform. The curing process would be vacuum bag compacting in an oven or autoclave.

9. Because of the multiple load path requirement of this product, it probably will be built predominantly from prepreg fabric. Ply splices must be strategically located to avoid radar distortion. Glass is the reinforcement of choice because of its radar transparency. Localized reinforcement plies will be required for the attachment areas. The drastic contour change at the nose would be difficult to mold from tape.

10. Automated fabric- and tape-laying equipment developed to date works best on convex, gentle-contoured surfaces. The radome contour does not fit the normal operating mode of automatic laydown equipment.

11. With the use of prepreg fabric and hand layup, there is some merit to laying up on a corresponding male form for easy access. The preform would then be transferred to the female mold.

12. Alternatively, since the product must be molded in a converging female shape that extends beyond the length of a person's arm, the tool could be mounted on a trunion suspension system. The mold could then be adjusted and positioned for ease of layup. This would allow the hand layup operator to keep both feet on the floor.

13. Mold materials could be steel, electroformed nickel, or fiberglass–epoxy composite. These materials have a coefficient of thermal expansion (CTE) compatible with that of the fiberglass–epoxy part. The female shape

FIGURE 7 Siamese winding.

could easily be electroformed in nickel or laid up with composites. Since only 50 radomes are needed, a fiberglass composite tool is an economical choice.

14. The female molded product would then be bagged and autoclave cured.

15. The cured product must then be trimmed and finished in the post-molding operation.

LOW TECHNOLOGY PROCESSING METHODS. Early composite fabrication techniques consisted exclusively of manual labor, with laminators laying fiber sheet onto a one-piece mold. These sheets were then soaked with resin containing a curing catalyst, a process known as wet layup. No pressure other than contact pressure was required. After several hours, the part hardened and was removed from the mold. Molding tools were made of plaster, wood, fiberglass composites, and handrafted metals. The tools needed only to be exposed to low temperature and pressure. With hand labor, the parts were trimmed, sanded, and further finished.

The layup process, as described above, is still used today, but with the use of prepreg fibers, the process has been enhanced. Vacuum bagging and autoclave pressurization, in the presence of heat, have created improved products with faster molding times.

HIGH TECHNOLOGY PROCESSING METHODS. Efforts to mechanize production have led to the development of automatic layup machines, filament winders, resin transfer molding, compression molding, pultrusion, thermoforming, and diaphragm forming. In every case, the material styles must be compatible with the process.

PRODUCTIVITY CHALLENGE. Composite processing methods are nearly as varied as composite part designs. No one method can be used to make all parts. It is therefore very difficult to conduct producibility studies comparing one method with another using a common composite configuration.

Selection of a process is very dependent on the geometric shape of the composite part and the degree of accuracy required in the fiber orientation. Rather than comparing one method with another for producing a common part, the process designer is well advised to (1) carefully consider the essential characteristics of the part design and (2) choose the best process that is likely to reach the required production rate, with the least tooling cost and the lowest cost raw material going into the process. Generally low production requirements utilize more expensive preprocessed materials and low cost tools, but have high manual labor costs. Conversely, parts with higher production rates have more tooling expenditures but use less expensive materials and less manual labor. The following actions are suggestions for improved producibility.

1. Where possible, use filaments or yarn to build the part. Textile preforms, developed by winding, knitting, weaving, or braiding, are possibilities. This would bypass the operation of building broadgoods.
2. Where possible, retain the drape characteristic as far into the manufacturing process as is feasible. This will require the addition of resin after the major shape has been established.
3. Use resin systems that remain uncatalyzed until they are needed for fiber penetration. This would eliminate the refrigeration/thawing requirement.
4. Use an intermediate-length fiber that is highly oriented within a thermosetting matrix. These materials, formed by heat stamping, could satisfy many of the secondary structural aircraft needs.
5. Use matched or net molding techniques. This would greatly reduce post-finishing operations, such as sanding and trimming. In some cases, this would eliminate the costly bagging operation.
6. Use thermoplastic matrix composites where feasible. Thermoplastics can be heated quickly, formed, and cooled rapidly. High pressure thermoforming, diaphragm forming, and injection molding are areas of promise.
7. Use the shortest length fiber feasible. Although continuous fiber produces higher strength and stiffness, a shorter fiber may be satisfactory for some applications.
8. Consider filament winding. Filament winding must continue to be developed, not only for winding and curing on the same mandrel, but as a preforming system.
9. Consider pultrusion. Pultrusion is an excellent system for high aspect ratio composites where constant cross sections are appropriate. Also, a pultruded, uncured, flat sheet product, for example, could be fed directly into a matched mold for final molding of three-dimensional contoured parts.
10. Standardize part functions where possible. For example, aircraft air ducts could be made more affordable if a universal ducting system were devised for multiple aircraft models.
11. Chose geometric shapes that work well in mechanized production systems. Avoid rapid contour changes, small radii, and negative draft angles.
12. Facilitate co-curing of the several elements making up an assembly, in preference to the separate molding and curing of each element, followed by secondary bonding.
13. Although autoclave processing is the most used system for producing structural composite parts, other more mechanized processes should be considered wherever possible.

Selecting Molding Tools for Composites

Mold tooling is the essential element of composite manufacture. The part can be no better than the tool from which it is molded. A brilliantly designed part will be substandard or dysfunctional when molded on an inferior tool. Likewise, the wrong type of mold, though well built, will be unsatisfactory. With an ill-chosen tool, the production rate may be too slow. Many issues must be considered when designing molding tools.

Mold Types

The two major types of tools used to mold composites are shown in Figure 8. Matched tools are made up of male/female mold sets. These tools mold two or more surfaces of the composite.

Single-faced tools mold only one surface of the composite. Ancillary components such as bleeders, breathers, vacuum bags, and hoses are required to complete the molding system.

Matched tools are operated in conjunction with hydraulic presses. Single-face molds require an oven or pressure vessel equipment.

Mold Design Issues

1. The mold must produce the required geometric shape and dimensions.
2. Temperature—Is the upper temperature achievable? Are the heating and cooling rates sufficient?
3. Structural stability during heat cycling.
4. Facilitation of use of lower cost composite materials when feasible.
5. Transportability and handling ease.
6. Facilitation of composite part release.
7. Generation of desired composite surface texture.
8. Which part surfaces should the mold establish?
9. Will the mold properly cure the composite?
10. Durability sufficient for the required production run.
11. Construction completion within the allotted lead time.
12. Alterability to accommodate design changes and repairability should damage occur.

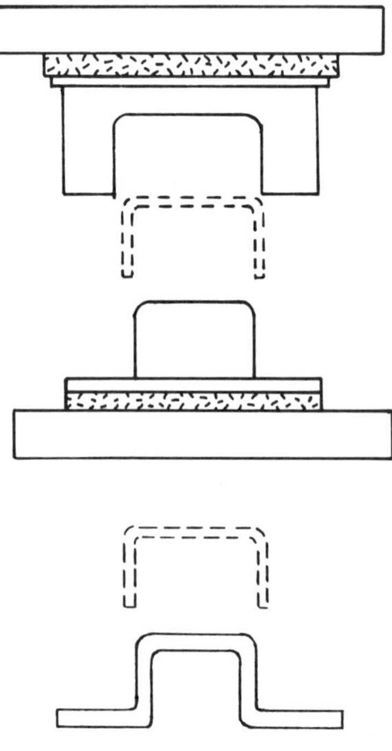

FIGURE 8 Mold types.

13. Low operating energy costs.
14. Ease of obtaining duplicate tools if required.
15. Suitability for the part size; this is especially important for very large parts.
16. Processing capability sufficient to produce the desired type and size of tool.
17. Ability to provide the required pressure on the composite at all locations.
18. Confidence of suitability for the task—limit risk taking to experimental molding situations.
19. When both heating and cooling are being supplied by the mold, the coefficient of thermal expansion (CTE) of the tooling material must match that of the composite.
20. If the CTEs do not match, dimensional compensations must be made in the mold to provide the correct part tolerance, and the mold generally must be oversized to account for composite part shrinkage.
21. Acquisition affordability for the production quantities required.

Tools that Mold-Control Multiple Faces of the Composite

For compression, resin transfer, and structural resin injection molding and for high pressure thermoforming, matched male/female mold sets are required. For prototype operations and small production runs, aluminum or composite molds are useful. However, for high quality and large production runs, the mold material of choice is machined steel.

Figure 9 depicts the salient features of a matched mold set. Temperature is established and controlled by circulating fluids or electric heaters. The mold base is generally made of an American Iron and Steel Institute (AISI) 1045 mild steel, while the mold male/female matched components are made of tool steels; AISI P20, 4130, and 420 stainless are suitable. The hardness of the tool steel during machining is typically 20 on the Rockwell C (Rc) scale. After machining, the mold faces are carburized to Rc 58–60. Flash chroming of the mold surfaces, when performed, will reduce mold wear, aid releasability of the composite part, and provide an improved part finish.

Hardened steel guide posts and bushings have a nominal 0.0013 mm total clearance. Nominal clearance of 0.0038 mm is provided at the shearing pocket.

The mold shown in Figure 9, with slight variations, can be used for compression, resin transfer, or structural resin injection molding and for high pressure thermoforming. In each of these cases, the mold would remain at a constant temperature. For compression molding, either bulk or sheet molding thermosetting compounds are used. These are loaded into the open mold in premeasured amounts before mold closing. Continuous fiber, either random or in oriented-ply stack-ups, can also be molded. The normal practice is to construct the fiber into a preform before mold loading. With a dry fiber preform, the resin is added before mold closing.

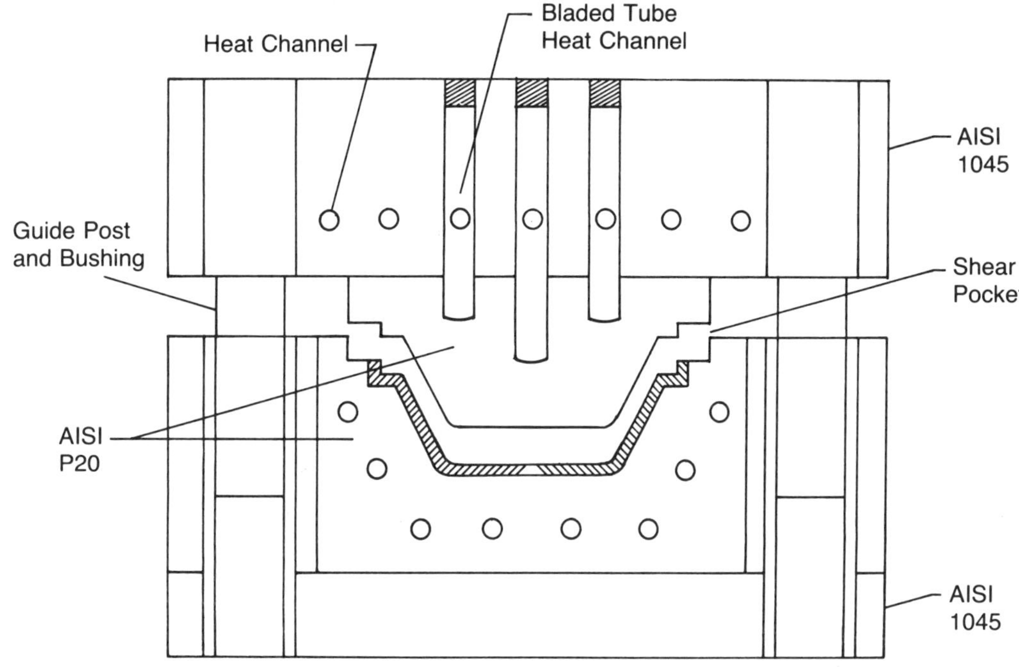

FIGURE 9 Matched mold set.

With resin transfer and structural resin injection molding, the fiber preform is placed into the open mold, and the thermosetting resin is transferred or injected under pressure after mold closing to impregnate the internal fiber bundle. Mold venting to permit the evacuation of air as it is displaced by the resin is accomplished around the mold perimeter and in some situations through the mold faces.

When forming thermoplastic composites, the materials are first placed in an auxiliary heater. After a proper melt, the materials are quickly transferred and compressed in the matched mold. Since the mold temperature is slightly below the chill point of the polymer, it facilitates both forming and cooling.

Since the mold in this discussion remains at a relatively constant temperature, CTE mismatches appear unimportant. Furthermore, shortness of cycle time can be maximized. The molder should avoid a sequence that requires heat-up and cool-down for each molding cycle.

While not shown in Figure 9, injector pins may be provided for automatic part removal when the mold is opened. With mechanized inserts, also not shown, parts with intricate features, such as undercuts, can be molded.

With matched mold tooling, it is possible to mold all composite faces, edges, bosses, and sometimes holes. The post-molding operations of trimming and drilling are virtually eliminated. The part is complete as molded, except for the deburring of the resin squeeze-out (flash), which occurs at the mating mold line.

There is considerable interest in what happens to alignment of oriented long fibers in high pressure thermoforming. Fiber orientation is relatively undisturbed for shallow and simple draws. For deep and complex draws, however, oriented long fibers are prone to breakage and misalignment. Because of these difficulties, oriented fibers 7–10 cm in length are used. While discontinuous fibers have less strength than continuous fiber, the discontinuous fiber will retain its orientation better when molded.

A specialized high pressure thermoforming technique uses only one hard tool surface. The second surface is a compliant mate made of silicone rubber.

DISADVANTAGES OF MATCHED MOLDS. While matched molds have advantages, as cited, they also have disadvantages. The double mold wall, while determining the composite wall thickness, does not assure uniform compaction. For example, the vertical portions of the molded part in Figure 9 will be less dense than the horizontal component. The reinforcement-to-resin ratio will also vary, especially in areas of rapid contour change.

It is very difficult to mold continuous-fiber parts with ply dropoffs or parts with local reinforcements. Even though carefully placed in the preform, local reinforcement will move as the mold closes.

Tools that Mold-Control One Face of the Composite

For many applications, a composite part does not need mold control on both faces. If two-faced mold control is not required, why pay for the added expense of a matched mold set? Furthermore, with a single-faced tool and a compliant pressure medium, uniform compaction can be obtained even though thicknesses vary.

ASSURANCE OF FIBER PLACEMENT. Most structural composites are made up of oriented prepreg lamina. These are laid on the mold ply by ply. Each ply, being tacky, will remain in position until the layup is completed. When compacted with a vacuum bag and autoclave pressure, the laminate receives pressure perpendicular and normal to the composite surface. Since there are no shearing forces, fiber placement is secure.

Maintaining desired orientation is critical, especially for tensile loading. Figures 10 and 11 show that strength and stiffness, respectively, are adversely affected as the fiber orientation deviates from the engineered norm.

HOMOGENEITY OF BAG-CURED LAMINATES. Uniformity of the resin-to-fiber ratio throughout the cured laminate is assured with the following practice: (a) Use a prepreg composite material that has the desired uniformity, (b) carefully lay down each ply tight against the mold face, and (c) use a proper bleeder/breather/vacuum bag installation. The bleeder will uniformly remove excess resin. The breather/bag combination will compact the composite fiber against the mold. Without compaction, fiber voids would result. Air or excess resin would then occupy the void space.

REMOVAL OF VOLATILES AND AIR FROM THE LAMINATE. To aid impregnation of fibers and to increase tack and drape, prepreg manufacturers add small quantities of solvent to the resin. While some solvents are reactive and become part of the ultimate solid, other solvents become expanding gases when heated. Since it is impossible to make intimate contact throughout the layup during ply laydown, air becomes trapped in the laminate. The resulting air pockets, if not removed, will expand when heated. Unremoved volatiles and air will cause voids and disbonding.

With the use of bleeders and breathers over the layup and with the action of the vacuum bag against the tool face, the mold becomes operational. In reality, the bag side becomes the mating tool. As atmospheric pressure is applied, all air and volatiles can escape through the breather system. This phenomenon is not fully operational with most matched male/female mold sets.

LIGHTWEIGHT MOLD CONSTRUCTION. With single-faced tools, the force of composite compaction is uniform. Importantly, the back side of the tool face receives the same uniform pressure. With this isopiestic condition, the tool face needs to be only thick enough to support its own contour and the layup. As will be discussed later in this section, a thin-faced tool greatly improves producibility.

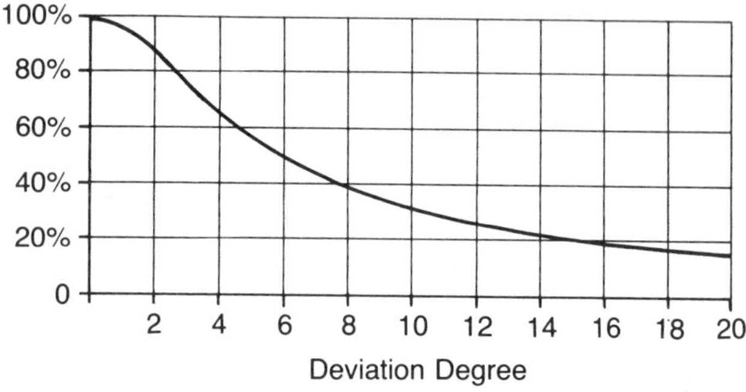

FIGURE 10 Typical tensile loss.

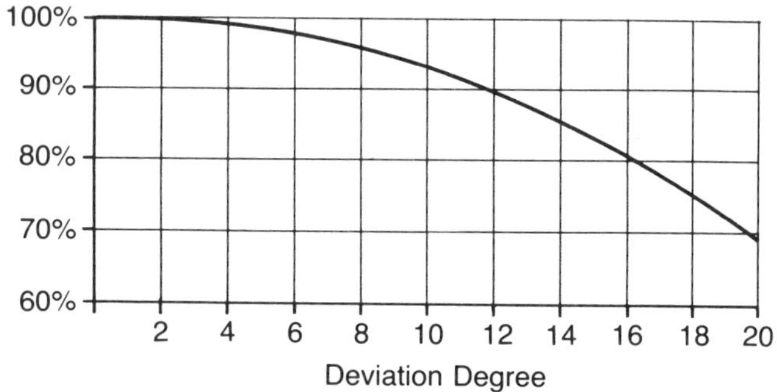

FIGURE 11 Typical stiffness loss.

Functions of a Tool Face (Fig. 12)

- Provision of required shape and dimension at processing temperature.
- Vacuum integrity.
- Hard and durable compaction surface.
- Imparting of desired surface finish.
- Heat transfer.
- Provision for composite part release after cure.

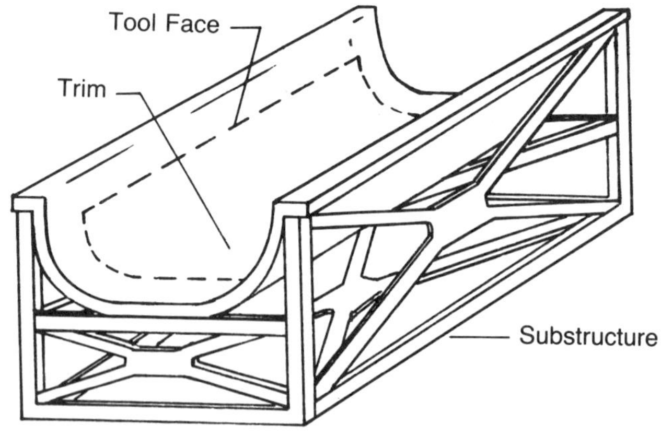

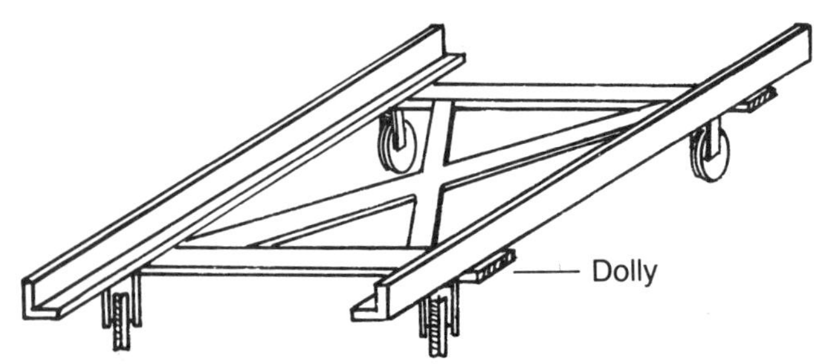

FIGURE 12 Single-faced mold tooling.

Functions of Tool Substructure (Fig. 12)

- Stiffening and stabilizing of tool face.
- Positioning the tool face for layup ease.
- Providing an interface for transport (forklift, hoist, or dolly)
- Caution: The substructure should be an open design to assure proper heat circulation to the tool face.

TOOL EFFICIENCY. Typically, single-face bag compacting tools are rotated through four work stations: (a) mold preparation, (b) mold layup and bagging, (c) oven or autoclave processing, and (d) demolding.

Each of the four stations has a unique environment and is located in a separate area of the production facility. In mold preparation, cleaning and polishing tools are used to remove resin flash and other contaminants from the tool face. When the tool face has been restored, liquid or paste parting solutions are applied to it. (Care must be taken to avoid storage or application of parting agents near and around composite materials and composite parts.) After proper drying of the parting materials, the molding tool is transported to a carefully controlled layup and bagging room. Cleanliness of room surfaces and atmospheric air is essential. Humidity and temperature levels are specified. When layup and bagging are completed, the assembly is moved into its cure station, oven or autoclave. Here the tool, composite part, and dolly move through the required thermal cycle. The last work station is the demolding area. Bagging, breathing, and caul sheets, as well as the composite part, are removed from the tool. The tool is now ready to start a new cycle of mold preparation, layup and bagging, heat processing, and demolding.

To ease transportation among the several areas and to facilitate heating and cooling, designers reduce the bulk and weight of the tools whenever possible. Also, in the interest of thermal efficiency, designers (a) construct the tool face and substructures out of materials with appropriate thermal diffusivity, and (b) construct the substructure so as to allow for free flow of oven- and autoclave-heated air to the back side of the tool face.

While thermal efficiency is important, durability of tooling must also be satisfactory. Since tools must be handled frequently, they should be tough and dependable. Often compromises must be made between thermal efficiency and durability of structure.

ANCILLARY TOOLING. When a single-faced tool is used for oven and autoclave processing, additional materials are needed to complete the system. These items, which significantly increase processing costs, are listed as fully expendable and semi-expendable supplies. When assessing the costs of expendable material, manufacturers must calculate acquisition costs of these supplies and also the labor charges for installation and removal of the bag-side expendables.

Fully Expendable	***Semiexpendable***
Parting agents	Vacuum hose
Peel plies	Thermal couples
Separator plies	Caul sheets
Bleeder plies	Vacuum exhaust valve
Breather plies	Rubber mandrels
Bagging film	Reusable rubber vacuum bags
Heating and cooling energy	
Vacuum sealant	
Flash breaker tape	
Resin dams	
Shrink and stretch tape	

MOLD TOOLING MATERIAL AND PROCESSING OPTIONS. The building of tools is preceded by enabling operations. These are:

1. Completion of the engineering drawing depicting the composite part.
2. Establishment of the production quantities desired.
3. Determination of the composite processing system to be employed.
4. Tool designing— Computer-aided design (CAD) allows for shrinkage, CTE differentials, and thermal stresses. Normally the tool geometry and dimensions *will not match* the part engineering drawing.

The material and processing choices for oven and autoclave, single-faced molds are many. The major options are listed in Table 1.

TABLE 1
Mold Material and Process Options

Metals	Polymer–Fiber Composites	Cementitious Inorganics	Monolithic Graphite
Aluminum			
Fabricating & machining	Epoxy–glass Laminating	Plasters Casting	Bulk graphite Machining
Nickel			
Electroform	Epoxy–graphite Laminating	Eutectic salts Casting	
Steel	(Linear)		
Fabricating & machining	Polyimide–graphite Laminating	Ceramics Casting	

COEFFICIENT OF THERMAL EXPANSION (CTE). Most materials expand when heated and contract when cooled. Aluminum, for example, expands 23×10^{-6} mm/mm·K.

When molding a composite, the mold must have the same CTE as the composite, or the mold dimension should be adjusted to compensate. With equivalent CTE or dimensional compensation, the composite part can be molded within dimensional tolerance. Table 2 lists typical CTE values for the more commonly used tooling materials. Note that fiberglass–epoxy, for example, closely matches the CTE of steel or nickel. With tooling of steel or nickel, no compensations are required in molding fiberglass–epoxy. An exact match is obtained when fiberglass–epoxy is molded with fiberglass–epoxy molds.

Preparatory Models Required

Regardless of the material and process chosen, a preparatory model(s) must be built before construction of tooling. Molding tools can be machined from bulk graphite, aluminum, or steel with tracer mills. The tracer operation requires the use of a carefully built model. This tracer model is typically built from plaster or wood.

With numerical control (N/C) and computer numerical control (CN/C), the model is a stored electronic model. From the electronic model, a cutter path can be programmed. When a multiple-axis milling machine is driven by the computer, a very accurate tool face of aluminum, steel, or bulk graphite can be achieved. Additional machining and fabrication would be required to obtain a uniform tool face thickness and to establish the substructure.

Should the designer choose nickel, composites, or ceramics for the tool face, coordinate data, full-scale loft lines, electronic models, or a three-dimension physical model(s) must be constructed. In these situations, a minimum of two steps is required to build the tool face. Frequently, however, three to five steps are required for tool-face construction. In the interest of cost reduction and achievement of desired dimensional tolerance, the tool builder is normally advised to minimize the number of preparatory steps. Nonetheless, model and pattern making will remain a necessary craft.

TABLE 2
Coefficient of Thermal Expansion

Materials	Expansion Coefficient mm/mm·K $\times 10^{-6}$
Hardwood	36 (across grain)
Plaster	16
Plastics model board	36
Cast epoxy	45
Graphite–epoxy (60–40%)	2
Fiberglass–epoxy (60–40%)	14
Aluminum	23
Steel	14
Nickel	13
Monolithic graphite	4
Ceramics	1

Model and Pattern Making

The craft of making models and doing the various preparatory steps required to build tools is called pattern making. A *master model* is essentially sculpted from stable and easily manipulated materials by hand and machine. From the master model, a *reverse replica(s)* is cast. Finally the tool face is either electroformed from nickel, cast of ceramic, laminated with a polymer–fiber composite, or laminated with a ceramic–fiber composite. A typical sequence of master model, reverse replica, and molding tool face is shown in Figure 13.

MASTER MODEL MATERIALS.

Hardwood. Although more modern modeling materials have been developed, wood continues to be used for some applications. Thin cross-sectional protrusions made of wood are more durable than those of plaster, plastics, or wax. Care must be taken to seal the completed wooden model properly to avoid shrinkage or expansion resulting from changing moisture environments. The taking of plaster splashes (reverse replicas) can be destructive to wood without extreme precautions in sealing. In an attempt to stabilize wood, it is frequently cut into thin planks that are glued together into a block. Each thin plank is of quarter-sawn stock. Wood is fairly stable in its length, but less stable in width and thickness. When taking reverse replicas from wood, temperatures above 65°C should be avoided.

Paper–phenolic laminates. Paper–phenolic laminates, nominally 19 mm thick, are bonded together to make the desired thickness for the model. With wood and metal cutting machines, these materials can be precision sculpted. Laminates of paper–phenolic have better stability than wood. Expansion and shrinkage factors, while minimal, are equal in all directions of the master block. When taking reverse replicas from the model, temperatures above 95°C should be avoided.

Plaster. Plaster is a highly versatile modeling medium. The dimensional stability and precision of plaster are satisfactory for most applications. Plaster, while curing, has a slight expansion factor, an advantage for modeling. Consequently, the model can be smoothed by sanding without falling under dimension.

Plaster can be machined after it has been poured as a liquid and then hardened; however, machining is a rare practice. shaping of fluid plaster is accomplished by *sweeping* and *screeding*. Sweeping requires a rigid metal template. The sweeping template is guided by hand against a mounted base template. Screeding of fluid plaster is done with a flexible steel blade. Vertically mounted

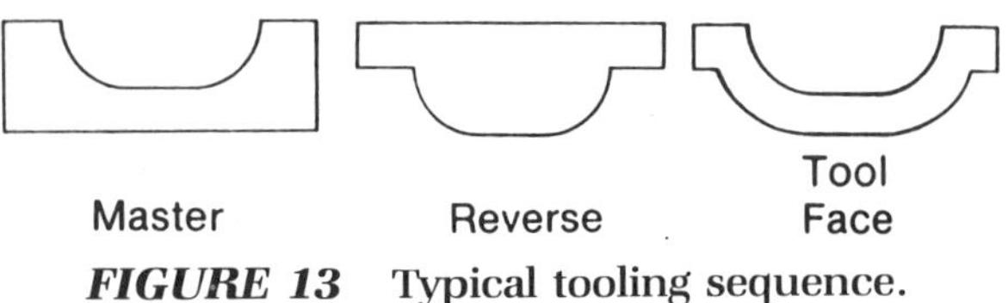

FIGURE 13 Typical tooling sequence.

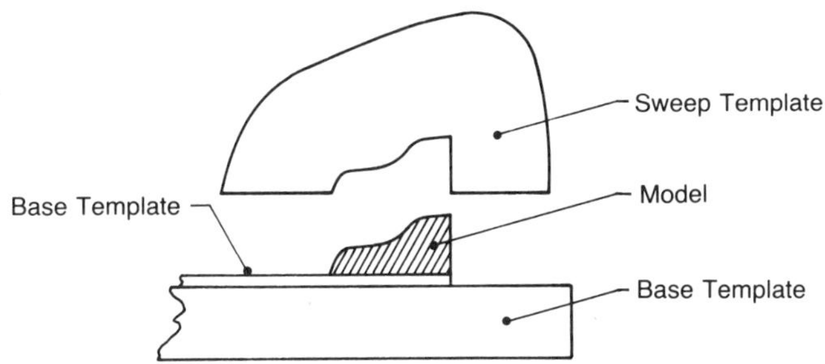

FIGURE 14 Sweeping plaster.

loft templates serve as guides for the flexed steel during the screeding process. See Figures 14 and 15.

Machined metals. Occasionally, master models are machined from aluminum or steel blocks. In these situations, a great deal of intricate detail is desired. Scaled aircraft models, for example, not only supply the tool maker with needed airfoil shapes, but can be used by flight dynamic engineers for water and wind tunnel testing.

Contemporary modeling stock. In the interest of higher precision, stability, and machineability, several alternatives to wood, paper–phenolic laminates, and plaster have been developed. *Monolithic graphite blocks* are easily machined and show virtually no change in dimension when cycled through heat environments. High heat cycling represents no difficulty, since the graphite was manufactured at a temperature in excess of 1000°C. A major difficulty, however, is its softness and lack of toughness. Because of its porous nature, vacuum integrity is attainable only with special sealer coats. *Plastics model planks, epoxy cast surfaces, uncured foam slabs,* and *wax blocks* are all useful at lower temperatures than are possible with monolithic graphite. Like graphite, all machine well. With plastics model planks and wax blocks, an oversized structure must first be created by gluing the planks together. These are mounted on a rigid platform base with special adhesives. The desired model is then N/C or CN/C machined. An automotive scale model made of wax, shown in Figure 16, illustrates several machining finishes. The right-hand hood, upper glass, and roof areas have the final machined finish.

Epoxy cast surfaces have some advantages over segmented plank build-ups. Their surface is continuous, without joints. A polymer–fiber laminate *armature* is first built to an approximate shape. By creating an offset gap between the armature and the plaster cover, space is provided for casting the epoxy surface (see Fig. 17). After resin cure and plaster removal, the epoxy surface is N/C machined to the precise contour.

To create a master model with uncured foam slabs, a composite armature is first built, as shown in Figure 17. The uncured foam slabs, roughly 1–2 cm thick, are contoured to the armature. The foam slabs have sufficient tack to adhere to the armature and to the adjacent foam segments. After room temperature cure, the surface is N/C machined to contour. Since the foam is made up of small spheres, which machine easily, an excellent surface finish can be obtained.

DIMENSIONAL COMPENSATION. With a few exceptions, master models reflect an exact three-dimensional shape as specified by the engineering drawing. Since several steps are required between model and composite part production, dimensional compensation may be required in the master model. Shrinkages or expansions, CTE mismatches, and thermal distortions must be calculated for every step.

REVERSE REPLICAS. When lifting exacting contours from master models, it is necessary to *splash* the master. The resulting product is a reverse replica. The lifting medium is plaster, plastic-faced plaster, cast ceramic, or a polymer–fiber layup. In the last case, a syntactic foam

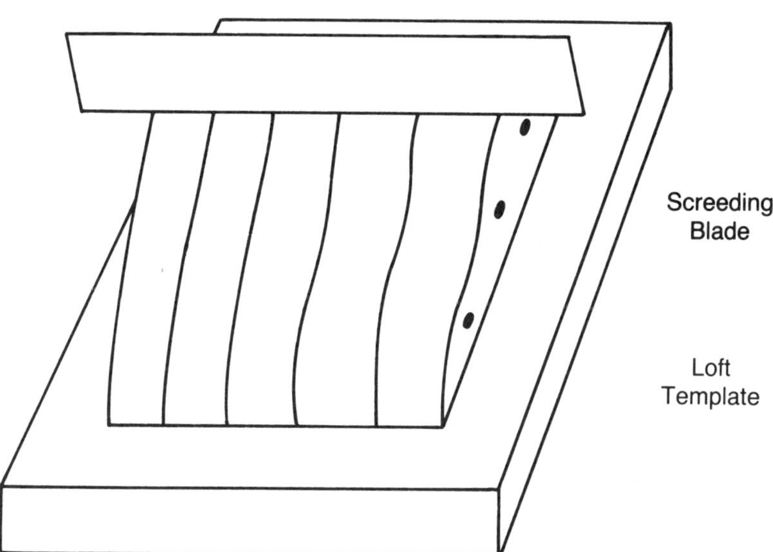

FIGURE 15 Screeding plaster.

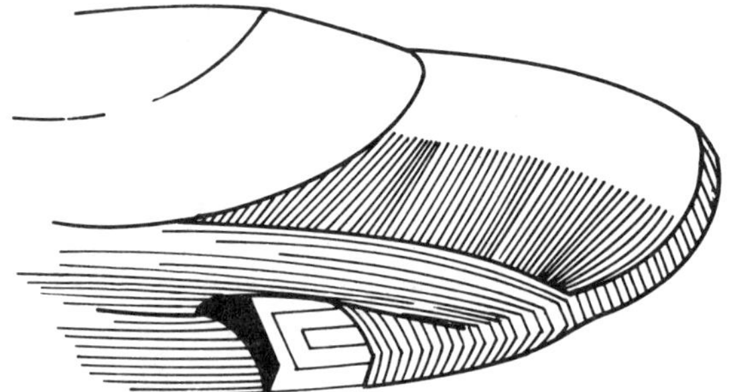

FIGURE 16 Machined wax model.

core can be used to help stiffen the laminate (see Fig. 18). Although splashing operations are accomplished at room temperature, slight heat is generated as the reverse replica material chemically hardens.

Since considerable cost and effort are expended in making the master model, great care must be taken to protect it when taking reverse replicas. Surfacing the master model with good parting agents is extremely important. Reverse replicas of plaster, ceramics, and plastic-faced plasters are generally used only once, while polymer–fiber laminates can be stored and readily reused if necessary.

Building the Tool Face and Substructure

POLYMER–FIBER COMPOSITE TOOLS. Epoxy–fiberglass and epoxy–graphite tools have gained wide acceptance in the aerospace industry. Experience in building polymer–fiber composite parts is easily transferred to building composite tools. Furthermore, the molding of fiberglass parts on fiberglass tools constitutes a CTE match. Likewise, a CTE match is created when aramid and graphite fiber composites are molded with graphite composite tools. Since composite tools are lightweight, they can be easily transported. Most polymer composite tools, especially polymer–graphite, are responsive to thermal cycling. Epoxy composite tools are limited to 175°C service.

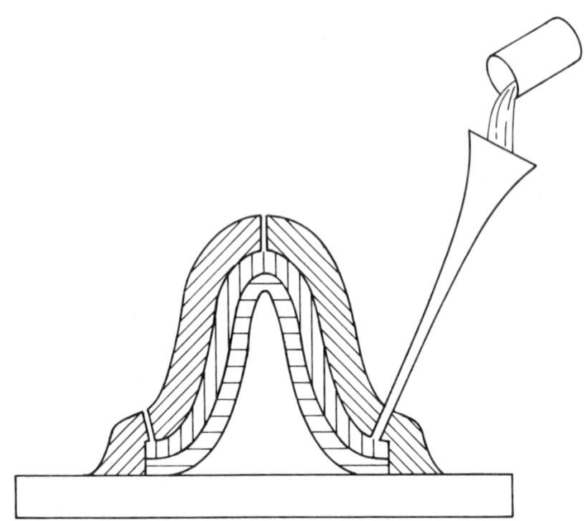

FIGURE 17 Composite armature with epoxy face.

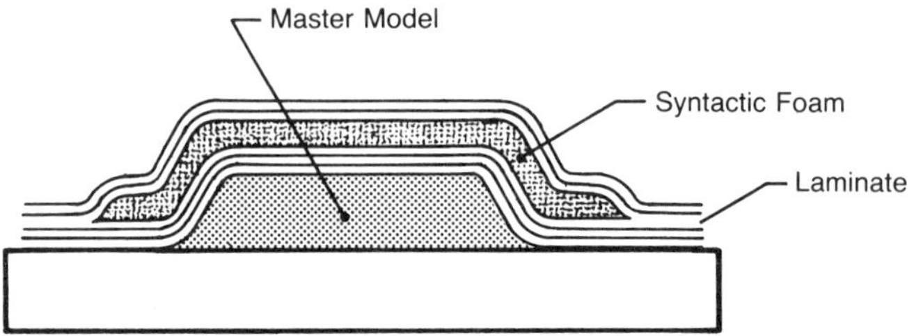

FIGURE 18 Laminate/foam splash.

When linear polyimide–graphite composites are used, processing at 400°C can be achieved. These tooling composites can be used for processing phenolics, bismaleimides, polyimides, and high temperature thermoplastics.

There are several ways of building a composite tool. The most prevalent system is to start with a swept plaster master or a machined master of carvable stock. A reverse replica of plastic-faced plaster or composite layup is then lifted from the master. When possible, the reverse replica should be built of the same materials that will be used for the final tool—a glass or graphite reverse is built for a glass or graphite tool respectively. During cure, compaction is achieved with a vacuum bag. After room temperature cure of the reverse replica composite face sheet, a substructure of similar materials is bonded to the tool face. The reverse replica, with the attached substructure, is then wedged from the master and postcured to 125°C in an oven. Parting agents are then applied to the reverse replica.

The desired tool face is now laid up ply by ply on the reverse replica, typically of preimpregnated fabric. Ply layup is done systematically so as to achieve a balanced and symmetric layup (see Table 3). Each ply is made up of numerous small pieces approximately 30 cm square. A tool face thickness of 51–64 mm is sufficient for most applications.

After the required plies have been applied, a bleeder/breather pack is placed over the laminate. This assembly

TABLE 3
Tool Face Layup Sequence (One Example)

Ply	Fabric Weave	Degrees Orientation	
1	Fine	0	
2	Fine	+45	
3	Coarse	90	
– – – – –	– – – – –	– – – – –	Vacuum debulk
4	Coarse	−45	
5	Coarse	0	– – Mid ply of symmetry
6	Coarse	−45	
7	Coarse	90	
8	Fine	+45	
9	Fine	0	

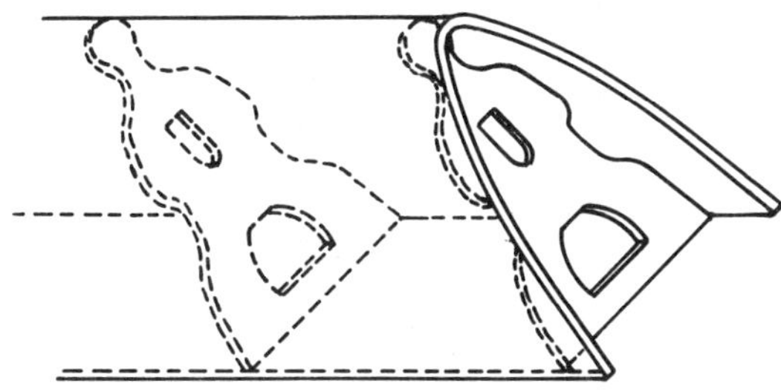

FIGURE 19 Fabricated and machined metal tool.

is then vacuum bagged and autoclave cured at a maximum temperature of 120°C and augmented pressure. Once again, after cure and debagging, a previously assembled substructure is attached to the tool face. After being removed by wedges, the tool is cleaned up and sent through a 175°C post-cure. With application of necessary parting agents, this tool is now ready for 175°C service (see Fig. 12).

If 400°C service from a composite tool is desired, a reverse replica of cast ceramic or machined monolithic graphite is required. In this case, the tool face is cured in an autoclave with temperature and pressure of 390°C and 15 atmospheres.

FABRICATED AND MACHINED METAL TOOLS. N/C-machined tools of aluminum and steel are fabricated from a rough-formed face sheet that is sufficiently thick to permit finished machining. Each piece of the substructure must be machined and then welded to the face sheet. The rough assembly must be stress-relieved before final machining.

The machining program must compensate for CTE mismatches between the metal tool and the composite to be molded. Also, a reasonable effort should be made to obtain a face sheet of uniform thickness. This will ensure consistent CTE and heat transfer throughout the tool (see Fig. 19).

FORMED METAL TOOLING. In situations requiring two-dimensional molding, formed metal tooling is a viable option. Thin metal face sheets can be rolled, brake formed, or stretch formed to the desired contour and size. These sheets are affixed to corresponding header plates made of the same metal (see Fig. 20).

ELECTROFORMED NICKEL TOOLING. Nickel tools are among the hardest tools used for composite manufacture. With face sheets of 5/mm thickness, their heat transfer is superior to that of a more massive steel tool. Nickel with dimensional compensations can be used to mold composites at cure temperatures of 425°C. An excellent finish can be obtained with electroformed nickel, mirroring the reverse replica finish.

Nickel tool faces are electroformed by placing a fiberglass–epoxy reverse replica in a plating tank. Before the replica is placed in the electroplating tank, an electrically conductive coating is placed on it. The process of forming the desired nickel thickness usually takes several weeks of tank time. Before the nickel is removed from the replica, a substantial steel substructure is mechanically attached. A completed tool is shown in Figure 21.

MONOLITHIC GRAPHITE TOOLING. Monolithic graphite blocks, as noted earlier, can be used for master models and reverse replicas. This material can also be used for the final molding tool. Bulk graphite is able to withstand temperatures high enough to mold the highest temperature polymer–fiber composites presently available. Monolithic graphite and graphite composites are CTE compatible. Monolithic graphite is easily machinable, but a good dust collection system is required.

Although the thermal conductivity of graphite is excellent, its thermal response is quite slow because of the thick sections required to offset its fragility. Graphite is also soft and easily damaged. Since the material lacks vacuum integrity, it is generally envelope bagged to ensure proper compaction of the composite being molded.

Substructures are usually made up of steel frames. Provision is made to allow the graphite to float, since its CTE differs from that of steel.

CERAMIC TOOLING. Ceramic tooling is derived from castable compounds that have virtually no CTE. With the addition of specified fillers, a regulated CTE can be achieved. Dimensional tolerances can be held with castable ceramics. Ceramics have a predictable but low shrinkage factor from liquid to cured castings.

For most ceramics, vacuum integrity is achieved by sealing the tool face surface with organic compounds. With some ceramics, vacuum integrity can be achieved as cast.

Ceramic tools are heavy, and their response to thermal cycling is slow. However, ceramics are capable of

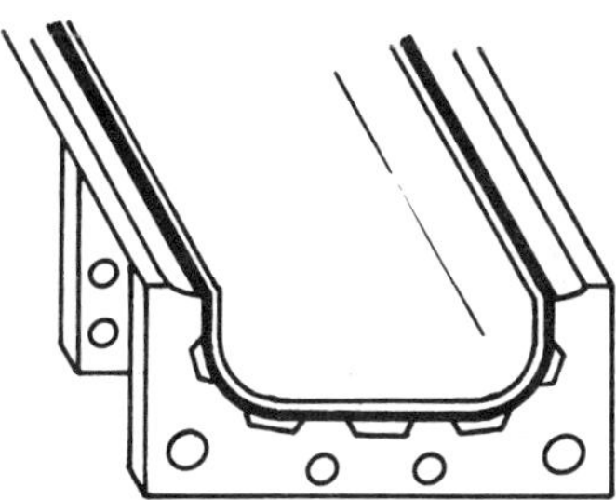

FIGURE 20 Formed face sheets and machined headers.

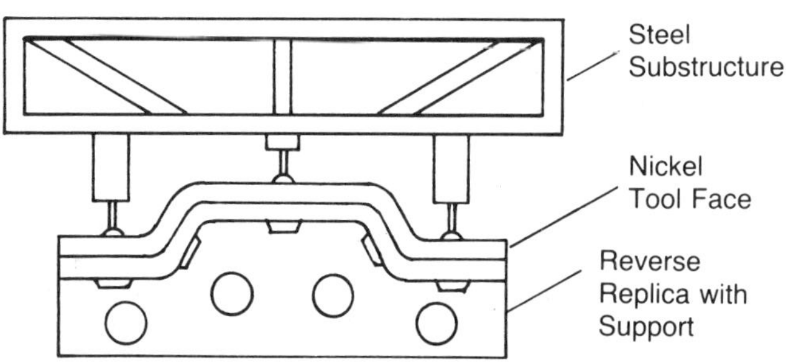

FIGURE 21 Electroformed nickel tool.

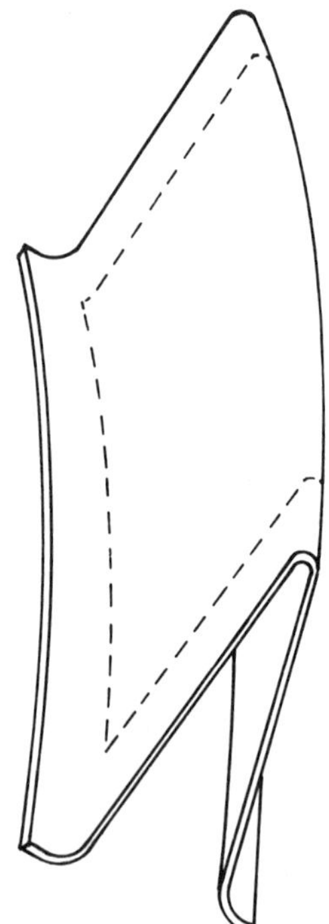

FIGURE 22 Ceramic tool face.

temperatures up to 1000°C. Ceramics should be well protected from shock and abuse, since they are brittle and relatively unrepairable when broken. As with monolithic graphite, a steel substructure, upon which the ceramic can float, is provided.

A mold shape served very well by ceramics is shown in Figure 22. This tool would be expensive if produced from steel or aluminum. The shape shown is difficult to lay up in composite. Monolithic graphite would be too fragile.

BREAKAWAY TOOLING. Ducts and other enclosed cylindrical shapes often require smooth interior surfaces. Some of these parts, because of their complex geometry, cannot be removed from normal tooling. Parts with these characteristics are built on male forms called *breakaways.* While many breakaways are actually broken out from the cured composite in small pieces, others are dissolved or melted out as a liquid slurry. Some breakaways are actually *take-apart mandrels.* These are built from detachable metal segments. After being removed from the composite, these pieces can be reassembled and held with mechanical fasteners for another molding operation. In Figure 23, the ball valve body, with its internal protruding segments, must be laid up on a partial take-apart component. Upon completion of the partial layup, the remaining tooling component is added. A final layup encompasses the entire take-apart tool.

Most breakaways are built from castable plasters or eutectic salts. Both plasters and eutectic salts are castable in multiple-piece female tools. Since plaster and salts are slow to respond to thermal cycling, these materials are cast hollow, where possible, to reduce their thermal mass. Plasters are normally cast in low temperature

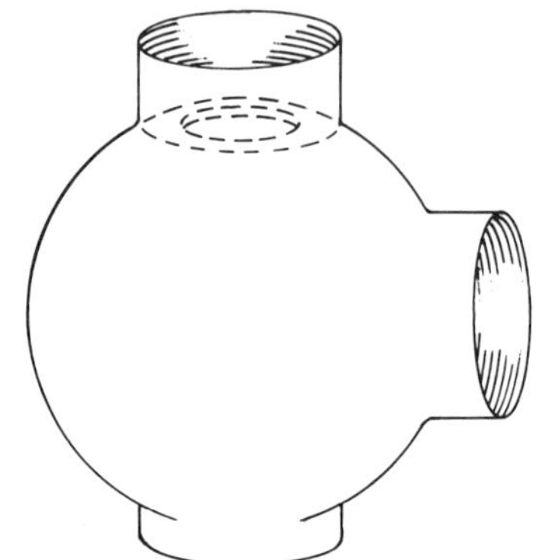

FIGURE 23 Composite ball valve body.

polymer composite tools, since no elevated temperatures are required. Eutectic salts are hot melt solutions that require aluminum or high temperature composite casting molds. With 500°C salts, for example, ceramic casting tools would be required. Eutectic salts are formulated for 150–540°C melt ranges. A cross section of a mandrel and mold is shown in Figure 24.

Filament winders frequently use wash-away mandrels built from bonded sand particles. Sand of the desired particle size is mixed with a water-soluble binder. Typical binders are polyvinyl alcohol and sodium silicate solutions. These are cast in female molds, where they solidify. After completion of the winding and cure operations, the mandrel is dissolved from the composite.

Cocure vs. Precure with Secondary Bonding

Whether to cocure or to precure with secondary bonding is an important decision that must be made before tool fabrication. The principle accepted by the composites trade is to cocure whenever feasible. Producibility is generally enhanced when cocure operations are achievable.

When cocuring, only one major molding tool and one cure cycle is required, whereas secondary bonding requires a bond tool in addition to the separate mold tools required to make the several elements. Preparing pre-

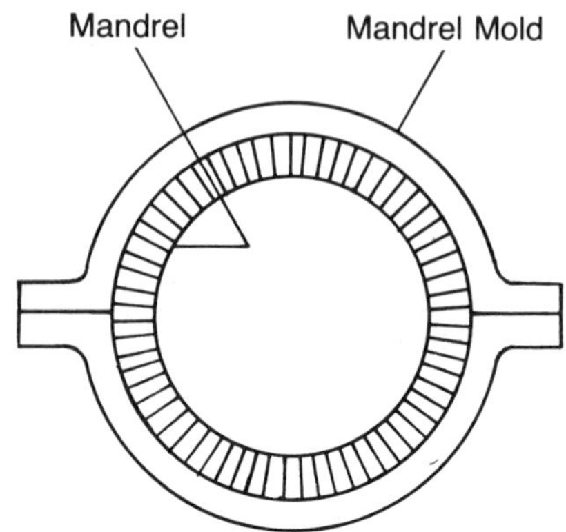

FIGURE 24 Mandrel and mold cross section.

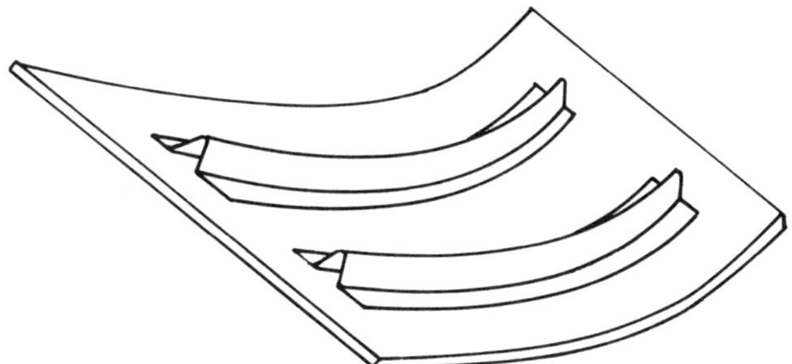

FIGURE 25 Stiffened panel.

cured elements for subsequent bonding is an additional and critical chore, a step which is bypassed when cocuring.

Cocuring often results in superior integration because all elements are free to flow and tightly compact. In contrast, mismatches among the separate tools and separate elements will create voids when secondary bonding is necessary. These voids are actually disbonds.

COCURE CASE STUDY—STIFFENED PANEL. The stiffened panel shown in Figure 25 may be cocured or precured with secondary bonding. Since cocuring offers producibility advantages, the following steps are suggested.

1. Develop a molding tool that mirrors the convex face of the panel.
2. Develop a split casting tool for casting the mandrels. The mandrels will mirror the inside shape of the stiffener. Breakaway or solid rubber mandrels may be used.
3. Lay up the panel face with prepreg fiber.
4. Position the two mandrels over the layup.
5. Lay up the stiffener members over each mandrel.
6. Bag as required and autoclave-cure.

COCURE CASE STUDY—SANDWICH PANEL. The sandwich panel shown in Figure 26 is made up of two composite face sheets, one sculptured honeycomb core, and a plastic cylinder that conceals and seals the honeycomb at the hole location.

The sandwich panel as shown cannot be cocured. Virtually no compressive forces can be exerted on the honeycomb edges perpendicular (90° angle) to the cell walls. The honeycomb edges would collapse completely under bag pressure.

This panel could be cocured if a structural foam were substituted for the honeycomb core. The foam could withstand the required autoclave pressures. With the foam substitution, the following cocure approach could be followed.

1. Develop a molding tool that mirrors the convex curved face. A male plug would be attached to this mold. The plug would establish the inside diameter of the hole.
2. Lay up the curved laminate with prepreg fiber.
3. Place the presculptured foam over the mold plug and rotate to proper position. Before placement of the foam, wrap the male plug with a thin layer of foaming adhesive.
4. Vacuum debulk the partial layup.
5. Continue the layup by placing prepreg over the sculpted foam. The layup extends beyond the foam insert to the first-stage layup.
6. Bag as required and autoclave-cure.

Conclusions

Mold design is an important process. The producibility of a composite is directly related to the level of tool technology chosen. Simple tools can be used to mold complex parts; however, the composite materials and labor

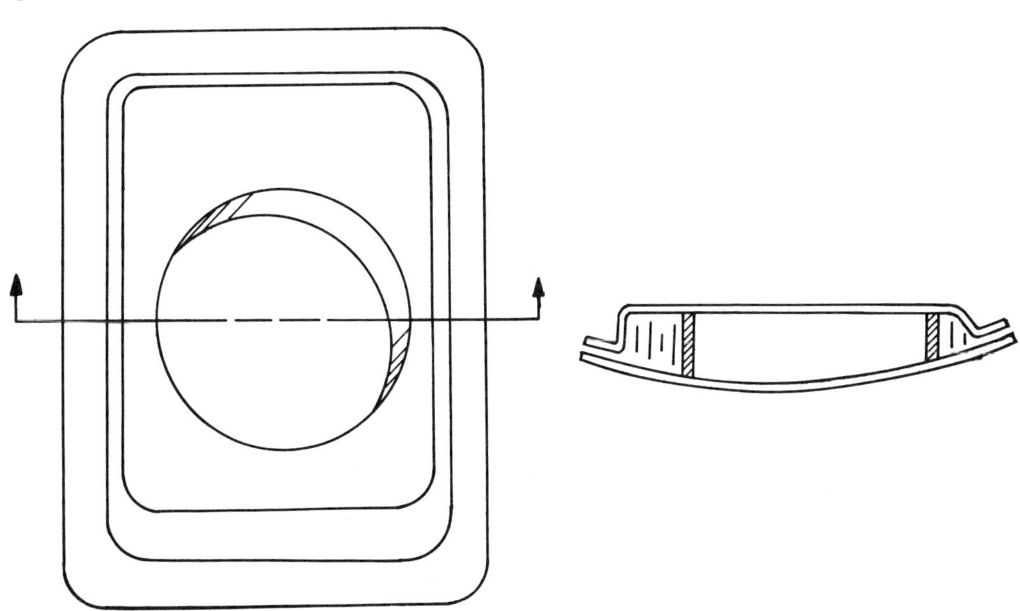

FIGURE 26 Sandwich panel.

are costly. The post-molding operations of trimming and finishing would be required. More sophisticated tools can produce parts that are virtually complete as molded. This type of high technology tool is affordable for only two markets:

1. High production products
2. Scientific products of national priority

There are many variables that should be considered when designing tools. The requirements are:

1. Temperature
   - Heating and cooling rates
   - Upper heat requirements
   - Thermal stability
   - Coefficient of thermal expansion
2. Durability
   - Compression, flexure, and tensile properties
   - Hardness and toughness
   - Maintenance of fit-up
   - Repairability
3. Surface finish
   - Establishment of desired texture
   - Releasibility of the molding
4. Part geometry and complexity
   - Male or female mold
   - Single- or multiple-piece mold
   - Breakaway mold
   - Cocure or precure molding
5. Composite material
   - Yarn, sheet goods, preconsolidated goods
   - Short, medium, or continuous fiber
   - Resin infusion before molding
   - Resin infusion while molding
6. Transportability
7. Lead time for tooling construction
8. Affordability

Selecting an appropriate tooling material is an important task. The choice of material is related to the intended use. Master models, reverse replicas, and heat cyclable tooling each require materials with special properties.

Finally, computer technology can be used to build data models that account for thermal stresses, CTE, and shrinkage factors. From such a model a cutter path can be developed. With cutter path data, the tool room can efficiently machine the master models or the desired molds.

D. Wayne Becker

Molding, Resin Transfer

Resin transfer molding (RTM) is now generally accepted as the best candidate for fully automatic production of complex composite parts. The rapid growth of injection molding processes in the past has been due to the successful application of automation techniques.

As attempts are made to increase the production of fiber-reinforced plastics to meet the increasing demand from new applications, currently popular composite fabrication techniques have been found lacking in several aspects. Compression molding techniques have high tooling costs and prohibitive press requirements. Filament winding processes cannot easily produce consistent part thicknesses and are limited to producing symmetric shapes. The pultrusion process is inherently limited to producing simple shapes with a constant cross section. Prepreg techniques require characteristically long cure cycles, utilize relatively expensive starting materials, and are labor-intensive.

RTM offers the promise of producing low cost composite parts with complex structures and larger net shapes. Relatively fast cycle times with good surface definition and appearance are easily achievable. The ability to consolidate parts allows the saving of considerable amount of time over conventional layup processes. Since RTM is not limited by the size of the autoclave or by pressure limitations, new tooling approaches can be utilized to fabricate large, complete structures. In some applications, RTM is the only way to manufacture certain complex structures.

In the RTM process, only the basic raw materials are required: a matrix resin and some fiber reinforcement. Fiber volumes of 55–65% can be achieved without appreciable difficulty. Chemical exposure of shop personnel is greatly reduced as a result of the closed-mold nature of the RTM process; the chemicals are handled inside process equipment.

Unfortunately, the development of the RTM process has not fulfilled its potential completely. The acceptance of RTM as a viable process has been resisted because of the general opinion that the process is wasteful of resin and is best suited for low tech applications. In addition, the RTM process is thought to be inherently difficult to automate.

Unresolved issues exist in RTM today, mainly in the areas of automation, preforming, tooling, mold flow analysis, and resin chemistries. The fragmented nature of the composites industry has compounded the difficulty of defining the RTM process so that these issues can be addressed. During the last four years, there have been rapid advances in RTM technology development that demonstrate the potential of the RTM process for producing advanced composite parts.

In the following sections, the resin transfer molding process will be discussed in detail, with emphasis placed on learning the basic steps. The discussion will cover variables such as mold complexity and materials of construction. Several methods of heating and cooling the mold will be suggested. Mold flow will be touched upon, as well as suggestions for gating.

Fiber reinforcements will be discussed, with emphasis placed on learning the characteristics of the various commercial types. Typical composite data will be shown for each general type of reinforcement.

The two basic types of resin delivery systems will be

discussed in detail. The relevant advantages and disadvantages of each will be summarized and discussed.

The physical properties of epoxy, vinyl ester, and polyester resin systems will be discussed, with a goal of understanding the advantages and disadvantages associated with the different types. A set of resin parameters will be given that can be used to screen new resin systems.

RTM Process Description

The actual process of RTM is very simple. A preform of dry fabric is laid in the mold, the mold is closed, and a pipe coming from the resin supply vessel is attached to the mold. A valve between the resin supply vessel and the mold is opened, and the resin is allowed to flow from the supply vessel to the mold. When the mold is full, as evidenced by resin issuing from other openings on the mold, the mold is sealed and the resin is then cured. The part is taken out after the mold has cooled sufficiently (Fig. 1).

In actual practice, the process is much more complicated. For example, in most cases the mold surface must be treated with a good release agent so that the composite can be removed without damage to itself or to the mold's face. The process of placing the reinforcement in the mold is always very involved, usually requiring several hours of labor unless a preform is available. The delivery of the resin system via a pumping device from the supply tanks to the mold is not a trivial matter. The successful bleeding of air from the mold requires judicious placement of the resin outlets so that the flow of resin does not trap air in the mold. The design of the mold has to address how the mold will be heated, what medium will be used, and how the mold will be cooled. The resin chemistry has to be selected such that the curing reaction is consistent with cycle time needs. The resin system has to accommodate the physical requirements of the RTM process. Finally, the mold design also has to anticipate the difficulty inherent in demolding cured thermoset resins.

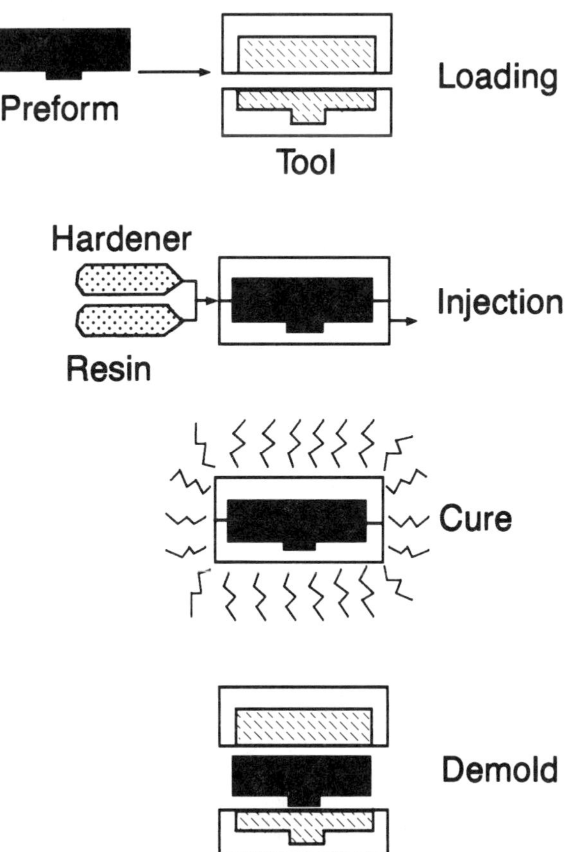

FIGURE 1 RTM process.

The following sections address these points in more detail. The first section deals with elements of mold design.

Mold Design Considerations

RTM mold design and construction are two of the most important factors in the successful use of the RTM process. Mold design may be broken down into five major areas: materials of construction, temperature control, inlet and outlet port design, gasketing (or sealing) scheme, and mold cavity design.

Materials of Construction

The most suitable material of construction for a mold is steel. Compared with aluminum or composite tooling, steel tooling provides much longer face life. Unfortunately, the disadvantages with steel tooling are the increased weight of the tool and the increased fabrication costs. Small tools (0.61 m × 0.61 m × 1.22 m) often weigh in excess of 454 kg. Tooling costs are high because of the amount of skilled labor needed to fabricate the mold.

Aluminum is generally considered satisfactory only for construction of prototype molds. The tooling costs are much lower, since the metal is softer and easier to machine. Because of the difficulties in welding aluminum, jacketing for the heating and cooling steps is not usually used. The molds are usually heated in an oven to cure the resin.

Another consideration in working with aluminum is that the metal galls very easily. This usually happens during the assembly of the mold if steel bolts are used. Aluminum molds also easily distort under moderate injection pressures if the mold is liquid-full.

Process development is currently taking place in adapting composite tooling to the needs of the RTM industry. Conventional construction techniques using steel frames to support composite mold faces are cost-effective for production runs of up to 5000 pieces. One complication with this type of mold construction is the "print through" effect on the working face. The composite skin will transmit the outline of the steel frame through to the face of the mold as a result of the effects of the elevated temperatures used during molding. This problem has been alleviated by using core materials in

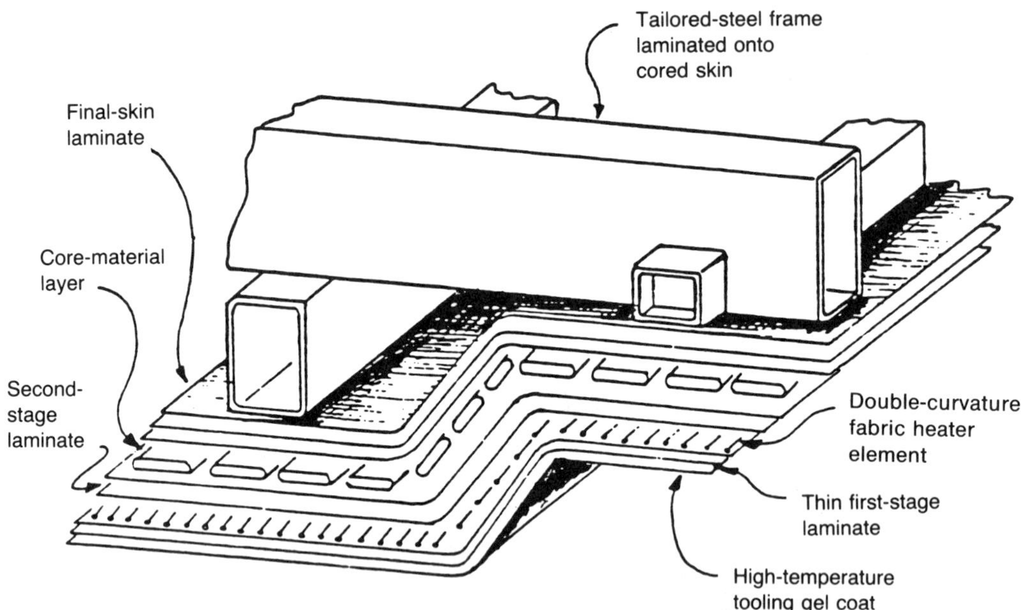

FIGURE 2 Composite mold construction.

the composite face to stiffen its cross section. This is shown in Figure 2.

Lengthening the life of the composite tool face is also desirable from a standpoint of lowering costs. The factors that cause deterioration of the mold face are temperature fatigue and/or attack by release agents. Electroplating the mold face with nickel metal is a new technique that will extend its face life by a factor of 10 times. This is a superior alternative to using low melt metal spray finishing techniques.

Heating and Cooling Design

The RTM mold has to be able to heat and cool the part during the fabrication cycle. Most resin systems cure faster at elevated temperatures. During demolding, lowering the temperature is sometimes helpful in removing the part. It also lowers the part's temperature below the glass transition temperature or softening point.

Normally, the mold is heated and cooled using either hot water or oil. The mold is constructed so as to allow the heating/cooling fluid to flow through passageways in its interior, similar to those in an engine block. The fluid is heated by conventional means, such as a gas- oil-fired heater. The fluid is cooled by chilled water using a conventional shell and tube heat exchanger.

When specifying the size of the heating unit, allowances must be made for the size of the mold to be heated. Large amounts of heat must be exchanged to raise even the smallest steel mold from room temperature to RTM temperature, and from there to the curing temperature. For example, for a 454 kg mold (0.3 m × 0.6 m × 0.6 m) with an assumed heat capacity of 1.26 kJ/kg·K, the heat required to raise the mold from 23°C to 57°C (the molding temperature) is 190,000 kJ. To raise the mold from 57°C to 135°C (the curing temperature) requires 316,500 kJ.

Putting this amount of heat into the mold cannot be accomplished instantaneously. For example, assuming a heat transfer coefficient of 28.85 w/m·K, a heat transfer area of 0.56 m^2, and a mean temperature difference of 38°C, the heat transfer rate is 21,000 kJ (Heat transfer coefficient × mold surface area × temperature difference). At this rate, it will take 9 minutes to heat the mold to the molding temperature, and 15 minutes to heat the mold to the curing temperature (Heat needed/Heat transfer rate).

For larger molds, the heating and cooling times will be longer if the heat transfer area does not increase in proportion to the weight. At some point, the production cycle time becomes limited by the rate at which heat can be added or removed, and becomes independent of the curing characteristics of the resin system.

Under development today is technology that allows the tool face to be heated by electric wires buried in the face. The construction of the mold face is such that the heat flows into the mold face and not outward toward the mold support structure. This is accomplished by use of a syntactic foam core that insulates the bulk of the mold from the tool face. This novel technology, if successful, would allow a more instantaneous transfer of heat where it will do the most good—at the mold face.

Injection Port and Vent Design

This part of the mold design is probably the most critical part of the process. The inlet port(s) must be correctly designed to allow the resin system to contact all of the reinforcement. Bypass of part of the reinforcement at the inlet of the mold usually results in a dry patch, since the resin system will not flow backwards. If a self-wetting

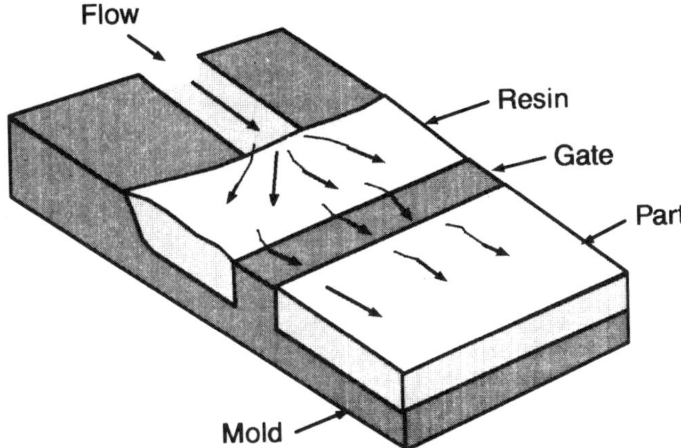

FIGURE 3 Typical gate construction.

fabric is used, sometimes the reinforcement will end up with complete wetout as a result of capillary flow from a resin-rich area to a dry area. The easiest way to ensure complete wetout initially is to gate the mold correctly.

Figure 3 shows the basic idea behind a resin gate. The resin backs up behind the weir and flows over it into the mold. The resin contacts the leading edge of the reinforcement as it makes its way into the part, thus wetting out all of the reinforcement as it flows through the mold.

Figure 4 shows several other gate configurations that have been used successfully in the past. If this figure is studied carefully, it can be seen that the factor common to each configuration is that the flow of resin is symmetrical about the vent ports, in a manner such that the volume of air left in the reinforcement is decreasing. This compression effect helps sweep the remaining air out of the part.

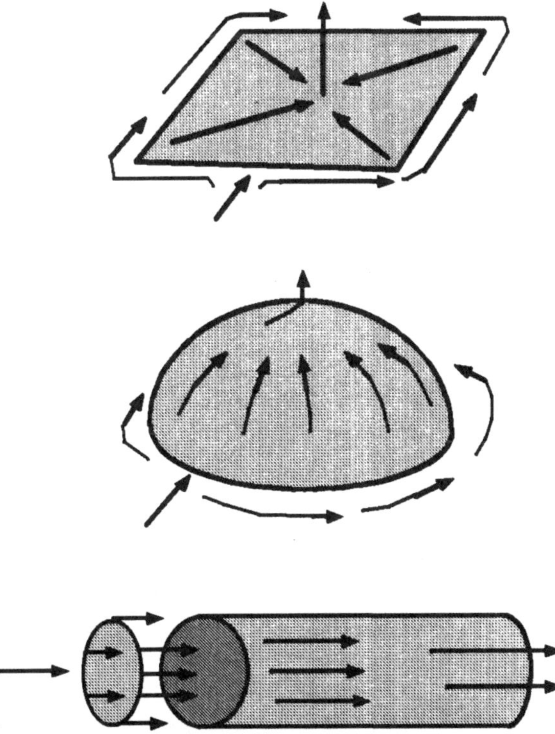

FIGURE 4 Desirable flow configurations.

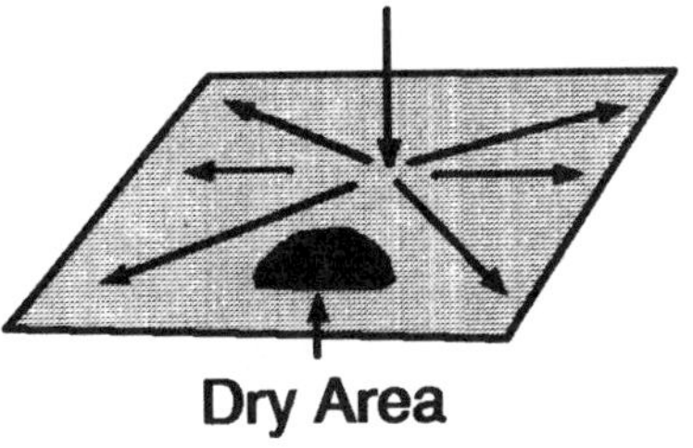

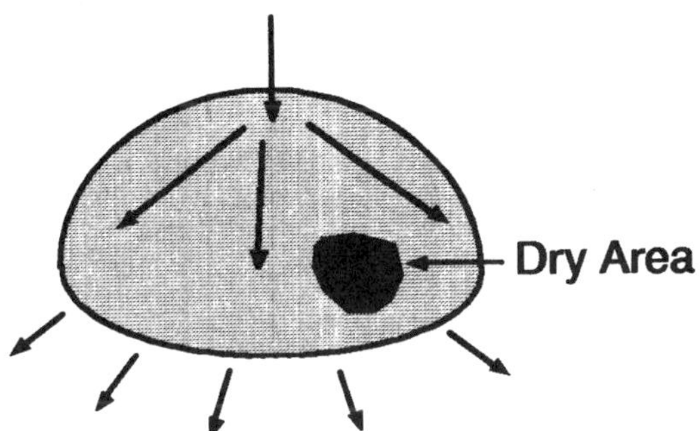

FIGURE 5 Undesirable flow configurations.

When the flow path is arranged in such a way that the resin flows into a configuration with increasing volume, there is a tendency to bypass areas of reinforcement. Figure 5 shows this situation. Particularly when there is reinforcement on either side of a core, there is the danger that slight imperfections in the core thickness will cause dry spots in the part. To overcome this problem, the resin must be introduced on either side of the core simultaneously. Holes may be drilled through the core to allow the resin system to flow to the other side. When this is done, the core "floats" on the wet reinforcement and equalizes itself.

Venting ports must be placed so as to draw the resin through sections of the part that are difficult to wet out. They are best placed at dead ends where the resin will not flow by itself. Venting ports must be capable of being sealed off after the resin has finished bleeding. This will allow pressure to build up in the mold, and force the resin to other sections of the part where the pressure drop is higher. All venting ports must be capable of being sealed after the bleeding operation is finished. This will allow the mold to be gelled under pressure, thus limiting further the chance for voids in the finished part.

Remember, the amount of air left in the part is usually a logarithmic function of the amount of resin bled through the part. The removal of entrapped air can be facilitated by using symmetry to design the inlet ports and outlet vents.

Sealing the Mold

Sealing the mold to achieve cavity pressures of 100 to 200 psig is an absolute necessity if the void content of the part is to be kept below 1%. The only practical way to accomplish this is to use silicon or Viton O-rings. Machining the face of the mold to close tolerances is prohibitively expensive. It is also usually impossible to maintain the mold absolutely flat to achieve a metal-tight seal.

O-ring design is well established. The slot has to be cut so that the O-ring can deform when the mold is closed and maintain a seal. Either square or round O-ring grooves can be used. The type of O-ring material used depends on the maximum temperature the O-ring will experience during the fabrication cycle and the type of solvent used to clean the mold. Normally, Viton can be used satisfactorily up to 250°F. Over 250°F, silicon rubber material can be used to temperatures approaching 350°F.

If help is needed in sealing around inlet or outlet tubes, Gortex Teflon sealant can be used. This product comes with a tacky edge that holds it to the mold. It is stable to over 350°F and can be compressed to a thin layer. This product is useful for making O-rings where grooves do not exist.

Mold Cavity Design

In planning the design of the mold cavity, the goal of the mold designer should be the consolidation of as many assembly steps as possible, without sacrificing part quality. One method is to incorporate as many subassemblies as possible into the main part. For example, instead of bonding terminals, bolts, and other such items onto a part after it is fabricated, it is usually more efficient to incorporate them into the part before the part is molded.

Another method is to fabricate portions of the composite part using bulk molding compound (BMC) before injecting the resin. Metal is removed from the mold cavity where the BMC part is desired. The cavity is then drilled with separate inlet and outlet ports for the BMC. The BMC is injected into its cavity, and stops when it encounters dry reinforcement. The resin is then injected into the dry reinforcement as is normally done during RTM. The mold is then put through a normal cure schedule, where the BMC concurs with the reinforced resin.

When designing molds for parts with rounded sides, use sliding mold sides to prevent the reinforcement from being pinched when the mold is closed. For parts with vertical sides, always allow several percent draft in the vertical dimension, or the part will be extremely hard to demold. Use jackscrews to push difficult sections of the part out of the mold.

Fiber Reinforcements

Another important factor in the RTM process is the proper choice of fiber reinforcement. The reinforcement carries 90% of the load in a normal composite and provides 90% to 95% of the stiffness. The fiber reinforcement in a composite can be directionally oriented to match the strength requirements of the part. Quasi-isotropic composites can be fabricated to duplicate the strength of aluminum, at a considerable weight savings.

Table 1 shows typical composite data for several styles of glass and graphite reinforcement. A wide range of strengths and moduli are obtained by mixing types and styles of reinforcement. Generally, glass reinforcements are used where stiffness requirements are in the range of 7000–14000 MPa. Graphite reinforcement is suitable for modulus requirements up to 138 GPa. Unidirectional reinforcement is subject to washing if proper precautions are not taken to prevent it. Usually preforms have to be

TABLE 1
Typical Composite Mechanical Data

	E-Glass[a]				Graphite[b,c]	
	Mat	8HS	Uni	Bi	8HS	Uni
Tensile						
Strength (MPa)	234	400	759	379	827	2000
Modulus (GPa)	13.8	27.6	31.0	17.2	69.0	144.8
Elongation (%)	2.5	1.4	2.4	2.2	1.2	1.2
Flexural						
Strength (MPa)	310	565	1034	483	731	2434
Modulus (GPa)	12.4	24.1	30.3	15.2	65.5	151.7
Compressive						
Strength (MPa)	248	427	503	290	662	1448
Short beam shear						
Strength (MPa)	37.2	54.5	41.4	31.0	57.9	117.2

[a] Resin wt % = 38; resin system was DERAKANE 530/BPO. (DERAKANE is a trademark of The Dow Chemical Company.)
[b] Resin wt % = 30, resin system was TACTIX 123 & TACTIX H31. (TACTIX is a trademark of The Dow Chemical Company.)
[c] Unidirectional data from Cofab literature.

used in order to hold the unidirectional material in place. Sometimes veil can be used to hold the layers and prevent washing.

The 8HS style of reinforcement is self-wetting. What this means is that a liquid will tend to adhere to the fabric and displace the air entrapped in the interstices of the fabric. The tendency of a fluid to wet a fabric is a function of the surface tension and density of the fluid and the dimensions of the voids in the fabric. The 8HS style of weave allows the individual tows of fiber to approach one another closely enough that the fabric becomes self-wetting. Ordinary fabrics are woven in a square weave pattern, and the interstices in the fabric cannot be made small enough to induce self-wetting.

This phenomenon can be used to help wetout by using the 8HS fabric as a single layer on top of the primary reinforcement in the part. The fluid will be drawn over the 8HS fabric by capillary action and will flow down into the main reinforcement under the influence of pump pressure.

Most bidirectional fabrics today are stitched together instead of woven. This reduces the stresses inherent in the woven roving design and leads to higher compressive strengths in the composite. These fabrics, however, tend not to be self-wetting.

The stitched fabric layups can be further bonded together into preforms, with or without a core. This makes loading of the mold much easier, of course, and quicker. The cores have to be cast from molds that allow for the thickness of the reinforcement. Typically, the core material is a urethane foam, cast directly in the mold.

If preforms are not used, then some means must be used to hold the layers of reinforcement together as they are built up on the tool surface. A tacky epoxy or vinyl ester resin, dissolved in acetone or other suitable solvent, can be used as a spot glue to hold the layers together. The tacky resin will be washed out during the resin injection cycle and will not interfere with the curing of the part.

Different sizings can be obtained on many reinforcements. Particularly on the 8HS-181 style of cloth, sizings can be tailored to the type of resin system. Different sizings are available that are compatible with epoxy, vinyl esters, or polyesters. The strength variation with type of sizing can be as much as 20%, so this factor needs to be considered in the choice of reinforcement.

Resin Delivery System

Once the fabric has been placed inside the mold, and the mold heated to the injection temperature, the next step is to begin injecting the resin system into the mold. The choice of the machine to accomplish this purpose is limited to one of several types available today, single- or multivalve.

Single-valve injection machines are merely one-pot systems that inject premixed resin–hardener systems. Usually these machines use air pressure to pump the resin mixture continuously into the mold cavity. The advantage of these machines is that they will tolerate a wide variation in resin system viscosities and mold back pressures without the need for priming and with very little waste of material. There is no restriction on the type of filler used in the resin, and any mix ratio can be used. Uneven catalyzation is eliminated, since resin system components are premixed. Furthermore, a linear cure profile is produced, since the resin and hardener have been mixed at the same time and not prior to injection. One disadvantage of this machine is that air is usually dissolved in the resin mixture at the time of injection. This can lead to an increased void content in the finished composite part.

There are very few moving mechanical parts in the single-valve machine. Therefore, the amount of periodic maintenance is reduced. Different resin systems can be tested without the need for extensive cleaning of the injection machine. Heating the system is easy, since there is only one pot to heat. This can lead to problems, however, if the resin system has to be left in the pot while problems are being worked out on the production line. If the problem cannot be fixed in a timely fashion, the resin mixture will have to be dumped to the waste container.

Multivalve injection machines are those with two or more supply tanks that have an injection valve for each tank. The valves are synchronized so that ingredient A is mixed with ingredient B in the desired ratio. The valves must be phased together so that they both start and stop at the same point in the injection cycle. These machines can be left idle for long periods of time, since their resin and hardener supply tanks are separate.

This factor can be a disadvantage if the resin and hardener have to be exchanged for another resin and hardener frequently. Waste disposal can become a problem because of the amount of material that must be dumped between system changes. The amount of cleaning solvent that must be used to flush the lines becomes a waste problem in itself.

Higher injection pressures are obtainable with multivalve injection machines, since they have pressure-multiplying valves to boost up the available line pressure. As designed fiber volumes increase in composite parts, higher back pressures (1050–1400 kPa) are required to adequately inject the parts. Single-valve pressure pots are hard pressed to handle this range of pressures.

Multivalve injection machines often induce flow patterns in composite parts as a result of the cycling of their injection valves between shots. These patterns take the form of waves or rings developing out from the resin system entry point. These rings, while seeming to be undesirable, do not appreciably affect the mechanical properties of the laminate.

Maintenance work must be performed periodically on multivalve injection machines because of the great number of close tolerance seals, check valves, pistons, and other such parts. These types of seals and pistons limit the types and amounts of filler that can be shot, because of the abrasive nature of these fillers.

Resin System Physical Properties

The parameters influencing the choice of the proper resin system for RTM are viscosity, pot life, tensile modulus, glass transition temperature, tensile elongation, and moisture absorbance. In considering a new resin system, the first three parameters must satisfy the following system criteria. Failure to meet these criteria usually means that the resin system is impractical for RTM.

System Criteria

Viscosity is probably the first parameter to determine in screening a resin system. The optimum viscosity for RTM is less than 500 cps for the duration of the injection cycle. Viscosities higher than this cause mold pressures that are too high to handle comfortably in conventional injection equipment. The temperature of the resin system can be raised to lower its viscosity, but its pot life may be adversely affected.

The next parameter to consider is the system pot life. The pot life is defined here as the time it takes the resin system's viscosity to reach 500 cps. Depending on the complexity of the part, from 5 to 60 minutes is required to complete most RTM injections.

The third parameter to consider is the value of the tensile modulus. The modulus must be over 2758 MPa or the composite compression strength will be less than the optimum value. A high tensile modulus is required to adequately support the fiber reinforcement and prevent premature buckling.

The remaining parameters determine the performance range of the resin system. The glass transition temperature should be as high as possible, of course. As a rule of thumb, the glass transition temperature should be at least 50°F, and preferably 100°F, higher than the use temperature. This is because of the effect of absorbed moisture. Absorbed moisture plasticizes the resin matrix and lowers the strength of the composite. The amount of moisture absorbed by the resin matrix should be small, $< 2\%$. This will limit the amount of mechanical performance degradation at elevated temperatures.

The tensile elongation is a measure of the amount of brittleness in a resin system. Ideally, the elongation should be 3–5% if sufficient damage tolerance is to be expected in a resin system.

System Characteristics

The physical characteristics of the resin system influence its selection for a RTM system. Of the various types of systems in use today for RTM, epoxy resins, vinyl ester resins, and polyester resins are the most widely used. Each system has its own unique chemistry and curing characteristics.

Epoxy resin systems can be cured with diamines or anhydrides. Since diamines offer the best hot-wet characteristics, those systems are used exclusively today. In Figure 6, a typical viscosity versus time and temperature relationship is shown. For a given temperature, the viscosity increases in an exponential fashion as time

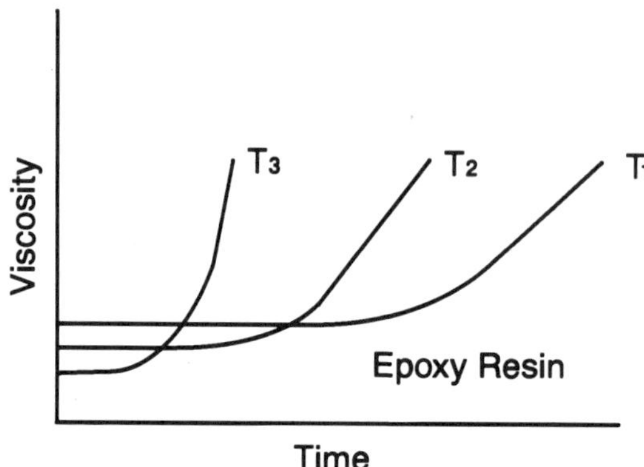

FIGURE 6 **Viscosity versus time for epoxy resins.**

passes. If the system temperature is increased, the initial viscosity of the system is lower, but the rate of change of viscosity with time is faster. The corresponding pot life of the resin system is decreased.

The glass transition temperature T_g for a thermoset resin system is also a function of temperature. Figure 7 shows this relationship, For a given temperature, the T_g increases exponentially with time until it reaches a constant value. As the curing temperature is raised, the T_g increases faster with time and reaches the steady-state value in a shorter amount of time. The steady-state value for T_g is related to the curing temperature, and usually approaches it unless the curing temperature is appreciably above the ultimate T_g.

Vinyl ester and polyester systems exhibit totally different behavior because of the nature of the curing reaction. These systems cure by a free radical mechanism. The peroxide catalyst is usually accelerated by cobalt naphthenate and/or a secondary amine. The viscosity versus time and temperature relationship is shown in Figure 8. At constant catalyst composition, the viscosity versus time relationship is essentially flat until just before the gel point. At this point, the viscosity increases dramatically and the system gels. If the temperature is increased, the time to gel shortens proportionately, and the initial viscosity is lower. The shape of the viscosity/time curve does not change appreciably as in the case of

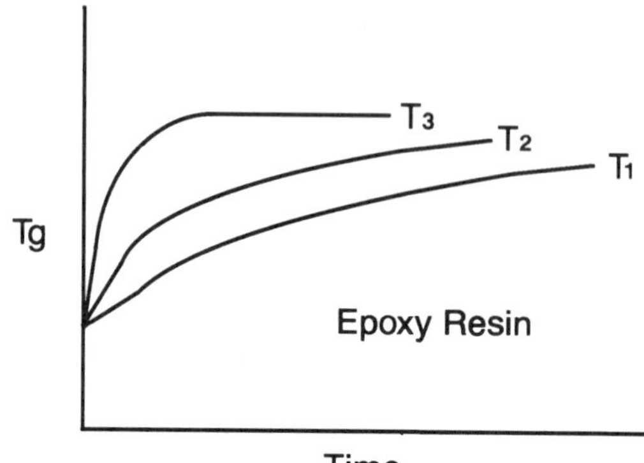

FIGURE 7 **Glass transition temperature versus time for epoxy resins.**

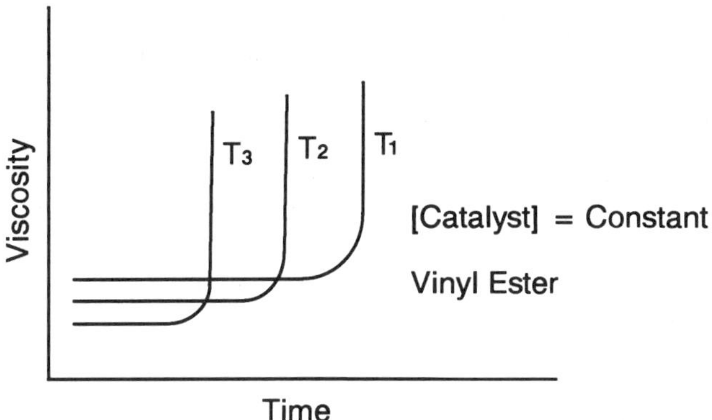

FIGURE 8 Viscosity versus time for vinyl ester resins.

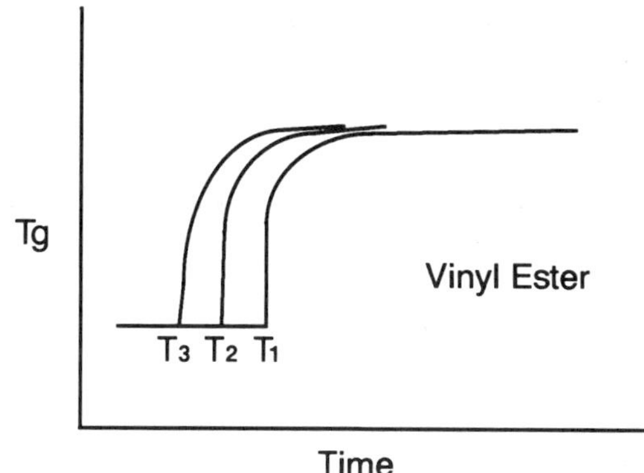

FIGURE 10 Glass transition temperature versus time for vinyl ester resins.

the epoxy–amine system. If the temperature is held constant, the gel point shortens as the catalyst concentration is increased (Fig. 9). The viscosity is essentially constant throughout the time in question.

The variation of T_g with curing temperature and time for vinyl ester systems is different from that observed for epoxy systems. Again, this is due to the nature of the curing reaction. Figure 10 shows typical T_g versus time and temperature relationships for a vinyl ester system. As the temperature is raised, the T_g versus time correlation is flat until the gel point. At that point, the T_g increases rapidly, leveling off at approximately 90% of the ultimate T_g. As the temperature of the system is raised, the gel point occurs earlier, and the steady-state value for T_g approaches the ultimate T_g of the system. The same kind of relationship is observed if the temperature is held constant and the amount of catalyst increased. One of the factors to keep in mind is that the curing reaction is highly exothermic, and an appreciable amount of heat is given off by the reaction. This helps drive the system's T_g up to close to that expected for the ultimate T_g.

Hot-Wet Considerations

One final topic to consider is the effect of the resin system on the hot-wet performance of the composite part.

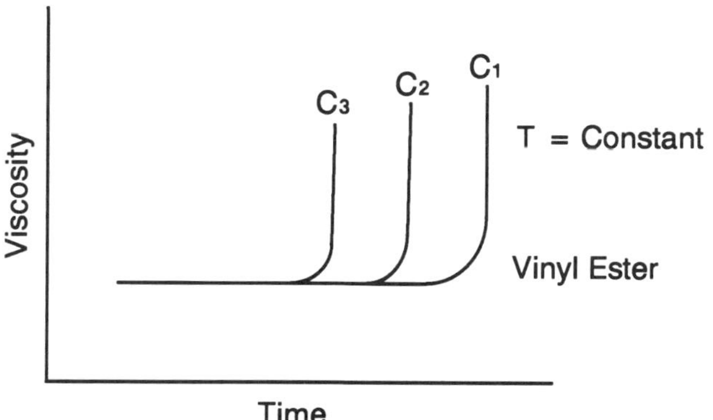

FIGURE 9 Viscosity versus time for vinyl ester resins.

Figures 11 and 12 show typical strength versus temperature and modulus versus temperature correlations for a generalized epoxy resin system. The strength of a typical composite laminate generally falls off gradually with increasing temperature until the ultimate T_g is approached. The strength at this point is essentially zero. Under wet conditions, the strength usually falls off at the same rate, but falls to zero at a lower temperature.

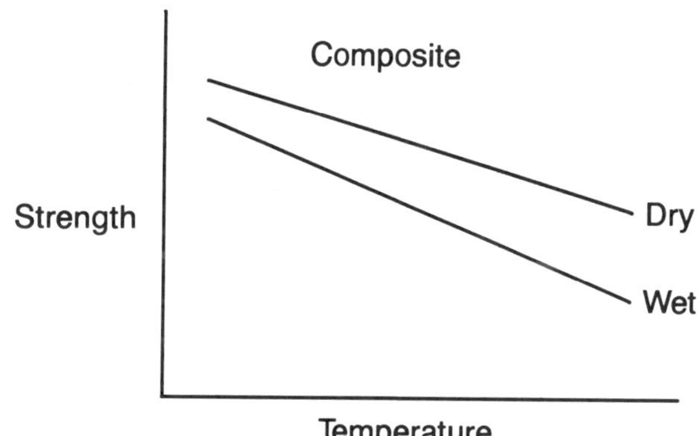

FIGURE 11 Strength versus temperature for typical composite.

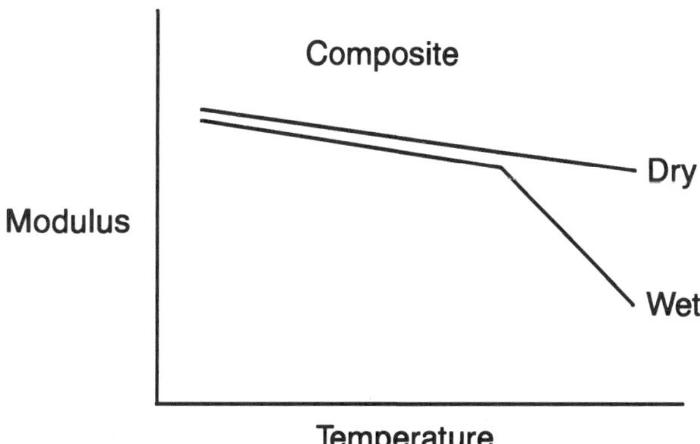

FIGURE 12 Modulus versus temperature for typical composite.

The modulus of a typical composite remains essentially constant until the temperature is close to the ultimate T_g. The modulus then falls off gradually as the temperature passes the ultimate T_g. In wet environments, the modulus falls off gradually as the temperature approaches the wet T_g. At the wet T_g, the modulus falls off very rapidly.

Conclusions

The concepts of resin transfer molding appear at first glance to be fairly simple in nature. As has been explained in this article, this is often not the case. After finishing this paper, it is hoped that the reader has a greater appreciation of the inherent complexity of the RTM process. The quality of parts produced by the RTM process can be equal to the quality of parts produced by the autoclave process using prepreg. Obviously, this is where the economic advantages exist.

The challenge for RTM in the future is to bring together the technologies of automation, preforming, and tooling development and join them with the existing resin chemistry technology to produce a fully automated RTM process.

Dexter White

Molding Short-Fiber Composites, Flow Processing

A short-fiber composite consists of a polymer matrix reinforced by fibers of much smaller length as compared with the overall dimensions of the fabricated structure. These reinforced polymers were developed to fill the mechanical property gap between the continuous-fiber laminates used as primary structures by the aircraft and aerospace industry and the neat polymers used in non-load-bearing applications. Although there is a notable increase in strength of short-fiber composites over the unreinforced polymers, because of the discontinuous nature of the fibers, they will always have lower strength than their continuous-fiber counterparts. However, short-fiber composites enjoy the advantage of providing stiffness levels comparable to those of continuous-fiber composites combined with ease of processing and the ability to be molded into complex shapes. Incorporation of fibers also lowers creep and improves dimensional stability. However, it does not follow that all mechanical and physical properties are enhanced. Strength, impact toughness, internal stresses, and fatigue properties intimately depend on the detailed local material microstructure. This structure depends on the deformation processes of the fiber suspension as it flows to produce the short-fiber composite part.

Short-fiber composites are handled as bulk suspensions and can be easily formed into complex-shaped structures by processes such as transfer, compression, and injection molding. The use of conventional molding techniques to produce large-scale composite hardware makes the manufacturing of these composites very efficient and inexpensive compared with manufacturing of continuous-fiber composites, which are produced by time-consuming processes in which fibers and/or plies are placed individually, rendering them unsuitable for high volume production. Thus, short-fiber composites have found their way into a market where moderate strength and low fabrication cost is desired and the stiffness dominates the design. Important industries for short-fiber composites include, but are not limited to, transportation, business machines, durable consumer items, and sporting goods.

The molding processes for all short-fiber systems involve a modest amount of flow through channels or mold cavities of different geometries. The complex microstructure and morphology that are produced depend on how well the fibers are dispersed, the range of fiber lengths and diameters, how the fibers interact with the mold walls and each other during flow, the heat transfer in the mold, the geometry of the mold, and the final microstructural configuration adopted by the reinforcements in the frozen-in discrete part. Composite properties are dominated by the microstructure of the fabricated part rather than the properties of the constituent materials. As the microstructure produced is related to the flow-processing route of the fiber-filled suspensions and the geometry of the mold, by exercising control over the flow processing, one can tailor the microstructure of the composite to the required application. The design of the mold influences flow processing and consequently the variations in the properties of the composite. To understand flow in the molding environment, one needs to study how it is influenced by mold geometry, processing conditions, and the composition and rheology of the fiber suspension. This will permit us to identify the proper processing environment for constructing a desirable material structure.

Current research is focusing on building theoretical and computational models so that components can be designed in a predictive manner without resorting to the cumbersome process of trial-and-error designs involving numerous prototypes and many repetitive tests. The ultimate goal is to manipulate the structure of the short-fiber composite through selection of processing conditions and techniques, materials, and mold design to produce a material that will meet a set of required specifications for product properties and likely performance in service.

Critical Issues

Molding of short-fiber-reinforced systems involves deformation and flow of fiber suspensions. In manufacturing short-fiber composite parts, critical issues such as physical and mechanical property control, shrinkage and warpage, appearance of the final product, use of new materials, and mold design are closely tied to the flow behavior of the suspensions.

Mechanical properties such as tensile strength and stiffness can vary considerably even in carefully controlled laboratory experiments. The primary sources of these variations are the inherent variation in the material and the fiber configurations (fiber orientation, fiber aspect ratio, and so on) in the final part. Some of the other probable causes of property variations can be attributed to defects during molding, such as weld lines, excessive voids, and weak interface between fibers and matrix. These variations complicate characterization of the mechanical and physical properties of the material and use of the laboratory data in product design. Recently, research is underway to study, understand, quantify, and predict the fiber configuration and structure during flow and separate their influence from the other causes of property variation.

To achieve dimensional stability for short-fiber composite parts, both shrinkage and warpage need to be predicted and controlled. It has been noted that higher loading of fibers in the resin reduces shrinkage problems. However, the variation in fiber orientation distributions leads to anisotropic shrinkage characteristics, which may result in highly warped structures. Severe tolerance problems may arise for tool makers and design engineers if the exact prediction of shrinkage and warpage is not incorporated into their calculations. Heating and cooling conditions for the mold have to be established to minimize these effects.

The surface finish of the part is important when the appearance of the product is crucial as well. Automobile doors and hoods are prime examples of composite parts where exterior finish is important. Fibers on the surface of the mold can cause surface waviness and sink marks. Appearance problems in compression molding of sheet molding compounds have been tackled by altering the composition of the polymer matrix, by combining different lengths of fibers reinforcements, and by adding a fiber-free coating to the part.

For the last decade, the plastics materials industry has tried to refine the grades of the existing polymers. The industry concentrated on producing polymers that were easy to process. However, the performance of these materials was limited. It was generally believed that polymers with excellent properties, such as fluoropolymers and polyaromatics, were difficult to process. The exploitation of liquid crystal polymers has changed the scenario. A new generation of materials, such as aromatic polyesters, combine good service properties with easy processing. A new generation of fiber-reinforced materials is altering the perception of the molding process. There is a need to study and characterize the rheological properties of these materials and to identify the underlying science that will help in designing the molding process and the operating conditions.

Tool or mold design plays a crucial role in the performance of the molded part. An ineffective design can lead to undesirable microstructure and weld lines, resulting in weaker and inadequate parts. A poor design will also cause warpage and surface finish problems. Construction of the mold is expensive and time-consuming. Hence predictive modeling that can establish a mold geometry with all its gating and vents to obtain a part with desired properties will be extremely useful and will lead to efficient use of short-fiber composites.

All these issues are related to the flow and rheology of fiber suspensions, transfer of heat, and/or the reaction that occurs in the mold. Analysis of these vital issues can be carried out during the molding process once we understand the rheology of the suspending fluid and the fiber suspensions, how fibers become oriented during flow, what causes fiber degradation, and the phenomena of fiber migration and clustering. This article will discuss fiber motion in flowing suspensions and review techniques for modeling and prediction of the microstructure in composite molding operations. Topics include the rheology of fiber suspensions, description and prediction of fiber orientation, fiber damage, and fiber clustering during processing operations.

Rheology of Fiber Suspensions

The rheology, or the science of deformation and flow, of fiber suspensions is extremely involved, and there are still many questions to be answered. As new materials are being used for the suspending medium (the matrix) for fibers, controlled rheological characterization becomes very important for processibility. The objective here is to present the state of our understanding of the complex modes of the rheological behavior in order to develop a general appreciation for the quality of the work on this subject and for its limitations. Fibers suspended in Newtonian as well as non-Newtonian fluids will be addressed. The central theme of this section will be to give a broad view of the bulk rheological behavior of fiber-filled suspensions and how the flow properties of the suspension are connected to processing conditions on one hand and the structure generated by the fibers on the other.

Experimental

To obtain a desired structure, better control of the processing of short-fiber composites can be achieved if the suspension can be characterized as Newtonian or non-Newtonian. For a non-Newtonian suspension, it is helpful to identify whether it exhibits one or more of the non-Newtonian properties, such as yield stress, time- and/or strain- or stress-dependent properties, shear thinning, shear thickening, and normal stresses. In general, the rheological behavior depends on fiber concentration, fiber aspect ratio, rheology of the suspending fluid, type of flow, and orientation of the fibers.

Most flow processes are complex because of the geometries of the boundaries and/or the motion imposed at the boundaries. However, the deformation of the fiber suspension is usually through a combination of shearing and elongation. One useful simplification is to study simple flows, such as shear and elongational flows, which provide the greatest utility and can be reproduced in a controlled laboratory atmosphere. No attempt will be made here to review the vast literature on the rheology of fiber suspensions; however, a short outline of the major ideas and results will be presented as pertinent to the processing of short-fiber composites.

Maschmeyer and Hill [1] reviewed the literature describing studies of the macroscopic rheological properties of rodlike suspensions in 1974. They indicated that most work focused on measurements of shear viscosity as a function of shear rate, fiber aspect ratio, and fiber volume fraction. They were not able to obtain even qualitative agreement among data obtained from various laboratories. Since then, research concerning the macroscopic rheology of fiber suspensions has proliferated. Experimental results have revealed many new phenomena as various researchers have tried to measure time-dependent shear and elongation properties [2,3], normal stress differences [2–4], and extensional viscosity [5–7]. Ganani and Powell [8] have compiled most of the experimental results and tried to correlate them based on their review. For concentrated suspensions (where the volume fraction of fibers is much greater than the ratio of diameter to length of the fibers, d/l), they were unable to find any correlations among the available data even for shear viscosity in-Newtonian suspending fluid. They reported that for suspensions in non-Newtonian fluids, which are typically encountered in the processing of composite materials, the situation was even less clear. Kamal and Mutel [9] have reviewed the various macroscopic rheological properties, such as the steady shear viscosity, normal stress, and dynamic shear, in Newtonian and non-Newtonian melts. From their documentation, it is clear that suspensions behave quite differently at low and high shear rates and that the rheological behavior is affected significantly by the suspending fluid, concentration of the fibers, surface treatment of the fibers, and other such factors. They have provided an extensive table summarizing the various studies about the rheology of suspensions in non-Newtonian media. Metzner [10] has systematically reviewed the rheology of suspensions in polymeric fluids and pointed out the effects of distribution of particle size, particle geometry, increasing deformation rate, and fluid elasticity on the rheological properties in simple flows.

Although all the studies have made an important contribution to the field, there still remain many open issues, and the conclusions still seem very system-specific. The major difficulty may be the use and the interpretation of the rheological data in the narrow gap viscometers. The ratio of the length of the suspended fibers to that of the characteristic gap is a crucial factor. This ratio should be small for any continuum assumptions about the suspension to be meaningful. Also, boundary conditions become difficult to define. Forces that arise as a result of shearing can cause the particles to move away from the walls, leaving a liquid-rich layer [11]. Furthermore, it is nontrivial to isolate the structure-property relationships in the fiber suspension rheology. The bulk properties of the suspension depend on the orientation distribution of the fibers in the fluid, which in turn depends on the rheological behavior. Hence, as the field of developing new molding materials keeps growing, the need for a standard rheological test to assess the processibility of these materials becomes more critical.

To summarize the experimental work, it is sufficient to point out that the molding process for short-fiber composite parts needs to be reevaluated. Various non-Newtonian effects will emerge as important when these bulk suspensions are examined. To be able to use these rheological properties to our advantage, we need to establish a method to characterize the suspension in simple fluids and be able to extend the procedure to complicated flow fields. Theoretical work is underway to perform parametric studies of the effects of suspension composition and structure on the rheological properties.

Theoretical Considerations

With regard to suspensions of rodlike particles, the theoretical study can be conveniently carried out by dividing the suspension into three regimes of concentration: dilute, semiconcentrated, and concentrated. By defining the boundaries of the semiconcentrated regime, one can establish the boundaries of each category. To keep the level of complexity to a minimum, the fibers are considered as rigid cylinders of length l and diameter d that occupy a fraction c of the total volume of the suspension. The concentration c and the fiber aspect ratio l/d are the dimensionless quantities that characterize the suspension. It has been shown [3,12] that the semiconcentrated limits are given by $(d/l)^2 < c < d/l$. For concentrations below $(d/l)^2$, the distance between a fiber and its nearest neighbor is greater than l, which defines the dilute regime. For concentrations above d/l, the spacing between the fibers is on the order of d and the suspension is said to be highly concentrated. For example, c has to be greater than 10% for the suspension with fibers of aspect ratio greater than 10 to be categorized in the concentrated regime. Composite melts are basically suspensions of 30–50% fibers with aspect ratios varying from 30 to 100. Thus, all composite melts definitely fall into the concentrated regime.

The bulk constitutive equations for a dilute suspension of rigid axisymmetric particles in Newtonian fluid are derived using a mechanistic approach [13–18]. The phenomenological theories of Erickson [19,20] and Hand [21] provide an alternative approach to predicting bulk properties. Dinh and Armstrong [3] systematically extended this work by deriving a stress constitutive equation for semiconcentrated fiber suspensions. Lipscomb et al. [22] showed that the flow of dilute fiber suspensions is qualitatively different from that of the suspending fluid, even when the shear viscosities are nearly the same. The continuum theory presented by them couples the flow and the fiber orientation field. They conclude, however, that there does not appear to be any internally consistent way of extending this theory to include long-range coordinate motions, which are prevalent in the concentrated fiber suspensions. Becraft [23] examined the liquid-crystalline theory of Doi [24] for its analogies to macroscopic suspensions of fibers. They applied the equations to a system of glass fibers in polypropylene.

The predictions met with reasonable agreement with their experimental shearing data from capillary and parallel plate geometry using two adjustable parameters.

More systematic and fundamental work is required to build a theory for rheological behavior of concentrated fiber suspensions, if one wants to provide a framework for predicting and controlling the microstructure during molding of short-fiber composites. The major difficulty is that the rheology is closely coupled with the anisotropic structure of the fluid as a result of the presence of the fibers. As the spacing between the fibers is of the order of magnitude of diameter d, there will be hydrodynamic interactions between the fibers and also direct fiber-fiber interactions that will reorient the fibers [25]. These sudden reorientations will also change the rheology. The fibers may also form aggregates, which will again influence the bulk properties. The basic concept is that when two fibers collide, this produces a change in the orientation of both fibers. The directional basis of these changes in orientation will depend on whether the suspension is orderly or completely random. Hence, knowledge about the concentration, aspect ratio, flow field, and suspending fluid is not sufficient to predict rheological behavior. We also need to know the structure (e.g., orientation distribution of fibers) during flow to predict the rheological properties.

Fiber Orientation

As the fiber suspension deforms and flows into the mold, one must know the fiber orientation structure not only to determine its rheological behavior but also to estimate its mechanical performance [26,27]. Flow and deformation of the suspension change the orientation of the fibers flowing in it. These orientations are subsequently frozen in as the material solidifies and become a key feature of the microstructure of the finished composite. If the fibers are randomly oriented, the mechanical and physical properties will be isotropic. If the fibers are aligned in one direction, the composite will be stiffer and stronger in that direction compared with any other direction. The distribution of orientations in a molded part could be quite diverse. One region may have random fiber orientation, while others may have preferred alignment in certain directions.

As pointed out earlier, knowing an average fiber orientation direction for the complete part is not sufficient because the local spatial orientation also plays an important part in determining properties such as strength and toughness. Hence, considerable research has been done with the intent of learning how to predict fiber orientation in molded parts. Mathematical models have been developed and incorporated into simulations of injection and compression molding with the emphasis on predicting the influence of tool geometry, processing conditions, and material properties on the final orientation pattern. The overall goal is to tailor the orientation microstructure to suit the desired application with minimum cost and maximum reliability [28].

This section will review the present state-of-the-art modeling of fiber orientation in molding short-fiber composites. Topics covered include characterizing orientation, fiber orientation mechanics for a collection of fibers, and incorporation of these ideas in manufacturing process models for injection and compression molding.

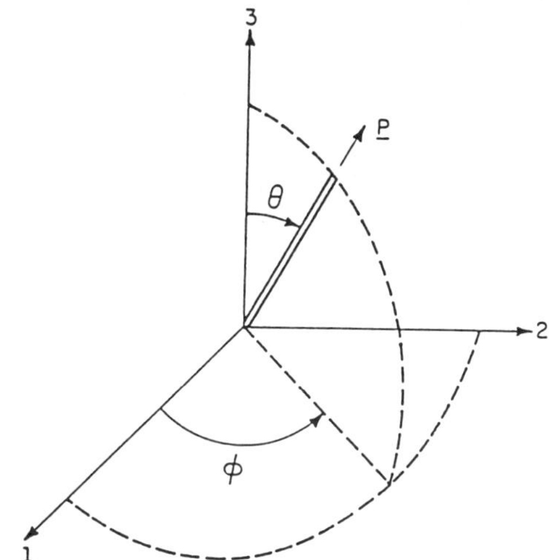

FIGURE 1 **Coordinate system used to define Ψ (θ,ϕ) and Ψ (p) in spherical coordinates.** (From Ref. 29.)

Characterizing Orientation

If we assume that all fibers are rigid rods or cylinders, a direction can be associated along the fiber length using a unit vector **p** relative to some reference coordinate frame. Equivalently, one can specify the orientation by the angles (θ,ϕ), as shown in Figure 1. The components of **p** are related to θ and ϕ:

$$p_1 = \sin\theta\cos\phi \quad (1a)$$

$$p_2 = \sin\theta\cos\phi \quad (1b)$$

$$p_3 = \cos\theta \quad (1c)$$

and are also interrelated, since the length of the vector is fixed:*

$$p_i p_i = 1 \quad (2)$$

The set of all possible directions of **p** corresponds to the unit sphere. For a certain class of molded parts where the fibers are much longer than mold thickness, most of the fibers will lie in one plane. This makes it possible to describe the orientation of a fiber with one angle ϕ or a two-dimensional vector **p**. The set of all possible directions for planar or two-dimensional orientation is a unit circle.

* Whenever cartesian tensor notation is used, summation over repeated indices is implied.

FIGURE 2 **Computer-redrawn plot of fibers from the radiograph of a rectangular plaque made from SMC.** (From Ref. 43.)

In a short-fiber composite, the fiber may be oriented in many different directions. Figure 2 is a redrawn image from a radiograph of a compression-molded plaque made from sheet molding compound; it depicts the complexity of the situation. It makes more practical sense to embody the idea of many directions to describe fiber orientation and call it *fiber orientation state* than to try to describe the direction of each individual fiber.

The most general description of the fiber orientation state is the probability distribution function. If we consider a small enough region of the composite, the distribution of fiber orientations in that region will represent the orientation state. The probability distribution function for orientation, also known as the orientation distribution function $\psi(\theta,\phi)$, is defined so that the probability of the fiber lying between angles θ and $\theta + d\theta$, ϕ and $d\phi$ is given by $\psi(\theta,\phi) \sin \theta \, d \, \theta \, d\phi$. The probability distribution function must satisfy two physical conditions. First, one end of the fiber is indistinguishable from the other end, so ψ must be periodic:

$$\psi(\theta,\phi) = \psi(\pi - \theta,\phi + \pi) \tag{3}$$

Second, every fiber must have some direction, so the integral over all possible directions or the orientation space must be equal to unity:

$$\int_0^{2\pi} \int_0^{\pi} \psi(\theta,\phi) \sin \theta \, d\theta \, d\phi = 1 \tag{4}$$

This is known as the normalization requirement. If the orientation statistics change with position, ψ is a function of x, y, z in addition to θ and ϕ.

The distribution function can be approximated by measuring the orientations of a large number of fibers selected from a region where the distribution function does not change drastically. The distribution function is a complete and unambiguous description of the fiber orientation state. The disadvantage of the distribution function for representing orientation is that it makes calculations to predict orientation in flowing suspensions very cumbersome. Note that ψ is a function of two variables (either θ and ϕ or two of the p_i) and is also a function of position in the part. For the moment, one can appreciate the fact that to represent this amount of information in a numerical simulation is a formidable task. This issue is addressed in detail in Ref. 29. A more compact description will be extremely useful.

To furnish easily interpreted measures of orientation, a number of orientation parameters have been defined. The most often used is the Hermans orientation parameter [30]. The parameter provides a scalar to represent the magnitude of orientation in axisymmetric cases. It is not possible to derive such orientation parameters for general cases of fiber orientation, nor is it possible to evaluate them from equations of change for flow-induced orientation.

A description that combines the generality of the distribution function and the compact nature of orientation parameters is a tensor description of orientation [29]. Second- and fourth-order orientation tensors can be defined as:

$$a_{ij} \equiv \langle p_i p_j \rangle \tag{5a}$$

$$a_{ijkl} \equiv \langle p_i p_j p_k p_l \rangle \tag{5b}$$

The angle brackets, $\langle \; \rangle$ denote the integral over all p weighted by ψ (p). (This notation expresses the average over all possible directions weighted by the orientation distribution function.) There are a number of physical interpretations of these tensors. They can be thought of as a generalization of orientation parameters, as moments of the distribution function, or as a series expansion of the distribution function. Advani and Tucker [29] provide a complete review. They have shown that these tensors are free from a priori assumptions about the shape of the distribution function and can be readily transformed from one coordinate frame of reference to another complying with the rules of tensor transformation. They are symmetric, that is:

$$a_{ij} = a_{ji} \tag{6}$$

The normalization condition such as Eq. (4) implies that the trace of a_{ij} is unity:

$$a_{ii} = 1 \tag{7}$$

The tensor description substitutes fewer number of scalar quantities for the distribution function to describe orientation. For example, for planar orientation, only two of the four components are independent. For the three-dimensional case, only five of the nine components are independent. Figure 3 shows examples of planar orientation states represented using the distribution function, orientation parameters, and orientation tensors. When a_{22} approaches unity, all fibers are along axis 2, and the

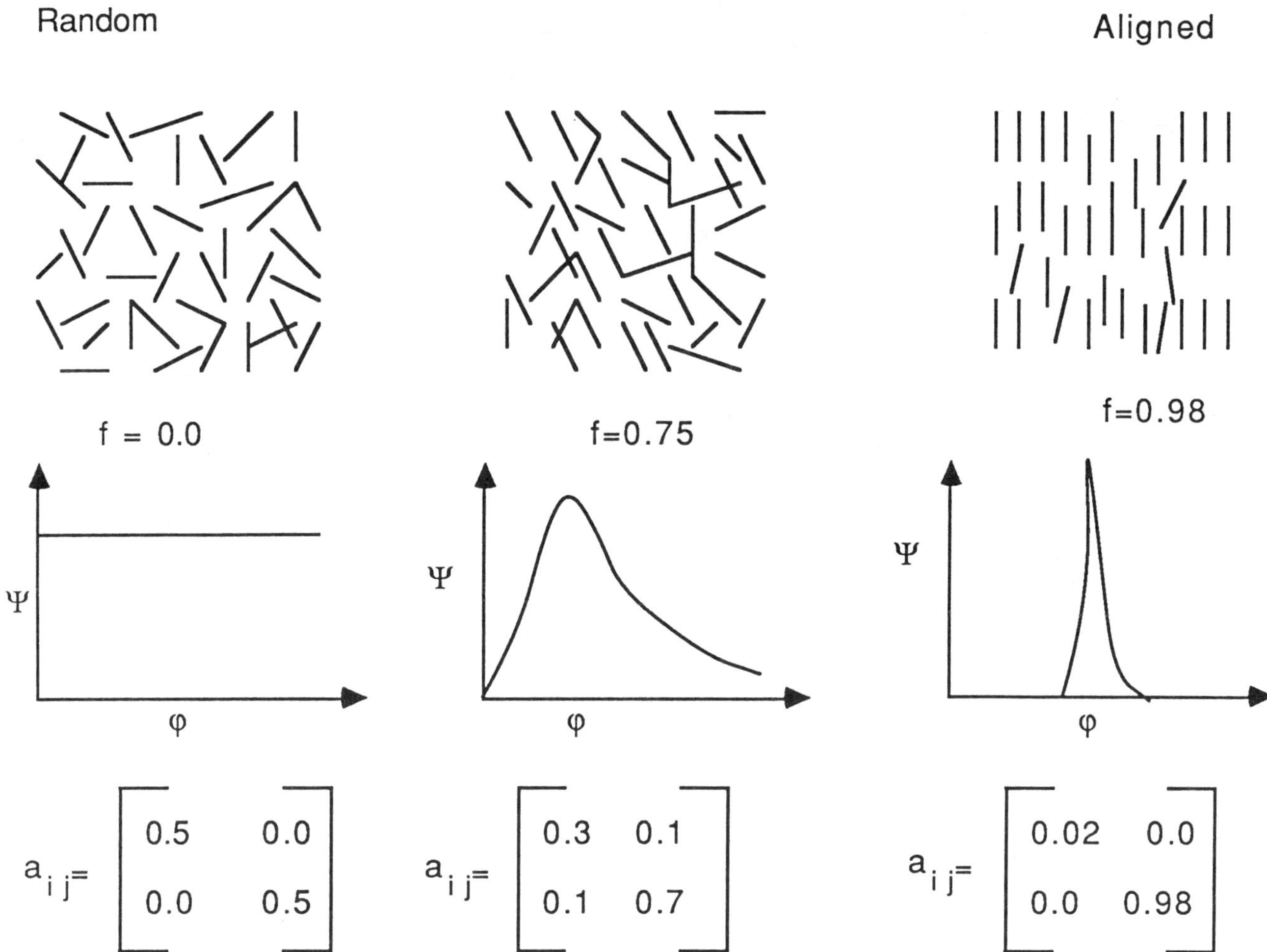

FIGURE 3 **Different representations of planar fiber orientation.**

distribution function approaches a Dirac delta function around that axis. When the fibers are randomly aligned, a_{11} and a_{22} are equal, and if the principal orientation is not along coordinate directions, a_{12} is nonzero. Note that orientation parameters erase vital information, such as the direction of the principal orientation if it is other than the coordinate direction.

Viscosity and elastic stiffness are normally fourth-order tensors, and one would normally require only a fourth-order orientation tensor to predict the effect of orientation on these fourth-rank properties [29]. By approximating the fourth-order tensor in terms of second-order tensors, it is possible to predict the effect of orientation on fourth-order properties using just the second-order orientation tensor [31].

Flow-Induced Fiber Orientation

For better performance, one wants to have as many fibers as possible in short-fiber composites. Most commercial composites are in the range of 10–50% fibers by volume. And since in these materials the fiber length-to-diameter ratios are always 10 or greater, all commercial materials fall under the concentrated suspension case.

Jeffery modeled accurately the motion of a single fiber immersed in a large body of incompressible Newtonian fluid [32]. However, fibers in concentrated suspensions behave somewhat differently. A number of researchers have observed fiber orientation in concentrated suspensions [25,27,33,34]. All of them reported fiber orientation behavior that is qualitatively similar to single-fiber motion. In elongational flows, fibers align along the direction of stretching, aligning with the streamlines in converging flows and normal to the streamlines in diverging flows. Shear flows tend to orient fibers in the flow direction. However, these workers do not observe perfect alignment of fibers, which the theory predicts. Fibers flowing in concentrated suspensions are so close to each other that they not only interact hydrodynamically but may physically collide with each other, causing more erratic motion that violates Jeffery's assumptions [35]. Good quantitative predictions of orientation, especially the amount of alignment and the breadth of alignment for realistic composites, will be useful for estimating the mechanical and physical properties more accurately. A model for the behavior of fibers in concentrated suspensions is required.

The dynamic behavior of a concentrated suspension

is quite complex, as the rheology is closely coupled with the fiber orientation structure, and unfortunately there is as yet no mechanistic model that includes the close coupling and the interactions between the fibers. This is an area that demands research and is actively being pursued [36,37]. So far, a phenomenological model proposed by Folgar and Tucker has proven useful [25]. They model the interaction between the fibers by introducing an additional term in the equation of motion for single-fiber motion. This term is similar to a diffusive term, and the effective diffusivity is made proportional to the strain rate, as interactions take place only when the suspension is deforming. A dimensionless "interaction coefficient" C_i term, typically of the order of 10^{-2}, served to match their experimental results [25,38]. The equation of motion can then be combined with the equation that conserves fibers in the orientation space [39] to produce the equation of change for fiber orientation in terms of distribution function and orientation tensors [25,29,38]. For example, the equation of change for the second-order tensor is given by:

$$\frac{Da_{ij}}{Dt} = -\frac{1}{2}(\omega_{ik}a_{kj} - a_{ik}\omega_{kj}) + \frac{1}{2}\lambda(\dot{\gamma}_{ik}a_{kj} + a_{ik}\dot{\gamma}_{kj} - 2\dot{\gamma}_{kl}a_{ijkl}) + 2C_I\dot{\gamma}(\delta_{ij} - \alpha a_{ij}) \tag{8}$$

where δ_{ij} is the unit tensor and α equals 3 for three-dimensional orientation and 2 for planar cases. Here, a cartesian tensor notation is used. ω_{ij} and γ_{ij} are the vorticity and the rate of deformation tensors, defined in terms of velocity gradients as:

$$\omega_{ij} = \frac{\partial v_j}{\partial x_i} - \frac{\partial v_i}{\partial x_j} \tag{9}$$

$$\dot{\gamma}_{ij} = \frac{\partial v_j}{\partial x_i} + \frac{\partial v_i}{\partial x_j} \tag{10}$$

The material derivative in Eq. (8) appears on the left-hand side because the fibers are convected with the fluid. This tensor form is particularly attractive for performing numerical computations.

Modeling Fiber Orientation Dynamics in Processing

After developing equations such as Eq. (8) that describe the nature of orientation at any point in the suspension using either the distribution function or orientation tensors, the goal is to solve for the orientation state at every point in the flow field as a function of time. The final orientation is the one that freezes in position as the suspension solidifies. The calculations generally involve numerical solutions of the governing equations, as the mold shape is complex and will have moving boundaries as the mold fills. (We will not address the flow of suspensions in confined geometries [22,28,40] where the boundaries of the flow domain do not change with time and are known a priori, for example, flow in an extrusion die.) Furthermore, flow and orientation calculations are coupled. Fiber orientation is determined by fluid deformation; the fibers are convected by the fluid, and the orientation is determined by the velocity gradients. One must know the flow field (velocities) to calculate the orientation. However, changes in the orientation will change the rheological properties as mentioned in the previous section, which affects the flow field. So one needs a constitutive equation that will include this effect [3,22]. One must then solve the constitutive equation and the equation of change for orientation such as Eq. (8) simultaneously. Including all these effects is an overwhelming task. Depending on the process, it may be possible to simplify the calculation to realistically model orientation in a particular process.

MODELING FOR COMPRESSION MOLDING. In compression molding, the mold cavity is always thin compared with the fiber length, and since the velocity does not vary across the thickness [41], the orientation state should not change through the thickness. This process environment makes it meaningful to solve for planar orientation in a two-dimensional flow field. Jackson, Advani, and Tucker [38] have modeled the flow and fiber orientation in compression-molded materials such as sheet molding compound (SMC). The flow of SMC in thin cavities is dominated by friction as the molding compound slips over the molding surface [41]. This material characteristic makes it possible to decouple flow and orientation because the rheology is unaffected by the structure of fibers.

Recently, Advani and Tucker [42] have developed a finite element calculation for fiber orientation in compression molding. It uses Folgar and Tucker's model [25] for orientation and a second-order tensor representation and treats the governing equation (Eq. 8) with conventional Galerkin's finite element method [43]. Their simulation can handle flat parts with otherwise arbitrary geometries. The program starts by specifying the mold geometry, the initial shape of the charge, and the initial fiber orientation in the charge. First the computer program finds the pressures and velocities inside the charge, then it calculates the orientation and stops the flow front in time. The procedure is repeated until the mold is full.

Sample results and comparison with experiments are shown in Figures 4 and 5. Figure 4 shows the flow front motion with time as predicted by the simulation and the comparison with the measured motion from the photographs. The agreement suggests that the flow kinematics of the mold filling are quite accurate. Figure 5 depicts the orientation distributions. The tensor notation of orientation is drawn by computing the principal axes and the principal values and drawing an ellipse with these as major and minor axes. A circle indicates a random orientation state, whereas an elongated ellipse indicates both the direction and amount of alignment. The only substantial errors are at the nodes that fill last and along the edges. Advani and Tucker [42] believe that the numerical method they used cannot calculate accurate velocities in these regions, resulting in poor predictions of fiber ori-

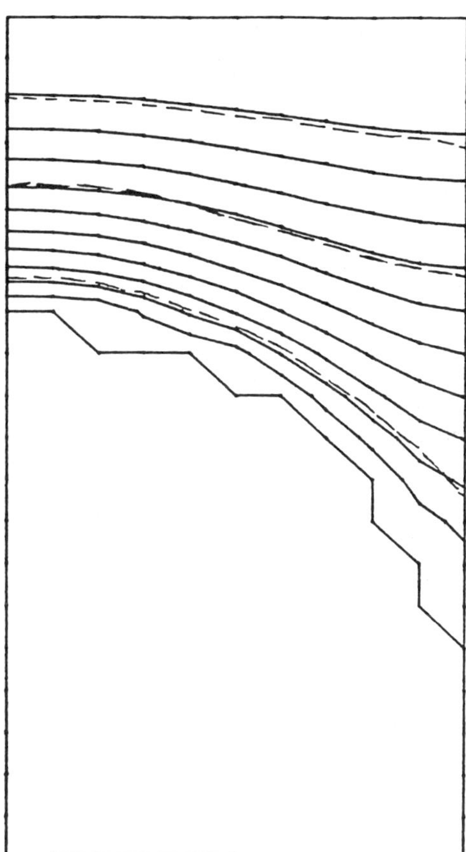

FIGURE 4 Flow fronts for the complex flow experiments. The solid lines are predictions and the dashed lines experimental. (From Ref. 42.)

FIGURE 5 Comparison of predicted and experimental fiber orientation states in the complex flow experiments. (From Ref. 42.)

entation. Otherwise the agreement is excellent. Their results support the assumptions they used to model flow and fiber orientation in compression molding.

Future work in this process that needs attention is refining the numerical schemes to predict accurate velocities at the boundaries and to represent the flow front more accurately. Also, the ability to model thin but nonplanar parts will be highly desirable. The behavior of fibers in similar materials, such as bulk molding compound (BMC) and dough molding compound (DMC), should also be explored. Prediction of fiber orientation in compression molding of thick sections will require a model that can predict the flow field [41] and adapt it to calculate not only fiber orientation in the plane but also its variation through the thickness.

MODELING FOR INJECTION MOLDING. Injection molding is more widely used to manufacture short-fiber composites. However, it is also more challenging to model. Several researchers have modeled the flow of neat polymers using the generalized Hele-Shaw formulation [44,45]. However, the addition of fibers to neat polymers introduces the complexity of flow being coupled with fiber orientation. Also, normally the fibers are much smaller compared with part thickness, and hence fiber orientation is three-dimensional. Skin/core effect has been observed by various researchers [46–50], which indicates variation in fiber orientation through the thickness of the part. One does understand this phenomenon qualitatively. Near the surface, the fibers are oriented in the flow direction as a result of shearing flow, and in the core they align in the transverse direction as a result of diverging flow. This shearing/stretching flow plays an important role in producing the core/shell structure. Injection speed, temperature, and material rheology all seem to affect the skin/core thickness [49,50]. Another phenomenon that contributes to this structure is the fountain flow effect, which has been observed in injection molding of neat polymers [51,52]. Thus, the flow field in injection molding is strongly affected by the interactions between viscous dissipation, temperature-dependent viscosity (which may be anisotropic as a result of the presence of fibers), convection of the hot polymer suspension, and conduction of heat to the cold mold. The presence and orientation of fibers may make the in-plane stresses significantly important, so the Hele-Shaw approximation may no longer be valid. Hence it will not be correct to use the flow field resulting from simulation of injection molding of neat polymers with Hele-Shaw approximation to predict fiber orientation in fiber-filled suspensions.

To date there have been only preliminary calculations describing fiber orientation in injection molding. No general simulation of the process is available because there are insufficient theories to describe the phenomena and also because of inadequate numerical methods. First, one would need a mechanistic theory that describes the orientation of fibers in concentrated fiber suspensions and how it modifies the rheology. It is not clear how to adopt quasi two-dimensional codes for unfilled polymers to describe fiber orientation in three dimensions [45,53]. Advances in numerical methods are needed to perform three-dimensional fluid flow and heat transfer calculations and to treat the moving flow front in three dimensions. The mold filling codes should calculate the flow field exactly, as precise velocities and velocity gradients are crucial to predicting fiber orientation accurately. This remains a pressing problem for simulation of fiber orientation in injection molding.

Fiber Degradation During Molding

One of the major concerns in producing fiber-reinforced parts by molding is fiber breakage during processing of these suspensions. Many experiments have established the fact that the fibers get damaged during processing and fiber length may be reduced by an order of magnitude [54–57]. This reduction in length can potentially reduce the reinforced efficiency of the fibers, substantially reducing the mechanical properties of the composite. If the intent is to maintain the fiber length, one should try to answer questions such as how fibers are broken during processing and what conditions in the materials or processing must be altered to avoid fiber damage. There are three common and likely mechanisms for fiber breakage: fiber-flow, fiber-fiber and fiber-wall interactions.

Fiber-Flow Interactions

Most flows in molding processes are a combination of extensional and shear deformations. In a purely elongational flow, fibers tend to align along the stretching direction and hence are under tension; they are rather unlikely to break under this mode. However, in shear flows, fibers rotate across the streamlines and may bend to their critical radius of curvature and buckle under viscous forces transmitted by the polymer.

The theory of fiber breakage in dilute suspensions, where the fiber-fiber interactions are absent, was first developed by Forgacs and Mason [55] and was verified by Salinas and Pittman [58]. The theory stated that the fluid shear stress τ required to buckle a straight fiber of aspect ratio l/d and modulus E is given by

$$\tau = E\frac{\ln(2l/d) - 1.75}{2(l/d)^4} \tag{11}$$

Experimental investigations of deformation of thread-like particles in shear flow by Forgacs and Mason [55] revealed that sufficient shear stress would induce buckling in the fibers, as predicted by the theory. Varying the fiber aspect ratio, fiber modulus, and/or shear stress will result in different conformations of the fiber during rotation, as illustrated in Figure 6. Since the threshold shear stresses for longer fibers are much lower than the stress levels induced during processing, longer fibers will break into shorter fibers until the surviving fibers are too short to buckle. This partly explains the common experience of injection molders that fibers end up the same length

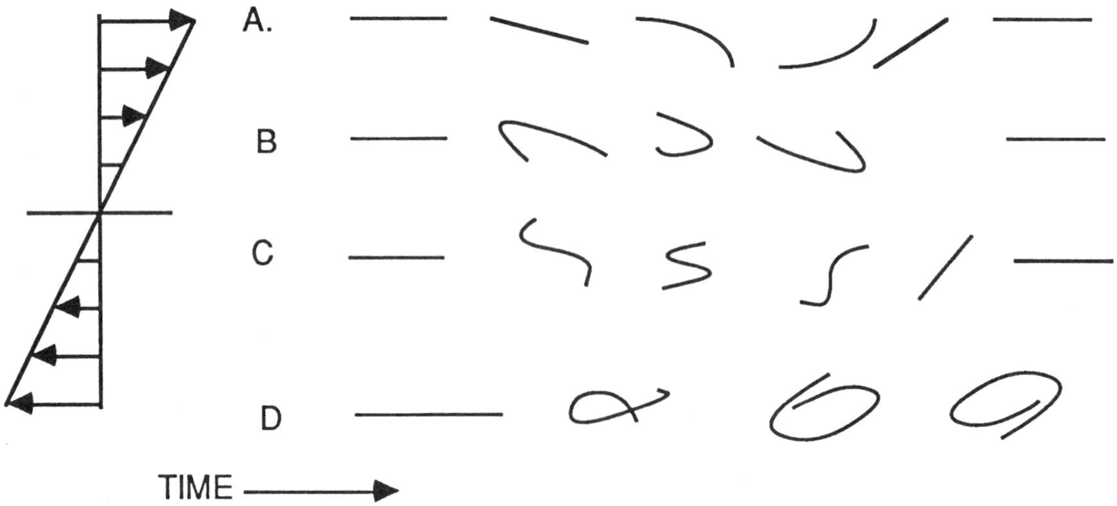

FIGURE 6 Types of fiber motion in simple shear velocity field. (*a*) Springy rotation; (*b*) single-ended snake orbit; (*c*) double-ended snake orbit; (*d*) coiled rotation. (After Ref. 55.)

after injection molding irrespective of the initial fiber lengths. The stress levels need to be reduced by at least one to two orders of magnitude during injection molding in order to retain longer fibers. In compression molding, it is possible to maintain the fiber lengths, as the stresses are lower because of slippage at the mold wall, resulting in squeezing flow [41]. In injection molding, the shearing stresses can be reduced by modifying the velocity profile through the thickness. A significant drop in the shear stresses can be achieved if one can change the velocity profile through the thickness from parabolic (in a Newtonian flow) to flat (plug flow). The velocity profile depends on many factors, such as fiber and filler concentration, rheology of the matrix, and heat transfer in the mold. The most damage is caused in the melting zone in injection molding [57], since this involves high shearing at temperatures just above the melting point, where the viscosity is still higher and consequently stress levels are at their highest. High performance fibers such as carbon and aramid, which have higher stiffness, are more likely to resist damage and make it possible to maintain longer fibers in discontinuous-fiber composites. Also a non-Newtonian matrix suspending the fibers may alleviate fiber breakage by either suppressing fiber rotation as a result of the elastic property in the matrix and/or by reducing shearing with a plug flow velocity profile (e.g., power law fluid behavior). One topic that needs attention is the fiber damage in a non-Newtonian matrix. Relevant questions that need to be answered are: What role does viscoelasticity play in fiber breakage? Do shear thinning fluids truly help in avoiding fiber damage? Further research in this area is clearly necessary.

Fiber-Fiber Interactions

Fiber-fiber interactions in concentrated suspensions can cause fiber overlaps, which will induce bending stresses in the fibers, resulting in breakage. The material will play an important role. For example, if the reinforcing fibers are glass, their abrasive surfaces during contact can induce stress concentration, which can lead to fiber damage. Not many studies have been carried out with carbon and aramid fibers. The effect of fiber volume fraction on breakage was studied by vonTurkovich and Erwin [57], who found that glass fibers in polystrene had the identical length distribution at the exit of an extruder for volume fractions ranging from 1 to 20%. Their results indicated that the fiber length was reduced by an order of magnitude after processing and that the average final length was insensitive to initial fiber length distribution. Higher glass fiber loadings (30–60%) were studied by Bailey and Kraft [60]. Their results show that fiber damage was reduced with higher fiber loadings. One could speculate on different reasons for the breakage, such as increasing the fiber volume fraction increases the abrasive environment, which will promote fiber damage. However, a higher concentration of fibers means that they cannot rotate as freely as before [25], thus reducing the breakage during the shearing flow. These opposing mechanisms make it difficult to arrive at any definite conclusion. Future research should try to investigate the effect of higher fiber loadings and the fiber material on fiber breakage.

Fiber-Wall Interactions

In both injection and compression molding, either the abrasive nature of the fibers or/and the fiber interaction with the mold geometry can cause damage. In injection molding, the screw geometry may play a crucial role in fiber length reduction as well. Wider channels have been shown to reduce breakage [59]. Also, the gate region of the mold may be a key factor in fiber degradation. Bailey and Kraft [60] studied long-fiber materials under various material and processing conditions in a plunger molding machine, a conventional injection molding machine, and an extruder. Mold configurations with a small gate and a generous gate were studied. Fiber loadings from 30 to 60% were used. Their study showed that a larger gate (2 mm × 1.5 mm) produced final fiber lengths with a mean of 0.99 mm, whereas a small gate (1mm × 1.5 mm) had a mean fiber length of only 0.49 mm. They also found that there was substantially more damage in the skin region than in the core. This could be attributed to a high shear rate near the mold surface coupled with fiber interactions with the mold geometry. Further study of the fiber–mold geometry relationship could aid the mold designer in selecting the gate size and mold geometry to minimize fiber damage with more confidence.

Fiber Clustering and Migration

In order to obtain satisfactory performance from a molded short-fiber composite part, the fibers should be uniformly distributed in the polymer matrix. It is necessary to form a homogeneous mixture of fibers and the suspending melt before the suspension is made to flow in the mold. In injection molding, this is accomplished by the screw. In compression and transfer molding, a material with fibers uniformly distributed must be used to constitute a homogeneous initial charge. However, during the molding process, the flow may cause the fibers to migrate and "clump" together, producing resin-rich areas with clusters or domains of fibers distributed nonuniformly in the composite. The segregation of the fibers from the matrix, also referred to as *phase separation* effect, results in an inhomogeneous mixture. The clusters of fibers are susceptible to microcracking, and the resin-rich areas are weak and more compliant than the composite part. Hence it is important to identify the cause of the segregation. Relevant questions are: How is it caused? What is its extent? How can we avoid it?

In 1836 Poiseuille [61] noticed that the blood corpuscles in the capillaries tend to keep away from the walls. Since then, many researchers have confirmed his observations that small particles in Newtonian fluids migrate along streamlines [62–65]. Flow in a cylinder viscometer or a tube gives rise to forces in the radial direction, which results in change in the particle distribution in that direction. For example, in Poiseuille flow, the forces act towards the tube axis, which results in particle concentration near the axis and a particle-free layer near the wall. Goettler [66] alluded to analogous phenomena in fiber-reinforced composites. A systematic parametric study was performed by Kubat and Szalanczi [67] in a spiral mold. They reported pronounced separation effects for low density polyethylene containing 25% glass spheres with diameters ranging from 50 to 100 μm. The difference in glass concentration was as much as 40% by weight between the tip and the center of the spiral. For melts with smaller spheres or fibers (aspect ratio 80), the effect was less dramatic. The change in concentration was only a few percent. Hegler and Mennig [68] also observed phase separation effects in injection-molded plaques. Again, the effects were more prominent with glass beads than with the fibers. However, they noted that the inhomogeneity increased with fiber concentration and increasing velocity gradients. Strong influence of the type of suspending fluid was also recorded. In fact, the migration towards the walls was observed for shear thinning fluids, whereas for Newtonian and viscoelastic fluids the migration of glass spheres across the streamlines was noted [64].

Inhomogeneity in a fabricated part is due to a combination of two factors: The initial distribution of fibers in the suspension may not have been uniform, and/or the flow may have caused migration of the fibers. The previous studies [67,68] reported separation effects for fibrous particles of an order of a few percent, and hence not very critical, in injection-molded parts. They had used pellets that were filamentized by extrusion compounding as their initial material. This operation of incorporating fibers in the resin produces a more evenly dispersed starting mixture (although there is considerable fiber degradation), and hence the migration effects will be only due to flow configuration. However, in compression molding, one starts with a material that is normally not homogeneous. McCullough [69] observed fiber clusters several orders of magnitude greater than the fiber lengths in compression-molded parts from SMC. Similarly, as shown in Figure 7, Casey and Advani [70] observed fiber clusters of the order of the fiber length in injection-molded parts that used granular material made by the pultrusion impregnation of continuous fibers, which were then chopped into pellets [71] to be used in molding machines. It was also noted that fiber concentration was much higher in the core than in the skin layer. The fiber volume fraction in the core was found to be nearly 50% by weight, as compared with 25% in the skin region [72]. Initial distribution of fibers seems to play an important role in clustering of fibers. Migration of particles during flow is important for fillers and spherical beads, and it is not very clear if it is as critical for fibers.

Is it possible to start with a nonhomogeneous mixture of fibers and the suspending melt and mold a homogeneous short-fiber composite? This will require us to create a flow field that will disperse the fibers uniformly in the matrix during processing of these short-fiber composites. To accomplish this task, we need to study the phenomena of fiber movement and migration. Not much has been done to analyze fiber migration phenomena quantitatively. Recently, Emerman [73] performed a semiquantitative analysis for spherical particles. He developed a set of rules for predicting the redistribution of particles in the melt. His calculations showed the effect of distance from the gate, filling time, and barrel temperature on the concentration distribution and compared favorably with his experiments. Kim [74] developed an analogous equation for ellipsoids in dilute suspensions. Monson and Sevick [75] used a Monte Carlo procedure to simulate cluster formation of an ellipsoid (aspect ratio < 5) suspension. Further work is required to minimize clustering of fibers during processing of composite materials.

FIGURE 7 **Fiber clusters in injection-molded part.** (From Ref. 70.)

Summary and Conclusions

As many new materials are being introduced as candidates for suspending mediums to handle bulk suspensions, controlled rheological characterization has become very important for estimating the processibility of these materials. Many system-specific studies have been performed that indicate that rheological behavior of fiber suspensions is markedly different from the behavior of the suspending fluids. Studies of simple flows have indicated that the suspension viscosity can differ by orders of magnitude at low and high shear rates and is affected significantly by the rheology of the suspending fluid, concentration, aspect ratio, and fiber orientation. To evaluate the dynamic rheological behavior in flow processes, we should seek a relation between suspension composition and structure and the rheological properties. More systematic work and investigation is needed to build this theory for rheological behavior of concentrated fiber suspensions in order to provide a framework for predicting and controlling the microstructure during the processing of short-fiber composite.

The fiber orientation field has also seen much progress. To characterize orientation, the use of a probability distribution function and a set of orientation tensors allows for the fact that fibers may orient in several directions in a given small region. These descriptions are used to model flow-induced orientation during mold filling, and it is now possible to predict the effect of flow on fiber orientation in complex geometries with useful accuracy. The exact effect of flow on fiber orientation can be modeled accurately for dilute suspensions. However, in concentrated fiber suspensions, the fiber behavior may be qualitatively similar but is quantitatively different. A phenomenological model exists to account for the concentration of the fibers, and work is underway to build rigorous theories for orientation in concentrated suspensions.

Numerical methods are used to model flow and fiber orientation in the complicated geometries typically encountered during the molding process. It is now possible to calculate the entire spatial orientation field using a tensor description of the orientation state as the mold fills. However, all calculations are in two-dimensional flow fields and treat decoupled flow and orientation fields. This development accurately models the orientation in the compression molding process for thin parts. The existing framework can be used to build a general simulation of the molding process if progress is made in the area of developing a theory to couple flow and orientation. In addition, advances in the field of numerical techniques will be needed.

Existing mold filling simulations use Hele-Shaw approximations, which may no longer be valid for mold filling of fiber-filled suspensions. Hence advances in numerical techniques are needed to build a simulation of mold filling that calculates accurate velocity and velocity gradient fields for meaningful fiber orientation calculations. Also, there is no existing method that can treat the movement of a three-dimensional flow front. This is an important issue if one needs to predict the fiber orientation near the flow front. More work remains to be done to adapt heat transfer analysis to minimize cycle time, control warpage and shrinkage, and avoid degradation of the polymer in mold filling simulations in neat polymers to processing of short-fiber composites.

Fiber damage during processing is mainly caused by fiber-fiber, fiber-wall, and fiber-flow interactions. Shearing plays an important role in fiber breakage. However, as the concentration of fibers is increased, the fiber-fiber and fiber-wall effects may become more prominent. No quantitative theory exists to predict fiber damage in molding. It is not very clear whether fiber clustering in a concentrated fiber suspension is due to fiber migration or to poor mixing. At present, there is no theory to characterize clusters in a composite. These issues have not received much attention, as short-fiber composites have been primarily used for moderate strength and stiffness applications. As short-fiber composites find uses in high performance applications, the need to minimize fiber degradation and to attain uniform distribution of fiber concentration will become crucial.

Although each of the above analyses tries to answer pertinent questions, greater potential would be realized if all the components could be coupled into a mold filling simulation to predict the microstructure. Various details could then be incorporated in mechanical and physical properties predictions to determine whether the part performs as required. This would provide the opportunity to construct the tooling and molds with confidence or to alter processing conditions and/or material parameters until the performance criteria are met. Such an approach to design and performance is under development, and will lead to enhanced quality and improved productivity of short-fiber composite parts.

Acknowledgements

The author would like to acknowledge the support provided by University of Delaware Research Foundation.

Suresh G. Advani

References

1. R. O. Maschmeyer and C. T. Hill, *Advances in Chemistry Series, 134,* 95 (1974).
2. A. Okagawa, R. G. Cox, and S. G. Mason, *J. Coll. Intf. Sci., 45,* 303 (1973).
3. S. M. Dinh and R. C. Armstrong, *J. Rheo., 28,* 207 (1984).
4. E. J. Hinch and L. G. Leal, *J. Fluid Mech., 52,* 683 (1972).
5. J. Mewis and A. B. Metzner, *J. Fluid Mech., 62,* 593 (1974).
6. C. B. Weinberger and J. D. Goddard, *Int. J. Mult. Flow, 1,* 465 (1974).
7. T. E. Kizior and F. A. Seyer, *Trans. Soc. Rheol., 18,* 271 (1974).
8. E. Ganani and R. L. Powell, *J. Compos. Mat., 19,* 194 (1985).
9. M. R. Kamal and A. Mutel, *J. Polym. Eng., 5,* 293 (1985).
10. A. B. Metzner, *J. Rheo., 29,* 739 (1985).
11. H. A. Barnes, M. F. Edwards and L. V. Woodcock, *Chem. Eng. Sci., 42,* 591. (1987).
12. M. Doi and S. F. Edwards, *J. Chem. Soc., Faraday Trans. II, 74,* 560 (1978).
13. G. K. Batchelor, *J. Fluid Mech., 41,* 545 (1970).
14. R. G. Cox and H. Brenner, *Chem. Eng. Sci., 26,* 65 (1971).
15. E. J. Hinch and L. G. Leal, *J. Fluid Mech., 52,* 683 (1972).
16. E. J. Hinch and L. G. Leal, *J. Fluid Mech., 57,* 753 (1973).
17. L. G. Leal and E. J. Hinch, *J. Fluid Mech., 46,* 685 (1971).
18. L. G. Leal and E. J. Hinch, *J. Fluid Mech., 55,* 745 (1972).
19. J. L. Erickson, *Arch. Rat. Mech. Anal., 4,* 231 (1960).
20. J. L. Erickson, *Arch. Rat. Kolloid Z., 173,* 117 (1960).
21. G. L. Hand, *J. Fluid Mech., 13,* 33 (1962).
22. G. G. Lipscomb, M. M. Denn, D. U. Hur, and D. V. Boger, *J. Non-Newtonian Fluid Mech., 26,* 297 (1988).
23. M. L. Becraft "Rheology of Concentrated Fiber Suspension," in Ph.D. Dissertation, University of Delaware (1988).
24. M. Doi, *J. Polym. Sci., 19,* 229 (1981).
25. F. Folgar and C. L. Tucker III, *J. Reinf. Plas. Compos., 3,* 98 (1984).

26. D. L. Denton, *Proc. 36th Ann. Conf. Reinf. Plast./Composite Institute, Society of Plastic Industries, Sect. 16-A,* 1981.
27. R. B. Pipes, R. L. McCullough, and D. G. Taggart, *Polym. Compos., 3,* 120 (1982).
28. L. A. Goettler, *Polym. Compos., 1,* 60 (1984).
29. S. G. Advani and C. L. Tucker III, *J. Rheo., 31,* 751 (1987).
30. P. H. Hermans, *Contributions to the Physics of Cellulose Fibers,* New York, Elsevier, 1946.
31. S. G. Advani and C. L. Tucker III, *J. Rheo., 34,* 367 (1990).
32. G. B. Jeffery, *Proc. Roy. Soc., Series A, 102,* 161 (1922).
33. M. J. Owen, D. H. Thomas, and M. S. Found, *Proc. 33rd Ann. Conf. Reinf. Plast./Composite Institute,* Society of Plastic Industries, *Section 20-B,* 1978, p. 1.
34. P. F. Bright, R. J. Crowson, and M. J. Folkes, *J. Mat. Sci., 13,* 2497 (1978).
35. C. L. Tucker, "Compression Molding of Polymers and Composites," in A. I. Isayev, Ed., *Injection and Compression Molding Fundamentals,* Marcel Dekker, New York, 1987, p. 481.
36. M. Doi, *J. Polym. Sci. Polym. Phys.* Ed., *19,* 229 (1981).
37. K. E. Evans and M. D. Ferrar, "Modeling Flow-Processed Composites," presented at 3rd International Conference on Fibre Reinforced Composites, 1988.
38. W. C. Jackson, S. G. Advani, and C. L. Tucker, *J. Compos. Mat., 26,* 539 (1986).
39. R. B. Bird, O. Hassager, R. C. Armstrong, and C. F. Curtiss, *Dynamics of Polymeric Liquids, Vol. 2, Kinetic Theory,* Wiley, New York, 1977.
40. R. C. Givler, M. J. Crochet, and R. B. Pipes, *J. Compos. Mat., 17,* 330 (1983).
41. M. R. Barone and D. A. Caulk, *J. Appl. Mech., 53,* 361 (1986).
42. S. G. Advani and C. L. Tucker, *SPE Tech. Papers, 34,* 687 (1988).
43. S. G. Advani, *Prediction of Fiber Orientation during Molding of Short Fiber Composites,* Ph.D. thesis, University of Illinois at Urbana Champaign, 1987.
44. C. A. Hieber and S. F. Shen, *J. Non-Newtonian Fluid Mech., 7,* 1 (1980).
45. S. Subbiah, D. Trafford, and S. Guceri, *Int. J. Heat Mass Transfer,* to appear.
46. M. Sanou, B. Chung, and C. Cohen, *Polym. Eng. Sci., 25,* 1008 (1985).
47. M. W. Darlington and P. L. McGinley, *J. Mat. Sci., 10,* 906 (1975).
48. M. W. Darlington, B. K. Gladwell, and G. R. Smith, *Polymer, 18,* 1269 (1977).
49. P. F. Bright, R. J. Crowson, and M. J. Folkes, *J. Mat. Sci., 13,* 2497 (1978).
50. M. W. Darlington and G. R. Smith, *Polym. Compos., 8,* 16 (1987).
51. Z. Tadmor, *J. Appl. Polym. Sci., 18,* 1753 (1974).
52. L. R. Schmidt, *Polym. Eng. Sci., 14,* 797 (1974).
53. V. W. Wang, C. A. Hieber, and K. K. Wang, *SPE Tech. Papers, 32,* 97 (1986).
54. R. J. Crowson and M. J. Folkes, *Polym. Eng. Sci, 20,* 934 (1980).
55. O. L. Forgacs and S. G. Mason, *J. Coll. Interf. Sci., 14,* 457 (1959).
56. L. Czarnecki and J. L. White, *J. Appl. Polym. Sci., 25,* 1217 (1980).
57. R. vonTurkovich and L. Erwin, *Polym. Eng. Sci., 21,* 743 (1983).
58. A. Salinas and J. F. T. Pittman, *Polym. Eng. Sci., 21,* 23 (1981).
59. K. O'Brien and J. F. Crincoli, "Improving the Properties of Reinforced Engineering Plastics by Employing Injecton Moldable Long Strand Variants," *RETEC Meeting,* 1987.
60. R. Bailey and H. Kraft, *Inter. J. Polym. Processing,* to appear.
61. J. L. M. Poiseuille, *Annales des Science Naturelles, 5,* 110 (1836).
62. G. Segre and A. Silberberg, *J. Fluid Mech., 14,* 115 (1962).
63. R. C. Jeffrey and J. R. A. Pearson, *J. Fluid Mech., 22,* 721 (1965).
64. A. Karnis and S. G. Mason, *Trans. Soc. Rheo., 10,* 571 (1966).
65. F. Gauthier, H. L. Goldsmith, and S. G. Mason, *Trans. Soc. Rheo., 15,* 297 (1971).
66. L. A. Goettler, *SPE Tech. Papers, 31,* 559 (1973).
67. J. Kubat and A. Szalanczi, *Polym. Eng. Sci., 14,* 833 (1974).
68. R. P. Hegler and G. Mennig, *Polym. Eng. Sci., 25,* 395 (1985).
69. R. McCullough, *Composites Design Guide,* Report 82-13, Center for Composite Materials, University of Delaware, 1982.
70. D. M. Casey and S. G. Advani, unpublished research.
71. J. B. Cattanach, G. Cuff, and F. N. Cogswell, *J. Polym. Eng., 6,* 345 (1986).
72. Jane Krosby, L & P, personal communication (1987).
73. S. H. Emerman, *Polym. Eng. Sci., 27,* 1105 (1987).
74. S. Kim, *Int. J. Multiphase Flow, 12,* 469 (1986).
75. E. M. Sevick, P. A. Monson, and J. M. Ottino, *J. Chem. Phys., 88,* 1198 (1988).

Molds

See Mold Fabrication